MICROPROCESSORS AND INTERFACING
Programming and Hardware

Douglas V. Hall

Gregg Division
McGRAW-HILL BOOK COMPANY
New York Atlanta Dallas St. Louis San Francisco
Auckland Bogotá Guatemala Hamburg Johannesburg Lisbon
London Madrid Mexico Montreal New Delhi Panama Paris
San Juan São Paulo Singapore Sydney Tokyo Toronto

MICROPROCESSORS AND INTERFACING: Programming and Hardware
INTERNATIONAL EDITION

1st Printing 1986

Sponsoring Editor: Paul Berk
Editing Supervisor: James Fields
Design and Art Supervisor/Cover Designer: Frances Conte Saracco
Production Supervisor: Priscilla Taguer
Text Designer: Susan Brorein

Library of Congress Cataloging in Publication Data

Hall, Douglas V., date.
Microprocessors and interfacing.
Includes index.
1. Microprocessors—Programming. 2. Microprocessors. 3. Computer interfaces. I. Title.
QA76.7.H2294 1986 005.26 86-156
ISBN 0-07-025526-1

When ordering this title use ISBN 0-07-100462-9

Printed and Bound in Singapore by Fong & Sons Printers Pte Ltd

CONTENTS

PREFACE

For the most part, *Microprocessors and Interfacing: Programming and Hardware* is based on a three-quarter series of microprocessor courses that my colleagues and I teach. The book is intended for students in electrical engineering programs, students in electronic engineering technician training programs, and people working in industry who want to upgrade their knowledge of microprocessors.

Before reading this book, you should have some basic knowledge of diodes, transistors, and digital circuitry. One of its aims is to teach you how to decipher and use manufacturer's literature; accordingly, many relevant parts of data sheets are shown. Because of the large number of actual devices discussed here, it was impossible to put the complete data sheets for all these devices in the appendixes. Therefore, I strongly suggest that you acquire or gain access to the latest edition of the Intel *Microsystem Components Handbook* so that, as you work your way through this book, you can refer to it if you need further information about a particular device. The bibliography lists other materials I have found useful.

I have chosen here to teach the programming, system connections, and interfacing of 16-bit microprocessors, which function as the "brains" of microcomputers such as the IBM PC. My experience as an engineer and as a teacher indicates that it is more productive to learn one microprocessor family very thoroughly, and from that strong base learn other families as needed. Therefore, this book concentrates on the Intel 8086/8088/80186/80188/80286/80386 family of microprocessors, rather than superficially covering the microprocessor families of several manufacturers.

I came into the world of electronics through the route of vacuum tubes. Therefore, my first tendency was to approach microprocessors from a hardware orientation. However, the more I designed with microprocessors and taught microprocessor classes, the more I became aware that the real essence of a microprocessor is what you can program it to do. For this reason the book begins with just a brief overview of the hardware of a computer. The next five chapters show how a microprocessor-based microcomputer can be programmed to do some real tasks.

The emphasis throughout is on writing assembly language programs in a top-down structured manner. The idea is to make programs easy to write, test, and debug. Experience has shown that the most successful approach to writing a program is to solve the problem first, and then simply implement the solution in the desired programming language.

The 8086 instructions are introduced in Chapters 2 through 5, as they are needed to solve simple programming problems. Chapter 6 contains a dictionary of all the 8086 and 80186 instructions. You can refer to this chapter to find further details about an instruction you want to use to do a particular operation in a program.

Chapter 7 discusses the hardware signals, timing, and system connections of a simple microcomputer. Chapter 7 also teaches a systematic approach to troubleshooting a malfunctioning 8086-based system. The remaining chapters show how the hardware and the programs work together. Troubleshooting a microprocessor-based system, for example, usually requires knowledge of both its hardware and programming, so I discuss here how diagnostic routines are written and used to find a problem.

Chapter 8 discusses how the 8086 responds to interrupts and how interrupt service procedures are written and used. Chapters 9 through 13 describe in detail how a microcomputer is interfaced with a wide variety of devices and systems. Also these chapters describe how the hardware and programs for microprocessor-based products are developed. Finally, Chapter 14 discusses operating system programs, and the 80286 and 80386 microprocessors that are designed to be used as the brains of multiuser microcomputer systems.

Program development for 16-bit microprocessors is somewhat tedious on hex-keypad-type development boards such as we used for 8-bit processors. Furthermore, industry does not usually develop microprocessor-based products in this manner. Therefore, for working with 16-bit processors I recommend a systems approach. A microprocessor development system, an IBM PC, or IBM PC-compatible computer, can be used to edit, assemble, link/locate, run, and debug 8086 assembly language programs. For programs that require external hardware, the object code for these programs can be downloaded to some prototype hardware such as an Intel SDK-86 development board. Chapter 13 contains a program that allows you to download object code programs from an IBM or IBM-compatible computer to an SDK-86 board. An available laboratory manual, written to accompany this book, shows you how to use the SDK-86 board and an IBM PC-compatible computer for assembly language programming and interfacing.

In the interfacing sections of this book I have tried to show as many circuits as possible that you can build, add to your microcomputer, and experiment with. Building and experimenting with real circuits will help you become fluent with microprocessors. The circuits in this book are intended just as starters. Hopefully you will grow far beyond what is shown here. If you have suggestions for improving this book or ideas that might clarify a point for someone else, please communicate with me.

I wish to express my profound thanks to the people who helped make this book a reality. Thanks to Pat Hunter, without whose cheerful encouragement I might not have made it through the book. She proofread and coded the manuscript, worked out the answers to the end-of-chapter problems to verify that they are solvable, and made many suggestions and contributions too numerous to mention. Thanks to Lee Campbell of Spokane Community College in Spokane, Washington, who meticulously worked his way through the manuscript and made many valuable suggestions. Thanks to Wayne J. Vyrostek of Westark College in Fort Smith, Arkansas, who reviewed the manuscript and contributed several valuable suggestions. Thanks to Intel Corporation for letting me use many drawings from their data books, so that this book could lead readers into the material they can use to continue their learning. Finally, thanks to my family and friends for their patience and support during the long effort of writing this book.

Douglas V. Hall

DEDICATION

To my students — Who grow beyond what I give and return to pull me into the future with them.

CHAPTER

1 Computer Number Systems, Codes, and Digital Devices

Before starting our discussion of microprocessors and microcomputers we need to make sure that some key concepts of the number systems, codes, and digital devices used in microcomputers are fresh in your mind. If the short summaries of these concepts in this chapter are not enough to refresh your memory, then it is a good idea to review them in a current digital text before going on in this book.

OBJECTIVES

At the conclusion of this chapter you should be able to:

1. Convert numbers between the following codes: binary, octal, hexadecimal, and BCD.
2. Define the terms bit, nibble, byte, word, most significant bit, and least significant bit.
3. Use a table to find the ASCII or EBCDIC code for a given alphanumeric character.
4. Perform addition and subtraction of binary, octal, hexadecimal, and BCD numbers.
5. Describe the operation of gates, flip-flops, latches, registers, ROMs, dynamic RAMs, static RAMs, and buses.
6. Describe how an arithemtic logic unit can be instructed to perform arithmetic or logical operations on binary words.

COMPUTER NUMBER SYSTEMS AND CODES

Review of Decimal System

To understand the structure of the binary number system, the first step is to review the familiar decimal or base-10 number system. Figure 1-1*a* shows a decimal number with the value of each place holder or digit expressed as a power of 10. The digits in the decimal number 5346.72 then tell you that you have 5 thousands, 3 hundreds, 4 tens, 6 ones, 7 tenths, and 2 hundredths. The number of symbols needed in any base number system is equal to the base number. In the decimal number system then, there are 10 symbols, 0 through 9. When the count in any digit position passes that of the highest value symbol, a carry of 1 is added to the next digit position and the other digit rolls back to zero. A car odometer is a good example of this.

A number system can be built using powers of any number as place holders or digits, but some bases are more useful than others. It is difficult to build electronic circuits which can store and manipulate 10 different voltage levels but relatively easy to build circuits which can handle two levels. Therefore, a *binary* or *base-2* number system is used.

The Binary Number System

Figure 1-1*b* shows the value of each digit in a binary number. Each binary digit represents a power of 2. A binary digit is often called a *bit*. Note that digits to the right of the *binary point* represent fractions used for numbers less than one. The binary system uses only two symbols, zero (0) and one (1). Therefore, in binary you count as follows: 0, 1, 10, 11, 100, 101, 110, 111, 1000, etc.

Binary numbers are often called *binary words* or just *words*. Binary words of certain numbers of bits have also acquired special names. A 4-bit binary word is

5 3 4 6 . 7 2

$10^3\ 10^2\ 10^1\ 10^0\quad 10^{-1}\ 10^{-2}$

(*a*)

1 0 1 1 0 . 1 1

$2^7\ 2^6\ 2^5\ 2^4\ 2^3\ 2^2\ 2^1\ 2^0\quad 2^{-1}\ 2^{-2}$

128 64 32 16 8 4 2 1 $\frac{1}{2}$ $\frac{1}{4}$

(*b*)

FIGURE 1-1 Digit values in decimal and binary. (*a*) Decimal. (*b*) Binary.

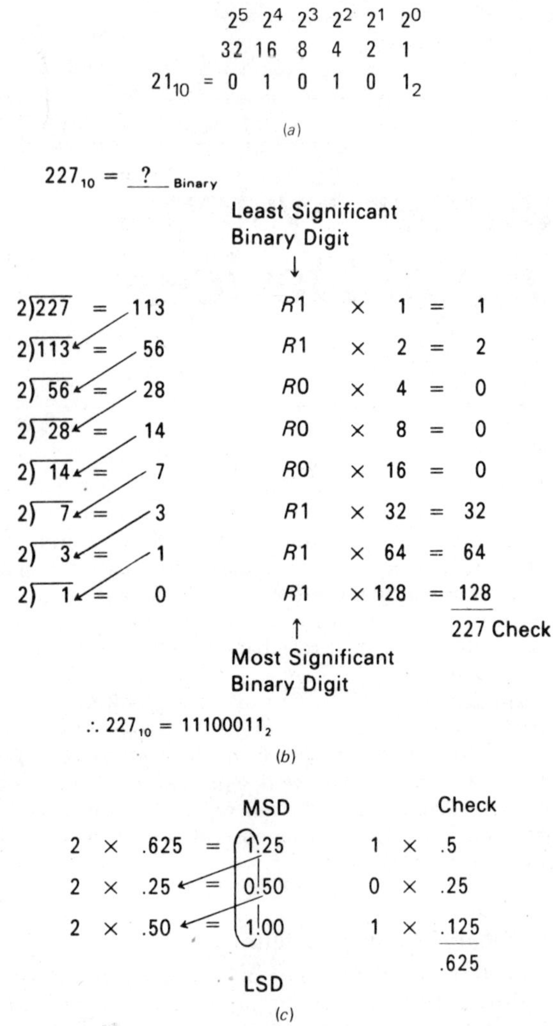

2^5 2^4 2^3 2^2 2^1 2^0
32 16 8 4 2 1
21_{10} = 0 1 0 1 0 1_2
(a)
227_{10} = ? Binary
Least Significant Binary Digit
2)227 = 113 R1 × 1 = 1
2)113 = 56 R1 × 2 = 2
2) 56 = 28 R0 × 4 = 0
2) 28 = 14 R0 × 8 = 0
2) 14 = 7 R0 × 16 = 0
2) 7 = 3 R1 × 32 = 32
2) 3 = 1 R1 × 64 = 64
2) 1 = 0 R1 × 128 = 128
227 Check
Most Significant Binary Digit
∴ 227_{10} = 11100011_2
(b)
MSD
Check
2 × .625 = 1.25 1 × .5
2 × .25 = 0.50 0 × .25
2 × .50 = 1.00 1 × .125
.625
LSD
(c)

FIGURE 1-2 Converting decimal to binary. *(a)* Digit value method. *(b)* Divide by 2 method. *(c)* Decimal fraction conversion.

called a *nibble*, and an 8-bit binary word is called a *byte*. A 16-bit binary word is often referred to just as a *word*, and a 32-bit binary word is referred to as a *doubleword*. The rightmost or *least-significant bit* of a binary word is usually referred to as the LSB. The leftmost or *most-significant bit* of a binary word is usually called the MSB.

To convert a binary number to its equivalent decimal number multiply each digit times the decimal value of the digit and just add these up. The binary number 101, for example, represents: $(1 \times 2^2) + (0 \times 2^1) + (1 \times 2^0)$ or $4 + 0 + 1 =$ decimal 5. For the binary number 10110.11 you have:

$$(1 \times 2^4) + (0 \times 2^3) + (1 \times 2^2) + (1 \times 2^1)$$
$$+ (0 \times 2^0) + (1 \times 2^{-1}) + (1 \times 2^{-2})$$
$$= 16 + 0 + 4 + 2 + 0 + 0.5 + 0.25$$
$$= \text{decimal } 22.75$$

To convert a decimal number to binary there are two common methods. The first (Figure 1-2*a*) is simply a reverse of the binary-to-decimal method above. For example, to convert the decimal number 21 (sometimes written as 21_{10}) to binary, first subtract the largest power of 2 that will fit in the number. For 21_{10} the largest power of 2 that will fit is 16 or 2^4. Subtracting 16 from 21 gives a remainder of 5. Put a one in the 2^4 digit position and see if the next lower power of 2 will fit in the remainder. Since 2^3 is 8 and 8 will not fit in the remainder of 5, put a zero in the 2^3 digit position. Then try the next lower power of 2. In this case the next is 2^2 or 4, which will fit in the remainder of 5. A 1 is, therefore, put in the 2^2 digit position. When 2^2 or 4 is subtracted from the old remainder of 5 a new remainder of 1 is left. Since 2^1 or 2 will not fit into this remainder, a zero is put in that position. A 1 is put in the 2^0 position because 2^0 is equal to 1 and this fits exactly into the remainder of 1. The result shows that 21_{10} is equal to 10101 in binary. The conversion process is somewhat messy to describe, but easy to do. Try converting 46_{10} to binary. You should get 101110.

Another method of converting a decimal number to binary is shown in Figure 1-2*b*. Divide the decimal number by 2 and write the quotient and remainder as shown. Divide this quotient and following quotients by 2 until the quotient reaches zero. The column of remainders will be the binary equivalent of the given decimal number. Note that the MSD is on the bottom of the column and the LSD is on the top of the column if you perform the divisions in order from the top to the bottom of the page. You can demonstrate that the binary number is correct by reconverting from binary to decimal as shown in the right-hand side of Figure 1-2*b*.

You can convert decimal numbers less than 1 to binary by successive multiplication by 2, and recording carries until the quantity to the right of the decimal point becomes zero, as shown in Figure 1-2*c*. The carries represent the binary equivalent of the decimal number, with the *most-significant bit* at the top of the column. Decimal 0.625 equals 0.101 in binary. For decimal values that do not convert exactly the way this one did (quantity to the right of the decimal never becomes zero), you can continue the conversion process until you get the number of binary digits desired.

At this point it is interesting to compare the number of digits required to express numbers in decimal with the number required to express them in binary. In decimal, one digit can represent 10^1 numbers, 0–9; two digits can represent 10^2 or 100 numbers, 0–99; and three digits can represent 10^3 or 1000 numbers, 0–999. In binary, a similar pattern exists. One binary digit can represent two numbers, 0–1; two binary digits can represent 2^2 or 4 numbers, 0–11; and three binary digits can represent 2^3 or 8 numbers. The pattern then is that N decimal digits can represent 10^N numbers and N binary digits can represent 2^N numbers. Eight binary digits can represent the 2^8 or 256 numbers, 0–255.

Octal

Binary is not a very compact code. This means that it requires many more digits to express a number than does, for example, decimal. Twelve binary digits can only

$$4096 \quad 512 \quad 64 \quad 8 \quad 1 \quad \tfrac{1}{8} \quad \tfrac{1}{64} \quad \tfrac{1}{512}$$

$$8^4 \quad 8^3 \quad 8^2 \, 8^1 \, 8^0 \, . \, 8^{-1} \, 8^{-2} \, 8^{-3}$$

(*a*)

$327_{\text{Decimal}} = \underline{\ ?\ }_{\text{Octal}} \qquad 327_D = 507_8$

				LSD			
$8\overline{)327}$	=	40	*R*	7	× 1	=	7
$8\overline{)40}$	=	5	*R*	0	× 8	=	0
$8\overline{)5}$	=	0	*R*	5	× 64	=	320
				MSD			327

(*b*)

Binary	101	011	111	. ← Binary Point
Octal	5	3	7	

(*c*)

FIGURE 1-3 Octal numbers. *(a)* Value of place holders. *(b)* Conversion of decimal to octal. *(c)* Conversion of binary to octal.

describe a number up to 4095_{10}. Computers require binary data, but people working with computers have trouble remembering the long binary words produced by the noncompact code. One solution to the problem is to use the *octal* or *base-8* code. As you can see in Figure 1-3*a*, the digits in this code represent powers of 8. The symbols then are 0 through 7. You can convert a decimal number to the octal equivalent number with the same trick you used to convert decimal to binary. Figure 1-3*b* shows the technique for decimal-to-octal conversion. Decimal 327 is equal to 507_8. Verification of this is shown by reconverting the octal to decimal in the second half of Figure 1-3*b*.

Since 8 is an integral power of 2, conversions from binary to octal, and octal to binary, are quite simple. If you have a binary number such as 101011111, then, starting from the binary point and moving to the left, mark off the binary digits in groups of three, as shown in Figure 1-3*c*. Each group of three binary digits is equal to one octal digit. For the example above, 111 is a 7, 011 is a 3, and 101 is a 5. Therefore, 101011111 binary is equal to 537_8.

You convert from octal to binary by replacing each octal digit with its 3-bit binary equivalent.

Hexadecimal

Some once-popular minicomputers, such as the PDP-8, have 12 parallel data lines. Four octal digits are an easy way to represent the binary data word on these 12 parallel lines. For example, 100001010111 binary is easily remembered or written as 4127 octal. Most microprocessors have 4-bit, 8-bit, 16-bit, or 32-bit data words. For these microprocessors, it is more logical to use a code which groups the binary digits in groups of four rather than three. *Hexadecimal* or base-16 code does this. Figure 1-4*a* shows the digit values for hexadecimal, which is often just called *hex*. Since hex is base-16, you have to have 16 possible symbols for each digit. The table of Figure 1-4*b* shows the symbols for hex code. After the decimal symbols 0 through 9 are used up, you use the letters A through F for values 10 through 15.

As mentioned above, each hex digit is equal to four binary digits. To convert the binary number 11010110

$$16^3 \quad 16^2 \, 16^1 \, 16^0 \, . \, 16^{-1} \quad 16^{-2} \quad 16^{-3}$$

$$4096 \quad 256 \quad 16 \quad 1 \quad \tfrac{1}{16} \quad \tfrac{1}{256} \quad \tfrac{1}{4096}$$

(*a*)

Dec		Hex
0	=	0
1	=	1
2	=	2
3	=	3
4	=	4
5	=	5
6	=	6
7	=	7
8	=	8
9	=	9
10	=	A
11	=	B
12	=	C
13	=	D
14	=	E
15	=	F

(*b*)

1101	0110_2	
D	6	HEX

(*c*)

$227_{10} = \underline{\ ?\ }_{\text{Hex}}$

$16\overline{)227}$	=	14	*R*3	× 1	=	3
$16\overline{)14}$	=	0	*R*E	× 16	=	224
						227

$227_{10} = E3_{16}$

(*d*)

FIGURE 1-4 Hexadecimal numbers. *(a)* Value of place holders. *(b)* Symbols. *(c)* Binary to hexadecimal conversion. *(d)* Decimal to hexadecimal conversion.

TABLE 1-1
COMMON NUMBER CODES

DECIMAL SYSTEM	BINARY	OCTAL	HEXA-DECIMAL	DECIMAL CODES				REFLECTED GRAY CODE	SEVEN-SEGMENT DISPLAY (1 = ON)							
				8421 BCD	2421	5421	EXCESS-3		*A*	*B*	*C*	*D*	*E*	*F*	*G*	DISPLAY
0	0000	0	0	0000	0000	0000	0011 0011	0000	1	1	1	1	1	1	0	0
1	0001	1	1	0001	0001	0001	0011 0100	0001	0	1	1	0	0	0	0	1
2	0010	2	2	0010	0010	0010	0011 0101	0011	1	1	0	1	1	0	1	2
3	0011	3	3	0011	0011	0011	0011 0110	0010	1	1	1	1	0	0	1	3
4	0100	4	4	0100	0100	0100	0011 0111	0110	0	1	1	0	0	1	1	4
5	0101	5	5	0101	1011	1000	0011 1000	0111	1	0	1	1	0	1	1	5
6	0110	6	6	0110	1100	1001	0011 1001	0101	1	0	1	1	1	1	1	6
7	0111	7	7	0111	1101	1010	0011 1010	0100	1	1	1	0	0	0	0	7
8	1000	10	8	1000	1110	1011	0011 1011	1100	1	1	1	1	1	1	1	8
9	1001	11	9	1001	1111	1100	0011 1100	1101	1	1	1	0	0	1	1	9
10	1010	12	A	0001 0000	0001 0000	0001 0000	0100 0011	1111	1	1	1	1	1	0	1	A
11	1011	13	B	0001 0001	0001 0001	0001 0001	0100 0100	1110	0	0	1	1	1	1	1	B
12	1100	14	C	0001 0010	0001 0010	0001 0010	0100 0101	1010	0	0	0	1	1	0	1	C
13	1101	15	D	0001 0011	0001 0011	0001 0011	0100 0110	1011	0	1	1	1	1	0	1	D
14	1110	16	E	0001 0100	0001 0100	0001 0100	0100 0111	1001	1	1	0	1	1	1	1	E
15	1111	17	F	0001 0101	0001 1011	0001 1000	0100 1000	1000	1	0	0	0	1	1	1	F

to hex, mark off groups of four, moving to the left from the binary point, as shown in Figure 1-4*c*. Then write the hex symbol for the value of each group of four. The 0110 group is equal to 6 and the 1101 group is equal to 13. Since 13 is D in hex, 11010110 binary is equal to D6 in hex. "H" is usually used after a number to indicate that it is a hexadecimal number. For example, D6 hex is usually written D6H. Eight bits require only two hex digits to represent them.

If you want to convert from decimal to hexadecimal, Figure 1-4*d* shows a familiar trick to do this. The result shows that 227_{10} is equal to E3H. As you can see, hex is an even more compact code than decimal. Two hexadecimal digits can indicate a number up to 255. Only four hex digits are needed to represent a 16-bit binary number.

To illustrate how hexadecimal numbers are used in digital logic, a service manual tells you that the 8-bit-wide data bus of an 8088A microprocessor should contain 3FH during a certain operation. Converting 3FH to binary gives the pattern of 1's and 0's (0011 1111) you would expect to find with your oscilloscope or logic analyzer on the parallel lines. The 3FH is simply a shorthand which is easier to remember and less prone to errors.

To convert from octal code to hex code, the easiest way is to write the binary equivalent of the octal and then convert the binary digits, four at a time, into the appropriate hex digits. Reverse the procedure to get from hex to octal.

BCD Codes

STANDARD BCD

In applications such as frequency counters, digital voltmeters, or calculators, where the output is a decimal display, a binary-coded decimal or BCD code is often used. The advantage of BCD for these applications is that information for each decimal digit is contained in a separate 4-bit binary word. As you can see in Table 1-1, the simplest BCD code uses the first 10 numbers of standard binary code for the BCD number 0 through 9. The hex codes A through F are invalid BCD codes. Each decimal digit then is individually represented by its 4-bit binary equivalent. Figure 1-5 illustrates this.

GRAY CODE

Gray code is another important binary code which is often used for encoding shaft position data from machines such as computer-controlled lathes. This code has the same possible combinations as standard binary, but as you can see in the 4-bit example in Table 1-1, they are arranged in a different order. Notice that only one

5	2	9	Decimal
0101	0010	1001	BCD

FIGURE 1-5 Decimal to BCD conversion.

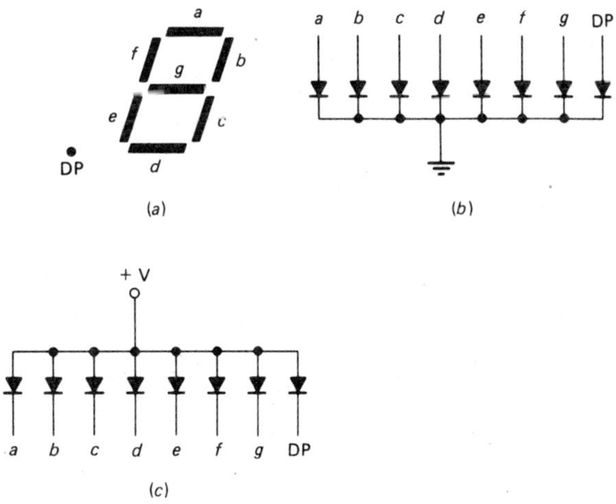

FIGURE 1-6 Seven-segment LED display. *(a)* Segment labels. *(b)* Schematic of common-cathode type. *(c)* Schematic of common-anode type.

binary digit changes at a time as you count up in this code.

If you need to construct a Gray-code table larger than that in Table 1-1, a handy way to do so is to observe the pattern of 1's and 0's and just extend it. The least-significant digit column starts with one 0 and then has alternating groups of two 1's and two 0's as you go down the column. The second-most-significant digit column starts with two 0's and then has alternating groups of four 1's and four 0's. The third column starts with four 0's, then has alternating groups of eight 1's and eight 0's. By now you should see the pattern. Try to figure out the Gray code for the decimal number 16. You should get 11000.

Seven-Segment Display Code

Since seven-segment displays such as that shown in Figure 1-6 are now so common in everything from calculators to gasoline pumps, the segment code for these has been included in Table 1-1. Some single seven-segment displays will display the last six numbers (10–15) of this code as the hexadecimal digits A–F. In Table 1-1, a 1 indicates that the segment is lit, which is true for displays such as the common-cathode light-emitting diode (LED) display in Figure 1-6*b*. For some displays, such as the common-anode LED display shown in Figure 1-6*c*, a low actually lights the segment, so you have to invert all the values.

Alphanumeric Codes

When communicating with or between computers you need a binary-based code which can represent letters of the alphabet as well as numbers. Common codes used for this have from 5 to 12 bits per word and are referred to as *alphanumeric codes*. To detect possible errors in

TABLE 1-2
COMMON ALPHANUMERIC CODES

ASCII SYMBOL	HEX CODE FOR 7-BIT ASCII	BCDIC SYMBOL	HEX CODE FOR EP BCDIC	EBCDIC SYMBOL	HEX CODE FOR EBCDIC	SELEC-TRIC SYMBOL	HEX CODE FOR SELEC-TRIC	HOL-LERITH SYMBOL	HOLES PUNCHED CODE FOR HOLLERITH
NUL	00			NUL	00			NUL	12 0 9 8 1
S0H	01			S0H	01			SCH	12 9 1
STX	02			STX	02			STX	12 9 2
ETX	03			ETX	03			ETX	12 9 3
E0T	04			E0T	37			ECT	9 7
ENQ	05			ENQ	2D			ENQ	0 9 8 5
ACK	06			ACK	2E			ACK	0 9 8 6
BEL	07			BEL	2F			BEL	0 9 8 7
BS	08			BS	16			BS	11 9 6
HT	09			HT	05			HT	12 9 5
LF	0A			LF	25			LF	0 9 5
VT	0B	‡	9A	VT	0B			VT	12 9 8 3
FF	0C			FF	0C			FF	12 9 8 4
CR	0D	ǂ	FF	CR	0D			CR	12 9 8 5
S0	0E			S0	0E			S0	12 9 8 6
S1	0F			S1	0F			S1	12 9 8 7
DLE	10			DLE	10			DLE	12 11 9 8 1
DC1	11			DC1	11			DC1	11 9 1
DC2	12			DC2	12			DC2	11 9 2
DC3	13			DC3	13			DC3	11 9 3
DC4	14			DC4	35			DC4	9 8 4
NAK	15			NAK	3D			NAK	9 8 5
SYN	16			SYN	32			SYN	9 2
ETB	17			E0B	26			ETB	0 9 6
CAN	18			CAN	18			CAN	11 9 8
EM	19			EM	19			EM	11 9 8 1
SUB	1A			SUB	3F			SUB	9 8 7
ESC	1B			BYP	24			ESC	0 9 7
FS	1C			FLS	1C			FS	11 9 8 4
GS	1D			GS	1D			GS	11 9 8 5
RS	1E			RDS	1E			RS	11 9 8 6
US	1F			US	1F			US	11 9 8 7
SP	20	SP	00	SP	40			SP	NO PNCH
!	21	!	6A	!	5A	½!	27	!	12 8 7
"	22	⧻	5F	"	7F	"	2D	"	8 7
#	23	#	4B	#	7B	#	7E	#	8 3
$	24	$	2B	$	5B	$	79	$	11 8 3
%	25	%	5C	%	6C	%	3D	%	0 8 4
&	26	&	30	&	50	&	7D		12
'	27	V	1D	'	7D	'	25		8 5
(	28	Blank	50	(	4D	(	38	(	12 8 5
)	29	△	6F	)	5D	)	39	)	11 8 5
*	2A	*	6C	*	5C	*	7C	*	11 8 4

(continued)

TABLE 1-2
COMMON ALPHANUMERIC CODES (*CONTINUED*)

ASCII SYMBOL	HEX CODE FOR 7-BIT ASCII	BCDIC SYMBOL	HEX CODE FOR EP BCDIC	EBCDIC SYMBOL	HEX CODE FOR EBCDIC	SELEC-TRIC SYMBOL	HEX CODE FOR SELEC-TRIC	HOL-LERITH SYMBOL	HOLES PUNCHED CODE FOR HOLLERITH
+	2 B			+	4 E	+	0 E	+	12 8 6
,	2 C	,	1 B	,	6 B	,	4 4	,	0 8 3
−	2 D			−	6 0	−	0 0	−	11
.	2 E	.	7 B	.	4 B	.	2 6	.	12 8 3
/	2 F	/	1 1	/	6 1	/	4 1	/	0 1
0	3 0	0	0 A	0	F 0	0	3 1	0	0
1	3 1	1	4 1	1	F 1	1	7 7	1	1
2	3 2	2	4 2	2	F 2	2	3 6	2	2
3	3 3	3	0 3	3	F 3	3	7 6	3	3
4	3 4	4	4 4	4	F 4	4	7 1	4	4
5	3 5	5	0 5	5	F 5	5	3 5	5	5
6	3 6	6	0 6	6	F 6	6	3 4	6	6
7	3 7	7	4 7	7	F 7	7	7 5	7	7
8	3 8	8	4 8	8	F 8	8	7 4	8	8
9	3 9	9	0 9	9	F 9	9	3 0	9	9
:	3 A	:	4 D	:	7 A	:	4 D	:	8 2
;	3 B	;	2 E	;	5 E	;	4 5	;	11 8 6
<	3 C	<	7 E	<	4 C			<	12 8 4
=	3 D	√	0 F	=	7 E	=	0 6	=	8 6
>	3 E	>	4 E	>	6 E			>	0 8 6
?	3 F	?	3 A	?	6 F	?	4 9	?	0 8 7
@	4 0	@	0 C	@	7 C	@	3 E	@	8 4
A	4 1	A	7 1	A	C 1	A	6 C	A	12 1
B	4 2	B	7 2	B	C 2	B	1 8	B	12 2
C	4 3	C	3 3	C	C 3	C	5 C	C	12 3
D	4 4	D	7 4	D	C 4	D	5 D	D	12 4
E	4 5	E	3 5	E	C 5	E	1 D	E	12 5
F	4 6	F	3 6	F	C 6	F	4 E	F	12 6
G	4 7	G	7 7	G	C 7	G	4 F	G	12 7
H	4 8	H	7 8	H	C 8	H	1 9	H	12 8
I	4 9	I	3 9	I	C 9	I	2 C	I	12 9
J	4 A	J	2 1	J	D 1	J	0 7	J	11 1
K	4 B	K	2 2	K	D 2	K	1 C	K	11 2
L	4 C	L	6 3	L	D 3	L	5 9	L	11 3
M	4 D	M	2 4	M	D 4	M	6 F	M	11 4
N	4 E	N	6 5	N	D 5	N	1 E	N	11 5
O	4 F	O	6 6	O	D 6	O	6 9	O	11 6
P	5 0	P	2 7	P	D 7	P	0 D	P	11 7
Q	5 1	Q	2 8	Q	D 8	Q	0 C	Q	11 8
R	5 2	R	6 9	R	D 9	R	6 D	R	11 9
S	5 3	S	1 2	S	E 2	S	2 9	S	0 2
T	5 4	T	5 3	T	E 3	T	1 F	T	0 3

(*continued*)

TABLE 1-2
COMMON ALPHANUMERIC CODES (*CONTINUED*)

ASCII SYMBOL	HEX CODE FOR 7-BIT ASCII	BCDIC SYMBOL	HEX CODE FOR EP BCDIC	EBCDIC SYMBOL	HEX CODE FOR EBCDIC	SELECTRIC SYMBOL	HEX CODE FOR SELECTRIC	HOLLERITH SYMBOL	HOLES PUNCHED CODE FOR HOLLERITH
U	5 5	U	1 4	U	E 4	U	5 E	U	0 4
V	5 6	V	5 5	V	E 5	V	6 E	V	0 5
W	5 7	W	5 6	W	E 6	W	2 8	W	0 6
X	5 8	X	1 7	X	E 7	X	5 F	X	0 7
Y	5 9	Y	1 8	Y	E 8	Y	0 9	Y	0 8
Z	5 A	Z	5 9	Z	E 9	Z	3 F	Z	0 9
[	5 B	[	7 D	[	A D	[	7 F	[	12 8 2
\	5 C	\	1 E	N L	1 5			\	0 8 2
]	5 D	]	2 D	]	D D			]	11 8 2
^	5 E	□	3 C	¬	5 F			^	11 8 7
—	5 F	—	6 0	—	6 D	—	0 8	—	0 8 5
`	6 0			R E S	1 4			`	8 1
a	6 1			a	8 1	a	6 4	a	12 0 1
b	6 2			b	8 2	b	1 0	b	12 0 2
c	6 3			c	8 3	c	5 4	c	12 0 3
d	6 4			d	8 4	d	5 5	d	12 0 4
e	6 5			e	8 5	e	1 5	e	12 0 5
f	6 6			f	8 6	f	4 6	f	12 0 6
g	6 7			g	8 7	g	4 7	g	12 0 7
h	6 8			h	8 8	h	1 1	h	12 0 8
i	6 9			i	8 9	i	2 4	i	12 0 9
j	6 A			j	9 1	j	0 7	j	12 11 1
k	6 B			k	9 2	k	1 4	k	12 11 2
l	6 C			l	9 3	l	5 1	l	12 11 3
m	6 D			m	9 4	m	6 7	m	12 11 4
n	6 E			n	9 5	n	1 6	n	12 11 5
o	6 F			o	9 6	o	6 1	o	12 11 6
p	7 0			p	9 7	p	0 5	p	12 11 7
q	7 1			q	9 8	q	0 4	9	12 11 8
r	7 2			r	9 9	r	6 5	r	12 11 9
s	7 3			s	A 2	s	2 1	s	11 0 2
t	7 4			t	A 3	t	1 7	t	11 0 3
u	7 5			u	A 4	u	5 6	u	11 0 4
v	7 6			v	A 5	v	6 6	v	11 0 5
w	7 7			w	A 6	w	2 0	w	11 0 6
x	7 8			x	A 7	x	5 7	x	11 0 7
y	7 9			y	A 8	y	0 1	y	11 0 8
z	7 A			z	A 9	z	3 7	z	11 0 9
{	7 B			{	8 B			{	12 0
|	7 C			|	4 F			|	12 11
}	7 D			}	9 B			}	11 0
~	7 E			¢	4 A			~	11 0 1
D E L	7 F			D E L	0 7			D E L	12 9 7

BCDIC: HEX DIGIT = PCBA; HEX DIGIT = $2^3 2^2 2^1 2^0$

SELECTRIC: $R_5 T_1 T_2$ = HEX DIGIT; $SR_{2A} R_2 R_1$ = HEX DIGIT

these codes, an additional bit, called a *parity bit*, is often added as the most-significant bit.

Parity is a term used to identify whether a data word has an odd or even number of 1's. If a data word contains an odd number of 1's, the word is said to have *odd parity*. The binary word 0110111 with five 1's has odd parity. The binary word 0110000 has an even number of 1's (two) so it has *even parity*.

In practice the parity bit may function as follows. The system that is sending a data word checks the parity of the word. If the parity of the data word is odd, the system will set the parity bit to a 1. This makes the parity of the data word plus parity bit even. If the parity of the data word is even, the sending system will reset the parity bit to a 0. This again makes the parity of the data word plus parity even. The receiving system checks the parity of the data word plus parity bit that it receives. If the receiving system detects odd parity in the received data word plus parity, it can assume an error occurred and tells the sending system to send the data again. The system is then said to be using even parity. The system could have been set up to use (maintain) odd parity in a similar manner.

The difficulty with this method of detecting errors introduced during transmission is that two errors introduced into a data word may keep the correct parity and, therefore, the parity checker won't indicate an error. Other, more complex methods, such as CRC and "Hamming codes" can be used to detect multiple errors in transmitted data, and even to correct errors. Some of these will be described in Chapters 12 and 13.

ASCII

Table 1-2 shows several alphanumeric codes. The first of these is *ASCII*, or American Standard Code for Information Interchange. This is shown in the table as a 7-bit code. With seven bits you can code up to 128 characters, which is enough for the full upper- and lower-case alphabet, numbers, punctuation marks, and control characters. The code is arranged so that if only upper-case letters, numbers, and a few control characters are needed, the lower six bits are all that is required. If a parity check is wanted, a parity bit is added to the basic 7-bit code in the MSB position. The binary word 1100 0100, for example, is the ASCII code for upper-case D with odd parity. Table 1-3 gives the meanings of the control character symbols used in the ASCII code table.

BCDIC

BCDIC code is the Binary Coded Decimal Interchange Code used with some computers. It uses seven bits plus a parity bit. The lower four bits are referred to as the *numeric bits*. The upper four bits contain a parity bit and three *zone* bits. The arrangement of these bits is shown at the bottom of Table 1-2. To save space in Table 1-2, the hex equivalent of the binary digits is used for the BCDIC code expressed with even parity.

EBCDIC

Another alphanumeric code commonly encountered in IBM equipment is the Extended Binary Coded Decimal Interchange Code or *EBCDIC*. This is an 8-bit code without parity. A ninth bit can be added for parity. To save space in Table 1-2, the eight binary digits of EBCDIC are represented with their 2-digit hex equivalent.

TABLE 1-3
DEFINITIONS OF CONTROL CHARACTERS

NUL	NULL	**DC2**	DIRECT CONTROL 2
SOH	START OF HEADING	**DC3**	DIRECT CONTROL 3
STX	START TEXT	**DC4**	DIRECT CONTROL 4
ETX	END TEXT	**NAK**	NEGATIVE ACKNOWLEDGE
EOT	END OF TRANSMISSION	**SYN**	SYNCHRONOUS IDLE
ENQ	ENQUIRY	**ETB**	END TRANSMISSION BLOCK
ACK	ACKNOWLEDGE	**CAN**	CANCEL
BEL	BELL	**EM**	END OF MEDIUM
BS	BACKSPACE	**SUB**	SUBSTITUTE
HT	HORIZONTAL TAB	**ESC**	ESCAPE
LF	LINE FEED	**FS**	FORM SEPARATOR
VT	VERTICAL TAB	**GS**	GROUP SEPARATOR
FF	FORM FEED	**RS**	RECORD SEPARATOR
CR	CARRIAGE RETURN	**US**	UNIT SEPARATOR
SO	SHIFT OUT		
SI	SHIFT IN		
DLE	DATA LINK ESCAPE		
DC1	DIRECT CONTROL 1		

SELECTRIC

Selectric is a 7-bit code used in the familiar IBM spinning ball typewriters and printers. Table 1-2 shows this code for reference also. Each bit position in the code controls an operation of the spinning ball.

From most-significant to least-significant bit, the meaning of the seven bits are: ROTATE 5, TILT 1, TILT 2, SHIFT, ROTATE 2A, ROTATE 2, and ROTATE 1. In addition to this 7-bit code, Selectrics have separate machine commands for space, return, backspace, tabs, bell, and index.

HOLLERITH

Hollerith is a 12-bit code used to encode data from those computer cards which threaten you with a fate worse than death if you "fold, spindle, or mutilate" them. Figure 1-7*b* shows a standard 12-row by 80-column card. The 12 data rows are referenced as, starting from the top, 12, 11, 0, 1, 2, 3, 4, 5, 6, 7, 8, 9. The top three rows are called *zone punches* and the bottom 10 rows are called *digit punches*. Note that the zero row is included in both categories. A punched hole represents a 1 and a data word is described by the 12 bits in a vertical column. The card in Figure 1-7*b* shows the Hollerith code for the numbers and letters printed across the top of the card. Table 1-3 shows the entire code and the punched-hole equivalent for each character. Since Hollerith code uses very few of the possible combinations for 12 bits, it is not very efficient. Therefore, it is usually converted to ASCII or EBCDIC for use.

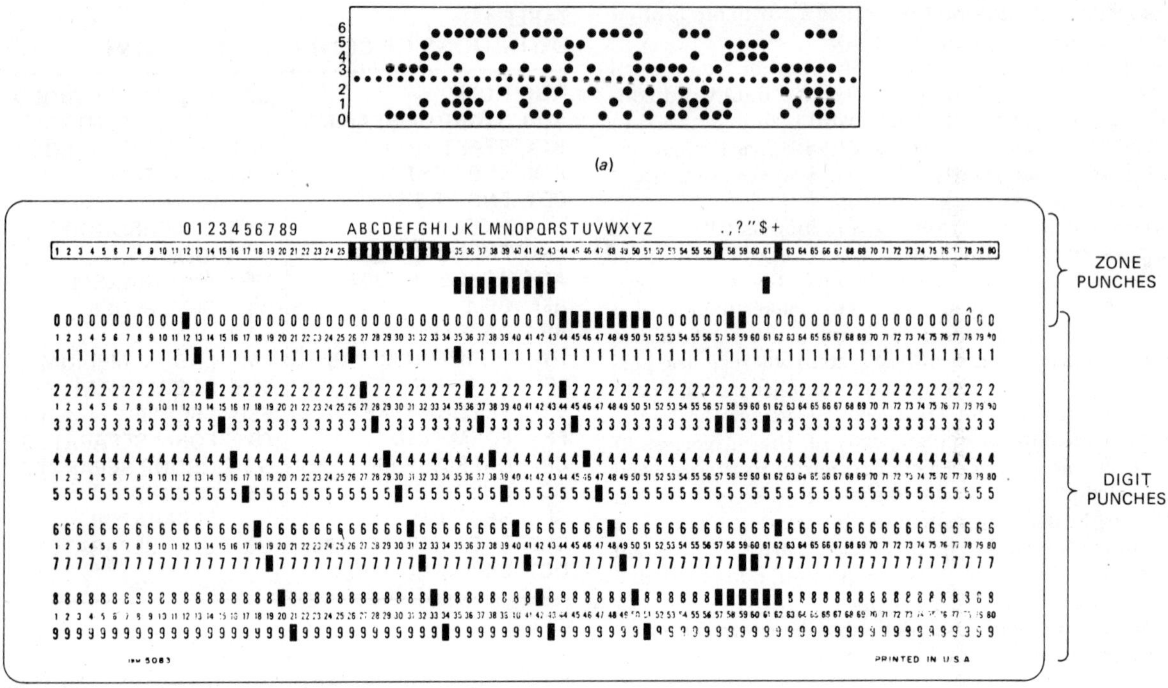

FIGURE 1-7 *(a)* ASCII punched paper tape; *(b)* Hollerith punched card.

ADDING AND SUBTRACTING BINARY, OCTAL, HEX, AND BCD NUMBERS

The previous section of this chapter reviewed common number systems and codes used with computers. This section reviews how to do computations in the previously described number systems.

Binary

ADDITION

Figure 1-8*a* shows the truth table for addition of two binary digits and carry in (C_{IN}) from addition of previous digits. Figure 1-8*b* shows the result of adding two 8-bit binary numbers together using these rules. Assuming that $C_{IN} = 1$, $1 + 0 + C_{IN}$ = a sum of 0 and a carry into the next digit, and $1 + 1 + C_{IN}$ = a sum of 1 and a carry into the next digit because the result in any digit position can only be a one or a zero.

2's COMPLEMENT BINARY

2's complement binary is a way of representing negative numbers in binary. When you handwrite a number which represents some physical quantity such as temperature, you can simply put a + sign in front of the number when you wish to indicate that the number is positive. You can write a − sign when you wish to indicate that the number is negative. If however, you want to store values such as temperatures, which can be positive or negative in a computer memory, there is a problem. Since the computer memory can only store 1's or

INPUTS			OUTPUTS	
A	B	C_{IN}	S	C_{OUT}
0	0	0	0	0
0	0	1	1	0
0	1	0	1	0
0	1	1	0	1
1	0	0	1	0
1	0	1	0	1
1	1	0	0	1
1	1	1	1	1

$S = A \oplus B \oplus C_{IN}$

$C_{OUT} = A \cdot B + C_{IN}(A \oplus B)$

(*a*)

```
    10011010
  + 11011100
 [1] 01110110
  ↑
  └─ Carry
```

(*b*)

FIGURE 1-8 Binary addition. *(a)* Truth table for 2 bits plus carry. *(b)* Addition of two 8-bit words.

0's, some way must be established to represent the sign of the number with a 1 or a 0.

The way to do this is to reserve the most significant bit of the data word as a *sign bit* and to use the rest of the bits of the data word to represent the size (magnitude) of the quantity. A computer that works with 8-bit words will use the MSB (bit 7) as the sign bit and the lower seven bits to represent the magnitude for the numbers. The usual convention is to represent a positive number with a 0 sign bit and a negative number with a 1 sign bit.

To make computations with signed numbers easier, the magnitude of negative numbers is represented in a special form called *2's complement*. The 2's complement of a binary number is formed by inverting each bit of the data word and adding one to the result. Some examples should help clarify all of this.

The number $+7_{10}$ is represented in 8-bit sign-and-magnitude form as 0000 0111. The sign bit is zero, which indicates a positive number. The magnitude of positive numbers is represented in straight binary, so 0000 0111 in the least-significant bits represents 7_{10}.

To represent -7_{10} in 8-bit 2's complement sign-and-magnitude form, start with the 8-bit code for +7, 0000 0111. Invert each bit to get 1111 1000. Then add 1 to get 1111 1001. This result is the correct representation of -7_{10}. Figure 1-9 shows some more examples of positive and negative numbers expressed in 8-bit sign-and-magnitude form. For practice, try generating each of these yourself to see if you get the same result as shown.

To reverse the above procedure and find the magnitude of a number expressed in sign-and-magnitude form, proceed as follows. If the number is positive, as indicated by the sign bit being a 0, then the least-significant 7 bits represent the magnitude directly in binary. If the number is negative, as indicated by the sign bit being a 1, then the magnitude is expressed in 2's complement. To get the magnitude of this negative number expressed in standard binary, invert each bit of the data word, including the sign bit and add one to the result. For example, given the word 1110 1011, invert each bit to get 0001 0100. Then add 1 to get 0001 0101. This equals 21_{10} so you know that the original numbers represent -21_{10}. Again, try reconverting a few of the numbers in Figure 1-9 for practice.

	Sign bit		
+ 7	0	0000111	
+ 46	0	0101110	
+105	0	1101001	
− 12	1	1110100	Sign and two's complement of magnitude
− 54	1	1001010	
−117	1	0001011	
− 46	1	1010010	

FIGURE 1-9 Positive and negative numbers represented with a sign bit and 2's complement.

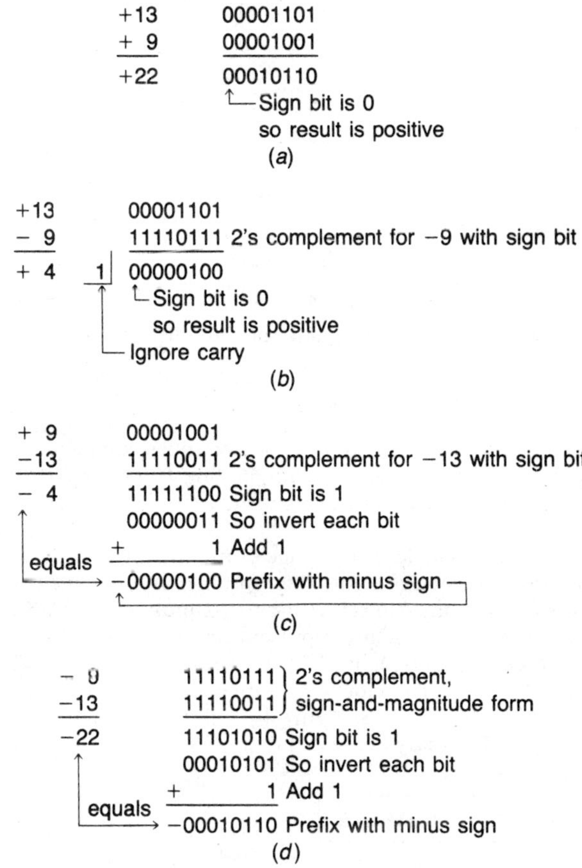

+13 00001101
+ 9 00001001
+22 00010110
Sign bit is 0 so result is positive
(a)
+13 00001101
− 9 11110111 2's complement for −9 with sign bit
+ 4 1 00000100
Sign bit is 0 so result is positive
Ignore carry
(b)
+ 9 00001001
−13 11110011 2's complement for −13 with sign bit
− 4 11111100 Sign bit is 1
00000011 So invert each bit
+ 1 Add 1
equals
−00000100 Prefix with minus sign
(c)
− 9 11110111
−13 11110011
2's complement, sign-and-magnitude form
−22 11101010 Sign bit is 1
00010101 So invert each bit
+ 1 Add 1
equals
−00010110 Prefix with minus sign
(d)

FIGURE 1-10 Addition of signed binary numbers. *(a)* +9 and +13. *(b)* −9 and +13. *(c)* +9 and −13. *(d)* −9 and −13.

Figure 1-10 shows some examples of addition of signed binary numbers of this type. Sign bits are added together just as the other bits are. Figure 1-10*a* shows the results of adding two positive numbers. The sign bit of the result is zero, so the result is positive. The second example, in Figure 1-10*b*, adds a −9 to a +13 or, in effect, subtracts 9 from 13. As indicated by the zero sign bit, the result of this, 4, is positive and in true binary form.

Figure 1-10*c* shows the result of adding a −13 to a smaller positive number, +9. The sign bit of the result is a 1. This indicates that the result is negative and the magnitude is in 2's complement form. To reconvert a 2's complement result to a signed number in true binary form:

1. Invert each bit, to produce 1's complement.
2. Add one.
3. Put a minus sign in front to indicate that the result is negative.

The final example in Figure 1-10*d* shows the results of adding two negative numbers. The sign bit of the result

01111111	+127
•	
•	
•	
00000001	+1
00000000	ZERO
11111111	−1
•	
•	
•	
10000001	−127
10000000	−128

FIGURE 1-11 Range of signed numbers that can be represented with 8 binary bits.

is a 1, and the result is negative and in 2's complement form. Again, inverting each bit, adding 1, and prefixing a minus sign will put the result in a more recognizable form.

Now let's consider the range of numbers that can be represented with eight bits in sign-and-magnitude form. Eight bits can represent a maximum of 2^8 or 256 numbers. Since we are representing both positive and negative numbers, half of this range will be positive and half negative. Therefore, the range then is 0 to +127 and from −1 to −128. Figure 1-11 shows the sign-and-magnitude binary representations for these values. If you like number patterns, you might notice that this scheme shifts the normal codes for 128 to 255 downward to represent −128 to −1.

If a computer is storing signed numbers as 16-bit words, then a much larger range of numbers can be represented. Since 16 bits gives 2^{16} or 65,536 possible values, the range for 16-bit sign-and-magnitude numbers is −32,768 to +32,767. Operations with 16-bit sign-and-magnitude numbers are done the same as was demonstrated above for 8-bit sign-and-magnitude numbers.

BINARY SUBTRACTION

There are two common methods for doing binary subtraction. These are the pencil method and the 2's complement add method. Figure 1-12*a* shows the truth table for binary subtraction of two binary digits A and B. Also included in the truth table is the effect of a borrow in, B_{IN}, from subtracting previous digits. Figure 1-12*b* shows an example of the "pencil" method of subtracting two 8-bit numbers. Using the truth table, this method is done the same way that you do decimal subtraction.

A second method of performing binary subtraction is by adding the 2's complement representation of the bottom number (subtrahend) to the top number (minuend). Figure 1-12*c* shows how this is done. First represent the top number in sign-and-magnitude form. Then form the 2's complement sign-and-magnitude represen-

INPUTS			OUTPUTS	
A	B	B_{IN}	D	B_{OUT}
0	0	0	0	0
0	0	1	1	1
0	1	0	1	1
0	1	1	0	1
1	0	0	1	0
1	0	1	0	0
1	1	0	0	0
1	1	1	1	1

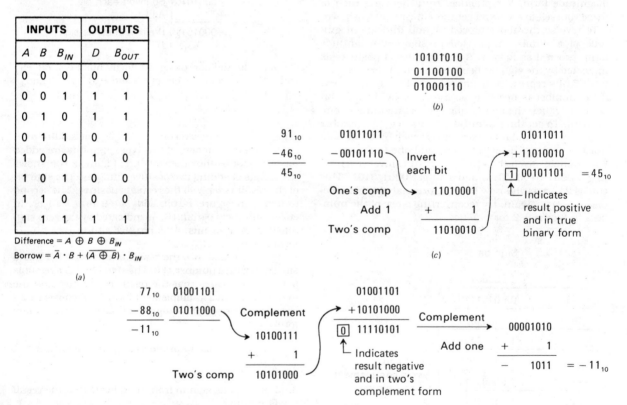

Difference $= A \oplus B \oplus B_{IN}$

Borrow $= \bar{A} \cdot B + (\overline{A \oplus B}) \cdot B_{IN}$

(*a*)

10101010
01100100
01000110

(*b*)

91_{10} − 46_{10} = 45_{10}

01011011
−00101110 → Invert each bit
One's comp 11010001
Add 1 + 1
Two's comp 11010010

01011011
+11010010
[1] 00101101 $= 45_{10}$
↑ Indicates result positive and in true binary form

(*c*)

77_{10} − 88_{10} = -11_{10}

01001101
01011000 → Complement → 10100111
+ 1
Two's comp 10101000

01001101
+10101000
[0] 11110101
↑ Indicates result negative and in two's complement form

Complement → 00001010
Add one + 1
− 1011 $= -11_{10}$

(*d*)

FIGURE 1-12 Binary subtraction. (*a*) Truth table for 2 bits and borrow. (*b*) Pencil method. (*c*) 2's complement positive result. (*d*) 2's complement negative result.

tation for the negative of the bottom number. Finally, add the two parts formed. For the example in Figure 1-12*c*, the sign of the result is a zero which indicates the result is positive and in true form. The final carry produced by the addition can be ignored. Figure 1-12*d* shows another example of this method of subtraction. In this case the bottom number is larger than the top number. Again, represent the top number in sign-and-magnitude form, produce the 2's complement sign-and-magnitude form for the negative of the bottom number, and add the two together. The sign bit of the result is a 1 for this example. This indicates that the result is negative and its magnitude is represented in 2's complement form. To get the result into a form that is more recognizable to you, invert each bit of the result, add 1 to it, and put a minus sign in front of it as shown in Figure 1-12*d*.

The examples shown use eight bits, but the process works for any number of bits. This method may seem awkward, but it is easy to do in a computer or microprocessor because it requires only the simple operations of inverting and adding.

BINARY MULTIPLICATION

There are several methods of doing binary multiplication. Figure 1-13 shows what is called the pencil method because it is the same as the way you learned to multiply decimal numbers. The top number or multiplicand is multiplied by the least-significant digit of the bottom number or multiplier. The partial product is written down. The top number is multiplied by the next digit of the multiplier. The resultant partial product is written down under the last, but shifted one place to the left. Adding all the partial products gives the total product. This method works well when doing multiplication by hand, but it is not practical for a computer because the type of shifts required make it awkward to implement.

One of the multiplication methods used by computers is repeated addition. To multiply 7 × 55, for example, the computer can just add up seven 55's. For large numbers, however, this method is slow. To multiply 786 × 253, for example, requires 252 add operations.

Most computers use an add-and-shift-right method. This method takes advantage of the fact that, for binary multiplication, the partial product can only be either the top number exactly if the multiplier digit is a 1, or a 0 if the multiplier digit is a 0. The method does the same thing as the pencil method except that the partial products are added as they are produced and the sum of the partial products is shifted right rather than each partial product being shifted left.

A point to note about multiplying numbers is the number of bits the product requires. For example, multiplying two 4-bit numbers can give a product with as many as 8 bits, and two 8-bit numbers can give a 16-bit product.

```
  11          1011   Multiplicand
×  9     ×    1001   Multiplier
----     ---------
  99          1011 ⎫
             0000  ⎪
            0000   ⎬ Partial products
           1011    ⎭
         ---------
           1100011   Product
```

FIGURE 1-13 Binary multiplication.

```
                01100     Quotient
Divisor   110 ) 1001000   Dividend          12
               -110                      6 ) 72
               ----
                 110
                -110
                ----
                   0
                         (a)

                110.01                    6.25
          100 ) 11001.00              4 ) 25.00
               -100
               ----
                 100
                -100
                ----
                  01 00
                         (b)
```

FIGURE 1-14 Binary division.

BINARY DIVISION

Binary division can also be performed in several ways. Figure 1-14 shows two examples of the pencil method. This is the same process as decimal long division. However, it is much simpler than decimal long division because the digits of the result (quotient) can only be 0 or 1. A division is attempted on part of the dividend. If this is not possible because the divisor is larger than that part of the dividend, a 0 is entered into the quotient. Another attempt is then made to divide using one more digit of the dividend. When a division is possible, a 1 is entered in the quotient. The divisor is then subtracted from the portion of the dividend used. The process is continued as with standard long division until all the dividend is used. As shown in Figure 1-14*b*, 0's can be added to the right of the binary point and division continued to convert a remainder to a binary equivalent.

Another method of division that is easier for computers and microprocessors to perform uses successive subtractions. The divisor is subtracted from the dividend and from each successive remainder until a borrow is produced. The desired quotient is 1 less than the number of subtractions needed to produce a borrow. This method is simple, but for large numbers it is slow.

For faster division of large numbers, computers use a subtract-and-shift-left method that is essentially the same process you go through with a pencil long division.

(a)			(b)
			Carry ↓
47_8		100 111	$^1 47_8$
$+36_8$	+	011 110	$+\ 36_8$
		1 000 101	$8_{10}\ 13_{10}$
		1 0 5_8	$1\ 0\ 5_8$

FIGURE 1-15 Octal addition. *(a)* Adding binary equivalents. *(b)* Direct octal addition.

Octal and Hexadecimal Addition and Subtraction

People working with computers or microprocessors often use octal or hexadecimal as a shorthand way of representing long binary numbers such as memory addresses. It is therefore useful to be able to add and subtract octal and hexadecimal numbers.

OCTAL ADDITION

Figure 1-15 shows two ways of adding the octal numbers 47 and 36. The first way is to convert both numbers to their binary equivalents. Remember, each octal digit represents three binary digits. These binary numbers are then added using the rules for binary addition from Figure 1-8*a*. The resultant binary sum is then converted back to octal.

The second method works directly with the octal form: 7 added to 6 gives 13, which is a carry to the next digit and a remainder of 5. The 5 is written down and the carry added to the next digit column. Then 4 plus 3 plus a carry gives 8, which is a carry with no remainder. The 0 is written down and the carry is added to the next digit column. This is the same process you use for decimal addition but a carry is produced any time the sum is 8 or greater, rather than 10.

HEXADECIMAL ADDITION

As shown in Figure 1-16 the same approaches can be used to add two hexadecimal numbers. For converting to binary, remember that each hex digit represents four binary digits. The binary numbers are added and the result is converted back to hexadecimal.

The second method works directly with the hex numbers. With hex addition, a carry is produced whenever

	(a)	(b)	
		Carry ↓	
7*A*	0111 1010	$7\ ^1$	A_{16}
+3*F*	+0011 1111	$+\ 3$	F_{16}
*B*9	1011 1001	11_{10}	25_{10}
		B_{16}	9_{16}

FIGURE 1-16 Hexadecimal addition.

the sum is 16 or greater. An A in hex is a 10 in decimal and an F is 15 in decimal. These add to give 25, which is a carry with a remainder of 9. The 9 is written down and the carry is added to the next digit column. Then 7 plus 3 plus a carry gives a decimal 11, or B in hex.

You may use whichever method seems easier to you and gives you consistently right answers. If you are doing a great deal of octal or hexadecimal arithmetic you might buy an electronic calculator specifically designed to do decimal, octal, and hexadecimal arithmetic.

OCTAL SUBTRACTION

Octal subtraction is shown in Figure 1-17. Since the least-significant digit of the top number is smaller than the least-significant digit of the bottom number, a borrow must be done. In octal subtraction, 8 is borrowed from the next digit position and added to the top number. The bottom number is then subtracted and the remainder written down. The process is continued until all digits are subtracted. If you are uncomfortable "borrowing 8's," you can just convert the number to decimal, subtract, and convert the result back to octal.

34_8	28_{10}
-17_8	-15_{10}
15_8	13_{10}

FIGURE 1-17 Octal subtraction.

HEXADECIMAL SUBTRACTION

Hexadecimal subtraction is similar to octal subtraction except that, when a borrow is needed, 16 is borrowed from the next-most-significant digit. Figure 1-18 shows this. It may help you to follow the example if you do partial conversions to decimal in your head. For example, 7 plus a borrowed 16 is 23. Subtracting B or 11 leaves 12 or C in hexadecimal. Then 3 from the 6 left after a borrow leaves 3, so the result is 3CH.

77_{16}	=	119_{10}
$-3B_{16}$	=	$-\ 59_{10}$
$3C_{16}$		60_{10}

FIGURE 1-18 Hexadecimal subtraction.

BCD Addition and Subtraction

In systems where the final result of a calculation is to be displayed, such as a calculator, it may be easier to work with numbers in a BCD format. These codes, as shown in Table 1-1, represent each decimal digit, 0 through 9, with a 4-bit binary word. The BCD words are the same as the binary equivalents for 0 through 9.

BCD ADDITION

BCD can have no digit-word with a value greater than 9. Therefore, a carry must be generated if the result of a

BCD addition is greater than 1001 or 9. Figure 1-19 shows three examples of BCD addition. The first, in Figure 1-19*a*, is very straightforward because the sum is less than 9. The result is the same as it would be for standard binary.

For the second example, in Figure 1-19*b*, adding BCD 7 to BCD 5 produces 1100. This is a correct binary result of 12 but it is an illegal BCD code. To convert the result to BCD format, a correction factor of 6 is added. The result of adding 6 is 0001 0010, which is the legal BCD code for 12.

Figure 1-19*c* shows another case where a correction factor must be added. The initial addition of 9 and 8 produces 0001 0001. Even though the lower four digits are less than 9, this is an incorrect BCD result because a carry out of bit 3 of the BCD digit-word was produced. This carry out of bit 3 is often called an *auxiliary carry*. Adding the correction factor of 6 gives the correct BCD result of 0001 0111 or 17.

To summarize, a correction factor of 6 must be added to the result if the result in the lower 4 bits is greater than 9 or if the initial addition produces a carry out of bit 3 of any BCD digit-word. This correction is sometimes called a *decimal adjust operation*.

The reason for the correction factor of 6 is that in BCD we want a carry into the next digit after 1001 or 9, but in binary a carry out of the lower four bits does not occur until after 1111 or 15, which is 6 more than 9.

	BCD	
35	0011 0101	
+23	+0010 0011	
58	0101 1000	

(*a*)

	BCD	
7	0111	
+ 5	+ 0101	
12	1100	Incorrect BCD
	+ 110	Add 6
	00010010	Correct BCD 12

(*b*)

	BCD	
9	1001	
+ 8	+ 1000	
17	00010001	Incorrect BCD
	110	Add 6
	00010111	Correct BCD 17

(*c*)

FIGURE 1-19 BCD addition. *(a)* No correction needed. *(b)* Correction needed because of illegal BCD result. *(c)* Correction needed because of carry out of BCD digit.

17	1 0111		
− 9	0 1001		
8	0 1110	Illegal BCD	
	− 110	Subtract 6	
	1000	$= 8_{10}$	

FIGURE 1-20 BCD subtraction.

BCD SUBTRACTION

Figure 1-20 shows a subtraction of BCD 17 (0001 0111) minus BCD 9 (0000 1001). The initial result, 0000 1110, is not a legal BCD number. Whenever this occurs in BCD subtraction, 6 must be *subtracted* from the initial result to produce the correct BCD result. For the example shown in Figure 1-20, subtracting 6 gives a correct BCD result of 0000 1000 or 8.

The correction factor of 6 must be subtracted from any BCD digit-word if that digit-word is greater than 1001, or if a borrow from the next higher digit occurred during the subtraction.

BASIC LOGIC GATES

Microcomputers such as those we discuss throughout this book often contain basic logic gates as "glue" between LSI (large scale integration) devices. For troubleshooting these systems it is important to be able to predict logic levels at any point directly from the schematic, rather than having to work your way through a truth table for each gate. This section should help refresh your memory of basic logic functions and help you remember how to quickly analyze logic gate circuits.

Inverting and Non-inverting Buffers

Figure 1-21 shows the schematic symbols and truth tables for simple buffers and logic gates. The first thing to remember about these symbols is that the shape of the symbol indicates the logic function performed by the device. The second thing to remember about these symbols is that a bubble or no bubble indicates the *assertion* level for an input or output signal. Let's review how modern logic designers use these symbols.

The first symbol for a *buffer* in Figure 1-21*a* has no bubbles on the input or output. Therefore, the input is active high and the output is active high. We read this symbol as follows. If the input, A, is asserted high, then the output, Y, will be asserted high. The rest of the truth table is covered by the assumption that if the A input is not asserted high, then the Y output will not be asserted high.

The next two symbols for a buffer each contain a bubble. The bubble on the output of the first of these indicates that the output is active low. The input has no bubble so it is active high. You can read the function of the device directly from the schematic symbol as follows. If the A input is asserted high, then the Y output will be

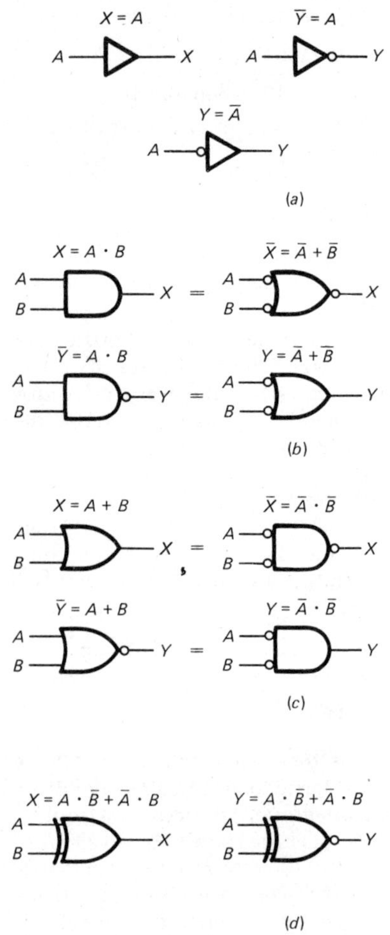

A	X	Y
0	0	1
1	1	0

A	B	X	Y
0	0	0	1
0	1	0	1
1	0	0	1
1	1	1	0

A	B	X	Y
0	0	0	1
0	1	1	0
1	0	1	0
1	1	1	0

A	B	X	Y
0	0	0	1
0	1	1	0
1	0	1	0
1	1	0	1

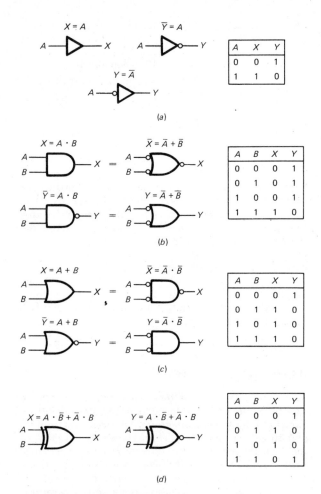

FIGURE 1-21 Buffers and logic gates. *(a)* Buffers. *(b)* AND—NAND. *(c)* OR—NOR. *(d)* Exclusive OR.

asserted low. This device then simply changes the assertion level of a signal. The output, Y, will always have a logic state which is the complement or inverse of that on the input, so the device is usually referred to as an *inverter*.

The second schematic symbol for an inverter in Figure 1-21*a* has the bubble on the input. We draw the symbol this way when we want to indicate that we are using the device to change an asserted-low signal to an asserted-high signal. For example, if we pass the signal $\overline{CS}$ through this device it becomes CS. The symbol tells you directly that if the input is asserted low, then the output will be asserted high. Now let's review how you express the functions of logic gates using this approach.

Logic Gates

Figure 1-21*b* shows the symbols and truth tables for simple logic gates. A symbol with a flat back and a round front indicates that the device performs the logical *AND* function. This means that the output will be asserted if the A input is asserted AND the B input is asserted. Again bubbles or no bubbles are used to indicate the assertion level of each input and output. The first AND symbol in Figure 1-21*b* has no bubbles so the inputs and the output are active high. The output then will be asserted high if the A input is asserted high AND the B input is asserted high. The bubble on the output of the second AND symbol in Figure 1-21*b* indicates that this device, commonly called a *NAND* gate, has an active low output. If the A input is asserted high and the B input is asserted high, then the Y output will be asserted low. Look at the truth table in Figure 1-21*b* to see if you agree with this.

Figure 1-21*c* shows the other two possible cases for the AND symbol. The first of these has bubbles on the inputs and on the outputs. If you see this symbol in a schematic, you should immediately see that the output will be asserted low if the A input is asserted low AND the B input is asserted low. The second AND symbol in Figure 1-21*c* has no bubble on the output, so the output will be asserted high if the A AND B inputs are both asserted low.

A logic symbol with a curved back indicates that the output of the device will be asserted if the A input is asserted OR the B input of the device is asserted. Again bubbles or no bubbles are used to indicate the assertion level for inputs and outputs. Note in Figure 1-21*b* and Figure 1-21*c* that each of the AND symbol forms has an equivalent OR symbol form. An AND symbol with active high inputs and an active high output, for example, represents the same device (a 74LS08 perhaps) as an OR symbol with active low inputs and an active low output. Use the truth table in Figure 1-21*b* to convince yourself of this. The bubbled-OR representation tells you that if one input is asserted low, the output will be low, regardless of the state on the other input. As we will show later in this chapter, this is often a useful way to think of the operation of an AND gate.

Figure 1-21*d* shows the symbol and truth table for an *exclusive* OR gate. The output of this device will be asserted if the A input is asserted OR if the B input is asserted, but the output will not be asserted if both A AND B are asserted.

You need to be able to read all of these symbols, because most logic designers will use the symbol that best describes the function they want a device to perform in a particular circuit.

Latches, Flip-flops, Registers, and Counters

THE D LATCH

A *latch* is a digital device that stores a one or a zero on its output. Figure 1-22*a* shows the schematic symbol and truth table for a D latch. The device functions as follows. If the *enable* input, E, is low, any data present on the D input will have no effect on the Q or $\overline{Q}$ outputs. This is indicated in the truth table by an *X* in the *D* column. If the enable input is high, a high or a low on the D input will be passed to the Q output. In other words, the Q output will follow the D input as long as the enable input is high. The $\overline{Q}$ output will contain the complement of the logic state on Q. When the enable input is made low again, the state on Q at that time will be latched there. Any changes on D will have no effect on Q until the enable input is made high again. When the enable input goes low, then, the state present on D just before the enable goes low will be stored on the Q output. Keep this operation in mind as you read about the D flip-flop in the next section.

THE D FLIP-FLOP

The first type of *flip-flop* to review is the D type. Figure 1-22*b* shows the schematic symbol and the truth table for a typical D flip-flop. Note that this device has a *clock* input, CK, in place of the enable input on the D latch. Also note the up arrows in the clock column of the truth table. These arrows are used to indicate that a one or zero on the D input will be copied to the Q output at the instant the clock input goes from low to high. In other words, the D flip-flop takes a snapshot of whatever state is on the D input when the clock goes high, and displays the photo on the Q output. If the clock input is low, a change on D will have no effect on the output. Likewise, if the clock input is high, a change on D will have no effect on the Q output. Contrast this operation with that of the D latch to make sure you understand the difference between the two devices.

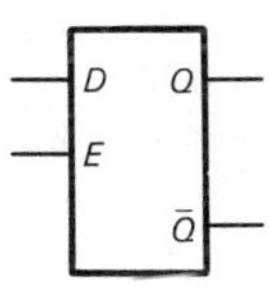

D	E	Q	Q̄
X	0	Q_N	$\bar{Q}_N$
0	1	0	1
1	1	1	0

(*a*)

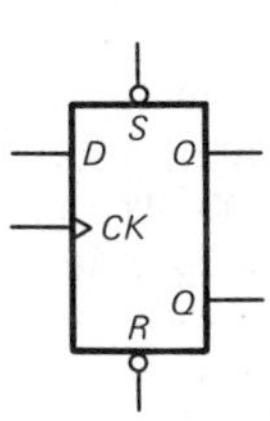

S	R	D	CK	Q	Q̄
1	1	1	↑	1	0
1	1	0	↑	0	1
1	1	X	0	Q_N	$\bar{Q}_N$
1	1	X	1	Q_N	$\bar{Q}_N$
0	1	X	X	1	0
1	0	X	X	0	1

(*b*)

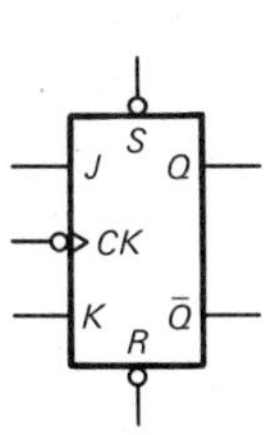

J	K	S	R	CK	Q	Q̄
0	0	1	1	↓	Q_N	$\bar{Q}_N$
0	1	1	1	↓	0	1
1	0	1	1	↓	1	0
1	1	1	1	↓	TOGGLE	
X	X	1	1	0	Q_N	$\bar{Q}_N$
X	X	1	1	1	Q_N	$\bar{Q}_N$
X	X	0	1	X	1	0
X	X	1	0	X	0	1

(*c*)

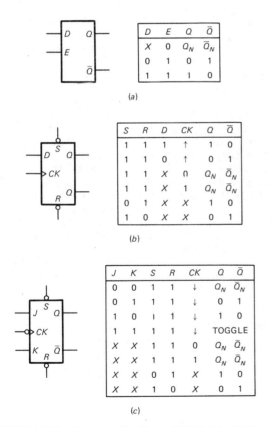

FIGURE 1-22 Latches and flip-flops. *(a)* **D** latch. *(b)* **D** flip-flop. *(c)* **J-K** flip-flop.

The D flip-flop in Figure 1-22*b* also has direct *set* (S) and *reset* (R) inputs. A flip-flop is considered *set* if its Q output is a one. It is *reset* if its Q output is a zero. The bubbles on the set and reset inputs tell you that these inputs are active low. The truth table for the D flip-flop in Figure 1-22*b* indicates that the set and reset inputs are *asynchonous*. This means that if the set input is asserted low, the output will be set, regardless of the state on the D and the clock inputs. Likewise, if the reset input is asserted low, the Q output will be reset, regardless of the state of the D and clock inputs. The *X*s in the *D* and *CK* columns of the truth table remind you that these inputs are "don't cares" if set or reset is asserted. The condition indicated by the asterisks (*) is a nonstable condition; that is, it will not persist when reset or clear inputs return to their inactive (high) level.

THE JK FLIP-FLOP

Figure 1-22*c* shows the schematic symbol and the truth table for a common JK flip-flop such as the 74LS76. The two data inputs, J and K make this device more versatile than a D flip-flop. The bubble on the clock input of the symbol and the downward arrows in the truth table indicate that the Q and Q̄ outputs will only change when the clock input goes from a high to a low. Changes on J or K will have no effect on the output if the clock input is low or if the clock input is high.

If J and K are both low when the CK input goes low, the outputs will remain the same as they were before the clock edge. This is indicated by Q_N and $\bar{Q}_N$ in the truth table. If J is low and K is high at the time of the clock edge, Q will become a zero. If J is high and K is low at the time of the clock edge, Q will become a one. If J and K are both high at the time of the clock edge, the Q output will *toggle*. This means that it will change to the opposite state of what it was before the clock edge. The JK flip-flop also has asynchronous set and reset inputs which function the same as those of the D flip-flop described previously.

REGISTERS

Flip-flops can be used individually or in groups to store binary data. A *register* is a group of D flip-flops connected in parallel as shown in Figure 1-23*a*. A binary word applied to the data inputs of this register will be transferred to the Q outputs when the clock input is made high. The binary word will remain stored on the Q outputs until a new binary word is applied to the D inputs and a low-to-high signal applied to the clock input. Other circuitry can read the stored binary word from the Q outputs at any time without changing its value.

If the Q output of each flip-flop in the register is connected to the D input of the next as shown in Figure 1-23*b*, then the register will function as a *shift register*. A one applied to the first D input will be shifted to the first Q output by a clock pulse. The next clock pulse will shift this one to the output of the second flip-flop. Each additional clock pulse will shift the one to the next flip-flop in the register. Some shift registers allow you to load a binary word into the register and shift the loaded

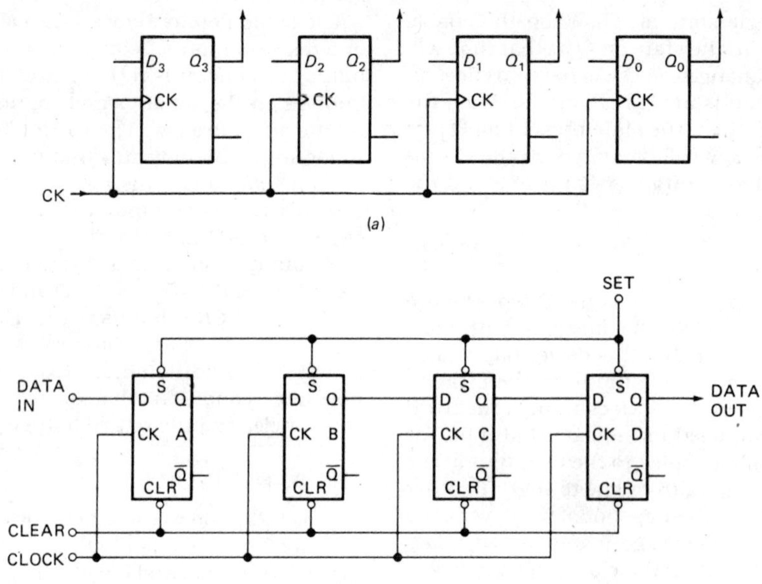

(b)

FIGURE 1-23 Registers. *(a)* Simple data storage. *(b)* Shift register.

word left or right when the register is clocked. As we will show later in this chapter, the ability to shift binary numbers is very useful.

COUNTERS

Flip-flops can also be connected in parallel to make *counters*. Figure 1-24 shows a schematic symbol and count sequence for a presettable 4-bit binary counter. The main point we want to review here is how a presettable counter functions, so there is no need to go into the internal circuitry of the device. If the reset input is asserted, the Q outputs will all be made zeros. After the reset signal is unasserted, each clock pulse will cause the binary count on the outputs to be incremented by one. As shown in Figure 1-24*b*, the count sequence will go from 0000 to 1111. If the outputs are at 1111, then the next clock pulse will cause the outputs to "roll over" to 0000 and a carry pulse to be sent out the carry output. This carry pulse can be used as the clock input for another counter.

Now, suppose that we want the counter to start counting from some number other than 0000. We can do this by applying the desired number to the four data inputs and asserting the load input. For example if we apply a binary 6, 0110, to the data inputs and assert the load input, this value will be transferred to the Q outputs. After the load signal is unasserted, the next clock signal will increment the Q outputs to 0111 or 7. Counters such as this can be connected in series (cascaded) to produce counters of any desired number of bits.

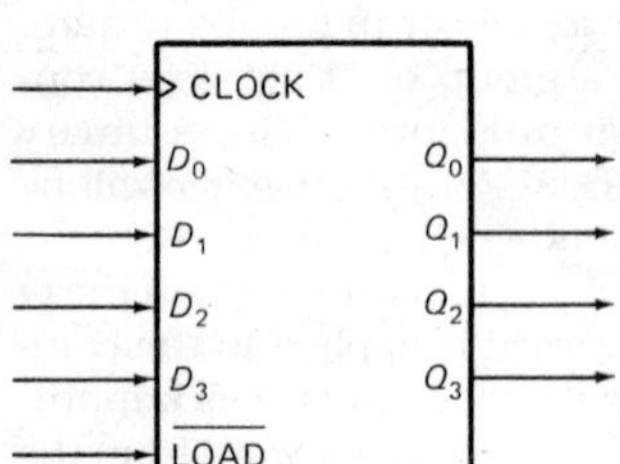

(a)

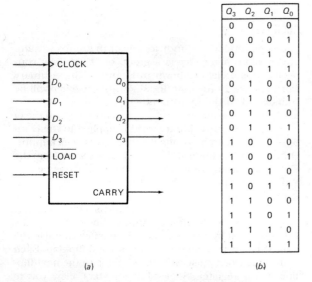

Q_3	Q_2	Q_1	Q_0
0	0	0	0
0	0	0	1
0	0	1	0
0	0	1	1
0	1	0	0
0	1	0	1
0	1	1	0
0	1	1	1
1	0	0	0
1	0	0	1
1	0	1	0
1	0	1	1
1	1	0	0
1	1	0	1
1	1	1	0
1	1	1	1

(b)

FIGURE 1-24 Four-bit, presettable binary counter. *(a)* Schematic symbol. *(b)* Count sequence.

ROMs, RAMs, and Buses

The next topics we need to review are the devices which store large numbers of binary words and how combinations of these devices can be connected together.

ROMs

The term ROM stands for *read-only memory*. There are several types of ROM that can be written to, read, erased, and written to with new data, but the main feature of ROMs is that they are *nonvolatile*. This means

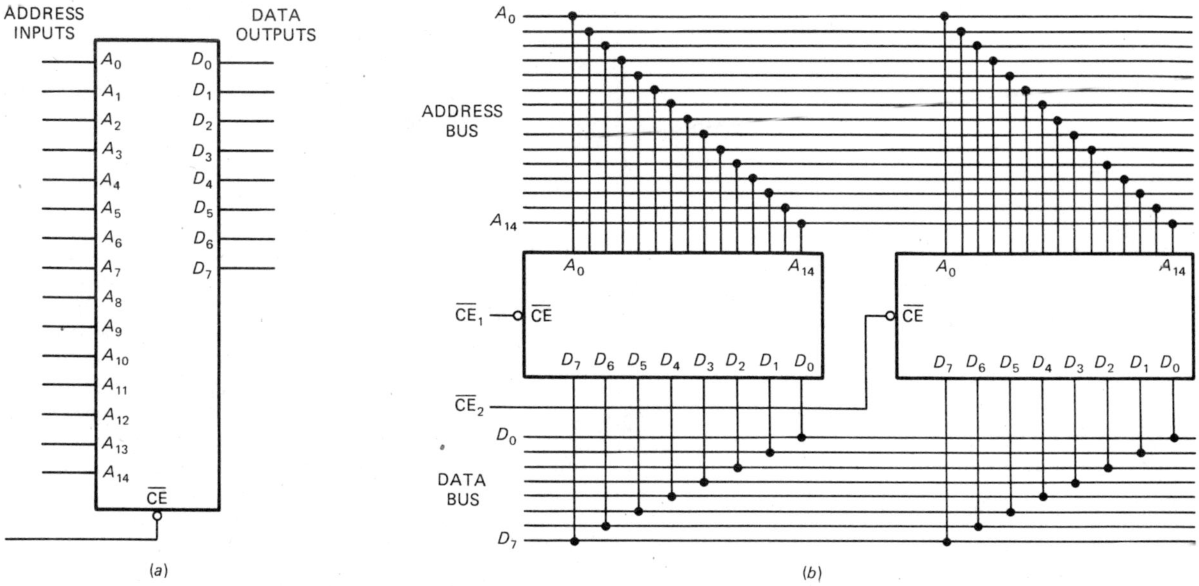

FIGURE 1-25 ROMs. *(a)* Schematic symbol. *(b)* Connection in parallel.

that the information stored in them is not lost when the power is removed from them.

Figure 1-25*a* shows the schematic symbol of a common ROM. As indicated by the eight *data* outputs, D0–D7, this ROM stores 8-bit data words. The data outputs are *three-state* outputs. This means that each output can be at a logic low state, a logic high state, or a high-impedance, floating state. In the high-impedance state an output is essentially disconnected from anything connected to it. If the $\overline{CE}$ input of the ROM is not asserted, then all of the outputs will be in the high-impedance state. Also, most ROMs switch to a lower-power-consumption condition if $\overline{CE}$ is not asserted. If the $\overline{CE}$ input is asserted, the device will be powered up, and the output buffers will be enabled. Therefore, the outputs will be at a normal logic low or logic high state. You will soon see why this is important if you don't happen to remember.

You can think of the binary words stored in the ROM as being in a long, numbered list. The number that corresponds to each stored word is called its *address*. In order to get a particular word onto the outputs of the ROM you have to do two things. You have to apply the address of that word to the address inputs, A0–A14, and you have to assert the $\overline{CE}$ input to turn on the outputs. Incidentally, you can tell the number of binary words stored in the ROM by the number of address inputs. The number of words is equal to 2^N where N is the number of address lines. The device in Figure 1-25*a* has 15 address lines, A0–A14, so the number of words is 2^{15} or 32,768. In a data sheet this device would be referred to as a 32K × 8 ROM. This means 32K addresses by 8 bits per address.

Now, let's see why we want three-state outputs on this ROM. Suppose that we want to store more than 32K data words. We can do this by connecting two or more ROMs in parallel as shown in Figure 1-25*b*. The address lines connect to each device to allow us to address one of the 32,768 words in each. A set of parallel lines used to send addresses or data to several devices in this way is called a *bus*. The data outputs of the ROMs are likewise connected in parallel so that any one of the ROMs can output data on the common data bus. If these ROMs had standard two state outputs, a serious problem would occur because each device would be trying to output an addressed word onto the data bus. The resulting argument between data outputs would probably destroy some of the outputs and give meaningless information on the data bus. Since the ROMs have three-state outputs, however, we can use external circuitry to make sure that only one ROM at a time has its outputs enabled. The very important principle here is that whenever several outputs are connected on a bus, the outputs should all be three-state, and only one set of outputs should be enabled at a time.

At the beginning of this section we mentioned that some ROMs can be erased and rewritten or reprogrammed with new data. Here's a summary of the different types of ROM.

Mask-programmed ROM—Programmed during manufacture; cannot be altered.

PROM—User programs by blowing fuses; cannot be altered except to blow additional fuses.

EPROM—Electrically programmable by user; erased by shining ultraviolet light on quartz window in package.

EEPROM—Electrically programmable by user; erased with electrical signals instead of ultraviolet light.

STATIC AND DYNAMIC RAMs

The name RAM stands for *random-access memory*, but since ROMs are also random access, the name probably

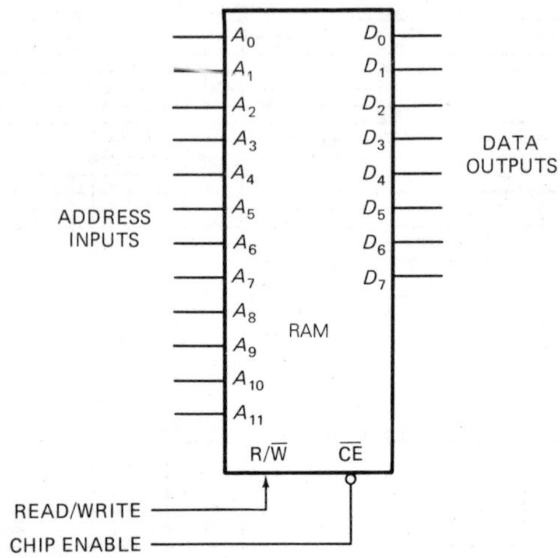

FIGURE 1-26 RAM schematic symbol.

should be read-write memory. RAMs are also used to store binary words. A *static* RAM is essentially a matrix of flip-flops. Therefore, we can write a new data word in a RAM location at any time by applying the word to the flip-flop data inputs and clocking the flip-flops. The stored data word will remain on the flip-flop outputs as long as the power is left on. This type of memory is *volatile* because data is lost when the power is turned off.

Figure 1-26 shows the schematic symbol for a common RAM. This RAM has 12 address lines, A0–A11, so it stores 2^{12} (4096) binary words. The eight data lines tell you that the RAM stores 8-bit words. When we are reading a word from the RAM these lines function as outputs. When we are writing a word to the RAM, these lines function as inputs. The *chip enable* input, $\overline{CE}$, is used to enable the device for a read or for a write. The R/$\overline{W}$ input will be asserted high if we want to read from the RAM or it will be asserted low if we want to write a word to the RAM. Here's how all these lines work for reading from and writing to the device.

To write to the RAM we apply the desired address to the address inputs, assert the $\overline{CE}$ input low to turn on the device, and assert the R/$\overline{W}$ input low to tell the RAM we want to write to it. We then apply the data word we want to store to the data lines of the RAM for a specified time. To read a word from the RAM we address the desired word, assert $\overline{CE}$ low to turn on the device, and assert R/$\overline{W}$ high to tell the RAM we want to read from it. For a read operation the output buffers on the data lines will be enabled and the addressed data word will be present on the outputs.

The static RAMs we have just reviewed store binary words in a matrix of flip-flops. In *dynamic* RAMs (DRAMs), binary 1's and 0's are stored as an electrical charge or no charge on a tiny capacitor. Since these tiny capacitors take up less space on a chip than a flip-flop would, a dynamic RAM chip can store many more bits than the same size static RAM chip. The disadvantage of dynamic RAMs is that the charge leaks off the tiny capacitors. The logic state stored in each capacitor must be *refreshed* every two milliseconds or so. A device called a *dynamic RAM refresh controller* can be used to refresh a large number of dynamic RAMs in a system. Some newer dynamic RAM devices contain built-in refresh circuitry so they appear static to external circuitry.

Arithmetic Logic Units

Previous sections of this chapter reviewed ANDing, ORing, exclusive ORing, adding, and subtracting of binary numbers. A device which can perform any of these functions and others on binary words is an *arithmetic logic* unit or ALU. Figure 1-27*a* shows a block diagram for the 74LS181 which is a 4-bit ALU. This device can perform any one of 16 logic functions or any one of 16 arithmetic functions on two 4-bit binary words. The function performed on the two words is determined by the logic level applied to the mode input, M, and by the 4-bit binary code applied to the select inputs, S0–S3.

Figure 1-27*b* shows the truth table for the 74LS181. In this truth table *A* represents the 4-bit binary word applied to the A0–A3 inputs and *B* represents the 4-bit binary word applied to the B0–B3 inputs. *F* represents the 4-bit binary word that will be produced on the F0–F3 outputs. If the mode input, M, is high, the device will perform one of 16 logic functions on the two words applied to the A and B inputs. For example, if M is high and we make S3 high, S2 low, S1 high, and S0 high, the 4-bit word on the A inputs will be ANDed with the 4-bit word on the B inputs. The result of this ANDing will appear on the F outputs. Each bit of the A word is ANDed with the corresponding bit of the B word to produce the result on F. Figure 1-27*c* shows an example of ANDing two words with this device. As you can see in this example, an output bit is high only if the corresponding bit is high in both the A word AND in the B word.

For another example of the operation of the 74LS181, suppose that the M input is high, S3 is high, S2 is high, S2 is high, and S0 is low. According to the truth table the device will now OR each bit in the A word with the corresponding bit in the B word and give the result on the corresponding F output. Figure 1-27*c* shows the result that will be produced by ORing two 4-bit words. Figure 1-27*c* also shows for your reference the result that would be produced by exclusive ORing these two 4-bit words together.

If the M input of the 74LS181 is low, then the device will perform one of 16 arithmetic functions on the A and B words. Again the result of the operation will be put on the F outputs. Several 74LS181s can be cascaded to operate on words longer than 4 bits. The ripple-carry input, $\overline{C}_n$, allows a carry from an operation on previous words to be included in the current operation. If the $\overline{C}_n$ input is asserted low, then a carry will be added to the results of the operation on A and B. For example if the M input is low, S3 is high, S2 is low, S1 is low, S0 is high, and $\overline{C}_n$ is low, the F outputs will have the sum of A plus B plus a carry.

The real importance of an ALU such as the 74LS181 is that it can be programmed with a binary instruction applied to its mode and select inputs to perform many different functions on two binary words applied to its

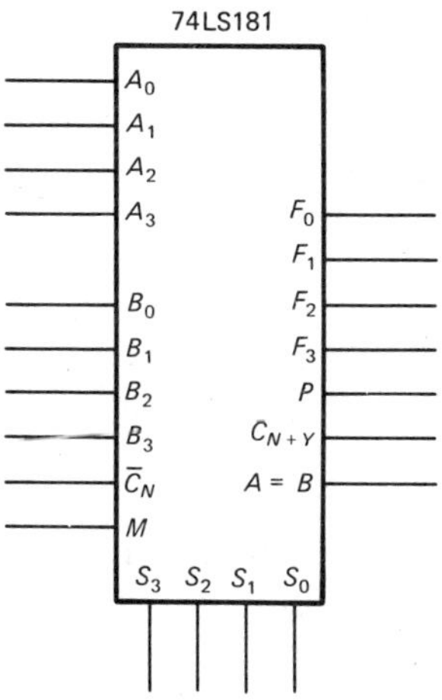

(a)

SELECTION				ACTIVE-HIGH DATA		
				$M = H$ LOGIC FUNCTIONS	$M = L$; ARITHMETIC OPERATIONS	
S3	S2	S1	S0		$\overline{C}_N = H$ (no carry)	$\overline{C}_N = L$ (with carry)
L	L	L	L	$F = \overline{A}$	$F = A$	$F = A$ PLUS 1
L	L	L	H	$F = \overline{A + B}$	$F = A + B$	$F = (A + B)$ PLUS 1
L	L	H	L	$F = \overline{A}B$	$F = A + \overline{B}$	$F = (A + \overline{B})$ PLUS 1
L	L	H	H	$F = 0$	$F =$ MINUS 1 (2's COMPL)	$F =$ ZERO
L	H	L	L	$F = \overline{AB}$	$F = A$ PLUS $A\overline{B}$	$F = A$ PLUS $A\overline{B}$ PLUS 1
L	H	L	H	$F = \overline{B}$	$F = (A + B)$ PLUS $A\overline{B}$	$F = (A + B)$ PLUS $A\overline{B}$ PLUS 1
L	H	H	L	$F = A \oplus B$	$F = A$ MINUS B MINUS 1	$F = A$ MINUS B
L	H	H	H	$F = A\overline{B}$	$F = A\overline{B}$ MINUS 1	$F = A\overline{B}$
H	L	L	L	$F = \overline{A} + B$	$F = A$ PLUS AB	$F = A$ PLUS AB PLUS 1
H	L	L	H	$F = \overline{A \oplus B}$	$F = A$ PLUS B	$F = A$ PLUS B PLUS 1
H	L	H	L	$F = B$	$F = (A + \overline{B})$ PLUS AB	$F = (A + \overline{B})$ PLUS AB PLUS 1
H	L	H	H	$F = AB$	$F = AB$ MINUS 1	$F = AB$
H	H	L	L	$F = 1$	$F = A$ PLUS A*	$F = A$ PLUS A PLUS 1
H	H	L	H	$F = A + \overline{B}$	$F = (A + B)$ PLUS A	$F = (A + B)$ PLUS A PLUS 1
H	H	H	L	$F = A + B$	$F = (A + \overline{B})$ PLUS A	$F = (A + \overline{B})$ PLUS A PLUS 1
H	H	H	H	$F = A$	$F = A$ MINUS 1	$F = A$

(b)

	A_3	A_2	A_1	A_0
A	1	0	1	0
	B_3	B_2	B_1	B_0
B	0	1	1	0
	F_3	F_2	F_1	F_0
$A \cdot B$	0	0	1	0
	F_3	F_2	F_1	F_0
$A + B$	1	1	1	0
	F_3	F_2	F_1	F_0
$A \oplus B$	1	1	0	0

(c)

FIGURE 1-27 Arithmetic logic unit (ALU). *(a)* Schematic symbol. *(b)* Truth table. *(c)* Sample AND, OR, XOR operations.

data inputs. In other words, instead of having to build a different circuit to perform each of these functions, we have one programmable device. We can perform any of the operations that we want in a computer with a sequence of simple operations such as those of the 74LS181. Therefore, an ALU is a very important part of the microprocessors and microcomputers which we discuss in the next chapter.

CHECKLIST OF IMPORTANT TERMS AND CONCEPTS IN THIS CHAPTER

If you do not remember any of the terms or concepts in this list, use the index to find them in the chapter.

Binary, bit, nibble, byte, word, double word

LSB, MSB, LSD, MSD

Octal, hexadecimal, standard BCD, Gray code

Seven-segment display code

Alphanumeric codes: ASCII, BCDIC, EBCDIC, Selectric, Hollerith

Parity bit, odd parity, even parity

Converting between binary, octal, hexadecimal, BCD

Arithmetic with binary, octal, hexadecimal, BCD

BCD decimal adjust operation

Signed numbers, sign bit

2's complement sign-and-magnitude form

Signal assertion level

Inverting and noninverting buffers

Symbols and truth tables for AND, NAND, OR, NOR, XOR logic gates.

D latch, D flip-flop, JK flip-flop

Register, shift register, binary counter

ROM: address lines, data lines, bus lines
  nonvolatile
  three-state
  cascaded outputs
  enable input

PROM, EPROM, EEPROM

RAM: static, dynamic
  volatile
  READ/WRITE input

ALU

REVIEW QUESTIONS AND PROBLEMS

1. Convert the following decimal numbers to binary:
   a. 22
   b. 76
   c. 500

2. Convert the following binary numbers to decimal:
   a. 1011
   b. 11010001
   c. 1110111001011001

3. Convert to following numbers to octal:
   a. 110101001 binary
   b. 11 decimal
   c. 111011101100 binary

4. Convert the following octal numbers to decimal:
   a. 314
   b. 74
   c. 43

5. Convert to hexadecimal:
   a. 53 decimal
   b. 756 decimal
   c. 01101100010 binary
   d. 11000010111 binary

6. Convert to decimal:
   a. D3H
   b. 3FEH
   c. 44H

7. Convert the following decimal numbers to BCD:
   a. 86
   b. 62
   c. 33

8. The L key is depressed on an ASCII-encoded keyboard. What pattern of 1's and 0's would you expect to find on the seven parallel data lines coming from the keyboard? What pattern would a carriage return, CR, give?

9. Define parity and describe how it is used to detect an error in transmitted data.

10. Show addition of:
    a. 10011_2 and 1011_2 in binary
    b. 37_{10} and 25_{10} in BCD
    c. 37_8 and 25_8 in octal
    d. 4AH and 77H

11. Express the following decimal numbers in 8-bit sign-and-magnitude form:
    a. +26
    b. −7
    c. −26
    d. −125

12. Show the subtraction, in binary, of the following decimal numbers using both the pencil method and the 2's complement addition method:
    a. 7 − 4
    b. 37 − 26
    c. 125 − 93

13. Show the multiplication of 1001 and 011 by the pencil method. Do the same for 11010 and 101.

14. Show the division of 1100100 by 1010 using the pencil method.

15. Perform the indicated operations on the following numbers:
    a. The octal numbers 27 + 16
    b. The octal numbers 132 − 45
    c. 3AH + 94H
    d. 17AH − 4CH
    e.
    $$\begin{array}{r} 0101\ 1001\ \text{BCD} \\ +\ \underline{0100\ 0010}\ \text{BCD} \end{array}$$
    f.
    $$\begin{array}{r} 0111\ 1001\ \text{BCD} \\ +\ \underline{0100\ 1001}\ \text{BCD} \end{array}$$
    g.
    $$\begin{array}{r} 0101\ 1001\ \text{BCD} \\ -\ \underline{0010\ 0110}\ \text{BCD} \end{array}$$
    h.
    $$\begin{array}{r} 0110\ 0111\ \text{BCD} \\ -\ \underline{0011\ 1001}\ \text{BCD} \end{array}$$

16. For the circuit in Figure 1-28
    a. Is the Y output active high or active low?
    b. Is the C signal active high or active low?
    c. What input conditions on A, B, and C will cause the Y output to be asserted?

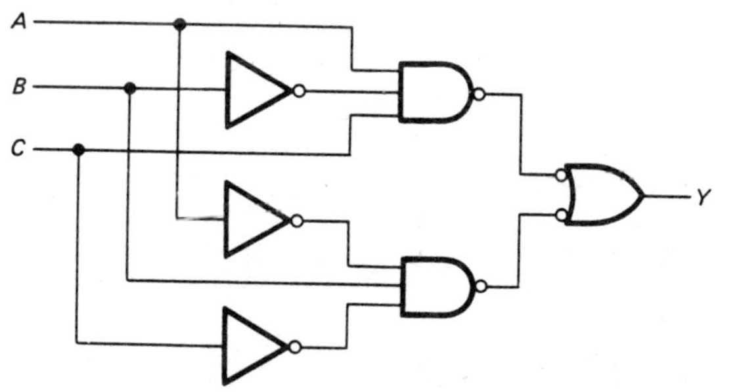

FIGURE 1-28 Circuit for Problem 1-16.

17. What is the main difference between a D latch and a D flip-flop?

18. The National Semiconductor INS8298 is a 65,536-bit ROM organized as 8192 words or bytes of 8 bits. How many address lines are required to address one of the 8192 bytes?

19. Why do most ROMs and RAMs have three-state outputs?

20. Using Figure 1-27, show the programming of the select and mode inputs the 74181 requires to perform the following arithmetic functions:

a. A + B
b. A − B − 1
c. A.B + A

21. Show the output word produced when the following binary words are ANDed with each other and when they are ORed with each other:

a. 1010 and 0111
b. 1011 and 1100
c. 11010111 and 111000
d. ANDing an 8-bit binary number with 1111 0000 is sometimes referred to as "masking" the lower 4 bits. Why?

CHAPTER

2 Computers, Microcomputers, and Microprocessors—An Introduction

We live in a computer oriented society and we are constantly bombarded with a multitude of terms relating to computers. Before getting started with the main flow of the book we will try to clarify some of these terms and to give an overview of computers and computer systems.

OBJECTIVES

At the conclusion of this chapter you should be able to:

1. Define the terms: microcomputer, microprocessor, hardware, software, firmware, time share, multitasking, distributed processing, and multiprocessing.
2. Describe how a microcomputer fetches and executes an instruction.
3. List the registers and other parts in the 8086/8088 execution unit and bus interface unit.
4. Describe the function of the 8086/8088 queue.
5. Demonstrate the way in which the 8086/8088 calculates memory addresses.

COMPUTERS

What is a Computer?

Figure 2-1 shows a block diagram for a simple computer. The major parts are the *central processing unit* or CPU, *memory*, and the *input and output* circuitry or I/O. Connecting these parts together are three sets of parallel lines called buses. The three buses are the *address bus*, the *data bus*, and the *control bus*.

MEMORY

The *memory* section usually consists of a mixture of RAM and ROM. It may also have magnetic floppy disks, magnetic hard disks, or laser optical disks. Memory has two purposes. The first purpose is to store the binary codes for the sequence of instructions you want the computer to carry out. When you write a computer program, what you are really doing is just writing a sequential list of instructions for the computer. The second purpose of the memory is to store the binary-coded data with which the computer is going to be working. This data might be the inventory records of a supermarket, for example.

INPUT/OUTPUT

The *input/output* or I/O section allows the computer to take in data from the outside world or send data to the outside world. Peripherals such as keyboards, video display terminals, printers, and modems are connected to the I/O section. These allow the user and the computer to communicate with each other. The actual physical devices used to interface the computer buses to external systems are often called *ports*. Ports in a computer function just as shipping ports do for a country. An *input port* allows data from a keyboard, an analog-to-digital (A/D) converter, or some other source to be read into the computer under control of the CPU. An *output port* is used to send data from the computer to some peripheral

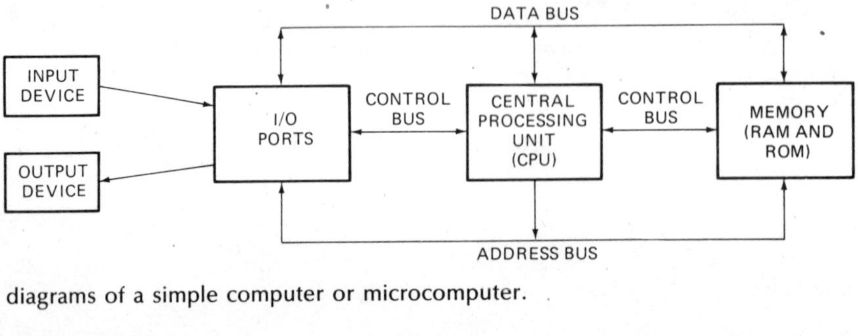

FIGURE 2-1 Block diagrams of a simple computer or microcomputer.

such as a video display terminal, a printer, or a digital-to-analog (D/A) converter. Physically, an input or output port is often just a set of parallel D flip-flops which let data pass through when they are enabled or clocked by a control signal from the CPU.

CENTRAL PROCESSING UNIT

The *central processing unit* or CPU controls the operation of the computer. It fetches binary-coded instructions from memory, decodes the instructions into a series of simple actions, and carries out these actions. The CPU contains an *arithmetic logic unit*, or ALU, which can perform add, subtract, OR, AND, invert, or exclusive-OR operations on binary words when instructed to do so. The CPU also contains an *address counter* which is used to hold the address of the next instruction or data to be fetched from memory, general-purpose registers which are used for temporary storage of binary data, and circuitry which generates the control bus signals.

ADDRESS BUS

The *address bus* consists of 16, 20, 24, or more parallel signal lines. On these lines the CPU sends out the address of the memory location that is to be written to or read from. The number of memory locations that the CPU can address is determined by the number of address lines. If the CPU has *N* address lines then it can directly address 2 to the *N* power memory locations. For example, a CPU with 16 address lines can address 2^{16} or 65,536 memory locations, a CPU with 20 address lines can address 2^{20} or 1,048,576 locations, and a CPU with 24 address lines can address 2^{24} or 16,777,216 locations. When the CPU reads data from or writes data to a port, the port address is also sent out on the address bus.

DATA BUS

The *data bus* consists of 8, 16, 32 or more parallel signal lines. As indicated by the double-ended arrows on the data bus line in Figure 2-1, the data bus lines are *bidirectional*. This means that the CPU can read data in on these lines from memory or from a port as well as send data out on these lines to a memory location or to a port. Many devices in a system will have their outputs connected to the data bus, but the outputs of only one device at a time will be enabled. Any device outputs connected on the data bus must be three-state so that they can be floated when the device is not in use.

CONTROL BUS

The *control bus* consists of 4–10 parallel signal lines. The CPU sends out signals on the control bus to enable the outputs of addressed memory devices or port devices. Typical control bus signals are *memory read*, *memory write*, *I/O read*, and *I/O write*. To read a byte of data from a memory location, for example, the CPU sends out the address of the desired byte on the address bus and then sends out a memory read signal on the control bus. The memory read signal enables the addressed memory device to output the byte of data onto the data bus where it is read by the CPU.

HARDWARE, SOFTWARE, AND FIRMWARE

When working around computers you hear the terms hardware, software, and firmware almost constantly. *Hardware* is the name given to the physical devices and circuitry of the computer. *Software* refers to the programs written for the computer. *Firmware* is the term given to programs stored in ROMs or in other devices which keep their stored information when the power is turned off.

Execution of a Three-Instruction Program

EXECUTION SEQUENCE

To give you a better idea of how the parts of a computer function together, we will now describe the actions a simple computer might go through to carry out (*execute*) a simple program. The three instructions of the program are:

1. Input a value from a keyboard connected to the port at address 05H.
2. Add 7 to the value read in.
3. Output the result to a display connected to the port at address 02H.

Figure 2-2*a* shows in diagram form the actions that the computer will perform to execute these three instructions.

For this example assume that the CPU fetches instructions and data from memory one byte at a time. Also assume that the binary codes for the instructions are in sequential memory locations starting at address 00100H. Figure 2-2*b* shows the binary codes that would be required in successive memory locations to execute this program on an 8086- or 8088-based microcomputer.

The first action a computer will do is to fetch the first instruction byte from memory. To do this the CPU sends out the address of the first instruction byte, in this case 00100H, to memory. This action is represented by line *1A* in Figure 2-2*a*. The CPU then sends out a memory read signal on the control bus (line *1B* in the figure). This causes the memory to output the first instruction byte (E4H) on the data bus as represented by line *1C*. The CPU reads in the byte from the data bus and *decodes* it. By decode we mean that the CPU determines from the binary code read in what actions it is supposed to take. In this case the CPU determines that the code read in represents an input instruction. Also from decoding this instruction byte, the CPU determines that it needs more information before it can carry out the instruction. The CPU must fetch from memory the input port address. To do this the CPU sends out the next sequential address (00101H) to memory as indicated by line *2A* in the figure. The CPU also sends out another memory read signal on the control bus (line *2B*). This enables the memory to put the addressed byte on the data bus (line *2C*). When the CPU reads in this second byte, 05H in this case, it has all the information it needs to execute the instruction.

FIGURE 2-2 *(a)* Execution of a three-step computer program. *(b)* Memory addresses and memory contents for a three-step program.

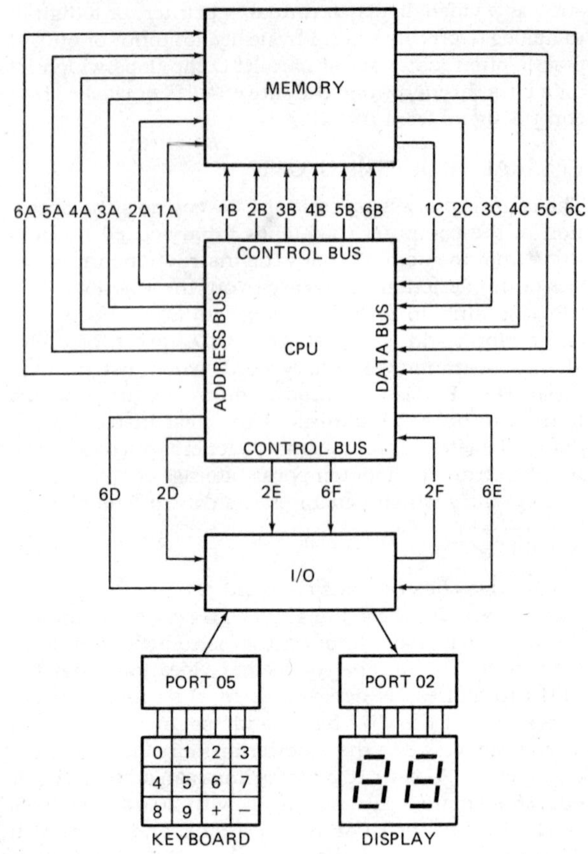

(a)

PROGRAM

1. Input a value from port 05.
2. Add 7 to this value.
3. Output the result to port 02.

SEQUENCE

1A CPU sends out address of first instruction to memory.
1B CPU sends out memory read control signal to enable memory.
1C Instruction byte sent from memory to CPU on data bus.
2A Address next memory location to get rest of instruction.
2B Send memory read control signal to enable memory.
2C Port address byte sent from memory to CPU on data bus.
2D CPU sends out port address on address bus.
2E CPU sends out input read control signal to enable port.
2F Data from port sent to CPU on data bus.
3A CPU sends address of next instruction to memory.
3B CPU sends memory read control signal to enable memory.
3C Instruction byte from memory sent to CPU on data bus.
4A CPU sends next address to memory to get rest of instruction.
4B CPU sends memory read control signal to enable memory.
4C Number 07H sent from memory to CPU on data bus.
5A CPU sends address of next instruction to memory.
5B CPU sends memory read control signal to enable memory.
5C Instruction byte from memory sent to CPU on data bus.
6A CPU sends out next address to get rest of instruction.
6B CPU sends out memory read control signal to enable memory.
6C Port address byte sent from memory to CPU on data bus.
6D CPU sends out port address on address bus.
6E CPU sends out data to port on data bus.
6F CPU sends out output write signal to enable port.

MEMORY ADDRESS	CONTENTS (BINARY)	CONTENTS (HEX)	OPERATION
00100H	11100100	E4	INPUT FROM
00101H	00000101	05	PORT 05H
00102H	00000100	04	ADD
00103H	00000111	07	07H
00104H	11100110	E6	OUTPUT TO
00105H	00000010	02	PORT 02

(*b*)

To execute the input instruction the CPU sends out the port address (05H) on the address bus (line *2D*) and sends out an I/O read signal on the control bus (line *2E*). The addressed port device then puts a byte of data on the data bus (line *2F*). The CPU reads in the byte of data and stores it in an internal register called the *accumulator*. This completes the first instruction.

Having completed the first instruction, the CPU must now fetch its next instruction from memory. To do this it sends out the next sequential address (00102H) on the address bus (line *3A*). The CPU then sends out a memory read signal on the control bus (line *3B*). This allows the memory to put the addressed byte (04H) on the data bus (line *3C*). The CPU reads in the instruction byte from the data bus and decodes it. From the instruction byte the CPU determines that it is supposed to add some number to the number stored in the accumulator. The CPU also determines from this instruction byte that it must go to memory again to get the number that it is supposed to add. To get the required byte, the CPU will send out the next sequential address (00103H) on the address bus (line *4A*) and a memory read signal on the control bus (line *4B*). The memory will then put the contents of the addressed byte (in this case the number 07H) on the data bus (line *4C*). The CPU will read in the byte on the data bus and add it to the contents of the accumulator as instructed. Assume the result of the addition is left in the accumulator. This completes the second instruction.

The CPU must now fetch its next instruction. To do this it sends out the next sequential address (00104H) on the address bus (line *5A*), sends out a memory read signal on the control bus (line *5B*), and reads in the addressed byte (E6H) from the data bus (line *5C*). From this byte the CPU determines that it is now supposed to do an output operation to a port. The CPU also determines that it must go to memory again to get the address of the port that it is supposed to output to. To do this it sends out the next sequential address (00105H) on the address bus (line *6A*), sends out a memory read signal on the control bus (line *6B*), and reads in the byte (02H) put on the data bus by the memory (line *6C*). The CPU now has all the information that it needs to execute the instruction. To output a data byte to a port, the CPU first sends out the address of the desired port on the address bus (line *6D*). Next it puts the data byte from the accumulator onto the data bus (line *6E*). The CPU then sends out an I/O write signal on the control bus (line *6F*). This signal enables the addressed output port device so the data from the data bus lines can pass through it. When the CPU removes the I/O write signal to proceed with the next instruction, the data output will remain latched on the output pins of the port device. Therefore, the computer does not have to keep outputting a value in order for it to remain there.

All of the steps described above may seem like a great deal of work just to input a value from a keyboard, add 7 to it, and output the result to a display. Even a simple computer, however, can run through all these steps in a few microseconds.

SUMMARY OF SIMPLE COMPUTER OPERATION

1. A simple computer CPU fetches instructions or reads data from memory (reads memory) by sending out an address on the address bus and a memory read signal on the control bus. The addressed instruction or data is sent from memory to the CPU on the data bus.
2. The CPU can write data in RAM by sending out an address on the address bus, sending out the data to be written on the data bus, and sending out a memory write signal on the control bus.
3. To read data from a port, the CPU sends the port address out on the address bus and sends an I/O read signal on the control bus. Data from the port comes into the CPU on the data bus.
4. To write data to a port, the CPU sends out the port address on the address bus, sends the data to be written to the port out on the data bus, and sends an I/O write signal out on the control bus.
5. A microcomputer fetches each program instruction in sequence, decodes the instruction, and executes it.

Types of Computers

MAINFRAMES

Computers come in a wide variety of sizes and capabilities. The largest and most powerful are often called *mainframes*. Mainframe computers may fill an entire room. They are designed to work at very high speeds with large data words, typically 64 bits or greater, and they have massive amounts of memory. Computers of this type are used for military defense control, business

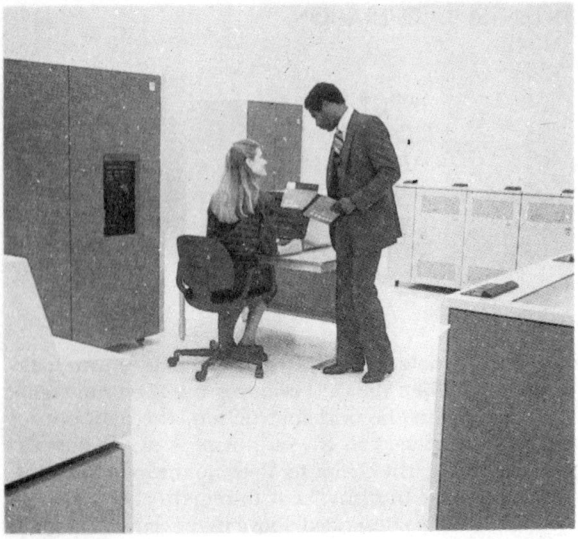

(a)

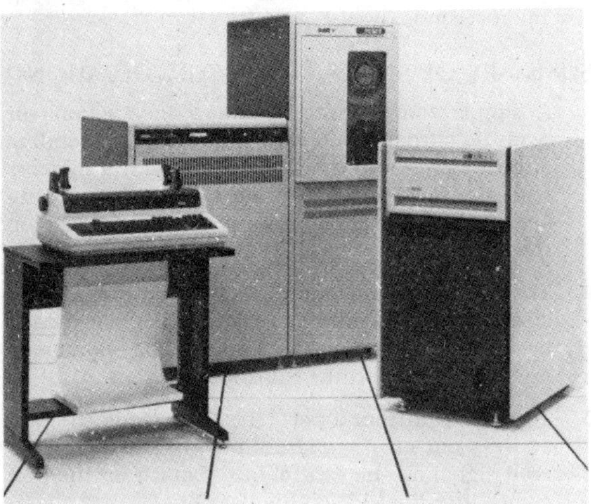

(b)

FIGURE 2-3 *(a)* Photograph of IBM mainframe computer. *(IBM Corp.)* *(b)* Photograph of DEC minicomputer. *(Digital Equipment Corp.)*

data processing (an insurance company, for example), and for creating computer graphics displays for science fiction movies. Examples of this type of computer are the IBM 4381, the Honeywell DPS8, and the CRAY X-MP/48. Figure 2-3*a* shows a photograph of an IBM 4381 mainframe.

MINICOMPUTERS

Scaled-down versions of mainframe computers are often called *minicomputers*. The main unit of a minicomputer usually fits in a single rack or box. A minicomputer runs more slowly, works directly with smaller data words (often 32-bit words), and does not have as much memory as a mainframe. Computers of this type are used for business data processing, industrial control (an oil refinery, for example), and scientific research. Examples of this type of computer are the Digital Equipment Corp. VAX 11/730 and the Data General MV/8000II. Figure 2-3*b* shows a photograph of a Digital Equipment Corp. VAX 11/730 minicomputer.

MICROCOMPUTERS

As the name implies, *microcomputers* are small computers. They range from small controllers that work directly with 4-bit words and can address a few thousand bytes of memory to larger units that work directly with 32-bit words and can address millions or billions of bytes of memory. Some of the more powerful microcomputers have all or most of the features of earlier minicomputers. Therefore, it has become very hard to draw a sharp line between these two types. One distinguishing feature of a microcomputer is that the CPU is usually a single integrated circuit called a *microprocessor*. Older books often used the terms microprocessor and microcomputer interchangeably, but actually the microprocessor is the CPU to which you add ROM, RAM, and ports to make a microcomputer. A later section in this chapter discusses the evolution of different types of microprocessors. Microcomputers are used in everything from smart sewing machines to computer-aided design systems. Examples of microcomputers are the Intel 8051 single-chip controller; the SDK-86, a single-board computer design kit; the IBM Personal Computer (PC); and the Apple Macintosh computer. Figure 2-4*a* shows a block diagram of the Intel 8051 single-chip microcontroller, Figure 2-4*b* shows the SDK-86 board, and Figure 2-4*c* shows the IBM PC. The purpose of this book is to teach you how microprocessors are connected with other components to build microcomputers, how the microcomputers are interfaced with peripheral components to build microcomputer systems, and how these systems are programmed. We use the IBM PC and the SDK-86 as example systems throughout this book. An available laboratory manual, written to accompany this book, shows you how to get started using the SDK-86 board and the IBM PC for assembly language programming.

SUMMARY OF IMPORTANT POINTS SO FAR

1. A computer or microcomputer consists of memory, a CPU, and some input/output circuitry.
2. These three parts are connected together by the address bus, the data bus, and the control bus.
3. The sequence of instructions, or program, for a computer is stored as binary numbers in successive memory locations.
4. The CPU fetches an instruction from memory, decodes the instruction to determine what actions must be done for the instruction, and carries out these actions.
5. Three types of computer are mainframes, minicomputers, and microcomputers.
6. The CPU in a microcomputer is called a microprocessor.

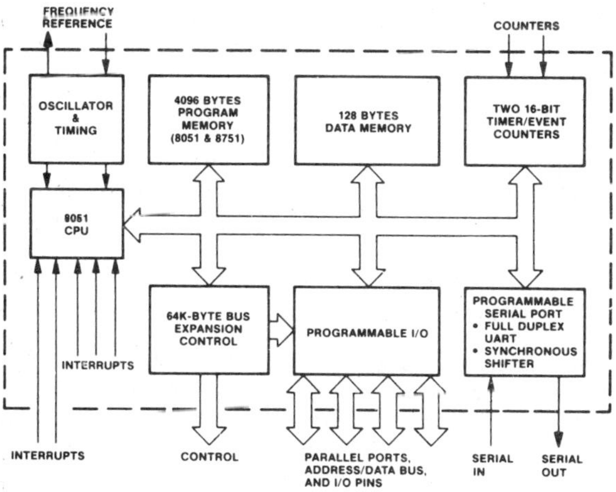

(a)

(b)

(c)

FIGURE 2-4 *(a)* Block diagram of Intel 8051 single chip microcomputer. *(Intel Corp.)* *(b)* Photograph of Intel SDK-86 board. *(Intel Corp.)* *(c)* Photograph of IBM PC. *(IBM Corp.)*

How Computers and Microcomputers are Used—An Example

The following sections are intended to give you an overview of how computers are interfaced with users to do useful work. These sections should help you understand many of the features designed into current microprocessors and where this book is heading.

COMPUTERIZING AN ELECTRONICS FACTORY—PROBLEM

Now, suppose that we want to "computerize" an electronics company. By this we mean that we want to make computer use available to as many people in the company as possible as cheaply as possible. We want the engineers to have access to a computer which can help them design circuits. People in the drafting department should have access to a computer which can be used for computer-aided drafting. The accounting department should have access to a computer for doing all of the financial bookkeeping. The warehouse should have access to a computer to help with inventory control. The manufacturing department should have access to a computer for controlling machines and testing finished products. The president, vice presidents, and supervisors should have access to a computer to help them with long range planning. Secretaries should have access to a computer for word processing. Sales people should have access to a computer to help them keep track of current pricing, product availability, and commissions. There are several ways to provide all the needed computer power. The next sections discuss some of the ways that are used to give people access to a computer.

BATCH PROCESSING

In the 1960s the available computers were very large and were kept in separate air-conditioned rooms. When programmers wanted to run their programs, they brought them to the computer room. Usually the program was in the form of a batch of punched cards. A computer operator would then run the program. A new programming job could not be started until the last one finished. Therefore, if a large job was being run, there might be a considerable wait before a programmer could get his or her job run. Also, if an error was found when the program ran, the programmer had to punch new cards, and either bribe the computer operator or put the corrected program cards on the bottom of the jobs-to-be-done pile. Needless to say, a system of this sort is not acceptable for computerizing our electronics company, because it only serves one user at a time and does not allow easy back-and-forth interaction between the computer and the user.

MULTIPROGRAMMING

An improvement over the basic batch system is a *multiprogramming* system. In this type of system several programs are put in the computer's memory at the same time. The computer runs one programming job until it reaches a point where it needs access to some slow peripheral device such as a printer. If the printer is not busy, the computer will print out the produced results. If the printer is busy, the data to be printed is stored on a magnetic disk. The computer can then start another programming job while it waits for the printer to become available. When the printer becomes available, the computer can print out the results from the first program, and then return to the second program. To further reduce the burden on the computer, some computers have separate circuitry that takes care of copying output data from magnetic disks to the printer. Multiprogramming improves the efficiency of the computer by keeping it busy more of the time, but it still does not allow the user to easily interact with the computer.

TIME-SHARE AND MULTITASKING SYSTEMS

A further improvement in computer access is *time-sharing*. Figure 2-5 shows a block diagram of one type of time-share system. Several video terminals are connected to the computer through direct wires or through telephone lines. The terminal can be on the user's desk or even in the user's home. The rate at which a user usually enters data is very slow as compared to the rate that a computer can process the data. Therefore, the computer can serve many users by dividing its time among them in small increments. In other words, the computer works on user #1's program for perhaps 20 milliseconds, works on user #2's program for 20 milliseconds, then works on user #3's program for 20 milliseconds, and so on until all all the users have had a turn. In a few milliseconds the computer will get back to user #1 again and repeat the cycle. To each user it will appear as if he or she has exclusive use of the computer because the computer processes data as fast as the user enters it. A time-share system such as this allows several users to interact with the computer at the same time. Each user can get information from or store information in the large memory attached to the computer. Each user can have an inexpensive printer attached to the terminal or can direct program or data output to a high-speed printer attached directly to the computer.

An airline ticket reservation computer might use a time-share system such as this to allow users from all over the country to access flight information and make reservations. A time-multiplexed or time-sliced system such as this can also allow a computer to control many machines or processes in a factory. A computer is much faster than the machines or processes. Therefore, it can check and adjust many pressures, temperatures, motor speeds, etc. before it needs to get back and recheck the first one. A system such as this is often called a *multitasking system* because it appears to be doing many tasks at the same time.

Now let's take another look at our problem of computerizing the electronics company. A time-share system seems to be a better idea than a batch system or even a multiprogramming system. We could put a powerful computer in some central location and run wires from it to video display terminals on users' desks. Each user could then run the program needed to do a particular task. The accountant can run a ledger program, the secretary can run a word processor program, etc. Each user can access the computer's large data memory. Incidentally, a large collection of data stored in a computer's

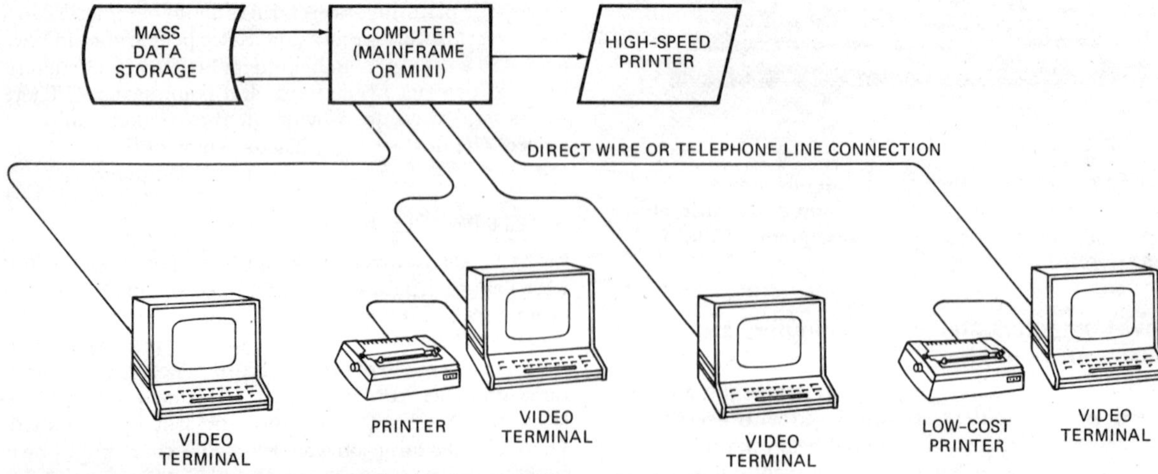

FIGURE 2-5 Block diagram of a computer time-share system.

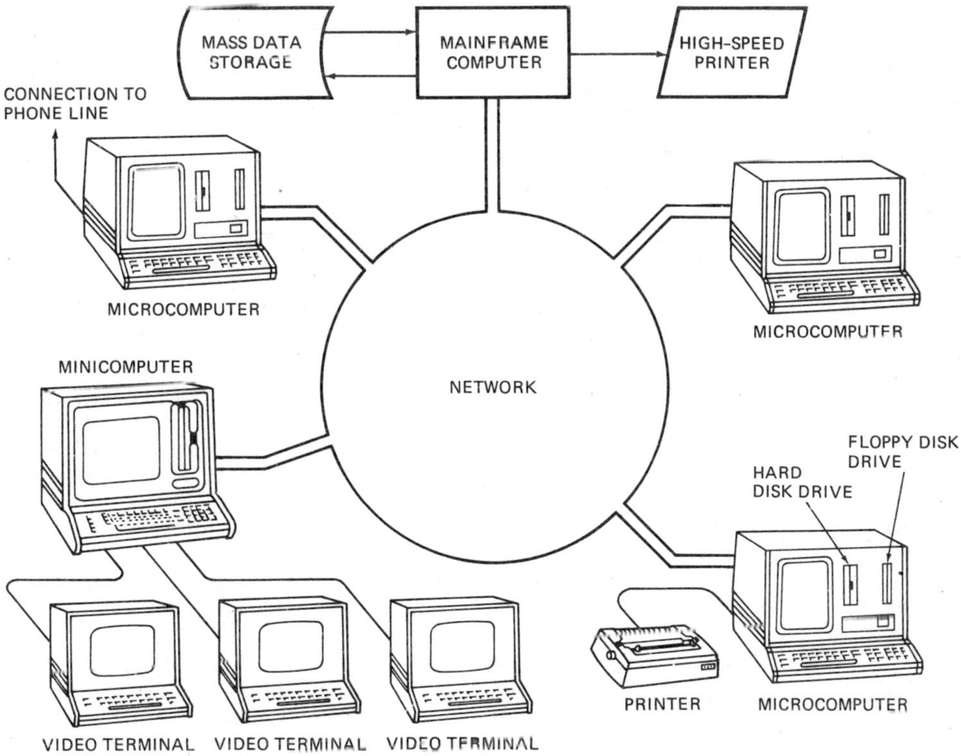

FIGURE 2-6 Block diagram of distributed processing computer system.

memory is often referred to as a *data base*. For a small company a system such as this might be adequate. However, there are at least two potential problems.

The first potential problem is "What happens if the computer is not working?" The answer to this question is that everything grinds to a halt. In a situation where people have become dependent upon the computer, not much gets done until the computer is up and running again. The old saying about putting all your eggs in one basket comes to mind here.

The second potential problem of the simple time share system is saturation. As the number of users increases, the time it takes the computer to do each user's task increases also. Eventually the computer's response time to each user becomes unreasonably long. People get very upset about the time they have to wait.

DISTRIBUTED PROCESSING OR MULTIPROCESSING

A partial solution for the two potential problems of a simple time-share system is to use a *distributed processor* system. Figure 2-6 shows a block diagram for such a system. The system has a powerful central computer with a large memory and a high-speed printer as does the simple time-share system decribed previously. However, in this system each user or group of users has a microcomputer instead of simply a video display terminal. In other words, each user station is an independent functioning microcomputer with a CPU, ROM, RAM, and probably magnetic or optical disk memory. This means that a person can do many tasks locally on the microcomputer without having to use the large computer at all. Since the microcomputers are connected to the large computer with a network, however, a user can access the computing power, memory, or other resources of the large computer when needed.

Distributing the processing around to multiple computers or processors in a system has several advantages. First, if the large computer goes down, the local microcomputers can continue working until they need to access the large computer for something. Second, the burden on the large computer is reduced greatly, because much of the computing is done by the local microcomputers. Finally, the distributed processor approach allows the system designer to use a local microcomputer best suited to the task it has to do.

COMPUTERIZED ELECTRONICS COMPANY OVERVIEW

Distributed processing seems to be the best way to go about computerizing our electronics factory. Engineers can each have a personal computer on their desk. With this they can use available programs to design and test circuits. They can access the large computer if they need data from its memory. Through the telephone lines, the engineer with a personal computer can access data in the memory of other computers all over the world. The drafting people can have personal computers for simple work, or large computer-aided design systems for more complex work. Completed work can be stored in the large computer memory. The accounting department can use personal computers with spread sheet programs to work with financial data kept in the memory of the large computer. The warehouse supervisor can likewise use a personal computer with an inventory program to keep personal records and those in the large computer's memory updated. Corporate officers can have personal computers tied into the network. They then can interact with any of the other systems on the network. Sales people can have portable personal computers that they can carry with them in the field. They can communicate with the main computer over the telephone lines using a modem. Secretaries doing word processing can use individual word processing units or personal computers. Since word processing is not a high intensity use for a computer, several video display terminals for word processing can be connected to a local microcomputer, and this local microcomputer can be connected to the large computer through the network. Users can also send messages to each other over the network. The specifics of a computer system such as this will obviously depend on the needs of the individual company for which the system is designed.

SUMMARY AND DIRECTION FROM HERE

The main concepts that you should take with you from this section are multiprogramming, time-sharing or multitasking, and distributed processing or multiprocessing. As you work your way through the rest of this book, keep an overview of the computerized electronics company in the back of your mind. The goal of this book is to teach you how all the parts of a system such as this work, how the parts are connected together, and how the system is programmed at different levels.

The first step toward this goal will be a quick look at the different types of microprocessors available. We then discuss a specific microprocessor, the Intel 8086, and the programming of a microcomputer built around a member of this microprocessor family, the IBM PC. Next we discuss the hardware connections and timing of this microcomputer. From there we show how the microcomputer is interfaced to a wide variety of peripheral devices. And finally we cycle back to our computerized electronics company, the networks it uses, and the system programs it requires.

Common Microprocessor Types

MICROPROCESSOR EVOLUTION

A common way of categorizing microprocessors is by the number of bits that their ALU can work with at a time. In other words, a microprocessor with a 4-bit ALU will be referred to as a 4-bit microprocessor, regardless of the number of address lines or the number of data bus lines that it has. The first microprocessor was the Intel 4004 produced in 1971. It contained 2300 PMOS transistors. The 4004 was a 4-bit device intended to be used with some other devices in making a calculator. Some logic designers, however, saw that this device could be used to replace PC boards full of combinational and sequential logic devices. Also, the ability to change the function of a system by just changing the programming, rather than redesigning the hardware, is very appealing. It was these factors that pushed the evolution of microprocessors.

In 1972 Intel came out with the 8008 which was capable of working with 8-bit words. The 8008, however, required 20 or more additional devices to form a functional CPU. In 1974 Intel announced the 8080, which had a much larger instruction set than the 8008 and only required two additional devices to form a functional CPU. Also, the 8080 used NMOS transistors, so it operated much faster than the 8008. The 8080 is referred to as a second-generation microprocessor.

Soon after Intel produced the 8080, Motorola came out with the MC6800, another 8-bit general-purpose CPU. The 6800 had the advantage that it required only a +5 V supply rather than the −5 V, +5 V, and +12 V supplies required by the 8080. For several years the 8080 and the 6800 were the top-selling 8-bit microprocessors. Some of their competitors were the MOS Technology 6502 used as the CPU in the Apple II microcomputer, and the Zilog Z80 used as the CPU in the Radio Shack TRS-80 microcomputer.

As designers found more and more applications for microprocessors, they pressured microprocessor manufacturers to develop devices with architectures and features optimized for doing certain types of tasks. In response to the expressed needs, microprocessors have evolved in three major directions during the last 10 years.

DEDICATED CONTROLLERS

One direction has been *dedicated controllers*. These devices are used to control "smart" machines such as microwave ovens, clothes washers, sewing machines, auto ignition systems, and metal lathes. Texas Instruments produced millions of their TMS-1000 family of 4-bit microprocessors for this type of application. In 1976 Intel introduced the 8048, which contains an 8-bit CPU, RAM, ROM, and some I/O ports all in one 40-pin package. Other manufacturers have followed with similar products. These devices are often referred to as *microcontrollers*. Some currently available devices in this category, the Intel 8051 and the Motorola MC6801, for example, contain programmable counters, a serial port (UART) as well as a CPU, ROM, RAM, and parallel I/O ports. A more recently introduced single-chip microcontroller, the Intel 8096, contains a 16-bit CPU, ROM, RAM, a UART, ports, timers, and a 10-bit analog-to-digital converter.

BIT-SLICE PROCESSORS

A second direction of microprocessor evolution has been bit-slice processors. For some applications general-purpose CPUs such as the 8080 and 6800 are not fast enough or their instruction sets are not suitable. For these applications several manufacturers produce devices which can be used to build a custom CPU. An example is the Advanced Micro Devices 2900 family of devices. This family includes 4-bit ALUs, multiplexers, sequencers, and other parts needed for custom-building a CPU. The term *slice* comes from the fact that these parts can be connected in parallel to work with 8-bit words, 16-bit words, or 32-bit words. In other words, a designer can add as many slices as needed for a particular application. The designer not only custom-designs the hardware of the CPU, but also custom-makes the instruction set for it using "microcode."

GENERAL-PURPOSE CPUs

The third major direction of microprocessor evolution has been toward general-purpose CPUs which give a microcomputer most or all of the computing power of earlier minicomputers. After Motorola came out with the MC6800, Intel produced the 8085, an upgrade of the 8080 requiring only a +5 V supply. Motorola then produced the MC6809 which has a few 16-bit instructions, but is still basically an 8-bit processor. In 1978 Intel came out with the 8086 which is a full 16-bit processor. Some 16-bit microprocessors, such as the National PACE and the Texas Instruments 9900 family of devices, were available previously, but the market apparently wasn't ready. Soon after Intel came out with the 8086, Motorola came out with the 16-bit MC68000, and the 16-bit race was off and running. The 8086 and the 68000 work directly with 16-bit words instead of with 8-bit words, they can address a million or more bytes of memory instead of the 64 Kbytes addressable by the 8-bit processors, and they execute instructions much faster than the 8-bit processors. Also these 16-bit processors have single instructions for functions that required a lengthy sequence of instructions on the 8-bit processors.

The evolution along this last path has continued on to 32-bit processors that work with giga (10^9) bytes or tera (10^{12}) bytes of memory. Examples of these devices are the Intel 80386, the Motorola MC68020, and the National 32032.

Since we could not possibly describe in this book the operation and programming of even a few of the available processors, we confine our discussions to primarily one group of related microprocessors. The family we have chosen is the Intel 8086, 8088, 80186, 80188, 80286 family. Members of this family are very widely used in personal computers, business computer systems, and industrial control systems. Our experience has shown that learning the programming and operation of one family of microcomputers very thoroughly is much more useful than looking at many processors superficially. If you learn one processor family well, you will most likely find it quite easy to learn another when you have to.

THE 8086, 8088, 80186, 80188, AND 80286 MICROPROCESSORS—INTRODUCTION

The Intel 8086 is a 16-bit microprocessor intended to be used as the CPU in a microcomputer. The term "16-bit" means that its arithmetic logic unit, internal registers, and most of its instructions are designed to work with 16-bit binary words. The 8086 has a 16-bit data bus, so it can read data from or write data to memory and ports either 16 bits or 8 bits at a time. The 8086 has a 20-bit address bus, so it can address any one of 2^{20} or 1,048,576 memory locations. Each of the 1,048,576 memory addresses of the 8086 represents a byte-wide location. Words will be stored in two consecutive memory locations. If the first byte of a word is at an even address, the 8086 can read the entire word in one operation. If the first byte of the word is at an odd address, the 8086 will read the first byte in one operation, and the second byte in another operation. Later we will discuss this in detail. The main point here is that if the first byte of a 16-bit word is at an even address, the 8086 can read the word in one operation.

The Intel 8088 has the same arithmetic logic unit, the same registers, and the same instruction set as the 8086. The 8088 also has a 20-bit address bus so it can address any one of 1,048,576 bytes in memory. The 8088, however, has an 8-bit data bus so it can only read data from or write data to memory and ports 8 bits at a time. The 8086, remember, can read or write either 8 or 16 bits at a time. To read a 16-bit word from two successive memory locations, the 8088 will always have to do two read operations. Since the 8086 and the 8088 are almost identical, any reference we make to the 8086 in the rest of the book will also pertain to the 8088 unless we specifically indicate otherwise. This is done to make reading easier. The Intel 8088, incidentally, is used as the CPU in the IBM Personal Computer and several compatible personal computers.

The Intel 80186 is an improved version of the 8086, and the 80188 is an improved version of the 8088. In addition to a 16-bit CPU the 80186 and 80188 each have programmable peripheral devices integrated in the same package. In a later chapter we will discuss these integrated peripherals. The instruction set of the 80186 and the 80188 is a *superset* of the instruction set of the 8086. The term superset means that all of the 8086 and 8088 instructions will execute properly on an 80186 or on an 80188, but the 80186 and the 80188 have a few additional instructions. In other words, a program written for an 8086 or for an 8088 is *upward-compatible* to an 80186 or to an 80188, but a program written for an 80186 or for an 80188 may not execute correctly on an 8086 or an 8088. In the instruction set descriptions in Chapter 6, we specifically indicate which instructions only work with the 80186 or 80188. The 80186 is used as the CPU in several personal computers.

The Intel 80286 is an advanced version of the 8086 specifically designed for use as the CPU in a multiuser or multitasking microcomputer. Programs written for an 8086 can be run on an 80286 operating in its *real address mode*. We discuss in Chapter 14 the operation and use of the 80286. The 80286 is the CPU used in the IBM PC/AT personal computer.

8086 INTERNAL ARCHITECTURE

The three-instruction program section of this chapter describes how a CPU sends out addresses, sends out

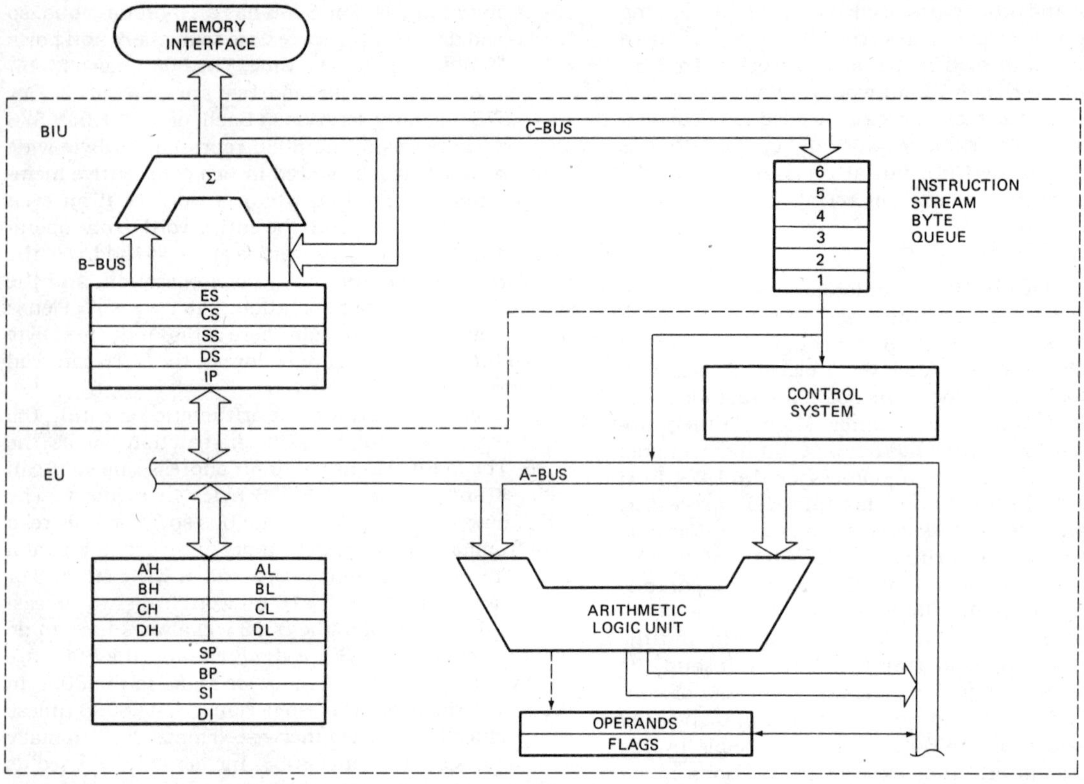

FIGURE 2-7 8086 internal block diagram. *(Intel Corp.)*

control signals, reads in instructions and data to internal registers, and sends out data to ports or memory. Before we can talk about how to write programs for the 8086, we need to discuss its specific internal features such as registers, instruction byte queue, and flags.

As shown by the block diagram in Figure 2-7, the 8086 CPU is divided into two independent functional parts, the *bus interface unit* or BIU, and the *execution unit* or EU. Dividing the work between these two units speeds up processing.

The Bus Interface Unit

The BIU sends out addresses, fetches instructions from memory, reads data from ports and memory, and writes data to ports and memory. In other words the BIU handles all transfers of data and addresses on the buses for the execution unit. The following sections describe the functional parts of the BIU.

THE QUEUE

To speed up program execution, the BIU fetches as many as six instruction bytes ahead of time from memory. The prefetched instruction bytes are held for the EU in a first-in-first-out group of registers called a *queue*. The BIU can be fetching instruction bytes while the EU is decoding an instruction or executing an instruction which does not require use of the buses. When the EU is ready for its next instruction, it simply reads the instruction from the queue in the BIU. This is much faster than sending out an address to the system memory and waiting for memory to send back the next instruction byte or bytes. The process is analogous to the way a bricklayer's assistant fetches bricks ahead of time and keeps a queue of bricks lined up so that the bricklayer can just reach out and grab a brick when necessary. Except in the cases of JUMP and CALL instructions where the queue must be dumped and then reloaded starting from a new address, this prefetch-and-queue scheme greatly speeds up processing. Fetching the next instruction while the current instruction executes is called *pipelining*.

SEGMENT REGISTERS

The BIU contains four 16-bit *segment registers*. They are: the *code segment* (CS) register, the *stack segment* (SS) register, the *extra segment* (ES) register, and the *data segment* (DS) register. These segment registers are used to hold the upper 16 bits of the starting addresses of four memory segments that the 8086 is working with at a particular time. The 8086 BIU sends out 20-bit addresses, so it can address any of 2^{20} or 1,048,576 bytes in memory. However, at any given time the 8086 only works with four, 65,536-byte (64 Kbyte) segments within this 1,048,576-byte (1 Mbyte) range. Figure 2-8

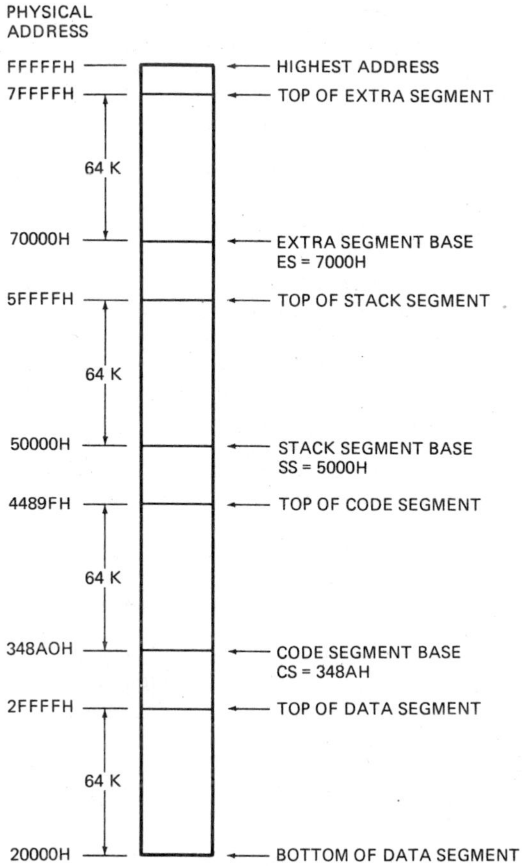

FIGURE 2-8 One way that four 64 Kbyte segments might be positioned within 1 Mbyte address space of 8086.

shows how these four segments might be positioned in memory at a given time. The four segments can be separated as shown, or, for small programs which do not need all 64 Kbytes in each segment, they can overlap. A minimum system, for example, might start all four segments at address 00000H.

To repeat then, a segment register is used to hold the upper 16 bits of the starting address for each of the segments. The code segment register, for example, holds the upper 16 bits of the starting address for the segment from which the BIU is currently fetching instruction code bytes. The BIU always inserts zeros for the lowest four bits (nibble) of the 20-bit starting address for a segment. If the code segment register contains 348AH, for example, then the code segment will start at address 348A0H. In other words, a 64 Kbyte segment can be located anywhere within the 1 Mbyte address space, but the segment will always start at an address with zeros in the lowest 4 bits. This constraint was put on the location of segments so that it is only necessary to store and manipulate 16-bit numbers when working with the starting address of a segment. The part of a segment starting address stored in a segment register is often called the *segment base*.

A *stack* is a section of memory set aside to store addresses and data while a *subprogram* executes. The stack segment register is used for the upper 16 bits of the starting address for the program stack. We will discuss the use and operation of a stack in detail later.

The extra segment register and the data segment register are used to hold the upper 16 bits of the starting addresses of two memory segments that are used for data.

INSTRUCTION POINTER

The next feature to look at in the BIU is the *instruction pointer* (IP) register. As discussed previously, the code segment register holds the upper 16 bits of the starting address of the segment from which the BIU is fetching instruction code bytes. The instruction pointer register holds the 16-bit address of the next code byte *within* this code segment. The value contained in the IP is often referred to as an *offset*, because this value must be offset from (added to) the segment base address in CS to produce the required 20-bit physical address. Figure 2-9*a* shows in diagram form how this works. The CS register points to the *base* or start of the current code segment. The IP contains the distance or offset from this base address to the next instruction byte to be fetched. Figure 2-9*b* shows how the 16-bit offset in IP is added to the 16-bit segment base address in CS to produce the 20-bit *physical* address. Notice that the two 16-bit numbers are not added directly in line. One way to describe this process is to say that the contents of the CS register are shifted left four bit positions before the contents of the IP are added to it. CS contains 348AH. When shifted left by four bit positions this produces 348A0H as the starting address of the code segment. The offset of 4214H in the IP is added to this base to give a 20-bit physical address of 38AB4H.

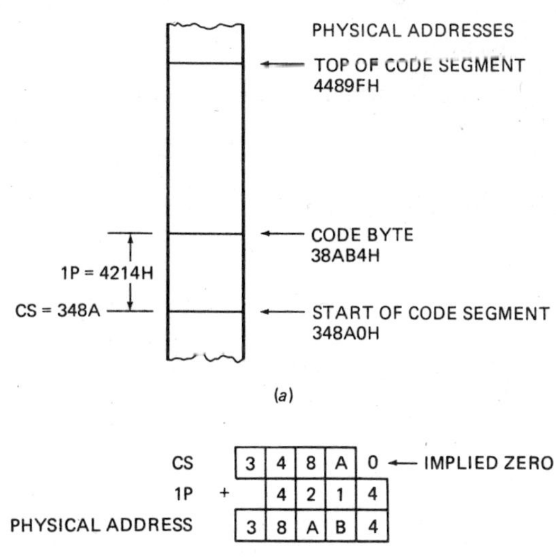

FIGURE 2-9 Addition of IP to CS to produce physical address of code byte. *(a)* Diagram. *(b)* Computation.

The 8086 20-bit physical addresses are often represented in a *segment base:offset form* rather than in the single number form. For the address of a code byte the alternative form will be CS:IP. For example, the address constructed in the preceding paragraph, 38AB4H, can also be represented as 348A:4214.

To summarize, then, the CS register contains the upper 16 bits of the starting address of the code segment in the 1 Mbyte address range of the 8086. The instruction pointer register contains a 16-bit offset which tells where in that 64 Kbyte code segment the next instruction byte will be fetched from. The actual physical address sent to memory is produced by shifting the contents of the CS register four bit positions left and adding the offset contained in IP.

As you will see in later sections, any time the 8086 accesses memory, the BIU produces the required 20-bit physical address by shifting the contents of one of the segment registers left four bit positions and adding to it a displacement or offset.

The Execution Unit

The execution unit of the 8086 tells the BIU where to fetch instructions or data from, decodes instructions, and executes instructions. The following sections describe the functional parts of the execution unit.

CONTROL CIRCUITRY, INSTRUCTION DECODER, AND ALU

Now take another look at the 8086 block diagram in Figure 2-7 to see what is contained in the execution unit. The EU contains *control circuitry* which directs internal operations. A *decoder* in the EU translates instructions fetched from memory into a series of actions which the EU carries out. The EU has a 16-bit *arithmetic logic unit* which can add, subtract, AND, OR, XOR, increment, decrement, complement, or shift binary numbers.

FLAG REGISTER

A *flag* is a flip-flop which indicates some condition produced by the execution of an instruction, or controls certain operations of the EU. A 16-bit *flag register* in the EU contains nine active flags. Figure 2-10 shows the location of the nine flags in the flag register. Six of the nine flags are used to indicate some *condition* produced by an instruction. For example, a flip-flop called the carry flag will be set to a one if the addition of two 16-bit binary numbers produces a carry out of the most significant bit position. If no carry out of the MSB is produced by the addition, then the carry flag will be a zero. The EU then effectively runs up a "flag" to tell you that a carry was produced.

The six conditional flags in this group are: the *carry flag* (CF), the *parity flag* (PF), the *auxiliary carry flag* (AF), the *zero flag* (ZF), the *sign flag* (SF), and the *overflow flag* (OF). The names of these flags should give you hints as to what conditions affect them. Certain 8086 instructions check these flags to determine which of two alternative actions should be done in executing the instruction.

The three remaining flags in the flag register are used to *control* certain operations of the processor. These flags are different from the six conditional flags described above in the way they get set or reset. The six conditional flags are set or reset by the EU on the basis of the results of some arithmetic or logic operation. The *control flags* are deliberately set or reset with specific instructions you put in the program. The three control flags are the *trap flag* (TF), which is used for single stepping through a program; the *interrupt flag* (IF), which is used to allow/prohibit the interruption of a program, and the *direction flag* (DF), which is used with string instructions.

Later we will discuss in detail the operation and use of the nine flags.

GENERAL-PURPOSE REGISTERS

Observe in Figure 2-7 that the EU has eight *general purpose registers* labeled AH, AL, BH, BL, CH, CL, DH, and DL. These registers can be used individually for temporary storage of 8-bit data. The AL register is also called the *accumulator*. It has some features that the other general-purpose registers do not have.

Certain pairs of these general-purpose registers can be used together to store 16-bit data words. The acceptable register pairs are AH and AL, BH and BL, CH and CL, and DH and DL. The AH-AL pair is referred to as the *AX register*, the BH-BL pair is referred to as the *BX register*, the CH-CL pair is referred to as the *CX register*,

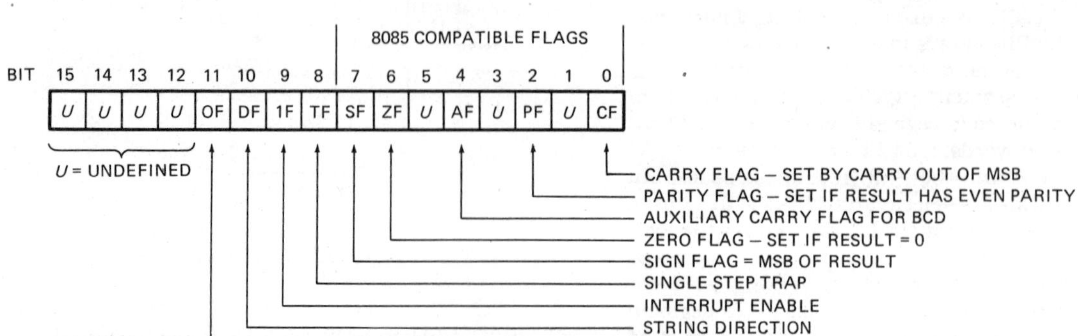

FIGURE 2-10 8086 flag register format. *(Intel Corp.)*

and the DH-DL pair is referred to as the *DX register*. For 16-bit operations, AX is called the accumulator.

The 8086 register set is very similar to those of the earlier generation 8080 and 8085 microprocessors. It was designed this way so that the many programs written for the 8080 and 8085 could easily be translated to run on the 8086 or the 8088. The advantage of using internal registers for the temporary storage of data is that, since the data is already in the EU, it can be accessed much more quickly than it could be accessed in external memory.

STACK POINTER REGISTER

A stack, remember, is a section of memory set aside to store addresses and data while a subprogram is executing. The 8086 allows you to set aside an entire 64 Kbyte segment as a stack. The upper 16 bits of the starting address for this segment is kept in the stack segment register. The *stack pointer* (SP) register contains the 16-bit offset from the start of the segment to the memory location where a word was most recently stored on the stack. The memory location where a word was most recently stored is called the *top of stack*. Figure 2-11*a* shows this in diagram form.

The physical address for a stack read or for a stack write is produced by adding the contents of the stack pointer register to the segment base address in SS. To do this the contents of the stack segment register are shifted four bit positions left and the contents of SP are added to the shifted result. Figure 2-11*b* shows an example. The 5000H in SS is shifted left four bit positions to give 50000H. When FFE0H in the SP is added to this, the resultant physical address for the top of the stack will be 5FFE0H. The physical address can be represented either as a single number, 5FFE0H, or it can be represented in SS:SP form as 5000:FFE0H.

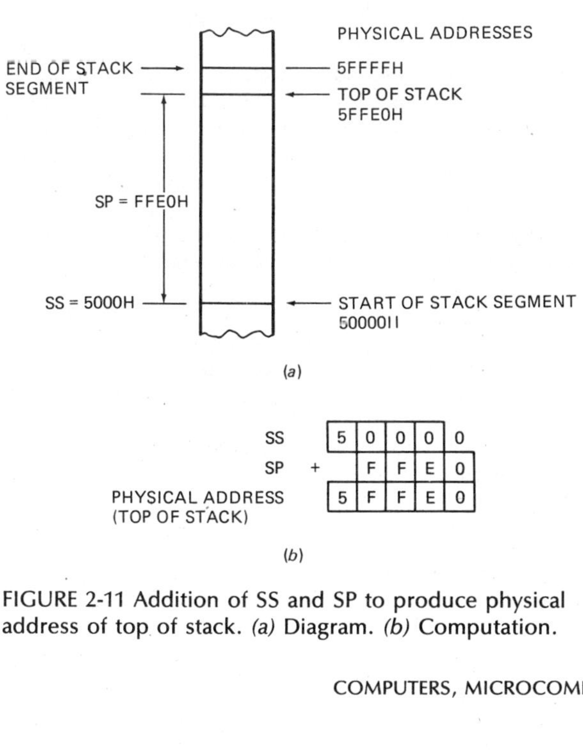

FIGURE 2-11 Addition of SS and SP to produce physical address of top of stack. *(a)* Diagram. *(b)* Computation.

The operation and use of the stack will be discussed in detail later as need arises.

OTHER POINTER AND INDEX REGISTERS

In addition to the stack pointer register, SP, the EU contains a 16-bit *base pointer* (BP) register. It also contains a 16-bit *source index* (SI) register and a 16-bit *destination index* (DI) register. These three registers can be used for temporary storage of data just as the general-purpose registers described above. However, their main use is to hold the 16-bit offset of a data word in one of the segments. SI, for example, can be used to hold the offset of a data word in the data segment. The physical address of the data in memory will be generated in this case by shifting the contents of the data segment register four bit positions to the left and adding the contents of SI to the result. A later section on addressing modes will discuss and show many examples of the use of these base and index registers.

INTRODUCTION TO PROGRAMMING THE 8086

Programming Languages

Now that you have an overview of the 8086 CPU, it is time to start you thinking about how it is programmed. To run a program, a microcomputer must have the program stored in binary form in successive memory locations. There are three language levels that can be used to write a program for a microcomputer.

MACHINE LANGUAGE

You can write programs as simply a sequence of the binary codes for the instructions you want the microcomputer to execute. The three-instruction program in Figure 2-2*b* is an example. This binary form of the program is referred to as *machine language* because it is the form required by the machine. However, it is difficult, if not impossible, for a programmer to memorize the thousands of binary instruction codes for a CPU such as the 8086. Also, it is very easy for an error to occur when working with long series of 1's and 0's. Using hexadecimal representation for the binary codes might help some, but there are still thousands of instruction codes to cope with.

ASSEMBLY LANGUAGE

To make programming easier many programmers write programs in *assembly language*. They then translate the assembly language program to machine language so it can be loaded into memory and run. Assembly language uses two-, three-, or four-letter *mnemonics* to represent each instruction type. A mnemonic is just a device to help you remember something. The letters in an assembly language mnemonic are usually initials or a shortened form of the English word(s) for the operation performed by the instruction. For example, the mnemonic for subtract is SUB, the mnemonic for exclusive OR is XOR, and the mnemonic for the instruction to copy data from one location to another is MOV.

LABEL FIELD	OP CODE FIELD	OPERAND FIELD	COMMENT FIELD
NEXT:	ADD	AL, 07H	; ADD CORRECTION FACTOR

FIGURE 2-12 Assembly language program statement format.

Assembly language statements are usually written in a standard form having four *fields*. Figure 2-12 shows an assembly language statement with the four fields indicated. The first field in an assembly language statement is the *label field*. A *label* is a symbol or group of symbols used to represent an address which is not specifically known at the time the statement is written. Labels are usually followed by a colon. Labels are not required in a statement, they are just inserted where they are needed. We will show later many uses of labels.

The *opcode field* of the instruction contains the mnemonic for the instruction to be performed. Instruction mnemonics are sometimes called *operation codes* or *opcodes*. The ADD mnemonic in the example statement in Figure 2-12 indicates that we want the instruction to do an addition. Chapter 6 describes the function of each 8086 instruction type and gives the opcodes for each.

The *operand field* of the statement contains the data, the memory address, the port address, or the name of the register on which the instruction is to be performed. *Operand* is just another name for the data item(s) acted on by an instruction. In the example instruction in Figure 2-12 there are two operands, AL and 07H, specified in the operand field. AL represents the AL register, and 07H represents the number 07H. This assembly language statement then says add the number 07H to the contents of the AL register. By Intel convention the result of the addition will be put in the register or the memory location specified *before* the comma in the operand field. For the example statement in Figure 2-12 then, the result will be left in the AL register. As another example, the assembly language statement, ADD BH, AL, when converted to machine language and run, will add the contents of the AL register to the contents of the BH register. The results will be left in the BH register.

Looking back at the example assembly language statement in Figure 2-12, observe the *comment field* which starts with a semicolon. This field is very important. Comments do not become part of the machine language program. You write *comments* in a program to remind you of the function that this instruction or group of instructions performs in the program.

To summarize why we use assembly language, let's look a little more closely at the assembly language ADD statement. The general format of the 8086 ADD instruction is:

ADD destination, source

The *source* can be a number written in the instruction, the contents of a specified register, or the contents of a memory location. The *destination* can be a specified register or a specified memory location. The source and the destination, however, cannot both be memory locations in an instruction.

A later section on 8086 addressing modes will show all of the ways in which the source of an operand and the destination of the result can be specified. The point here is that the single mnemonic, ADD, together with a specified source and a specified destination can represent a great many 8086 instructions in an easily understandable form.

The question that may occur to you at this point is, "If I write a program in assembly language, how do I get it translated into machine language which can be loaded into the microcomputer and executed?" There are two answers to this question. The first method of doing the translation is by working out the binary code for each instruction a bit at a time using the templates given in the manufacturer's data books. We will show you how to do this in the next chapter. It is a tedious and error-prone task. The second method of doing the translation is with an *assembler*. An assembler is a program which can be run on a personal computer or *microcomputer development system*. It reads the assembly language instructions and generates the correct binary code for each. For developing all but the simplest assembly language programs, an assembler and other program development tools are essential. We will introduce you to these program development tools in the next chapter and describe their use throughout the rest of this book.

HIGH LEVEL LANGUAGES

Another way of writing a program for a microcomputer is with a *high level language* such as BASIC, FORTRAN, or Pascal. These languages use program statements which are even more English-like than those of assembly language. Each high level statement may represent many machine code instructions. An *interpreter program* or a *compiler program* is used to translate higher level language statements to machine codes which can be loaded into memory and executed. Programs can usually be written faster in high level languages than in assembly language because the high level language works with bigger building blocks. However, programs written in a high level language and interpreted or compiled execute slower than the same programs written in assembly language. Programs that involve a lot of hardware control, such as robots and factory control systems, or programs that must run as quickly as possible are usually best written in assembly language. Programs that manipulate massive amounts of data, such as insurance company records, are usually best written in a high level language. The decision of which language to use has recently been made more difficult by the fact that current assemblers allow the use of many high level language features, and the fact that some current high level languages provide assembly language features.

OUR CHOICE

Throughout this book we will use mostly assembly language because we will be working very closely with hardware interfacing. Before we start teaching you assembly language programming in the next chapter, however, we want to give you an introduction to how the 8086 accesses data.

How the 8086 Accesses Immediate and Register Data

In a previous discussion of the 8086 BIU we described how the 8086 accesses code bytes using CS and IP. We also described how the 8086 accesses the stack using SS and SP. Before we can teach you assembly language programming techniques, we need to discuss some of the different ways that an 8086 can access the data that it operates on. The different ways that a processor can access data are referred to as its *addressing modes*. In assembly language statements the addressing mode is indicated in the instruction. We will use the 8086 MOV instruction to illustrate some of the 8086 addressing modes.

The MOV instruction has the format:

```
MOV destination, source
```

When executed, this instruction copies a word or a byte from the specified source location to the specified destination location. The source can be a number written directly in the instruction, a specified register, or a memory location specified in one of 24 different ways. The destination can be a specified register or a memory location specified in any one of 24 different ways. The source and the destination cannot both be memory locations in an instruction.

IMMEDIATE ADDRESSING MODE

Suppose that in a program you need to put the number 437BH in the CX register. The MOV CX, 437BH instruction can be used to do this. When it executes, this instruction will put the *immediate* hexadecimal number 437BH in the 16-bit CX register. This is referred to as *immediate addressing mode* because the number to be loaded into the CX register will be put in two memory locations immediately following the code for the MOV instruction. This is similar to the way the port address was put in memory immediately after the code for the input instruction in the three-instruction program in Figure 2-2*b*.

A similar instruction, MOV CL, 48H could be used to load the 8-bit immediate number 48H into the 8-bit CL register. You can also write instructions to load an 8-bit immediate number into an 8-bit memory location or to load a 16-bit number into two consecutive memory locations, but we are not yet ready to show you how to specify these.

REGISTER ADDRESSING MODE

Register addressing mode means that a register is the source of an operand for an instruction. The instruction MOV CX, AX, for example, copies the contents of the 16-bit AX register into the 16-bit CX register. Remember that the destination location is specified in the instruction before the source. Also note that the contents of AX are just *copied* to CX, not actually moved. In other words, the previous contents of CX are written over, but the contents of AX are not changed. For example, if CX contains 2A84H and AX contains 4971H before the MOV CX, AX instruction executes, then after the instruction executes CX will contain 4971H and AX will still contain 4971H. You can MOV any 16-bit register to any 16-bit register, or you can MOV any 8-bit register to any 8-bit register. However, you cannot use an instruction such as MOV CX, AL because this is an attempt to copy a *byte-type* operand (AL) into a *word-type* destination (CX). The byte in AL would fit in CX, but the 8086 would not know which half of CX to put it in. If you try to write an instruction like this and you are using a good assembler, the assembler will tell you that the instruction contains a *type error*. To copy the byte from AL to the high byte of CX you can use the instruction MOV CH, AL. The instruction MOV CL, AL will copy the byte from AL to CL, the low byte of CX.

How the 8086 Accesses Data in Memory

OVERVIEW OF MEMORY ADDRESSING MODES

The addressing modes described in the following sections are used to specify the location of an operand in memory. A previous section described how the 8086 produces the physical address for instruction codes by adding an offset in the instruction pointer to the code segment base in the CS register. Remember that the contents of CS are shifted four bit positions left before the contents of IP are added. Another previous section described how the 8086 accesses stack locations by adding an offset in the stack pointer register to the stack segment base in the stack segment register. Here again the contents of the stack segment register are shifted four bit positions left before the contents of the stack pointer are added.

To access data in memory the 8086 must also produce a 20-bit physical address. It does this by adding a 16-bit value called the *effective address* to one of the four segment bases. The effective address (EA) represents the *displacement* or *offset* of the desired operand from the segment base. In most cases, any of the segment bases can be specified, but the data segment is the one most often used. Figure 2-13*a* shows in graphic form how the EA is added to the data segment base to point to an operand in memory. Figure 2-13*b* shows how the 20-bit physical address is generated by the BIU. The starting address for the data segment in Figure 2-13*b* is 20000H so the data segment register will contain 2000H. The BIU shifts the 2000H four bit positions left and adds the effective address, 437AH, to the result. The 20-bit physical address sent out to memory by the BIU will then be 2437AH. The physical address can be represented either as a single number, 2437AH, or in the segment base: offset form as 2000:437AH.

The execution unit calculates the effective address for an operand using information you specify in the in-

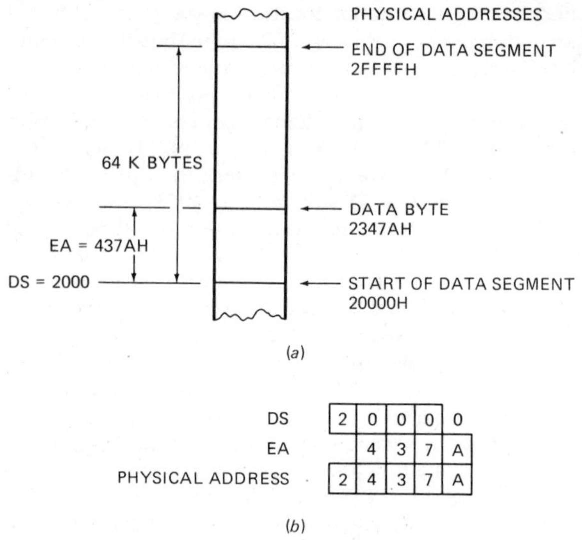

FIGURE 2-13 Addition of data segment register and effective address to produce physical address of data byte. *(a)* Diagram. *(b)* Computation.

struction. You can tell the EU to use a number in the instruction as the effective address, to use the contents of a specified register as the effective address, or to compute the effective address by adding a number in the instruction to the contents of one or two specified registers. The following section describes one way you can tell the execution unit to calculate an effective address. In later chapters we show other ways of specifying the effective address. We also show how the addressing modes this provides are used to solve some common programming problems.

DIRECT ADDRESSING MODE

For the simplest memory addressing mode the effective address is just an 8- or 16-bit number written directly in the instruction. The instruction MOV CL, [437AH] is an example. The square brackets around the 437AH are shorthand for "the contents of the memory location(s) at a displacement from the segment base of." When executed, this instruction will copy the contents of the memory location, at a displacement of 437AH from the data segment base, into the CL register. The actual 20-bit physical memory address will be produced by shifting the data segment base in DS four bits left and adding the effective address 437AH to the result. Figure 2-13*b* shows how the operation is done. This addressing mode is called *direct* because the displacement of the operand from the segment base is specified directly in the instruction. The displacement in the instruction will be added to the data segment base in DS unless you use a *segment override prefix* to tell the BIU to add it to some other segment base. We will discuss the segment overide prefix later.

Another example of this addressing mode is the instruction MOV BX, [437AH]. When executed, this instruction copies a word from memory into the BX register. Since each memory address of the 8086 represents a byte of storage, the word must come from two memory locations. The byte at a displacement of 437AH from the data segment base will be copied into BL. The contents of the next higher address, displacement 437BH, will be copied into the BH register. The 8086 will automatically access the required number of bytes in memory for a given instruction.

The previous two examples showed how the direct addressing mode can be used to specify the source of an operand. It can also be used to specify the destination of an operand. The instruction MOV [437AH], BX, for example, will copy the contents of the BX register to two memory locations in the data segment. The contents of BL will be copied to the memory location at a displacement of 437AH. The contents of BH will be copied to the memory location at a displacement of 437BH.

NOTE: When you are *hand-coding* progams using direct addressing of the form shown above, make sure to put in the square brackets to remind you how to code the instruction. If you leave the brackets out of an instruction such as **MOV CX, [437AH]**, you will code it as if it were the instruction as **MOV CX, 437AH**. This will load the immediate number 437AH into CX, rather than load a word from memory at a displacement of 437AH. Also note that if you are writing an instruction using direct addressing such as this for an *assembler*, you must use a form such as **MOV BL, DS:BYTE PTR[437AH]** to give the assembler all the information it needs. As will be shown in the next chapter, when using an assembler, we usually use a name to represent the direct address rather than the actual numerical offset.

A FEW WORDS ABOUT SEGMENTATION

At this point you may be wondering why Intel designed the 8086 family devices to use *memory segmentation*. At least two reasons come to mind. First, by working with only 64 Kbyte segments of memory at a time, the 8086 only has to work with 16-bit effective addresses to access any location in the segment. In other words, because of the segmentation scheme the 8086 only has to manipulate and store 16-bit address components. The second reason has to do with the type of microcomputer in which an 8086 family CPU is likely to be used. A previous section of this chapter described briefly the operation of a time-share microcomputer system. In a time-share system several users share a CPU. The CPU works on one user's program for perhaps 20 milliseconds, then works on the next user's program for 20 milliseconds. After working 20 milliseconds for each of the other users, the CPU comes back to working on the first user's program again. Each time the CPU switches from one user's program to the next it must access a new section of code and new sections of data. Segmentation makes this switching quite easy. Each user's program can be assigned a separate set of logical segments for its code and data. The user's program will contain offsets or displacements from these segment bases. To change from one user's program to a second user's program all that has to be done is to reload the four segment registers with the segment base addresses assigned to the second user's program. In other words, segmentation makes it easy to keep users' programs and data separate from each other, and segmentation makes it easy to switch from one user's program to another user's program.

IMPORTANT TERMS AND CONCEPTS FROM THIS CHAPTER

If you do not remember any of the terms or concepts in the following list, use the index to find them in the chapter.

Microcomputer, microprocessor

Hardware, software, firmware

Time-share

Multitasking computer system

Distributed processing system

Multiprocessing

CPU

Memory, RAM, ROM

I/O ports

Address, data, and control buses

Control bus signals

ALU

Segmentation

BIU
  Instruction byte queue, pipelining
  ES, CS, SS, DS registers, IP register

EU
  AX, BX, CX, DX registers, flag register
  ALU, SP, BP, SI, DI registers

Machine language

Assembly language
  Mnemonic, opcode, operand, label, comment,

Assembler

High level language

Compiler

Immediate address mode, register address mode, direct address mode

Effective address

REVIEW QUESTIONS AND PROBLEMS

1. Describe the sequence of signals that occurs on the address bus, the control bus, and the data bus when a computer fetches an instruction.

2. Describe the main advantages of a distributed processing computer system over a simple time-share system.

3. What determines whether a microprocessor is considered an 8-bit, 16-bit, or 32-bit device?

4. *a.* How many address lines does an 8086 have?
   b. How many memory addresses does this number of address lines allow the 8086 to access directly?
   c. At any given time, the 8086 works with four segments in this address space. How many bytes are contained in each segment?
   d. Why was the 8086 designed with this segmentation of the address space?

5. What is the main difference between the 8086 and the 8088?

6. *a.* Describe the function of the 8086 queue.
   b. How does the queue speed up process operation?

7. *a.* If the code segment for an 8086 program starts at address 70400H, what number will be in the CS register?
   b. Assuming this same code segment base, what physical address will a code byte be fetched from if the instruction pointer contains 539CH?

8. What physical address is represented by:
   a. 4370:561EH
   b. 7A32:0028H

9. What is the advantage of using a CPU register for temporary data storage over using a memory location?

10. If the stack segment register contains 3000H and the stack pointer register contains 8434H, what is the physical address of the top of the stack?

11. *a.* What is the advantage of using assembly language instead of writing a program directly in machine language?
    b. Describe the operation an 8086 will perform when it executes ADD AX, BX.

12. What types of programs are usually written in assembly language?

13. Describe the operation that an 8086 will perform when it executes each of the following instructions:
    a. MOV BX, 03FFH
    b. MOV AL, 0DBH
    c. MOV DH, CL
    d. MOV BX, AX

14. Write the 8086 assembly language statement which will perform the following operations:
   a. Load the number 7986H into the BP register.
   b. Copy the BP register contents to the SP register.
   c. Copy the contents of the AX register to the DS register.
   d. Load the number F3H into the AL register.

15. If the 8086 execution unit calculates an effective address of 14A3H and DS contains 7000H, what physical address will the BIU produce?

16. If the data segment register, DS, contains 4000H, what physical address will the instruction MOV AL, [234BH] read?

17. If the 8086 data segment register contains 7000H, write the instruction that will copy the contents of DL to address 74B2CH.

18. Describe the difference between the instructions MOV AX, 2437H and MOV AX, [2437H].

CHAPTER

3 8086 Family Assembly Language Programming—Introduction

The last chapter showed you the format for 8086 assembly language programs and introduced you to a few 8086 instructions. Developing a program, however, requires more than just writing down a series of instructions. When you want to build a house, it is a good idea to first develop a complete set of plans for the house. With the plans you can see if the house has the rooms you need, if the rooms are efficiently placed, and if the house is structured so that you can easily add on to it if you have more kids. We have all probably seen examples of what happens when someone attempts to build a house by just putting pieces together without a plan.

Likewise, when you write computer programs it is a good idea to start by developing a detailed plan or outline. A good outline helps you to break a large and seemingly overwhelming programming job down into small modules which can easily be written, tested, and debugged. The more time you spend organizing your programs the less time it will take you to write and debug them. You should *never* start writing an assembly language program by just writing down instructions! In this chapter we show you how to develop assembly language programs in a systematic way.

OBJECTIVES

At the conclusion of this chapter you should be able to:

1. Write a task list, flowchart, or pseudocode for a simple programming problem.
2. Write, code or assemble, and run a very simple assembly language program.
3. Describe the use of program development tools such as editors, assemblers, linkers, locators, debuggers, and emulators.
4. Properly document assembly language programs.

PROGRAM DEVELOPMENT STEPS

Defining the Problem

The first step in writing a program is to think very carefully about the problem that you want the program to solve. In other words, ask yourself many times, "What do I really want this program to do?" If you don't do this, you may write a great program that works, but does not do what you need it to do. As you think about the problem it is a good idea to write down exactly what you want the program to do and the order in which you want the program to do it. At this point you do not write down program statements, you just write the operations you want in general terms. An example for a simple programming problem might be:

1. Read temperature from sensor
2. Add correction factor of +7
3. Save result in a memory location

For a program as simple as this, the three actions desired are very close to the eventual assembly language statements. For more complex problems, however, we develop a more extensive outline before writing the assembly language statements. The next section shows you some of the common ways of representing program operations in a program outline.

Representing Program Operations

The formula or sequence of operations used to solve a programming problem is often called the *algorithm* of the program. The following sections show you several ways of representing the algorithm for a program or program segment.

SEQUENTIAL TASK LISTS

Some programmers use just a *sequential list of the tasks* such as that in the preceding section to show the algorithm for their programs. To give you a better idea of this form, we will show another slightly different example. Suppose that, instead of taking in one data sample from the temperature sensor, we want to take in a data sample every hour for 24 hours, add 7 to each sample, and put each corrected value in a memory location. We could write a task list for this problem as:

1. Read data sample from temperature sensor.

2. Add 7 to value read in.

3. Store corrected value in memory location.

4. Wait one hour.

5. Read next sample from temperature sensor.

6. Add 7 to value read in.

7. Store corrected value in next memory location.

.
.
.
.
.

97. Read last data sample from temperature sensor.

98. Add 7 to value read in.

99. Store corrected value in next memory location.

As you can see, this direct form is not a very compact or efficient way of representing the operation of the program. A more efficient way of writing the sequential task list for this program is:

Read a data sample from temperature sensor.

Add 7 to the value read in.

Store corrected value in memory location.

Wait one hour.

24 samples yet?
  No, read next sample and process.
  Yes, done.

The last three lines indicate that we want the program to do the read, add, store, and wait operations 24 times. Carefully written sequential task lists are often quite close to the assembly language statements that will implement them, so you may find them useful. As you determine hardware details, such as port addresses for the system on which the program is to run, you can add this information to the appropriate task statement. The next section shows you a more graphic way of representing the algorithm of a program or program segment.

FLOWCHARTS

If you have done any previous programming in BASIC or in FORTRAN, you are probably familiar with *flowcharts*. Flowcharts use graphic shapes to represent different types of program operations. The specific operation desired is written in the graphic symbol. Figure 3-1 shows some of the common flowchart symbols. Plastic templates are available to help you draw these symbols if you decide to use them for your programs.

Figure 3-2 shows a flowchart for a program to read in 24 data samples from a temperature sensor at 1-hour intervals, add 7 to each, and store each result in a memory location. A *racetrack-shaped* symbol labeled START is used to indicate the *beginning* of the program. A *parallelogram* is used to represent *input* or *output operations*. In the example we use it to indicate reading data from the temperature sensor. A *rectangular box symbol* is used to represent *simple operations* other than input and output operations. The box containing "add 7" in Figure 3-2 is an example.

A *rectangular box with double lines at each end* is often used to represent a *subroutine* or *procedure* that will be written separately from the main program. When a set of operations must be done several times throughout a program, it is usually more efficient to write the series of operations once as a separate *subprogram* and then just use or "call" this subprogram as it is needed. For example, suppose that there are several times in a program where you need to compute the square root of a number. Instead of writing the series of instructions for computing a square root each time you need it in the program, you can write the instruction sequence once as a subprogram and set it aside in some location in memory. You can then call this subprogram each time you need to compute a square root. In the flowchart in Figure 3-2 we use the double-ended box to indicate that the "wait 1 hour" operation will be programmed as a subroutine. Incidentally, the terms *subprogram*, *subroutine*, and *procedure* all have the same meaning. Chapter 5 shows how procedures are written and used.

A *diamond-shaped* box is used in flowcharts to represent a *decision* point or crossroad. Usually it indicates that some condition is to be checked at this point in the program and, if the condition is found to be *true*, one set

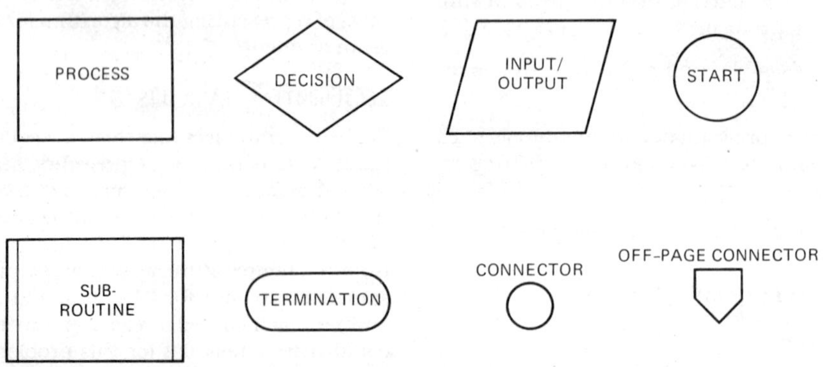

FIGURE 3-1 Flowchart symbols.

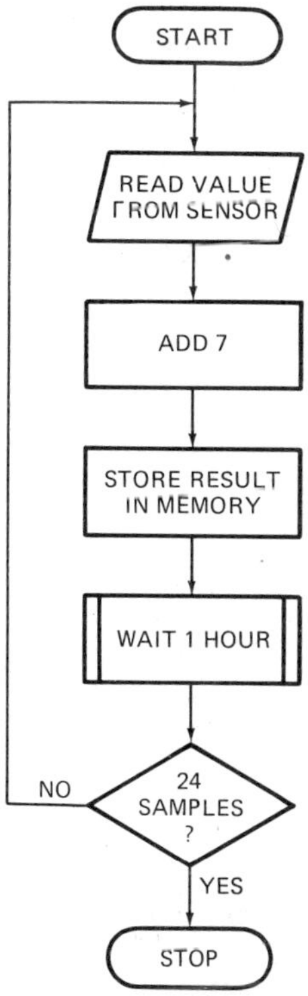

FIGURE 3-2 Flowchart for program to read in 24 data samples from a port, correct each value, and store each in a memory location.

of actions is to be done. If the condition is found to be *false*, then another set of actions is to be done. In the example flowchart in Figure 3-2 the condition to be checked is whether 24 samples have been read in and processed. If 24 samples have not been read in and processed, the arrow labeled NO in the flowchart indicates that we want the computer to jump back and execute the read, add, store, and wait steps again. If 24 samples have been read in, the arrow labeled YES in the flowchart of Figure 3-2 indicates that all the desired operations have been done. The *racetrack-shaped* symbol at the bottom of the flowchart indicates the *end* of the program.

The two additional flowchart symbols in Figure 3-1 are *connectors*. If a flowchart column gets to the bottom of the paper, but all of the program has not been represented, you can put a small circle with a letter in it at the bottom of the column. You then start the next column at the top of the same paper with a small circle containing the same letter. If you need to continue a flowchart to another page, you can end the flowchart on the first page with the five-sided off-page connector symbol containing a letter or number. You then start the flowchart on the next page with an off-page connector symbol containing the same letter or number.

For simple programs and program sections, flowcharts are a graphic way of showing the operational flow of the program. We will show flowcharts for many of the program examples throughout this book. Flowcharts, however, have several disadvantages. First, you can't write much information in the little boxes. Second, flowcharts do not present information in a very compact form. For more complex problems, flowcharts tend to become spread out over many pages. They are very hard to follow back and forth between pages. Third, and most important, with flowcharts the overall structure of the program tends to get lost in the details. The following section describes a more clearly *structured* and *compact* method of representing the algorithm of a program or program segment.

STRUCTURED PROGRAMMING AND PSEUDOCODE OVERVIEW

In the early days of computers a single brilliant person might write even a large program single-handedly. The main concerns in this case were, "Does the program work?" and "What do we do if this person leaves the company?" As the number of computers increased and the complexity of the programs being written increased, large programming jobs were usually turned over to a team of programmers. In this case the compatibility of parts written by different programmers became an important concern. During the 1970s it became obvious to many professional programmers that, in order for team programming to work, a systematic approach and standardized tools were absolutely necessary.

One suggested systematic approach is called *top-down design*. In this approach a large programming problem is first broken down into major *modules*. The top level of the outline shows the relationship and function of these modules. This top level then presents a one-page overview of the entire program. Each of the major modules is broken down into still smaller modules on following pages. The division is continued until the steps in each module are clearly understandable. Each programmer can then be assigned a module or set of modules to write for the program. Also, those who want to learn about the program later can start with the overview and work their way down to the level of detail they need. This approach is the same as drawing the complete plans for a house before starting to build it.

The opposite of top-down design is *bottom-up design*. In this approach each programmer starts writing low-level modules and hopes that all the pieces will eventually fit together. When completed, the result should be similar to that produced by the top-down design. Many modern programming teams use a combination of the two techniques. They do the top-down design and then build, test, and link modules starting from the smallest and working upward.

The development of standard programming tools was helped by the discovery that any desired program operation could be represented by three basic types of operation. The first type of operation is *sequence* which means simply doing a series of actions. The second basic type of operation is *decision* or *selection*, which means choosing between two alternative actions. The third basic type of operation is *repetition* or *iteration*, which means repeating a series of actions until some condition is or is not present.

On the basis of this observation, the suggestion was made that all programmers use a set of three to seven standard *structures* to represent all of the operations in their programs. Actually, only three structures, SEQUENCE, IF–THEN–ELSE, and WHILE–DO, are required to represent any desired program action, but three or four more structures derived from these often make programs clearer. If you have previously written programs in a structured language such as Pascal, then these structures are probably already familiar to you. Figure 3-3 uses flowchart symbols to represent the commonly used structures so you can more easily visualize their operations. In actual program documentation, however, English-like statements called *pseudocode* are used rather than the space-consuming flowchart symbols. Figure 3-3 also shows the pseudocode format and an example for each structure.

Each structure has only *one entry point* and *one exit point*. The output of one structure is connected to the input of the next structure. Program execution then proceeds through a series of these structures.

Any structure can be used within another. An IF–THEN–ELSE structure, for example, can contain a sequence of statements. Any place that the term *statement(s)* appears in Figure 3-3, one of the other structures could be substituted for it. The term "statement(s)" can also represent a subprogram or procedure that is called to do a series of actions.

STANDARD PROGRAMMING STRUCTURES

The structure shown in Figure 3-3*a* is an example of a simple sequence. In this structure the actions are simply written down in the desired order. An example is:

Read temperature from sensor.

Add correction factor of +7.

Store corrected value in memory.

Figure 3-3*b* shows an IF–THEN–ELSE example of the decision operation. This structure is used to direct operation to one of two different actions based on some condition. An example is:

```
IF temperature less than 70 degrees THEN
  Turn on heater
ELSE
  Turn off heater
```

The example says that if the temperature is below the thermostat setting, we want to turn the heater on. If the temperature is equal to or above the thermostat setting, we want to turn the heater off.

The IF–THEN structure shown in Figure 3-3*c* is the same as the IF–THEN–ELSE except that one of the paths contains no action. An example of this is:

```
IF hungry THEN
  Get food.
```

The assumption for this example is that if you are not hungry, you will just continue on with your next task.

The WHILE–DO structure in Figure 3-3*d* is one form of repetition. It is used to indicate that you want to do some action or sequence of actions as long as some condition is present. This structure represents a *program loop*. The example in Figure 3-3*d* is:

```
WHILE money lasts DO
  Eat supper out.
  Go to movie.
  Take a taxi home.
```

This example shows a sequence of actions you might do each evening until you ran out of money. Note that, in this structure, the condition is checked *before* the action is done the first time. You certainly would want to check how much money you have before eating out.

Another useful structure, derived from the WHILE–DO structure, is the REPEAT–UNTIL structure shown in Figure 3-3*e*. You use this structure to indicate that you want the program to repeat some action or series of actions until some condition is present. A good example of the use of this structure is the programming problem we used in the discussion of flowcharts. The example is:

```
REPEAT
  Get data sample from sensor.
  Add correction of +7.
  Store result in a memory location.
  Wait one hour.
UNTIL 24 samples taken.
```

Compare the space required by the pseudocode representation for the desired action with the space required by the flowchart representation shown in Figure 3-2. The space advantage of pseudocode should be obvious.

As indicated previously, the REPEAT–UNTIL structure is derived from the WHILE–DO. In other words, any problem that can be represented by a REPEAT–UNTIL can also be represented by a properly written WHILE–DO. The example in Figure 3-3*e* could be written as:

```
WHILE NOT 24 samples DO
  Read data sample from temperature sensor.
  Add correction factor of +7.
  Store result in memory location.
  Wait one hour.
```

Note that the REPEAT–UNTIL structure indicates that the condition is first checked *after* the statement(s) is performed. In other words, a REPEAT–UNTIL structure indicates that the action or series of actions will always be done at least once. If you don't want this to happen, then use the WHILE–DO which indicates that the condition is checked before any action is taken. As we will show later, the structure you use makes a difference in the actual assembly language program you write to implement it.

The WHILE–DO and REPEAT–UNTIL structures contain a simple IF–THEN–ELSE decision operation. However, since this decision is an *implied* part of these two structures, we don't indicate the decision separately in them.

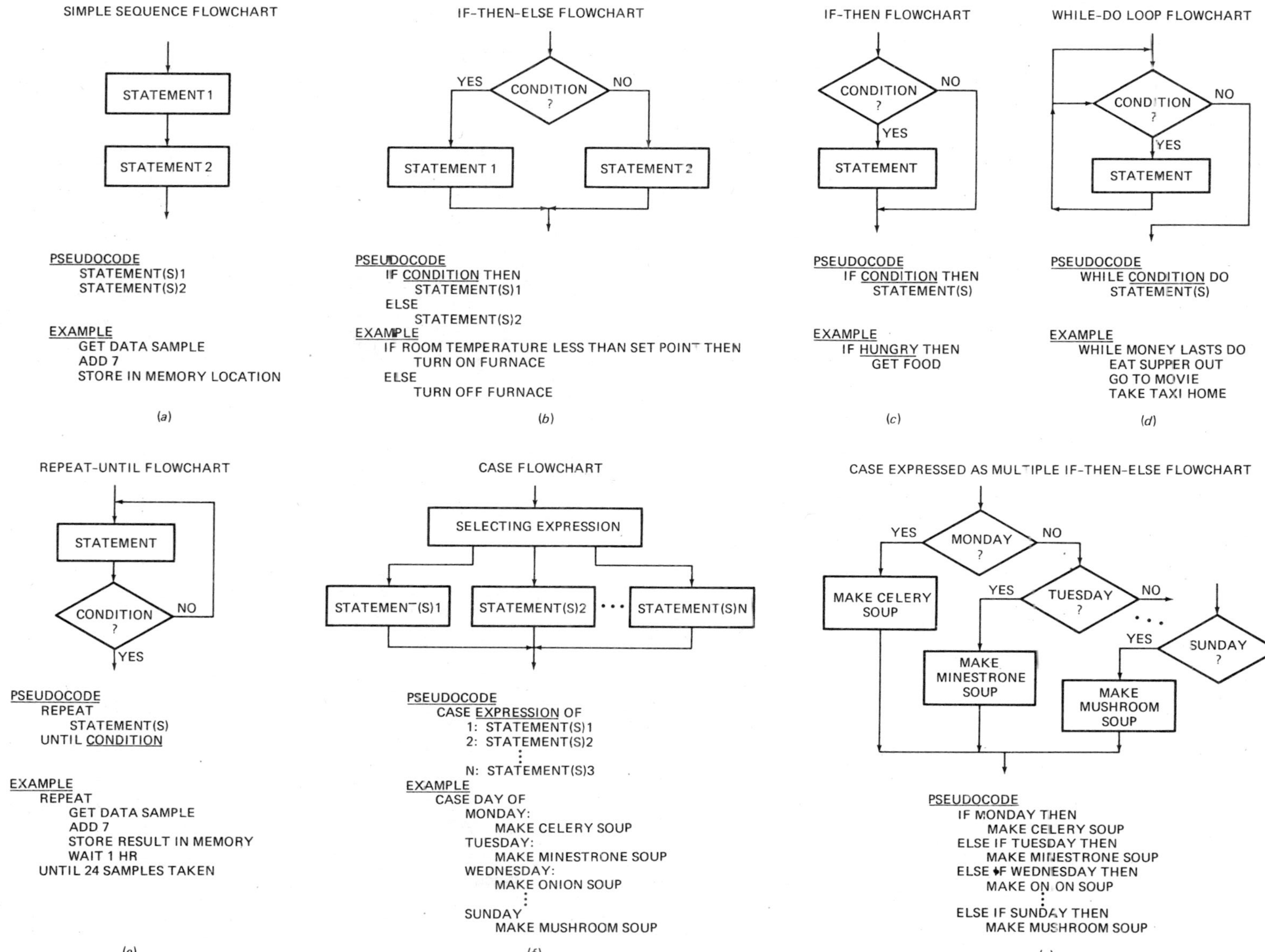

FIGURE 3-3 Standard program structures. *(a)* SEQUENCE. *(b)* IF-THEN-ELSE. *(c)* IF-THEN. *(d)* WHILE-DO. *(e)* REPEAT-UNTIL. *(f)* CASE. *(g)* CASE expressed as multiple IF-THEN-ELSE.

Another form of the repetition operation that you might see in high level language programs is the FOR–DO loop. This structure has the form:

```
FOR count = 1 TO n DO
  statement
  statement
```

In assembly language we usually implement this type of operation with a REPEAT–UNTIL structure, so we have not included a sample of it.

The CASE structure shown in Figure 3-3*f* is a compact way of representing a choice among several alternative actions. The choice is determined by testing some quantity. The example in Figure 3-3*f* best shows how this is used. This everyday example describes the desired actions for a cook in a restaurant. The pseudocode is just a summary of the thinking the cook might go through. The cook or the computer checks the value of the variable called "day" and selects the appropriate actions for that day. Each of the indicated actions, such as "Make celery soup," is itself a sequence of actions which could be represented by the structures we have described.

The CASE structure is really just a compact way to represent a complex IF–THEN–ELSE structure. To illustrate this, Figure 3-3*g* also shows how the soup cook example can be represented as a series of IF–THEN–ELSE structures. Note that, in this example, the last IF–THEN has no ELSE after it because all of the possible days have been checked. You can, if you want, add the final ELSE to the IF–THEN–ELSE chain to send an error message if the data does not match any of the choices. The CASE structure does contain the final ELSE, however. The CASE form is more compact for documentation purposes and some high-level languages such as Pascal allow you to implement it directly. However, the IF–THEN–ELSE structure gives you a much better idea of how you write an assembly language program section to choose between several alternative actions.

Throughout the rest of this book we show you how to use these structures to represent program actions and how to implement these structures in assembly language.

SUMMARY OF PROGRAM STRUCTURE REPRESENTATION FORMS

Writing a successful program does not consist of just writing down a series of instructions. You must first think carefully about what you want the program to do and how you want the program to do it. Then you must represent the structure of the program in some way that is very clear to you and to anyone else who might have to work on the program. If the structure is well developed, it is usually not a difficult step to write the actual programming language statements that implement it.

One way of representing program operations is with a sequential task list. For initial thinking and simple programming problems this technique works well. For more complex programming problems, a sequential list may become very messy because it has little real structure or standardization. Another way of representing program operations is with flowcharts. Flowcharts are a very graphic representation, and they are useful for short program segments, especially those that deal directly with hardware. However, flowcharts use a great deal of space. Consequently, the flowchart for even a moderately complex program may take up several pages. It often becomes difficult to follow program flow back and forth between pages. Also, since there are no agreed-upon structures, a poor programmer can write a flowchart which jumps all over the place and is even more difficult to follow. The term "logical spaghetti" comes to mind here.

A third way of representing the operations you want in a program is with a top-down design approach and standard program structures. The overall program problem is first broken down into major functional modules. Each of these modules is broken down into smaller and smaller modules until the steps in each module are obvious. The algorithms for the whole program and for each module are each expressed with a standard structure. Only three basic structures, SEQUENCE, IF–THEN–ELSE, and WHILE–DO, are needed to represent any needed program action or series of actions. However, other useful structures such as IF–THEN, REPEAT–UNTIL, FOR–DO, and CASE can be derived from these basic three. A structure can contain another structure of the same type or one of the other types. Each structure has only one entry point and one exit point. These programing structures may seem restrictive, but using them usually results in program representations which are easy to understand and for which it is easy to write the programs. A program written in a structured manner is easier to debug and much more understandable to someone else who has to work on it. Furthermore, a program representation developed with structured programming techniques can be implemented easily in assembly language or in a high-level language such as Modula II or C.

Finding the Right Instruction

After you get the structure of a program worked out and written down, the next step is to determine the instruction statements required to do each part of the program. Since the examples in this book are based on the 8086 family of microprocessors, now is a good time to give you an overview of the instructions the 8086 has for you to use.

You do not usually learn a new language by studying its dictionary from cover to cover. It is more productive to first learn a few very useful words and learn how to put together simple sentences. You can then learn more words as you need them to express more complex thoughts. Chapter 6 contains a dictionary of all of the 8086 instructions with detailed descriptions and examples for each. You can use this as a reference as you write programs. Here we simply list the 8086 instructions in *functional* groups with single-sentence descriptions so that you can see the types of instructions that are available to you. As you read through this section, do not expect to understand all of the instructions. When you start writing programs, you will probably use this section to determine the type of instruction and Chapter 6 to get the instruction details as you need them. After you have written a few programs, you will remember most of the basic instruction types and will be able to just look up an instruction in Chapter 6 to get any additional details you need. Chapter 4 shows you in detail how to use the *move*, *arithmetic*, *logical*, *jump*, and *string* instructions. Chapter 5 shows how to use the *call* instructions and the *stack*.

As you skim through the following overview of the 8086 instructions, see if you can find the instructions needed to do the "read temperature sensor value from a port, add +7, and store result in memory" example program.

DATA TRANSFER INSTRUCTIONS

General-purpose byte or word transfer instructions:

MNEMONIC	DESCRIPTION
MOV	Copy byte or word from specified source to specified destination.
PUSH	Copy specified word to top of stack.
POP	Copy word from top of stack to specified location.
PUSHA	(80186/80188 ONLY) Copy all registers to stack.
POPA	(80186/80188 ONLY) Copy words from stack to all registers.
XCHG	Exchange bytes or exchange words.
XLAT	Translate a byte in AL using a table in memory.

Simple input and output port transfer instructions:

IN	Copy a byte or word from specified port to accumulator.
OUT	Copy a byte or word from accumulator to specified port.

Special address transfer instructions:

LEA	Load effective address of operand into specified register.
LDS	Load DS register and other specified register from memory.
LES	Load ES register and other specified register from memory.

Flag transfer instructions:

LAHF	Load (copy to) AH with the low byte of the flag register.
SAHF	Store (copy) AH register to low byte of flag register.
PUSHF	Copy flag register to top of stack.
POPF	Copy word at top of stack to flag register.

ARITHMETIC INSTRUCTIONS

Addition instructions:

ADD	Add specified byte to byte, or specified word to word.
ADC	Add byte + byte + carry flag or word + word + carry flag.
INC	Increment specified byte or specified word by one.
AAA	ASCII adjust after addition.
DAA	Decimal (BCD) adjust after addition.

Subtraction instructions:

SUB	Subtract byte from byte, or word from word.
SBB	Subtract byte and carry flag from byte, or word and carry flag from word.
DEC	Decrement specified byte or specified word by one.
NEG	Negate—invert each bit of a specified byte or word and add 1 (form 2's complement).
CMP	Compare two specified bytes or two specified words.
AAS	ASCII adjust after subtraction.
DAS	Decimal (BCD) adjust after subtraction.

Multiplication instructions:

MUL	Multiply unsigned byte by byte or unsigned word by word.
IMUL	Multiply signed byte by byte or signed word by word.
AAM	ASCII adjust after multiply.

Division instructions:

DIV	Divide unsigned word by byte, or unsigned double word by word.
IDIV	Divide signed word by byte, or signed double word by word.
AAD	ASCII adjust before division.
CBW	Fill upper byte of word with copies of sign bit of lower byte.
CWD	Fill upper word of double word with sign bit of lower word.

BIT MANIPULATION INSTRUCTIONS

Logical instructions:

NOT	Invert each bit of a byte or word.
AND	AND each bit in byte or word with the corresponding bit in another byte or word.
OR	OR each bit in a byte or word with the corresponding bit in another byte or word.
XOR	Exclusive OR each bit in a byte or word with the corresponding bit in another byte or word.
TEST	AND operands to update flags, but don't change operands.

Shift instructions:

SHL/SAL	Shift bits of word or byte left, put zero(s) in LSB(s).
SHR	Shift bits of word or byte right, put zero(s) in MSB(s).
SAR	Shift bits of word or byte right, copy old MSB into new MSB.

Rotate instructions:

ROL	Rotate bits of byte or word left, MSB to LSB and to CF.
ROR	Rotate bits of byte or word right, LSB to MSB and to CF.
RCL	Rotate bits of byte or word left, MSB to CF and CF to LSB.
RCR	Rotate bits of byte or word right, LSB to CF and CF to MSB.

STRING INSTRUCTIONS

NOTES A *string* is a series of bytes or a series of words in sequential memory locations. A string often consists of ASCII character codes. In the list a "/" is used to separate different mnemonics for the same instruction. Use the mnemonic which most clearly describes the function of the instruction in a specific application. A "B" in a mnemonic is used to specifically indicate that a string of bytes is to be acted upon. A "W" in the mnemonic is used to indicate that a string of words is to be acted upon.

REP	An instruction prefix. Repeat following instruction until CX = 0.
REPE/REPZ	An instruction prefix. Repeat instruction until CX = 0 or ZF ≠ 1.
REPNE/REPNZ	An instruction prefix. Repeat until CX = 0 or ZF = 1.
MOVS/MOVSB/MOVSW	Move byte or word from one string to another.
COMPS/COMPSB/COMPSW	Compare two string bytes or two string words.
INS/INSB/INSW	(80186/80188) Input string byte or word from port.
OUTS/OUTSB/OUTSW	(80186/80188) Output string byte or word to port.
SCAS/SCASB/SCASW	Scan a string. Compare a string byte with byte in AL or a string word with word in AX.
LODS/LODSB/LODSW	Load string byte into AL or string word into AX.
STOS/STOSB/STOSW	Store byte from AL or word from AX into string.

PROGRAM EXECUTION TRANSFER INSTRUCTIONS

These instructions are used to tell the 8086 to start fetching instructions from some new address, rather than continuing in sequence.

Unconditional transfer instructions:

CALL	Call a procedure (subprogram), save return address on stack.
RET	Return from procedure to calling program.
JMP	Go to specified address to get next instruction.

Conditional transfer instructions:

NOTE A "/" is used to separate two mnemonics which represent the same instruction. Use the mnemonic which most clearly describes the decision condition in a specific program. These instructions are often used after a compare instruction. The terms *below* and *above* refer to unsigned binary numbers. Above means larger in magnitude. The terms *greater than* or *less than* refer to signed binary numbers. Greater than means more positive.

JA/JNBE	Jump if above/Jump if not below nor equal.
JAE/JNB	Jump if above or equal/Jump if not below.
JB/JNAE	Jump if below/Jump if not above nor equal.

JBE/JNA	Jump if below or equal/Jump if not above.
JC	Jump if carry flag (CF) = 1.
JE/JZ	Jump if equal/Jump if zero flag (ZF) = 1.
JG/JNLE	Jump if greater/Jump if not less than nor equal.
JGE/JNL	Jump if greater than or equal/Jump if not less than.
JL/JNGE	Jump if less than/Jump if not greater than nor equal.
JLE/JNG	Jump if less than or equal/Jump if not greater than.
JNC	Jump if no carry (Jump if carry flag = 0).
JNE/JNZ	Jump if not equal/Jump if not zero (zero flag = 0).
JNO	Jump if no overflow (Jump if overflow flag = 0).
JNP/JPO	Jump if not parity/Jump if parity odd (PF = 0).
JNS	Jump if not sign (Jump if sign flag = 0).
JO	Jump if overflow flag = 1.
JP/JPE	Jump if parity/Jump if parity even (PF = 1).
JS	Jump if sign flag = 1.

Iteration control instructions:

These instructions can be used to execute a series of instructions some number of times. Here mnemonics separated by a "/" represent the same instruction. Use the one that best fits the specific application.

LOOP	Loop through a sequence of instructions until CX = 0.
LOOPE/LOOPZ	Loop through a sequence of instructions while zero flag = 1 and CX ≠ 0.
LOOPNE/LOOPNZ	Loop through a sequence of instructions while zero flag = 0 and CX ≠ 0.
JCXZ	Jump to specified address if CX = 0.

If you aren't tired of instructions, continue skimming through the rest of the list. Don't worry if the explanation is not clear to you because we will explain these instructions in detail in later chapters.

Interrupt instructions:

INT	Interrupt program execution, call service procedure.
INTO	Interrupt program execution if overflow flag = 1.
IRET	Return from interrupt service procedure to main program.

High-level language interface instructions:

ENTER	(80186/80188 ONLY) Enter procedure.
LEAVE	(80186/80188 ONLY) Leave procedure.
BOUND	(80186/80188 ONLY) Check if effective address within specified array bounds.

PROCESSOR CONTROL INSTRUCTIONS

Flag set/clear instructions:

STC	Set carry flag (CF) to 1.
CLC	Clear carry flag (CF) to 0.
CMC	Complement the state of the carry flag (CF).
STD	Set direction flag (DF) to 1 (decrement string pointers).
CLD	Clear direction flag (DF) to 0.
STI	Set interrupt enable flag to 1 (enable INTR input).
CLI	Clear interrupt enable flag to 0 (disable INTR input).

External hardware synchronization instructions:

HLT	Halt (do nothing) until interrupt or reset.
WAIT	Wait (do nothing) until signal on the TEST pin is low.
ESC	Escape to external coprocessor such as 8087 or 8089.
LOCK	An instruction prefix. Prevents another processor from taking bus while the adjacent instruction executes.

No operation instruction:

NOP	No action except fetch and decode.

Now that you have glanced through an overview of the 8086 instruction set, let's see if you found the instructions needed to implement the "read sensor, add +7, and store result in memory" example program. The IN instruction can be used to read the temperature value from an A/D converter connected to a port. The ADD instruction can be used to add the correction factor of +7 to the value read in. Finally, the MOV instruction can be used to copy the result of the addition to a memory location. A major point here is that breaking the programming problem down into a sequence of steps makes it easy to find the instruction or small group of instructions that will perform each step. The next section shows you how to write the actual program using these instructions.

Writing a Program

INITIALIZATION INSTRUCTIONS

After finding the instructions needed to do the main part of your program, there are a few additional instructions you need to determine before you actually write your program. The purpose of these additional instructions is to *initialize* various parts of the system such as segment registers, flags, and programmable port devices. Segment registers, for example, must be loaded with the upper 16 bits of the address in memory where you want the segment to begin. For our "read temperature sensor, add +7, and store result in memory" example program, the only part we need to initialize is the data segment register. The data segment register must be initialized so that we can copy the result of the addition to a location in memory. If, for example, we want to store data in memory starting at address 00100H, then we want the data segment register to contain the upper 16 bits of this address, 0010H. The 8086 does not have an instruction to move a number directly into a segment register. Therefore, we move the desired number into one of the 16-bit general-purpose registers, and then copy it to the desired segment register. Two MOV instructions will do this.

If you are using the stack in your program, then you must include an instruction to load the stack pointer register with the offset of the top of the stack. Most microcomputer systems contain several programmable peripheral devices such as ports, timers, and controllers. You must include instructions which send control words to these devices to tell them the function you want them to perform. Also, you usually want to include instructions which set or clear the control flags such as the interrupt enable flag and the direction flag.

The best way to approach the initialization task is to make a checklist of all the registers, programmable devices, and flags in the system you are working on. Then you can mark the ones you need for a specific program and determine the instructions needed to initialize each part. An initialization list for an 8086-based system, such as the SDK-86 prototyping board, might look like the following.

INITIALIZATION LIST

Data segment register

Stack segment register

Extra segment register

Stack pointer register

Base pointer register

Source index register

Destination index register

8255 programmable ports

8259A priority interrupt controller

8254 programmable counter

8251A programmable serial port

Initialize data variables

Reset/clear direction flag and interrupt enable flag

As you can see, the list can become quite lengthy even though we have not included all of the devices a system might commonly have. Note that initializing the code segment register is absent from this list. The code segment register gets loaded with the correct starting value by the system command you use to run the program. Now let's see how you put all of these parts together to make a program.

A STANDARD PROGRAM FORMAT

In this section we show you the form your programs should have if you are going to construct the machine codes for each instruction by *hand*. A later section of this chapter will show you the additional parts you need to add to the program if you are going to use an *assembler* to produce the binary codes for the instructions.

To help you format your programs, *assembly language coding sheets* such as that shown in Figure 3-4 are available. The *address* column is used for the address or the offset of a code byte or data byte. The actual code bytes or data bytes are put in the *data/code* column. A *label* is a name which represents an address referred to in a jump or call instruction. A label is put in the *label* column. It is followed by a colon (:) if it is used by a jump or call instruction in the same code segment. The *opcode* column contains the mnemonics for the instructions. The *operand* column contains the registers, memory locations, or data acted upon by the instructions. A *comment* column gives you space to describe the function of the instruction for future reference.

Figure 3-4 shows how the instructions for the "read temperature, add +7, store result in memory" program can be written in sequence on a coding sheet. We will discuss here the operation of these instructions to the extent needed. If you want more information about any of these, detailed descriptions of the *syntax* (assembly language grammar) and operation of each of these instructions can be found in Chapter 6.

The first line at the top of coding form in Figure 3-4 does not represent an instruction. It simply indicates that we want to set aside a memory location to store the result. This location must be in available RAM so that we can write to it. Address 00100H is an available RAM location on an SDK-86 prototyping board, for example. Next, we decide where in memory we want to start putting the code bytes for the instructions of the program. Again, on an SDK-86 prototyping board, address 00200H and above is available RAM, so we chose to start the program at address 00200H.

The first operation we want to do in the program is to initialize the data segment register. As discussed previously, two MOV instructions are used to do this. The MOV AX, 0010H instruction, when executed, will load

PROGRAMMER D. V. HALL SHEET 1 OF 1

PROGRAM TITLE READ TEMPERATURE & CORRECT DATE: 1/1/86

ABSTRACT: This program reads in a temperature value from a sensor connected to port 05H, adds a correction factor of +7 to the value read in, and then stores the result in a reserved memory location.

PROCEDURES: None called.

REGISTERS USED: Ax

FLAGS AFFECTED: All conditional

PORTS: Uses 05 as input port

MEMORY: 00100H–DATA; 00200H–0020CH, CODE

ADDRESS	DATA or CODE	LABELS	MNEM.	OPERAND(S)	COMMENTS
00100	XX				Reserve memory location to store
00101					result. This location will be loaded
00102					with a data byte as read in
00103					& corrected by the program.
00104					XX means "don't care" about
00105					contents of location.
00106					
00107					
00108					
00109					
0010A					
0010B					
0010C					
0010D					
0010E					Code starts here
0010F					Note break in address
200	B8		MOV	AX, 0010H	Initialize DS to point to start of
01	10				memory set aside for storing data
02	00				
03	8E		MOV	DS, AX	
04	D8				
05	E4		IN	AL, 05H	Read temperature from
06	05				port 05H
07	04		ADD	AL, 07H	Add correction factor
08	07				of +07
09	A2		MOV	[0000], AL	Store result in reserved
0A	00				memory
0B	00				
0C	CC		INT	3	Stop, wait for command
0D					from user
0E					
0F					

FIGURE 3-4 Assembly language program on standard coding form.

the upper 16 bits of the address we chose for data storage into the AX register. The MOV DS, AX instruction will copy this number from the AX register to the data segment register. Now we get to the instructions that do what we started out to do. The IN AL, 05H instruction will copy a data byte from port 05H to the AL register. The ADD AL, 07 instruction will add 07H to the AL register and leave the result in the AL register. The MOV [0000], AL instruction will copy the byte in AL to a memory location at a displacement of 0000H from the data segment base. In other words, AL will be copied to a physical address computed by shifting the data segment base in DS, 0010H, four bit positions left and adding the displacement, 0000H, contained in the instruction. The data will then be copied to physical address 00100H in memory. This is an example of the direct addressing mode described near the end of the previous chapter.

The INT 3 instruction at the end of the program functions as a *breakpoint*. In most 8086 systems, when the 8086 executes this instruction it will cause the 8086 to stop executing the instructions of your program and return control to the *monitor* or *system program*. You can then use *system commands* to look at the contents of registers and memory locations, or run another program. Without an instruction such as this at the end of the program, the 8086 would fetch and execute the code bytes for your program, and then it would go on fetching meaningless bytes from memory and trying to execute them as if they were code bytes.

The next major section of this chapter will show you how to construct the binary codes for these and other 8086 instructions so that you can assemble and run the programs on a development board such as the SDK-86. First, however, we want to use Figure 3-4 to make an important point about writing assembly language programs.

DOCUMENTATION

In a previous section of this chapter we stressed the point that you should do a lot of thinking and carefully write down the algorithm for a program before you start writing instruction statements. You should also document the program itself so that its operation is clear to you and to anyone else who needs to understand it.

Each page of the program should contain the name of the program, the page number, the name of the programmer, and perhaps a version number. Each program or procedure should have a heading block which contains an *abstract* describing what the program is supposed to do, which procedures it calls, which registers it uses, which ports it uses, which flags it affects, the memory used, and any other information which will make it easier for another programmer to interface with the program.

Comments should be used generously to describe the specific function of an instruction or group of instructions. Not every statement needs an individual comment. Comments should not just repeat the instruction mnemonic.

We cannot overemphasize the importance of clear, concise documentation in your programs. Experience has shown that even a short program that you wrote a month ago and forgot to put comments on may not be at all understandable to you now.

CONSTRUCTING THE MACHINE CODES FOR 8086 INSTRUCTIONS

This section shows you how to construct the binary codes for 8086 instructions. Most of the time you will probably use an assembler do this for you, but it is useful to understand how the codes are constructed. If you have an 8086-based prototyping board such as the Intel SDK-86 available, knowing how to hand-code instructions will enable you to code, enter, and run simple programs as you work your way through the 8086 instruction set examples in the next chapters.

Instruction Templates

To code the instructions for 8-bit processors such as the 8085, all you have to do is look up the hexadecimal code for each instruction on a one-page chart. For the 8086 the process is not quite as simple. Here's why. There are 32 ways to specify the source of the operand in an instruction such as MOV CX, source. The source of the operand can be any one of eight 16-bit registers, or a memory location specified by any one of 24 memory addressing modes. Each of the 32 possible instructions requires a different binary code. If CX is made the source rather than the destination then there are 32 ways of specifying the destination. Each of these 32 possible instructions requires a different binary code. There are then 64 different codes for MOV instructions using CX as a source or as a destination. Likewise, another 64 codes are required to specify all of the possible MOVs using CL as a source or a destination, and 64 more are required to specify all of the possible MOVs using CH as a source or a destination. The point here is that, because there is such a large number of possible codes for the 8086 instructions, it is impractical to list them all in a simple table. Instead, we use a *template* for each basic instruction type and fill in bits within this template to indicate the desired addressing mode, data type, etc. In other words, we build up the instruction codes on a bit-by-bit basis.

Different Intel literature shows the code templates for the 8086 instructions in two slightly different formats. One format is shown at the end of the 8086 data sheet in Appendix A. The second format is shown along with the 8086 instruction timings in Appendix B. We will start by showing you how to use the templates shown in the 8086 data sheet. As a first example of how to use these templates we will build the code for the IN AL, 05H instruction from our example program. Figure 3-5*a* shows the template for this instruction. Note that two bytes are required for the instruction. The upper 7 bits of the first byte tell the 8086 that this is an "input from a fixed port" instruction. The bit labeled W in the template is used to tell the 8086 whether you want to input a byte or input a word. If you want the 8086 to input a byte from

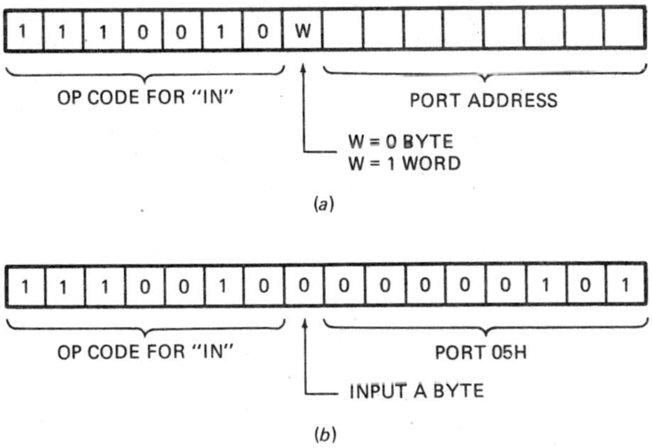

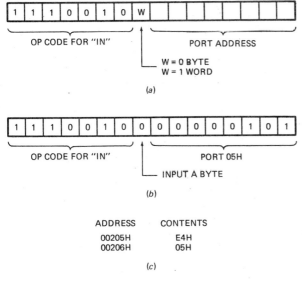

ADDRESS	CONTENTS
00205H	E4H
00206H	05H

(c)

FIGURE 3-5 Coding template for 8086 IN (fixed-port) instruction. *(a)* Template. *(b)* Example. *(c)* Hex codes in sequential memory locations.

an 8-bit port to AL, then make the W bit a 0. If you want the 8086 to input a word from a 16-bit port to the AX register, then make the W bit a 1. The 8-bit port address, 05H or 00000101 binary, is put in the second byte of the instruction. When the program is loaded into memory to be run, the first instruction byte will be put in one memory location and the second instruction byte will be put in the next. Figure 3-5*c* shows this in hexadecimal form as E4H, 05H.

To further illustrate how these templates are used, we will show here several examples with the simple MOV instruction. We will then construct the codes for the example program in Figure 3-4. Other examples will be shown as needed in the following chapters. Figure 3-6 shows the coding template or format for 8086 instructions which MOV data from a register to a register, from a register to a memory location, or from a memory location to a register. Note that at least two code bytes are required for the instruction.

The upper 6 bits of the first byte are an opcode which indicates the general type of instruction. Look in the table in Appendix A to find the 6-bit opcode for this MOV register/memory to /from register instruction. You should find it to be 100010. The W bit in the first word is used to indicate whether a byte or a word is being moved. If you are moving a byte, make this bit a 0. If you are moving a word, make this bit a 1. In this instruction, one operand must always be a register, so 3 bits in the second byte are used to indicate which register is involved. The 3-bit codes for each register are shown at the end of the table in Appendix A and in Figure 3-7*a*. Look in one of these places to find the code for the CL register. You should get 001. The D bit in the first byte of the instruction code is used to indicate whether the data is being moved *to* the register identified in the REG field of the second byte or *from* that register. If you are moving data *to* the register identified in the REG field, make the D bit a 1. If you are moving data *from* that register, make the D bit a 0.

Now remember that in this instruction one operand must be a register. The 2-bit field labeled MOD and the 3-bit field labeled R/M in the second byte of the instruction code are used to specify the desired addressing mode for the other operand. Figure 3-8 shows the MOD and R/M bit patterns for each of the 32 possible addressing modes.

If the other operand in the instruction is also one of the eight registers, then put in 11 for the MOD bits in the code template. Put the 3-bit code for that register in the R/M bits in the code template.

For the case where the other operand is a memory location there are 24 ways of specifying how the execution unit should compute the effective address of the operand in memory. Remember from Chapter 2 that the effective address can be specified directly in the instruction, it can be contained in a register, or it can be the sum of one or two registers and a displacement. Figure 3-8 shows the MOD and R/M codes for each of the 24 ways of specifying an effective address. The MOD code indicates whether the address specification in the in-

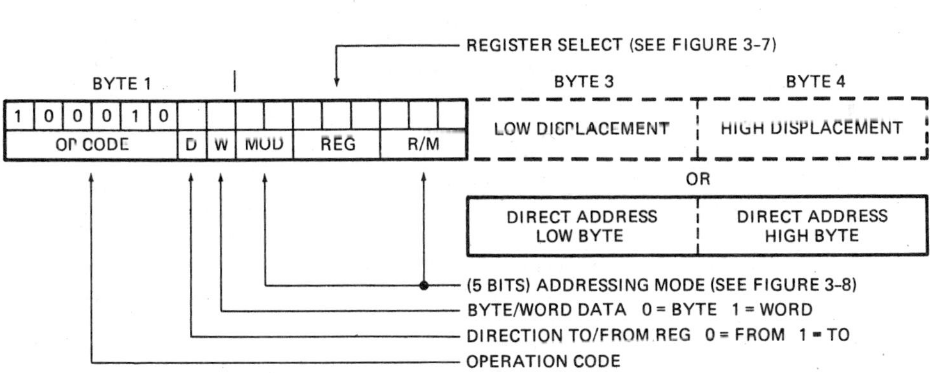

FIGURE 3-6 Coding template for 8086 instructions which MOV data between registers or between a register and a memory location.

```
REG FIELD BIT ASSIGNMENTS
IF W = 1   6-BIT REGISTER

    REGISTER CODE

      AX    ØØØ

      CX    ØØ1

      DX    Ø1Ø

      BX    Ø11

      SP    1ØØ

      BP    1Ø1

      SI    11Ø

      DI    111

IF W = Ø    8-BIT REGISTER

    REGISTER CODE

      AL    ØØØ

      CL    ØØ1

      DL    Ø1Ø

      BL    Ø11

      AH    1ØØ

      CH    1Ø1

      DH    11Ø

      BH    111
```

(a)

```
    SEGREG    CODE

      ES       ØØ

      CS       Ø1

      SS       1Ø

      DS       11
```

(b)

FIGURE 3-7 Instruction codes for 8086 registers. *(a)* General purpose, pointers, and index. *(b)* Segment registers.

struction contains a displacement. The R/M code indicates which register(s) contain part(s) of the effective address. Here's how it works:

1. If the specified effective address contains no displacement as in the instruction MOV CX, [BX] or in the instruction MOV [BX][SI], DX, then make the MOD bits 00, and chose the R/M bits which correspond to the register(s) containing the effective address. For example, if an instruction contains [BX], the 3-bit R/M code is 111. For an instruction which contains [BX][SI], the R/M code is 000. Note that for direct addressing where the displacement of the operand from the segment base is specified directly in the instruction, MOD is 00 and R/M is 110. For an instruction using direct addressing the low byte of the direct address is put in as a third instruction code byte of the instruction, and the high byte of the direct address is put in as a fourth instruction code byte.
2. If the effective address specified in the instruction contains a displacement less than 256 along with a reference to the contents of a register, as in the instruction MOV CX, 43H[BX], then code in MOD as 01, and chose the R/M bits which correspond to the register(s) which contain the part(s) for the effective address. For the instruction MOV CX, 43H[BX], MOD will be 01 and R/M will be 111. Put the 8-bit value of the displacement in as the third byte of the instruction.
3. If the expression for the effective address contains a displacement which is too large to fit in 8 bits, as in the instruction MOV DX, 4527H[BX], then put in 10 for MOD, and chose the R/M bits which correspond to the register(s) which contain the part(s) for the effective address. For the instruction MOV DX, 4527H[BX] the R/M bits are 111. The low byte of the displacement is put in as a third byte of the instruction. The high byte of the displacement is put in as a fourth byte of the instruction. The examples which follow should help clarify all of this for you.

MOV Instruction Coding Examples

All of the examples in this section use the MOV instruction template in Figure 3-6. As you read through these examples, it is a good idea to keep track of the bit-by-bit development on a separate paper for practice.

CODING MOV SP, BX

This instruction will copy a word from the BX register to the SP register. Consulting the table in Appendix A, you find that the 6-bit opcode for this instruction is 100010. Make the W bit a 1 because you are moving a word. The D bit for this instruction may be somewhat confusing, however. Since two registers are involved, you can think of the move as *to* SP, or *from* BX. Actually, it does not matter which you assume as long as you are consistent in coding the rest of the instruction. If you think of the instruction as moving a word *to* SP, then make the D bit

R/M \ MOD	00	01	10	11	
				W = 0	W = 1
000	(BX) + (SI)	(BX) + (SI) + d8	(BX) + (SI) + d16	AL	AX
001	(BX) + (DI)	(BX) + (DI) + d8	(BX) + (DI) + d16	CL	CX
010	(BP) + (SI)	(BP) + (SI) + d8	(BP) + (SI) + d16	DL	DX
011	(BP) + (DI)	(BP) + (DI) + d8	(BP) + (DI) + d16	BL	BX
100	(SI)	(SI) + d8	(SI) + d16	AH	SP
101	(DI)	(DI) + d8	(DI) + d16	CH	BP
110	d16 (direct address)	(BP) + d8	(BP) + d16	DH	SI
111	(BX)	(BX) + d8	(BX) + d16	BH	DI
	MEMORY MODE			REGISTER MODE	

d8 = 8-bit displacement d16 = 16-bit displacement

FIGURE 3-8 MOD and R/M bit patterns for 8086 instructions. The effective address (EA) produced by these addressing modes will be added to the data-segment base to form the physical address except for those cases where BP is used as part of the EA. In that case the EA will be added to the stack-segment base to form the physical address. You can use a segment override prefix to indicate that you want the EA to be added to some other segment base.

a 1, and put 100 in the REG field to represent SP. The MOD field will be 11 to represent register addressing mode. Make the R/M field 011 to represent the other register, BX. The resultant code for the instruction MOV SP,BX will be 10001011 11100011. Figure 3-9*a* shows the meaning of all of these bits.

If you change the D bit to a 0 and swap the codes in the REG and R/M fields, you will get 10001001 11011100, which is another equally valid code for the instruction. Figure 3-9*b* shows the meaning of the bits in this form. This second form, incidentally, is the form that the Intel 8086 Macroassembler produces.

CODING MOV CL, [BX]

This instruction will copy a byte to CL from the memory location whose effective address is contained in BX. The effective address will be added to the data segment base in DS to produce the physical address.

To find the 6-bit opcode for byte one of the instruction, consult the table in Appendix A. You should find that this code is 100010. Make the D bit a 1 because data is being moved *to* register CL. Make the W bit a 0 because the instruction is moving a byte into CL. Next you need to put the 3-bit code which represents register CL in the REG field of the second byte of the instruction code. The codes for each register are shown in Figure 3-7. In this figure you should find that the code for CL is 001. Now, all you need to determine is the bit patterns for the MOD and R/M fields. Again use the table in Figure 3-8 to do this. To use the table, first find the box containing the desired addressing mode. The box containing [BX], for example, is in the lower left corner of the table. Read the required MOD-bit pattern from the top of the column. In this case, MOD is 00. Then read the required R/M-bit pattern at the left of the box. For this instruction you should find R/M to be 111. Assembling all of these bits together should give you 10001010 00001111 as the binary code for the instruction MOV CL, [BX]. Figure 3-9*c* summarizes the meaning of all the bits in this result.

MOV 43H[SI], DH

This instruction will copy a byte from the DH register to a memory location. The effective address of the memory location will be computed by adding the indicated displacement of 43H to the contents of the SI register. The actual physical address will be produced by shifting the contents of the data segment base in DS 4 bits left and adding this effective address to the result.

The 6-bit opcode for this instruction is again 100010. Make the D bit a 0 because you are moving *from* a register. Make the W bit a 0 because you are moving a byte. Put 110 in the REG field to represent the DH register. The R/M field will be 100 because SI contains part of the effective address. Make the MOD field 01 because the displacement contained in the instruction, 43H, will fit in one byte. If the specified displacement had been a number larger than FFH, then MOD would have been 10. Putting all these pieces together gives 10001000 01110100 for the first two bytes of the instruction code. The specified displacement, 43H or 01000011 binary is put after these two as a third instruction byte. Figure 3-9*d* shows this. If an instruction specifies a 16-bit displacement, then the low byte of the displacement is put

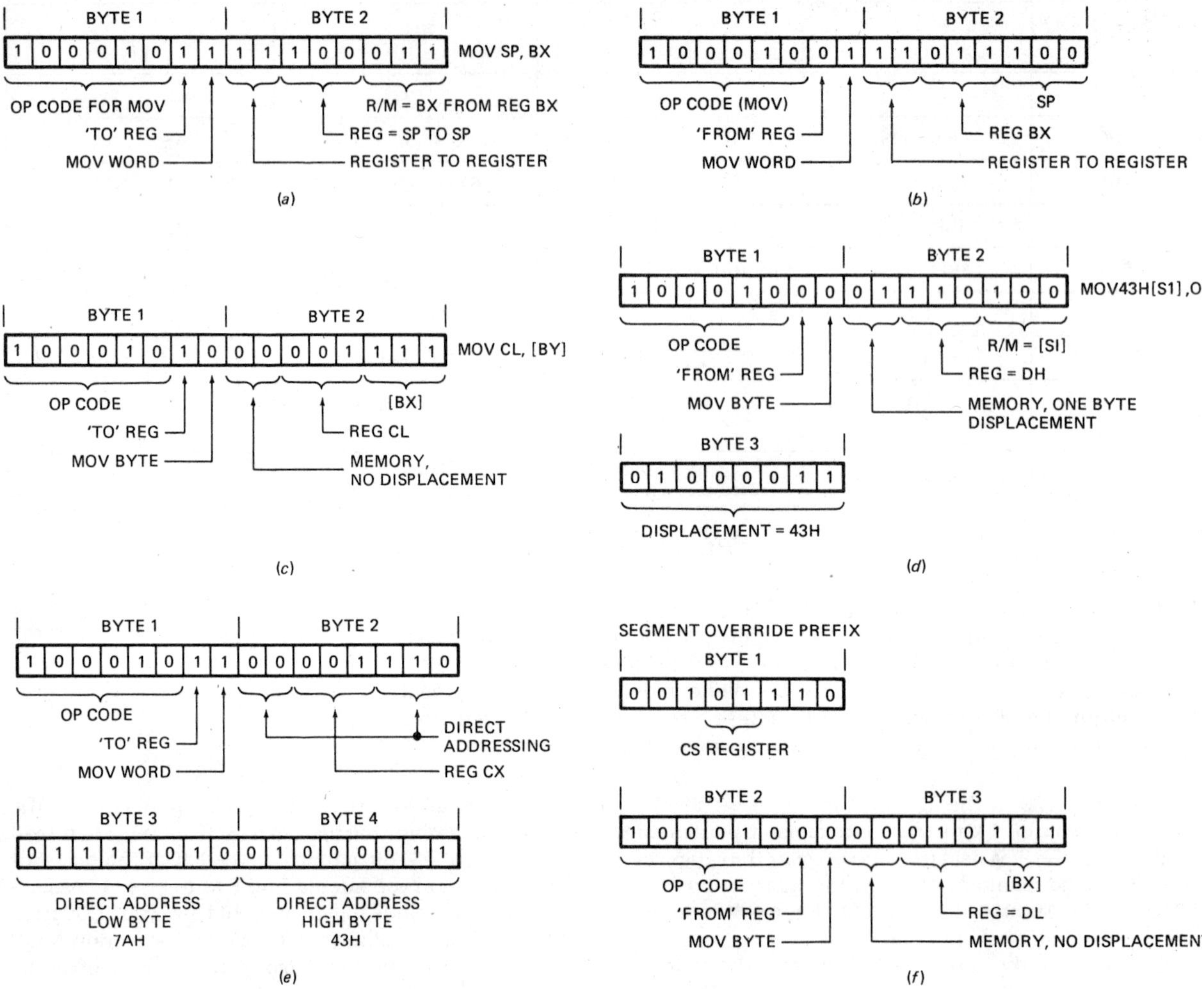

FIGURE 3-9 MOV instruction coding examples. *(a)* MOV SP, BX. *(b)* MOV SP, BX alternative. *(c)* MOV CL, [BX]. *(d)* MOV 43H [SI], DH. *(e)* MOV CX, [437AH]. *(f)* MOV CS:[BX], DL.

in as byte three of the instruction code, and the high byte of the displacement is put in as byte four of the instruction code.

CODING MOV CX, [437AH]

This instruction copies the contents of two memory locations into CX. The direct address or displacement of the first memory location from the start of the data segment is 437AH. The physical memory address will be produced by shifting the contents of the data segment register, DS, 4 bit positions left and adding this direct address to the result.

The 6-bit opcode for this instruction is again 100010. Make the D bit a 1 and the W bit a 1 because you are moving a word to CX. Put 001 in the REG field to represent the CX register, and then consult Figure 3-8 to find the MOD and R/M codes. In the first column of the figure you should find a box labeled "direct addressing," which is the mode specified by this instruction. For direct addressing you should find MOD to be 00 and R/M to be 110. The first two code bytes for the instruction then are 10001011 00001110. These two bytes will be followed by the low byte of the direct address, 7AH (01111010 binary). The high byte of the direct address 43H (01000011 binary) will be put after that. The instruction will be coded into four successive memory addresses as 8BH, 0EH, 7AH, and 43H. Figure 3-9*e* spells this out in detail.

CODING MOV CS:[BX], DL

This instruction copies a byte from the DL register to a memory location. The effective address for the memory location is contained in the BX register. Normally an effective address in BX will be added to the data segment base in DS to produce the physical memory address. In this instruction, CS: indicates that we want the BIU to add the effective address to the code segment base in CS to produce the physical address. The CS: is called a *segment override prefix*.

When an instruction containing a segment override prefix is coded, an 8-bit code for the segment override prefix is put in memory before the code for the rest of the instruction. The code byte for the segment override prefix has the format 001XX110. You insert a 2-bit code in place of the X's to indicate which segment base you want the effective address to be added to. The codes for these 2 bits are as follows: 00 = ES, 01 = CS, 10 = SS, and 11 = DS. The segment override prefix byte for CS then is 00101110. For practice, code out the rest of this instruction. Figure 3-9*f* shows the result you should get and how the code for the segment override prefix is put before the other code bytes for the instruction.

Coding the Example Program

Again, as you read through this section follow the bit-by-bit development of the instruction codes on a separate paper for practice.

MOV AX, 0010H

This instruction will move the immediate word 0010H into the AX accumulator. The simplest code template to use for this instruction is listed in the table in Appendix A under the MOV "immediate to register" heading. The format for it is 1011 W REG, data byte low, data byte high. Make the W bit a 1 because you want to move a word. Consult Figure 3-7 to find the code for the AX register. You should find this to be 000. Put this 3-bit code in the REG field of the instruction code. The completed instruction code byte is 10111000. Put the low byte of the immediate number, 10H, in as the second code byte. Then put the high byte of the immediate data, 00H, in as the third code byte.

MOV DS, AX

This instruction copies the contents of the AX register into the data segment register. The template to use for coding this instruction is found in the table in Appendix A under the heading "Register/memory to segment register." The format for this template is 10001110 MOD 0 segreg R/M . Segreg represents the 2-bit code for the desired segment register. These codes are also found in the table at the end of Appendix A. The segreg code for the DS register is 11. Since the other operand is a register, MOD should be 11. Put the 3-bit code for the AX register, 000, in the R/M field. The resultant codes for the two code bytes should then be 10001110 11011000.

IN AL, 05H

This instruction copies a byte of data from port 05H to the AL register. The coding for this instruction was described in a previous section. The code for the instruction is 11100100 00000101.

ADD AL, 07H

This instruction adds the immediate number 07H to the AL register and puts the result in the AL register. The simplest template to use for coding this instruction is found in the table in Appendix A under the heading "ADD—Immediate to accumulator." The format is 0000010W, data byte, data byte. Since we are adding a byte, the W bit should be a 0. The immediate data byte we are adding will be put in the second code byte. The third code byte will not be needed because we are only adding a byte.

MOV [0000], AL

This instruction copies the contents of the AL register to a memory location. The direct address or displacement of the memory location from the start of the data segment is 0000H. The code template for this instruction is found in the table in Appendix A under the heading "MOV—Accumulator to memory." The format for the instruction is 1010001W, address low byte, address high byte. Since the instruction moves a byte, the W bit should be a 0. The low byte of the direct address is written in as the second instruction code byte, and the high byte of the direct address is written in as the third instruction code byte. The codes for these 3 bytes then will be 10100010 00000000 00000000.

INT 3

In most 8086 systems this instruction causes the 8086 to stop executing instructions and do nothing but wait for the user to tell it what to do next. According to the format table in the appendix, the code for this instruction is the single byte 11001100 or CCH.

SUMMARY OF HAND CODING THE EXAMPLE PROGRAM

Figure 3-4 shows the example program with all the instruction codes in sequential order as you would write them so that you could load the program into memory and run it. Codes are in HEX to save space.

A Look at Another Coding Template Format

As we mentioned previously, Intel literature shows the 8086 instruction coding templates in two different forms. The preceeding sections have shown you how to use the templates found in the 8086 data sheet in Appendix A. Now let's take a brief look at the second form shown along with the instruction clock cycles in Appendix B.

The only difference between the second form for the templates and the form we discussed previously is that the D and W bits are not individually identified. Instead, the complete opcode bytes are shown for each version of an instruction. For example, in Appendix B the opcode byte for the MOV memory, register 8 instruction is shown as 88H, and the opcode byte for the MOV memory, register 16 instruction is shown as 89H. The only difference between the two codes is that the W bit is a 0 for the 8-bit move and the W bit is a 1 for the 16-bit move. One important point to make about using the templates in Appendix B is that for operations which involve two registers, the 3-bit code for the source register is put in the REG field of the MOD/RM instruction byte. The 3-bit code for the destination register is put in the R/M field of the MOD/RM instruction byte. The instruction MOV BX, CX, for example, is coded out as 10001001 11001011, or 89H CBH. You can use whichever set of templates you find easier to use.

A Few Words About Hand Coding

If you have to hand-code 8086 assembly language programs, here are a few tips to make your life easier. First, check your algorithm very carefully to make sure that it really does what it is supposed to do. Second, initially write down just the assembly language statements and comments for your program. You can check the table in the appendix to determine how many bytes each instruction takes so you know how many blank lines to leave between instruction statements. You may find it helpful to insert three or four NOP instructions after every nine or ten instructions. The NOP instruction doesn't do anything but kill time. However, if you accidentally leave out an instruction in your program, you can replace the NOPs with the needed instruction. This way you don't have to rewrite the entire program after the missing instruction.

After you have written down the instruction statements, recheck very carefully to make sure you have the right instructions to implement your algorithm. Then, work out the binary codes for each instruction and write them in the appropriate places on the coding form.

Hand coding is laborious for long programs. When writing long programs, it is much more efficient to use an assembler. The next section of this chapter shows you how to write your programs so you can use an assembler to produce the machine codes for the instructions.

WRITING PROGRAMS FOR USE WITH AN ASSEMBLER

If you have an 8086 assembler available, you should learn to use it as soon as possible. Besides doing the tedious task of producing the binary codes for your instruction statements, an assembler also allows you to refer to data items by name rather than by numerical addresses. As you should soon see, this greatly reduces the work you have to do and makes your programs much more readable. In this section we show you how to write your programs so that you can use an assembler on them. The assemblers used for the programs in this book were the Intel 8086/8088/80186/80188 Macro Assembler and the Microsoft Macro Assembler for the IBM Personal Computer. If you are using another assembler, some features may be slightly different so consult the manual for it.

Program Format

The best way to approach this section seems to be to show you a simple, but complete, program written for an assembler and explain the function of the various parts. By now you are probably tired of the "read temperature, add +7, and store result in memory" program, so we will use another example.

Figure 3-10 shows an 8086 assembly language program which multiplies two 16-bit binary numbers to give a 32-bit binary result. If you have a development system or a computer with an 8086 assembler to work on, this is a good program for you to key in, assemble, and run to become familiar with the operation of your system. If you are working on a prototyping board such as the SDK-86, you can construct the binary codes for each of the instructions, load the program into the on-board RAM, and run it. In any case, you can use the structure of this example program as a model for your own programs.

In addition to program instructions, the example program in Figure 3-10 contains directions to the assembler. These directions to the assembler are commonly called *assembler directives* or *pseudo operations*. A section at the end of Chapter 6 lists and describes for your reference a large number of the available assembler directives. Here we will discuss the basic assembler directives you need to get started writing programs. We will introduce more of these directives as we need them in the next two chapters.

SEGMENT and ENDS Directives

The SEGMENT and ENDS directives are used to identify a group of data items or a group of instructions that you want to be put together in a particular segment. These directives are used in the same way that parentheses are used to group like terms in algebra. A group of data statements or a group of instruction statements contained between SEGMENT and ENDS directives is called a *logical segment*. When you set up a logical segment, you give it a name of your choosing. In the example program the statements DATA_HERE SEGMENT and DATA_HERE ENDS set up a logical segment named DATA_HERE. There is nothing sacred about the name DATA_HERE. We simply chose this name to help us remember that this logical segment contains data statements. The statements CODE_HERE SEGMENT and CODE_HERE ENDS in the example program set up a logical segment named CODE_HERE which contains instruction statements. The Intel and the IBM 8086 macro assemblers, incidentally, allow you to use names and labels of up to 31 characters. You can't use spaces in a name, but you can use an underscore as shown to separate words in a name. Also, you can't use instruction mnemonics as segment names or labels. Throughout the rest of the program you will refer to a logical segment by the name that you give it when you define it.

```
PAGE ,132              ; Makes listing file lines 132 characters wide
;8086 program
;ABSTRACT              : This program multiplies the two 16-bit words in
                       ; the memory locations called MULTIPLICAND and
                       ; MULTIPLIER. The result is stored in the memory
                       ; location called PRODUCT
;PORTS USED            : None
;PROCEDURES USED: None
;REGISTERS USED : CS, DS, DX and AX

DATA_HERE       SEGMENT

MULTIPLICAND    DW      204AH            ; first word here
MULTIPLIER      DW      3B2AH            ; second word here
PRODUCT         DW      2 DUP(0)         ; result here

DATA_HERE       ENDS


CODE_HERE       SEGMENT
                ASSUME CS : CODE_HERE, DS : DATA_HERE

                MOV AX, DATA_HERE        ; initialize DS register
                MOV DS,AX

                MOV AX, MULTIPLICAND     ; get one word
                MUL MULTIPLIER           ; multiply by second word
                MOV PRODUCT, AX          ; store low word of result
                MOV PRODUCT+2, DX        ; store high word of result
                INT 3                    ; wait for command from user

CODE_HERE       ENDS
                END
```

FIGURE 3-10 Assembly language source program to multiply two 16-bit binary numbers to give a 32-bit result.

A logical segment is not usually given a physical starting address when it is declared. After the program is assembled, and perhaps linked with other assembled program modules, it is then assigned the physical address where it will be loaded in memory to be run.

Data and Addresses Naming Directives—EQU, DB, DW, and DD

Programs work with three general categories of data: constants, variables, and addresses. The value of a constant does not change during the execution of the program. The number 7 is an example of a constant you might use in a program. A variable is the name given to a data item which can change during the execution of a program. The current temperature of an oven is an example of a variable. Addresses are referred to in many instructions. You may, for example, load an address into a register or jump to an address.

Constants, variables, and addresses used in your programs can be given names. This allows you to refer to them by name rather than having to remember or calculate their value each time you refer to them in an instruction. In other words, if you give names to constants, variables, and addresses the assembler can use these names to find the desired data item or address when you refer to it in an instruction. Specific directives are used to give names to constants and variables in your programs. Labels are used to give names to addresses in your programs.

THE EQU DIRECTIVE

The EQU or *equate* directive is used to assign a name to constants used in your programs. The statement CORRECTION_FACTOR EQU 07H, in a program such as our previous example, would tell the assembler to insert the value 07H every time that it finds the name CORRECTION_FACTOR in a program statement. In other words, when the assembler reads the statement ADD AL, CORRECTION_FACTOR, it will automatically code the instruction as if you had written it ADD AL, 07H. Here's the advantage of using an EQU directive to declare constants at the start of your program. Suppose that you use the correction factor of +07H 23 times in your program. Now the company you work for changes brands of temperature sensor and the new correction factor is +09H. If you used the number 07H in the 23 instructions which contain this correction factor, then you have to go through the entire program, find each instruction that uses the correction factor, and update the value. Murphy's law being what it is, you are likely to miss one or two of these, and the program won't work correctly. If you used an EQU at the start of your program and then referred to CORRECTION_FACTOR by name in the 23 instructions, then all you do is change the value in the EQU statement from 07H to 09H and reassemble the program. The assembler automatically inserts the new value of 09H in all 23 instructions.

NOTE In large programs consisting of modules assembled separately, constants must be declared in each module. The assembler has no way to remember an EQU value from one module when it assembles another module.

DB, DW, AND DD DIRECTIVES

The DB, DW, and DD directives are used to assign names to variables in your programs. The DB directive after a name specifies that the data is of *type byte*. The program statement OVEN_TEMPERATURE DB 27, for example declares a variable of type byte and gives it the name OVEN_TEMPERATURE. DW is used to specify that the data is of *type word* (16 bits), and DD is used to specify that the data is of *type double word* (32 bits). If a number is written after the DB, DW, or DD, the data item will be *initialized* with that value when the program is loaded from disk into RAM. The statement CONVERSION_FACTORS DB 27H,48H,32H,69H will declare a data item of 4 bytes and initialize the 4 bytes with the specified 4 values. If we don't care what a data item is initialized to then we can indicate this with a "?", as in the statement TARE_WEIGHT DW ?. Note that data variables which are changed during the operation of a program should also be initialized with program instructions so that the program can be rerun from the start without reloading it to initialize the variables. Figure 3-10 shows three more examples of naming and initializing data items.

The first example, MULTIPLICAND DW 204AH, declares a data word named MULTIPLICAND, and initializes that data word with the value 204AH. What this means is that the assembler will set aside two successive memory locations and assign the name MULTIPLICAND to the first location. As you will see, this allows us to access the data in these memory locations by name. The MULTIPLICAND DW 204AH statement also indicates that when the final program is loaded into memory to be run, these memory locations will be loaded with (initialized to) 204AH. Actually, since this is an Intel microprocessor, the first address in memory will contain the low byte of the word, 4AH, and the second memory address will contain the high byte of the word, 20H.

The second data declaration example in Figure 3-10, MULTIPLIER DW 3B2AH, sets aside storage for a word in memory and gives the starting address of this word the name MULTIPLIER. When the program is loaded, the first memory address will be initialized with 2AH, and the second memory location with 3BH.

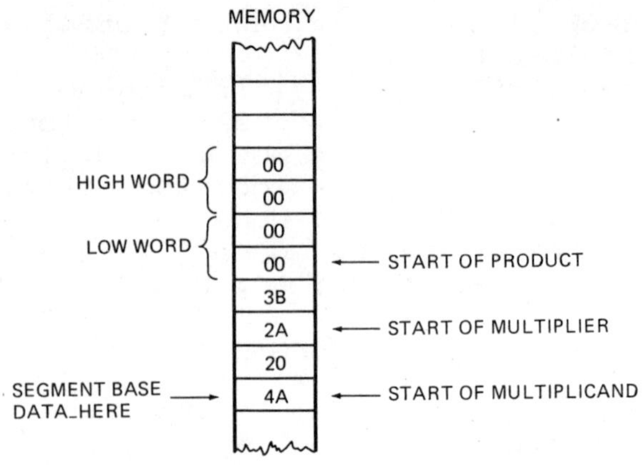

FIGURE 3-11 Data arrangement in memory for multiply program.

The third data declaration example in Figure 3-10, PRODUCT DW 2 DUP(0), sets aside storage for two words in memory and gives the starting address of the first word the name PRODUCT. The DUP(0) part of the statement tells the assembler to initialize the two words to all zeros. When we multiply two 16-bit binary numbers, the product can be as large as 32 bits. Therefore, we must set aside this much space to store the product. We could have used the DD directive to declare PRODUCT as a double word, but, since in the program we move the result to PRODUCT one word at a time, it is more convenient to declare PRODUCT as 2 words.

Figure 3-11 shows how the data for MULTIPLICAND, MULTIPLIER, and PRODUCT will actually be arranged in memory starting from the base of the DATA_HERE segment. The first byte of MULTIPLICAND, 4AH, will be at a displacement of zero from the segment base, because MULTIPLICAND is the first data item declared in the logical segment DATA_HERE. The displacement of the second byte of MULTIPLICAND is 0001. The displacement of the first byte of MULTIPLIER from the segment base is 0002H, and the displacement of the second byte of MULTIPLIER is 0003H. These are the displacements that we would have to figure out for each data item if we were not using names to refer to them.

If the logical segment DATA_HERE is eventually put in ROM or EPROM, then MULTIPLICAND will function as a constant, because it cannot be changed during program execution. However, if DATA_HERE is eventually put in RAM then MULTIPLICAND can function as a variable because a new value could be written in those memory locations during program execution.

Types of Numbers Used in Data Statements

All of the previous examples of DB, DW, and DD declarations use hexadecimal numbers as indicated by an "H" after the number. You can, however, put in a number in any one of several other forms. For each form you must tell the assembler which form you are using.

BINARY

For example, when you use a binary number in a statement, you put a B after the string of 1's and 0's to let the assembler know that you want the number to be treated as a binary number. The statement TEMP_MAX DB 01111001B is an example. If you want to put in a negative binary number, write the number in its 2's complement sign-and-magnitude form.

OCTAL

To indicate that you want a number to be evaluated as base-8 or octal, put a Q after the string of octal digits. The statement OLD_COMPUTER DW 7341Q is an example.

DECIMAL

The assembler treats a number with no identifying letter after it as a decimal number. In other words, if you forget to put an H after a number that you want the assembler to treat as hexadecimal, the assembler will treat it as a decimal number. The assembler automatically converts a decimal number in a statement to binary so the value can be loaded into memory. The statement TEMPERATURE_MAX DB 49 is an example. If you indicate a negative number in a data declaration statement, the assembler will convert the number to its 2's complement sign-and-magnitude form. For example, given the statement TEMP_MIN DB −20, the assembler will insert the value 11101100, which is the 2's complement representation for −20 decimal.

NOTE You can put a D after decimal values if you want to more clearly indicate that the value is decimal.

HEXADECIMAL

As shown in several previous examples, a hexadecimal number is indicated by an H after the hexadecimal digits. The statement MULTIPLIER DW 3B2AH is an example.

ASCII

ASCII characters can be put in data declaration statements by enclosing them in single quotation marks. The statement BOY_1 DB 'ALBERT', for example, tells the assembler to set aside six memory locations named BOY_1. It also tells the assembler to put the ASCII code for A in the first memory location, the ASCII code for L in the second, the ASCII code for B in the third, etc. The assembler will automatically determine the ASCII codes for the letters or numbers within the quotes.

NOTE ASCII can only be used with the DB directive.

DECIMAL REAL AND HEXADECIMAL REAL

These two types are used to represent noninteger numbers such as 3.14159. We will discuss how these are used in Chapter 11.

Accessing Named Data with Program Instructions

Now that we have shown you how the data structure is set up, let's look at how program instructions access this data. Temporarily skipping over the first two instructions in the CODE_HERE section of the program in Figure 3-10, find the instruction MOV AX, MULTIPLICAND. This instruction, when executed, will copy a word from memory to the AX register. When the assembler reads through this program the first time, it will automatically calculate the displacement of each of the named data items from the segment base DATA_HERE. Referring to Figure 3-11 you can see that the displacement of MULTIPLICAND from the segment base is 0000. This is because MULTIPLICAND is the first data item declared in the segment. The assembler, then, will find that the displacement of MULTIPLICAND is 0000H. When the assembler reads the program the second time to produce the binary codes for the instructions, it will insert this displacement as part of the binary code for the instruction MOV AX, MULTIPLICAND. Since we know that the displacement of MULTIPLICAND is 0000, we could have written the instruction as MOV AX, [0000]. However, there would be a problem if we later changed the program by adding another data item before MULTIPLICAND in DATA_HERE. The displacement of MULTIPLICAND would be changed. Therefore, we would have to remember to go through the entire program and correct the displacement in all instructions that access MULTIPLICAND. If you use a name to refer to each data item as shown, the assembler will automatically calculate the correct displacement of that data item for you and insert this displacement each time you refer to it in an instruction.

To summarize how this works, then, the instruction MOV AX, MULTIPLICAND is an example of direct addressing where the direct address or displacement within a segment is represented by a name. For instructions such as this, the assembler will automatically calculate the displacement of the named data item from the start of the segment and insert this value as part of the binary code for the instruction. When the instruction executes, the BIU will add the displacement contained in the instruction to the data segment base in DS. (Remember, the contents of DS are shifted 4 bit positions left before the displacement is added.) This addition produces the 20-bit physical address needed to address the data named MULTIPLICAND in memory.

The next instruction in the program in Figure 3-10 is another example of direct addressing using a named data item. The instruction MUL MULTIPLIER multiplies the word named MULTIPLIER in DATA_HERE times the word in the AX register. The low word of the result is left in the AX register, and the high word of the result is left in the DX register. When the assembler reads through this program the first time, it will find the displacement of MULTIPLIER in DATA_HERE is 0002H. When it reads through the program the second time it inserts this displacement as part of the binary code for the MUL instruction. When the MUL MULTIPLIER instruction executes, the BIU will add the displacement contained in the instruction to the data segment base in DS to address MULTIPLIER in memory.

The next instruction, MOV PRODUCT, AX, in the program in Figure 3-10 copies the low word of the result from AX to memory. The low byte of AX will be copied to a memory location named PRODUCT. The high byte of

AX will be copied to the next higher address which we can refer to as PRODUCT + 1.

The following instruction in the program, MOV PRODUCT + 2, DX, copies the high word of the multiplication result from DX to memory. When the assembler reads this instruction, it will add the indicated "2" to the displacement it calculated for PRODUCT and insert the result as part of the binary code for the instruction. Therefore, when the instruction executes, the low byte of DX will be copied to memory at a displacement of PRODUCT + 2. The high byte of DX will be copied to a memory location which we can refer to as PRODUCT + 3. Figure 3-11 shows how the two words of the product are put in memory. Note that the lower byte of a word is always put in the lower memory address.

This example program should show you that if you are using an assembler, names are a very convenient way of specifying the direct address of data in memory.

Naming Addresses—Labels

Names representing addresses are called *labels*. They are written in the label field of an instruction statement or a directive statement. One major use of labels is to represent the destination for jump and call instructions. Suppose, for example, we want the 8086 to jump back to some previous instruction over and over. Instead of computing the numerical address that we want to jump to, we put a label in front of the instruction we want to jump to and write the jump instruction as JMP label. Here is a specific example.

```
NEXT: IN    AL, 05H ; Get data sample form port 05H
      .             ; Process data value read in.
      .
      .
      JMP NEXT      ; Get next data value and process
```

If you use a label to represent an address as shown in this example, the assembler will automatically calculate the address that needs to be put in the code for the jump instruction. The next two chapters show many examples of the use of labels with jump and call instructions.

Another example of using a name to represent an address is in the SEGMENT directive statement. The name DATA_HERE in the statement DATA_HERE SEGMENT, for example, represents the starting address of a segment named DATA_HERE. Later we show you how we use this name to initialize the data segment register. We will now discuss some other parts of the example program that you will need to use in your programs.

The ASSUME Directive

An 8086 program may have several logical segments which contain code, several that contain data, and several that can serve as a stack. However, at any given time the 8086 works directly with only four physical segments; a *code segment*, a *data segment*, a *stack segment*, and an *extra segment*. The ASSUME directive tells the assembler which logical segment to use for each of these physical segments at a given time.

In Figure 3-10, for example, the statement ASSUME CS:CODE_HERE, DS:DATA_HERE tells the assembler that the logical segment CODE_HERE contains the instruction statements for the program and should be treated as a code segment. It also tells the assembler that it should treat the logical segment DATA_HERE as the data segment for this program. In other words, the DS:DATA_HERE part of the statement tells the assembler that, for any instruction which refers to data in the data segment, that data will be found in the logical segment DATA_HERE. The ASSUME . . . DS:DATA_HERE, for example, tells the assembler that a named data item such as MULTIPLICAND is contained in the logical segment called DATA_HERE. Given this information, the assembler can construct the binary codes for the instruction. The displacement of MULTIPLICAND from the start of DATA_HERE will be inserted as part of the instruction by the assembler.

If you are using an assembler, you must use an ASSUME statement in your program. Also, if you are using the stack segment and the extra segment in your program, you must include terms in the statement to tell the assembler the name of the logical segment to use for the stack and the name of the logical segment to use for the extra segment. These additional terms might look like: SS:STACK_HERE, ES:EXTRA_HERE. As we will show later, you can put another ASSUME directive later in the program to tell the assembler to use different logical segments from that point on.

If the ASSUME directive is not completely clear to you at this point, don't worry. We show many more examples of its use thoroughout the rest of the book. We introduced the ASSUME directive here because you need to put it in your programs for most 8086 assemblers. You can use the assume statement in Figure 3-10 as a model of how to write this directive for your programs.

Initializing Segment Registers

The ASSUME directive tells the assembler the names of the logical segments to use as code segment, data segment, stack segment, and extra segment. The assembler uses displacements from the start of the specified logical segment to code out instructions. When the instructions are executed, the displacements in the instructions will be added to segment registers to produce the actual physical addresses. The assembler, however, cannot directly load the segment registers with the starting physical addresses of the segments.

The segment registers, other than the code segment register, must be initialized by program instructions before they are used to access data. The first two instructions of the example program in Figure 3-10 show how this is done for the data segment register. DATA_HERE in the first instruction represents the upper 16 bits of the starting address you give the segment DATA_HERE. Since the 8086 does not allow us to move this immediate number directly into the data segment register, we must first load it into one of the general-purpose registers and then copy it into the data segment register. MOV AX, DATA_HERE loads the upper 16 bits of the segment starting address into the AX register. MOV DS, AX copies this value from AX to the data segment register. This is the same operation we described for hand coding the example program in Figure 3-4, except that here we use the segment name instead of a number to refer to the segment base address. In this example we used the AX register to pass the value, but any 16-bit register other than a segment register can be used. If you are hand coding your programs, you can just insert the upper 16 bits of the 20-bit segment starting address in place of DATA_HERE in the instruction. For example, if in your particular system you decide to locate DATA_HERE at address 00300H, DS should be loaded with 0030H. If you are using an assembler, you can use the segment name to refer to its base address as shown in the example.

If you use the stack segment and the extra segment in a program, the stack segment register and the extra segment register must also be initialized by program instructions in the same way.

When the assembler reads through your assembly language program, it calculates the displacement of each named variable from the start of the logical segment that contains it. The assembler also keeps track of the displacement of each instruction code byte from the start of a logical segment. The CS:CODE_HERE part of the ASSUME statement in Figure 3-10 tells the assembler to calculate the displacements of the following instructions from the start of the logical segment CODE_HERE. In other words, it tells the assembler that, when this program is run, the code segment register will contain the upper 16-bits of the address where the logical segment CODE_HERE was located in memory. The instruction byte displacements that the assembler is keeping track of are the values that the 8086 will put in the instruction pointer, IP, to fetch each instruction byte.

There are several ways that the CS register can be loaded with the code segment base address and the instruction pointer can be loaded with the displacement of the instruction byte to be fetched next. The first way is with the command you give your system to execute a program starting at a given address. A typical command of this sort is G = 0010:0000 <CR>. (<CR> means "press the return key.") This command will load CS with 0010 and load IP with 0000. The 8086 will then fetch and execute instructions starting from address 00100, the address produced when the BIU shifts CS and adds IP. The other ways of loading CS and IP will be discussed in later sections.

The END Directive

The END directive, as the name implies, tells the assembler to stop reading. Any instructions or statements that you write after an END directive will be ignored.

ASSEMBLY LANGUAGE PROGRAM DEVELOPMENT TOOLS

Introduction

For all but the very simplest assembly language programs you will probably want to use some type of *microcomputer development system* and *program development tools* to make your work easier. These systems usually contain several hundred Kbytes of RAM, a keyboard and video display, floppy and/or hard disk drives, a printer, and an emulator. Figure 3-12 shows an Applied Microsystems ES 1800 16-bit emulator which can be added to an IBM PC/AT or compatible computer to produce a complete 8086/80186/80286 development system. The following sections give you an introduction to several common program development tools which you use with these systems. Most of these tools are programs which you run to perform some function on the program you are writing. You will have to consult the manuals for your system to get the specific details for it, but this section should give you an overview of the steps involved in developing an assembly language microcomputer program using a system. An accompanying lab manual steps you through the use of all these tools with the SDK-86 board and the IBM Personal Computer.

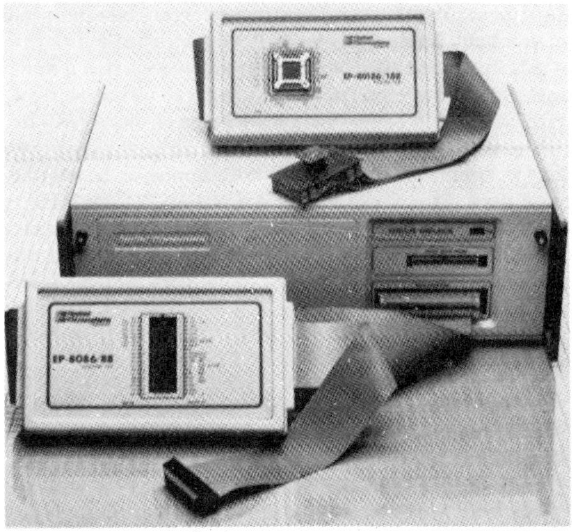

FIGURE 3-12 Applied Microsystems ES 1800 16-bit emulator. *(Applied Microsystems Corp.)*

Editor

An *editor* is a program which, when run on a system, lets you type in the assembly language statements for your program. Examples of editors are ALTER which runs on Intel systems, EDLIN which runs on IBM PCs, and Wordstar which runs on most systems. The main function of an editor is to help you construct your assembly language program in just the right format so that the assembler will translate it correctly to machine language. Figure 3-10 shows an example of the format you should use when typing in your program. This form of your program is called the *source program*. The actual position of each field on a line is not important, but you must put the fields of each statement in the correct order, and you must leave at least one blank between fields. Whenever possible, we like to line the fields up in columns so that it is easier to read the program.

As you type in your program, the editor stores the ASCII codes for the letters and numbers in successive

RAM locations. If you make a typing error the editor will let you back up and correct it. If you leave out a program statement, the editor will let you move everything down and insert the line. This is much easier than working with pencil and paper, even if you type as slowly as I do.

When you have typed in all of your program, you then copy it from memory to a file on a floppy or hard magnetic disk. This file, such as the one in Figure 3-10, is called a *source file*. If you later find that your program contains errors, you can use the editor to load the source file back into RAM and make the needed corrections in the source program.

Assembler

An *assembler* program is used to translate assembly language mnemonics to the correct binary code for each instruction. The assembler will read the source file of your program from the disk where you saved it after editing. An assembler usually reads your source file more than once. On the first pass through the source program, the assembler finds everything. It determines the displacement of named data items and the offset of labels, and puts this information in a *symbol table*. On a second pass through the source program, the assembler produces the binary code for each instruction and assigns addresses to each.

The assembler generates two files on the floppy or hard disk. The first file is called the *object file*. The object file contains the binary codes for the instructions and information about the addresses of the instructions. This file contains the information that will eventually be loaded into memory and executed. The second file generated by the assembler is called the *assembler list file*. Figure 3-13 shows the assembler list file for the source program in Figure 3-10. This file contains the assembly language statements, the binary codes for each instruction, and the offset for each instruction. You usually send this file to a printer so that you will have a printout of the entire program to work with when you are testing and troubleshooting the program. The assembler listing will also indicate any typing or syntax (assembly language grammar) errors you made in typing in your source program.

NOTE The assembler will not tell you if you made a programming error. You usually have to run the program to find these. To correct the errors indicated on the listing, you use the editor to reedit your source program and save the corrected source program on disk. You then reassemble the corrected source program. It may take several times through the edit-assemble loop before you get all of the syntax errors out of your source program.

Now let's take a look at some of the information given on the assembler listing. The left-most column in the listing gives the offsets of data items from the start of the data segment and the offsets of code bytes from the start of the code segment. Note that the assembler does not generate absolute physical addresses. A linker or locator will do this later. Also note that the MOV AX, DATA_HERE statement is assembled with some blanks after the basic instruction code because the start of DS is not known at the time the program is assembled.

The trailer section of the listing in Figure 3-13 gives some additional information about the segments and names used in the program. The statement CODE_HERE 0014 PARA NONE, for example, tells you that the segment CODE_HERE is 14H bytes long and will be located at a physical address whose lower 4 bits are 0000. The statement MULTIPLIER . . . . L WORD 0002 DATA_HERE tells you that MULTIPLIER is a variable of type word and that it is located at an offset of 0002 in the segment DATA_HERE.

Linker

A *linker* is a program used to join together several object files into one large object file. When writing large programs it is usually much more efficient to divide the large program into smaller *modules*. Each module can be individually written, tested, and debugged. When all of the modules work they can be linked together to form a large functioning program. Also, the object modules for useful programs, a square root program, for example, can be kept in a *library file* and linked into other programs as needed.

The linker produces a *link file* which contains the binary codes for all the combined modules. The linker also produces a *link map* file which contains the address information about the linked files. The linker, however, does not assign absolute addresses to the program, it only assigns relative addresses starting from zero. This form of the program is said to be *relocatable*, because it can be put anywhere in memory to be run. If you are going to run your program on a system such as the IBM PC, you can just load the link file into memory and run it. If you are going to run your program on a system such as the Intel Series IV, then you must use a *locator program* to assign absolute addresses to the linker file.

Locator

A *locator* is a program used to assign the specific addresses of where the object code is to be loaded into memory. A locator program that comes with the IBM PC Disk Operating System (DOS) is called EXE2BIN. Here's how you proceed if you want to produce a program with absolute addresses that you can download to an SDK-86 from an IBM PC. First build a source (.ASM) file using the EDLIN or perhaps the WORDSTAR editor. Assemble the source file with the IBM PC Macroassembler (MASM) to produce the .OBJ file. Use the LINK program to produce a relocatable .EXE file. Then use the EXE2BIN program to give your program an absolute starting address such as 0010:0000H. Finally, use the SDKDMP program from Chapter 13 to download the .BIN file produced by EXE2BIN and run it. In some systems a single program performs both the link and the locate functions.

Debugger

If your program requires no external hardware or requires only hardware accessible directly from your system, then you can use a *debugger* to run and debug your program. A debugger is a program which allows you to load your object code program into system mem-

```
The IBM Personal Computer MACRO Assembler 03-06-85      PAGE    1-1
                                PAGE ,132       ; Makes listing file lines 132 characters wide
                                ;8086 program
                                ;ABSTRACT       : This program multiplies the two 16-bit words in
                                                ; the memory locations called MULTIPLICAND and
                                                ; MULTIPLIER. The result is stored in the memory
                                                ; location called PRODUCT
                                ;PORTS USED     : None
                                ;PROCEDURES USED: None
                                ;REGISTERS USED : CS, DS, DX and AX

 0000                           DATA_HERE       SEGMENT

 0000  204A                     MULTIPLICAND    DW      204AH           ; first word here
 0002  3B2A                     MULTIPLIER      DW      3B2AH           ; second word here
 0004    02 [                   PRODUCT         DW      2 DUP(0)        ; result here
             0000
                  ]
 0008                           DATA_HERE       ENDS

 0000                           CODE_HERE       SEGMENT
                                                ASSUME CS : CODE_HERE, DS : DATA_HERE

 0000  B8 ---- R                                MOV AX, DATA_HERE       ; initialize DS register
 0003  8E D8                                    MOV DS,AX

 0005  A1 0000 R                                MOV AX, MULTIPLICAND    ; get one word
 0008  F7 26 0002 R                             MUL MULTIPLIER          ; multiply by second word
 000C  A3 0004 R                                MOV PRODUCT, AX         ; store low word of result
 000F  89 16 0006 R                             MOV PRODUCT+2, DX       ; store high word of result
 0013  CC                                       INT 3                   ; wait for command from user

 0014                           CODE_HERE       ENDS
                                                END

The IBM Personal Computer MACRO Assembler 03-06-85      PAGE    Symbols-1

Segments and groups:

                N a m e                 Size    align   combine class

CODE_HERE. . . . . . . . . . . .        0014    PARA    NONE
DATA_HERE. . . . . . . . . . . .        0008    PARA    NONE

Symbols:
                N a m e                 Type    Value   Attr

MULTIPLICAND . . . . . . . . . .        L WORD  0000    DATA_HERE
MULTIPLIER . . . . . . . . . . .        L WORD  0002    DATA_HERE
PRODUCT. . . . . . . . . . . . .        L WORD  0004    DATA_HERE       Length =0002

Warning Severe
Errors  Errors
0       0
```

FIGURE 3-13 Assembler listing for example program in Figure 3-10.

ory, execute the program, and troubleshoot or "debug" it. The debugger allows you to look at the contents of registers and memory locations after your program runs. It allows you to change the contents of registers and memory locations and rerun the program. Some debuggers allow you stop execution after each instruction so you can check or alter memory and register contents. A debugger also allows you to set a *breakpoint* at any point in your program. When you run the program the system will execute instructions up to this breakpoint and stop. You can then examine register and memory contents to see if the results are correct at that point. If the results are correct, you can move the breakpoint to a later point in the program. If the results are not correct, you can check the program up to that point to find out why they are not correct. The debugger tools can help you isolate a problem in your program. Once you find the problem, you can then cycle back and correct the algorithm if necessary. You then use the editor to correct your source program, reassemble the corrected source program, relink, and run the program again.

Microprocessor prototyping boards such as the SDK-86 contain a debugger program in ROM. On boards such as this the debugger is commonly called a *monitor program* because it lets you monitor program activity. The SDK-86 monitor program, for example, lets you enter and run programs, single step through programs, examine register and memory contents, and insert breakpoints. The DEBUG program, used with the IBM PC, allows you to do the same functions and also has a trace function which shows you the contents of all the registers after each instruction executes.

Emulator

Another way to run your program is with an *emulator*. An emulator is a mixture of hardware and software. It is usually used to test and debug the hardware and software of an external system such as the prototype of a microprocessor-based instrument. Part of the hardware of an emulator is a multiwire cable which connects the host system to the system being developed. A plug at the end of the cable is plugged into the prototype in place of its microprocessor. Through this connection the software of the emulator allows you to download your object-code program into RAM in the system being tested and run it. As with a debugger, an emulator allows you to load and run programs, examine and change the contents of registers, examine and change the contents of memory locations, and insert breakpoints in the program. The emulator also takes a "snapshot" of the contents of registers, activity on the address and data bus, and the state of the flags as each instruction executes. The emulator stores this *trace data*, as it is called, in a large RAM. You can do a printout of the trace data to see the results that your program produced on a step-by-step basis.

Another powerful feature of an emulator is the ability to use either system memory or the memory on the prototype for the program you are debugging. In a later chapter we discuss in detail the use of an emulator in developing a microprocessor-based product.

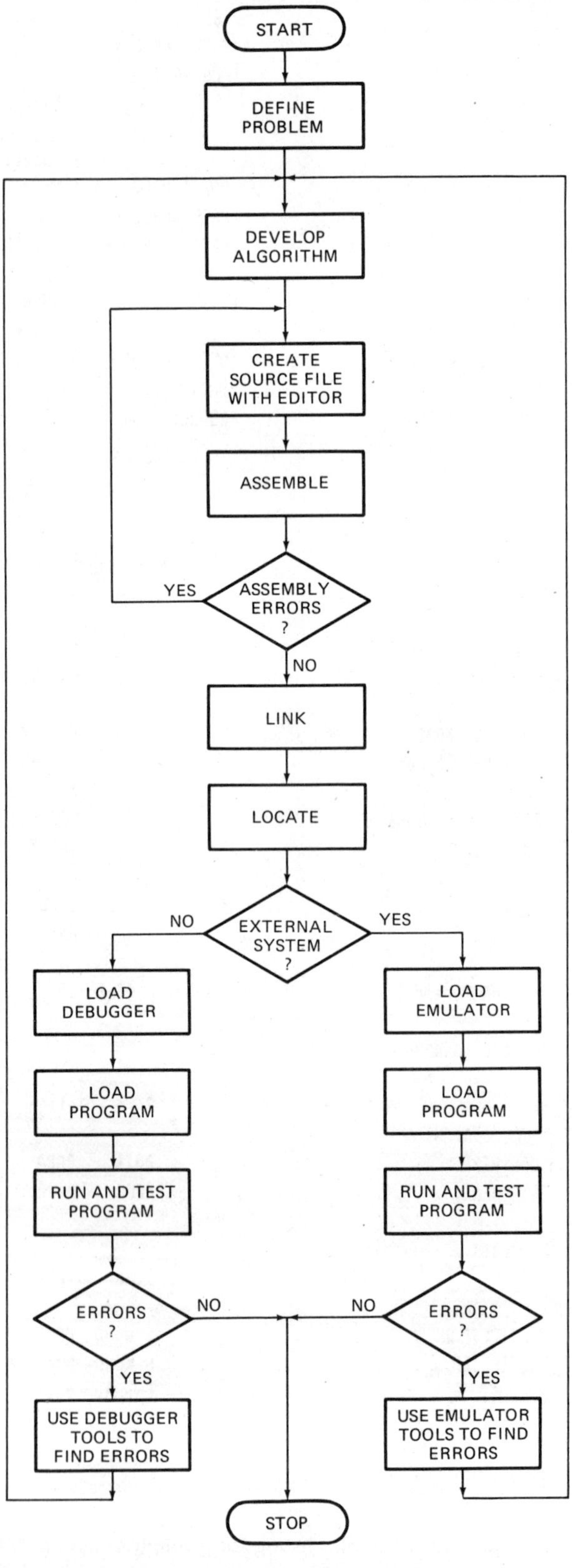

FIGURE 3-14 Program development algorithm.

Summary of the Use of Program Development Tools

Figure 3-14 shows in diagram form the order in which you will use the program development tools we have described. The first and most important step is to think out very carefully what you want the program to do and how you want the program to do it. Next, use an editor to create the source file for your program. Assemble the source file with the assembler. If the assembler list file indicates any errors in your program, use the editor to correct these errors. Cycle through the edit-assemble loop until the assembler tells you on the listing that it found no errors. If your program consists of several modules, then use the linker to join their object modules together into one large object module.

NOTE On some systems such as the IBM PC you must use the linker even if your program has only one module.

Now, if your system requires it, use the locate program to specify where in memory you want your program to be put. Your program is now ready to be loaded into memory and run. If your program does not interact with any external hardware other than that connected directly to the system, then you can use the system debugger to run and debug your program. If your program is intended to work with external hardware such as the prototype of a microprocessor-based instrument, then you will probably use an emulator to run and debug your program. We will be discussing and showing the use of these program development tools throughout the rest of this book, but this section should give you an overview.

CHECKLIST OF IMPORTANT TERMS AND CONCEPTS IN THIS CHAPTER

If you do not remember any of the terms or concepts in the following list, use the index to find them in the chapter.

Algorithm

Sequential task list

Flowcharts and flowchart symbols

Structured programming

Pseudocode

Top-down and bottom-up design

Sequence, repetition, and decision operations

IF—THEN—ELSE, IF—THEN, WHILE—DO, REPEAT—UNTIL, and CASE structures

8086 instruction types

Mnemonics

Initialization list

Standard program format

Documentation

Instruction template: W-bit, MOD, R/M, D-bit

Segment-override prefix

Assembler

Assembler directives: SEGMENT, ENDS, END, DB, DW, DD, EQU, ASSUME

Named data items

Development tools

Editor

Linker: library file, link files, link map, relocatable

Locator

Debugger, monitor program

Emulator, trace data

REVIEW QUESTIONS AND PROBLEMS

1. List the major steps in developing an assembly language program.

2. What is the main advantage of a top-down design approach to solving a programming problem?

3. Why is it necessary to develop a detailed algorithm for a program before writing down any assembly language instructions?

4. *a.* What are the three basic structure types used when writing programs?
   b. What is the advantage of using only these structures when writing the algorithm for a program?

5. A program is like a recipe. Use a flowchart or pseudocode to show the algorithm for the following recipe. The operations in it are sequence and repetition. Instead of implementing the resulting algorithm in assembly language, implement it in your microwave and use the result to help you get through the rest of the book.

   Peanut Brittle:

1	cup sugar	1 teaspoon butter
0.5	cup white corn syrup	1 teaspoon vanilla
1	cup unsalted peanuts	1 teaspoon baking soda

   i. Put sugar and syrup in 1.5 quart casserole (with handle) and stir until thoroughly mixed.

ii. Microwave at HIGH setting for 4 minutes.

iii. Add peanuts and stir until thoroughly mixed.

iv. Microwave at HIGH setting for 4 minutes. Add butter and vanilla, stir until well mixed and microwave at HIGH setting for 2 more minutes.

v. Add baking soda and gently stir until light and foamy. Pour mixture onto nonstick cookie sheet and let cool for 1 hour. When cool, break into pieces. Makes 1 pound.

6. Use a flowchart or pseudocode to show the algorithm for a program which gets a number from a memory location, subtracts 20H from it, and outputs 01H to port 3AH if the result of the subtraction is greater than 25H.

7. Given the register contents in Figure 3-15, answer the following questions:
   a. What physical address will the next instruction be fetched from?
   b. What is the physical address for the top of the stack?

8. Describe the operation and results of each of the following instructions, given the register contents shown in Figure 3-15. Include in your answer the physical address or register that each instruction will get its operands from and the physical address or register that each instruction will put the result. Use the instruction descriptions in Chapter 6 to help you. Assume that the instructions below are independent, not sequential unless listed together under a letter.
   a. MOV AX, BX
   b. MOV CL, 37H
   c. INC BX
   d. MOV CX, [246BH]
   e. MOV CX, 246BH
   f. ADD AL, DH
   g. MUL BX
   h. DEC BP
   i. DIV BL
   j. SUB AX, DX
   k. OR CL, BL
   l. NOT AH
   m. ROL BX, 1
   n. AND AL, CH
   o. MOV DS, AX
   p. ROR BX, CL
   q. AND AL, 0FH
   r. MOV AX, [BX]
   s. MOV [BX][SI], CL

9. See if you can spot the grammatical (SYNTAX) errors in the following instructions (use Chapter 6 to help you):
   a. MOV BH, AX
   b. MOV DX, CL
   c. ADD AL, 2073H
   d. MOV 7632H, CX
   e. IN BL, 04H

10. Show the results that will be in the affected registers or memory locations after each of the following groups of instructions execute. Assume that each group of instructions starts with the register and memory contents shown in Figure 3-15. (Use Chapter 6.)
   a. ADD BL, AL
      MOV [0004], BL
   b. MOV CL, 04
      ROR DI, CL
   c. MOV BX, 000AH
      MOV AL, [BX]
      SUB AL, CL
      INC BX
      MOV [BX], AL
   d. ADD AL, BH
      DAA

ES	6000		DATA SEGMENT	
CS	4000		C	D7
SS	7000		B	9A
DS	5000		A	7C
IP	43E8		9	DB
	AH	AL	8	C3
AX	42	35	7	B2
	BH	BL	6	49
BX	07	5A	5	21
	CH	CL	4	89
CX	00	04	3	71
	DH	DL	2	22
DX	33	02	1	4A
			5000H	3B
SP	0000			
BP	2468			
SI	4C00			
DI	7D00			

FIGURE 3-15 8086 register and memory contents for Problems 7, 8, and 9.

11. Write the 8086 instruction which will perform the indicated operation. Use the instruction overview in this chapter and the detailed descriptions in Chapter 6 to help you.
   a. Copies AL to BL
   b. Loads 43H into CL
   c. Increments the contents of CX by one
   d. Copies SP to BP
   e. Adds 07H to DL
   f. Multiplies AL times BL
   g. Copies AX to a memory location at offset 245AH in the data segment
   h. Decrements SP by one
   i. Rotates the most significant bit of AL into the least-significant bit position
   j. Copies DL to a memory location whose offset is in BX
   k. Masks the lower 4 bits of BL

l. Sets the most significant bit of AX to a one but does not affect the other bits

m. Inverts the lower 4 bits of BL but does not affect the other bits.

12. Construct the binary code for each of the following 8086 instructions.

a. MOV BL, AL

b. MOV [BX], CX

c. ADD BX, 59H[DI]

d. SUB [2048], DH

e. XCHG CH, ES:[BX]

f. ROR AX, 1

g. OUT DX, AL

h. AND AL, OFH

i. NOP

j. IN AL, DX

13. Describe the function of each assembler directive and instruction statement in the following short program.

```
;     pressure read program
DATA_HERE SEGMENT
      PRESSURE  DB   0          ;storage for pressure
DATA_HERE ENDS
PRESSURE_PORT  EQU  04H         ;pressure sensor connected
                                ;to port 04H
CORRECTION_FACTOR   EQU  07H    ;current correction factor, 07
CODE_HERE SEGMENT
     ASSUME    CS:CODE_HERE, DS:DATA_HERE
     MOV  AX, DATA_HERE
     MOV  DS, AX
     IN   AL, PRESSURE_PORT
     ADD  AL, CORRECTION_FACTOR
     MOV  PRESSURE, AL
CODE_HERE ENDS
          END
```

14. Describe how an assembly language program is developed and debugged using system tools such as editors, assemblers, linkers, locators, emulators, and debuggers.

15. Write the pseudocode representation for the flowchart in Figure 3-14.

CHAPTER

4 8086 Assembly Language Programming Techniques—Part 1

The purposes of this chapter are to show you how some of the standard program structures described in the last chapter are implemented in 8086 assembly language, how these structures are used to solve some common programming problems, and how some of the 8086 instructions work.

OBJECTIVES

At the conclusion of this chapter you should be able to:

1. Write flowcharts or pseudocode for simple programming problems.
2. Implement WHILE—DO and REPEAT—UNTIL program structures in 8086 assembly language.
3. Describe the operation of selected data transfer, arithmetic, logical, jump, loop, and string instructions.
4. Use based and indexed addressing modes to access data in your programs.
5. Describe a systematic approach to debugging a simple assembly language program using debugger, monitor, or emulator tools.

MORE PRACTICE WITH SIMPLE SEQUENCE PROGRAMS

Finding the Average of Two Numbers

DEFINING THE PROBLEM AND WRITING THE ALGORITHM

A common need in programming is to find the average of two numbers. Suppose, for example, we know the maximum temperature of a day and the minimum temperature of a day, and we want to determine the average temperature. The sequence of steps we go through to do this might look something like the following.

Add maximum temperature and minimum temperature.

Divide sum by two to get average temperature.

This sequence doesn't look much like an assembly language program and it shouldn't. The algorithm at this point should be general enough that it could be implemented in any programming language, or on any machine. Once you are reasonably sure of your algorithm, then you can start thinking about the architecture and instructions of the specific microcomputer on which you plan to run the program. Now let's show you how we get from the algorithm to the assembly language program for it.

SETTING UP THE DATA STRUCTURE

One of the first things for you to think about in this process is the data that the program will be working with. You need to ask yourself questions such as:

1. Will the data be in memory or in registers?
2. Is the data of type byte, type word, or perhaps type doubleword?
3. How many data items are there?
4. Does the data represent only positive numbers, or does it represent positive and negative (signed) numbers?
5. For more complex problems you might ask how the data is structured. For example, is the data in an array or in a record?

Let's assume for this example that the data is all in memory, the data is of type byte, and that the data represents only positive numbers in the range 0 to 0FFH. The top part of Figure 4-1, between the DATA_HERE SEGMENT and the DATA_HERE ENDS directives, shows how you might set up the data structure for this program. It is very similar to the data structure for the multiply example in the last chapter. In the logical segment called

```
;8086 PROGRAM
;ABSTRACT        : this program averages two temperatures
;                  name  HI_TEMP and LO_TEMP and puts the
;                  result in the memory location AV_TEMP.
;REGISTERS USED: DS, CS, AX, BX
;PORTS USED      : None used
;PROCEDURES      : None used

DATA_HERE        SEGMENT

HI_TEMP          DB        92H       ; max temp storage
LO_TEMP          DB        52H       ; low temp storage
AV_TEMP          DB        ?         ; put average here

DATA_HERE        ENDS

CODE_HERE        SEGMENT
                 ASSUME CS : CODE_HERE, DS : DATA_HERE

         MOV     AX, DATA_HERE       ; initialize data segment
         MOV     DS, AX
         MOV     AL, HI_TEMP         ; get first temperature
         ADD     AL, LO_TEMP         ; add second to it
         MOV     AH, 00H             ; clear all of AH register
         ADC     AH, 00H             ; put carry in LSB of AH
         MOV     BL, 02H             ; load divisor in BL register
         DIV     BL                  ; divide AX by BL
                                     ; quotient in AL,
                                     ; remainder in AH
         MOV     AV_TEMP, AL         ; copy result to memory

CODE_HERE        ENDS
                 END
```

FIGURE 4-1 8086 program to average two temperatures.

DATA_HERE, HI_TEMP is declared as a variable of type byte and initialized with a value of 92H. In an actual application, the value in HI_TEMP would probably be put there by another program which reads the output from a temperature sensor. The statement LO_TEMP DB 52H declares a variable of type byte and initializes it with the value 52H. The statement AV_TEMP DB ? sets aside a byte location to store the average temperature, but does not initialize the location to any value. When the program executes, it will write a value to this location.

INITIALIZATION CHECKLIST

Now that you have the data structure set up, let's start thinking about the instructions that we can use to perform the actions we want on this data. Although it does not show in the algorithm, we know from a discussion in Chapter 3 that we should start the program with a list of initialization instructions. Start by putting this checklist at the top of the paper. At this point you may not know exactly which parts on the checklist will have to be initialized, but the presence of the list will remind you that it has to be done. For this example program the only part you have to initialize is the data segment register.

CHOOSING INSTRUCTIONS

Next look at the major actions that you want the program to perform other than moving data from one place to another. You want the program to add two byte-type numbers together, so scan through the instruction groups in Chapter 3 to determine which 8086 instruction will do this for you. The ADD instruction is the obvious choice in this case. Now find and read the detailed discussion of this instruction in Chapter 6. From this discussion you can determine how the instruction works and see if it will do the necessary job. From the discussion of the ADD instruction you should find that the ADD instruction has the format ADD destination, source. A byte from the specified source is added to a byte in the specified destination, or a word from the specified source is added to a word in the specified destination. (Note that you cannot directly add a byte to a word.) The result in either case is put in the specified destination. The source can be an immediate number, a register, or a memory location. The destination can be a register or a memory location. The source and the destination cannot both be memory locations in an instruction. This means that you have to move one of the operands from memory to a register before you can do the ADD. Another point to consider here is that if you add two 8-bit numbers, the sum can be larger than 8 bits. Adding F0H and 40H, for example, gives 130H. The 8-bit destination will contain 30H, and the carry will be held in the carry flag. What this means is that you must collect the parts of the result in a location large enough to hold all 9 bits. A 16-bit register is a good choice. To summarize then, you need to move one of the numbers you want to add into a register such as AL, add the other number from memory to it, and move any carry produced by the addition to the upper half of the 16-bit register containing the result (which is in AL).

Now let's see how you can do this with program instructions. Take a look now at the first six instruction statements of the example program in Figure 4-1. As explained in the last chapter, the first two instructions, MOV AX, DATA_HERE and MOV DS, AX, are required to initialize the data segment register. These instructions load the DS register with the upper 16 bits of the starting address for the data segment. If you are using an assembler, you can use the name DATA_HERE in the instruction to refer to this address. If you are not using an assembler, then just put the hex for the upper 16 bits of the address in the MOV AX, DATA_HERE instruction in place of the name.

The next instruction in the example program in Figure 4-1, MOV AL, HI_TEMP, copies one of the temperatures from a memory location to the AL register. The name HI_TEMP in the instruction represents the direct address or displacement of the variable in the logical segment DATA_HERE. The ADD AL, LO_TEMP instruction adds a byte from memory to the contents of the AL register. The result of the addition (sum) is left in the AL register.

Now that we have done the addition, the next thing to do is get the carry bit where we want it. We would like to get the contents of the carry flag into the least significant bit of the AH register. The MOV AH, 00H instruction clears all of the bits of AH to 0's. The ADC AH, 00H instruction adds the immediate number 00H plus the contents of the carry flag to the contents of the AH register. The result will be left in the AH register. Since we cleared AH to all 0's before the add, what we are really adding is 00H + 00H + CF. The result of all this is that the carry bit ends up in the least-significant bit of AH, which is what we set out to do.

The next major action in our algorithm is to divide the sum of the two temperatures by two. Look at the instruction groups in the last chapter to see if the 8086 has a divide instruction. You should find that it has two divide instructions, DIV and IDIV. DIV is for dividing unsigned numbers, and IDIV is used for dividing signed binary numbers. Since in this example we are dividing unsigned binary numbers, look up the DIV instruction in Chapter 6 to find out how it works. The DIV instruction can be used to divide a 16-bit number in AX by a specified byte in a register or in a memory location. After the division an 8-bit quotient is left in the AL register and an 8-bit remainder is left in the AH register. The DIV instruction can also be used to divide a 32-bit number in the DX and AX registers by a 16-bit number from a specified register or memory location. In this case a 16-bit quotient is left in the AX register, and a 16-bit remainder is left in the DX register. In either case there is a problem if the quotient is too large to fit in the indicated destination. In a later chapter we discuss what to do about this problem. Fortunately, for the example here the data is such that the problem will not arise.

As you can see, we already have the sum of the two temperatures already positioned in the AX register, ready for the DIV operation. Before we can do the DIV operation, however, we have to get the divisor, 02H, into a register or memory location to satisfy the requirements of the DIV instruction. A simple way to do this is with the MOV BL, 02H instruction, which loads the immediate number 02H into the BL register. Now we can do the divide operation with the instruction DIV BL. The 8-bit quotient from the division will be left in the AL register. All we have left to do is to copy the quotient to the memory location we set aside for the average temperature. The instruction MOV AV_TEMP, AL will copy AL to this memory location. Take another look at Figure 4-1 to see how these instructions are added on to the previous instructions.

NOTE: We could have used the remainder in AH to round off the average temperature, but that would have made the program more complex than desired for this example.

SUMMARY OF CONVERTING AN ALGORITHM TO ASSEMBLY LANGUAGE

A first step in converting an algorithm to assembly language is to set up and declare the data structure that the algorithm will be working with. Then write down the instructions required for initialization at the start of the code section. Next determine the instructions required to implement the major actions in the algorithm, and decide how the data must be positioned for these instructions. Finally, insert the MOV or other instructions required to get the data in the correct position.

A Few Comments about the 8086 Arithmetic Instructions

The 8086 has instructions to add, subtract, multiply, and divide. It can operate on signed or unsigned binary numbers, BCD numbers, or numbers represented in ASCII. Rather than put a lot of arithmetic examples at this point in the book, we show arithmetic examples with each arithmetic instruction description in Chapter 6. The description of the MUL instruction in Chapter 6, for example, shows how unsigned binary numbers are multiplied. Also we show other arithmetic examples as needed throughout the rest of the book. If you need to do some arithmetic operations on the 8086 there are a few instructions in addition to the basic add, subtract, multiply, and divide instructions that you need to look up in Chapter 6.

If you are adding BCD numbers, you need to also look up the Decimal Adjust for Addition (DAA) instruction. If you are subtracting BCD numbers, then you need to look up the Decimal Adjust for Subtraction (DAS) instruction. If you are working with ASCII numbers, then you need to look up the ASCII Adjust after Addition (AAA) instruction, the ASCII Adjust after Subtraction (AAS) instruction, the ASCII Adjust after Multiply (AAM) instruction, and the ASCII Adjust before Division (AAD) instruction.

Converting Two ASCII Number Codes to Packed BCD

DEFINING THE PROBLEM AND WRITING THE ALGORITHM

If you type a 9 on an ASCII-encoded computer terminal keyboard, the 8 bit ASCII code sent to the computer will be 00111001 binary, or 39H. If you type a 5 on the keyboard, the code sent to the computer will be 00110101 binary or 35H, the ASCII code for 5. The ASCII codes for the numbers 0 through 9 are 30H through 39H. As you can see, the lower nibble of the ASCII codes contains the 4-bit BCD code for the number represented by the ASCII code. For many applications we want to convert the ASCII code coming in from the terminal to its simple BCD equivalent. We can do this by simply replacing the 3 in the upper nibble of the byte with four 0's. For example, suppose we read in 00111001 binary or 39H, the ASCII code for 9. If we replace the upper 4 bits with 0's, we are left with 00001001 binary or 09H. The lower 4 bits contain 1001 binary, the BCD code for 9. Numbers represented as one BCD digit per byte are referred to as *unpacked BCD*. If two BCD digits are put in a byte, this form is referred to as *packed BCD*. Figure 4-2 shows examples of ASCII, unpacked BCD, and packed BCD. When we want to store BCD numbers in memory the packed form is obviously more efficient because it has two BCD digits in each byte memory location. The problem we are going to work on here is how to convert two numbers from ASCII code form to unpacked BCD and then pack the two BCD digits into one byte. Figure 4-2 shows the steps in numerical form.

The algorithm for this problem can be stated simply as:

Convert first ASCII number to unpacked BCD.

Convert second ASCII number to unpacked BCD.

Move first BCD nibble to upper nibble position in byte.

Pack two BCD nibbles in one byte.

Now let's see how you can implement this algorithm in 8086 assembly language.

THE DATA STRUCTURE AND INITIALIZATION LIST

For this example program let's assume that the first ASCII code entered is in the BL register, and the second ASCII code entered is in the AL register. Since we are not using memory for data in this program, we do not need to declare a data segment. Also then we do not need to initialize the data segment register. In a real application this program would probably be a procedure or a part of a larger program.

MASKING WITH THE AND INSTRUCTION

The first operation in the algorithm is to convert a number in ASCII form to its unpacked BCD equivalent. This is done by replacing the upper 4 bits of the ASCII byte with four 0's. The 8086 AND instruction can be used to do this operation. Remember from basic logic or from the review in Chapter 1 that, when a 1 or a 0 is ANDed with a 0, the result is always a zero. ANDing a bit with a 0 is called *masking* that bit, because the previous state of the bit is hidden or masked. To mask 4 bits in a word, then, all you do is AND each bit you want to mask with a 0. A bit ANDed with a 1, remember, is not changed.

```
ASCII                          5     0011 0101  =  35H
ASCII                          9     0011 1001  =  39H

UNPACKED BCD                   5     0000 0101  =  05H
UNPACKED BCD                   9     0000 1001  =  09H

UNPACKED BCD                   5     0101 0000  =  50H
moved to upper nibble
PACKED BCD                     59    0101 1001  =  59H
```

FIGURE 4-2 ASCII, UNPACKED BCD, and PACKED BCD examples.

ASCII 5	0 0 1 1	0 1 0 1
MASK	0 0 0 0	1 1 1 1
RESULT	0 0 0 0	0 1 0 1

FIGURE 4-3 Effects of ANDing with 1's and 0's.

According to the description of the AND instruction in Chapter 6, the instruction has the format AND destination, source. The instruction ANDs each bit of the specified source with the corresponding bit of the specified destination and puts the result in the specified destination. The source can be an immediate number, a register, or a memory location specified in one of those 24 different ways. The destination can be a register or a memory location. The source and the destination must both be bytes, or they must both be words. The source and the destination cannot both be memory locations in an instruction.

For this example the first ASCII number is in the BL register, so we can just AND an immediate number with this register to mask the desired bits. The upper 4 bits of the immediate number should be 0's because these correspond to the bits we want to mask in BL. The lower 4 bits of the immediate number should be 1's because we want to leave these bits unchanged. The immediate number then should be 00001111 binary or 0FH. The instruction to convert the first ASCII number is AND BL, 0FH. When this instruction executes, it will leave the desired unpacked BCD in BL. Figure 4-3 shows how this will work for an ASCII number of 35H initially in BL.

For the next action in the algorithm we want to perform the same operation on a second ASCII number in the AL register. The instruction AND AL, 0FH will do this for us. After this instruction executes AL will contain the unpacked BCD for the second ASCII number.

MOVING A NIBBLE WITH THE ROTATE INSTRUCTION

The next action in the algorithm is to move the 4 BCD bits in the first unpacked BCD byte to the upper nibble position in the byte. We need to do this so that the 4 BCD bits are in the correct position for packing with the second BCD nibble. Take another look at Figure 4-2 to help you visualize this. What we are effectively doing here is swapping or exchanging the top nibble with the bottom nibble of the byte. If you check the instruction groups in Chapter 3 you will find that the 8086 has an exchange instruction, XCHG, which can be used to swap two bytes or to swap two words. The 8086 does not have a specific instruction to swap the nibbles in a byte. However, if you think of the operation that we need to do as shifting or rotating the BCD bits four bit positions to the left, this will give you a good idea which instruction will do the job for you. The 8086 has a wide variety of rotate and shift instructions. For now let's look at the rotate instructions. There are two instructions, ROL and RCL, which rotate the bits of a specified operand to the left. Figure 4-4 shows in diagram form how these two work. For ROL each bit in the specified register or memory location is rotated one bit position to the left. The bit that was the MSB is rotated around into the LSB position. The old MSB is also copied to the carry flag. For the RCL instruction each bit of the specified register or memory location is also rotated one bit position to the left. However, the bit that was in the MSB position is moved to the carry flag, and the bit that was in the carry flag is moved into the LSB position. As indicated by the C in the middle of the mnemonic, the carry flag is in the rotated loop when the RCL instruction executes.

In the example program we really don't want the contents of the carry flag rotated into our operand, so the ROL instruction seems to be the one we want. If you consult the ROL instruction description in Chapter 6, you will find that the instruction has the format ROL destination, count. The destination can be a register or a memory location. It can be a byte location or a word location. The count can be the immediate number 1 specified directly in the instruction, or the count can be a number previously loaded into the CL register. The instruction ROL AL,1 for example will rotate the contents of AL one bit position to the left. We could repeat this instruction four times to produce the shift of four bit positions we need for our BCD packing problem. However, there is an easier way to do it. We first load the CL register with the number of times we want to rotate AL. The instruction MOV CL, 04H will do this. Then we use the instruction ROL BL, CL to do the rotation. When it executes, this instruction will automatically rotate BL the number of bit positions loaded into CL. Note that for the 80186 you can write the single instruction ROL BL, 04H to do this job.

Now that we have determined the instructions needed to mask the upper nibbles and the instructions necessary to move the first BCD digit into position, the only thing left is to pack the upper nibble in BL and the lower nibble in AL into the same byte.

COMBINING BYTES OR WORDS WITH THE ADD OR THE OR INSTRUCTION

You can't use a standard MOV instruction to combine two bytes into one as we need to do here. The reason is that the MOV instruction copies an operand from a specified source to a specified destination. The previous

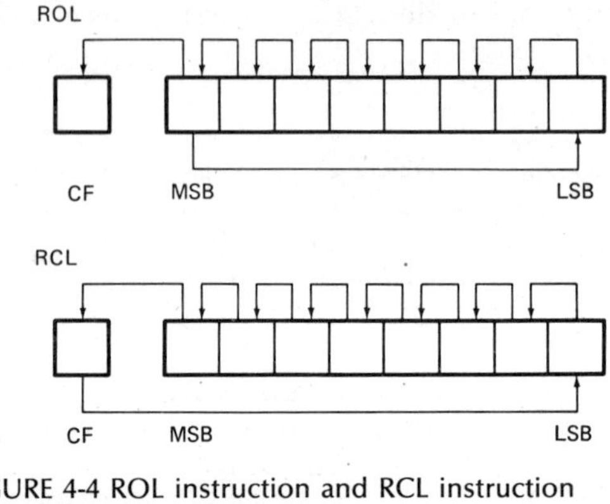

FIGURE 4-4 ROL instruction and RCL instruction operations for byte operands.

contents of the destination are lost. You can, however, use an ADD or an OR instruction to pack the two BCD nibbles.

As described in the previous program example, the ADD instruction adds the contents of a specified source to the contents of a specified destination and leaves the result in the specified destination. For the example program here, the instruction ADD AL, BL can be used to combine the two BCD nibbles. Take a look at Figure 4-2 to help you visualize this addition.

If you look up the OR instruction in Chapter 6, you will find that it has the format OR destination, source. This instruction ORs each bit in the specified source with the corresponding bit in the specified destination. The result of the ORing is left in the specified destination. Remember from basic logic or the review in Chapter 1 that ORing a bit with a 1 always produces a result of 1. ORing a bit with a 0 leaves the bit unchanged. To set a bit in a word to a 1 then, all you have to do is OR that bit with a word which has a 1 in that bit position and 0's in all the other bit positions. This is similar to the way the AND instruction is used to clear bits in a word to 0's. See the OR instruction description in Chapter 6 for examples of this.

For the example program here we use the instruction OR AL, BL to pack the two BCD nibbles. Bits ORed with 0's will not be changed. Bits ORed with 1's will become or stay 1's. Again look at Figure 4-2 to help you visualize this operation.

SUMMARY OF BCD PACKING PROGRAM

Figure 4-5 shows the complete program to produce a packed BCD byte from two ASCII bytes. Work your way through this to make sure you understand how each part works. In this program we use the AND instruction to zero (mask) unwanted bits in the ASCII bytes. Any bit ANDed with a 0 will become or remain a zero. Any bit ANDed with a 1 will remain the same. We use the ROL instruction to rotate a nibble from the lower nibble position to the higher nibble position. In this case the ROR instruction would also accomplish the same result. Finally, we use the OR instruction to combine the two BCD nibbles in one byte. Any bit ORed with a 1 will become or remain a 1. Any bit ORed with a 0 will remain the same.

FLAGS, JUMPS, AND WHILE—DO IMPLEMENTATION

Introduction

The real power of a computer comes from its ability to repeat a sequence of instructions *as long as* some condition exists, repeat a sequence of instructions *until* some condition exists, or choose between two or more sequences of actions based on some condition. *Flags* indicate whether some condition is present or not. *Jump* instructions are used to tell the computer what sequence of actions to take based on the condition indicated by the flags. In this section we first discuss the 8086 conditional flags and the 8086 jump instructions. Then we show with examples how the WHILE—DO structure is implemented and used.

The 8086 Conditional Flags

The 8086 has six *conditional flags*. They are the *carry* flag, the *parity* flag, the *auxiliary carry* flag, the *zero*

```
;8086 PROGRAM
;ABSTRACT        : Program to produce a packed BCD byte from
;                  two ASCII-encoded digits.
;                  The first ASCII digit (5) is located in AL
;                  The second ASCII digit (9)is located in BL
;                  The result (packed BCD) to be left in AL
;REGISTERS USED: CS, AL, BL, CL
;PORTS USED      : None
;PROCEDURES      : None used

CODE_HERE         SEGMENT
                  ASSUME CS : CODE_HERE
        MOV       AL, '5'           ; load first ASCII digit into AL
        MOV       BL, '9'           ; load second ASCII digit into BL
        AND       AL, 0FH           ; mask upper 4 bits of first digit
        AND       BL, 0FH           ; mask upper 4 bits of second digit
        MOV       CL, 04H           ; load CX for 4 rotates required
        ROL       AL, CL            ; rotate AL 4 bit positions
        ADD       AL, BL            ; combine nibbles, result in AL

CODE_HERE         ENDS
                  END
```

FIGURE 4-5 8086 assembly language program to produce packed BCD from two ASCII characters.

flag, the *sign* flag, and the *overflow* flag. Chapter 1 shows numerical examples of the conditions indicated by these flags. Here we review these conditions and show how some of the important 8086 instructions affect these flags.

THE CARRY FLAG WITH ADD, SUBTRACT, AND COMPARE INSTRUCTIONS

If the addition of two 8-bit numbers produces a sum greater than 8 bits, the carry flag will be set to a 1 to indicate a carry into the next bit position. Likewise, if the addition of two 16-bit numbers produces a sum greater than 16 bits then the carry flag will be set to a 1 to indicate that a final carry was produced by the addition.

During subtraction the carry flag functions as a borrow flag. If the bottom number in a subtraction is larger than the top number, then the carry/borrow flag will be set to indicate that a borrow was needed to perform the subrtraction.

The 8086 compare instruction has the format CMP destination, source. The source can be an immediate number, a register, or a memory location. The destination can be a register or a memory location. The comparison is done by subtracting the contents of the specified source from the contents of the specified destination. Flags are updated to reflect the result of the comparison, but neither the source nor the destination is changed. If the source operand is greater than the specified destination operand, then the carry/borrow flag will be set to indicate that a borrow was needed to do the comparison (subtraction). If the source operand is the same size as or smaller than the specified destination operand, then the carry/borrow flag will not be set after the compare. If the two operands are equal, the zero flag will be set to a 1 to indicate that the result of the compare (subtraction) was all 0's. Here's an example and summary of this for your reference.

CMP BX, CX

condition	CF	ZF
CX > BX	1	0
CX < BX	0	0
CX = BX	0	1

The compare instruction is very important because it allows you to easily determine whether one operand is greater than, less than, or the same size as another operand.

THE PARITY FLAG

Parity is a term used to indicate whether a binary word has an even number of 1's or an odd number of 1's. A binary number with an even number of 1's is said to have even parity. The 8086 parity flag will be set to a 1 after an instruction if the lower 8 bits of the destination operand has an even number of 1's. Probably the most common use of the parity flag is to determine if ASCII data sent to a computer over phone lines or some other communications link contains any errors. A later chapter will describe this use of parity.

THE AUXILIARY CARRY FLAG

This flag has significance in BCD addition or BCD subtraction. If a carry is produced when the least-significant nibbles of 2 bytes are added, the auxiliary carry flag will be set. In other words, a carry out of bit 3 sets the auxiliary carry flag. Likewise, if the subtraction of the least-significant nibbles requires a borrow, the auxiliary carry/borrow flag will be set. The auxiliary carry/borrow flag is only used by the DAA and the DAS instructions. Consult the DAA and the DAS instruction descriptions in Chapter 6 and the BCD operation examples section of Chapter 1 for further discussion of BCD operations.

THE ZERO FLAG WITH INCREMENT, DECREMENT, AND COMPARE INSTRUCTIONS

As the name implies, this flag will be set to a 1 if the result of an arithmetic or logic operation is zero. For example, if you subtract two numbers which are equal, the zero flag will be set to indicate that the result of the subtraction was zero. If you AND two words together and the result contains no 1's, the zero flag will be set to indicate that the result was all 0's.

Besides the more obvious arithmetic and logic instructions which affect the zero flag, there are a few other very useful instructions which also do. One of these is the compare instruction, CMP, which we discussed with the carry flag previously. As shown there, the zero flag will be set to a 1 if the two operands compared are equal.

Another important instruction which affects the zero flag is the decrement instruction, DEC. This instruction will decrement or, in other words, subtract one from, a number in a specified register or memory location. If, after decrementing, the contents of the register or memory location are zero, the zero flag will be set. Here's a preview of how this is used. Suppose that we want to repeat a sequence of actions nine times. To do this we first load a register with the number 09H, and execute the sequence of actions. We then decrement the register and look at the zero flag to see if the register is down to zero yet. If the zero flag is not set, then we know that the register is not yet down to zero, so we tell the 8086, with a jump instruction, to go back and execute the sequence of instructions again. The following sections will show many specific examples of how this is done.

The increment instruction, INC destination, also affects the zero flag. If an 8-bit destination containing FFH or a 16-bit destination containing FFFFH is incremented, the result in the destination will be all 0's. The zero flag will be set to indicate this.

THE SIGN FLAG—POSITIVE AND NEGATIVE NUMBERS

When you need to represent both positive and negative numbers for an 8086, you use 2's complement sign-and-magnitude form as described in Chapter 1. In this form the most significant bit of the byte or word is used as a sign bit. A 0 in this bit indicates that the number is positive. A 1 in this bit indicates that the number is negative. The remaining 7 bits of a byte or the remaining 15 bits of a word are used to represent the magnitude of the number. For a positive number the magnitude will be in standard binary form. For a negative number the magnitude will be in 2's complement form. After an arithmetic or logic instruction executes, the sign flag will be a copy of the most significant bit of the destination byte or the destination word. In addition to its use with signed arithmetic operations, the sign flag can be used to determine if an operand has been decremented beyond zero. Decrementing 00H, for example, will give FFH. Since the MSB of FFH is a 1, the sign flag will be set.

THE OVERFLOW FLAG

This flag will be set if the result of a signed operation is too large to fit in the number of bits available to represent it. To remind you of what overflow means, here is an example. Suppose you add the 8-bit signed number 01110101 (+117 decimal) and the 8-bit signed number 00110111 (+55 decimal). The result will be 10101100 (+172 decimal) which is the correct binary result in this case, but is too large to fit in the 7 bits allowed for the magnitude in an 8-bit signed number. For an 8-bit signed number, a 1 in the most significant bit indicates a negative number. The overflow flag will be set after this operation to indicate that the result of the addition has overflowed into the sign bit.

The 8086 Unconditional Jump Instruction

INTRODUCTION

Jump instructions can be used to tell the 8086 to start fetching its instructions from some new location. Figure 4-6 shows in diagram form how a jump instruction affects the program execution flow. The 8086 remember, computes the physical address to fetch the next code byte from by adding the offset in the instruction pointer to the code segment base in CS. Jump instructions change the number in the instruction pointer register, and in some cases they also load a new number into the code segment register. The 8086 JMP instruction always causes a jump to occur. This is referred to as an *unconditional* jump. The 8086 also has a large collection of conditional jump instructions which cause a jump based on whether some condition is present or not. In this section we discuss how the unconditional jump instruction operates. In a later section we discuss the operation of the conditional jump instructions.

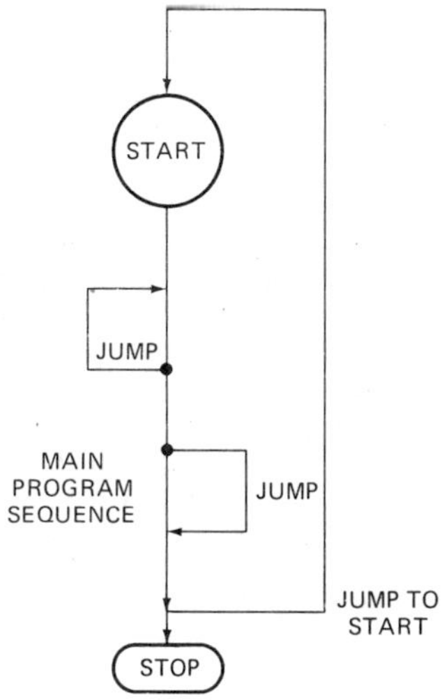

FIGURE 4-6 Change in program flow that can be caused by jump instructions.

UNCONDITIONAL JUMP INSTRUCTION TYPES—OVERVIEW

The 8086 unconditional jump instruction, JMP, has five different types. Figure 4-7 shows the names and instruction coding templates for these five types. We will first summarize how these five work to give you an overview, and then we will describe in detail the two types you need for your programs at this point. The JMP instruction description in Chapter 6 shows examples of each of the five types.

JMP = Jump

Within segment or group, IP relative

Opcode	DispL	DispH

Opcode	Clocks	Operation
E9	15	IP ← IP + Disp16
EB	15	IP ← IP + Disp8 (Disp8 sign-extended)

Within segment or group, Indirect

Opcode	mod 100 r/m	mem-low	mem-high

Opcode	Clocks	Operation
FF	11	IP ← Reg16
FF	18+EA	IP ← Mem16
FF	18+EA	IP ← Mem16

Inter-segment or group, Direct

Opcode	offset-low	offset-high	seg-low	seg-high

Opcode	Clocks	Operation
EA	15	CS ← segbase IP ← offset

Inter-segment or group, Indirect

Opcode	mod 101 r/m		

Opcode	Clocks	Operation
FF	24+EA	CS ← segbase IP ← offset

FIGURE 4-7 8086 unconditional JMP instructions *(Intel Corp.)*.

THE DIRECT WITHIN-SEGMENT NEAR JMP INSTRUCTION

This instruction can cause the next instruction to be fetched from anywhere in the current code segment. A jump to an address in the same segment as the jump instruction is commonly called an *intrasegment* or a *near* jump. To produce the new instruction fetch address this instruction adds a 16-bit signed displacement contained in the instruction to the contents of the instruction pointer register. A signed 16-bit displacement means that the jump can be to a location anywhere from +32,767 to −32,768 bytes from the current instruction pointer location. A positive displacement usually means you are jumping ahead in the program, and a negative displacement usually means you are jumping "backward" in the program.

THE DIRECT WITHIN-SEGMENT SHORT-TYPE JMP INSTRUCTION

This instruction is a special case of a near jump. This JMP instruction produces the new instruction fetch address by adding a signed 8-bit displacement, contained in the instruction, to the contents of the instruction pointer register. With an 8-bit signed displacement the jump can be to a location anywhere from +127 to −128 bytes from the current instruction pointer location.

THE INDIRECT WITHIN-SEGMENT JMP INSTRUCTION

This instruction replaces the contents of the instruction pointer register with the contents of a specified 16-bit register or the contents of a specified memory location. The MOD—R/M byte in the second byte position of the coding template for this instruction indicates that the register or memory location can be specified in any of the 32 register and memory addressing modes shown in Figure 3-8. Since this type JMP is to an address in the same code segment as the JMP instruction, it is another example of a near jump.

THE DIRECT INTERSEGMENT-TYPE JMP

This instruction causes a jump to another code segment. A jump to another code segment is often referred to as an *intersegment* or *far* jump. In order to get to another segment, you have to change the contents of both the instruction pointer and the code segment registers. As shown in Figure 4-7, for this type instruction the new value for the instruction pointer is written in as bytes 2 and 3 of the instruction code. The new value for the code segment register is written in as bytes 4 and 5 of the instruction code. Note that in each case, the low byte is written before the high byte.

THE INDIRECT INTERSEGMENT JMP

This instruction also causes a far (to another code segment) JMP. Therefore, both the instruction pointer register and the code segment register contents have to be changed. For this type instruction the new values are taken from four memory locations. The new value for IP will be written in the first two memory locations, low byte first, and the new value for CS will be written in the next two memory locations, low byte first. Again, the MOD—R/M byte in the second byte position of the instruction code template indicates that the first memory address can be specified in any one of the 24 memory addressing modes shown in Figure 3-8.

DIRECT WITHIN-SEGMENT NEAR AND DIRECT WITHIN-SEGMENT SHORT JMP EXAMPLES

Suppose that in a program you want to keep executing an instruction or group of instructions over and over again. Figure 4-8 shows how the JMP instruction can be used to do this. In this program the label BACK followed by a colon is used to give a name to the address we want to jump back to. When the assembler reads this label it will make an entry in its symbol table as to where it found the label. Then when the assembler reads the JMP instruction and finds the name BACK, it will be able to calculate the displacement from the jump instruction to the label. This displacement will be part of the code for the instruction. Even if you are not using an assembler, you should use labels to indicate jump destinations so that you can easily see them. The NOP instruction used in the program in Figure 4-8 does nothing except fill space. We used it in this example to represent the instructions that we want to loop through over and over. We also use it to represent the instructions after the JMP—BACK loop. Actually, the way this program is written the 8086 will never get to the instructions after the JMP instruction. Can you see why? The answer is that once the 8086 gets into the JMP—BACK loop, the only ways it can get out are if the power is turned off, an interrupt occurs, or the system is reset. In most programs one of the instructions we have represented with a NOP would be a conditional jump instruction which would get execution out of the loop when the specified condition occurred.

Now let's see how the binary code for the JMP instruction in Figure 4-8 is constructed. The jump is to a label in the same segment so this narrows our choices down to the first three types of JMP instruction shown in Figure 4-7. For several reasons it is best to use the direct-type JMP instruction whenever possible. This narrows our choices down to the first two types in Figure 4-7. The choice between these two is determined by whether you need a 1-byte displacement to reach the JMP destination address, or whether you need a 2-byte displacement to reach the JMP destination. Since for our example program the destination address is within the range of −128 to +127 bytes from the instruction after the JMP instruction, we can use the direct within-segment short type of JMP. According to Figure 4-7 the instruction template for this instruction is 11101011 (EBH) followed by some displacement. Here's how you calculate the displacement to put in the instruction.

NOTE: An assembler automatically does this for you, but you should still learn how it is done to help you in troubleshooting.

The numbers in the left column of Figure 4-8 represent the offset of each code byte from the code segment

```
The IBM Personal Computer MACRO Assembler 10-17-84     PAGE    1-1

                                   page,132
                                   ;8086 program
                                   ;ABSTRACT      : This program illustrates a "backwards" jump
                                   ;REGISTERS USED: CS, AL
                                   ;PORTS USED    : None
                                   ;PROCEDURES    : None used

0000                               CODE_HERE      SEGMENT
                                                  ASSUME  CS : CODE_HERE

0000  04 03                        BACK:    ADD   AL, 03H         ; add 3 to total
0002  90                                    NOP                   ; dummy instructions
0003  90                                    NOP                   ; to represent those
0004  90                                    NOP                   ; instructions jumped
0005  90                                    NOP                   ; back over
0006  EB F8                                 JMP   BACK            ; loop back through
                                                                  ; series of instructions
0008  90                                    NOP                   ; dummy instructions to
0009  90                                    NOP                   ; represent continuation
                                                                  ; after loop
000A                               CODE_HERE      ENDS
                                                  END
```

FIGURE 4-8 Program demonstrating "backward" JMP.

base. These are the numbers that will be in the instruction pointer as the program executes. After the 8086 fetches an instruction byte it automatically increments the instruction pointer to point to the next instruction byte. The displacement in the instruction then will be added to the offset of the next in-line instruction after the JMP instruction. For the example program in Figure 4-8 the displacement in the JMP instruction will be added to offset 0008H, which is in the instruction pointer after the JMP instruction executes. What this means is that when you are counting the number of bytes of displacement, you always start counting from the address of the instruction immediately after the JMP instruction. For the example program we want to jump from offset 0008H back to offset 0000H. This is a displacement of −8H.

You can't, however, write the displacement in the instruction as −8H. Negative displacements must be expressed in 2's complement, sign-and-magnitude form. We showed how to do this in Chapter 1. First, write the number as an 8-bit positive binary number. In this case that is 00001000. Then, invert each bit of this, including the sign bit, to give 11110111. Finally, add 1 to that result to give 11111000 binary or F8H which is the correct 2's complement representation for −8H. As shown in the assembler listing for the program in Figure 4-8, the two code bytes for this JMP instruction then are EBH and F8H.

To summarize this example then, a label is used to give a name to the destination address for the jump. This name is used to refer to the destination address in the JMP instruction. Since the destination in this example is within the range of −128 to +127 bytes from the address after the JMP instruction, the instruction can be coded as a direct within-segment short-type JMP. The displacement is calculated by counting the number of bytes from the next address after the JMP instruction to the destination. If the displacement is negative (backward in the program), then it must be expressed in 2's complement form before it can be written in the instruction code template.

Now let's look at another simple example program, in Figure 4-9, to see how you can jump ahead over a group of instructions in a program. Here again we use a label to give a name to the address that we want to JMP to. We also use NOP instructions to represent the instructions that we want to skip over and the instructions that continue after the JMP. Now let's see how this JMP instruction is coded.

When the assembler reads through the source file for this program it will find the label "THERE" after the JMP mnemonic. At this point the assembler has no way of knowing whether it will need 1 byte or 2 bytes to represent the displacement to the destination address. The assembler plays it safe by reserving 2 bytes for the displacement. Then the assembler reads on through the rest of the program. When the assembler finds the specified label, it calculates the displacement from the instruction after the JMP instruction to the label. If the assembler finds the displacement to be outside the range of −128 bytes to +127 bytes, then it will code the instruction as a direct within-segment near JMP with 2 bytes of displacement. If the assembler finds the displacement to be within the −128 to +127 byte range,

```
The IBM Personal Computer MACRO Assembler 10-17-84      PAGE    1-1

                                 page,132
                                 ;8086 program
                                 ;ABSTRACT        : This program illustrates a "forwards" jump
                                 ;REGISTERS USED : CS, AX
                                 ;PORTS USED      : None
                                 ;PROCEDURES USED: None

0000                             CODE_HERE      SEGMENT
                                                ASSUME  CS : CODE_HERE

0000  EB 05 90                          JMP     THERE            ; skip over a series
                                                                 ; of instructions
0003  90                                NOP                      ; dummy instructions
0004  90                                NOP                      ; to represent those
0005  90                                NOP                      ; instructions skipped
0006  90                                NOP                      ; over
0007  B8 0000                    THERE: MOV     AX, 0000H        ; zero accumulator before addition
000A  90                                NOP                      ; dummy instructions to
000B  90                                NOP                      ; represent continuation
                                                                 ; of execution
000C                             CODE_HERE      ENDS
                                                END
```

FIGURE 4-9 Program demonstrating "forward" JMP.

then it will code the instruction as a direct within-segment short-type JMP with a 1-byte displacement. In the latter case the assembler will put the code for a NOP instruction, 90H, in the third byte it had reserved for the JMP instruction. The instruction codes for the JMP THERE instruction in Figure 4-9 demonstrate this. As shown in the instruction template in Figure 4-7, EBH is the basic opcode for the direct within-segment short JMP. The 05H represents the displacement to the JMP destination. Since we are jumping forward in this case, the displacement is a positive number. The 90H in the next memory byte is the code for a NOP instruction. The displacement is calculated from the offset of this instruction, 0002H, to the offset of the destination label, 0007H. The difference of 05 between these two is the displacement you see coded in the instruction.

If you are hand coding a program such as this, you will probably know how far it is to the label and you can leave just 1 byte for the displacement if that is enough. If you are using an assembler and you don't want to waste the byte of memory or the time it takes to fetch the extra NOP instruction, you can write the instruction as JMP SHORT label. The SHORT operator is a promise to the assembler that the destination will not be outside the range of −128 to +127 bytes. Trusting your promise, the assembler then only reserves 1 byte for the displacement.

SUMMARY OF UNCONDITIONAL JMPS

The 8086 has five types of unconditional JMP instructions. The types you will probably use most often in your programs are the direct within-segment near and the direct within-segment short. A label followed by a colon is used to give the destination address a name for both of these JMP types. For the direct within-segment near type, a 16-bit displacement contained in the instruction is added to the contents of the instruction pointer to produce the destination address. This type of jump can be to an address in the range of −32,768 bytes to +32,767 bytes from the current IP contents. The direct within-segment short JMP instruction adds an 8-bit displacement contained in the instruction to the IP to produce the destination address. For this type JMP the destination can be in the range of −128 bytes to +127 bytes from the current instruction pointer contents. The displacement for both of these JMP types is counted from the address of the instruction after the JMP instruction to the address of the destination instruction. A jump ahead in the program is usually represented by a positive displacement. A jump backward in the program is usually represented by a negative displacement which is coded in the instruction in its 2's complement sign-and-magnitude form. Note that if you are making a JMP from an address near the start of a 64 Kbyte segment to an address near the end of the segment, you may not be able to get there with a jump of +32,767. The way you get there is to JMP backwards around to the desired destination address. An assembler will automatically do this for you.

One advantage of the direct near- and short-type JMPs is that the destination address is specified *relative* to the address of the instruction after the JMP instruction.

Since the JMP instruction in this case does not contain an absolute address or offset, the program can be loaded anywhere in memory and it will still run correctly. A program which can be loaded anywhere in memory to be run is said to be *relocatable*. You should try to write your programs so that they are relocatable.

The indirect within-segment type of JMP instruction replaces the contents of the instruction pointer with a 16-bit value from a register or memory location specified in the instruction. The direct intersegment far-type JMP loads IP with a new value contained in bytes 2 and 3 of the instruction code, and it loads CS with a new value from bytes 4 and 5 of the instruction code. The intersegment indirect far-type JMP loads IP and CS with new values read from a memory location specified in the instruction.

The 8086 Conditional Jump Instructions

As we stated previously, much of the real power of a computer comes from its ability to choose between two courses of action depending on whether some condition is present or not. In the 8086 the six conditional flags indicate the conditions that are present after an instruction. The 8086 conditional jump instructions look at the state of a specified flag(s) to determine whether a jump should be made or not. Figure 4-10 shows the mnemonics for the 8086 conditional jump instructions. Next to each mnemonic is a brief explanation of the mnemonic. Note that the terms "above" and "below" are used when you are working with unsigned binary numbers. The 8-bit unsigned number 11000110 is above the 8-bit unsigned number 00111001, for example. The terms "greater" and "less" are used when you are working with signed binary numbers. The 8-bit signed number 00111001 is greater (more positive) than the 8-bit signed number 11000110 which represents a negative number. Also shown in Figure 4-10 is an indication of the flag conditions that will cause the 8086 to do the jump. If the specified flag conditions are not present, the 8086 will just continue on to the next instruction in sequence. In other words, if the jump condition is not met, the conditional jump instruction will effectively function as a NOP. Suppose, for example, we have the instruction JC SAVE, where SAVE is the label at the destination address. If the carry flag is set, this instruction will cause the 8086 to jump to the instruction at the SAVE: label. If the carry flag is not set, the instruction will have no effect other than taking up a little processor time.

All conditional jumps are *short-type* jumps. This means that the destination label must be in the same code segment as the jump instruction. Also, the destination address must be in the range of −128 bytes to +127 bytes from the address of the instruction after the jump instruction. As we show in later examples, this limit on the range of unconditional jumps is important to be aware of as you write your programs.

The conditional jump instructions are usually used after arithmetic or logic instructions. Very commonly they are used after compare instructions. For this case the compare instruction syntax and the conditional jump instruction syntax are such that a little trick makes it very easy to see what will cause a jump to occur. Here's the trick. Suppose that you see the instruction sequence

MNEMONIC	CONDITION TESTED	"JUMP IF . . ."
JA/JNBE	(CF or ZF)=0	above/not below nor equal
JAE/JNB	CF=0	above or equal/not below
JB/JNAE	CF=1	below/not above nor equal
JBE/JNA	(CF or ZF)=1	below or equal/not above
JC	CF=1	carry
JE/JZ	ZF=1	equal/zero
JG/JNLE	((SF xor OF) or ZF)=0	greater/not less nor equal
JGE/JNL	(SF xor OF)=0	greater or equal/not less
JL/JNGE	(SF xor OF)=1	less/not greater nor equal
JLE/JNG	((SF xor OF) or ZF)=1	less or equal/not greater
JNC	CF=0	not carry
JNE/JNZ	ZF=0	not equal/not zero
JNO	OF=0	not overflow
JNP/JPO	PF=0	not parity/parity odd
JNS	SF=0	not sign
JO	OF=1	overflow
JP/JPE	PF=1	parity/parity equal
JS	SF=1	sign

Note: "above" and "below" refer to the relationship of two unsigned values;
"greater" and "less" refer to the relationship of two signed values.

FIGURE 4-10 8086 conditional JMP instructions *(Intel Corp.)*.

```
CMP BL, DH

JAE HEATER_OFF
```

in a program, and you want to determine what these instructions do. The CMP instruction compares the byte in the DH register with the byte in the BL register and sets flags according to the result. A previous section showed you how the carry and zero flags are affected by a compare instruction. According to Figure 4-10 the JAE instruction says "Jump if above or equal" to the label HEATER_OFF. The question now is, will it jump if BL is above DH, or will it jump if DH is above BL. You could determine how the flags will be affected by the comparison and use Figure 4-10 to answer the question. However, an easier way is to mentally read parts of the compare instruction between parts of the jump instruction. If you read the example sequence as "Jump if BL is above or equal to DH," the meaning of the sequence is immediately clear. As you write your own programs, thinking of a conditional sequence in this way should help you to choose the right conditional jump instruction. The next sections show you how we use conditional and unconditional jump instructions to implement some of the standard program structures and solve some common programming problems.

WHILE—DO Implementation and Example

Remember from the discussion in Chapter 3 that the WHILE—DO structure has the form:

```
WHILE some condition is present DO
  Action
  Action
```

An important point about this structure is that the condition is checked *before* any action is done. In industrial control applications of microprocessors there are many cases where we want to do this. The following very simple example will show you how to implement this structure in 8086 assembly language.

DEFINING THE PROBLEM AND WRITING THE ALGORITHM

Suppose that in controlling a chemical process we want to bring the temperature of a solution up to 100°C before going on to the next step in the process. If the solution temperature is below 100°, we want to turn on a heater and wait for the temperature to reach 100°. If the solution temperature is at or above 100°, then we want to go on with the next step in the process. The WHILE—DO structure fits this problem because we want to check the condition (temperature) before we turn on the heater. We don't want to turn on the heater if the temperature is already high enough because we might overheat the solution.

Figure 4-11 shows a flowchart and the pseudocode of an algorithm for this problem. The first step in the algorithm is to read in the temperature from a sensor connected to a port. The temperature read in is then compared with 100°. These two parts represent the condition-checking part of the structure. If the temperature is at or above 100°, execution will exit the structure and do the next mainline action, turn off the heater. If the heater is already off, it will not do any harm to turn it off again. If the temperature is less than 100°, the heater is turned on and the temperature rechecked. Execution will stay in this loop while the temperature is below 100°. Incidentally, it will not do any harm to turn the heater on if it is already on. When the temperature reaches 100°, execution will exit the structure and go on to the next mainline action, turn off the heater.

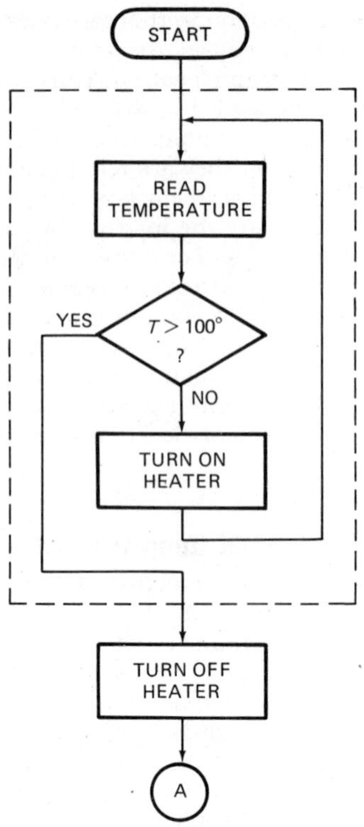

FLOWCHART

(*a*)

```
READ TEMPERATURE
WHILE TEMPERATURE < 100° DO
    TURN HEATER ON
TURN HEATER OFF
```

PSEUDOCODE

(*b*)

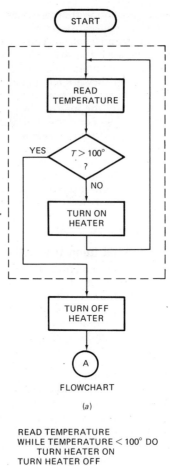

FIGURE 4-11 Flowchart and pseudocode for heater control problem.

IMPLEMENTING THE ALGORITHM IN ASSEMBLY LANGUAGE

Figure 4-12 shows one way to write the assembly language for this example. We have assumed for this example that the temperature sensor inputs an 8-bit binary value for the Celsius temperature to port FFF8H. We have also assumed that the heater control output is connected to the most significant bit of port FFFAH. (Incidentally, these port addresses are two of the available

```
The IBM Personal Computer MACRO Assembler 02-16-85      PAGE    1-1

                              page, 132
                              ;8086 program
                              ;ABSTRACT      : program turns heater off if temperature equals
                              ;                100 degrees or more, and to turn the heater on
                              ;                if the temperature is below 100 degrees.
                              ;REGISTERS USED: CS, DX, AL
                              ;PORTS USED    : FFF8H - for temperature data input
                              ;                FFFAH - MSB for heater control output
                              ;PROCEDURES    : None used

0000                          CODE_HERE      SEGMENT
                                             ASSUME  CS : CODE_HERE

                                                                  ; initialize port FFFA as an output port
0000  BA FFFE                                MOV     DX, 0FFFEH   ; point DX to port contol register
0003  B0 99                                  MOV     AL, 99H      ; control word to set up port FFFA as an output
0005  EE                                     OUT     DX, AL       ; send control word to port

0006  BA FFF8                 TEMP_IN:       MOV     DX, 0FFF8H   ; read in temperature data
0009  EC                                     IN      AL, DX
000A  3C 64                                  CMP     AL, 100      ; if temp >= 100
000C  73 08                                  JAE     HEATER_OFF   ; go turn heater off

000E  B0 80                                  MOV     AL, 80H      ; load code for heater on
0010  BA FFFA                                MOV     DX, 0FFFAH   ; point DX to output port
0013  EE                                     OUT     DX, AL       ; turn heater on
0014  EB F0                                  JMP     TEMP_IN      ; go & read temp again
0016  B0 00                   HEATER_OFF:    MOV     AL, 00       ; load code for heater off
0018  BA FFFA                                MOV     DX, 0FFFAH   ; point DX to output port
001B  EE                                     OUT     DX, AL       ; turn heater off

001C                          CODE_HERE      ENDS
                                             END
```

(*a*)

```
The IBM Personal Computer MACRO Assembler 02-16-85      PAGE    1-1

                              page, 132
                              ;8086 PROGRAM
                              ;ABSTRACT      : program to turn heater off if temperature
                              ;                equals 100 degrees or more, and to turn the
                              ;                heater on if the temperature is below 100 degrees.
                              ;REGISTERS USED: CS, DX, AL
                              ;PORTS USED    : FFF8H - for temperature data input
                              ;                FFFAH _ MSB for heater control output
                              ;PROCEDURES    : None used

0000                          CODE_HERE      SEGMENT
                                             ASSUME  CS : CODE_HERE
```

FIGURE 4-12 Assembly language program for heater control problem. *(a)* First approach. *(b)* Improved version.

```
                                                                  ; initialize port FFFA as an output port
0000  BA FFFE                        MOV    DX, 0FFFEH            ; point DX to port contol register
0003  B0 99                          MOV    AL, 99H               ; control word to set up port FFFA as an output
0005  EE                             OUT    DX, AL                ; send control word to port

0006  BA FFF8         TEMP_IN:       MOV    DX, 0FFF8H            ; point DX at input port
0009  EC                             IN     AL, DX                ; read in temperature data
000A  3C 64                          CMP    AL, 100
000C  72 03                          JB     HEATER_ON             ; if temp < 100 go
                                                                  ; turn heater ON
000E  EB 09 90                       JMP    HEATER_OFF            ; temp >= 100 go
                                                                  ; turn heater OFF
0011  B0 80           HEATER_ON:     MOV    AL, 80H               ; load code for heater ON
0013  BA FFFA                        MOV    DX, 0FFFAH            ; point DX at output port
0016  EE                             OUT    DX, AL                ; turn heater ON
0017  EB ED                          JMP    TEMP_IN               ; read temp again

0019  B0 00           HEATER_OFF:    MOV    AL,  00               ; load code for heater OFF
001B  BA FFFA                        MOV    DX, 0FFFAH            ; point DX at output port
001E  EE                             OUT    DX, AL                ; turn heater OFF

001F                  CODE_HERE      ENDS
                                     END
```

(b)

FIGURE 4-12 *(continued).*

ports, P2A and P2B, on an SDK-86 board.) A 1 sent to the MSB of port FFFAH turns the heater on.

The 8086 has two types of input instruction, *fixed* port and *variable* port. The fixed port instruction has the format IN AL, port or IN AX, port. The term "port" in these represents an 8-bit port address to be put directly in the instruction. The instruction IN AX, 07H, for example, will copy a word from port 07H to the AX register. With an 8-bit port address you can address any one of 256 possible ports. The port address is fixed, however. The program cannot change the port address as it executes.

For the variable-port input instruction, the address of the desired port is put in the DX register. The input instruction for this type then has the format IN AL, DX or IN AX, DX. If you load DX with FFF8H and then do an IN AL, DX as in Figure 4-12*a*, the 8086 will copy a byte of data from port FFF8H to the AL register. The variable-port type instruction has two major advantages. First, up to 65,536 different ports can be specified with the 16-bit port address in DX. Second, the port address can be changed as a program executes by simply putting a different number in DX. This is handy in a case where you want the computer to be able to input from 15 different terminals, for example. Instead of writing 15 different input programs, you can write one input program which changes the contents of DX to input from different terminals.

The 8086 also has a fixed-port output instruction and a variable-port output instruction. The fixed-port output instruction has the form OUT port, AL or OUT port, AX. Here again the term *port* represents an 8-bit port address written in the instruction. OUT 05, AL, for example, will copy the contents of the AL register to port 05H. For the variable port output instruction the 16-bit port address is put in the DX register. The output instruction format for this type is OUT DX, AL or OUT DX, AX. If you load DX with FFFAH and then do an OUT DX, AL instruction as in Figure 4-12*a*, the 8086 will copy the contents of the AL register to port FFFAH.

Most common devices used as ports for microcomputers can be used for input or output. When the power is first applied to these devices they are in the input mode. If you want to use any of these devices as output ports, you must send the device a control word which switches the device to output mode. Chapter 9 and later chapters will describe in detail how you initialize programmable port devices, but to give you an introduction we show you here how to initialize one of the ports in an 8255 device on an SDK-86 microcomputer for use as an output port. To specify the function of one of these programmable devices you send a control word to a register inside the device. You can find the control word format for each type of device in the manufacturer's data book. For one of the 8255s on an SDK-86 board, the address of the control register in the device is FFFEH. The instruction MOV DX, 0FFFEH points DX at this address. The control word needed to make port P2B of this 8255 an output, and P2A and P2C inputs, is 99H. (In Chapter 9 we show how we determined this control word.) We load this control word into AL with MOV AL, 99H and send it to the 8255 control register with OUT DX, AL. Now we can output a byte to port P2B of this device any time we need to in the program. The actual address of this port P2B on the SDK-86 board is FFFAH. It is to this address that we will output a byte to turn the heater on or off.

After we input the data from the temperature sensor in Figure 4-12*a* we compare the value read with 100

(64H). The JAE instruction after the compare can be read as "jump to the label HEATER_OFF if AL is above or equal to 100." Note that we used the Jump if Above or Equal instruction rather than a Jump if Equal instruction. Can you see why? To see the answer, visualize what would happen if we had used a JE instruction and the temperature of the solution were 101°. On the first check the temperature would not be equal to 100° so the 8086 would turn on the heater. The heater would not get turned off until meltdown.

If the heater temperature is below 100°, we turn on the heater by loading a 1 in the most significant bit of AL and outputting this value to the most significant bit of port FFFAH. We then do an unconditional JMP back to check the temperature again.

When the temperature is at or above 100°, we load a 0 in the most significant bit of AL and output this to port FFFAH to turn off the heater. Note that the action of turning off the heater is outside the basic WHILE—DO structure. The WHILE-DO structure is shown by the dotted box in the flowchart in Figure 4-11*a* and by the indentation in the pseudocode in Figure 4-11*b*.

SOLVING A POTENTIAL PROBLEM OF CONDITIONAL JUMP INSTRUCTIONS

In the example program in Figure 4-12*a* we used the conditional jump instruction JAE to help implement the WHILE—DO structure. Conditional jump instructions have a potential problem which you should become aware of at this point. *All the conditional jump instructions are short-type jumps.* This means that a conditional jump can only be to a location within the range of −128 to +127 bytes from the instruction after the conditional jump instruction. This limit on the range of the jump posed no problem for the example program in Figure 4-12*a* because we were only jumping to a location 8 bytes ahead in the program. Suppose, however, that the instructions for turning on the heater required 220 bytes of memory. The HEATER_OFF label would then be outside the range of the JAE instruction.

Figure 4-12*b* shows how you can change the instructions slightly to solve the problem without changing the basic WHILE—DO overall structure. In this example we read the temperature in as before and compare it to 100 (64H). We then use the Jump if Below instruction to jump to the program section which turns on the heater. This instruction, together with the CMP instruction, says jump to the label HEATER_ON if AL is below 100. If the temperature is at or above 100, the JB instruction will act like a NOP, and the 8086 will go on to the JMP HEATER_OFF instruction. Changing the conditional jump instruction and writing the program in this way means that the destination for the conditional jump instruction is always just two instructions away. Therefore, you know that the destination will always be reachable. Except for very time-critical program sections, you should always write conditional jump instruction sequences in this way so that you don't have to worry about the potential problem. The disadvantages of this approach are the time and memory space required by the extra JMP instruction.

REPEAT—UNTIL IMPLEMENTATION AND EXAMPLES

Remember from the discussion in Chapter 3 that the REPEAT—UNTIL structure has the form

```
REPEAT
  Action
  Action
  .
  .
  .
UNTIL some condition is present
```

An important point about this structure is that the action or series of actions is done once before the condition is checked. Compare this with the WHILE—DO structure.

The following examples will show you how you can implement the REPEAT—UNTIL with 8086 assembly language and introduce you to some more assembly language programming techniques.

Waiting for a Strobe Signal

DEFINING THE PROBLEM AND WRITING THE ALGORITHM

Many systems that interface with a microcomputer output data on parallel-signal lines and then output a separate signal to indicate that valid data is on the parallel lines. The data-ready signal is often called a *strobe*. An example of a strobed data system such as this is an ASCII-encoded computer-type keyboard. Figure 4-13 shows how the parallel data lines and the strobe line from such a keyboard are connected to ports of a micro-

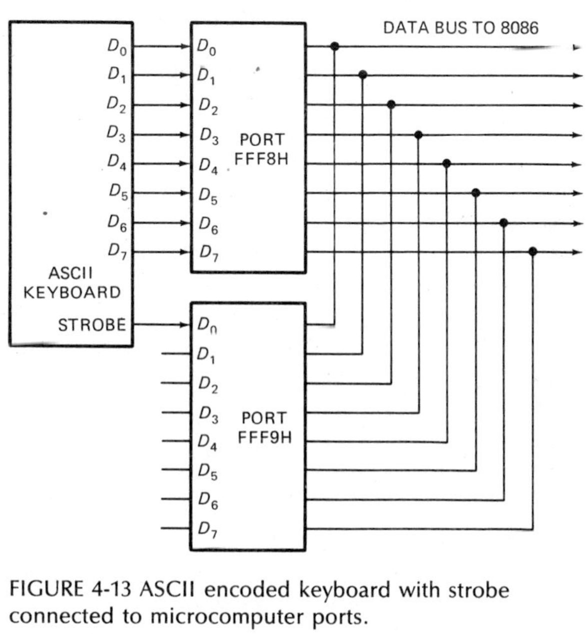

FIGURE 4-13 ASCII encoded keyboard with strobe connected to microcomputer ports.

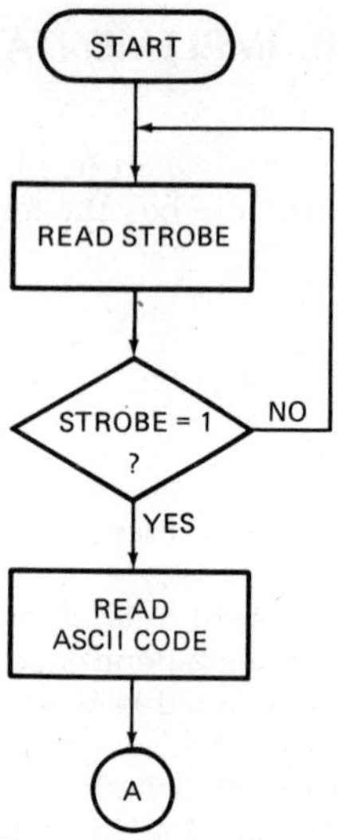

FLOWCHART

(*a*)

```
REPEAT
     READ KEYPRESSED STROBE
UNTIL STROBE = 1
READ ASCII CODE FOR KEY PRESSED
```

PSEUDOCODE

(*b*)

computer. When a key is pressed on the keyboard, circuitry in the keyboard detects which key is pressed and sends the ASCII code for that key out on the eight data lines connected to port FFF8H. After the data has had time to settle on these lines, the circuitry in the keyboard sends out a key-pressed strobe which lets you know that the data on the eight lines is valid. We have connected this strobe line to the least-significant bit of port FFF9H. A strobe can be an active high signal or an active low signal. For the example here, assume that the strobe signal goes high when a valid ASCII code is on the parallel data lines.

If we want to read the data from this keyboard, we can't do it at just any time. We must wait for the strobe to go high so that we know that the data we read will be valid. Basically what we have to do is look at the strobe signal and test it over and over until it goes high. Figure 4-14*a* shows how we can represent this operation with a flowchart, and Figure 4-14*b* shows the pseudocode. We want to repeat the read-strobe-and-test loop until the strobe is found to be high. Then we want to exit the loop and read in the ASCII code byte. The basic REPEAT-UNTIL structure is shown by the indentation in the pseudocode. Note that the read ASCII data action is not part of this structure and is therefore not indented.

```
The IBM Personal Computer MACRO Assembler 02-18-85      PAGE    1-1

                              page, 132
                              ;8086 PROGRAM
                              ;ABSTRACT      : program to read ASCII code when a strobe
                              ;                signal is sent from a keyboard
                              ;REGISTERS USED: CS, DX, AL
                              ;PORTS USED    : FFF9H - strobe signal input port
                              ;              : FFF8H - ASCII data input port
                              ;PROCEDURES    : None used

0000                          CODE_HERE      SEGMENT
                                             ASSUME  CS : CODE_HERE

0000  BA FFF9                                MOV     DX, 0FFF9H    ; point DX at strobe port
0003  EC                      LOOK_AGAIN:    IN      AL, DX        ; read keyboard strobe
0004  24 01                                  AND     AL, 01        ; mask extra bits and
                                                                   ; set flags
0006  74 FB                                  JZ      LOOK_AGAIN    ; strobe low, keep looking
0008  BA FFF8                                MOV     DX, 0FFF8H    ; point DX at data port
000B  EC                                     IN      AL, DX        ; read in ASCII code

000C                          CODE_HERE      ENDS
                                             END
```

(*c*)

FIGURE 4-14 Flowchart, pseudocode, and assembly language for reading ASCII code when a strobe is present. *(a)* Flowchart. *(b)* Pseudocode. *(c)* Assembly language program.

IMPLEMENTING THE ALGORITHM WITH ASSEMBLY LANGUAGE

Figure 4-14*c* shows the 8086 assembly language to implement this algorithm. To read in the key-pressed strobe signal, we first load the address of the port to which it is connected into the DX register. Then we use the variable port input instruction, IN AL, DX, to read the strobe data to AL. This input instruction copies a byte of data from port FFF9H to the AL register. However, we only care about the least-significant bit of the byte, because that is the one the strobe is connected to. We would like to find out if this bit is a 1. We will show you three ways to do it.

The first way, shown in Figure 4-14*c*, is to AND the byte in AL with the immediate number 01H. Remember that a bit ANDed with a 0 becomes a 0 (is masked). A bit ANDed with a 1 is not changed. If the least-significant bit is a 0, then the result of the ANDing will be all 0's. The zero flag, ZF, will be set to a 1 to indicate this. If the least-significant bit is a 1, the zero flag will not be set to a 1 because the result of the ANDing will still have a 1 in the least-significant bit. The Jump if Zero instruction, JZ, will check the state of the zero flag and, if it finds the zero flag set, will jump to the label LOOK_AGAIN. If the JZ instruction finds the zero flag not set (indicating that the LSB was a one), it passes execution on to the instructions which read in the ASCII data.

Another way to check the least-significant bit of the strobe word is with the TEST instruction instead of the AND instruction. The 8086 TEST instruction has the format TEST destination, source. The TEST instruction ANDs the contents of the specified source with the contents of the specified destination and sets flags according to the result. However, the TEST instruction does not change the contents of either the source or the destination. The AND instruction, remember, puts the result of the ANDing in the specified destination. The TEST instruction is useful if you want to set flags without changing the operands. In the example program in Figure 4-14*c* the AND AL, 01H instruction could be replaced with the TEST AL, 01H instruction.

Still another way to check the least-significant bit of the strobe byte is with a rotate instruction. If we rotate the least-significant bit into the carry flag, we can use a Jump if Carry or Jump if Not Carry instruction to control the loop. For this example program we can use either the ROR instruction or the RCR instruction. Assuming that we choose the ROR instruction, the check and jump instruction sequence would look like this:

```
LOOK_AGAIN: IN AL, DX
            ROR AL,1         ; Rotate LSB into
                               carry
            JNC LOOK_AGAIN;  If LSB = 0, keep
                               looking
```

For your programs you can use the way of checking a bit that seems easiest in a particular situation.

To read the ASCII data we first have to load the port address, FFF8H, into the DX register. We then use the variable port input instruction IN AL, DX to copy the ASCII data byte from the port to the AL register.

The main purpose of the preceding section was to show you how you can use a conditional jump instruction to make the 8086 REPEAT a series of actions UNTIL the flags indicate that some condition is present. The following section shows another example of implementing the REPEAT—UNTIL structure. This example also shows you how a register-based addressing mode is used to access data in memory.

Operating on a Series of Data Items in Memory

In many programming situations we want to perform some operation on a series of data items stored in successive memory locations. We might, for example, want to read in a series of data values from a port and put the values in successive memory locations. A series of data values of the *same type* stored in successive memory locations is often called an *array*. Each value in the array is referred to as an *element* of the array. For our example program here we want to add an inflation factor of 03H to each price in an 8-element array of prices. Each price is stored in a byte location as packed BCD (two BCD digits per byte). The prices then are in the range of 1 cent to 99 cents. Figure 4-15*a* and *b* shows a flowchart and the pseudocode for the operations that we want to perform. Follow through whichever form you feel more comfortable with.

We read one of the BCD prices from memory, add the inflation factor to it, and adjust the result to keep it in BCD format. The new value is then copied back to the array, replacing the old value. After that, a check is made to see if all of the prices have been operated on. If they haven't, then we loop back and operate on the next price. The two questions that may occur to you at this point are, "How are we going to indicate in the program which price we want to operate on, and how are we going to know when we have operated on all of the prices?" To indicate which price we are operating on at a particular time, we use a register as a pointer. To keep track of how many prices we have operated on we use another register as a counter. The example program in Figure 4-15*c* shows one way in which our algorithm for this problem can be implemented in assembly language.

The example program in Figure 4-15*c* uses several assembler directives. Let's review the function of these before describing the operation of the program instructions. The ARRAYS_HERE SEGMENT and the ARRAYS_HERE ENDS directives are used to set up a logical segment containing the data definitions. The CODE_HERE SEGMENT and the CODE_HERE ENDS directives are used to set up a logical segment which contains the program instructions. The ASSUME CS:CODE_HERE, DS:ARRAYS_HERE directive tells the assembler to use CODE_HERE as the code segment and use ARRAYS_HERE for all references to the data segment. The END directive lets the assembler know that it has reached the end of the program. Now let's discuss the data structure for the program.

The statement, COST DB 20H,28H,15H,26H,19H, 27H,16H,29H, in the program tells the assembler to set aside successive memory locations for an 8-element array of bytes. The array is given the name COST. When the assembled program is loaded into memory to be run, the eight memory locations will be loaded with the eight

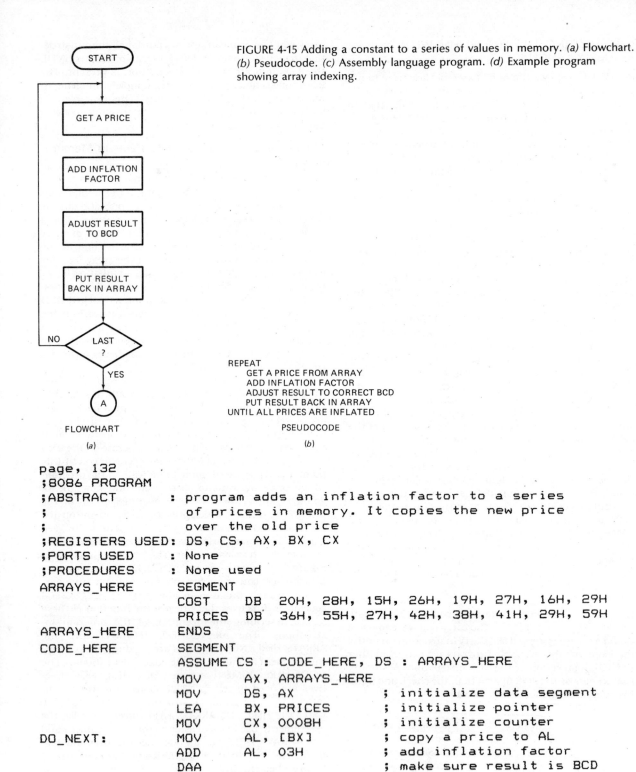

FIGURE 4-15 Adding a constant to a series of values in memory. *(a)* Flowchart. *(b)* Pseudocode. *(c)* Assembly language program. *(d)* Example program showing array indexing.

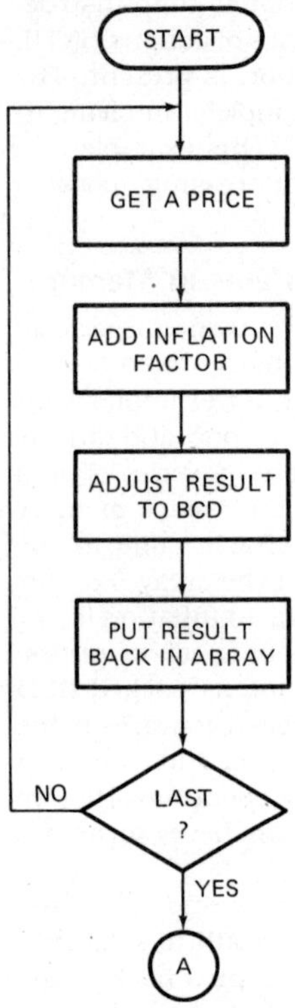

FLOWCHART

(a)

```
REPEAT
    GET A PRICE FROM ARRAY
    ADD INFLATION FACTOR
    ADJUST RESULT TO CORRECT BCD
    PUT RESULT BACK IN ARRAY
UNTIL ALL PRICES ARE INFLATED
```

PSEUDOCODE

(b)

```
page, 132
;8086 PROGRAM
;ABSTRACT        : program adds an inflation factor to a series
;                  of prices in memory. It copies the new price
;                  over the old price
;REGISTERS USED: DS, CS, AX, BX, CX
;PORTS USED      : None
;PROCEDURES      : None used
ARRAYS_HERE      SEGMENT
                 COST    DB  20H, 28H, 15H, 26H, 19H, 27H, 16H, 29H
                 PRICES  DB  36H, 55H, 27H, 42H, 38H, 41H, 29H, 59H
ARRAYS_HERE      ENDS
CODE_HERE        SEGMENT
                 ASSUME CS : CODE_HERE, DS : ARRAYS_HERE
                 MOV     AX, ARRAYS_HERE
                 MOV     DS, AX            ; initialize data segment
                 LEA     BX, PRICES        ; initialize pointer
                 MOV     CX, 0008H         ; initialize counter
DO_NEXT:         MOV     AL, [BX]          ; copy a price to AL
                 ADD     AL, 03H           ; add inflation factor
                 DAA                       ; make sure result is BCD
                 MOV     [BX], AL          ; copy result back to memory
                 INC     BX                ; point to next price
                 DEC     CX                ; decrement counter
                 JNZ     DO_NEXT           ; if not last, go get next
CODE_HERE        ENDS
                 END
```

(c)

```
page ,132
;8086 PROGRAM
;ABSTRACT          : Program adds a profit factor to each element in
;                  : a COST array and puts the result in an array
;                  : called PRICES
;REGISTERS USED    : DS, CS, AX, BX, CX
;PORTS USED        : None
;PROCEDURES        : None used

ARRAYS_HERE        SEGMENT
                   COST    DB  20H, 28H, 15H, 26H, 19H, 27H, 16H, 29H
                   PRICES  DB  8 DUP(0)
ARRAYS_HERE        ENDS

PROFIT             EQU     15H                    ; profit = 15 cents

CODE_HERE          SEGMENT
                   ASSUME  CS:CODE_HERE, DS:ARRAYS_HERE

                   MOV   AX, ARRAYS_HERE          ; initialize data segment
                   MOV   DS, AX
                   MOV   CX, 0008H                ; initialize counter
                   MOV   BX, 0000H                ; initialize pointer
DO_NEXT:           MOV   AL, COST[BX]             ; point to element in COST
                   ADD   AL, PROFIT               ; add the profit to COST
                   DAA                            ; decimal adjust result
                   MOV   PRICES[BX], AL           ; store result in PRICES
                   INC   BX                       ; point to next element
                                                  ; in the arrays
                   DEC   CX                       ; decrement the counter
                   JNZ   DO_NEXT                  ; if not last, do again
CODE_HERE          ENDS
                   END
```

(*d*)

FIGURE 4-15 *(continued)*.

values specified in the DB statement. The statement, PRICES DB 36H,55H,27H,42H,38H,41H,29H,59H, sets up another 8-element array of bytes and gives it the name PRICES. The eight memory locations will be loaded with the specified values when the assembled program is loaded into memory. Figure 4-16 shows how these two arrays will be arranged in memory. Note that the name of the array represents the displacement or offset of the first element of the array from the start of the data segment.

The first two instructions, MOV AX, ARRAYS_HERE and MOV DS, AX initialize the data segment register as was described for the example program in Figure 3-10. The LEA mnemonic in the next instruction stands for load effective address. An effective address, remember, is the number of bytes from the start of a segment to the desired data item. The instruction LEA BX, PRICES loads the displacement of the first element of PRICES into the BX register. A displacement contained in a register is usually referred to as an *offset*. If you take another look at the data structure for this program in Figure 4-16 you should see that the offset of PRICES is 0008H. Therefore, the LEA BX, PRICES instruction will load BX with

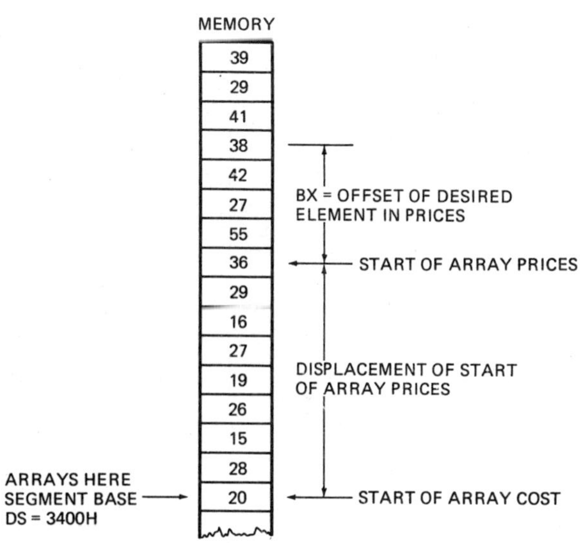

FIGURE 4-16 Data arrangement in memory for "inflate prices" program.

0008H. We say that BX is a *pointer* to an element in PRICES. We will soon show you how this pointer is used to indicate which price we want to operate on at a given time in the program. The next instruction, MOV CX, 0008H, loads the CX register with the number of prices in the array. We use this register as a counter to keep track of how many prices we have operated on. After we operate on each price, we decrement the counter by one. When the counter reaches zero, we know that we have operated on all of the prices.

The MOV AL, [BX] instruction copies one of the prices from memory to the AL register. Remember, the 8086 produces the physical address for accessing data in memory by shifting the contents of a segment register four bit positions left and adding an effective address, EA, to the result. A section in Chapter 3 showed you how the effective address could be specified directly in the instruction with either a name or a number. The instructions MOV AX, MULTIPLICAND and MOV AX, DS:WORD PTR [0000H] are examples of this addressing mode. For the instruction MOV AL, [BX] the effective address is contained in the BX register where we put it with the LEA instruction above. The first time this instruction executes, BX will contain 0008H, the effective address or offset of the first price in the array. Therefore, the first price will be copied into AL. To produce the physical memory address the 8086 will shift the contents of the data segment register four bit positions left and add this 0008H to the result.

The next instruction ADD AL, 03H adds the immediate number 03H to the contents of the AL register. The binary result of the addition will be left in AL. We want the prices in the array to be in BCD form, so we have to make sure the result is adjusted to be a legal BCD number. For example, if we add 03 to 29 the result in AL will be 2C. Most people would not understand this as a price so we have to adjust the result to the desired BCD number. The Decimal Adjust after Addition instruction DAA will automatically make this adjustment for us. DAA will adjust the 2CH by adding six to the lower nibble and the carry produced to the upper nibble. The result of this in AL will be 32H which is the result we want from adding 03 to 29. The DAA instruction only works on the AL register. For further examples of DAA operation, consult the DAA instruction description in Chapter 6.

The INC BX instruction adds 1 to the number in BX. BX now contains the effective address or offset of the next price in the array. We like to say that BX now points to the next element in the array. The DEC CX instruction decrements the count we set up in the CX register by 1. If CX contains 0 after this decrement, the zero flag will be set to a 1. The JNZ DO_NEXT checks the zero flag. If it finds the zero flag set, it just passes execution out of the structure to the next mainline instruction. If it finds the zero flag not set, the JNZ instruction will cause a jump to the label DO_NEXT. Execution will repeat the sequence of instructions between the label and the JNZ instruction until CX is counted down to zero. Each time through the loop, BX will be incremented to point to the next price in the array.

Using a pointer to access data items in memory is a powerful technique that you will want to use in your programs. Figure 4-15*d* shows another example. Here we want to add a profit of 15 cents to each element of an array called COST and put the result in the corresponding element of an array called PRICES. We first initialize BX as a pointer to the first element of each array with MOV BX, 000H. The instruction MOV AL, COST[BX] will copy the first cost value into AL. The effective address for this instruction will be produced by adding the displacement represented by the name COST to the contents of BX. Likewise, the instruction MOV PRICES[BX], AL copies the result of the addition to the first element of PRICES. When BX is incremented, COST[BX] and PRICES[BX] will each access the next element in the array. A programmer familiar with higher level lan-

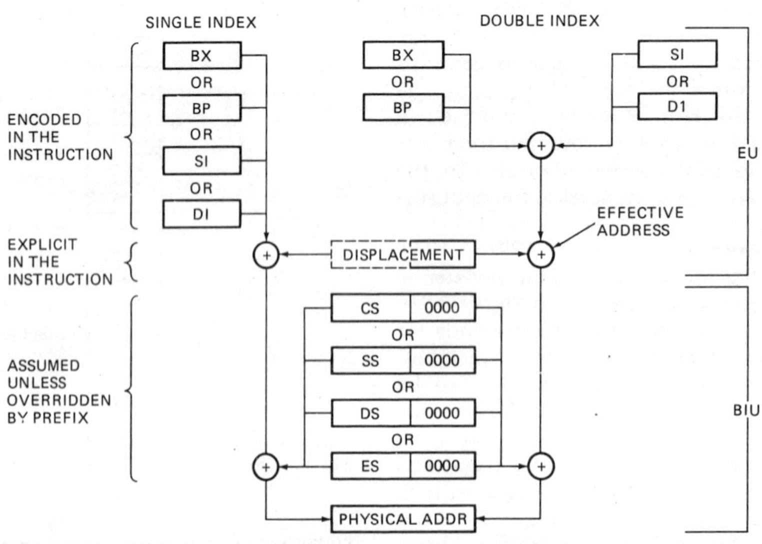

FIGURE 4-17 Summary of 8086 addressing modes.

```
SEGMENT BASE
        .
        .             Name PATIENTS represents displacement of
        .             start of array of records from segment base

PATIENTS                ; array of patient records starts here

                                RECORD 1
                                TV N. BEER
                                1324 Down Street
                                Portland, OR 97219
                                2/15/45
                                247 lb
                                $327.56

BX holds offset of ---------> RECORD 2
desired record in array         IM A. RUNNER
                                13733 S.W. Knaus Rd
                                Lake Oswego, OR 97304
                                6/30/41
                                145 lb
SI holds offset of ---------> $0.00
desired field in record
                                RECORD 3
```

FIGURE 4-18 Use of double indexed addressing mode.

guages would probably say that BX is being used as an array index. The 8086 has several registers which can be used to index or to point to data in memory.

Figure 4-17 summarizes all the ways you can tell the 8086 to calculate an effective address and a physical address for accessing data in memory. In all cases the physical address is generated by adding an effective address to one of the segment bases, CS, SS, DS, or ES. The effective address can be a direct displacement specified directly in the instruction as, for example, MOV AX, MULTIPLIER. The effective address or offset can be specified to be in a register, as in the instruction MOV AL, [BX]. Also the effective address can be specified to be the contents of a register plus a displacement in the instruction. The instruction MOV AX, PRICES[BX] is an example. For this example, PRICES represents the displacement of the start of the array from the segment base and BX represents the number of the element in the array that we want to access. The effective address of the desired element then is the sum of these two.

For working with more complex data structures such as records, you can tell the 8086 to compute an effective address by adding the contents of BX or BP plus the contents of SI or DI plus an 8-bit or a 16-bit displacement contained in the instruction. The instruction MOV AL, PATIENTS [BX][SI] is an example of this addressing mode. Figure 4-18 shows an example of why you might want an addressing mode such as this to access the balance due field in some medical records in memory. We will illustrate the use of some of these more complex addressing modes in later chapters.

When BX, SI, or DI is used to contain all or part of the effective address, the physical address will be produced by adding the effective address to the data segment base in DS. When BP is used to contain all or part of the effective address, the physical address will be produced by adding the effective address to the stack segment base in SS. For any of these four, you can use a segment override prefix to tell the 8086 to add the effective address to some other segment base. The instruction MOV AL, CS:[BX] tells the 8086 to produce a physical memory address by adding the offset in BX to the code segment base instead of to the data segment base. An exception to this is that with a special group of instructions called *string instructions* an offset in DI will always be added to the extra segment base in ES to produce the physical address.

Summary of REPEAT—UNTIL Implementation

The preceding sections have shown two examples of implementing the REPEAT—UNTIL structure. In the first example we repeated a series of actions until a condition was found to be present. Specifically, we kept looking and testing until we found a strobe signal high. In the second. We used a conditional jump instruction to check the condition of a flag and make the decision whether to repeat the series of actions or not.

In the second REPEAT—UNTIL example we introduced the concept of using a register as a pointer to a data element in an array. We also showed in this example how to make a program repeat a sequence of instructions a specific number of times. To do this we load the desired number of repeats in a register or memory location. Each time we execute the series of instructions, we decrement this counter by one. When the count in the register is decremented to zero, the zero flag will be set. Again we use a conditional jump instruction to check this flag and to decide whether to repeat the instruction sequence in the loop again.

The need for performing a sequence of actions a specified number of times in a program is so common that some programming languages use a specific structure to express it. This structure, derived from the basic WHILE—DO, is called the FOR—DO. It has the form

FOR count = 1 to count = *n* DO
  Action
  Action

where *n* is the number of times we want to do the sequence of actions. In assembly language you will usually implement this by loading *n* into a register and counting it down as shown in Figure 4-15c.

The common need to repeat a sequence of actions a specified number of times also led the designers of the 8086 to give it a group of instructions which make this easier for you. These instructions are the LOOP instructions which we discuss in the next section.

The 8086 LOOP Instructions

INSTRUCTION OPERATION AND EXAMPLES

The LOOP instructions have the format LOOP label. These instructions combine two operations in each instruction. The first operation is to decrement the CX register by one. The second operation is to check the CX register and, in some cases, also the zero flag to decide whether to do a jump to the specified label. As with the previously described conditional jump instructions, the LOOP instructions can only do short jumps. This means that the destination label must be in the range of −128 bytes to +127 bytes from the instruction after the LOOP instruction. Figure 4-19 summarizes the LOOP instructions. Instruction mnemonics separated by a "/" represent the same instruction. NE in the mnemonics stands for not equal, and NZ in the mnemonics stands for not zero. Also shown in the figure is the condition(s) checked by each instruction to decide if it should do the jump.

LOOP	Loop until CX = 0
LOOPE/LOOPZ	Loop if zero flag set and CX < > 0
LOOPNE/LOOPNZ	Loop if zero flag not set and CX < > 0
JCXZ	Jump if CX = 0

FIGURE 4-19 8086 LOOP instructions.

The basic LOOP-label instruction will decrement the CX register by 1 and jump to the specified label if the CX register is not 0. The instruction LOOP DO_NEXT, for example, could be used in place of the DEC CX and the JNZ DO_NEXT instructions in the program in Figure 4-15c.

The LOOP instructions decrement the CX register, but do not affect the zero flag. This leaves the zero flag available for other tests. The LOOPE/LOOPZ label instruction will decrement the CX register by one and jump to the specified label if CX is not equal to zero *and* if the zero flag is set to a one. In other words program execution will exit from the repeat loop if CX has been decremented down to zero *or* the zero flag is not set. This instruction might be used after a compare instruction, for example, to continue a sequence of operations for a specified number of times or until compared values were no longer equal.

The LOOPNE/LOOPNZ label instruction decrements the CX register by one. If CX is not zero *and* the zero flag is not set, this instruction will cause a jump to the specified label. In other words, execution will exit from the loop if CX is equal to zero *or* the zero flag is set. This instruction is useful when you want to execute a sequence of instructions a fixed number of times or until two values are equal. An example might be a program to read data from a disk. We typically write this type of program so that it attempts to read the data until the checksums are equal or until 10 unsuccessful attempts have been made to read the disk.

Another instruction often listed with the LOOP instructions is the JCXZ label instruction. This instruction does not affect the CX register. It simply does a short jump to the specified label if the CX register is zero. The JCXZ instruction checks the CX register directly, it does not check the zero flag.

In summary then, the LOOP instructions are useful for implementing the REPEAT—UNTIL structure for those special cases where we want to do a series of actions a fixed number of times *or* until the zero flag changes state. LOOP instructions incorporate two operations in each instruction; therefore, they are somewhat more efficient than single instructions to do the same job. The 8086 string instructions, which we discuss in a later section of this chapter, incorporate even more operations in single instructions. Some of the string instructions can implement an entire REPEAT—UNTIL structure with a single instruction. In the next section, we introduce you to instruction timing and show how the LOOP instruction can be used to produce a delay between the execution of instructions.

INSTRUCTION TIMING AND DELAY LOOPS

The rate at which 8086 instructions are executed is determined by a crystal-controlled clock with a frequency of a few megahertz. Each instruction takes a certain number of clock cycles to execute. The MOV register, register instruction, for example, requires 2 clock cycles to execute and the DAA instruction requires 4 clock cycles. The JNZ instruction requires 16 clock cycles if it does the jump and only 4 clock cycles if it doesn't do the jump. A table in Appendix B shows the number of clock cycles required by each instruction. Using the numbers in this table you can calculate how long it takes to execute an instruction or series of instructions. For example, if we are running an 8086 with a 5-MHz clock, then each clock cycle takes ⅕ MHz or 0.2 μs. An instruction which takes 4 clock cycles then will take 4 clock cycles x 0.2 μs/clock cycle or 0.8 μs to execute.

A common programming problem is the need to introduce a delay between the execution of two instructions. For example, we might want to read a data value from a port, wait 1 ms, and then read the port again. A later chapter will show how you can use interrupts to mark off time intervals. Here we show you how to use a program loop to do it.

The basic principle is to execute an instruction or series of instructions over and over until the desired time has elapsed. Figure 4-20*a* shows a program we might use to do this. The MOV CX, N instruction loads the CX register with the number of times we want to repeat the delay loop. Just ahead we show you how to calculate this number for a desired amount of delay. The NOP instructions next in the program are not required. The KILL_TIME label could be right in front of the LOOP instruction. In this case, only the LOOP instruction would be repeated. We put the NOPs in to show you how you can get more delay by extending the time it takes to execute the loop. The LOOP KILL_TIME instruction will decrement CX and, if it is not down to zero yet, do a jump to the label KILL_TIME. The program then will execute the two NOP instructions and the LOOP instruction over and over until CX is counted down to zero. The number in CX will determine how long this takes. Here's how you determine the value to put in CX for a given amount of delay.

First you calculate the number of clock cycles needed to produce the desired delay. If you are running your 8086 with a 5-MHz clock, then the time for each clock cycle then is ⅕ MHz or 0.2 μs. Now, suppose that you want to create a delay of 1 ms or 1000 μs with a delay loop. If you divide the 1000 μs desired by the 0.2 μs per clock cycle, you get the number of clock cycles required to produce the desired delay. For this example then you need a total of 5000 (1000/0.2) clock cycles to produce the desired delay.

The next step is to write the number of clock cycles required for each instruction next to that instruction as shown in Figure 4-20*a*. Then look at the program to determine which instructions get executed only once. The number of clock cycles for these instructions will only contribute to the total once. Instructions which only enter once in the calculation are often called *overhead*. We will represent the number of cycles of overhead with the symbol C_o. In Figure 4-20*a* the only instruction which executes just once is MOV CX, N, which takes 4 clock cycles. For this example then, C_o is 4.

Now determine how many clock cycles are required for the loop. The two NOPs in the loop require a total of 6 clock cycles. The LOOP instruction requires 17 clock cycles if it does the jump back to KILL_TIME, but it requires only 5 clock cycles when it exits the loop. The jump takes longer because the instruction byte queue has to be reloaded starting from the new address. For all but the very last time through the loop it will require 17 clock cycles for the LOOP instruction. Therefore, you can use 17 as the number of cycles for the LOOP instruction and compensate later for the fact that for the last time it uses 12 cycles less. For the example program the number of cycles per loop, C_L, is 6 + 17 or 23. The total number of clock cycles delayed by the loop is equal to the number of times the loop executes multiplied by

```
                                       Clock Cycles
             MOV   CX, N               4           = C
KILL TIME:   NOP                       3              O
             NOP                       3           = C
             LOOP KILL_TIME            17/5           L
```

(*a*)

$$C_T = C_O + N(C_L) - 12$$

$$N = \frac{C_T - C_O + 12}{C_L} = \frac{5000 - 4 + 12}{23} = 218 = 0D9H$$

(*b*)

FIGURE 4-20 Delay loop program and calculations. *(a)* Program. *(b)* Calculations.

the time per loop. To be somewhat more accurate you can subtract the 12 cycles that were not used when the last LOOP instruction executed. The total number of clock cycles required for the example program to execute is $C_O + N(C_L) - 12$. Set this equal to the number of clock cycles of delay you want, 5000 for this example, and solve the result for *N*. Figure 4-20*b* shows how this is done. The resultant value for *N* is 218 decimal or 09DH. This is the number of times you want the loop to repeat, so this is the value of *N* that you will load into CX before entering the loop.

With the simple relationship shown in Figure 4-20*b*, you can determine the value of *N* to put in a delay loop you write, or you can determine the time a delay loop written by someone else will take to execute.

If you can't get a long enough delay by counting down a single register or memory location, you can nest delay loops. An example of this nesting is:

```
                          ;number of states
        MOV  BX,COUNT1    ;4
CNTDN1: MOV  CX, COUNT2   ;4 x COUNT1
CNTDN2: LOOP CNTDN2       ;((17xCOUNT2)-12)COUNT1
        DEC  BX           ;2(COUNT1)
        JNZ  CNTDN1       ;16(COUNT1)-12
```

The principle here is to load CX with COUNT2 and count CX down COUNT1 times. To determine the number of states that this program section will take to execute, observe that the LOOP instruction will execute COUNT2 times for each time CX is loaded with COUNT1. The total number of states then is COUNT1 times the number of states for the last four instructions plus 4, for the MOV BX, COUNT1 instruction. The best way to approach getting values for the two unknowns, COUNT1 and COUNT2, is to choose a value such as FFFFH for COUNT2 and then solve for the value of COUNT1. A couple of tries should get reasonable values for both COUNT1 and COUNT2.

Delay loops are a very common use of the REPEAT—UNTIL structure. The next section describes the 8086 string instructions which are often used in REPEAT—UNTIL structures.

The 8086 String Instructions

INTRODUCTION AND OPERATION

A string is a series of bytes or words stored in successive memory locations. Often a string consists of a series of ASCII character codes. When you use a word processor or text-editor program, you are actually creating a string of this sort as you type in characters. One important feature of a word processor is the ability to move a sentence or group of sentences from one place in the text to another. Doing this involves moving a string of ASCII characters from one place in memory to another. The 8086 Move String instruction, MOVS, allows you to do this very easily. Another important feature of most word processors is the ability to search through the text looking for a given word or phrase. The 8086 Compare String instruction, CMPS, allows you to do this easily. Let's see how these string instructions work.

MOVING A STRING

Suppose that we have a string of ASCII characters in successive memory locations starting at offset 2000H in the data segment, and we want to move this string to an offset of 2400H in the data segment. Figure 4-21*a* shows the basic pseudocode for this operation. When we start thinking about how we can implement this algorithm in assembly language, several points come to mind. We need a pointer to the source string to keep track of which string element we are moving at a given time. This is the same reason we needed a pointer in the price-fixing program in Figure 4-15*c*. We use the source index register for this pointer. SI will hold the offset of the byte that we are moving at a given time. We also need a pointer to the location where we are moving string elements to. The destination index register DI is used to hold the offset of the location where a byte is being moved to at a given time. Another need is for a counter to keep track of how many string bytes have been moved so we can determine when we have moved all of the string. We use the CX register as a counter for string operations. Having these pieces in mind we can expand the pseudocode for the problem as shown in Figure 4-21*b*. We often describe an algorithm in general terms at first and then expand sections as needed to help us see how the algorithm is implemented in a specific language. In the expanded version in Figure 4-21*b*, you can see that we need to initialize the two pointers and the counter. The REPEAT—UNTIL loop consists of moving a byte, incrementing the pointers to point to the source and destination for the next byte, and decrementing the counter so we can see if all of the bytes have been moved.

As it turns out, the single 8086 string instruction, MOVSB, will perform all of the actions in the REPEAT—UNTIL loop. In other words the MOVSB instruction will copy a byte from the location pointed to by the SI register to a location pointed to by the DI register. It will then automatically increment SI to point to the next source location, increment DI to point to the next destination location. Actually, as we will show you soon, we can specify whether we want SI and DI to increment or decrement. If we add a special prefix called the *repeat* prefix in front of the MOVSB instruction, the MOVSB instruction will be repeated and CX decremented until CX is counted down to zero. In other words it will repeat the MOVSB instruction until the entire string is copied to the destination location.

Figure 4-21*c* shows the program instructions to molve our string of bytes. The first three instructions in the program initialize the data segment register and the extra segment register. After that we load the SI register with 2000H so that it points to the start of the source string. We then load the DI register with 2400H so that it points to the first address of the desired destination. Actually, for string instructions, the offset in DI is added to the extra segment to produce the physical address.

However, if DS and ES are initialized with the same value as we did with the first three instructions in the program, then SI and DI will both be added to the same segment base. Next we load the CX register with the number of bytes in the string we are moving. CX will function as a counter to keep track of how many string bytes have been moved at any given time. Finally, we make the direction flag a 0 with the Clear Direction Flag instruction, CLD. This will cause both SI and DI to be automatically incremented after a string byte is moved. If the direction flag is set with the STD instruction, then SI and DI will be automatically decremented after each string byte is moved. Now when the Move String Byte instruction, MOVSB, executes, a byte pointed to by SI will be copied to the location pointed to by DI. SI and DI will be automatically incremented to point to the next source and to the next destination. The count register will be automatically decremented. The MOVSB instruction by itself will just copy one byte and update SI and DI. However, with the repeat prefix, REP, in front of the MOVSB instruction as shown, CX will be decremented and the instruction will execute over and over again until the CX register is counted down to zero. When the program is coded, the 8-bit code for the REP prefix, 11110010, is put in the memory location before the code for the MOVSB instruction, 10100100. After the MOVSB instruction is finished, SI will be pointing to the location after the last source string byte, DI will be pointing to the location after the last destination address, and CX will be zero.

The MOVSW instruction can be used to move a string of words. Depending on the state of the direction flag, SI and DI will automatically be incremented or decremented by two after each move. CX will be decremented by one after each word move with the REP prefix so CX should be initialized with the number of words in the string.

```
REPEAT
       MOVE BYTE FROM SOURCE STRING TO DESTINATION STRING
UNTIL ALL BYTES MOVED
```

(a)

```
INITIALIZE SOURCE POINTER, SI
INITIALIZE DESTINATION POINTER, DI
INITIALIZE COUNTER, CX

REPEAT
      COPY BYTE FROM SOURCE TO DESTINATION
      INCREMENT SOURCE POINTER, SI
      INCREMENT DESTINATION POINTER, DI
      DECREMENT COUNTER, CX
UNTIL CX = 0
```

(b)

```
CODE_HERE SEGMENT
      ASSUME  CS:CODE_HERE, DS:STRINGS_HERE, ES:STRINGS_HERE
            MOV   AX, 0000
            MOV   DS, AX              ; initialize DS
            MOV   ES, AX              ; initialize ES
            MOV   SI, 2000H           ; initialize SI
            MOV   DI, 2400H           ; initialize DI
            MOV   CX, 0080H           ; initialize counter -
                                      ; 128 bytes in string
            CLD
REP         MOVSB
```

(c)

FIGURE 4-21 Program for moving a string from one location to another in memory. *(a)* First-version pseudocode. *(b)* Expanded-version pseudocode. *(c)* Assembly language.

STRING INSTRUCTIONS OVERVIEW

A section in Chapter 3 shows a list of the string instructions with short descriptions of their operations. Take a look at this list to give you an overview of this group of instructions, and then go on to the second string instruction example which follows. Consult the detailed descriptions of the individual instructions in Chapter 6 for further information and short program examples for each.

USING THE COMPARE STRING BYTE TO CHECK A PASSWORD

For this program example suppose that we want to compare a password entered by a person who wants to use the computer with the correct password stored in memory. If the passwords do not match, we want to sound an alarm. If the passwords match, we want to continue on with the mainline program. Figure 4-22 shows how we might represent the algorithm for this with a flowchart and with pseudocode. Note that we want to terminate the REPEAT—UNTIL when either the compared bytes do not match, or when we are at the end of the string. We then use an IF—THEN structure to sound the alarm if the compared strings were not equal at any point. If the strings match, the IF—THEN just directs execution on to the main program.

To implement this algorithm in assembly language, we probably would first expand the basic structures as we did for the previous string example in Figure 4-21. Figure 4-22c shows how we might do this expansion. The first action in the expanded algorithm is to initialize the port device for output. We need to have an output port because we will turn on the alarm by outputting a 1 to the alarm control circuit. You can see that we need a pointer to each string and a counter to keep track of how many string elements have been compared. If you use SI and DI for the pointers and CX for the counter, then the 8086 Compare String Bytes instruction, CMPSB, will implement all of the actions between REPEAT and UNTIL. If we put the correct repeat prefix in front of this instruction, the single instruction statement will implement the entire REPEAT—UNTIL structure.

Figure 4-23 reviews some old concepts, introduces a few new ones, and shows how this program can be done in assembly language. First let's look at the data structure for this program. The statement PASSWORD DB 'FAILSAFE' sets aside 8 bytes of memory and gives the first memory location the name PASSWORD. This statement also initializes the eight memory locations with the ASCII codes for the letters FAILSAFE. The single quotes around FAILSAFE tell the assembler to put the ASCII codes for the letters of this word in successive memory locations. For FAILSAFE the ASCII codes will be 46H, 41H, 49H, 4CH, 53H, 41H, 46H, 45H. The statement INPUT_WORD DB 8 DUP(?) will set aside eight memory locations and assign the name INPUT_WORD to the first location. The DUP(?) in the statement tells the assembler not to initialize these eight locations. We assume that another program section will load these locations with ASCII codes read from the keyboard.

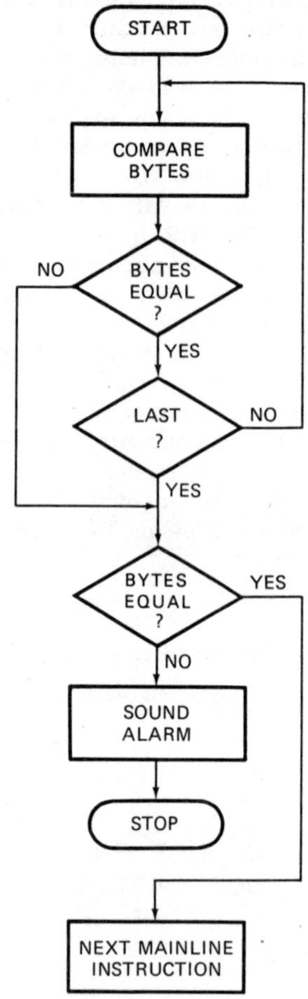

(a)

```
REPEAT
    COMPARE SOURCE BYTE WITH DESTINATION BYTE
UNTIL (BYTES NOT EQUAL) OR (END OF STRING)
IF BYTES NOT EQUAL THEN
    SOUND ALARM
    STOP
DO NEXT MAINLINE INSTRUCTION
```

(b)

```
INITIALIZE PORT DEVICE FOR OUTPUT
INITIALIZE SOURCE POINTER — SI
INITIALIZE DESTINATION POINTER — DI
INITIALIZE COUNTER — CX
REPEAT
    COMPARE SOURCE BYTE WITH DESTINATION BYTE
    INCREMENT SOURCE POINTER
    INCREMENT DESTINATION POINTER
    DECREMENT COUNTER
UNTIL (STRING BYTES NOT EQUAL) OR (CX = 0)
IF STRING BYTES NOT EQUAL THEN
    SOUND ALARM
    STOP
DO NEXT MAINLINE INSTRUCTION
```

(c)

FIGURE 4-22 Flowchart and pseudocode for comparing strings program. *(a)* Flowchart. *(b)* Initial pseudocode. *(c)* Expanded pseudocode.

Now let's look at the code segment section of the program. The ASSUME statement tells the assembler that the instructions will be in the segment CODE_HERE. It also tells the assembler that any references to the data segment or to the extra segment will mean the segment DATA_HERE. We have to tell the assembler what to assume about the extra segment, because with string instructions an offset in DI is added to the extra segment base to produce the physical address.

The next three MOV statements in the program initialize the data and extra segment registers. Since we initialize DS and ES with the same values, both SI and DI will point to locations in the segment DATA_HERE. The next three instructions initialize port P2B of an SDK-86 board as an output port.

LEA SI, PASSWORD loads the effective address or offset of the start of the FAILSAFE string into the SI register. Since PASSWORD is the first data item in the segment DATA_HERE, SI will be loaded with 0000H. LEA DI, INPUT_WORD loads the effective address or offset of the start of the INPUT_WORD string into the DI register. Since the offset of INPUT_WORD is 0008H, DI will be loaded with this value. The MOV CX, 08H statement initializes CX with the number of bytes in the string. The clear direction flag instruction tells the 8086 to automatically increment SI and DI after two string bytes are compared.

The CMPSB instruction will compare the byte pointed to by SI with the byte pointed to by DI and set the flags according to the result. It will also increment the pointers, SI and DI, to point to the next string elements, and decrement the counter, CX, to indicate that two string elements have been compared. The REPE prefix in front of this instruction tells the 8086 to repeat the CMPSB instruction if the compared bytes were equal *and* CX is not decremented down to zero yet. When the instruction is coded, the code for this prefix, 11110011, will be put in memory before the code for the CMPSB instruction, 10100110.

If the zero flag is not set when execution leaves the repeat loop then we know that the two strings are not equal. This means that the password entered was not valid so we want to sound an alarm. The JNE SOUND_ALARM will check the zero flag and, if it is not set, do a jump to the specified label. If the zero flag is set, indicating a valid password, then execution falls through to the JMP OK instruction. This JMP instruction simply jumps over the instructions which sound the alarm and stops the computer.

For this example we assume that the alarm control is connected to the least-significant bit of port FFFAH and that a 1 output to this bit turns on the alarm. The MOV AL, 01 instruction loads a 1 in the LSB of AL. The MOV DX, 0FFFAH instruction points DX at the port that the alarm is connected to and the OUT DX, AL instruction copies this byte to port FFFAH. Finally, the HLT instruction stops the computer. An interrupt or reset will be required to get it started again.

As the preceding examples show, the string instructions make it very easy to implement some commonly needed REPEAT—UNTIL algorithms. Some of the programming problems at the end of the chapter will give you practice with these. The next section here gives you some hints on how to debug the programs that you write.

DEBUGGING ASSEMBLY LANGUAGE PROGRAMS

So far in this book we have tried to show you the tools and techniques used to write assembly language programs. By now you should be writing some programs of your own, so we need to give you a few hints on how to debug your programs if they don't work correctly the first time you try to run them.

The first technique you use when you hit a difficult-to-find problem in either hardware or software is the *Five Minute Rule*. This rule says "You get 5 minutes to freak out and mumble about changing vocations, then you have to cope with the problem in a systematic manner." What this means is step back from the problem, collect your wits, and think out a systematic series of steps to find the solution. We have seen many technicians waste a lot of valuable time randomly poking and probing to try to find the cause of a problem. Here is a list of additional techniques you may find useful in writing and debugging your programs.

1. Very carefully define the problem you are trying to solve with the program and work out the best algorithm you can.

2. If the program consists of several parts, write, test and debug each part individually then add parts one at a time.

3. If a program or program section does not work, first recheck the algorithm to make sure it really does what you want it to. You might have someone else look at it also. Another person may quickly spot an error you have overlooked 17 times.

4. If the algorithm seems correct, check to make sure that you have used the correct instructions to implement the algorithm. It is very easy to accidentally switch the operands in an instruction. You might, for example, write down the instruction MOV AX, DX when the instruction you really want is MOV DX, AX. Sometimes it helps to work out on paper the effect that a series of instructions will have on some sample numbers. These predictions on paper can later be compared with the actual results produced when the program section runs.

5. If you are hand coding your programs, this is the next place to check. It is very easy to get a bit wrong when you construct the 8086 instruction codes. Also remember, when constructing instruction codes which contain addresses or displacements, that the low byte of the address or displacement is coded in before the high byte.

```
page, 132
;8086 PROGRAM
;ABSTRACT        : This program inputs a password and sounds an
;                  alarm if the password is incorrect
;REGISTERS USED: CS, DS, ES, AX, DX, CX, SI, DI
;PORTS USED      : FFFAH - alarm output port
;PROCEDURES      : None used

DATA_HERE           SEGMENT
        PASSWORD            DB    'FAILSAFE'
        INPUT_WORD          DB    8 DUP(?)      ; space for user input
DATA_HERE           ENDS

CODE_HERE           SEGMENT
                    ASSUME CS:CODE_HERE, DS:DATA_HERE, ES:DATA_HERE

                    MOV     AX, DATA_HERE      ; initialize data &
                    MOV     DS, AX             ; extra segments
                    MOV     ES, AX

                    MOV     DX, 0FFFEH         ; set up port
                    MOV     AL, 99H            ; as an output port
                    OUT     DX, AL

                    LEA     SI, PASSWORD       ; load source pointer
                    LEA     DI, INPUT_WORD     ; load destination pointer
                    MOV     CX, 08H            ; counter = password length
                    CLD                        ; increment DI & SI
REPE                CMPSB                      ; compare the two strings
                    JNE     SOUND_ALARM        ; not equal, sound alarm
                    JMP     OK                 ; equal - continue
SOUND_ALARM:        MOV     AL, 01             ; to sound alarm, send
                    MOV     DX, 0FFFAH         ; a 1 to the output
                    OUT     DX, AL             ; port whose address is
                    HLT                        ; in DX and HALT.
OK:                 NOP                        ; rest of program for user
                                               ; whose password = FAILSAFE
CODE_HERE           ENDS
                    END
```

FIGURE 4-23 Assembly language program for comparing strings.

6. If you don't find a problem in the algorithm, instructions, or coding, now is the time to use debugger, monitor, or emulator tools to help you localize the problem. You could use these tools right from the start, but by doing this it is easy to get lost in chasing bits and not see the bigger picture of what is causing the program to fail. For short program sections, the debugger or monitor *trace* and *single-step* functions may help you determine where the program is not doing what you want it to do. The IBM PC Debugger Trace command displays the contents of the registers after each instruction executes. After you run to a breakpoint then you can use the dump memory command to examine the contents of the memory. The SDK-86 board's Single Step command executes one instruction and then stops execution. You can then use the Examine Register and Examine Memory commands to see if registers and memory contain the correct data at that point. If the results are correct at that point you can use the trace or single step command to execute the next instruction. Once you have localized the problem to one or two instructions, it is usually not too hard to find out what is wrong. See the accompanying laboratory manual instructions for using these functions.

7. For longer programs, the single-step approach can be somewhat tedious. Using breakpoints is often a faster technique to narrow the source of a problem down to a small region. Most debuggers, monitors,

and emulators allow you to specify both a starting address and an ending address in their "GO" command. The SDK-86 monitor GO command, for example, has the format: GO address, breakpoint address. The GO command for the IBM PC debugger has the format: G = address address. When you give these commands, execution will start at the address specified first in the command and stop when it reaches the address specified in the second position in the command. After the program runs to a breakpoint you can use the examine register and examine memory commands to check the results at that point. Here's how we use breakpoints.

Instead of running the entire program, specify a breakpoint so that execution stops some distance into the program. You can then check to see if the results are correct at this point. If they are, you can run the program again with the breakpoint at a later address and check the results at that point. If the results are not correct, you can move the breakpoint to an earlier point in the program, run it again, and check if the results in registers and memory are correct.

Suppose, for example, you write a program such as the price-fixing program in Figure 4-15*c* and it does not give the correct results. The first place to put a breakpoint might be at the address of the ADD AL, 03 instruction. Incidentally, the instruction at the address where you put the breakpoint does not get executed in most systems. After the program runs to this breakpoint, you check to see if the data segment register, pointer register and counter register were correctly initialized. You can also see if the first price got copied into AL. If the program works correctly to this point, you can run it again with the breakpoint at the address of the JNZ DO_NEXT instruction. After the program executes to this breakpoint you can check AL to see if the addition and decimal adjustment produced the results you predicted. If the 8086 is working at all it will almost always do operations such as this correctly, so recheck your predictions if you disagree with it. You can check the pointer in BX to see if it is pointing at the next price, and you can check the count in CX to see if it has been decremented as it should be. Also you can check to see if the adjusted price got put back in memory. If you have not found the problem by now, the problem may be in the JNZ DO_NEXT instruction. Perhaps you accidentally put the DO_NEXT label next to the ADD AL, 03H instruction instead of next to the MOV AL, [BX] instruction. Or, if you are hand coding, perhaps you calculated the displacement for the JNZ instruction incorrectly.

It helps your frustration level if you make a game of thinking where to put breakpoints to track down the little bug that is messing up your program. With a little practice you should soon develop an efficient debugging algorithm of your own using the specific tools available on your system.

CHECKLIST OF IMPORTANT TERMS AND CONCEPTS IN THIS CHAPTER

If you do not remember any of the terms or concepts in the following list, use the index to find them in the chapter.

Defining a problem

Setting up a data structure

Making an initialization checklist

Masking and moving nibbles using AND and OR instructions

Packed and unpacked BCD numbers

Conditional flags: CF, PF, AF, ZF, SF, OF

Jump instructions:
- Unconditional
- Direct and indirect within-segment near jumps
- Direct and indirect within-segment short jumps
- Direct and indirect intersegment jumps

Relocatable

Conditional jumps

Fixed- and variable-port input/output

Based and indexed addressing modes

Loop instruction

Delay loop clock cycles

String instructions

Debugging—breakpoints, trace, single step

REVIEW QUESTIONS AND PROBLEMS

1. Describe the operation and results of each of the following instructions, given the register contents shown in Figure 4-24. Include in your answer the physical address or register that each instruction will get its operands from and the physical address or register that each instruction will put the result in. Use the instruction descriptions in Chapter 6 to help you. Assume that the instructions below are independent, not sequential unless listed together under a letter.

a. ROL AX, CL

b. IN AL, DX

c. MOV CX, [BX]

d. ADD AX, [BX][SI]

e. JMP 023AH

f. JMP BX

g.
```
NEXT:   MOV  AL, [BX]
        ADD  AL, 02
        DEC  CL
        JNZ  NEXT
```

h.
```
            MOV  CX, 3FC2H
COUNT_DOWN: LOOP COUNT_DOWN
```

i.
```
    MOV CX, 100 ;length of STRING_1
    MOV SI, OFFSET STRING_1
    MOV DI, OFFSET STRING_2
    CLD
REP MOVSB
```

2. Construct the binary codes for the instructions of Questions 1*a* through 1*f*.

3. Predict the state of the six 8086 conditional flags after each of the following instructions or group of instructions executes. Use the register contents shown in Figure 4-24. Assume all flags are reset before the instructions execute. Use the detailed instruction descriptions in Chapter 6 to help you.
   a. MOV AL, AH *c*. ADD CL, DH
   b. ADD BL, CL *d*. OR CX, BX

4. See if you can find any errors in the following instructions or groups of instructions.
   a.
   ```
   CNTDOWN: MOV  BL, 72H
            DEC  BL
            JNZ  CNTDOWN
   ```
   b. REP ADD AL, 07 *d*. ADD CX, AL
   c. JMP BL *e*. DIV AX, BX

5. *a*. Write an algorithm for a program which adds a byte number from one memory location to a byte from the next memory location, puts the sum in a third memory location, and saves the state of the carry flag in the least-significant bit of a fourth memory location. Mask the upper 7 bits of the memory location where the carry is stored.
   b. Write an 8086 assembly language program for this algorithm. *HINTS*: Set up data declarations similar to those in Figure 3-10. Use a rotate instruction to get the carry flag state into the LSB of a register or memory location.
   c. What additional instructions would you have to add to this program so that it correctly adds 2 BCD bytes?

For each of the following programming problems, draw a flowchart or write the pseudocode for an algorithm to solve the problem. Then write an 8086 assembly language program to implement the algorithm. If you have an 8086 system available, enter and assemble your source program, then load the object code for the program into memory so you can run and test it. If the program does not work correctly, use the approach described in the last section of this chapter to help you debug it.

6. Convert a packed BCD byte to two ASCII characters for the two BCD digits in the byte. For example, given a BCD byte containing 57H (01010111 binary), produce the two ASCII codes 35H and 37H.

7. Compute the average of 4 bytes stored in an array in memory.

8. Compute the average of any number of bytes in an array in memory. The number of bytes to be added is in the first byte of the array.

9. Add a 5-byte number in one array to a 5-byte number in another array. Put the sum in another array. Put the state of the carry flag in byte 6 of the array that contains the sum. The first value in each array is the least-significant byte of that number. *HINT*: See Figure 4-15*d*.

10. An 8086-based process control system outputs a measured Fahrenheit temperature to a display on its front panel. You need to write a short program which converts the Fahrenheit temperature to Celsius so that the system can be sold in Europe. The relationship between Fahrenheit and Celsius is: $C = (F - 32)5/9$. The Fahrenheit temperature will always be in the range of 50° to 250°. Round the Celsius value to the nearest degree.

11. An ASCII keyboard outputs parallel ASCII + parity to port FFF8H of an SDK-86 board. The keyboard also outputs a strobe to the least-significant bit (D0) of port FFFAH. (See Figure 4-13.) When you press a key, the keyboard outputs the ASCII code for the pressed key on the eight parallel lines and outputs a strobe pulse high for 1 ms. You want to poll the strobe over and over until you find it high. Then you want to read in the ASCII code, mask the parity bit (D7), and store the ASCII code in an array in memory. Next you want to poll the strobe over and over again until you find it low. When you find the strobe has gone low, check to see if you have read in 10 characters yet. If not, then go back and wait for the strobe to go high again. If 10 characters have been read in, stop.

12. *a*. Write a delay loop which produces a delay of 500 μs on an 8086 with an 8-MHz clock.

	AH	AL			BH	BL
AX	A4	07		BX	24	B3
	CH	CL			DH	DL
CX	00	02		DX	FF	FA

SP = FFFF		CS = 2000	
BP = 0009		DS = 3000	
SI = 4200		SS = 4000	
DI = 4300		ES = 3000	

FIGURE 4-24 Figure for Chapter 4 problems.

b. Write a short program which outputs a 1-kHz square wave on D0 of port FFFAH. The basic principle here is to output a high, wait 500 μs (0.5 ms), output a low, wait 500 μs and output a high, etc. Remember that, before you can output to a port device, you must first initialize it as in Figure 4-12*a*. If you connect a buffer such as that shown in Figure 8-22 and a speaker to D0 of the port, you will be able to hear the tone produced.

13. *a.* Move a string containing your name in the form "Charlie T. Tuna" from one string location in memory to a new string location named NEW_HOME which is just above the initial location.

b. Move the string containing your name up four addresses in memory. Consider whether the pointers should be incremented or decremented after each byte is moved in order to keep any needed byte from being written over. *HINT*: Initialize DI with the value of SI + 4.

14. Scan a string of 80 characters, looking for a carriage return (0DH). If a carriage return is found, put the length of the string up to the carriage return in AL. If no carriage return is found, put 50H (80 decimal) in AL.

15. Given a string containing your name in the form "Charlie T. Tuna", put the characters in a second string called LAST_FIRST in the order "Tuna Charlie T".

CHAPTER

5 IF—THEN—ELSE STRUCTURES, PROCEDURES, AND MACROS

The last chapter showed you how quite a few of the 8086 instructions work, and how jump instructions are used to implement WHILE—DO and REPEAT—UNTIL structures. A section of this chapter shows how IF—THEN—ELSE structures are also implemented with jump instructions. The major point of this chapter, however, is to show you how to write and use subprograms called *procedures*. A final section of the chapter shows you how to write and use assembler MACROs.

OBJECTIVES

At the conclusion of this chapter you should be able to:

1. Write 8086 assembly language programs to solve IF—THEN, IF—THEN—ELSE, and multiple IF—THEN—ELSE type programming problems.
2. Write an 8086 assembly language program which calls a near procedure.
3. Write an 8086 assembly language program which calls a far procedure.
4. Describe how a stack is initialized and used in 8086 assembly language programs which call procedures.
5. Write and use an assembler MACRO.

IF—THEN, IF—THEN—ELSE, AND MULTIPLE IF—THEN—ELSE PROGRAMS

IF—THEN Programs

Remember from Chapter 2 that the IF—THEN structure has the format:

IF condition THEN
  action
  action

This structure says that IF the stated condition is found to be true, the series of actions following THEN will be executed. If the condition is false, then execution will skip over the actions after the THEN and proceed on with the next mainline instruction.

The simple IF—THEN is implemented with a conditional jump instruction. In some cases an instruction to set flags is needed before the conditional jump instruction. Figure 5-1*a* shows, with a program fragment, one way to implement the simple IF—THEN structure. In this program we first compare BX with AX to set the required flags. If the zero flag is set after the comparison, indicating that AX is equal to BX, the JE instruction will cause execution to jump to the MOV CL, 07H instruction labeled THERE. If AX is not equal to BX, then the three NOP instructions after the JE instruction will be executed before the MOV CL, 07H instruction.

The implementation in Figure 5-1*a* will work well for a short sequence of instructions after the conditional jump instruction. However, if the sequence of instructions is lengthy, there is a potential problem. Remember from the discussion of conditional jumps in the last chapter that a conditional jump can only be to a location in the range of −128 bytes to +127 bytes from the address after the conditional jump instruction. A long sequence of instructions after the conditional jump instruction may put the label out of range of the conditional jump instruction. If you are absolutely sure that the destination label will not be out of range, then use the instruction sequence shown in Figure 5-1*a* to implement an IF—THEN structure. If you are not sure if the destination will be in range, Figure 5-1*b* shows an instruction sequence that will always work. In this sequence the conditional jump instruction only has to jump over the JMP instruction. The JMP instruction used to get to the label THERE can jump to anywhere in the code segment, or even to another code segment. Note that you have to change the conditional jump instruction from JE to JNE in this second version. The price

```
        CMP     AX, BX   ; compare to set flags
        JE      THERE    ; if equal then skip correction
        NOP
        NOP              ; NOPs represent correction
        NOP              ; instructions
THERE:  MOV     CL, 07H  ; load count
```

(a)

```
        CMP     AX, BX   ; compare to set flags
        JNE     FIX      ; if not equal do correction
        JMP     THERE
FIX:    NOP              ; NOPs represent correction
        NOP              ; instructions
        NOP
THERE:  MOV     CL, 07H  ; load count
```

(b)

FIGURE 5-1 IF—THEN implementations. *(a)* Conditional jump destinations closer than ±128 bytes. *(b)* Conditional jump destinations further than ±128 bytes.

you pay for not having to worry whether the destination is in range is an extra jump instruction.

By now you are probably thinking that this IF—THEN structure looks very familiar. It should, because a simple IF—THEN is part of the WHILE—DO and REPEAT—UNTIL structures. If you look back at the programs in the last chapter, you should see several examples of simple IF—THEN. One example is the instruction sequence in Figure 4-23 which turns on an alarm if two compared strings are not equal. We cycled through the simple IF—THEN again here as a lead-in to the IF—THEN—ELSE discussed next.

IF—THEN—ELSE Programs

The IF—THEN—ELSE structure is used to indicate a choice between two alternative courses of action. Figure 3-3*b* shows the flowchart and pseudocode for this structure. Basically the structure has the format:

```
IF condition THEN
  action
  action
ELSE
  action
  action
```

This is a different situation than the simple IF—THEN, because here either one series of actions or another series of actions is done before the program goes on with the next mainline instruction. An example will show how we implement this structure.

Suppose that in the computerized factory we discussed in Chapter 2 we have an 8086 microcomputer which controls a printed-circuit-board-making machine. Part of the job of this 8086 is to check a temperature sensor and turn on a green lamp or a yellow lamp depending on the value of the temperature it reads in. If the temperature is below 30°C, we want to turn on a yellow lamp to tell the operator that the solution is not up to temperature. If the temperature is greater than or equal to 30°C, we want to light a green lamp. With a system such as this the operator can visually scan all the lamps on the control panel until all green lamps are lit. When all the lamps are green, the operator can push the GO button to start making boards. The reason that we have the yellow lamp is to let the operator know that this part of the machine is working, but that the temperature is not yet up to 30°C.

Figure 5-2 shows with flowcharts and with pseudocode two ways we can represent the algorithm for this problem. The difference between the two is simply a matter of whether we make the decision based on the temperature being below 30°C, or we make the decision based on the temperature being above or equal to 30°C. The two approaches are equally valid, but your choice determines which conditional jump instruction you choose. Figure 5-3*a* shows the 8086 assembly language implementation of the algorithm in Figure 5-2*a*.

For this program segment, assume that we read the temperature in from an analog-to-digital converter connected to input port FFF8H. Also assume that the control for the yellow lamp is connected to bit 0 of port FFFAH, and the control for the green lamp is connected to bit 1 of port FFFAH. A 1 sent to a bit position of port FFFAH turns on the lamp connected to that line. After we read the data in from the port, we compare it with our set point value of 30°C. If the input value is below 30°C, then we jump to the instructions which turn on the yellow lamp. If the temperature is above or equal to 30°C, we jump to the instructions which turn on the green lamp. Note that we have implemented this algorithm in such a way that the JB instruction will always be able to reach the label YELLOW.

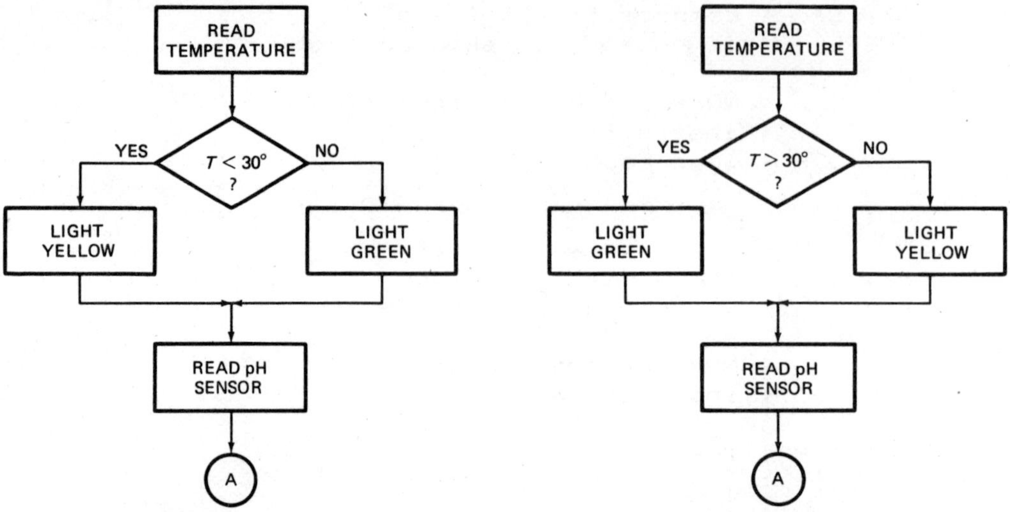


```
READ TEMPERATURE
IF TEMPERATURE < 30° THEN
    LIGHT YELLOW LIGHT
ELSE
    LIGHT GREEN LIGHT
READ pH SENSOR
```

(a)

```
READ TEMPERATURE
IF TEMPERATURE > 30° THEN
    LIGHT GREEN LIGHT
ELSE
    LIGHT YELLOW LIGHT
READ pH SENSOR
```

(b)

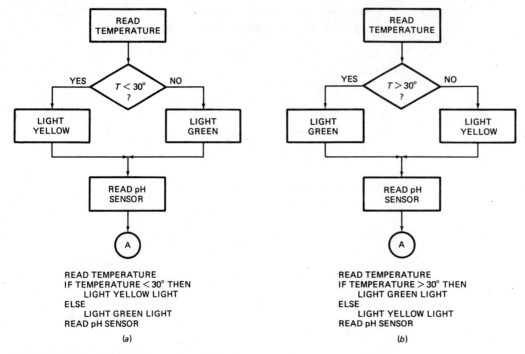

FIGURE 5-2 Flowcharts and pseudocode for two ways of expressing algorithm for printed-circuit-board-making machine. *(a)* Temperature below 30° test. *(b)* Temperature above 30° test.

```
The IBM Personal Computer MACRO Assembler 02-19-85      PAGE    1-1

                        PAGE ,132
                        ;8086 program section for PC board making machine
                        ;
                        ;ABSTRACT:      This program section reads the temperature of a cleaning
                        ;               bath solution and lights one of two lamps according to
                        ;               the temperature read. If the temp is below 30 degrees
                        ;               Celsius, a yellow lamp will be turned on. If the temp
                        ;               is above or equal to 30 degrees, a green lamp willbe
                        ;               turned on.
                        ;REGISTERS USED: CS, AL, DX
                        ;PORTS USED     : FFF8H as a temperature input
                        ;                 FFFAH as lamp control output (yellow=bit 0, green=bit 1)
                        ;PROCEDURES     : None used
0000                    CODE_HERE       SEGMENT
                                        ASSUME CS:CODE_HERE

                        ;intialize port FFFAH as an output port
0000  BA FFFE                           MOV DX, 0FFFEH          ; Point DX to port control register
0003  B0 99                             MOV AL, 99H             ; load control word to set up output port
0005  EE                                OUT DX, AL              ; send control word to control register
                        ;initialization complete

0006  BA FFF8                           MOV DX, 0FFF8H          ; point DX at input port
0009  EC                                IN  AL, DX              ; read temp from sensor on input port
000A  3C 1E                             CMP AL, 30              ; compare temp with 30 degrees C
```

FIGURE 5-3 Assembly language program segments for printed-circuit-board-making machine. *(a)* Below 30° version. *(b)* Above 30° version.

```
000C  72 03                          JB  YELLOW          ; if temp < 30 go light yellow lamp
000E  EB 0A 90                       JMP GREEN           ; else go light green lamp
0011  B0 01          YELLOW:         MOV AL, 01H         ; load code to light yellow lamp
0013  BA FFFA                        MOV DX, 0FFFAH      ; point DX at output port
0016  EE                             OUT DX, AL          ; send code to light yellow lamp
0017  EB 07 90                       JMP EXIT            ; go to next mainline instruction
001A  B0 02          GREEN:          MOV AL, 02H         ; load code to light green lamp
001C  BA FFFA                        MOV DX, 0FFFAH      ; point DX at output port
001F  EE                             OUT DX, AL          ; send code to light green lamp
0020  BA FFFC        EXIT:           MOV DX, 0FFFCH      ; next mainline instruction

0023                 CODE_HERE       ENDS
                                     END
```

(*a*)

```
IBM Personal Computer MACRO Assembler  Version 2.00  Page 1-1 11-07-85

                     PAGE ,132
                     ;8086 program section for PC board making machine
                     ;
                     ;ABSTRACT:       This program section reads the temperature of a cleaning
                     ;                bath solution and lights one of two lamps according to
                     ;                the temperature read. If the temp is below 30 degrees
                     ;                Celsius, a yellow lamp will be turned on. If the temp
                     ;                is above or equal to 30 degrees, a green lamp will be
                     ;                turned on.
                     ;REGISTERS USED: CS, AL, DX
                     ;PORTS USED     : FFF8H as a temperature input
                     ;                 FFFAH as lamp control output (yellow=bit 0, green=bit 1)
                     ;PROCEDURES     : None used

0000                 CODE_HERE       SEGMENT
                                     ASSUME CS:CODE_HERE

                     ;intialize port FFFAH as an output port
0000  BA FFFE                        MOV DX, 0FFFEH      ; Point DX to port control register
0003  B0 99                          MOV AL, 99H         ; load control word to set up output port
0005  EE                             OUT DX, AL          ; send control word to control register
                     ;initialization complete

0006  BA FFF8                        MOV DX, 0FFF8H      ; point DX at input port
0009  EC                             IN  AL, DX          ; read temp from sensor on input port
000A  3C 1E                          CMP AL, 30          ; compare temp with 30 degrees C
000C  73 03                          JAE GREEN           ; if temp >= 30, go light green lamp
000E  EB 0A 90                       JMP YELLOW          ; else go light yellow lamp
0011  B0 02          GREEN:          MOV AL, 02H         ; load code to light green lamp
0013  BA FFFA                        MOV DX, 0FFFAH      ; point DX at output port
0016  EE                             OUT DX, AL          ; send code to light green lamp
0017  EB 07 90                       JMP EXIT            ; go to next mainline instruction
001A  B0 01          YELLOW:         MOV AL, 01H         ; load code to light yellow lamp
001C  BA FFFA                        MOV DX, 0FFFAH      ; point DX at output port
001F  EE                             OUT DX, AL          ; send code to light yellow lamp
0020  BA FFFC        EXIT:           MOV DX, 0FFFCH      ; next mainline instruction

0023                 CODE_HERE       ENDS
                                     END
```

(*b*)

FIGURE 5-3 (*continued*)

To actually turn on a lamp, we load a 1 in the appropriate bit of the AL register with a MOV instruction and send the byte to the lamp control port, FFFAH. For example, the instruction sequence MOV AL, 01H—OUT DX, AL will light the yellow lamp by sending a 1 to bit 0 of port FFFAH.

Figure 5-3*b* shows another equally valid assembly language program segment to solve our problem. This one uses a Jump if Above or Equal instruction, JAE, at the decision point and switches the order of the actions. This program more closely follows the second algorithm statement in Figure 5-2*b*. Perhaps you can see from these examples why two programmers may write very different programs to solve even very simple programming problems.

Multiple IF—THEN—ELSE Implementation

In the preceding section we showed how to implement and use the IF—THEN—ELSE structure which chooses between two alternative courses of action. In many situations we want a computer to choose one of several alternative actions based on the value of some variable read in or on a command code entered by a user. To choose one alternative from several we can *nest* IF—THEN—ELSE structures. The result has the form:

```
IF condition THEN
    action
    action
ELSE IF condition THEN
        action
        action
      ELSE
         action
         action
```

It is important to note in this structure that the last ELSE is part of the IF—THEN just before it. Figure 3-3*g* showed a flowchart and pseudocode for a "soup cook" example using this structure. The soup cook example, however, is too messy to implement here. Therefore, while the printed-circuit-board machine from the last section is still fresh in your mind, we will expand that example to show you how a multiple IF—THEN—ELSE is implemented.

Suppose that we want to have three lamps on our printed-circuit-board-making machine. We want a yellow lamp to indicate that the temperature is below 30°C, a green lamp to indicate that the temperature is above or equal to 30°C but below 40°C, and a red lamp to indicate that the temperature is at or above 40°C. Figure 5-4 shows three ways to indicate what we want to do here. The first way in Figure 5-4*a* simply indicates the desired action next to each temperature range. You may find this form very useful in visualizing problems where the alternatives are based on the range of a variable. Don't miss the ASCII-to-hexadecimal problem at the end of the chapter for some practice with this. Once you get the problem defined in this list form, you can easily convert it to a flowchart or pseudocode. When writing the flowchart or the pseudocode, it is best to start at one end of the overall range and work your way to the other. For example, in the flowchart in Figure 5-4*b* we start by checking if the temperature is below 30°. If the temperature is not below 30° then it must be above or equal to 30° and you do not have to do another test to determine this. You then check if the temperature is below 40°. If the temperature is above or equal to 30°, but below 40°, then you know that the temperature is in the green lamp range. If the temperature is not below 40°, then you know that the temperature must be above or equal to 40°. In other words, two carefully chosen tests will direct execution to one of the three alternatives.

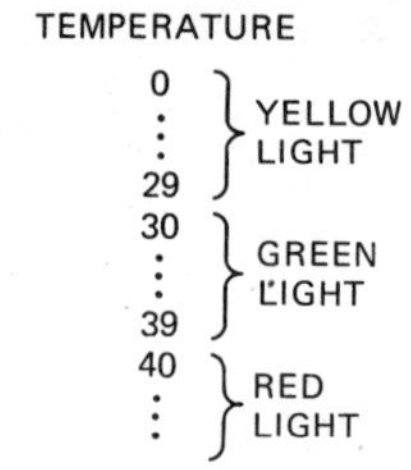

(*a*)

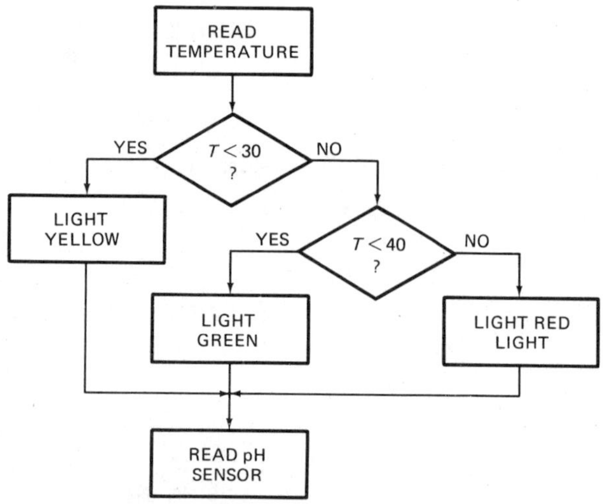

(*b*)

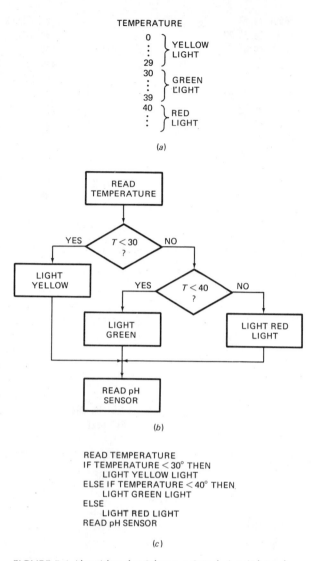

```
READ TEMPERATURE
IF TEMPERATURE < 30° THEN
    LIGHT YELLOW LIGHT
ELSE IF TEMPERATURE < 40° THEN
    LIGHT GREEN LIGHT
ELSE
    LIGHT RED LIGHT
READ pH SENSOR
```

(*c*)

FIGURE 5-4 Algorithm for 3-lamp printed-circuit-board-making machine. *(a)* Condition list. *(b)* Flowchart. *(c)* Pseudocode.

Figure 5-5 shows how we can write a program for this algorithm in 8086 assembly language. In the program we first initialize port FFFAH as an output port. We then

```
The IBM Personal Computer MACRO Assembler 02-19-85      PAGE    1-1

                                PAGE ,132
                                ;8086 program section for PC board making machine
                                ;
                                ;ABSTRACT:      This program section reads the temperature of a cleaning
                                ;               bath solution and lights one of 3 lamps according to
                                ;               the temperature read. If the temp is below 30 degrees
                                ;               Celsius, a yellow lamp will be turned on. If the temp
                                ;               is >= 30 and < 40 degrees, a green lamp will be turned on
                                ;               Temps >= 40 degrees will turn on a red lamp.
                                ;REGISTERS USED: CS, AL, DX
                                ;PORTS USED    : FFF8H as a temperature input
                                ;                FFFAH as lamp control output (yellow = bit 0,
                                ;                              green = bit 1, red  = bit 2)
                                ;PROCEDURES    : None used

0000                            CODE_HERE       SEGMENT
                                                ASSUME CS:CODE_HERE

                                ;intialize port FFFAH as an output port
0000  BA FFFE                                   MOV DX, 0FFFEH          ; Point DX to port control register
0003  B0 99                                     MOV AL, 99H             ; load control word to set up output port
0005  EE                                        OUT DX, AL              ; send control word to control register
                                ;initialization complete

0006  BA FFF8                                   MOV DX, 0FFF8H          ; point DX at input port
0009  EC                                        IN  AL, DX              ; read temp from sensor on input port
000A  BA FFFA                                   MOV DX, 0FFFAH          ; point DX at output port
000D  3C 1E                                     CMP AL, 1EH             ; compare temp with 30 degrees C
000F  72 0A                                     JB  YELLOW              ; if temp < 30 go light yellow lamp
0011  3C 28                                     CMP AL, 28H             ; compare with 40 degrees
0013  72 0C                                     JB  GREEN               ; if temp < 40 go light green lamp
0015  B0 04                     RED:            MOV AL, 04H             ; temp >= 40 so load code to light red lamp
0017  EE                                        OUT DX, AL              ; send code to light red lamp
0018  EB 0A 90                                  JMP EXIT                ; go to next mainline instruction
001B  B0 01                     YELLOW:         MOV AL, 01H             ; load code to light yellow lamp
001D  EE                                        OUT DX, AL              ; send code to light yellow lamp
001E  EB 04 90                                  JMP EXIT                ; go to next mainline instruction
0021  B0 02                     GREEN:          MOV AL, 02H             ; load code to light green lamp
0023  EE                                        OUT DX, AL              ; send code to light green lamp
0024  BA FFFC                   EXIT:           MOV DX, 0FFFCH          ; next mainline instruction
0027  EC                                        IN  AL, DX              ; read ph sensor
0028                            CODE_HERE       ENDS
                                                END
```

FIGURE 5-5 Assembly language program for 3-lamp printed-circuit-board-making machine.

read in the temperature from an A/D converter connected to port FFF8H. We compare the temperature read in with the first set point value, 30° (1EH). If the temperature is below 30°, the jump if below, JB, instruction will cause a jump to the label YELLOW. If the jump is not taken, we know the temperature is above or equal to 30° so we go on to the CMP AL, 28H instruction to see if the temperature is below the second set point, 40° (28H). The JB GREEN instruction will cause a jump to the label GREEN, if the temperature is less than 40° (28H). If the jump is not taken, we know that the temperature must be at or above 40°C so we just go ahead and turn on the red lamp.

For this program we assume that the lines which control the three lamps are connected to port FFFAH. The yellow lamp is connected to bit 0, the green is connected to bit 1, and the red is connected to bit 2. We turn on a lamp by outputting a 1 to the the appropriate bit of port

FFFAH. The instruction sequence MOV AL, 02H—OUT DX, AL, for example, will turn on the green lamp by sending a 1 to bit 1 of port FFFAH.

SUMMARY OF IF—THEN—ELSE IMPLEMENTATION

Conditional jump instructions and instructions which set flags for them are used to implement IF—THEN—ELSE structures. A single IF—THEN—ELSE structure is used to choose one of two alternative series of actions. IF—THEN—ELSE structures can be linked to choose one of three or more alternative series of actions. As shown in Figure 3-3*g*, linked IF—THEN—ELSE structures are one way to implement the CASE structure. The algorithm for the printed-circuit-board machine lamps program in the preceding section example could have been expressed as:

```
CASE temperature OF
< 30              : light yellow lamp
≤ 30 and < 40     : light green lamp
≤ 40              : light red lamp
```

This CASE structure would be implemented in the same way as the program in Figure 5-5. However, expressing the algorithm for the problem as linked IF—THEN—ELSE structures makes it much easier to see how to implement the algorithm in assembly language. Later we show you another way to implement some CASE situations using a *jump table*.

In many programs where we want to choose between two or more alternative series of actions, each of the series of actions is quite lengthy. In this case we write each series of actions as a *subprogram* and CALL this subprogram when it is needed. The next major section of this chapter shows you how to write and use subprograms, or *procedures*, as they are often called.

WRITING AND USING PROCEDURES

Introduction

Whenever we have a series of instructions that we want to execute several times in a program, we write the series of instructions as a separate subprogram. We can then CALL this subprogram each time we want to execute that series of instructions. This saves us from having to write the series of instructions over and over each time we want it to execute in the program. This subprogram is usually called a *subroutine* or a *procedure*. To be consistent with the Intel literature we will use the term *procedure* when referring to called subprograms.

There is another major reason for using procedures in programs. Recall from Chapter 2 the *top-down design* approach to solving a programming problem. In this approach the problem is carefully defined, and then the overall job is broken down into modules. Each of these modules is broken down into smaller modules. The process is continued until the algorithm for each module is clearly obvious. Figure 5-6 shows how this hierarchy of modules can be represented in diagram form. A diagram such as this is often called a *hierarchial chart*. The point of all this is to break a large problem down into manageable-size pieces which can be individually written, tested, and debugged. The individual modules are usually written as procedures and called from a mainline program which implements the highest level of the hierarchy. This approach has the added advantage that a person can read the mainline program to get an overview of what the program does, and then work down into the procedures to see the amount of detail needed at a particular point. Now that you know what procedures are used for, we will give you an overview of how they work.

Figure 5-7*a* shows in diagram form how program execution goes from the mainline to a procedure and back to the mainline. A CALL instruction in the mainline loads the instruction pointer and in some cases also the code segment register with the starting address of the procedure. The next instruction fetched then will be the first instruction of the procedure. At the end of the procedure a return instruction, RET, sends execution back to the next instruction after the CALL in the mainline program. The RET instruction does this by loading the instruction pointer, and, if necessary, the code segment register with the address of the next instruction after the CALL instruction. As shown in Figure 5-7*b*, a procedure can call another procedure. This is called *nesting* procedures. Nested procedures are used to implement the hierarchy of modules we described in the preceding

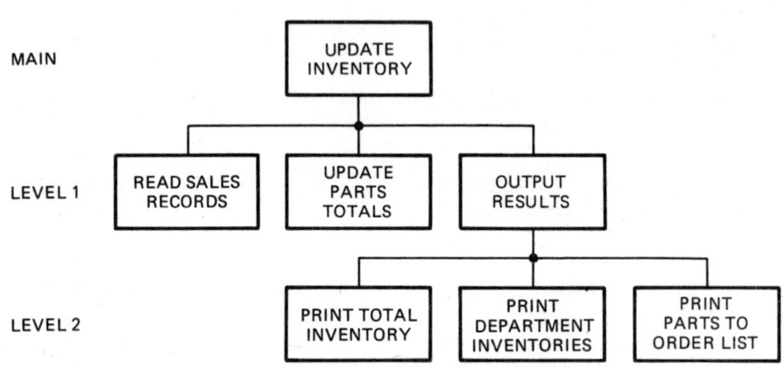

FIGURE 5-6 Hierarchical chart for inventory update program.

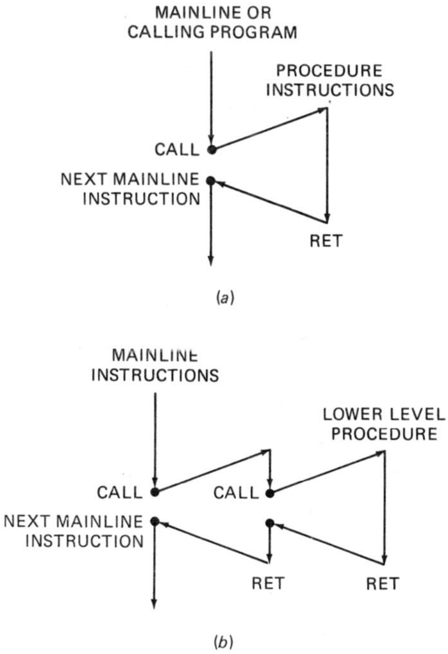

FIGURE 5-7 Program flow to and from procedures. *(a)* Single procedures. *(b)* Nested procedures.

paragraph. In the case of nested procedures, a RET instruction at the end of the lower level procedure returns execution to the higher level procedure. A second RET instruction at the end of the higher level procedure returns execution to the mainline program.

The question that may occur to you at this point is, "If a procedure can be called from anywhere in a program, how does the RET instruction know where to return execution to?" The answer to this question is that when a CALL instruction executes, it automatically stores the return address in a special section of memory called *the stack*. A later section will introduce you to how the 8086 stack works. For now let's take a closer look at the 8086 CALL and RET instructions.

The 8086 CALL and RET Instruction

THE CALL INSTRUCTION OVERVIEW

The 8086 CALL instruction performs two operations when it executes. First, it stores the address of the instruction after the CALL instruction on the stack. This address is called the *return* address because it is the address that execution will return to after the procedure executes. If the CALL is to a procedure in the same code segment, then only the instruction pointer contents will be saved on the stack. If the CALL is to a procedure in another code segment, both the instruction pointer and the code segment register contents will be saved on the stack.

The second operaton of the CALL instruction is to change the contents of the instruction pointer and, in some cases, the code segment register to contain the starting address of the procedure. Figure 5-8*a* shows the coding formats for the four forms of the 8086 CALL instruction. The differences between these four forms are in the way they tell the 8086 to get the starting address for the procedure.

CALL = Call

Within segment or group, IP relative

Opcode	DispLow	DispHigh

Opcode	Clocks	Operation
E8	19	IP ← IP+Disp16—(SP) ← return link

Within segment or group, Indirect

Opcode	mod 010 r/m		

Opcode	Clocks	Operation
FF	16	IP ← Reg16—(SP) ← return link
FF	21+EA	IP ← Mem16—(SP) ← return link
FF	21+EA	IP ← Mem16—(SP) ← return link

Inter-segment or group, Direct

Opcode	offset-low	offset-high	seg-low	seg-high

Opcode	Clocks	Operation
9A	28	CS ← segbase IP ← offset

Inter-segment or group, Indirect

Opcode	mod 011 r/m	mem-low	mem-high

Opcode	Clocks	Operation
FF	37+EA	CS ← segbase IP ← offset

(a)

RET = Return from Subroutine

Opcode

Opcode	Clocks	Operation
C3	8	intra-segment return
CB	18	inter-segment return

Return and add constant to SP

Opcode	DataL	DataH

Opcode	Clocks	Operation
C2	12	intra-segment ret and add
CA	17	inter-segment ret and add

(b)

FIGURE 5-8 8086 CALL and RET instruction formats *(Intel Corp.)*. *(a)* CALL. *(b)* RET.

DIRECT WITHIN-SEGMENT NEAR CALLS

The first form, direct within-segment near call, tells the 8086 to produce the starting address of the procedure by adding a 16-bit signed displacement contained in the instruction to the contents of the instruction pointer. This is the same process as we described for the direct within-segment near JMP instruction in Chapter 4. With this instruction the starting address of the procedure can be anywhere in the range of −32,768 bytes to +32,767 bytes from the address of the instruction after the CALL. If you are hand coding a program, you calculate the displacement by counting from the address of the instruction after the CALL to the starting address of the procedure. If the procedure is in memory before the CALL instruction, then the displacement will be negative. In this case you represent the displacement in 16-bit, 2's complement sign-and-magnitude form just as you do for backward JMP instructions. If you are using an assembler, the assembler will automatically calculate the displacement from the instruction after the CALL to a label at the start of the procedure.

THE INDIRECT WITHIN-SEGMENT CALL

The indirect within-segment CALL instruction is also a near call. When this form of CALL executes, the instruction pointer is replaced with a 16-bit value from a specified register or memory location. As indicated by the MOD—R/M byte in the coding template, the source of the value can be any of the eight 16-bit registers or a memory location specified by any one of the 24 addressing modes shown in Figure 3-8. This form of CALL instruction can be used to choose one of several procedures based on a computed value. The instruction CALL BP, for example, will do a near call to the offset contained in BP. In other words the value in BP will be put in the instruction pointer. The instruction CALL WORD PTR [BX] will get the new value for the instruction pointer from a memory location pointed to by BX.

THE DIRECT INTERSEGMENT FAR CALL

The direct intersegment far CALL is used when the procedure is in another segment. If the procedure is in another segment, you have to change both the instruction pointer and the code segment register to get to it. For this form of the CALL instruction, the new value for the instruction pointer is written in as bytes 2 and 3 of the instruction code. Note that the low byte of the new IP value is written before the high byte. The new value for the code segment register is written in as bytes 4 and 5 of the instruction code. Again the low byte is written before the high byte. A program example later in this chapter shows you how to write your programs so that an assembler can find a procedure label in another segment.

THE INDIRECT INTERSEGMENT FAR CALL

This form of the CALL instruction replaces the instruction pointer and the code segment register contents with two 16-bit values from memory. Since two 16-bit values are needed, the values cannot come from a register. The MOD—R/M byte in the instruction is used to specify the addressing mode for the memory location where the 8086 goes to get the new values. The first word from memory is put in the instruction pointer, and the second word from memory is put in the code segment register. The instruction CALL DWORD PTR [BX] will compute a new value for IP from [BX] and [BX + 1] and a new value for CS from [BX + 2] and [BX + 3]. In other words it does a far call to an address contained in 4 bytes pointed to by BX in the data segment.

THE 8086 RET INSTRUCTION

As we described in the previous section, when the 8086 does a near CALL it saves the instruction pointer value for the instruction after the CALL on the stack. A return instruction, RET, at the end of the procedure copies this value from the stack back to the instruction pointer. This then returns execution to the mainline program.

When the 8086 does a far CALL it saves the contents of both the instruction pointer and the code segment register on the stack. An RET instruction at the end of the procedure copies these values from the stack back into the IP and CS registers to return execution to the mainline program. Obviously we need one form of the RET instruction to handle returns from near procedures, and another form of the instruction to handle returns from far procedures. Actually the 8086 has four forms of the RET instruction. Figure 5-8*b* shows the coding templates for these four.

The simple within-segment form of RET copies a word from the top of the stack to the instruction-pointer register. This is the instruction form you will usually use to return from a near procedure. The within-segment adding immediate to SP form is also used to return from a near procedure. When this form executes, however, it will copy the word at the top of the stack to the instruction pointer and also add an immediate number contained in the instruction to the contents of SP. Later, we will show you what this form is used for.

The intersegment form of the RET instruction is used to return from far procedures. When this form of the RET instruction executes, it will copy the word from the top of the stack to the instruction pointer. It will then increment the stack pointer by two and copy the next word from the stack to the code segment register. The intersegment adding immediate to SP form of the instruction also copies a new value for IP and a new value for CS from the stack. However, it also adds a 16-bit immediate number contained in the instruction code to SP.

Throughout the preceding discussions of the CALL and RET instructions we have talked about writing words to the stack and copying these words back to the instruction pointer and/or code segment register. Now we will show you how to set up a stack in your programs.

The 8086 Stack

The *stack* is a section of memory you set aside for storing return addresses. The stack is also used to save the contents of registers for the mainline program while a procedure executes. A third use of the stack is to hold data or addresses that will be acted upon by a procedure.

The 8086 lets you set aside up to an entire 64 Kbyte segment of memory as a stack. Remember from the block diagram in Figure 2-7 that the 8086 contains a stack segment register and a stack pointer register. The stack segment register is used to hold the upper 16 bits of the starting address you give to the stack segment. If you decide to start the stack segment at 70000H, for example, the stack segment register will contain 7000H. The stack pointer register is used to hold the offset of the last word written on the stack. The 8086 produces the physical address for a stack location by shifting the contents of the stack segment register four bit positions to the left and adding the contents of the stack pointer to the result. Figure 2-11 shows a numerical example of this.

If you are going to call procedures or use the stack in some other way in your program, then you need to initialize both the stack segment register and the stack pointer register. Figure 5-9 shows the pieces you need to add to your programs to declare a stack segment, and to initialize SS and SP. We have shown in Figure 5-9 how you should format all this for an assembler. If you are not using an assembler, then you should use the same format, but put the desired numbers in place of the names we have used.

The STACK_HERE SEGMENT STACK and STACK_HERE ENDS statements in Figure 5-9 are used to declare a logical segment that will be used for the stack. The STACK directive tells the assembler that this segment will be used as a last-in, first-out stack.

NOTE: If you are going to use the IBM program EXE2BIN on your programs so that you can download them to an SDK-86, omit the STACK directive here. The linker will then give you an error message, WARNING—NO STACK SEGMENT, which you can ignore.

We don't need all 64 Kbytes of the logical segment in our programs so we tell the assembler to set aside 40 decimal or 28H words of storage in this logical segment with the DW 40 DUP(0) statement. By limiting the stack to near the size actually needed, this segment can be overlapped with other logical segments to save on the amount of physical memory required for a program. In other words, there is no use having a larger stack set aside than you are going to need.

Now, when we store addresses or data in these stack locations, we start at the highest location and fill toward the bottom. This is opposite to the way you put instruction code bytes in memory. In the case of instruction codes you start at the lowest address in a code segment and fill toward the top. Since we start writing to the highest location in the stack first, we need a name attached to this location so we can access it by name. The statement STACK_TOP LABEL WORD in Figure 5-9 gives the name STACK_TOP to the next even address after the 40 words we set aside for the stack. We will explain later why we want the name at the address after the actual stack. The WORD in this statement indicates that writes to and reads from the stack will be done as words. Figure 5-10 shows in diagram form how this example stack would be arranged in memory.

We arbitrarily choose to start the stack segment at address 70000H for this example, and we set a stack length of 40 words with the DW 40 DUP(0) statement. Since each memory address represents a byte, these 40 words will occupy the 80 addresses 70000H to 7004FH as shown in Figure 5-10. The label STACK_TOP is associated with address 70050H, the next address after the stack.

The next program addition you need to look at is in the ASSUME statement. Note that we have added the term SS:STACK_HERE, to tell the assembler that any reference in the program to the stack means the segment STACK_HERE. This term tells the assembler that SS will contain the starting address of STACK_HERE, but

```
;8086 Program fragment showing the intialization of the stack
;       segment and the stack pointer

STACK_HERE      SEGMENT STACK
        DW      40      DUP(0)
STACK_TOP       LABEL   WORD
STACK_HERE      ENDS

CODE_HERE       SEGMENT
                ASSUME CS:CODE_HERE, SS:STACK_HERE

                MOV     AX, STACK_HERE  ; initialize stack segment
                MOV     SS, AX          ; register
                LEA     SP, STACK_TOP   ; intialize stack pointer
                :                       ; continue with program
                :
CODE_HERE       ENDS
                END
```

FIGURE 5-9 Required program additions when a stack is used.

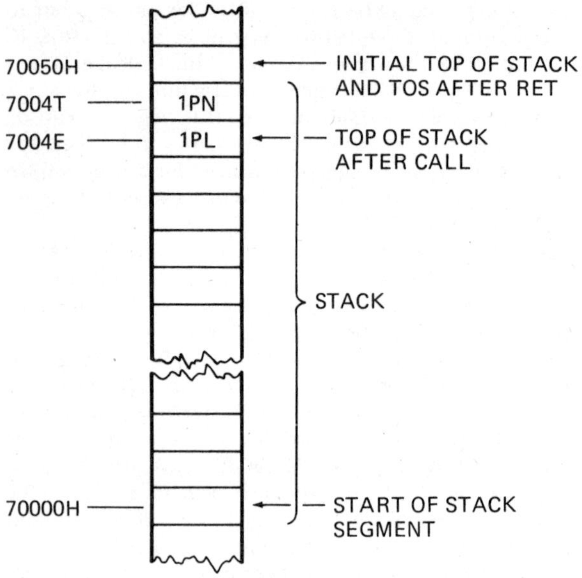

FIGURE 5-10 Stack diagram showing how the return address is pushed on the stack by CALL.

it does not load this value in the SS register. Loading the SS register must be done with program instructions, just as we do with the data segment register and the extra segment register. Again, we can't load an immediate number directly into a segment register, so we load the starting address of the segment into a register and then copy it into the stack segment register. The MOV AX, STACK_HERE and the MOV SS, AX instructions do this. Now all we have to do is initialize the stack pointer. We want to initialize SP so that the first word written to the stack goes to the highest location in the memory we set aside for the stack. All of the instructions which directly write a word to the stack decrement the stack pointer by two before writing the word. Therefore, we want the stack pointer to be initially loaded with the next even address above the actual stack. We gave this location the name STACK_TOP. Therefore, we can use the LEA SP, STACK_TOP instruction to initialize the stack pointer. We could also have used the instruction MOV SP, OFFSET STACK_TOP to initialize the stack pointer.

The next section shows how the pieces we have discussed are put in an example program which calls a near procedure. We also use this example to show you how the stack functions during a procedure call and return.

A Near Procedure Call and Return Example

Previous sections introduced you to the 8086 CALL and RET instructions and showed you how to set up a stack. Here we use a program example to show you how procedures are written and to dig more deeply into how the stack operates.

DEFINING THE PROBLEM AND WRITING THE ALGORITHM

Delay loops such as that shown in Figure 4-20 are often written as procedures so that they can be called from anywhere in a program. Suppose that we want to have a program which reads 100 data words from a port at 1-ms intervals, masks the upper 4 bits of each word, and puts each result in an array in memory. Before you read on, see if you can write a flowchart or pseudocode for this problem. Now compare your results with those in Figure 5-11*a* or *b*. Hopefully you recognized this problem as a REPEAT—UNTIL situation.

The next step is to expand the algorithm to take into account the specific architectural features of the 8086 that we will use to implement the algorithm. Figure 5-11*c* shows one way to do this expansion. We know that

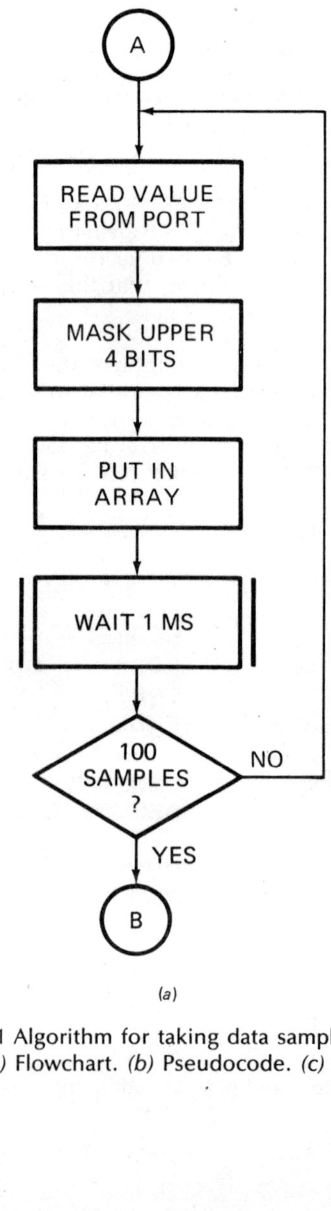

(*a*)

FIGURE 5-11 Algorithm for taking data samples at 1-ms intervals. *(a)* Flowchart. *(b)* Pseudocode. *(c)* Pseudocode expanded.

```
DATA SAMPLES PROGRAM
REPEAT
        GET DATA SAMPLE FROM PORT
        MASK UPPER 4 BITS
        PUT IN ARRAY
        WAIT 1 ms
UNTIL   100 samples taken
```

(b)

```
DATA SAMPLES PROGRAM

INITIALIZE POINTER TO ARRAY [SI]
INITIALIZE COUNTER, BX

REPEAT
        READ PORT
        MASK UPPER 4 BITS
        PUT IN ARRAY [SI]
        INCREMENT POINTER SI
        DECREMENT COUNTER BX
        WAIT_1MS PROCEDURE
UNTIL  COUNTER = 0

WAIT_1MS  PROC
          LOAD COUNT VALUE
          REPEAT
             DECREMENT COUNT VALUE
          UNTIL  COUNT = 0
```

(c)

FIGURE 5-11 *(continued)*

we need a pointer to the array and a counter to keep track of how many values we have put in the array. Therefore we initialize these at the start. After we read in each value and put it in the array, we increment the pointer so that it points to the next location in the array. We then decrement the counter to indicate that we have taken another sample, and call the WAIT_1MS procedure. Note that the algorithm for the procedure is done separately from that for the main program. As we discussed in the introduction to procedures, the flow of the mainline program is clearer if much of the detail is put in separate procedures. Upon returning from the delay procedure we repeat the series of instructions if our sample counter is not yet down to zero.

For the delay procedure we simply load a number in a register or memory location and decrement the number until it is down to zero. Note that even this expanded algorithm is general enough that it could be implemented on almost any microprocessor.

THE 8086 ASSEMBLY LANGUAGE PROGRAM

Figure 5-12 shows the assembly language program for our expanded algorithm. This program reviews some of the concepts from previous chapters and demonstrates some new ones from this chapter. The program is a little longer than our previous examples, but don't let this overwhelm you. A large part of the program is simply initializing everything. Read through this program and see how much of it you can remember and/or figure out before you read our explanations in the following paragraphs. Deciphering a program written by someone else is an important skill to develop.

At the start of the program we declare a logical segment for data with the DATA_HERE SEGMENT—DATA_HERE ENDS statements. The statement PRESSURES DW 100 DUP(0) in this segment sets aside 100 words of memory to store the values read in from a pressure sensor. This statement also initializes these 100 words to all 0's. It really doesn't matter what values are initially in these locations, because the program is going to write values in them. However, we like to initialize arrays such as this to all 0's so that during debugging we can tell if the program wrote any values to these locations.

Next we declare a logical segment to be used for the stack with the STACK_HERE SEGMENT STACK and the STACK_HERE ENDS statements. The statement DW 40 DUP(0) sets up a stack length of 40 words and initializes these words to all 0's. Again we really don't care what value these words have initially because we will be writing values there as we call procedures. The statement STACK_TOP LABEL WORD gives a name to the next even address after the highest address in the stack we have set up. As described in the previous section, we can then access this location by name when we initialize the stack pointer.

Now let's work our way through the main program and the procedure in the code segment. We have to tell the assembler which logical segments are being used for code, data, and stack in the program. The ASSUME CS:CODE_HERE, DS:DATA_HERE, SS:STACK_HERE statement does this. The ASSUME statement, however, does not actually initialize the segment registers. We have to do this with program instructions. The MOV AX,DATA_HERE and MOV DS,AX instructions initialize the data segment register. The MOV AX, STACK_HERE and MOV SS,AX instructions initialize the stack segment register. The stack pointer register must be initialized to point to the next even address after the memory space we set aside for the stack. The MOV SP, OFFSET STACK_TOP statement will do this. The OFFSET operator, remember, tells the assembler to calculate the distance from the start of a segment to the specified name and put this number in the specified register. We set aside 40 words for the stack so the offset of the label STACK_TOP will be 80 decimal or 0050H. This number is twice the number of words because each 8086 address represents a byte. The 0050H is the number that you would put in the instruction, if you were hand coding the program.

Up to this point most of what we have done is essentially housekeeping chores. Now we get started on the actual algorithm for our initially stated problem. The statement LEA SI, PRESSURES initializes the SI register as a pointer to the first location in the array PRESSURES. It loads the effective address or offset of the first

```
IBM Personal Computer MACRO Assembler  Version 2.00 Page 1-1 11-07-85
                                PAGE ,132
                                ;8086 Program
                                ;ABSTRACT:      This program takes in data samples from a port at 1 ms
                                ;               intervals, masks the upper 4 bits of each sample, and
                                ;               puts each masked sample in successive locations in an array.
                                ;REGISTERS USED:CS, SS, DS, AX, BX, CX, DX, SI, SP
                                ;PORTS USED:    0FFF8H - input port for data samples
                                ;PROCEDURES:    WAIT_1MS

 = FFF8                         PRESSURE_PORT   EQU  0FFF8H

 0000                           DATA_HERE       SEGMENT
 0000     64 [                  PRESSURES  DW  100  DUP(0)      ; set up array of 100 words
               0000
                    ]
 00C8                           DATA_HERE       ENDS

 0000                           STACK_HERE      SEGMENT STACK
 0000     28 [                             DW   40  DUP(0)      ; set stack length of 40 words
               0000
                    ]
 0050                           STACK_TOP       LABEL   WORD
 0050                           STACK_HERE      ENDS

 0000                           CODE_HERE       SEGMENT
                                                ASSUME  CS:CODE_HERE, DS:DATA_HERE, SS:STACK_HERE

 0000  B8 ---- R                START:     MOV AX, DATA_HERE            ; initialize data segment register
 0003  8E D8                               MOV DS, AX
 0005  B8 ---- R                           MOV AX, STACK_HERE           ; initialize stack segment register
 0008  8E D0                               MOV SS, AX
 000A  BC 0050 R                           MOV SP, OFFSET STACK_TOP     ; intialize stack pointer to top of stack

 000D  8D 36 0000 R                        LEA SI, PRESSURES            ; point SI to start of array
 0011  BB 0064                             MOV BX, 100                  ; load BX with number of samples
 0014  BA FFF8                             MOV DX, PRESSURE_PORT        ; Point DX at input port
 0017                           NEXT_VALUE:
 0017  ED                                  IN  AX, DX                   ; read data from port
 0018  25 0FFF                             AND AX, 0FFFH                ; mask upper 4 bits
 001B  89 04                               MOV [SI], AX                 ; store data word in array
 001D  E8 0026 R                           CALL WAIT_1MS                ; delay of 1 ms
 0020  46                                  INC SI                       ; point SI at next location in array
 0021  46                                  INC SI
 0022  4B                                  DEC BX                       ; decrement sample counter
 0023  75 F2                               JNZ NEXT_VALUE               ; repeat until 100 samples done
 0025  90                       STOP:      NOP

 0026                           WAIT_1MS   PROC NEAR
 0026  B9 23F2                             MOV CX, 23F2H                ; load delay constant into CX
 0029  E2 FE                    HERE:      LOOP HERE                    ; loop until CX = 0
 002B  C3                                  RET
 002C                           WAIT_1MS   ENDP
 002C                           CODE_HERE  ENDS
                                           END
```

FIGURE 5-12 Assembly language program to read in 100 samples of data at 1-ms intervals.

```
IBM Personal Computer MACRO Assembler  Version 2.00  Page Symbols-1 11-07-85


Segments and Groups:

                N a m e                 Size    Align   Combine Class

CODE_HERE. . . . . . . . . . . .        002C    PARA    NONE
DATA_HERE. . . . . . . . . . . .        00C8    PARA    NONE
STACK_HERE . . . . . . . . . . .        0050    PARA    STACK

Symbols:

                N a m e                 Type    Value   Attr

HERE . . . . . . . . . . . . . .        L NEAR  0029    CODE_HERE
NEXT_VALUE . . . . . . . . . . .        L NEAR  0017    CODE_HERE
PRESSURES. . . . . . . . . . . .        L WORD  0000    DATA_HERE       Length =0064
PRESSURE_PORT. . . . . . . . . .        Number  FFF8
STACK_TOP. . . . . . . . . . . .        L WORD  0050    STACK_HERE
START. . . . . . . . . . . . . .        L NEAR  0000    CODE_HERE
STOP . . . . . . . . . . . . . .        L NEAR  0025    CODE_HERE
WAIT_1MS . . . . . . . . . . . .        N PROC  0026    CODE_HERE       Length =0006

50092 Bytes free

Warning Severe
Errors  Errors
0       0
```

FIGURE 5-12 (*continued*)

word in PRESSURES into SI. For our example here PRESSURES is the first data item in the segment so the value loaded into SI will be 0000H. We chose to use the BX register as a sample counter, so we use the statement MOV BX, 100 to initialize BX with the number of samples we want to take and store. Finally, we are going to get to some action.

As indicated by the PRESSURE_PORT EQU 0FFF8H statement at the top of the program, the pressure sensor is connected to port FFF8H. Since this port address is larger than FFH, we have to use the variable port input instruction. For this input instruction we first load the port address in the DX register with the MOV DX, PRESSURE_PORT instruction, and then read the data word in with the IN AX, DX instruction. Notice how much more understandable it makes a program when we use a name such as PRESSURE_PORT in an instruction rather than 0FFF8H, the numerical port address. If you are working with an assembler, use EQU statements to give names to constants in your program.

When we get the pressure value into AX, we mask out the upper 4 bits with the AND AX, 0FFFH instruction. The reason why we want to do this is that the analog-to-digital converter that the pressure sensor is connected to is a 12-bit unit. The upper 4 bits of the 16-bit port are not connected to anything and may pick up random-noise signals. To prevent noise signals on the upper 4 bits from getting put in memory with our data, we mask these bits out by ANDing them with 0's. The instruction MOV [SI], AX will copy the data word from the AX register to the memory location pointed to by SI in the data segment.

To produce the desired delay between samples we CALL the WAIT_1MS procedure. This is a direct within segment CALL because the procedure is contained in the same code segment as the CALL instruction.

We use the PROC and ENDP directives to "bracket" the assembly language statements of the procedure. Putting a name in front of these directives allows us to call the procedure by name. For the example in Figure 5-12 we gave the procedure the name WAIT_1MS to remind us of the function of the procedure. To produce the desired delay we load a number into the CX register with the MOV CX, 23F2H instruction and count the number down to 0 with the LOOP HERE instruction. The LOOP instruction, remember, decrements CX by 1 and jumps to the specified label if CX is not yet down to 0. Since we put the label on the LOOP instruction, the LOOP instruction will simply execute over and over until CX reaches 0. The RET instruction at the end of the procedure will return execution to the next instruction after the CALL in the mainline of the program. Since this procedure is in the same code segment as the mainline program, only the instruction pointer has to be changed to get back to the mainline. The CALL instruction copied the desired value for the instruction pointer to the stack before going to the procedure. The RET instruction will copy this value from the stack back to the instruction pointer. If you are hand coding a program such as this, make sure to use the correct form of the RET instruction. After we briefly discuss the rest of the mainline program, we will show you what happens to the stack and stack pointer as the return address is copied to and from the stack.

Now, back in the mainline we need to get ready for reading the next data value. First we want to get SI pointed to the location where we want to put the next data word. Since each address represents a byte, and we are storing words, we have to increment the pointer by two to point to the next storage location. Two INC SI instructions do this. You could use the single instruction ADD SI, 02H to do the same job. After updating the pointer we decrement the sample counter in BX with the DEC BX instruction. If BX is not yet counted down to 0, the JNZ NEXT_VALUE instruction will cause the 8086 to read in and process another value from the port. If BX is 0, indicating that all 100 samples have been taken, execution goes on to the next mainline instruction after JNZ. Now let's see what happens to the stack and the stack pointer during all of this.

More Stack Operation and Use

STACK OPERATION DURING A CALL AND RET

To show how the stack operates during a CALL and RET we will use some specific numbers with the example program in Figure 5-12. Suppose that for the program we start the stack at address 70000H. The stack segment register then will be initialized with 7000H. We set a stack length of 40 decimal or 28H words with the DW 40 DUP(0) statement. These 28H words will occupy the 50H memory locations from 70000H to 7004FH. Figure 5-13*a* shows this in diagram form. Now remember, when we write words to the stack, we put the first word at the highest address. For our example here the first word will be written at addresses 7004EH and 7004FH. As we write other words to the stack they are written at lower addresses. In other words the stack fills from the top down. We use the stack pointer to keep track of where the last word was written to the stack. The location pointed to by the stack pointer at any time is called the *top of the stack*. In the program we initialized the stack pointer to offset 0050H, the next even address above our actual stack, with the MOV SP, OFFSET TOP_STACK instruction.

After the 8086 fetches the CALL instruction from the instruction-byte queue in the BIU it automatically increments the instruction pointer to 0020H, the offset of the next instruction after the CALL. You can see this if you look at the first column of the program listing in Figure 5-12. The instruction pointer then contains the address we want execution to return to after the procedure is completed. When the near CALL instruction in our example program executes, the 8086 first decrements the stack pointer by two. Then it copies the return address

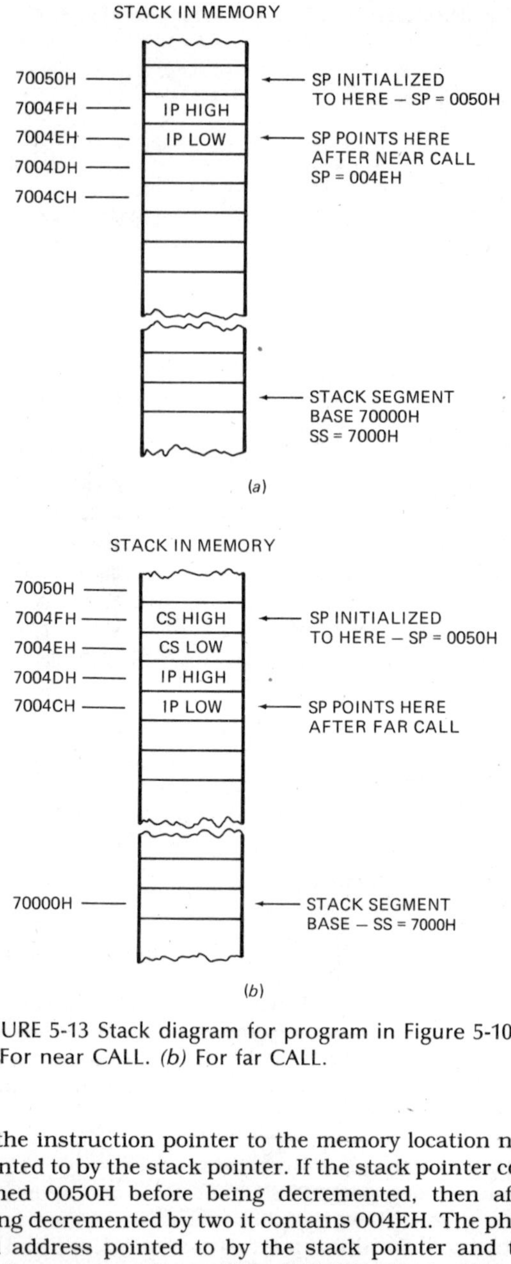

FIGURE 5-13 Stack diagram for program in Figure 5-10. *(a)* For near CALL. *(b)* For far CALL.

in the instruction pointer to the memory location now pointed to by the stack pointer. If the stack pointer contained 0050H before being decremented, then after being decremented by two it contains 004EH. The physical address pointed to by the stack pointer and the stack segment register will be 7004EH. The low byte of the instruction pointer will be copied to address 7004EH and the high byte of the instruction pointer will be copied to address 7004FH. This follows the Intel convention of putting the lower byte of a word at the lower address in memory. Figure 5-13*a* shows these two bytes labeled as IP LOW and IP HIGH. After the CALL instruction executes, the stack pointer is left pointing to offset 004EH. This location is now the top of the stack or TOS.

When the RET instruction at the end of the procedure in the example program executes, the 8086 copies the return address from the top of the stack to the instruction pointer. Since the top of the stack was at offset 004EH, the word from addresses 7004EH and 7004FH will be copied to the instruction pointer. After it copies the word from the top of the stack to the instruction pointer, the 8086 increments the stack pointer by two. For our example here it will increment the stack pointer from 004EH to 0050H. The stack pointer is now back where it was before the CALL instruction executed. Note that the return address is still present in memory because the RET instruction simply copied it to the instruction pointer and incremented the stack pointer over it.

When the 8086 executes a far CALL instruction it decrements the stack pointer by two and copies the contents of the code segment register to the stack. It then decrements the stack pointer by two again and copies the offset of the next mainline instruction from the instruction pointer to the stack. To help you visualize this Figure 5-13*b* shows how these would be written to the stack assuming the same stack starting addresses that we used for the previous example. As you can see from this figure, after a far CALL the top of the stack will be four addresses lower than it was before the CALL.

A far RET used at the end of a far procedure will copy a word from the top of the stack to the instruction pointer and increment the stack pointer by two. It will then copy the word from the new top of the stack to the code segment register. The next instruction will then be fetched from the physical address after the far CALL instruction. The top of the stack will be back to where it was before the CALL and RET.

As we mentioned previously the stack is also used to save the contents of registers while a procedure executes and to hold data that the procedure is to act on. The next section shows you how we do this.

USING PUSH AND POP TO SAVE REGISTER CONTENTS

In the example program in Figure 5-12 we used the BX register to keep track of how many data samples we had taken in. After each data sample was taken in we decremented the BX register and used the JNZ instruction to determine whether to take another sample or to exit. We would like to have used the CX register to keep track of the number of samples taken so that we could have used a single LOOP instruction instead of the DEC BX and JNZ label instructions. The reason that we couldn't use CX for this in the program is because CX is used in the procedure. Any value we put in CX in the mainline program would be written over by the MOV CX, 23F2H instruction in the procedure. It is very common to want to use registers both in the mainline program and in a procedure without the two uses interfering with each other. The PUSH and POP instructions make this very easy to do.

The PUSH register/memory instruction decrements the stack pointer by two and copies the contents of the specified 16-bit register or memory location to memory at the new top of stack location. The PUSH CX instruction, for example, will decrement the stack pointer by two and copy the contents of the CX register to the stack where the stack pointer now points. This instruction then can be used to save the contents of CX while a procedure executes. The next question is, how do we get the saved value back when we want it?

The POP register/memory instruction copies a word from the top of the stack to the specified 16-bit register or memory location and increments the stack pointer by two. The POP CX instruction, for example, will copy a word from the top of the stack to the CX register and increment the stack pointer by two. After a POP the stack pointer will point to the next word on the stack.

You can PUSH any of the 16-bit general purpose registers, AX, BX, CX, and DX; any of the base or pointer registers, BP, SP, SI, and DI; any of the segment registers, CS, DS, SS, and ES; or even a word from a memory location specified by one of those 24 memory addressing modes in Figure 3-8. A separate instruction, PUSHF, decrements the stack pointer by two and copies the flag word to the stack. The 80186 and 80188 PUSHA instruction copies AX, CX, DX, BX, SP, BP, SI and DI to the stack.

You can POP a word from the stack to any of the registers except CS, and you can POP a word from the stack to a memory location specified in any one of those 24 ways. The POPF instruction copies a word from the stack to the flag register and increments the stack pointer by two. The 80186 and 80188 POPA instruction copies words from the stack to the DI, SI, BP, BX, DX, CX, and AX registers. Note that the POPA instruction does not return a value to the SP register.

When you PUSH several registers on the stack you have to remember to POP them off in the reverse order that you pushed them on. This is because the stack functions in a *last-in—first-out* manner. An everyday example of this type of operation is the spring-loaded plate stacks seen in some restaurants. The last plate pushed on the stack is the first one popped off. Figure 5-14*a* should help you visualize how this works for the 8086. It shows a sequence of PUSH instructions you might use to save registers and flags at the start of a near procedure called MULTO. Figure 5-14*b* shows how the PUSH instructions will put the contents of these registers on the stack. The first entry in the stack is the copy of the instruction pointer put there by the CALL instruction that called the procedure. Following this are the flag word and the words from registers AX, BX, and CX. After all of these are pushed on the stack, the stack pointer is left pointing at the location in the stack where CX was pushed.

When we want to restore the saved values to the registers and flags at the end of the procedure we first POP CX because it was the last register pushed on the stack. After CX is popped the stack pointer will be left pointing at the location where BX is stored. Therefore, we POP BX next. We continue popping until all of the registers and the flags are restored. The RET instruction then copies the return address from the stack to the instruction pointer to return execution to the main program. It is very important to keep the number of pushes equal to the number of pops or in some other way keep the stack balanced so that the RET instruction finds the correct word to put in the instruction pointer.

Some programmers like to push and pop registers in the mainline or calling program rather than in the procedure as we did in Figure 5-14*a*. This approach has the advantage that you can push only those registers that

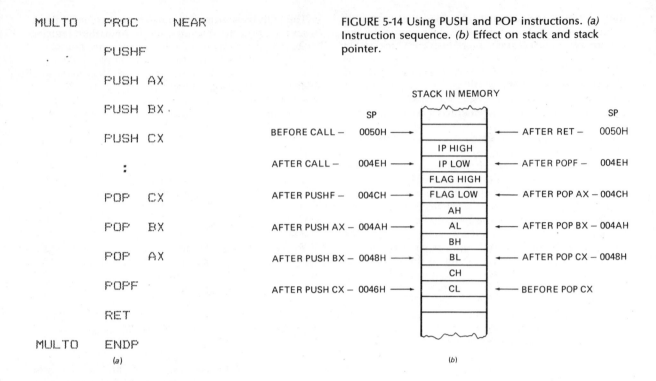

```
MULTO   PROC    NEAR
        PUSHF
        PUSH AX
        PUSH BX
        PUSH CX
          :
        POP  CX
        POP  BX
        POP  AX
        POPF
        RET
MULTO   ENDP
```

(a)

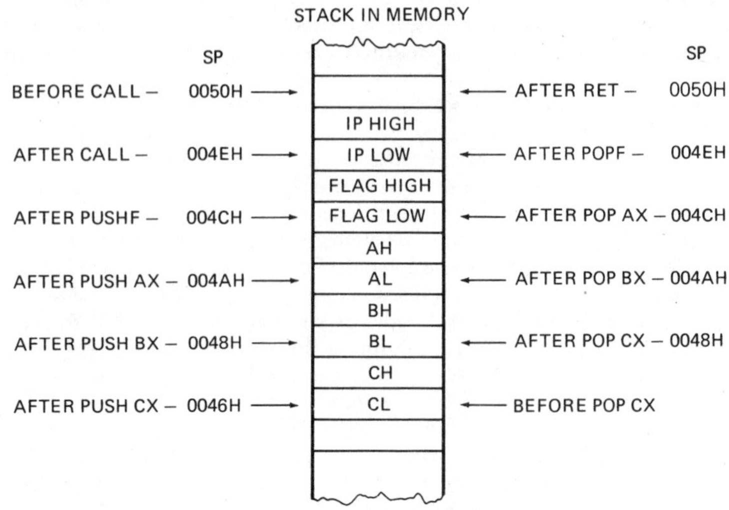

(b)

FIGURE 5-14 Using PUSH and POP instructions. *(a)* Instruction sequence. *(b)* Effect on stack and stack pointer.

you care about saving each time you call the procedure. The disadvantages of this approach are that the pushes and pops clutter up the mainline program, and you may decide to use another register at some point in the program and forget to add a push for it. We like to push the flags and any registers used in a procedure directly in the procedure. This way we always know that the procedure can be called from anywhere in the program without losing the contents of any registers. Another advantage of this approach is that you only have to write the pushes and pops once. A disadvantage is that in a situation where all the pushes are not needed, the procedure may take a little longer to run.

Passing Parameters to and from Procedures

Often when we call a procedure we want to make some data values or addresses available to the procedure. Likewise we often want a procedure to make some processed data values or addresses available to the main program. These address or data values passed back and forth between the mainline and the procedure are commonly called *parameters*. There are three major ways of passing parameters to and from a procedure. Parameters can be passed in *registers*, they can be passed in *dedicated memory locations*, or they can be passed in *stack locations*. In the following sections we use three versions of a simple program to show you how each of these methods work.

DEFINING THE PROGRAMMING PROBLEM

A common programming need is to convert a packed BCD number such as 4596 to its binary or hexadecimal equivalent. The hexadecimal equivalent of BCD 4596 is 11F4H, for example. There are several ways to do this conversion, but to us the easiest is based on using the

```
4596  =    (4 x 1000) + (5 x 100) + (9 x 10) + (6 x 1)

1000  = 03E8H   therefore    4000 = 4 x 3E8H  = 0FA0H
 100  = 0064H   therefore     500 = 5 x 064H  = 01F4H
  10  = 000AH   therefore      90 = 9 x 00AH  = 005AH
   1  = 00001H  therefore       6 = 6 x 001H  = 0006H

                                        4596  = 11F4H
```

FIGURE 5-15 BCD-to-HEX or -BINARY algorithm.

value of each placeholder in the BCD number. Figure 5-15 shows the names and values for each digit in a 4-digit BCD number such as 4596. When we write a number such as this it means that we have a total of 4 thousands + 5 hundreds + 9 tens + 6 units. To determine the value of this number in hexadecimal we just multiply the number in each digit position by the value of that digit position in hexadecimal and add up the results. The right-hand side of Figure 5-15 shows how this works. The units position has a value of 1 in hex so multiplying this by 6 units gives 0006H. The tens position has a value of 0AH. Multiplying this value by 9, the number of tens, gives 005AH. The hex value of the hundreds position is 64H. When we multiply this value by 5, the number of hundreds, we get 01F4H. When we multiply the hex value of the thousands position, 03E8H, by 4 (the number of thousands), we get 0FA0H. Adding up the results for the 4 digits gives 11F4H which is the hex equivalent of 4596 BCD. You can use this method to convert a BCD number with any number of digits to its binary equivalent, but to conserve space here we will do it for just a 2-digit BCD number.

The algorithm for this program then is the simple sequence of operations:

Separate nibbles

Save lower nibble (don't need to multiply by one)

Multiply upper nibble by 0AH

Add lower nibble to result of multiplication

We want to implement this program as a procedure which can be called from anywhere in a mainline program. For our first version we pass the BCD number to the procedure in a register.

PASSING PARAMETERS IN REGISTERS

Figure 5-16 shows our first version of a procedure to convert a 2-digit packed BCD number to its hex (binary) equivalent. The BCD number is passed to the procedure in the AL register and the hex equivalent is passed back to the calling program in the AL register. We start the procedure by pushing the flag register and the other registers we use in the procedure. Notice that we don't need to push and pop the AX register because we are using it to pass a value to the procedure and expecting the procedure to pass a different value back to the calling program in it.

Hopefully the function of the rest of the instructions in the procedure are reasonably clear from the comments with them. We first make a copy of the BCD in AL so we have two copies to work on. We then mask the upper nibble of one and save it in BL. Since multiplying this nibble by one would not change its value, we are done with it for now. We mask the lower nibble of the other copy of the BCD and rotate this nibble into the lower nibble position of the byte so we can multiply it correcty. When we multiply this nibble by the digit weight of 0AH, the result is left in the AX register. However, since the result can never be greater than 8 bits, we can disregard the contents of AH. Finally, we add the lower nibble we saved in BL to the result in AL to get the hex total. The desired result is left in AL. Before returning to the main program we pop the registers we pushed at the start of the procedure.

USING GENERAL MEMORY TO PASS PARAMETERS

For cases where we only have to pass a few parameters to and from a procedure, registers are a convenient way to do it. However, in cases where we need to pass a large number of parameters to a procedure or in cases where we don't want to use registers, we use memory. This memory may be a dedicated section of general memory or part of the stack. The following example shows a very simple case using dedicated memory locations.

Figure 5-17*a* shows a fragment of a program that uses another version of our BCD_TO_HEX procedure. In this version the number to be converted is stored in a dedicated memory location named BCD_INPUT and the hex result is returned from the procedure to a dedicated memory location called HEX_VALUE.

In the procedure we first push the flags and all of the registers used in the procedure. We then copy the BCD number into AL with the MOV AL, BCD_INPUT instruction. From here on the procedure is the same as the previous version until we reach the point where we want to pass the hex result back to the calling program. Here we use the MOV HEX_VALUE, AL instruction to copy the result to the dedicated memory location we set aside for it. To complete the procedure we pop the flags and registers, and return to the main program.

The approach used in Figure 5-17*a* works in this case, but it has a severe limitation. Can you see what it is? The limitation is that this procedure will always look to the memory location named BCD_INPUT to get its data and always put its result in the memory location called HEX_VALUE. In other words, the way it is written we can't easily use this procedure to convert a BCD number in some other memory location.

PASSING PARAMETERS USING POINTERS

A parameter passing method which overcomes the disadvantage of using data item names directly in a procedure is to pass the procedure a pointer to the desired data. Figure 5-17*b* shows one way to do this. In the main program before we call the procedure we use the MOV SI, OFFSET BCD_INPUT instruction to set up the SI register as a pointer to the memory location BCD_INPUT. We also use the MOV DI, OFFSET HEX_VALUE instruction to set up the DI register as a pointer to the memory location named HEX_VALUE. In the procedure the MOV AL, [SI] instruction will copy the byte pointed to by SI into AL. Likewise, the instruction MOV [DI], AL instruction later in the procedure will copy the byte from AL to the memory location pointed to by DI.

This second approach which actually uses a combination of registers and memory is more versatile because you can pass the procedure pointers to data anywhere in memory. You can pass pointers to individual values or pointers to arrays or strings. If you don't want to use

```
;8086 PROGRAM FRAGMENT BCD TO HEX CONVERSION
;ABSTRACT:          Program fragment that uses a procedure to convert
;                   BCD numbers to HEX (binary). It shows how to use
;                   AL register to pass parameters to the procedure
;Not shown   -      SS contains segment base for STACK_HERE
;Not shown   -      Initialization of segment registers
;PORTS USED       : none
;PROCEDURES USED: BCD_HEX

DATA_HERE          SEGMENT
                   BCD_INPUT       DB ?        ; storage for BCD value
                   HEX_VALUE       DB ?        ; storage for binary value
DATA_HERE          ENDS

CODE_HERE          SEGMENT WORD
                   ASSUME CS:CODE_HERE, DS:DATA_HERE, SS:STACK_HERE
                   :
                   :
         MOV       AL, BCD_INPUT
         CALL      BCD_HEX
         MOV       HEX_VALUE, AL              ; store the result
                   :


;PROCEDURE:        BCD_HEX
;                  Converts BCD numbers to HEX (binary), uses
;                  registers to pass parameters to the procedure
;SAVES:            All registers used except AH

BCD_HEX            PROC     NEAR
                   PUSHF             ; save flags
                   PUSH     BX       ; and registers
                   PUSH     CX
;start conversion
                   MOV      AH, AL   ; save copy of BCD in AH
                   AND      AH, OFH  ; separate and save lower
                   MOV      BL, AH   ; BCD digit
                   AND      AL, OFOH ; separate upper nibble
                   MOV      CL, 04   ; move upper BCD digit to low
                   ROR      AL, CL   ; nibble position for multiply
                   MOV      BH, OAH  ; load conversion factor in BH
                   MUL      BH       ; upper BCD digit in AL * OAH in BH
                                     ; result in AX
                   ADD      AL, BL   ; add lower BCD to result of MUL
                                     ; final result in AL
;end conversion, restore registers
                   POP      CX
                   POP      BX
                   POPF
                   RET
BCD_HEX            ENDP
CODE_HERE          ENDS
                   END
```

FIGURE 5-16 Example program passing parameters in registers.

```
;8086 PROGRAM FRAGMENT - BCD to HEX CONVERSION
;ABSTRACT:          Program fragment that uses a procedure to convert BCD
;                   numbers to HEX (binary). It shows how to use dedicated
;                   memory locations to pass parameters to a procedure.
;not shown -        SS contains segment base for STACK_HERE
;not shown -        Initialization of segment registers
;PORTS USED         : None
;PROCEDURES USED: BCD_HEX

DATA_HERE           SEGMENT
                    BCD_INPUT       DB ?   ; storage for BCD value
                    HEX_VALUE       DB ?   ; storage for binary value
DATA_HERE           ENDS
CODE_HERE           SEGMENT
                    ASSUME CS:CODE_HERE, DS:DATA_HERE, SS:STACK_HERE
                    :
                    :                       ; intitialization
                    CALL BCD_HEX
                    :

;PROCEDURE:         BCD_HEX
;ABSTRACT :         Converts BCD numbers to HEX, uses dedicated
;                   memory locations for data
;SAVES:             All registers used

BCD_HEX             PROC     NEAR
                    PUSH     AX
                    PUSHF                        ; save flags
                    PUSH     BX                  ; and registers
                    PUSH     CX
;get BCD value from named memory location
                    MOV      AL, BCD_INPUT
;do conversion
                    MOV      AH, AL    ; save copy of BCD in AH
                    AND      AH, 0FH   ; separate and save lower
                    MOV      BL, AH    ; BCD digit
                    AND      AL, 0F0H  ; separate upper nibble
                    MOV      CL, 04    ; move upper BCD digit to low
                    ROR      AL, CL    ; nibble position for multiply
                    MOV      BH, 0AH   ; load conversion factor in BH
                    MUL      BH        ; upper BCD digit in AL * 0AH in BH
                                       ; result in AX
                    ADD      AL, BL    ; add lower BCD to result of MUL
                                       ; final result in AL
;end of conversion now store Hex value in named memory location
                    MOV      HEX_VALUE, AL
                    POP      CX          ; restore flags and
                    POP      BX          ; registers
                    POPF
                    POP      AX
                    RET
BCD_HEX             ENDP
CODE_HERE           ENDS
                    END
```

(a)

FIGURE 5-17 Example program passing parameters in named memory locations. *(a)* Named memory location only. *(b)* More versatile approach using pointers to named memory locations.

```
;8086 PROGRAM FRAGMENT - BCD to HEX CONVERSION
;ABSTRACT:          Program fragment that uses a procedure to convert BCD
;                   numbers to HEX (binary). It shows how to use pointers
;                   to pass parameters to procedure.
;not shown -        SS contains segment base for STACK_HERE
;not shown -        Initialization of segment registers
;PORTS USED         : None
;PROCEDURES USED: BCD_HEX

DATA_HERE           SEGMENT
                    BCD_INPUT       DB ?   ; storage for BCD value
                    HEX_VALUE       DB ?   ; storage for binary value
DATA_HERE           ENDS
CODE_HERE           SEGMENT
                    ASSUME CS:CODE_HERE, DS:DATA_HERE, SS:STACK_HERE
                    :
                    :
;put pointer to BCD in SI and pointer to HEX storage in DI
                    MOV     SI, OFFSET BCD_INPUT
                    MOV     DI, OFFSET HEX_VALUE
                    CALL BCD_HEX
                    :

;PROCEDURE:         BCD_HEX
;ABSTRACT :         Converts BCD numbers to HEX. Uses pointers
;                   to get data parameters
;SAVES:             All registers used
BCD_HEX             PROC    NEAR
                    PUSH    AX                  ; save registers and flags
                    PUSHF
                    PUSH    BX
                    PUSH    CX
;byte in DS pointed to by SI is moved to AL
                    MOV     AL, [SI]
;do conversion
                    MOV     AH, AL    ; save copy of BCD in AH
                    AND     AH, 0FH   ; separate and save lower
                    MOV     BL, AH    ; BCD digit
                    AND     AL, 0F0H  ; separate upper nibble
                    MOV     CL, 04    ; move upper BCD digit to low
                    ROR     AL, CL    ; nibble position for multiply
                    MOV     BH, 0AH   ; load conversion factor in BH
                    MUL     BH        ; upper digit * 0AH,result in AX
                    ADD     AL, BL    ; add lower BCD to result of MUL
                    MOV     [DI], AL  ; move HEX value result in AL
                                      ; to DS location pointed to by DI
                    POP     CX        ; restore registers and flags
                    POP     BX
                    POPF
                    POP     AX
                    RET
BCD_HEX             ENDP
CODE_HERE           ENDS
                    END
```

(*b*)

FIGURE 5-17 (*continued*)

registers to pass the pointers, you can use memory locations dedicated specifically to holding the pointers. In that case the procedure will first fetch the pointer and then use it to access the desired data.

For many of your programs you will probably use registers or a combination of registers and general memory to pass parameters to procedures. However, for more complex programs, such as those which allow several users to time-share a system, we often use the stack to pass parameters to and from procedures.

PASSING PARAMETERS USING THE STACK

To pass parameters to a procedure using the stack we push them on the stack somewhere in the mainline program before we call the procedure. Instructions in the procedure then read these parameters from the stack. Likewise, parameters to be passed back to the calling program are written to the stack by instructions in the procedure and read off the stack by instructions in the mainline. A simple example will best show you how this works.

Figure 5-18*a* shows a version of our BCD-to-hex procedure which uses the stack for passing the BCD number to the procedure and for passing the hex value back to the calling program. To save space here we assume that previous instructions in the mainline set up a stack segment, initialized the stack segment register, and initialized the stack pointer. We also assume that previous instructions in the mainline have left the BCD number in AL. Now in the mainline fragment in Figure 5-18*a* we copy AX to the stack with the PUSH AX instruction. In a more complex example the BCD number or a pointer to it would probably be put on the stack by a different mechanism, but the important point for now is that the parameter is on the stack for the procedure to access. The CALL instruction in the mainline decrements the stack pointer by two, copies the return address on the stack, and loads the instruction pointer with the starting address of the procedure. PUSH instructions at the start of the procedure save the flags and all of the registers used in the procedure on the stack. Before discussing any more instructions, let's take a look at the contents of the stack after these pushes.

Figure 5-18*b* shows how the values pushed on the stack will be arranged. Note that the BCD value is in the stack at a higher address than the return address. After the registers are pushed on the stack the stack pointer is left pointing to the stack location where BP is stored. Now, the question is, how can we easily access the parameter that seems buried in the stack? One way is to add 12 to the stack pointer with an ADD SP, 0CH instruction so the stack pointer points to the word we want from the stack. A POP AX instruction could then be used to copy the desired word from the stack to AX. However, for a variety of reasons which we will explain later, we would like to be able to access the parameter without changing the contents of the stack pointer.

The design of the 8086 makes it very easy to use the base pointer register to do this. Remember from Chapter 2 that an offset in the BP register will be added to the stack segment register to produce a physical memory address. In other words the BP register can act as a second pointer to a location in the stack. This is how we use it in our example program here. In the procedure we copy the contents of the stack pointer register to the BP register with the MOV BP, SP instruction. BP then points to the same location as the stack pointer. Now we use the MOV AX, [BP + 12] instruction to copy the desired word from the stack to AX. The 8086 will produce the effective address for this instruction by adding the displacement of 12, specified in the instruction, to the contents of the BP register. The 0042H in BP gives an effective address of 004EH. As you can see in Figure 5-18*b* the effective address produced will be that of the desired parameter. Note that this instruction does not change the contents of BP. BP can then be used to access other parameters on the stack by simply specifying a different displacement in the instruction used to access the parameter.

Once we have the BCD number copied from the stack into AL, the instructions which convert it to hex are the same as those in the previous versions. When we want to put the hex value back in the stack to return it to the calling program, we again use BP as a pointer to the stack. The instruction MOV [BP +12], AX will copy AX to a stack location 12 addresses higher than that where BP is pointing. This of course is the same location we used to pass the BCD number to the procedure. After we pop the registers and return to the calling program, the registers will all have the values they had before the CALL instruction executed. AX will contain the original BCD number and the stack pointer will be pointing to the hex value now at the top of the stack. In the mainline we can now pop this hex value into a register with an instruction such as POP CX.

Whenever you are using the stack to pass parameters it is very important to keep track of what you have pushed on the stack and where the stack pointer is at each point in a program. We have found that diagrams such as the one in Figure 5-18*b* are very helpful in doing this. One potential problem to watch for when using the stack to pass parameters is *stack overflow*. Stack overflow means that the stack fills up and overflows the memory space you set aside for it. To see how this can easily happen if you don't watch for it, consider the following. Suppose that we use the stack to pass four word parameters to a procedure, but that we only pass one word parameter back to the calling program on the stack. Figure 5-19 shows a stack diagram for this situation. Before a CALL instruction the four parameters to be passed to the procedure are pushed on the stack. During the procedure the parameter to be returned is put in the stack location previously occupied by the fourth input parameter. After the RET instruction at the end of the procedure executes, the stack pointer will be left pointing at this value. Now assume we pop this value into a register. The POP instruction will copy the value to a register and increment the stack pointer by two. The stack pointer now points to the third word we pushed to pass to the procedure. In other words the stack pointer is six addresses lower than it was when we started this process. Now suppose that we call this procedure many times in the course of the mainline pro-

```
;8086 PROGRAM
;ABSTRACT:          Program fragment that uses a procedure to convert
;                   BCD numbers to HEX (binary). It shows how to use
;                   the stack to pass parameters to procedure.
;not shown -        SS contains segment base for STACK_HERE
;not shown -        Initialization of segment registers
;PROCEDURES USED: BCD_HEX

DATA_HERE           SEGMENT
                    BCD_INPUT       DB  ? ; storage for BCD value
                    HEX_VALUE       DB  ? ; storage for binary value
DATA_HERE           ENDS
CODE_HERE           SEGMENT
                    ASSUME CS:CODE_HERE, DS:DATA_HERE, SS:STACK_HERE
                    :                         ; initialize segments
                    :                         ; move BCD_VALUE into AL
                    PUSH     AX               ; put BCD_VALUE on stack
                    CALL     BCD_HEX          ; convert to binary
;program continues here with result of conversion on the top of stack
                    :
;PROCEDURE:         BCD_HEX
;ABSTRACT :         converts BCD numbers to HEX (binary).
;                   Takes its parameters from stack
;SAVES:             All registers used and flags
BCD_HEX             PROC     NEAR
                    PUSH     AX                  ; save registers and flags
                    PUSHF
                    PUSH     BX
                    PUSH     CX
                    PUSH     BP                  ; save BP
                    MOV      BP, SP              ; copy SP into BP
                    MOV      AX, [BP+12]         ; copy BCD # from stack to AX
;do conversion
                    MOV      AH, AL              ; save copy of BCD in AH
                    AND      AH, OFH             ; separate and save lower
                    MOV      BL, AH              ; BCD digit
                    AND      AL, OFOH            ; separate upper nibble
                    MOV      CL, 04              ; move upper BCD digit to low
                    ROR      AL, CL              ; nibble position for multiply
                    MOV      BH, OAH             ; load conversion factor in BH
                    MUL      BH                  ; upper digit*OAH,result in AX
                    ADD      AL, BL              ; add lower BCD to result of MUL
                                                 ; final result in AL
;end of conversion now move HEX value from AL to location onto the stack
                    MOV      [BP+12], AX
                    POP      BP                  ; restore registers
                    POP      CX                  ; and flags and return
                    POP      BX
                    POPF
                    POP      AX
                    RET
BCD_HEX             ENDP
CODE_HERE           ENDS
                    END
```

(a)

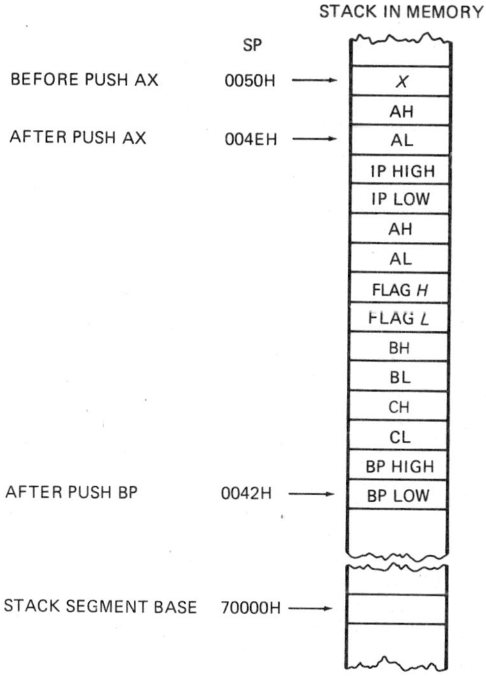

(b)

FIGURE 5-18 Example program passing parameters on the stack. *(a)* Assembly language program. *(b)* Stack diagram.

gram. Each time we push four words on the stack but only pop one word off, the stack pointer will be left six addresses lower than it was before the process. The top of the stack will keep getting moved downward. When the stack pointer gets down to 0000H, the next push will roll it around to FFFEH and write a word at the very top of the 64 Kbyte stack segment. If you overlapped segments as you usually do in a small system, the word may get written in a memory location that you are using for data or your program code and your data or code will be lost! This is what we mean by the term stack overflow.

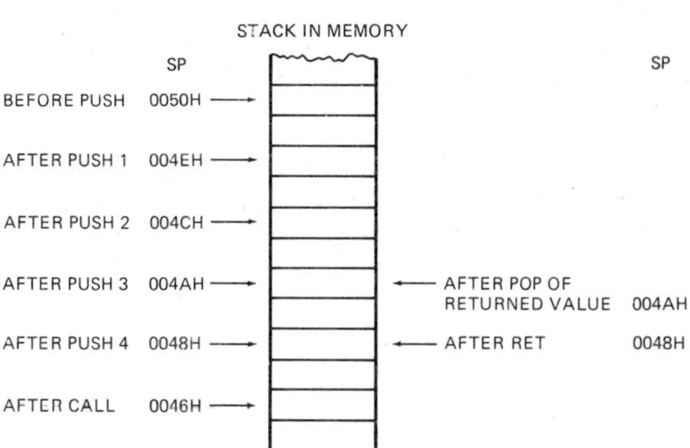

FIGURE 5-19 Stack diagram showing cause of stack overflow.

The cure for this potential problem is to use your stack diagrams to help you keep the stack balanced. You need to keep the number of pops equal to the number of pushes or in some other way make sure the stack pointer gets back to its initial location.

For this example we could use an ADD SP, 06H instruction after the POP instruction to get the stack pointer back up the additional six addresses to where it was before we pushed the four parameters on the stack.

For other cases such as this the 8086 RET instruction has two forms which help you to keep the stack balanced. Remember from a previous section of this chapter that the 8086 has four forms of the RET instruction. The regular near RET instruction copies the return address from the stack to the instruction pointer and increments the stack pointer by 2. The regular far RET instruction copies the return IP and CS values from the stack to IP and CS, and increments the stack pointer by 4. The other two RET instruction forms perform the same functions respectively, but they also add a number specified in the instruction to the stack pointer. The near RET 6 instruction, for example, will first copy a word from the stack to the instruction pointer and increment the stack pointer by 2. It will then add 6 more to the stack pointer. This is a quick way to skip the stack pointer up over some old parameters on the stack.

SUMMARY OF PASSING PARAMETERS TO AND FROM PROCEDURES

You can pass parameters between a calling program and a procedure using registers, dedicated memory locations, or the stack. The method you choose depends largely on the specific program. There are no hard rules, but here are a few guidelines. For simple programs with just a few parameters to pass, registers are usually the easiest to use. For passing arrays or other data structures to and from procedures you can use registers to pass pointers to the start of these data structures. As we explained previously, passing pointers to the procedure is a much more versatile method than having the procedure access the data structure directly by name.

For procedures in a multiuser-system program, procedures that will be called from a high level language program, or procedures that call themselves, parameters should be passed on the stack. When writing programs which pass parameters on the stack you should use stack diagrams such as the one in Figure 5-18*b* to help you keep track of where everything is in the stack at a particular time. The following section will give you some additional guidance as to when to use the stack to pass parameters, and it will give you some additional practice following the stack and stack pointer as a program executes.

Reentrant and Recursive Procedures

The terms *reentrant* and *recursive* are often used in microprocessor manufacturers' literature, but seldom illustrated with examples. Here we try to give these terms some meaning for you. You should make almost all of the procedures you write reentrant, so read that section carefully. You will seldom have to write a recursive procedure so the main points to look for in that section are the definition of the term and the operation of the stack as a recursive procedure operates.

REENTRANT PROCEDURES

The 8086 has a signal input which allows a signal from some external device to interrupt the normal program execution sequence and call a specified procedure. In our electronics factory, for example, a temperature sensor in a flow-solder machine could be connected to the interrupt input. If the temperature gets too high, the sensor sends an interrupting signal to the 8086. The 8086 will then stop whatever it is doing and go to a procedure which takes whatever steps are necessary to cool down the solder bath. This procedure is called an *interrupt service procedure*. Chapter 8 discusses 8086 interrupts and interrupt service procedures in great detail, but it is appropriate to introduce the concept here.

Now, suppose that the 8086 was in the middle of executing a multiply procedure when the interrupt signal occurred, and that we also need to use the multiply procedure in the interrupt service subroutine. Figure 5-20 shows the program execution flow we want for this situation. When the interrupt occurs, execution goes to the interrupt service procedure. The interrupt service procedure then calls the multiply procedure when it needs it. The RET instruction at the end of the multiply procedure returns execution to the interrupt service procedure. A special return instruction at the end of the interrupt service procedure returns execution to the multiply procedure where it was executing when the interrupt occurred. The multiply procedure must be written such that it can be interrupted, used, and "reentered" without losing or writing over anything. A procedure which can function in this way is said to be reentrant.

To be reentrant a procedure must first of all push the flags and all registers used in the procedure. Also, to be reentrant a program should use only registers or the

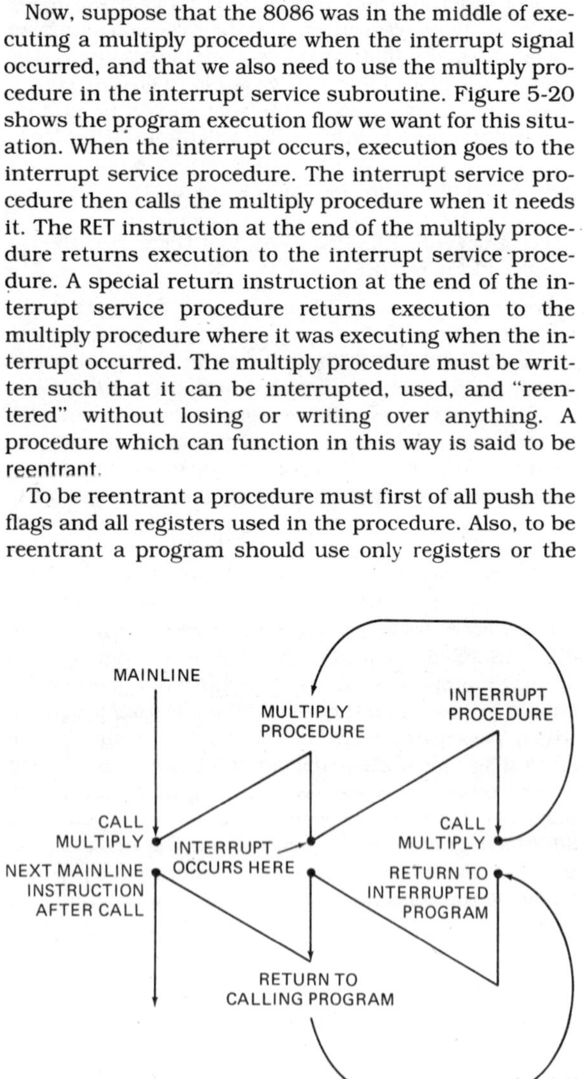

FIGURE 5-20 Program execution flow for reentrant procedure.

stack to hold parameters. To see why this second point is necessary, let's take another look at the program in Figure 5-17*a*. This program uses the named variables BCD_INPUT and HEX_VALUE. The procedure BCD_TO_HEX accesses these two directly by name. Now, suppose that the 8086 is in the middle of executing the BCD_TO_HEX procedure and an interrupt occurs. Further suppose that the interrupt service procedure loads some new value in the memory location named BCD_INPUT, and calls the BCD_TO_HEX procedure again. The initial value in BCD_INPUT has now been written over. If the interrupt occurred before the first execution of the procedure had a chance to read this value in, the value will be lost forever. When execution returns to BCD_TO_HEX after the interrupt service procedure, the value used for BCD_INPUT will be that put there by the interrupt service routine instead of the desired initial value. There are several ways we can handle the parameters so that the procedure BCD_TO_HEX is reentrant.

The first is to simply pass the parameters in registers as we did in the program in Figure 5-16. If this form of the procedure is called by an interrupt service procedure, all of the variables will be saved by push instructions at the start of the procedure and they will be restored by pop instructions before returning to complete the first execution.

A second method of making the BCD_TO_HEX procedure reentrant is to pass pointers to the data items in registers as we did in the program in Figure 5-17*b*. Again, anything in registers will be saved by push instructions and restored by pop instructions when the procedure is called by the interrupt service routine.

Usually at this point someone remembers that the 8086 allows you to push the contents of a memory location on the stack and asks, "Why can't I just save the contents of BCD_INPUT on the stack with a PUSH BCD_INPUT instruction?" You can do this, but if an interrupt occurs before this instruction occurs, you still have the problem.

The third way to make the BCD_TO_HEX procedure reentrant is by passing parameters on the stack as we did in the version in Figure 5-18. In this version the mainline pushes the BCD number on the stack and then calls the procedure. The procedure pushes registers on the stack and accesses the BCD number relative to where the stack pointer ends up. If an interrupt occurs, the interrupt service procedure will push on the stack the BCD number it wishes to convert and call BCD_TO_HEX. This BCD number will be pushed on the stack at a different location from the first BCD number that was pushed. Since everything is saved on the stack no matter where the interrupt occurs, the first execution of the procedure will produce correct results when it is reentered.

If you are writing a procedure that you may want to call from a program in a high-level language such as Pascal, PL/M, or C, then you should definitely use the stack for passing parameters because that is how these languages do it. Check the manual for the high-level language to determine the parameter passing conventions for that language.

RECURSIVE PROCEDURES

A recursive procedure is a procedure which calls itself. This seems simple enough, but the question you may be thinking is, "Why would we want a procedure to call it self?" The answer is that certain types of problems, such as choosing the next move in a computer chess program, can best be solved with a recursive procedure. Recursive procedures are used to work with complex data structures called *trees*. It is unlikely that you will have to write a recursive procedure because most of the programming problems that you are likely to encounter can be solved with a simple WHILE—DO or REPEAT UNTIL approach. You should, however, know what the term means when you encounter it. For those of you who wish to know more about how a recursive procedure works, we have included an example in the following sections.

Most of the examples of recursive procedures that we could think of are too complex to show here. Therefore, to show you how recursion works, we have chosen a simple problem which could be solved without recursion.

RECURSIVE PROCEDURE EXAMPLE—ALGORITHM

The problem we have chosen to solve is to compute the factorial of a given number in the range of 1 to 9. The factorial of a number is the product of the number and all of the positive integers less than the number. For example, 5 factorial is equal to 5 × 4 × 3 × 2 × 1. The word factorial is often represented with "!". You can therefore write 5 factorial as 5!.

What we want to do here is write a recursive procedure which will compute the factorial of a number, *N*, which we pass to it on the stack, and pass the factorial back to the calling program on the stack. The basic algorithm can be expressed very simply as: IF N = 1 THEN factorial = 1, ELSE factorial = N × (factorial of N − 1). This says that if the number we pass to the procedure is 1, the procedure should return the factorial of 1 which is 1. If the number we pass is not 1, then the procedure should multiply this number by the factorial of the number minus one. Now here's where the recursion comes in. Suppose we pass a 3 to the procedure. When the procedure is first called it has the value of 3 for *N*, but it does not have the value for the factorial of *N* − 1

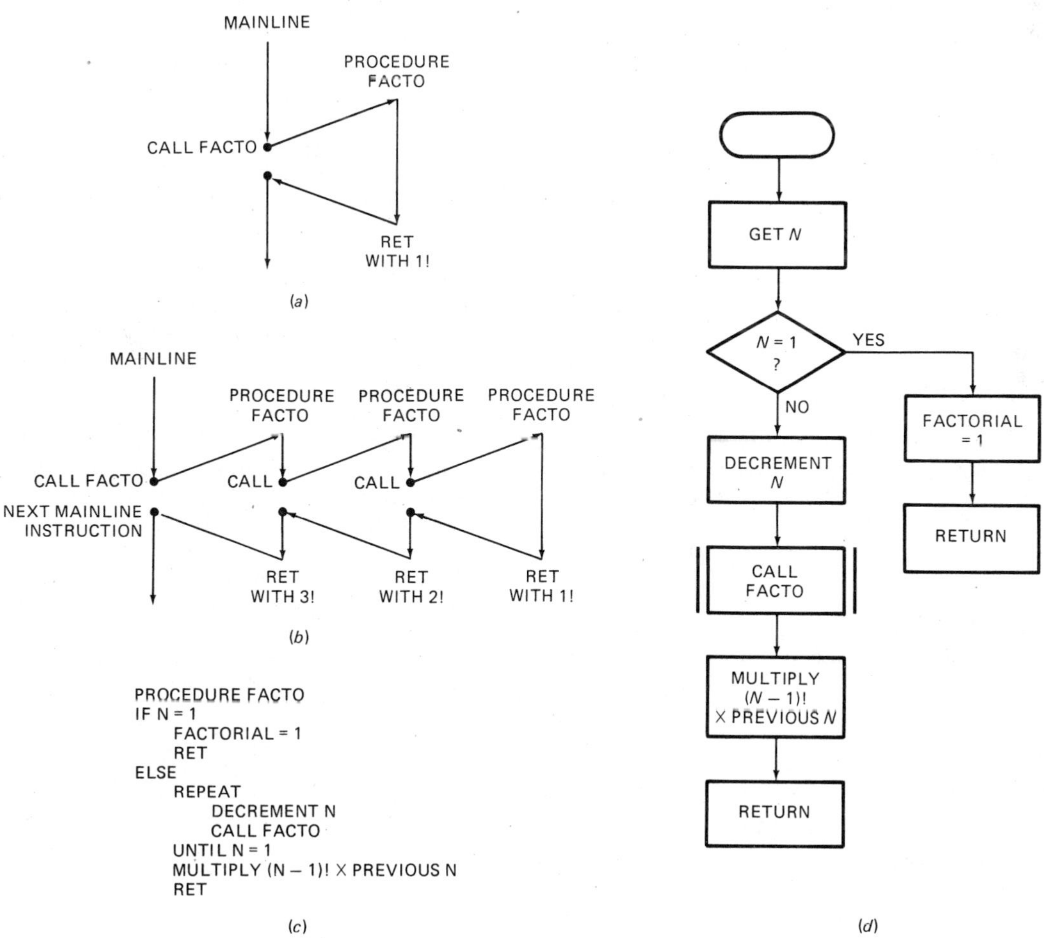

FIGURE 5-21 Algorithm for program to compute factorial for a number *N* between 1 and 9. *(a)* Flow diagram for *N* = 1. *(b)* Flow diagram for *N* = 3. *(c)* Pseudocode. *(d)* Flowchart.

that it needs to do the multiplication indicated in the algorithm. The procedure solves this problem by calling itself to compute the needed factorial of $N - 1$. It calls itself over and over until the factorial of $N - 1$ that it has to compute is the factorial of 1.

Figure 5-21 shows in several ways how we can represent this process. In the program flow diagram in Figure 5-21*a* you can see that if the value of N passed to the procedure is 1 then the procedure simply loads 1 in the stack location reserved for $N!$ and returns to the calling program. Figure 5-21*b* shows the program flow that will occur when the number passed to the procedure is some number other than one. If we call the procedure with an N of 3, the procedure will call itself to compute $N - 1!$ or 2! It will then call itself again to compute the value of the next $N - 1$ factorial or 1!. Since 1! is 1 the procedure will return this value to the program that called it. In this case the program that called it was a previous execution of the same procedure that needed this value to compute 2! Given this value it will compute 2! and return the value to the program that called it. Here again the program that called it was a previous execution of the same procedure that needed 2! to compute the factorial of 3. Given the factorial of 2 this call of the procedure can now compute the factorial of 3 and return to the program that called it. For the example here the return now will be to the mainline program.

Figure 5-21*c* shows how we can represent this algorithm in slightly expanded pseudocode. Use the program flow diagram in Figure 5-21*b* to help you see how execution continues after the return when $N = 1$ and $N = 3$. Can you see that if N is initially 1 the first return will return execution to the instruction following CALL FACTO in the mainline? If the initial N was 3, for example, this return will return execution back to the instruction after the call in the procedure. Likewise, the return after the multiply can send execution back to the next instruction after the call or back to the mainline if the final result has been computed.

Figure 5-21*d* shows a flowchart for this algorithm. Note that the flowchart shows the same ambiguity about where the return operations send execution to.

ASSEMBLY LANGUAGE RECURSIVE FACTORIAL PROCEDURE

Figure 5-22*a* shows an 8086 assembly language procedure which computes the factorial of a number in the range of 1 to 9. To save space we have not included instructions to return an error message if the number passed to the procedure is out of this range. Figure 5-22*b* shows, with a stack diagram, how the stack will be affected if this procedure is called with an N of 3. When working your way through a recursive procedure or any procedure which uses the stack extensively, a stack diagram such as this is absolutely necessary to keep track of everything.

The first parts of the program are housekeeping chores we described in some previous examples. We start the mainline program by declaring a stack segment and setting aside a stack of 200 words with a label at the top of the stack. The first three instructions in the code segment of the mainline program initialize the stack segment register and the stack pointer register. The SUB SP,04 instruction after this will decrement the stack pointer register by 4. In other words we skip the stack pointer down over 2 words in the stack. These two word locations will be used to pass the computed factorial from the procedure back to the mainline program. Next we load the number whose factorial we want into AX and push the value on the stack where the procedure will access it. Now we are ready to call the procedure. We have given the procedure the name FACTO with the FACTO PROC NEAR and FACTO ENDP directives. The procedure is near because it is in the same code segment as the instruction which calls it.

At the start of the procedure we save the flags and all the registers used in the procedure on the stack. Take a look at Figure 5-22*b* to see what is on the stack at this point. Note that the value of N is buried 10 addresses up the stack from where the stack pointer was left after BP was pushed. To access this buried value we first copy SP to BP with the MOV BP, SP instruction so that BP points to the top of the stack. We can then use the expression [BP + 10] to refer to the address where N was pushed. The MOV AX, [BP + 10] instruction will copy N from the stack to AX. If the value of N read in is 1 then the factorial is 1. We want to put 00000001H in the stack locations we reserved for the result, restore the registers and return to the mainline program. Follow this path through the program in Figure 5-22*a*. Note how the MOV WORD PTR [BP + 12], 0001H instruction is used to load a value to a location buried in the stack. The WORD PTR directives tell the assembler that you want to move a word to the specified memory location. Without these directives the assembler will not know whether to code the instruction for moving a byte or for moving a word. The MOV WORD PTR [BP + 14], 0000H instruction is

```
;8086 PROGRAM
;ABSTRACT          : This program computes the factorial of a
;                    number between 1 and 9
;PORTS USED        : None
;PROCEDURES USED: FACTO

STACK_HERE          SEGMENT STACK
        DW          200 DUP(0)      ; set aside 200 words for stack
STACK_TOP           LABEL   WORD    ; assign name to word above stack top
STACK_HERE          ENDS
```

```
NUMBER  EQU      03
CODE_HERE        SEGMENT
                 ASSUME CS:CODE_HERE, SS:STACK_HERE
        MOV      AX, STACK_HERE  ; initialize stack segment register
        MOV      SS, AX
        MOV      SP, OFFSET STACK_TOP ; initialize stack pointer
        SUB      SP, 0004H         ; make space in stack for factorial
                                   ; to be returned
        MOV      AX, NUMBER        ; put number to be passed on stack
        PUSH     AX
        CALL     FACTO             ; compute factorial of number
        POP      AX                ; get result
        NOP                        ; simulate next mainline instructions
        NOP
        NOP
;PROCEDURE: FACTO
;ABSTRACT : Recursive procedure that computes the factorial of
;           a number. It takes its parameter from the stack and
;           returns the result on the stack.
;SAVES    : all registers used
FACTO   PROC     NEAR
        PUSHF                      ; save flags and registers on stack
        PUSH     AX
        PUSH     DX
        PUSH     BP
        MOV      BP, SP            ; point BP at top of stack
        MOV      AX,[BP+10]        ; copy number from stack to AX
        CMP      AX, 0001H         ; if number not equal 1 then go on
        JNE      GO_ON             ;  and compute factorial
        MOV WORD PTR [BP+12], 0001H      ; else load factorial of one in
        MOV WORD PTR [BP+14], 0000H      ; stack and return to mainline
        JMP      EXIT
GO_ON:  SUB      SP, 0004H         ; make space in stack for preliminary
                                   ;  factorial
        DEC      AX                ; decrement number now in AX
        PUSH     AX                ; save number - 1 on stack
        CALL     FACTO             ; compute factorial of number - 1
        MOV      BP, SP            ; point BP at top of stack
        MOV      AX, [BP+2]        ; last (N-1)! from stack to AX
        MUL WORD PTR [BP+16]       ; multiply by previous N
        MOV      [BP+18], AX       ; copy new factorial to stack
        MOV      [BP+20], DX
        ADD      SP, 0006H         ; point stack pointer at pushed register
EXIT:   POP      BP                ; restore registers
        POP DX
        POP AX
        POPF
        RET
FACTO   ENDP
CODE_HERE ENDS
        END
```

(a)

FIGURE 5-22 Recursive procedure to calculate factorial of number between 1 and 9. *(a)* Assembly language. *(b)* Stack diagram showing contents of stack for $N = 3$.

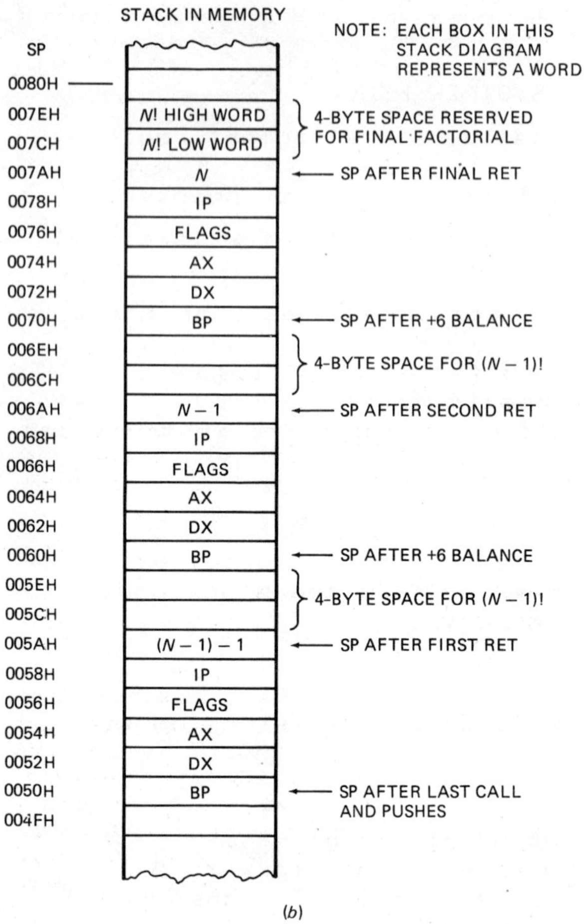

FIGURE 5-22*b* Stack diagram showing contents of stack for N = 3.

likewise used to move a word value to the other word location reserved in the stack for the factorial.

Now let's see what happens if the number passed to FACTO is a 3. The CMP AX, 0001H instruction and the JNE GO_ON instructions determine that *N* is not 1 and send execution to the SUB SP, 04H instruction. According to the algorithm we are going to find the value of *N*! by multiplying *N* times the value of $(N - 1)!$. We will be calling FACTO again to find the value of $(N - 1)!$. The SUB SP, 04H instruction skips the stack pointer down over four addresses in the stack. The value of $(3 - 1)!$ will be returned in these locations. We then decrement *N* by one and push the value of $N - 1$ on the stack where FACTO will access it.

When we call FACTO now to compute the value of $(N - 1)!$ the registers and flags will again be pushed on the stack. Take another look at Figure 5-22*b* to see what is on the stack at this point. The value of $N - 1$ that we need is again buried 10 addresses up in the stack. This is no problem because the MOV BP, SP and MOV AX, [BP + 10] instructions will allow us to access the value. We started with $N = 3$ for this example, so the value of $N - 1$ that we read in at this point is equal to 2. Since this value is not 1 execution will again go to the label GO_ON. The SUB SP, 04 instruction will again skip the stack pointer down over four addresses to leave space for $(2 - 1)!$ to be returned by FACTO. We decrement *N* by 1 to get $N - 1$, which is now 1. We push this value on the stack and call FACTO to compute the factorial of 1.

After pushing all the registers on the stack FACTO reads this 1 from the stack with the MOV AX, [BP + 10] instruction. When the CMP AX, 0001H instruction in FACTO finds that the number passed to it is 1, FACTO loads a factorial value of 1 in the four memory locations we most recently set aside for a returned factorial. The MOV WORD PTR [BP + 12], 0001 and the MOV WORD PTR [BP + 14], 0000 instructions do this. Look at the stack diagram in Figure 5-22*b* to see where these four locations are in the stack. FACTO will then do a return to the next instruction after the CALL instruction that called it.

Now in this case FACTO was called from a previous execution of FACTO so the return will be to the MOV BP,SP instruction after CALL FACTO. The MOV BP, SP instruction points BP at the top of the stack so that we can access data on the stack without affecting the stack pointer. The MOV AX, [BP + 2] instruction after this copies the low word of the last computed $(N - 1)!$ from the stack to AX so that we can multiply it by *N*. We only need the lower word of the two we set aside for the factorial, because for an *N* of eight or less, only the lower word will contain data. Restricting the allowed range of *N* for this example means that we only have to do a 16-bit by 16-bit multiply. We could increase the allowed range of *N* by simply setting aside larger spaces in the stack for factorials and including instructions to multiply larger numbers. In this example the MUL WORD PTR [BP + 16] instruction multiplies the $(N - 1)!$ in AX by the previous *N* from the stack. The low word of the product is left in AX and the high word of the product is left in DX. The MOV [BP + 18], AX and the MOV [BP + 20], DX instructions copy these two words to the stack locations we reserved for the next factorial. Now take a look at the stack diagram in Figure 5-22*b* to see where these two words get put and where the stack pointer is at this time. The next operation we would like to do in the program is pop the registers and return. As you can see from Figure 5-22*b*, however, the stack pointer is now pointing at some old data on the stack, not at the first register we want to pop. To get the stack pointer pointing where we want it, we add six to it with the ADD SP,06H instruction. Then we pop the registers and return.

To see where we are returning to, take another look at Figure 5-22*b*. We are returning with 2! in the stack so we still have one more computation to produce the desired 3!. Therefore, the return is again to the MOV BP, SP instruction after CALL in FACTO. The instructions after this will multiply 2! times 3 to produce the desired 3!, and copy 3! to the stack as described in the preceding paragraph. The ADD SP,06H instruction will again adjust the stack pointer so that we can pop the registers and return. Since we have done all the required computations, this time the return will be to the mainline program. The desired result, 3!, will be in the memory locations we reserved for it in the stack. We can access this result with a BP addressing mode when we need the value in the mainline.

If you work your way through the flow of the stack and the stack pointer in this example program, you should have a good understanding of how the stack is used.

Writing and Calling FAR Procedures

INTRODUCTION AND OVERVIEW

A far procedure is one which is located in a different segment from the CALL instruction which calls it. To get to the starting address of a far procedure the 8086 must change the contents of both the code segment register and the contents of the instruction pointer. Therefore, if you are hand coding a program which calls a far procedure, make sure to use one of the intersegment forms of the CALL instruction to do this. You might, for example, use the direct intersegment CALL instruction. If you look at the coding template for this instruction you will see that the destination instruction pointer value is coded in as bytes 2 and 3 of the instruction, and the destination code segment register value is coded in as bytes 4 and 5 of the instruction. Likewise, at the end of a far procedure, both the contents of the code segment register and the contents of the instruction pointer must be popped off the stack to return to the calling program. Make sure to use one of the intersegment forms of the RET instruction to do this.

If you are using an assembler to assemble a program containing a far procedure, there are a few additional directives you have to give the assembler. The following sections show you how to put these needed additions in your programs. The first case we will describe is one where the procedure is in the same assembly module, but it is in a segment with a different name than the segment that contains the CALL instruction.

ACCESSING A PROCEDURE IN ANOTHER SEGMENT

Suppose that in a program we want to put all of the mainline in one logical segment and we want to put several procedures in another logical segment to keep them separate. Figure 5-23 shows some program fragments which illustrate this situation. For this example our mainline instructions are in a segment named CODE_HERE. A procedure called MULTIPLY_32 is in a segment named PROCEDURES_HERE. Since the procedure is in a different segment from the CALL instruction we must change the contents of the code segment register to access it. Therefore, the procedure is far.

We let the assembler know that the procedure is far by using the word FAR in the MULTIPLY_32 PROC FAR statement. When the assembler finds that the procedure is declared as far, it will automatically code the CALL instruction as an intersegment CALL.

Now the remaining thing we have to do, so that the program gets assembled correctly, is to make sure that the assembler uses the right code segment for each part of the program. We use the ASSUME statement to do this. At the start of the mainline we use the statement ASSUME CS:CODE_HERE to tell the assembler to compute the offsets of the following instructions from the segment base named CODE_HERE. At the start of the procedure we use the ASSUME CS:PROCEDURES_HERE

```
CODE_HERE SEGMENT

          ASSUME     CS:CODE_HERE, DS:DATA_HERE, SS:STACK_HERE

             :

          CALL MULITPLY_32

CODE_HERE ENDS



PROCEDURES_HERE      SEGMENT

  MULTIPLY_32        PROC FAR

          ASSUME     CS:PROCEDURES_HERE

             :

  MULTIPLY_32        ENDP

PROCEDURES_HERE      ENDS
```

FIGURE 5-23 Program additions needed for a far procedure.

statement to tell the assembler to compute the offsets for the instructions in the procedure. The assembler will then compute these offsets starting from the segment base named PROCEDURES_HERE.

When the assembler finally codes out the CALL instruction, it will put the value of PROCEDURES_HERE in for CS in the instruction. It will put the offset of the first instruction of the procedure in PROCEDURES_HERE as the IP value in the instruction.

To summarize then, if a procedure is in another segment you must declare it far with the FAR directive. Also you must put an ASSUME statement in the procedure to tell the assembler what segment base to use for the instructions in the procedure.

ACCESSING A PROCEDURE AND DATA IN A SEPARATE ASSEMBLY MODULE

As we have discussed previously, the best way to write a large program is as a series of modules. Each module can be individually written, assembled, tested, and debugged. Working modules can then be linked together. The previous section showed you how to access a procedure in the same assembly module, but in a different segment from the CALL instruction. Here we show you how to write your programs so that they can access data or procedures in another assembly module.

In order for a linker to be able to access data or a procedure in another assembly module correctly, there are four major types of information that you must give the assembler. We will give you an overview of these four and then show with a program example how you actually write them.

In the assembly module which contains the calling program you must use the EXTRN directive to tell the assembler the names of any procedures or data items that are in other assembly modules. Also, in the module that contains the calling program you must use the PUBLIC directive to tell the assembler any labels or data items that will be accessed from another assembly module.

In the assembly module which contains the procedure you must likewise use the EXTRN directive to tell the assembler the names of any labels or data items that it must look for in another assembly module. Also in the assembly module that contains the procedure you must use the PUBLIC directive to tell the assembler that a label or data item will be accessed from another assembly module.

PROBLEM DEFINITION AND ALGORITHM DISCUSSION

The procedure in the following example program was written to solve a small problem we encountered when writing the program for a microprocessor-controlled medical instrument. Here's the problem.

In the program we add up a series of values read in from an A/D converter. The sum is an unsigned number of between 24 and 32 bits. We needed to scale this value by dividing it by 10. This seems easy because the 8086 DIV instruction will divide a 32-bit unsigned binary number by a 16-bit binary number. The quotient from the division, remember, is put in AX and the remainder is put in DX. However, if the quotient is larger than 16 bits as it will be for our scaling, the quotient will not fit in AX. In this case the 8086 will automatically respond in the same way that it would if you tried to divide a number by zero. We will discuss the details of this response in Chapter 8. For now it is enough to say that we don't want the 8086 to make this response. The simple solution we came up with is to do the division in two steps in such a way that we get a 32-bit quotient and a 16-bit remainder.

Our algorithm is a simple sequence of actions very similar to the way we were taught to do long division. We will first describe how this works with decimal numbers and then we will show how it works with 32-bit and 16-bit binary numbers. Figure 5-24*a* shows an example of long division of the decimal number 433 by the decimal number 9. The 9 won't divide into the 4, so we put a 0 or nothing in this digit position of the quotient. We then see if 9 divides into 43. It fits 4 times, so we put a 4 in this digit position of the quotient and subtract 4 × 9 from the 43. The remainder of 7 now becomes the high digit of the 73, the next number we try to divide the 9 into. After we find that the 9 fits 8 times and subtract 9 × 8 from the 73, we are left with a final remainder of 1. Now let's see how we do this with large binary numbers.

As shown in Figure 5-24*b* we first divide the 16-bit divisor into a 32-bit number made up of a word of all 0's and the high word of the dividend. This division gives us the high word of the quotient and a remainder. The remainder becomes the high word of the dividend for the next division, just as it did for the decimal division.

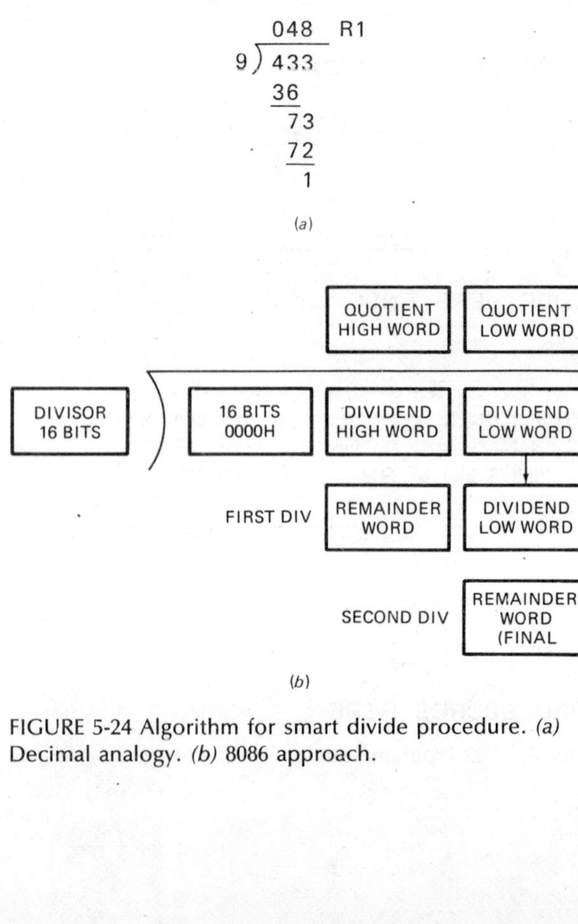

FIGURE 5-24 Algorithm for smart divide procedure. *(a)* Decimal analogy. *(b)* 8086 approach.

We move the low word of the original dividend in as the low word of this dividend and divide by the 16-bit divisor again. The 16-bit quotient from this division is the low word of the 32-bit quotient we want. The 16-bit final remainder can be used to round off the quotient or be discarded, depending on the application.

THE ASSEMBLY LANGUAGE PROGRAM

Figure 5-25*a* shows the mainline of a program which calls the procedure shown in Figure 5-25*b* which implements our division algorithm. We wrote these two as separate assembly modules so that we could show you what you need to add to each module in order for the modules to be linkable. Let's look closely at these added parts before we discuss the actual division procedure.

The first added part of the program to look at is in the statement DATA_HERE SEGMENT WORD PUBLIC. The word PUBLIC in this statement tells the assembler that the contents of this segment will be added to the contents of a segment with the same name in another assembly module when the two modules are linked. In other words, if two or more assembly modules have PUBLIC segments named DATA_HERE, their contents will be pulled together in successive memory locations when the program modules are linked. You should then declare a segment PUBLIC anytime you want it to be linked with other segments of the same name in other modules.

The next addition to look at is the statement PUBLIC DIVISOR in the mainline module in Figure 5-25*a*. This statement is necessary to tell the assembler that the data item named DIVISOR will be accessed from some other assembly module or modules. Essentially what we are doing here is telling the assembler to put the offset of DIVISOR in a special table where it can be accessed when the program modules are linked. Whenever you want a named data item or a label to be accessible from another assembly module you must declare it as PUBLIC. Note in the table at the end of the assembler listing that DIVISOR is global. This is the assembler's way of telling you that it can be accessed from other modules by the linker.

The other side of this coin is that, when you need to access a label or a named data item in another module, you must use the EXTRN directive to tell the assembler that the label or data item is not in the present module. In the example program the statement EXTRN SMART_DIVIDE:FAR tells the assembler that we will be accessing a label or procedure of type far in some other assembly module. For this example we will be accessing our procedure, SMART_DIVIDE. We enclose the EXTRN statement with the PROCEDURES_HERE PUBLIC and the PROCEDURES_HERE ENDS statements to tell the assembler that the procedure SMART_DIVIDE is located in the segment PROCEDURES_HERE. There are some cases where these statements are not needed, but we have found that bracketing the EXTRN statement with SEGMENT—ENDS directives in this way is the best way to make sure that the linker can find everything when it links modules. As you can see in the table at the end of the assembler listing in Figure 5-25*a*, SMART_DIVIDE is identified as an external label of type far, found in a segment named PROCEDURES_HERE.

Now let's see how we handle EXTRN and PUBLIC in the procedure module. The procedure accesses the data item named DIVISOR which is defined in the mainline module. Therefore, we must use the statement EXTRN DIVISOR:WORD to tell the assembler that DIVISOR, a data item of type word, will be found in some other module. Furthermore we enclose the EXTRN statement with the DATA_HERE SEGMENT PUBLIC and DATA_HERE ENDS statements to tell the assembler that DIVISOR will be found in a segment named DATA_HERE.

NOTE: If we had needed to also access DIVIDEND we could have written the EXTRN statement as EXTRN DIVISOR:WORD, DIVIDEND:WORD. To add more terms, just separate them with a comma.

The procedure SMART_DIVIDE must be accessible from other modules so we declare it public with the PUBLIC SMART_DIVIDE statement in the procedure module. If we needed to make other labels or data items public, we could have listed them separated by commas after PUBLIC SMART_DIVIDE. An example is PUBLIC SMART_DIVIDE, EXIT.

Now that we have explained the use of PUBLIC and EXTRN, let's work our way through the rest of the program. At the start of the mainline the ASSUME statement tells the assembler which logical segments to use as code, data, and stack. We then initialize the data segment, stack segment, and stack pointer registers as described in previous example programs. Now before calling the SMART_DIVIDE procedure we copy the dividend and divisor from memory to some registers. The dividend and the divisor are passed to the procedure in these registers. As we explained in a previous section, if we pass parameters to a procedure in registers, the procedure does not have to refer to specific named memory locations. The procedure is then more general and can more easily be called from any place in the mainline. However, in this example we referenced the named memory location, DIVISOR, from the procedure to show you how it can be done using the EXTRN and PUBLIC directives. The procedure is of type far so when we call it, both the code segment register and the instruction pointer contents will be changed.

In the procedure we first check to see if the divisor is zero with a CMP DIVISOR, 0 instruction. If the divisor is zero the JE instruction will send execution to the label ERROR_EXIT. There we set the carry flag with STC as an error indicator and return to the mainline program. If the divisor is not zero, then we go on with the division. To understand how we do the division, remember that the 8086 DIV instruction divides the 32-bit number in DX and AX by the 16-bit number in a specified register or memory location. It puts a 16-bit quotient in AX and a 16-bit remainder in DX. Now, according to our algorithm in Figure 5-24*b* we want to put 0000H in DX and the high word of the dividend in AX for our first DIV operation. MOV BX, AX saves a copy of the low word of the dividend for future reference. MOV AX, DX copies the high word of the dividend into AX where we want it, and MOV DX, 0000H puts all 0's in DX. After the first DIV

```
The IBM Personal Computer MACRO Assembler 02-19-85      PAGE    1-1
                                    PAGE ,132
                                    ;8086 Program
                                    ;ABSTRACT:      This program divides a 32-bit number by a 16-bit number
                                    ;               to give a 32-bit quotient and a 16-bit remainder.
                                    ;REGISTERS USED:CS, DS, SS, AX, SP, BX, CX, DX
                                    ;PROCEDURES:    Calls SMART_DIVIDE which is a far procedure
                                    ;PORTS USED:    None
0000                                DATA_HERE       SEGMENT WORD    PUBLIC
0000  403B 8C72                             DIVIDEND        DW      403BH, 8C72H    ; dividend 8C72403BH
0004  5692                                  DIVISOR         DW      5692H           ; 16-bit divisor
0006                                DATA_HERE       ENDS

0000                                MORE_DATA       SEGMENT WORD
0000       02 [                             QUOTIENT        DW      2 DUP(0)
                  0000
                        ]

0004  0000                                  REMAINDER       DW      0
0006                                MORE_DATA       ENDS

0000                                STACK_HERE      SEGMENT STACK
0000       64 [                                             DW      100 DUP(0)      ; stack of 100 words
                  0000
                        ]

00C8                                        TOP_STACK       LABEL   WORD            ; name pointer to top of stack
00C8                                STACK_HERE      ENDS

                                    PUBLIC  DIVISOR

0000                                PROCEDURES_HERE SEGMENT PUBLIC                  ; let assembler know that SMART_DIVIDE
                                            EXTRN   SMART_DIVIDE : FAR              ; is a label of type FAR and is located
0000                                PROCEDURES_HERE ENDS                            ; in the segment PROCEDURES_HERE

0000                                CODE_HERE       SEGMENT WORD    PUBLIC
                                                    ASSUME  CS:CODE_HERE, DS:DATA_HERE, SS:STACK_HERE

0000  B8  ---- R                    START:          MOV     AX, DATA_HERE           ; initialize data segment
0003  8E D8                                         MOV     DS, AX
0005  B8  ---- R                                    MOV     AX, STACK_HERE          ; initialize stack segment
0008  8E D0                                         MOV     SS, AX
000A  BC 00C8 R                                     MOV     SP, OFFSET TOP_STACK    ; initialize stack pointer
                                    ; load low word of dividend in AX, high word of dividend in DX, divisor in CX
000D  A1 0000 R                                     MOV     AX, DIVIDEND            ; load low  word
0010  8B 16 0002 R                                  MOV     DX, DIVIDEND + 2        ; load high word
0014  8B 0E 0004 R                                  MOV     CX, DIVISOR             ; load divisor
0018  9A 0000 ---- E                                CALL    SMART_DIVIDE
                                    ; quotient returned in DX:AX, remainder returned in BX, carry set if result invalid
001D  73 03                                         JNC     SAVE_ALL                ; carry = 0, result valid
001F  EB 13 90                                      JMP     STOP                    ; carry set, don't save result
                                            ASSUME DS:MORE_DATA                     ; change data segment
0022  1E                            SAVE_ALL:       PUSH    DS                      ; save old DS
```

FIGURE 5-25 Assembly language program to divide a 32-bit number by a 16-bit number and return a 32-bit quotient. *(a)* Mainline program module. *(b)* Procedure module.

```
The IBM Personal Computer MACRO Assembler 02-19-85      PAGE    1-2

 0023  BB  ---- R                        MOV     BX, MORE_DATA          ; load new data segment
 0026  8E DB                             MOV     DS, BX
 0028  A3 0000 R                         MOV     QUOTIENT, AX           ; store low  word of quotient
 002B  89 16 0002 R                      MOV     QUOTIENT + 2, DX       ; store high word of quotient
 002F  89 0E 0004 R                      MOV     REMAINDER, CX          ; store remainder
                                ASSUME DS:DATA_HERE
 0033  1F                                POP     DS                     ; restore initial DS
 0034  90                  STOP:         NOP
 0035                      CODE_HERE     ENDS
                                         END

Segments and groups:

              N a m e               Size    align   combine class

CODE_HERE. . . . . . . . . . . .    0035    WORD    PUBLIC
DATA_HERE. . . . . . . . . . . .    0006    WORD    PUBLIC
MORE_DATA. . . . . . . . . . . .    0006    WORD    NONE
PROCEDURES_HERE. . . . . . . . .    0000    PARA    PUBLIC
STACK_HERE . . . . . . . . . . .    00C8    PARA    STACK

Symbols:

              N a m e               Type    Value   Attr

DIVIDEND . . . . . . . . . . . .    L WORD  0000    DATA_HERE       Global
DIVISOR. . . . . . . . . . . . .    L WORD  0004    DATA_HERE       Global
QUOTIENT . . . . . . . . . . . .    L WORD  0000    MORE_DATA       Length =0002
REMAINDER. . . . . . . . . . . .    L WORD  0004    MORE_DATA
SAVE_ALL . . . . . . . . . . . .    L NEAR  0022    CODE_HERE
SMART_DIVIDE . . . . . . . . . .    L FAR   0000    PROCEDURES_HERE External
START. . . . . . . . . . . . . .    L NEAR  0000    CODE_HERE
STOP . . . . . . . . . . . . . .    L NEAR  0034    CODE_HERE
TOP_STACK. . . . . . . . . . . .    L WORD  00C8    STACK_HERE

Warning Severe
Errors  Errors
0       0
```

(a)

```
The IBM Personal Computer MACRO Assembler 01-01-80      PAGE    1-1

                           PAGE ,132
                           ;8086 procedure called SMART DIVIDE
                           ;ABSTRACT:       This procedure divides a 32-bit number by a 16-bit number
                           ;                to give a 32-bit quotient and a 16-bit remainder. The
                           ;                parameters are passed to and from the procedure in the
                           ;                following way:
                           ;                Dividend : low word in AX, high word in DX
                           ;                Divisor  : word in CX
                           ;                Quotient : low word in AX, high word in DX
                           ;                Remainder: in CX
                           ;                Carry    : carry set if try to divide by zero
                           ;USES:           AX, BX, CX, DX, BP, FLAGS
```

```
                         ;the following block tells the assembler that the divisor is a word
                         ;variable found in the external segment named DATA_HERE
 0000                    DATA_HERE       SEGMENT PUBLIC
                                  EXTRN   DIVISOR:WORD
 0000                    DATA_HERE       ENDS

                         ;the next statement makes SMART_DIVIDE available to other modules
                         PUBLIC          SMART_DIVIDE

 0000                    PROCEDURES_HERE SEGMENT PUBLIC
 0000                    SMART_DIVIDE    PROC    FAR
                                         ASSUME  CS:PROCEDURES_HERE, DS:DATA_HERE
 0000  83 3E 0000 E 00                   CMP     DIVISOR, 0      ; check for illegal divide
 0005  74 17                             JE      ERROR_EXIT      ; divisor = 0 so exit
 0007  8B D8                             MOV     BX, AX          ; save low order of dividend
 0009  8B C2                             MOV     AX, DX          ; position high word for 1st divide
 000B  BA 0000                           MOV     DX, 0000H       ; zero DX
 000E  F7 F1                             DIV     CX              ; AX/CX, quotient in AX, remainder in DX
 0010  8B E8                             MOV     BP, AX          ; save high order of final result
 0012  8B C3                             MOV     AX, BX          ; get back low order of dividend
 0014  F7 F1                             DIV     CX              ; AX/CX, quotient in AX, remainder in DX
 0016  8B CA                             MOV     CX, DX          ; pass remainder back in CX
 0018  8B D5                             MOV     DX,BP           ; pass high order result back in DX
 001A  F8                                CLC                     ; clear carry to indicate valid result
 001B  EB 02 90                          JMP     EXIT            ; finished
 001E  F9                ERROR_EXIT:     STC                     ; set carry to indicate divide by zero
 001F  CB                EXIT:           RET
 0020                    SMART_DIVIDE    ENDP
 0020                    PROCEDURES_HERE ENDS
                                         END


The IBM Personal Computer MACRO Assembler 01-01-80      PAGE    Symbols-1

Segments and groups:

                N a m e                 Size    align   combine class

DATA_HERE. . . . . . . . . . . .        0000    PARA    PUBLIC
PROCEDURES_HERE. . . . . . . . .        0020    PARA    PUBLIC

Symbols:

                N a m e                 Type    Value   Attr

DIVISOR. . . . . . . . . . . . .        V WORD  0000    DATA_HERE       External
ERROR_EXIT . . . . . . . . . . .        L NEAR  001E    PROCEDURES_HERE
EXIT . . . . . . . . . . . . . .        L NEAR  001F    PROCEDURES_HERE
SMART_DIVIDE . . . . . . . . . .        F PROC  0000    PROCEDURES_HERE Global  Length =0020

Warning Severe
Errors  Errors
0       0
```

(*b*)

FIGURE 5-25 (*continued*)

instruction executes, AX will contain the high word of the 32-bit quotient we want as our final answer. We save this in BP with the MOV BP,AX instruction so that we can use AX for the second DIV operation.

The remainder from the first DIV operation was left in the DX register. As shown by the diagram in Figure 5-24*b*, this is right where we want it for the second DIV operation. All we have to do now, before we do the second DIV operation, is to get the low word of the original dividend back into AX with the MOV AX, BX instruction. After the second DIV instruction executes, the 16-bit quotient will be in AX. This word is the low word of our desired 32-bit quotient. We just leave this word in AX to be passed back to the mainline. The DX register was left with the final remainder. We copy this remainder to CX with the MOV CX, DX instruction to be passed back to the mainline program. After the first DIV operation we saved the high word of our 32-bit quotient in BP. We now use the MOV DX, BP instruction to copy this word back to DX where we want it to be when we return to the mainline. You really don't have to shuffle the results around the way we did with these last three instructions, but we like to pass parameters to and from procedures in as systematic a way as possible so that we can more easily keep track of everything. After the shuffling we clear the carry flag with CLC before returning.

Back in the mainline we check the carry flag with the JNC instruction. If the carry flag is set we know that the divisor was 0, no division was done, and there is no result to put in memory. If the carry flag is not set then we know that a valid 32-bit quotient was returned in DX and AX and a 16-bit remainder was returned in CX. We now want to copy this quotient and this remainder to some named memory locations we set aside for them. If you look at some earlier lines in the program, you will see that the memory locations called QUOTIENT and REMAINDER are in a segment called MORE_DATA. At the start of the mainline we tell the assembler to ASSUME that we will be using DATA_HERE as the data segment. Now, however, we want to access some data items in MORE_DATA using DS. To do this we have to do two things. First we have to tell the assembler to ASSUME DS:MORE_DATA. Second, we have to load the segment base of MORE_DATA into DS. In our program we save the old value of DS by pushing it on the stack. We do this so that we can easily reload DS with the base address of DATA_HERE later in the program. The MOV BX, MORE_DATA and MOV DS,BX instructions load the base address of MORE_DATA into DS. The three MOV instructions after this copy the quotient and the remainder into the named memory locations.

Finally in the program we point DS back at DATA_HERE so that later instructions can access data items in the DATA_HERE segment. To do this we first tell the assembler to ASSUME DS:DATA_HERE. We then POP the base address of DATA_HERE off the stack into DS. As you write more complex programs you will often want to access different segments at different times in the program. We wrote this example to show you how to do it. When you do change segments, make sure to change both the ASSUME and the actual contents of the segment register.

Writing and Debugging Programs Containing Procedures

The most important point in writing a program containing procedures is to approach the overall job very systematically. We carefully work out the overall structure of the program and break it down into modules which can easily be written as procedures. We then write the mainline program so that we know what each procedure has to do and how parameters can be most easily be passed to each procedure. To test this mainline we simulate each procedure with a few instructions which simply pass test values back to the mainline. Some programmers refer to these "dummy" procedures as *stubs*. If the structure of the mainline seems reasonable, we then develop each procedure and replace the dummy with it. The advantage of this approach is that you have a structure to hang the procedures on. If you write the procedures first, you have the messy problem of trying to write a mainline to connect all the pieces together.

Now, suppose that you have approached a program as we suggested, and the program doesn't work. After you have checked the algorithm and instructions, you should check that the number of PUSH and POP instructions are equal for each call and return operation. If none of the checks turns up anything, you can use the system debugging tools to track down the problem. Probably the best tools to help you localize a problem to a small area are breakpoints. Run the program to a breakpoint just before a CALL instruction to see if the correct parameters are being passed to the procedure. Put a breakpoint at the start of the procedure to see if execution ever gets to the procedure. Move the breakpoint to a later point in the procedure to determine if the procedure found the parameters passed from the mainline. Use a breakpoint just before the RET instruction to see if the procedure produced the correct results and put these results in the correct locations to pass them back to the mainline program. Inserting breakpoints at key points in your program is much more effective in locating a problem than random poking and experimenting.

WRITING AND USING ASSEMBLER MACROS

Macros and Procedures Compared

Whenever we need to use a group of instructions several times throughout a program there are two ways we can avoid having to write the group of instructions each time we want to use it. One way is to write the group of instructions as a separate procedure. We can then just CALL the procedure whenever we need to execute that group of instructions. A big advantage of using a procedure is that the machine codes for the group of instructions in the procedure only have to be put in memory once. Disadvantages of using a procedure are the need for a stack, and the overhead time required to call the procedure and return to the calling program.

When the repeated group of instructions is too short or not appropriate to be written as a procedure, we use a macro. A macro is a group of instructions we bracket and give a name to at the start of our program. Each time we "call" the macro in our program, the assembler will insert the defined group of instructions in place of the "call." In other words the macro call is like a shorthand expression which tells the assembler, "Every time you see a macro name in the program, replace it with the group of instructions defined as that macro at the start of the program." An important point here is that the assembler generates machine codes for the group of instructions each time the macro is called. Replacing the macro with the instructions it represents is commonly called "expanding" the macro. Since the generated machine codes are right *in-line* with the rest of the program, the processor does not have to go off to a procedure and return. Therefore, using a macro avoids the overhead time involved in calling and returning from a procedure. A disadvantage of generating in-line code each time a macro is called is that this may make the program take up more memory. The examples which follow should help you see how to define and call macros. For these examples we use the syntax of the IBM PC macro assembler, MASM, written by Microsoft Corporation. If you are developing your programs on some other machine, consult the assembly language programming manual for your machine to find the macro definition and calling formats for it.

Defining and Calling a Macro Without Parameters

For our first example suppose that we are writing an 8086 program which has many complex procedures. At the start of each procedure we want to save the flags and all of the registers by pushing them on the stack. At the end of each procedure we want to restore the flags and all of the registers by popping them off the stack. Each procedure would normally contain a long series of PUSH instructions at the start and a long series of POP instructions at the end. Typing in these lists of push and pop instructions is tedious and prone to errors. We could write a procedure to do the pushing and another procedure to do the popping. However, this adds more complexity to the program and is therefore not appropriate. Two simple macros will solve the problem for us.

Here's how we write a macro to save all the registers.

```
PUSH_ALL MACRO
  PUSHF
  PUSH AX
  PUSH BX
  PUSH CX
  PUSH DX
  PUSH BP
  PUSH SI
  PUSH DI
  PUSH DS
  PUSH ES
  PUSH SS
ENDM
```

The PUSH_ALL MACRO statement identifies the start of the macro and gives the macro a name. The ENDM identifies the end of the macro.

Now, to call the macro in one of our procedures we simply put in the name of the macro just as we would an instruction mnemonic. The start of a procedure which does this might look like this.

```
BREATH_RATE  PROC FAR
ASSUME CS:PROCEDURES_HERE, DS:PATIENT_PARAMETERS
PUSH_ALL                          ; macro call
MOV AX, PATIENT_PARAMETERS        ; Initialize data
                                  ; segment reg
MOV DS, AX
        .
        .
        .
```

When the assembler assembles this program section it will replace PUSH_ALL with the instructions that it represents and insert the machine codes for these instructions in the object code version of the program. The assembler listing tells you which lines were inserted by a macro call by putting a + in each program line inserted by a macro call. As you can see from the example here, using a macro makes the source program much more readable because the source program does not have the long series of push instructions cluttering it up.

The preceding example showed how a macro can be used as simple shorthand for a series of instructions. The real power of macros, however, comes from being able to pass parameters to them when you call them. The next section shows you how and why this is done.

Passing Parameters to Macros

Most of us have received computer printed letters of the form:

> Dear MR. HALL,
>
> We are pleased to inform you that you may have won up to $1,000,000 in the *Reader's Weekly* sweepstakes. To find out if you are a winner MR. HALL, return the gold card to *Reader's Weekly* in the enclosed envelope before OCTOBER 22, 1986. You can take advantage of our special offer of three years of *Reader's Weekly* for only $24.95 by putting an X in the YES box on the gold card. If you do not wish to take advantage of this offer, which is one third off the newstand price, mark the no box on the gold card.
>
> Thank you,

A letter such as this is an everyday example of the macro with parameters concept. The basic letter "macro" is written with dummy words in place of the addressee's name, the reply date, and the cost of a three-year subscription. Each time the macro which prints the letter is called, new values for these parameters are passed to the macro. The result is a "personal" looking letter.

In assembly language programs we likewise can write a generalized macro with dummy parameters. Then when we call the macro we can pass it the actual parameters needed for the specific application.

Suppose, for example, we are writing a word processor program. A frequent need in a word processor program is to move strings of ASCII characters from one place in memory to another. The 8086 MOVS instruction is intended to do this. Remember from the discussion of the string instructions in Chapter 4, however, that in order for the MOVS instruction to work correctly, you first have to load SI with the offset of the source start, DI with the offset of the destination start, and CX with the number of bytes or words to be moved. We can define a macro to do all of this as follows.

```
    MOVE_ASCII MACRO NUMBER, SOURCE,
                                   DESTINATION
    MOV CX, NUMBER      ; Number of characters
                        ; to be moved in CX
    LEA SI, SOURCE      ; Point SI at ASCII source
    LEA DI, DESTINATION ; Point DI at ASCII
                        ; destination
REP MOVSB               ; Copy ASCII string to
                        ; new location
    ENDM
```

The words NUMBER, SOURCE, and DESTINATION in this macro are called dummy variables. When we call the macro, values from the calling statement will be put in the instructions in place of the dummies. If, for example, we call this macro with the statement: MOVE_ASCII 03DH, BLOCK_START, BLOCK_DEST, the assembler will expand the macro as follows.

```
    MOV CX, 03DH        ; Number of characters to be
                        ; moved in CX
    LEA SI, BLOCK_START ; Point SI at ASCII destination
    LEA DI, BLOCK_DEST  ; Point DI at ASCII destination
REP MOVSB               ; Copy ASCII string to new
                        ; location
```

We do not have space here to show you very much of what you can do with macros. Read through the assembly language programming manual for your system to find more details about working with macros.

Summary of Procedures vs. Macros

PROCEDURE

Accessed by CALL and RET mechanism during program execution.

Machine code for instructions only put in memory once.

Parameters passed in registers, memory locations, or stack.

MACRO

Accessed during assembly with name given to macro when defined.

Machine code generated for instructions each time called.

Parameters passed as part of statement which calls macro.

IMPORTANT TERMS AND CONCEPTS FROM THIS CHAPTER

If you do not remember any of the terms in the following list, use the index to help you find them in the chapter for review.

Procedure

Subprogram

CALL

RET

Nested procedures

Direct intersegment far CALL

Indirect intersegment far CALL

Direct intersegment near CALL

Indirect intersegment near CALL

Stack: top of stack, stack pointer

PUSH, POP

Parameter, parameter passing

Near and far procedures

Stack overflow

Reentrant and recursive procedures

Interrupt

Interrupt service procedure

Separate assembly modules

Macro

REVIEW QUESTIONS AND PROBLEMS

1. In order to avoid hand keying programs into an SDK-86 board we wrote a program to send machine code programs from an IBM PC to an SDK-86 board through a serial link. As part of this program we had to convert each byte of the machine code program to ASCII codes for the two nibbles in the byte. In other words, a byte of 7AH has to be sent as 37H, the ASCII code for 7, and 41H, the ASCII code for A.

Once you separate the nibbles of the byte, this conversion is a simple IF—THEN—ELSE situation. Write an algorithm and assembly language program section which does the needed conversion.

2. A common problem when reading a series of ASCII characters from a keyboard is the need to filter out those codes which represent the hex digits 0–9 and A–F, and convert these ASCII codes to the hex digits they represent. For example, if we read in 34H, the ASCII code for 4, we want to mask the upper 4 bits to leave 04, the 8-bit hex code for 4. If we read in 42H, the ASCII code for B, we want to add 09 and mask the upper 4 bits to leave 0B, the 8-bit code for hex B. If we read in an ASCII code that is not in the range of 30H–39H or 41H–46H, then we want to load an error code of FFH instead of the hex value of the entered character. Figure 5-26 shows the desired action next to each range of ASCII values. Write an algorithm and an assembly language program which implements these actions. *HINT*: a nested IF—THEN—ELSE structure might be useful.

3. Show the 8086 instruction or group of instructions which will:
   a. Initialize the stack segment register to 4000H and the stack pointer register to 8000H.
   b. Call a near procedure named FIXIT.
   c. Save BX and BP at the start of a procedure and restore them at the end of the procedure.
   d. Return from a procedure and automatically increment the stack pointer by 8.

4. *a.* Use a stack map to show the effect of each of the following instructions on the stack pointer and on the contents of the stack.

```
        MOV  SP,4000H
        PUSH  AX
        CALL MULTO
        POP AX
MULTO   PROC NEAR
        PUSHF
        PUSH BX
          .
          .
          .
        POP  BX
        POPF
        RET
MULTO   ENDP
```

   b. What effect would it have on the execution of this program if the POPF instruction in the procedure was accidentally left out? Describe the steps you would take in tracking down this problem if you did not notice it in the program listing.

5. Show the binary codes for the following instructions.
   a. CALL BX
   b. CALL WORD PTR [BX]
   c. The instruction which will call a procedure which is 97H addresses higher in memory than the instruction after a call instruction.
   d. An instruction which returns execution from a far procedure to a mainline program and increments the stack pointer by 4.

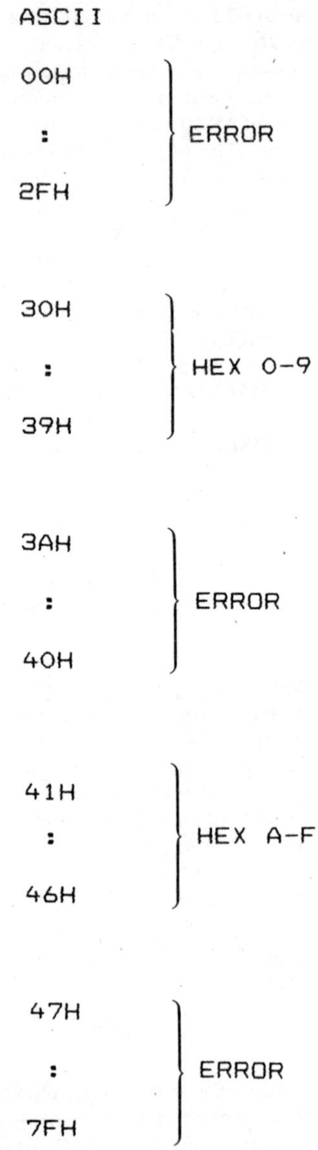

FIGURE 5-26 ASCII chart for Problem 5-2.

6. *a.* List three methods of passing parameters to a procedure. Give the advantage and disadvantage of each method.
   b. Define the term "reentrant" and explain how you must pass parameters to a procedure so that it is reentrant.

7. *a.* Write a procedure which produces a delay of

3.33 ms when run on an 8086 with a 5-MHz clock.

b. Write a mainline program which uses this procedure to output a square wave on bit D0 of port FFFAH.

8. Write a procedure which converts a 4-digit BCD number passed in AX to its binary equivalent. Use the algorithm in Figure 5-15.

9. The 8086 MUL instruction allows you to multiply a 16-bit number by a 16-bit binary number to give a 32-bit result. In some cases, however, you may need to multiply a 32-bit number by a 32-bit number to give a 64-bit result. With the MUL instruction and a little adding you can easily do this. Figure 5-27 shows in diagram form how to do it. Each letter in the diagram represents a 16-bit number. The principle is to use MUL to form partial products and add these partial products together as shown. Write an algorithm for this multiplication and then write the 8086 assembly language program for the algorithm.

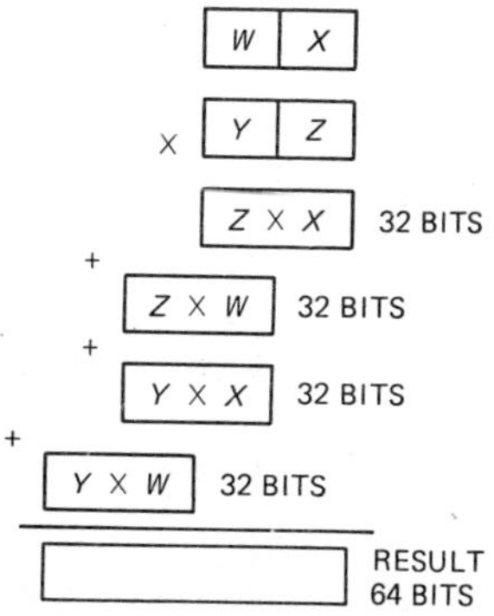

FIGURE 5-27 32-bit by 32-bit multiply method for Problem 5-9.

10. Calculating the factorial of a number which we did with a recursive procedure in Figure 5-22*a* can easily be done with a simple REPEAT UNTIL structure of the form:

```
IF N=1 THEN
     FACTORIAL = 1
ELSE
     FACTORIAL = 1
     REPEAT
          FACTORIAL = FACTORIAL × N
          DECREMENT N
     UNTIL N = 0
```

Write an 8086 procedure which implements this algorithm for an *N* between 1 and 8.

11. Write an assembler macro which will restore, in the correct order, the registers saved by the macro PUSH_ALL in this chapter.

12. *a.* Show how you would tell the assembler to make the label BINADD available to other assembly modules.

b. Show how you would tell the assembler to look for a byte type data item named CONVERSION_FACTOR in a segment named FIXUPS.

CHAPTER

6 8086 Instruction Descriptions and Assembler Directives

This chapter consists of two major sections. The first section is a dictionary of all of the 8086/8088/80186/80188 instructions. For each instruction we give a detailed description of its operation, the correct syntax for the instruction, and the flags affected by the instruction. Also, numerical examples are shown for those instructions where they are appropriate. The binary coding templates for the instructions are shown alphabetically in a table in the appendix. Putting the codes together in a table makes it easier to find codes if you are hand coding a program.

The second major section of this chapter is a dictionary of commonly used 8086 assembler directives. The directives described here are those defined for the Intel 8086 macro assembler and the IBM macro assembler, MASM. If you are using some other assembler, it probably has similar capabilities, but the names may be different.

You will probably use this chapter mostly as a reference to get the details of an instruction or directive as you write programs of your own or decipher someone else's programs. However, you should skim through the chapter at least once to give yourself an overview of the material contained here. You should not try to absorb all of this chapter at once. Most of the instructions described here are used and discussed in various example programs throughout the book. Therefore, we have included references to the appropriate sections in the instruction descriptions here.

INSTRUCTION DESCRIPTIONS

AAA INSTRUCTION—ASCII Adjust for Addition—AAA

Numerical data coming into a computer from a terminal is usually in ASCII code. In this code the numbers 0–9 are represented by the ASCII codes 30H–39H. The 8086 allows you to add the ASCII codes for two decimal digits without masking off the "3" in the upper nibble of each. The AAA instruction is then used to make sure the result is the correct unpacked BCD. A simple numerical example will show how this works.

EXAMPLE:

```
ADD AL, BL  ; AL = 00110101,  ASCII 5
            ; BL = 00111001,  ASCII 9
            ; AL = 01101110,
            ; 6EH - Incorrect Temporary Result
AAA         ; AL = 00000101.  Unpacked BCD for 4.
            ; Carry = 1 to indicate correct answer is
            ; 14 decimal.
```

NOTES: OR AL with 30H to get 34H, the ASCII code for 4, if you want to send the result back to a CRT terminal. The one in the carry can be rotated into the low nibble of a register, ORed with 30H to give the ASCII code for 1, and then sent to the terminal.
The AAA instruction only works on the AL register.
The AAA instruction correctly updates the AF and the CF, but the OF, PF, SF, and ZF are left undefined.

AAD INSTRUCTION—BCD-to-Binary Convert before Division—AAD

The mnemonic for this instruction might tempt you to call it ASCII Adjust for Division. However, the instruction actually works with unpacked BCD. You must mask out the 3 in the upper nibble of the ASCII codes for decimal digits before you can use AAD. AAD converts two unpacked BCD digits in AH and AL to the equivalent binary number in AL. This adjustment must be made before dividing the two unpacked BCD digits in AX by an unpacked BCD byte. After the division AL will then contain the unpacked BCD quotient and AH will contain the unpacked BCD remainder. The PF, SF, and ZF are updated. The AF, CF, and OF are undefined after AAD.

EXAMPLE:

```
            ; AX = 0607   unpacked BCD for 67 decimal
            ; CH = 09H
AAD         ; Adjust to binary before division
            ; AX = 0043 = 43H = 67 decimal
DIV CH      ; Divide AX by unpacked BCD in CH
            ; AL = quotient = 07 unpacked BCD
            ; AH = remainder = 04 unpacked BCD
            ; PF = 0 SF = 0 ZF = 0
```

NOTE: If an attempt is made to divide by 0, or if the quotient is greater than 09, the 8086 will do a type 0 interrupt. Interrupts are explained in Chapter 8.

AAM INSTRUCTION—BCD Adjust after Multiply—AAM

The mnemonic for this instruction may tempt you to call it ASCII Adjust for Multiply. In truth, however, the 8086 does not allow you to multiply the ASCII codes for decimal digits directly. Before you can multiply two ASCII digits, you must first mask the upper 4 bits of each. This leaves unpacked BCD (one BCD digit per byte) in each byte. After the two unpacked BCD digits are multiplied, the AAM instruction is used to adjust the product to two unpacked BCD digits in AX.

AAM only works after the multiplication of two unpacked BCD bytes. It only works on an operand in AL. The PF, SF, and ZF are updated by AAM. The AF, CF, and OF are undefined after AAM.

EXAMPLE:

```
            ; AL = 00000101 = unpacked BCD 5
            ; BH = 00001001 = unpacked BCD 9
MUL BH      ; AL × BH Result in AX
            ; AX = 00000000 00101101 = 002DH
AAM         ; AX = 00000100 00000101
            ; which is unpacked BCD for 45.
            ; If ASCII codes for the result are
            ; desired,
            ; use the next instruction.
OR AX,3030H ; Put 3 in upper nibble of each byte.
            ; AX = 00110100 00110101 = ASCII
            ; codes for 45.
```

AAS INSTRUCTION—ASCII Adjust for Subtraction—AAS

Numerical data coming into a computer from a terminal is usually in ASCII code. In this code the numbers 0–9 are represented by the ASCII codes 30H–39H. The 8086 allows you to subtract the ASCII codes for two decimal digits without masking the "3" in the upper nibble of each. The AAS instruction is then used to make sure the result is the correct unpacked BCD. Some simple numerical examples will show how this works.

EXAMPLE:

```
            ;           (a)
            ; AL = 0011 1001 = ASCII 9
            ; BL = 0011 0101 = ASCII 5
SUB AL, BL  ; (9 – 5) Results:
            ; AL = 0000 0100 = BCD 04
            ; CF = 0
AAS         ; Results:
            ; AL = 0000 0100 = BCD 04
            ; CF = 0 no borrow required
```

```
            ;           (b)
            ; AL = 0011 0101 = ASCII 5
            ; BL = 0011 1001 = ASCII 9
SUB AL, BL  ; (5 – 9) Results:
            ; AL = 1111 1100 = –4
            ; in 2's complement
            ; CF = 1
AAS         ; Results;
            ; AL = 0000 0100 = BCD 04
            ; CF = 1 borrow needed
```

The AAS instruction leaves the correct unpacked BCD result in the low nibble of AL and resets the upper nibble of AL to all 0's. If you want to send the result back to a CRT terminal, you can OR AL with 30H to produce the correct ASCII code for the result. If multiple-digit numbers are being subtracted, the CF can be taken into account by using the SBB instruction when subtracting the next digits.

NOTES: The AAS instruction only works on the AL register. The AAS instruction correctly updates the AF and the CF, but the OF, PF, SF, and the ZF are left undefined.

ADC INSTRUCTION—Add with carry—ADC destination, source

ADD INSTRUCTION—Add—ADD destination, source

These instructions add a number from some source to a number from some destination and put the result in the specified destination. The add with carry instruction, ADC, also adds the status of the carry flag into the result. The source may be an immediate number, a register, or a memory location as specified by any of the 24 addressing modes shown in Figure 3-8. The destination may be a register or a memory location specified by any one of the 24 addressing modes in Figure 3-8. The source and the destination in an instruction cannot both be memory locations. The source and the destination must be of the same type. In other words, they must both be byte locations, or they must both be word locations. If you want to add a byte to a word, you must copy the byte to a word location and fill the upper byte of the word with zeroes before adding. Flags affected: AF, CF, OF, PF, SF, ZF.

EXAMPLES (CODING):

```
ADD AL, 74H     ; Add immediate number 74H
                ; to contents of AL

ADC CL, BL      ; Add contents of BL plus
                ; carry status to
                ; contents of CL.
                ; Result in CL.

ADD DX, BX      ; Add contents of BX to
                ; contents of DX
```

```
ADD DX,[SI]           ; Add word from memory at
                      ; offset [SI] in DS
                      ; to contents of DX

ADC AL, PRICES[BX]    ; Add byte from Effective
                      ; Address PRICES[BX]
                      ; plus carry status to
                      ; contents of AL

ADD PRICES[BX], AL    ; Add contents of AL to
                      ; contents of memory
                      ; location at Effective
                      ; Address PRICES[BX]
```

EXAMPLES (NUMERICAL):

```
                      ; Addition of Unsigned
                      ; numbers
ADD CL, BL            ; CL = 01110011 = 115
                      ; decimal
                      ; + BL = 01001111 =
                      ; 79 decimal
                      ; Result in CL
                      ; CL = 11000010 = 194
                      ; decimal
                      ; Addition of Signed numbers
ADD CL, BL            ; CL = 01110011 = +115
                      ; decimal
                      ; + BL = 01001111 = + 79
                      ; decimal
                      ; Result in CL
                      ; CL = 11000010 = −62
                      ; decimal.
    ; Incorrect because result too large to fit in 7 bits.
```

FLAG RESULTS:

CF = 0 No carry out of bit 7

PF = 0 Result has odd parity

AF = 1 Carry was produced out of bit 3

ZF = 0 Result in destination was not zero

SF = 1 Copies most-significant bit of result; indicates negative result if you are adding signed numbers

OF = 1 Set to indicate that the result of the addition was too large to fit in the lower 7 bits of the destination used to represent the magnitude of a signed number. In other words the result was greater than +127 decimal so the result overflows into the sign bit position and incorrectly indicates that the result is negative. If you are adding two signed 16-bit values, the OF will be set if the magnitude of the result is too large to fit in the lower 15 bits of the destination.

NOTES: The PF is only meaningful for an 8-bit result. The AF is only set by a carry out of bit 3. Therefore, the DAA instruction cannot be used after word additions to convert the result to correct BCD.

AND INSTRUCTION—AND corresponding bits of two operands—AND destination, source

This instruction ANDs each bit in a source byte or word with the same number bit in a destination byte or word. The result is put in the specified destination. The contents of the specified source will not be changed. The result for each bit position will follow the truth table for a two-input AND gate. In other words, a bit in the specified destination will be a one only if that bit is a one in both the source and the destination operands. Therefore, a bit can be masked (reset) by ANDing it with 0.

The source operand can be an immediate number, the contents of a register, or the contents of a memory location specified by one of the 24 addressing modes shown in Figure 3-8. The destination can be a register or a memory location. The source and the destination cannot both be memory locations in the same instruction. The CF and OF are both 0 after AND. The PF, SF, and ZF are updated by AND. AF is undefined. Note that PF only has meaning for an 8-bit operand.

EXAMPLES (CODING):

```
AND BH, CL       ; AND byte in CL with byte in BH,
                 ; result in BH

AND BX, 00FFH    ; AND word in BX with immediate
                 ; 00FFH.
                 ; Mask upper byte, leave
                 ; lower unchanged.

AND CX, [SI]     ; AND word at offset [SI] in
                 ; data segment
                 ; with word in CX
                 ; register. Result in CX register.
```

EXAMPLE (NUMERICAL)

```
                 ; BX = 10110011 01011110
AND BX, 00FFH    ; Mask out upper 8 bits of BX.
                 ; Result BX = 00000000 01011110.
    ; CF = 0 OF = 0 PF = 0 SF = 0 ZF = 0
```

BOUND—80186/80188 ONLY—Check if Array Operation Out of Bounds

When performing some operation on an array of data in memory the BOUND instruction can be used to make sure that data values outside the array are not being operated on. To use this instruction the offset of the lowest element in the array (lower bound) is loaded in two memory addresses. The offset of the highest element in the array (upper bound) is loaded into the next two memory addresses. The offset of the array element currently being worked on is loaded into a general-purpose register such as BX. When the BOUND instruction executes, it will compare the value in BX with the lower and upper bounds in the two memory locations. If the offset in BX is less than the lower bound or greater than the upper bound, then the 80186/80188 will do a type 5 interrupt. Refer to Chapter 8 for a thorough discussion of interrupts. BOUND affects no flags.

EXAMPLE:

L_ARRAY1 EQU 15 ; Length of ARRAY1 = 15 bytes
MOV BOUND_STORE, OFFSET ARRAY1 ; Store offset
; of lowest element
; Now store offset of highest element of the array.
MOV BOUND_STORE+2, OFFSET ARRAY1+L_ARRAY1
; Assume BX contains offset of array element
; currently being operated upon and then
; generate type 5 interrupt if trying to
; operate on an element which is out of bounds
BOUND BX, BOUND_STORE

CALL INSTRUCTION—Call a procedure

The CALL instruction is used to transfer execution to a subprogram or procedure. There are two basic types of CALLs, *near* and *far*. A near CALL is a call to a procedure which is in the same code segment as the CALL instruction. When the 8086 executes a near CALL instruction it decrements the stack pointer by two and copies the offset of the next instruction after the CALL on the stack. This offset saved on the stack is referred to as the return address, because this is the address that execution will return to after the procedure executes. A near CALL instruction will also load the instruction pointer with the offset of the first instruction in the procedure. A RET instruction at the end of the procedure will return execution to the instruction after the CALL by copying the offset saved on the stack back to IP.

A far CALL is a call to a procedure which is in a different segment from that which contains the CALL instruction. When the 8086 executes a far CALL it decrements the stack pointer by two and copies the contents of the CS register to the stack. It then decrements the stack pointer by two again and copies the offset of the instruction after the CALL to the stack. Finally, it loads CS with the segment base of the segment which contains the procedure and IP with the offset of the first instruction of the procedure in that segment. A RET instruction at the end of the procedure will return execution to the next instruction after the CALL by restoring the saved values of CS and IP from the stack.

EXAMPLES

; Direct within-segment (near or intrasegment)

CALL MULTO ; MULTO is the name of the procedure. The assembler determines displacement of MULTO from the instruction after the CALL and codes this displacement in as part of the instruction.

; Indirect within-segment near or intrasegment

CALL BX ; BX contains the offset of the first instruction of the procedure. Replaces contents of IP with contents of register BX.

CALL WORD PTR [BX] ; Offset of first instruction of procedure is in two memory addresses in DS. Replaces contents of IP with contents of word memory location in DS pointed to by BX.

; Direct to another segment-far or intersegment

CALL SMART_DIVIDE ; SMART_DIVIDE is the name of the procedure. The procedure must be declared FAR with SMART_DIVIDE PROC FAR at its start (see section in Chapter 5). The assembler will determine the code segment base for the segment which contains the procedure and the offset of the start of the procedure in that segment. It will put these values in as part of the instruction code.

; Indirect to another segment-far or intersegment

CALL DWORD PTR[BX] ; New values for CS and IP are fetched from four memory locations in DS. The new value for CS is fetched from [BX] and [BX + 1], the new IP is fetched from [BX + 2] and [BX + 3].

CBW INSTRUCTION—Convert signed Byte to signed Word—CBW

This instruction copies the sign of a byte in AL to all the bits in AH. AH is then said to be the sign extension of AL. The CBW operation must be done before a signed byte in AL can be divided by another signed byte with the IDIV instruction. CBW affects no flags.

EXAMPLE:

; AX = 00000000 10011011 = −155 decimal
CBW; Convert signed byte in AL to signed word in AX.
; Result in AX = 11111111 10011011
; = −155 decimal.

For further examples of the use of CBW, see the IDIV instruction description.

CLC INSTRUCTION—Clear the carry flag, CF—CLC

This instruction resets the carry flag to zero. No other flags are affected.

EXAMPLE:

CLC ; Clear carry flag

CLD INSTRUCTION—Clear direction flag—CLD

This instruction resets the direction flag to zero. No other flags are affected. If the direction flag is reset, SI and DI will automatically be incremented when one of the string instructions such as MOVS, CMPS, or SCAS executes. Consult the string instruction descriptions for examples of the use of the direction flag.

EXAMPLE:

CLD ; Clear direction flag so that string pointers
; autoincrement

CLI INSTRUCTION—Clear interrupt flag—CLI

This instruction resets the interrupt flag to zero. No other flags are affected. If the interrupt flag is reset, the 8086 will not respond to an interrupt signal on its INTR input. The CLI instruction, however, has no effect on the nonmaskable interrupt input, NMI.

CMC INSTRUCTION—Complement the carry flag—CMC

If the carry flag, CF, is a zero before this instruction, it will be set to a one after the instruction. If the carry flag is one before this instruction, it will be reset to a zero after the instruction executes. CMC affects no other flags.

EXAMPLE:

```
CMC   ; Invert the carry flag
```

CMP INSTRUCTION—Compare byte or word—CMP destination, source

This instruction compares a byte from the specified source with a byte from the specified destination, or a word from a specified source with a word from a specified destination. The source can be an immediate number, a register, or a memory location specified by one of the 24 addressing modes shown in Figure 3-8. The destination can also be an immediate number, a register, or a memory location. However, the source and the destination cannot both be memory locations in the same instruction. The comparison is actually done by subtracting the source byte or word from the destination byte or word. The source and the destination are not changed, but the flags are set to indicate the results of the comparison. The AF, OF, SF, ZF, PF, and CF are updated by the CMP instruction. For the instruction CMP CX, BX the CF, ZF, and SF will be left as follows:

	CF	ZF	SF	
CX = BX	0	1	0	; Result of subtraction is 0
CX > BX	0	0	0	; No borrow required ; so CF = 0
CX < BX	1	0	1	; Subtraction required ; borrow so CF = 1

EXAMPLES:

```
CMP AL, 01H          ; Compare immediate number
                     ; 01H with byte in AL

CMP BH, CL           ; Compare byte in CL with
                     ; byte in BH

CMP CX, TEMP_MIN     ; Compare word at
                     ; displacement TEMP_MIN
                     ; in DS
                     ; with word in CX

CMP TEMP_MAX, CX     ; Compare CX with word at
                     ; displacement TEMP_MAX
                     ; in data segment

CMP PRICES[BX],49H   ; Compare immediate 49H
                     ; with byte at offset
                     ; [BX] in array PRICES
```

NOTE: The compare instructions are often used with the conditional jump instructions described in a later section. Having the compare instructions formatted the way they are makes this use very easy to understand. For example, given the instruction sequence:

```
CMP BX, CX
JAE TARGET
```

you can mentally read it as jump to target if BX is above or equal to CX. In other words, just mentally insert the first operand after the J for jump and the second operand after the condition.

CMPS/CMPSB/CMPSW—Compare string bytes or string words

A string is a series of the same type of data items in sequential memory locations. The CMPS instruction can be used to compare a byte in one string with a byte in another string or to compare a word in one string with a word in another string. SI is used to hold the offset of a byte or word in the source string and DI is used hold the offset of a byte or a word in the other string. The comparison is done by subtracting the byte or word pointed to by DI from the byte or word pointed to by SI. The AF, CF, OF, PF, SF, and ZF flags are affected by the comparison, but neither operand is affected. After the comparison SI and DI will automatically be incremented or decremented to point to the next elements in the two strings. If the direction flag has previously been set to a one with an STD instruction, then SI and DI will automatically be decremented by one for a byte string or by two for a word string. If the direction flag has been previously reset to a zero with a CLD instruction, then SI and DI will automatically be incremented after the compare. They will be incremented by one for byte strings and by two for word strings.

The string pointed to by DI must be in the extra segment. The string pointed to by SI is assumed to be in the data segment, but you can use a segment override prefix to tell the 8086 to add the offset in SI to CS, SS, or ES.

The CMPS instruction can be used with a REP, REPE, or REPNE prefix to compare all of the elements of a string. To see how this is done, read the discussion of strings in Chapter 4 and the example program in Figure 4-23.

EXAMPLES:

```
                 ; Point SI at source string
                 ; Point DI at destination string
MOV   SI, OFFSET FIRST_STRING
MOV   DI, OFFSET SECOND_STRING
CLD              ; DF cleared so SI and DI will
                 ; autoincrement after compare
CMPS  FIRST_STRING, SECOND_STRING
; The assembler uses names to determine whether
; strings were declared as type byte or as type word.

  MOV CX, 100 ; Put number of string
              ; elements in CX
              ; Point SI at source of string
              ; and DI at destination of string
  MOV SI, OFFSET FIRST_STRING
  MOV DI, OFFSET SECOND_STRING
```

```
STD          ; DF set so SI and DI will auto-
             ; decrement after compare
REPE CMPSB   ; Repeat the comparison of string
             ; bytes until end
             ; of string or until compared
             ; bytes are not equal
```

NOTE: CX functions as a counter which the REPE prefix will cause to be decremented after each compare. The B attached to CMPS tells the assembler that the strings are of type byte. With this addition you don't have to put in the string names as we did in the previous example. If you want to tell the assembler that the strings are of type word, write the instruction as CMPSW.

CWD INSTRUCTION—Convert Signed Word to Signed Doubleword—CWD

CWD copies the sign bit of a word in AX to all the bits of the DX register. In other words it extends the sign of AX into all of DX. The CWD operation must be done before a signed word in AX can be divided by another signed word with the IDIV instruction. CWD affects no flags.

EXAMPLE:

```
      ; DX = 00000000 00000000
      ; AX = 11110000 11000111 = -3897 decimal
CWD   ; Convert signed word in AX to signed double
      ; word in DX:AX.
      ; Result DX = 11111111 11111111
      ; AX = 11110000 11000111 = -3897 decimal
```

For a further example of the use of CWD see the IDIV instruction description.

DAA INSTRUCTION—Decimal Adjust Accumulator—DAA

This instruction is used to make sure the result of adding two packed BCD numbers is adjusted to be a legal BCD number. DAA only works on AL. If the lower nibble in AL after an addition is greater than 9 or the AF was set by the addition, then the DAA instruction will add 6 to the lower nibble in AL. If the result in the upper nibble of AL is now greater than 9 or if the carry flag was set by the addition or correction, then the DAA instruction will add 60H to AL. A couple of simple examples should clarify how this works.

EXAMPLES:

```
                   (a)
             ; AL = 0101 1001 = 59 BCD
             ; BL = 0011 0101 = 35 BCD
ADD AL, BL   ; AL = 1000 1110 = 8EH
DAA
             ; add 0110 Because 1110 > 9
             ; AL = 1001 0100 = 94 BCD

                   (b)
             ; AL = 1000 1000 = 88 BCD
             ; BL = 0100 1001 = 49 BCD
ADD AL, BL   ; AL = 1101 0001, AF = 1
DAA
             ; add 0110 Because AF = 1
             ; AL = 1101 0111 = D7H
             ; 1101 > 9 so add 0110 0000
             ; AL = 0011 0111 = 37 BCD, CF = 1
```

NOTES: The DAA instruction updates the AF, CF, PF, and ZF. The OF is undefined after a DAA instruction.

A decimal UP counter can be implemented using the DAA instruction as follows:

```
MOV COUNT, 00H   ; Initialize count in memory
                 ; location to 0
   .
   .
MOV AL, COUNT    ; Bring count into AL to work on
ADD AL, 01H      ; Can also count up by 2, by 3, or
                 ; by some other number using the
                 ; ADD instruction
DAA              ; Decimal adjust the result
MOV COUNT, AL    ; Put decimal result back in
                 ; memory store
```

DAS INSTRUCTION—Decimal Adjust after Subtraction—DAS

This instruction is used after subtracting two packed BCD numbers to make sure the result is correct packed BCD. DAS only works on the AL register. If the lower nibble in AL after a subtraction is greater than 9 or the AF was set by the subtraction, then the DAS instruction will subtract 6 from the lower nibble of AL. If the result in the upper nibble is now greater than 9 or if the carry flag was set, the DAS instruction will subtract 60 from AL. A couple of simple examples should clarify how this works.

EXAMPLES:

```
             ;        (a)
             ; AL = 1000 0110 = 86 BCD
             ; BH = 0101 0111 = 57 BCD
SUB AL, BH   ; AL = 0010 1111 = 2FH, CF = 0
DAS          ; Subtract 0000 0110 (-06H).
             ; Because 1111 in low nibble > 9
             ; AL = 0010 1001 = 29 BCD

             ;        (b)
             ; AL = 0100 1001 = 49 BCD
             ; BH = 0111 0010 = 72 BCD
SUB AL, BH   ; AL = 1101 0111 = D7H CF = 1
DAS          ; Subtract 0110 0000 (-60H).
             ; Because 1101 in upper nibble > 9
             ; AL = 0111 0111 = 77 BCD CF = 1
             ; CF = 1 means borrow was needed
```

NOTES: The DAS instruction updates the AF, CF, SF, PF, and ZF. The OF is undefined after DAS.

A decimal down counter can be implemented using the DAS instruction as follows:

```
MOV AL, COUNT   ; Bring count into AL to work on
SUB AL, 01H     ; Decrement. Can also count down
                ; by 2, 3,
                ; etc. using SUB instruction.
DAS             ; Keep results in BCD format
MOV COUNT, AL   ; Put new count back in memory
```

DEC INSTRUCTION—Decrement destination register or memory—DEC destination

This instruction subtracts one from the destination word or byte. The destination can be a register or a memory location specified by any one of the 24 addressing modes shown in Figure 3-8. The AF, OF, PF, SF, and ZF are updated, but the CF is not affected. This means that if an 8-bit destination containing 00H or a 16-bit destination containing 0000H is decremented, the result will be FFH or FFFFH with no carry (borrow).

EXAMPLES:

DEC CL ; Subtract one from contents of CL register

DEC BP ; Subtract one from contents of BP register

DEC BYTE PTR [BX] ; Subtract one from byte at offset [BX] in DS. The BYTE PTR directive is necessary to tell the assembler to put in the correct code for decrementing a byte in memory, rather than decrementing a word. The instruction essentially says "Decrement the byte in memory pointed to by [BX]."

DEC WORD PTR [BP] ; Subtract one from a word at offset [BP] in SS. The WORD PTR directive tells the assembler to put in the code for decrementing a word pointed to by the contents of BP. An offset in BP will be added to the SS register contents to produce the physical address.

DEC TOMATO_CAN_COUNT ; Subtract one from byte or word named TOMATO_CAN_COUNT in DS. If TOMATO_CAN_COUNT was declared with a DB then the assembler will code this instruction to decrement a byte. If TOMATO_CAN_COUNT declared with DW, then the assembler will code this instruction to decrement a word.

DEC HERD_COUNT [BX] ; Decrement word or byte at offset [BX] in array HERD_COUNT. Array is in DS. This instruction will decrement a byte if HERD_COUNT was declared with a DB. It will decrement a word if HERD_COUNT was declared with a DW directive.

DIV INSTRUCTION—Unsigned divide—DIV source

This instruction is used to divide an unsigned word by a byte, or to divide an unsigned double word (32 bits) by a word.

When dividing a word by a byte, the word must be in the AX register. After the division AL will contain an 8-bit result (quotient) and AH will contain an 8-bit remainder. If an attempt is made to divide by 0 or the quotient is too large to fit in AL (greater than FFH), the 8086 will automatically do a type 0 interrupt. Interrupts are explained in Chapter 8.

When a double word is divided by a word, the most-significant word of the double word must be in DX and the least-significant word of the double word must be in AX. After the division AX will contain the 16-bit result (quotient), and DX will contain a 16-bit remainder. Again, if an attempt is made to divide by 0 or the quotient is too large to fit in AX (greater than FFFFH), the 8086 will do a type 0 interrupt.

For a DIV the dividend (numerator) must always be in AX or DX and AX, but the source of the divisor (denominator) can be a register or a memory location specified by any one of the 24 addressing modes shown in Figure 3-8. If the divisor does not divide an integral number of times into the dividend, the quotient is truncated, not rounded. The example below will illustrate this. All flags are undefined after a DIV instruction.

If you want to divide a byte by a byte, you must first put the dividend byte in AL and fill AH with all 0's. The SUB AH, AH instruction is a quick way to do this. Likewise, if you want to divide a word by a word, put the dividend word in AX, and fill DX with all 0's. The SUB DX, DX instruction does this quickly.

EXAMPLES (CODING):

```
DIV BL   ; Word in AX/byte in BL.
         ; Quotient in AL, remainder in AH.

DIV CX   ; Double word in DX and AX/word in CX.
         ; Quotient in AX, remainder in DX.
```

DIV SCALE[BX] ; AX/(byte at effective address SCALE[BX]) if SCALE[BX] is of type byte or (DX and AX)/(word at effective address SCALE[BX]) if SCALE[BX] is of type word.

EXAMPLE (NUMERICAL)

```
         ; AX = 37D7H = 14295 decimal
         ; BH = 97H = 151 decimal
DIV BH   ; AX/BH
         ; Quotient in AL = 5EH = 94 decimal.
         ; Remainder in AH = 65H = 101 decimal.
```

Since the remainder is greater than half of the divisor, the actual quotient is closer to 5FH than to the 5EH produced. However, as indicated above, the quotient is always truncated to the next lower integer rather than rounded to the closest integer. If you want to round the quotient, you can compare the remainder with (divisor/2), and then add one to the quotient if the remainder is greater than the (divisor/2).

ENTER—80186/80188 ONLY—Enter Procedure

This instruction is used at the start of an assembly language procedure which is intended to be called from a high-level language program such as those written in Pascal or C. Its main functions are to save space on the stack for variables used in the procedure, and to determine pointers to data areas in the stack used by lower-level procedures (subprocedures). Refer to the Intel literature for further explanation of this instruction if you need it.

ESC INSTRUCTION
Escape—ESC

This instruction is used to pass instructions to a coprocessor such as the 8087 math coprocessor which shares the address and data bus with an 8086. Instructions for the coprocessor are represented by a 6-bit code imbedded in the escape instruction. As the 8086 fetches instruction bytes, the coprocessor also catches these bytes from the data bus and puts them in its queue. However, the coprocessor treats all of the normal 8086 instructions as NOPs. When the 8086 fetches an ESC instruction, the coprocessor decodes the instruction and carries out the action specified by the 6-bit code specified in the instruction. In most cases the 8086 treats the ESC instruction as a NOP. In some cases the 8086 will access a data item in memory for the coprocessor. A section in Chapter 11 describes the operation and use of the ESC instruction.

HLT INSTRUCTION
Halt processing—HLT

The HLT instruction will cause the 8086 to stop fetching and executing instructions. The 8086 will enter a halt state. The only ways to get the processor out of the halt state are with an interrupt signal on the INTR pin, an interrupt signal on the NMI pin, or a reset signal on the RESET input. See Chapter 7 for further details about the halt state.

IDIV INSTRUCTION
Divide by signed byte or word—IDIV source

This instruction is used to divide a signed word by a signed byte, or to divide a signed double word (32 bits) by a signed word.

When dividing a signed word by a signed byte, the word must be in the AX register. After the division, AL will contain the signed result (quotient), and AH will contain the signed remainder. The sign of the remainder will be the same as the sign of the dividend. If an attempt is made to divide by 0, the quotient is greater than 127 (7FH), or the quotient is less than −127 (81H), the 8086 will automatically do a type 0 interrupt. Interrupts are discussed in Chapter 8. For the 80186 this range is −128 to +127.

NOTE: When dividing a signed double word by a signed word, the most-significant word of the dividend (numerator) must be in the DX register and the least-significant word of the dividend must be in the AX register. After the division AX will contain a signed 16-bit quotient and DX will contain a signed 16-bit remainder. The sign of the remainder will be the same as the sign of the dividend. Again, if an attempt is made to divide by 0, the quotient is greater than +32,767 (7FFFH), or the quotient is less than −32,767 (8001H), the 8086 will automatically do a type 0 interrupt.

NOTE: For the 80186 this range is −32,768 to +32,767.

The dividend for IDIV must always be in AX or DX and AX but the source of the divisor can be a register or a memory location specified by any one of the 24 addressing modes shown in Figure 3-8. If the divisor does not divide into the dividend the quotient will be truncated, not rounded. The example below will illustrate this. All flags are undefined after an IDIV.

If you want to divide a signed byte by a signed byte, you must first put the dividend byte in AL and fill AH with copies of the sign bit from AL. In other words, if AL is positive (sign bit = 0) then AH should be filled with 0's. If AL is negative (sign bit = 1), the AH should be filled with 1's. The 8086 Convert Byte to Word instruction, CBW, does this by copying the sign bit of AL to all the bits of AH. AH is then said to contain the "sign extension of AL." Likewise, if you want to divide a signed word by a signed word, you must put the dividend word in AX and extend the sign of AX to all the bits of DX. The 8086 Convert Word to Doubleword instruction, CWD, will copy the sign bit of AX to all the bits of DX.

EXAMPLES (CODING)

```
IDIV BL              ; Signed word in AX/signed byte
                     ; in BL

IDIV BP              ; Signed double word in DX and
                     ; AX/signed word in BP

IDIV BYTE PTR [BX]   ; AX/byte at offset [BX] in DS

MOV AL, DIVIDEND     ; Position byte dividend
CBW                  ; Extend sign of AL into AH
IDIV DIVISOR         ; Divide by byte divisor
```

EXAMPLES (NUMERICAL)

```
; Example showing a signed word divided by a signed
; byte
; AX = 00000011 10101011 = 03ABH = 939 decimal
; BL = 11010011 = D3H = −2DH = −45 decimal

IDIV BL   ; Quotient in AL = 11101100
          ; AH = ECH = −14H = −20 decimal
          ; Remainder in AH = 00100111
          ; AH = 27H = +39 decimal
```

NOTE: Quotient is negative because positive was divided by negative. Remainder has same sign as dividend (positive).

```
; Example showing a signed byte divided by a signed
; byte
          ; AL = 11001010 = −26H = −38 decimal
          ; CH = 00000011 = +3H = +3 decimal
CBW       ; Extend sign of AL through AH,
          ; AX = 11111111 11001010
IDIV CH   ; Divide AX by CH
          ; AL = 11110100 = −0CH = −12 decimal
          ; AH = 11111110 = −2H = −2 decimal
```

Although the quotient is actually closer to 13 (12.666667) than to 12, the 8086 truncates it to 12 rather than rounding it to 13. If you want to round the quotient, you can compare the magnitude of the remainder with the (dividend/2) and add one to the quotient if the remainder is greater than (dividend/2). Note that the sign of the remainder is the same as the sign of the dividend (negative). All flags are undefined after IDIV.

IMUL INSTRUCTION—Multiply signed number—IMUL source

This instruction multiplies a signed byte from some source times a signed byte in AL, or a signed word from some source times a signed word in AX. The source can be another register or a memory location specified by any one of the 24 addressing modes shown in Figure 3-8. When a byte from some source is multiplied by AL, the signed result (product) will be put in AX. A 16-bit destination is required because the result of multiplying two 8-bit numbers can be as large as 16 bits. When a word from some source is multiplied by AX, the result can be as large as 32 bits. The high-order (most-significant) word of the signed result is put in DX and the low-order (least-significant) word of the signed result is put in AX. If the magnitude of the product does not require all of the bits of the destination, the unused bits will be filled with copies of the sign bit. If the upper byte of a 16-bit result or the upper word of a 32-bit result contains only copies of the sign bit (all 0's or all 1's), then the CF and the OF will both be 0's. If the upper byte of a 16-bit result or the upper word of a 32-bit result contains part of the product, the CF and OF will both be 1's. You can use the status of these flags to determine whether the upper byte or word of the product needs to be kept. The AF, PF, SF, and ZF are undefined after IMUL.

If you want to multiply a signed byte by a signed word, you must first move the byte into a word location and fill the upper byte of the word with copies of the sign bit. If you move the byte into AL you can use the 8086 Convert Byte to Word instruction, CBW, to do this. CBW extends the sign bit from AL into all the bits of AH. Once you have converted the byte to a word, you can do word times word IMUL. The result will be in DX and AX.

EXAMPLES(CODING):

```
IMUL BH    ; Signed byte in AL times signed
           ; byte in BH, result in AX

IMUL AX    ; AX times AX, result in DX and AX

; Example showing a signed byte multiplied by a
; signed word

MOV CX, MULTIPLIER      ; Load signed word
                        ; multiplier in CX
MOV AL, MULTIPLICAND    ; Load byte multiplicand
                        ; into AL
CBW                     ; Extend sign of AL into AH
IMUL CX                 ; Word multiply;
                        ; result in DX and AX
```

EXAMPLES(NUMERICAL):

```
           ;             (a)
           ; AL = 01000101 = 69 decimal
           ; BL = 00001110 = 14 decimal
IMUL BL    ; AX = +966 decimal
           ; AX = 00000011 11000110
           ; MSB = 0, positive result
           ; magnitude in true form.
           ; SF = 0, CF = 1, OF = 1

           ;             (b)
           ; AL = 11100100 = -28 decimal
           ; BL = 00111011 = +59 decimal
IMUL BL    ; AX = -1652 decimal
           ; AX = 11111001 10001100
           ; MSB = 1, negative result
           ; magnitude in 2's complement.
           ; SF = 1, CF = 1, OF = 1
```

IMUL—80186/80188 ONLY—Integer (signed) Multiply Immediate—IMUL destination register, source, immediate byte or word

This version of the IMUL instruction functions the same as the IMUL instruction described in the preceding section except that this version allows you to multiply an immediate byte or word by a byte or word in a specified register and put the result in a specified general-purpose register. If the immediate number is a byte, it will be automatically sign-extended to 16 bits. The source of the other operand for the multiplication can be a register or a memory location specified by any one of the addressing modes shown in Figure 3-8. Since the result is put in a 16-bit general-purpose register, only the lower 16 bits of the product are saved!

EXAMPLE:

```
IMUL CX, BX, 07H   ; Multiply contents of BX times
                   ; 07H and put

                   ; lower 16 bits of result in CX
```

IN INSTRUCTION—IN accumulator, port

The IN instruction will copy data from a port to the accumulator. If an 8-bit port is read, the data will go to AL. If a 16-bit port is read, the data read will go to AX. The IN instruction has two possible formats, fixed port and variable port.

For the fixed port type the 8-bit address of a port is specified directly in the instruction.

EXAMPLES:

```
IN AL, 0C8H      ; Input a byte from port 0C8H to AL

IN AX, 34H       ; Input a word from port 34H to AX

A_TO_D EQU 4AH
IN AX, A_TO_D    ; Input a word from port 4AH to AX
```

For the variable-port type IN instruction, the port address is contained in the DX register. Since DX is a 16-bit register, the port address can be any number between 0000H and FFFFH. Therefore, up to 65,536 ports are addressable in this mode. The DX register, however, must always be loaded with the port address before the IN instruction.

EXAMPLES:

MOV DX,0FF78H	; Initialize DX to point to port
IN AL, DX	; Input a byte from 8 bit port ; 0FF78H to AL
IN AX, DX	; Input a word from 16-bit port ; 0FF78H to AX

The variable-port IN instruction has the advantage that the port address can be computed or dynamically determined in the program. Suppose, for example, that an 8086-based computer needs to input data from 10 terminals, each having its own port address. Instead of having separate routines to input data from each port, we can write one general input subroutine and simply pass the address of the desired port to the subroutine in DX. The IN instructions do not change any flags.

INC INSTRUCTION—INCREMENT destination

The INC instruction adds 1 to the indicated destination. The destination can be a register or memory location specified by any one of the 24 ways shown in Figure 3-8. The AF, OF, PF, SF, and ZF are affected (updated) by this instruction. Note that the carry flag, CF, is not affected. This means that if an 8-bit destination containing FFH or a 16-bit destination containing FFFFH is incremented, the result will be all 0's with no carry.

EXAMPLES:

INC BL ; Add 1 to contents of BL register

INC CX ; Add 1 to contents of CX register

INC BYTE PTR [BX] ; Increment byte at offset of BX in data segment. The BYTE PTR directive is necessary to tell the assembler to put in the right code to indicate that a byte in memory, rather than a word, is to be incremented. The instruction essentially says "increment the byte pointed to by the contents of BX."

INC WORD PTR [BX]; Increment the word at offset of [BX] and [BX + 1] in the data segment. In other words, increment the word in memory pointed to by BX.

INC MAX_TEMPERATURE ; Increment byte or word named MAX TEMPERATURE in data segment. Increment byte if MAX_TEMPERATURE declared with DB. Increment word if MAX_TEMPERATURE declared with DW.

INC PRICES [BX] ; Increment element [BX] in array PRICES. Increment a word if PRICES was defined as an array of words with a DW directive. Increment a byte if PRICES was defined as an array of bytes with a DB directive.

NOTE: The PTR operator is not needed in the last two examples because the assembler knows the type of the operand from the DB or DW used to declare the named data initially.

INS/INSB/INSW—80186/80188 ONLY—Input String from Port—INS destination string, DX

INS copies a byte or a word from a port to a memory location in the extra segment pointed to by DI. The address of the port to be copied from must be put in DX before this instruction executes. If the direction flag is cleared when this instruction executes, DI will automatically be incremented by one for a byte operation, and incremented by two for a word operation after the data is copied from the port. If the direction flag is set, DI will automatically be decremented by one for a byte operation and decremented by two for a word operation after data is copied from the port. When used with the REP prefix or as part of a loop, the INS instruction can input a block of data directly to a series of memory locations without having the data go through AL or AX as it does with the regular IN instruction.

When using the INS instruction you must in some way tell the assembler whether you want to input bytes or input words. There are two ways to do this. The first is to use the name of the destination string in the instruction statement as in the statement INS BUFFER, DX. The assembler will code the instruction for a byte input if BUFFER was declared with a DB and it will code the instruction for a word input if BUFFER was declared with a DW. The second way to tell the assembler whether to code the instruction for a byte or for a word input is to add a B or a W to the basic instruction mnemonic. For example, INSB DX, tells the assembler to code the instruction for copying a byte from a port pointed to by DX to a memory location pointed to by DI in the extra segment. INS affects no flags.

EXAMPLE:

CLD	; Clear direction flag to ; autoincrement DI
MOV DI, OFFSET BUF	; Point DI at input buffer
MOV DX, 0FFF8H	; Load DX with ; port address
MOV CX, LENGTH BUF	; Load number of bytes ; to be read in CX
REP INSB DX	; Copy bytes from port ; until buffer full

INT INSTRUCTION—Interrupt program execution—INT type

This instruction causes the 8086 to call a far procedure in a manner similar to the way in which the 8086 responds to an interrupt signal on its INTR or NMI inputs. Part of the response is to do an indirect far call to a procedure which responds to that particular interrupt. The term type in the instruction refers to a number between 0 and 255 which identifies the interrupt. When an 8086 executes an INT type instruction, it will:

1. Decrement the stack pointer by two and push the flags on the stack.
2. Decrement the stack pointer by two and push the contents of CS on the stack.
3. Decrement the stack pointer by two and push the offset of the next instruction after the INT number instruction on the stack.
4. Get a new value for IP from an absolute memory address of 4 times the type specified in the instruction. For an INT 8 instruction, for example, the new IP will be read from address 00020H.
5. Get a new value for CS from an absolute memory address of 4 times the type specified in the instruction plus 2. For an INT 8 instruction, for example, the new value of CS will be read from address 00022H.
6. Reset both the IF and the TF. Other flags are not affected.

Chapter 8 further describes the use of this instruction.

EXAMPLES:

INT 35 ; New IP from 0008CH, new CS from 0008EH

INT 3 ; This is a special form which has the single byte code of CCH. Many systems use this as a breakpoint instruction. New IP from 0000CH, new CS from 0000EH.

INTO INSTRUCTION—Interrupt on overflow

If the overflow flag, OF, is set, this instruction will cause the 8086 to do an indirect far call to a procedure you write to handle the overflow condition. Before doing the call the 8086 will:

1. Decrement the stack pointer by 2 and push the flags on the stack.
2. Decrement the stack pointer by 2 and push CS on the stack.
3. Decrement the stack pointer by 2 and push the offset of the next instruction after the INTO instruction on the stack
4. Reset the TF and the IF, other flags are not affected.

To do the call the 8086 will read a new value for IP from address 00010H and a new value of CS from address 00012H.

Chapter 8 further describes the 8086 interrupt system.

EXAMPLE:

INTO ; Call interrupt procedure if OF = 1

IRET INSTRUCTION—Interrupt return—IRET

When the 8086 responds to an interrupt signal or to an interrupt instruction, it pushes the flags, the current value of CS, and the current value of IP on the stack. It then loads CS and IP with the starting address of the procedure which you write for the response to that interrupt. The IRET instruction is used at the end of the interrupt service procedure to return execution to the interrupted program. To do this return the 8086 copies the saved value of IP from the stack to IP, the stored value of CS from the stack to CS, and the stored value of the flags back to the flag register. Flags will have the values they had before the interrupt, so any flag settings from the procedure will be lost unless they are specifically saved in some way.

NOTE: The RET instruction should not normally be used to return from interrupt procedures because it does not copy the flags from the stack back to the flag register. See Chapter 8 for further discussion of interrupts and the use of IRET.

EXAMPLE:

IRET

JA/JNBE INSTRUCTION—Jump if above/Jump if not below nor equal

These two mnemonics represent the same instruction. The terms "above" and "below" are used when referring to the magnitude of unsigned numbers. The number 0111 is above the number 0010. If, after a compare or some other instruction which affects flags, the zero flag and the carry flag are both 0, this instruction will cause execution to jump to a label given in the instruction. If CF and ZF are not both 0, the instruction will have no effect on program execution. The destination label for the jump must be in the range of −128 bytes to +127 bytes from the address of the instruction after the JA. JA/JNBE affects no flags. For further explanation of conditional jump instructions, see Chapter 4.

EXAMPLES:

```
CMP AX, 4371H      ; Compare by subtracting 4371H
                   ; from AX
JA RUN_PRESS       ; Jump to label RUN_PRESS if AX
                   ; above 4371H

CMP AX, 4371H      ; Compare by subtracting 4371H
                   ; from AX
JNBE RUN_PRESS     ; Jump to label RUN_PRESS if AX
                   ; not below nor
                   ; equal to 4371H
```

JAE/JNB/JNC INSTRUCTIONS
Jump if above or equal/Jump if not below/Jump if no carry

These three mnemonics represent the same instruction. The terms "above" and "below" are used when referring to the magnitude of unsigned numbers. The number 0111 is above the number 0010. If, after a compare or some other instruction which affects flags, the carry flag is 0, this instruction will cause execution to jump to a label given in the instruction. If CF is 1, the instruction will have no effect on program execution. The destination label for the jump must be in the range of −128 bytes to +127 bytes from the address of the instruction after the JA. JAE/JNB/JNC affects no flags. For further explanation of conditional jump instructions, see Chapter 4.

EXAMPLES:

```
CMP AX, 4371H     ; Compare by subtracting 4371H
                  ; from AX
JAE RUN_PRESS     ; Jump to label RUN_PRESS if AX
                  ; is above or equal to 4371H

CMP AX, 4371H     ; Compare by subtracting 4371H
                  ; from AX
JNB RUN_PRESS     ; Jump to label RUN_PRESS if AX
                  ; not below 4371H

ADD AL, BL        ; Add two bytes
JNC OK            ; Result within acceptable range,
                  ; continue
```

JB/JC/JNAE INSTRUCTIONS
Jump if below/Jump if carry/Jump if not above nor equal

These three mnemonics represent the same instruction. The terms "above" and "below" are used when referring to the magnitude of unsigned numbers. The number 0111 is above the number 0010. If, after a compare or some other instruction which affects flags, the carry flag is a 1, this instruction will cause execution to jump to a label given in the instruction. If CF is 0, the instruction will have no effect on program execution. The destination label for the jump must be in the range of −128 bytes to +127 bytes from the address of the instruction after the JB. JB/JC/JNAE affects no flags. For further explanation of conditional jump instructions, see Chapter 4.

EXAMPLES:

```
CMP AX, 4371H     ; Compare by subtracting 4371H
                  ; from AX
JB RUN_PRESS      ; Jump to label RUN_PRESS if
                  ; AX below 4371H

ADD BX, CX        ; Add two words
JC ERROR_FIX      ; Jump to label ERROR_FIX if
                  ; CF = 1

CMP AX, 4371H     ; Compare by subtracting 4371H
                  ; from AX
JNAE RUN_PRESS    ; Jump to label RUN_PRESS if
                  ; AX not above nor
                  ; equal to 4371H
```

JBE/JNA INSTRUCTIONS
Jump if below or equal/Jump if not above

These two mnemonics represent the same instruction. The terms "above" and "below" are used when referring to the magnitude of unsigned numbers. The number 0111 is above the number 0010. If, after a compare or some other instruction which affects flags, either the zero flag or the carry flag is 1, this instruction will cause execution to jump to a label given in the instruction. If CF and ZF are both 0, the instruction will have no effect on program execution. The destination label for the jump must be in the range of −128 bytes to +127 bytes from the address of the instruction after the JBE. JBE/JNA affects no flags. For further explanation of conditional jump instructions, see Chapter 4.

EXAMPLES:

```
CMP AX, 4371H     ; Compare by subtracting 4371H
                  ; from AX

JBE RUN_PRESS     ; Jump to label RUN_PRESS if AX
                  ; is below or equal to 4371H

CMP AX, 4371H     ; Compare by subtracting 4371H
                  ; from AX
JNA RUN_PRESS     ; Jump to label RUN_PRESS if AX
                  ; not above 4371H
```

JCXZ INSTRUCTION
Jump if the CX register is zero

This instruction will cause a jump to a label given in the instruction if the CX register contains all 0's. If CX does not contain all 0's, execution will simply proceed to the next instruction. Note that this instruction does not look at the zero flag when it decides whether to jump or not. The destination label for this instruction must be in the range of −128 to +127 bytes from the address of the instruction after the JCXZ instruction. JCXZ affects no flags.

EXAMPLE:

```
; If CX already 0 skip over the process

            JCXZ SKIP_LOOP
LOOP:       SUB [BX], 07H     ; Subtract 7 from
                              ; data value
            INC BX            ; point to
                              ; next value
            LOOP COUNT        ; Loop until CX = 0
SKIP_LOOP:                    ; next instruction
```

JE/JZ INSTRUCTIONS—Jump if equal/Jump if zero

These two mnemonics represent the same instruction. If the zero flag is set, then this instruction will cause execution to jump to a label given in the instruction. If the zero flag is not 1, then execution will simply go on to the next instruction after JE or JZ. The destination label for the JE/JZ instruction must be in the range of −128 to +127 bytes from the address of the instruction after the JE/JZ instruction. JE/JZ affects no flags.

EXAMPLES:

```
AGAIN: CMP BX, DX      ; Compare by subtracting DX
                       ; from BX
       JE DONE         ; Jump to label DONE
                       ; if BX = DX
       SUB BX, AX      ; Else subtract AX
       INC CX          ; Increment counter
       JMP AGAIN       ; Check again
DONE:  MOV AX, CX      ; Copy count to AX

IN AL, 8FH             ; Read data from port 8FH
SUB AL, 30H            ; Subtract minimum value
JZ START_MACHINE       ; Jump to label if result of
                       ; subtraction was 0
```

JG/JNLE INSTRUCTION—Jump if greater/Jump if not less than nor equal

These two mnemonics represent the same instruction. The terms "greater" and "less" are used to refer to the relationship of two signed numbers. Greater means more positive. The number 00000111 is greater than the number 11101010, because the second number is negative. This instruction is usually used after a compare instruction. The instruction will cause a jump to a label given in the instruction if the zero flag is 0 and the carry flag is the same as the overflow flag. The destination label must be in the range of −128 bytes to +127 bytes from the address of the instruction after the JG/JNLE instruction. If the jump is not taken, execution simply goes on to the next instruction after the JG or JNLE instruction. JG/JNLE affects no flags.

EXAMPLES:

```
CMP BL, 39H            ; Compare by subtracting 39H
                       ; from BL
JG SHORT_LABEL         ; Jump to label if BL more
                       ; positive than 39H

CMP BL, 39H            ; Compare by subtracting
                       ; 39H from BL
JNLE SHORT_LABEL       ; Jump to label if BL not less
                       ; than nor equal to 39H
```

JGE/JNL INSTRUCTION—Jump if greater than or equal/Jump if not less than

These two mnemonics represent the same instruction. The terms "greater" and "less" are used to refer to the relationship of two signed numbers. Greater means more positive. The number 00000111 is greater than the number 11101010, because the second number is negative. This instruction is usually used after a compare instruction. The instruction will cause a jump to a label given in the instruction if the sign flag is equal to the overflow flag. The destination label must be in the range of −128 bytes to +127 bytes from the address of the instruction after the JGE/JNL instruction. If the jump is not taken, execution simply goes on to the next instruction after the JGE or JNL instruction. JGE/JNL affects no flags.

EXAMPLES:

```
CMP BL, 39H            ; Compare by subtracting 39H
                       ; from BL
JGE SHORT_LABEL        ; Jump to label if BL more
                       ; positive than 39H
                       ; or equal to 39H

CMP BL, 39H            ; Compare by subtracting 39H
                       ; from BL
JNL SHORT_LABEL        ; Jump to label if BL not less
                       ; than 39H
```

JL/JNGE INSTRUCTION—Jump if less than/Jump if not greater than nor equal

These two mnemonics represent the same instruction. The terms "greater" and "less" are used to refer to the relationship of two signed numbers. Greater means more positive. The number 00000111 is greater than the number 11101010, because the second number is negative. This instruction is usually used after a compare instruction. The instruction will cause a jump to a label given in the instruction if the sign flag is not equal to the overflow flag. The destination label must be in the range of −128 bytes to +127 bytes from the address of the instruction after the JL/JNGE instruction. If the jump is not taken, execution simply goes on to the next instruction after the JL or JNGE instruction. JL/JNGE affects no flags.

EXAMPLES:

```
CMP BL, 39H            ; Compare by subtracting 39H
                       ; from BL
JL SHORT_LABEL         ; Jump to label if BL more
                       ; negative than 39H

CMP BL, 39H            ; Compare by subtracting 39H
                       ; from BL
JNGE SHORT_LABEL       ; Jump to label if BL not more
                       ; positive than 39H
                       ; or BL not equal to 39H
```

JLE/JNG INSTRUCTIONS—Jump if less than or equal/Jump if not greater

These two mnemonics represent the same instruction. The terms "greater" and "less" are used to refer to the relationship of two signed numbers. Greater means more positive. The number 00000111 is greater than the number 11101010, because the second number is negative. This instruction is usually used after a compare instruction. The instruction will cause a jump to a label given in the instruction if the zero flag is set, or if the sign flag is not equal to the overflow flag. The destination label must be in the range of −128 bytes to +127 bytes from the address of the instruction after the JLE/JNG instruction. If the jump is not taken, execution simply goes on to the next instruction after the JLE/JNG instruction. JLE/JNG affects no flags.

EXAMPLES:

```
CMP BL, 39H          ; Compare by subtracting 39H
                     ; from BL
JLE SHORT_LABEL      ; Jump to label if BL more
                     ; negative than 39H or
                     ; equal to 39H
CMP BL, 39H          ; Compare by subtracting 39H
                     ; from BL
JNG SHORT_LABEL      ; Jump to label if BL not more
                     ; positive than 39H
```

JMP INSTRUCTION—Unconditional jump to specified destination

This instruction will always cause the 8086 to fetch its next instruction from the location specified in the instruction rather than from the next location after the JMP instruction. If the destination is in the same code segment as the JMP instruction, then only the instruction pointer will be changed to get to the destination location. This is referred to as a near jump. If the destination for the jump instruction is in a segment with a name different from that containing the JMP instruction, then both the instruction pointer and the code segment register contents will be changed to get to the destination location. This is referred to as a far jump. The JMP instruction affects no flags. Refer to a section in Chapter 4 for a detailed discussion of the different forms of the unconditional JMP instruction.

EXAMPLES:

JMP CONTINUE ; Fetch next instruction from address at label CONTINUE. If label is in same segment, an offset coded as part of the instruction will be added to the instruction pointer to produce the new fetch address. If the label is in another segment then IP and CS will be replaced with values coded in as part of the instruction. This type of jump is referred to as direct because the displacement of the destination or the destination itself is specified directly in the instruction.

JMP BX ; Replace the contents of IP with the contents of BX. BX must first be loaded with the offset of the destination instruction in CS. This is a near jump. It is also referred to as an indirect jump because the new value for IP comes from a register rather than from the instruction itself as in a direct-type jump.

JMP WORD PTR [BX] ; Replace IP with a word from a memory location pointed to by BX in DS. This is an indirect near jump.

JMP DWORD PTR [SI] ; Replace IP with word pointed to by SI in DS. Replace CS with word pointed to by SI + 2 in DS. This is an indirect far jump.

JNA/JBE INSTRUCTION—Jump if not above/Jump if below or equal

Please refer to the discussion of this instruction under the heading JBE.

JNAE/JB INSTRUCTION—Jump if not above or equal/Jump if below

Please refer to the discussion of this instruction under the heading JB.

JNB/JNC/JAE INSTRUCTION—Jump if not below/Jump if no carry/Jump if above or equal

Please refer to the discussion of this instruction under the heading JAE.

JNBE/JA INSTRUCTION—Jump if not below or equal/Jump if above

Please refer to the discussion of this instruction under the heading JA.

JNC/JNB/JAE INSTRUCTION—Jump if not carry/Jump if not below/Jump if above or equal

Please refer to the discussion of this instruction under the heading JAE.

JNE/JNZ INSTRUCTION—Jump if not equal/Jump if not zero

These two mnemonics represent the same instruction. If the zero flag is 0, then this instruction will cause execution to jump to a label given in the instruction. If the zero flag is 1, then execution will simply go on to the next instruction after JNE or JNZ. The destination label for the JNE/JNZ instruction must be in the range of −128 to +127 bytes from the address of the instruction after the JNE/JNZ instruction. JNE/JNZ affects no flags.

EXAMPLES:

```
AGAIN:  IN AL, 0F8H       ; Read data value from port
        CMP AL, 72        ; Compare by subtracting
                          ; 72 from AL
        JNE AGAIN         ; Jump to label AGAIN if
                          ; AL not equal 72
        IN AL, 0F9H       ; Read next port when
                          ; AL = 72
        MOV BX, 2734H     ; Load BX as counter
```

```
NEXT 1: ADD AX, 0002H   ; Add count factor to AX
        DEC BX          ;
        JNZ NEXT 1:     ; Repeat until BX = 0
```

JNG/JLE INSTRUCTION—Jump if not greater/Jump if less than or equal

Please refer to the discussion of this instruction under the heading JLE.

JNGE/JL INSTRUCTION—Jump if not greater than nor equal/Jump if less than

Please refer to the discussion of this instruction under the heading JL.

JNL/JGE INSTRUCTION—Jump if not less than/Jump if greater than or equal

Please refer to the discussion of this instruction of this instruction under the heading JGE.

JNLE/JG INSTRUCTION—Jump if not less than nor equal to/Jump if greater than

Please refer to the discussion of this instruction under the heading JG.

JNO INSTRUCTION—Jump if no overflow

The overflow flag will be set if the result of some signed arithmetic operation is too large to fit in the destination register or memory location. The JNO instruction will cause the 8086 to jump to a destination given in the instruction if the overflow flag is not set. The destination must be in the range of −128 bytes to +127 bytes from the address of the instruction after the JNO instruction. If the overflow flag is set, execution will simply continue with the next instruction after JNO. JNO affects no flags.

EXAMPLE:

```
      ADD AL, BL   ; Add signed bytes in AL and BL
      JNO DONE     ; Process done if no overflow
      MOV AL, 00H  ; Else load error code in AL
DONE: OUT 24H,AL   ; Send result to display
```

JNP/JPO INSTRUCTION—Jump if no parity/Jump if parity odd

If the number of 1's left in a data byte after an instruction which affects the parity flag is odd, then the parity flag will be 0. The JNP/JPO instruction will cause execution to jump to a specified destination address if the parity flag is 0. The destination address must be in the range of −128 bytes to +127 bytes from the address of the instruction after the JNP/JNO instruction. If the parity flag is set, execution will simply continue on to the instruction after the JNO/JPO instruction. The JNO/JPO instruction affects no flags.

EXAMPLE:

```
IN AL, 0F8H        ; Read ASCII character
                   ; from UART
OR AL, AL          ; Set flags
JPO ERR_MESSAGE    ; Even parity expected, send
                   ; error message if parity
                   ; found odd
```

JNS INSTRUCTION—Jump if not signed (Jump if positive)

This instruction will cause execution to jump to a specified destination if the sign flag is 0. Since a 0 in the sign flag indicates a positive signed number, you can think of this instruction as saying "jump if positive." If the sign flag is set, indicating a negative signed result, execution will simply go on to the next instruction after JNS. The destination for the jump must be in the range of −128 bytes to +127 bytes from the address of the instruction after the JNS. JNS affects no flags.

EXAMPLE:

```
DEC AL     ; Decrement counter
JNS REDO   ; Jump to label REDO if counter has not
           ; decremented to FFH
```

JNZ/JNE INSTRUCTION—Jump if not zero/Jump if not equal

Please refer to the discussion of this instruction under the heading JNE.

JO INSTRUCTION—Jump if overflow

The overflow flag will be set if the result of some signed arithmetic operation is too large to fit in the destination register or memory location. The JO instruction will cause the 8086 to jump to a destination given in the instruction if the overflow flag is set. The destination must be in the range of −128 bytes to +127 bytes from the address of the instruction after the JO instruction. If the overflow flag is not set, execution will simply continue with the next instruction after JO. JO affects no flags.

EXAMPLE:

```
ADD AL, BL     ; Add signed bytes in AL and BL
JO ERROR       ; Jump to label ERROR if overflow
               ; from add
MOV SUM, AL    ; Else put result in memory location
               ; named SUM
```

JP/JPE INSTRUCTION—Jump if parity/Jump if parity even

If the number of 1's left in a data word after an instruction which affects the parity flag is even, then the parity flag will be set. The JP/JPE instruction will cause execution to jump to a specified destination address if the parity flag is set. If the parity flag is 0, execution will simply continue on to the instruction after the JP/JPE instruction. The destination address must be in the range of −128 bytes to +127 bytes from the address of the instruction after the JP/JPE instruction. The JP/JPE instruction affects no flags.

EXAMPLE:

```
IN AL, 0F8H         ; Read ASCII character
                    ; from UART
OR AL, AL           ; Set flags
JPE ERR_MESSAGE     ; Odd parity expected, send
                    ; error message if parity
                    ; found even
```

JPE/JP INSTRUCTION—Jump if parity even/Jump if parity

Please refer to the discussion of this instruction under the heading JP.

JPO/JNP INSTRUCTION—Jump if parity odd/Jump if no parity

Please refer to the discussion of this instruction under the heading JNP.

JS INSTRUCTION—Jump if signed (Jump if negative)

This instruction will cause execution to jump to a specified destination if the sign flag is set. Since a 1 in the sign flag indicates a negative signed number, you can think of this instruction as saying "jump if negative" or "jump if minus." If the sign flag is 0, indicating a positive signed result, execution will simply go on to the next instruction after JS. The destination for the jump must be in the range of −128 bytes to +127 bytes from the address of the instruction after the JS. JS affects no flags.

EXAMPLE:

```
ADD BL, DH     ; Add signed byte in DH to signed
               ; byte in BL
JS TOO_COLD    ; Jump to label TOO_COLD if result
               ; of addition is negative number
```

JZ/JE INSTRUCTION—Jump if zero/Jump if equal

Please refer to the discussion of this instruction under the heading JE.

LAHF INSTRUCTION—Copy low byte of flag register to AH

The lower byte of the 8086 flag register is the same as the flag byte for the 8085. LAHF copies these 8085 equivalent flags to the AH register. They can then be pushed on the stack along with AL by a PUSH AX instruction. An LAHF instruction followed by a PUSH AX instruction has the same effect as the 8085 PUSH PSW instruction. The LAHF instruction was included in the 8086 instruction set so that the 8085 PUSH PSW instruction could easily be simulated on an 8086. LAHF changes no flags.

EXAMPLE: LAHF

LDS INSTRUCTION—Load register and DS with words from memory—LDS register, memory address of first word

This instruction copies a word from two memory locations into the register specified in the instruction. It then copies a word from the next two memory locations into the DS register. LDS is useful for pointing SI and DS at the start of a string before using one of the string instructions. LDS affects no flags.

EXAMPLES:

LDS BX, [4326] ; Copy contents of memory at displacement 4326H in DS to BL, contents of 4327H to BH. Copy contents at displacement of 4328H and 4329H in DS to DS register.

LDS SI, STRING_POINTER ; Copy contents of memory at displacements STRING_POINTER and STRING_POINTER+1 in DS to SI register. Copy contents of memory at displacements STRING_POINTER+2 and STRING_POINTER+3 in DS to DS register. DS:SI now points at start of desired string.

LEA INSTRUCTION—Load Effective Address—LEA register, source

This instruction determines the offset of the variable or memory location named as the source and puts this offset in the indicated 16-bit register. LEA changes no flags.

EXAMPLES:

```
LEA BX, PRICES          ; Load BX with offset of
                        ; PRICES in DS
LEA BP, SS:STACK_TOP    ; Load BP with offset of
                        ; STACK_TOP in SS
LEA CX, [BX][DI]        ; Load CX with
                        ; EA = (BX) + (DI)
```

A program example will better show the context in which this instruction is used. If you look at the program in Figure 4-15c you will see that PRICES is an array of bytes in a segment called ARRAYS_HERE. The program gets a byte from this array with the instruction LEA BX, PRICES. This will load the displacement of the first element of PRICES directly into BX. The instruction MOV AL, [BX] can then be used to bring an element from the array into AL. After one element in the array is processed, BX is incremented to point to the next element.

LEAVE—80186/80188 ONLY—Leave procedure

The LEAVE instruction is used at the end of an assembly language procedure which is intended to be called from a high level language program such as Pascal or C. An ENTER instruction at the start of such a procedure sets aside stack space for variables used in the procedure and in subprocedures. The main function of the LEAVE instruction is to increment SP and BP up over this reserved space so they have the values they had before the ENTER instruction. In other words, LEAVE restores SP and BP to the values they had at the start of the procedure. A RET instruction can then be used to return execution to the calling program. Leave affects no flags. Refer to the Intel literature for further explanation of this instruction if you need it.

LES INSTRUCTION—Load register and ES with words from memory—LES register, memory address of first word

This instruction loads new values into the specified register and into the ES register from four successive memory locations. The word from the first two memory locations is copied into the specified register, and the word from the next two memory locations is copied into the ES register. LES can be used to point DI and ES at the start of a string before a string instruction is executed. LES affects no flags.

EXAMPLES:

LES BX, [789AH] ; Contents of memory at displacements 789AH and 789BH in DS copied to BX. Contents of memory at displacements 789CH and 789DH in DS copied to ES register.

LES DI, [BX] ; Copy contents of memory at offset [BX] and offset [BX + 1] in DS to DI register copy contents of memory at offsets [BX + 2] and [BX + 3] to ES register.

LOCK INSTRUCTION—Assert bus lock signal

Many microcomputer systems contain several microprocessors. Each microprocesor has its own local buses and memory. The individual microprocessors are connected together by a system bus so that each can access system resources such as disk drives or memory. Each microprocessor only takes control of the system bus when it needs to access some system resource. The LOCK prefix allows a microprocessor to make sure that another processor does not take control of the system bus while it is in the middle of a critical instruction which uses the system bus. The LOCK prefix is put in front of the critical instruction. When an instruction with a LOCK prefix executes, the 8086 will assert its bus lock signal output. This signal is connected to an external bus controller device which then prevents any other processor from taking over the system bus. LOCK affects no flags.

EXAMPLE:

LOCK XCHG SEMAPHORE, AL ; The XCHG instruction requires two bus accesses. The LOCK prefix prevents another processor from taking control of the system bus between the two accesses.

LODS/LODSB/LODSW INSTRUCTION—Load string byte into AL or Load string word into AX

This instruction copies a byte from a string location pointed to by SI to AL, or a word from a string location pointed to by SI to AX. If the direction flag is cleared (0), SI will automatically be incremented to point to the next element of the string. For a string of bytes SI will be incremented by one. For a string of words SI will be incremented by two. If the direction flag, DF, is set (1), SI will be automatically decremented to point to the next string element. For a byte string SI will be decremented by one, and for a word string SI will be decremented by two. LODS affects no flags.

EXAMPLE:

```
CLD   ; Clear direction flag so SI is autoincremented
MOV SI, OFFSET SOURCE_STRING   ; Point SI at
                               ; string
LODS SOURCE_STRING
```

NOTE: Assembler uses name of string to determine whether string is of type byte or of type word. Instead of using the string name to do this, you can use the mnemonic LODSB to tell the assembler that the string is of type byte or the mnemonic LODSW to tell the assembler that the string is of type word.

LOOP INSTRUCTION—Loop to specified label until CX = 0

This instruction is used to repeat a series of instructions some number of times. The number of times the instruction sequence is to be repeated is loaded into the count register. Each time the LOOP instruction executes, CX is automatically decremented by one. If CX is not 0, execution will jump to a destination specified by a label in the instruction. If CX = 0 after the autodecrement, execution will simply go on to the next instruction after LOOP. The destination address for the jump must be in the range of −128 bytes to +127 bytes from the address of the instruction after the LOOP instruction. LOOP affects no flags. See Chapter 4 for further discussion and examples of the LOOP instruction.

EXAMPLE:

```
      MOV BX, OFFSET PRICES  ; Point BX at
                             ; first element in array
      MOV CX, 40     ; Load CX with number of
                     ; elements in array
NEXT: MOV AL, [BX]   ; Get element from array
      ADD AL, 07H    ; Add correction factor
      DAA            ; Decimal adjust result
      MOV [BX],AL    ; Put result back in array
      LOOP NEXT      ; Repeat until all elements
                     ; adjusted
```

LOOPE/LOOPZ INSTRUCTION—Loop while CX not = 0 and ZF = 1

LOOPE and LOOPZ are two mnemonics for the same instruction. This instruction is used to repeat a group of instructions some number of times or until the zero flag becomes 0. The number of times the instruction sequence is to be repeated is loaded into the count register, CX. Each time the LOOP instruction executes, CX is automatically decremented by one. If CX is not 0 and the zero flag is set, execution will jump to a destination specified by a label in the instruction. If CX is 0 after the autodecrement or if the zero flag is not set, execution will simply go on to the next instruction after LOOPE/LOOPZ. In other words, the two ways to exit the loop are CX = 0 or ZF = 0. The destination address for the jump must be in the range of −128 bytes to +127 bytes from the address of the instruction after the LOOPE/LOOPZ instruction. LOOPE/LOOPZ affects no flags. See Chapter 4 for further discussion and examples of the LOOPE/LOOPZ instruction.

EXAMPLE:

```
      MOV BX, OFFSET ARRAY  ; Point BX at start
      DEC BX                ; of array
      MOV CX, 100           ; Put number of
                            ; array elements
                            ; in CX
NEXT: INC BX                ; Point to next
                            ; element in array
      CMP [BX], OFFH        ; Compare array
                            ; element with
                            ; FFH
      LOOPE NEXT
```

NOTE: The next element is checked if the element equals FFH and the element was not the last one in the array. If CX = 0 and ZF = 1 on exit, all elements were equal to FFH. If CX is not equal to 0 on exit from loop, then BX points to next element after first byte that was not FFH.

LOOPNE/LOOPNZ INSTRUCTION—Loop while CX is not 0 and ZF = 0

LOOPNE and LOOPNZ are two mnemonics for the same instruction. This instruction is used to repeat a group of instructions some number of times or until the zero flag becomes a 1. The number of times the instruction sequence is to be repeated is loaded into the count register, CX. Each time the LOOPNE/LOOPNZ instruction executes, CX is automatically decremented by one. If CX is not 0 and the zero flag is not set, execution will jump to a destination specified by a label in the instruction. If CX is 1 after the autodecrement or if the zero flag is set, execution will simply go on to the next instruction after LOOPNE/LOOPNZ. In other words, the two ways to exit the loop are CX = 0 or ZF = 1. The destination address for the jump must be in the range of −128 bytes to +127 bytes from the address of the instruction after the LOOPNE/LOOPNZ instruction. LOOPNE/LOOPNZ affects no flags. See Chapter 4 for further discussion and examples of the LOOPNE/LOOPNZ instruction.

EXAMPLE:

```
      MOV BX, OFFSET ARRAY  ; Point BX at start
      DEC BX                ; of array
      MOV CX, 100           ; Put number of
                            ; array elements
                            ; in CX
NEXT: INC BX                ; Point to next
                            ; element in array
      CMP [BX], ODH         ; Compare array
      LOOPNE NEXT           ; element with
                            ; ODH
```

NOTE: The next element in the array is checked if the element was not equal ODH and the element was not the last one in array. If CX = 0 and ZF = 0 on exit, ODH was not found in the array. If CX does not equal 0 on exit from loop, then BX points to next element after the first element containing ODH.

LOOPNZ/LOOPNE INSTRUCTION—Loop while CX is not 0 and ZF = 0

Please see the discussion of this instruction under the heading LOOPNE.

LOOPZ/LOOPE INSTRUCTION—Loop while CX is not 0 and ZF = 1

Please see the discussion of this instruction under the heading LOOPE.

MOV INSTRUCTION—MOV destination, source

The MOV instructions transfer a word or byte of data from some source to a destination. The destination can be a register or a memory location. The source can be a register, a memory location, or an immediate number. The source and destination in an instruction cannot both be memory locations. The source and destination in a MOV instruction must both be of type byte, or they must both be of type word. MOV instructions do not affect any flags.

EXAMPLES:

```
MOV CX, 037AH    ; Put the immediate number
                 ; 037AH in CX
MOV BL, [437AH]  ; Copy byte from offset 437AH
                 ; in DS to BL
MOV AX, BX       ; Copy contents of register BX
                 ; to AX
MOV DL, [BX]     ; Copy byte from memory address
                 ; to DL
                 ; Offset of memory address
                 ; in DS is in BX
MOV DS, BX       ; Copy word from BX to data
                 ; segment register
```

MOV RESULTS [BP], AX ; Copy AX to two memory locations, AL to first location, AH to second. EA of first memory location is the sum of displacement represented by RESULTS and contents of BP. Physical address = EA + SS.

MOV CS:RESULTS [BP], AX ; Same as the above instruction, but Physical Address = EA + CS because of segment override prefix, CS.

MOVS/MOVSB/MOVSW INSTRUCTION—Move string byte or string word—MOVS destination, source

This instruction copies a byte or word from a location in the data segment to a location in the extra segment. The offset of the source byte or word in the data segment must be in the SI register. The offset of the destination in the extra segment must be contained in the DI register. For multiple byte or multiple word moves the number of elements to be moved is put in the CX register so that it can function as a counter. After the byte or word is moved SI and DI are automatically adjusted to point to the next source and the next destination. If the direction flag is 0, then SI and DI will be incremented by 1 after a byte move and they will be incremented by 2 after a word move. If the DF is a 1, then SI and DI will be decremented by 1 after a byte move and they will be decremented by 2 after a word move. MOVS affects no flags.

When using the MOVS instruction you must in some way tell the assembler whether you want to move a string as bytes or as words. There are two ways to do this. The first way is to indicate the names of the source and the destination strings in the instruction as, for example, MOVS STRING_DUMP, STRING_CREATE. The assembler will code the instruction for a byte move if STRING_DUMP and STRING_CREATE were declared with a DB. It will code the instruction for a word move if they were declared with a DW. Note that this reference to the source and destination strings does not load SI and DI. This must be done with separate instructions. The second way to tell the assembler whether to code the instruction for a byte or word move is to add a "B" or a "W" to the MOVS mnemonic. MOVSB, for example, says move a string as bytes. MOVSW says move a string as words.

EXAMPLE:

```
    CLD   ; Clear Direction Flag to autoincrement SI
          ; and DI
    MOV AX, 00H
    MOV DS, AX        ; Initialize data segment
                      ; register to 0
    MOV ES, AX        ; Initialize extra segment
                      ; register to 0
    MOV SI, 2000H     ; Load offset of start of source
                      ; string into SI
    MOV DI, 2400H     ; Load offset of start of
                      ; destination into DI
    MOV CX, 04H       ; Load length of string in CX
                      ; as counter
REP MOVSB             ; Decrement CX and
                      ; MOVSB until CX = 0
```

After move SI will be one greater than offset of last byte in source string. DI will be one greater than offset of last byte of destination string. CX will be 0.

MUL INSTRUCTION—Multiply unsigned bytes or words—MUL source

This instruction multiplies an unsigned byte from some source times an unsigned byte in the AL register, or an unsigned word from some source times an unsigned word in the AX register. The source can be a register or a memory location specified by any one of the 24 addressing modes shown in Figure 3-8. When a byte is multiplied by the contents of AL, the result (product) is put in AX. A 16-bit destination is required because the result of multiplying an 8-bit number by an 8-bit number can be as large as 16 bits. The most-significant byte of the result is put in AH and the least-significant byte of the result is put in AL. When a word is multiplied by the contents of AX, the product can be as large as 32 bits. The most-significant word of the result is put in the DX register, and the least-significant word of the result is put in the AX register. If the most-significant byte of a 16-bit result or the most-significant word of a 32-bit result is 0, the CF and the OF will both be 0's. Checking these flags then allows you to detect and perhaps discard unnecessary leading 0's in a result. The AF, PF, SF, and ZF are undefined after a MUL instruction.

If you want to multiply a byte by a word, you must first move the byte to a word location such as an extended register and fill the upper byte of the word with all 0's.

NOTE: You cannot use the 8086 Convert Byte to Word instruction, CBW, to do this. The CBW instruction fills the upper byte of AX with copies of the MSB of AL. If the number in AL is 80H or greater, CBW will fill the upper half of AX with 1's instead of with 0's. Once you get the byte converted correctly to a word with 0's in the upper byte, you can then do a word times word multiply. The 32-bit result will be in DX and AX.

EXAMPLES:

```
MUL BH    ; AL times BH, result in AX

MUL CX    ; AX times CX, result high word in DX,
          ; low word in AX

MUL BYTE PTR [BX]   ; AL times byte in DS pointed
                    ; to by [BX]
```

MUL CONVERSION_FACTOR [BX] ; Multiply AL times byte at effective address CONVERSION_FACTOR [BX] if it was declared as type byte with DB. Multiply AX times word at effective address CONVERSION_FACTOR [BX], if it was declared as type word with DW.

```
; Example showing a byte multiplied by a word
MOV AX, MULTIPLICAND_16   ; Load 16-bit
                          ; multiplicand in AX
MOV CL, MULTIPLIER_8      ; Load 8-bit multiplier
                          ; in CL
MOV CH, 00H               ; Set upper byte of CX
                          ; to all 0's
MUL CX                    ; AX times CX, 32-bit
                          ; result in DX and AX
```

NEG INSTRUCTION—Form 2's complement—NEG destination

This instruction replaces the number in a destination with the 2's complement of that number. The destination can be a register or a memory location specified by any one of the addressing modes shown in Figure 3-8. This instruction forms the 2's complement by subtracting the original word or byte in the indicated destination from zero. You may want to try this with a couple of numbers to convince yourself that it gives the same result as the invert each bit and add one algorithm. As shown in some examples below, the NEG instruction is useful for changing the sign of a signed word or byte. An attempt to NEG a byte location containing −128 or a word location containing −32,768 will produce no change in the destination contents because the maximum positive signed number in 8 bits is +127 and the maximum positive signed number in 16 bits is +32,767. The OF will be set to indicate that the operation could not be done. The NEG instruction updates the AF, CF, SF, PF, ZF, and OF.

EXAMPLES (CODING)

```
NEG AL    ; Replace number in AL with its
          ; 2's complement

NEG BX    ; Replace word in BX with its
          ; 2's complement

NEG BYTE PTR [BX]   ; Replace byte at offset [BX] in
                    ; DS with its 2's complement

NEG WORD PTR [BP]   ; Replace word at offset [BP] in
                    ; SS with its 2's complement
```

Note that the BYTE PTR and WORD PTR directives are required in the last two examples to tell the assembler whether to code the instruction for a byte operation or a word operation. The [BP] reference by itself does not indicate the type of the operand.

NOP INSTRUCTION—Perform no operation

This instruction simply uses up three clock cycles and increments the instruction pointer to point to the next instruction. NOP affects no flags. The NOP instruction can be used to increase the delay of a delay loop as shown in Figure 4-20. It can also be used to hold a place in a program for instructions that will be added later.

NOT INSTRUCTION—Invert each bit of operand—NOT destination

The NOT instruction inverts each bit (forms the 1's complement of) the byte or word at the specified destination. The destination can be a register or a memory location specified by any one of the addressing modes shown in Figure 3-8. No flags are affected by the NOT instruction.

EXAMPLES:

```
NOT BX              ; Complement contents of BX
                    ; register

NOT BYTE PTR [BX]   ; Complement memory byte at
                    ; offset [BX] in
                    ; data segment
```

OR INSTRUCTION—Logically OR corresponding bits of two operands—OR destination, source

This instruction ORs each bit in a source byte or word with the corresponding bit in a destination byte or word. The result is put in the specified destination. The contents of the specified source will not be changed. The result for each bit will follow the truth table for a two-input OR gate. In other words, a bit in the destination will become a one if that bit is a one in the source operand OR that bit is a one in the original destination operand. Therefore, a bit in the destination operand can be set to a one by simply ORing that bit with a one in the same bit of the source operand. A bit ORed with zero is not changed.

The source operand can be an immediate number, the contents of a register, or the contents of a memory location specified by one of the 24 addressing modes shown in Figure 3-8. The destination can be a register or a memory location. The source and the destination cannot both be memory locations in the same instruction. The CF and OF are both zero after OR. The PF, SF, and ZF are updated by the OR instruction. AF is undefined after OR. Note that PF only has meaning for an 8-bit operand.

EXAMPLES (CODING):

```
OR AH, CL      ; CL ORed with AH, result in AH.
               ; CL not changed.
OR BP, SI      ; SI ORed with BP, result in BP.
               ; SI not changed.
OR SI, BP      ; BP ORed with SI, result in SI.
               ; BP not changed.
OR BL, 80H     ; BL ORed with immediate 80H.
               ; Set MSB of BL to a 1.
```

OR CX, TABLE[BX][SI] ; CX ORed with word from effective address TABLE[BX][SI] in data segment. Word in memory is not changed.

EXAMPLE (NUMERICAL):

```
               ; CX = 00111101 10100101
OR CX, 0FF00H  ; OR CX with immediate FF00H.
               ; Result in CX = 11111111 10100101
               ; note upper byte now all 1's
               ; CF = 0 OF = 0 PF = 1 SF = 1
               ; ZF = 0
```

OUT INSTRUCTION—Output a byte or word to a port—OUT port, accumulator AL or AX

The OUT instruction copies a byte from AL or a word from AX to the specified port. The OUT instruction has two possible forms, fixed-port and variable port.

For the fixed-port form the 8-bit port address is specified directly in the instruction. With this form any one of 256 possible ports can be addressed.

EXAMPLES:

```
OUT 3BH, AL    ; Copy the contents of AL to port 3BH
OUT 2CH, AX    ; Copy the contents of AX to port 2CH
```

For the variable port form of the OUT instruction, the contents of AL or AX will be copied to the port at an address contained in DX. Since DX is a 16 bit register, the port address contained there can be any number between 0000H and FFFFH. Therefore, up to 65,536 possible ports can be addressed in this mode. The DX register must always be loaded with the desired port address before this form of the OUT instruction is used. The advantage of the variable port form of addressing is described within the discussion of the IN instruction.

EXAMPLES:

```
MOV DX, 0FFF8H ; Load desired port address in DX
OUT DX, AL     ; Copy contents of AL to port FFF8H
OUT DX, AX     ; Copy contents of AX to port FFF8H
```

NOTE: The OUT instruction does not affect any flags.

OUTS/OUTSB/OUTSW—80186/80188 ONLY—Output String to Port—OUTS port, source string

OUTS copies a byte or a word from a string location pointed to by SI to a port whose address is contained in DX. The address of the port to be copied to must be put in DX before this instruction executes. If the direction flag is cleared when this instruction executes, SI will automatically be incremented by one for a byte operation, and incremented by two for a word operation after the data is copied to the port. If the direction flag is set, SI will automatically be decremented by one for a byte operation and decremented by two for a word operation after data is copied to the port. When used with the REP prefix or as part of a loop the OUTS instruction can output a block of data directly from a series of memory locations to a port without having the data go through AL or AX as it does with the regular OUT instruction.

When using the OUTS instruction you must in some way tell the assembler whether you want to output bytes or output words. There are two ways to do this. The first is to use the name of the source string in the instruction statement as in the statement OUTS DX, BUFFER. The assembler will code the instruction for a byte output if BUFFER was declared with a DB and it will code the instruction for a word output if BUFFER was declared with a DW. The second way to tell the assembler whether to code the instruction for a byte or for a word input is to add a B or a W to the basic instruction mnemonic. OUTSB DX, for example, tells the assembler to code the instruction for copying a byte from a string location pointed to by SI to a port whose address is in DX. SI normally points to a location in the data segment, but you can use a segment override prefix to point it to a location in some other segment. OUTS affects no flags.

EXAMPLE:

```
      CLD   ; Clear direction flag, autoincrement DI
      MOV DI, OFFSET BUFFER   ; Point DI at
                              ; output buffer
      MOV DX, 0FFF8H   ; Load DX with
                       ; port address
      MOV CX, 100      ; Load number of
                       ; bytes to be output in CX
REP OUTSB DX           ; Copy bytes to port
                       ; until buffer empty
```

POP INSTRUCTION—POP destination

The POP instruction copies a word from the stack location pointed to by the stack pointer to a destination specified in the instruction. The destination can be a general-purpose register, a segment register, or a memory location. The data in the stack is not changed. After the word is copied to the specified destination, the stack pointer is automatically incremented by 2 to point to the next word on the stack. No flags are affected by the POP instruction.

NOTE: POP CS is illegal.

EXAMPLES:

```
POP DX          ; Copy a word from top of stack to
                ; DX, SP = SP + 2

POP DS          ; Copy a word from the top of the
                ; stack to DS.
                ; Increment SP by 2

POP TABLE [BX]  ; Copy a word from the top of the
                ; stack to memory in DS
                ; with EA = TABLE + [BX]
```

POPA INSTRUCTION—80186/80188 ONLY—Pop all Registers from Stack

POPA restores four pointer and index registers and four general-purpose registers that were saved on the stack with a PUSHA instruction. After the saved value is copied from the stack to the appropriate register, the stack pointer is incremented by two. Register contents are popped off the stack in the following order: DI, SI, BP, SP, BX, DX, CX, AX. POPA affects no flags.

POPF INSTRUCTION—Pop word from top of stack to flag register

This instruction copies a word from the two memory locations at the top of the stack to the flag register and increments the stack pointer by two. The stack segment register is not affected. All flags are affected.

PUSH INSTRUCTION—PUSH source

The PUSH instruction decrements the stack pointer by two and copies a word from some source to the location in the stack segment where the stack pointer then points. The source of the word can be a general-purpose register, a segment register, or memory. The stack segment register and the stack pointer must be initialized before this instruction can be used. PUSH can be used to save data on the stack so it will not be destroyed by a procedure. It can also be used to put data on the stack so that a procedure can access it there as needed. No flags are affected by this instruction. Refer to Chapter 5 for further discussion of the stack and the PUSH instruction.

EXAMPLES:

```
PUSH BX          ; Decrement SP by 2, copy BX to stack

PUSH DS          ; Decrement SP by 2, copy DS to stack

PUSH AL          ; Illegal, must push a word

PUSH TABLE [BX]  ; Decrement SP by 2, copy word
                 ; from memory
                 ; at EA = TABLE + [BX]
                 ; in DS to stack
```

PUSH INSTRUCTION—80186/80188 ONLY—PUSH Immediate

This version of the PUSH instruction allows you to store an immediate byte or word given in the instruction on the stack. If an immediate byte is specified, the byte will be sign-extended to a word before the PUSH is done, because all stack pushes are word operations. The stack pointer will be decremented by two before the word is pushed on the stack. PUSH affects no flags.

EXAMPLE:

```
PUSH 437AH  ; Decrement SP by 2 and write
            ; number 437AH on stack.
```

PUSHA INSTRUCTION—80186/80188 ONLY—Push all Registers on Stack

PUSHA copies the contents of the four general-purpose registers and the contents of four pointer and index registers to memory locations in the stack. The stack pointer is decremented by two before each register is pushed on the stack. The registers are pushed in the following order: AX, CX, DX, BX, SP, BP, SI, DI. The value pushed for SP is the value that SP had before AX was pushed. PUSHA affects no flags. PUSHA can be used at the start of a procedure to save the contents of these eight registers. A POPA instruction at the end of the procedure can be used to restore the original contents of the registers before returning to the program which called the procedure.

PUSHF INSTRUCTION—Push flag register on the stack

This instruction decrements the stack pointer by two and copies the word in the flag register to the memory location(s) pointed to by the stack pointer. The stack segment register is not affected. No flags are changed.

EXAMPLE: PUSHF

RCL INSTRUCTION—Rotate operand around to the left through CF—RCL destination, count

This instruction rotates all of the bits in a specified word or byte some number of bit positions to the left. The operation is circular because the MSB of the operand is rotated into the carry flag and the bit in the carry flag is rotated around into the LSB of the operand. See the diagram below.

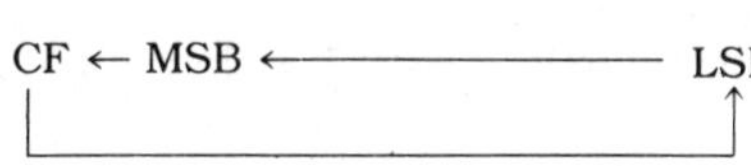

The "C" in the middle of the mnemonic should help you remember that CF is in the rotated loop and it should help distinguish this instruction from the ROL instruction. For multibit rotates CF will contain the bit most recently rotated out of the MSB.

The destination operand can be in a register or in a memory location specified by any one of the 24 addressing modes shown in Figure 3-8. If you want to rotate the operand one bit position, you can specify this by putting a 1 in the count position of the instruction. To rotate more than one bit position, load the desired number in the CL register and put "CL" in the count position of the instruction.

NOTE: The 80186 and the 80188 allow you to specify a rotate of up to 32 bit positions with either an immediate number in the instruction or with a number in CL.

RCL affects only the CF and OF. After RCL the CF will contain the bit most recently rotated out of the MSB. The OF will be a 1 after a single bit RCL if the MSB was changed by the rotate. OF is undefined after a multibit rotate.

The RCL instruction is a handy way to move the CF into the LSB of a register or memory location to save it after addition.

EXAMPLES (CODING)

```
RCL DX, 1          ; Word in DX 1 bit left, MSB to CF
                   ; CF to LSB

MOV CL, 4          ; Load number of bit positions to
                   ; rotate in CL

RCL SUM[BX], CL    ; Rotate byte or word at effective
                   ; address SUM[BX] 4 bits left.
                   ; Original bit 4 now in CF,
                   ; original CF now in bit 3.
```

EXAMPLES (NUMERICAL)

```
             ; CF = 0 BH = 10110011
RCL BH, 1    ; Byte in BH 1 bit left. MSB to CF,
             ; CF to LSB.

             ; CF = 1 BH = 01100110
             ; OF = 1 because MSB changed

             ; CF = 1 AX =00011111 10101001

MOV CL, 2    ; Load CL for rotating two bit positions.
RCL AX, CL   ; Rotate AX two bit positions.
             ; CF = 0  AX = 01111110 10100110
             ; OF undefined
```

RCR INSTRUCTION—Rotate operand around to the right through CF—RCR destination, count

This instruction rotates all of the bits in a specified word or byte some number of bit positions to the right. The operation is circular because the LSB of the operand is rotated into the carry flag and the bit in the carry flag is rotated around into the MSB of the operand. See the diagram below.

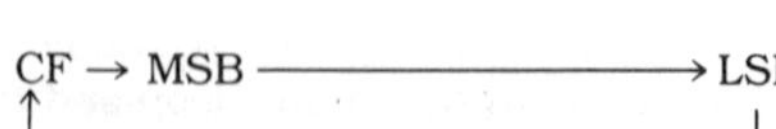

The "C" in the middle of the mnemonic should help you remember that CF is in the rotated loop and it should help distinguish this instruction from the ROR instruction. For multibit rotates CF will contain the bit most recently rotated out of the LSB.

The destination operand can be in a register or in a memory location specified by any one of the 24 addressing modes shown in Figure 3-8. If you want to rotate the operand one bit position, you can specify this by putting a 1 in the count position of the instruction. To rotate more than one bit position, load the desired number in the CL register and put "CL" in the count position of the instruction.

NOTE: The 80186 and the 80188 allow you to specify a rotate of up to 32 bit positions with either an immediate number in the instruction or with a number in CL.

RCR affects only the CF and OF. After RCR the CF will contain the bit most recently rotated out of the MSB. The OF will be a 1 after a single bit RCR if the MSB was changed by the rotate. OF will be undefined after multibit rotates.

EXAMPLES (CODING)

```
RCR BX, 1             ; Word in BX right 1 bit.
                      ; CF to MSB, LSB to CF

MOV CL, 04H           ; Load CL for rotating four bit
                      ; positions

RCR BYTE PTR [BX]     ; Rotate byte at offset [BX] in
                      ; data segment four bit
                      ; positions right.

                      ; CF = original bit 3.
                      ; Bit 4 = original CF.
```

EXAMPLES (NUMERICAL)

```
              ; CF = 1 BL = 00111000

RCR BL, 1     ; Byte in BL one bit position right.
              ; LSB to CF.

              ; CF = 0,BL = 10011100
              ; OF = 1 because MSB changed to 1

              ; CF = 0
              ; WORD PTR [BX] = 01011110 00001111

MOV CL, 02H                 ; Load CL for rotate two
                            ; bit positions

RCR WORD PTR [BX], CL       ; Rotate word at offset [BX]
                            ; in data segment 2
                            ; bits right.

                            ; CF = original bit 1.
                            ; Bit 14 = original CF

                            ; WORD PTR [BX] =
                            ; 10010111 10000011
```

REP/REPE/REPZ/REPNE/REPNZ—(Prefix) Repeat string instruction until specified conditions exist

REP is a prefix which is written before one of the string instructions. It will cause the CX register to be decremented and the string instruction to be repeated until CX = 0. The instruction REP MOVSB, for example, will continue to move string bytes until the length of the string which was loaded into CX is counted down to zero.

REPE and REPZ are two mnemonics for the same prefix. They stand for Repeat if Equal and Repeat if Zero, respectively. You can use whichever prefix makes the operation clearer to you in a given program. REPE or REPZ is often used with the Compare String instruction or with the Scan String instruction. REPE or REPZ will cause the string instruction to be repeated as long as the compared bytes or words are equal (ZF = 1), AND CX is not yet counted down to zero. In other words there are two conditions that will stop the repetition: CX = 0 or string bytes or words NOT equal.

EXAMPLE:

REPE CMPSB ; Compare string bytes until end of string or until string bytes not equal. See the discussion of the CMPS instruction for a more detailed example of the use of REPE.

REPNE and REPNZ are also two mnemonics for the same prefix. They stand for Repeat if Not Equal and Repeat if Not Zero, respectively. REPNE or REPNZ is often used with the Scan String instruction. REPNE or REPNZ will cause the string instruction to be repeated until the compared bytes or words are not equal (ZF = 0), OR until CX = 0 (end of string).

EXAMPLE:

REPNE SCASW ; Scan a string of words until a word in the string matches the word in AX or until all of the string has been scanned. See the discussion of SCAS for a more detailed example of the use of this prefix.

The string instruction used with the prefix determines which flags are affected. See the individual instructions for this information. Also see Chapter 5 for further examples of the REP instruction with string instructions.

NOTE: Interrupts should be disabled when multiple prefixes are used, such as LOCK, segment override, and REP with string instructions on the 8086/8088. This is because, during an interrupt response, the 8086 can only remember the prefix just before the string instruction. The 80186/80188 will correctly remember all of the prefixes and start up correctly after an interrupt.

RET INSTRUCTION—Return execution from procedure to calling program

The RET instruction will return execution from a procedure to the next instruction after the CALL instruction in the calling program. If the procedure is a near procedure (in the same code segment as the CALL instruction), then the return will be done by replacing the instruction pointer with a word from the top of the stack. The word from the top of the stack is the offset of the next instruction after the CALL. This offset was pushed on the stack as part of the operation of the CALL instruction. The stack pointer will be incremented by two as the return address is popped off the stack.

If the procedure is a far procedure (in a different code segment from the CALL instruction which calls it), then the instruction pointer will be replaced by the word at the top of the stack. This word is the offset part of the return address put there by the CALL instruction. The stack pointer will then be incremented by two. The code segment register is then replaced with a word from the new top of the stack. This word is the segment part of the return address that was pushed on the stack by a far call operation. After the code segment word is popped off the stack the stack pointer is again incremented by two.

A RET instruction can be followed by a number, for example, RET 6. In this case the stack pointer will be incremented by an additional six addresses after the IP or the IP and CS are popped off the stack. This form is used to increment the stack pointer up over parameters passed to the procedure on the stack.

The RET instruction affects no flags.

Please refer to Chapter 5 for further discussion of the CALL and RET instructions.

ROL INSTRUCTION—Rotate all bits of operand left, MSB to LSB—ROL destination, count

This instruction rotates all the bits in a specified word or byte to the left some number of bit positions. The operation can be thought of as circular, because the data bit rotated out of the MSB is circled back into the LSB. The data bit rotated out of the MSB is also copied to the CF during ROL. In the case of multiple bit rotates, CF will contain a copy of the bit most recently moved out of the MSB. See the diagram below.

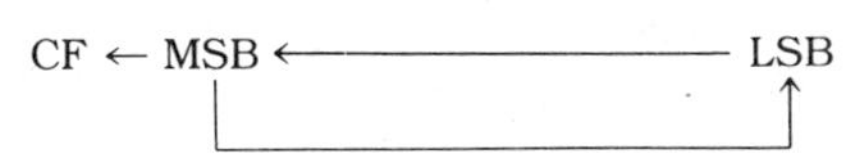

The destination operand can be in a register or in a memory location specified by any one of the 24 addressing modes shown in Figure 3-8. If you want to rotate the operand one bit position, you can specify this by putting a 1 in the count position of the instruction. To rotate more than one bit position, load the desired number in the CL register and put "CL" in the count position of the instruction.

NOTE: The 80186 and the 80188 allow you to specify a rotate of up to 32 bit positions with either an immediate number in the instruction or with a number in CL.

ROL affects only the CF and OF. After ROL the CF will contain the bit most recently rotated out of the MSB. The OF will be a 1 after ROL if the MSB was changed by the rotate.

The ROL instruction can be used to swap the nibbles in a byte or to swap the bytes in a word. It can also be used to rotate a bit into CF where it can be checked and acted upon by the conditional jump instructions, JC (Jump if Carry) or JNC (Jump if No Carry).

EXAMPLES (CODING)

```
ROL AX,1         ; Word in AX one bit position left,
                 ; MSB to LSB and CF

MOV CL, 04H      ; Load number of bits to rotate in CL

ROL BL, CL       ; Rotate BL four bit positions
                 ; (swap nibbles)

ROL FACTOR[BX], 1  ; MSB of word or byte at
                   ; effective address

                   ; FACTOR[BX]
                   ; in data segment one bit

                   ; position left into CF
JC ERROR           ; Jump if CF = 1 to error routine
```

EXAMPLES (NUMERICAL)

```
                 ; CF = 0   BH = 10101110
ROL BH, 1        ; CF = 1   BH = 01011101   OF = 1

                 ; CL = 8   Set for 8-bit rotate
                 ; BX = 01011100 11010011
ROL BX, CL       ; Rotate BX 8 times left (swap bytes)
                 ; CF = 0, BX = 11010011 01011100
                 ; OF = ?
```

ROR INSTRUCTION—Rotate all bits of operand right, LSB to MSB—ROR destination, count

This instruction rotates all of the bits of the specified word or byte some number of bit positions to the right. The operation is described as a rotate rather than a shift because the bit moved out of the LSB is rotated around into the MSB. To help visualize the operation, think of the operand as a loop with the LSB connected around to the MSB. The data bit moved out of the LSB is also copied to the CF during ROR. See diagram below. In the case of multiple bit rotates the CF will contain a copy of the bit most recently moved out of the LSB.

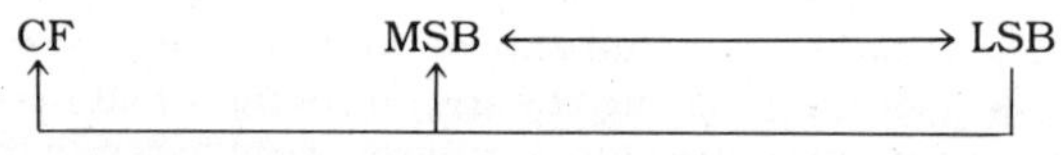

The destination operand can be in a register or in a memory location specified by any one of the 24 addressing modes shown in Figure 3-8. If you want to rotate the operand one bit position, you can specify this by putting a 1 in the count position of the instruction. To rotate more than one bit position, load the desired number in the CL register, and put "CL" in the count position of the instruction.

NOTE: The 80186 and the 80188 allow you to specify a rotate of up to 32 bit positions with either an immediate number or with a number in CL.

ROR affects only the CF and the AF. After ROR the CF will contain the bit most recently rotated out of the LSB. The OF will be a 1 after ROR if the MSB is changed by the rotate.

The ROR instruction can be used to swap the nibbles in a byte or to swap the bytes in a word. It can also be used to rotate a bit into the CF where it can be checked and acted upon by the conditional jump instructions, JC (Jump if Carry) or JNC (Jump if No Carry).

EXAMPLES (CODING)

```
ROR BL, 1        ; Rotate all bits in BL right one bit
                 ; position

                 ; LSB to MSB and to CF

MOV CL, 08H      ; Load number of bit positions to be
                 ; rotated in CL

ROR WORD PTR [BX], CL   ; Rotate word at offset

                 ; [BX] in data segment eight bit

                 ; positions right (swaps bytes in word)
```

EXAMPLES (NUMERICAL)

```
                 ; CF = 0, BX = 00111011 01110101
ROR BX,1         ; Rotate all bits of BX one bit
                 ; position right

                 ; CF = 1, BX = 10011101 10111010

                 ; CF = 0, AL = 10110011 OF = 1
MOV CL, 04H      ; Load CL for rotate four bit positions
ROR AL, CL       ; Rotate all bits of AL around four
                 ; bit positions right

                 ; CF = 0   AL = 00111011   OF = ?
```

SAHF INSTRUCTION—Copy AH register to low byte of flag register

The lower byte of the 8086 flag register corresponds exactly to the 8085 flag byte. SAHF replaces this 8085 equivalent flag byte with a byte from the AH register. SAHF is used with the POP AX instruction to simulate the 8085 POP PSW instruction. As described under the heading LAHF, an 8085 PUSH PSW instruction will be translated to an LAHF—PUSH AX sequence to run on an 8086. An 8085 POP PSW instruction will be translated to a POP AX— SAHF sequence to run on an 8086. SAHF changes the flags in the lower byte of the flag register.

EXAMPLE: SAHF

SAL/SHL INSTRUCTION—Shift operand bits left, put zero in LSB(s)—SAL/SHL destination, count

SAL and SHL are two mnemonics for the same instruction. This instruction shifts each bit in the specified destination some number of bit positions to the left. As a bit is shifted out of the LSB position, a 0 is put in the LSB position. The MSB will be shifted into the CF. In the case of multiple bit shifts, CF will contain the bit most recently shifted in from the MSB. Bits shifted into CF previously will be lost. See diagram below.

CF ← MSB ←———————— LSB ← 0

The destination operand can be a byte or a word. It can be in a register or in a memory location specified by any one of the 24 addressing modes shown in Figure 3-8.

If the desired number of shifts is one, this can be specified by putting a 1 in the count position of the instruction. For shifts of more than 1 bit position the desired number of shifts is loaded into the CL register and CL is put in the count position of the instruction. The advantage of the CL way is that the number of shifts can be dynamically calculated as the program executes.

NOTE: The 80186 and the 80188 allow you to specify a shift of up to 32 bit positions with either an immediate number in the instruction or with a number in CL.

The flags are affected as follows: CF contains the bit most recently shifted in from MSB. For a count of one OF will be 1 if the CF and the current MSB are not the same. For multiple bit shifts, the OF is undefined. The SF and the ZF will be updated to reflect the condition of the destination. The PF will only have meaning if the destination is AL. AF is undefined.

The SAL or SHL instruction can also be used to multiply an unsigned binary number by a power of two. Shifting a binary number one bit position to the left and putting a 0 in the LSB multiplies the number by 2. Shifting the number two bit positions multiplies it by 4. Shifting the number three bit positions multiplies it by 8, etc. For this specific type of multiply the SAL method is faster than using MUL, but you must make sure that the result does not become too large for the destination.

EXAMPLES (CODING)

```
SAL BX, 1        ; Shift word in BX 1 bit position left,
                 ; 0 in LSB

MOV CL, 02H      ; Load desired number of shifts in CL
SAL BP, CL       ; Shift word in BP left (CL) bit
                 ; positions, 0's in
                 ; 2 least-significant bits

SAL BYTE PTR [BX], 1   ; Shift byte at offset [BX] in
                       ; data segment one bit
                       ; position left, 0 in LSB

; Example of SAL instruction's use to help pack BCD
IN AL, COUNTER_DIGIT   ; Unpacked BCD from
                       ; counter to AL

MOV CL, 04H      ; Set count for four bit positions

SAL AL, CL       ; Shift BCD to upper nibble, 0's in
; lower nibble. Ready to OR another BCD digit into
; lower nibble of AL.
```

EXAMPLE (NUMERICAL)

```
           ; CF = 0,  BX = 11100101 11010011
SAL BX,1   ; Shift BX register contents one bit left
           ; CF = 1, BX = 11001011 10100110
           ; OF = 0  PF = ?  SF = 1  ZF = 0
```

SAR INSTRUCTION—Shift operand bits right, new MSB = old MSB—SAR destination, count

This instruction shifts each bit in the specified destination some number of bit positions to the right. As a bit is shifted out of the MSB position, a copy of the old MSB is put in the MSB position. In other words the sign bit is copied into the MSB. The LSB will be shifted into CF. In the case of multiple bit shifts, CF will contain the bit most recently shifted in from the LSB. Bits shifted into CF previously will be lost. See diagram below.

MSB → MSB ————————→ LSB → CF

The destination operand can be a byte or a word. It can be in a register or in a memory location specified by any one of the 24 addressing modes shown in Figure 3-8.

If the desired number of shifts is one, this can be specified by putting a 1 in the count position of the instruction. For shifts of more than one bit position the desired number of shifts is loaded into the CL register and CL is put in the count position of the instruction.

NOTE: The 80186 and the 80188 allow you to specify a shift of up to 32 bit positions with either an immediate number in the instruction or with a number in CL.

The flags are affected as follows: CF contains the bit most recently shifted in from the LSB. For a count of one the OF will be a 1 if the two MSBs are not the same. After a multibit SAR the OF will be 0. The SF and the ZF will be updated to show the condition of the destination. PF will only have meaning for an 8-bit destination. AF will be undefined after SAR.

The SAR instruction can be used to divide a signed byte or word by a power of two. Shifting a binary number right one bit position divides it by 2. Shifting a binary number right two bit positions divides it by 4. Shifting it three positions divides it by 8, etc. For unsigned numbers a 0 is put in the MSB after the old MSB is shifted right. (See discussion of SHR instruction.) For signed binary numbers the sign bit must be copied into the new MSB as the old sign bit is shifted right. This is necessary to retain the correct sign in the result. SAR shifts the operand right and copies the sign bit into the MSB as required for this operation. Using SAR to do a divide by 2, however, gives slightly different results than using the IDIV instruction to do the same job. IDIV always truncates a signed result toward zero. For example, an IDIV of 7 by 2 gives 3 and an IDIV of −7 by 2 gives −3. SAR always truncates a result in a downward direction. Using SAR to divide 7 by 2 gives 3, but using SAR to divide −7 by 2 gives −3.

EXAMPLES (CODING)

```
SAR DI, 1    ; Shift word in DI one bit position right,
             ; new MSB = old MSB

MOV CL, 02H   ; Load desired number of shifts in CL
SAR WORD PTR [BP], CL   ; Shift word at offset [BP]
; in stack segment right two bit positions. Two MSBs
; are now copies of original MSB.
```

EXAMPLES (NUMERICAL)

```
           ; AL = 00011101 = +29 decimal CF = 0
SAR AL, 1  ; Shift signed byte in AL right
           ; to divide by 2.
           ; AL = 00001110 = +14 decimal. CF = 1
           ; OF = 0  PF = 0  SF = 0  ZF = 0
           ; BH = 11110011 = −13 decimal
SAR BH, 1  ; Shift signed byte in BH right to
           ; divide by 2
           ; BH = 11111001 = −7 decimal CF = 1
           ; OF = 0  PF = 1  SF = 1  ZF = 0
```

SBB INSTRUCTION—Subtract with borrow—SBB destination, source

SUB INSTRUCTION—Subtract—SUB destination, source

These instructions subtract the number in the indicated source from the number in the indicated destination and put the result in the indicated destination. For subtraction the carry flag, CF, functions as a borrow flag. The carry flag will be set after a subtraction if the number in the specified source is larger than the number in the specified destination. In other words, the carry/borrow flag will be set if a borrow was required to do the subtraction. The Subtract instruction, SUB, subtracts just the contents of the specified source from the contents of the specified destination. The Subtract with Borrow instruction, SBB, subtracts the contents of the source and the contents of the CF from the indicated destination. The source may be an immediate number, a register, or a memory location specified by any of the 24 addressing modes shown in Figure 3-8. The destination may be a register or a memory location. The source and the destination cannot both be memory locations in the same instruction. The source and the destination must both be of type byte or they must both be of type word. If you want to subtract a byte from a word, you must first move the byte to a word location such as an extended register and fill the upper byte of the word with 0's. The AF, CF, OF, PF, SF, and ZF are updated by the SUB instruction.

EXAMPLES (CODING):

```
SUB CX, BX        ; Subtract contents of BX from
                  ; contents of CX Leave result in CX
SBB CH, AL        ; Subtract contents of AL and
                  ; contents of CF
                  ; from contents of CH. Result in CH.
SUB AX, 3427H     ; Subtract immediate number
                  ; 3427H from AX
SBB BX, [3427H]   ; Subtract word at displacement
                  ; 3427H in DS
                  ; and contents of CF from BX.
SUB PRICES [BX], 04H  ; Subtract 04 from byte at
effective address PRICES[BX] if PRICES declared with
DB. Subtract 04 from word at effective address
PRICES[BX] if PRICES declared with DW.
SBB CX, TABLE[BX] ; Subtract word from effective
                  ; address TABLE[BX] and
                  ; status of CF from CX
SBB TABLE[BX], CX ; Subtract CX and status of CF
                  ; from word in memory at
                  ; effective address TABLE[BX]
```

EXAMPLES(NUMERICAL):

```
; Example subtracting unsigned numbers
             ; CL = 10011100 = 156 decimal
             ; BH = 00110111 = 55 decimal
SUB CL, BH   ; Subtract BH from CL. Result:
             ; CL = 01100101 = 101 decimal
             ; CF = 0, AF = 0, PF = 1, OF = 0,
             ; SF = 0, ZF = 0

; Two examples subtracting Signed numbers
             ;        (a)
             ; CL = 00101110 =+46 decimal
             ; BH = 01001010 =+74 decimal
SUB CL, BH   ; Results:
             ; CL = 11100100 = −28 decimal
             ; CF = 1, borrow required
             ; AF = 0  PF = 1  ZF = 0
             ; SF = 1, result negative
             ; OF = 0, magnitude of result
             ; fits in 7 bits

             ;        (b)
             ; CL = 10100010 = −95 decimal
             ; BH = 01001100 = +76 decimal
SUB CL, BH   ; Results:
             ; CL = 01010101 = +85 decimal
             ; CF = 1, borrow required
             ; AF = 0  PF = 1  ZF = 0
             ; SF = 0, result positive !
             ; OF = 1, invalid result
```

The overflow flag being set indicates that the magnitude of the expected result, −171 decimal, is too large to fit in the 7 bits used for the magnitude in an 8-bit signed number. If the Interrupt on Overflow instruction, INTO, has been executed, this error will cause the 8086 to perform a software interrupt procedure. Part of this procedure is a user-written subroutine to handle the error.

NOTES: The SBB instruction allows you to subtract two multibyte numbers because any borrow produced by subtracting less-significant bytes is included in the result when the SBB instruction executes. Although the examples above were for 8-bit numbers to save space, the principles are the same for 16-bit numbers. For 16-bit signed numbers, however, the SF is a copy of bit 15, and the least-significant 15 bits of the number are used to represent the magnitude. Also, the PF and the AF only function for the lower 8 bits.

SCAS/SCASB/SCASW INSTRUCTION—Scan string byte or a string word

SCAS compares a string byte with a byte in AL or a string word with word in AX. The instruction affects the flags, but it does not change either the operand in AL(AX) or the operand in the string. The string to be scanned must be in the extra segment and DI must contain the offset of the byte or the word to be compared. After SCAS executes, DI will be automatically incremented or decremented to point to the next element in the string. For byte strings DI will be incremented or decremented by one, and for word strings DI will be incremented or decremented by two. If the direction flag is cleared (0), then DI will be incremented after SCAS. If the direction flag is set (1), then DI will be decremented after SCAS. SCAS affects the AF, CF, OF, PF, SF, and ZF. This instruction is often used with a repeat prefix to find the first occurrence of a specified byte or word in a string.

EXAMPLE:

```
; Scan a text string of 80 characters for a carriage
; return
        MOV AL, 0DH   ; Byte to be scanned for into AL
MOV DI, OFFSET TEXT_STRING   ; Offset of string
                             ; to DI

        MOV CX, 80    ; CX used as element counter
        CLD           ; Clear DF so DI autoincrements
REPNE SCAS TEXT_STRING   ; Compare byte in
                         ; string with byte in AL
```

NOTE: Scanning is repeated while the bytes are not equal and it is not end of the string. If a carriage return 0DH is found, ZF = 1 and DI will point at the next byte after the carriage return in string. If a carriage return is not found then CX = 0 and ZF = 0. The assembler uses the name of the string to determine whether the string is of type byte or type word. Instead of using the name you can tell the assembler directly the type of the string by using the mnemonic SCASB for a byte string and SCASW for a word string.

SHL/SAL INSTRUCTION—Shift operand bits left, put zero in LSB(s)—SHL/SAL destination, count

SAL and SHL are two mnemonics for the same instruction. Please refer to the discussions of this instruction under the heading SAL/SHL.

SHR INSTRUCTION—Shift operand bits right, put zero in MSB(s)—SHR destination, count

This instruction shifts each bit in the specified destination some number of bit positions to the right. As a bit is shifted right out of the MSB position, a 0 is put in its place. The bit shifted out of the LSB position goes to the CF. In the case of a multiple bit shift, CF will contain the bit most recently shifted in from the LSB. Bits shifted into CF previously will be lost. See diagram below.

0 → MSB ——————————→ LSB → CF

The destination operand can be a byte or a word. It can be in a register or in a memory location specified by any one of the 24 addressing modes shown in Figure 3-8.

If the desired number of shifts is one, this can be specified by putting a 1 in the count position of the instruction. For shifts of more than one bit position the desired number of shifts is loaded into the CL register, and CL is put in the count position of the instruction.

NOTE: The 80186 and 80188 allow you to specify a shift of up to 32 bit positions with either an immediate number in the instruction or a number in CL.

The flags are affected by SHR as follows: CF contains the bit most recently shifted in from the LSB. For a count of one, OF will be a 1 if the two MSBs are not both 0's. For multiple bit shifts, OF is meaningless. The SF and ZF will be updated to show the condition of the destination. PF will only have meaning for the lower eight bits of destination. AF is undefined.

The SHR instruction can be used to divide an unsigned binary number by a power of two. Shifting a binary number one bit position to the right and putting 0 in the MSB divides the number by two. Shifting the number two bit positions divides it by 4. Shifting it three bit positions divides it by 8, etc. When an odd number is divided with this method, the result will be truncated. In other words, dividing 7 by 2 will give a result of 3.

EXAMPLES (CODING)

```
SHR BP, 1   ; Shift word in BP one bit position right,
            ; 0 in MSB

MOV CL, 03H           ; Load desired number of shifts
                      ; into CL

SHR BYTE PTR [BX]     ; Shift byte at offset [BX] in
                      ; data segment
                      ; 3 bits right.
                      ; 0's in 3 MSBs.

; Example of SHR Used to Help Unpack Two BCD
; Digits in AL to BH and BL

MOV BL, AL    ; Copy packed BCD to BL
AND BL, 0FH   ; Mask out upper nibble. Low BCD
              ; digit now in BL.

MOV CL, 04H   ; Load count for shift in CL
SHR AL, CL    ; Shift AL four bit positions right and
              ; put 0's in upper 4 bits
MOV BH, AL    ; Copy upper BCD nibble to BH
```

EXAMPLES (NUMERICAL)

```
          ; SI = 10010011 10101101   CF = 0
SHR SI,1  ; Shift contents of SI register right one bit
          ; position, SI = 01001001 11010110

          ; CF = 1   OF = 1   PF = ?   SF = 0
          ; ZF = 0
```

STC INSTRUCTION—Set the carry flag to a one

STC does not affect any other flags.

STD INSTRUCTION—Set the direction flag to a one

STD is used to set the direction flag to a one so that SI and/or DI will automatically be decremented to point to the next string element when one of the string instructions executes. If the direction flag is set, SI and/or DI will be decremented by one for byte strings, and by two for word strings. STD affects no other flags. Please refer to Chapter 5 and the discussion of the REP prefix in this chapter for examples of the use of this instruction.

STI INSTRUCTION—Set interrupt flag (IF)

Setting the interrupt flag to a one enables the INTR interrupt of the 8086 after the next instruction after STI. An interrupt signal on this input will then cause the 8086 to interrupt program execution, push the return address and flags on the stack, and execute an interrupt service procedure. An IRET instruction at the end of the interrupt service procedure will restore the flags which were pushed on the stack, and return execution to the interrupted program. Because STI does not allow the INTR input to be enabled until the instruction after STI executes, the instruction sequence STI—IRET will return execution to the interrupted program before another interrupt will be recognized. This is important to keep the stack from filling up in systems which have many different interrupts. STI does not affect any other flags.

Please refer to Chapter 8 for a thorough discussion of interrupts.

EXAMPLE:

```
STI    ; Enable interrupts after next instruction
IRET   ; Return from interrupt service procedure.
       ; Interrupts enabled after return.
```

STOS/STOSB/STOSW INSTRUCTION—Store byte or word in string

The STOS instruction copies a byte from AL or a word from AX to a memory location in the extra segment. In effect it replaces a string element with a byte from AL or a word from AX. DI is used to hold the offset of the memory location in the extra segment. After the copy, DI is automatically incremented or decremented to point to the next string element in memory. If the direction flag, DF, is cleared, then DI will automatically be incremented by one for a byte string or incremented by two for a word string. If the direction flag is set, DI will be automatically decremented by one for a byte string or decremented by two for a word string. STOS does not affect any flags.

EXAMPLES:

```
MOV DI, OFFSET TARGET_STRING   ; Point DI at
                               ; destination string
STOS TARGET_STRING             ; Assembler uses string
```
name to determine whether string is of type byte or type word. If byte string, then string byte replaced with contents of AL. If word string, then string word replaced with contents of AX.

```
MOV DI, OFFSET TARGET_STRING   ; Point DI at
                               ; destination string

STOSB                          ; "B" added to STOS
```
mnemonic directly tells assembler to replace byte in string with byte from AL. STOSW would tell assembler directly to replace a word in the string with a word from AX.

SUB INSTRUCTION—Subtract—SUB destination, source

Please refer to the discussion of this instruction under the heading SBB.

TEST INSTRUCTION—AND operands to update flags—TEST destination, source

This instruction ANDs the contents of a source byte or word with the contents of the specified destination word. Flags are updated, but neither operand is changed. The TEST instruction is often used to set flags before a conditional jump instruction.

The source operand can be an immediate number, the contents of a register, or the contents of a memory location specified by one of the 24 addressing modes shown in Figure 3-8. The destination operand can be from a register or from a memory location. The source and the destination cannot both be memory locations in an instruction. The CF and OF are both 0's after TEST. The PF, SF, and ZF will be updated to show the results of the ANDing. AF will be undefined. PF only has meaning for the lower eight bits of the destination.

EXAMPLES (CODING)

```
TEST AL, BH        ; AND BH with AL, no result.
                   ; Update PF, SF, ZF.

TEST CX, 0001H     ; AND CX with immediate number
                   ; 0001H no result stored.
                   ; Update PF, SF, ZF.

TEST BP, [BX][DI]  ; AND word at offset [BX][DI] in
                   ; data segment with word in BP,
                   ; no results stored.
                   ; Update PF, SF, and ZF.

; Example of a Polling Sequence Using TEST

AGAIN: IN AL, 2AH     ; Read port with strobe
                      ; connected to LSB
       TEST AL, 01H   ; AND immediate 01H with AL
                      ; to test if LSB of AL is 1 or 0.
                      ; ZF = 1 if LSB of result is 0.
                      ; No result stored.
       JZ AGAIN       ; Read port again if LSB = 0
```

EXAMPLES (NUMERICAL)

```
              ; AL = 01010001
TEST AL, 80H  ; AND immediate 80H with AL to test
              ; if MSB of AL is 1 or 0.
              ; ZF = 1 if MSB of AL =0.
```

```
; AL = 01010001 (Unchanged)
; PF = 0   SF = 0
; ZF = 1 because ANDing produced 00
```

WAIT INSTRUCTION—Wait for test signal or interrupt signal

When this instruction executes, the 8086 enters an idle condition where it is doing no processing. The 8086 will stay in this idle state until a signal is asserted on the 8086 TEST input pin, or until a valid interrupt signal is received on the INTR or the NMI interrupt input pins. If a valid interrupt occurs while the 8086 is in this idle state, the 8086 will return to the idle state after the interrupt service procedure executes. It returns to the idle state because the address of the WAIT instruction is the address pushed on the stack when the 8086 responds to the interrupt request. WAIT affects no flags. The WAIT instruction is used to synchronize the 8086 with external hardware such as the 8087 math coprocessor. In Chapter 11 we describe how this works.

XCHG INSTRUCTION—XCHG destination, source

The XCHG instruction exchanges the contents of a register with the contents of another register or the contents of a register with the contents of a memory location(s). The instruction cannot directly exchange the contents of two memory locations. A memory location can be specified as the source or as the destination by any of the 24 addressing modes summarized in Figure 3-8. The source and destination must both be words, or they must both be bytes. The segment registers cannot be used in this instruction. No flags are affected by this instruction.

EXAMPLES:

```
XCHG AX, DX            ; Exchange word in AX
                       ; with word in DX

XCHG BL, CH            ; Exchange byte in BL
                       ; with byte in CH

XCHG AL, PRICES [BX]   ; Exchange byte in AL with
                       ; byte in memory at
                       ; EA = PRICES [BX] in DS
```

XLAT/XLATB INSTRUCTION—Translate a byte in AL

The XLAT instruction replaces a byte in the AL register with a byte from a lookup table in memory. Before the XLAT instruction can be executed the lookup table containing the values for the new code must be put in memory, and the offset of the starting address of the lookup table must be loaded in BX. To point to the desired byte in the lookup table the XLAT instruction adds the byte in AL to the offset of the start of the table in BX. It then copies the byte from the address pointed to by (BX + AL) back into AL. XLAT changes no flags. The section "Converting One Keyboard Code to Another" in Chapter 9 should clarify the use of the XLAT instruction.

EXAMPLE:

```
; 8086 routine to convert ASCII code byte to EBCDIC
; equivalent
; ASCII code byte is in AL at start. EBCDIC code in
; AL at end.
    MOV BX, 2800H   ; Point BX at start of EBCDIC
                    ; table in DS

    XLAT            ; Replace ASCII in AL with
                    ; EBCDIC from table
```

The XLAT instruction can be used to convert any code of 8 bits or less to any other code of 8 bits or less.

XOR INSTRUCTION—Exclusive OR corresponding bits of two operands—XOR destination, source

This instruction exclusive ORs each bit in a source byte or word with the same number bit in a destination byte or word. The result replaces the contents of the specified destination. The contents of the specified source will not be changed. The result for each bit position will follow the truth table for a two-input exclusive OR gate. In other words, a bit in the destination will be set to a 1 if that bit in the source and that bit in the original destination were not the same. A bit exclusive-ORed with a 1 will be inverted. A bit exclusive-ORed with a 0 will not be changed. Because of this you can use the XOR instruction to selectively invert or not invert bits in an operand.

The source operand can be an immediate number, the contents of a register, or the contents of a memory location specified by any one of the addressing modes shown in Figure 3-8. The destination can be a register or a memory location. The source and destination cannot both be memory locations in the same instruction. The CF and OF are both 0 after XOR. The PF, SF, and ZF are updated. AF is undefined after XOR.

NOTE: PF only has meaning for an 8-bit operand.

EXAMPLES (CODING):

```
XOR CL, BH    ; Byte in BH exclusive ORed with byte
              ; in CL. Result in CL. BH not changed.

XOR BP, DI    ; Word in DI exclusive ORed with word
              ; in BP. Result in BP. DI not changed.

XOR WORD PTR [BX], 00FFH   ; Exclusive OR
              ; immediate number 00FFH
              ; with word at offset [BX] in data
              ; segment. Result in memory
              ; location [BX].
```

EXAMPLE (NUMERICAL)

```
              ; BX = 00111101 01101001
              ; CX = 00000000 11111111
```

```
XOR BX, CX   ; Exclusive OR CX with BX,
             ; result in BX
             ; BX = 00111101 10010110
             ; Note bits in lower byte are inverted
             ; CF = 0  OF = 0  PF = 1  SF = 0
             ; ZF = 0  AF = ?
```

ASSEMBLER DIRECTIVES

The words defined in this section are directions to the assembler, they are not instructions for the 8086. The assembler directives described here are those for the Intel 8086 macro assembler and the IBM macro assembler, MASM. If you are using some other assembler, consult the manual for it to find the corresponding directives.

ASSUME

The ASSUME directive is used to tell the assembler the name of the logical segment it should use for a specified segment. The statement ASSUME CS:CODE_HERE, for example tells the assembler that the instructions for a program are in a logical segment named CODE_HERE. The statement ASSUME DS:DATA_HERE, tells the assembler that for any program instruction which refers to the data segment it should use the logical segment called DATA_HERE. If, for example, the assembler reads the statement MOV AX, [BX] after it reads this ASSUME, it will know that the memory location referred to by [BX] is in the logical segment DATA_HERE. You must tell the assembler what to assume for any segment you use in a program. If you use a stack in your program you must tell the assembler the name of the logical segment you have set up as a stack with a statement such as ASSUME SS:STACK_HERE. For a program with string instructions which use DI, the assembler must be told what to assume for the extra segment with a statement such as ASSUME ES:STRING_DESTINATION. For further discussion of the ASSUME directive refer to the appropriate section of Chapter 3.

DB—Define Byte

The DB directive is used to define a byte-type variable, or to set aside one or more storage locations of type byte in memory. The statement CURRENT_TEMPERATURE DB 42H, for example, tells the assembler to reserve 1 byte of memory for a variable named CURRENT_TEMPERATURE and to put the value 42H in that memory location when the program is loaded into memory to be run. Refer to Chapter 3 for further discussion of the DB directive and to Chapter 4 for a discussion of how you can access variables named with a DB in your programs. Here are a few more examples of DB statements.

```
PRICES DB 49H, 98H, 29H   ; Declare array of 3
                          ; bytes named PRICES
                          ; and initialize 3 bytes
                          ; as shown

NAME_HERE DB 'THOMAS'   ; Declare array of
                        ; 6 bytes and initialize
                        ; with ASCII codes for
                        ; letters in THOMAS
```

TEMPERATURE_STORAGE DB 100 DUP(?) ; Set aside 100 bytes of storage in memory and give it the name TEMPERATURE_STORAGE, but leave the 100 bytes uninitialized. Program instructions will load values into these locations.

PRESSURE_STORAGE DB 20H DUP(0) ; Set aside 20H bytes of storage in memory, give it the name PRESSURE_STORAGE, and put 0 in all 20H locations.

DD—Define Doubleword

The DD directive is used to declare a variable of type doubleword or to reserve memory locations which can be accessed as type doubleword. The statement ARRAY_POINTER DD 25629261H, for example, will define a doubleword named ARRAY_POINTER, and initialize the doubleword with the specified value when the program is loaded into memory to be run. The low word, 9261H, will be put in memory at a lower address than the high word. A declaration of this type is often used with the LES or LDS instruction. The instruction LES DI, ARRAY_POINTER, for example, will copy the low word of this doubleword, 9261H, into the DI register, and the high word of the doubleword, 2562H, into the extra segment register.

DQ—Define Quadword

This directive is used to tell the assembler to declare a variable 4 words in length, or to reserve 4 words of storage in memory. The statement BIG_NUMBER DB 243598740192A92BH, for example, will declare a variable named BIG_NUMBER, and initialize the 4 words set aside with the specified number when the program is loaded into memory to be run. The statement STORAGE DQ 100 DUP(0) reserves 100 quad words of storage and initializes them all to 0 when the program is loaded into memory to be run.

DT—Define Ten bytes

DT is used to tell the assembler to define a variable which is 10 bytes in length, or to reserve 10 bytes of storage in memory. The statement PACKED_BCD DT 1234567890 will declare an array named PACKED_BCD which is 10 bytes in length. It will initialize the 10 bytes with the values 1234567890 when the program is loaded into memory to be run. This directive is often used when declaring data arrays for the 8087 math coprocessor discussed in Chapter 11. The statement RESULTS DT 20H DUP(0) will declare an array of 20H blocks of 10 bytes each and initialize all 320 bytes to 00 when the program is loaded into memory to be run.

DW—Define Word

The DW directive is used to tell the assembler to define a variable of type word, or to reserve storage locations of type word in memory. The statement MULTIPLIER DW 437AH, for example, declares a variable of type word named MULTIPLIER. The statement also tells the assembler that the variable MULTIPLIER should be initialized with the value 437AH when the program is loaded into memory to be run. Refer to Chapter 3 for further discussion of the DW directive and how you can access variables named with a DW in your programs. Here are a few more examples of DW statements.

THREE_LITTLE_WORDS DW 1234H, 3456H, 5678H ; Declare array of three words and initialize with specified values

STORAGE DW 100 DUP(0) ; Reserve an array of 100 words of memory and initialize all 100 words with 0000. Array is named STORAGE.

STORAGE DW 100 DUP(?) ; Reserve 100 words of storage in memory and give it the name STORAGE, but leave the words uninitialized.

END—End Program

The END directive is put after the last statement of a program to tell the assembler that this is the end of the program module. The assembler will ignore any statements after an END directive, so you should make sure to only use one END directive at the very end of your program module. A carriage return is required after the END directive.

ENDP—End Procedure

This directive is used along with the name of the procedure to indicate the end of a procedure to the assembler. This directive, together with the procedure directive, PROC, is used to "bracket" a procedure. Here's an example.

```
SQUARE_ROOT PROC    ; Start of procedure
                    ; Procedure instruction
                    ; statements
SQUARE_ROOT ENDP    ; End of procedure
```

Chapter 5 shows more examples and describes how procedures are written and called.

ENDS—End Segment

This directive is used with the name of a segment to indicate the end of that logical segment. ENDS is used with the SEGMENT directive to "bracket" a logical segment containing instructions or data. Here's an example.

```
CODE_HERE SEGMENT    ; Start of logical segment
                     ; containing code
                     ; Instruction statements
CODE_HERE ENDS       ; End of segment named
                     ; CODE_HERE
```

EQU—Equate

EQU is used to give a name to some value or symbol. Each time the assembler finds the given name in the program it will replace the name with the value or symbol you equated with that name. Suppose, for example, you write the statement CORRECTION_FACTOR EQU 03H at the start of your program and later in the program you write the instruction statement ADD AL, CORRECTION_FACTOR. When it codes this instruction statement, the assembler will code it as if you had written the instruction ADD AL, 03H. The advantage of using EQU in this manner is that if CORRECTION_FACTOR is used 27 times in a program, and you want to change the value, all you have to do is change the EQU statement and reassemble the program. The assembler will automatically put in the new value each time it finds the name CORRECTION_FACTOR. If you had used 03H instead of the EQU approach, then you would have to try to find and change all 27 instructions yourself. Here are some more examples.

```
DECIMAL_ADJUST EQU DAA    ; Create clearer
                          ; mnemonic for DAA
STRING_START EQU [BX]     ; Give name to [BX]
```

EVEN—Align on Even Memory Address

As the assembler assembles a section of data declarations or instruction statements, it uses a location counter to keep track of how many bytes it is from the start of a segment at any time. The EVEN directive tells the assembler to increment the location counter to the next even address if it is not already at an even address. The 8086 can read a word from memory in one bus cycle if the word is at an even address. If the word starts on an odd address, the 8086 must do two bus cycles to get the 2 bytes of the word. Therefore, a series of words can be much more quickly read if they are on even addresses. When EVEN is used in a data segment, the location counter will simply be incremented to the next even address if necessary. When EVEN is used in a code segment, the location counter will be incremented to the next even address if necessary. A NOP instruction will be inserted in the location incremented over. Here's an example which shows why you might want to use EVEN in a data segment.

```
DATA_HERE SEGMENT
; Declare array of 9 bytes. Location
; counter will point to 0009 after
; assembler reads next statement.
SALES_AVERAGES DB 9 DUP(?)
EVEN   ; Increment location counter to 000AH
INVENTORY_RECORDS DW 100 DUP(0)
   ; Array of 100 words starting
   ; on even address for quicker read.
DATA_HERE ENDS
```

EXTRN

The EXTRN directive is used to tell the assembler that the names or labels following the directive are in some other assembly module. For example, if you want to call a procedure which is in a program module assembled at a different time from that which contains the CALL instruction, you must tell the assembler that the procedure is external. The assembler will then put information in the object code file so that the linker can connect the two modules together. For a reference to an external named variable you must specify the type of the variable as in the statement EXTRN DIVISOR:WORD. For a reference to a label you must specify whether the label is near (in a code segment with the same name), or far (in a code segment with a different name). The statement EXTRN SMART_DIVIDE:FAR tells the assembler that SMART_DIVIDE is a label of type far in another assembly module. Names or labels referred to as external in one module must be declared public with the PUBLIC directive in the module where they are defined.

EXTRN statements should usually be bracketed with SEGMENT—ENDS directives which identify the segment in which the external name or label will be found. Here's an example of how to do this.

```
PROCEDURES_HERE SEGMENT

    EXTRN SMART_DIVIDE:FAR   ; Found in segment
                             ; PROCEDURES_HERE

PROCEDURES_HERE ENDS
```

Refer to Chapter 5 for a thorough discussion of the use of the EXTRN and the PUBLIC directives.

GROUP—Group-Related Segments

The GROUP directive is used to tell the assembler to group the logical segments named after the directive into one logical group segment. This allows the contents of all of the segments to be accessed from the same group segment base. The assembler sends a message to the linker and/or locator telling it to link the segments so that the segments are physically in the same 64 Kbyte segment. Here's an example of the GROUP directive: SMALL_SYSTEM GROUP CODE_HERE, DATA_HERE, STACK_HERE

An appropriate ASSUME statement to follow this would be: ASSUME CS:SMALL_SYSTEM, DS:SMALL_SYSTEM, SS:SMALL_SYSTEM

LABEL

As the assembler assembles a section of data declarations or instruction statements, it uses a location counter to keep track of how many bytes it is from the start of a segment at any time. The LABEL directive is used to give a name to the current value in the location counter. The LABEL directive must be followed by a term which specifies the type you want associated with that name. If the label is going to be used as the destination for a jump or a call, then the label must be specified as type near or as type far. If the label is going to be used to reference a data item, then the label must be specified as type byte, type word, or type double word. Here's how we use the LABEL directive for a jump address.

```
ENTRY_POINT LABEL FAR   ; Can jump to here from
                        ; another segment

NEXT: MOV AL, BL        ; Cannot do a far jump
                        ; directly to a label
                        ; with a colon
```

Here's how we use the LABEL directive for a data reference.

```
STACK_HERE SEGMENT STACK

    DW 100 DUP(0)       ; Set aside 100 words
                        ; for stack

STACK_TOP LABEL WORD    ; Give name to next
                        ; location after last
                        ; word in stack
STACK_HERE ENDS
```

To initialize stack pointer then, MOV SP, OFFSET STACK_TOP.

LENGTH—Not implemented in IBM MASM

LENGTH is an operator which tells the assembler to determine the number of elements in some named data item such as a string or array. When the assembler reads the statement MOV CX, LENGTH STRING1, for example, it will determine the number of elements in STRING1 and code this number in as part of the instruction. When the instruction executes then, the length of the string will be loaded into CX. If the string was declared as a string of bytes, LENGTH will produce the number of bytes in the string. If the string was declared as a word string, LENGTH will produce the number of words in the string.

NAME

The NAME directive is used to give specific names to each assembly module when programs consisting of several modules are written. The statement NAME PC_BOARD, for example, might be used to name an assembly module which contains the instructions for controlling a printed-circuit-board-making machine.

OFFSET

OFFSET is an operator which tells the assembler to determine the offset or displacement of a named data item (variable) from the start of the segment which contains it. This operator is usually used to load the offset of a variable into a register so that the variable can be accessed with one of the indexed addressing modes. When the assembler reads the statement MOV BX, OFFSET PRICES, for example, it will determine the offset of the variable PRICES from the start of the segment in which PRICES is defined and code this displacement in as part of the instruction. When the instruction executes, this computed displacement will be loaded into BX. An instruction such as ADD AL, [BX] can then be used to add a value from PRICES to AL.

ORG—Originate

As the assembler assembles a section of data declarations or instruction statements, it uses a location counter to keep track of how many bytes it is from the start of a segment at any time. The location counter is automatically set to 0000 when the assembler starts reading a segment. The ORG directive allows you to set the location counter to a desired value at any point in the program. The statement ORG 2000H tells the assembler to set the location counter to 2000H, for example.

A "$" is often used to symbolically represent the current value of the location counter. The $ actually represents the next available byte location where the assembler can put a data or code byte. The $ is often used in ORG statements to tell the assembler to make some change in the location counter relative to its current value. The statement ORG $ + 100 tells the assembler to increment the value of the location counter by 100 from its current value. A statement such as this might be used in a data segment to leave 100 bytes of space for future use.

PROC—Procedure

The PROC directive is used to identify the start of a procedure. The PROC directive follows a name you give the procedure. After the PROC directive the term NEAR or the term FAR is used to specify the type of the procedure. The statement SMART_DIVIDE PROC FAR, for example, identifies the start of a procedure named SMART_DIVIDE and tells the assembler that the procedure is far (in a segment with a different name from that which contains the instruction which calls the procedure). The PROC directive is used with the ENDP directive to "bracket" a procedure. Refer to the ENDP discussion for an example of this. Also refer to Chapter 5 for a thorough discussion of how procedures are written and called.

PTR—Pointer

The PTR operator is used to assign a specific type to a variable or to a label. It is necessary to do this in any instruction where the type of the operand is not clear. When the assembler reads the instruction INC [BX], for example, it will not know whether to increment the byte pointed to by BX or increment the word pointed to by BX. We use the PTR operator to clarify how we want the assembler to code the instruction. The statement INC BYTE PRT [BX] tells the assembler that we want to increment the byte pointed to by BX. The statement INC WORD PTR [BX] tells the assembler that we want to increment the word pointed to by BX. The PTR operator assigns the type specified before PTR to the variable specified after PTR.

The PTR operator can be used to override the declared type of a variable. Suppose, for example, that we have declared an array of words with the statement WORDS DW 437AH, B972H, 7C41H. Normally we would access the elements in this array as words. However, if we want to access a byte in the array, we can do it with an instruction such as MOV AL, BYTE PTR WORDS.

We also use the PTR operator to clarify our intentions when we use indirect jump instructions. The statement JMP [BX], for example, does not tell the assembler whether to code the instruction for a near jump or for a far jump. If we want to do a near jump we write the instruction as JMP WORD PTR [BX]. If we want to do a far jump we write the instruction as JMP DWORD PTR [BX]. Please refer to Chapter 3 for further discussion of the 8086 jump instructions.

PUBLIC

Large programs are usually written as several separate modules. Each module is individually assembled, tested and debugged. When all the modules are working correctly, their object code files are linked together to form the complete program. In order for the modules to link together correctly, any variable name or label referred to in other modules must be declared public in the module where it is defined. The PUBLIC directive is used to tell the assembler that a specified name or label will be accessed from other modules. An example is the statement PUBLIC DIVISOR, DIVIDEND which makes the two variables, DIVISOR and DIVIDEND, available to other assembly modules.

If an instruction in a module refers to a variable or label in another assembly module, the assembler must be told that it is external with the EXTRN directive. Refer to the discussion of the EXTRN directive to see how this is done.

SEGMENT

The SEGMENT directive is used to indicate the start of a logical segment. Preceding the SEGMENT directive is the name you want to give the segment. The statement CODE_HERE SEGMENT, for example, indicates to the assembler the start of a logical segment called CODE_HERE. The SEGMENT and ENDS directives are used to "bracket" a logical segment containing code or data. Refer to the ENDS directive for an example of how this is done.

Additional terms are often added to a SEGMENT directive statement to indicate some special way in which we want the assembler to treat the segment. The statement CODE_HERE SEGMENT WORD tells the assembler that we want this segment located on the next available word address when the segments are located and given absolute addresses. Without this WORD addition the segment will be located on the next available paragraph (16-byte) address which might waste as much as 15 bytes of memory. The statement CODE_HERE SEGMENT

PUBLIC tells the assembler that this segment will be put together with other segments named CODE_HERE from other assembly modules when the modules are linked together.

SHORT

The SHORT operator is used to tell the assembler that only a 1-byte displacement is needed to code a jump instruction. If the jump destination is after the jump instruction in the program, the assembler will automatically reserve 2 bytes for the displacement. Using the short operator saves 1 byte of memory by telling the assembler that it only needs to reserve 1 byte for this particular jump. In order for this to work the destination must be in the range of −128 bytes to +127 bytes from the address of the instruction after the jump. The statement JMP SHORT NEARBY_LABEL is an example of the use of SHORT.

TYPE

The TYPE operator tells the assembler to determine the type of a specified variable. The assembler actually determines the number of bytes in the type of the variable. For a byte-type variable the assembler will give a value of 1. For a word-type variable the assembler will give a value of 2, and for a doubleword-type variable it will give a value of 4. The TYPE operator can be used in an instruction such as ADD BX, TYPE WORD_ARRAY, where we want to increment BX to point to the next word in an array of words.

CHAPTER

7 8086 System Connections, Timing, and Troubleshooting

In Chapter 2 we showed that a microcomputer consists of a CPU, memory, and ports. We also showed in Chapter 2 that these parts are connected together by three major buses: the address bus, the control bus, and the data bus. For Chapters 3 through 6, however, we made little mention of the hardware of a microcomputer because we were mostly concerned in these chapters with how a microcomputer is programmed. In this chapter we come back to take a closer look at the hardware of a microcomputer.

OBJECTIVES

At the conclusion of this chapter you should be able to:

1. Draw a diagram showing how RAMs, ROMs, and ports are added to an 8086 CPU to make a simple microcomputer.
2. Describe how addresses sent out on the 8086 data bus are demultiplexed.
3. Describe the signal sequence on the buses as a simple 8086-based microcomputer fetches and executes an instruction.
4. Describe how address decoding circuitry gives a specific address to each device in a system and makes sure only one device is enabled at a time.
5. Calculate the required access time for a memory device or port to work correctly in an 8086 microcomputer system.
6. List a series of steps you might take to troubleshoot a malfunctioning microcomputer system that once worked.

8086 HARDWARE OVERVIEW

In previous chapters we worked with what is often called the *programmer's model* of the 8086. This model shows features, such as internal registers, number of address lines, and number of data lines, that we need in order to be able to program the device. Now we will look at the hardware model of the 8086 so that we can show how a microcomputer system is built around it. We will also discuss in this chapter the hardware connections for an 8088. A later chapter will show the hardware connections for the 80186 and 80286 microprocessors.

To get started, let's take a look at the pin diagram for the 8086 in Figure 7-1. Don't be overwhelmed by all of those pins with strange mnemonics next to them. You don't need to learn the detailed functions of all of these at once. We describe and show the use of these different pins throughout the next few chapters as needed. When you later need to refresh your memory of the function of a particular pin, consult the index to find the section where that particular pin or signal is described in detail. For reference, the complete data sheet showing all of the pin descriptions is shown in the appendix.

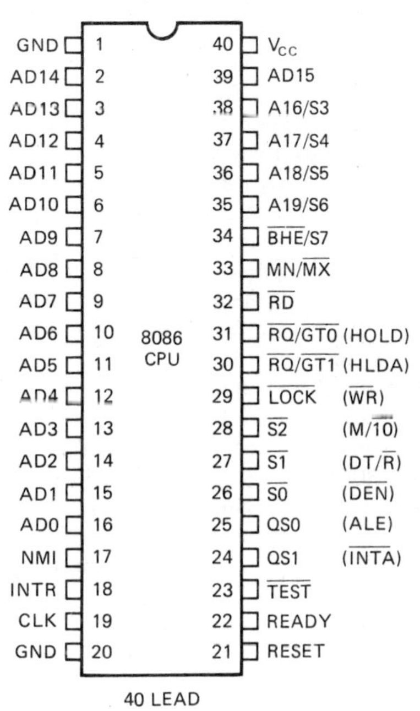

FIGURE 7-1 8086 pin diagram. *(Intel Corporation)*

Note first in Figure 7-1 that V_{CC} is on pin 40 and ground on pins 1 and 20. Next find the clock input labeled CLK on pin 19. An 8086 requires a clock signal from some external, crystal-controlled clock generator to synchronize internal operations in the processor. Different versions of the 8086 have maximum clock frequencies ranging from 5 MHz to 10 MHz.

Now look for the address and data bus lines. Remember from previous chapters that the 8086 has a 20-bit address bus and a 16-bit data bus. A look at Figure 7-1, however, does not immediately reveal these 36 lines. The reason is that the designers multiplexed the lower 16 address lines out on the data bus to minimize the number of pins needed. The 8086 could then be put in a 40-pin package. In other words, the data bus lines, labeled AD0 through AD15 in Figure 7-1, are used at the start of a machine cycle to send out addresses, and later in the machine cycle they are used to send or receive data. The 8086 sends out a signal called *address latch enable*, or ALE, on pin 25 to let external circuitry know that an address is on the data bus. Later we will discuss in detail how this works. The upper 4 bits of an address are sent out on the lines labeled A16/S3 through A19/S6. The double mnemonic on these pins indicates that address bits A16 through A19 are sent out on these lines during the first part of a machine cyle and status information, which identifies the type of operation to be done in that cycle, is sent out on these lines during a later part of the cycle.

Having found the address bus and the data bus, now look for the control bus lines. Some of the control bus lines on a microprocessor usually have mnemonics such as $\overline{RD}$, $\overline{WR}$, and M/$\overline{IO}$. Pin 32 of the 8086 in Figure 7-1 is labeled $\overline{RD}$. This signal will be asserted low when the 8086 is reading data from memory or from a port. Pin 29 has a label $\overline{WR}$ next to it. However, pin 29 also has a label $\overline{LOCK}$ next to it, because this pin has two functions. The function of this pin and the functions of the other pins between 24 and 31 depend on the *mode* in which the 8086 is operating.

The operating mode of the 8086 is determined by the logic level applied to the MN/$\overline{MX}$ input, pin 33. If pin 33 is assserted high, then the 8086 will function in *minimum mode*, and pins 24 through 31 will have the functions shown in parentheses next to the pins in Figure 7-1. Pin 29, for example, will function as $\overline{WR}$ which will go low any time the 8086 writes to a port or to a memory location. Pin 28 will function as M/$\overline{IO}$. The 8086 will assert this signal high if it is reading from or writing to a memory location, and it will assert this signal low if it is reading from or writing to a port. The $\overline{RD}$, $\overline{WR}$, and M/$\overline{IO}$ signals form the heart of the control bus for a minimum mode 8086 system. The 8086 is operated in minimum mode in systems where it is the only microprocessor on the system buses. Later in this chapter we discuss in detail the operation of a minimum mode system.

If the MN/$\overline{MX}$ pin is asserted low, then the 8086 is in *maximum mode*. In this mode pins 24 through 31 will have the functions described by the mnemonics next to the pins in Figure 7-1. In this mode the control bus signals ($\overline{S0}$, $\overline{S1}$, $\overline{S2}$) are sent out in encoded form on pins 26, 27, and 28. An external bus controller device decodes these signals to produce the control bus signals required for a system which has two or more microprocessors sharing the same buses. In Chapter 11 we discuss how a maximum mode 8086 system operates.

Here's a brief introduction to the functions of a few more of the 8086 pins. First note pin 21, the RESET input. If this input is made high, the 8086 will, no matter what it is doing, reset its DS, SS, ES, IP, and flag registers to all 0's. It will set its CS register to FFFFH. When the RESET signal is removed from pin 21, the 8086 will then fetch its next instruction from physical address FFFF0H. This address is produced in the 8086 Bus Interface Unit (BIU) by shifting the FFFFH in the CS register 4 bits left and adding the 0000H in the instruction pointer to it. The first instruction you want to execute after a reset is put at this address, FFF0H. An example would be the first instruction of a monitor program such as the one on the SDK-86.

Next notice that the 8086 has two interrupt inputs, *nonmaskable interrupt* (NMI) input on pin 17 and the *interrupt* (INTR) input on 18. A signal can be applied to one of these inputs to cause the 8086 to interrupt the program it is executing and go execute a specified procedure. We might, for example, connect a temperature sensor from a steam boiler to an interrupt input on an 8086. If the boiler gets too hot, then it will assert the interrupt input. This will cause the 8086 to stop executing its current program and go execute a procedure to turn off the fuel supply to the boiler. At the end of the procedure we can return to executing the interrupted program. Chapter 8 describes in detail the operation and uses of interrupts.

Now that you have an overview of most of the major pins on an 8086, we will take a closer look at what is happening on the buses during a read operation and during a write operation.

Basic Signal Flow on 8086 Buses

Figure 7-2 shows, in timing diagram form, the activities on the 8086 buses during simple read and write operations. Don't be overwhelmed by all of the lines on this diagram. Their meaning should become clear to you as we work our way through the diagram.

8086 BUS ACTIVITIES DURING A READ MACHINE CYCLE

The first line to look at in Figure 7-2 is the *clock waveform* at the top. This represents the crystal-controlled clock signal sent to the 8086 from an external clock generator device as shown in the top left of Figure 7-3. One cycle of this clock is referred to as a *state*. A state is measured from the 50 percent point on the falling edge of one clock pulse to the 50 percent point on the falling edge of the next clock pulse. T1 in the figure is a state. Each basic bus operation such as reading a byte from memory or writing a word to a port requires some number of states. The group of states required for a basic bus operation is called a *machine cycle*. The total time it takes the 8086 to fetch and execute an instruction is

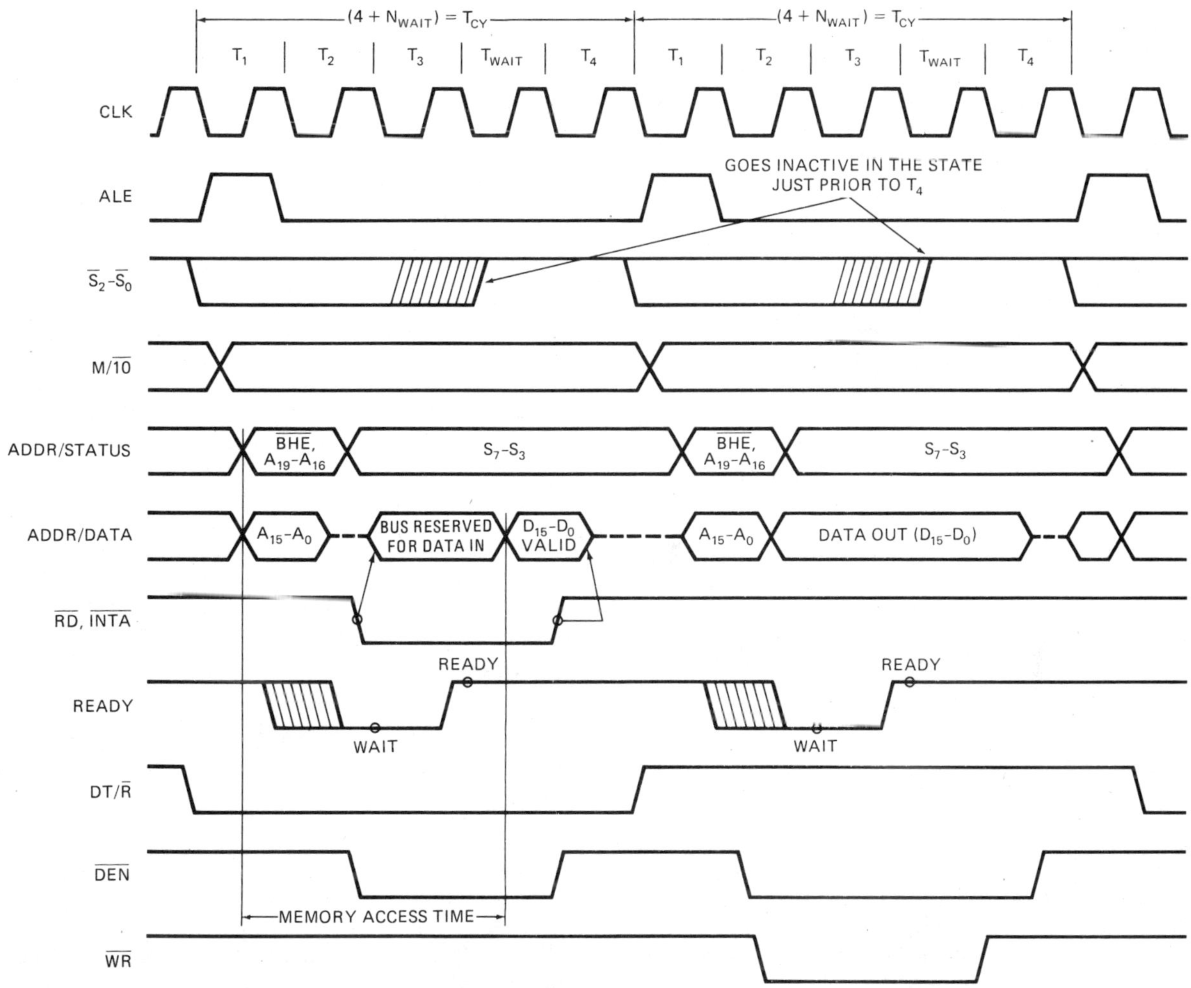

FIGURE 7-2 Basic 8086 system timing. *(Intel Corporation)*

called an *instruction cycle*. An instruction cycle consists of one or more machine cycles. To summarize this, then, an instruction cycle is made up of machine cycles, and a machine cycle is made up of states. What we are going to examine here are the activities that occur on the buses during a read machine cycle.

During T1 of a read machine cycle an 8086 first asserts the M/$\overline{IO}$ signal. It will assert this signal high if it is going to do a read from memory during this cycle, and it will assert M/$\overline{IO}$ low if it is going to do a read from a port during this cycle. The timing diagram in Figure 7-2 shows two waveforms for the M/$\overline{IO}$ signal, because the signal may be going low or going high for a read cycle. The point where the two waveforms cross indicates the time at which the signal becomes valid for this machine cycle. Likewise, in the rest of the timing diagram, crossed lines are used to represent the time when information on a line or group of lines is changed. Incidentally, the best way to analyze a timing diagram such as this one is to think of time as a vertical line moving from left to right across the diagram. With this technique you can easily see the sequence of activities on the signal lines as you move your imaginary time line across the waveforms.

After asserting M/$\overline{IO}$, the 8086 sends out a high on the *address latch enable signal*, ALE. This signal is connected to the enable input (STB) of the 8282 latches as shown in Figure 7-3. As you can also see in Figure 7-3, the data inputs of these latches are connected to the 8086 AD0–AD15, A16–A19, and $\overline{BHE}$ (*bus high enable*) lines. After the 8086 asserts ALE high, it sends out on these lines the address of the memory location that it wants to read. Since the latches are enabled by ALE being high, this address information passes through the latches to their outputs. The 8086 then makes the ALE output low. This disables the latches and holds the address information latched on the latch outputs. The address information on the latch outputs can now be used to select the desired memory or port location.

Observe in the timing diagram in Figure 7-2 how the activity on the ADDR/DATA lines is represented. The first point at which the two waveforms cross represents

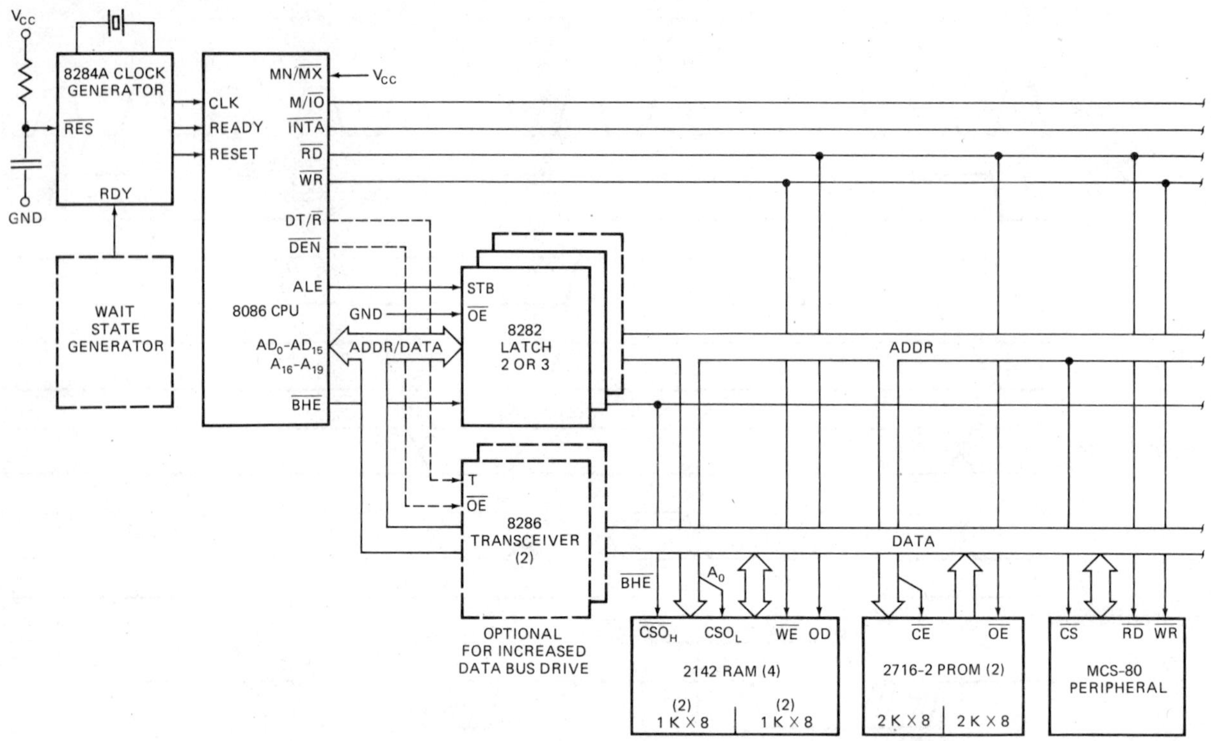

FIGURE 7-3 Basic 8086 minimum mode system. *(Intel Corporation)*

the time at which the 8086 has put a valid address on these lines. These two waveforms DO NOT indicate that all 16 lines are going high or going low at this point. Again, the crossed lines indicate the time at which a valid address is on the bus.

Since the address information is now held on the latches, the 8086 does not need to send it out anymore. Therefore, as shown by a dashed line in Figure 7-2, the 8086 floats the AD0–AD15 lines so that they can be used to input data from memory or from a port. At about the same time the 8086 also removes the BHE and A16–A19 information from the upper lines and sends out some status information on those lines.

The 8086 is now ready to read data from the addressed memory location or port, so near the end of state T2 the 8086 asserts its RD signal low. As you will see in a later section of the chapter, this signal is used to enable the addressed memory device or port device. When enabled the addressed device will put a byte or word of data on the data bus. In other words, asserting the RD signal low causes the addressed device to put data on the data bus. This cause-and-effect relationship is shown on the timing diagram in Figure 7-2 by an arrow going from the falling edge of RD to the "bus reserved for data in" section of the ADDR/DATA waveforms. The bubble on the tail of the arrow always is put on the signal transition or level that causes some action, and the point of the arrow always indicates the action caused. Arrows of this sort are only used to indicate the effect a signal from one device will have on another device. They are not usually used to indicate signal cause and effect within a device.

Now, referring to Figure 7-2 again, find the section of the AD0–AD15 waveform marked off as memory access time near the bottom of the diagram. The addressed memory location or port must put valid data on the data bus before the end of this indicated time interval. Suppose, for example, that we are addressing a ROM. ROMs typically have an access time of a few hundred nanoseconds. In other words, after we apply an address to a ROM, it will be a few hundred nanoseconds before we will see valid data on the outputs of the ROM. If the access time for a ROM in a system is longer than the maximum memory access time specified for the 8086, then the 8086 will not get valid data when it addresses that ROM. A later section of this chapter shows you how to calculate whether a particular ROM, RAM, or port device has a short enough access time to work properly in a given 8086 system. For now, however, we just need you to understand the concept so we can show you one way that an 8086 can accommodate a slow device.

If you look at the pin diagram for the 8086 in Figure 7-1, you should find an input labeled READY. If this pin is high the 8086 is "ready" and operates normally. If the READY input is made low at the right time in a machine cycle, the 8086 will insert one or more *WAIT* states between T3 and T4 in that machine cycle. The timing diagram in Figure 7-2 shows an example of this. An external hardware device is set up to pulse READY low before the rising edge of the clock in T2. After the 8086 finishes

T3 of the machine cycle, it enters a WAIT state. During a WAIT state the signals on the buses remain the same as they were at the start of the WAIT state. The address of the addressed memory location is held on the output of the latches so it does not change. As you can see from the timing diagram in Figure 7-2, the control bus signals, M/$\overline{IO}$ and $\overline{RD}$, also do not change during the WAIT state, Twait. If the READY input is made high again during T3 or during the WAIT state as shown in Figure 7-2, then after one WAIT state the 8086 will go on with the regular T4 of the machine cycle. What we have done by inserting the WAIT state is to freeze the action on the buses for one clock cycle. This gives the addressed device an extra clock cycle time to put out valid data. If, for example, we want to use a slower (cheaper) ROM in a system, we can add a simple circuit which pulses the READY input low each time that ROM is addressed. Note in Figure 7-3 that a READY input signal is usually passed through the 8284 clock generator IC so that the signal actually applied to the 8086 is synchronized with the system clock. Incidentally, a cross-hatched section on a waveform indicates that the signal may be changed at any time during that time interval.

If the 8086 READY input is still low at the end of a WAIT state, then the 8086 will insert another WAIT state. The 8086 will continue inserting WAIT states until the READY input is made high again. In later chapters we show more applications using the READY input to insert WAIT states in a machine cycle.

Another look now at Figure 7-2 will show you that there are still two waveforms we haven't discussed yet. These two are the $\overline{DEN}$ signal and the DT/$\overline{R}$ signal. The $\overline{DEN}$ signal is used to enable bidirectional buffers on the data bus. Figure 7-3 shows how buffers such as 8286s are connected on the data bus in a system. For a very small system these buffers are not needed, but as more devices are added to a system they become necessary. Here's why. Most of the devices such as ROMs and RAMs used around microprocessors have MOS inputs, so on a dc basis they don't require much current. However, each input or output added to the system data bus, for example, acts like a capacitor of a few picofarads connected to ground. In order to change the logic state on these inputs from low to high, all of this added capacitance must be charged. To change the logic state to a low, the capacitance must be discharged. If we add more than a few devices on the data bus lines, the 8086 outputs cannot supply enough current drive to charge and discharge the circuit capacitance rapidly. Therefore, we add high-current drive buffers to do the job.

We must be able to float the outputs of buffers used on the data bus so that they do not interfere with other activities on these lines. For example, we certainly don't want data bus buffer outputs enabled onto the data bus while the 8086 is putting out the lower 16 bits of an address on these lines. The *data enable signal*, $\overline{DEN}$, from the 8086 will enable the data bus buffers when it is asserted low. Buffers used on the data bus must also be bidirectional, because we both send data out on the data bus and read data in on the data bus. The *data transmit/receive signal*, DT/$\overline{R}$, from the 8086 is used to specify the direction in which the buffers are enabled. When DT/$\overline{R}$ is asserted high, the buffers will, if enabled by $\overline{DEN}$, transmit data from the 8086 to ROM, RAM, or ports. When DT/$\overline{R}$ is asserted low, the buffers, if enabled by $\overline{DEN}$, will allow data to come in from ROM, RAM, or ports to the 8086.

Now let's look back at Figure 7-2 to see how $\overline{DEN}$ and DT/$\overline{R}$ function during a read machine cycle. During T1 of the machine cycle the 8086 asserts DT/$\overline{R}$ low to put the data buffers in the receive mode. Then, after the 8086 finishes using the data bus to send out the lower 16 address bits, it asserts $\overline{DEN}$ low to enable the data bus buffers. The data put on the data bus by an addressed port or memory will then be able to come in through the buffer to the 8086.

We can summarize the activities on the buses during an 8086 read machine cycle as follows. The 8086 asserts M/$\overline{IO}$ high if the read is to be from memory and asserts M/$\overline{IO}$ low if the read is going to be from a port. At about the same time, the 8086 asserts ALE high to enable some external latches. It then sends out $\overline{BHE}$ and, on the lines AD0–A19, the desired address. The 8086 then pulls the ALE line low to latch the address information in the external latches. After the 8086 is through using lines AD0–AD15 for an address, it removes the address from these lines and puts the lines in the input mode (floats them). It then asserts its $\overline{RD}$ signal low. The $\overline{RD}$ signal going low turns on the addressed memory or port which then puts the desired data on the data bus. To complete the cycle the 8086 brings the $\overline{RD}$ line high again. This causes the addressed memory or port to turn off, thereby floating the bus again. If the 8086 READY input is made low before or during T2 of a machine cycle, the 8086 will insert WAIT states as long as the READY input is low. When READY is made high the 8086 will continue on with T4 of the machine cycle. WAIT states can be used to give slow devices additional time to put out valid data. If a system is large enough to need data bus buffers, then the 8086 $\overline{DEN}$ signal will be asserted low to enable the buffers, and the 8086 DT/$\overline{R}$ signal will be asserted high to set the buffers for output or asserted low to set the buffers for input.

8086 BUS ACTIVITIES DURING A WRITE MACHINE CYCLE

Now that we have analyzed the 8086 bus activities for a read machine cycle, let's take a look at the timing diagram for a write machine cycle in the right-hand half of Figure 7-2. Most of this diagram should look very familiar to you because it is very similar to the read cycle.

During T1 of a write machine cycle the 8086 asserts M/$\overline{IO}$ low if the write is going to be to a port and it asserts M/$\overline{IO}$ high if the write is going to be to memory. At about the same time the 8086 raises ALE high to enable the address latches. The 8086 then outputs $\overline{BHE}$ and the address that it will be writing to on AD0–A19. When reading from or writing to a port, lines A16–A19 will always be low, because the 8086 only sends out 16-bit port addresses. After this address has had time to pass through the latches, the 8086 brings ALE low again to latch the address on the outputs of the latches. In addition to holding the address, these latches also function as buffers for the address lines. After the address information is latched, the 8086 removes the address information from AD0–AD15 and puts the desired data on the data bus. It then asserts its $\overline{\text{WR}}$ signal low. The $\overline{\text{WR}}$ signal is used to turn on the memory or port where the data is to be written. After the addressed memory or port has had time to accept the data from the data bus, the 8086 raises the $\overline{\text{WR}}$ signal line high again and floats the data bus.

If the READY input is made low by external hardware before or during T2 of the machine cycle, the 8086 will insert a WAIT state after T3. If the READY input is made high before the end of the WAIT state, the 8086 will go on with state T4 as soon as it finishes the WAIT state. If the READY input is still low just before the end of the WAIT state, the 8086 will insert another WAIT state. It will continue to insert WAIT states until READY is made high. During a WAIT state the logic levels on the buses are held constant. Therefore, if we have a memory or port device which needs more time to absorb the data from the data bus, we can use some external hardware to pulse the READY line low each time this device is addressed. Pulling the READY line low will cause the 8086 to insert one or more WAIT states in the machine cycle, thus giving the addressed device more time to absorb the data.

If the system is large enough to need buffers on the data bus, then DT/$\overline{\text{R}}$ will be connected to the buffers. During a write cycle the 8086 asserts DT/$\overline{\text{R}}$ high to put the buffers in the transmit mode. When the 8086 asserts $\overline{\text{DEN}}$ low to enable the buffers, data output from the 8086 will pass through the buffers to the addressed port or memory location.

Work your way across the timing diagrams for the read and write machine cycles in Figure 7-2 until you feel that you understand the sequence of activities that occur. Understanding this well will make later sections easier to understand.

FIGURE 7-4 Intel SDK-86 microprocessor development board. *(Intel Corporation)*

Analyzing a Minimum-Mode System, the SDK-86

The previous sections showed how a clock generator, address latches, and data bus buffers are connected to an 8086 to form what we might call the minimum-mode CPU group. As shown in Figure 7-3 this group of ICs generates the address bus, data bus, and control bus signals needed for an 8086 minimum-mode system. In this major section of the chapter we discuss how this CPU group is connected with ROM, RAM, ports, and other devices to form a system. The system we use for this discussion is the *Intel SDK-86 system design kit*, a readily available 8086-based unit suitable for building the prototypes of small microcomputer-based instruments.

Figure 7-4 shows a photograph of an SDK-86 board. From the photograph you can see that, in addition to the microcomputer ICs, the board has a hexadecimal keypad, some seven-segment displays, and a large open area for adding more ROM, RAM, ports, or other circuitry. A monitor program in ROM on the board allows you to enter, execute, and debug machine code programs using the on-board hex keypad or an external CRT terminal connected to the serial port on the board. The board comes with 2 Kbytes of RAM and sockets where you can add another 2 Kbytes. The board also has six 8-bit parallel ports which you can program to be inputs or outputs. To get a better idea of the hardware functions on the board and the devices used to implement these functions, let's look at the detailed block diagram of the SDK-86 in Figure 7-5.

Whenever you are approaching a system that is new to you, it is a good idea to carefully study the detailed block diagram of the system before you start digging into the actual schematics. The schematics for even a small system such as this are often spread over many pages. Without the overview that the block diagram gives, it is very difficult for you to see how all of the schematic pieces fit together.

The first parts to look at in Figure 7-5 are the 8086 CPU and the 8284 *clock generator*. Note that the 8284 has a 14.7456-MHz crystal connected to it. According to the data sheet for the the 8284, the frequency of the crystal connected to the 8284 will be divided by three to produce the clock signal sent to the 8086. Therefore, the actual 8086 clock frequency for this board will be 4.915 MHz. Another clock signal called PCLK is also produced by the 8284. This signal is used as a general-purpose clock signal throughout the system. The hardware RST signal and the RDY signal are also passed through the 8284 to synchronize them with the clock signal before they are sent to the 8086. As you can see in Figure 7-5, considerable circuitry is connected to the RDY input so that several conditions can cause a WAIT state to be inserted in a machine cycle. The structure labeled W27 through W34 above the WAIT state generator in Figure

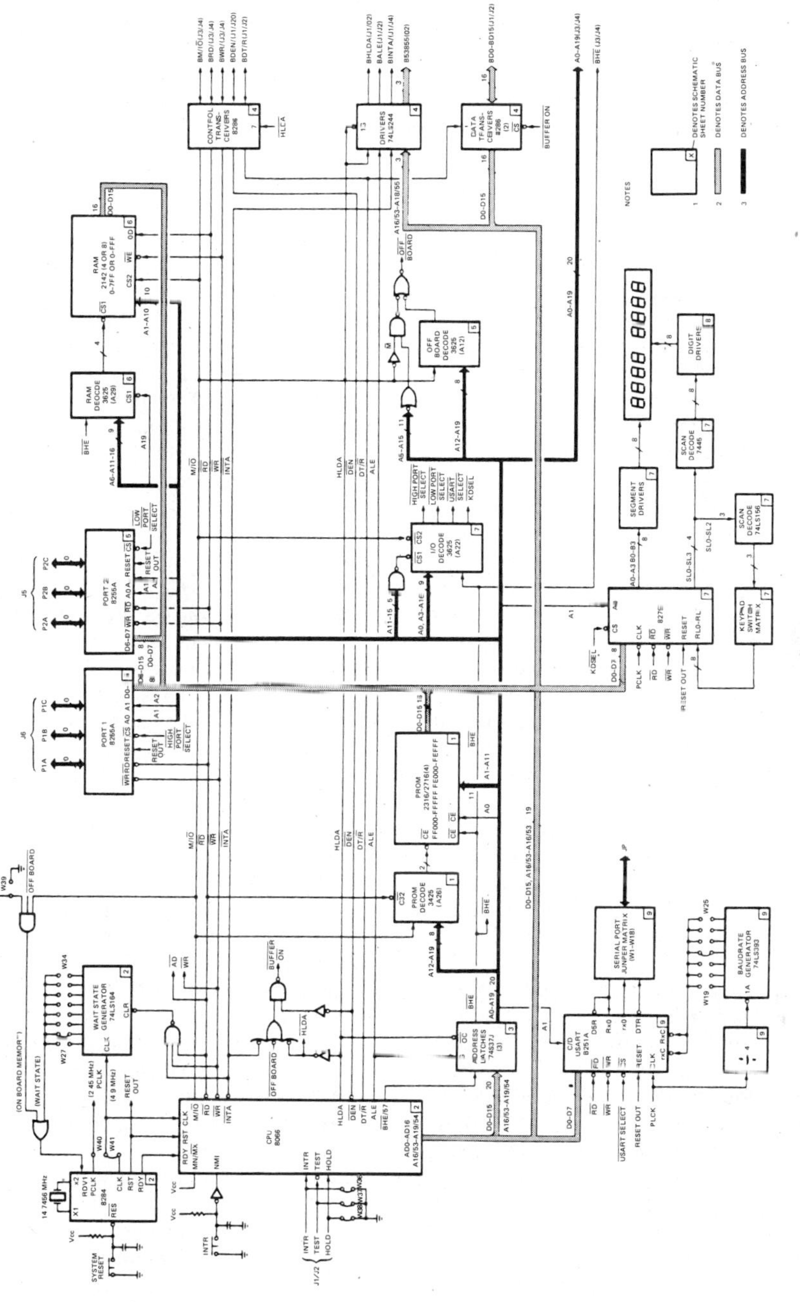

FIGURE 7-5 Detailed block diagram of SDK-86 board. *(Intel Corporation)*

7-5 represents wire wrap pins which can be jumpered to specify the number of WAIT states desired in a machine cycle. We will discuss this in detail later.

By this time you may have noticed that the symbols for the 8284, 8086, and WAIT state generator each have a small box containing a 2 in their lower right corner. This number tells you that the detailed schematic for these parts will be found on sheet number two of the set of schematics. Figure 7-6 shows the complete schematic set for the SDK-86 board, so you can check this out if you wish.

The next parts to look for in the block diagram of the SDK-86 are the *address latches* which you know are needed to grab address information during T1 of a machine cycle. The box just below the 8086 in the diagram indicates that three 74S373s are used for address latches. AD0–AD15, A16–A19, and $\overline{BHE}$ are connected to the inputs of these latches. As expected, ALE is used to enable the latches. The information held on the output of the latches after ALE goes low is A0–A19 and $\overline{BHE}$. The /20 after A0–A19 on the output of the latches indicates that there are 20 lines in this group. A heavy black line is used to distinguish the demultiplexed address bus from the data bus.

Next, follow the address lines to the right on the diagram to find the ROM in the system. The box labeled PROM indicates that four 2316 or 2716 devices are used for EPROM in the system. Each of these devices holds 2 Kbytes of memory. Also indicated in the PROM box in the diagram are the absolute addresses where these devices are located. Two of the EPROMs occupy the address space from FE000H–FEFFFH, and the other two occupy the address space from FF000H–FFFFFH. The 3625 PROM decoder connected to these EPROMs has two related purposes. The first is to produce a signal which turns on the desired EPROM when you send out an address in the range assigned to that device. The second purpose is to make sure that only one device is outputting onto the data bus at a time. We discuss in detail later how address decoders are connected to give a desired address to a particular device in a system. Note that the enable input, $\overline{CS2}$, of the decoder PROM is connected to the $\overline{RD}$ signal from the 8086. The result of this is that the PROM decoder will only be enabled if the 8086 is doing a read operation. Can you see why you would not want an EPROM to be turned on if you accidentally sent out an address in its range during a write operation? The answer is that attempting to write to the outputs of an EPROM can burn out both the ROM and buffer outputs. The "*A26*" in the PROM decoder box of the block diagram, incidentally, indicates that the 3625 IC will be numbered *XA*26 on the schematic sheet where it is found.

Follow the address bus to the upper right corner of the block diagram in Figure 7-5 to find how RAM is implemented in this system. The board comes with 2 Kbytes of static RAM contained in four 2142s, but there are sockets for another four 2142s. The initial four devices occupy the address space from 00000H–007FFH. If four more 2142s are added, they will be in the address space 00800H–00FFFH. Another 3625 is used here as a RAM decoder. As with the PROM decoder, the purposes of this device are to turn on a memory device which corresponds to the address sent out on the address bus, and to make sure that only one device at a time is outputting data on a data bus line. The 8086 can read or write a byte, or it can read or write a word. Therefore, 16 data lines are connected to the RAM block.

Now let's find the *system ports* in the block diagram in Figure 7-5. Two 8255As at the top of the page give the system *programmable* parallel ports. The term programmable in this case means that as part of your program, you send the 8255A a *control byte*. The control byte tells the 8255A whether you want a particular group of lines on the device to function as outputs or as inputs. In Chapter 9 we show you how to make up and send these control words. The two 8255As in this system can be used individually to input or output parallel bytes. They can also be used together to input or output words. For byte input or output operations, only one of the devices will be turned on by asserting its $\overline{CS}$ input low. For word input or output operations, both 8255As will be turned on by asserting their $\overline{CS}$ inputs low. The high byte of a word to be output, for example, will then be sent to one of the ports in the PORT 1 device. The low byte of the word to be output will go to the corresponding port in the PORT 2 device. To be more specific, if the high byte of an output word goes to port P1A, then the low byte of that word will go to port P2A. In a later section on address decoding, we show how the addresses work out for these ports.

Most systems need a serial port so they can communicate with CRT terminals, modems, and other devices which require data to be sent and received in serial form. As shown in the lower left corner of Figure 7-5 the SDK-86 uses an 8251A as a *serial port*. The letters USART on this device stand for *universal synchronous/asynchronous receiver transmitter*, which is quite a mouthful. Chapter 13 discusses the initialization and use of the 8251A. For now, just think of this device as two back-to-back shift registers. One shift register accepts a parallel byte from the system data bus and shifts it out the TxD output in serial form. The other shift register shifts in serial data from the RxD input and converts it to parallel bytes which can be read by the 8086 on the system data bus. The 8251A has only eight data inputs, so data can only be written to or read from the 8251A a byte at a time. Therefore, only the lower 8 bits of the data bus are connected to it. Each of the shift registers in the 8251A requires a clock signal with a frequency of 16 or 64 times the rate at which you want to shift data bits in or out. The clock for the transmit shift register is called TxC and the clock for the receive shift register is called RxC on the block diagram. These are tied together because you usually want to send and receive data at the same rates. The clock for these inputs is produced by dividing down the 2.45-MHz PCLK signal from the 8284 clock generator. Wire wrap jumper pins, W19–W25, allow you to select the desired TxC and RxC frequency from a divider chain in the 74LS393 *baud rate generator*. *Baud rate* is a way of specifying the rate at which data bits are shifted in or out of a serial device. Baud rate for a device such as the 8251A is defined as one over the time per bit. If the time per bit is 3.33 ms, for example, then the baud rate is 300 baud. Common baud rates for serial data transmission are 300, 600, 1200, 2400, 9600, and 19,200.

The final port to discuss here is the 8279 in the bottom center of the SDK-86 block diagram (Figure 7-5). The 8279 is a *specialized input/output device* which has two major functions. The first function is to scan the hex keypad, detect when a key is pressed, debounce the signal from a pressed key, and store the code for the pressed key in an internal RAM where it can be read by the 8086. The second major function of the 8279 is to refresh the multiplexed display on the eight 7-segment LED displays. Seven-segment codes for the digits to be displayed are sent to a RAM in the 8279. The 8279 then

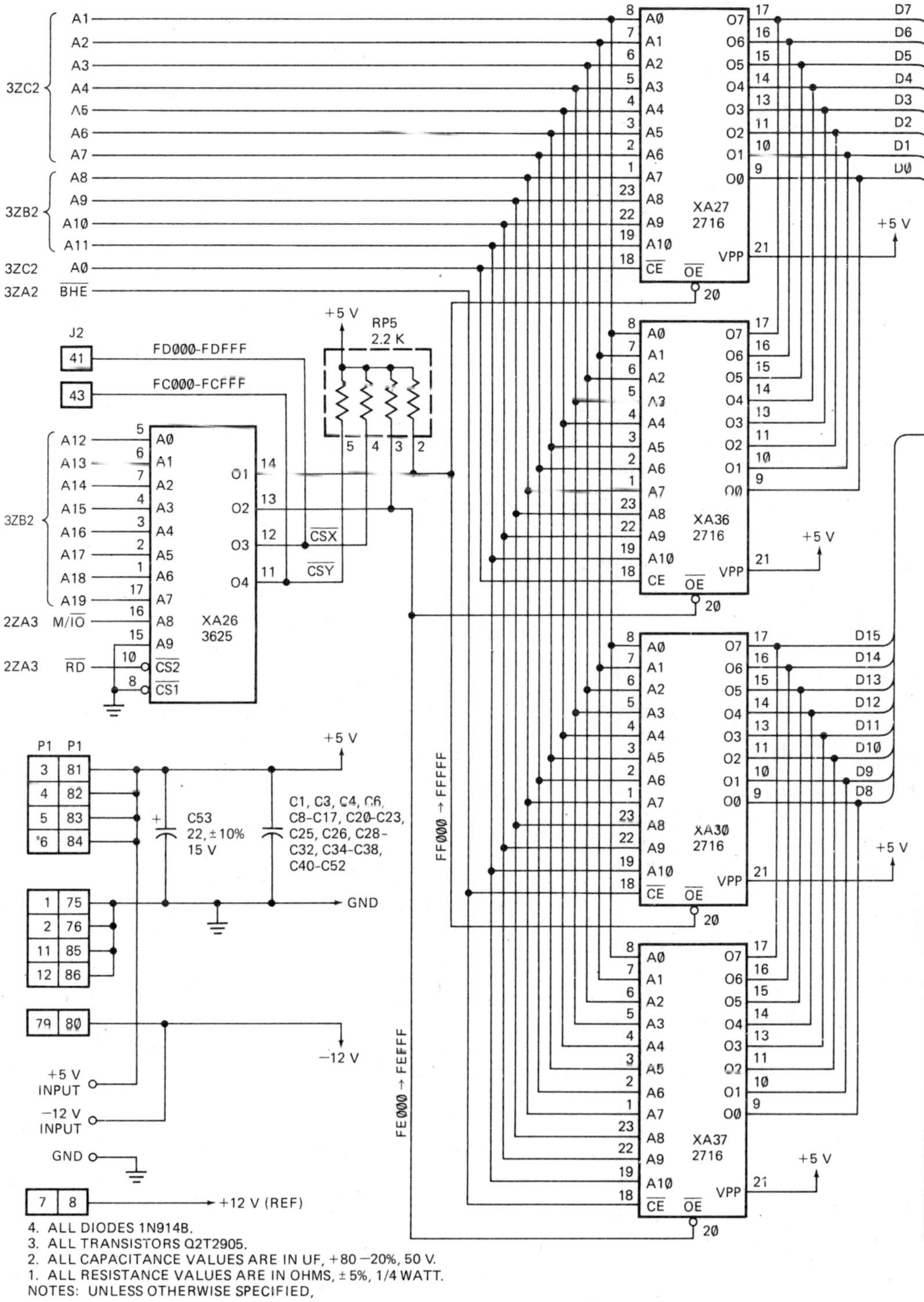

IC	+5 V VCC	GND
8Ø86	4Ø	1, 2Ø
8255A	26	7
8251A	26	4
8284	18	9
8279	4Ø	2Ø
2716	24	12
2142	2Ø	1Ø
3625	18	9
8286	2Ø	1Ø
74LSØØ	14	7
74LSØ4	14	7
74LS1Ø	14	7
74LS14	14	7
74LS2Ø	14	7
74S3Ø	14	7
74LS74	14	7
74LS156	16	8
74LS164	14	7
74LS244	2Ø	1Ø
74LS393	14	7
7445	16	8
74S133	16	8
74S373	2Ø	1Ø
ULN2ØØ3A	9	8

FIGURE 7-6 SDK-86 complete schematics. *(Intel Corporation)*

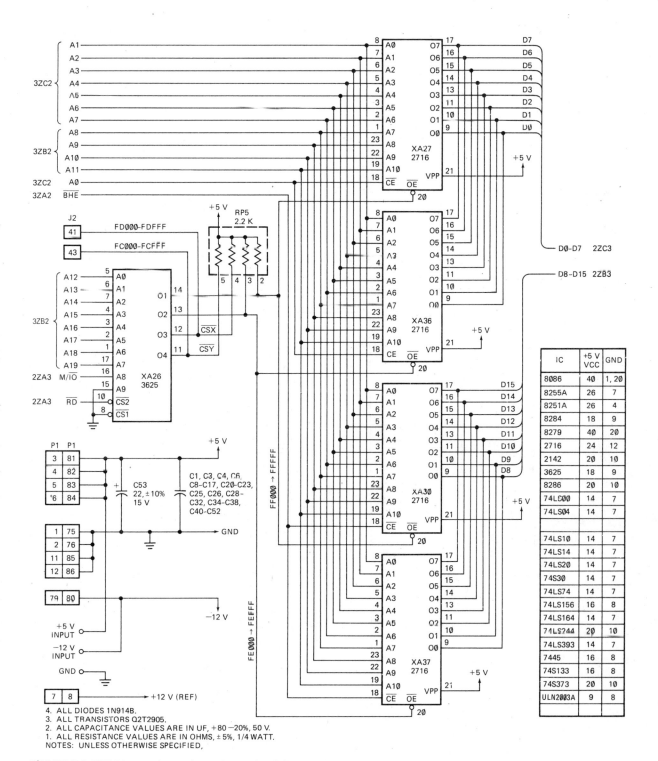

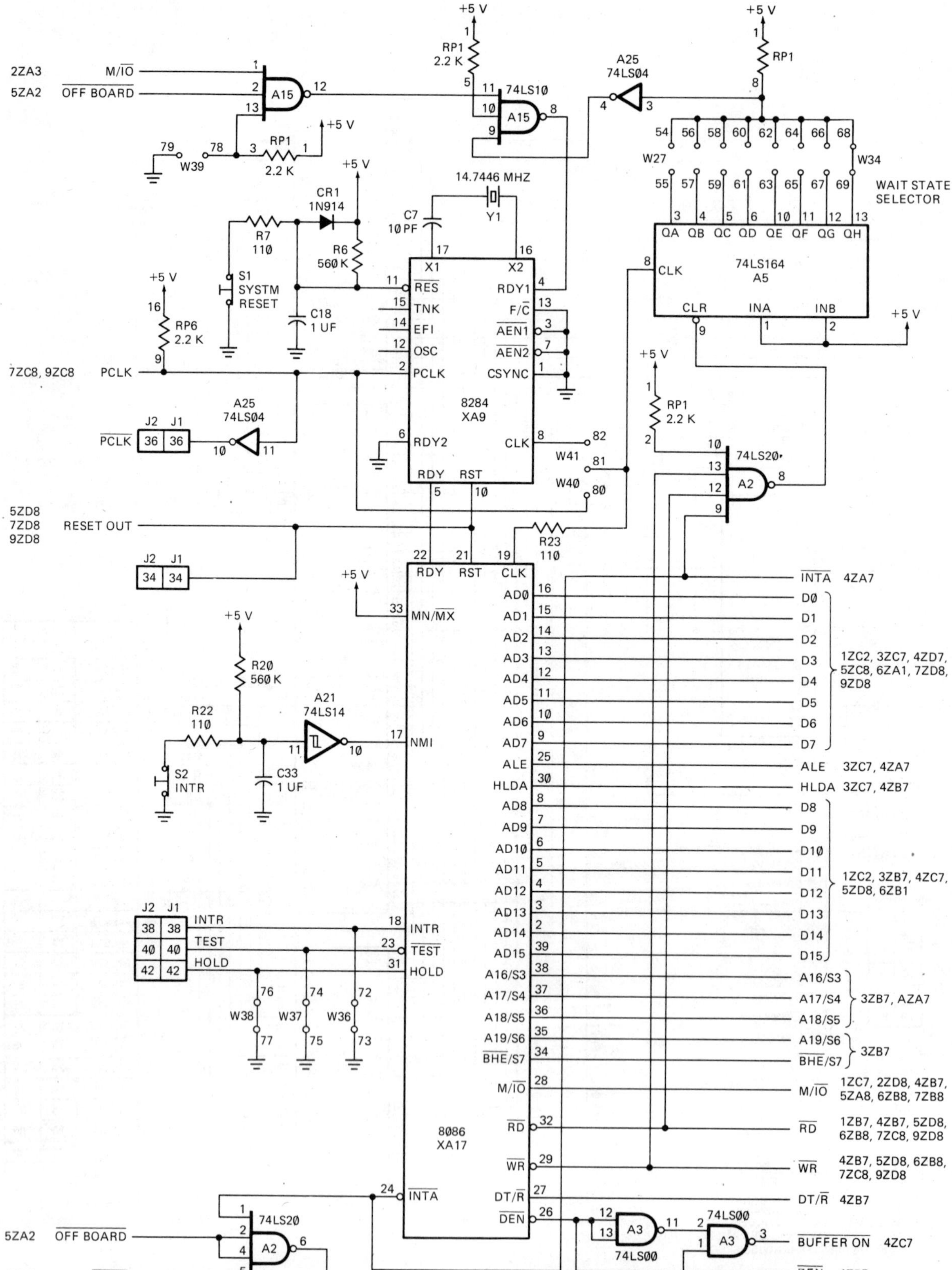

FIGURE 7-6 (*continued*)

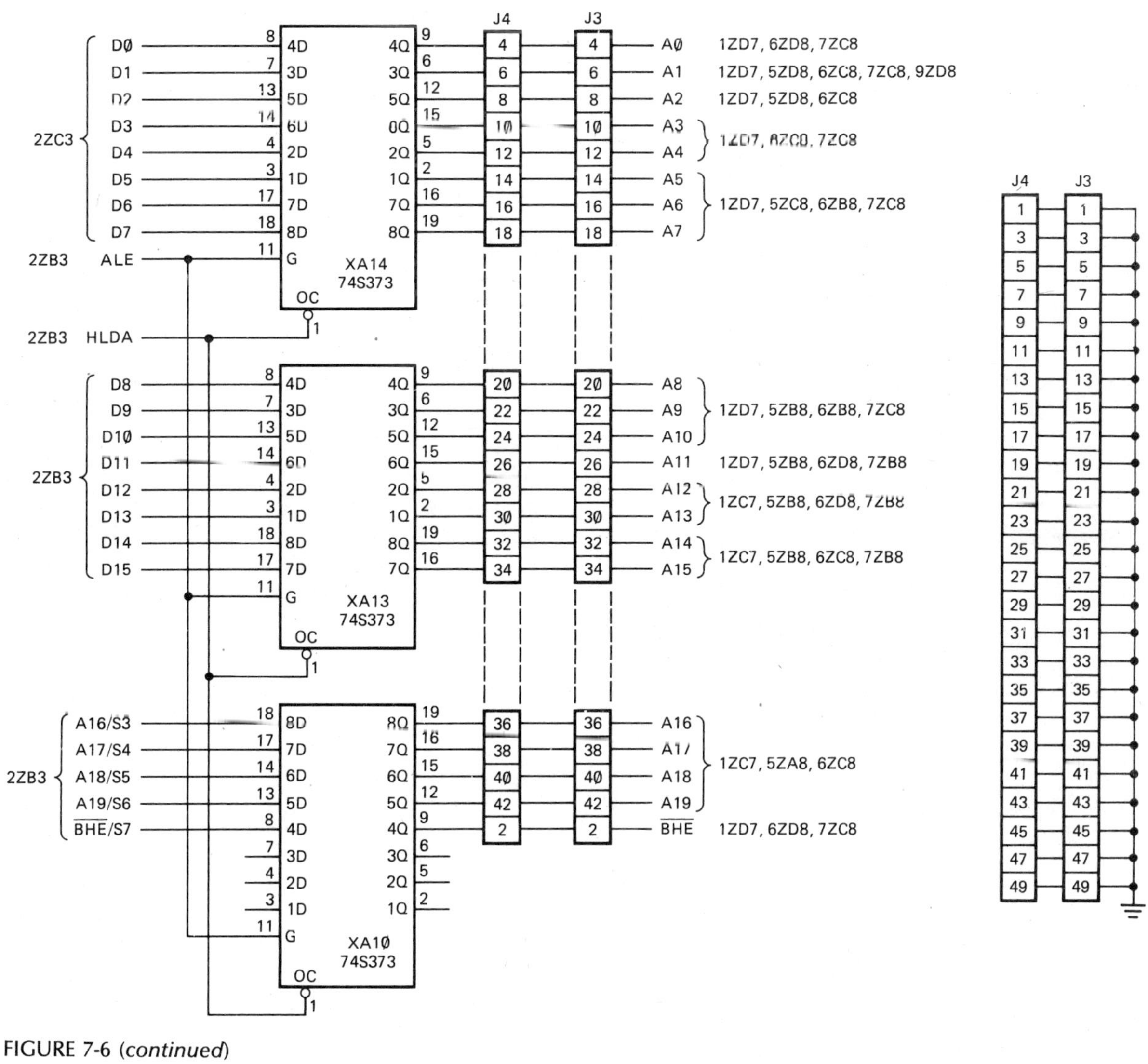

FIGURE 7-6 (*continued*)

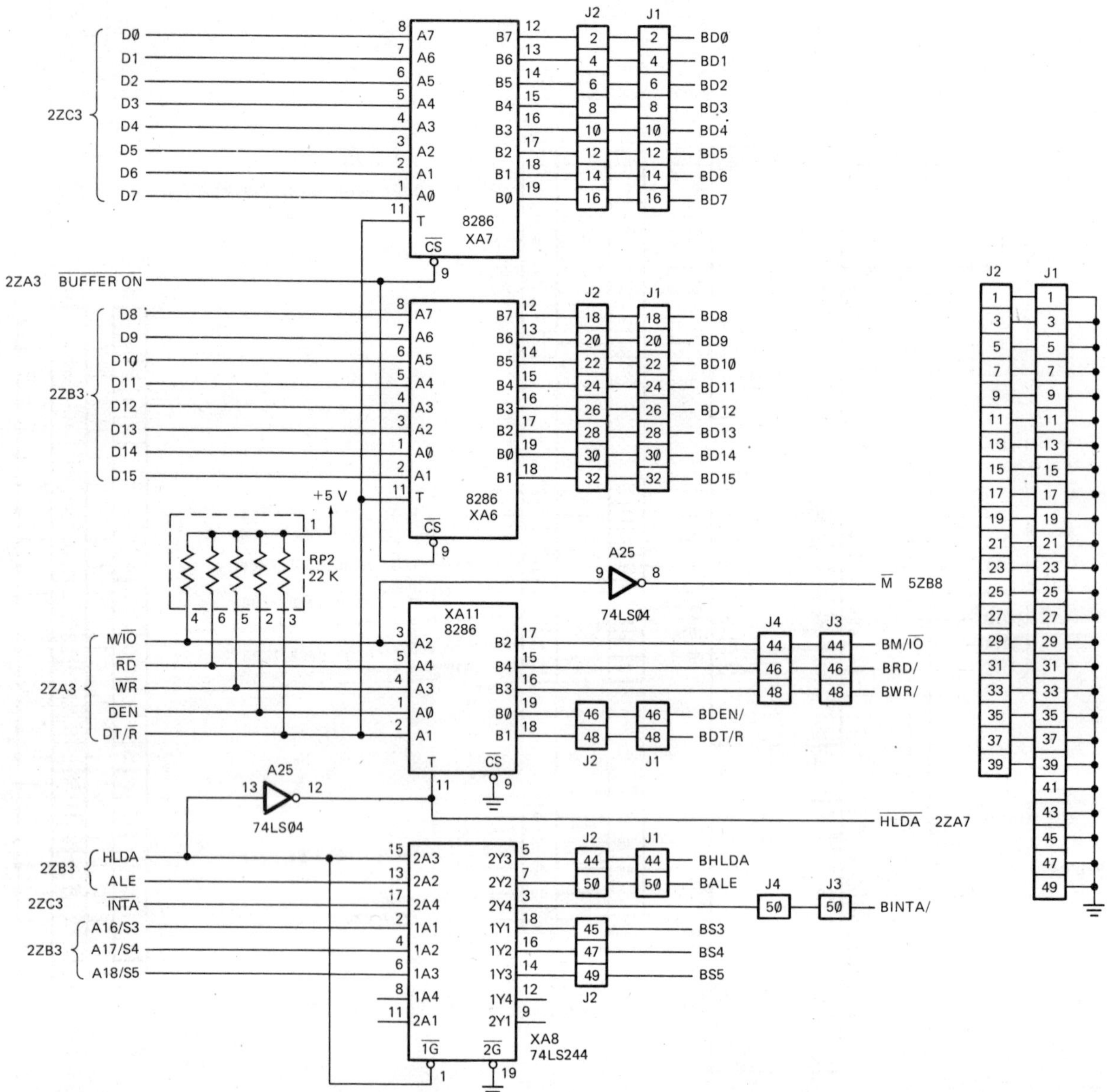

FIGURE 7-6 *(continued)*

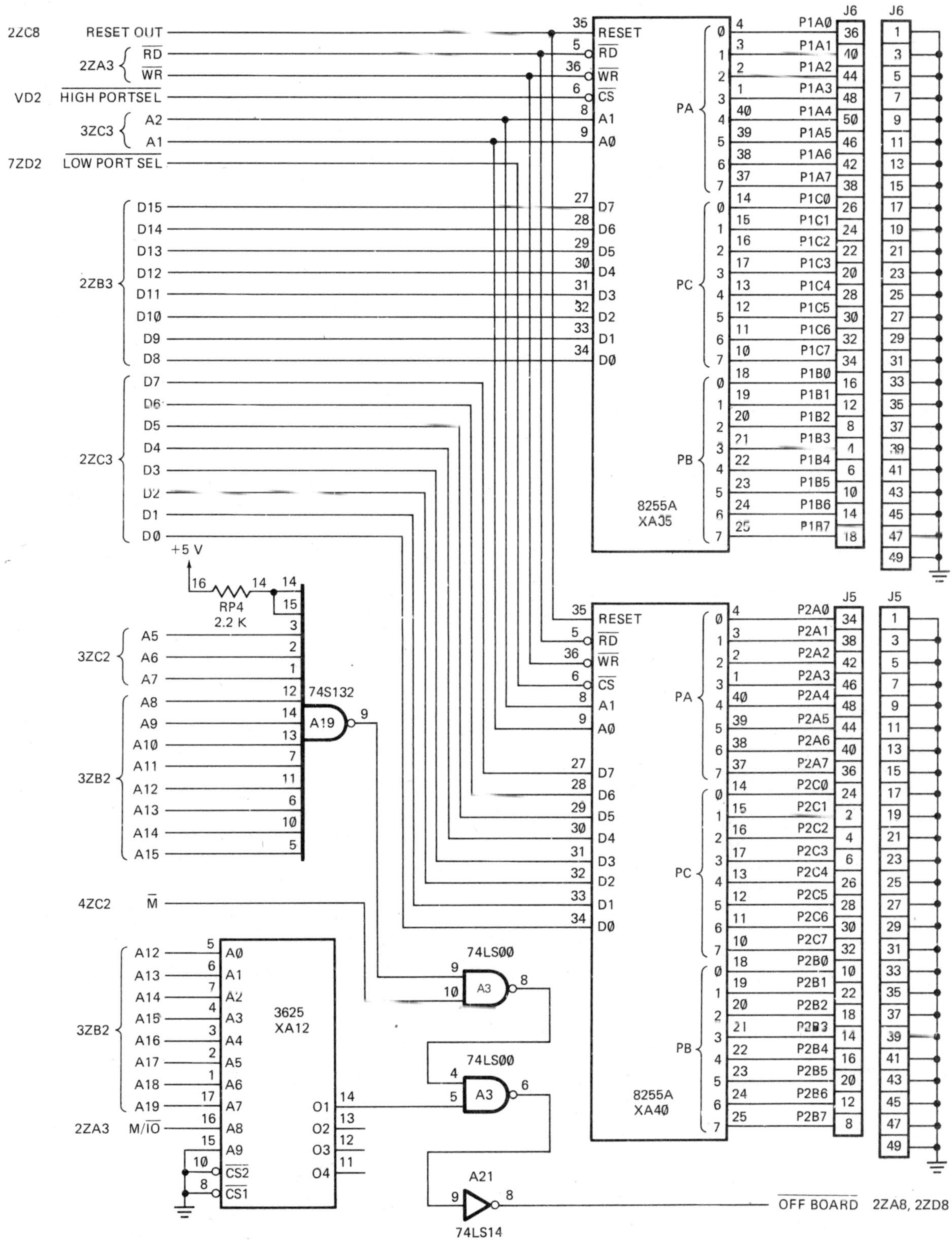

FIGURE 7-6 (*continued*)

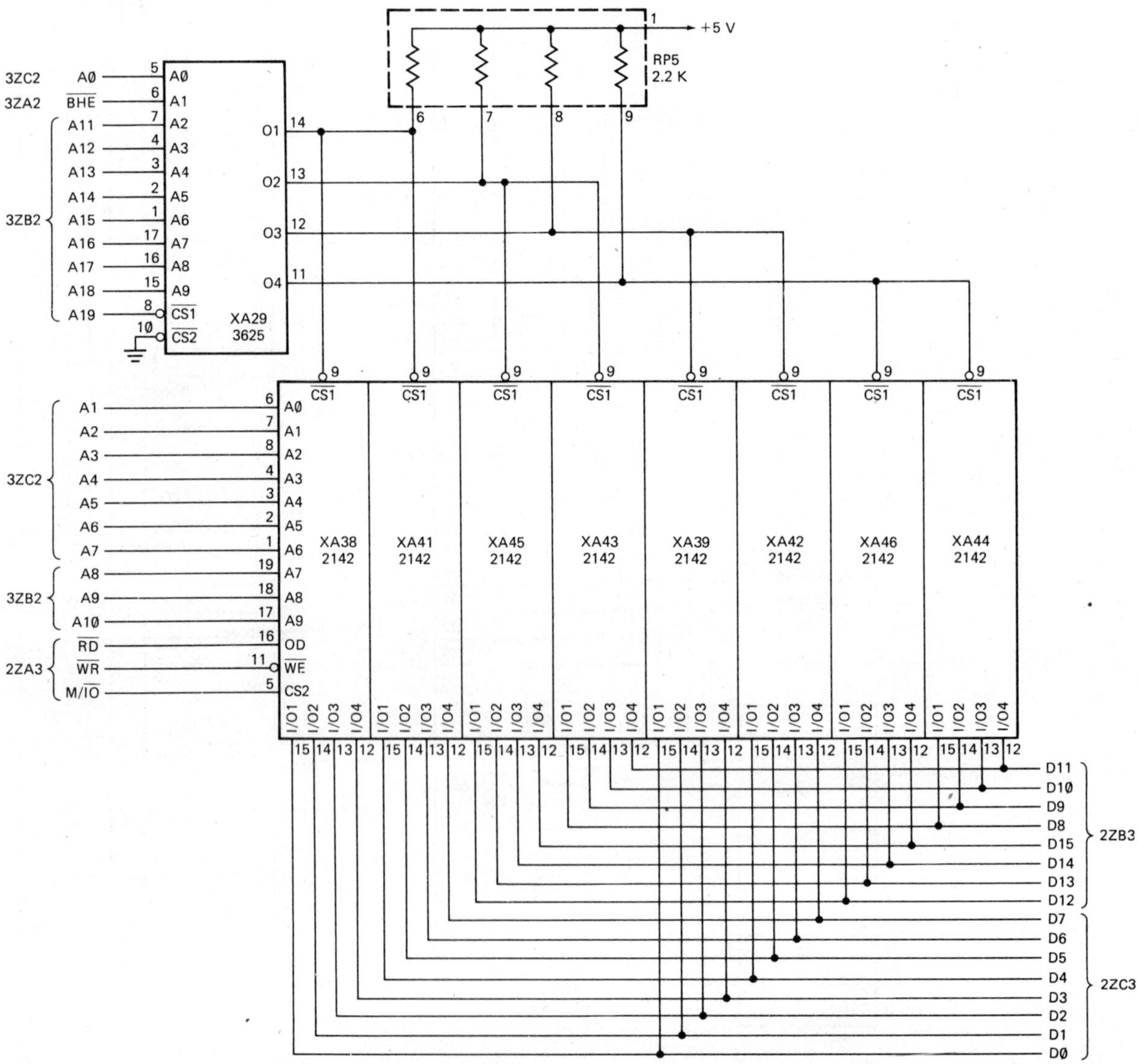

FIGURE 7-6 (*continued*)

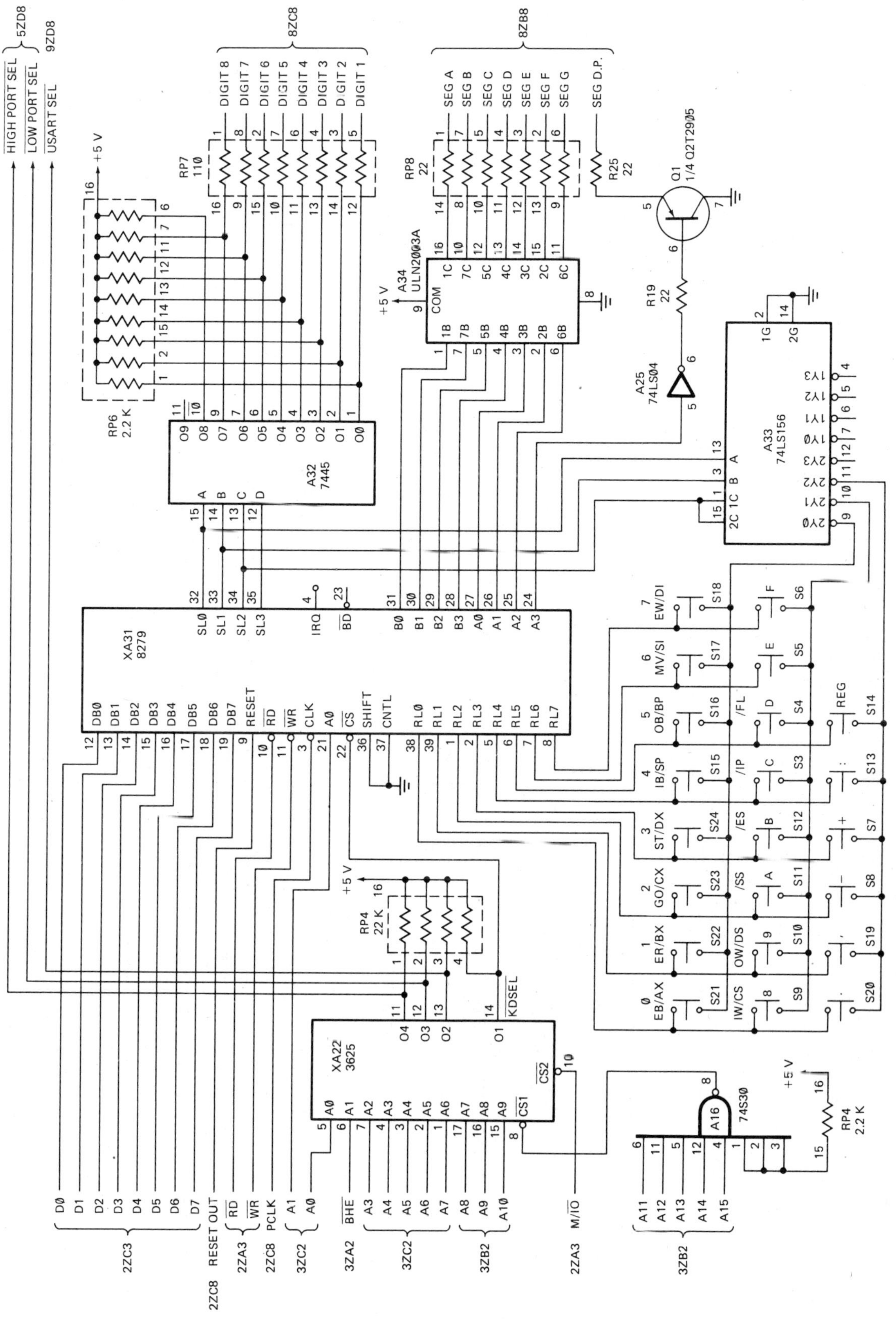

FIGURE 7-6 (continued)

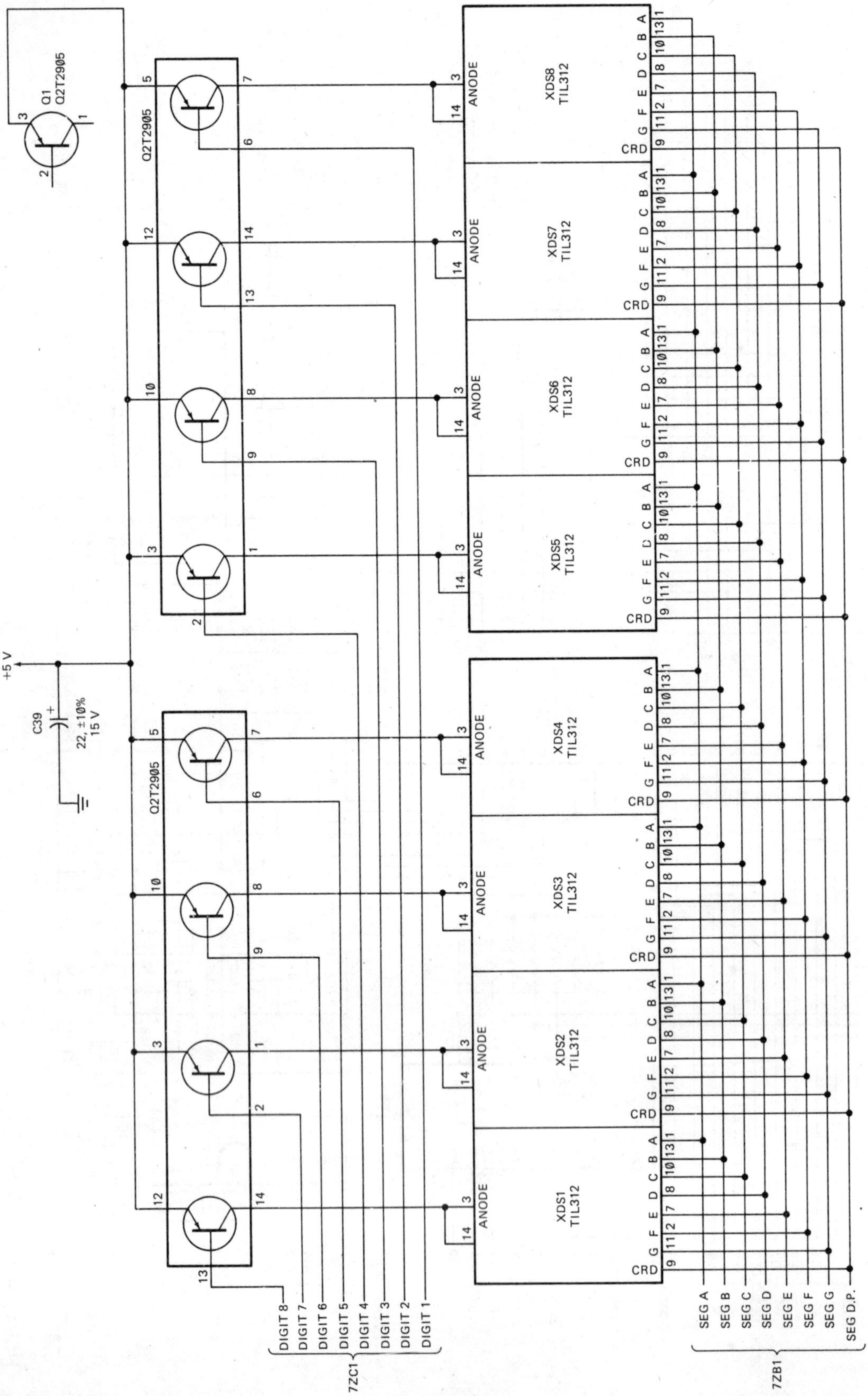

FIGURE 7-6 (*continued*)

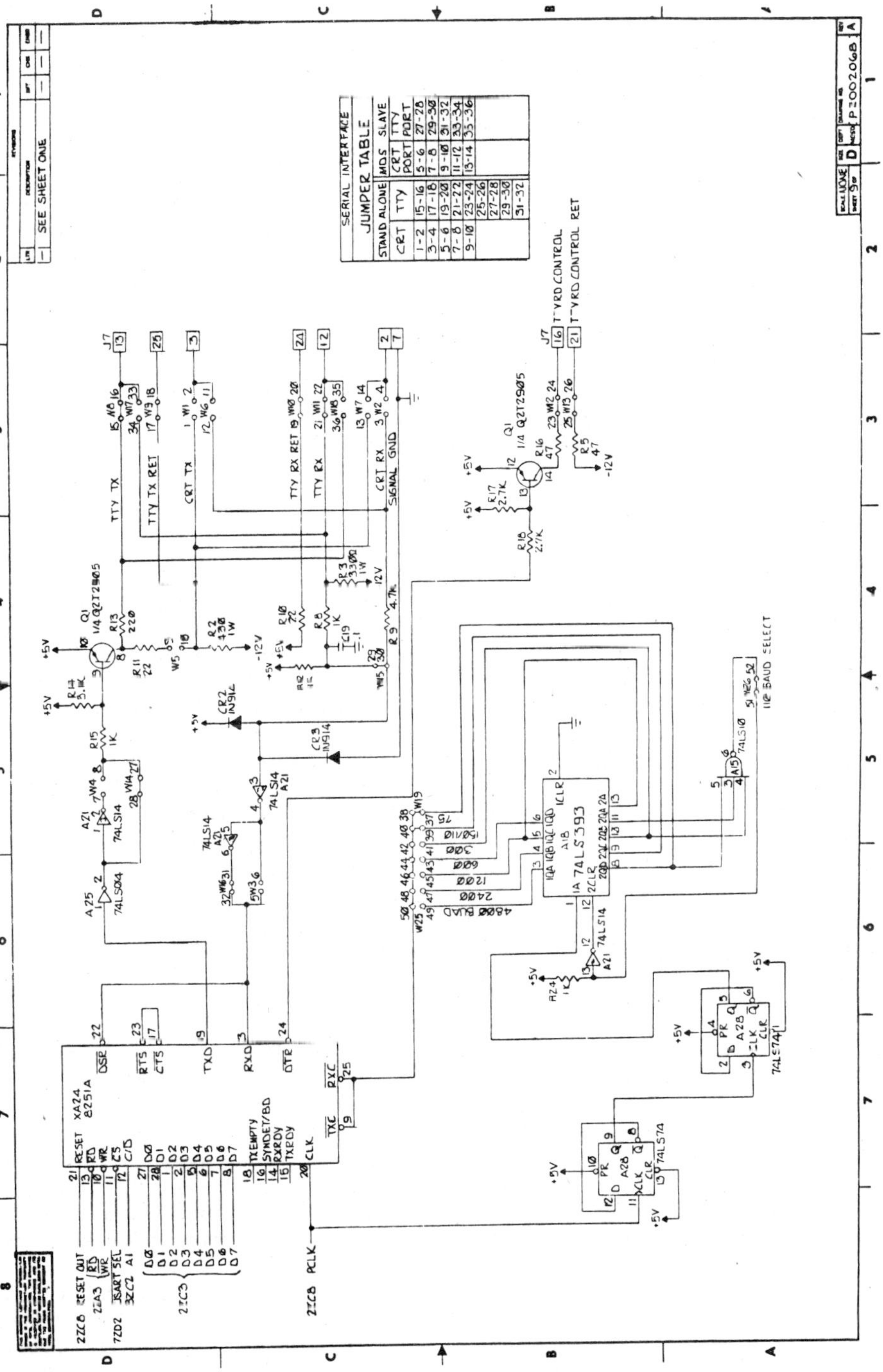

FIGURE 7-6 (*continued*)

automatically sends out the code for one digit and turns on that digit. After a millisecond or so the 8279 sends out the seven-segment code for the next digit and turns on that digit. The process is continued until all digits have been lit, and then the 8279 cycles back to the first digit again. In Chapter 9 we discuss in detail how you use an 8279. The main point for now is that this device takes care of scanning a keyboard and refreshing a display so that you don't have to do these operations as part of your program.

Now that you have an overview of the ports in this system, see if you can find in the block diagram the decoder which selects an addressed port. You should find the 3625 labeled *A22* about in the center of the block diagram. We discuss later how this device produces the port select signals from a port address sent out by the 8086.

The final part of the SDK-86 block diagram to take a look at is the buffers along the right-hand edge. The purpose of these devices is to buffer the data and control bus lines so that they can drive additional ROM, RAM, or ports that you might add to the expansion area of the board. Note that the address lines are already buffered by the 74S373 address latches.

A First Look at the SDK-86 Schematics

Now that you have seen an overview of the SDK-86, the next step is to take a first look at Figure 7-6, which shows the actual schematics for the board. At first these many pages of schematics may seem overwhelming to you, but if you use the *5-minute freak-out rule* and then approach the schematics one part at a time, you should have no trouble understanding them. The schematics simply show greater detail for each of the parts of the block diagram that we discussed in the preceding sections of the chapter.

At this point we want to make clear that it is not the purpose of this chapter to make you an expert on the circuit connections of an SDK-86 board. We use parts of these schematics to demonstrate some major concepts such as address decoding and to show how the parts are connected together to form a small but real system. Even if you do not have an SDK-86 board, you can learn a great deal from these schematics about how an 8086 system functions. Multipage schematics such as these are typical for any microprocessor-based board or product, so you need to get used to working with them.

Before getting started on the next major concept, we will discuss some of the symbols used on most microprocessor system schematics. The first thing that we want to look at in the schematics are the numbers across the top and bottom of each and the letters along the sides of each. These are called *zone coordinates*. You use these coordinates to identify the location of a part or connection on the schematic just as you might use similar coordinates on a road map to help you locate Bowers Avenue. For example, on sheet 1 of the schematics find the lines labeled A1–A7 in the upper left corner. Next to these lines you should see *3ZC2*. This indicates that these address lines come from zone *C2* on sheet 3. To see what the lines actually connect to, first find schematic sheet 3. Then move across the row of the schematic labeled *C* until you come to the column labeled 2. This zone is small enough that you should easily be able to find where these lines come from. The zone coordinates next to these lines on sheet 3 indicate the other schematic sheets and zones that these lines go to. For practice, try finding where a few more lines connect from and to.

The next points to look at on the schematics are the numbers on the ICs. In addition to a part number such as 2716, each IC has a number of the form *XA*36. This second number is used to help locate the IC on the printed circuit board. The number is usually silk-screened on the board next to the corresponding IC. Usually IC numbers are sequential and start from the upper left corner of the component side of a board. There may be several 2716s on the board, but only one will be labeled *XA*36.

Other devices often found on microprocessor boards are *resistor packs*. You can find an example in zone *C5* of schematic sheet 1. As you can see from the schematic, this device contains four 2.2-kΩ resistors. Resistor packs may physically be thin, vertical, rectangular wafers, or they may be in packages similar to small ICs. The advantages of resistor packs are that they take up less printed circuit board space and that they are easier to install than individual resistors.

Some other symbols to look at in the schematics are the structures with labels such as J2 and P1. You can find examples of these in zones *C7* and *B7* of schematic sheet 1. These symbols are used to indicate *connectors*. The number in the rectangular box specifies the pin number on the connector that a signal goes to. The letter P stands for *plug*. A connector is considered a plug if it plugs into something else. In the case of the SDK-86, the connector labeled P1 is the printed circuit board edge connector. The letter J next to a connector stands for *jack*. A connector is considered a jack if something else plugs into it. On the SDK-86 board the jacks J1–J6 are 50-pin connectors that you can plug ribbon cable connectors into. These jacks allow the address bus, data bus, control bus, and parallel ports to be connected to additional circuitry.

One more point to notice on the SDK-86 schematics is the capacitors on the power supply inputs shown in zone *B6* of sheet 1. As you can see there, the schematic shows a large number of 0.1-μF capacitors in parallel with a 22-μF capacitor. Most systems have *filtering* such as this on their power lines. You may wonder what is the use of putting all of these small capacitors in parallel with one which is obviously many times larger. The point of this is that the large capacitor filters out or *bypasses* low-frequency noise on the power lines, and the small capacitors, spread around the board, bypass high-frequency noise on the power supply lines. Noise is produced on the power supply lines by devices switching from one logic state to another. If this noise is not filtered out with bypass capacitors, it may become large enough to disturb system operation.

Glance through the SDK-86 schematics to get an idea of where various parts are located and to see what additional information you can pick up from the notes on them. In the next section of this chapter we discuss how microcomputer systems address memory and ports. As part of the discussion we cycle back to these schematics to see how the SDK-86 does it.

ADDRESSING MEMORY AND PORTS IN MICROCOMPUTER SYSTEMS

Address Decoder Concept

While discussing the block diagram of the SDK-86 board earlier in this chapter, we mentioned that the 3625 devices on the board serve as *address decoders*. One function of an address decoder is to produce a signal which enables the ROM, RAM, or port device that you want enabled for a particular address. A second, related function of an address decoder is to make sure that only one device at a time is enabled to put data on the data bus lines.

It seems that every microcomputer system does address decoding in a different way from every other system. Therefore, instead of memorizing the method used in one particular system, it is important that you understand the concept of address decoding. You can then figure out any system you have to work on.

A SYSTEM ROM DECODER

To start, look at Figure 7-7. This figure shows how eight EPROMs can be connected in parallel on a common address bus and common data bus. From just looking at the schematic you can see that these EPROMs output bytes of data because each has eight outputs connected to the system data bus. The number of address lines connected to each device gives you an indication of how many bytes are stored in it. Each EPROM has 12 address lines (A0–A11) connected to it. Therefore, the number of bytes stored in the device is 2^{12} or 4096. If you have trouble with this, think of how many bits a counter has to have to count the 4096 states from 0 to 4095 decimal, or 0000H to 0FFFFH.

Note that each 2732 in Figure 7-7 has a Chip Select ($\overline{CS}$) input. When this input is asserted low the addressed byte in a device will be output on the data bus. To get meaningful data from the EPROMs we need to make sure that the $\overline{CS}$ input of only one device at a time is low. In the circuit in Figure 7-7 this is done by the 74LS138. If the 74LS138 is enabled by making its $\overline{G2A}$ and $\overline{G2B}$ inputs low and its G1 input high, then only one output of the device will be low at a time. The output that will be low is determined by the 3-bit address applied to the C, B, and A select inputs. For example, if CBA is 000, then the Y0 output will be low, and all the other outputs will be high. ROM 0 will be selected. If CBA is 001, the Y1 output will be low and the ROM 1 will be selected. If CBA is 111, then Y7 will be low, and only ROM 7 will be enabled. Now let's see what address range each of these ROMs will have in the system.

To determine the addresses of ROMs, RAMs, and ports in a system, a good approach in many cases is to use a worksheet such as that in Figure 7-8. To make one of these worksheets you start by writing the address bits and the binary weight of each address bit across the top of the paper as shown in the figure. To make it easier to convert binary addresses to hex, it helps if you mark off the address lines in groups of four as shown. Next, draw vertical lines which mark off the three address lines that connect to the decoder select inputs (C, B, and A). For the decoder in Figure 7-7 address lines A14, A13, and A12 are connected to the C, B, and A inputs of the decoder,

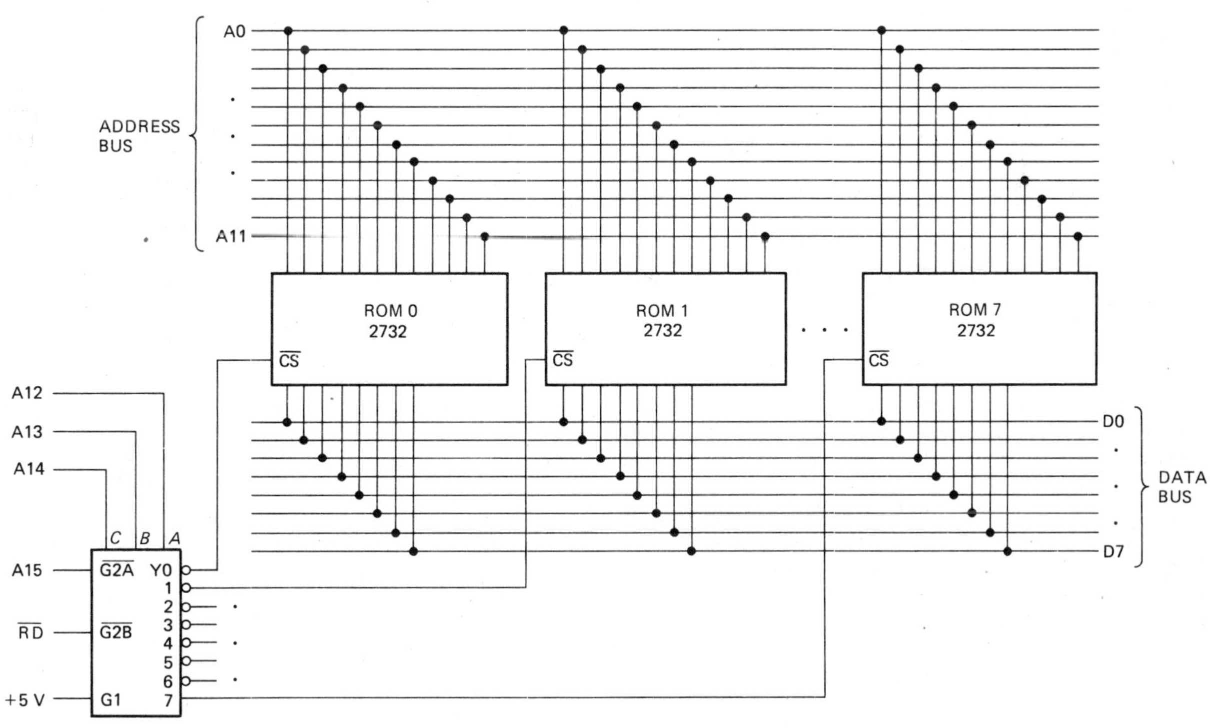

FIGURE 7-7 Parallel ROMs with decoder.

		HEX DIGIT				HEX DIGIT				HEX DIGIT				HEX DIGIT				
		2^{15} A15	2^{14} A14	2^{13} A13	2^{12} A12	2^{11} A11	2^{10} A10	2^{9} A9	2^{8} A8	2^{7} A7	2^{6} A6	2^{5} A5	2^{4} A4	2^{3} A3	2^{2} A2	2^{1} A1	2^{0} A0	HEX EQUIVALENT ADDRESS
BLOCK 1	START	0	0	0	0	0	0	0	0	0	0	0	0	0	0	0	0	= 0000
	END	0	0	0	0	1	1	1	1	1	1	1	1	1	1	1	1	= 0FFF
BLOCK 2	START	0	0	0	1	0	0	0	0	0	0	0	0	0	0	0	0	= 1000
	END	0	0	0	1	1	1	1	1	1	1	1	1	1	1	1	1	= 1FFF
BLOCK 3	START	0	0	1	0	0	0	0	0	0	0	0	0	0	0	0	0	= 2000
	END	0	0	1	0	1	1	1	1	1	1	1	1	1	1	1	1	= 2FFF
BLOCK 4	START	0	0	1	1	0	0	0	0	0	0	0	0	0	0	0	0	= 3000
	END	0	0	1	1	1	1	1	1	1	1	1	1	1	1	1	1	= 3FFF
BLOCK 5	START	0	1	0	0	0	0	0	0	0	0	0	0	0	0	0	0	= 4000
	END	0	1	0	0	1	1	1	1	1	1	1	1	1	1	1	1	= 4FFF
BLOCK 6	START	0	1	0	1	0	0	0	0	0	0	0	0	0	0	0	0	= 5000
	END	0	1	0	1	1	1	1	1	1	1	1	1	1	1	1	1	= 5FFF
BLOCK 7	START	0	1	1	0	0	0	0	0	0	0	0	0	0	0	0	0	= 6000
	END	0	1	1	0	1	1	1	1	1	1	1	1	1	1	1	1	= 6FFF
BLOCK 8	START	0	1	1	1	0	0	0	0	0	0	0	0	0	0	0	0	= 7000
	END	0	1	1	1	1	1	1	1	1	1	1	1	1	1	1	1	= 7FFF

DECODER ADDRESS INPUTS (A14, A13, A12)

FIGURE 7-8 Address decoder worksheet showing address decoding for eight 2732s in Figure 7-7.

respectively. Then write under each address bit the logic level that must be on that line to address the first location in the first EPROM. To address the first location in any of the EPROMs, the A0 through A11 address lines must all be low, so put a 0 under each of these address bits on the worksheet. To enable EPROM 0, the select inputs of the decoder must be all 0's. Since address lines A14, A13, and A12 are connected to these select inputs, they must then all be 0's to enable EPROM 0. Write a 0 under each of these address bits on the worksheet. Since address line A15 is connected to the $\overline{G2A}$ enable input of the decoder, it must be asserted low in order for the decoder to work at all. Write a 0 under the A15 bit on your worksheet. Note that the $\overline{RD}$ signal from the microprocessor control bus is connected to the $\overline{G2B}$ enable input of the decoder. The decoder then will only be enabled during a read operation. This is done to make sure that data cannot accidentally be written to ROM. The G1 enable input of the decoder is permanently asserted by tying it to +5 V because we don't need it for anything else in this circuit.

You can now read the starting address of EPROM 0 directly from the worksheet as 0000H. The highest address in EPROM 0 is that address where A0–A11 are all 1's. If you put a 1 under each of these bits as shown on the worksheet, you can see that the ending address for EPROM 0 is 0FFFH. Remember that A12–A14 have to be low to select EPROM 0. A15 has to be low to enable the decoder. The address range of EPROM 0 is said to be 0000H to 0FFFH, a 4 Kbyte block.

Now let's use the worksheet to determine the address range for EPROM 1. EPROM 1 is enabled when A15 is 0, A14 is 0, A13 is 0, and A12 is 1. For the first address in EPROM 1 address lines A0–A11 must all be low. Therefore, the starting address of EPROM 1 is 1000H. Its ending address, when A0–A11 are all 1's, is 1FFFH. If you look at the worksheet in Figure 7-8 you should see that the address ranges for the other six EPROMs in the system are 2000H to 2FFFH, 3000H to 3FFFH, 4000H to 4FFFH, 5000H to 5FFFH, 6000H to 6FFFH, and 7000H to 7FFFH. In this system then we use address lines A14, A13, and A12 to select one of eight EPROMs in the overall address range of 0000H to 7FFFH. Some people like to think of address lines A14, A13, and A12 as "counting off" 4096-byte blocks of memory. If you think of the address lines as the outputs of a 16-bit counter, you can see how this works. The end address for each EPROM has all 1's in address bits A0–A11. When you increment the address to access the next byte in memory, these bits all go to 0, and a 1 rolls over into bits A14, A13, and A12. This increments the count in these 3 bits by one and enables the next highest 4096-byte EPROM. The count in these bits goes from binary 000 to 111.

A SYSTEM RAM DECODER

The system in Figure 7-7 contains only ROM. In most systems we want to have ROM, RAM, and ports. To give you more practice with basic address decoding we will show you now how we can add a decoder for RAM to the system.

Suppose that we want to add eight 2K × 8 RAMs to the system, and we want the first RAM to start at address 8000H, just above the EPROMs which end at address 7FFFH.

To start, make another worksheet such as the one in Figure 7-8. Addressing one of the 2048 bytes (2^{11}) in each RAM requires 11 address lines, A0–A10. These lines will be connected directly to each RAM, so draw a vertical line on the worksheet to indicate this. Since we want to select one of eight RAM devices, we can use another 74LS138 such as we used for the EPROMs. We

HEX DIGIT				HEX DIGIT				HEX DIGIT				HEX DIGIT				HEX EQUIVALENT ADDRESS	START OF BLOCK
A15	A14	A13	A12	A11	A10	A9	A8	A7	A6	A5	A4	A3	A2	A1	A0		
1	0	0	0	0	0	0	0	0	0	0	0	0	0	0	0	= 8000H	1
1	0	0	0	1	0	0	0	0	0	0	0	0	0	0	0	= 8800H	2
1	0	0	1	0	0	0	0	0	0	0	0	0	0	0	0	= 9000H	3
1	0	0	1	1	0	0	0	0	0	0	0	0	0	0	0	= 9800H	4
1	0	1	0	0	0	0	0	0	0	0	0	0	0	0	0	= A000H	5
1	0	1	0	1	0	0	0	0	0	0	0	0	0	0	0	= A800H	6
1	0	1	1	0	0	0	0	0	0	0	0	0	0	0	0	= B000H	7
1	0	1	1	1	0	0	0	0	0	0	0	0	0	0	0	= B800H	8

DECODER ADDRESS INPUTS (A13, A12, A11)

FIGURE 7-9 Address decoder worksheet for eight 2 Kbyte RAMs starting at address 8000H.

want to select 2048-byte blocks of memory, so address line A11 will be connected to the A input of the decoder, A12 will be connected to the B input of the decoder, and A13 will be connected to the C input of the decoder. Under these 3 address bits on the worksheet list the 3-bit binary count sequence from 000 to 111 as we did in Figure 7-8. All we have left to decide is what to connect to the enable inputs of the decoder. We want the block of RAM selected by the outputs of this decoder to start at address 8000H. For this, address A15 is high and A14 is low. The G1 enable input of the decoder is active high so we connect it to the A15 address line. This input will then be asserted when A15 is high. We connect A14 to $\overline{G2A}$ of the decoder so that this input will be asserted when A14 is low. Because we don't need to use it in this circuit, we simply tie the $\overline{G2B}$ input of the decoder to ground so that it will be asserted all the time. Note that we don't connect the $\overline{RD}$ signal to an enable input on a RAM decoder, because we want to enable the RAMs for both read and write operations. Figure 7-9 shows the address decoder worksheet for the 74LS138 connections that we have just described.

Now, if you put a 1 under A15 on your worksheet and a 0 under A14, you will be able to quickly determine the address range for each of the RAMs. The first RAM will start at address 8000H. The ending address for this RAM will be at the address where bits A0–A10 are all 1's. If you put 1's under these bits on your worksheet, you should see that the ending address for the first RAM is 87FFH. For practice, work out the hexadecimal addresses for each of the other seven RAMs. When you finish, compare your results with those in Figure 7-9. The eight RAMs occupy the address space from 8000H to BFFFH.

A SYSTEM PORT DECODER

Figure 7-10*a* shows how another 74LS138 can be connected in our system to produce chip select signals for some port devices. Make another address decoder worksheet and see if you can figure out the system address that corresponds to each of these decoder outputs. Check your results with those in Figure 7-10*b*. First note that A15 and A14 must be high to enable the decoder, and write 1's under these bits on your worksheet. Then notice that A13 and A12 must be low to enable the

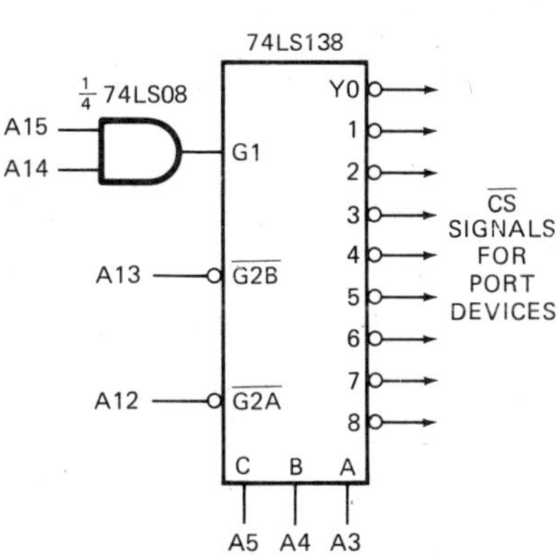

(*a*)

HEX DIGIT				HEX DIGIT				HEX DIGIT				HEX DIGIT				HEX PORT DEVICE ADDRESS			
A15	A14	A13	A12	A11	A10	A9	A8	A7	A6	A5	A4	A3	A2	A1	A0				
1	1	0	0	X	X	X	X	X	X	0	0	0	0	0	0	C	0	0	0
										0	0	1				C	0	0	8
										0	1	0				C	0	1	0
										0	1	1				C	0	1	8
										1	0	0				C	0	2	0
										1	0	1				C	0	2	8
										1	1	0				C	0	3	0
1	1	0	0	X	X	X	X	X	X	1	1	1				C	0	3	8

DECODER SELECT INPUTS (A5, A4, A3)

(*b*)

FIGURE 7-10 Adding a port device decoder. *(a)* Schematic for 74LS138 connections. *(b)* Address decoder worksheet.

decoder, and write a 0 under each of these bits on your worksheet. Finally determine which three address lines are connected to the select inputs of the decoder, and write the binary count sequence from 000 to 111 under these bits. For this port decoder, address lines A3, A4, and A5 will be connected to the decoder select inputs. Address lines A0, A1, and A2 will be connected directly to the port devices to address individual ports and control registers in the devices. This is the same idea as connecting the lower address lines directly to a ROM so that we can address one of the bytes stored there.

Address lines A6–A11 are not connected to the port devices or to the decoder, so they have no effect on selecting a port. We don't care then whether these bits are 1's or 0's. As you will see, these "don't care" bits mean that there are many addresses which will turn on one of the port devices. To give the simplest address for each device, however, we assume that each of these don't care bits is 0. Write 0's under each of these bits on your worksheet. You should now see that the address C000H will cause the Y0 output of the decoder to be asserted. The address C008H will cause the Y1 output of the decoder to be asserted. Using address lines A3, A4, and A5 on the decoder select inputs then leaves eight address spaces for each port device.

To see that any one of several different addresses can select one of these port devices, replace the 0 you put under A6 on the first line of your worksheet with a 1. This represents a system address of C040H. A15 and A14 are 1's and A13, A12, A5, A4, and A3 are 0's, for this address. Therefore, this address will also cause the Y0 output of the decoder to be asserted. You can try other combinations of 1's and 0's on A6–A11 if you need to further convince yourself that these bits don't matter when addressing ports. Again, we usually use 0's for these bits to give the simplest address.

Using a decoder which translates memory addresses to chip select signals for port devices is called *memory-mapped I/O*. In this system a port will be written to or read from in the same way as any other memory location. In other words, if this were an 8088 system, you would use an instruction such as MOV AL, DS:BYTE PTR 0C000H to read a byte of data from the first port to the AL register, instead of using the MOV DX, 0C000H and IN AL, DX instructions. The advantage of memory-mapped I/O is that any instruction which references memory can be used to input data from or output data to ports. In a system such as this, for example, the single instruction ADD AL, DS:BYTE PTR [0C000H] could be used to input a byte of data from the port at address C000H and add the byte to the AL register. The disadvantage of memory-mapped I/O is that some of the system memory address space is used up for ports and is therefore not available for memory.

You can use memory-mapped I/O with any microprocessor, but some microprocessors such as those of the 8086 family allow you to set up separate address spaces for input ports and for output ports. You access ports in these separate address spaces directly with the IN and OUT instructions. Having separate address spaces for input and output ports is called *direct I/O*. The advantage of direct I/O is that none of the system memory space is used for ports. The disadvantage is that only the specialized IN and OUT instructions can be used to input or output data.

In a later section of this chapter we show how direct I/O is done with the 8086, but first we will discuss how the 8086 addresses memory.

8086 and 8088 Addressing and Address Decoding

8086 MEMORY BANKS

The 8086 has a 20-bit address bus, so it can address 2^{20} or 1,048,576 addresses. Each address represents a stored byte. As you know from previous chapters, when you write a word to memory with an instruction such as MOV DS:WORD PTR[437AH], BX, the word is actually written into two consecutive memory addresses. Assuming that DS contains 0000, the low byte of the word is written into the specified memory address, 0437AH, and the high byte of the word is written into the next higher address, 0437BH. To make it possible to read or write a word with one machine cycle, the memory for an 8086 is set up as two "banks" of up to 524,288 bytes each. Figure 7-11*a* shows this in diagram form.

One memory bank contains all the bytes which have even addresses such as 00000, 00002, and 00004. The data lines of this bank are connected to the lower eight data lines, D0–D7, of the 8086. The other memory bank contains all of the bytes which have odd addresses such as 00001, 00003, and 00005. The data lines of this bank are connected to the upper eight data lines, D8–D15, of the 8086. Address line A0 is used as part of the enabling for memory devices in the lower bank. An addressed memory device in this bank will be enabled when address line A0 is low, as it will be for any even address. Address lines A1–A19 are used to select the desired memory device in the bank and to address the desired byte in that device.

Address lines A1–A19 are also used to select a desired memory device in the upper bank and to address the desired byte in that bank. An additional part of the enabling for memory devices in the upper bank is a separate signal called *bus high enable*, or $\overline{\text{BHE}}$. $\overline{\text{BHE}}$ is sent out from the 8086 at the same time as an address is sent out. An external latch, strobed by ALE, grabs the $\overline{\text{BHE}}$ signal and holds it stable for the rest of the machine cycle. The $\overline{\text{BHE}}$ signal will be asserted low if a *byte* is being accessed at an odd address, or if a *word* at an even address is being accessed. Figure 7-11*b* shows you what will be on the $\overline{\text{BHE}}$ and A0 lines for different types of memory accesses.

If you read a byte from or write a byte to an even address such as 00000H, A0 will be asserted low and $\overline{\text{BHE}}$ will be high. The lower bank will be enabled and the upper bank will be disabled. A byte will be transferred to or from the addressed location in the low bank on D0–D7. For an instruction such as MOV AH, DS:BYTE PTR [0000], the 8086 will automatically transfer the byte of

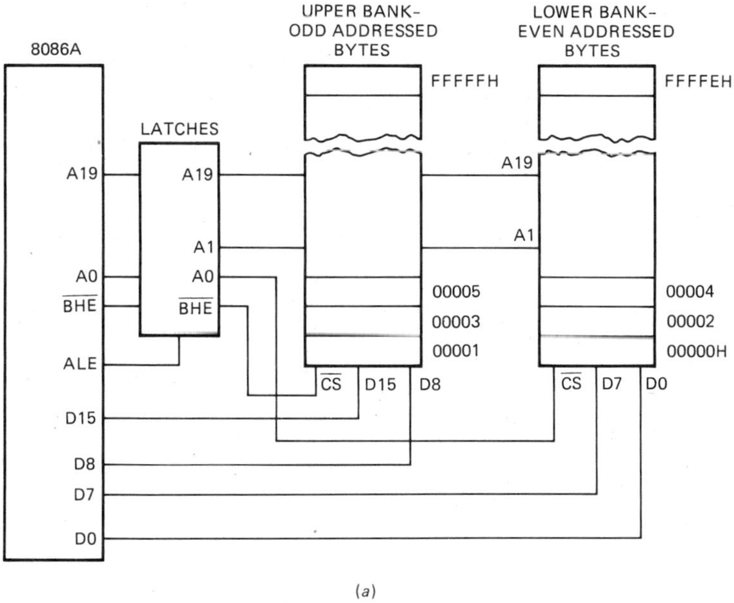

ADDRESS	DATA TYPE	$\overline{BHE}$	A0	BUS CYCLES
0000	BYTE	1	0	ONE
0000	WORD	0	0	ONE
0001	BYTE	0	1	ONE
0001	WORD	0	1	FIRST
		1	0	SECOND

(b)

FIGURE 7-11 8086 memory banks. *(a)* Block diagram. *(b)* Signals for byte and word operations.

data from the lower data bus lines to AH, the upper byte of the AX register. You just write the instruction and the 8086 takes care of getting the data in the right place.

Now, if the DS register contains 0000H and you use an instruction such as MOV AX, DS:WORD PTR [0000] to read a word from memory into AX, both A0 and $\overline{BHE}$ will be asserted low. Therefore, both banks will be enabled. The low byte of the word will be transferred from address 00000H to the 8086 on D0–D7. The high byte of the word will be transferred from address 00001H to the 8086 on D8–D15. The 8086 memory, remember, is set up in banks so that words, which have their low byte at an even address, can be transferred to or from the 8086 in one bus cycle. When programming an 8086, then, it is important to start an array of words on an even address for most efficient operation. If you are using an assembler, the EVEN directive is used to do this.

When you use an instruction such as MOV AL, DS:BYTE PTR [0001] to access just a byte at an odd address, A0 will be high and $\overline{BHE}$ will be asserted low. Therefore, the low bank will be disabled, and the high bank will be enabled. The byte will be transferred from memory address 00001H in the high bank to the 8086 on lines D8–D15. The 8086 will automatically transfer the byte of data from the higher eight data lines to AL, the low byte of the AX register. Note that address 00001H is actually the first location in the upper bank.

The final case in Figure 7-11*b* is that where you want to read a word from or write a word to an odd address. The instruction MOV AX, DS:WORD PTR [0001H] copies the low byte of a word from address 00001 to AL and the high byte from address 00002H to AH. In this case the 8086 requires two machine cycles to copy the two bytes from memory. During the first machine cycle the 8086 will output address 00001H and assert $\overline{BHE}$ low. A0 will be high. The byte from address 00001H will be read into the 8086 on lines D8–D15 and put in AL. During the second machine cycle the 8086 will send out address 00002H. A0 will be low, but $\overline{BHE}$ will be high. The second byte will be read into the 8086 on lines D0–D7 and put in AH. Note that the 8086 automatically takes care of getting a byte to the correct register regardless of which data lines the byte comes in on.

The main reason that the A0 and $\overline{BHE}$ signals function the way they do is to prevent the writing of an unwanted byte into an adjacent memory location when the 8086 writes a byte. To understand this, think what would happen if both memory banks were turned on for all write operations, and you wrote a byte to address 00002 with the instruction MOV DS:BYTE PTR [0002], AL. The data from AL would be written to address 00002 as desired. However, since the upper bank is also enabled, the random data on D8–D15 would be written into address 00003. The 8086 then is designed so that $\overline{BHE}$ is high during this byte write. This disables the upper bank of memory and prevents the random data on D8–D15 from being written to address 00003.

Now that you have an overview of address decoding and of the 8086 memory banks, let's look at some examples of how all of this is put together in a small system.

ROM ADDRESS DECODING ON THE SDK-86

Sheet 1 of the SDK-86 schematics in Figure 7-6 shows the circuit connections for the EPROMs and EPROM decoder. The 2716 EPROMs there are 2K × 8 devices.

Two of the EPROMs have their eight data outputs connected in parallel to system data lines D0–D7. These two EPROMs then give 4 Kbytes of storage in the lower memory bank. The other two EPROMs have their data outputs connected in parallel on system data lines D8–D15 to give 4 Kbytes of storage in the upper bank of ROM. Eleven address lines are needed to address the 2 Kbytes in each device. Therefore, system address lines A1–A11 are connected to each EPROM. Remember that we can't use A0 for this because, as we described in the last section, it is used in enabling the lower bank.

A 2716 has two enable inputs, $\overline{CE}$ and $\overline{OE}$. In order for the 2716 to output an addressed byte, both of these enable inputs must be asserted low. The $\overline{CE}$ inputs of the two devices in the lower bank are connected to system address line A0, so the $\overline{CE}$ inputs of these devices will be asserted if A0 is a 0. The $\overline{CE}$ inputs of the two 2716s in the upper bank are connected to the $\overline{BHE}$ line. The $\overline{CE}$ inputs of these devices then will be asserted whenever $\overline{BHE}$ is asserted low. To summarize, then, the two devices labeled *XA*27 and *XA*36 form the lower bank of EPROM and the two devices labeled *XA*30 and *XA*37 form the upper bank of EPROM in this system. To see how the $\overline{OE}$ enable input of each of these devices gets asserted and to determine the address that each device will have in the system you need to look next at the 3625 address decoder labeled *XA*26.

A 3625 is a 1K × 4 bipolar PROM which performs the same function that a 74LS138 performs in Figures 7-7 and 7-10. Since a 3625 has open collector outputs, a pull-up resistor to +5 V is required on each output. The dotted box around the four resistors on the schematic indicates the four are all contained in one package, resistor pack 5 (*RP*5). The 3625 translates an address to a signal which is used as part of the enabling of the desired device. Using a PROM as an address decoder, however, is for several reasons much more powerful than using a simple decoder such as the 74LS138. In the first place, the 3625 is programmable, which means that you can move the memory devices to new addresses in memory by simply programming a new PROM. Secondly, the large number of inputs on the PROM allow you to select a specific area of memory without using external gates. If, for example, you wanted the $\overline{G2A}$ input of a 74LS138 to be asserted if A11–A15 were all high, you would have to use an external NAND gate to detect this condition. With a PROM you can just make this condition part of the truth table you use to burn the PROM.

Now, to analyze any decoder circuit, first determine what signals are required to enable the decoder. The $\overline{CS1}$ enable input of the 3625 EPROM decoder is tied to ground so it is permanently enabled. The $\overline{CS2}$ enable input is tied to the $\overline{RD}$ signal from the 8086, so that the decoder will only be enabled if the 8086 is doing a read operation. As explained previously, you don't want to accidentally enable a ROM if you send out a wrong address during a write operation.

The next step in analyzing a decoder circuit using a PROM is to consult the manufacturer's manual for the system. You need to do this because, for a PROM, the relationship between the inputs and the outputs cannot be determined directly from the schematic.

Figure 7-12 shows the truth table for the PROM from the SDK-86 manual. This truth table is very similar to the address decoder worksheet that we used in previous sections of the chapter. From the truth table you can see that in order for the O1 output to be asserted low, M/$\overline{IO}$ has to be high. This is reasonable since this decoder is enabling memory devices. Address lines A12–A19 also have to be high in order for the O1 output of the PROM to be asserted low. Since the upper eight address bits must all be 1's for the O1 output to be asserted, the lowest address which will cause this is FF000H. Refer to sheet 1 of the SDK-86 schematics in Figure 7-6 to see that the O1 output of the decoder PROM connects to the $\overline{OE}$ enable inputs of two of the 2716 EPROMs, *XA*27 and *XA*30. Note also on the schematic, or remember from a previous discussion, that the other enable input of *XA*27, $\overline{CE}$, is connected to system address line A0. The *XA*27 EPROM then will be enabled whenever the 8086 does a memory read from an even address (A0 = 0) in the range FF000H to FFFFFH. Now let's look at the *XA*30 EPROM.

The $\overline{CE}$ enable input of *XA*30 is connected to the system $\overline{BHE}$ line. As shown in Figure 7-11, $\overline{BHE}$ will be asserted low whenever the 8086 accesses a byte at an odd address or a word at an even address. The *XA*30 EPROM then will be enabled when the 8086 reads a byte from an odd address in the range FF000H to FFFFFH. *XA*30 will also be enabled when the 8086 asserts both A0 and $\overline{BHE}$ low to read a word that starts on an even address in the range FF000H to FFFFFH.

These EPROMs are put at this high address in memory on the SDK-86 board because, after a RESET, the 8086 goes to address FFFF0H to get its first instruction. Since we want the SDK-86 to execute its monitor program after we press the RESET button, we locate the

PROM INPUTS				PROM OUTPUTS*				PROM ADDRESS BLOCK SELECTED
M/$\overline{IO}$	A14–A19	A13	A12	O4	O3	O2	O1	
1	1	1	1	1	1	1	0	FF000H–FFFFFH
1	1	1	0	1	1	0	1	FE000H–FEFFFH
1	1	0	1	1	0	1	1	FD000H–FDFFFH ($\overline{CSX}$)
1	1	0	0	0	1	1	1	FC000H–FCFFFH ($\overline{CSY}$)
ALL OTHER STATES				1	1	1	1	NONE

FIGURE 7-12 Truth table for an SDK-86 (A26) ROM decoder PROM.

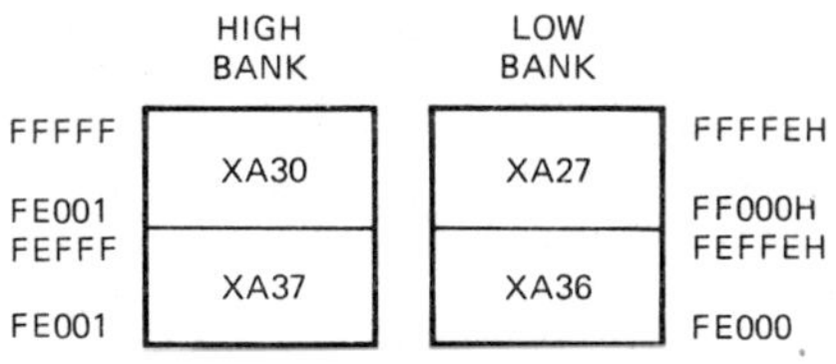

FIGURE 7-13 ROM memory map for SDK-86 board.

EPROM containing the monitor program such that this address is in it. The four SDK-86 EPROMs actually contain two monitor programs. One monitor in devices *XA*27 and *XA*30 allows you to use the hex keypad for entering and running programs. The other monitor in devices *XA*36 and *XA*37 allows you to use an external CRT terminal to enter and run programs. Using sheet 1 of the schematic and the PROM truth table in Figure 7-12, see if you can determine the address range for the *XA*36 and *XA*37 EPROMS.

The $\overline{OE}$ enable inputs of the *XA*36 and *XA*37 devices are connected to the O2 output of the address decoder PROM. According to the truth table for the PROM, the O2 output will be asserted low if M/$\overline{IO}$ is high, A14–A19 are high, A13 is high, and A12 is low. The lowest address that will assert the O2 output of the PROM then is FE000H, and the highest address that will assert the O2 output low is FEFFFH. Therefore the address range for *XA*36 and *XA*37 is FE000H—FEFFFH. Since A0 must also be low to enable the *XA*36 EPROM, this device contains the even-addressed bytes in this range. Since $\overline{BHE}$ must also be low to enable the *XA*37 EPROM, this device contains the odd-addressed bytes in the range FE000H—FEFFFH. A memory map such as the one in Figure 7-13 is a convenient way to summarize where each device is located in the system address space. Note that the 3625 ROM decoder has two unused outputs which can be used as part of the enabling for EPROMs you add to the prototyping section of the board.

RAM ADDRESS DECODING ON THE SDK-86

To give you another example of memory address decoding in a real system, we now discuss the RAM decoding of the SDK-86 board. Sheet 6 of the SDK-86 schematics in Figure 7-6 shows the circuit for the system RAM and RAM decoder. Let's look at this schematic to see what we can learn from it.

First take a look at the input and output lines on the 2142 static RAM devices. From the fact that each device has four data I/O lines you can conclude that the devices store 4-bit words. The fact that each device has 10 address inputs, A0–A9, indicates that each one stores 2^{10} or 1024 of these 4-bit words. To store bytes, two 2142s are enabled in parallel. Devices *A*38 and *A*41, for example, are enabled together to store bytes from the lower eight data lines,and devices *A*43 and *A*45 are enabled together to store bytes from the upper eight data lines. Note next that the control bus signals $\overline{RD}$, $\overline{WR}$, and M/$\overline{IO}$ are connected to all of the 2142s. $\overline{RD}$ is connected to the *output disable*, OD, pin on the 2142s. When the $\overline{RD}$ signal is high or when the device is not enabled, the output buffers will be disabled. During a read operation the $\overline{RD}$ signal is asserted low. If a 2142 is enabled and its OD input is low, the output buffers will be turned on so that an addressed word is output onto the data bus.

$\overline{WR}$ from the 8086 is connected to the *write enable*, $\overline{WE}$, input of the 2142s. If a 2142 is enabled, data on the data bus will be written into the addressed location in the RAM when the 8086 asserts $\overline{WR}$ low.

The 2142s have two *enable* inputs, $\overline{CS1}$ and CS2. The M/$\overline{IO}$ signal from the 8086 is connected to the CS2 input of all of the 2142s. Since the CS2 input is active high, it will be asserted whenever the 8086 is doing a memory operation. The $\overline{CS1}$ inputs of the 2142s are connected in pairs to the outputs of a 3625 PROM which functions as an address decoder.

In order to assert any of its outputs and enable some RAM, the 3625 must itself be enabled. Since the $\overline{CS2}$ enable input of the PROM is tied to ground, it is permanently enabled. The $\overline{CS1}$ enable input will be asserted when system address line A19 is low. To determine any more information about this PROM you need to look at the truth table for the device. Before we go on to that, however, note that A0 and $\overline{BHE}$ are connected to two of the address inputs on the 3625 PROM. Knowing what you do about 8086 memory banks, why do you think we want A0 and $\overline{BHE}$ to be part of what determines the outputs for this decoder? If you don't have the answer to this question, a look at the truth table for the device in Figure 7-14 should help you.

According to the third line of the truth table/address decoder worksheet in Figure 7-14, the O1 output of the PROM will be asserted low if A12–A18 are low, A11 is low, $\overline{BHE}$ is high, and A0 is low. The O1 output then will be

PROM INPUTS				PROM OUTPUTS*				BYTE(S) SELECTED (ADDRESS BLOCK)
A12-A18	A11	$\overline{BHE}$	A0	O4	O3	O2	O1	
0	0	0	0	1	1	0	0	BOTH BYTES (0H-07FFH)
0	0	0	1	1	1	0	1	HIGH BYTE (0H-07FFH)
0	0	1	0	1	1	1	0	LOW BYTE (0H-07FFH)
0	1	0	0	0	0	1	1	BOTH BYTES (0800H-0FFFH)
0	1	0	1	0	1	1	1	HIGH BYTE (0800H-0FFFH)
0	1	1	0	1	0	1	1	LOW BYTE (0800H-0FFFH)
ALL OTHER STATES				1	1	1	1	NONE

FIGURE 7-14 Truth table for an SDK-86 (A29) RAM decoder PROM.

asserted for even system addresses starting with 00000H. A low on the O1 output will enable the *A*38 and *A*41 RAMs which are connected to the lower half of the data bus. These two devices are part of the lower bank of RAM.

Next look at the second line of the PROM truth table in Figure 7-14. From this line you should see that the O2 output of the PROM will be asserted low if A12–A18 are low, A11 is low, BHE is low, and A0 is high. The O2 output will then be asserted for odd system addresses starting with 0001H. A low on the O2 output will enable the *A*43 and *A*45 RAMs which are connected on the upper half of the data bus. These two devices are part of the upper bank of RAM.

Now, suppose we want to write a 16-bit word to RAM at an even address. To do this we want both O1 and O2 to be asserted low so that both the lower bank RAMs and the upper bank RAMs are enabled. According to the first line of the PROM truth table in Figure 7-14, O1 and O2 will both be asserted low if BHE and A0 are both low. Remember from Figure 7-11 that BHE and A0 will both be low whenever you write a word to an even address or read a word from an even address. This last case gives the answer to the question we asked earlier about why A0 and BHE are connected to the address decoder PROM inputs. The two inputs are required to tell the PROM decoder to assert both O1 and O2 for a word read or write operation.

The address range for the *XA*38, *XA*41, *XA*43, and *XA*45 RAMs is 00000H to 007FFH. Another look at the PROM truth table in Figure 7-14 should show you that RAMS *XA*39, *XA*42, *XA*44, and *XA*46 contain 2K more bytes in the range 00800H to 00FFFH. Again, both banks of this additional RAM will be enabled if A0 and BHE are both low, as they are for reading or writing a word to an even address.

SDK-86 PORT ADDRESSING AND PORT DECODING

In a previous section of this chapter we described *memory-mapped input/output*. In a system with memory-mapped I/O, port devices are addressed and selected by decoders as if they were memory devices. The main advantage of memory-mapped I/O is that any instruction which refers to memory can theoretically be used to read from or write to a port. The single instruction ADD BH, DS:BYTE PTR [437AH] could be used to read a byte from a port and add the byte read in to the BH register. The disadvantage of memory-mapped I/O is that the ports occupy part of the system memory space. This space is then not available for storing data or instructions.

To avoid having to use part of the system memory space for ports, the 8086 family microprocessors have a separate address space for ports. Having a separate address space for ports is called *direct I/O*, because this separate address space is accessed directly with the IN and the OUT instructions.

Remember from previous chapters that the 8086 IN and OUT instructions each have two forms, *fixed* port and *variable* port. For fixed-port instructions an 8-bit port address is written as part of the instruction. The instruction IN AL, 38H, for example, copies a byte from port 38H to the AL register. For variable-port input or output operations, the 16-bit port address is first loaded into the DX register with an instruction such as MOV DX, 0FFF8H. The instruction IN AL, DX is then used to copy a byte from port FFF8H to the AL register. MOV DX, 0038H followed by IN AL, DX has the same effect as IN AL, 38H.

Whenever the 8086 uses the IN or OUT instructions to access a port, the port address is sent out directly from the 8086 on lines AD0–AD15. None of the segment registers has any effect on the address for an IN or OUT instruction. The 8086 always outputs 0's on lines A16–A19 during an IN or OUT instruction. Since the 8086 outputs a 16-bit address for direct I/O operations, it can address any one of 2^{16} or 65,536 input ports, and any one of 65,536 output ports. An 8086 system which uses direct I/O is designed so that the separate address space for ports is enabled when the M/IO signal from the 8086 is low. Remember that the M/IO signal being high was one of the enabling conditions for the SDK-86 ROM and RAM decoders we discussed in previous sections. The M/IO signal will be low during any direct input or output operation. The RD signal from the 8086 will also be low during an IN operation, so this signal is used to enable an addressed port device for input. The WR signal from the 8086 will be low along with M/IO during an OUT operation, so this signal is used to enable an addressed port for output.

For an example of how direct I/O ports are addressed and selected in a real system, we will again look at the SDK-86 schematics in Figure 7-6, sheet 7. Here another 3625 PROM (*XA*22) is used to produce the chip select signals for four I/O devices. The O1 output of the PROM is used to enable the 8279 keyboard/display interface device. A section of Chapter 9 discusses in detail the operation of this device. The O2 output of the PROM is used to enable the 8251A USART shown on sheet 9 of the schematics. The 8251A allows communication with other systems in serial form. A section in Chapter 13 discusses the operation of this device. The O3 and O4 outputs are connected to two 8255A parallel port devices shown on sheet 5 of the schematics. These devices can be enabled individually to input or output bytes. They can also be enabled together to input or output words. A later section in this chapter shows you how to tell each port in these devices whether you want it to be an input or an output.

Take a look now at the 3625 decoder PROM to see if you can determine what conditions enable it. You should find that the CS2 enable input of the PROM will be asserted when M/IO is low as it is during an input or output operation. Furthermore, you should see that the CS1 input will be asserted when A11–A15 are all high. Now, to see what addresses cause each of the PROM outputs to be asserted, refer to the address decoder worksheet for the PROM in Figure 7-15*a*. From this figure you can see that to assert the O1 output low, A5–A15 have to be high, A4 has to be low, and A3 has to be high and A0 has to be low. BHE can be either a high or a low. Note, however, that only the lower eight data lines, D0–

PROM INPUTS						PROM OUTPUTS*			
A11-A15	A5-A10	A4	A3	$\overline{BHE}$	A0	O4 HIGH PORT SELECT	O3 LOW PORT SELECT	O2 USART SELECT	O1 KDSEL
1	1	0	1	0	0	1	1	1	0
1	1	0	1	1	0	1	1	1	0
1	1	1	0	0	0	1	1	0	1
1	1	1	0	1	0	1	1	0	1
1	1	1	1	0	0	0	0	1	1
1	1	1	1	0	1	0	1	1	1
1	1	1	1	1	0	1	0	1	1
ALL OTHER STATES						1	1	1	1

(a)

PORT ADDRESS	PORT FUNCTION
0000 to FFDF	OPEN
FFE8	READ/WRITE 8279 DISPLAY RAM OR READ 8279 FIFO
E9	
EA	READ 8279 STATUS OR WRITE 8279 COMMAND
EB	
EC	RESERVED
ED	
EE	RESERVED
FFEF	
FFF0	READ/WRITE 8251A DATA
F1	
F2	READ 8251A STATUS OR WRITE 8251A CONTROL
F3	
F4	RESERVED
F5	
F6	RESERVED
FFF7	
FFF8	READ/WRITE 2855A PORT P2A
F9	READ/WRITE 8255A PORT P1A
FA	READ/WRITE 8255A PORT P2B
FB	READ/WRITE 8255A PORT P1B
FC	READ/WRITE 8255A PORT P2C
FD	READ/WRITE 8255A PORT P1C
FE	WRITE 8255A P2 CONTROL
FFFF	WRITE 8255A P1 CONTROL

(b)

FIGURE 7-15 Truth table and map for SDK-86 port decoder. *(a)* Truth table. *(b)* Map.

D7, are connected to the 8279. Therefore, data must be sent to or read from the 8279 at an even byte address. In other words data must be sent as a byte to an even address or as the lower byte of a word to an even address.

The system base address for this device then is FFE8H. System address line A1 is connected to the 8279 to select one of two internal addresses in the device. A1 low selects one internal address and A1 high selects the other internal address. A1 low gives system address FFE8H, and A1 high gives system address FFEAH. These are then the two addresses for the 8279 in this system.

According to the worksheet in Figure 7-15*a*, the O2 output of the decoder PROM will be asserted low when A4–A15 are high, and A3 and A0 are low. $\overline{BHE}$ can be either a low or a high, but, since only the lower eight data lines are connected to the 8251A USART, data must be sent to or read from the device as bytes at an even address. Again system address line A1 is used to select one of two internal addresses in the 8251A (Figure 7-6, sheet 6). A1 low selects one internal address and A1 high selects the other internal address. The two system addresses for this device then are FFF0H and FFF2H.

Now, before discussing the O3 and O4 outputs of the decoder PROM, we will take a brief look at the two 8255 parallel port devices they enable. These devices are shown on sheet 5 of the schematics in Figure 7-6. Each of these devices contains three 8-bit parallel ports and a control register. System address lines A1 and A2 are used to address the desired port or register in the device just as lower address lines are used to address the desired internal location in a memory device. Note that the lower eight data lines, D0–D7, are connected to the *XA*40 device, and the upper eight data lines are connected to the *XA*35 device. This is done so that you have several input or output possibilities. You can read a byte from or write a byte to an even-addressed port in device

*XA*40. You can read a byte from or write a byte to an odd-addressed port in device *XA*35. You can read a word from or write a word to a 16-bit port made up from an 8-bit port from device *XA*40 and an 8-bit port from device *XA*35. To input or output a word both devices have to be enabled. Now let's look at the decoder truth table to determine what addresses enable the various ports in these devices.

The *XA*40 device will be enabled by the O3 output of the decoder PROM if address lines A3–A15 are high and A0 is low. A1 and A2 are used to select internal ports of the 8255A. Let's assume that these two bits are 0 for the first address in the device. To select the A port in the *XA*40 8255A, address lines A1 and A2 have to be low. The system address that will enable this device and select the A port within it is FFF8H. Other values of A2 and A1 will select one of the other ports or the control register in this device. Figure 7-15*b* shows the system addresses for the ports and control register in this 8255. Note that the ports in this device (*XA*40) are identified as port 2A, port 2B, and port 2C. These all have even addresses because A0 must be low for this device to be selected.

The *XA*35 8255A which contains port 1A, port 1B, and port 1C will be enabled by the O4 output of the decoder PROM if A3–A15 are high and the $\overline{\text{BHE}}$ line is low. If this 8255A is being enabled for a byte read or write, then the A0 line will also be high. A2 and A1 are again used to address one of the ports or the control register within the 8255A. A2 = 0 and A1 = 1 will select port 1A in this 8255A. As shown in Figure 7-15*b*, then, the system address for port 1A is FFF9H. Port 1B will be accessed with a system address of FFFBH, port 1C will be accessed with a system address of FFFDH, and the internal control register will be accessed with a system address of FFFFH.

Note in the truth table in Figure 7-15*a* that the 3625 PROM decoder will enable a port device only when the specific address assigned to that device is sent out by the 8086. This is sometimes called *complete decoding* because all of the address lines play a part in selecting a device and one of its internal ports or registers. As we show in a later section, adding another decoder to produce enable signals for more port devices is very easy in a system which uses this complete decoding.

THE SDK-86 "OFF-BOARD" DECODER

Before we show you how another port decoder can be added to the SDK-86, we need to briefly discuss the operation of the *off-board circuitry* on sheet 5 of the SDK-86 schematics. The purpose of this circuitry is to produce the signal $\overline{\text{OFF BOARD}}$ whenever the 8086 sends out a memory or port address which does not correspond to a device decoded on the board. The $\overline{\text{OFF BOARD}}$ signal will be asserted low if pin 4 of the A3 NAND gate is low or if pin 5 of the *A*3 NAND gate is low. According to the truth table for the *XA*12 PROM in Figure 7-16, the O1 output will be low if the the 8086 is doing a memory operation and the address sent out is not in one of the ranges decoded for the on-board RAM or ROM.

In order for pin 4 of the *A*3 NAND gate to be low, both pin 9 and pin 10 of the *A*3 NAND gate must be high. Pin 10 will be high if the 8086 is doing an input or output operation (IO/$\overline{\text{M}}$ from the 8286 inverting buffer equals 1). Pin 9 of the *A*3 NAND gate will be high if any one of the *A*19 NAND gate inputs is low. Since system address lines A5–A15 are connected to the inputs of the 74LS133 NAND gate, the signal to pin 9 of *A*3 will be high for any address less than FFE0H. In other words, pin 4 of the *A*3 NAND gate will be asserted low for any I/O operation in an address range not selected by the *XA*22 port decoder.

The $\overline{\text{OFF BOARD}}$ signal produced by the previously discussed PROM and logic gates is connected to an input of a NAND gate labeled *A*2 on sheet 2 of the schematics. If $\overline{\text{OFF BOARD}}$ is asserted low, or $\overline{\text{INTA}}$ is asserted low, or $\overline{\text{HLDA}}$ is asserted low, the output of this gate will be high. For now all we are interested in is the fact that if $\overline{\text{OFF BOARD}}$ is asserted low, a high will be applied to pin 1 of the *A*3 NAND gate in zone *A*4 of the schematic. If the $\overline{\text{DEN}}$ signal from the 8086 is also asserted low, the signal labeled $\overline{\text{BUFFER ON}}$ will be asserted low. The $\overline{\text{DEN}}$ signal from the 8086 will be asserted whenever the 8086 reads in data from a memory location or a port, or when it writes data to a memory location or port. The $\overline{\text{BUFFER ON}}$ signal produced here is used to enable the 8286 data bus buffers (*XA*6 and *XA*7) shown on sheet 4 of the schematics. Now here's the point of all this.

In the next chapter we show you how to add another I/O decoder and some other devices to the prototyping area of an SDK-86 board. To drive these additional devices, the address, data and control buses must all be be buffered. The address bus on the SDK-86 board is buffered by the 74S373 address latches shown on sheet 3 of the schematics. Data bus and control bus buffers are not needed to drive the ROM, RAM, and port devices that come with the SDK-86 board. To read data from or write data to external devices, however, the data bus is buffered by two 8286s shown as *XA*7 and *XA*6 on sheet 4 of the SDK-86 schematics. These two buffers are turned

PROM INPUTS									PROM OUTPUT (O1)	CORRESPONDING ADDRESS BLOCK
M/$\overline{\text{IO}}$	A19	A18	A17	A16	A15	A14	A13	A12		
1	0	0	0	0	0	0	0	0	1 (INACTIVE)	0H–0FFFH (ON-BOARD RAM)
1	1	1	1	1	1	1	1	0	1 (INACTIVE)	FE000H–FEFFFH (ON-BOARD PROM)
1	1	1	1	1	1	1	1	1	1 (INACTIVE)	FF000H–FFFFFH (ON-BOARD PROM)
ALL OTHER STATES									0 (ACTIVE)	01000H–FDFFFH (OFF-BOARD)

FIGURE 7-16 SDK-86 "off-board" decoder PROM truth table.

on when the $\overline{\text{BUFFER ON}}$ signal, described in the preceding paragraph, is asserted low. The 8286 buffers are bidirectional. When these buffers are enabled, the Data Transmit/Receive signal, DT/$\overline{\text{R}}$, from the 8086 will determine in which direction the buffers are pointed. If DT/$\overline{\text{R}}$ is high, the buffers will be enabled to write data to some external device. If DT/$\overline{\text{R}}$ is low, the buffers will be enabled to read data in from some external device.

The control bus signals are buffered by an 8286 labeled *XA*11 and a 74LS244 labeled *XA*8 on sheet 4 of the SDK-86 schematics. These buffers are permanently enabled to send out the control bus signals except during a HOLD state which we will explain later.

THE SDK-86 WAIT STATE GENERATOR CIRCUITRY

Now that you know how the $\overline{\text{OFF BOARD}}$ signal is produced on the SDK-86 board, we can explain the operation of the *WAIT state generator circuitry* shown on sheet 2 of the schematics.

In a previous section of the chapter we mentioned that if the RDY input of the 8086 is asserted low, the 8086 will insert one or more WAIT states in the machine cycle it is currently executing. Figure 7-2 shows how a WAIT state is inserted in an 8086 machine cycle. During a WAIT state the information on the buses is held constant. Whatever was on the buses at the start of the WAIT state remains throughout the WAIT state. The main purpose of inserting one or more WAIT states in a machine cycle is to give an addressed memory device or I/O device more time to accept or output data. In the next major section of the chapter we show you how to determine whether a WAIT state is needed for a given device with a given 8086 clock frequency. For now, however, let's just see how the circuitry on the SDK-86 board causes the 8086 to insert a selected number of WAIT states.

WAIT states are inserted by pulling the RDY1 input of the 8284 clock generator IC low (Figure 7-6, sheet 2, zone *C6*). The 8284 internally synchronizes the RDY1 input signal with the clock signal and sends the resultant signal to the RDY input of the 8086. For the SDK-86 the RDY1 input will be asserted low if all three inputs of the *A*15 NAND gate shown in zone *D5* of the schematic are high. Pin 10 of this device is tied to +5 V, so it is permanently high. Pin 11 of *A*15 will be high if any of the inputs of the NAND gate in zone *D7* are asserted low. Pin 1 of this gate will be low whenever the 8086 does an input or output operation. Pin 2 of this gate will be low whenever the 8086 accesses a port or memory location which is not decoded on the board. In other words, with these connections the selected number of WAIT states will be inserted in each machine cycle when the 8086 does a read from or a write to an on-board I/O device, or when the 8086 does a read to or a write from any device not decoded on the board. If jumper W39 is installed on pin 13 of *A*15, pin 11 of *A*15 will always be high. The selected number of WAIT states selected by the W27–W34 jumpers will be inserted for all read and write operations.

The desired number of wait states to be inserted is selected by putting a jumper between two pins in the W27–W34 matrix shown in zone *D3* (sheet 2) of the schematic. If a jumper is installed in the W27 position, for example, no WAIT states will be inserted. If a jumper is installed in the W28 position, one WAIT state will be inserted. The pattern continues to jumper W34 which will cause seven WAIT states to be inserted in each machine cycle. Here's how the WAIT state generator itself works.

The 74LS164 WAIT state generator is an 8-bit shift register. At the start of a machine cycle the $\overline{\text{RD}}$, $\overline{\text{WR}}$, and $\overline{\text{INTA}}$ signals from the 8086 are all high. These three signals being high will cause the NAND gate in zone *C4* to assert the clear input, CLR, of the shift register. The outputs of the shift register will then all be low. One of these lows will be coupled through a jumper and an inverter to pin 9 of the *A*15 NAND gate we discussed previously. This high on pin 9 together with a high on pin 11 will cause the RDY1 input of the 8284 to be pulled low. However, wait states will not be inserted unless RDY1 remains low long enough. Now, when $\overline{\text{RD}}$, $\overline{\text{WR}}$, or $\overline{\text{INTA}}$ goes low in the machine cycle, the $\overline{\text{CLR}}$ input of the 74LS164 shift register will go high, and the shift register will function normally. The highs on the INA and INB inputs will be loaded onto the QA output on the next positive edge of the clock. If the wait state jumper is in the W27 position, then this high on the QA output will, through the inverter and NAND gate, cause the RDY1 input of the 8284 to go high again. For this case the RDY1 input goes high soon enough that no WAIT states are inserted.

The high loaded into the 74LS164 shift register is shifted one stage to the right by each successive clock pulse. When the high reaches the jumper connected to the *A*25 inverter, it will cause the RDY1 input of the 8284 to go high. The 8086 will then exit from a WAIT state on the next clock pulse. The number of WAIT states inserted in a machine cycle is determined by how many stages the high has to be shifted before it reaches the installed jumper.

To summarize all of this, the 8086 will insert the selected number of WAIT states in any machine cycle which accesses any device not addressed on the board, or any I/O device on the board. If jumper W39 is inserted, the selected number of WAIT states will be inserted for any on-board or off-board access. The purpose of inserting wait states is to give the addressed device more time to accept or output data.

How the 8088 Microprocessor Accesses Memory and Ports

Now that we have shown in detail how the 8086 accesses memory and port devices, we can show you how the 8088 does it.

In Chapter 2 we mentioned that the 8088 is the CPU used in the IBM PC and the IBM PC/XT. The instruction set of the 8088 is identical to that of the 8086, and the registers of the two are the same. There are two major differences between the two devices. First, the 8088 instruction byte queue is only 4 bytes long instead of 6. Second, and more important, the 8088 memory is not divided into two banks as the 8086 memory is. The 8088 only has an 8-bit data bus, AD0–AD7. All of the memory devices and ports in an 8088 system are con-

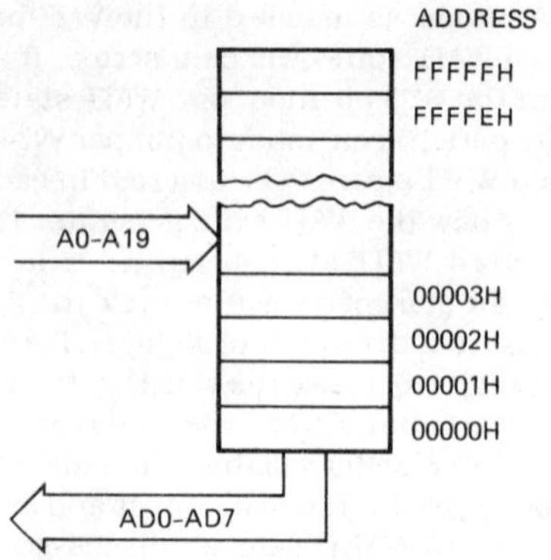

FIGURE 7-17 8088 memory structure.

nected onto these eight lines. The 8088 memory then functions as a single bank of up to 1,048,576 bytes. Figure 7-17 shows this structure. This single bank structure means that an 8088 cannot read a word from or write a word to memory in one machine cycle as the 8086 can. The 8088 can only read or write bytes, so the 8088 must always do 2 machine cycles to read or write a word. Address lines A0–A19 are used with some decoders to select a desired byte in memory. The 8088 does not produce the $\overline{BHE}$ signal, because it is not needed.

Most of the available memory devices and I/O devices were designed for 8-bit microprocessors which have 8-bit data buses. The 8088 was designed with an 8-bit data bus so that it would interface more easily with 8-bit memory devices and I/O devices. For example, in an 8088 system a simple 74LS138 can be used for a port device decoder as we showed in Figure 7-10*a*. An 8086 system requires a more complex decoder such as the 3625 PROM in Figure 7-6 (sheet 7), because the decoder has to take into account the states of A0 and $\overline{BHE}$.

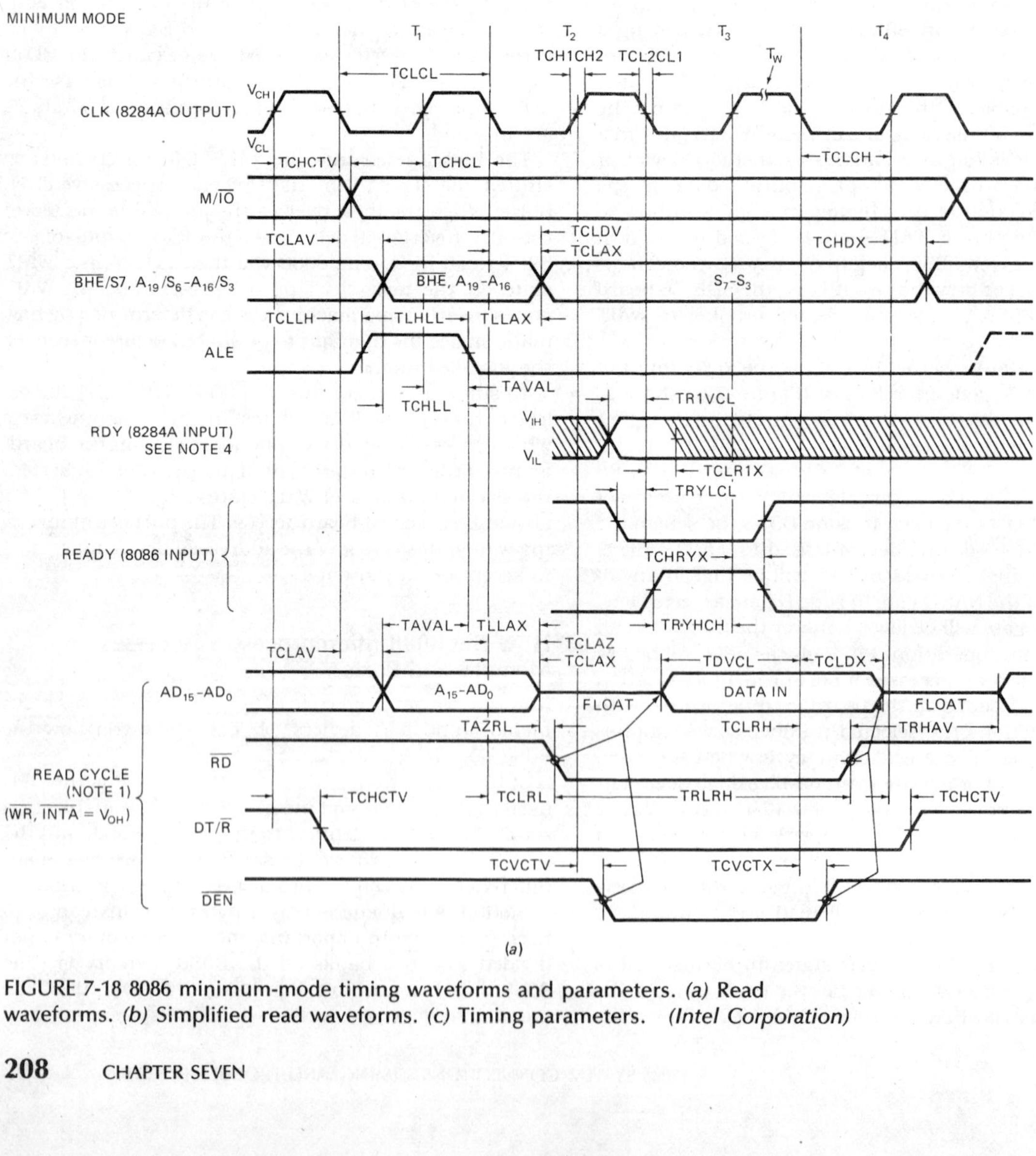

FIGURE 7-18 8086 minimum-mode timing waveforms and parameters. *(a)* Read waveforms. *(b)* Simplified read waveforms. *(c)* Timing parameters. *(Intel Corporation)*

8086 TIMING PARAMETERS

In previous sections of this chapter we used generalized timing waveforms such as that in Figure 7-2. These diagrams are sufficient to show the sequence of activities on the 8086 buses. However, they are not detailed enough to determine, for example, whether a memory device is fast enough to work in a given 8086 system. To allow you to make precise timing calculations, manufacturer's data books give detailed timing waveforms and lists of timing parameters for each microprocessor. Complete timing information for the 8086 is contained in the data sheet in the appendix. Figure 7-18 shows some of these for the 8086 operating in minimum mode.

As you look at Figure 7-18 remember the *5-minute freak-out rule*. Most of the time there are only a very few of these parameters that you need to worry about. In most systems, for example, you don't need to worry about the clock signal parameters, because an 8284 clock generator and a crystal will be used to produce the clock signal. The frequency of the clock signal from an 8284 is always one-third the resonant frequency of the crystal connected to it. The 8284 is designed to guarantee the correct clock period, clock time low, clock time high, etc. as long as the correct suffix number part is used. The 8284A, for example, can be used in an 8-MHz system, but a faster part, the 8284A-1 must be used for a system where a 10-MHz clock is desired.

The edges of the clock signal cause operations in the 8086 to occur; therefore, as you can see in Figure 7-18*a*, the clock waveform is used as a reference for other times. The timing values for when the 8086 puts out $M/\overline{IO}$, addresses, ALE, and control signals, for example, are all specified with reference to an appropriate clock edge.

As we mentioned earlier, one of the main things you use these diagrams and parameters for is to find out whether a particular memory or port device is fast enough to work in a system with a given clock frequency. Here's an example of how you do this.

If you look in zone C5 of sheet 2 of the SDK-86 schematics, you will see that if jumper W41 is installed, the 8086 will receive a 4.9-MHz clock signal from the 8284. If jumper W40 is installed, the 8086 will receive the 2.45-MHz PCLK signal from the 8284. Now, suppose that you want to determine whether the 2716 EPROMs on the SDK-86 board will work correctly with no WAIT states if you install jumper W41 to run the 8086 with the 4.9-MHz clock.

First you look up the access times for the 2716

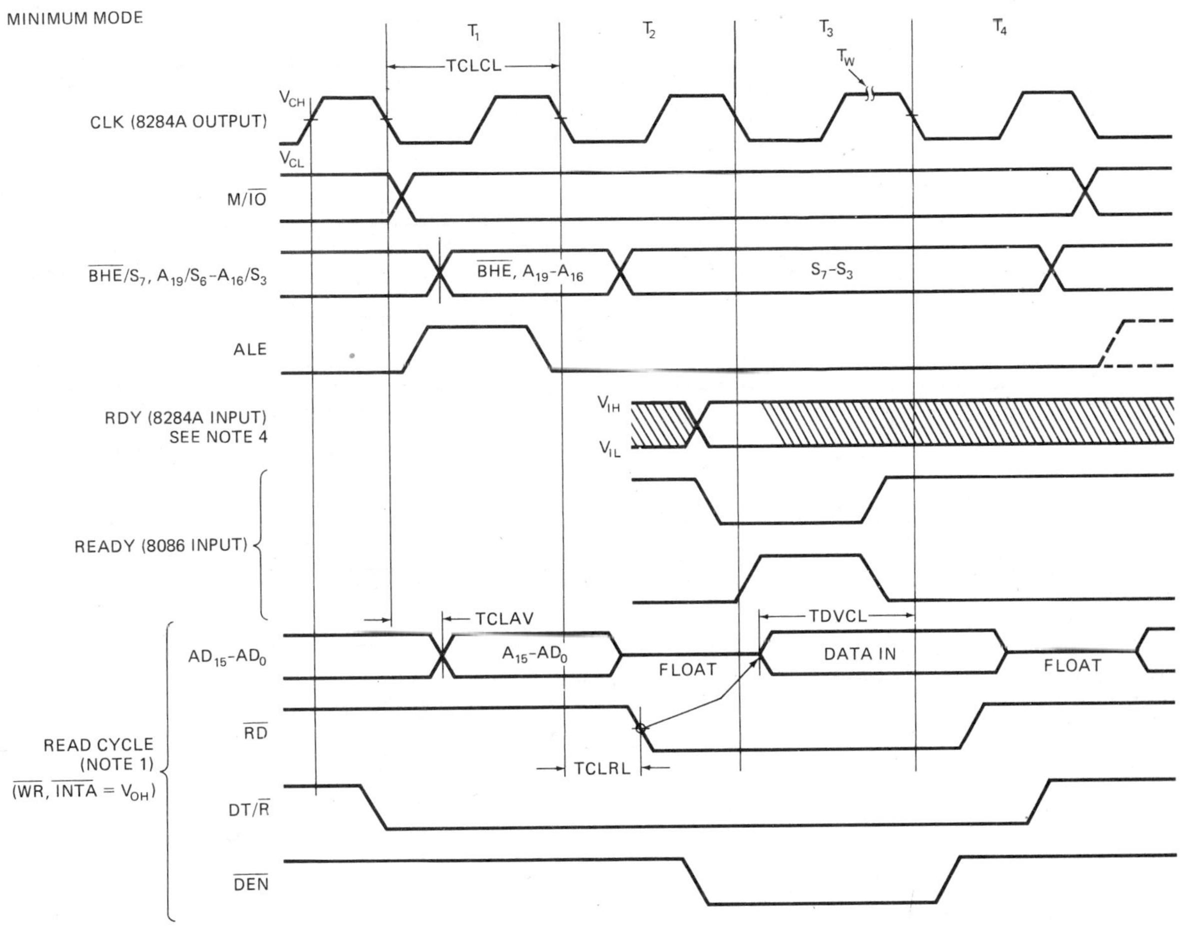

FIGURE 7-18 (*continued*)

EPROM in the appropriate data book. According to an Intel data book, the 2716 has a maximum address to output access time, t_{ACC}, of 450 ns. This means that if the 2716 is already enabled and its output buffers turned on, it will put valid data on its outputs no more than 450 ns after an address is applied to the address inputs. The 2716 data sheet also gives a chip enable to output access time, t_{CE}, of 450 ns. This means that if an address is already present on the address inputs of the 2716, and the output buffers are already enabled, the 2716 will put valid data on its outputs no later than 450 ns after the $\overline{CE}$ input is asserted low. A third parameter given for the 2716 in the data book is an output enable to output time, t_{OE}, of 120 ns maximum. This means that if the device already has an address on its address inputs, and its $\overline{CE}$ input is already asserted, valid data

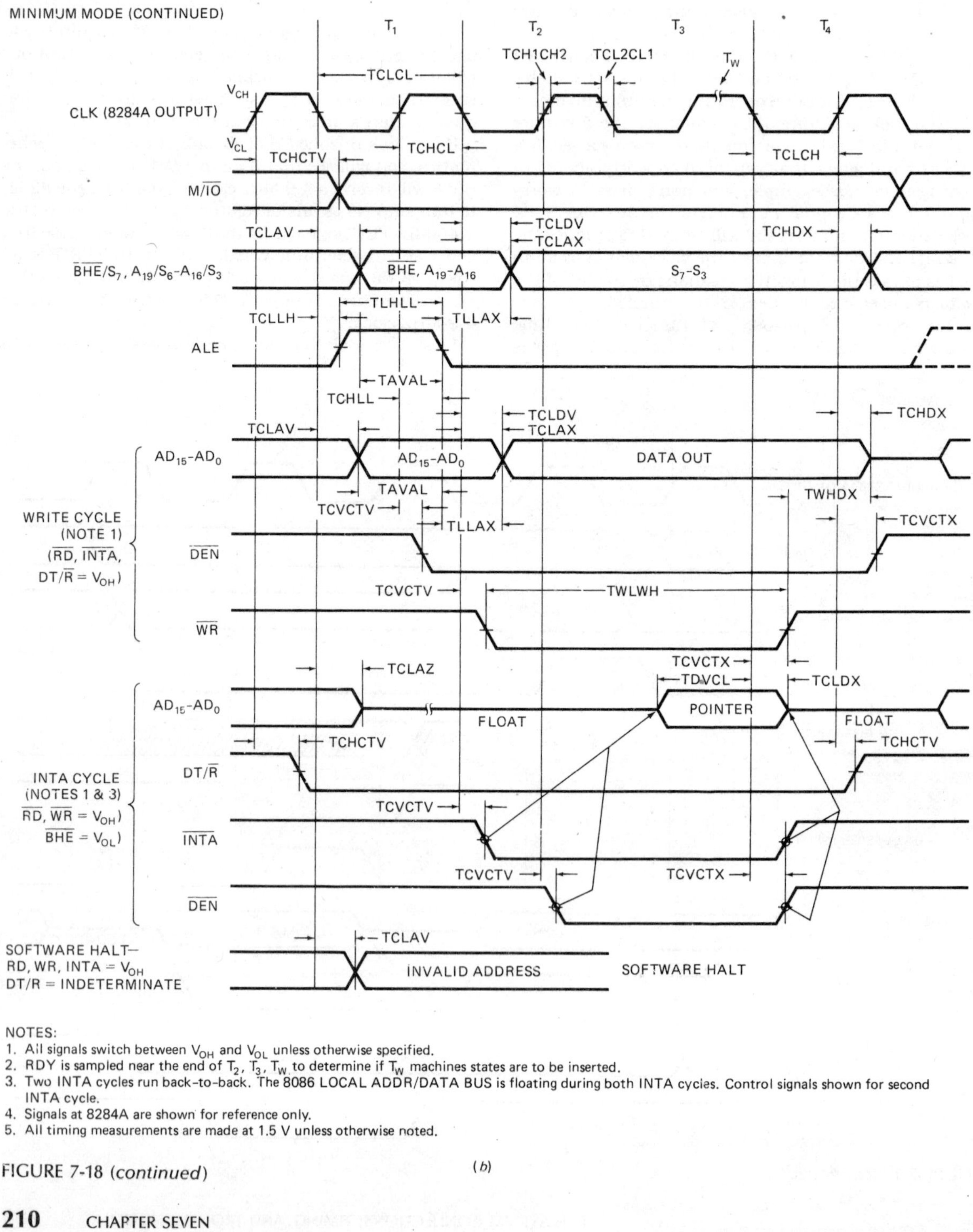

NOTES:
1. All signals switch between V_{OH} and V_{OL} unless otherwise specified.
2. RDY is sampled near the end of T_2, T_3, T_W to determine if T_W machines states are to be inserted.
3. Two INTA cycles run back-to-back. The 8086 LOCAL ADDR/DATA BUS is floating during both INTA cycles. Control signals shown for second INTA cycle.
4. Signals at 8284A are shown for reference only.
5. All timing measurements are made at 1.5 V unless otherwise noted.

FIGURE 7-18 *(continued)* (*b*)

MINIMUM COMPLEXITY SYSTEM TIMING REQUIREMENTS

SYMBOL	PARAMETER	8086 MIN.	8086 MAX.	8086-(Preliminary) MIN.	8086-(Preliminary) MAX.	8086-2 MIN.	8086-2 MAX.	UNITS	TEST CONDITIONS
TCLCL	CLK Cycle Period	200	500	100	500	125	500	ns	
TCLCH	CLK Low Time	118		53		68		ns	
TCHCL	CLK High Time	69		39		44		ns	
TCH1CH2	CLK Rise Time		10		10		10	ns	From 1.0 V to 3.5 V
TCL2CL1	CLK Fall Time		10		10		10	ns	From 3.5 V to 1.0 V
TDVCL	Data in Setup Time	30		5		20		ns	
TCLDX	Data in Hold Time	10		10		10		ns	
TR1VCL	RDY Setup Time into 8284A (See Notes 1, 2)	35		35		35		ns	
TCLR1X	RDY Hold Time into 8284A (See Notes 1, 2)	0		0		0		ns	
TRYHCH	READY Setup Time into 8086	118		53		68		ns	
TCHRYX	READY Hold Time into 8086	30		20		20		ns	
TRYLCL	READY Inactive to CLK (See Note 3)	−8		−10		−8		ns	
THVCH	HOLD Setup Time	35		20		20		ns	
TINVCH	INTR, NMI, $\overline{TEST}$ Setup Time (See Note 2)	30		15		15		ns	
TILIH	Input Rise Time (Except CLK)		20		20		20	ns	From 0.8 V to 2.0 V
TIHIL	Input Fall Time (Except CLK)		12		12		12	ns	From 2.0 V to 0.8 V
TCLAV	Address Valid Delay	10	110	10	50	10	60	ns	*C_L = 20 − 100 pF for all 8086 Outputs (In addition to 8086 selfload)
TCLAX	Address Hold Time	10		10		10		ns	
TCLAZ	Address Float Delay	TCLAX	80	10	40	TCLAX	50	ns	
TLHLL	ALE Width	TCLCH-20		TCLCH-10		TCLCH-10		ns	
TCLLH	ALE Active Delay		80		40		50	ns	
TCHLL	ALE Inactive Delay		85		45		55	ns	
TLLAX	Address Hold Time to ALE Inactive	TCHCL-10		TCHCL-10		TCHCL-10		ns	
TCLDV	Data Valid Delay	10	110	10	50	10	60	ns	
TCHDX	Data Hold Time	10		10		10		ns	
TWHDX	Data Hold Time After WR	TCLCH-30		TCLCH-25		TCLCH-30		ns	
TCVCTV	Control Active Delay 1	10	110	10	50	10	70	ns	
TCHCTV	Control Active Delay 2	10	110	10	45	10	60	ns	
TCVCTX	Control Inactive Delay	10	110	10	50	10	70	ns	
TAZRL	Address Float to READ Active	0		0		0		ns	
TCLRL	$\overline{RD}$ Active Delay	10	165	10	70	10	100	ns	
TCLRH	$\overline{RD}$ Inactive Delay	10	150	10	60	10	80	ns	
TRHAV	$\overline{RD}$ Inactive to Next Address Active	TCLCL-45		TCLCL-35		TCLCL-40		ns	
TCLHAV	HLDA Valid Delay	10	160	10	60	10	100	ns	
TRLRH	$\overline{RD}$ Width	2TCLCL-75		2TCLCL-40		2TCLCL-50		ns	
TWLWH	$\overline{WR}$ Width	2TCLCL-60		2TCLCL-35		2TCLCL-40		ns	
TAVAL	Address Valid to ALE Low	TCLCH-60		TCLCH-35		TCLCH-40		ns	
TOLOH	Output Rise Time		20		20		20	ns	From 0.8 V to 2.0 V
TOHOL	Output Fall Time		12		12		12	ns	From 2.0 V to 0.8 V

NOTES:
1. Signal at 8284A shown for reference only.
2. Setup requirement for asynchronous signal only to guarantee recognition at next CLK.
3. Applies only to T2 state. (8 ns into T3).

(*c*)

FIGURE 7-18 (*continued*)

will appear on the output pins at most 120 ns after the $\overline{\text{OE}}$ pin is asserted low.

Now that you have these three parameters for the 2716, the next step is to check if each one of these times is short enough for the device to work with a 4.9-MHz 8086. In other words, does the 2716 put out valid data soon enough after it is addressed and enabled to satisfy the requirements of the 8086? To determine this you need to look at both the 8086 timing parameters and how the 2716 is addressed and enabled on the SDK-86 board.

To make it easier for you to find the important parameters for these calculation, we show in Figure 7-18*b* a simplified version of the timing diagram in Figure 7-18*a*. You should try to mentally do this simplification whenever you are faced with a timing diagram. As shown by the timing diagram in Figure 7-18*b* the 8086 sends out M/$\overline{\text{IO}}$, $\overline{\text{BHE}}$, and an address during T1 of the machine cycle. Note on the AD15–AD0 line of the timing diagram that the 8086 outputs this information within a time labeled TCLAV after the falling edge of the clock at the start of T1. TCLAV stands for *time from clock low to address valid.* According to the data sheet shown in Figure 7-18*c* in the 8086 column, the maximum value of this time is 110 ns. Now look further to the right on the AD15–AD0 lines. You should see that valid data must arrive at the 8086 from memory a time TDVCL before the falling edge of the clock at the end of T3. TDVCL stands for *time data must be valid before clock goes low.* The data sheet gives a value of 30 ns for this parameter.

The time between the end of the TCLAV interval (time clock low to address valid) and the start of the TDVCL interval is the time available for getting the address to the memory, and for the t_{ACC} of the memory device. You can determine this time by subtracting TCLAV and TDVCL from the time for 3 clock cycles. With a 4.9-MHz clock each clock cycle will be 204 ns. Three clock cycles then total 612 ns. Subtracting a TCLAV of 110 ns and a TDVCL of 30 ns leaves 472 ns available for getting the address to the 2716 and for its t_{ACC}. To help you visualize these times, Figure 7-19*a* shows this operation in simplified diagram form.

If you look at sheets 1 and 3 of the SDK-86 schematics you should see that the $\overline{\text{BHE}}$ signal and the A0–A11 address information goes from the 8086 through the 74S373 latches to get to the 2716s. The propagation delay of the 74S373s then must be subtracted from the 472 ns to determine how much time is actually available for the t_{ACC} of the 2716. The maximum delay of a 74S373 is 12 ns. As shown in Figure 7-19*a*, subtracting this from the 472 ns leaves 460 ns for the t_{ACC} of the 2716. Now, as we told you in a previous paragraph, the 2716 has a maximum t_{ACC} of 450 ns. Since this 450 ns is less than the 460 ns available, you know that the t_{ACC} of the 2716 is acceptable for the SDK-86 operating with a 4.9-MHz clock. You still, however, must check if the values of t_{CE} and t_{OE} for the 2716 are acceptable.

If you look at sheet 1 of the SDK-86 schematics, you should see that the $\overline{\text{CE}}$ inputs of the 2716s are connected to either A0 or to $\overline{\text{BHE}}$. The timing for these signals is the same as that for the addresses in the preceding section. As shown in Figure 7-19*a* the time available for t_{CE} of the 2716 will be 460 ns. Since the maximum t_{CE} of the 2716 is 450 ns, you know that this parameter is also acceptable for an SDK-86 operating with a 4.9-MHz clock.

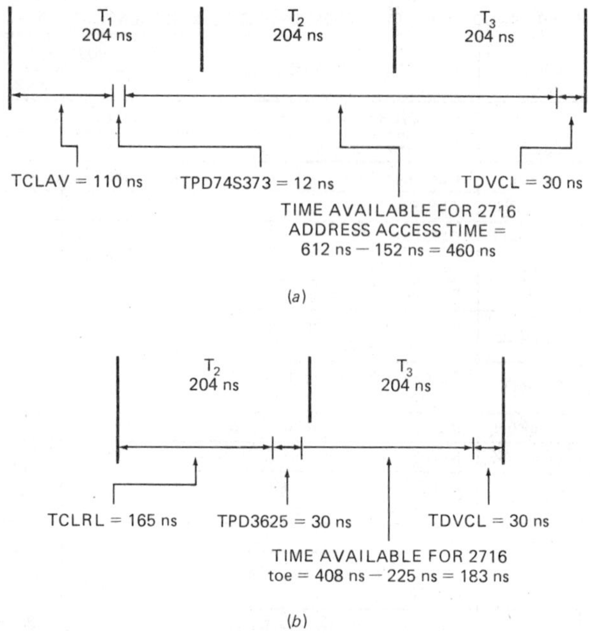

FIGURE 7-19 Calculations of 8086 times available for 2716 EPROM access. *(a)* Time for t_{ACC} and t_{RD}. *(b)* Time for t_{OE}.

The final parameter to check is t_{OE} of the 2716. According to sheet 1 of the SDK-86 schematics, the $\overline{\text{OE}}$ signals for the 2716s are produced by the 3625 decoder. The signals coming to this decoder are A12–A19, M/$\overline{\text{IO}}$, and $\overline{\text{RD}}$. Look at the 8086 timing diagram in Figure 7-18*b* to see if you can determine which of these signals arrives last at the 3625. You should find that addresses and M/$\overline{\text{IO}}$ are sent out during T1, but $\overline{\text{RD}}$ is not sent out until T2. As indicated by the arrow from the falling edge of the $\overline{\text{RD}}$ signal, $\overline{\text{RD}}$ going low causes the address decoder to send an $\overline{\text{OE}}$ signal to the 2716 EROMs. Since $\overline{\text{RD}}$ is sent out so much later than addresses, it will be the limiting factor for timing. $\overline{\text{RD}}$ going low and the EPROM returning valid data must occur within the time of states T2 and T3. Now, according to the timing diagram, $\overline{\text{RD}}$ is sent out from the 8086 within a time TCLRL after the falling edge of the clock at the start of T2. From the data sheet the maximum value of TCLRL is 165 ns. As we discussed before, the 8086 requires that valid data arrive on AD0–AD15 from memory a time TDVCL before the falling edge of the clock at the end of T3. The minimum value of TDVCL from the data sheet is 30 ns. The time between the end of the TCLRL interval and the start of the TDVCL interval is the time available for the $\overline{\text{OE}}$ signal to get produced and for the $\overline{\text{OE}}$ signal to turn on the memory. To determine the actual time available for these operations, first compute the time for states T2 and T3. For a 4.9-MHz clock each clock cycle or state will be 204 ns, so the two together total 408 ns. Then subtract the TCLRL of 165 ns and the TDVCL of 30 ns. As shown by the simple diagram in Figure 7-19*b*, this leaves 213 ns available for the decoder delay and the t_{OE} of the 2716. Checking a data sheet for the 3625 would show you that it has a maximum $\overline{CS2}$ to output delay of 30 ns. Subtract this from the available 213 ns to see how much time is left for the t_{OE} of the 2716. The result of this subtraction is 183 ns. As we indicated in a previous paragraph, the 2716 has a maximum t_{OE} of 120 ns. Since this time is considerably less than the 183 ns available, the 2716 has an acceptable t_{OE} value for operating on the SDK-86 board with a 4.9-MHz clock.

You have now checked all three 2716 parameters and found that all three are acceptable for an SDK-86 operating on a 4.9-MHz clock. No wait states need to be inserted when these devices are accessed, so jumper W39 in zone *D7* on sheet 2 of the schematics can be left out. As discussed in a previous section, installing jumper W39 will cause the selected number of WAIT states to be inserted for all memory or I/O operations.

Here's a final point about calculating the time available for t_{ACC}, t_{CE}, and t_{OE} of some device in a system. Suppose that you want to add another pair of 2716 EPROMs in the prototyping area of the SDK-86 board, and you want to enable the outputs of these added devices with the O3 output of the 3625 ROM decoder on sheet 1 of the schematics. The timing for these added devices will be the same as for the previously discussed 2716s except that the data from the added devices must come back through the 8286 buffers shown on sheet 4 of the SDK-86 schematics. According to an 8286 data sheet, these buffers have a maximum delay of 30 ns. This 30 ns must be subtracted from the times available for t_{ACC}, t_{CE}, and t_{OE}. If you look back at our calculations of the time available for t_{ACC} in Figure 7-19*a*, for example, you will see that we ended up with 460 ns available for t_{ACC}. Subtracting the 30 ns of buffer delay from this leaves only 430 ns, which is considerably less than the maximum t_{ACC} of 450 ns for the 2716. This tells you that, because of the buffer delay, the added 2716's are not fast enough to operate on an SDK-86 board with a 4.9-MHz clock and no WAIT states. To take care of this problem, the SDK-86 is designed so that any access to a memory or I/O device "off board" will cause the selected number of WAIT states to be inserted in the machine cycle. For our example here, selecting one WAIT state with jumper W28 on sheet 2 will give another 204 ns for the data to get from the 2716s to the 8086. This is more than enough time to compensate for the buffer delay, so the added 2716s will work correctly.

TROUBLESHOOTING A SIMPLE 8086-BASED MICROCOMPUTER

Now that you have some knowledge of the software and the hardware of a microcomputer system, we can start teaching you how to troubleshoot a simple microcomputer system such as an SDK-86 board. For this section assume that the microcomputer or microprocessor-based instrument previously worked. Later sections of this book will describe how the prototype of a microprocessor-based instrument is developed.

The following sections describe a series of steps that we have found effective in troubleshooting various microcomputer systems. The first point to impress on your mind about troubleshooting a microcomputer is that a systematic approach is almost always more effective than random poking, probing, and hoping. You don't, for example, want to spend 2 hours troubleshooting a system and finally find that the only problem is that the power supply is putting out only 3 V instead of 5 V. Use the list of steps below or a list of your own each time you have to troubleshoot a microcomputer.

1. Identifying the Symptoms

Make a list of the symptoms that you find or those that a customer describes to you. Find out, for example, whether the symptom is present immediately or whether the system must operate for a while before it shows up. If someone else describes the symptoms to you, check them yourself, or have that person demonstrate the symptoms to you. This allows you to check if the problem is with the machine or with how the person is attempting to use the machine.

2. Making a Careful Visual and Tactile Inspection

This step is good for preventive maintenance as well for finding a current problem. Check for components that have been or are excesively hot. When touching components to see if any are too hot, do it gently, because a bad IC can get hot enough to give a nasty burn if you keep your finger on it too long.

Check to see that all ICs are firmly seated in their sockets and that the ICs have no bent pins. Vibration can cause ICs to work loose in their sockets. A bent pin may make contact for a while, but after heating, cooling, and vibration, it may no longer make contact. Also, inexpensive IC sockets may oxidize with age and no longer make good contact.

Check for broken wires and loose connectors. A thin film of dust, etc. may form on printed circuit board edge connectors and prevent them from making dependable contact. The film can be removed by gently rubbing the edge connector fingers with a clean, nonabrasive pencil eraser. If the microcomputer has ribbon cables, check to see if they have been moved around or stressed. Ribbon cables usually have small wires that are easily broken. If you suspect a broken conductor in a ribbon cable, you can later make an electrical check to verify your suspicions.

3. Checking the Power Supply

From the manual for the microcomputer determine the power supply voltages. Check the supply voltage(s) directly on the appropriate pins of some ICs to make sure the voltage is actually getting there. Check with a scope to make sure the power supplies do not have excessive noise or ripple. One microcomputer that we were called on to troubleshoot had very strange symptoms caused by 2-V peak-to-peak ripple on the 5-V supply.

4. Signal Roll Call

The next step is to make a quick check of some key signals around the CPU of the microcomputer. If the problem is a bad IC, this can help point you toward the one that is bad. First, check if the clock signal is present and at the right frequency. If not, perhaps the clock generator IC is bad. If the microcomputer has a clock, but doesn't seem to be doing anything, use an oscilloscope to check if the CPU is putting out control signals such as $\overline{RD}$, $\overline{WR}$, and ALE. Also, check the least-significant data bus line to see if there is any activity on the buses. If there is no activity on these lines, a common cause is that the CPU is stuck in a wait, hold, halt, or reset condition by the failure of some TTL devices. To check this out, use the manual to help you predict what logic level should be on each of the CPU input control signals for normal operation. The RDY input of the 8086, for example, should be high for normal operation. If an external logic gate fails and holds RDY low, the 8086 will go on inserting WAIT states forever, and the buses will be held constant. If the 8086 HOLD input is held high or the RST input is held high, the 8086 buses will be floating. Connecting a scope probe to these lines will pull them to ground, so you will see them as lows.

If there is activity on the buses, use an oscilloscope to see if the CPU is putting out control signals such as $\overline{RD}$ and $\overline{WR}$. Also check with your oscilloscope to see if select signals are being generated on the outputs of the ROM, RAM, and port decoders as the system attempts to run its monitor or basic program. If no select signals are being produced, then the address decoder may be bad or the CPU may not be sending out the correct addresses.

After a little practice you should be able to work through the previously described steps quite quickly. If you have not located the problem at this point, the next step for a system with its ICs in sockets is to systematically substitute known good ICs for those in the nonworking system.

5. Systematically Substituting ICs

The easiest case of substitution is that where you have two identical microcomputers, one that works and one that doesn't, and the ICs of both units are in sockets. For this case you can use the working system to test the ICs from the nonworking system. The trick here is to do this in such a way that you don't end up with two systems that do not work! Here's how you do it.

First of all, DO NOT REMOVE OR INSERT ANY ICs WITH THE POWER ON! Now, with the power off, remove the CPU from the good system and put it in a piece of conductive foam. Plug the CPU from the bad system into the now empty socket on the good board and turn on the power. If the good system still works, then probably the CPU is good. Turn off the power and put the CPU back in the bad system. If the good system does not work with the CPU from the bad system, then the CPU is probably bad. Remove it from the good system and bend the pins so that you know it is bad. If the CPU seems bad, you can try replacing it with the CPU you removed from the good system. If you do this, however, it is important to keep track of which IC came from where. To do this we like to mark each IC from the good system with a wide-tip, water-soluble marking pen. We can then rebuild the good system by simply putting back all the marked ICs. The marks on the ICs can easily be removed with a damp cloth.

The procedure from here on is to keep testing ICs from the bad system until you find all of the bad ICs. Make sure to turn the power off before you remove or insert any ICs. Be aware that more than one IC may be bad. It is not unusual, for example, that an AC power-line surge will wipe out several devices in a system. We usually work our way out from the CPU to address latches, buffers, decoders, and memory devices. Often the specific symptoms point you to the problem group of ICs without your having to test every IC in the system. If, for example, the system accesses ROM, but doesn't access RAM, suspect the RAM decoder. If a system uses buffers on the buses, suspect these devices. Buffers are high-current devices and they often fail.

6. Troubleshooting a System with Soldered-in ICs

The approach described in the preceding paragraphs works well if the system ICs are all in sockets and you have two identical systems. However, since sockets add to the cost and unreliability of a system, many small systems put only the CPU and ROMs in sockets. This makes your troubleshooting work harder, but not impossible.

Again, if you have two identical systems, one that works and one that doesn't work, you can attempt to run the monitor or basic system program on each and compare signals on the two. A missing or wrong signal may point you to the bad IC or ICs.

If the system works enough to read some instructions from ROM and execute them, you can replace the monitor or basic system ROM with one that contains diagnostic programs which test RAM and I/O devices. A RAM test routine, for example, might attempt to write all 1's to each RAM location, and then read the memory location to see if the the data was written correctly to that location. If the data read back is not correct, the diagnostic program can stop and in some way indicate the address that it could not write to. If a write of all 1's is successful, then the test routine will try to write all 0's to each memory location. A port test routine might initialize a port for output, and then write alternating 1's and 0's to the port over and over again. With an oscilloscope you can then see if the port device is getting enabled and if the data is getting to the output of the port device. Another port test routine might try to read a byte of data in from a port over and over so that you can again see if the device is getting enabled and if the data is getting through the device to the system data bus. The technique of using program routines to test hardware is a very important one that you will use many times when you are working with microcomputer systems.

Now, suppose that you have localized the problem to a few ICs that are soldered in. If the problem is one that occurs when the unit gets hot, you might try spraying some Freon cold spray on the ICs, one at a time, to see if you can determine which one has a problem. If this does not find the bad IC or the problem is not heat-related, what you do next is to replace these ICs one at a time until the system works correctly. The point we want to stress here is that cost of these few ICs is probably much less that the cost of the time it would take you to determine just which IC is bad, if you do not have specialized test equipment.

To remove an IC from a printed circuit board, DO NOT attempt to desolder pins with a hand-held solder "slurper." Modern multilayer printed circuit boards are quite fragile, and these tools can slip and knock a trace right off the board. Instead, use cutters with narrow tips to cut all the leads of the IC next to the body. Since you are going to throw it out anyway, you don't care if you destroy the IC. With the body of the IC out of the way, you can then gently heat each pin individually and use needle nose pliers to remove it from the PC board. If the hole fills with solder, heat it gently and insert a small wooden toothpick until the solder cools. After you replace each IC, power up the system and see if it now works.

The techniques described in the preceding sections will enable you to troubleshoot many microcomputer systems with a minimum of test equipment. However, specialized test equipment is available to speed up the process and help find complex problems. The following sections describe two of these instruments.

7. Using a Logic Analyzer to Troubleshoot a Microcomputer System

A *logic analyzer* is an instrument which allows you to see the signals on 16 to 64 signal lines at once. With a logic analyzer you can, for example, see the signals on the address bus, data bus, and control bus of a microcomputer. Figure 7-20 shows a picture of a relatively low-cost logic analyzer, the Tektronix 318. Instruments such as this are themselves controlled by internal microprocessors. Small clipleads plug into a pod shown at the bottom of the drawing to get parallel data signals into the analyzer. The model shown only has 16 parallel data inputs. In addition, a scope-type probe can be used to send in serial data such as that sent out from a UART to a modem.

Figure 7-21 shows a functional block diagram of a simple logic analyzer. Since logic analyzers are used to detect and display only 1's and 0's, a comparator is put on each input. The reference input of the comparator is set for the logic threshold of the devices in the system. The signals out of the comparators to the rest of the analyzer are then clear-cut 1's or 0's.

The analyzer takes "snapshots" of the logic levels on each of the data inputs and stores these samples in an

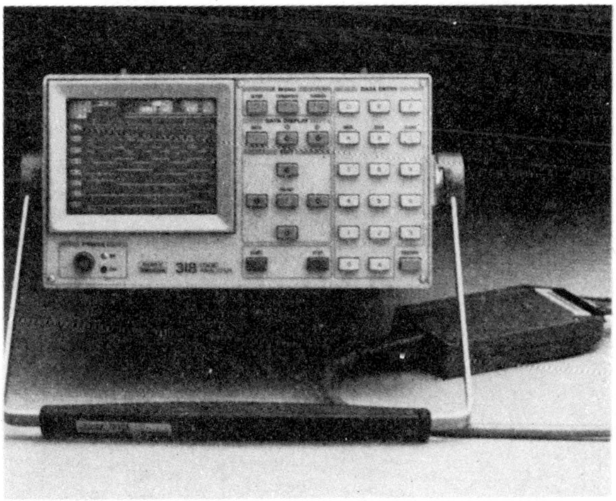

FIGURE 7-20 Tektronix 318 logic analyzer. *(Copyright 1983, Tektronix Inc.)*

internal RAM. Different analyzers store between 256 and 1024 samples for each input channel. A *clock* signal tells the analyzer how often to take samples. As shown by the block diagram in Figure 7-21, an internally produced signal or some external signal can be used to clock the analyzer. If you are using an analyzer to look at 8086 address and data lines, for example, you could use ALE as a clock signal. The analyzer will then take a sample each time the 8086 puts out an address and pulses ALE. The samples stored in the analyzer memory will then represent the sequence of addresses output by the 8086 after some specified *trigger.* As another example, you could clock the analyzer on $\overline{RD}$ from an 8086. After a specified trigger the analyzer will take a sample each time the 8086 does a read operation. In this case the samples stored in the analyzer memory will represent the sequence of data words read in from memory or from ports.

To make precise timing measurements with an analyzer a clock signal from an internal, crystal-controlled oscillator is used. In this case the analyzer will take a sample each time a pulse from the internal clock oscillator occurs. If, for example, you choose an internal clock frequency of 50 MHz, the analyzer will take a sample every 20 ns.

If the analyzer is receiving either an internal or an external clock, it will be continuously taking samples of the input data and storing these samples in the internal RAM. A *trigger* signal tells the analyzer when to display the samples stored in the RAM. As shown by the block diagram in Figure 7-21 some external signal can be used to trigger the analyzer, or the trigger signal can come from a word recognizer in the analyzer. A word recognizer compares the binary word on the input signal lines with a word you set with switches or a keyboard. When the two words match, the word recognizer sends out a trigger signal.

Since the analyzer is continuously taking samples, you can set the analyzer for a *pretrigger* display, a *center*

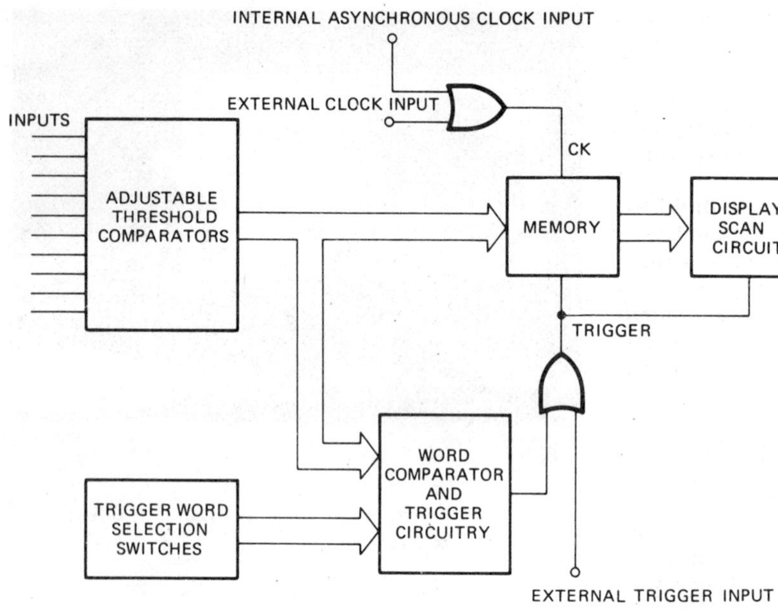

FIGURE 7-21 Logic analyzer block diagram.

trigger display, or a *posttrigger* display. For an analyzer that displays 256 samples, pretrigger means that the display will show the 256 samples that were taken just before the trigger occurred. For center trigger mode, 128 samples taken before the trigger and 128 samples taken after the trigger will be displayed. Posttrigger mode means that the analyzer will take 256 more samples after the trigger and display them.

Figure 7-22 shows some of the formats in which a logic analyzer can display the samples stored in its RAM. The series of displayed data samples is often called a *trace*. The timing diagram format in Figure 7-22*a* is most useful when making time measurements with an internal clock. A binary listing such as that in Figure 7-22*b* is useful for seeing the actual pattern of 1's and 0's on signal lines, but a hexadecimal listing such as that in Figure 7-22*c* makes it easier to recognize if a microcomputer is putting out addresses in the right sequence.

Some analyzers, such as the Tektronix 318, allow you take a series of samples from a functioning system, store these samples in a second memory in the analyzer, and then compare these samples with a series of samples taken from a nonfunctioning system. We have found this feature quite helpful in troubleshooting malfunctioning instruments which have poor documentation.

As we mentioned previously, the 318 can also be used to display a sequence of serial data as shown in Figure 7-22*c*. Note that in this format the analyzer shows the binary, hexadecimal, and the equivalent ASCII code for each of the data bytes taken in.

To summarize then, a logic analyzer takes samples of the signals present on its data inputs each time a clock pulse occurs and stores these samples in an internal RAM. A system signal such as ALE may be used as a clock source. In this case the analyzer is said to be operating in synchronous mode. For precise timing measurements an internal, crystal controlled clock source is used. The group of samples that are actually displayed on the screen is determined by a trigger signal. The trigger signal may come from some external source, or it may be produced by a word recognizer when it finds a specified data word on the parallel signal lines. Now that you have an overview of how a logic analyzer works, here's a few hints on how to use one for troubleshooting an 8086 microcomputer.

Connect the analyzer data inputs to the address and data bus lines from the CPU. For an 8086, connect the external clock input of the analyzer to the 8086 ALE pin. Look at an 8086 timing diagram such as the one in Figure 7-2 to see at which edge of the ALE signal valid addresses are present on the buses. Set the analyzer to clock on this edge. Set the analyzer to trigger on address FFFF0H, the first address output by the 8086 after a reset. Set the analyzer display format for posttrigger display. Tell the analyzer to do a trace and press the 8086 system reset button. The display on the analyzer should show you the sequence of addresses output after a reset. If you have one, use the system monitor listing to see if the displayed sequence is correct. If the sequence is not correct, look for address bits that should change, but don't. The cause of this problem may be the CPU or an address buffer. A common failure mode for buffers is that an input or an output will short to V_{cc} or to ground. This prevents that line from changing.

If the address sequence seems reasonable, connect the analyzer external clock input to the 8086 $\overline{RD}$ pin. Set the analyzer to clock on the positive edge of this signal. Set the format for posttrigger display. Tell the analyzer to do a trace and push the system reset button. The display on the analyzer should show the data transferred

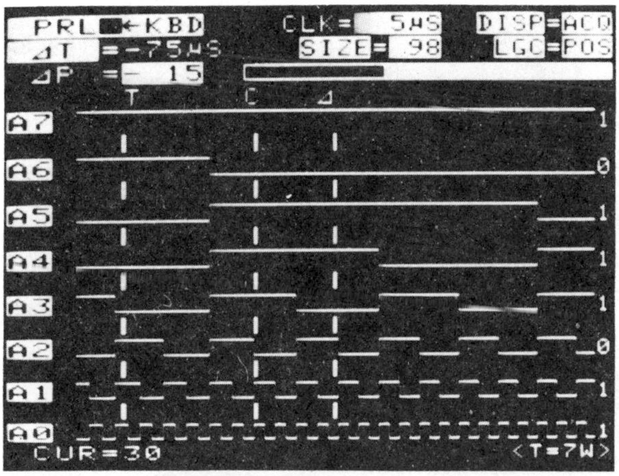

(a)

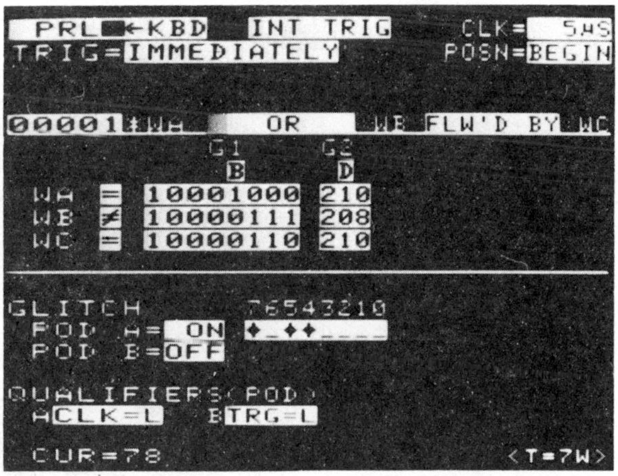

(b)

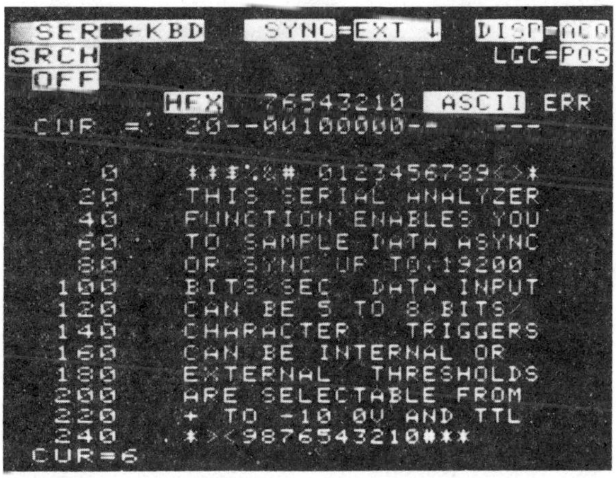

(c)

FIGURE 7-22 Logic analyzer display formats. *(a)* Parallel timing display. *(b)* Parallel state display. *(c)* Serial data display. *(Tektronix Inc.)*

on AD0–AD15. Again, use the monitor program listing to see if instruction bytes are coming in correctly. To help with this, some analyzers allow you to display the instruction mnemonic that corresponds to the bytes read in. If the data sequence is not correct, again check for stuck bits.

Another important logic analyzer feature you should learn to use is the clock qualifier input. If you switch on this input, the analyzer will only accept clock signals when a specified logic level is present on it. You can use the clock qualifier input to, for example, do a trace of only data read from ports. To do this in an 8086 minimum mode system, you connect the data inputs to the data bus, the $\overline{RD}$ signal to the analyzer clock input, and the 8086 M/$\overline{IO}$ signal to the analyzer clock qualifier input. You set the clock qualifier input to respond to a low. The analyzer will then take samples only when $\overline{RD}$ pulses and M/$\overline{IO}$ are low, as they will be during reads from ports.

We obviously can't describe here all of the ways to use a logic analyzer. If you have one, consult the manual for it to learn some of the finer points of its use. Also, the lab manual that is available for use with this book has some exercises to help you gain more skill with an analyzer. The point here was to show you how to use the analyzer as a "window" into what's going on in a system. By carefully choosing what signals you look at, what signal you clock on, and what word you trigger on, you can often solve difficult problems. For this reason, a logic analyzer is a valuable tool when developing a new microcomputer-based product. However, it is important for you to have a perspective of when to use an analyzer in troubleshooting simple systems that previously worked. Most of the time you can use the techniques described in previous sections to find and fix a problem in less time than it would take you to connect up the logic analyzer and figure out the trace display. If you have an analyzer, however, don't hesitate to use it when the simple techniques don't seem to be getting you anywhere.

8. Other Microcomputer Troubleshooting Equipment

A logic analyzer is a very powerful troubleshooting tool, but to use it effectively you need some detailed knowledge and a program listing for the system that you are trying to troubleshoot. If you are working as a repair technician and have to repair several different types of microcomputer systems with poor documentation to work from, most analyzers then are not too useful. To make your life easier in this case, "smart" instruments such as the Fluke 9010A microsystem troubleshooter have been created.

As you can see from the picture of the 9010A in Figure 7-23, it has a keyboard, a display, and an "umbilical" cable with an IC plug on the end. The unit also contains a minicasette tape recorder. For troubleshooting, the 9010A is used as follows.

The microprocessor in a fully functioning unit is re-

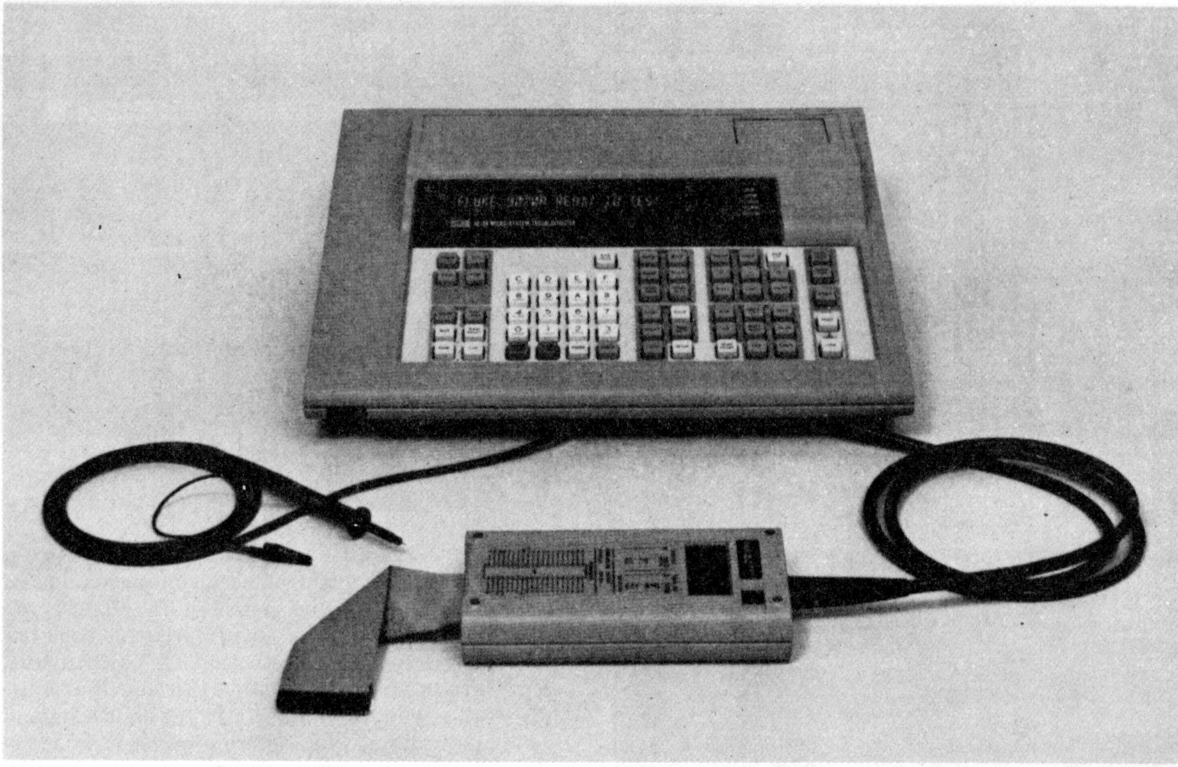

FIGURE 7-23 Fluke 9010A microsystem troubleshooter. *(John Fluke Mfg. Co., Inc.)*

moved, and the plug at the end of the cable is inserted in its place. The learn function of the 9010A is then executed. This function finds and maps ROM, RAM, and I/O registers that can be written into and read from. It also computes signatures (checksums) for blocks of ROM. All of these parameters are stored in the 9010A's RAM and/or on a minicassette tape. The microprocessor on a malfunctioning unit is then removed and the plug at the end of the umbilical cable inserted in its place. An automatic test function is then executed. In this mode the 9010A tests the buses, RAM, ROM, ports, power supply, and clock on the malfunctioning system. Any problem found, such as stuck nodes or adjacent trace short-circuits, is indicated on the display. The results of this test give some good hints as to the source of the problem. Because of its built-in intelligence, the 9010A can be programmed to do other tests as well.

The point of an instrument such as the 9010A is that with it you do not have to be intimately familiar with the programming language and hardware details of a simple microcomputer system in order to troubleshoot it.

IMPORTANT TERMS AND CONCEPTS FROM THIS CHAPTER

If you do not remember any of the terms or concepts in the following list, use the index to find them in the chapter.

Pin functions of 8086:
V_{cc}, $\overline{RD}$, $\overline{WR}$, CLK, ALE, M/$\overline{IO}$, $\overline{LOCK}$, MN/$\overline{MX}$, RESET, NMI, INTR, $\overline{BHE}$, $\overline{DEN}$, DT/$\overline{R}$

8086 RESET response

Maximum and minimum mode of 8086

8086 timing diagram interpretation

State, instruction cycle, machine cycle, wait state, READY signal

Bus activities during read/write

Bidirectional buffer

General functions: 3625, 8284, 8255A, 8251A, 8279, 2716, 2142

SDK-86 schematic: zones, plugs, jacks, wire wraps, resistor packs

Address decoding: ROM decoding, RAM decoding, port decoding

Memory-mapped I/O

Direct I/O

8086 memory banks

Timing parameters: t_{ACC}, t_{CL}, t_{OE}, t_{CE}, TCLAV, TCLRL, TDVCL

8086 typical clock frequencies

Troubleshooting steps for a simple 8086-based microcomputer

Logic analyzer: clock signal, trigger, trace

REVIEW QUESTIONS AND PROBLEMS

1. From what point on the clock waveform is the start of an 8086 state measured?

2. Why are latches required on the AD0–AD15 bus in an 8086 system?

3. What is the purpose of the ALE signal in an 8086 system?

4. Describe the sequence of events on the 8086 data/address bus, the ALE line, the M/$\overline{IO}$ line, and the $\overline{RD}$ line as the 8086 fetches an instruction word.

5. What logic levels will be on the 8086 $\overline{RD}$, $\overline{WR}$, and M/$\overline{IO}$ lines when the 8086 is doing a write to a memory location? A read from a port?

6. What is the major difference between an 8086 operating in minimum mode and an 8086 operating in maximum mode?

7. Describe the response an 8086 will make when its RESET (RST) input is asserted high.

8. Why are buffers often needed on the address, data, and control buses in a microcomputer system?

9. *a.* How is an 8086 entered into a WAIT state?
   b. At what point in a machine cycle does an 8086 enter a WAIT state?
   c. What information is on the buses during a WAIT state?
   d. How long is a WAIT state?
   e. How many WAIT states can be inserted?
   f. Why would you want the 8086 to insert a WAIT state?

10. What are the functions of the 8086 DT/$\overline{R}$ and $\overline{DEN}$ signals?

11. What does an arrow going from a transition on one signal waveform to a transition on another tell you?

12. How are wire wrap jumpers indicated on a schematic?

13. What is the meaning of /8 on a signal line on a schematic?

14. Describe the two purposes of address decoders in microcomputer systems.

15. A memory device has 15 address lines connected to it and 8 data outputs. What size words and how many words does the device store?

16. Briefly describe the function of the 8255, 8251A, and 8279 devices in the SDK-86 microcomputer system.

17. A group of signal lines on a schematic have the label *2ZB3* next to them. What is the meaning of this label?

18. What is the difference between a connector identified with a "J" and a connector identified with a "P"?

19. Describe the purpose of the many small capacitors connected between V_{cc} and ground on microcomputer printed circuit boards.

20. A 74LS138 decoder has its three SELECT inputs connected to A12, A13, and A14 of the system address bus. It has $\overline{G2A}$ connected to A15, $\overline{G2B}$ connected to $\overline{RD}$, and G1 connected to +5 V. Use an address decoder worksheet to determine what eight ROM address blocks the decoder outputs will select. Why is $\overline{RD}$ used as one of the enables on a ROM decoder?

21. Show a memory map for the ROMs in Problem 20.

22. Use an address decoder worksheet to help you draw a circuit to show how another 74LS138 can be connected to select one of eight 1 Kbyte RAMs starting at address 8000H.

23. Why are there actually many addresses that will select one of the port devices connected to the port decoder in Figure 7-10*a*.

24. Describe memory-mapped I/O and direct I/O. Give the main advantage and main disadvantage of each.

25. *a.* Why is the 8086 memory set up as two byte-wide banks?
    b. What logic levels would you find on $\overline{BHE}$ and A0 when an 8086 is writing a byte to address 04274H? Writing a word to 04274H?
    c. Describe the 8086 bus operations required to write a word to address 04373H.

26. How does the circuitry on the SDK-86 make sure that you cannot accidentally write a byte or word to ROM?

27. Why is some ROM put at the top of the address space in an 8086 system?

28. *a.* Show the truth table you would use for a 3625 PROM decoder to produce $\overline{CS1}$ signals for 4K × 8 RAMs in an 8086 system. Assume the first RAM starts at address 00000H. Don't forget A0 and $\overline{BHE}$.
    b. Draw the circuit connections for the 3625 decoder PROM and for two of the 4K × 8 RAMs.

29. Use sheets 5 and 7 of the SDK-86 schematics to help you determine for the SDK-86 what logic levels will be on $\overline{BHE}$, A0–A19, M/$\overline{IO}$, $\overline{RD}$, and $\overline{WR}$ when a word is read from ports FFF8H and FFF9H. Are these ports memory-mapped or direct? What instruction(s) would you use to do this read operation?

30. *a.* How is the $\overline{OFF\ BOARD}$ signal produced on the SDK-86 board?
    b. Describe the purpose of the $\overline{OFF\ BOARD}$ signal.

31. Describe how the 8088 memory is configured. Why doesn't the 8088 need a $\overline{BHE}$ signal?

32. By referring to the 8086 timing diagrams in Figure 7-18*a* and parameters in Figure 7-18*c* determine for the 8086-2:
   a. The maximum clock frequency.
   b. The time between CLOCK going low and $\overline{RD}$ going low.
   c. The time for which memory must hold data on the data bus after CLOCK goes low at the start of T4.
   d. The time that the lower 16 address bits remain on the data bus after ALE goes low.

33. The 27128-25 is a 16K × 8 EPROM with a t_{ACC} of 250 ns maximum, a t_{CE} of 250 ns maximum, and a t_{OE} of 100 ns maximum. Will this device work correctly without WAIT states in an 8-MHz 8086-2 system with circuit connections such as those in the SDK-86 schematics? Assume the address latches have a propagation delay of 12 ns and the decoder has a delay of 30 ns.

34. List the major steps you would take to troubleshoot a microcomputer system such as the SDK-86 which previously worked. Assume all ICs are in sockets.

35. Why is it important to check power supplies with an oscilloscope?

36. Describe how you can keep from mixing up ICs from a good system with those from a bad system when substituting.

37. Write an 8086 routine to test the system RAM in addresses 00200H–07FFFH.

38. Write a test routine to output alternating 1's and 0's to port FFFAH over and over. With this routine running you could check with an oscilloscope to see if the port device is getting enabled and is outputting data.

39. Describe the symptoms that an SDK-86 would show for each of the following problems.
   a. Pin 8 of *A*15 in zone *D*5 of schematic sheet 2 is stuck low.
   b. The reset key is stuck on.
   c. None of the outputs of *XA*29 in zone *D*7 of schematic sheet 6 ever goes low.
   d. Pin 6 of *A*3 in zone *A*5 of schematic sheet 5 is stuck low.

40. Draw a block diagram of a simple logic analyzer and briefly describe how it operates. Include in your answer the function of the clock and the function of the trigger.

41. What do you use for a logic analyzer clock when you want to make detailed timing measurements.

42. On what signal and what edge of that signal would you clock a logic analyzer and on what word would you trigger to see in an 8086 system:
   a. The sequence of addresses output after a RESET?
   b. The sequence of instructions read in after a RESET? (Assume the first instruction word is FAH.)
   c. Both the addresses sent out and the words read in?
   d. Most logic analyzers have a clock qualifier input. If this input is used, the logic analyzer will not respond to a clock signal unless a specified logic level is on the qualifier input. You might, for example, connect the M/$\overline{IO}$ to the clock qualifier input and set it for a high to see a trace of data read from memory. What clock qualifier would you use to see a trace of only data read in from ports?

43. How is it possible for a logic analyzer to display data that occurred before the trigger?

CHAPTER

8 Interrupts and Interrupt Service Procedures

Most microprocessors allow normal program execution to be interrupted by some external signal or by a special instruction in the program. When a microprocessor is interrupted, it stops executing its current program and calls a procedure which "services" the interrupt. At the end of the interrupt service procedure, execution is usually returned to the interrupted program. This chapter shows you how the 8086 family members respond to interrupts, how to write interrupt service procedures, and how interrupts are used in a variety of applications.

OBJECTIVES

At the conclusion of this chapter you should be able to:

1. Describe the interrupt response of an 8086 family processor.
2. Initialize an 8086 interrupt vector (pointer) table.
3. Write interrupt service procedures.
4. Describe the operation of an 8254 programmable counter/timer and write the instructions necessary to initialize an 8254 for a specified application.
5. Describe the operation of an 8259A priority interrupt controller and write the instructions needed to initialize an 8259A for a specified application.

8086 INTERRUPTS AND INTERRUPT RESPONSES

Overview

An 8086 interrupt can come from any one of three sources. One source is from an external signal applied to the *nonmaskable interrupt* (NMI) input pin, or the *interrupt* (INTR) input pin. An interrupt caused by a signal applied to one of these inputs is referred to as a *hardware interrupt*.

A second source of an interrupt is execution of the interrupt instruction, INT. This is referred to as a *software interrupt*.

The third source of an interrupt is from some condition produced in the 8086 by the execution of an instruction. An example of this is the divide by zero interrupt. Program execution will automatically be interrupted if you attempt to divide an operand by zero. *Conditional interrupts* such as this are also referred to as software interrupts.

At the end of each instruction cycle the 8086 checks to see if any interrupts have been requested. If an interrupt has been requested, the 8086 responds to the interrupt by stepping through the following series of major actions.

1. It decrements the stack pointer by two and pushes the flag register on the stack.
2. It disables the INTR interrupt input by clearing the interrupt flag in the flag register.
3. It resets the trap flag in the flag register.
4. It decrements the stack pointer by two and pushes the current code segment register contents on the stack.
5. It decrements the stack pointer again by two and pushes the current instruction pointer contents on the stack.
6. It does an indirect far jump to the start of the procedure you wrote to respond to the interrupt.

To summarize these steps, then, the 8086 pushes the flag register on the stack, disables the single step and the INTR input, and does essentially an indirect far call to the interrupt service procedure. Figure 8-1 shows this in diagram form. Note that an IRET instruction at the end of the interrupt service procedure returns execution to the main program.

Now remember from Chapter 5 that when the 8086 does a far call to a procedure, it puts a new value in the code segment register and a new value in the instruction pointer. For an indirect call the 8086 gets the new values for CS and IP from four memory addresses. Likewise, when it responds to an interrupt the 8086 goes to memory locations to get the CS and IP values for the

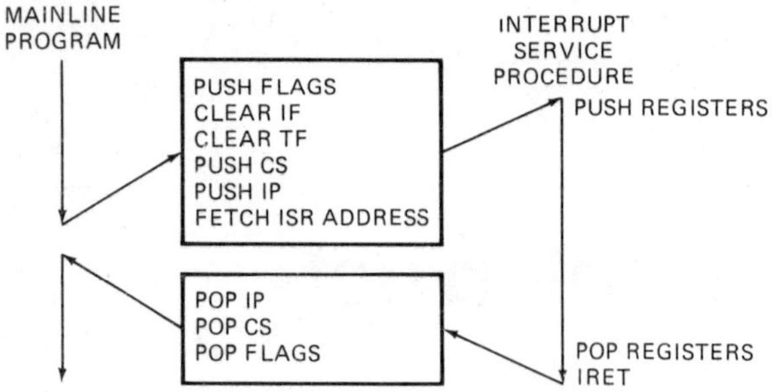

FIGURE 8-1 8086 interrupt response.

start of the interrupt service procedure. In an 8086 system the first 1 Kbyte of memory from 00000H to 003FFH is set aside as a table for storing the starting addresses of interrupt service procedures. Since 4 bytes are required to store the CS and IP values for each interrupt service procedure, the table can hold the starting addresses for up to 256 interrupt procedures. The starting address of an interrupt service procedure stored in this table is often called the *interrupt vector* or the *interrupt pointer*, and the table itself is then referred to as the *interrupt vector table* or the *interrupt pointer table*.

Figure 8-2 shows how the 256 interrupt pointers are arranged in the memory table. Each doubleword interrupt pointer is identified by a number from 0 to 255. Intel calls this number the *type* of the interrupt. The lowest five types are dedicated to specific interrupts such as the divide by zero interrupt and the nonmaskable interrupt which we explain in detail later. The next 27 interrupt types, from 5 to 31, are reserved by Intel for use in future microprocessors. The upper 224 interrupt types, from 32 to 255, are available for you to use for hardware or software interrupts.

When the 8086 responds to an interrupt, it automatically goes to the specified location in the interrupt pointer table to get the starting address of the interrupt service procedure. The 8086, however, does not automatically load the starting address in the pointer table. As we will show later, you have to do this with instructions at the start of your program. Note that the new value for the instruction pointer is put in as the low word of the pointer, and the new value for the code segment register is put in as the high word of the pointer.

Now that you have an overview of how the 8086 responds to interrupts, we can show in detail how one of these interrupts works.

An 8086 Interrupt Response Example—Type 0

Probably the easiest 8086 interrupt to understand is the divide-by-zero interrupt, identified as *type 0* in Figure 8-2. We'll use a type 0 interrupt to show you in detail how an 8086 interrupt works, and how to write a procedure to service an interrupt.

First of all let's refresh your memory about how the 8086 DIV and IDIV instructions work. The 8086 DIV instruction allows you to divide a 16-bit unsigned binary number in AX by an 8-bit unsigned number from a specified register or memory location. The 8-bit result (quotient) from this division will be left in the AL register. The 8-bit remainder will be left in the AH register. The DIV instruction also allows you to divide a 32-bit unsigned binary number in DX and AX by a 16-bit number in a specified register or memory location. The 16-bit quotient from this division is left in the AX register, and the 16-bit remainder is left in the DX register. The 8086 IDIV instruction, in the same manner, allows you to divide a 16-bit signed number in AX by an 8-bit signed number in a specified register, or a 32-bit signed number in DX and AX by a 16-bit signed number from a specified register or memory location.

If the quotient from dividing a 16-bit number is too large to fit in AL or the quotient from dividing a 32-bit number is too large to fit in AX, the result of the division will be meaningless. A special case of this is where an attempt is made to divide a 32-bit number or a 16-bit number by zero. The result of dividing by zero is infinity (actually undefined), which is somewhat too large to fit in AX or AL. Whenever the quotient from a DIV or IDIV operation is too large to fit in the result register, the 8086 will do a type 0 interrupt.

The type 0 response proceeds as follows. The 8086 first decrements the stack pointer by two and copies the flag register to the stack. It then clears the IF and the TF. Next it saves the return address on the stack. To do this the 8086 decrements the stack pointer by two, pushes the CS value of the return address on the stack, decrements the stack pointer by two again, and pushes the IP value of the return address on the stack. The 8086 then gets the starting address of the interrupt service procedure from the type 0 locations in the interrupt pointer table. As you can see in Figure 8-2 it gets the new value for CS from addresses 00002H and 00003H, and the new value for IP from addresses 00000H and 00001H. After the starting address of the procedure is loaded into CS and IP, the 8086 then fetches and executes the first instruction of the procedure.

At the end of the interrupt service procedure an IRET instruction will be used to return execution to the interrupted program. The IRET instruction pops the stored value of IP off the stack and increments the stack pointer by two. It then pops the stored value of CS off the stack and increments the stack pointer again by two. Finally it restores the flags by popping off the stack the values stored during the interrupt response and increments the stack pointer by two. Remember from the previous paragraph that during its interrupt response, the 8086 disables the INTR and single-step interrupt by clearing IF and TF. Now, if the INTR input and/or the trap interrupt were enabled before the interrupt, they will be enabled upon return to the interrupted program. The reason for this is that flags from the interrupted program were pushed on the stack before IF and TF were cleared by the 8086 in its interrupt response. To summarize, then, IRET returns execution to the interrupted program and restores the IF and the TF to the state they were in before the interrupt. Now that we have described the type 0 response, we can show you how to write a program to handle this interrupt.

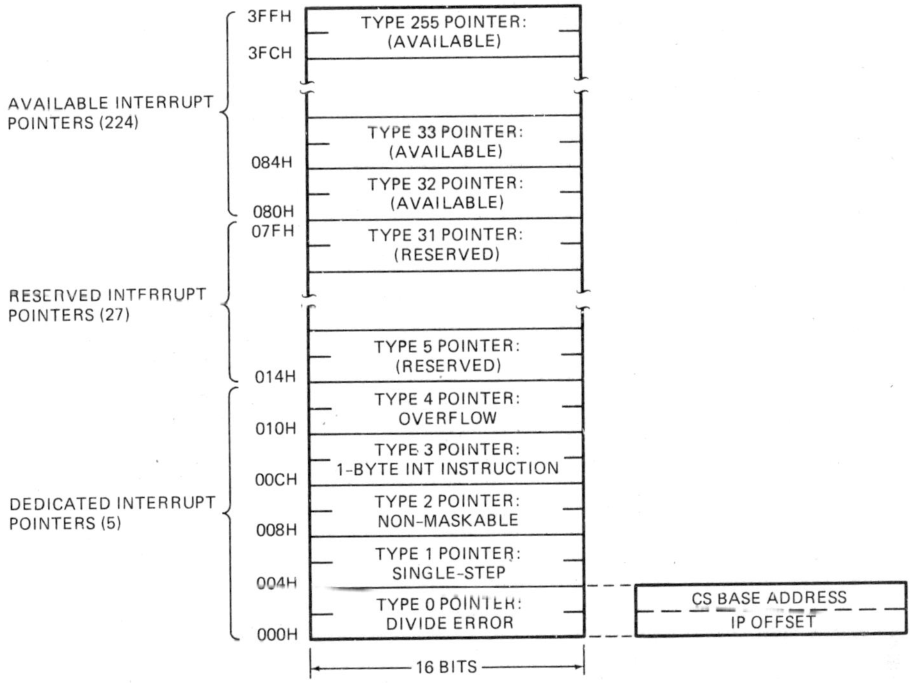

FIGURE 8-2 8086 interrupt pointer table.

An 8086 Interrupt Program Example

DEFINING THE PROBLEM AND WRITING THE ALGORITHM

In the last chapter we were mucking around mostly in hardware, so instead of jumping directly into the program, let's use this example to review how you go about writing any program.

As described in Chapter 3, you start by carefully defining the problem that you want the program to solve or the operations that you want it to perform. Part of this step is to determine the amount and types of data that the program is to work with.

For the example program here we have four word-sized hexadecimal values stored in memory. We want to divide each of these values by a byte-type scale factor to give a byte-type scaled value. If the result of the division is valid, we want to put the scaled value in an array in memory. If the result of the division is invalid (too large to fit in the 8-bit result register), we want to put 0 in the array for that scaled value. Figure 8-3 shows the algorithm for this program in pseudocode. As shown in Figure 8-3*a*, the mainline part of this program gets each 16-bit value from memory in turn and divides that value by the 8-bit scale factor. If the result of the division is too large to fit in the quotient register, AL, then the 8086 will do a type 0 interrupt immediately after the divide instruction finishes.

Figure 8-3*b* shows the algorithm for our type 0 interrupt service procedure. The main function of this procedure is to set a *flag* which will be checked by the mainline program. The flag in this case is not one of the flags in the 8086 flag register. The flag here is a bit in a memory location we set aside for this purpose. In the actual program we give this memory location the name BAD_DIV_FLAG. At the end of the interrupt service procedure we return to the interrupted mainline program.

After the division in the mainline program we check to see if the result of the division is valid. If the result is valid, we store it in the correct place in the scaled values array in memory. If the result is not valid, we leave zero in that place in the scaled values array. The way we actually make the decision whether a result is valid or not is to check the BAD_DIV_FLAG. If the result of the division was too large, then the 8086 will have done a type zero interrupt, and our interrupt service procedure will have set the BAD_DIV_FLAG to a one. If the result of the division is valid, then the 8086 will not do the interrupt, and the BAD_DIV_FLAG will be zero.

The sequence of operations is repeated until all of the values have been scaled. We use a register to keep track of which input value is being operated on at a particular time.

WRITING THE INITIALIZATION LIST

After you have worked out the data structure and the algorithm for a program, the next step is to make an initialization list such as the one shown in Chapter 3. Here is a list for this program.

1. Initialize the interrupt pointer table. In other words, the starting address of our type 0 interrupt service routine must be put in locations 00000H and 00002H.

```
INITIALIZATION LIST

REPEAT
     get INPUT_VALUE
     divide by scale factor
     IF result valid THEN
          store result as scaled value
     ELSE store zero
UNTIL all values scaled
```

(a)

```
Save registers
Set error flag
Restore registers
Return to mainline
```

(b)

FIGURE 8-3 Algorithm for divide by zero program example. *(a)* Mainline program. *(b)* Interrupt service procedure.

2. Set up the data segment where the values to be scaled, the scale factor, the scaled values, and the BAD_DIV_FLAG will be put.
3. Initialize the data segment register to point to the base address of the data segment containing the values to be scaled.
4. Set up a stack to store the return address, since we are essentially calling a procedure.
5. Initialize the stack segment and the stack pointer registers.
6. Initialize a pointer to the start of the data to be scaled, a counter to keep track of how many values we have scaled, and a pointer to the start of the array where we want to put the scaled values.

Once you have the algorithm and the initialization list for a program, the next step is to start writing the instructions for the program, so now let's look at the assembly language program for this problem.

ASSEMBLY LANGUAGE PROGRAM AND INTERRUPT PROCEDURE

Figure 8-4 shows our 8086 assembly language program for the mainline and for the type 0 interrupt service procedure. You can use many of the parts of these when you write your own interrupt programs. To help refresh your memory of the PUBLIC and the EXTRN directives, we have written the mainline program and the interrupt service procedure as two separate assembly modules. Remember, if you are not using an assembler, you can just substitute the actual offsets or numbers for the names used in the example program.

At the start of the mainline program in Figure 8-4*a*, we declare a segment named DATA_HERE for the data that the program will be working with. The WORD directive tells the locator to start this segment on the next available even address. The PUBLIC directive in this statement identifies the segment name as public so it can be referred to in other assembly modules. The input values are words, so we use a DW directive to declare these four values. The scaled values will be bytes, so we use the DB directive to set aside four locations for these. The DUP(0) in the statement initializes the 4-byte locations to all 0's. As the program executes, the results will be written into these locations. SCALE_FACTOR DB 09H sets aside a byte location for the number that we are going to be dividing the input values by. The advantage of using a DB to declare the scale factor, rather than an EQU directive, is that with a DB the value of scale factor can be held in RAM where it can be changed dynamically in the program as needed. If you use a statement such as SCALE_FACTOR EQU 09H to set a value, you have to reassemble the program to change the value.

Part of the 8086 interrupt response is essentially a far call to the interrupt service procedure. In any program that calls a procedure we have to set up a stack to store the return address and parameters passed to and from the procedure. The next section of the program declares a stack segment called STACK_HERE. It also establishes a pointer to the next location above the stack with the statement TOP_STACK LABEL WORD. Remember from the examples in Chapter 5 that this label is used to initialize the stack pointer to the next location after the top of the stack.

The next two parts of the program are necessary because we wrote the main program and the interrupt service procedure as two separate assembly modules. When the assembler reads through a source program, it makes a *symbol table* which contains the segment and offset of all of the names and labels used in the program. The statement PUBLIC BAD_DIV_FLAG tells the assembler to identify the name BAD_DIV_FLAG as public. This means that when the object module for this program is linked with some other object module that declares BAD_DIV_FLAG as EXTRN, the linker will be allowed to

```
page ,132
;8086 PROGRAM
;ABSTRACT:      Program scales some data values by division.
;PROCEDURES:    Uses BAD_DIV, a type 0 interrupt service procedure
;PORTS USED:    None

DATA_HERE       SEGMENT WORD    PUBLIC
        INPUT_VALUES    DW  0035H, 0855H, 2011H, 1359H
        SCALED_VALUES   DB 4 DUP(0)
        SCALE_FACTOR    DB 09H
        BAD_DIV_FLAG    DB 0
DATA_HERE       ENDS
STACK_HERE      SEGMENT STACK
                        DW  100 DUP (0)         ; Set up stack of 100 words
        TOP_STACK  LABEL  WORD                  ; Pointer to top of stack
STACK_HERE      ENDS

PUBLIC  BAD_DIV_FLAG            ; Make flag available to other modules

INT_PROC_HERE   SEGMENT WORD    PUBLIC
        EXTRN  BAD_DIV:FAR              ; Let assembler know procedure BAD_DIV is
INT_PROC_HERE   ENDS                    ;  in  another assembly module

CODE_HERE       SEGMENT WORD    PUBLIC
        ASSUME CS:CODE_HERE, DS:DATA_HERE, SS:STACK_HERE
START:  MOV  AX, STACK_HERE             ; Initialize stack segment register
        MOV  SS, AX
        MOV  SP, OFFSET TOP_STACK       ; Initialize stack pointer
        MOV  AX, DATA_HERE              ; Initialize data segment register
        MOV  DS, AX
;store the address for the BAD_DIV routine at address 0000:0000
;address 00000-00003 is where type 0 interrupt gets interrupt
;service procedure address. CS at 00002 & 00003, IP at 00000 & 00001
        MOV  AX, 0000
        MOV  ES, AX
        MOV  WORD PTR ES:0002, SEG BAD_DIV
        MOV  WORD PTR ES:0000, OFFSET BAD_DIV
        MOV  SI, OFFSET INPUT_VALUES    ; Initialize pointer for input values
        MOV  BX, OFFSET SCALED_VALUES   ; Point BX at start of result array
        MOV  CX, 0004H                  ; Initialize data value counter
NEXT:   MOV  AX, [SI]                   ; Bring a value to AX for divide
        DIV  SCALE_FACTOR               ; Divide by scale factor
        CMP  BAD_DIV_FLAG, 01H          ; Check if divide produced invalid result
        JNE  OK                         ; No, go save scaled value
        MOV  BYTE PTR [BX], 00          ; Yes, load 0 as scaled value
        JMP  SKIP
OK:     MOV  [BX], AL                   ; Save scaled value
SKIP:   MOV  BAD_DIV_FLAG, 0            ; Reset BAD_DIV_FLAG before doing next
        ADD  SI, 02H                    ; Point at location of next input value
        INC  BX                         ; Point at location for next result
        LOOP NEXT                       ; Repeat until all values done
STOP:   NOP
CODE_HERE       ENDS
        END     START
```

(a)

FIGURE 8-4 8086 assembly language program for divide by zero example. *(a)* Mainline. *(b)* Interrupt service procedure.

```
age , 132

;8086 PROCEDURE TO SERVICE DIVIDE BY ZERO INTERRUPT (TYPE 0)

; This procedure sets the LSB of a memory location called BAD_DIV_FLAG,
;  enables INTR, and returns execution to the interrupted program with
;  registers and flags unchanged.

DATA_HERE           SEGMENT WORD PUBLIC
       EXTRN  BAD_DIV_FLAG:BYTE         ; Let assembler know BAD_DIV_FLAG
DATA_HERE           ENDS                ;  is in another assembly module

PUBLIC  BAD_DIV                         ; Make procedure BAD_DIV available to
                                        ;  other assembly modules

INT_PROC_HERE       SEGMENT WORD PUBLIC ; Set up a segment for all
                                        ;  interrupt service procedures
BAD_DIV  PROC FAR                       ; Procedure for type 0 interrupt
       ASSUME CS:INT_PROC_HERE, DS:DATA_HERE

       PUSH  AX                ; Save AX of interrupted program
       PUSH  DS                ; Save DS of interrupted program
       MOV   AX, DATA_HERE     ; Load DS value needed here
       MOV   DS, AX
       MOV   BAD_DIV_FLAG, 01  ; Set LSB of BAD_DIV_FLAG byte
       POP   DS                ; Restore DS of interrupted program
       POP   AX                ; Restore AX of interrupted program
       IRET                    ; Return to next instruction in
                               ;  interrupted program

BAD_DIV  ENDP
INT_PROC_HERE       ENDS

        END
```

(b)

FIGURE 8-4 (*continued*)

make the connection. Some programmers say that the PUBLIC directive "exports" a name or label.

The other end of this export operation is to "import" labels or names that are defined in other assembly modules. The statement EXTRN BAD_DIV:FAR in our example program, for example, tells the assembler that BAD_DIV is a label of type FAR and that BAD_DIV is defined in some other assembly module. The INT_PROC_HERE SEGMENT WORD PUBLIC and INT_PROC_HERE ENDS statements tell the assembler that BAD_DIV is defined in a segment named INT_PROC_HERE. When the assembler reads these statements it will make an entry in its symbol table for BAD_DIV, and identify it as external. When the object module for this program is linked with the object module for the program where BAD_DIV is defined, the linker will fill in the proper values for the CS and IP of BAD_DIV.

For the actual instructions of our mainline program we declare a code segment with the statement CODE_HERE SEGMENT WORD PUBLIC. The WORD in this statement tells the linker/locator to locate this segment on the first available even address. The PUBLIC in this statement tells the linker that this segment can be joined together (concatenated) with segments of the same name from other assembly modules.

As usual at the start of the code segment we use an ASSUME statement to tell the assembler what logical segments to use for code, data, and stack. After this comes the hopefully familiar instructions for initializing the stack segment register, the stack pointer register, and the data segment register.

The next four instructions are needed to place the address of the BAD_DIV procedure in the type 0 location in the interrupt pointer table. The code segment address for BAD_DIV is stored at 00002 and 00003 and the address of the offset of BAD_DIV at 00000 and 00001. It is necessary to load the interrupt procedure addresses in this way if you are using an SDK-86 board, or the

MASM and Link programs on an IBM PC-type machine. This is because the linker overrides any ORG directives which makes it difficult to put programs at absolute addresses.

Next we initialize SI as a pointer to the first input value with the statement MOV SI, OFFSET INPUT_VALUES. The statement MOV BX, OFFSET SCALED_VALUES then initializes BX as a pointer to the first of the locations we set aside for the 8-bit scaled results.

To keep track of how many values have been scaled we set up the CX register as a counter. The statement MOV CX, 0004H initializes the counter with the number of values we want to scale. This register will be decremented after each input value is scaled. When CX = 0, we know that all values have been scaled.

Finally everything is initialized, and we get to the operations we set out to do. The statement MOV AX, [SI] copies an input value from memory to the AX register where it has to be for the divide operation. The DIV SCALE_FACTOR instruction divides the number in AX by 09H, the value we assigned to SCALE_FACTOR previously with a DB directive. The 8-bit quotient from this division will be put in AL and the 8-bit remainder will be put in AH. If the quotient is too large to fit in AL, then the 8086 will automatically do a type 0 interrupt. For our program here, the 8086 will push the flags on the stack, reset the IF and TF, and push the return address on the stack. It will then go to addresses 0000H and 0002H to get the IP and CS values for the start of BAD_DIV, the procedure we wrote to service a type 0 interrupt. It will then execute the BAD_DIV procedure. Now let's look at the procedure in Figure 8-4*b* and see how it works.

The BAD_DIV procedure starts by letting the assembler know that the name BAD_DIV_FLAG represents a variable of type byte, and that this variable is defined in a segment called DATA_HERE in some other (EXTRN) assembly module. We also tell the assembler that the label BAD_DIV should be made available to other assembly modules (PUBLIC).

Next we declare a logical segment called INT_PROC_HERE. We could have put this procedure in the segment CODE_HERE with the mainline program. However, in system programs where there are many interrupt service procedures, a separate segment is usually set aside for them. What we are doing here, then, is to show you an overall structure that we will fill in as we work our way through the rest of the book.

The statement BAD_DIV PROC FAR identifies the actual start of the procedure, and tells the assembler that both the CS and IP values for this procedure must be saved. The ASSUME statement at the start of the procedure then tells the assembler the names of the segments to use for code and data for this procedure.

Now, an important operation to do at the start of any interrupt service procedure is to push on the stack any registers that you are going to use in the procedure. You can then restore these registers by popping them off the stack just before returning to the interrupted program. The interrupted program will then resume with its registers as they were before the interrupt. In our procedure here we save AX and DS. Since we use the same same data segment, DATA_HERE, in the mainline and in the procedure, you may wonder why we saved DS. The point is that an interrupt service procedure should be written so that it can be used at any point in a program. By saving the DS value of the interrupted program, this interrupt service procedure can be used in a program section that does not use DATA_HERE as its data segment.

The ASSUME statement tells the assembler the name of the segment to use as a data segment, but remember that it does not load the DS register with a value for the start of that segment. The instructions MOV AX, DATA_HERE and MOV DS, AX do this in our procedure.

Finally, we get to the whole point of this procedure with the MOV BAD_DIV_FLAG, 01 instruction. This instruction simply sets the least-significant bit of the memory location we set aside with a DB directive at the start of the mainline program. Note that in order to access this variable by name you have to let the assembler know that it is external, and you have to make sure that the DS register contains the segment base for the segment in which BAD_DIV_FLAG is located.

To complete the procedure we pop the saved registers off the stack and return to the interrupted program. The IRET instruction, remember, is different from the regular RET instruction in that it pops the flag register and the return address off the stack. Note in the source program in Figure 8-4*b* that if you are using an assembler, the procedure must be "closed" with an ENDP directive, and the segment must as usual be closed with an ENDS directive.

Now let's look back in the mainline to see what it does with this BAD_DIV_FLAG. Immediately after the DIV instruction, the mainline checks to see if the BAD_DIV_FLAG is set by comparing it with 01. If the BAD_DIV_FLAG was not set by the type 0 interrupt service procedure, then a jump is made to the MOV [BX], AL instruction. This instruction copies the result of the division in AL to the memory location in SCALED_VALUES pointed to by BX. If BAD_DIV_FLAG was set by a type 0 interrupt, then zero is put in the memory location in SCALED_VALUES and a jump will be made to the MOV BAD_DIV_FLAG, 00 instruction which resets the BAD_DIV_FLAG. Since this jump passes over the MOV [BX], AL instruction, the invalid result of the division will not be copied into one of the locations in SCALED_VALUES.

After putting the scaled value or zero in the array and resetting the flag, we get ready to operate on the next input value. The ADD SI, 02 instruction increments SI by two so that it points to the next 16-bit value in INPUT_VALUES. The INC BX instruction points BX at the next 8-bit location in SCALED_VALUES. The LOOP instruction after these automatically decrements the CX register by one, and, if CX is not then zero, it causes the 8086 to jump to the specified label, NEXT.

The preceding section has shown you how to set up an interrupt pointer table, how to write an interrupt service procedure, and how the 8086 responds to a type 0 interrupt. Now we can discuss some of the other types of 8086 interrupts.

8086 Interrupt Types

The preceding sections used the type 0 interrupt as an example of how the 8086 interrupts function. In this section we discuss in detail the different ways an 8086 can be interrupted, and how the 8086 responds to different types of interrupts. We discuss these in order, starting with type 0, so that you can easily find a particular discussion when you need to refer back to it. However, as you read though this section you should not attempt to learn all of the details of all of the kinds of interrupts at once. Read through all of the kinds to get an overview, and then focus on the details of the hardware-caused NMI interrupt, the software interrupts produced by the INT instruction, and the hardware interrupt produced by applying a signal to the INTR input pin.

DIVIDE-BY-ZERO INTERRUPT—TYPE 0

As we described in the preceding section, the 8086 will automatically do a type 0 interrupt if the result of a DIV operation or an IDIV operation is too large to fit in the destination register. For a type 0 interrupt the 8086 pushes the flag register on the stack, resets the IF and TF, and pushes the return address (CS and IP) on the stack. It then gets the CS value for the start of the interrupt service procedure from address 00002H in the interrupt pointer table, and the IP value for the start of the procedure from address 00000H in the interrupt pointer table.

Since the 8086 type 0 response is automatic and cannot be disabled in any way, you have to account for it in any programs where you use the DIV or IDIV instructions. One way to do this is to in some way make sure the result will never be too large for the result register. We showed one way to do this in the example program in Figure 5-25*b*. In that example you may remember we first make sure the divisor is not zero, and then we do the division in several steps so that the result of the division will never be too large.

Another way to account for the 8086 type 0 response is to simply write an interrupt service procedure which takes the desired action when an invalid division occurs. The advantage of this approach is that you don't have the overhead of a more complex division routine in your mainline program. The 8086 automatically does the checking and only does the interrupt procedure if there is a problem. Remember that when using any interrupts with the 8086 you must in some way load the starting address of the interrupt service procedure in the interrupt pointer table.

SINGLE-STEP INTERRUPT—TYPE 1

In a section of Chapter 3 on debugging assembly language programs we discussed the use of the single-step feature present in some monitor programs and debugger programs. When you tell a system to single-step, it will execute one instruction and stop. You can then examine the contents of registers and memory locations. If they are correct, you can tell the system to go on and execute the next instruction. In other words, when in single-step mode, a system will stop after it executes each instruction, and wait for further direction from you. The 8086 trap flag and type 1 interrupt response make it quite easy to implement a single-step feature in an 8086-based system.

If the 8086 trap flag is set, the 8086 will automatically do a type 1 interrupt after each instruction executes. When the 8086 does a type 1 interrupt it pushes the flag register on the stack, resets the TF and IF, and pushes the CS and IP values for the next instruction on the stack. It then gets the CS value for the start of the type 1 interrupt service procedure from address 00006H, and it gets the IP value for the start of the procedure from address 00004H.

The tasks involved in implementing single step then are: set the trap flag, write an interrupt service procedure which saves all registers on the stack where they can later be examined or perhaps displayed on the CRT, and load the starting address of the type 1 interrupt service procedure into addresses 00004H and 00006H. The actual single-step procedure will depend very much on the system that it is to be implemented on. We do not have space here to show you the different ways to do this. We will, however, show you how the trap flag is set or reset, because this is somewhat unusual.

The 8086 has no instructions to directly set or reset the trap flag. These operations are done by pushing the flag register on the stack, changing the trap-flag bit to what you want it to be, and then popping the flag register back off the stack. Here is the instruction sequence to set the trap flag.

```
PUSHF                ; Push flags on stack
MOV BP,SP            ; Copy SP to BP for use as index
OR [BP+0], 0100H     ; Set TF bit
POPF                 ; Restore flag register
```

To reset the trap flag, simply replace the OR instruction in the above sequence with the instruction AND [BP+0], 0FEFFH.

NOTE: We have to use [BP + 0] because BP cannot be used as a pointer without a displacement. See Figure 3-8.

The trap flag is reset when the 8086 does a type 1 interrupt, so the single-step mode will be disabled during the interrupt service procedure.

NONMASKABLE INTERRUPT—TYPE 2

The 8086 will automatically do a *type 2* interrupt response when it receives a low-to-high transition on its NMI input pin. When it does a type 2 interrupt the 8086 will push the flags on the stack, reset TF and IF, and push the CS value and the IP value for the next instruction on the stack. It will then get the CS value for the start of the type 2 interrupt service procedure from address 0000AH, and the IP value for the start of the procedure from address 00008H.

The name *nonmaskable* given to this input pin on the 8086 means that the type 2 interrupt response cannot be disabled (masked) by any program instructions. Because this input cannot be intentionally or accidentally disabled, we use it to signal the 8086 that some condition in an external system must be taken care of.

We could, for example, have a pressure sensor on a large steam boiler connected to the NMI input. If the pressure goes above some preset limit the sensor will send an interrupt signal to the 8086. The type 2 interrupt service procedure we write for this case can turn off the fuel to the boiler, open a pressure relief valve, and sound an alarm.

Another common use of the type 2 interrupt is to save program data in the case of a system power failure. Some external circuitry detects when the ac power to the system fails and sends an interrupt signal to the NMI input. Because of the large filter capacitors in most power supplies, the dc system power will remain for perhaps 50 ms after the ac power is gone. This is more than enough time for a type 2 interrupt service procedure to copy program data to some RAM which has a battery backup power supply. When the ac power returns, program data can be restored from the battery-backed-up RAM and the program can resume execution where it left off. A practice problem at the end of the chapter gives you a chance to write a simple procedure for this task.

BREAKPOINT INTERRUPT—TYPE 3

The type 3 interrupt is produced by execution of the INT 3 instruction. The main use of the type 3 interrupt is to implement a breakpoint function in a system. In Chapter 4 we described the use of breakpoints in debugging assembly language programs. Hopefully you have been using them in debugging your programs. When you insert a breakpoint the system executes the instructions up to the breakpoint, and then goes to the breakpoint procedure. Unlike the single-step feature which stops execution after each instruction, the breakpoint feature executes all the instructions up to the inserted breakpoint and then stops execution.

When you tell most 8086 systems to insert a breakpoint at some point in your program, they actually do it by temporarily replacing the instruction byte at that address with CCH, the 8086 code for the INT 3 instruction. When the 8086 executes this INT 3 instruction it pushes the flag register on the stack, resets TF and IF, and pushes the CS and IP values for the next mainline instruction on the stack. The 8086 then gets the CS value of the start of the type 3 interrupt service procedure from address 0000EH and the IP value for the procedure from address 0000CH. A breakpoint interrupt service procedure usually saves all of the register contents on the stack. Depending on the system, it may then send the register contents to the CRT display and wait for the next command from the user, or in a simple system it may just return control to the user. In this case an examine register command can be used to check if the register contents are correct at that point in the program.

OVERFLOW INTERRUPT—TYPE 4

The 8086 overflow flag, OF, will be set if the signed result of an arithmetic operation on two signed numbers is too large to be represented in the destination register or memory location. For example, if you add the 8-bit signed number 01101100 (108 decimal) and the 8-bit signed number 01010001 (81 decimal), the signed result will be 10111101 (189 decimal). This is the correct result if we were adding unsigned binary numbers, but it is not the correct signed result. For signed operations the 1 in the most-significant bit of the result indicates that the result is negative and in 2's complement form. The result then actually represents −67 decimal, which is obviously not the correct result for adding +108 and +89.

There are two major ways to detect and respond to an overflow error in a program. One way is to put the Jump if Overflow instruction, JO, immediately after the arithmetic instruction. If the overflow flag is set as a result of the arithmetic operation, execution will jump to the address specified in the JO instruction. At this address you can put an error routine which responds in the way you want to the overflow.

The second way of detecting and responding to an overflow error is to put the *Interrupt on Overflow* instruction, INTO, immediately after the arithmetic instruction in the program. If the overflow flag is not set when the 8086 executes the INTO instruction, the instruction will simply function as an NOP. However, if the overflow flag is set, indicating an overflow error, the 8086 will do a *type 4* interrupt after it executes the INTO instruction.

When the 8086 does a type 4 interrupt, it pushes the flag register on the stack, resets the TF and IF, and pushes the CS and IP values for the next instruction on the stack. It then gets the CS value for the start of the interrupt service procedure from address 00012H and the IP value for the procedure from address 00010H. Instructions in the interrupt service procedure then perform the desired response to the error condition. The procedure might, for example, set a "flag" in a memory location as we did in the BAD_DIV procedure in Figure 8-4*b*. The advantage of using the INTO and type 4 interrupt approach is that the error routine is easily accessible from any program.

SOFTWARE INTERRUPTS—TYPE 0—255

The 8086 INT instruction can be used to cause the 8086 to do any one of the 256 possible interrupt types. The desired interrupt type is specified as part of the instruction. The instruction INT 32, for example, will cause the 8086 to do a *type 32* interrupt response. The 8086 will push the flag register on the stack, reset the TF and IF, and push the CS and IP values of the next instruction on the stack. It will then get the CS and IP values for the start of the interrupt service procedure from the interrupt pointer table in memory. The IP value for any interrupt type is always at an address of 4 times the interrupt type, and the CS value is at a location two addresses higher. For a type 32 interrupt, then, the IP value will be put at 4 × 32 or 128 decimal (80H), and the CS value will be put at address 82H in the interrupt pointer table.

Software interrupts produced by the INT instruction have many uses. In a previous section we discussed the use of the INT 3 instruction to insert breakpoints in programs for debugging. Another use of software interrupts is to test various interrupt service procedures. You could, for example, use an INT 0 instruction to send execution to a divide-by-zero interrupt service procedure without having to run the actual division program. As another example, you could use an INT 2 instruction to send execution to an NMI interrupt service procedure. This allows you to test the NMI procedure without needing to apply an external signal to the NMI input of the 8086.

Another important use of software interrupts is to call desired procedures from many different programs in a system. The BIOS in the IBM PC is a good example of this. The IBM PC has in its ROMs a collection of procedures. Each procedure performs some specific function such as reading a character from the keyboard, writing some characters to the CRT, or reading some information from a disk. This collection of procedures is referred to as the *B*asic *I*nput *O*utput *S*ystem or BIOS. The BIOS procedures are called with INT instructions. You can read the BIOS section of the IBM PC technical reference manual to get all of the details of these if you need them, but here's an example of how you might use one of them.

Suppose that, as part of an assembly language program that your are writing to run on an IBM PC, you want to send some characters to the printer. Figure 8-5 is a program which shows how to do this.

Note that the DX, AH, and AL registers are used to pass parameters to the procedure. Also note that the procedure is used for two different operations: initializing the printer port and sending a character to the printer. The operation performed by the procedure is determined by the number passed to the procedure in the AH register. AH = 1 means initialize the printer port, AH = 0 means print the character in AL, and AH = 2 means read the printer status and return it in AH. If an attempt to print a character was not successful for some reason such as the printer not being turned on, not being selected, or being busy, 01 is returned in AH.

The main advantage of calling procedures in this way is that you don't need to worry about the absolute address where the procedure actually resides or about trying to link the procedure into your program. All you have to know is the interrupt type for the procedure and the format for the parameters you need to pass to the procedure. We show some other examples of using BIOS procedures in later chapters.

INTR INTERRUPTS—TYPE 0—255

The 8086 INTR input allows some external signal to interrupt execution of a program. Unlike the NMI input, however, INTR can be masked (disabled) so that it cannot cause an interrupt. If the interrupt flag, IF, is cleared, then the INTR input is disabled. The IF can be cleared at any time with the *clear interrupt* instruction, CLI. If the interrupt flag is set, the INTR input will be enabled. The IF can be set at any time with the *set interrupt* instruction, STI.

When the 8086 is reset, the interrupt flag is automatically cleared. Before the 8086 can respond to an interrupt signal on its INTR input you have to set the IF with an STI instruction. The 8086 was designed this way so that ports, timers, registers, etc. can be initialized before enabling the INTR input. In other words this allows you to get the 8086 ready to handle an interrupt before letting an interrupt in, just as you might want to get yourself ready in the morning with a cup of coffee before turning on the telephone and having to cope with the interrupts it produces.

The interrupt flag is also automatically cleared as part of the response of an 8086 to an interrupt. This is done for two reasons. First, it prevents a signal on the INTR input from interrupting a higher priority interrupt service procedure in progress. You can, however, set the IF with an STI instruction at the start of the procedure if you want an INTR input signal to be able to interrupt a procedure in progress.

The second reason for automatically disabling the INTR input at the start of an INTR interrupt service procedure is to make sure that a signal on the INTR input does not cause the 8086 to continuously interrupt itself. The INTR input is activated by a high level. In other words, whenever the INTR input is high and INTR is enabled, the 8086 will be interrupted. If INTR were not disabled during the first response, the 8086 would be continuously interrupted, and never get to the actual interrupt service procedure. Since the INTR is level-activated, the interrupt signal must remain present until it is recognized by the 8086.

The IRET instruction at the end of the interrupt service procedure restores the flags to the condition they were in before the procedure by popping the flag register off the stack. This will reenable the INTR input. If a high level signal is still present on the INTR input, it will cause the 8086 to be interrupted again. If we do not want the 8086 to be interrupted again by the same input signal, we have to use external hardware to make sure the signal is made low again before we reenable INTR with the STI instruction, or before the end of the INTR service procedure.

When the 8086 responds to an INTR interrupt signal, its response is somewhat different from its response to other interrupts. The main difference is that for an INTR interrupt, the interrupt type is sent to the 8086 from an external hardware device such as the 8259A *priority interrupt controller* which we discuss later in this chapter. An 8086 INTR response proceeds as follows.

The 8086 first does two interrupt acknowledge machine cycles, as shown in Figure 8-6. The purpose of these two machine cycles is to get the interrupt type from the external device. At the start of the first interrupt acknowledge machine cycle the 8086 floats the data bus lines, AD0–AD15. It then sends out an interrupt acknowledge pulse on its $\overline{INTA}$ output pin. This pulse essentially tells the external device, "get ready." During the second interrupt acknowledge machine cycle the 8086 sends out another pulse on its $\overline{INTA}$ output pin. In response to this second $\overline{INTA}$ pulse the external device puts the interrupt type (number) on the lower eight lines of the data bus where it is read by the 8086.

Once the 8086 receives the interrupt type, it pushes the flag register on the stack, clears TF and IF, and pushes the CS and IP values of the next instruction on

```
PAGE, 132
;8086 PROGRAM
;ABSTRACT       : This program sends a string of characters to a
;                 printer from the IBM PC
;REGISTERS USED : CS, SS, DS, BX, AX, CX, DX
;PORTS USED     : printer port 0
;PROCEDURES USED: Calls BIOS printer IO procedure INT 17

STACK_HERE      SEGMENT STACK
        DW      200 DUP(0)    ; set aside 200 words for stack
STACK_TOP       LABEL   WORD  ; assign name to word above stack top
STACK_HERE      ENDS

CHAR_COUNT      EQU     27

DATA_HERE       SEGMENT
                MESSAGE         DB      'HELLO THERE, HOW ARE YOU?'
                MESSAGE_END     DB      0DH, 0AH   ; return & line feed
DATA_HERE       ENDS
CODE_HERE       SEGMENT
                ASSUME CS:CODE_HERE, SS:STACK_HERE, DS:DATA_HERE

        MOV     AX, STACK_HERE  ; initialize stack segment register
        MOV     SS, AX
        MOV     SP, OFFSET STACK_TOP ; initialize stack pointer

        MOV     AX, DATA_HERE   ; initialize data segment
        MOV     DS, AX
        MOV     AH, 01          ; initialize printer port
        MOV     DX, 0           ; to use printer port 0
        INT     17H             ; call procedure to intitialize printer port
        LEA     BX, MESSAGE     ; get to start of message
        MOV     CX, CHAR_COUNT  ; set up a count variable
 AGAIN:
        MOV     AH, 0           ; code to tell procedure to send character
        MOV     AL, [BX]        ; load character to be sent into AL
        INT     17H             ; send character to printer
        CMP     AH, 01H         ; if character not printed then AH =1
        JNE     NEXT
NOT_RDY:STC                     ; set carry to indicate message not sent
        JMP     EXIT            ; leave loop
NEXT:   CLC                     ; clear carry flag to show character is sent
        INC     BX              ; address of next character
        LOOP    AGAIN           ; send the next character
EXIT:   NOP
CODE_HERE ENDS
        END
```

FIGURE 8-5 8086 assembly language program for outputting characters to a printer.

the stack. It then uses the type it read in from the external device to get the CS and IP values for the interrupt service procedure from the interrupt pointer table in memory. The IP value for the procedure will be put at an address equal to 4 times the type number, and the CS value will be put at an address equal to 4 times the type number plus 2, just as is done for the other interrupts.

The advantage of having an external device insert the desired interrupt type is that the external device can "funnel" interrupt signals from many sources into the INTR input pin on the 8086. When the 8086 responds with $\overline{INTA}$ pulses, the external device can send to the 8086 the interrupt type that corresponds to the source of the interrupt signal. As you will see later the external

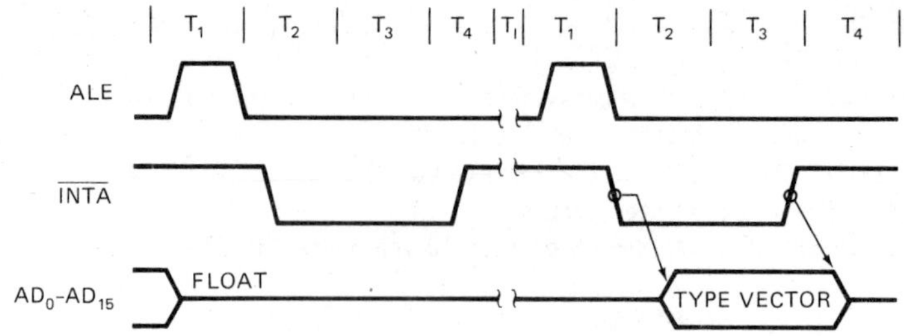

FIGURE 8-6 8086 interrupt acknowledge machine cycles.

device can also prevent an argument if two or more sources send interrupt signals at the same time.

PRIORITY OF 8086 INTERRUPTS

As you read through the preceding discussions of the different interrupt types, the question may have occurred to you, "What happens if two or more interrupts happen at the same time?" The answer to this question is that the highest priority interrupt will be serviced first, and then the next highest priority interrupt will be serviced. Figure 8-7 shows the priorities of the 8086 interrupts as shown in the Intel data book. Some examples will show you what these priorities actually mean.

As a first example, suppose that the INTR input is enabled, the 8086 receives an INTR signal during execution of a divide instruction, and the divide operation produces a divide-by-zero interrupt. Since the internal interrupts such as divide error, INT, and INTO have higher priority than INTR, the 8086 will do a divide error (type 0) interrupt response first. Part of the type 0 interrupt response is to clear the IF. This disables the INTR input and prevents the INTR signal from interrupting the higher priority type 0 interrupt service procedure. An IRET instruction at the end of the type 0 procedure will restore the flags to what they were before the type 0 response. This will reenable the INTR input and the 8086 will do an INTR interrupt response. A similar sequence of operations will occur if the 8086 is executing an INT or INTO instruction and a high level signal arrives at the INTR input.

As a second example of how this priority works, suppose that a rising-edge signal arrives at the NMI input while the 8086 is executing a DIV instruction, and that the division operation produces a divide error. Since the 8086 checks for internal interrupts before it checks for an NMI interrupt, the 8086 will push the flags on the stack, clear TF and IF, push the return address on the stack, and go to the start of the divide error (type 0) service procedure. However, because the NMI interrupt request is not disabled, the 8086 will then do an NMI (type 2) interrupt response. In other words, the 8086 will push the flags on the stack, clear TF and IF, push the return address on the stack, and go execute the NMI interrupt service procedure. When the 8086 finishes the NMI procedure, it will return to the divide error procedure, finish executing that procedure, and then return to the mainline program.

To finish our discussion of 8086 interrupt priorities, let's see how the single step (TRAP or type 1) interrupt fits in. If the trap flag is set, the 8086 will do a type 1 interrupt response after every mainline instruction. When the 8086 responds to any interrupt, however, part of its response is to clear the trap flag. This disables the single-step function, so the 8086 will not normally single-step through the instructions of the interrupt service procedure. In actuality, if the 8086 is in single-step mode when it enters an interrupt service procedure, it will execute the single-step procedure once before it executes the called interrupt procedure. The trap flag can be set again in the single-step procedure if single stepping is desired in the interrupt service procedure.

Now that we have shown you the different types of 8086 interrupts and how the 8086 responds to each, we will show you a few examples of how the 8086 hardware interrupts are used. Other applications of interrupts will be shown throughout the rest of the book.

INTERRUPT	PRIORITY
DIVIDE ERROR, INT n, INTO	HIGHEST
NMI	
INTR	
SINGLE-STEP	LOWEST

FIGURE 8-7 Priority of 8086 interrupts. *(Intel Corporation)*

HARDWARE INTERRUPT APPLICATIONS

Hardware and Software Considerations When Using Interrupts

HARDWARE

Whenever you are going to do some task with an interrupt, there are some important hardware points for you to consider. Among these are:

1. How many interrupt inputs does the microprocessor have?
2. Do these inputs require active high, active low, or edge-active signals to assert them?
3. Do the interrupt inputs have priorities?
4. Is external hardware required to insert a restart instruction or interrupt type, or is this done automatically when the CPU responds to the interrupt?

SOFTWARE

Among the software considerations when you are going to use an interrupt are the following:

1. What instructions are required to unmask/enable the interrupt input you want to use.
2. How are the stack and stack pointer initialized?
3. Does the CPU automatically save flags and register contents when it responds to the interrupt, or do you have to use push instructions at the start of the routine to do this?
4. How can data required by the interrupt service procedure be accessed no matter where in the main program the interrupt occurs?
5. What instructions are required at the end of the procedure to restore main program flags and registers, enable interrupts, and return to the interrupted mainline program.

SIMPLE INTERRUPT DATA INPUT

One of the most common uses of interrupts is to relieve a CPU of the burden of polling. Back in Chapter 4 we showed you how ASCII characters can be read in from an encoded keyboard on a polled basis. Figure 4-13 shows the circuit connections, and Figure 4-14 shows the algorithm and program for this. To refresh your memory, polling works as follows.

The strobe or data ready signal from some external device is connected to an input port line on the microcomputer. The microcomputer uses a program loop to read and test this port line over and over until the data ready signal is found to be asserted. The microcomputer then exits the polling loop and reads in the data from the external device. Data can also be output on a polled basis.

The disadvantage of polled input or output is that while the microcomputer is polling the strobe or data ready signal, it can not easily be doing other tasks. In systems where the microcomputer must be doing many tasks, polling is a waste of time, so interrupt input and output is used. In this case the data ready or strobe signal is connected to an interrupt input on the microcomputer. The microcomputer then goes about doing its other tasks until it is interrupted by a data ready signal from the external device. An interrupt service procedure can read in or send out the desired data and, when finished, return execution to the interrupted program.

For our example here we will connect the keypressed strobe to the NMI interrupt input of the 8086 on an SDK-86. The NMI input is usually reserved for responding to a power failure or some other catastropic condition. However, since we are not expecting any catastropic conditions to befall our SDK-86, we choose to use this input because it does not require an external hardware device to insert the interrupt type as does the INTR input.

Sheet 2 of the SDK-86 schematics in Figure 7-6 shows the circuitry normally connected to the NMI input. This circuitry is designed so that you can cause an NMI interrupt by pressing a key labeled INTR on the hex keypad.

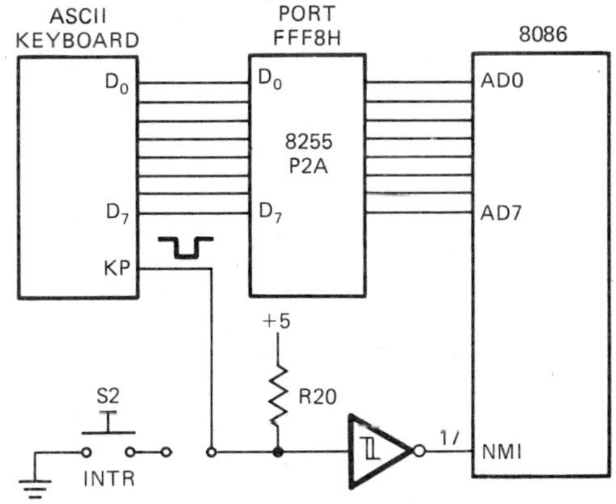

FIGURE 8-8 Circuit modifications for SDK-86 **NMI** input.

When this key is pressed, the input of the 74LS14 inverter will be made low, and the output of the inverter will go high. The low-to-high transition on the NMI input causes the 8086 to automatically do an NMI (type 2) interrupt response.

Figure 8-8 shows how we modified circuitry for our example here. We removed *R*22, a 110-Ω resistor, and *C*33, a 1-μF capacitor, so the keypad switch can no longer cause an interrupt. We then connected an active low strobe line from an ASCII-encoded keyboard directly to the input of *A*21, the 74LS14 inverter. When a key on the ASCII keyboard is pressed, the keyboard circuitry will send out the ASCII code for the pressed key on its eight parallel data lines and it will assert the keypressed strobe line low. The keypressed strobe going low will cause the NMI input of the 8086 to be asserted high. This will cause the 8086 to do a type 2 interrupt. Now let's look at the hardware and software considerations for this interrupt example.

The hardware considerations for this example are quite simply answered. The NMI input requires a low-to-high transition, and, with the circuit connections shown in Figure 8-8, this will be produced when a key on the ASCII keyboard is pressed. Since we are only using one interrupt here, we are not concerned about priorities. In response to its NMI input being asserted, the 8086 automatically does a type 2 interrupt response. No external hardware is needed for the interrupt type.

The software considerations require a little more thought, but their answers are very similar to those for the divide by zero example in a previous section. At the start of the mainline we need to load address 00008H with the IP value for the start of the type 2 procedure, and address 0000AH with the CS value for the start of the procedure. Since any interrupt response uses the stack, we need to set up a stack. Assuming that we are going to read in the ASCII characters from the keyboard and put them in an array in memory, we need to set up a data segment for the array. In the actual code section of the mainline we need to initialize the data segment register, the stack segment register, and the stack pointer register. Figure 8-9*a* shows the instructions for doing all this. Another important thing to do in the start of the mainline program is to initialize a pointer to the start of

```
page ,132
;8086 PROGRAM TO READ CHARACTERS FROM A KEYBOARD
;ABSTRACT:   The mainline of this procedure initializes the interrupt
;            table with the address of the procedure that reads the
;            characters from a keyboard on an interrupt basis.
;PROCEDURES: Uses KEYBOARD
;PORTS USED: None in mainline, FFF8H for keyboard input in KEYBOARD
;
DATA_HERE       SEGMENT WORD    PUBLIC
        ASCII_STRING    DB      100 DUP(0)      ; store for characters
        ASCII_POINTER   DW      OFFSET ASCII_STRING
        CHARCNT         DB      100             ; read 100 characters
        KEYDONE         DB      0               ; =1 if characters all read
DATA_HERE       ENDS
STACK_HERE      SEGMENT STACK
                        DW  100 DUP (0)         ; Set up stack of 100 words
        TOP_STACK  LABEL  WORD                  ; Pointer to top of stack
STACK_HERE      ENDS

PUBLIC  ASCII_POINTER, CHARCNT, KEYDONE         ; Make available to other modules
EXTRN   KEYBOARD:FAR                            ; Procedure in another assembly module

CODE_HERE       SEGMENT WORD    PUBLIC
        ASSUME CS:CODE_HERE, DS:DATA_HERE, SS:STACK_HERE
START:  MOV  AX, STACK_HERE             ; Initialize stack segment register
        MOV  SS, AX
        MOV  SP, OFFSET TOP_STACK       ; Initialize stack pointer
        MOV  AX, DATA_HERE              ; Initialize data segment register
        MOV  DS, AX
;store the address for the KEYBOARD routine at address 0000:0008
;address 00008-0000B is where type 2 interrupt gets interrupt
;service procedure address. CS at 0000A & 0000B, IP at 00008 & 00009
        MOV  AX, 0000
        MOV  ES, AX
        MOV  WORD PTR ES:000AH, SEG KEYBOARD
        MOV  WORD PTR ES:0008H, OFFSET KEYBOARD
;simulate larger program.
HERE:   JMP  HERE

CODE_HERE       ENDS
        END
```

(a)

FIGURE 8-9 Reading characters from an ASCII keyboard on interrupt basis. *(a)* Initialization and mainline. *(b)* Interrupt service procedure.

the array where the ASCII characters will be put as they are read in. The statement ASCII_POINTER DW OFFSET ASCII_STRING in the data segment in Figure 8-9*a* sets aside a word location in memory and initializes that location with the offset of the start of the array we declared to put the ASCII characters in. In the procedure we get this pointer, use it to store a character, and increment it to point to the next location in the array. Since this pointer is stored in a named memory location, it can be accessed easily by the procedure, no matter when the interrupt occurs in the mainline program.

The HERE: JMP HERE instruction at the end of the mainline program simulates a complex mainline program that the 8086 might be executing. The 8086 will execute this instruction over and over until an interrupt occurs. When an interrupt occurs the 8086 will service the interrupt and then return to execute the HERE: JMP HERE instruction over and over again until the next interrupt. Note that if we had connected the interrupt signal to the 8086 INTR interrupt input instead of the NMI input, we would have had to enable the INTR input with an STI instruction before the HERE: JMP HERE.

Figure 8-9*b* shows the interrupt service procedure for this example. The comments for the procedure express its algorithm fairly clearly. After saving AX, BX, CX, and DX on the stack, we check to see if all characters have

```
;8086 READ KEYBOARD ON INTERRUPT BASIS PROCEDURE
;ABSTRACT  : This procedure reads in ASCII characters from an
;          : encoded keyboard on an interrupt basis and stores them
;          : in a buffer in memory
;SAVES     : all registers used
;PORTS USED: input port FFF8H for the keyboard input


DATA_HERE       SEGMENT WORD    PUBLIC
      EXTRN   ASCII_POINTER:WORD, CHARCNT:BYTE, KEYDONE:BYTE
DATA_HERE       ENDS


 PUBLIC  KEYBOARD


 CODE_HERE       SEGMENT WORD    PUBLIC
 KEYBOARD        PROC            FAR
        ASSUME CS:CODE_HERE, DS:DATA_HERE


        STI             ; enable 8086 INTR so higher priority
                        ; interrupts can be recognized
        PUSH AX         ; save registers
        PUSH BX
        PUSH CX
        PUSH DX
        CMP  CHARCNT, 00        ; see if all characters read in
        JZ   EXIT               ; leave if all done
        MOV  BX, ASCII_POINTER  ; get pointer to buffer
        MOV  DX, 0FFF8H         ; point at keyboard port
        IN   AL, DX             ; Read ASCII code
        AND  AL, 7FH            ; Mask parity bit
        MOV  [BX], AL           ; Write character to buffer
        INC  ASCII_POINTER, BX  ; point to next buffer location
        DEC  CHARCNT            ; Check if 100 characters yet
        JNZ  NOTDONE            ; No, clear carry to indicate
        MOV  KEYDONE, 01H       ; Yes, set flag to indicate done
        JMP  EXIT
NOTDONE:MOV  KEYDONE, 00H       ; No, clear keydone flag
EXIT:   POP  DX                 ; restore registers
        POP  CX
        POP  BX
        POP  AX
        IRET
KEYBOARD        ENDP
CODE_HERE       ENDS
                END
```

(*b*)

FIGURE 8-9 (*continued*)

been read. If CHARCNT is zero, then we do not read in any characters. If CHARCNT is not zero, we copy the array pointer from its named memory location, ASCII_POINTER, to BX. We then read in the ASCII character from the port that the keyboard is connected to and mask the parity bit of the ASCII character. The MOV [BX], AL instruction next copies the ASCII character to the memory location pointed to by BX. To get the pointer ready for the read and store operation, we increment the stored pointer with the INC ASCII_POINTER instruction. Finally, our work done, we restore DX, CX, BX, and AX, and return to the mainline program.

Sitting in a HERE: JMP HERE loop waiting for an interrupt signal may not seem like much of an improvement over polling the keypressed strobe. However, in a more realistic program the 8086 would be doing many other tasks between keyboard interrupts. With polling the 8086 would not easily be able to do this.

Using Interrupts for Counting and Timing

COUNTING

As a simple example of the use of an interrupt input for counting, suppose that we are using an 8086 to control a printed-circuit-board-making machine in our computerized electronics factory. Further suppose that we want to detect each finished board as it comes out of the machine and to keep a count of finished boards so that we can compare this count with the number of boards fed in. This way we can determine if any boards were lost in the machine.

To do this count on an interrupt basis, all we have to do is detect when a board passes out of the machine and send an interrupt signal to an interrupt input on the 8086. The interrupt service procedure for that input can simply increment the board count stored in a named memory location.

To detect a board coming out of the machine we use an infrared LED, a phototransistor, and two conditioning gates as shown in Figure 8-10. The LED is positioned over the track where the boards come out, and the phototransistor is positioned below the track. When no board is between the LED and the phototransistor, the light from the LED will strike the phototransistor and turn it on. The collector of the phototransistor will then be low, as will the NMI input on the 8086. When a board passes between the LED and the phototransistor, the light will not reach the phototransistor, and it will turn off. Its collector will go high, and so will the signal to the NMI input of the 8086. The 74LS14 Schmitt trigger inverters are necessary to turn the slow rise-time signal from the phototransistor collector into a signal which meets the rise-time requirements of the NMI input on the 8086. When the 8086 receives the low-to-high signal on its NMI input, it will automatically do a type 2 interrupt response. As we mentioned above, all the type 2 interrupt service procedure has to do in this case is increment the board count in a named memory location and return to running the machine. This same technique can be used to count people going into a stadium, cows coming in from the pasture, or just about any thing else you might want to count.

USING AN INTERRUPT INPUT FOR TIMING APPLICATIONS

In Chapter 4 we showed how a delay loop could be used to set the time between microcomputer operations. In the example there we used a delay loop to let us take in data samples at 1-ms intervals. The obvious disadvantage of a delay loop is that while the microcomputer is stuck in the delay loop, it cannot easily be doing other useful work. In many cases a delay loop would be a waste of the microcomputer's valuable time. For most microcomputer timing, an interrupt approach is much more efficient.

Suppose, for example, that in our 8086-controlled printed circuit board machine we need to check the pH of a solution approximately every 4 minutes. If we used a delay loop to count off the 4 minutes, either the 8086 wouldn't be able to do much else, or we would have some difficult calculations to figure out at what points in the program to go check the pH.

To solve this problem, all we have to do is connect a simple 1-Hz pulse source to an interrupt input as shown in Figure 8-11. This 555 timer circuit is not very accurate, but it is inexpensive, and it is good enough for this application. We connect the timer output to the 8086 NMI input as you might do to demonstrate this concept on an SDK-86 board. The 555 timer will send an interrupt signal to the 8086 NMI input approximately once every second. If we simply count the number of NMI interrupts that occur, we will then know how many seconds have passed.

Here's how the programming is done for this application. In the mainline we set aside a memory location for

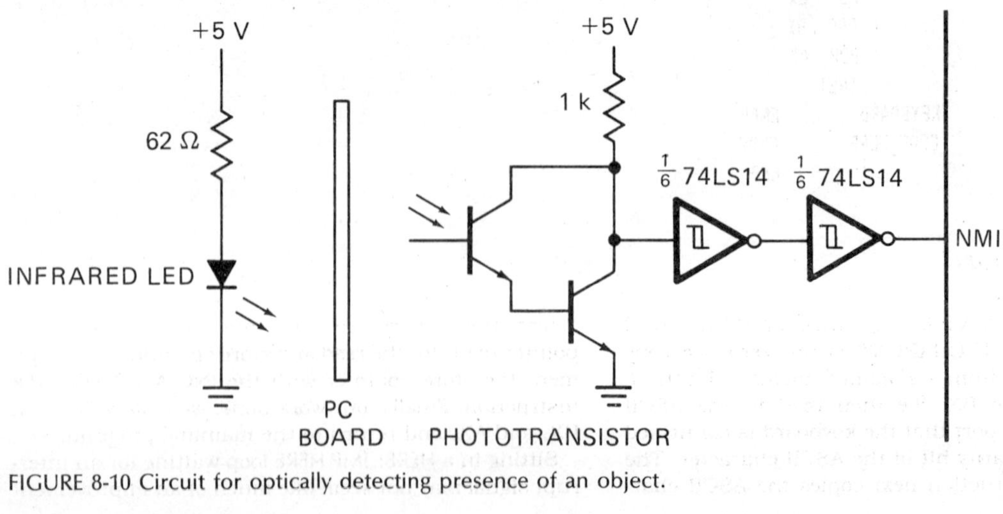

FIGURE 8-10 Circuit for optically detecting presence of an object.

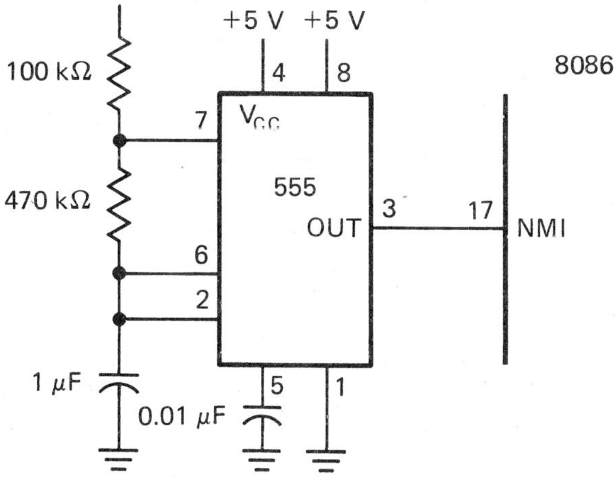

FIGURE 8-11 Inexpensive 1-Hz pulse source for interrupt timing.

the seconds count and initialize that location to the number of seconds that we want to count off. In this case we want 4 minutes, which is 240 decimal or F0H seconds. Each time the 8086 receives an interrupt from the 555 timer, it executes the interrupt service procedure for the NMI interrupt. In this procedure we decrement the seconds count in the named memory location and test to see if the count is down to zero yet. If the count is zero, we know that 4 minutes have elapsed, so we reload the seconds count memory location with F0H and call the procedure which reads the pH of the solution and takes appropriate action if the pH is not correct. If the seconds count is not zero, execution simply returns to the mainline program until the next interrupt from the 555 or from some other source occurs. To help you visualize how this works, Figure 8-12 shows the algorithm for this mainline and procedure. The advantage of this interrupt approach is that the interrupt service procedure only takes a few microseconds of the 8086's time once every second. The rest of the time the 8086 is free to run the mainline program.

USING AN INTERRUPT TO PRODUCE A REAL-TIME CLOCK

Another application using a 1-Hz interrupt input might be to generate a real-time clock of seconds, minutes, and hours. The time from this clock can then be displayed and/or printed out on timecards, etc. To generate the clock a 1-Hz signal is applied to an interrupt input. A seconds count, a minutes count, and an hours count are kept in three successive memory locations. When an interrupt occurs, the seconds count is incremented by one. If the seconds count is not equal to 60, then execution is simply returned to the mainline program. If the seconds count is equal to 60 then the seconds count is reset to zero, and the minutes count is incremented by one. If the minutes count is not 60 then execution is simply returned to the mainline. If the minutes count is 60 then the minutes count is reset to zero, and the

```
INITIALIZE
     INTERRUPT POINTER TABLE
     STACK AND STACK SEGMENT POINTER
     DATA SEGMENT
     SECONDS COUNT TO 240 DECIMAL
WAIT FOR INTERRUPT
```

(a)

```
SAVE REGISTERS
DECREMENT SECONDS COUNT
IF SECONDS COUNT = 0 THEN
     RELOAD SECONDS COUNT WITH 240 DECIMAL
     CALL pH READ PROCEDURE
     RESTORE REGISTERS
     RETURN TO MAINLINE
ELSE RESTORE REGISTERS
     RETURN TO MAINLINE
```

(b)

FIGURE 8-12 Algorithm for pH read at 4-minute intervals. *(a)* Initialization and mainline. *(b)* Interrupt service procedure.

hours count is incremented by one. If the hours count is not 13, then execution is simply returned to the mainline. If the hours count is equal to 13 then it is reset to 1 and execution returned to the mainline. A problem at the end of the chapter asks you to write the algorithm and program for this real-time clock.

The interrupt service routine for the real-time clock can easily be modified to also keep track of other time measurements such as the 4-minute timer shown in the preceding example. In other words, the single interrupt service routine can be used to keep track of several different time intervals. By counting a different number of interrupts or applying a different frequency signal to the interrupt input, this technique can be used to time many different tasks in a microcomputer system.

GENERATING AN ACCURATE TIME BASE FOR TIMING INTERRUPTS

The 555 timer that we used for the 4-minute timer described above was accurate enough for that application, but for many applications, it is not. For more precise timing we usually use a signal derived from a crystal-controlled oscillator such as the processor clock signal. The processor clock signal is stable, but it is obviously too high in frequency to drive a processor interrupt input directly. Therefore, it is divided down with an external counter device to an appropriate frequency for the interrupt input. Most microcomputers have a counter device such as the Intel 8253 or 8254, which can be programmed with instructions to divide an input frequency by any desired number. Besides acting as programmable frequency dividers, these devices have many important uses in microcomputer systems. Therefore, the next section describes how an 8254 operates, how an 8524 can easily be added to an SDK-86 board, and how an 8254 is used in a variety of interrupt applications. Also in the next section we use the 8254 discussion to show you the general procedure for initializing any of the programmable peripheral devices we discuss in later chapters.

A Software-Programmable Timer/Counter, the Intel 8253 and 8254

Because of the many tasks that they can be used for in microcomputer systems, programmable timer/counters are very important for you to learn about. As you read through the following sections, pay particular attention to the applications of this device in systems and the general procedure for initializing a programmable device such as the 8254. Read lightly through the discussions of the different counter modes to become aware of the types of problems that the device can solve for you. You can later dig into the details of these discussions when you have a specific problem to solve.

Another important point to make to you here is that the discussions of various devices throughout the rest of this book are not intended to replace the manufacturers' data sheets for the devices. Many of the programmable peripheral devices we discuss are so versatile that they require almost a small book for each to describe all the details of their operations. The discussions here are intended to introduce you to the devices, show you what they can be used for, and show you enough details about them that you can do some real jobs with them. After you become familiar with using a device in some simple applications, you can read the data sheets to learn further "bells and whistles" that the devices have.

Basic 8253 and 8254 Operation

The Intel 8253 and 8254 each contain three 16-bit counters which can be programmed to operate in several different modes. The 8253 and 8254 devices are pin-for-pin compatible, and they are nearly identical in function. The major differences are:

1. The maximum input clock frequency for the 8253 is 2.6 MHz, the maximum clock frequency for the 8254 is 8 MHz (10 MHz for the 8254-2).
2. The 8254 has a *read-back* feature which allows you to latch the count in all of the counters and the status of the counter at any point. The 8253 does not have this read-back feature.

To simplify reading of this section we will refer only to the 8254. However, you can assume that the discussion also applies to the 8253 except where we specifically state otherwise.

As shown by the block diagram of the 8254 in Figure 8-13, the device contains three 16-bit counters. In some ways these counters are similar to the TTL presettable counters we reviewed in Chapter 1. The big advantage of these counters, however, is that you can load a count in them, start them, and stop them with instructions in your program. The device is then said to be software

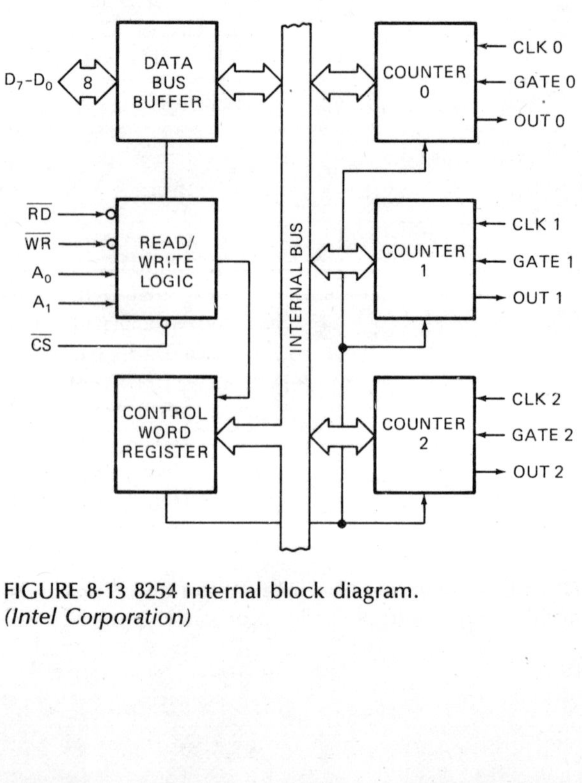

FIGURE 8-13 8254 internal block diagram. *(Intel Corporation)*

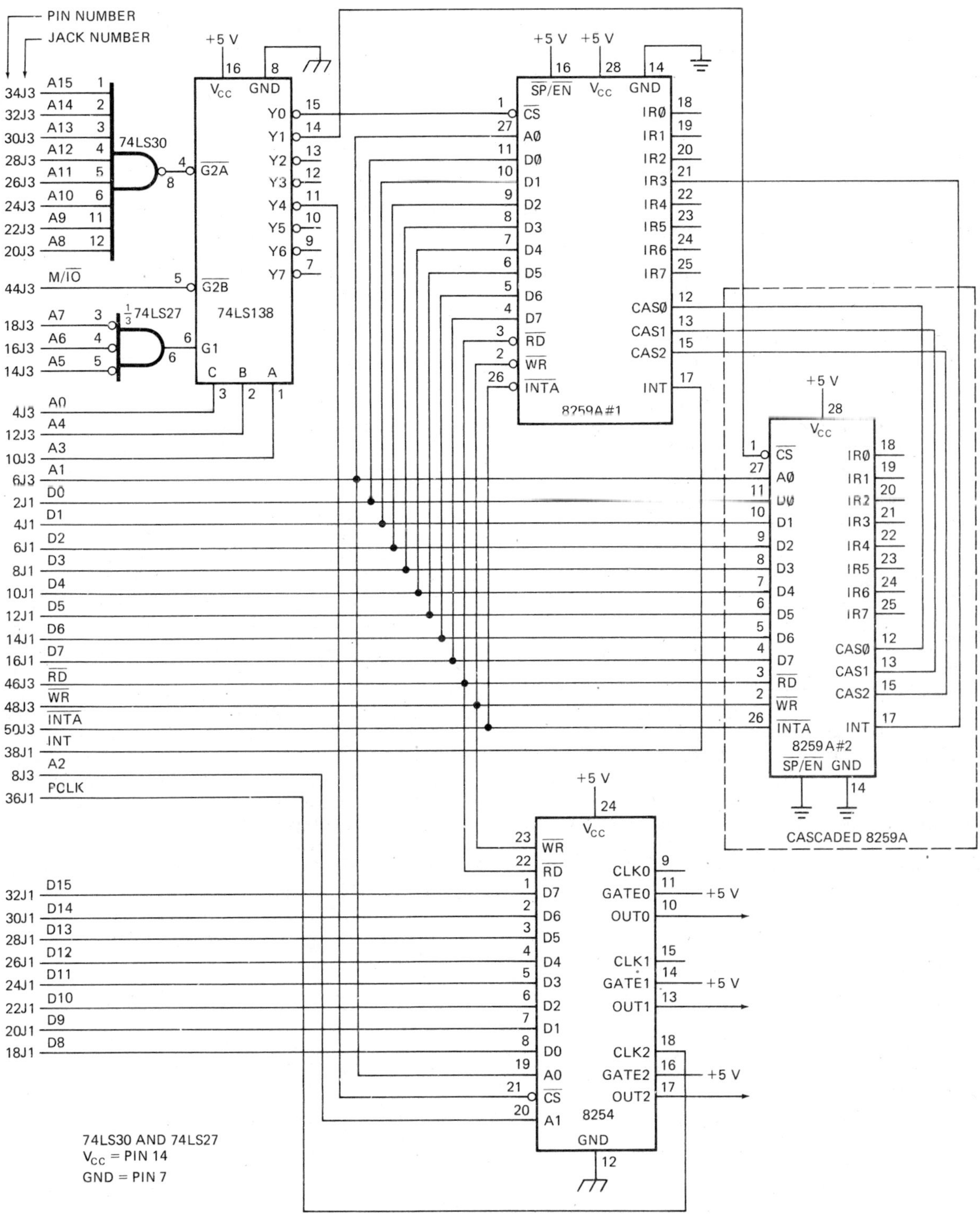

FIGURE 8-14 Circuit showing how to add an 8254 and 8259A(s) to an SDK-86 board.

programmable. To program the device you send count bytes and control bytes to the device just as you would send data to a port device.

If you look along the left side of the block diagram in Figure 8-13 you will see the signal lines used to interface the device to the system buses. A little later we show how these are actually connected in a real system. The main points for you to note about the 8254 at the moment are that it has an 8-bit interface to the data bus, it has a $\overline{\text{CS}}$ input which will be asserted by an address decoder when the device is addressed, and that it has two address inputs, A0 and A1, to allow you to address one of the three counters or the control word register in the device.

The right side of the 8254 block diagram in Figure 8-13 shows the counter inputs and outputs. You can apply a signal of any frequency from dc to 8 MHz (2.6 MHz for the 8253) to the counter clock inputs, labeled CLK in the diagram. The GATE inputs on each counter allow you to start or stop that counter with an external hardware signal. If the GATE input of a counter is high (1), then the counter is enabled for counting. If the GATE input is low, the counter is disabled. The resultant frequency or pulse from each counter appears on its OUT pin. Now let's see how a programmable peripheral device such as the 8254 is connected in a system.

SYSTEM CONNECTIONS FOR AN 8254 TIMER/COUNTER

An 8254 is a very useful device to have in a microcomputer system, but, in order to keep the cost down, the SDK-86 was not designed with one on the board. For a real example of how an 8254 is connected in a system, we show you here how to add one to an SDK-86 board. If you use wire-wrap headers for connectors J1 and J3, the circuitry shown can easily be wire-wrapped on the prototyping area of the SDK-86 board. Install the WAIT state jumper to insert one WAIT state, as explained in Chapter 7. A WAIT state is needed because of the added delay of the decoders and buffers.

Figure 8-14 shows the circuit connections for adding an 8254 and an 8259A to an SDK-86 board. We discuss the 8259A priority interrupt controller (PIC) in the last major section of this chapter. Analyzing the circuit in Figure 8-14 should help refresh your memory on address decoding.

The 74LS138 in Figure 8-14 is used to produce chip select ($\overline{\text{CS}}$) signals for the 8254, the 8259A, and any other I/O devices we might want to add. We will look first at the circuitry around this device to determine the system base address which selects each device.

In order for any of the outputs of the 74LS138 to be asserted, the G1, $\overline{\text{G2A}}$, and $\overline{\text{G2B}}$ enable inputs must all be asserted. The G1 input will be asserted (high) if system address lines A5, A6, and A7 are all low. The $\overline{\text{G2A}}$ input will be asserted (low) if system address lines A8–A15 are all high. As shown by the truth table in Figure 8-15, these two inputs then will be asserted for a system base address of FF00H. The $\overline{\text{G2B}}$ input of the 74LS138 will be asserted (low) if the M/$\overline{\text{IO}}$ line is low, as it will be for a port read or write operation.

Now, remember from Chapter 7, that only one of the Y outputs of the 74LS138 will ever be asserted at a time. The output asserted is determined by the 3-bit binary code applied to the A, B, and C select inputs. In the circuit in Figure 8-14 we connected system address line A0 to the C input, address line A4 to the B input, and address line A3 to the A select input. The truth table in Figure 8-15 shows the system base addresses that will enable each of the 74LS138 Y outputs. As you will see a little later, system address lines A1 and A2 are used to select internal parts of the 8254 and 8259.

We connected A0 to the C input so that half of the Y outputs will be selected for even addresses and half of the Y outputs will be selected by odd addresses. We did this so that we can equalize loading on the two halves of the data bus as we add peripheral devices such as the 8254 and 8259A. To see how this works, note that the peripheral devices have only eight data lines. For an odd-addressed device we connect these data lines to the upper eight system data lines, and for an even-addressed device we connect these to the lower eight system data lines. By alternating between odd and even selected outputs as we add peripheral devices, we equalize loading on the bus as desired.

As shown by the truth table in Figure 8-15, the system base address of the added 8254 is FF01H. Other connec-

A8–A15	A5–A7	A4	A3	A2	A1	A0	M/$\overline{\text{IO}}$	Y OUTPUT SELECTED	SYSTEM BASE ADDRESS	DEVICE
1	0	0	0	X	X	0	0	0	F F 0 0	8259A #1
1	0	0	1	X	X	0	0	1	F F 0 8	8259A #2
1	0	1	0	X	X	0	0	2	F F 1 0	
1	0	1	1	X	X	0	0	3	F F 1 8	
1	0	0	0	X	X	1	0	4	F F 0 1	8254
1	0	0	1	X	X	1	0	5	F F 0 9	
1	0	1	0	X	X	1	0	6	F F 1 1	
1	0	1	1	X	X	1	0	7	F F 1 9	
ALL OTHER STATES								NONE		

FIGURE 8-15 Truth table for 74LS138 address decoder in Figure 8-14.

tions to the 8254 are the system $\overline{RD}$ and $\overline{WR}$ lines used to enable the 8254 for reading or writing; eight data lines, used to send control bytes, status bytes, and count values between the CPU and the 8254; and system address lines A1 and A2, used to select the control register or one of the three counters in the 8254. Next we will show you how to initialize an 8254 to do some useful work for you.

INITIALIZING A PROGRAMMABLE PERIPHERAL DEVICE—THE 8254

When the power is first turned on, programmable peripheral devices such as the 8254 are usually in *undefined states.* Before you can use them for anything you have to initialize them in the *mode* you need for your specific application. Initializing these devices is not usually difficult, but it is very easy to make errors if you do not do it in a very systematic way. To initialize any programmable peripheral device you should work your way through the following series of steps.

1. Determine the system base address for the device. You do this from the address decoder circuitry or the address decoder truth table. From the truth table in Figure 8-15 the system base address of the 8254 in our example here is FF01H.

2. Use the device data sheet to determine the internal addresses for each of the control registers, ports, timers, status registers, etc. in the device. Figure 8-16*a* shows the internal addresses for the three counters and the control word register for the 8254. A0 in this table represents the A0 input of the device and A1 represents the A1 input of the device. Note in the schematic in Figure 8-14 that we connected system address line A1 to the A0 input, and system address line A2 to the A1 input of the 8254. Among other reasons, we did this because, as described before, we wanted to use system address line A0 as an input to the address decoder.

3. Add each of the internal addresses to the system base address to determine the system address of each of the parts of the device. You need to do this so that you know to what address to send control words, timer values, etc. Figure 8-16*b* shows the system addresses for the three timers and the control register of the 8254 we added to the SDK-86 board. Note that the addresses are all odd.

4. Look in the data sheet for the device for the format of the control word(s) that you have to send to the device to initialize it. For different devices, incidentally, the control word(s) may be referred to as command words or mode words. To initialize the 8254 you send a control word to the control register for each counter that you want to use. Figure 8-17 shows the format for the 8254 control word.

A_1	A_2	SELECTS
0	0	COUNTER 0
0	1	COUNTER 1
1	0	COUNTER 2
1	1	CONTROL WORD REGISTER

(*a*)

SYSTEM ADDRESS	8254 PART
F F 0 1	COUNTER 0
F F 0 3	COUNTER 1
F F 0 5	COUNTER 2
F F 0 7	CONTROL REG

(*b*)

FIGURE 8-16 8254 internal addresses and system addresses. *(a)* Internal. *(b)* System.

D_7	D_6	D_5	D_4	D_3	D_2	D_1	D_0
SC1	SC0	RW1	RW0	M2	M1	M0	BCD

SC – SELECT COUNTER:

SC1	SC0	
0	0	SELECT COUNTER 0
0	1	SELECT COUNTER 1
1	0	SELECT COUNTER 2
1	1	READ-BACK COMMAND (SEE READ OPERATIONS)

RW – READ/WRITE:

RW1	RW0	
0	0	COUNTER LATCH COMMAND (SEE READ OPERATIONS)
0	1	READ/WRITE LEAST SIGNIFICANT BYTE ONLY.
1	0	READ/WRITE MOST SIGNIFICANT BYTE ONLY.
1	1	READ/WRITE LEAST SIGNIFICANT BYTE FIRST, THEN MOST SIGNIFICANT BYTE.

M – MODE:

M2	M1	M0	
0	0	0	MODE 0 – INTERRUPT ON TERMINAL COUNT
0	0	1	MODE 1 – HARDWARE ONE-SHOT
X	1	0	MODE 2 – PULSE GENERATOR
X	1	1	MODE 3 – SQUARE WAVE GENERATOR
1	0	0	MODE 4 – SOFTWARE TRIGGERED STROBE
1	0	1	MODE 5 – HARDWARE TRIGGERED STROBE

BCD:

0	BINARY COUNTER 16-BITS
1	BINARY CODED DECIMAL (BCD) COUNTER (4 DECADES)

NOTE: DON'T CARE BITS (X) SHOULD BE 0 TO INSURE COMPATIBILITY WITH FUTURE INTEL PRODUCTS.

FIGURE 8-17 8254 control word format. *(Intel Corporation)*

5. Construct the control word required to initialize the device for your specific application. You construct this control word on a bit-by-bit basis. We have found it helpful to actually draw the eight little boxes as shown at the top of Figure 8-17 so that we don't miss any bits. (An easy way to draw the eight boxes is to draw a long rectangle, divide it in half, divide each resulting half in two, and finally divide each resulting quarter in two.) To help keep track of the meaning of each bit of a control word, write under each bit the meaning of that bit. A little later we show you how to do this for an 8254 control word. Documentation of this sort is very valuable when you are trying to debug a program or modify an old program for some new application.

6. Finally, send the control word(s) you have made up to the control register address for the device, and send the starting count to the counter registers. Now, Let's take a closer look at the 8254 control word format now to see how you make up one of these words.

A separate control word must be sent for each counter that you want to use in the device. However, according to Figure 8-16*a*, the 8254 has only one control register address. The trick here is that the control words for all three counters are sent to the same address in the device. You use the upper two bits of each control word to tell the 8254 which counter you want that control word to initialize. For example, if you are making up a control word for counter 0 in the 8254 you make the SC1 bit of the control word a 0, and the SC0 bit a 0. Later we will explain the meaning of the read-back command specified by a 1 in each of these bits.

Next let's look at the bit labeled BCD in the control word. The 16-bit counters in the 8254 are down-counters. This means that the number in a counter will be decremented by each clock pulse. You can program the 8254 to count down a loaded number in BCD (decimal) or in binary. If you make the D0 bit of the control word a 0, then the counter will treat the loaded number as a pure binary number. For this case the largest number that you can load in is FFFFH. If you make the D0 bit of the control word a 1, then the largest number you can load in the counter is 9999H, and the counter will count a loaded number down in decimal (BCD). Actually, because of the way the 8254 counts, the "largest" number you can load in for both cases is 0000, but thinking of FFFFH and 9999H makes it easier to remember the difference between the two modes.

Now let's take a brief look at the mode bits (M2, M1, and M0) in the control word format in Figure 8-17. The binary number you put in these bits specifies the effect that the gate input will have on counting and the waveform that will be produced on the OUT pin. For example, if you specify mode 3 for a counter by putting 011 in these 3 bits, the counter will be put in a square-wave mode. In this mode the output will be high for the first half of the loaded count and low for the second half of the loaded count. When the count reaches zero, the original count is automatically reloaded and the countdown repeated. The waveform on the OUT pin in this mode will then be a square wave with a frequency equal to the input clock frequency divided by the count you wrote to the counter. A little later we will discuss and show applications for some of the six different modes. First let's finish looking at the control word bits and see how you send the control word and a count to the device.

The RW1 and RW0 bits of the control word are used to specify how you want to write a count to a counter or to read the count from a counter. If you want to load a 16-bit number into a counter, you put 1's in both of these bits in the control word you send for that counter. After you send the control word, you send the low byte of the count to the counter address, and then send the high byte of the count to the counter address. In a later paragraph we show an example of the instruction sequence to do this. In cases where you only want to load a new value in the low byte of a counter, you can send a control word with 01 in the RW bits, and then send the new low byte to the counter. Likewise, if you only want to load a new high byte value in the counter, you can send a control word with 10 in the RW bits, and then send only the new high byte to the counter.

You can read the number in one of the counters at any time. The usual way to do this is to first latch the current count in some internal latches by sending a control word with 00 in the RW bits. Send another control word with 01, 10 or 11 in the RW bits to specify how you want to read out the bytes of the latched count. Then read the count from the counter address.

Now, for a specific example, suppose that we want to use counter 0 of the 8254 in Figure 8-14 to produce a stable 78.6-kHz square-wave signal for a UART clock by dividing down the 2.45-MHz PCLK signal available on the SDK-86 board. To do this we first connect the SDK-86 PCLK signal to the CLK input of counter 0 and tie the GATE input of counter high to enable it for counting. To produce 78.6 kHz from 2.45 MHz we have to divide by 32 decimal, so this is the value that we will eventually load into counter 0. First, however, we have to determine the system addresses for the device, make up the control word for counter 0, and send the control word.

As shown in Figure 8-16*b* the system address for the control register of this 8254 is FF07H. This is where we will send the control word. For our control word we want to select counter 0, so we make the SC1 and SC0 bits both 0's. We want the counter to operate in square-wave mode. This is mode 3, so we make the mode bits of the control word 011. Since we want to divide by 32 decimal, we tell the counter to count down in decimal by making the BCD bit of the control word a 1. This makes our life easier, because we don't have to convert the 32 to binary or hex. Finally we have to decide how we want to load the count into the counter. Since the count that we need to load in is less than 99, we only have to load the lower byte of the counter. According to Figure 8-17, the RW1 bit should be a 0 and the RW0 bit a 1 for a write to only the lower byte (LSB). The complete control word then is 00010111 in binary. Here are the instructions to send the control word and count to counter 0 of the 8254 in Figure 8-14. Note how the bits of the control word are documented.

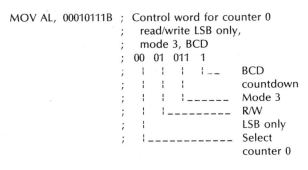

```
MOV AL, 00010111B ; Control word for counter 0
                  ;   read/write LSB only,
                  ;   mode 3, BCD
                  ; 00 01 011  1
                  ;  |  |   |  |__  BCD
                  ;  |  |   |       countdown
                  ;  |  |   |______ Mode 3
                  ;  |  |_________  R/W
                  ;  |              LSB only
                  ;  |____________  Select
                  ;                 counter 0

MOV DX, 0FF07H    ; Point at 8254 control register
OUT DX, AL        ; Send control word
MOV AL, 32H       ; Load lower byte of count
MOV DX, 0FF01H    ; Point to counter 0 count register
OUT DX, AL        ; Send count to count register
```

Note that since we set the RW bits of the control word for read/write LSB only, we do not have to include instructions to load the MSB of the counter. Programmed in this way the 8254 will automatically load 0's in the upper byte of the counter.

If you need to load a count that is larger than 1 byte, make the RW bits in the control word both 1's. Send the lower byte of the count as shown above. Then send the high byte of the count to the count register by adding the instructions:

```
MOV AL, HIGH_BYTE_OF_COUNT; Load MSB of count
OUT DX, AL    ; Send MSB to count register
```

Note that the high byte of the count is sent to the same address that the low byte of the count was sent. For each counter that you want to use in an 8254, you repeat the above series of six or eight instructions with the control word and count for the mode that you want. Before going on with this chapter, review the six initialization steps shown at the start of this section to make sure these are firmly fixed in your mind. In the next section we discuss and show some applications of the different modes that an 8254 counter can be operated in, but we do not have space there to show all of the steps for each of the modes.

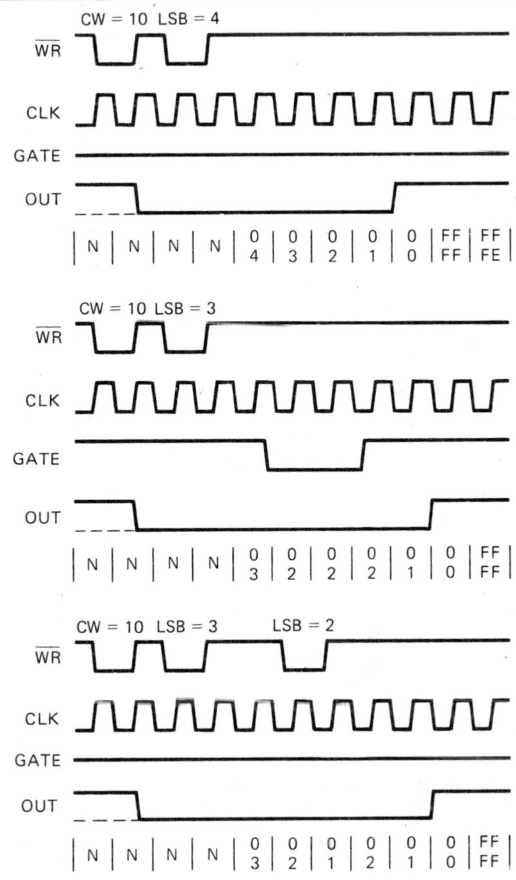

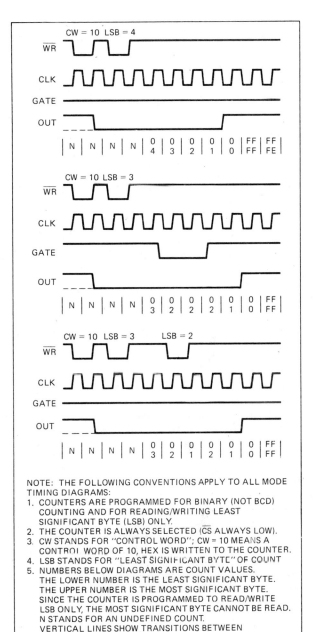

NOTE: THE FOLLOWING CONVENTIONS APPLY TO ALL MODE TIMING DIAGRAMS:

1. COUNTERS ARE PROGRAMMED FOR BINARY (NOT BCD) COUNTING AND FOR READING/WRITING LEAST SIGNIFICANT BYTE (LSB) ONLY.
2. THE COUNTER IS ALWAYS SELECTED ($\overline{CS}$ ALWAYS LOW).
3. CW STANDS FOR "CONTROL WORD"; CW = 10 MEANS A CONTROL WORD OF 10, HEX IS WRITTEN TO THE COUNTER.
4. LSB STANDS FOR "LEAST SIGNIFICANT BYTE" OF COUNT
5. NUMBERS BELOW DIAGRAMS ARE COUNT VALUES. THE LOWER NUMBER IS THE LEAST SIGNIFICANT BYTE. THE UPPER NUMBER IS THE MOST SIGNIFICANT BYTE. SINCE THE COUNTER IS PROGRAMMED TO READ/WRITE LSB ONLY, THE MOST SIGNIFICANT BYTE CANNOT BE READ. N STANDS FOR AN UNDEFINED COUNT. VERTICAL LINES SHOW TRANSITIONS BETWEEN COUNT VALUES.

MODE 0

FIGURE 8-18 8254 MODE 0 example timing diagrams. *(Intel Corporation)*

8254 Counter Modes and Applications

As we mentioned previously, an 8254 counter can be programmed to operate in any one of six different modes. The Intel data book uses timing diagrams such as those in Figure 8-18 to show how a counter functions in each of these modes. Since all of these waveforms may not be totally obvious to you at first glance, we will work our way through some of these to show you how to interpret them. We will also show some uses of the different counter modes.

MODE 0—INTERRUPT ON TERMINAL COUNT

First read the Intel notes at the bottom of Figure 8-18, then take a look at the top set of waveforms in the figure. For this first example the GATE input is held high so the counter is always enabled for counting. The first dip in the waveform labeled $\overline{WR}$ represents the control word for the counter being written to the 8254. CW = 10 over this dip indicates that the control word written is 10H. According to the control word format in Figure 8-17, this means that counter 0 is being initialized for binary counting, mode 0, and a read/write of only the LSB. After the control word is written to the control register, the output pin of counter 0 will go low. The next dip in the $\overline{WR}$ waveform represents a count of 4 being written to the count register of counter 0. Before this count can be counted down, it must be transferred from the count register to the actual counter. If you look at the count values shown under the OUT waveform in the timing diagram, you should see that the count of 4 is transferred into the counter by the next clock pulse after $\overline{WR}$ goes high. Each clock pulse after this will decrement the count by one. When the counter transitions to zero, the OUT pin will go high. If you write a count N to a counter in mode 0, the OUT pin will go high after $N + 1$ clock pulses have occurred. Note that the counter decrements from 0000 to FFFFH on the next clock pulse unless you load some new count into the counter. If the OUT pin is connected to an active high interrupt input of the processor, then the processor will be interrupted when the counter reaches zero (terminal count).

The second set of waveforms in Figure 8-18 shows that if the GATE input is made low, the counter value will be held. When the GATE input is made high again, the counter continues to decrement by one for each clock pulse. The third set of waveforms in Figure 8-18 shows that if a new count is written to a counter, the new count will be loaded into the counter on the next clock pulse. Following clock pulses will decrement the new count until it reaches zero.

As an example of what you can use this mode for, suppose that as one of its jobs we want to use an available 8086 to control some parking lot signs around our electronics factory. The main parking lot can hold 1000 cars. When it gets full, we want to turn on a sign which directs people to another available lot. To detect when a car enters the lot we can use an optical sensor such as the one shown in Figure 8-10. Each time a car passes through, this circuit will produce a pulse. We could connect the signal from this sensor to an interrupt input, and have the processor count interrupts as we did for the printed-circuit-board-making machine in a previous example. However, the less we burden the processor with trivial tasks such as this, the more it is available to do complex work for us. Therefore, we let a counter in an 8254 count cars and only interrupt the 8086 when it has counted 1000 cars.

We connect the output from the optical sensor circuit to the CLK input of, say, counter 1 of an 8254. We tie the GATE input of counter 1 to +5 V so it will be enabled for counting. We connect the OUT pin of counter 1 to an interrupt input on the 8086.

In the mainline program we initialize counter 1 for mode 0, BCD counting, and read/write LSB then MSB with a control word of 01110011 binary. We want the counter to produce an interrupt after 1000 pulses from the sensor, so we will send a count of 999 decimal to the counter. The reason that we want to send 999 instead of 1000 is that, as shown in Figure 8-18, the OUT pin will go high $N + 1$ clock pulses after the count value is written to the counter. Since we initialized the counter for read/write LSB then MSB, we send 99H and then 09H to the address of counter 1. Note that we initialized the counter for BCD counting, so we can just send the count value as a BCD number instead of having to convert it to hex.

The service procedure for this interrupt will contain instructions which turn on the parking-lot-full sign, close off the main entrance, and return to the mainline program. For this example we don't worry that the counter decrements from 0000 to FFFFH, because, after we shut the gate, the counter will not receive any more interrupts.

MODE 1—HARDWARE RETRIGGERABLE ONE-SHOT

The basic principle of a *one-shot* is that when a signal is applied to the trigger input of the device, its output will be asserted. After a fixed amount of time the output will automatically return to its unasserted state. For a TTL one-shot such as the 74LS122, the time that the output is asserted is determined by the time constant of a resistor and a capacitor connected to the device. For an 8254 counter in one-shot mode the time that the output is

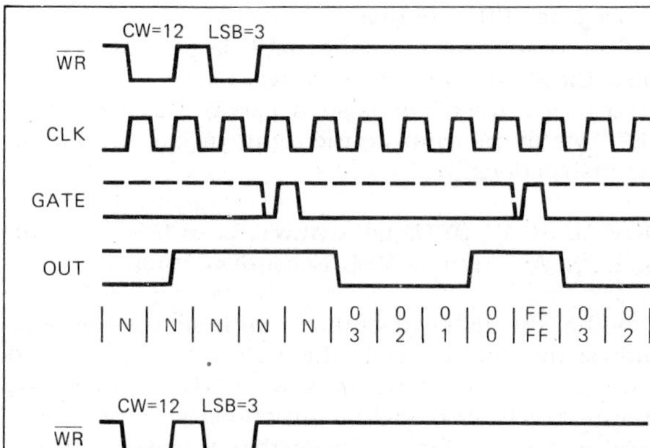

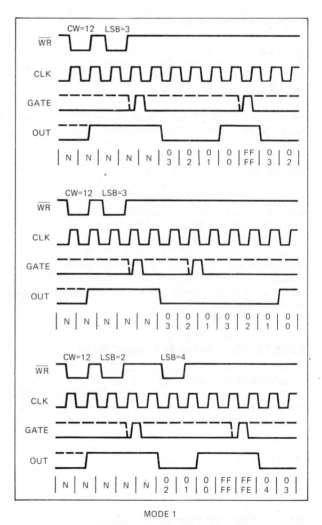

FIGURE 8-19 8254 MODE 1 example timing diagrams. *(Intel Corporation)*

asserted low is determined by the frequency of an applied clock and a count loaded into the counter. The advantage of the 8254 approach is that the output pulse width can be changed under program control, and if a crystal-controlled clock is used, the output pulse width can be very accurately specified.

Figure 8-19 shows some example timing waveforms for an 8254 counter in mode 1. Let's take a look at the top set of waveforms. Again the first dip in the $\overline{WR}$ waveform represents the control word of 12H being sent to the 8254. Use Figure 8-17 to help you determine how this control word initializes the device. You should find that a control word of 12H programs counter 0 for binary count, mode 1, read/write LSB then MSB. When the control word is written to the 8254, the OUT pin goes high.

The second dip in the $\overline{WR}$ waveform represents writing a count to the counter. Note that, because the GATE input is low, the counter does not start counting down immediately when the count is written as it does in mode 0. For mode 1 the GATE input functions as a trigger input. When the GATE/trigger input is made high, the count will be transferred from the count register to the actual counter on the next clock pulse. Each following clock pulse will decrement the counter by one. When the counter reaches zero, the OUT pin will go high again. In other words, if we load a value of *N* in the counter, and we trigger the device by making the GATE input high, the OUT pin will go low for a time equal to *N* clock cycles. The output pulse width is then *N* times the period of the signal applied to the CLK input. Incidentally, the dashed sections of the GATE waveforms in Figure 8-19 mean that the GATE/trigger input signal can go low again anytime during that time interval.

The second set of waveforms in Figure 8-19 demonstrate what is meant by the term *retriggerable*. If another trigger pulse comes before the previously loaded count has been counted down to zero, the original count will be reloaded on the next clock pulse. The countdown will then start over and continue until another trigger occurs or until the count reaches zero. If trigger pulses continue to come before the count is decremented to zero, the OUT pin will remain low.

The bottom set of waveforms in Figure 8-19 show that if you write a new count to a count register while the OUT pin is low, the new count will not be loaded into the counter and counted down until the next trigger pulse occurs.

For an example of the use of mode 1 we will show you how to make a circuit which produces an interrupt signal if the ac power fails. This circuit could be connected to the NMI input of an 8086 to vector to a procedure which saves parameters in battery-backed RAM when the ac power fails. Figure 8-20 shows a circuit which uses an optical coupler (LED and a phototransistor packaged together) to produce logic level pulses at power-line frequency. The 74LS14 inverters sharpen the edges of these pulses so that they can be applied to the GATE/trigger input of an 8254. For a 60-Hz line frequency, a pulse will be produced every 16.66 ms. Now what we want to do here is to load the counter with a value such that the counter will always be retriggered by the power-line pulses before the countdown is completed. As shown by the second set of waveforms in Figure 8-19, the OUT pin will then stay low and not send an interrupt signal to the NMI input of the 8086. If the ac power fails, no more pulses come in to the 8254 trigger input. The trigger input will be left high, and the countdown will be completed. The 8254 OUT pin will then go high and interrupt the 8086.

To determine the counter value for this application, just calculate the number of input clock pulses required to produce a countdown time longer than 16.66 ms, for example, 20 ms. If we use the 2.4576-MHz PCLK signal on an SDK-86 board, 20 ms requires 49,152 cycles of PCLK, so this is the number we would load in the 8254 counter. Since this number is too large to load in as a BCD count, we load it in as C000H, and in the control word we tell the 8254 to count the number down in binary.

MODE 2—TIMED INTERRUPT GENERATOR

In a previous section we described how a real-time clock of seconds, minutes, and hours could be kept in three memory locations by counting interrupts from a 1-Hz

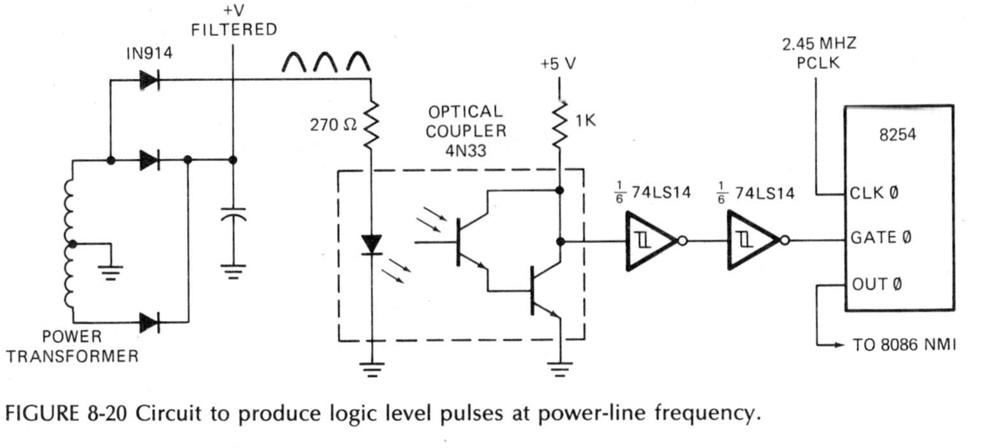

FIGURE 8-20 Circuit to produce logic level pulses at power-line frequency.

pulse source. We also described how the 1-Hz interrupts could be used to measure off other time intervals. The difficulty with using a 1-Hz interrupt signal is that the maximum resolution of any time measurement is 1 second. In other words, if you use a 1-Hz signal, you can only measure times to the nearest second. To improve the resolution of time measurements, most microcomputer systems use a higher frequency signal such as 1 kHz for a real-time clock interrupt. With a 1-kHz interrupt signal the time resolution is then 1 ms. An 8254 counter operating in mode 2 can be used to produce a stable 1-kHz signal by dividing down the processor clock signal.

Figure 8-21 shows the waveforms for an 8254 counter operating in mode 2. Let's look at the top set of waveforms first. The two dips in the $\overline{WR}$ waveform represent a control word and the LSB of a count being written to the count register. The next clock pulse after the count is written will transfer the count from the count register to the actual counter. Since the GATE input is high, succeeding clock pulses will count down this value until it reaches one. When the count reaches one, the OUT pin, which was previously high, will go low. The falling edge of the next clock pulse will cause the OUT pin to go high again and the original count to be loaded into the counter again. Successive clock pulses will cause the countdown and load cycle to repeat over and over. If the counter is loaded with a number *N*, the OUT pin will go low for one clock cycle every *N* input clock pulses. The frequency of the output waveform then will be equal to the input clock frequency divided by *N*.

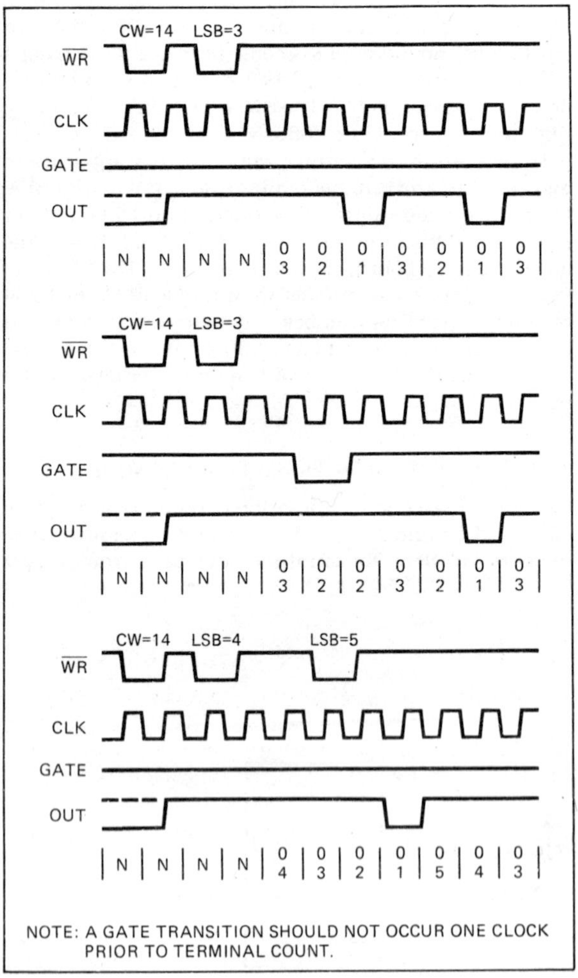

MODE 2

FIGURE 8-21 8254 MODE 2 example timing waveforms. *(Intel Corporation)*

Now, for a specific example, suppose that we want to produce a 1-kHz signal for a real-time clock from an 8-MHz processor clock signal. To do this we connect the processor clock signal to the CLK input on one of the 8254 counters and tie the GATE input of that counter high. We initialize that counter for BCD counting, mode 2, and read/write LSB then MSB. Since we want to divide the 8 MHz by 8000 decimal to get 1 kHz, we then write 00H to the counter as the LSB, and 80H to the counter as the MSB.

A question that may occur to you at this point is "How do I count seconds if the interrupts are coming in every millisecond?" The answer to the question is that you set aside a memory location as a milliseconds counter and initialize that location with 1000 decimal (3E8H). The interrupt service procedure decrements this count each time an interrupt occurs and checks to see if the count is down to zero yet. If the count is not zero, then execution is simply returned to the mainline. If the count is down to zero, 1000 interrupts or 1 second has passed. Therefore, the milliseconds counter location is reloaded with 3E8H, and the seconds-minutes-hours procedure is called to update the count of seconds. In a similar way the 1-kHz interrupt service procedure can measure off several different time intervals that are multiples of 1 ms.

The middle set of mode 2 waveforms in Figure 8-21 demonstrates that if the GATE input is made low while the counter is counting, counting will stop. If the GATE input is made high again, the original count will be reloaded into the counter by the next clock pulse. Succeeding clock pulses will decrement the loaded count.

The bottom set of mode 2 waveforms in Figure 8-21 show that if a new count is written to the count register, this new count will not be transferred to the counter until the the previously loaded count has been decremented to one.

MODE 3—SQUARE-WAVE MODE

If an 8254 counter is programmed for mode 3 and an even number is written to its count register, the waveform on the OUT pin will be a *square wave*. The frequency of the square wave will be equal to the frequency of the input clock divided by the number written to the count register. If an odd number is written to a counter programmed for mode 3, the output waveform will be high for one more clock cycle than it is low, so the waveform will not be quite symmetrical. Figure 8-22 shows some example waveforms for mode 3. By now these waveforms should look quite familiar to you.

The top set of waveforms show that after a control word is written to the control register and a count is written to the count register, the count is transferred to the counter on the next clock pulse. As shown by the count sequence under the OUT waveform, each additional clock pulse decrements the counter by 2. When the count is down to 2, the OUT pin goes low and the original count is reloaded. The OUT pin stays low while the loaded count is again counted down by twos. When the count is down to 2, the OUT pin goes high again and the original count is again loaded into the counter. The cycle then repeats.

The center set of waveforms in Figure 8-22 shows what happens if an odd number is written to the count register. As you can see from this waveform, the number of clock cycles for each waveform is still equal to the number loaded into the count register. However, as we mentioned above, the clock is high for one more clock cycle than it is low.

The bottom set of waveforms in Figure 8-22 shows that counting stops if the gate is made low at any time. After the GATE input is made high again, the original count will be loaded by the next clock pulse.

Mode 3 can be used for any case where you want a

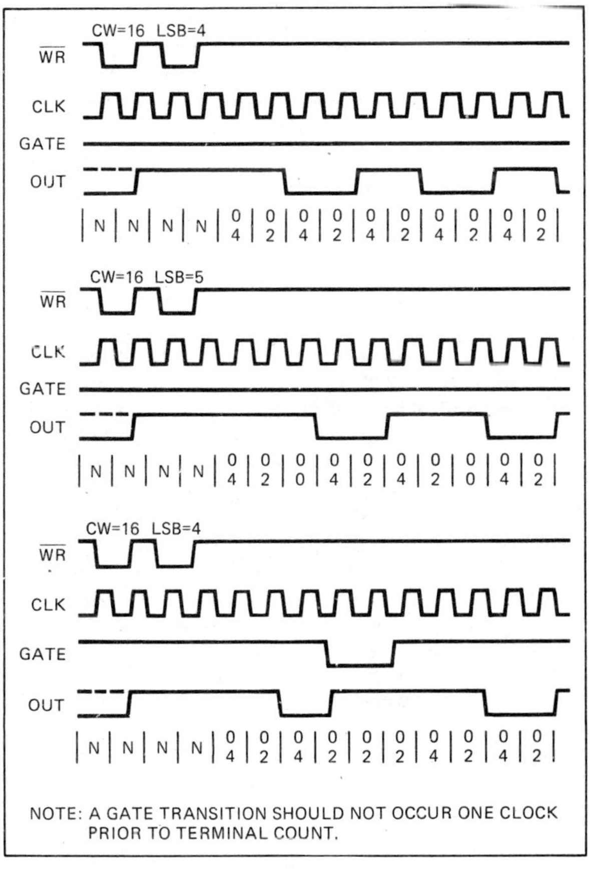

FIGURE 8-22 8254 MODE 3 example timing waveforms. *(Intel Corporation)*

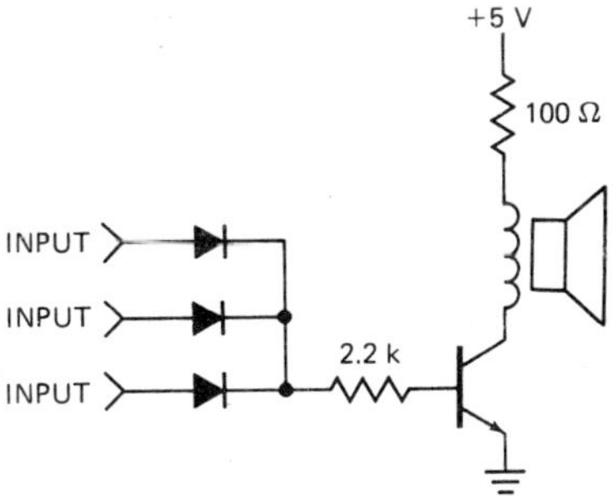

FIGURE 8-23 Audio speaker buffer for 8254 timer output or port.

repetitive square-wave-type signal. In a previous section we showed how an 8254 counter operating in mode 3 can be used to generate the baud rate clock for a USART such as the 8251A. Mode 3 could also be used to generate interrupt pulses for a real-time clock as we described for mode 2.

Another use of 8254 counters operating in mode 3 is as programmable audio tone generators. For this application a high-frequency clock such as the 2.4576-MHz PCLK signal on an SDK-86 board is connected to the counter CLK input, the GATE input is tied high, and the OUT pin is connected to an audio buffer such as that shown in Figure 8-23. This simple buffer allows the outputs of several counters to be added together if desired, and supplies the current required to drive a small speaker.

As an example of this application, suppose that you want to produce a tone that is a musical A of 440 Hz from the 2.4576-MHz PCLK signal. Dividing the PCLK signal by 5585 will give the desired 440 Hz. Therefore, you simply send a control word which programs a counter for mode 3, read/write LSB then MSB, and BCD counting. You then write the LSB of 85H and the MSB of 55H to the counter. If you want to change the frequency, all you have to do is write a new count to the count register. With a few programmable counters and some relatively simple programming you can play your favorite songs.

MODE 4—SOFTWARE-TRIGGERED STROBE

This mode and mode 5 are often confused with mode 1, but there is an obvious difference. Mode 1 is used to produce a low-going pulse that is *N* clock pulses wide. If you look at the top set of waveforms for mode 4 in Figure 8-24, you should see that mode 4 produces a low-going pulse *after* $N + 1$ clock pulses. For mode 4 the output pulse is low for the time of one input clock pulse and then returns high. In other words, in mode 4 a counter produces a low-going strobe pulse $N + 1$ clock cycles after a count is written to the count register. Mode 4 is referred to as *software-triggered* because it is the writing of the count to the count register that starts the process. Note that after the loaded count is counted down, the counter decrements to FFFFH and then continues to decrement from there.

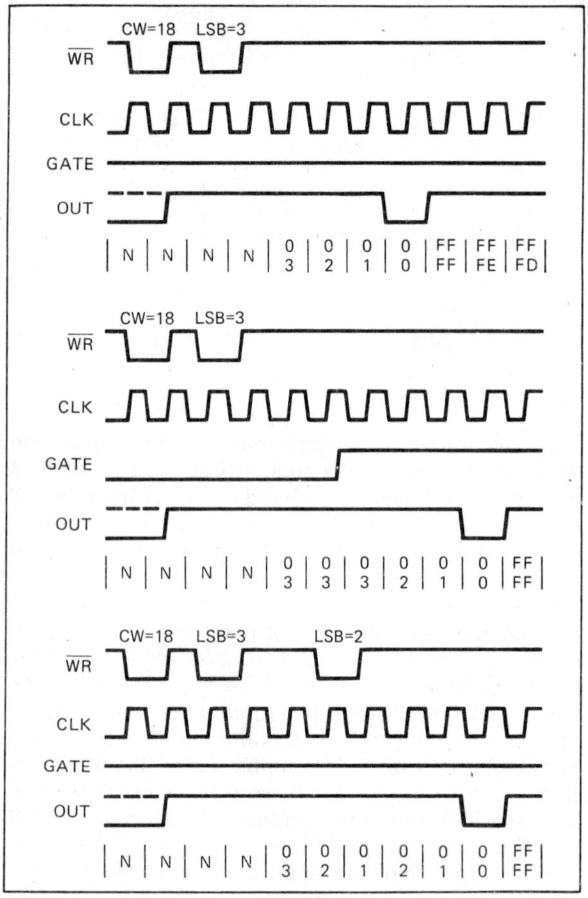

MODE 4

FIGURE 8-24 8254 MODE 4 example timing waveforms. *(Intel Corporation)*

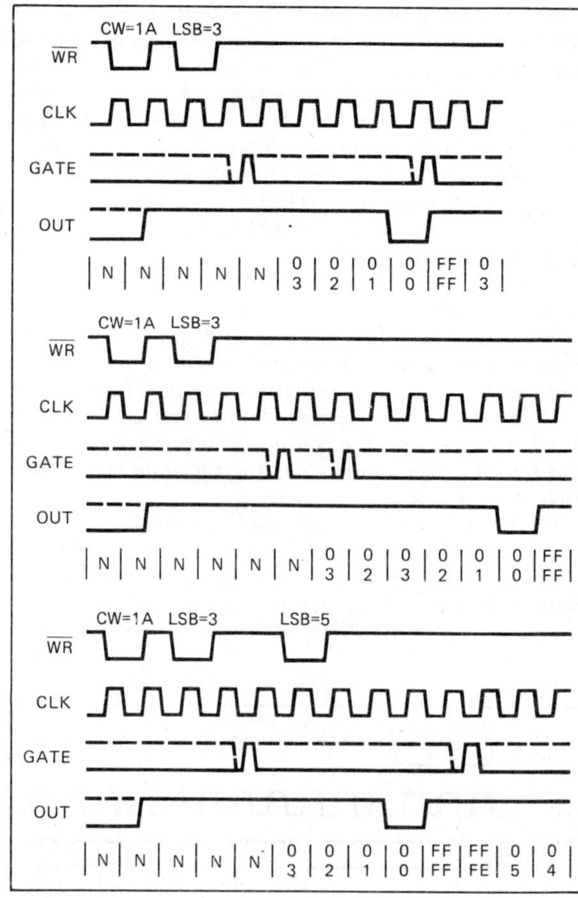

MODE 5

FIGURE 8-25 8254 MODE 5 example timing waveforms. *(Intel Corporation)*

Mode 4 can be used in a case where you want to send out some parallel data on a port, and then after some delay send out a strobe signal to let the receiving system know that the data is available.

MODE 5—HARDWARE-TRIGGERED STROBE

Mode 5 is used where we want to produce a low-going strobe pulse some programmable time interval after a rising-edge trigger signal is applied to the GATE input. This mode is very useful when you want to delay a rising edge signal by some amount of time.

Figure 8-25 shows some example waveforms for a counter operating in mode 5. For a start let's look at the top set of waveforms. As usual we write a control word and the desired count to a counter. As shown by the count sequence under the OUT waveform, however, the count does not get transferred to the counter until the the GATE (trigger) input is made high. When the trigger input is made high the count will be transferred to the counter on the next clock pulse. Succeeding clock pulses will decrement the counter. When the counter reaches zero, the OUT pin will go low for one clock pulse time. The OUT pin will go low $N + 1$ clock pulses after the trigger input goes high.

The second set of waveforms in Figure 8-25 shows that if another trigger pulse occurs during the countdown time, the original count will be reloaded on the next clock pulse and the countdown will start over. The OUT pin will remain high until the count is finally counted down. If trigger pulses continue to come before the countdown is completed, the OUT pin will continue to stay high. Therefore you can use a counter in mode 5 to produce a power fail signal as we showed in the previous discussion of mode 1. Note that for mode 5, however, the OUT pin will be high if the power is on and go low when the power fails.

The bottom set of waveforms in Figure 8-25 shows that if a new count is written to a counter, the new count will not be loaded into the counter until a new trigger pulse occurs.

USING A NONSYSTEM CLOCK WITH 8254 IN MODES 2 AND 3

If you are applying a signal which is not derived from the system clock to the CLK input of an 8254 (not 8253), then a small note in the Intel data sheet indicates that the GATE input of a counter must be pulsed low just after the count is written to the counter. An easy way to do this is to connect the GATE input of the counter to an otherwise unused output port pin. You can then pulse the GATE by outputting a low and then outputting a high to that port pin.

READING THE COUNT FROM AN 8254 COUNTER

For many counter applications we want to be able to read the current count in the counter. Suppose, for example, that we are using an 8254 counter to count the cars coming into a parking lot as we did in our example for mode 0 above. In that case we used the counter to produce an interrupt when the parking lot was full, so we could shut the gate. Now further suppose that as part of a traffic flow study we want to find out how many cars have come into the lot by 7:30 a.m. An interrupt-driven real-time clock procedure can, at 7:30 a.m., call a procedure which reads in the current count from the counter. Since the counter was initially loaded with 1000 decimal and is being counted down as cars come in, we can simply subtract the current count from 1000 to determine how many cars have come in.

The counters in an 8254 have latches on their outputs. When you read the count from a counter, what you are actually reading is the data on the outputs of these latches. These latches are normally enabled during counting so that the latch outputs just follow the counter outputs. If you try to read the count while the counter is counting, the count may change between reading the LSB and the MSB. This may give you a strange count. To read a correct count, then, you must in some way stop the counting or latch the current count on the output of the latches. There are three major ways of doing this.

The first is to stop counting by turning off the clock signal or making the GATE input low with external hardware. This method has the disadvantages that it requires external hardware and that a clock pulse which occurs while the clock is disabled will obviously not be counted.

The second way of reading a stable value from a counter is to latch the current count with a counter latch command, and then read the latched count. A counter is latched by sending a control word to the control register address in the 8254. If you look at the format for the 8254 control word in Figure 8-17 you should see that a counter latch command is specified by making the RW1 and RW0 bits both 0. The SC1 and SC0 bits specify which counter we want to latch. The lower 4 bits of the control word are "don't cares" for a counter latch command word so we usually make them 0's for simplicity. As an example, here is the sequence of instructions you would use to latch and read the LSB and MSB from counter 1 of the 8254 in Figure 8-14. We assume that the counter was already programmed for read/write LSB then MSB when the device was initialized. If the counter was programmed for only LSB or only MSB, then only that byte can be read.

```
MOV AL, 01000000B ; Counter latch command
                  ; for counter 1
MOV DX, 0FF07H    ; Point at 8254 control register
OUT DX, AL        ; Send latch command
MOV DX, 0FF03H    ; Point at counter 1 address
IN  AL, DX        ; Read LSB of latched count
IN  AH, DX        ; Read MSB of latched count
                  ; Count now in AX
```

When a counter latch command is sent, the latched count is held until it is read. When the count is read from the latches, the latch outputs return to following the counter outputs.

The third method of reading a stable count from a counter is to latch the count with a read-back command. This method is available in the 8254, but not in the 8253. It is essentially an enhanced version of the counter latch command approach described in the preceding paragraphs.

Figure 8-26 shows the format for the 8254 counter read-back command word. It is sent to the same address that other control words are for a particular 8254. The 1's in bits D7 and D6 identify this as a read-back command word. To latch the count on a counter you put a 0 in bit D5 of the control word and put a 1 in the bit position that corresponds to that counter in the control word. The advantage of this control word is that you can latch one, two, or all three counters by putting 1's in the appropriate bits. Once a counter is latched, the count is read as shown in the example program above. After being read, the latch outputs return to following the counter outputs.

If a read-back command word with bit D4 = 0 is sent to an 8254, the status of one or more counters will be latched on the output latches. Consult the Intel data sheet for further information on this latched status.

The preceding sections have shown how 8254 counters can be used to do a wide variety of tasks around microcomputers. Many of these applications produce an interrupt signal which must be connected to an interrupt input on the microprocessor. In the next section we show how a *priority interrupt controller* device, the Intel 8259A, is used to service multiple interrupts.

A0, A1 = 11 $\overline{CS}$ = 0 $\overline{RD}$ = 1 $\overline{WR}$ = 0

D_7	D_6	D_5	D_4	D_3	D_2	D_1	D_0
1	1	$\overline{COUNT}$	$\overline{STATUS}$	CNT 2	CNT 1	CNT 0	0

D_5: 0 = LATCH COUNT OF SELECTED COUNTERS(S)
D_4: 0 = LATCH STATUS OF SELECTED COUNTER(S)
D_3: 1 = SELECT COUNTER 2
D_2: 1 = SELECT COUNTER 1
D_1: 1 = SELECT COUNTER 0
D_0: RESERVED FOR FUTURE EXPANSION; MUST BE 0

FIGURE 8-26 8254 read-back control word format.

Multiple Interrupts and the 8259A Priority Interrupt Controller

Previous sections of this chapter show how interrupts can be used for a variety of applications. In a small system, for example, we might read ASCII characters in from a keyboard on an interrupt basis; count interrupts from a timer to produce a real-time clock of seconds, minutes, and hours; and detect several emergency or job-done conditions on an interrupt basis. Each of these interrupt applications requires a separate interrupt input. If we are working with an 8086, we have a problem here because the 8086 has only two interrupt inputs, NMI and INTR. If we save NMI for a power failure interrupt, this leaves only one interrupt input for all the other applications. For applications where we have interrupts from multiple sources such as this we use an external device called *a priority interrupt controller* (PIC) to "funnel" the interrupt signals into an interrupt input on the processor. In this section we show how a common PIC, the Intel 8259A, is connected in an 8086 system, how it is initialized, and how it is used to handle interrupts from multiple sources.

8259A OVERVIEW AND SYSTEM CONNECTIONS

To show you how an 8259A functions in an 8086 system we first need to review how the 8086 INTR input works. Remember from a discussion earlier in this chapter that if the 8086 interrupt flag is set and the INTR input receives a high signal, the 8086 will:

1. Push the flags on the stack.
2. Clear the IF and TF.
3. Push the return address on the stack.
4. Put the data bus in the input mode.
5. Send out two interrupt acknowledge pulses on its $\overline{\text{INTA}}$ pin. The $\overline{\text{INTA}}$ pulses tell some external hardware device such as an 8259A to send the desired interrupt type to the 8086.
6. When the 8086 receives the interrupt type from the external device, it will multiply that interrupt type by 4 to produce an address in the interrupt pointer table.
7. From that address and the three following addresses the 8086 gets the IP and CS values for the start of the interrupt service procedure. Once these values are loaded into CS and IP, the 8086 will then execute the interrupt service procedure.

Now if you look at the internal block diagram of the 8259A in Figure 8-27, I think you will be able to start seeing how it fits into the INTR operation. First notice the 8-bit data bus and control signal pins in the upper left corner of the diagram. The data bus allows the 8086 to send control words to the 8259A and read a status word from the 8259A. The $\overline{\text{RD}}$ and $\overline{\text{WR}}$ inputs control these transfers when the device is selected by asserting its chip select ($\overline{\text{CS}}$) input low. The 8-bit data bus also allows the 8259A to send interrupt types to the 8086. Next notice the eight interrupt inputs labeled IR0–IR7 on the right side of the diagram. If the 8259A is properly enabled, a interrupt signal applied to any one of these inputs will cause the 8259A to assert its INT output pin high. If this pin is connected to the INTR pin of an 8086 and if the 8086 interrupt flag is set, then this high signal will cause the previously described INTR response.

The $\overline{\text{INTA}}$ input of the 8259A is connected to the $\overline{\text{INTA}}$ output of the 8086. The 8259A uses the first $\overline{\text{INTA}}$ pulse from the 8086 to do some activities which depend on the mode that it is programmed in. When it receives the sec-

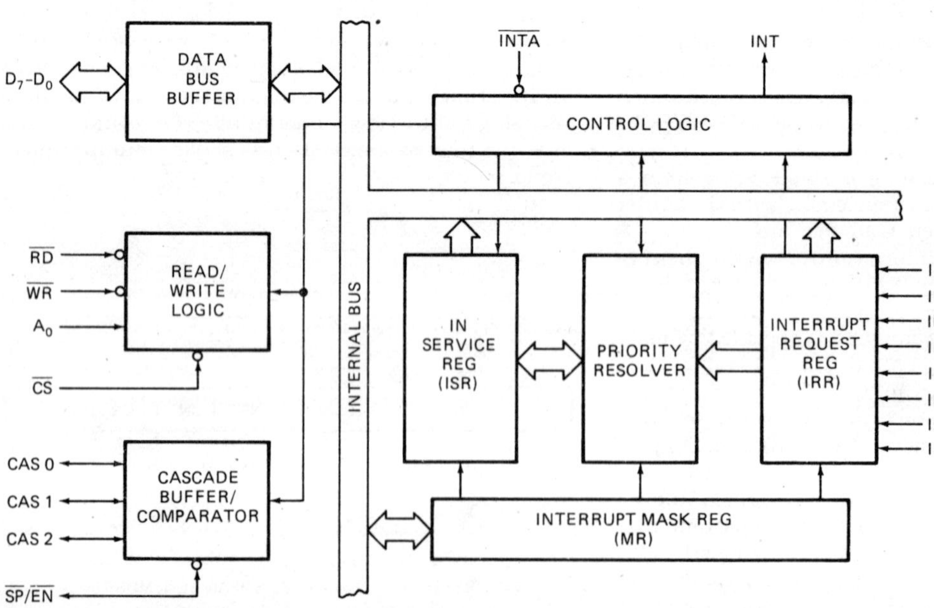

FIGURE 8-27 8259A internal block diagram. *(Intel Corporation)*

ond $\overline{INTA}$ pulse from the 8086, the 8259A outputs an interrupt type on the 8-bit data bus as shown in Figure 8-6. The interrupt type that it sends to the 8086 is determined by the IR input that received an interrupt signal and by a number you send the 8259A when you initialize it. The point here is that the 8259A "funnels" interrupt signals from up to eight different sources into the 8086 INTR input, and it sends the 8086 a specified interrupt type for each of the eight interrupt inputs.

At this point the question may occur to you, "What happens if interrupt signals appear at, for example, IR2 and IR4 at the same time?" In the *fixed priority mode* that the 8259A is usually operated in, the answer to this question is quite simple. In this mode the IR0 input has the highest priority (most important), the IR1 input the next highest, and so on down to IR7 which has the lowest priority. What this means is that if two interrupt signals occur at the same time, the 8259A will service the one with the highest priority first, assuming that both inputs are unmasked (enabled) in the 8259A.

Now let's look again at the block diagram of the 8259A in Figure 8-27 so we can explain in more detail how the device will respond to multiple interrupt signals. In the block diagram note the four boxes labeled *interrupt request register* (IRR), *interrupt mask register* (IMR), *in-service register* (ISR), and *priority resolver*. The operation of these four functional blocks is quite logical.

The interrupt mask register is used to disable (mask) or enable (unmask) individual interrupt inputs. Each bit in this register corresponds to the interrupt input with the same number. You unmask an interrupt input by sending a command word with a 0 in the bit position that corresponds to that input.

The interrupt request register keeps track of which interrupt inputs are asking for service. If an interrupt input is unmasked, and has an interrupt signal on it, then the corresponding bit in the interrupt request register will be set.

The in-service register keeps track of which interrupt inputs are currently being serviced. For each input that is currently being serviced, the corresponding bit will be set in the in-service register. An example will show how the priority resolver acts as a judge in the middle of all this.

Suppose that IR2 and IR4 are unmasked and that an interrupt signal comes in on the IR4 input. Since IR4 is unmasked, bit 4 of the interrupt request register will be set. The priority resolver will detect that this bit is set and see if any action needs to be taken. To do this it checks the bits in the in-service register (ISR) to see if a higher priority input is being serviced. If a higher priority input is being serviced as indicated by a bit being set for that input in the ISR, then the priority resolver will take no action. If no higher priority interrupt is being serviced, then the priority resolver will activate the circuitry which sends an interrupt signal to the 8086. When the 8086 responds with $\overline{INTA}$ pulses, the 8259A will send the interrupt type that we specified for the IR4 input when we initialized the device. The 8086 will use the type number it receives to find and execute the interrupt service procedure we wrote for the IR4 interrupt.

Now, suppose that while the 8086 is executing the IR4 service procedure, an interrupt signal arrives at the IR2 input of the 8259A. Since we assumed for this example that IR2 was unmasked, bit 2 of the interrupt request register will be set. The priority resolver will detect that this bit in the IRR is set and make a decision whether to send another interrupt signal to the 8086. To make the decision, the priority resolver looks at the in-service register. If a higher priority bit in the ISR is set, then a higher priority interrupt is being serviced. The priority resolver will wait until the higher priority bit in the ISR is reset before sending an interrupt signal to the 8086 for the new interrupt input. If the priority resolver finds that the new interrupt has a higher priority than the highest priority interrupt currently being serviced, it will set the appropriate bit in the ISR and activate the circuitry which sends a new INT signal to the 8086. For our example here, IR2 has a higher priority than IR4 so the priority resolver will set bit 2 of the ISR and activate the circuitry, which sends a new INT signal to the 8086. If the 8086 INTR input was reenabled with an STI instruction at the start of the IR4 service procedure, as shown in Figure 8-28*a*, then this new INT signal will interrupt the 8086 again. When the 8086 sends out a

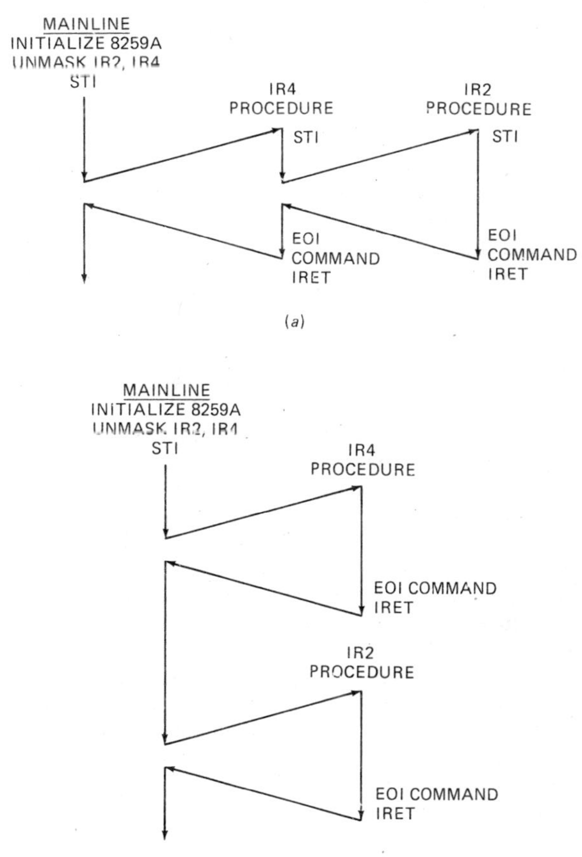

FIGURE 8-28 8259A and 8086 program flow for IR4 interrupt followed by IR2 interrupt. *(a)* Response with INTR enabled in IR4 procedure. *(b)* Response with INTR not enabled in IR4 procedure.

second $\overline{\text{INTA}}$ pulse in response to this interrupt, the 8259A will send it the type number for the IR2 service procedure. The 8086 will use the received type number to find and execute the IR2 service procedure.

At the end of the IR2 procedure we send the 8259A a command word that resets bit 2 of the in-service register so that lower priority interrupts can be serviced. After that, an IRET instruction at the end of the IR2 procedure sends execution back to the interrupted IR4 procedure. At the end of the IR4 procedure we send the 8259A a command word which resets bit 4 of the in-service register so that lower priority interrupts can be serviced. An IRET instruction at the end of the IR4 procedure returns execution to the mainline program. This all sounds very messy, but it is really just a special case of nested procedures. Incidentally, if the IR4 procedure did not reenable the INTR input with an STI instruction, as shown in Figure 8-28*b*, the 8086 would not respond to the IR2-caused INT signal until it finished executing the IR4 procedure. We can't describe all of the possible cases, but the main point here is that the 8086 and the 8259A can be programmed to respond to interrupt signals from multiple sources in almost any way you want them to. Now, before we show you how to initialize and write programs for an 8259A, we will show you more about how it is connected in microcomputer systems.

8259A SYSTEM CONNECTIONS AND CASCADING

Figure 8-14 shows how an 8259A can be added to an SDK-86 board. As shown by the truth table in Figure 8-15, the 74LS138 address decoder will assert the $\overline{\text{CS}}$ input of the 8259A when an I/O base address of FF00H is on the address bus. The A0 input of the 8259A is used to select one of two internal addresses in the device. This pin is connected to system address line A1, so the system addresses for the two internal addresses are FF00H and FF02H. The eight data lines of the 8259A are always connected to the lower half of the 8086 data bus because the 8086 expects to receive interrupt types on these lower eight data lines. $\overline{\text{RD}}$ and $\overline{\text{WR}}$ are connected to the system $\overline{\text{RD}}$ and $\overline{\text{WR}}$ lines. $\overline{\text{INTA}}$ from the 8086 is connected to $\overline{\text{INTA}}$ on the 8259A. The interrupt request signal, INT, from the 8259A is connected to the INTR input of the 8086. The multipurpose $\overline{\text{SP}}/\overline{\text{EN}}$ pin is just tied high because we are only using one 8259A in this system. Since we are not cascading any slave 8259As on the IR inputs, the cascade lines (CAS0, CAS1, and CAS2) can be left open. The eight IR inputs are available for interrupt signals. Unused IR inputs should be tied to ground so that a noise pulse cannot accidentally cause an interrupt. In a later section we will show you how to initialize this 8259A, but first we need to show you how more than one 8259A can be added to a system.

The dashed box on the right side of Figure 8-14 shows how another 8259A could be added to the SDK-86 system to give 15 interrupt inputs. If needed, an 8259A could be connected to each of the eight IR inputs of the original 8259A to give a total of 64 interrupt inputs. Note that since the 8086 has only one INTR input, only one of the 8259A INT pins is connected to the 8086 INTR pin. The 8259A connected directly into the 8086 INTR pin is referred to as the *master*. The INT pin from the other 8259A connects into an IR input on the master. This secondary or *cascaded* device is referred to as a *slave*. Note that the $\overline{\text{INTA}}$ signal from the 8086 goes to both the master and to the slave devices.

Each 8259A has its own addresses so that command words can be written to it and status bytes read from it. For the cascaded 8259A in Figure 8-14, the two system I/O addresses will be FF08H and FF0AH.

The cascade pins (CAS0, CAS1, and CAS2) from the master are connected to the corresponding pins of the slave. For the master these pins function as outputs, and for the slave device they function as inputs. A further difference between the master and the slave is that on the slave the $\overline{\text{SP}}/\overline{\text{EN}}$ pin is tied low to let the device know that it is a slave.

Briefly, here is how the master and the slave work when the slave receives an interrupt signal on one of its IR inputs. If that IR input is unmasked on the slave and if that input is a higher priority than any other interrupt level being serviced in the slave, then the slave will send an INT signal to the IR input of the master. If that IR input of the master is unmasked and if that input is a higher priority than any other IR inputs currently being serviced, then the master will send an INT signal to the 8086 INTR input. If the 8086 INTR is enabled, the 8086 will go through its INTR interrupt procedure and sends out two $\overline{\text{INTA}}$ pulses to both the master and the slave. The slave ignores the first interrupt acknowledge pulse. When the master receives the first $\overline{\text{INTA}}$ pulse, it outputs a 3-bit slave identification number on the CAS0, CAS1, and CAS2 lines. (Each slave in a system is assigned a 3-bit ID as part of its initialization.) Sending the 3-bit ID number enables the slave. When the slave receives the second $\overline{\text{INTA}}$ pulse from the 8086, the slave will send the desired interrupt type number to the 8086 on the eight data lines.

If an interrupt signal is applied directly to one of the IR inputs on the master, the master will send the desired interrupt type to the 8086 when it receives the second $\overline{\text{INTA}}$ pulse from the 8086.

Now that we have given you an overview of how an 8259A operates and how 8259As can be cascaded, the initialization command words for the 8259A should make some sense to you.

INITIALIZING AN 8259A

Earlier in this chapter, when we showed you how to initialize an 8254, we listed a series of steps you should go through to initialize any programmable device. To refresh your memory of these very important steps we will work quickly through them again for the 8259A.

The first step in initializing any device is to find the system base address for the device from the schematic or from a memory map for the system. In order to have a specific example here, we will use the 8259A shown in Figure 8-14. The base address for the 8259A in this system is FF00H.

The next step is to find the internal addresses for the device. For an 8259A the two internal addresses are se-

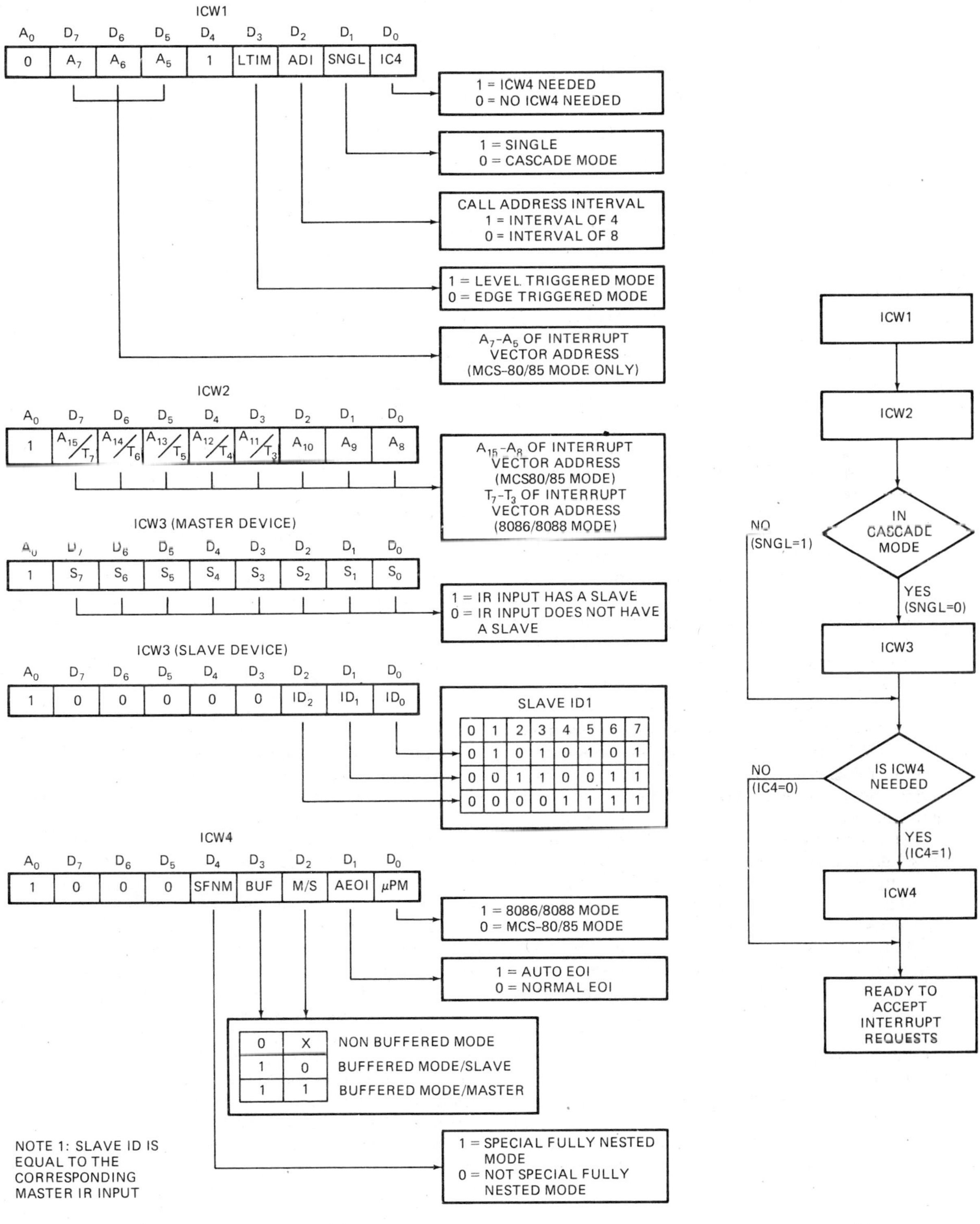

FIGURE 8-29 8259A initialization command word formats and sending order. *(a)* Formats. *(b)* Sending order and requirements. *(Intel Corporation)*

lected by a high or a low on the A0 pin. In the circuit in Figure 8-14 the A0 pin is connected to system address line A1, so the internal addresses correspond to 0 and 2.

Next you add the internal addresses to the base address for the device to get the system address for each internal part of the device. The two system addresses for this 8259A then are FF00H and FF02H.

Now look at Figure 8-29*a* for the format of the command words that must be sent to this device to initialize it. The sight of all of these command words may seem overwhelming at first, but taken one at a time they are quite straightforward. To help you see which initialization command words (ICWs) are needed for various 8259A applications, Figure 8-29*b* shows this in flowchart form. According to this flowchart an ICW1 and an ICW2 must be sent to any 8259A in the system. If the system has any slave 8259As (cascade mode) then an ICW3 must be sent to the master, and a different ICW3 must be sent to the slave. If the system is an 8086, or if you want to specify certain special conditions, then you have to send an ICW4 to the master and to each slave. Now let's look at the formats for the different ICWs.

The first thing to notice about the the ICW formats in Figure 8-29*a* is that the bit labeled A0 on the left end of each of these is not part of the actual command word. This bit tells you the internal address that the control word must be sent to. The A0 = 0 next to ICW1, for example, tells you that ICW1 must be sent to internal address 0, which for our 8259A corresponds to system address FF00H.

The next step in the initialization procedure is to make up the control words. The least-significant bit of ICW1 tells the 8259A whether it needs to look for an ICW4 or not. Since we are using the device in an 8086 system we need to send ICW4. Therefore we make bit D0 a 1. We only want to use one 8259A for now, so we make bit D1 a 1. When used with an 8086, bit D2 is a don't care, so we make it a 0. Bit D3 is used to specify level-triggered mode or edge-triggered mode. In level-triggered mode, service will be requested whenever a high level is present on an IR input. In edge-triggered mode, a signal on an IR input must go from low to high and stay high until serviced. We usually use the edge-triggered mode so that a signal such as a square wave will not cause multiple interrupts. Making bit D3 a 0 does this. Bit D4 has to be a 1. For operation in an 8086 system, bits D5, D6, and D7 are don't cares, so we make them 0's for simplicity. The ICW1 for our example here then is 00010011.

In an 8086 system ICW2 is used to tell the 8259A the type number to send in response to an interrupt signal on the IR0 input. In response to an interrupt signal on some other IR input, the 8259A will automatically add the number of the IR input to this base number and send the result to the 8086 as the type number for that input. Because 8086 interrupt types 0–31 are either dedicated or reserved, type 32 (decimal) is the lowest type number available for us to use. If we send the 8259A an ICW2 of 00100000 binary or 32 decimal, the 8259A will send this number as the type to the 8086 in response to an IR0 interrupt. For an IR1 input the 8259A will send 00100001 binary or 33 decimal and so on for the other IR inputs. In any ICW2 you send the 8259A, the lowest three bits must always be 0's, because the 8259A automatically supplies these bits to correspond to the number of the IR input.

Since we are not using a slave in our example, we don't need to send an ICW3. If you are using a slave 8259A in a system, you have to send an ICW3 to the master to tell it which IR inputs have slaves. The master has to be told this so that it knows for which IR input signals it has to send out a slave ID number on the CAS0, CAS1, and CAS2 lines. You have to send an ICW3 to a slave 8259A to give it an ID number. The ID number you give a slave is equal to the IR input of the master that its INT output is connected to. When the master sends out an ID number on the CAS lines, the slave will recognize its ID number and output the desired type number to the 8086 when it receives an $\overline{\text{INTA}}$ pulse.

For our example here, the only reason we need to send an ICW4 is to let the 8259A know that it is operating in an 8086 system. We do this by making bit D0 of the word a 1. Another interesting bit in this command word is D1, the automatic end-of-interrupt bit. If this bit is set in ICW4, the 8259A will automatically reset the in-service register bit for the interrupt input that is being responded to when the second interrupt acknowledge pulse is received. The effect of this is that the 8259A will then be able to respond to an interrupt signal on a lower priority IR input. In other words, a lower priority interrupt input could then interrupt a higher priority procedure. Since we don't want automatic end of interrupt, the ICW4 for our example here is 00000001.

In addition to the initialization command words shown in Figure 8-29*a*, the 8259A has a second set of command words called *operation command words* or OCWs. These are shown in Figure 8-30. An OCW1 must be sent to an 8259A to unmask any IR inputs that you want it to respond to. For our example here let's assume that we only want to use IR2 and IR3. Since a 0 in a bit position of OCW1 unmasks the corresponding IR input, we put 0's in these two bits and 1's in the rest of the bits. Our OCW1 then is 111110011.

OCW2 is mainly used to reset a bit in the in-service register. This is usually done at the end of the interrupt service procedure, but it can be done at any time in the procedure. The effect of resetting the ISR bit for an interrupt level is that once the bit is reset, the 8259A can then respond to interrupt signals of lower priority. In small systems we usually use the nonspecific end-of-interrupt command word. The OCW2 for this is 00100000. When the 8259A receives this OCW it will automatically reset the in-service register bit for the IR input currently being serviced. If you want to reset a specific ISR bit, you can send the 8259A an OCW2 with 011 in bits D7, D6, and D5, and the number of the ISR bit you want to reset in the lowest 3 bits of the word. You can also use OCW2 to tell the 8259A to rotate the priorities of the IR inputs so that after an IR input is serviced, it drops to the lowest priority. If you are interested, consult the Intel data sheet for more information on this and on the use of OCW3.

Now that we have made up the required ICWs and OCWs the next step is to write the instructions to send these command words to the 8259A.

Figure 8-31 shows an 8086 assembly language program which shows how to initialize an 8259A and combines many of the concepts of this chapter. You can use this program as a pattern for writing programs which service several interrupts. The purpose of this program is to initialize the SDK-86 system in Figure 8-14 for generating a real-time clock of seconds, minutes, and hours from a 1-kHz interrupt signal, and for reading ASCII codes from a keyboard on an interrupt basis. This program assumes that the 2.4576-MHz PCLK signal on the board is connected to the CLK input of 8254 counter 0, the GATE input of the 8254 counter 0 is tied high, and the OUT pin of counter 0 is connected to the IR0 input of

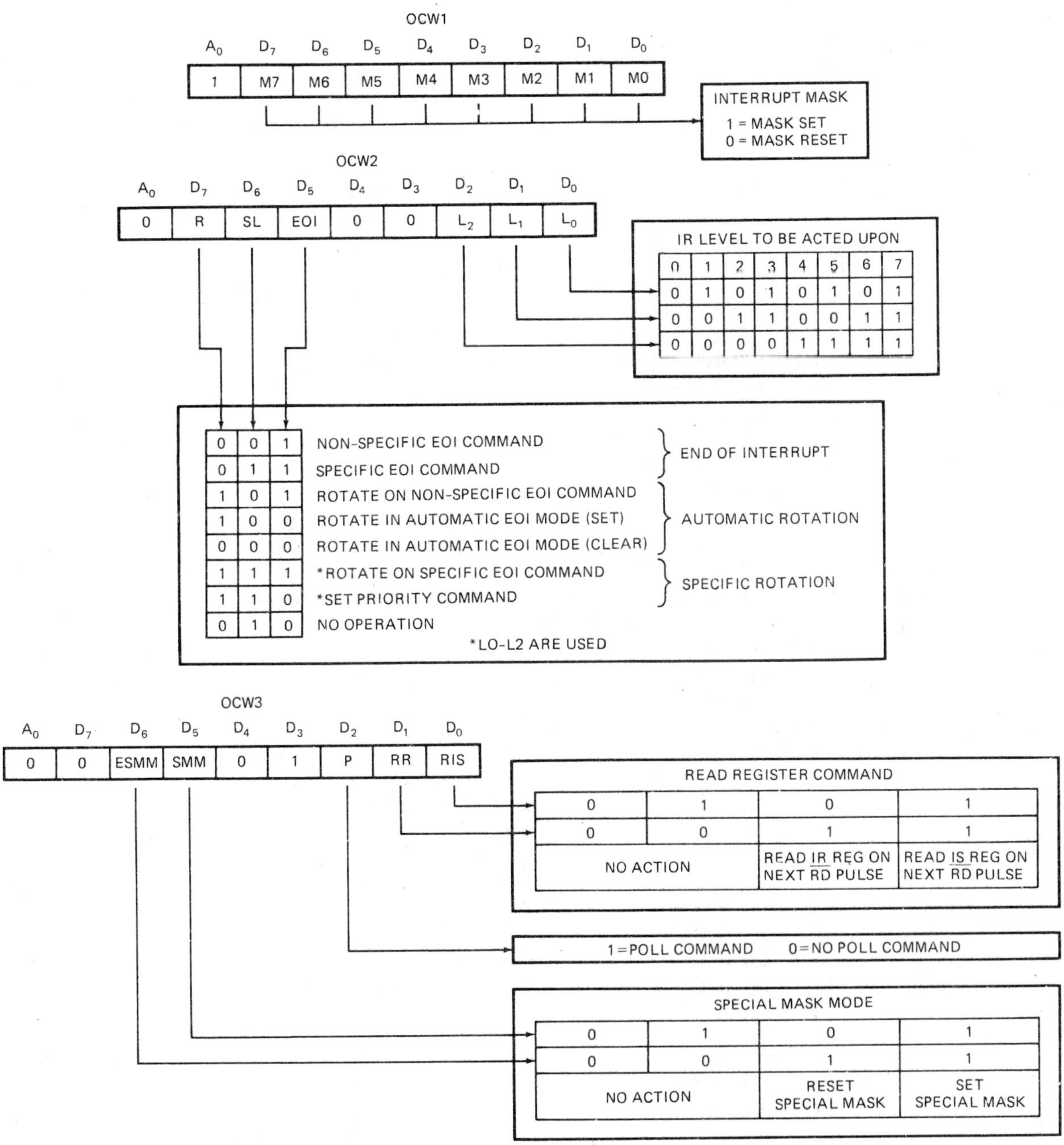

FIGURE 8-30 8259A operational command words. *(Intel Corporation)*

```
page ,132
;8086 PROGRAM FRAGMENT TO SHOW INITIALIZATION OF INTERRUPT JUMP TABLE,
;            8259A, AND COUNTER 0 OF 8254.

AINT_TABLE      SEGMENT WORD    PUBLIC
        TYPE_64         DW      2 DUP(0)        ;reserve space for clock proc addr
        TYPE_65         DW      2 DUP(0)        ;not used in this program
        TYPE_66         DW      2 DUP(0)        ;reserve space for keyboard proc addr
AINT_TABLE      ENDS

DATA_HERE       SEGMENT WORD    PUBLIC
        SECONDS         DB      0
        MINUTES         DB      0
        HOURS           DB      0
        INT_COUNT       DW      03E8H           ;1 kHz interrupt counter
        KEY_BUF         DB      100     DUP(0)  ;Buffer for 100 ASCII chars
DATA_HERE       ENDS

STACK_HERE      SEGMENT                         ; no STACK directive, because
                        DW      100     DUP(0)  ;   will be using EXE2BIN
        TOP_STACK       LABEL   WORD
STACK_HERE      ENDS

CODE_HERE       SEGMENT PUBLIC
        ASSUME CS:CODE_HERE, DS:AINT_TABLE, SS:STACK_HERE
;initialize stack segment register, stack pointer,data segment
        MOV     AX, STACK_HERE
        MOV     SS, AX
        MOV     SP, OFFSET TOP_STACK
        MOV     AX, AINT_TABLE
        MOV     DS, AX
;define the addresses for the interrupt service procedures
        MOV     TYPE_64+2, SEG CLOCK            ; put in clock proc addr
        MOV     TYPE_64,   OFFSET CLOCK
        MOV     TYPE_66+2, SEG KEYBOARD         ; put in keyboard proc addr
        MOV     TYPE_66,   OFFSET KEYBOARD
;initialize data segment
        ASSUME DS:DATA_HERE
        MOV     AX, DATA_HERE
        MOV     DS, AX
;initialize 8259A
        MOV     AL, 00010011B           ; edge triggered, single, ICW4
        MOV     DX, 0FF00H              ; point at 8259A control
        OUT     DX, AL                  ; send ICW1
        MOV     AL, 01000000B           ; type 64 is first 8259A type
        MOV     DX, 0FF02H              ; point at ICW2 address
        OUT     DX, AL                  ; send ICW2
        MOV     AL, 00000001B           ; ICW4, 8086 mode
        OUT     DX, AL                  ; send ICW4
        MOV     AL, 11111010B           ; OCW1 to unmask IR0 and IR2
        OUT     DX, AL                  ; send OCW1

;initialize 8254 counter 0 for 1 kHz output
; 8254 command word for counter 0, LSB then MSB, square wave, BCD
        MOV     AL, 00110111B
        MOV     DX, 0FF07H              ; point at 8254 control addr
```

FIGURE 8-31 Assembly language program showing initialization of 8086, 8259A, and 8254 for real-time clock and keyboard interrupt procedures.

```
        OUT     DX, AL                  ; send counter 0 command word
        MOV     AL, 58H                 ; Load LSB of count
        MOV     DX, 0FF01H              ; point at counter 0 data addr
        OUT     DX, AL                  ; send LSB of count
        MOV     AL, 24H                 ; load MSB of count
        OUT     DX, AL                  ; send MSB of count
;enable interrupt input of 8086
        STI
HERE:   JMP     HERE                    ; wait for interrupt

CLOCK   PROC    FAR
;       :                               ; clock procedure intructions
        MOV     AL, 00100000B           ; OCW2 for non-specific EOI
        MOV     DX, 0FF00H              ; address for OCW2
        OUT     DX, AL                  ; send OCW2 for end of interrupt
        IRET
CLOCK   ENDP

KEYBOARD PROC   FAR
;       :                               ; keyboard proc intructions
        MOV     AL, 00100000B           ; OCW2 for non-specific EOI
        MOV     DX, 0FF00H              ; address for OCW2
        OUT     DX, AL                  ; send OCW2 for end of interrupt
        IRET
KEYBOARD ENDP
CODE_HERE ENDS
        END
```

FIGURE 8-31 *(continued)*

the 8259A. The program further assumes that the keypressed strobe from the ASCII keyboard is connected to the IR2 input of the 8259A.

In the program we first declare a segment called AINT_TABLE to reserve space for the pointers to the interrupt procedures. The statement TYPE_64 DW 2 DUP(0), for example, sets aside a word space for the offset of the the type 64 procedure and a word for the segment base of the procedure. Because of the way the IBM MASM, LINK, and EXE2BIN programs work, it is necesssary to do a little trick to get the AINT_TABLE segment located at absolute address 0000:0100H where it must be for the program to work correctly. The trick is simply to give the segment which sets aside space for these pointers a name which alphabetically comes before the names of the other segments in your program, just as we named our segment here AINT_TABLE. When you MASM and LINK your program, the result is a relocatable object code (.EXE) program which the computer will load into any convenient location to run. The .EXE form of the program will not get put at the required absolute locations. To solve this problem you process your .EXE program with a program called EXE2BIN. When run, this program prompts you to put in a segment fixup (absolute starting address) for your program. If you give a segment fixup value of 0010H, EXE2BIN will produce a .BIN file which will load at absolute address 0000:0100H. Since AINT_TABLE is alphabetically first, it will be located starting at this address, which is the correct absolute address for the 8086 type 64 interrupt.

The next thing we do in our program is to declare a data segment and set aside some memory locations for seconds count, minutes count, hours count, and 100 characters read in from the keyboard. After the data segment we set up a stack segment.

At the start of the mainline program we initialize the stack segment register, the stack pointer register, and the data segment register. We will be using interrupt type 64 for a real-time clock and type 66 will point at the start of the procedure that reads ASCII codes from the keyboard. We will not be using a type 65 interrupt in this program. The next four instructions are needed to place the addresses of the clock and keyboard procedures in the type 64 and type 66 locations in the interrupt pointer table. Later we initialize the 8259A so that type 64 corresponds to its IR0 input and type 66 corresponds to its IR2 interrupt. We then ASSUME DS:DATA_HERE and initialize DATA_HERE as the data segment.

For the example here we have chosen type 64 to correspond to an IR0 interrupt, so the needed ICW2 will be 01000000. We then initialize the 8259A with the command words we worked out above and this new ICW2. Note that those command words shown with a 0 as the A0 bit in Figures 8-29 and 8-30 are sent to system address FF00H and those command words shown with a 1 as the A0 bit are sent to system address FF02H.

The next section of the mainline program initializes counter 0 of the 8254 for mode 2, BCD countdown, and read/write LSB then MSB. To produce a 1-kHz signal from the 2.4576-MHz PCLK we then write a count of 2458 to counter 0. This will not give exactly 1 kHz, but it is as close as we can get with this particular input clock frequency. The PCLK frequency for this board was chosen to make baud rate clock frequencies come out exact, not a 1-kHz real-time clock. Larger systems usually have two or more crystal-controlled oscillators to accommodate both.

Finally, after the timer is initialized, we enable the 8086 INTR input with the STI instruction so that the 8086 can respond to INT signals from the 8259A, and wait for an interrupt with the HERE: JMP HERE instruction.

For the two interrupt service procedures we show just the skeletons and the end of interrupt instructions. We leave it to you to write the actual procedures. Note that the procedures must be declared as FAR so that the assembler will load both the IP and the CS in the interrupt pointer table. Remember from a previous discussion that when the 8259A responds to an IR signal, it sets the corresponding bit in the ISR. This bit must be reset at some time during or at the end of the interrupt service procedure so that the priority resolver can respond to future interrupts of the same or lower priority. At the end of our procedures here we do this by sending an OCW2 to the 8259A. The OCW2 of 00100000 that we send tells the 8259A to reset the ISR bit for the IR level that is currently being serviced. This is a nonspecific end of interrupt (EOI) instruction.

This chapter has introduced you to interrupts and some interrupt applications. The following chapters will show you more of this, because much of the interfacing discussed there is done on an interrupt basis.

IMPORTANT TERMS AND CONCEPTS FROM THIS CHAPTER

If you do not remember any of the terms in the following list, use the index to help you find them in the chapter for review.

Interrupt—INTR

Nonmaskable interrupt—NMI

Software interrupts—INT instruction

Interrupt service procedure

Interrupt vector, interrupt pointer

Interrupt vector table, interrupt pointer table

Interrupt type

Divide by zero interrupt—type 0

Single-step interrupt—type 1

Nonmaskable interrupt—type 2

Breakpoint interrupt—type 3

Overflow interrupt—type 4

Software interrupts—type 0–255

INTR interrupts—type 0–255

BIOS

Edge-activated interrupt input

Level-activated interrupt input

Interrupt priority

Hardware interrupts

Software programmable

Programmable timer/counter devices—8253, 8254

Internal addresses

Control words, command words, mode words

8259A priority interrupt controller
- Fixed priority
- In-service register—ISR
- Priority resolver
- Interrupt request register—IRR
- Interrupt mask register—IMR

REVIEW QUESTIONS AND PROBLEMS

1. List and describe in general terms the steps an 8086 will take when it responds to an interrupt.
2. Describe the purpose of the 8086 interrupt pointer table.
3. What addresses in the interrupt pointer table are used for a type 2 interrupt?
4. The starting address for a type 4 interrupt service procedure is 0010:0082. Show where and in what order this address should be placed in the interrupt jump table.
5. Address 00080H in the interrupt jump table contains 4A24H, and address 00082H contains 0040H. To what interrupt type do these locations correspond? What is the starting address for the interrupt service procedure?
6. Briefly describe the condition(s) which cause the 8086 to perform each of the following types of interrupts: type 0, type 1, type 2, type 3, type 4.
7. Why is it necessary at the start of an interrupt service procedure to PUSH all registers used in the procedure and to POP them at the end of the procedure?
8. Why must you use an IRET instruction rather than the regular RET instruction at the end of an interrupt service procedure?

9. Show the assembler directive and instructions you would use to initialize the interrupt pointer table locations for a type 0 procedure called DIV_0_ERROR and a type 2 procedure called POWER_FAIL.

10. Describe the main use of the 8086 type 1 interrupt. Show the assembly language instructions necessary to set the 8086 trap flag.

11. In a system which has battery-backed RAM for saving data in case of a power failure, the stack is often put in the battery-backed RAM. This makes it easy to save registers and critical program data. Assume that the battery-backed RAM is in the address range of 08000H-08FFFH. Write an 8086 power failure interrupt service procedure which:

    Sets an external battery-backed flip-flop connected to bit 0 of port 28H to indicate that a power failure has occurred.

    Saves all registers on the stack.

    Saves the stack pointer value for the last entry at location 8000H.

    Saves the contents of memory locations 00100H–003FFH after the saved stack pointer value at the start of the battery backed memory. (A string instruction might be useful for this.)

    Halts.

    When the power comes back on, the start-up routine can check the power fail flip-flop. If the flip-flop is set, the start-up procedure can copy the saved data back into its operating locations, initialize the stack segment register, and then get the saved SP value from address 08000H. Using this value it can restore the pushed registers and return execution to where the power fail interrupt occurred. This is called a "warm start." If we don't want it to do a warm start, we can reset the flip-flop with an external RESET key so the system does a start from scratch or "cold start."

12. Why is the 8086 INTR input automatically disabled when the 8086 is RESET? How is the 8086 INTR input enabled to respond to interrupts? What instruction can be used to disable the INTR input? Why is the INTR input automatically disabled as part of the response to an INTR interrupt? How does the INTR input automatically get reenabled at the end of an INTR interrupt service procedure?

13. Describe the response an 8086 will make if it receives an NMI interrupt signal during a division operation which produces a divide by zero interrupt.

14. The data outputs of an 8-bit analog-to-digital converter are connected to bits D0–D7 of port FFF9H and the end-of-conversion signal from the A/D converter is connected to the NMI input of an 8086. Write a simple mainline program and an interrupt service procedure which reads in a byte of data from the converter. If the MSB of the data is a 0, indicating the value is in range, add the byte to a running total kept in two successive memory locations. If the MSB of data is 1, showing that the value is out of range, ignore the input. After 100 samples have been totaled, divide by 100 to get the average, store this average in another reserved memory location, and reset the total to zero.

15. Write the algorithm and the program for an interrupt service procedure which turns on an LED connected to bit D0 of port FFFAH on for 25 seconds and off for 25 seconds. The procedure should also turn a second LED connected to bit D1 of port FFFAH on for 1 minute and off for 1 minute. Assume that a 1-Hz interrupt signal is connected to the NMI input of an 8086, and that a high on a port bit turns on the LED connected to it.

16. Write the algorithms for a mainline program and an interrupt service procedure which generate a real-time clock of seconds, minutes, and hours in three memory locations using a 1-Hz signal applied to the NMI input of an 8086. Then write the assembly language programs for the mainline and the procedure. If you are working on an SDK-86 board, there is a procedure in Figure 9-33 that you can add to your program to display the time on the data and address field LEDs of the board. You can use this procedure without needing to understand the details of how it works. To display a word on the data field, simply put the word in the CX register, put 00H in AL, and call the procedure. To display a word on the address field, put the word in CX, 01H in AL, and call the procedure. *HINT*: Clear carry before incrementing a count in AL so that DAA works correctly.

17. In Chapter 5 we discussed using breakpoints to debug programs containing procedures. List the sequence of locations where you would put breakpoints in the example program in Figure 8-9 to debug it if it did not work when you loaded it into memory.

18. Suppose that we add another 8254 to the SDK-86 add-on circuitry shown in Figure 8-14, and that the $\overline{CS}$ input of the new 8254 is connected to the Y5 output of the 74LS138 decoder.

    a. What will be the system base address for this added 8254?
    b. To which half of the 8086 data bus should the eight data lines from this 8254 be connected?
    c. What will be the system addresses for the three counters and the control word register in this 8254?
    d. Show the control word you would use to initialize counter 1 of this device for read/write LSB then MSB, mode 3, and BCD countdown.
    e. Show the sequence of instructions you would use to write this control word and a count of 0356 to the counter.
    f. Assuming the GATE input is high, when does the counter start counting down in mode 3?

g. Assuming initialization as in parts *d* and *f*, and that a 712-kHz signal is applied to the CLK input of counter 1 in mode 3, describe the frequency, period, and duty cycle of the waveform that will be on the OUT pin of the counter.

h. Describe the effect that a control word of 10010000 sent to this 8254 will have.

19. Show the instructions you would use to initialize counter 2 of the 8254 in Figure 8-14 to produce a 1.2-ms-wide STROBE pulse on its OUT pin when it receives a trigger input on its GATE input.

20. Show the instructions needed to latch and read a 16-bit count from counter 1 of the 8254 in Figure 8-14.

21. Describe the sequence of actions that an 8259A and an 8086, as connected in Figure 8-14, will take when the 8259A receives an interrupt signal on its IR2 input. Assume only IR2 is unmasked in the 8259A and that the 8086 INTR input has been enabled with an STI instruction.

22. Describe the use of the CAS0, CAS1, and CAS2 lines in a system with a cascaded 8259A.

23. Describe the response that an 8259A will make if it receives an interrupt signal on its IR3 and IR5 inputs at the same time. Assume fixed priority for the IR inputs. What response will the 8259A make if it is servicing an IR5 interrupt and an IR3 interrupt signal occurs

24. Why is it necessary to send an end-of-interrupt (EOI) command to an 8259A at some time in an interrupt service routine?

25. Show the sequence of command words and instructions that you would use to initialize an 8259A with a base address of FF10H as follows:

edge-triggered; only one 8259A; 8086 system; interrupt type 40 corresponds to IR0 input; normal EOI; nonbuffered mode, not special fully nested mode; IR1 and IR3 unmasked.

CHAPTER

9 Digital Interfacing

The major goal of this chapter and the next is to show you much of the interface circuitry and software needed to control a complex machine such as our printed-circuit-board-making machine or a medical instrument with a microprocessor. We try to show enough detail in each topic so that you can build and experiment with some real circuits and programs. Perhaps you can use some of these to control appliances around your house or solve some problems at work.

OBJECTIVES

At the conclusion of this chapter you should be able to:

1. Describe simple input and output, strobed input and output, and handshake input and output.
2. Initialize a programmable parallel port device such as the 8255A for simple input or output and for handshake input or output.
3. Interpret the timing waveforms for handshake input and output operations.
4. Describe how phonemes are sent to a speech synthesizer on a handshake basis.
5. Describe how parallel data is sent to a printer on a handshake basis.
6. Show the hardware connections and the programs that can be used to interface keyboards to a microcomputer.
7. Show the hardware connections and the programs that can be used to interface alphanumeric displays to a microcomputer.
8. Describe how an 8279 can be used to refresh a multiplexed LED display and scan a matrix keyboard.
9. Initialize an 8279 for a given display and keyboard format.
10. Show the circuitry used to interface high-power devices to microcomputer ports.
11. Describe the hardware and software needed to control a stepper motor.

PROGRAMMABLE PARALLEL PORTS AND HANDSHAKE INPUT/OUTPUT

Throughout the program examples in the preceding chapters, we have used port devices to input parallel data to the microprocessor and to output parallel data from the microprocessor. Most of the available port devices such as the 8255A on the SDK-86 board contain two or three ports which can be programmed to operate in one of several different modes. The different modes allow you to use the device for many common types of parallel data transfer. First we will discuss some of these common methods of transferring parallel data, and then we will show how the 8255A is initialized and used in a variety of I/O operations.

Methods of Parallel Data Transfer

SIMPLE INPUT AND OUTPUT

When you need to get digital data from some simple switch such as a thermostat into a microprocessor, all you have to do is connect the switch to an input port line and read the port. The thermostat data is always present and ready, so you can read it at any time.

Likewise, when you need to output data to a simple display device such as an LED, all you have to do is connect the input of the LED buffer on an output port pin and output the logic level required to turn on the light. The LED is always there and ready, so you can send data to it at any time. The timing waveform in Figure 9-1*a* represents this situation. The crossed lines on the waveform represent the time at which a new data byte becomes valid on the output lines of the port. The absence of other waveforms indicates that this output operation is not directly dependent on any other signals.

SIMPLE STROBE I/O

In many applications valid data is only present on an external device at a certain time and it must be read in

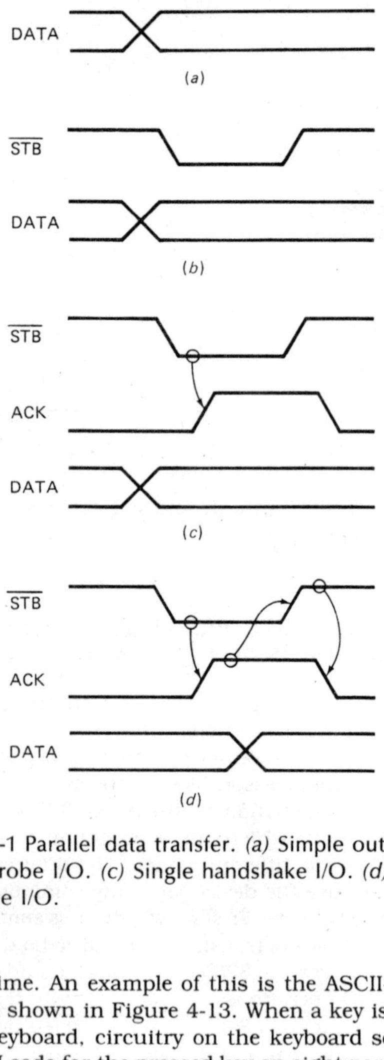

FIGURE 9-1 Parallel data transfer. *(a)* Simple output. *(b)* Simple strobe I/O. *(c)* Single handshake I/O. *(d)* Double handshake I/O.

at that time. An example of this is the ASCII-encoded keyboard shown in Figure 4-13. When a key is pressed on the keyboard, circuitry on the keyboard sends out the ASCII code for the pressed key on eight parallel data lines. The keyboard circuitry then sends out a strobe signal on another line to indicate that valid data is present on the eight data lines. As shown in Chapter 3, you can connect this strobe line to an input port line and poll it to determine when you can input valid data from the keyboard. Another alternative, described in Chapter 8, is to connect the strobe line to an interrupt input on the processor and have an interrupt service routine read in the data when the processor receives an interrupt. The point here is that this transfer is time-dependent. You can only read in data when a strobe pulse tells you that the data is valid.

Figure 9-1*b* shows the timing waveforms which represent this type of operation. The sending device, such as a keyboard, outputs parallel data on the data lines and then outputs an $\overline{STB}$ signal to let you know that valid data is present.

For low rates of data transfer, such as from a keyboard to a microprocessor, a simple strobe transfer works well. However, for high-speed data transfer this method does not work because there is no signal which tells the sending device when it is safe to send the next data byte. In other words the sending system might send data bytes faster than the receiving system could read them. To prevent this problem a *handshake* data transfer scheme is used.

SINGLE HANDSHAKE I/O

Figure 9-1*c* shows some example timing waveforms for a *handshake data transfer* from a peripheral device to a microprocessor. The peripheral outputs some parallel data and sends an $\overline{STB}$ signal to the microprocessor. The microprocessor detects the asserted $\overline{STB}$ signal on a polled or interrupt basis and reads in the byte of data. The microprocessor then sends an acknowledge signal, ACK, to the peripheral to indicate that the peripheral can send the next byte of data. From the viewpoint of the microprocessor, this operation is referred to as a handshake or strobed input.

These same waveforms might represent a handshake output from a microprocessor to a parallel printer. In this case the microprocessor outputs a character to the printer and asserts an $\overline{STB}$ signal to the printer to tell the printer, "Here is a character for you." When the printer is ready, it answers back with the ACK signal to tell the microprocessor, "I got that one, send me another." We will show you much more about printer interfacing in a later section.

The point of this handshake scheme is that the sending device or system cannot send the next data byte until the receiving device or system indicates with an ACK signal that it is ready to receive the next byte.

DOUBLE HANDSHAKE DATA TRANSFER

For data transfers where even more coordination is required between the sending system and the receiving system, a *double handshake* is used. Figure 9-1*d* shows some example waveforms for a double handshake input from a peripheral to a microprocessor. Perhaps it will help you to follow these waveforms by thinking of them as a conversation between two people. In these waveforms each signal edge has meaning. The sending device asserts its $\overline{STB}$ line low to ask, "Are you ready?" The receiving system raises its ACK line high to say, "I'm ready." The peripheral device then sends the byte of data and raises its $\overline{STB}$ line high to say, "Here is some valid data for you." After it has read in the data the receiving system drops its ACK line low to say, "I have the data, thank you, and I await your request to send the next byte of data."

For a handshake output of this type, from a microprocessor to a peripheral, the waveforms are the same but the microprocessor sends the $\overline{STB}$ signal and the data, and the peripheral sends the ACK signal. In a later section we show how this type of handshake is used to transfer *phoneme* bytes from a microprocessor to a speech-synthesizer device.

For handshake data transfer, a microprocessor can determine when it is time to send the next data byte on a polled or on an interrupt basis. We usually use the interrupt approach because it makes better use of the processor's time. The $\overline{STB}$ or ACK signals for these handshake transfers can be produced on a port pin by instructions in the program. This method, however, tends to use too much processor time. Therefore, port devices such as the 8255A have been designed so that they can be programmed to automatically manage the handshake operation. For example, the 8255A can be programmed to automatically receive a $\overline{STB}$ signal from a peripheral, send an interrupt signal to the processor, and send the ACK signal to the peripheral at the proper times. The following sections show you how to connect, initialize, and use an 8255A for a variety of applications.

8255A Internal Block Diagram and System Connections

Figure 9-2 shows the internal block diagram of the 8255A. Along the right side of the diagram you can see that the device has 24 input/output lines. Port A can be used as an 8-bit input port or as an 8-bit output port. Likewise, port B can be used as an 8-bit input port or as an 8-bit output port. Port C can be used as an 8-bit input or output port, two 4-bit ports, or to produce handshake signals for ports A and B. We will discuss the different modes for these lines in detail a little later.

Along the left side of the diagram you see the usual signal lines used to connect the device to the system buses. Eight data lines allow you to write data bytes to a port or the control register and to read bytes from a port or the status register under the control of the $\overline{RD}$ and $\overline{WR}$ lines. The address inputs, A0 and A1, allow you to selectively access one of the three ports or the control register. The internal addresses for the device are: port A–00, port B–01, port C–10, control–11. Asserting the $\overline{CS}$ input of the 8255A enables it for reading or writing. The $\overline{CS}$ input will be connected to the output of the address decoder circuitry to select the device when it is addressed.

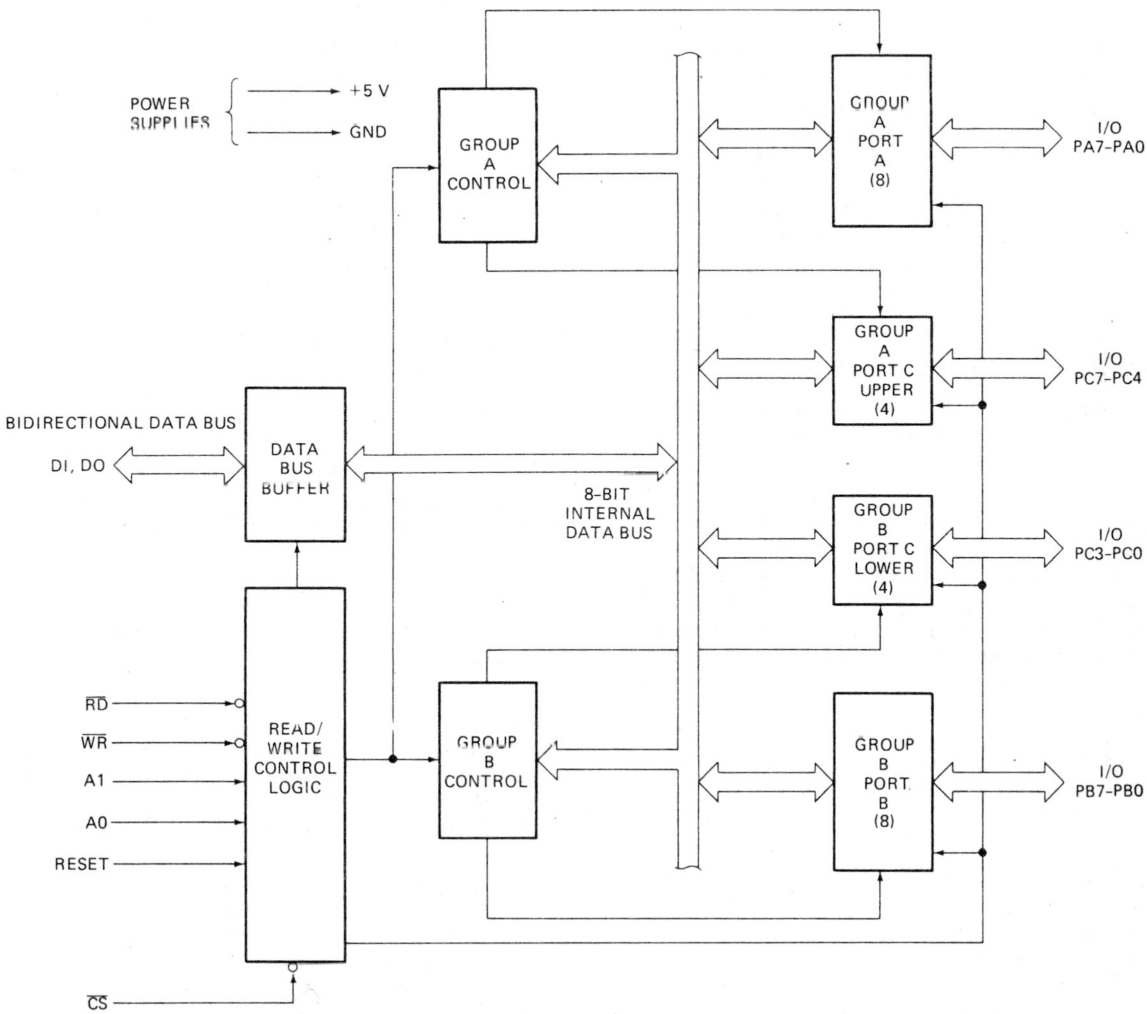

FIGURE 9-2 Internal block diagram of 8255A programmable parallel port device. *(Intel Corporation)*

The RESET input of the 8255A is connected to the system reset line so that, when the system is reset, all of the port lines are initialized as input lines. This is done to prevent destruction of circuitry connected to port lines. If port lines were initialized as outputs after a power-up or reset, the port might try to output into the output of a device connected to the port. The possible argument between the two outputs might destroy one or both of them. Therefore all of the programmable port devices initialize their port lines as inputs when reset.

We discussed in Chapter 7 how two 8255As can be connected in an 8086 system. Take a look at Figure 7-6 (sheet 5) to refresh your memory of these connections. Note that one of the 8255As is connected to the lower half of the 8086 data bus, and the other 8255A is connected to the upper half of the data bus. This is done so that a byte can be transferred by enabling one device, or a word can be transferred by enabling both devices at the same time. According the truth table for the I/O port address decoder in Figure 7-15, the *XA*40 8255A on the lower half of the data bus will be enabled for a base address of FFF8H, and the *XA*35 8255A will be enabled for a base address of FFF9H.

Another point to notice in Figure 7-6 is that system address line A1 is connected to the 8255A A0 inputs, and system address line A2 is connected to the 8255A A1 inputs. With these connections the system addresses for the three ports and the control register in the *XA*40 8255A will be FFF8H, FFFAH, FFFCH, and FFFEH as shown in Figure 7-15. Likewise, the system addresses for the three ports and the control register of the *XA*35 8255A are FFF9H, FFFBH, FFFDH, and FFFFH.

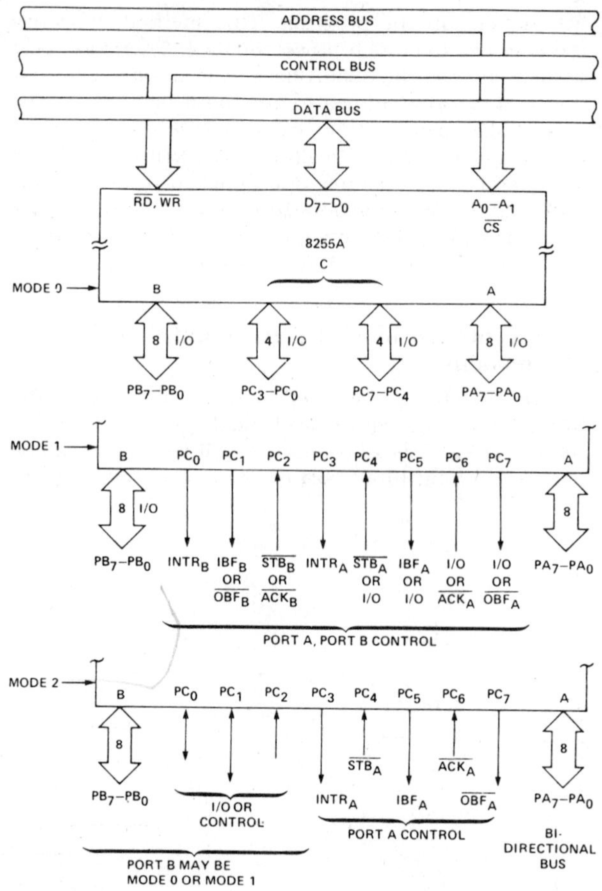

FIGURE 9-3 Summary of 8255A operating modes. *(Intel Corporation)*

8255A Modes and Initialization

Figure 9-3 summarizes the different modes in which the ports of the 8255A can be initialized.

MODE 0

When you want to use a port for simple input or output without handshaking, you initialize that port in MODE 0. If both port A and port B are initialized in MODE 0, then the two halves of port C can be used together as an additional 8-bit port, or they can be used individually as two 4-bit ports. When used as outputs the port C lines can be individually set or reset by sending a special control word to the control register address. Later we will show you how to do this. The two halves of port C are independent, so one half can be initialized as input, and the other half initialized as output.

MODE 1

When you want to use port A or port B for a handshake (strobed) input or output operation such as we discussed in previous sections, you initialize that port in MODE 1. In this mode some of the pins of port C function as handshake lines. Pins PC0, PC1, and PC2 function as handshake lines for port B if it is initialized in MODE 1. If port A is initialized as a handshake (MODE 1) input port, then pins PC3, PC4, and PC5 function as handshake signals. Pins PC6 and PC7 are available for use as input lines or output lines. If port A is initialized as a handshake output port, then port C pins PC3, PC6, and PC7 function as handshake signals. Port C pins PC4 and PC5 are available for use as input or output lines. Since the 8255A is often used in this mode, we show several examples in the following sections.

MODE 2

Only port A can be initialized in MODE 2. In MODE 2, port A can be used for *bidirectional handshake* data transfer. This means that data can be output or input on the same eight lines. The 8255A might be used in this mode to extend the system bus to a slave microprocessor or to transfer data bytes to and from a floppy disk controller board. If port A is initialized in MODE 2, then pins PC3–PC7 are used as handshake lines for port A. The other three pins of port C can be used for I/O if port B is in MODE 0. The three pins will be used for port B handshake lines if port B is initialized in MODE 1. After you work your way through the MODE 1 examples in the following sections you should have little difficulty understanding the discussion of MODE 2 in the Intel data sheet if you encounter it in a system.

Constructing and Sending 8255A Control Words

Figure 9-4 shows the formats for the two 8255A control words. The MSB of the control word tells the 8255A which control word you are sending it. You use the *mode definition control word* format in Figure 9-4*a* to tell the device what modes you want the ports to operate in. For the mode definition control word you put a 1 in the MSB. You use the *bit set/reset control word* format in Figure 9-4*b* when you want to set or reset the output on a pin of port C, or when you want to enable the interrupt output signals for handshake data transfers. The MSB is 0 for this control word. Both control words are sent to the control register address for the 8255A.

As usual, making up a control word consists of figuring out what to put in the eight little boxes one bit at a time. As an example for this device, suppose that you want to initialize the 8255A (*XA*40) in Figure 7-6 as follows: Port B MODE 1 input, port A MODE 0 output, port C upper as inputs, and bit 3 of port C as output. Figure 9-5*a* shows the control word which will program the 8255A in this way. The figure also shows how you should document any control words you make up for use in your programs. Using Figure 9-4*a*, work your way through this word to make sure you see why each bit has the value it does.

As we said previously, the control register address for the *XA*40 8255A is FFFEH. To send a control word then you load the control word in AL with a MOV AL, 10001110B instruction, point DX at the port address with the MOV DX, 0FFFEH instruction, and send the control word to the 8255A control register with the OUT DX, AL instruction.

As an example of how to use the bit set/reset control word, suppose that you want to output a 1 to (set) bit 3 of port C, which was initialized as an output with the mode set control word above. To set or reset a port C output pin you use the bit set/reset control word shown in Figure 9-4*b*. Make bit D7 a 0 to identify this as a bit set/reset control word and put a 1 in bit D0 to specify that you want to set a bit of port C. Bits D3, D2, and D1 are used to tell the 8255A which bit you want to act on. For this example you want to set bit 3, so you put 011 in these three bits. For simplicity and compatibility with future products, make the other three bits of the control word 0's. The result of 00000111B, with proper documentation, is shown in Figure 9-5*b*.

To send this control word to the 8255A simply load it into AL with the MOV AL, 00000111B instruction, point DX at the control register address with the MOV DX, 0FFFEH instruction if DX is not already pointing there, and send the control word with the OUT DX, AL instruction. As part of the application examples in the following sections, we will show you how you know which bit in port C to set to enable the interrupt output signal for handshake data transfer.

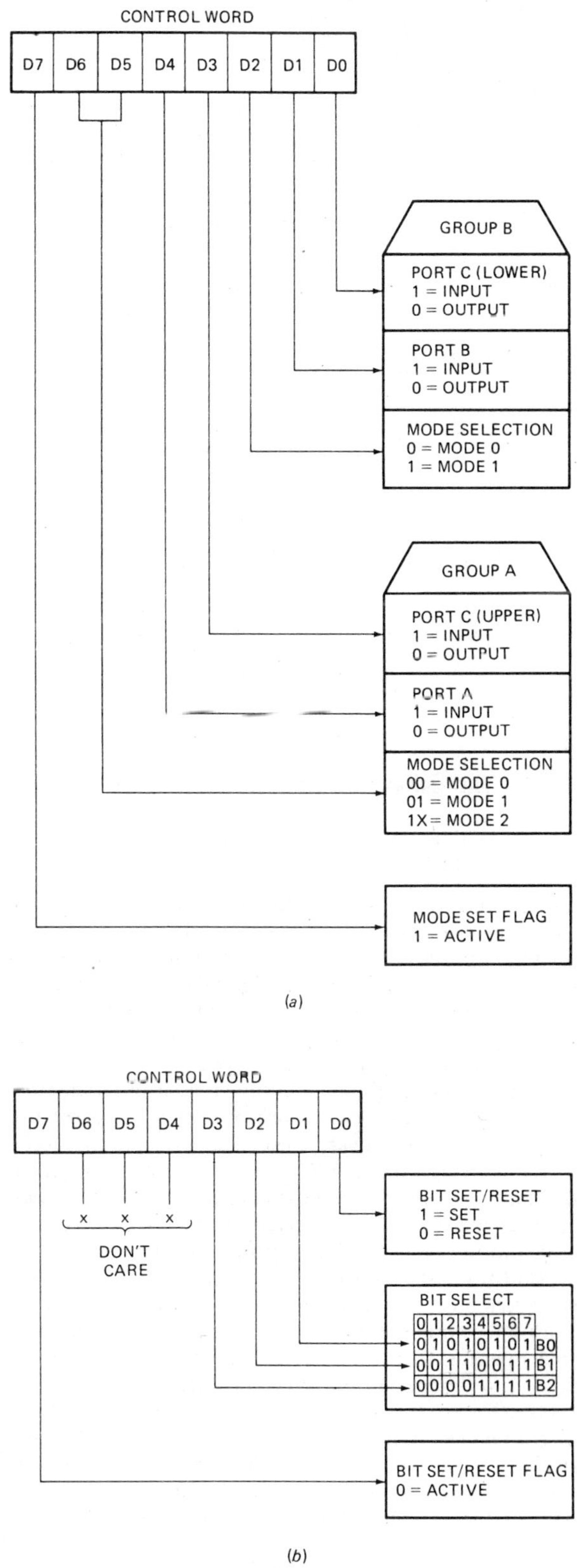

FIGURE 9-4 8255A control word formats. *(a)* Mode set control word. *(b)* Port C bit set/reset control word.

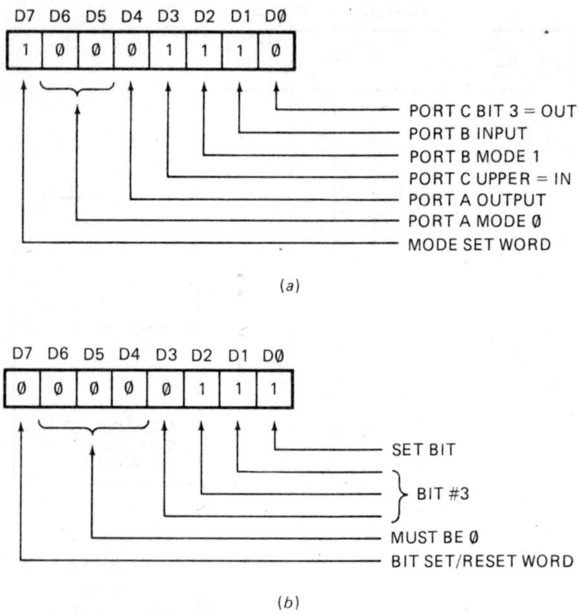

FIGURE 9-5 Control word examples for 8255A. *(a)* Mode set control word. *(b)* Port C bit set/reset control word to set bit 3.

8255A Handshake Application Examples

INTERFACING TO A MICROCOMPUTER-CONTROLLED LATHE

All of the machines in the machine shop of our computer-controlled electronics factory operate under microcomputer control. One example of the machines here is a lathe which makes bolts from long rods of stainless steel. The cutting instructions for each type of bolt that we need to make are stored on a ¾-in wide paper or metal tape. Each instruction is represented by a series of holes in the tape. A tape reader pulls the tape through an optical or mechanical sensor to detect the hole patterns and convert these to an 8-bit parallel code. The microcomputer reads the instruction codes from the tape reader on a handshake basis and sends the appropriate control instructions to the lathe. The microcomputer must also monitor various conditions around the lathe. It must, for example, make sure the lathe has cutting lubricant oil, is not out of material to work on, and is not jammed up in some way. Machines that operate in this way are often referred to as *computer numerical control* or *CNC machines*.

Figure 9-6 shows in diagram form how an 8255A might be used to interface a microcomputer to the tape reader and lathe. In the next chapter we will show you some of the actual circuitry needed to interface the port pins of the 8255A to the sensors and the high-power motors of the lathe. For now we want to talk about initializing the 8255A for this application and analyze the timing waveforms for the handshake input of data from the tape reader.

First you want to make up the control word to initialize the 8255A in the correct modes for this application. To do this start by making a list showing how you want each port pin or group of pins to function. Then put in the control word bits that implement those pin functions. For our example here:

Port A needs to be initialized for handshake input (MODE 1), because instruction codes have to be read in from the tape reader on a handshake basis.

Port B needs to initialized for simple output (MODE 0). No handshaking is needed here because this port is being used to output simple on or off control signals to the lathe.

Port C, bits PC0, PC1, and PC2 are used for simple input of sensor signals from the lathe.

Port C, bits PC3, PC4, and PC5 function as the handshake signals for the data transfer from the tape reader connected to port A.

Port C, bit PC6 is used for output of the STOP/GO signal to the tape reader.

Port C, bit PC7 is not used for this example.

Figure 9-7 shows the control word to initialize the 8255A for these pin functions. This word will be sent to the control register address as shown above. Now let's talk about how the program for this machine might operate, and how the handshake data transfer actually takes place.

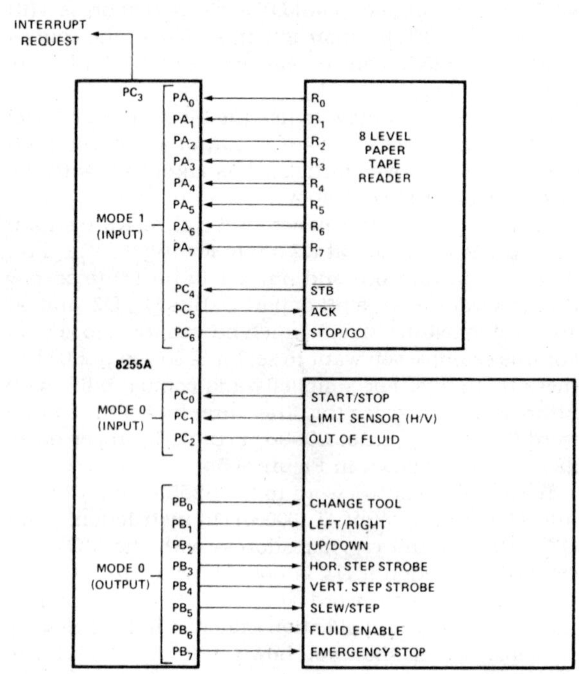

FIGURE 9-6 Interfacing a microprocessor to a tape reader and lathe.

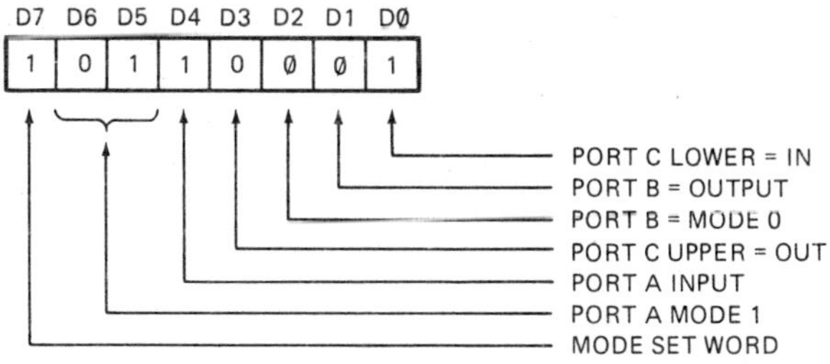

FIGURE 9-7 Control word to initialize 8255A for interface with tape reader and lathe.

After initializing everything, the system would probably read port C bits PC0, PC1, and PC2 to check if the lathe was ready to operate. For any 8255A mode you read port C by simply doing an input from the port C address. Then the microprocessor would output a start command to the tape reader on bit PC6. This is done with a bit set/reset command. Assuming you want to reset bit PC6 to start the tape reader, the bit set/reset control word for this is 00001100. When the tape reader receives the go command, it will start the handshake data transfer to the 8255A. Let's work our way through the timing waveforms in Figure 9-8 to see how the data transfer takes place.

The tape reader starts the process by sending out a byte of data to port A on its eight data lines. The tape reader then asserts its $\overline{STB}$ line low to tell the 8255A that a new byte of data has been sent. In response the 8255A raises its input buffer full (IBF) signal on PC5 high to tell the tape reader that it is ready for the data. When it detects the IBF signal at a high level, the tape reader raises its $\overline{STB}$ signal high again. The rising edge of the $\overline{STB}$ signal has two effects on the 8255A. It first latches the data byte in the input latches of the 8255A. Once the data is latched, the tape reader can remove the data byte in preparation for sending the next data byte. This is shown by the dashed section on the right side of the data waveform in Figure 9-8. Secondly, if the interrupt signal output has been enabled, the rising edge of the $\overline{STB}$ signal will cause the 8255A to output an interrupt request signal to the microprocessor on bit PC3.

The processor's response to the interrupt request will be to go to an interrupt service procedure which reads in the byte of data latched in port A. When the $\overline{RD}$ signal from the microprocessor goes low for this read of port A, the 8255A will automatically reset its interrupt request signal on PC3. This is done so that a second interrupt cannot be caused by the same data byte transfer. When the processor raises its $\overline{RD}$ signal high again at the end of the read operation, the 8255A automatically drops its IBF signal on PC5 low again. IBF going low again is the signal to the tape reader that the data transfer is complete, and that it can send the next byte of data. The time between when the 8255A sends the interrupt request signal and when the processor reads the data byte from port A depends on when the processor gets around to servicing that interrupt. The point here is that this time doesn't matter. The tape reader will not send the next byte of data until it detects that the IBF signal has gone low again. The transfer cycle will then repeat for the next data byte.

After the processor reads in the lathe control instruction byte from the tape reader, it will decode this instruction, and output the appropriate control byte to the lathe on port B of the 8255A. The tape reader then sends the next instruction byte. If the instruction tape is made into a continuous loop, the lathe will keep making the specified parts until it runs out of material. The unused bit of port C, PC7, could be connected to a mechanism which loads in more material so the lathe can continue.

Before we go on there is one more point we have to make about initializing the 8255A for this microcomputer-controlled lathe application. In order for the handshake input data transfer from the tape reader to work correctly the interrupt request signal from bit PC3 has to be enabled. This is done by sending a bit set/reset control word for the appropriate bit of port C. Figure 9-9 shows the port C bit that must be set to enable the interrupt output signal for each of the 8255A handshake modes. For the example here port A is being

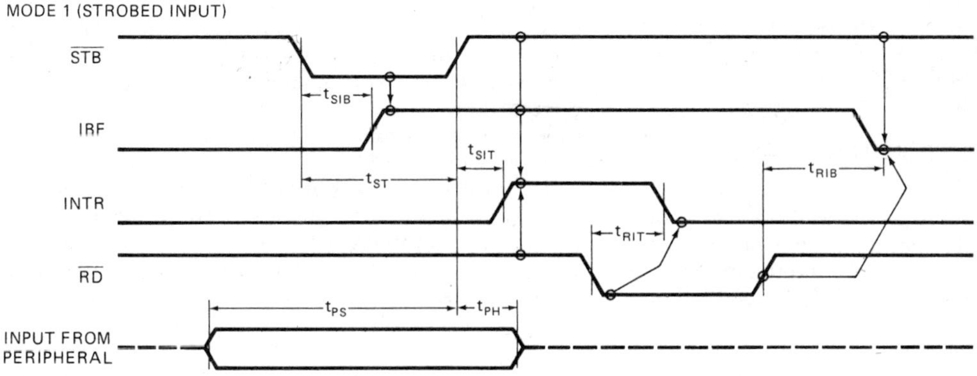

FIGURE 9-8 Timing waveforms for 8255 handshake data input from a tape reader.

FOR MODE 1	PORT C INTERRUPT SIGNAL PIN NUMBER	TO ENABLE INTERRUPT REQUEST SET PORT C BIT
PORT A IN	PC3	PC4
PORT B IN	PC0	PC2
PORT A OUT	PC3	PC6
PORT B OUT	PC0	PC2

FOR MODE 2	PORT C INTERRUPT SIGNAL PIN NUMBER	TO ENABLE INTERRUPT REQUEST SET PORT C BIT
PORT A IN	PC3	PC4
PORT A OUT	PC3	PC6

FIGURE 9-9 Port C bits to set to enable interrupt request outputs for handshake modes.

used for handshake input. According to Figure 9-9, port C bit PC4 must be set to enable the interrupt output for this operation. The bit set/reset control word to do this is 00001001. This bit set/reset control word will be sent to the control address of the 8255A.

Handshake data transfer from the tape reader to the 8255A can be stopped by disabling the 8255A interrupt output on port C pin PC3. This is done by resetting bit PC4 with a bit set/reset control word of 00001000. You will later see another example of the use of this interrupt enable/disable process in Figure 9-17.

As another example of 8255A interrupt output enabling, suppose that you are using port B as a handshake output port. According to Figure 9-9 you need to set port C bit PC2 to enable the 8255A interrupt output signal. The bit set/reset control word to do this is 00000101.

The microcomputer-controlled lathe we have described here is a small example of automated manufacturing. The advantage of this approach is that it relieves humans of the drudgery of standing in front of a machine continually making the same part, day after day. Hopefully society can find more productive use for the human time made available.

A SPEECH SYNTHESIZER INTERFACE—8255A HANDSHAKE OUTPUT

Many microprocessor-based products now recognize spoken commands and speak to you. In Chapter 12 we discuss in detail several methods of producing human speech under microprocessor control. For our example here we chose the *Votrax SC-01A phoneme speech synthesizer* because it is relatively inexpensive, it is easy to program, and it interfaces easily with a microprocessor port on a handshake basis. You may want to build up the circuit shown here and give your programs a voice. The circuit can be connected to one of the 8255As on an SDK-86 board if you have one of these available.

SC-01A OPERATION AND CIRCUIT CONNECTIONS

Figure 9-10*a* shows how an SC-01A speech synthesizer IC can be connected to an 8255A. The SC-01A uses *phonemes* to produce speech. Phonemes are the individual sounds in words. By linking phonemes, you can produce any word. To produce words, phrases, or even sentences, the microcomputer simply has to output a series of phonemes to the SC-01A with the proper timing. A 6-bit binary code sent to the P0–P5 inputs of the SC-01A determines which of its 64 phonemes it will output. An additional two bits sent to the SC-01A's I1 and I2 inputs determine the inflection of the sounded phoneme. A table in the appendix shows the 64 phoneme codes and the phoneme sequence for some example words. To sound a phoneme you send the phoneme and inflection codes for that phoneme and then assert the STB input of the SC-01 high. The SC-01A will then assert its acknowledge/request ($\overline{A}$/R) line low to tell you that it received the phoneme, and it will sound the phoneme. The time required to sound a phoneme ranges from 47 to 250 ms. When the SC-01A finishes sounding the phoneme, it will raise its $\overline{A}$/R line high again to indicate that it is ready for the next phoneme. The variable time it takes to sound a phoneme means that you have to send phonemes to the SC-01A on a handshake basis. You could poll the $\overline{A}$/R line to determine when the SC-01A is ready for the next phoneme, but because of the relatively long time between requests, it is much more reasonable to service the device on an interrupt basis. An 8255A port operating in MODE 1 easily manages the required STB, $\overline{A}$/R, and interrupt signals, so these lines are con-

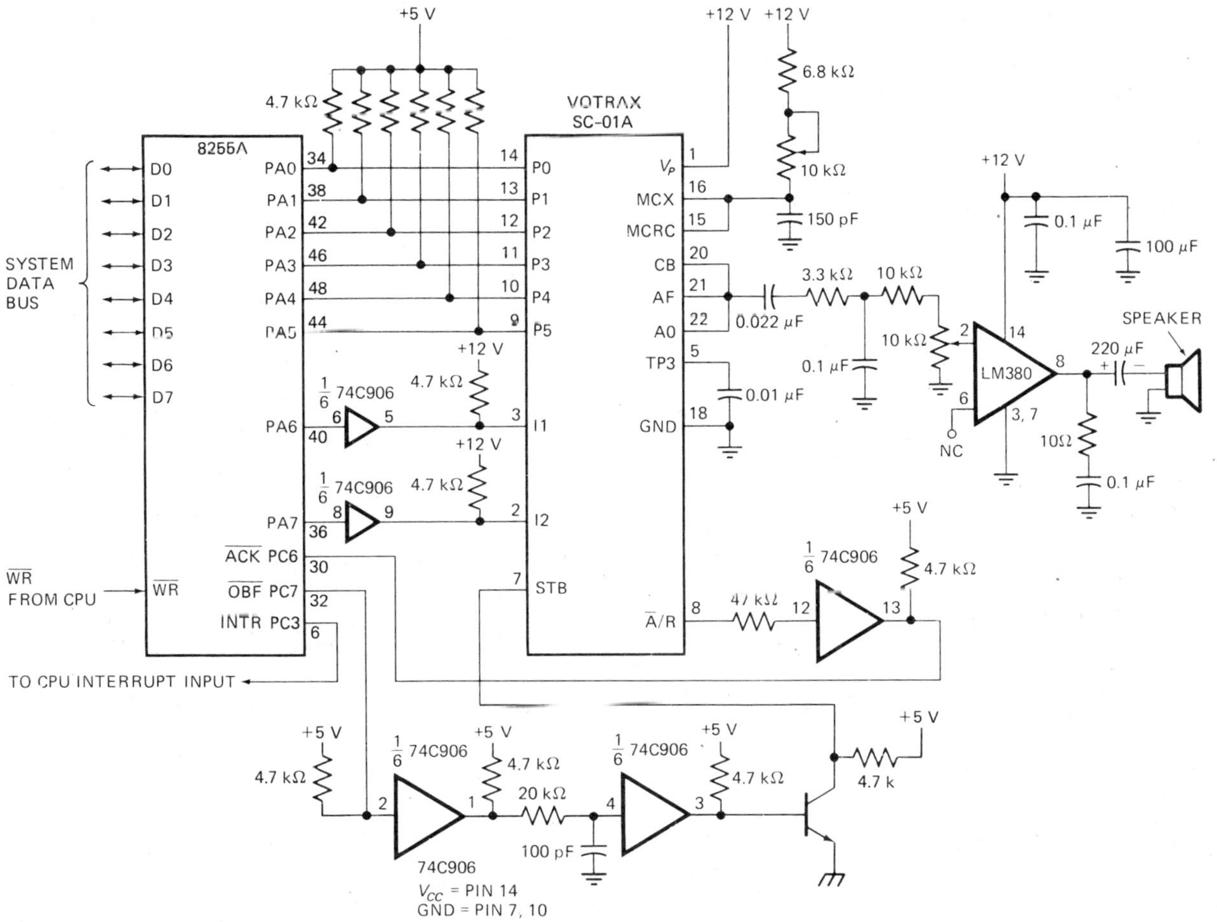

(a)

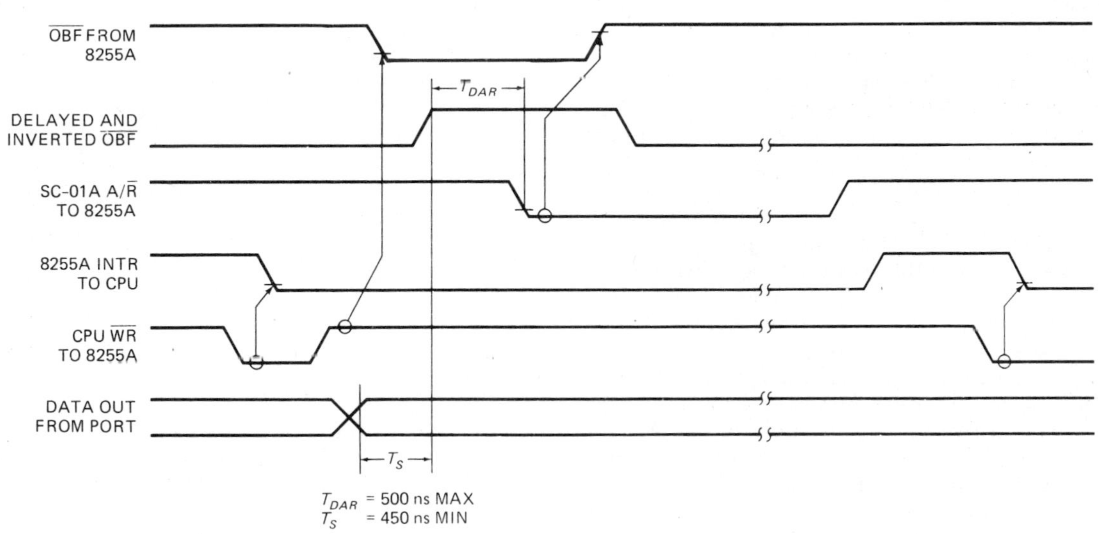

(b)

FIGURE 9-10 *(a)* Connection of a Votrax SC-01A speech synthesizer to an 8255A for handshake output of phonemes. *(b)* Timing waveforms for transfer of a phoneme from 8255A to SC-01A on handshake basis. *(Votrax Incorporated)*

nected to the appropriate bits of port C for this mode. Before we go on to the 8255A operation and timing waveforms, here are a few more points about the circuit connections.

The LM380 in Figure 9-10*a* is an audio amplifier which amplifies the signal from the SC-01A so that it can drive a speaker. The resistors and capacitors connected to pins 15 and 16 of the SC-01A determine the internal clock frequency. This clock frequency determines the pitch of the phoneme. Adjust the 10-kΩ potentiometer to get a frequency of about 680 kHz on pin 16 or until you like the pitch of the sounded phonemes. The 74C906 open drain CMOS buffers, between the 8255A PA6–PA7 pins and the I1–I2 pins, convert the 0–5 V range signals from the 8255A to the 0–12 V range signals required by the SC-01A inputs. Likewise, the 74C906 buffer on the A/R output of the SC-01A converts the 0–12 V range signal from the SC-01A to the 0–5 V range signal required by the 8255A. The STB signal to the SC-01A must come at least 450 ns after the phoneme and inflection codes arrive at the device. The 20-kΩ resistor and the 100-pF capacitor between the two 74C906 buffers on the STB line produce the required delay for this signal. The transistor after the second buffer inverts the OBF signal from the 8255A so it has the correct polarity for the SC-01A STB input. It is often necessary to "massage" the handshake strobe signal so that it meets the timing requirements of the receiving device. In our next application example, a printer interface, we show you another way to do this.

PHONEME TRANSFER TIMING WAVEFORMS

Figure 9-10*b* shows the timing waveforms for a handshake output data transfer to the SC-01A. Here's how this works.

When the SC-01 is first powered up it raises its A/R output high to indicate that it is ready for a phoneme. This causes the 8255A to send an interrupt signal to the processor. In response to the interrupt request the processor does an interrupt service procedure which writes a phoneme and an inflection code to port A of the 8255A. The left edge of the waveforms in Figure 9-10*b* represents the start of the phoneme write operation. During this write operation the WR from the 8086 will go low. When the 8255A detects this low, it will automatically reset its interrupt request output on pin PC3. A little later you will see how this was set. Now, when the WR signal from the 8086 goes high, the phoneme and inflection codes will be present on the output of the 8255A. WR being at a high state causes the 8255A to automatically assert its output buffer full (OBF) signal low on bit PC7. This signal, inverted and delayed 450 ns by the buffer circuit, arrives at the STB input of the SC-01A. This signal edge says to the SC-01A, "Here is a phoneme for you." In response, the SC-01A drops its A/R output low to say, "I got the phoneme, thank you." When this falling edge arrives at the 8255A ACK input on bit PC6, the 8255A automatically raises its OBF signal high again. This edge essentially asks the SC-01A, "May I send you another phoneme?" After the SC-01A finishes sounding the phoneme (47–250 ms later) it raises its A/R line high again to say, "Send me the next phoneme." When the 8255A ACK input receives the rising edge of this A/R signal, it automatically raises the interrupt request signal on pin PC3 high if that signal has been enabled. If the 8086 interrupt input being used is enabled, the 8086 will go and execute the interrupt service routine that writes a phoneme to port A of the 8255A. Writing a phoneme to the 8255A will cause the 8255A interrupt request output on PC3 to be automatically reset. The handshake sequence then repeats for this phoneme.

8255A INITIALIZATION FOR HANDSHAKE OUTPUT

In order to have specific addresses let's assume the SC-01A is connected to the 8255A, *XA*40, on an SDK-86 board. As shown in Figure 7-15, the port addresses for this device are port P2A—FFF8H, port P2B—FFFAH, port P2C—FFFCH, and P2 control register—FFFEH. Now let's make up the mode control word to send to the 8255A.

For the mode control word we make bit D7 a 1. We want to use port A as a handshake port, so we initialize it in MODE 1 by putting 0 in bit D6 and 1 in bit D5. To initialize port A for output, we put a 0 in bit D4. The other bits in this control word would be determined by the use of port B and the remaining pins of port C. If you are not using these, just make these bits 0's. Figure 9-11*a* shows the resultant control word. We send this mode control word to the control register at address FFFEH.

Since we want to do the handshake data transfer on an interrupt basis, we have to send another control word to enable the interrupt request signal on pin PC3. According to Figure 9-8 we enable this interrupt request

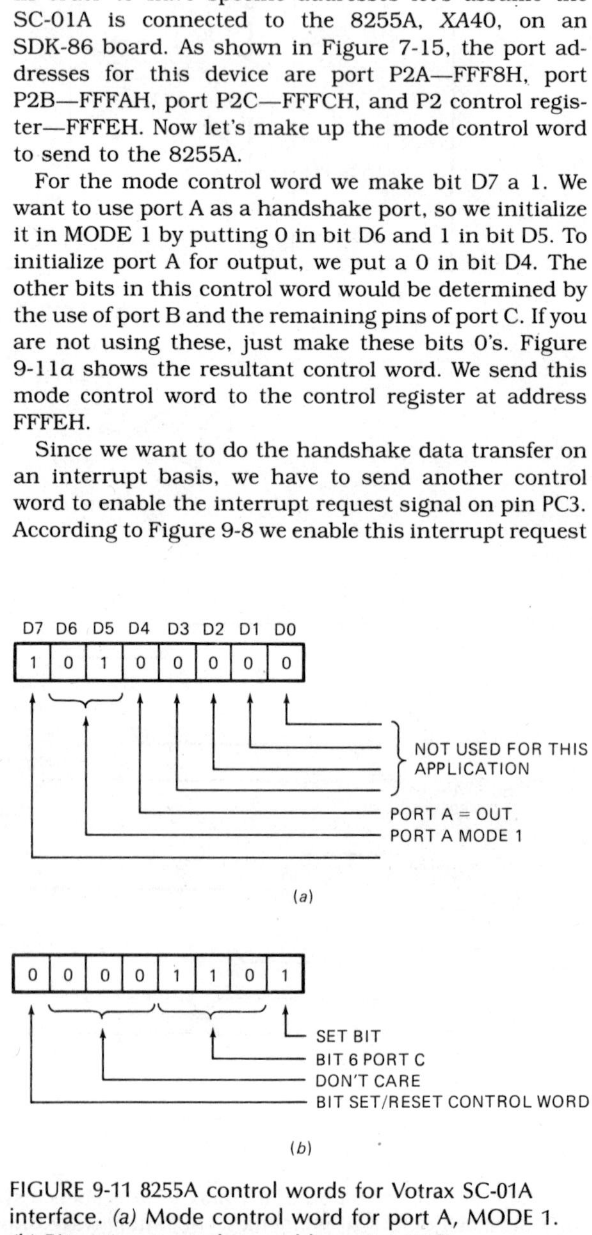

FIGURE 9-11 8255A control words for Votrax SC-01A interface. *(a)* Mode control word for port A, MODE 1. *(b)* Bit set/reset word to enable port A INTR.

by setting bit PC6. Figure 9-11*b* shows the bit set/reset control word to set bit PC6. This control word is also output to the control register at address FFFEH.

PROGRAM NOTES FOR SC-01A MAINLINE AND INTERRUPT PROCEDURE

The major tasks you have to do for the mainline here are:

1. Set up a table containing the sequence of phoneme codes you want to send. Make the last code in the table the no-sound phoneme, FFH, so that you can easily determine when all of the phoneme codes have been sent. As you read out the codes from the table you can then compare each with this *sentinel* value to see if you have reached the end of the table.
2. Initialize, in a memory location, a pointer to the start of the phoneme table.
3. Initialize the interrupt pointer table to point to the start of the interrupt service procedure.
4. Enable and unmask the interrupt input you are using.
5. Initialize the 8255A and enable the 8255A interrupt request output. When the SC-01A is ready for the next phoneme, the 8255A will send an interrupt signal to the 8086.

The 8086 interrupt service procedure must get the table pointer from memory, use the pointer to get the next phoneme from the table, and send the phoneme to the 8255A. The service procedure should then compare the phoneme code to the sentinel value of FFH. If the phoneme code is equal to FFH, then the procedure can simply return to the interrupted program. If the code is not FFH, then the procedure should increment the table pointer to point to the next phoneme, store the pointer back in memory, and do an IRET.

We leave the actual assembly language program for you to write as an exercise at the end of the chapter.

PARALLEL PRINTER INTERFACE—ANOTHER HANDSHAKE OUTPUT EXAMPLE

For most common printers such as the IBM PC printer, the Epson FX-80, and the NEC 8023, the data to be printed is sent to the printer as ASCII characters on eight parallel lines. The printer receives the characters to be printed and stores them in an internal RAM buffer. When the printer detects a carriage return character (ODH), it prints out the first row of characters from the print buffer. When the printer detects a second carriage return it prints out the second row of characters. The process continues until all the desired characters have been printed.

Transfer of the ASCII codes from a microcomputer to a printer must be done on a handshake basis because the microcomputer can send characters much faster than the printer can print them. The printer must in some way let the microcomputer know that its buffer is full, and that it cannot accept any more characters until it prints some out. A common standard for interfacing with parallel printers is the *Centronics Parallel Standard*, named for the company that developed it. In the following sections we show you how a Centronics parallel interface works, and how to implement it with an 8255A.

Centronics Interface Pin Descriptions and Circuit Connections

Centronics-type printers usually have a 36-pin interface connector. Figure 9-12 shows the pin assignments and descriptions for this connector as it is used in the IBM PC printer and the Epson printers. Some manufacturers use one or two pins differently so consult the manual for your specific printer before connecting it up as we show here.

Thirty-six pins may seem like a lot of pins just to send ASCII characters to a printer. The large number of lines is caused by the fact that each data and signal line has its own individual ground return line. For example, as shown in Figure 9-12, pin 2 is the LSB of the data character sent to the printer, and pin 20 is the ground return for this signal. The reason for the individual ground returns is to reduce the chance of picking up electrical noise in the lines. If you are making an interface cable for a parallel printer, these ground return lines should only be connected together and to ground at the microcomputer end of the cable as shown in Figure 9-14. While we are talking about grounds, note that pin 16 is listed as logic ground and pin 17 is listed as chassis ground. In order to prevent large noise currents from flowing in the logic ground wire, these wires should only be connected together in the microcomputer. (This precaution is necessary whenever you connect any external device or system to a microcomputer.)

The rest of the pins on the 36-pin connector fall into two categories, signals sent to the printer to tell it what operation to do and signals from the printer that indicate its status. The major control signals to the printer are $\overline{\text{INIT}}$ on pin 31, which tells the printer to perform its internal initialization sequence and $\overline{\text{STROBE}}$ on pin 1, which tells the printer, "Here is a character for you." Two additional input pins, pin 14 and pin 36 are usually taken care of inside the printer.

The major status signals output from the printer are:

1. The $\overline{\text{ACKNLG}}$ signal on pin 10 which, when low, indicates that the data character has been accepted and the printer is ready for the next character.
2. The BUSY signal on pin 11, which is high if, for some reason such as being out of paper, the printer is not ready to receive a character.
3. The PE signal on pin 12, which goes high if the out-of-paper switch in the printer is activated.
4. The SLCT signal on pin 13 which goes high if the printer is selected for receiving data.
5. The $\overline{\text{ERROR}}$ signal on pin 32 which goes low for a variety of problem conditions in the printer.

SIGNAL PIN NO.	RETURN PIN NO.	SIGNAL	DIRECTION	DESCRIPTION
1	19	$\overline{\text{STROBE}}$	IN	STROBE pulse to read data in. Pulse width must be more than 0.5 μs at receiving terminal. The signal level is normally "high"; read-in of data is performed at the "low" level of this signal.
2	20	DATA 1	IN	These signals represent information of the 1st to 8th bits of parallel data respectively. Each signal is at "high" level when data is logical "1" and "low" when logical "0."
3	21	DATA 2	IN	
4	22	DATA 3	IN	
5	23	DATA 4	IN	
6	24	DATA 5	IN	
7	25	DATA 6	IN	
8	26	DATA 7	IN	
9	27	DATA 8	IN	
10	28	$\overline{\text{ACKNLG}}$	OUT	Approximately 5 μs pulse; "low" indicates that data has been received and the printer is ready to accept other data.
11	29	BUSY	OUT	A "high" signal indicates that the printer cannot receive data. The signal becomes "high" in the following cases: 1. During data entry. 2. During printing operation. 3. In "offline" state. 4. During printer error status.
12	30	PE	OUT	A "high" signal indicates that the printer is out of paper.
13	–	SLCT	OUT	This signal indicates that the printer is in the selected state.
14	–	$\overline{\text{AUTO FEED XT}}$	IN	With this signal being at "low" level, the paper is automatically fed one line after printing. (The signal level can be fixed to "low" with DIP SW pin 2-3 provided on the control circuit board.)
15	–	NC		Not used.
16	–	OV		Logic GND level.
17	–	CHASIS-GND	–	Printer chasis GND. In the printer, the chassis GND and the logic GND are isolated from each other.
18	–	NC	–	Not used.
19-30	–	GND	–	"Twisted-Pair Return" signal; GND level.
31	–	$\overline{\text{INIT}}$	IN	When the level of this signal becomes "low" the printer controller is reset to its initial state and the print buffer is cleared. This signal is normally at "high" level, and its pulse width must be more than 50 μs at the receiving terminal.
32		$\overline{\text{ERROR}}$	OUT	The level of this signal becomes "low" when the printer is in "Paper End" state, "Offline" state and "Error" state.
33	–	GND	–	Same as with pin numbers 19 to 30.
34	–	NC	–	Not used.
35				Pulled up to +5 Vdc through 4.7 k-ohms resistance.
36	–	$\overline{\text{SLCT IN}}$	IN	Data entry to the printer is possible only when the level of this signal is "low." (Internal fixing can be carried out with DIP SW 1-8. The condition at the time of shipment is set "low" for this signal.)

Notes: 1. "Direction" refers to the direction of signal flow as viewed from the printer.
2. "Return" denotes "Twisted-Pair Return" and is to be connected at signal-ground level. When wiring the interface, be sure to use a twisted-pair cable for each signal and never fail to complete connection on the return side. To prevent noise effectively, these cables should be shielded and connected to the chassis of the system unit.
3. All interface conditions are based on TTL level. Both the rise and fall times of each signal must be less than 0.2 μs.
4. Data transfer must not be carried out by ignoring the $\overline{\text{ACKNLG}}$ or BUSY signal. (Data transfer to this printer can be carried out only after confirming the $\overline{\text{ACKNLG}}$ signal or when the level of the BUSY signal is "low.")

FIGURE 9-12 Pin connections and descriptions for Centronix-type parallel interface to IBM PC and EPSON FX-100 printers. *(IBM Corporation)*

Figure 9-13 shows the timing waveforms for transferring data characters to an IBM printer using the basic handshake signals. Here's how this works.

Assuming the printer has been initialized, you first check the BUSY signal pin to see if the printer is ready to receive data. If this signal is low, indicating the printer is ready (not busy), you send an ASCII code on the eight parallel data lines. After at least 0.5 μs you assert the $\overline{\text{STROBE}}$ signal low to tell the printer a character has been sent. The $\overline{\text{STROBE}}$ signal going low causes the printer to assert its BUSY signal high. After a minimum time of 0.5 μs the $\overline{\text{STROBE}}$ signal can be raised high again. Note that the data must be held valid on the data lines for at least 0.5 μs after the $\overline{\text{STROBE}}$ signal is made high.

When the printer is ready to receive the next character, it asserts its $\overline{\text{ACKNLG}}$ signal low for about 5 μs. The rising edge of the $\overline{\text{ACKNLG}}$ signal tells the microcomputer that it can send the next character. The rising edge of the $\overline{\text{ACKNLG}}$ signal also resets the BUSY signal from the printer. BUSY being low is another indication that the printer is ready to accept the next character. Some systems use the $\overline{\text{ACKNLG}}$ signal for the handshake, and some systems use the BUSY signal. Now let's see how you can do this handshake printer interface with an 8255A.

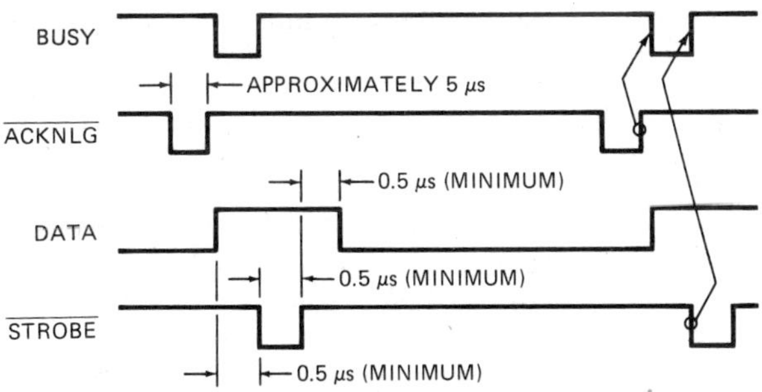

FIGURE 9-13 Timing waveforms for transfer of a data character to a Centronix-type parallel printer such as the IBM-PC or Epson printer. *(IBM Corporation)*

8255A CONNECTIONS AND INITIALIZATION

Figure 9-14 shows the circuit for connecting the Centronics parallel printer signals to an 8255A. We show here the pin connections for the J6 connector on the SDK-86 board so you can easily add this interface if you are working with one.

For this interface circuitry, 74LS07 open-collector buffers are used on the signal and data lines from the 8255A, because the 8255A outputs do not have enough current drive to charge and discharge the capacitance of the connecting cable fast enough. Pull-up resistors for the open-collector outputs of the 74LS07s are built into the printer.

Port B is used for the handshake output data lines. Therefore, as shown in Figure 9-3, bit PC0 functions as the interrupt request output to the 8086. The $\overline{ACKNLG}$ signal from the printer is connected to the 8255A $\overline{ACK}$ input on bit PC2. The $\overline{OBF}$ signal from the 8255A does not have the right timing parameters for this handshake, so PC1 is left unconnected. For this the $\overline{STROBE}$ input of the printer is connected to bit PC4. The $\overline{STROBE}$ signal will be generated by a bit set/reset of this pin.

The four printer status signals are connected to port A so the program can read them in, determine the condi-

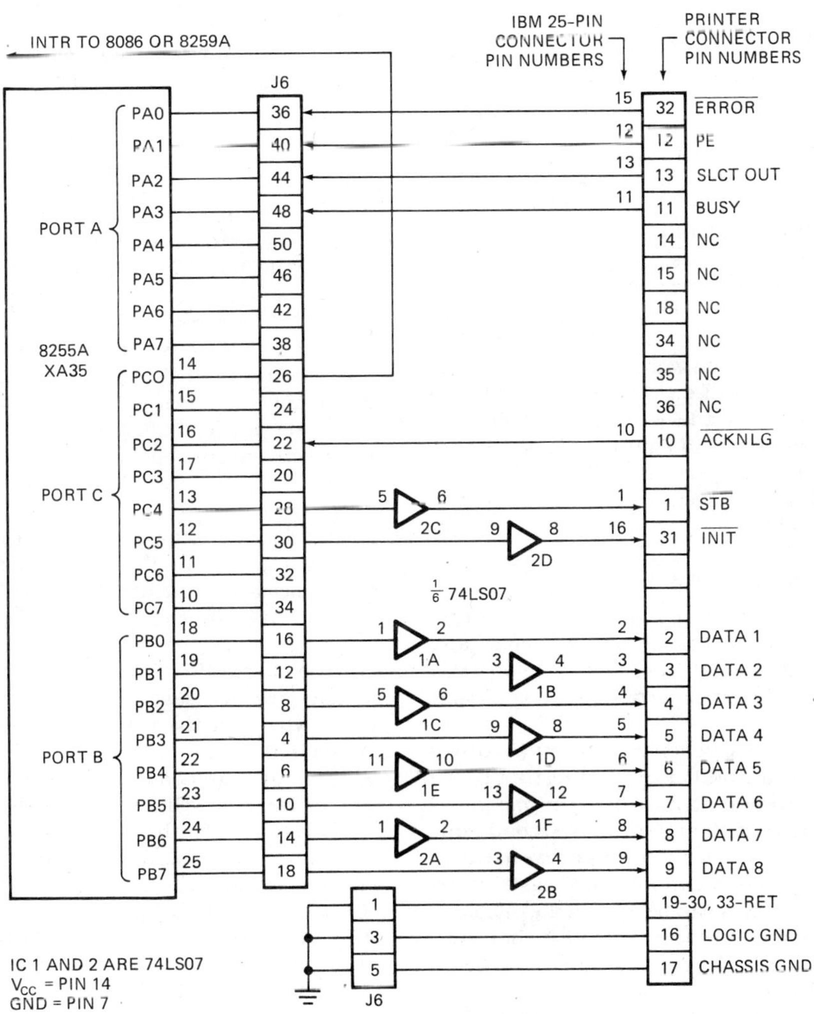

FIGURE 9-14 Circuit for interfacing Centronix-type parallel input printer to 8255A on SDK-86 board.

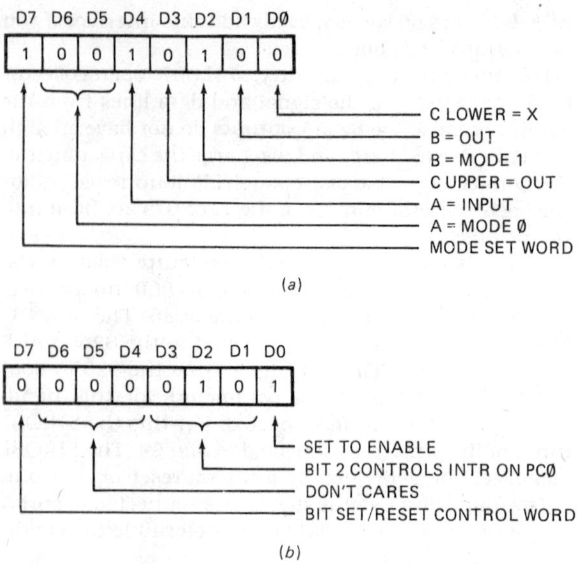

FIGURE 9-15 8255A control words for printer interface. *(a)* Mode control word. *(b)* Bit set/reset control word.

tion of the printer, and send the appropriate messages to the CRT if the printer is not ready.

Finally the INIT input of the printer is connected to bit PC5 so that the printer can be reinitialized under program control. Now let's look at the 8255A control words for this application.

Figure 9-15*a* shows the mode control word to initialize port B in MODE 1 output, port A for MODE 0 input, and the upper 4 bits of port C as outputs. Figure 9-15*b* shows the bit set/reset control word necessary to enable the interrupt request signal on bit PC0 for the handshake. The addresses for the 8255A, *XA*35, on the SDK-86 board are, as shown in Figure 7-15, port P1A—FFF9H, port P1B—FFFBH, port P1C—FFFDH, and control P1—FFFFH. For that system, then, both control words are output to FFFFH.

THE PRINTER DRIVER PROGRAM

Procedures which input data from or output data to peripheral devices such as disk drives, modems, and printers are often called *I/O drivers*. Here we show you one way to write the driver procedure for our parallel printer interface.

The first point to consider when writing any I/O driver is whether to do it on a polled or on an interrupt basis. For the parallel Centronics interface here the maximum data transfer rate is about 1000 characters/second. This means that there is a little less than 1 ms between transfers. If characters are sent on an interrupt basis, many other program instructions can be executed while waiting for the interrupt request to send the next character. Also, when the printer buffer gets full, there will be an even longer time that the processor can be working on some other job while waiting for the next interrupt. This is another illustration of how interrupts allow the computer to do several tasks "at the same time." For our example here assume that the interrupt request from PC0 of the 8255A is connected to the IR6 interrupt input of the 8259A shown in Figure 8-14 so that a clock interrupt, a keyboard interrupt, and the printer interrupt can all be serviced in turn.

Figure 9-16*a* shows the steps you need in the mainline to initialize everything and "call" the printer driver to send a string of ASCII characters to the printer.

At the start of the mainline some named memory locations are set aside to store parameters needed for transfer of data to the printer. The memory locations set aside for passing information between the mainline and the driver procedure are often called a *control block*. In the control block a named location is set aside for a pointer to the address of the ASCII character that is currently being sent. Another memory location is set aside to store the number of characters to be sent. The number in this location will function as a counter so you know when you have sent all of the characters in the buffer. Instead of using this counter approach to keep track of how many characters have been sent, the *sentinel method* we described for handshake output to the SC-01A in Figure 9-10 could have been used. With the sentinel approach you put a *sentinel* character in memory after the last character to be sent out. MSDOS, for example, uses a $ (24H) as a sentinel character for some of its drivers. As you read each character in from memory, you compare it with the sentinel value. If it matches, you know all of the characters have been sent. The sentinel approach and the counter approach are both widely used, so you should be familiar with both.

To get the hardware ready to go, you need to initialize the 8259A and unmask the IR inputs of the 8259A that are used. The 8086 INTR input must also be enabled. Next the 8255A must be initialized by sending it the mode control word shown above. A bit set/reset control word is then sent to the 8255A to make the STROBE signal to the printer high because this is its unasserted level. To make sure the printer is internally initialized, you pulse the INIT line to the printer low for a few microseconds.

When you are actually ready to print some characters in a program, you first read the printer status from port A and check if the printer is selected, not out of paper, and not busy. In a more complete program you could send a specific error message to the display indicating the type of error found. The program here just sends a general error message. If no printer error condition is found, you load the starting address of the string of ASCII characters into the control block location you set aside for this, and load the number of characters to be sent in the reserved location in the control block. Finally, you enable the interrupt request pin on the 8255A. Note that you do not enable this interrupt until you are actually ready to send data. A high on the ACKNLG line from the printer causes the 8255A to output an interrupt request signal. This interrupt request signal goes through the 8259A to the processor and causes it to go to the interrupt service procedure.

Figure 9-16*b* shows the algorithm for the procedure which services this interrupt and actually sends the characters to the printer. After pushing some registers the 8086 INTR input is enabled so that higher-priority interrupts can interrupt this procedure. The string address pointer is then read in from the control block and used to read a character in from the memory buffer to AL. The character in AL is then output to port B of the 8255A.

From here on the program follows the timing diagram in Figure 9-13. After sending the character the program waits at least 0.5 μs, asserts the $\overline{\text{STROBE}}$ input low, waits at least another 0.5 μs and raises the $\overline{\text{STROBE}}$ line high again. The data byte will be latched on the port B output pins until the next character is sent, so the data hold parameter in the timing diagram is satisfied. Sending of the character is now complete, so the next step is to get ready to send another character.

To do this the buffer pointer is incremented by one, and the incremented value is placed back in the control block location. The character counter in the control block is then decremented. If the character counter is not down to zero, there are more characters to send so the EOI command is simply sent to the 8259A, everything popped off the stack, and execution returned to the interrupted program. If the character counter is down to zero, all of the characters have been sent, so the interrupt request output of the 8255A is disabled with a bit set/reset control word to prevent further interrupt

```
Mainline algorithm for printer driver
Initialization
    set up control block
        word for storing pointer to ASCII string
        word for number of characters in string
    initialization control words to 8259A
    unmask 8259A IR6 and any other IR inputs used
    unmask 8Ø86 INTR input
    mode set word to 8255A
    send STROBE high to printer
    initialize printer (pulse init low)
To send ASCII string
    read printer status from port
    if error then
        send error message
        exit
    set print done status bit
    load starting address of string into pointer store
    load length of string into character counter
    enable 8255A INTR output
    wait for interrupt
```

(a)

FIGURE 9-16 Algorithm for printer mainline and interrupt-based printer driver procedure. *(a)* Mainline steps. *(b)* Printer driver procedure steps.

```
Printer Driver Procedure Algorithm

    save registers

    enable 8Ø86 INTR for higher priority interrupts

    get pointer to string

    get ASCII character from buffer

    send character to printer

    wait 1 usec

    send STROBE low

    wait 1 usec

    send STROBE high

    increment pointer to string

    put pointer back in pointer store

    decrement character count

    if character count = Ø then

         disable 8255A interrupt request output

    send EOI command to 8259A

    restore registers

    return from interrupt procedure
```

(*b*)

FIGURE 9-16 (*continued*)

requests from there. This interrupt source will remain disabled until you want to send another buffer of characters to the printer. Execution then exits from the procedure by sending an EOI command to the 8259A, popping registers, and doing an IRET.

Figure 9-17 shows the pertinent parts of the mainline program and the printer driver procedure. The preceding discussion of the algorithms and the comments with the instructions should make most of these reasonably clear if you work your way through them one step at a time. You have seen many of the pieces in previous programs. One part of the program that we do want to expand and clarify is the generation of the $\overline{STROBE}$ signal with bit PC3.

In the speech synthesizer example in a preceding section we used external hardware to "massage" the $\overline{OBF}$ signal from the 8255A so it matched the timing and polarity requirements of the receiving device. Here we generate the strobe directly under software control.

In the mainline we make the $\overline{STROBE}$ signal on PC4 high by sending a bit set/reset control word of 00001001 to the control register of the 8255A. In the printer driver procedure a character is sent to the printer with the OUT DX, AL instruction. According to the timing diagram in Figure 9-13 we then want to wait at least 0.5 μs before asserting the $\overline{STROBE}$ signal low. This is automatically done in the program because the instructions required to assert the strobe low take longer than 0.5 μs. The MOV AL, 00001000B instruction requires 4 clock cycles, and the OUT DX, AL instruction requires 8 clock cycles to execute. Assuming a 5-MHz clock (0.2-μs period), these two instructions take 2.4 μs to execute, which is more than required.

Again referring to the timing diagram in Figure 9-13,

```
page ,132
;8086 Printer-driver program
;
;ABSTRACT:      This program sets up the 8259A and the 8255A on an SDK-86
;               board so that a message in a buffer can be sent to a
;               printer. The mainline sets up a control block and
;               initializes all variables
;PORTS USED:    SDK-86 ports P1A - FFF9H used to input status of printer
;                            P1B - FFFBH used to output a character
;                            P1C - for handshake signals for port B
;PROCEDURES:    PRINT_IT used to output characters

A_INT_TABLE     SEGMENT WORD
        TYPE_64_69      DW      12 DUP(0)       ; reserved for IR0-IR5
        TYPE_70         DW      2  DUP(0)       ; IR6 interrupt
        TYPE_71         DW      2  DUP(0)       ; IR7 interrupt - not used
A_INT_TABLE     ENDS

DATA_HERE       SEGMENT WORD PUBLIC
        MESSAGE_1       DB      'This is the message from the printer driver!'
                        DB      0DH, 0AH, 0DH    ; return & line feed for printer
        PRINT_DONE      DB      0
        POINTER         DW      00      ; storage for pointer to MESSAGE_1
        COUNTER         DB      0       ; counter for length of MESSAGE_1
        PRINTER_ERROR   DB      0
DATA_HERE       ENDS

PUBLIC  PRINT_DONE, POINTER, COUNTER, MESSAGE_1
EXTRN   PRINT_IT:FAR
MESSAGE_LENGTH  EQU     47              ; length of MESSAGE_1

STACK_HERE      SEGMENT
                        DW      30 DUP(0)
        STACK_TOP       LABEL   WORD
STACK_HERE      ENDS

CODE_HERE       SEGMENT WORD PUBLIC
        ASSUME CS:CODE_HERE, DS:A_INT_TABLE, SS:STACK_HERE
;initialize stack and data segments
        MOV     AX, STACK_HERE
        MOV     SS, AX
        MOV     SP, OFFSET STACK_TOP
        MOV     AX, A_INT_TABLE
        MOV     DS, AX
;set up interrupt table and put in address for printer interrupt subroutine
        MOV     TYPE_70+2, SEG PRINT_IT
        MOV     TYPE_70,   OFFSET PRINT_IT
;set up data segment
ASSUME DS:DATA_HERE
        MOV     AX, DATA_HERE
        MOV     DS, AX
;initialize 8259A and unmask IR6
        MOV     DX, 0FF00H              ; point at 8259A control address
```

FIGURE 9-17 8086 assembly language instructions for mainline and printer driver procedure. *(a)* Mainline. *(b)* Procedure.

```
        OUT     DX, AL                  ; send ICW1
        MOV     DX, 0FF02H              ; point at ICW2 address
        MOV     AL, 01000000B           ; type 64 is first 8259A type
        OUT     DX, AL                  ; send ICW2
        MOV     AL, 00000001B           ; ICW4, 8086 mode
        OUT     DX, AL                  ; send ICW4
        MOV     AL, 10111111B           ; OCW1 to unmask IR6
        OUT     DX, AL                  ; send OCW1
;initialize 8255A,
; Port A: mode-0 i/p. Port B: mode-1 o/p. Unused port C bits: o/p
        MOV     DX, 0FFFFH              ; control address for 8255A
        MOV     AL, 10010100B           ; control word for above conditions
        OUT     DX, AL                  ; send control word
        STI                             ;unmask 8086 INTR interrupt
;send strobe high to printer with bit set on PC4
        MOV     AL, 00001001B
        OUT     DX, AL
;initialize printer-pulse INIT low on PC5
        MOV     AL, 00001010B
        OUT     DX, AL
;read printer status from port A, status OK - AL = XXXX0101
;PA3-BUSY=0, PA2-SLCT=1, PA1-PE=0, PA0-ERROR=1
        MOV     PRINTER_ERROR, 0        ; printer OK so far
        MOV     DX, 0FFF9H              ; point at port A
        IN      AL, DX                  ; get status of printer
        AND     AL, 0FH                 ; upper four bits are not used
        CMP     AL, 00000101B           ; is status OK
        JZ      SEND_IT                 ; send it if OK
;printer not ready, try once more after waiting 20 ms.
        MOV     CX, 16EAH               ; load count for 20ms
PAUSE:  LOOP    PAUSE                   ; and wait
        IN      AL, DX                  ; repeat steps to read status
        AND     AL, 0FH
        CMP     AL, 00000101B
        JZ      SEND_IT                 ; is printer ready yet?
        MOV     PRINTER_ERROR, 01       ; set error code
        JMP     FIN                     ; not ready so terminate send
;set up pointer to message storage and say print not done yet
SEND_IT:MOV     AX, OFFSET MESSAGE_1
        MOV     POINTER, AX
        MOV     PRINT_DONE, 00
        MOV     COUNTER, MESSAGE_LENGTH
;enable 8255A interrupt request output on PC0 by setting PC2
        MOV     DX, 0FFFFH              ; point at port control addr
        MOV     AL, 00000101B           ; bit set word for PC0 intr
        OUT     DX, AL
;wait for an interrupt from the printer
WT:     JMP     WT
FIN:    NOP

CODE_HERE       ENDS
                END
```

(*a*)

FIGURE 9-17 (*continued*)

```
PAGE ,132
;8086 Procedure for printer driver program
;ABSTRACT:       This procedure outputs a character from a buffer to
;                a printer. If no characters are left in the buffer
;                then the interrupt to the 8086 on IR6 of the 8259A
;                is disabled
;PROCEDURES:     None used
;PORTS:          Uses SDK-86 board Port P1B (FFFBH) to output characters
;                Port P1C bits for handshake signals and printer intr
;REGISTERS USED:AX, BX, DX Flags, destroys no registers

PUBLIC  PRINT_IT
DATA_HERE       SEGMENT PUBLIC
        EXTRN   COUNTER:BYTE, POINTER:WORD
        EXTRN   MESSAGE_1:BYTE, PRINT_DONE:BYTE
DATA_HERE       ENDS

CODE_HERE       SEGMENT WORD PUBLIC
PRINT_IT        PROC    FAR
                ASSUME  CS:CODE_HERE, DS:DATA_HERE
                PUSHF                   ; save registers
                PUSH    AX
                PUSH    BX
                PUSH    DX
                STI                     ; enable higher interrupts
                MOV     DX, 0FFFBH      ; point at port B
                MOV     BX, POINTER     ; load pointer to message
                MOV     AL, [BX]        ; get a character
                OUT     DX, AL          ; send the character to printer
;send printer strobe on PC4 low then high
                MOV     DX, 0FFFFH      ; point at port control addr
                MOV     AL, 00001000B   ; strobe low control word
                OUT     DX, AL
                MOV     AL, 00001001B   ; strobe high control word
                OUT     DX, AL
;increment pointer and decrement counter
                INC     BX
                MOV     POINTER, BX
                DEC     COUNTER
                JNZ     NEXT            ; wait for next character?
;no more characters-disable 8255A int request on PC0 by bit reset of PC2
                MOV     AL, 00000100B   ; bit reset word for PC0 interrupt
                OUT     DX, AL
                MOV     PRINT_DONE, 0
NEXT:           MOV     AL, 00100000B   ; OCW2 for non-specific EOI
                MOV     DX, 0FF00H      ; point at 8259A control addr
                OUT     DX, AL
                POP     DX              ; restore registers
                POP     BX
                POP     AX
                POPF
                IRET
PRINT_IT        ENDP
CODE_HERE       ENDS
                END
```

FIGURE 9-17 *(continued)* (*b*)

the $\overline{STROBE}$ time low must also be at least 0.5 μs. The MOV AL, 00001001B instruction takes 4 clock cycles and the OUT DX, AL instruction takes 8 clock cycles. With a 5-MHz clock this totals to 2.4 μs, which again is more than enough time for $\overline{STROBE}$ low. In this case creating the $\overline{STROBE}$ signal with software does not use much of the processor's time, so this is an efficient way to do it.

A FEW MORE POINTS ABOUT THE 8255A

Before leaving our discussion of the 8255A we want to show you a little more about how port C is used.

Any bits of port C which are programmed as inputs can be read by simply doing a read from the port C address. You can mask out any unwanted bits of the word read in. If port A and/or port B is programmed in a handshake mode, then some of the bits of a byte read in from port C represent *status information* about the handshake signals. Figure 9-18 shows the meaning of the bits read from port C for port A and/or port B in MODE 1. Here's how you read this diagram. If port B is initialized as a handshake (MODE 1) input port, then bits D0, D1, and D2 read from port C represent the status of the port B handshake signals. Bit D2 will be high if the port B interrupt request output has been enabled. Bit D2 is a copy of the level on the input buffer full (IBF) pin. Bit D3 is a copy of the interrupt request output, so it will be high if port B is requesting an interrupt.

In our previous application examples, we showed how to do handshake data transfer on an interrupt basis to make maximum use of the CPU time. However, in applications where the CPU has nothing else to do while waiting to, for example, read in the next character from some device, then you can save one interrupt input by reading data from the 8255A on a polled basis. To do this for a handshake input operation on port B you simply loop through reading port C and checking bit D1 over and over until you find this bit high. The IBF pin being high means that the input data byte has been latched into the 8255A and can now be read. The timing

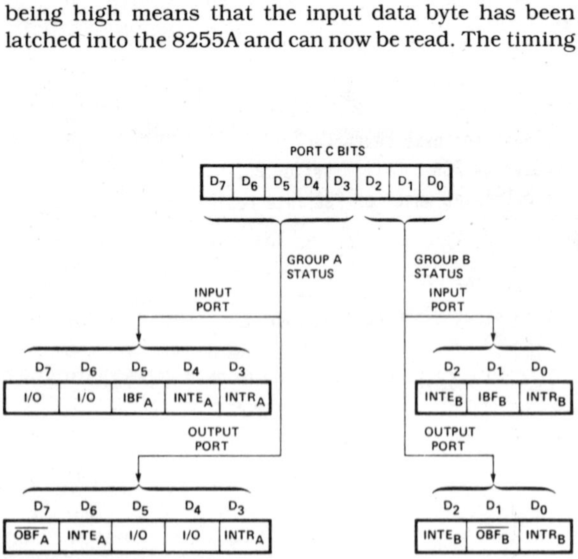

FIGURE 9-18 8255A status word format for MODE 1 input and output operations.

waveforms for this case are the same as those in Figure 9-9, except that you are not using the interrupt request output from the 8255A.

Port C bits not used for handshake signals and programmed as outputs can be written to by sending bit set/reset control words to the control register. Technically, bits PC0–PC3 can also be written to directly at the port C address, but we have found it safer to just use the bit set/reset control word approach to write to all leftover port C bits programmed as outputs.

INTERFACING A MICROPROCESSOR TO KEYBOARDS

Keyboard Types

When you press a key on your computer you are activating a switch. There are many different ways of making these switches. Here's an overview of the construction and operation of some of the most common types.

MECHANICAL KEYSWITCHES

In mechanical switch keys, two pieces of metal are pushed together when you press the key. The actual switch elements are often made of a phosphor-bronze alloy with gold plating on the contact areas. The keyswitch usually contains a spring to return the key to the nonpressed position and perhaps a small piece of foam to help damp out bouncing. Mechanical switches are relatively inexpensive, but they have several disadvantages. First, they suffer from *contact bounce*. A pressed key may make and break contact several times before it makes solid contact. Second, the contacts may become oxidized or dirty with age so they no longer make a dependable connection. Higher-quality mechanical switches typically have a rated lifetime of about 1 million keystrokes.

MEMBRANE KEYSWITCHES

These switches are really just a special type of mechanical switch. They consist of a three-layer plastic or rubber sandwich as shown in Figure 9-19*a*. The top layer has a conductive line of silver ink running under each row of keys. The middle layer has a hole under each key position. The bottom layer has a conductive line of silver ink running under each column of keys. When you press a key you push the top ink line through the hole to contact the bottom ink line. The advantage of membrane keyboards is that they can be made as very thin, sealed units. They are often used on cash registers in fast food restaurants, on medical instruments, and in other messy applications. Lifetime of membrane keyboards varies over a wide range.

CAPACITIVE KEYSWITCHES

As shown in Figure 9-19*b*, a capacitive keyswitch has two small metal plates on the printed-circuit board and another metal plate on the bottom of a piece of foam. When you press the key, the movable plate is pushed closer to the fixed plate. This changes the capacitance between the fixed plates. Sense amplifier circuitry de-

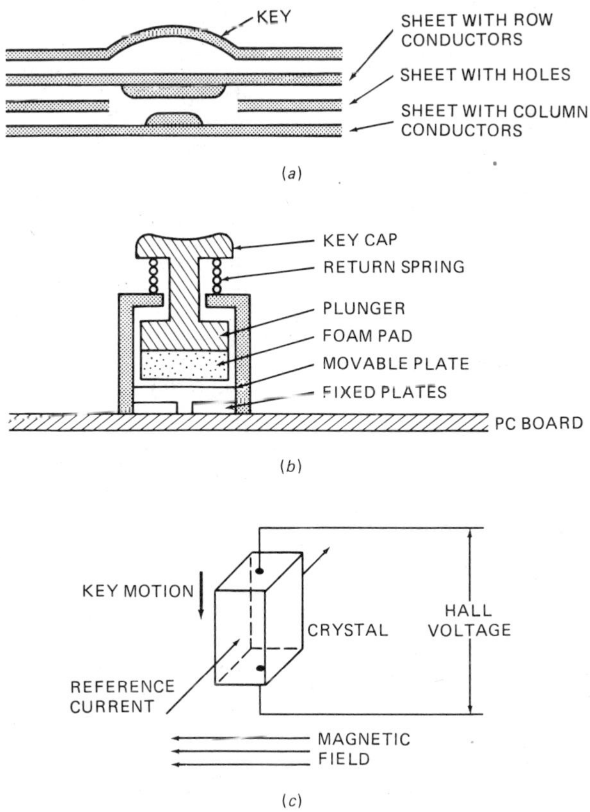

FIGURE 9-19 Key switch types. *(a)* Membrane. *(b)* Capacitive. *(c)* Hall effect.

tects this change in capacitance and produces a logic-level signal that indicates a key has been pressed. The big advantage of a capacitive switch is that it has no mechanical contacts to become oxidized or dirty. A small disadvantage is the specialized circuitry needed to detect the change in capacitance. Capacitive keyswitches typically have a rated lifetime of about 20 million keystrokes.

HALL EFFECT KEYSWITCHES

This is another type of switch which has no mechanical contact. It takes advantage of the deflection of a moving charge by a magnetic field. Figure 9-19*c* shows you how this works. A reference current is passed through a semiconductor crystal between two opposing faces. When a key is pressed, the crystal is moved through a magnetic field which has its flux lines perpendicular to the direction of the current flow in the crystal. (Actually it is easier to move a small magnet past the crystal.) Moving the crystal through the magnetic field causes a small voltage to be developed between two of the other opposing faces of the crystal. This voltage is amplified and used to indicate that a key is pressed. Hall effect keyboards are more expensive because of the more complex switch mechanisms, but they are very dependable, and have typical rated lifetimes of 100 million or more keystrokes.

Keyboard Circuit Connections and Interfacing

In most keyboards the keyswitches are connected in a matrix of rows and columns as shown in Figure 9-20*a*. We will use simple mechanical switches for our examples here, but the principle is the same for other types of switches. Getting meaningful data from a keyboard such as this requires doing three major tasks. They are:

1. Detect a keypress.
2. Debounce the keypress.
3. Encode it (Produce a standard code for the pressed key).

The three tasks can be done with hardware, software, or a combination of the two, depending on the application. We will first show you how they can be done with software as might be done in a microprocessor-based grocery scale where the microprocessor is not pressed for time. Later we describe some hardware devices which do these tasks.

Software Keyboard Interfacing

CIRCUIT CONNECTIONS AND ALGORITHM

Figure 9-20*a* shows how a hexadecimal keypad can be connected to a couple of microcomputer ports so our three tasks can be done as part of a program. The rows of the matrix are connected to four output port lines. The column lines of the matrix are connected to four input port lines. When no keys are pressed, the column lines are held high by the pull-up resistors to +5 V. The main principle here is that pressing a key connects a row to a column. If a low is output on a row and a key in that row is pressed, then the low will appear on the column which contains that key and can be detected on the input port. If you know the row and the column of the pressed key, you then know which key was pressed, and can make up any code you want to represent that key. Figure 9-20*b* shows a flowchart for a procedure to detect, debounce, and produce the hex code for a pressed key.

The first step is to output 0's to all of the rows. Next the columns are read and and checked over and over until the columns are all high. This is done to make sure a previous key has been released before looking for the next one. In standard keyboard terminology this is called *two-key lockout*. Once the columns are found to be all high the program enters another loop which waits until a low appears on one of the columns, indicating a key has been pressed. This loop does the detect task for us. A simple 20-ms delay procedure then does the debounce task.

After the debounce time another check is made to see if the key is still pressed. If the columns are all high, then no key is pressed and the initial detection was just a noise pulse or a light brushing past a key. If any of the columns are still low, then the assumption is made that it is a valid keypress.

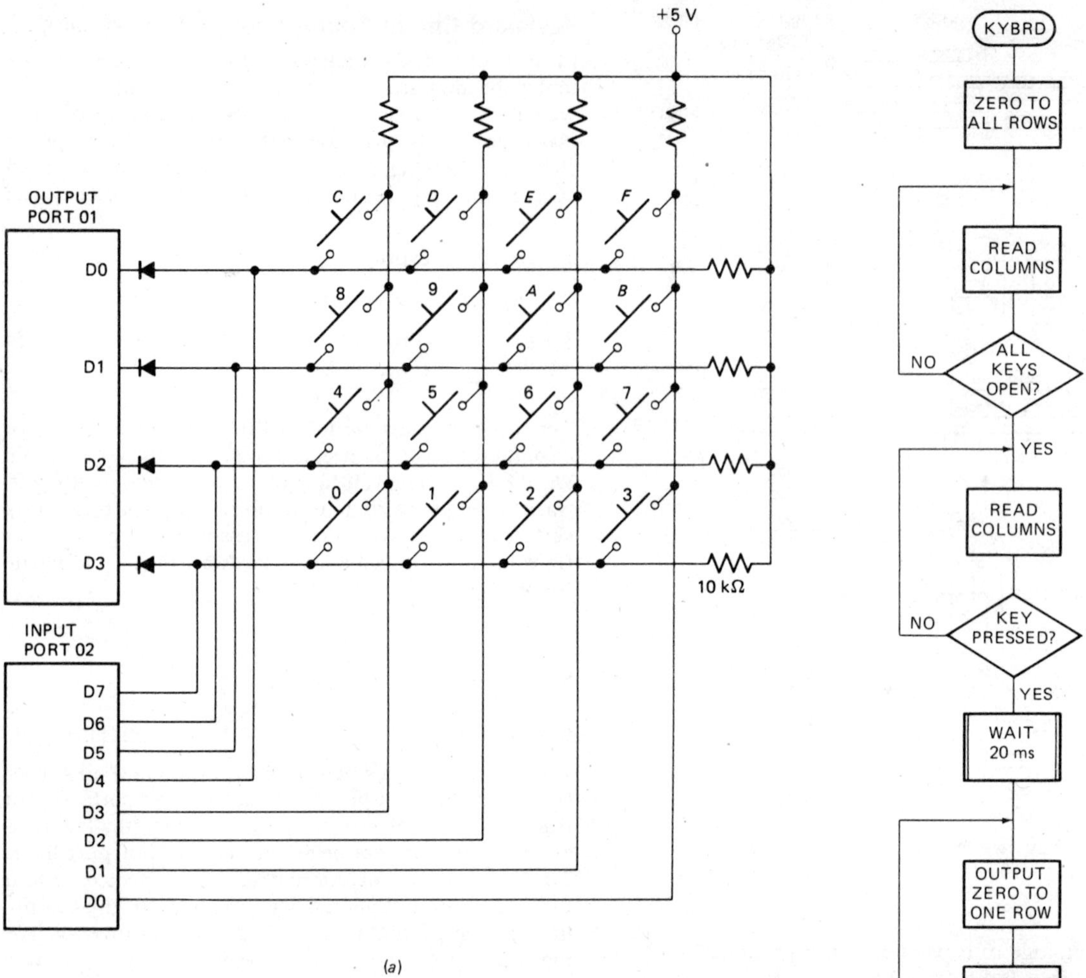

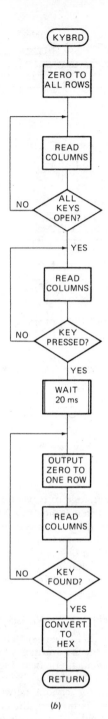

FIGURE 9-20 Detecting a matrix keyboard key-press, debouncing it, and encoding it with a microcomputer. *(a)* Port connections. *(b)* Flowchart for procedure.

The final task is to determine the row and column of the pressed key, and convert this row and column information to the hex code for the pressed key. To get the row and column information, a low is output to one row and the columns are read. If none of the the columns are low, the pressed key is not in that row, so the low is rotated to the next row and the columns are checked again. The process is repeated until a low on a row produces a low on one of the columns. The pressed key is in the row which is low at that time. The byte read in from the input port will contain a 4-bit code which represents the row of the pressed key and a 4-bit code which represents the column of the pressed key. As we show later, this row-column code can easily be converted to hex using a lookup table.

Figure 9-21 shows the assembly language program for this procedure. The detect, debounce, and row-detect parts of the program follow the flowchart very closely and should be easy for you to follow. Work your way down through these parts until you reach the ENCODE label, then continue with the discussion here.

CODE CONVERSION

There are two major ways of converting one code to another in a program. The ENCODE portion of this program uses a *compare* technique, which is important for you to learn, so we will discuss this portion in detail. In a later section on keyboard interfacing with hardware

```
page ,132
;8086 Program to scan and decode a 16 switch keypad
;ABSTRACT :  This program initializes the ports below and then
;            calls a procedure to input an 8-bit value from
;            a 16-switch keypad and encode it.
;PORTS    :  SDK-86 board Port P1A - FFF9H as output, P1B - FFFBH as input
;ROUTINES : Calls KEYBRD to scan and decode 16-switch keypad
;REGISTERS: Uses DS, SS, SP, AX, DX

DATA_HERE       SEGMENT WORD    PUBLIC
TABLE   DB   77H,  7BH,  7DH,  7EH, 0B7H, 0BBH, 0BDH, 0BEH
;             0     1     2     3     4     5     6     7
        DB  0D7H, 0DBH, 0DDH, 0DEH, 0E7H, 0EBH, 0EDH, 0EEH
;             8     9     A     B     C     D     E     F
DATA_HERE       ENDS
STACK_HERE      SEGMENT
                        DW      30      DUP(0)  ; set up stack of 30 words
        TOP_STACK       LABEL   WORD            ; pointer to top of stack
STACK_HERE      ENDS

CODE_HERE       SEGMENT WORD    PUBLIC
                ASSUME  CS:CODE_HERE, DS:DATA_HERE, SS:STACK_HERE
;initialize segment registers
START:          MOV     AX, STACK_HERE
                MOV     SS, AX
                MOV     SP, OFFSET TOP_STACK
                MOV     AX, DATA_HERE
                MOV     DS, AX
; initialize ports, load DX with port control address
                MOV     DX, 0FFFFH
;mode set word, port A as output, mode 0, port B & C input ports, mode 0
                MOV     AL, 10001011B           ; code 8BH
                OUT     DX, AL                  ; Send control word
                CALL    KEYBRD
                NOP
                NOP
;program will continue here with other tasks


;PROCEDURE KEYBRD
;ABSTRACT :  procedure gets a code from a 16-switch keypad and decodes
;            it. It returns the code for the keypress in AL and AH=00. If there
;            is an error in the keypress then it returns AH=01.
;PORTS    : Uses SDK-86 ports P1A - FFF9H as output and P1B - FFFBH as input
;INPUTS   : Keypress from port
;OUTPUTS  : Keypress code or error message in AX
;ROUTINES : None used
;REGISTERS: Destroys AX


KEYBRD  PROC    NEAR
                PUSHF                   ; save registers used
                PUSH    BX
                PUSH    CX
```

FIGURE 9-21 Assembly language instructions for keyboard detect, debounce, and encode procedure.

```
                PUSH    DX
                MOV     AL, 00              ; send 0's to all rows
                MOV     DX, 0FFF9H          ; load output address
                OUT     DX, AL              ; send 0's
;Read columns
                MOV     DX, 0FFFBH          ; load i/p port address
WAIT_OPEN:      IN      AL, DX
                AND     AL, 0FH             ; mask row bits
                CMP     AL, 0FH             ; wait until no keys pressed
                JNE     WAIT_OPEN
;Read columns for keypress
WAIT_PRESS:     IN      AL, DX              ; read columns
                AND     AL, 0FH             ; mask row bits
                CMP     AL, 0FH             ; see if keypressed
                JE      WAIT_PRESS
;Debounce keypress
                MOV     CX, 16EAH           ; delay of 20 ms
DELAY:          LOOP    DELAY
                IN      AL, DX              ; read columns
                AND     AL, 0FH
                CMP     AL, 0FH             ; see if key still pressed
                JE      WAIT_PRESS
;Initialize a row mask with bit 0 low
                MOV     AL, 0FEH
                MOV     CL, AL              ; save mask
NEXT_ROW:       MOV     DX, 0FFF9H          ; put a low on one row
                OUT     DX, AL
                MOV     DX, 0FFFBH          ; read columns & check for low
                IN      AL, DX
                AND     AL, 0FH             ; mask out row code
                CMP     AL, 0FH             ; check for low in a column
                JNE     ENCODE              ; found column, now encode it
                ROL     CL, 01              ; rotate mask
                MOV     AL, CL              ; move mask
                JMP     NEXT_ROW            ; look at next row
;Encode the row/column information
ENCODE:         MOV     BX, 000FH           ; set up BX as a counter
                IN      AL, DX              ; read row and column from port
TRY_NEXT:       CMP     AL, TABLE[BX]       ; compare row/col code with table
                JZ      DONE
                DEC     BX                  ; point at next table entry
                JNS     TRY_NEXT
                MOV     AH, 01              ; pass an error code in AH
                JMP     EXIT
DONE:           MOV     AL, BL              ; hex code for key in AL
                MOV     AH, 00              ; valid-code key in AH
EXIT:           NOP
                POP     DX
                POP     CX
                POP     BX
                POPF
                RET
KEYBRD          ENDP
CODE_HERE       ENDS
                END     START
```

FIGURE 9-21 (*continued*)

we will show you the other major code conversion technique which we call *add and point.*

After the row which produces a low on one of the columns is found, execution jumps to the label ENCODE. The IN AL, DX instruction here reads the row and column codes in from the input port. This 8-bit code read in represents the key pressed. All that has to be done now is to convert this 8-bit code to the simple hex code for the key pressed. For example if you press the D key, you want to exit from the procedure with 0DH in AL.

The conversion is done with the lookup table declared with DBs at the top of Figure 9-21. This table contains the 8-bit keypressed codes for each of the 16 keys. Note that the codes are put in the table in order for the hex code they represent. The principle of the conversion technique we use here is to compare the row and column code read in with each of the values in the table until a match is found. We use a counter to keep track of how far down the table we have to go to find a match for a particular input code. When a match is found, the counter will contain the hex code for the key pressed.

In the program in Figure 9-21 we use the BX register as the counter and as a pointer to one of the codes in the table. To start we load a count of 000FH in BX with the MOV BX, 000FH instruction. The CMP AL, TABLE[BX] after this compares the code at offset [BX] in the table with the row and column code in AL. BX contains 000FH and the code in the table at this offset is the row and column code for the F key. If we get a match on this first compare, we know the F key was pressed, and BL contains the hex code for this key. The hex code in BL is copied to AL to pass it back to the calling program. AH is loaded with 00H to tell the calling program this was a valid keypress, and a return made to the calling program.

If we don't get a match on the first compare, we decrement BX to point to the code for the E key in the table and do another compare. If a match occurs this time, the E key was the key pressed, and the hex code for that key, 0EH, is in BL. If we don't get a match on this compare, we cycle through the loop until we get a match or until the row and column code for the pressed key has been compared with all of the values in the table. As long as the value in BX is zero or above after the DEC BX instruction, the Jump if Not Sign instruction, JNS TRY_NEXT, will cause execution to go back to the compare instruction. If no match is found in the table, BX will decrement from 0 to FFFFH. Since the sign bit is a copy of the MSB of the result after the DEC instruction, the sign bit will then be set. Execution will fall through to an instruction which loads an error code of 01H in AH. We then return to the calling program. The calling program will check AH on return to determine if the contents of AL represent the code for a valid keypress.

ERROR TRAPPING

The concept of detecting some error condition such as "no match found" is called *error trapping*. Error trapping is a very important part of real programs. Even in this simple program, think what might happen with no error trap if two keys in the same row were pressed at exactly the same time. A column code with two lows in it would be produced. This would not match any of the row and column codes in the table. After all of the values in the table were checked, BX would be decremented to FFFFH and AL would then be compared with a value off in memory at offset FFFFH. The cycle would continue until, by chance, the value in a memory location matched the row and column code in AL. The contents of BL at that point would be passed back to the calling routine. The chances are 1 in 256 that this would be the correct value. Since these are not very good odds, it is advisable to put error traps in your programs wherever there is a chance for it to go off to "never-never land" in this way. The error/no-error code can be passed back to the calling program in a register as shown, in a dedicated memory location, or on the stack.

Keyboard Interfacing With Hardware

The previous section described how you can connect a keyboard matrix to a couple of microprocessor ports, and perform the three interfacing tasks with program instructions. For systems where the CPU is too busy to be bothered doing these tasks in software, an external device is used to do them. One example of an MOS device which can do this is the General Instruments AY-5-2376, which can be connected to the rows and columns of a keyboard switch matrix. The AY-5-2376 independently detects a keypress by cycling a low down through the rows and checking the columns just as we did in software. When it finds a key pressed, it waits a debounce time. If the key is still pressed after the debounce time, the AY-5-2376 produces the 8-bit code for the pressed key and sends it out to, for example, a microcomputer port on eight parallel lines. To let the microcomputer know that a valid ASCII code is on the data lines, the AY-5-2376 outputs a strobe pulse. The microcomputer can detect this strobe pulse and read in the ASCII code on a polled basis as we showed in Figure 4-14, or it can detect the strobe pulse on an interrupt basis as we showed in Figure 8-9. With the interrupt method the microcomputer doesn't have to pay any attention to the keyboard until it receives an interrupt signal, so this method uses very little of the microcomputer's time.

The AY-5-2376 has a feature called *two-key rollover.* This means that if two keys are pressed at nearly the same time, each key will be detected, debounced, and converted to ASCII. The ASCII code for the first key and a strobe signal for it will be sent out, then the ASCII code for the second key and a strobe signal for it will be sent out. Compare this with two-key lockout which we described previously in the software method of keyboard interfacing.

CONVERTING ONE KEYBOARD CODE TO ANOTHER USING XLAT

Suppose that you are building up a simple microcomputer to control the heating, watering, lighting, and ventilation of your greenhouse. As part of the hardware, you buy a high-quality, fully encoded keyboard at the local electronics surplus store for a few dollars. When you get the keyboard home you find that it works perfectly, but that it outputs EBCDIC codes instead of the ASCII codes that you want. Here's how you use the 8086 XLAT instruction to easily solve this problem.

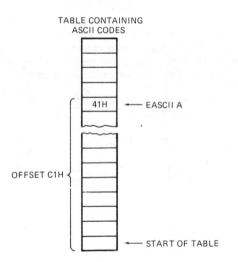

First look at Table 1-2 which shows the ASCII and EBCDIC codes. The job you have to do here is convert each input EBCDIC input code to the corresponding ASCII code. One way to do this is the compare technique described previously for the hex-keyboard example. For that method you first put the EBCDIC codes in a table in memory in the order shown in Table 1-2, and set up a register as a counter and pointer to the end of the table. Then enter a loop which compares the EBCDIC character in AL with each of the EBCDIC codes in the table until a match is found. The counter is decremented after each compare so that when a match is found, the count register contains the desired ASCII code. This compare technique works well, but for this conversion it will, on the average, have to do 64 compares before a match is found. The compare technique then is often too time-consuming for long tables. The XLAT method is much faster.

The first step in the XLAT method is to make up in memory a table which contains all of the ASCII codes. You can use the DB assembler directive to do this. Since EBCDIC code is an 8-bit code, the table will require 256 memory locations. The trick here is to put each ASCII code in the table at a displacement equal to the value of the EBCDIC character from the start of the table. For example, the EBCDIC code for uppercase A is C1H, so at offset C1H in the table you put the ASCII code for uppercase A, 41H, as shown in Figure 9-22.

To do the actual conversion, you simply load the BX register with the offset of the start of the table, load the EBCDIC character to be converted in the AL register,

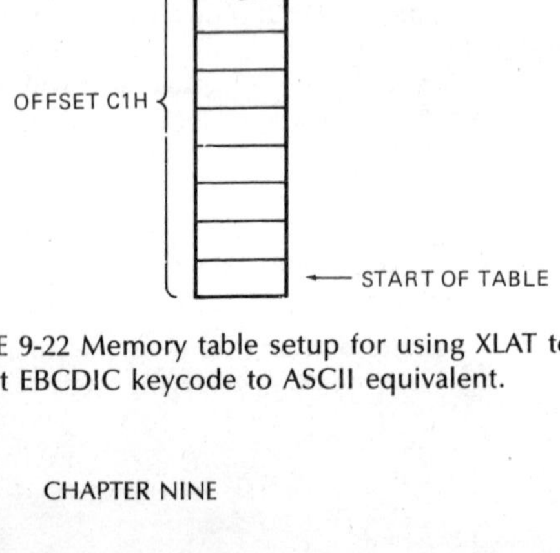

FIGURE 9-22 Memory table setup for using XLAT to convert EBCDIC keycode to ASCII equivalent.

and do the XLAT instruction. When the 8086 executes the XLAT instruction, it internally adds the EBCDIC value in AL to the starting offset of the table in BX. Because of the way the table is made up, the result of this addition will be a pointer to the desired ASCII value in the table. The 8086 uses this pointer to copy the desired ASCII character from the table to AL.

The advantage of this technique is that, no matter where in the table the desired ASCII value is, the conversion only requires execution of two loads and one XLAT instruction. The question may occur to you at this point, "If this method is so fast, why didn't we use it for the hex keypad conversion described earlier?" The answer is that since the row and column code from the hex keypad is an 8-bit code, the lookup table for the XLAT method would require 256 memory locations. Of these 256 memory locations, only 16 would actually be used. This would be a waste of memory, so the compare method is a better choice. It is important for you to become familiar with both code conversion methods, so that you can use the one that best fits a particular application.

DEDICATED MICROPROCESSOR KEYBOARD ENCODERS

Most computers and computer terminals now use detached keyboards with built-in encoders. Instead of using a hardware encoder device such as the AY-5-2376 these keyboards use a dedicated microprocessor. Figure 9-23 shows the encoder circuitry for the IBM PC capacitive-switch matrix keyboard. The 8048 microprocessor used here contains an 8-bit CPU, a ROM, some RAM, three ports, and a programmable timer/counter. A program stored in the on-chip ROM performs the three keyboard tasks and sends the code for a pressed key out to the computer. To cut down the number of connecting wires, the key code is sent out in serial form rather than in parallel form. Some keyboards send data to the computer in serial form using a beam of infrared light instead of a wire.

Note in Figure 9-23 the sense amplifier to detect the change in capacitance produced when a key is pressed. Also note that the 8048 uses a tuned *LC* circuit rather than a more expensive crystal to determine its operating clock frequency.

One of the major advantages of using a dedicated microprocessor to do the three keyboard tasks is programmability. Special function keys on the keyboard can be programmed to send out any code desired for a particular application. By simply plugging in an 8048 with a different lookup table in ROM, the keyboard can be changed from outputting ASCII characters to outputting some other character set.

The IBM keyboard, incidentally, does not send out ASCII codes, but instead sends out a hex "scan" code for each key when it is pressed and a different scan code when that key is released. This double-code approach gives the system software maximum flexibility because a program command can be implemented either when a key is pressed or when it is released.

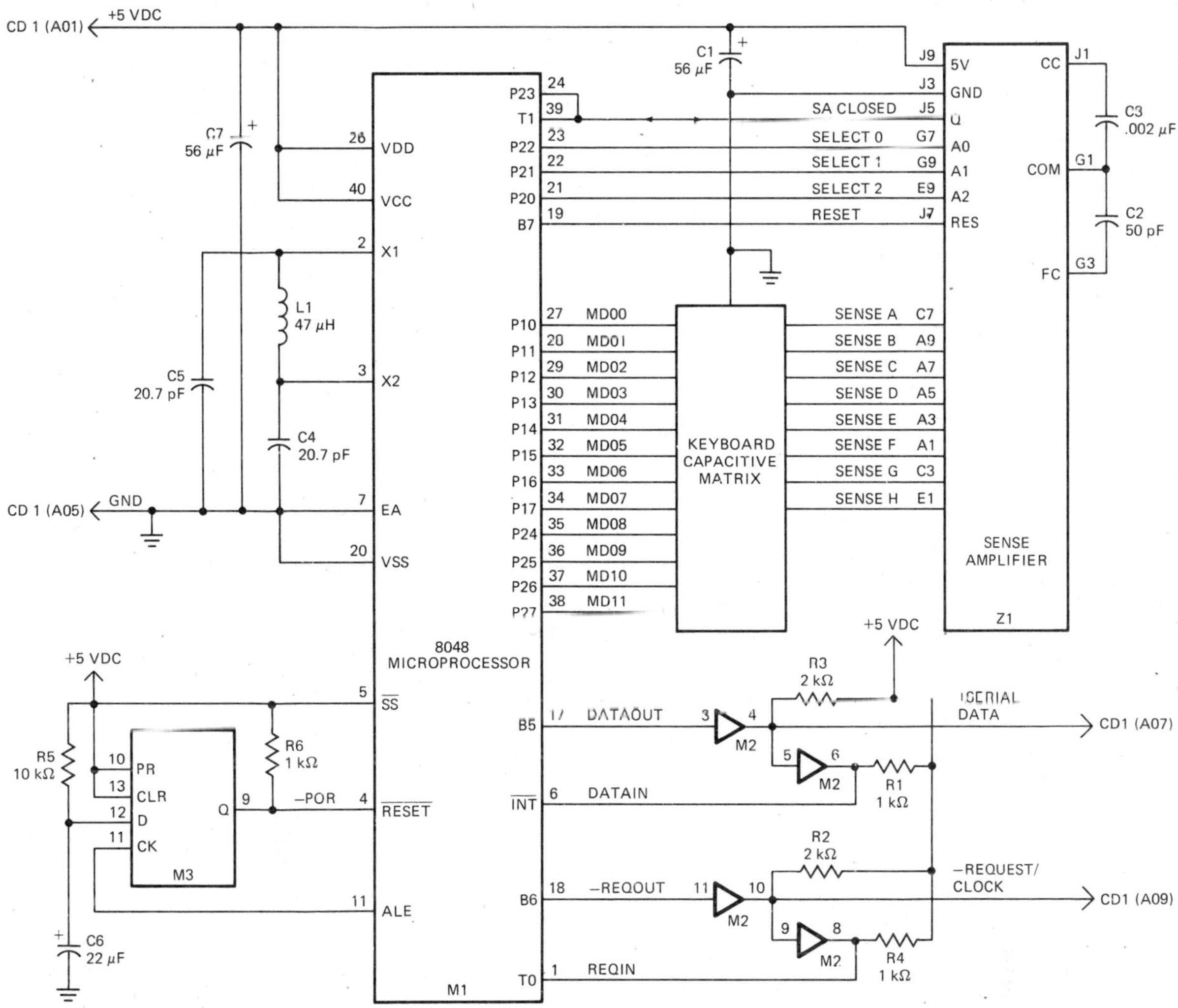

FIGURE 9-23 IBM PC keyboard scan circuitry using a dedicated microprocessor. *(IBM Corporation)*

INTERFACING TO ALPHANUMERIC DISPLAYS

Many microprocessor-controlled instruments and machines need to display letters of the alphabet and numbers to give directions or data values to users. In systems where a large amount of data needs to be displayed, a CRT is usually used to display the data. In a later chapter we show you how to interface a microcomputer to a CRT. In systems where only a small amount of data needs to be displayed, simple digit-type displays are often utilized. There are several technologies used to make these digit-oriented displays, but we only have space here to discuss the two major types. These are *light-emitting diodes* (LEDs) and *liquid-crystal displays* (LCDs). LCD displays use very low power, so they are often used in portable, battery-powered instruments. LCDs however, do not emit their own light, they simply change the reflection of available light. Therefore, for an instrument that is to be used in dim light conditions you have to include a light source for the LCDs, or use LEDs which emit their own light. Starting with LEDs, the following sections show you how to interface these two types of displays to microcomputers.

Interfacing LED Displays to Microcomputers

Alphanumeric LED displays are available in three common formats. For displaying only numbers and hexadecimal letters, simple seven-segment displays such as that shown in Figure 1-6*a* are used.

To display numbers and the entire alphabet, 18-segment displays such as that shown in Figure 9-24*a*, or 5 by 7 dot-matrix displays such as that shown in Figure 9-24*b* can be used. The seven-segment type is the

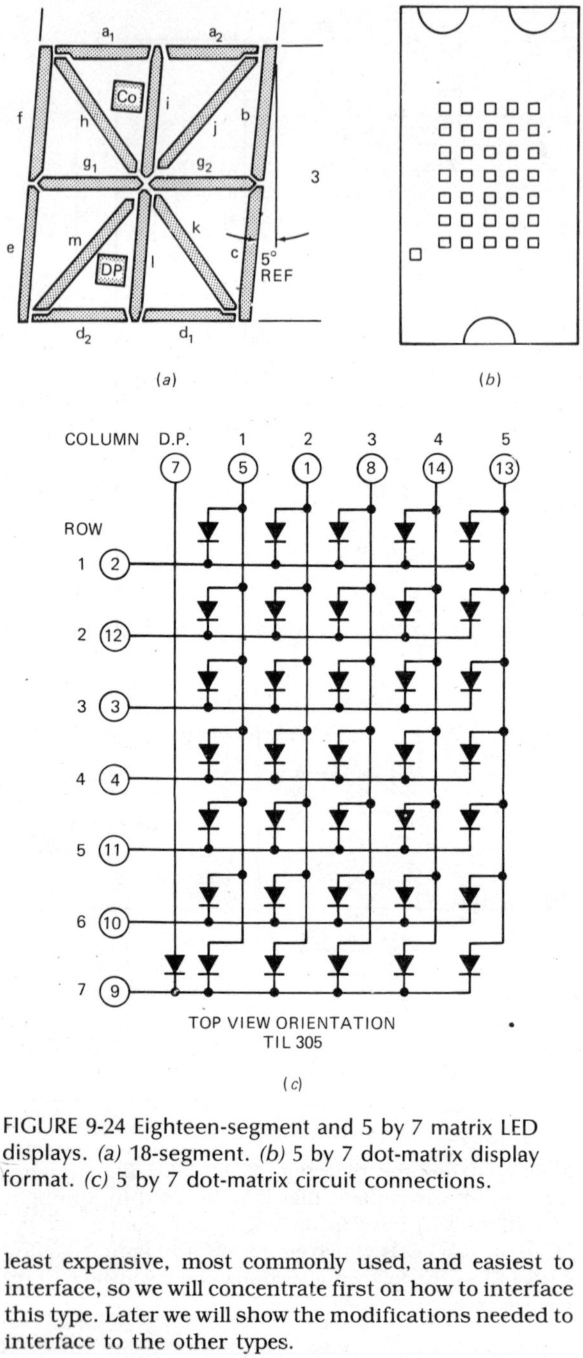

FIGURE 9-24 Eighteen-segment and 5 by 7 matrix LED displays. *(a)* 18-segment. *(b)* 5 by 7 dot-matrix display format. *(c)* 5 by 7 dot-matrix circuit connections.

least expensive, most commonly used, and easiest to interface, so we will concentrate first on how to interface this type. Later we will show the modifications needed to interface to the other types.

STATIC AND MULTIPLEXED DISPLAY CIRCUITS

Figure 9-25 shows a circuit you might use to drive a single, seven-segment, common-anode display. For a common-anode display, a low is applied to a segment to turn it on. When a BCD code is sent to the inputs of the 7447, it outputs lows on the segments required to display the number represented by the BCD code. This circuit connection is referred to as a *static display* because current is being passed through the display at all times. Note that current-limiting resistors are required in series with each segment. Here's how you calculate the value of these resistors.

Each segment requires a current of between 5 and 30 mA to light. Let's assume you want a current of 20 mA. The voltage drop across the LED when it is lit is about 1.5 V. The output low voltage for the 7447 is a maximum of 0.4 V at 40 mA, so assume that it is about 0.2 V at 20 mA. Subtracting these two voltage drops from the supply voltage of 5 V leaves 3.3 V across the current-limiting resistor. Dividing 3.3 V by 20 mA gives a value of 168 Ω for the current-limiting resistor. The voltage drops across the LED and the output of the 7447 are not exactly predictable and the exact current through the LED is not critical as long as we don't exceed its maximum rating. Therefore, a standard value of 150 Ω is reasonable.

The circuit in Figure 9-25 works well for driving just one or two LED digits. However, there are problems if you want to drive, for example, eight digits. The first problem is power consumption. For worst-case calculations, assume that all eight digits are displaying the digit 8 so all seven segments are lit. Seven segments times 20 mA per segment gives a current of 140 mA per digit. Multiplying this by 8 digits gives a total current of 1120 mA or 1.12 A for the the eight digits! A second problem of the static approach is that each display digit requires a separate 7447 decoder, each of which uses, perhaps, another 13 mA. The current required by the decoders and the LED displays might be several times the current required by the rest of the circuitry in the instrument.

To solve the problems of the static display approach, we use a *multiplex* method. A circuit example is the easiest way to explain to you how this multiplexing works. Figure 9-26 shows a circuit you can add to a couple of microcomputer ports to drive some common-anode LED displays in a multiplexed manner. Note that the circuit

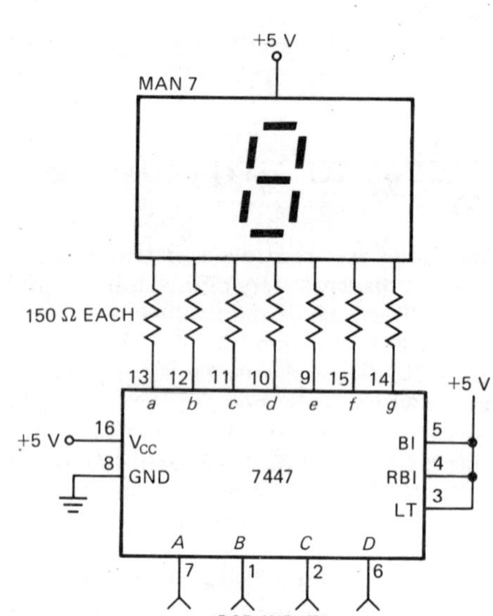

FIGURE 9-25 Circuit for driving single seven-segment LED display with 7447.

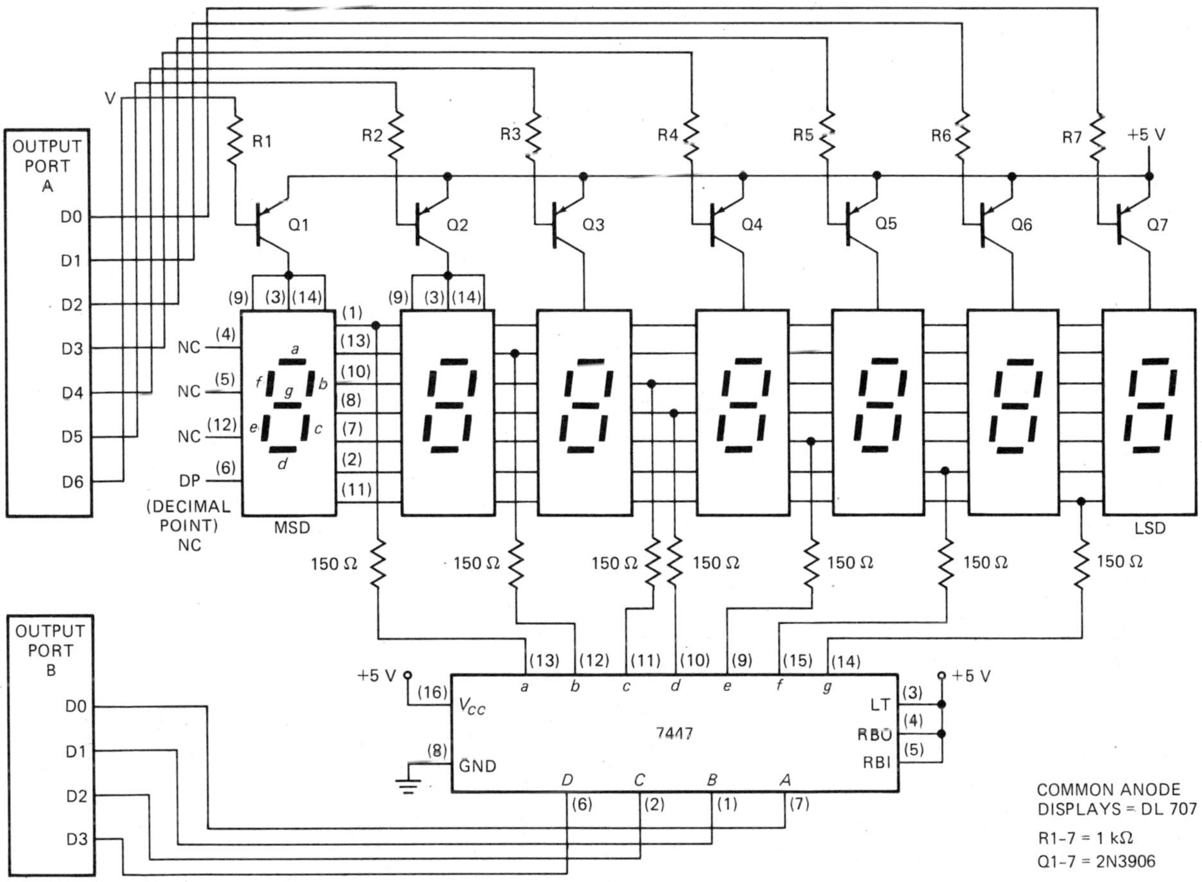

FIGURE 9-26 Circuit for multiplexing seven-segment displays with a microcomputer.

has only one 7447 and that the segment outputs of the 7447 are bused to the segment inputs of all of the digits. The question that may occur to you on first seeing this is, "Aren't all of the digits going to display the same number?" The answer is that they would if all of the digits were turned on at the same time. The trick of multiplexing displays is that the segment information is sent out to all of the digits on the common bus, but only one display digit is turned on at a time. The PNP transistor in series with the common-anode of each digit acts as an on and off switch for that digit. Here's how the multiplexing process works.

The BCD code for digit 1 is first output from port B to the 7447. The 7447 outputs the corresponding seven-segment code on the segment bus lines. The transistor connected to digit 1 is then turned on by outputting a low to that bit of port A. (Remember, a low turns on a PNP transistor.) All of the rest of the bits of port A should be high to make sure no other digits are turned on. After 1 or 2 ms, digit 1 is turned off by outputting all highs to port A. The BCD code for digit 2 is then output to the 7447 on port B, and a word to turn on digit 2 is output on port A. After 1 or 2 ms, digit 2 is turned off and the process repeated for digit 3. The process is continued until all of the digits have had a turn. Then digit 1 and the following digits are lit again in turn. We leave it to you as an exercise at the end of the chapter to write a procedure which is called on an interrupt basis every 2 ms to keep these displays refreshed with some values stored in a table.

With 8 digits and 2 ms per digit you get back to digit 1 every 16 ms or about 60 times a second. This refresh rate is fast enough that, to your eye, the digits will each appear to be lit all of the time. Refresh rates of 40 to 200 times a second are acceptable.

The immediately obvious advantages of multiplexing the displays are that only one 7447 is required, and only one digit is lit at a time. We usually increase the current per segment to between 40 and 60 mA for multiplexed displays so that they will appear as bright as they would if not multiplexed. Even with this increased segment current, multiplexing gives a large saving in power and parts.

NOTE: If you are calculating the current-limiting resistors for multiplexed displays, make sure to check the data sheet for the maximum current rating for the displays you are using.

A disadvantage of the software multiplexing approach shown here is that it puts an additional burden on the CPU. Also, if the CPU gets involved in doing some lengthy task which cannot be interrupted to refresh the display, only one digit of the display will be left lit. An alternative approach to interfacing multiplexed displays to a microcomputer is to use a *dedicated display controller* such as the Intel 8279, which independently keeps displays refreshed and scans a matrix keyboard. In the next section we show you how an 8279 is connected in a circuit, discuss how the 8279 operates, and show you how to initialize an 8279.

Display and Keyboard Interfacing with the 8279

8279 CIRCUIT CONNECTIONS AND OPERATION OVERVIEW

Sheets 7 and 8 of the SDK-86 schematics in Figure 7-6 show the circuit connections for the keypad and the multiplexed seven-segment displays. First let's look at the display circuitry on sheet 8. The displays here are common-anode and each digit has a PNP transistor switch between its anode and the +5-V supply. A logic low is required to turn on one of these switches. Note the 22-μF capacitor between +5 V and ground at the top of the schematic. This is necessary to filter out transients caused by switching the large currents to the LEDs off and on. The segments of each digit are all connected on a common bus. Since these are common-anode displays, a low is needed to turn on a segment. Now let's look at sheet 7 in Figure 7-6 to see how these displays are driven.

The drive for the digit-switch transistors comes from a 7445 BCD to decimal decoder. This device is also known as a one-of-ten-low decoder. When a 4-bit BCD code is applied to the inputs of this device, the output corresponding to that BCD number will go low. For example, when the 8279 outputs 0100 or BCD 4, the 7445 output labeled O4 will go low. In the mode used for this circuit the 8279 outputs a continuous count sequence from 0000 to 1111 over and over. This causes a low to be stepped from output to output of the 7445 in ring counter fashion, turning on each LED digit in turn. Only one output of the 7445 will ever be low at a time, so only one LED digit will be turned on at a time.

The segment bus lines for the displays are connected to the A3–A0 and B3–B0 outputs of the 8279 through some high-current buffers in the ULN2003A. Note that the 22-Ω current-limiting resistors in series with the segment lines are much smaller in value than those we calculated for the static circuit in Figure 9-25. There are two reasons for this. First, there is an additional few tenths of a volt drop across the transistor switch on each anode. Second, when multiplexing displays we pass a higher current through the displays so that they appear as bright as they would if not multiplexed. Here's how the 8279 keeps these displays refreshed.

When you want to display some letters or numbers you write the seven-segment codes for the letters or numbers that you want displayed to a 16-byte RAM inside the 8279. The 8279 then automatically cycles through the process we described previously for sending these codes in sequence to the displays. Figure 9-27 shows the operation in timing diagram form. The 8279 first outputs the binary number for the first digit to the 7445 on the SL0–SL3 lines (Figure 7-6, sheet 7) to turn on the first one of the digit-driver transistors. The lines S0 and S1 in Figure 9-27 represent the SL0 and SL1 lines from the 8279. The 8279 then outputs the seven-segment code for the first digit on the A3–A0 and B3–B0 lines. This will light the first digit with the desired pattern. After 490 μs the 8279 outputs on the A and B lines a code which turns off all of the segments. For the circuit in Figure 7-6, sheet 7, this blanking code will be all zeros (00H). The display is blanked here to prevent "ghosting" of information from one digit to the next when the digit strobe is switched to the next digit. While the displays are blanked, the 8279 sends out the BCD code for the next digit to the 7445 to enable the digit-2 driver transistor. It then sends out the seven-segment code for digit 2 on the A and B lines. This then lights the desired pattern on digit 2. After 490 μs the 8279 blanks the display again and goes on to digit 3. The 8279 steps through all of the digits and then returns to digit 1 and repeats the cycle. Since each digit requires about 640 μs, the 8279 gets back to digit 1 after about 5.1 ms for an 8-digit display and back to digit 1 after about 10.3 ms for a 16-digit display. The time it takes to get back to a digit again is referred to as the *scan time*.

The point is that once you load the seven-segment codes into the internal RAM in the 8279, it automatically keeps the displays refreshed without you having to do anything else in the program. As we will show later, the 8279 can be connected and initialized to refresh a wide variety of displays.

The 8279 can also automatically perform the three tasks for interfacing to a matrix keyboard. Remember from previous discussions that the three tasks involve putting a low on a row of the keyboard matrix and checking the columns of the matrix. If any keys are pressed in that row, a low will be present on the column which contains the key, because pressing a key shorts a row to a column. If no low is found on the columns the low is stepped to the next row and the columns checked again. If a low is found on a column, then, after a debounce time, the column is checked again. If the keypress was valid, a compact code representing the key is constructed. Take a look at the circuit on sheet 7 of Figure 7-6 to see how an 8279 can be connected to do this.

When connected as shown in Figure 7-6, sheet 7, the 74LS156 functions as a one-of-eight-low decoder. In other words, if you apply 011, the binary code for 3, to its inputs, the 74LS156 will output a low on its 2Y3 output. Now remember from the discussion of 8279 display refreshing, that the 8279 is outputting a continuous counting sequence from 0000 to 1111 on its SL0–SL3 lines. This count sequence applied to the inputs of the 74LS156 will cause it to step a low along its outputs. The 74LS156 then puts a low on one row of the keyboard at a time.

The column lines of the keyboard are connected to the

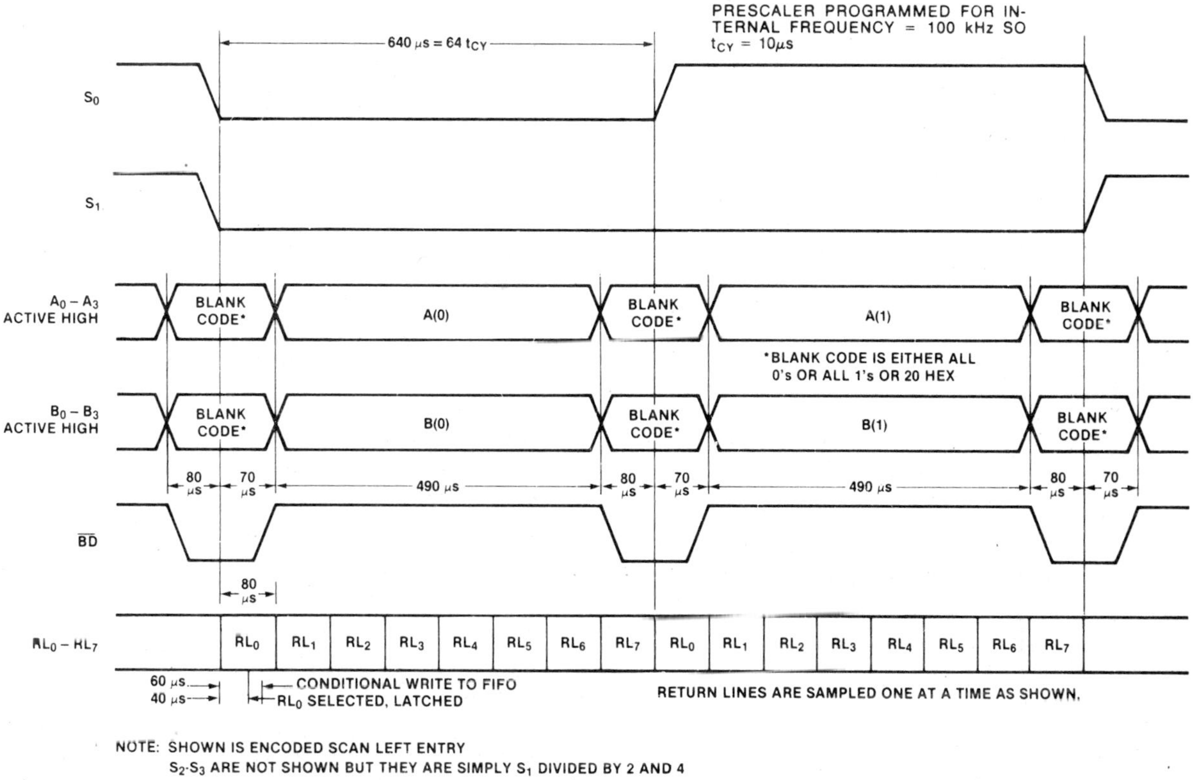

FIGURE 9-27 8279 display refresh timing and keyboard scan timing. *(Intel Corporation)*

return lines, RL0–RL7 of the 8279. As a low is put on each row by the scan-line count and the 74LS156, the 8279 checks these return lines one at a time to see if any of them are low. The bottom line of the timing waveforms in Figure 9-27 shows when the return lines are checked. If the 8279 finds any of the return lines low, indicating a keypress, it waits a debounce time of about 10.3 ms and checks again. If the keypress is still present, the 8279 produces an 8-bit code which represents the key pressed. Figure 9-28 shows the format for the code produced. Three bits of this code represent the number of the row in which it found the pressed key, and another 3 bits represent the column of the pressed key. For interfacing to full typewriter keyboards the shift and control keys are connected to pins 36 and 37 respectively of the 8279. The upper 2 bits of the code produced represent the status of these two keys.

After the 8279 produces the 8-bit code for the pressed key it stores the word in an internal 8-byte *FIFO* RAM. The term FIFO stands for first-in–first-out, which means that when you start reading codes from the FIFO, the first code you read out will be that for the first key pressed. The FIFO can store the codes for up to eight pressed keys before overflowing.

When the 8279 finds a valid keypress, it does two things to let you know about it. It asserts its interrupt request pin, IRQ, high, and it increments a FIFO count in an internal status register. You can connect the IRQ output to an interrupt input and detect when the FIFO has a character for you on an interrupt basis, or you can simply check the count in the status word to determine when the FIFO has a code ready to be read. The point here is that, once the 8279 is initialized, you don't need to pay any attention to it until you want to send some new characters to be displayed, or until it notifies you that it has a valid keypressed code for you in its FIFO. Now that you have an overview of how the 8279 functions, we will show you how to initialize an 8279 to do all of these wondrous things and more.

INITIALIZING AND COMMUNICATING WITH AN 8279

As we have shown before, the first step in initializing a programmable device is to determine the system base address for the device, the internal addresses, and the system addresses for the internal parts. As an example

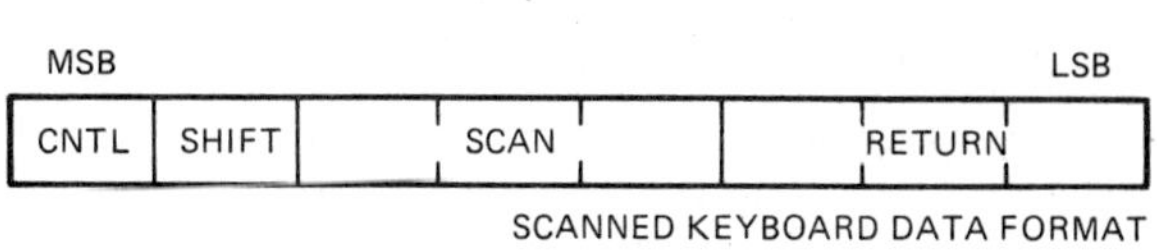

FIGURE 9-28 Format for data word produced by 8279 keyboard encoding.

Write Display RAM

Code:

1	0	0	AI	A	A	A	A

The CPU sets up the 8279 for a write to the Display RAM by first writing this command. After writing the command with $A_0 = 1$, all subsequent writes with $A_0 = 0$ will be to the Display RAM. The addressing and Auto-Increment functions are identical to those for the Read Display RAM. However, this command does not affect the source of subsequent Data Reads; the CPU will read from whichever RAM (Display or FIFO/Sensor) which was last specified. If, indeed, the Display RAM was last specified, the Write Display RAM will, nevertheless, change the next Read location.

Display Write Inhibit/Blanking

Code:

				A	B	A	B
1	0	1	X	IW	IW	BL	BL

The IW Bits can be used to mask nibble A and nibble B in applications requiring separate 4-bit display ports. By setting the IW flag (IW = 1) for one of the ports, the port becomes marked so that entries to the Display RAM from the CPU do not affect that port. Thus, if each nibble is input to a BCD decoder, the CPU may write a digit to the Display RAM without affecting the other digit being displayed. It is important to note that bit B_0 corresponds to bit D_0 on the CPU bus, and that bit A_3 corresponds to bit D_7.

If the user wishes to blank the display, the BL flags are available for each nibble. The last Clear command issued determines the code to be used as a "blank." This code defaults to all zeros after a reset. Note that both BL flags must be set to blank a display formatted with a single 8-bit port.

Clear

Code:

1	1	0	C_D	C_D	C_D	C_F	C_A

The C_D bits are available in this command to clear all rows of the Display RAM to a selectable blanking code as follows:

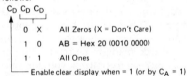

C_D	C_D C_D	
	0 X	All Zeros (X = Don't Care)
	1 0	AB = Hex 20 (0010 0000)
	1 1	All Ones

Enable clear display when = 1 (or by C_A = 1)

During the time the Display RAM is being cleared (~160 μs), it may not be written to. The most significant bit of the FIFO status word is set during this time. When the Display RAM becomes available again, it automatically resets.

If the C_F bit is asserted ($C_F = 1$), the FIFO status is cleared and the interrupt output line is reset. Also, the Sensor RAM pointer is set to row 0.

C_A, the Clear All bit, has the combined effect of C_D and C_F; it uses the C_D clearing code on the Display RAM and also clears FIFO status. Furthermore, it resynchronizes the internal timing chain.

End Interrupt/Error Mode Set

Code:

1	1	1	E	X	X	X	X

X = Don't care.

For the sensor matrix modes this command lowers the IRQ line and enables further writing into RAM. (The IRQ line would have been raised upon the detection of a change in a sensor value. This would have also inhibited further writing into the RAM until reset).

For the N-key rollover mode — if the E bit is programmed to "1" the chip will operate in the special Error mode. (For further details, see Interface Considerations Section.)

Keyboard/Display Mode Set

Code:

MSB							LSB
0	0	0	D	D	K	K	K

Where DD is the Display Mode and KKK is the Keyboard Mode.

DD

0	0	8 8-bit character display — Left entry
0	1	16 8-bit character display — Left entry*
1	0	8 8-bit character display — Right entry
1	1	16 8-bit character display — Right entry

For description of right and left entry, see Interface Considerations. Note that when decoded scan is set in keyboard mode, the display is reduced to 4 characters independent of display mode set.

KKK

0	0	0	Encoded Scan Keyboard — 2 Key Lockout*
0	0	1	Decoded Scan Keyboard — 2-Key Lockout
0	1	0	Encoded Scan Keyboard — N-Key Rollover
0	1	1	Decoded Scan Keyboard — N-Key Rollover
1	0	0	Encoded Scan Sensor Matrix
1	0	1	Decoded Scan Sensor Matrix
1	1	0	Strobed Input, Encoded Display Scan
1	1	1	Strobed Input, Decoded Display Scan

Program Clock

Code:

0	0	1	P	P	P	P	P

All timing and multiplexing signals for the 8279 are generated by an internal prescaler. This prescaler divides the external clock (pin 3) by a programmable integer. Bits PPPPP determine the value of this integer which ranges from 2 to 31. Choosing a divisor that yields 100 kHz will give the specified scan and debounce times. For instance, if Pin 3 of the 8279 is being clocked by a 2 MHz signal, PPPPP should be set to 10100 to divide the clock by 20 to yield the proper 100 kHz operating frequency.

Read FIFO/Sensor RAM

Code:

0	1	0	AI	X	A	A	A

X = Don't Care

The CPU sets up the 8279 for a read of the FIFO/Sensor RAM by first writing this command. In the Scan Keyboard Mode, the Auto-Increment flag (AI) and the RAM address bits (AAA) are irrelevant. The 8279 will automatically drive the data bus for each subsequent read ($A_0 = 0$) in the same sequence in which the data first entered the FIFO. All subsequent reads will be from the FIFO until another command is issued.

In the Sensor Matrix Mode, the RAM address bits AAA select one of the 8 rows of the Sensor RAM. If the AI flag is set (AI = 1), each successive read will be from the subsequent row of the sensor RAM.

Read Display RAM

Code:

0	1	1	AI	A	A	A	A

The CPU sets up the 8279 for a read of the Display RAM by first writing this command. The address bits AAAA select one of the 16 rows of the Display RAM. If the AI flag is set (AI = 1), this row address will be incremented after each following read *or write* to the Display RAM. Since the same counter is used for both reading and writing, this command sets the next read *or write* address and the sense of the Auto-Increment mode for both operations.

FIGURE 9-29 8279 command word formats and bit descriptions. *(Intel Corporation)*

here we will use the 8279 on sheet 7 of Figure 7-6. Figure 7-15*b* shows that the system base address for this device is FFE8H. The 8279 has only two internal addresses which are selected by the logic level on its A0 input, pin 21. If the A0 input is low when the 8279 is selected, then the 8279 is enabled for reading data from it or writing data to it. A0 being high selects the internal control/status registers. For the circuit on sheet 7 of Figure 7-6, the A0 input is connected to system address line A1. Therefore, the data address for this 8279 is FFE8H and the control/status address is FFEAH.

After you have figured out the addresses for a device, the next step is to look at the format for the control word(s) you have to send to the device to make it operate in the mode you want. Figure 9-29 shows the format for the 8279 control words as they appear in the Intel data book. After you use up your 5-minute "freak-out" time we will help you decipher these.

A question that may occur to you when you see all of these control words is, "If the 8279 only has one control register address, how am I going to send it all of these different control words?" The answer to this is that all of the control words are sent to the same control register address, FFEAH for this example. The upper three bits of each control word tell the 8279 which control word is being sent. A pattern of 010 in the upper three bits of a control word, for example, identifies that control word as a "Read FIFO/Sensor RAM" control word.

The first control word you send to initialize the 8279 is the *keyboard/display mode set* word. Keep Figure 9-29 handy as we discuss this and the other control words. The bits labeled DD in the control word specify first of all whether you have 8 digits or 16 digits to refresh. If you have eight or fewer displays, make sure to initialize for 8 digits so the 8279 doesn't spend half of its time refreshing nonexistent displays. The DD bits in this control word also specify the order in which the characters in the internal 16-byte display RAM will be sent out to the digits. In the left entry mode, the seven-segment code in the first address of the internal display RAM will be sent to the leftmost digit of the display. If you want to display the letters AbCd on the 4 leftmost digits of an 8-digit display, then you put the seven-segment codes for these letters in the first four locations of the display RAM as shown in Figure 9-30*a*. Codes put in higher addresses in the display RAM will be displayed on following digits to the right. In the right entry mode, the first code sent to the display RAM is put in the lowest address. This character will be displayed on the rightmost digit of the display. If a second character is written to the display RAM it will be put in the second location in the RAM as shown in Figure 9-30*b*. On the display, however, the new character will be displayed on the rightmost digit, and the previous character will be shifted over to the second position from the right. This is the way that the displays of most calculators function as you enter numbers.

Now let's look at the KKK bits of the mode-set control word. The first choice you have to make here if you are using the 8279 with a keyboard is whether you want *encoded scan* or *decoded scan*. You know that for scanning a keyboard or turning on digit drivers, you need a pattern of stepping lows. In encoded mode the 8279 puts out a binary count sequence on its SL0–SL3 scan lines, and an external decoder such as the 7445 is used to produce the stepping lows. If you only have 4 digits to refresh, you can program the 8279 in decoded mode. In this mode the 8279 directly outputs stepping lows on the four scan lines. The second choice you have to make for this control word is whether you want *two-key lockout*, or *N-key rollover*. In the two-key mode, one key must be released before another keypress will be detected and processed. In the *N*-key rollover mode, if two keys are pressed at nearly the same time, both keypresses will be detected and debounced and their codes put in the FIFO in the order the keys were pressed.

In addition to being used to scan a keyboard, the 8279 can also be used to scan a matrix of switch sensors such as the metal strips and magnetic sensors you see on store windows and doors. In sensor matrix mode the 8279 scans all of the sensors and stores the condition of up to 64 switches in the FIFO RAM. If the condition of any of the switches changes, an IRQ signal is sent out to the processor. An interrupt service procedure can then sound an alarm and turn the dogs loose. The return lines of the 8279 can also function as a strobed input port in much the same way as the 8255A.

The SDK-86 initializes the 8279 for eight-character display, left entry, encoded scan, two-key lockout. See if you can determine the mode-set control word for these conditions. You should get 00000000.

The next control word you have to send the 8279 is

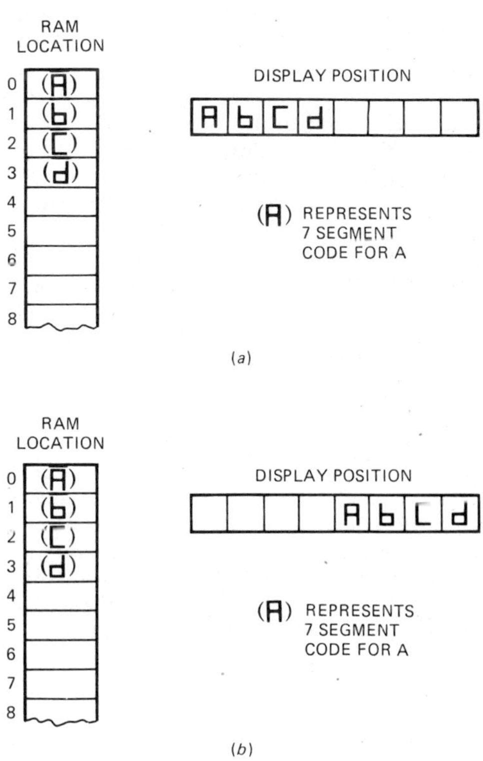

FIGURE 9-30 8279 RAM and display location relationships. *(a)* Left entry. *(b)* Right entry.

the *program-clock word*. The 8279 requires an internal clock frequency of about 100 kHz. A programmable divider in the 8279 allows you to apply some available frequency such as the 2.45-MHz PCLK signal to its clock input and divide this frequency down to the needed 100 kHz. The lower 5 bits of the program clock control word simply represent the binary number you want to divide the applied clock by. For example, if you want to divide the input clock frequency by 24, you send a control word with 001 in the upper 3 bits and 11000 in the lower 5 bits.

The final control word needed for basic initialization is the *clear* word. You need to send this word to tell the 8279 what code to send to the segments to turn them off while the 8279 is switching from one digit to the next. (Refer to Figure 9-27 and its discussion.) In addition to telling the 8279 what blanking character to use during refresh, this control word can be used to clear the display RAM and/or the FIFO at any time. For now we are only concerned with the first function. The lower two bits labeled C_D in the control word in Figure 9-29 specify the desired blanking code. The required code will depend on the hardware connections in a particular system. For the SDK-86 a high from the 8279 turns on a segment, so the required blanking code is all 0's. Therefore you can put 0's in the two C_D bits. The resultant control word is 11000000.

The three control words described so far take care of the basic initialization. However, before you can send codes to the internal display RAM, you have to send the 8279 a *write-display-RAM* control word. This word tells the 8279 that data later sent to the data address should be put in the display RAM, and it tells the 8279 where in the display RAM to put the data byte sent in. Refer to Figure 9-29 for the format of this word. The 8279 has an internal 4-bit pointer to the display RAM. You use the lower 4 bits of this control word to initialize the pointer to the location where you want to write a data byte in the RAM. If you want to write a data byte to the firstlocation in the display RAM, you put 0000 in these bits. If you put a 1 in the auto increment bit, labeled AI in the figure, the internal pointer will be automatically incremented to point to the next RAM location after each data byte is written. To start loading characters in the first location in the RAM and select auto increment, then, the control word is 10010000.

Figure 9-31 shows the sequence of instructions to send the control words we have developed here to the 8279 on the SDK-86 board. Also shown are some instructions to send a seven-segment code to the first location in the display RAM. Note that the control words are all sent to the control address, FFEAH, and the character going to the display RAM is sent to the data address, FFE8H. Also note from sheet 7 of Figure 7-6 that the D0 bit of the byte sent to the display RAM corresponds to segment output B0, and D7 of the byte sent to the display corresponds to segment output A3. This is important to know when you are making up a table of seven-segment codes to send to the 8279.

You now know how to initialize an 8279 and send characters to its display RAM. Two additional points we need to show you are how to read keypressed codes from the FIFO, and how to read the status word. In order to read a code from the FIFO you first have to send a *read FIFO/sensor RAM* control word to the 8279 control address. Figure 9-29 shows the format for this word. For a read of the FIFO RAM, the lower 5 bits of the control word are don't cares, so you can just make them zeros. You send the resultant control word, 01000000, to the control register address and then do a read from the data address. The bottom section of Figure 9-31 shows this.

Now, suppose that the processor receives an interrupt signal from the 8279 indicating that one or more valid keypresses have occurred. The question then comes up, "How do I know how many codes I should read from the FIFO?" The answer to this question is that you read the status register from the control register address before you read the FIFO. Figure 9-32 shows the format for this status word. The lowest 3 bits of the status word indicate the number of valid characters in the FIFO. You can load this number into memory location and count it down as you read in characters. Incidentally, if more than 8 characters have been entered in the FIFO, only the last 8 will be kept. The error-overrun bit, labeled O in the status word, will be set to tell you characters have been lost.

Characters can be read from the 8279 on a polled basis as well as on an interrupt basis. To do this you simply read and test the status word over and over again until bit 0 of the status word becomes a 1. The SDK-86 uses this method to tell when the FIFO holds a keypressed code.

SDK-86 DISPLAY DRIVER PROCEDURE

Figure 9-33 shows an 8086 assembly language procedure to display the contents of the CX register on SDK-86 LED displays. This procedure assumes the 8279 has already been initialized by the SDK-86 monitor program, or as shown in the first part of Figure 9-31. If AL is zero when this procedure is called, the contents of CX will be displayed on the data field LEDs. If AL is not zero then the contents of CX will be displayed on the address field LEDs. There are two main points for you to see in this procedure. The first is the sending of the write-display-RAM control word to the 8279 so we can write to the desired locations in the display RAM. Note that, for the data field, we write a control word of 90H which tells the 8279 to put the next data word sent into the first location in the display RAM. Since the 8279 is initialized for left entry, the first location should correspond to the leftmost display digit. However, if you look at sheet 8 of the SDK-86 schematics you will see that digit 1 (leftmost as far as the 8279 is concerned) is actually the rightmost on the board. This means that, for the SDK-86, the position of a seven-segment code in the display RAM corresponds to its position in the display starting from the right! All you have to do is send the seven-segment code for a number you want to display in a particular digit position to the corresponding location in the display RAM.

The next part of the display procedure to take a close look at is the instructions which convert the four hex

```
INITIALIZATION

MOV   DX, 0FFEAH          ; Point at 8279 control address
MOV   AL, 00000000B       ; Mode set word for left entry,
                          ; encoded scan, 2-key lockout
OUT   DX, AL              ; Send to 8279
MOV   AL, 00111000B       ; Clock word for divide by 24
OUT   DX, AL
MOV   AL, 11000000B       ; Clear display char is all zeros
OUT   DX, AL

SEND SEVEN SEGMENT CODE TO DISPLAY RAM

MOV   AL, 10010000B       ; Write display RAM, first location,
                          ;  auto increment
MOV   DX, 0FFEAH          ; Point at 8279 control address
OUT   DX, AL              ; Send control word
MOV   DX, 0FFE8H          ; Point at 8279 data address
MOV   AL, 6FH             ; Seven segment code for 9
OUT   DX, AL              ; Send to display RAM
MOV   AL, 5BH             ; Seven segment code for 2
OUT   DX, AL              ; Send to display RAM
; .
; .
; .

READ KEYBOARD CODE FROM FIFO

MOV   AL, 01000000B       ; Control word for read FIFO RAM
MOV   DX, 0FFEAH          ; Point at 8279 control address
OUT   DX, AL              ; Send control word
MOV   DX, 0FFE8H          ; Point at 8279 data address
IN    AL, DX              ; Read FIFO RAM
```

FIGURE 9-31 8086 instructions to initialize SDK-86 8279, write to display RAM, and read FIFO RAM.

nibbles in the CX register to the corresponding seven-segment codes for sending to the display RAM. To do this we first shuffle and mask to get each nibble into a byte by itself. We then use a lookup table and the XLAT instruction to do the actual conversion. Note that when making up seven-segment codes for the SDK-86 board, a high turns on a segment, bit D0 of a display RAM byte represents the "a" segment, bit D6 represents the "g" segment, and bit D7 represents the decimal point. Work your way through this section as a review of using XLAT.

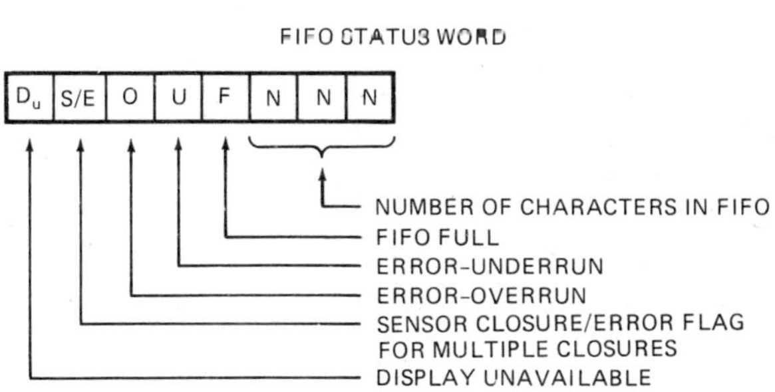

FIGURE 9-32 8279 status word format.

INTERFACING TO 18-SEGMENT AND DOT-MATRIX LED DISPLAYS

In the preceding examples we used an 8279 to refresh some seven-segment displays. The seven-segment codes for each digit were stored in successive locations in the display RAM. To display ASCII codes on 18-segment LED displays you can store the ASCII codes for each digit in the display RAM. (Remember that the A lines are driven from the upper nibble of the display RAM and the B lines are driven by the lower nibble). An external ROM is used to convert the ASCII codes to the required 18-segment codes and send them to the segment drivers. Strobes for each digit driver are produced just as they are for the seven-segment displays in Figure 7-6. The refreshing of each digit then proceeds just as it does for the seven-segment displays.

```
PAGE ,132
;8086 PROCEDURE TO DISPLAY DATA ON SDK-86 LEDs
;ABSTRACT:  This procedure will display a 4-digit hex or BCD number
;           passed in the CX register on LEDs of the SDK-86
;INPUTS:    Data in CX, control in AL.
;           AL =  00H  data displayed in data-field of LEDs
;           AL <> 00H, data displayed in address field of LEDs.
;PORTS:     Uses none
;PROCEDURES:Uses none
;REGISTERS: saves all registers and flags
PUBLIC  DISPLAY
DATA_HERE SEGMENT       WORD    PUBLIC
SEVEN_SEG       DB      3FH, 06H, 5BH, 4FH, 66H, 6DH, 7DH, 07H
                ;        0    1    2    3    4    5    6    7
                DB      7FH, 6FH, 77H, 7CH, 39H, 5EH, 79H, 71H
                ;        8    9    A    b    C    d    E    F
DATA_HERE ENDS
CODE_HERE       SEGMENT WORD    PUBLIC
                ASSUME CS:CODE_HERE, DS:DATA_HERE
DISPLAY PROC    FAR
        PUSHF                       ; save flags
        PUSH DS                     ; save caller's DS
        PUSH AX                     ; save registers
        PUSH BX
        PUSH CX
        PUSH DX
        MOV  BX, DATA_HERE          ; init DS as needed for procedure
        MOV  DS, BX
        MOV  DX, 0FFEAH             ; point at 8279 control address
        CMP  AL, 00H                ; see if data field required
        JZ   DATFLD                 ; yes, load control word for data field
        MOV  AL, 94H                ; no, load address-field control word
        JMP  SEND                   ; go send control word
DATFLD: MOV  AL, 90H                ; load control word for data field
SEND:   OUT  DX, AL                 ; send control word to 8279
        MOV  BX, OFFSET SEVEN_SEG   ; pointer to seven-segment codes
        MOV  DX, 0FFE8H             ; point at 8279 display RAM
        MOV  AL, CL                 ; get low byte to be displayed
        AND  AL, 0FH                ; mask upper nibble
        XLATB                       ; translate lower nibble to 7-seg code
        OUT  DX, AL                 ; send to 8279 display RAM
        MOV  AL, CL                 ; get low byte again
        MOV  CL, 04                 ; load rotate count
        ROL  AL, CL                 ; Move upper nibble into low position
        AND  AL, 0FH                ; Mask upper nibble
        XLATB                       ; translate 2nd nibble to 7-seg code
        OUT  DX, AL                 ; send to 8279 display RAM
        MOV  AL, CH                 ; Get high byte to translate
        AND  AL, 0FH                ; Mask upper nibble
        XLATB                       ; Translate to 7-seg code
        OUT  DX, AL                 ; send to 8279 display RAM
        MOV  AL, CH                 ; get high byte to fix upper nibble
        ROL  AL, CL                 ; move upper nibble into low position
        AND  AL, 0FH                ; mask upper nibble
```

FIGURE 9-33 Procedure to display contents of CX register on SDK-86 LED displays.

```
        XLATB                 ; translate to 7-seg code
        OUT  DX, AL           ; 7-seg code to 8279 display RAM
        POP  DX               ; restore all registers and flags
        POP  CX
        POP  BX
        POP  AX
        POP  DS
        POPF
        RET
DISPLAY ENDP
CODE_HERE ENDS
        END
```

FIGURE 9-33 (*continued*)

Refreshing 5 by 7 dot-matrix LED displays is a little more complex, because instead of lighting an entire digit, you have to refresh one row or one column at a time in each digit. Think of how you might do this for one 5 by 7 matrix which has its row drivers connected to one port and its column drivers connected to another port. To display a letter on this matrix you send out the code for the first column to the row drivers and send a code to the column drivers to turn on that column. After a millisecond or so you turn off the first column, send out the code for the second column, and light the second column. You repeat the process until all of the columns have been refreshed and then cycle back to column 1 again. You could use additional ports to drive additional digits, but the number of ports required soon gets too large. To reduce the number of ports required, inexpensive external latches can be used to hold the row codes for each digit. You then write the row codes for the first columns of all the digits to these latches. The columns of all the digits are connected in parallel, so when you output a code to turn on the first column, the first column of each digit will be lit with the code stored in its row latch. The process is repeated for each column until all columns are refreshed, and then started over again.

To further simplify interfacing multidigit dot-matrix LED displays to a microcomputer, Beckman Instruments, Hewlett-Packard, and several other companies make large integrated display/driver devices which require you to send only a series of ASCII codes for the characters you want displayed and one or two strobe signals for each character sent.

Liquid Crystal Display Operation and Interfacing

LCD OPERATION

Liquid crystal displays are created by sandwiching a thin (10- to 12-micron) layer of a liquid-crystal fluid between two glass plates. A transparent, electrically conductive film or backplane is put on the rear glass sheet. Transparent sections of conductive film in the shape of the desired characters are coated on the front glass plate. When a voltage is applied between a segment and the backplane, an electric field is created in the region under the segment. This electric field changes the transmission of light through the region under the segment film.

There are two commonly available types of LCD: *dynamic scattering*, and *field effect*. The dynamic scattering type scrambles the molecules where the field is present. This produces an etched-glass-looking light character on a dark background. Field effect types use polarization to absorb light where the electric field is present. This produces dark characters on a silver-gray background.

Most LCDs require a voltage of 2 or 3 V between the backplane and a segment to turn on the segment. You can't, however, just connect the backplane to ground and drive the segments with the outputs of a TTL decoder as we did the static LED display in Figure 9-25! The reason for this is that LCDs rapidly and irreversibly deteriorate if a steady dc voltage of more than about 50 mV is applied between a segment and the backplane. To prevent a dc buildup on the segments, the segment-drive signals for LCDs must be square waves with a frequency of 30–150 Hz. Even if you pulse the TTL decoder, it still will not work because the output low voltage of TTL devices is greater than 50 mV. CMOS gates are often used to drive LCDs.

Figure 9-34*a* shows how two CMOS gate outputs can be connected to drive an LCD segment and backplane. Figure 9-34*b* shows typical drive waveforms for the backplane and for the on and the off segments. The off (in this case unused) segment receives the same drive signal as the backplane. There is never any voltage between them, so no electric field is produced. The waveform for the on segment is 180° out of phase with the backplane signal, so the voltage between this segment and the backplane will always be +V. The logic for this is quite simple, because you only have to produce two signals, a square wave and its complement. To the driving gates the segment-backplane sandwich appears as a somewhat leaky capacitor. The CMOS gates can easily supply the current required to charge and discharge this small capacitance.

Older and/or inexpensive LCD displays turn on and off too slowly to be multiplexed in the way we do with LED displays. At 0°C some LCDs may require as much as 0.5 seconds to turn on or off. To interface to these types we use a nonmultiplexed driver device. Newer LCDs can turn on and off faster. To reduce the number of connect-

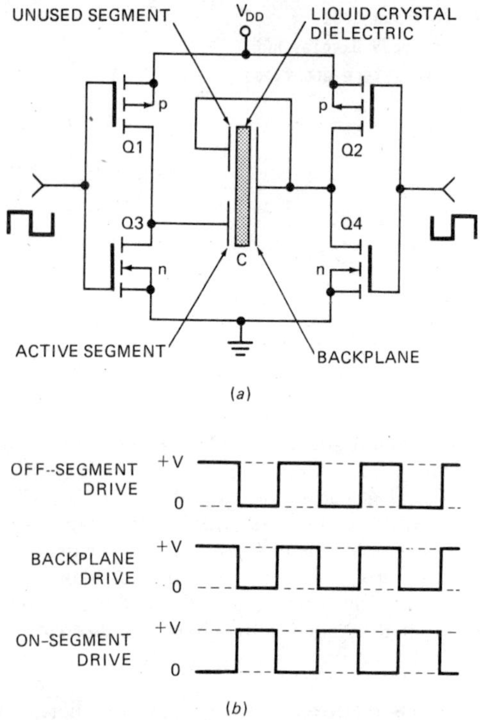

FIGURE 9-34 LCD drive circuit and drive waveforms. *(a)* CMOS drive circuits. *(b)* Segment and backplane drive waveforms.

ing wires when interfacing to these, we use a *triplex* technique. The following sections show you brief examples of each of these.

INTERFACING A MICROCOMPUTER TO NONMULTIPLEXED LCD DISPLAYS

Figure 9-35 shows how an Intersil ICM7211M can be connected to drive a 4-digit, nonmultiplexed, seven-segment LCD display such as you might buy from your local electronics surplus store. The 7211M inputs can be connected to port pins or directly to microcomputer buses as shown. For our example here we have connected the $\overline{CS}$ inputs to the Y2 output of the 74LS138 port decoder that we showed you how to add to an SDK-86 board in Figure 8-14. According to the truth table in Figure 8-15, the device will then be addressable as ports with a base address of FF10H. SDK-86 system address line A2 is connected to the digit-select input (DS2) and system address line A1 is connected to the DS1 input. This gives digit 4 a system address of FF10H. Digit 3 will be addressed at FF12H, digit 2 at FF14H, and digit 1 at FF16H. The data inputs are connected to the lower four lines of the SDK-86 data bus. The oscillator input is left open.

To display a character on one of the digits, you simply put the 4-bit hex code for that digit in the lower 4 bits of the AL register, and output it to the address of that digit. The ICM 7211M converts the 4-bit hex code to the required seven-segment code. The rising edge of the $\overline{CS}$ input signal causes the seven-segment code to be latched in the output latches for the addressed digit. An internal oscillator automatically generates the segment and backplane drive waveforms shown in Figure 9-34*b*.

INTERFACING TO TRIPLEXED LCD DISPLAYS

For many microcomputer-based instruments we want to display letters as well as numbers. To do this we usually use 18-segment digits such as the one shown in Figure 9-24*a*. For 18-segment LED digits we simply bus all of the segment inputs and multiplex the displays as described previously. Current LCD digits, however, cannot be multiplexed in the same way because of their slow switching response time. To reduce the connections required for a set of LCD digits, a compromise approach called *triplexing* has been devised. For triplexing, each digit is built as a matrix of six rows and three columns. Each digit has a 6-bit latch to hold the 6-bit row code for the segments in a column. The row codes are sent to all of the latches and the columns of each digit turned on. After a period of time the row codes for the second column are sent out to the latches. The first column is turned off and the second column turned on. After a period of time the row codes for the third columns are sent out to digits and the third columns turned on. At any given time one of the three columns in each display is activated, which is where the term triplexing comes from. Since only three columns ever need to be refreshed, no matter how many digits are connected, the switching rates are much lower than they are for the LED multiplexing method. The Intersil ICM7233 is an example of a device which contains all of the circuitry needed to drive four triplexed, 18-segment LCD digits. It can be connected directly to a microcomputer bus as we showed for the ICM7211M in Figure 9-35. To display a series of characters all you have to do is output a 6-bit ASCII code for each character to the appropriate digit address in the device. A demonstration kit containing two 7233s, eight 18-segment LCD displays and a PC board is available from Intersil, if you want to add this type of display to something you are building.

INTERFACING MICROCOMPUTER PORTS TO HIGH-POWER DEVICES

As shown for the 8255A in Figure 9-36, the output pins on programmable port devices can typically source only a few tenths of a milliampere from the +5-V supply and sink only 1 or 2 mA to ground. If you want to control some high-power devices such as lights, heaters, solenoids, and motors with your microcomputer, you need to use interface devices between the port pins and the high-power device. This section shows you a few of the commonly used devices and techniques.

INTEGRATED CIRCUIT BUFFERS

One approach to buffering the outputs of port devices is with TTL buffers such as the 7406 hex inverting and 7407 hex noninverting. In Figure 9-14 for example, we

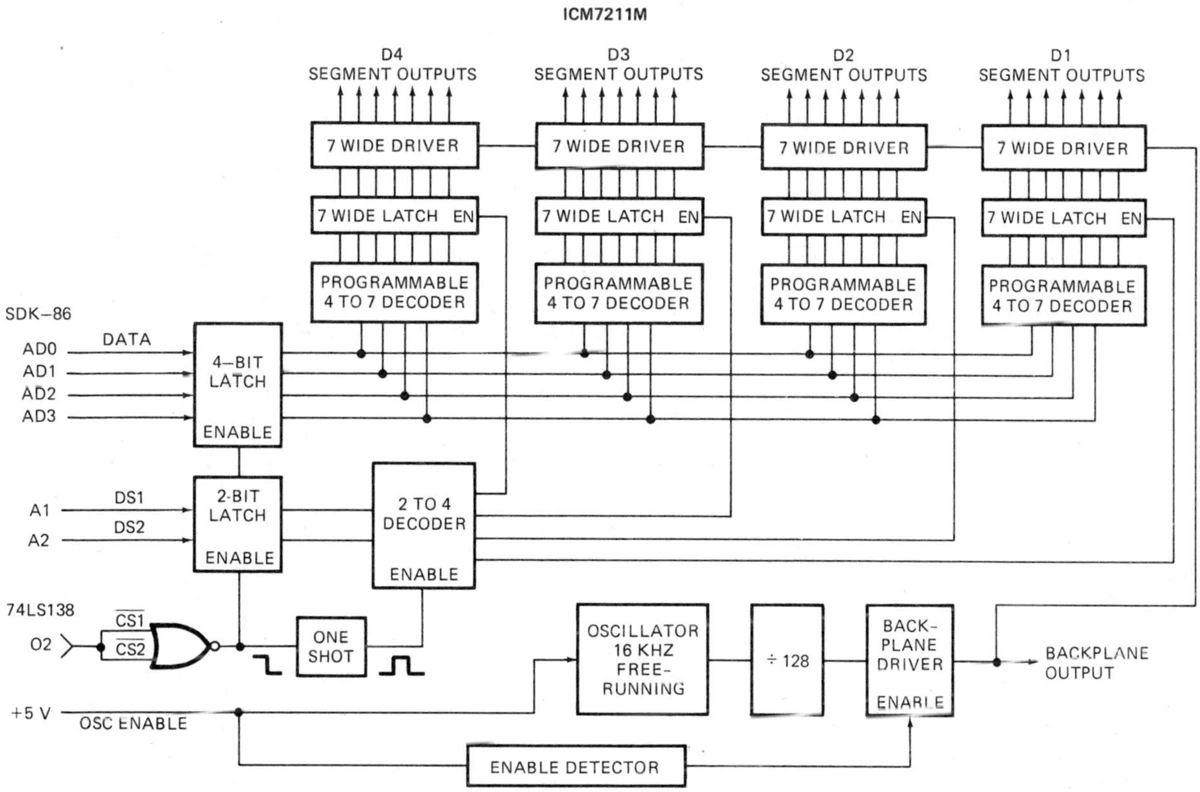

FIGURE 9-35 Circuit for interfacing four LCD digits to an SDK-86 bus using Intersil ICM7211M.

show 74LS07 buffers on the lines from ports to a printer. In an actual circuit the 8255A outputs to the computer-controlled lathe in Figure 9-6 should also have buffers of this type. The 74LS06 and 74LS07 have open collector outputs, so you have to connect a pull-up resistor from each output to +5 V. Each of the buffers in a 74LS06 or 74LS07 can sink as much as 40 mA to ground. You could then drive an LED with each output by simply connecting the LED and a current-limiting resistor in series between the buffer output and +5 V.

Buffers of this type have the advantage that they come six to a package, and they are easy to apply. For cases where you only need a buffer on one or two port pins, you may use discrete transistors.

D.C. CHARACTERISTICS
$T_A = 0°C$ to $70°C$, $V_{CC} = +5\ V \pm 5\%$; GND = 0V

SYMBOL	PARAMETER	MIN.	MAX.	UNIT	TEST CONDITIONS
V_{IL}	INPUT LOW VOLTAGE	–0.5	0.8	V	
V_{IH}	INPUT HIGH VOLTAGE	2.0	V_{CC}	V	
V_{OL} (DB)	OUTPUT LOW VOLTAGE (DATA BUS)		0.45	V	$I_{OL} = 2.5$ mA
V_{OL} (PER)	OUTPUT LOW VOLTAGE (PERIPHERAL PORT)		0.45	V	$I_{OL} = 1.7$ mA
V_{OH} (DB)	OUTPUT HIGH VOLTAGE (DATA BUS)	2.4		V	$I_{OH} = -400\ \mu A$
V_{OH} (PER)	OUTPUT HIGH VOLTAGE (PERIPHERAL PORT)	2.4		V	$I_{OH} = -200\ \mu A$
I_{DAR}[1]	DARLINGTON DRIVE CURRENT	–1.0	–4.0	mA	$R_{EXT} = 750\ \Omega$; $V_{EXT} = 1.5$ V
I_{CC}	POWER SUPPLY CURRENT		120	mA	
I_{IL}	INPUT LOAD CURRENT		±10	μA	$V_{IN} = V_{CC}$ TO 0V
I_{OFL}	OUTPUT FLOAT LEAKAGE		±10	μA	$V_{OUT} = V_{CC}$ TO 0V

NOTE 1: AVAILABLE ON ANY 8 PINS FROM PORT B AND C.

FIGURE 9-36 8255A dc operating characteristics.

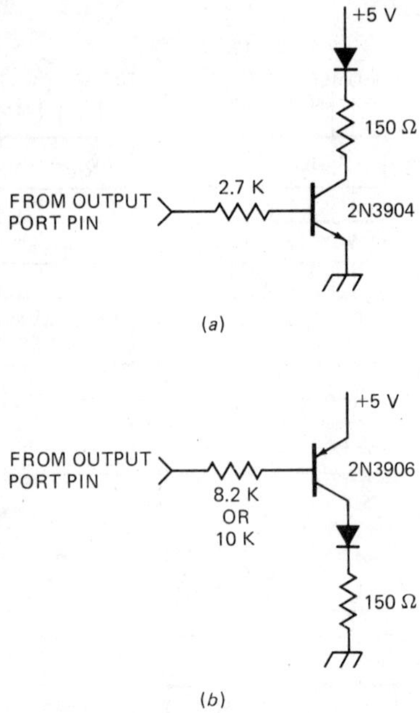

FIGURE 9-37 Transistor buffer circuits for driving LED from 8255A port pin. *(a)* NPN. *(b)* PNP.

TRANSISTOR BUFFERS

Figure 9-37 shows some transistor circuits you can connect to microprocessor port lines to drive LEDs or small dc lamps. We will show you how to quickly determine the parts values to put in these circuits for your particular application. First determine how much current you need to flow through the LED, lamp, or other device. For our example here, suppose that we want 20 mA to flow through an LED. Next determine whether you want a logic high on the output port pin to turn on the device or whether you want a logic low to turn on the device. If you want a logic high to turn on the LED, then use the NPN circuit. Now look through your transistor collection to find an NPN transistor which can carry the required current, has a collector-to-emitter breakdown voltage (V_{BCEO}) greater than the applied supply voltage, and can dissipate the power generated by the current flowing through it. We usually keep some inexpensive 2N3904 NPNs and some 2N3906 PNPs on hand for low-current switch applications such as this. Some alternatives are the 2N2222 NPN and the 2N2907 PNP. When you decide what transistor you are going to use, look up its current gain, h_{FE}, on a data sheet. If you don't have a data sheet, assume a value of 50 for the current gain of small-signal transistors such as these. Remember, current gain, or beta as it is commonly called, is the ratio of collector current to the base current needed to produce that current. To produce a collector current of 20 mA in a transistor with a beta of 50 then requires a base current of 20 mA/50 or 0.4 mA. To drive this buffer transistor, then, the output port pin only has to supply the 0.4 mA.

A look at the V_{OH}(PER) specification of the 8255A in Figure 9-36 shows that an 8255A peripheral port pin can only source 200 μA (0.2 mA) of current and still maintain a legal TTL-compatible output voltage of 2.4 V! When you see this specification, you may at first think the port output will not be able to drive the transistor. However, the fact is that the outputs can source more than 0.2 mA, but if they source more than 0.2 mA, the output high voltage will drop below 2.4 V. You don't care about the output high voltage dropping below 2.4 V except in the unlikely case that you are trying to drive a logic gate input off the same port pin as the transistor. The I_{DAR} specification in Figure 9-36 indicates that port B and port C pins can source at least 1.0 mA, but when doing so the output voltage may be as low as 1.5 V. Let's assume an output voltage of 2.0 V for calculating the value of our current-limiting resistor, R_b. The value of this resistor is not very critical as long as it lets through enough base current to drive the transistor. The base of the NPN transistor will be at about 0.7 V when the transistor is conducting, and the output port pin will be at least 2.0 V. Dividing the 1.3 V across R_b by the desired current of 0.4 mA gives an R_b value of 3.25 kΩ. A 2.7-kΩ or 3.3-kΩ resistor will work fine here.

For the PNP circuit in Figure 9-37*b* the output port pin can easily supply the needed drive current. The V_{OL}(PER) specification in Figure 9-36 shows that an output pin can sink at least 1.7 mA and still have an output low voltage no greater than 0.45 V. R_b in Figure 9-37*b* has about 4 V across it. Dividing this voltage by the required 0.4 mA gives an R_b value of 10 kΩ.

When you need to switch currents larger than about 50 mA on and off with an output port line, a single transistor does not have enough current gain to do this dependably. One solution to this problem is to connect two transistors in a Darlington configuration. Figure 9-38 shows how we might do this to drive a small solenoid-controlled valve which controls the flow a chemical into our printed-circuit-board-making machine, or a small solenoid in the print heads of a dot-matrix printer. For the case of the printer solenoid, when a current is passed through the coil of the solenoid, a print wire is

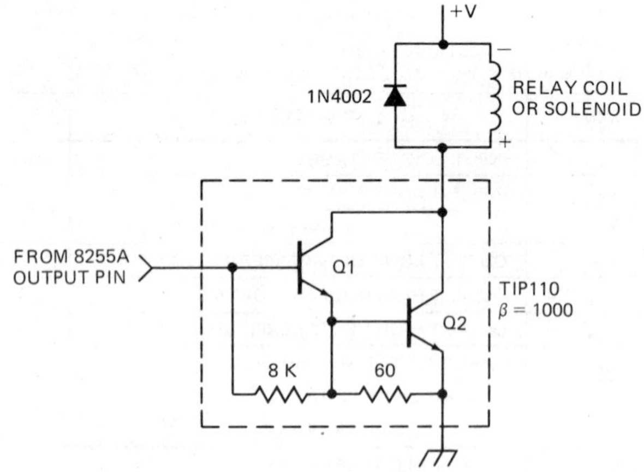

FIGURE 9-38 Darlington transistor used to drive relay coil or solenoid.

forced out. The print wire hits the ribbon against the paper and produces a dot on the paper.

The dotted lines around the two transistors in Figure 9-38 indicate that both devices are contained in the same package. Here's how this configuration works. The output port pin supplies base current to transistor Q1. This base current produces a collector current beta times as large in Q1. The collector current of Q1 becomes the base current of Q2 and is amplified by the current gain of Q2. The result of all this is that the device acts like a single transistor with a current gain of (beta Q1 × beta Q2) and a base-emitter voltage of about 1.4 V. The internal resistors help turn off the transistors. The TIP110 device we show here has a minimum beta of 1000 at 1 A. If we assume that we need 400 mA to drive the solenoid, then the worst-case current that must be supplied by the output port pin is about 400 mA/1000 or 0.4 mA, which it can easily do. If the drive current required for the Darlington is too high for the port output, you can add a resistor from the transistor base to +5 V to supply the added current. The output can easily sink the added current when the output is in the low state. Also another transistor could be added as a buffer between the output pin and the Darlington input. Note that since the V_{BE} of the Darlington is about 2 V, no R_b is needed here. Now let's check out the power dissipation.

According to the data sheet for the TIP110, it comes in a TO-220 package which can dissipate up to 2 W at an ambient temperature of 25°C with no heat sink. With 400 mA flowing through the device it will have a collector-emitter saturation voltage of about 2 V. Multiplying the current of 400 mA times the voltage drop of 2 V gives us a power dissipation of 0.8 W for our circuit here. This is well within the limits for the device. A rule of thumb that we like to follow is, if the calculated power dissipation for a device such as this is more than half of its 25°C no-heat-sink rating, mount the device on the chassis or a heat sink to make sure it will work on a hot day. If mounted on the appropriate heat sink the device will dissipate 50 W at 25°C.

One more important point to mention about the circuit in Figure 9-38 is the reverse-biased diode connected across the solenoid coil. You must remember to put in this diode whenever you drive an inductive load such as a solenoid, relay, or motor. Here's why. The basic principle of an inductor is that it fights against a change in the current through it. When you apply a voltage to the coil by turning on the transistor, it takes a while for the current to start flowing. This does not cause any major problems. However, when you turn off the transistor, the collapsing magnetic field in the inductor keeps the current flowing for a while. This current cannot flow through the transistor because it is off. Instead this current develops a voltage across the inductor with the polarity shown by the + and − signs on the coil in Figure 9-38. This induced voltage, sometimes called inductive "kick," will usually be large enough to break down the transistor if you forget to put in the diode. When the coil is conducting, the diode is reverse-biased so it doesn't conduct. However, as soon as the induced voltage reaches 0.7 V this diode turns on and supplies a return path for the induced current. The voltage across the inductor is clamped at 0.7 V, which saves the transistor.

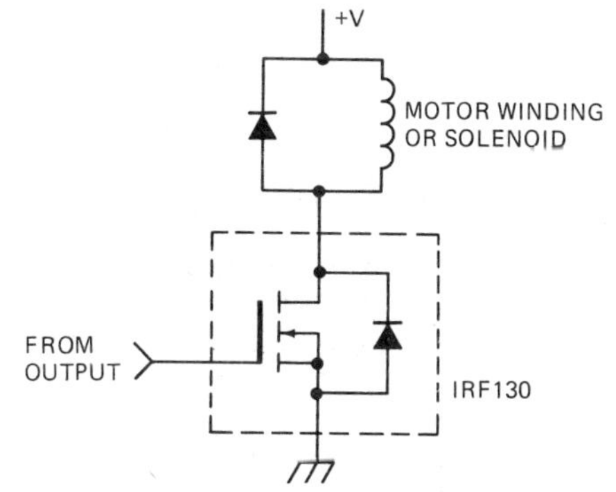

FIGURE 9-39 Power MOSFET circuit for driving solenoid or motor winding.

Figure 9-39 shows how a power MOSFET transistor can be used to drive a solenoid, relay, or motor winding. Power MOSFETS are several times more expensive than bipolar Darlingtons, but they have the advantage that they only require a voltage to drive them. The Motorola IRF130 shown here, for example, only requires a maximum gate voltage of 4 V for a drain current of 8 A. Note that a diode is required across the coil here also.

INTERFACING TO AC POWER DEVICES

To turn 110-V, 220-V, or 440-V ac devices on and off under microprocessor control we usually use *mechanical* or *solid-state relays*. The control circuitry for both of these types of relay is electrically isolated from the actual switch. This is very important because if the 110 V ac line gets shorted to the V_{cc} line of a microcomputer, it usually bakes most of the microcomputer's ICs. Figure 9-40*a* shows a picture of a mechanical relay. This relay has both normally open and normally closed contacts. When a current is passed through the coil of the relay, the switch arm is pulled down, opening the top contacts and closing the bottom set of contacts. The contacts are rated for a maximum current of 25 A, so this relay could be used to turn on a 1 or 2-horsepower motor or a large electric heater in one of the machines in our electronics factory. When driven from a 12-V supply, the coil requires a current of about 170 mA. The circuit shown in Figure 9-38 could easily drive this relay coil from a microcomputer port line.

Mechanical relays, sometimes called contactors, are available to switch currents from milliamps up to several thousand amps. Mechanical relays, however, have several serious problems. When the contacts are opened and closed, arcing takes place between the contacts. This causes the contacts to oxidize and pit just as the ignition points in your car do with age. As the contacts become oxidized they make a higher-resistance contact and may get hot enough to melt. Another disadvantage of mechanical relays is that they can switch on or off at any point in the ac cycle. Switching on or off at a high voltage point in the ac cycle can cause a large amount of electrical noise called electromagnetic interference

(a)

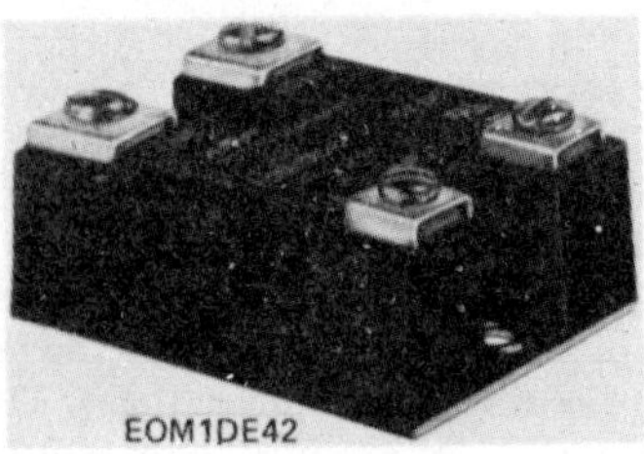

(b)

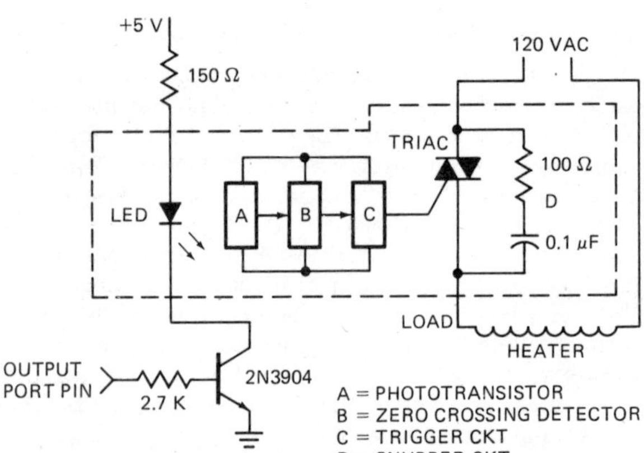

(c)

FIGURE 9-40 Relays for switching large currents. *(a)* Mechanical. *(b)* Solid-state. *(Potter and Brumfield)*. *(c)* Internal circuitry for solid-state relay.

(EMI). The solid-state relays discussed next avoid these problems to a large extent.

Figure 9-40*b* shows a picture of a solid-state relay which is rated for 25 A at 25°C if mounted on a suitable heat sink. Figure 9-40*c* shows a block diagram of the circuitry in the device and its connection from an output port to an ac load.

The input circuit is essentially an LED and a current-limiting resistor. To turn the device on you simply turn on the buffer transistor which pulls the required 11 mA through the internal LED. The light from the LED is focused on a phototransistor connected to the actual output-control circuitry. Since the only connection between the input and the output is a beam of light, there is several thousand volts of isolation between the input circuitry and the output circuitry.

The actual switch in a solid-state relay is a triac which conducts in either direction when triggered. The zero-voltage detector makes sure that the triac is only triggered when the ac line voltage is very close to one of its zero voltage crossing points. If you output a signal to turn on the relay, the relay will not actually turn on until the next time the ac line voltage crosses zero. Triacs automatically turn off when the current through them drops below a small value called the holding current. If the control signal is on, the trigger circuitry will automatically retrigger the triac for each half cycle. If you send a signal to turn off the relay, it will actually turn off the next time the ac current drops to zero. Zero-point switching eliminates most of the EMI that would be caused by switching the triac on at high-voltage points in the ac cycle.

Solid-state relays then have the advantages that they produce less EMI, they have no mechanical contacts to arc, and they are easily driven from microcomputer ports. Their disadvantages are that they are more expensive than an equivalent mechanical relay and there is a voltage drop of a couple of volts across the triac when it is on. Another potential problem with solid-state relays occurs when driving a large inductive load such as a motor. Remember from basic ac theory that the voltage waveform leads the current waveform in an ac circuit with inductance. A triac turns off when the current through it drops to near zero. In an inductive circuit the voltage waveform may be at several tens of volts when the current is at zero. When the triac is conducting it has perhaps 2 V across it. When the triac turns off, the voltage across the triac will quickly jump to several tens of volts. This large *dV/dT* may possibly turn on the triac at a point you don't want it turned on. To keep the voltage across the triac from changing too rapidly, an *RC snubber* circuit is connected across the triac as shown in Figure 9-40*c*. A system in the next chapter uses a solid-state relay to control an electrical heater.

INTERFACING A MICROCOMPUTER TO A STEPPER MOTOR

A unique type of motor useful for moving things in small increments is the stepper motor. If you have a dot-matrix printer such as the Epson FX-80, look inside and you will probably see one small stepper motor which is used to advance the paper to the next line position, and another small stepper motor which is used to move the print head to the next character position. While you are in there, you might look for a small device containing an LED and a phototransistor which detects when the print head is in the "home" position. Stepper motors are also used to position the read/write head over the desired track of a hard or floppy disk, and to position the pen on X-Y plotters.

Instead of rotating smoothly around and around as

most motors do, stepper motors rotate or "step" from one fixed position to the next. Common step sizes range from 0.9° to 30°. A stepper motor is stepped from one position to the next by changing the currents through the fields in the motor. The two common field connections are referred to as two-phase and four-phase. We will discuss *four-phase steppers* here because their drive circuitry is much simpler.

Figure 9-41 shows a circuit you can use to interface a small four-phase stepper such as the Superior Electric MO61-FD302, IMC Magnetics Corp. Tormax 200, or a similar, nominal 5-V unit to four microcomputer port lines. If you build up this circuit, bolt some small heat sinks on the MJE2955 transistors and mount the 10-W resistors where you aren't likely to touch them.

Since the 7406 buffers are inverting, a high on an output-port pin turns on current to a winding. Figure 9-41*b* shows the switching sequence to step a motor such as this clockwise, as you face the motor shaft, or counterclockwise. Here's how this works. Suppose that SW1 and SW2 are turned on. Turning off SW2 and turning on SW4 will cause the motor to rotate one step of 1.8° clockwise. Changing to SW4 and SW3 on will cause the motor to rotate another 1.8° clockwise. Changing to SW3 and SW2 on will cause another step. After that, changing to SW2 and SW1 on again will cause another step clockwise. You can repeat the sequence until the motor has rotated as many steps clockwise as you want. To step the motor counterclockwise, you simply work through the switch sequence in the reverse direction. In either case the motor will be held in its last position by the current through the coils. Figure 9-41*c* shows the switch sequence that can be used to rotate the motor half-steps of 0.9° clockwise or counterclockwise.

A close look at the switch sequence in Figure 9-41*b* shows an interesting pattern. To take the first step clockwise from SW2 and SW1 being on, the pattern of 1's and 0's is simply rotated one bit position around to the right. The 1 from SW1 is rotated around into bit 4. To take the next step the switch pattern is rotated one more bit position. To step counterclockwise the switch pattern is rotated left one bit position for each step desired. Suppose that you initially load 00110011 into AL and output this to the switches. Duplicating the switch pat-

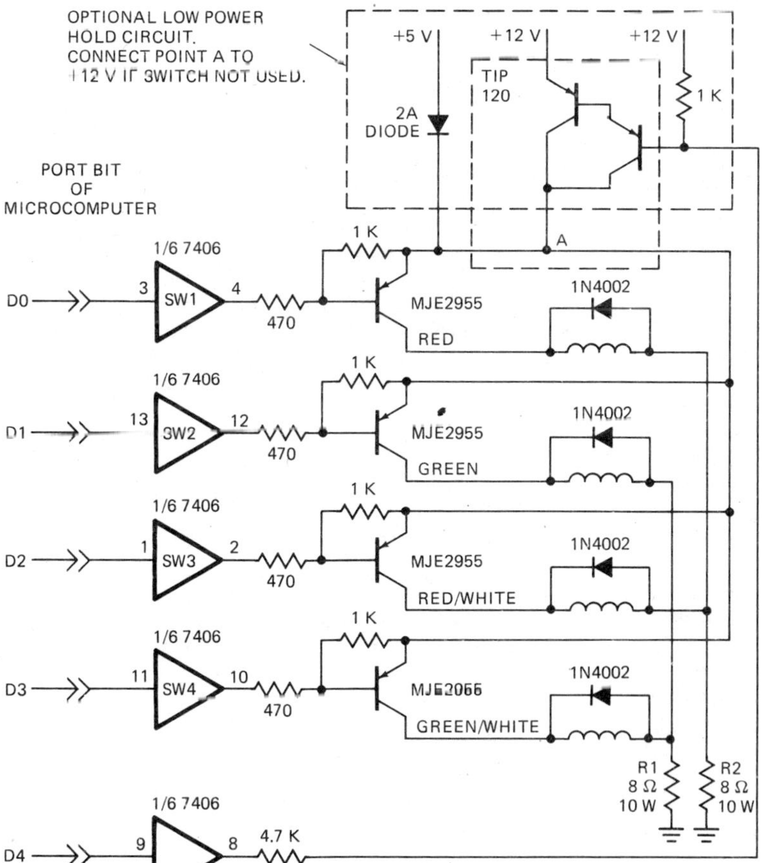

(*a*)

STEP	SWITCH			
	SW4	SW3	SW2	SW1
1	0	0	1	1
2	1	0	0	1
3	1	1	0	0
4	0	1	1	0
1	0	0	1	1

CW ↓ ↑ CCW

1 = SWITCH ON

(*b*)

EIGHT-STEP INPUT SEQUENCE
(HALF-STEP MODE)

STEP	SW4	SW3	SW2	SW1
1	OFF	OFF	ON	ON
2	OFF	OFF	OFF	ON
3	ON	OFF	OFF	ON
4	ON	OFF	OFF	OFF
5	ON	ON	OFF	OFF
6	OFF	ON	OFF	OFF
7	OFF	ON	ON	OFF
8	OFF	OFF	ON	OFF
1	OFF	OFF	ON	ON

(*c*)

FIGURE 9-41 Four-phase stepper motor interface circuit and stepping waveforms. *(a)* Circuit. *(b)* Full-step drive signal order. *(c)* Half-step drive signal order.

tern in the upper half of AL will make stepping easy. To step the motor clockwise, you just rotate this pattern right one bit position and output it to the switches. To step counterclockwise, you rotate the switch pattern left one bit position and output it. After you output one step code you must wait a few milliseconds before you output another step command, because the motor can only step so fast. Maximum stepping rates for different types of steppers vary from a few hundred steps per second to several thousand steps per second. To achieve high stepping rates the stepping rate is slowly increased to the maximum, then it is decreased as the desired number of steps is approached.

As a stepper motor steps to a new position it tends to oscillate around the new position before settling down. A common software technique to damp out this oscillation is to first send the pattern to step the motor toward the new position. When the motor has rotated part of the way to the new position, a word to step the motor backward is output for a short time. This is like putting the brakes on. The step forward word is then sent again to complete the step to the next position. The timing for the damping command must be determined experimentally for each motor and load.

Before we go on, here are a couple of additional points about the circuit in Figure 9-41*a*, in case you want to add a stepper to your robot or some other project. First of all, don't forget the clamp diodes across each winding to save the transistors from inductive kick. Second, we need to explain the function of the current-limiting resistors, *R*1 and *R*2. The motor we used here has a nominal voltage rating of 5.5 V. This means that we could have designed the circuit to operate with a voltage of about 6.5 V on the emitters of the driver transistors (5.5 V for the motor plus 1 V for the drop across the transistor). For low stepping rates, this would work fine. However, for higher stepping rates and more torque while stepping, we use a higher supply voltage and current-limiting resistors as shown. The point of this is that by adding series resistance, we decrease the *L*/*R* time constant. This allows the current to change more rapidly in the windings. For the motor we used, the current per winding is 0.88 A. Since only one winding on each resistor is ever on at a time, 6.5 V/0.88 A gives a resistor value of 6.25 Ω. To be conservative we used 8-Ω, 10-W resistors. The optional transistor switch and diode connection to the +5-V supply are used as follows. When not stepping, the switch to +12 V is off so the motor is held in position by the current from the +5 V supply. Before you send a step command, you turn on the transistor to +12 V to give the motor more current for stepping. When stepping is done you turn off the switch to +12 V, and drop back to the +5 V supply. This cuts the power dissipation.

In small printers such as the IBM PC parallel printer, a dedicated microprocessor is used to control the various operations in the printer. In this case the microprocessor has plenty of time to control the print-head and line-feed stepper motors in software as we described above. For applications where the main microcomputer is too busy to be bothered with controlling a stepper directly, a simple one-chip microcomputer or a device such as the Cybernetic Microsystems CY525 stepper controller is used.

OPTICAL MOTOR SHAFT ENCODERS

In order to control the machines in our electronics factory, the microcomputers in these machines often need information about the position, direction of rotation, and speed of rotation of various motor shafts. The microcomputer, of course, needs this information in digital form. The circuitry which produces this digital information from each motor for the microcomputer is called a *shaft encoder.* There are two basic types of shaft encoder, *absolute* and *incremental*. Here's how these two types work.

Absolute Encoders

Absolute encoders attach a binary-coded disk such as the one shown in Figure 9-42 on the rotating shaft. Light sections of the disk are transparent, and dark sections are opaque. An LED is mounted on one side of each track and a phototransistor is mounted on the other side, opposite the LED. Outputs from the four phototransistors will produce one of the binary codes shown in Figure 9-42. The phototransistor outputs can be conditioned with Schmitt-trigger buffers and connected to a microcomputer port. Each code represents an absolute angular position of the shaft in its rotation. With a 4-bit disk, 360° is divided up into 16 parts, so the position of the shaft can be determined to the nearest 22.5°. With an 8-bit disk the position of the disk can be determined to the nearest 360°/256, or 1.4°.

Observe that the codes in Figure 9-42 do not follow a normal binary count sequence. The codes here follow a sequence called *Gray code.* Using Gray code reduces the size of the largest possible error in reading the shaft

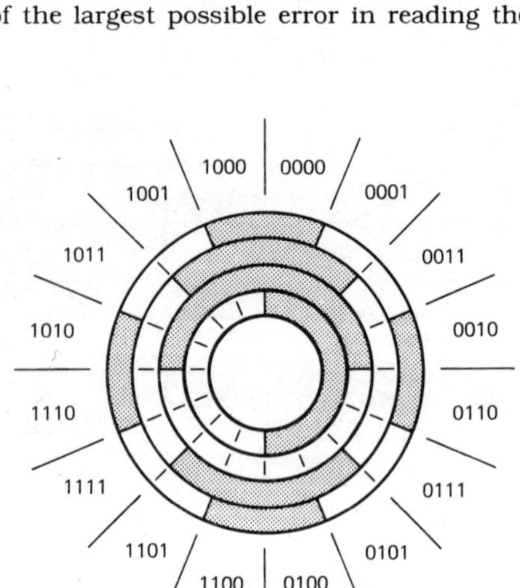

FIGURE 9-42 Gray code optical-encoder disk used to determine angular position of a rotating shaft.

position to the value of the least-significant bit. If the disk used straight binary code, the largest possible error would be the value of the most significant bit. Look at the parallel listings of binary and Gray codes in Table 1-1 to help you see why this is the case.

To start, assume we did have a binary disk and the disk was rotating from position 0111 (7) to position 1000 (8). Now suppose that the detectors pick up the change to 000 on the least-significant 3 bits, but don't pick up the change to 1 on the most-significant bit. The output code would then be 0000 instead of the desired 1000. This is an error equal to the value of the MSB. Now, while this is fresh in your mind, look across the table at the same position change for the Gray code encoder. The Gray code for position 7 is 0100 and the Gray code for position 8 is 1100. Note that only one bit changes for this transition. If you look at the Gray code table closely, you will see that this is the case for all of the transitions. What this means is that if a detector fails to pick up the new bit value during a transition, the resulting code will always be the code for the preceding position. This represents an error equal to the value of the LSB.

If you need to construct a Gray code table for more than 4 bits, a handy method is to observe the pattern of 1's and 0's in Table 1-1, and just extend it. The least-significant-bit column starts with a 0, and then has alternating groups of two 1's and two 0's. The second-most-significant column starts with two 0's and then has alternating groups of four 1's and four 0's. The third column starts with four 0's and has alternating groups of eight 1's and eight 0's. By now you should see the pattern. Try to figure out the Gray code for the decimal number 16. You should get 11000.

Absolute encoding using a Gray code disk has the advantage that each position is represented by a specific code which can be directly read in by the microcomputer. Disadvantages are the multiple detectors needed, the multiple lines required, and the difficulty keeping track of position during multiple rotations.

Incremental Encoders

An incremental encoder produces a pulse for each increment of shaft rotation. Figure 9-43 shows the Rhino XR-2 robot arm, which uses incremental encoders to determine the position and direction of rotation for each of its motors. For this encoder, a metal disk with two tracks of slotted holes is mounted on each motor shaft. An LED is mounted on one side of each track of holes and a phototransistor is mounted opposite the LED on the other side of the disk. Each phototransistor produces a train of pulses as the disk is rotated. The pulses are passed through Schmitt trigger buffers to sharpen their edges.

The two tracks of slotted holes are 90° out of phase with each other as shown at the top of Figure 9-44. Therefore, as the disk is rotated, the waveforms shown at the bottom of Figure 9-44 will be produced by the phototransistors for rotation in one direction. Rotation in the other direction will shift the phase of the waveforms 180° so that the B waveform leads the A waveform by 90° instead of lagging it by 90°. Now the question is, "How do you get position, speed, and direction information from these waveforms?"

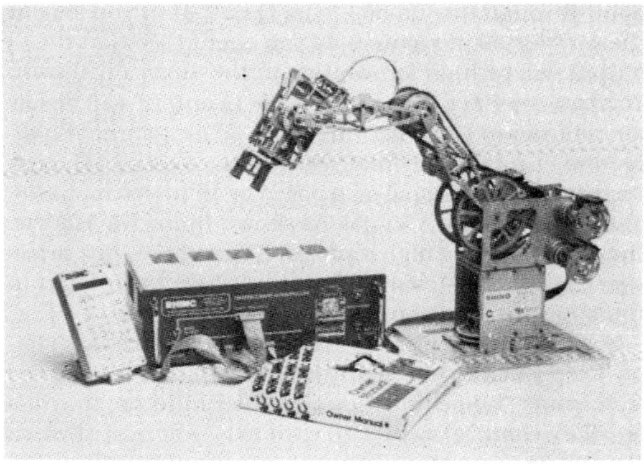

FIGURE 9-43 Rhino XR robotics system. *(Rhino Robots Incorporated)*

You can determine the speed of rotation by simply counting the number of pulses in the time between two interrupts as we described in Chapter 8. Each track has six holes so six pulses will be produced for each revolution. Some simple arithmetic will give you the speed in revolutions per minute (rpm).

You can determine the direction of rotation with hardware or with software. For the hardware approach connect the A signal to the D input of a D flip-flop, and the B signal to the clock input of the flip-flop. The rising edge of the B signal will clock the level of the A signal at that

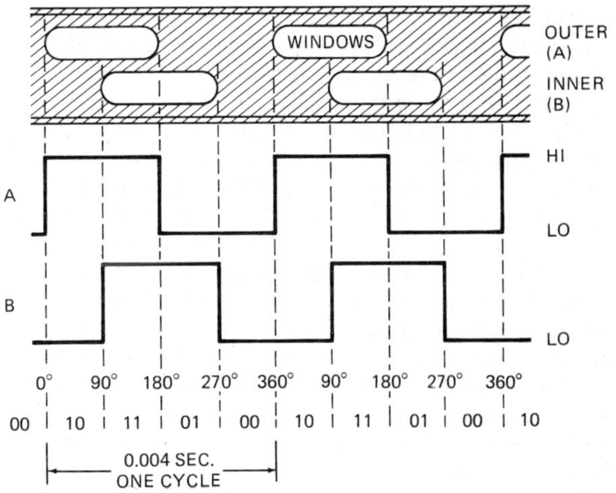

FIGURE 9-44 Optical-encoder disk slot pattern and output waveforms.

point through the flip-flop to its Q output. If you look at the waveforms in Figure 9-44 you should see that the Q output will be high for rotation in the direction shown. You can convince yourself that the Q output will be low for rotation in the other direction. To determine the direction of rotation more directly, you can detect the rising edge of the B signal on a polled or an interrupt basis, and then read the A signal. As shown in the waveforms, the A signal being high represents rotation in one direction, and the A signal being low represents rotation in the opposite direction.

To determine the position of the motor shaft you simply keep track of how many holes the motor has moved from some "home" position. On the Rhino robot arm a small mechanical switch on each axis is activated when the arm is in its starting or "home" position. When you turn on the power, the motor controller/driver box automatically moves the arm to this home position. To move the arm to some new position you calculate the number of holes each motor must rotate to get the arm to that position. For each motor you then send the controller a command which tells it which direction to rotate that motor and how many holes to rotate it. The controller will drive the motor the specified number of holes in the specified direction. If you then manually rotate the encoder wheel or some heavy load moves the arm and rotates the encoder disk, the controller will detect the change in position of the disk and drive the motor back to its specified position. This is an example of digital *feedback control*, which is easily done with a microcomputer. The Rhino controller uses an 8748 single-chip microcomputer to interpret and carry out the commands you send it. Commands are sent to the controller in the serial ASCII form described at the start of Chapter 13.

Incidentally, you may wonder at this point why the designers of the Rhino arm did not use stepper motors such as those we described in a previous section. The answers are: stepper motors are much more expensive than the simple dc motors used; if a stepper motor is forced back a step there is no way to know about it and correct for it unless you have an external encoder. Also the dc motor-encoder approach better demonstrates the method used in large commercial robots.

In the Rhino robot arm each motor drives its section of the arm through a series of gears. Gearing the motor down reduces the force that the motor has to exert, and makes the exact position of the motor shaft less critical. Therefore, for the Rhino, six sets of slots in the encoder disk are sufficient. However, for applications where a much more accurate indication of shaft position is needed, a self-contained shaft encoder such as the Hewlett-Packard HEDS-5000 is attached to the motor shaft. These encoders have two track-encoder disks with 500 tiny radial slits per track. The waveforms produced are same as those shown for the Rhino encoder in Figure 9-44, but at a much higher frequency for the same motor speed.

Optical encoders in their many different forms are an important part of a large number of microcomputer-controlled machines.

IMPORTANT TERMS AND CONCEPTS FROM THIS CHAPTER

If you do not remember any of the terms in the following list, use the index to help you find them in the chapter for review.

Simple input and output

Simple strobe I/O

Single handshake I/O

Double handshake data transfer

8255A initialization of ports A, B, and C
- MODE 0, MODE 1, MODE 2
- Mode definition control word
- Set/reset control word

Computer numerical control (CNC) machines

VOTRAX SC-01A speech synthesizer
- Phoneme

Centronix parallel standard
- I/O driver
- Control block
- Counters and sentinels

Keyswitches—mechanical, capacitive, Hall effect

Debounce keypress

Two-key lockout, two-key rollover

Code conversion
- Compare
- Add and point, XLAT

Error trapping

LED interfacing
- Static display
- Multiplexed display
- Dedicated display controller
- Scan time

8279
- FIFO
- Encoded and decoded scan
- Keyboard/display mode-set control word
- Clear control word
- Write-display control word

LCD interfacing
- Dynamic scattering
- Field effect
- Backplane
- Triplexing

Relays
- Mechanical

Solid state
Electromagnetic interference
Zero-point switching
RC-snubber circuit

Four-phase stepper motor
Shaft encoders—absolute and incremental
Digital feedback control

REVIEW QUESTIONS AND PROBLEMS

1. Why must data be sent to a printer on a handshake basis?

2. For the double handshake data transfer in Figure 9-1*d*
   a. Indicate which signal is asserted by the sender and which signal is asserted by the receiver.
   b. Describe the meaning of each of the signal transitions.

3. Why are the port lines of programmable port devices automatically put in the input mode when the device is first powered up or reset?

4. An 8255A has a system base address of FFF9H. What are the system addresses for the three ports and the control register for this 8255A?

5. *a.* Show the mode-set control word needed to initialize an 8255A as follows:
   Port A—handshake input
   Port B—handshake output
   Port C—bits PC6 and PC7 as outputs
   b. Show the bit set/reset control word needed to initialize the port A interrupt request and the port B interrupt request.
   c. Show the assembly language instructions you would use to send these control words to the 8255A in Question 4.
   d. Show the additional instruction you need if you want the handshake to be done on an interrupt basis through the IR3 input of the 8259A in Figure 8-14.
   e. Show the instructions you would use to put a high on port C bit PC6 of this device.

6. Describe the exchange of signals between the tape reader, 8255A, and 8086 in Figure 9-6 as a byte of data is transferred from the tape reader to the microprocessor.

7. Why is it more efficient to send phonemes to the SC-01A speech synthesizer in Figure 9-10*a* on an interrupt basis than on a polled basis?

8. If you have an SC-01A speech IC connected to your system as shown in Figure 9-10*a*, write the mainline program and the interrupt procedure to send phonemes to the SC-01A. The mainline can terminate with the HERE: JMP HERE instruction so that it simply waits for interrupts from the 8255A. Use the phoneme table in the appendix to help you make up the table of phonemes for the message "self-test complete."

9. When connecting peripheral devices such as printers, terminals, etc. to a computer, why is it very important to connect the logic ground and the chassis ground together only at the computer?

10. Describe the function and direction of the following signals in a Centronics parallel-printer interface.
    a. $\overline{\text{STROBE}}$
    b. $\overline{\text{ACKNLG}}$
    c. BUSY
    d. $\overline{\text{INIT}}$

11. Modify the printer driver procedure in Figure 9-17 so that it stops sending characters to the printer when it finds a sentinel character of 03H, instead of using the counter approach.

12. Would the software method of generating the $\overline{\text{STROBE}}$ signal to the printer in Figure 9-17 still work if you try to run the program with an 8-MHz 8086?

13. Show the instructions you would use to read the status byte from the 8255A in Question 5.

14. Describe the three major tasks needed to get meaningful information from a matrix keyboard.

15. Describe how the "compare" method of code conversion in Figure 9-21 works.

16. Why is "error trapping" necessary in real programs? Describe how the error trap in the program in Figure 9-21 works.

17. Assume the rows of the circuit shown in Figure 9-45 are connected to ports FFF8H and the 74148 is connected to port FFFAH of an SDK-86 board. The 74148 will output a low on its $\overline{\text{GS}}$ output if a low is applied to any of its inputs. The way the keyboard is wired, the $\overline{\text{A2}}$, $\overline{\text{A1}}$, and $\overline{\text{A0}}$ outputs will have a 3-bit binary code for the column in which a low appears. Use the algorithm and discussion of Figure 9-21 to help you write a procedure which detects a keypress, debounces the keypress, and determines the row number and column number of the pressed key. The procedure should then combine the row code, column code, shift bit and control bit into a single byte in the form: control, shift, row code, column code. The XLAT instruction can then be used to convert this code byte to ASCII for return to the calling program. *HINT*: Use DB directive to make up table of ASCII codes.

    Why is the XLAT approach more efficient than the compare technique for this case?

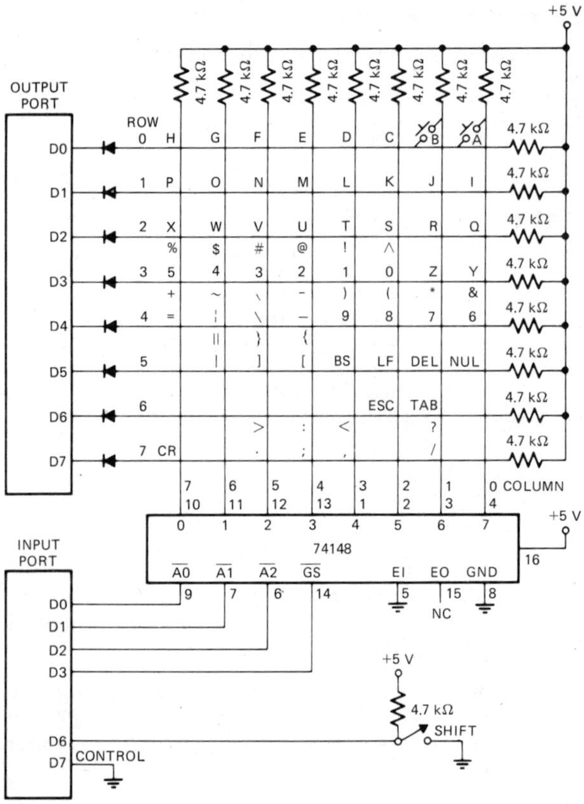

FIGURE 9-45 Interface circuitry for unencoded matrix keyboard for Problem 9-17.

NOTE: For test purposes, the keyboard matrix can be simulated by building the diodes, resistors, and 74148 on a prototyping board and using a jumper wire to produce a "keypress."

18. *a.* Calculate the value of the current-limiting resistor needed in series with each segment of a seven-segment display driven by a 7447, if you want 40 mA per segment.

b. Approximately how much current is being pulsed through each LED segment on the SDK-86 board?

19. *a.* Write the algorithm for a procedure which refreshes the multiplexed LED displays shown in Figure 9-26. Assume the procedure will be called every 2 ms by an interrupt signal to IR4 of an 8259A.

b. Write the assembly language instructions for the display refresh procedure. Since this procedure is called on an interrupt basis, all display parameters should be kept in named memory locations. If you have time, you can add the circuitry shown in Figure 9-26 to your microcomputer so you can test your program.

20. Figure 9-46 shows a circuit for an 8 by 8 matrix of LEDs that you can add to a couple of ports on your microcomputer to produce some interesting displays. The principle here is to output a 1 to port B for each LED you want turned on in the top row and then output a 1 to the D0 bit of port A to turn on that row. After 2 ms, you output the pattern you want in the row to port B and a 1 to bit 1 of port A to turn on the second row. The process is repeated until all rows are done and then started over.

The row patterns can be kept in a table in memory. If you want to display a sequence of letters, you can display the contents of one table for a few seconds, then switch to another table containing the second letter. Using the rotate instruction, you can produce some scrolled displays. *HINT*: The wiring required to build the LED matrix can be reduced by using an IC 5 by 7 dot-matrix LED display such as the Texas Instruments TIL305.

Write the algorithm and program for an interrupt procedure (called every 2 ms) to refresh these displays.

21. You are assigned the job of fixing several SDK-86 boards with display problems. For each of the problems listed below, describe a possible cause of the problem and tell where you would look with an oscilloscope to check out your theory. Use the circuit on sheet 7 of Figure 7-6 to help you.

a. The segment never lights.

b. The leftmost digit of the data field never lights.

c. All of the displays show dim "eights."

22. *a.* Show the command words and assembly language instructions necessary to initialize an 8279 at address 80H and 82H as follows:
16-character display, left entry, encoded-scan keyboard, *N*-key rollover.
1-MHz input clock divided to 100 kHz.
Blanking character FFH.

b. Show the 8279 instructions necessary to write 99H to the first location in the display RAM and autoincrement the display RAM pointer.

c. Show the assembly language instructions necessary to read the first byte from the 8279 FIFO RAM.

d. Determine the seven-segment codes you would have to send to the SDK-86 8279 to display the letters HELP on the data field display. Remember that D0 of the byte sent = B0 and D7 of the byte sent = A3.

e. Show the sequence of instructions you can send to the 8279 of the SDK-86 board to blank the entire display.

23. Write a procedure which polls the LSB of the 8279 status register on the SDK-86 board until it finds a key pressed, then reads the keypressed code from the FIFO RAM to AL and returns.

24. Why must the backplane and segment-line signals be pulsed for LCD displays?

25. Draw a circuit you could attach to an 8255A port B pin to drive a 1-A solenoid valve from a +12-V supply. You want a high on the port pin to turn on the solenoid.

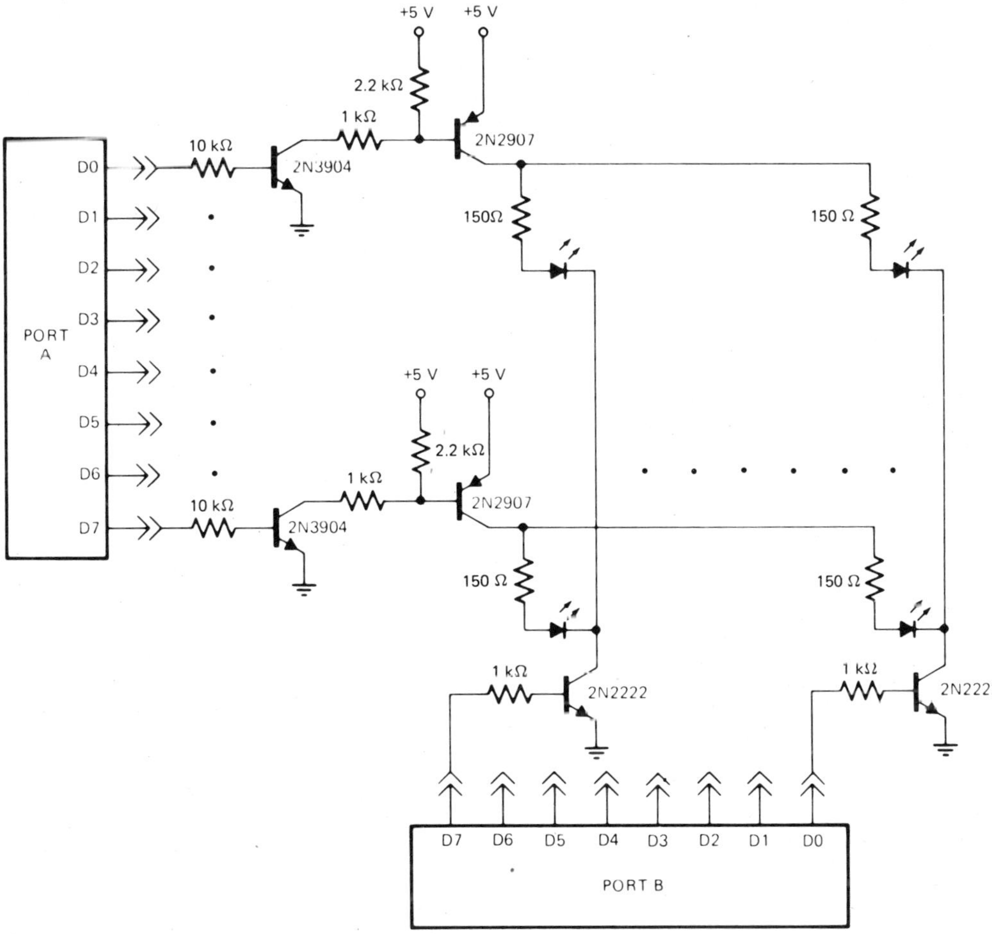

FIGURE 9-46 Eight by eight LED matrix circuitry for Problem 9-20.

26. Why must reverse-biased diodes always be placed across inductive devices when you are driving them with a transistor?

27. What are the major advantages and disadvantages of mechanical relays and solid-state relays.

28. *a.* How is electrical isolation between the control input and the output circuitry achieved in a solid-state relay?

b. Describe the function of the zero-crossing detector used in better-quality solid-state relays.

c. Why is a "snubber circuit" required across the triac of a solid-state relay when you are driving inductive loads?

29. Write the algorithm and the program for an 8086 procedure to drive the stepper motor shown in Figure 9-41. Assume the desired direction of rotation is passed to the procedure in AL (AL = 1 is clockwise, AL = 0 is counterclockwise) and the number of steps is passed to the procedure in CX. Also assume full-step mode as shown in Figure 9-41*b*. Don't forget to delay 20 ms between step commands!

30. *a.* Why is Gray code, rather than straight binary code, used on many absolute-position shaft encoders?

b. If a Gray-code wheel has six tracks and each track represents 1 binary bit, what is its angular resolution?

31. *a.* Look at the encoder disk on the Rhino arm in Figure 9-43. Do the waveforms in Figure 9-44 represent clockwise or counterclockwise rotation of the motor shaft as seen from the gear end of the motor, which is what you care about.

b. Assume the A signal shown in Figure 9-44 is connected to bit D0 and the B signal is connected to bit D1 of port FFF8H. Write a procedure which determines the direction of rotation and passes a 1 back in AL for clockwise and a 0 back in AL for counterclockwise rotation.

c. The dc motors, such as those on the RHINO arms, are rotated clockwise by passing a current through them in one direction and rotated counterclockwise by passing a current through them in the opposite direction. Assume you have a motor controller that responds to a 2-bit control word as follows:

00 = hold   01 = rotate clockwise
11 = hold   10 = rotate counter-clockwise

Write the algorithm and program for a procedure to rotate a motor. The number of holes is passed to the procedure in CX; the direction of rotation is determined by the value in AL. AL = 1 is clockwise, AL = 0 is counterclockwise.

CHAPTER

10 Analog Interfacing and Industrial Control

In order to control the machines in our electronics factory, medical instruments, or automobiles with a microcomputer we need to determine the values of variables such as pressure, temperature, and flow. There are usually several steps in getting an electrical signal which represents the values of these variables, and converting the electrical signals to a digital form that the microcomputer understands.

The first step involves a *sensor* which converts the physical pressure, temperature, or other variable to a proportional voltage or current. The signals from most sensors are quite small, so they must next be amplified, and perhaps filtered. This is usually done with some type of operational-amplifier (op-amp) circuit. The final step is to convert the signal to digital form with an analog-to-digital (A/D) converter. In this chapter we review some op-amp circuits commonly used in these steps, show the interface circuitry for some common sensors, and discuss the operation and interfacing of A/D converters. We also discuss the interfacing of D/A converters and show how all of these pieces are put together in a microcomputer-based scale and a machine-control system.

OBJECTIVES

At the conclusion of this chapter you should be able to:

1. Recognize several common op-amp circuits, describe their operation, and predict the voltages at key points in each.
2. Describe the operation and interfacing of several common sensors used to measure temperature, pressure, flow, etc.
3. Draw circuits showing how to interface D/A converters with any number of bits to a microcomputer.
4. Define D/A data-sheet parameters such as resolution, settling time, accuracy, and linearity.
5. Describe briefly the operation of flash, successive approximation, and ramp A/D converters.
6. Draw circuits showing how A/D converters of various types can be interfaced to a microcomputer.
7. Write programs to control A/D and D/A converters.
8. Describe how feedback is used to control variables such as pressure, temperature, flow, motor speed, etc.
9. Describe the operation of a "time slice" factory control system.

REVIEW OF OPERATIONAL-AMPLIFIER CHARACTERISTICS AND CIRCUITS

Basic Operational Amplifier Characteristics

Figure 10-1*a* shows the schematic symbol for an op amp. Here are the important points for you to remember about the basic op amp. First, the pins labeled +V and −V represent the power supply connections. The voltages applied to these pins will usually be +15 V and −15 V, or +12 V and −12 V. The op amp also has two signal inputs. The amplifier amplifies the difference in voltage between these two inputs by 100,000 or more. The input labeled with a − sign is called the inverting input and the one labeled with the + sign is called the noninverting input. The + and − on these inputs has nothing to do with the power-supply voltages. These signs indicate the phase relationship between a signal applied to that input and the result that signal produces on the output. If for example, the noninverting input is made more positive than the inverting input, the output will move in a positive direction, which is in phase with the applied input signal. Now let's see how much the output changes for a given input signal, and see how an op amp is used as a comparator.

Op-amp Circuits and Applications

OP AMPS AS COMPARATORS

We said previously that the op amp amplifies the difference in voltage between its inputs by 100,000 or more.

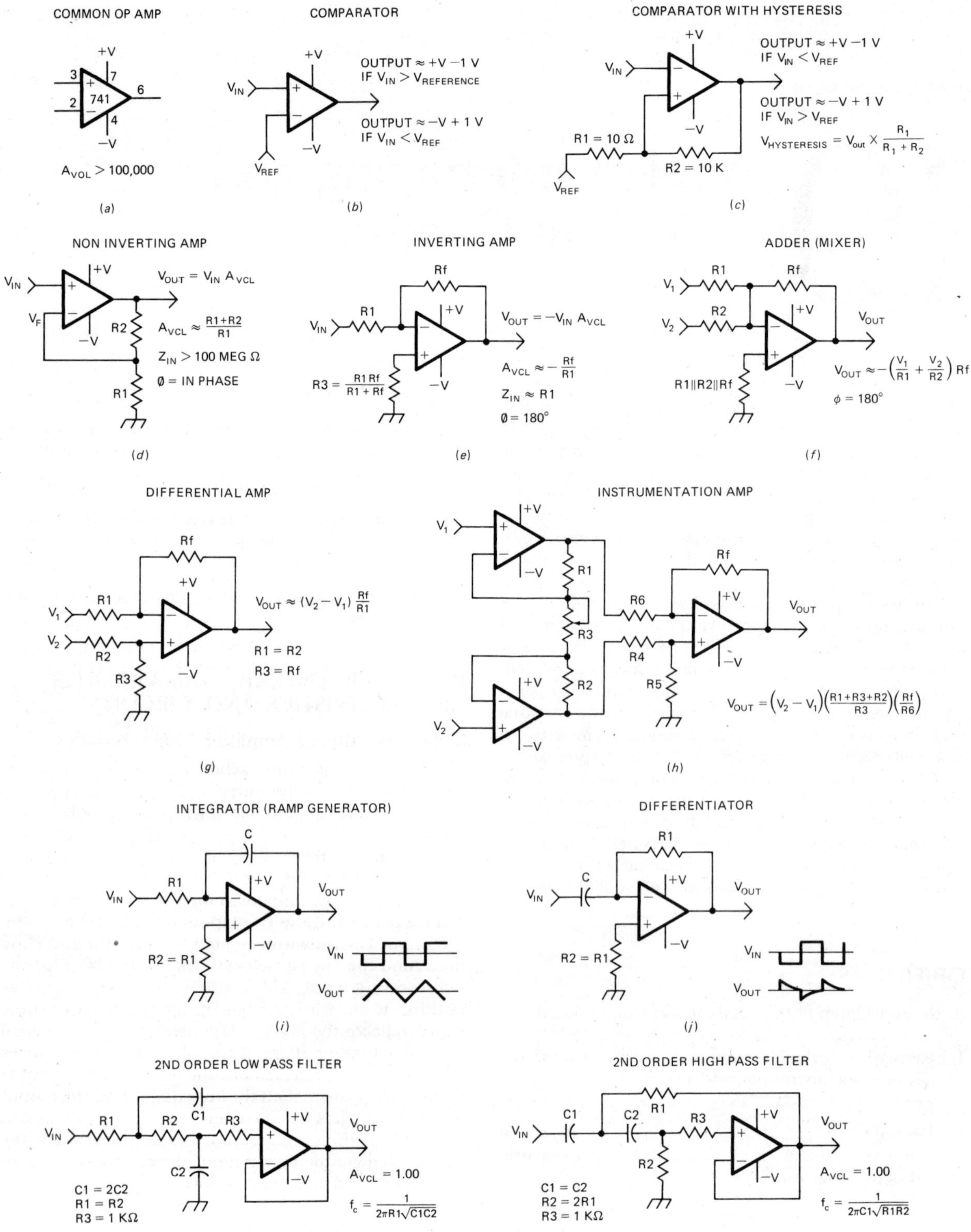

FIGURE 10-1 Overview of commonly used op-amp circuits. *(a)* Common op amp. *(b)* Comparator. *(c)* Comparator with hysteresis. *(d)* Noninverting amp. *(e)* Inverting amp. *(f)* Adder (mixer). *(g)* Differential amp. *(h)* Instrumentation amp. *(i)* Integrator (ramp generator). *(j)* Differentiator. *(k)* Second-order low-pass filter. *(l)* Second-order high-pass filter.

(The number is variable with temperature and from device to device.) Suppose that you power an op amp with +15 V and −15 V, tie the inverting input of the op amp to ground, and apply a signal of +0.01 V dc to the noninverting input. The op amp will attempt to amplify this signal by 100,000 and produce the result on its output. An input signal of 0.01 V times a gain of 100,000 predicts an output voltage of 100 V. The op-amp output, however, can only go positive to a voltage a volt or two less than the positive supply voltage, perhaps 13 V, so this is as far as it goes. Now suppose that you apply a signal of −0.01 V to the noninverting input. The output will now try to go to −100 V as fast as it can. The output, however, can only go to about −13 V, so it stops here.

In this circuit the op amp effectively compares the input voltage with the voltage on the inverting input and gives a high or low output depending on the result of the comparison. If the input is more than a few microvolts above the reference voltage on the inverting input, the output will be high (+13 V). If the input voltage is a few microvolts more negative than the reference voltage, the output will be low (−13 V). An op amp used in this way is called a *comparator*. Figure 10-1*b* shows how a comparator is usually labeled. The reference voltage applied to the inverting input does not have to be ground (0 V). An input voltage can be compared to any voltage within the input range specified for the particular op amp.

As you will see throughout this chapter, comparators have many applications. We might, for example, connect a comparator to a temperature sensor on the boiler in our electronics factory. When the voltage from the temperature sensor goes above the voltage on the reference input of the comparator, the output of the comparator will change state and send an interrupt signal to the microprocessor controlling the boiler. Commonly available comparators such as the LM319 have TTL-compatible outputs which can be connected directly to microcomputer ports or interrupt inputs.

Figure 10-1*c* shows another commonly used comparator circuit. Note in this circuit that the reference signal is applied to the noninverting input, and the input voltage is applied to the inverting input. This connection simply inverts the output state from those in the previous circuit. Note also in Figure 10-1*c* the positive-feedback resistors from the output to the noninverting input. This feedback gives the comparator a characteristic called *hysteresis*. Hysteresis means that the output voltage changes at a different input voltage when the input is going in the positive direction than it does when the input voltage is going in a negative direction. If you have a thermostatically controlled furnace in your house you have seen hysteresis in action. The furnace, for example, may turn on when the room temperature drops to 65°F, and then not turn off until the temperature reaches 68°F. Hysteresis is the difference between the two temperatures. Without this hysteresis the furnace would rapidly be turning on and off if the room temperature were near 68°F. Another situation where hysteresis saves the day is the case where you have a slowly changing signal with noise on it. Hysteresis prevents the noise from causing the comparator output to oscillate as the input signal gets close to the reference voltage.

To determine the amount of hysteresis in a circuit such as that in Figure 10-1*c*, assume $V_{REF} = 0$ V and $V_{OUT} = 13$ V. A simple voltage-divider calculation will tell you that the noninverting input is at about 13 mV. The voltage on the inverting input of the amplifier will have to go more positive than this before the comparator will change states. Likewise, if you assume V_{OUT} is −13 V, the noninverting input will be at about −13 mV, so the voltage on the inverting input of the amplifier will have to go below this to change the state of the output. The hysteresis of this comparator is +13 mV and −13 mV.

NONINVERTING AMPLIFIER OP-AMP CIRCUIT

When operating in open-loop mode (no feedback to the inverting input), an op amp has a very high, but unpredictable, gain. This is acceptable for use as a comparator, but not for use as a predictable amplifier. Figure 10-1*d* shows one way negative feedback is added to an op amp to produce an amplifier with stable, predictable gain. First of all notice that the input signal in this circuit is applied to the noninverting input, so the output will be in phase with the input. Second, note that a fraction of the output signal is fed back to the inverting input. Now, here's how this works.

To start assume that V_{IN} is 0 V, V_{OUT} is 0 V, and the voltage on the inverting input is 0. Now, suppose that you apply a +0.01-V dc signal to the noninverting input. Since the 0.1-V difference between the two inputs will be amplified by 100,000, the output will head towards 100 V as fast as it can. However, as the output goes positive, some of the output voltage will be fed back to the inverting input through the resistor divider. This feedback to the inverting input will decrease the difference in voltage between the two inputs. To make a long story short, the circuit quickly reaches a predictable balance point where the voltage on the inverting input (V_F) is very, very close to the voltage on the noninverting input (V_{IN}). For a 1.0-V dc output this equilibrium voltage difference might be about 10 μV. If you assume that the voltages on the two inputs are equal, then predicting the output voltage for a given input voltage is simply a voltage divider problem. $V_{OUT} = V_{IN}(R1 + R2)/R1$. If $R2$ is 99 kΩ and $R1$ is 1 kΩ, then $V_{OUT} = V_{IN} \times 100$. For a 0.01-V input signal the output voltage will be 1.00 V. The closed-loop gain, A_{VCL}, for this circuit is equal to the simple resistor ratio, $(R1 + R2)/R1$.

To see another advantage of feeding some of the output signal back to the inverting input, let's see what happens when the load connected to the output of the op amp changes and draws more current from the output. The output voltage will temporarily drop because of the increased load. Part of this drop will be fed back to the inverting input, increasing the difference in voltage between the two inputs. This increased difference will cause the op amp to drive its output to correct for the increased load. Feedback which causes an amplifier to oppose a change on its output is called *negative feedback*. Because of the negative feedback, then, the op amp will work day and night to keep its output stabilized and its two inputs at nearly the same voltage! This is probably the most important point you need to know to analyze or troubleshoot an op-amp circuit with negative feedback. Draw a box around this point in your mind so you don't forget it.

The noninverting circuit we have just discussed is used mostly as a *buffer*, because it has a very high *input impedance* (Z_{IN}), and will therefore not load down a sensor or some other device you connect to its input. Bipolar-transistor-input op amps will have an input impedance greater than 100 MΩ. Some op amps such as the National LF356 have a FET input stage so their input impedance is 10^{12} Ω.

INVERTING AMPLIFIER OP-AMP CIRCUIT

Figure 10-1*e* shows a somewhat more versatile amplifier circuit using negative feedback. Note that in this circuit, the noninverting input is tied to ground with a resistor and the signal you want to amplify is sent to the inverting input through a resistor. The output signal will therefore be 180° out of phase with the input signal. Resistor R_f supplies the negative feedback which keeps the two inputs at nearly the same voltage. Since the noninverting input is tied to ground, the op amp will sink or source current to hold the inverting input also at zero volts. In this circuit the inverting input point is referred to as *virtual ground* because the op amp holds it at ground. The voltage gain of this circuit is also determined by the ratio of two resistors. The A_{VCL} for this circuit at low frequencies is equal to $-R_f/R1$. You can derive this for yourself by just thinking of the two resistors as a voltage divider with V_{IN} at one end, zero volts in the middle, and V_{OUT} on the other end. The minus sign in the gain expression simply indicates that the output is inverted from the input. The input impedance (Z_{IN}) of this circuit is approximately $R1$ because the op amp holds one end of this resistor at zero volts.

One additional characteristic we need to refresh in your mind about op-amp circuits before going on to other op-amp circuits is *gain-bandwidth product*. As we indicated previously, an op amp may have an open-loop dc gain of 100,000 or more. At higher frequencies the gain decreases until, at some frequency, the open-loop gain drops to 1. Figure 10-2*a* shows an open-loop voltage gain versus frequency graph for a common op amp such as a 741. The frequency at which the gain is 1 is referred to on data sheets as the *unity-gain bandwidth* or the *gain-bandwidth product*. A common value for this is 1 MHz. The bandwidth of an amplifier circuit with negative feedback times the low-frequency closed-loop gain will be equal to this value. For example, if an op amp with a gain-bandwidth product of 1 MHz is used to build an amplifier circuit with a closed-loop gain of 100, the bandwidth of the circuit (f_c) will be about 1 MHz/100 or 10 kHz, as shown in Figure 10-2*b*.

OP-AMP ADDER CIRCUIT

Figure 10-1*f* shows a commonly used variation of the inverting amplifier described in the previous section. This circuit adds together or mixes two or more input signals. Here's how it works.

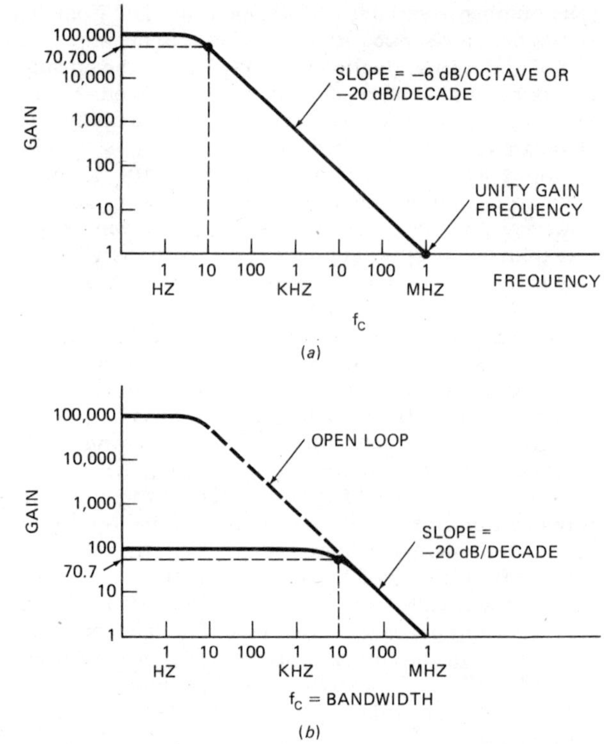

FIGURE 10-2 *(a)* Open-loop gain versus frequency response of 741 op amp. *(b)* Gain versus frequency response of 741 op-amp circuit with closed-loop gain of 100.

Remember from the previous discussion that in an inverting circuit, the op amp holds the inverting input at virtual ground. For the circuit here the input voltage, V_1, produces a current through $R1$ to this point. The input voltage, V_2, causes a current through $R2$ to this point. The two currents add together at the virtual ground, which is commonly called the *summing point* for this circuit. The sum of the two currents is pulled through resistor R_f by the op amp to hold the inverting input at zero volts. The output voltage is then equal to the sum of the currents times the value of R_f, or $(V_1/R1 + V_2/R2) \times R_f$. A circuit such as this is used to "mix" audio signals and to sum binary-weighted currents in a D/A converter. An adder circuit can have several inputs.

SIMPLE DIFFERENTIAL-INPUT AMPLIFIER CIRCUIT

As we will show later, many sensors have two output signal lines with a dc voltage of several volts on each signal line. The dc voltage present on both signal leads is referred to as a *common-mode signal*. The actual signal you need to amplify from these sensors is the few millivolts difference between them. If you try to use a standard inverting or noninverting amplifier circuit to do this, the large dc voltage will be amplified along with the small difference voltage you need to amplify. Figure 10-1*g* shows a simple circuit which, for the most part, solves this problem without using coupling capacitors to block the dc. The analysis of this circuit is beyond the space we have here, but basically the resistors on the noninverting input hold this input at a voltage near the common-mode voltage. The amplifier holds the inverting input at the same voltage. If the resistors are matched carefully, the result is that only the difference in voltage between V_2 and V_1 will be amplified. The output will be a single line which contains only the amplified difference. We say that the common-mode signal has been *rejected*.

AN INSTRUMENTATION AMPLIFIER CIRCUIT

Figure 10-1*h* shows an op-amp circuit used in applications needing a greater rejection of the common-mode signal than is provided by the simple differential circuit in Figure 10-1*g*. The first two op amps in this circuit remove the common-mode voltage and the last op amp converts the result from a differential signal to a signal referenced to ground. Instrumentation amplifier circuits such as this are available in single packages.

AN OP-AMP INTEGRATOR CIRCUIT

Figure 10-1*i* shows an op-amp circuit that can be used to produce linear voltage ramps. A dc voltage applied to the input of this circuit will cause a constant current of $V_{IN}/R1$ to flow into the virtual-ground point. This current flows onto one plate of the capacitor. In order to hold the inverting input at ground, the op-amp output must pull the same current from the other plate of the capacitor. The capacitor is then getting charged by the constant current $V_{IN}/R1$. Basic physics tells you that the voltage across a capacitor being charged by a constant current is a *linear ramp*. Note that because of the inverting amplifier connection, the output will ramp negative for a positive input voltage. Also note that some provision must be made to prevent the amplifier output from ramping into *saturation*.

The circuit is called an integrator because it produces an output voltage proportional to the integral or "sum" of the current produced by an input voltage over a period of time. The waveforms in Figure 10-1*i* show the circuit response for a pulse-input signal.

AN OP-AMP DIFFERENTIATOR CIRCUIT

Figure 10-1*j* shows an op-amp circuit which produces an output signal proportional to the rate of change of the input signal. With the input voltage to this circuit at zero or some other steady dc voltage, the output will be at zero. If a new voltage is applied to the input, the voltage across the capacitor cannot change instantly, so the inverting input will be pulled away from zero volts. This will cause the op amp to drive its output in a direction to charge the capacitor and pull the inverting input back to zero. The waveforms in Figure 10-1*j* show the circuit response for a pulse-input signal. The time required for the output to return to zero is determined by the time constant of *R*1 and *C*.

OP-AMP ACTIVE FILTERS

In many control applications we need to filter out unwanted low-frequency or high-frequency noise from the signals read in from sensors. This could be done with simple *RC* filters, but *active filters* using op amps give much better control over filter characteristics. There are many different filter configurations using op amps. The main points we want to refresh here are the meanings of the terms *low-pass* filter, *high-pass* filter, and *bandpass* filter; and how you identify the type when you find one in a circuit you are analyzing.

A low-pass filter amplifies or passes through low frequencies, but at some frequency determined by circuit values, the output of the filter starts to decrease. The frequency at which the output is down to 0.707 of the low-frequency value is called the *critical frequency* or *break point*. Figure 10-3*a* shows a graph of gain versus frequency for a low-pass filter with the critical frequency (f_c) labeled. Note that above the critical frequency the gain drops off rapidly. For a first-order filter such as a single *R* and *C*, the gain decreases by a factor of 10 for each increase of 10 times in frequency (−20 dB/decade). For a second-order filter the gain decreases by a factor of 100 for each increase of 10 times in frequency.

Figure 10-1*k* shows a common circuit for a second-order low-pass filter. The way you recognize this as a low-pass filter is to look for a dc path from the input to the noninverting input of the amplifier. If the dc path is present, as it is in Figure 10-1*k*, you know that the amplifier can amplify dc and low frequencies. Therefore, it is a low-pass filter with a response such as that shown in Figure 10-3*a*.

For contrast look at the circuit for the second-order high-pass filter in Figure 10-1*l*. Note that in this circuit the dc component of an input signal cannot reach the noninverting input because of the two capacitors in series with that input. Therefore, this circuit will not amplify dc and low-frequency signals. Figure 10-3*b* shows the graph of gain versus frequency for a high-pass filter such as this. Note that the gain-bandwidth product of the op amp limits the high-frequency response of the circuit.

For the low-pass circuit in Figure 10-1*k*, the gain for the flat part of the response curve is 1, or unity, because the output is fed back directly to the inverting input. At the critical frequency, f_c, the gain will be 0.707, and above this frequency the gain will drop off. The critical frequency for the circuit is determined by the equation next to the circuit. The equation assumes that *R*1 and *R*2 are equal, and that the value of *C*1 is twice the value of *C*2. *R*3 is simply a damping resistor. The positive feedback supplied by *C*1 is the reason the gain is only down to 0.707 at the critical frequency rather than down to 0.5 as it would be if we simply cascaded two simple *RC* circuits.

For the high-pass filter, the gain for the flat section of the response curve is also one. Assuming that the two capacitors are equal and the value of *R*2 is twice the value of *R*1, the critical frequency is determined by the formula shown next to Figure 10-1*l*. Again, *R*3 is for damping.

A low-pass filter can be put in series with a high-pass filter to produce a bandpass filter which lets through a desired range of frequencies. There are also many different single amplifier circuits which will pass or reject a band of frequencies.

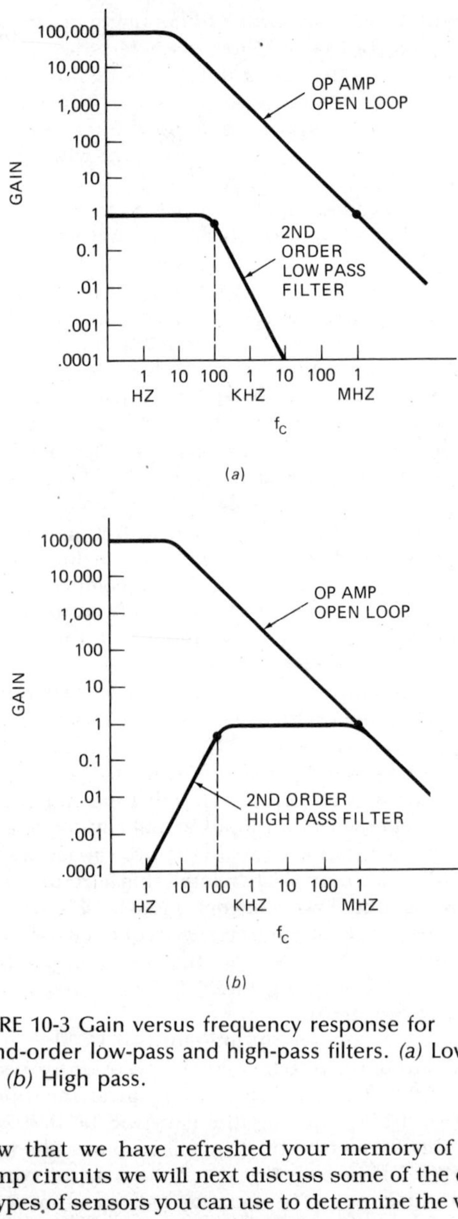

FIGURE 10-3 Gain versus frequency response for second-order low-pass and high-pass filters. *(a)* Low pass. *(b)* High pass.

Now that we have refreshed your memory of basic op-amp circuits we will next discuss some of the different types of sensors you can use to determine the values of temperatures, pressures, position, etc.

SENSORS AND TRANSDUCERS

It would take a book many times the size of this one to describe the operation and applications of all of the different types of available sensors and transducers. What we want to do here is introduce you to a few of these and show how they can be used to get data for microcomputer-based instruments in, for example, our electronics factory.

Light Sensors

One of the simplest light sensors is a light-dependent resistor such as the Clairex CL905 shown in Figure 10-4*a*. A glass window allows light to fall on a zig-zag pattern of cadmium sulfide or cadmium selenide whose resistance depends on the amount of light present. The resistance of the CL905 varies from about 15 MΩ when in the dark to about 15 kΩ when in a bright light. Photoresistors such as this do not have a very fast response time and are not stable with temperature, but they are inexpensive, durable, and sensitive. For these reasons they are usually used in applications where a measurement of the amount of light need not be precise. The devices on top of streetlights which turn them on when it gets dark, for example, contain a photoresistor, a transistor driver, and a mechanical relay as shown in Figure 10-4*b*. As it gets dark, the resistance of the photoresistor goes up. This increases the voltage on the base of the transistor until, at some point, it turns on. This turns on the transistor driving the relay, which in turn switches on the lamp.

Another device used to sense the amount of light present is a photodiode. If light is allowed to fall on a specially constructed silicon diode, the reverse leakage current of the diode increases linearly as the amount of light falling on it increases. A circuit such as that shown in Figure 10-5 can be used to convert this small leakage current to a proportional voltage. Note that in this circuit a negative reference voltage is applied to the noninverting input of the amplifier. The op amp will then produce this same voltage on its inverting input, reverse biasing the photodiode. The op amp will pull the photodiode leakage current through R_f to produce a proportional voltage on the output of the amplifier. For a typical photodiode such as the HP 5082-4203 shown, the reverse leakage current varies from near 0 μA to about 100 μA, so with the 100-kΩ R_f, an output voltage of about 0 V to 10 V will be produced. The circuit will work

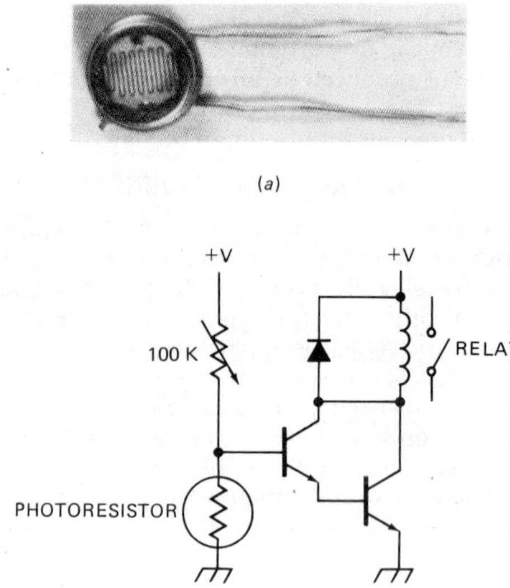

FIGURE 10-4 *(a)* Cadmium sulfide photocell. *(Clairex Electronics)*. *(b)* Light-controller relay circuit using a photocell.

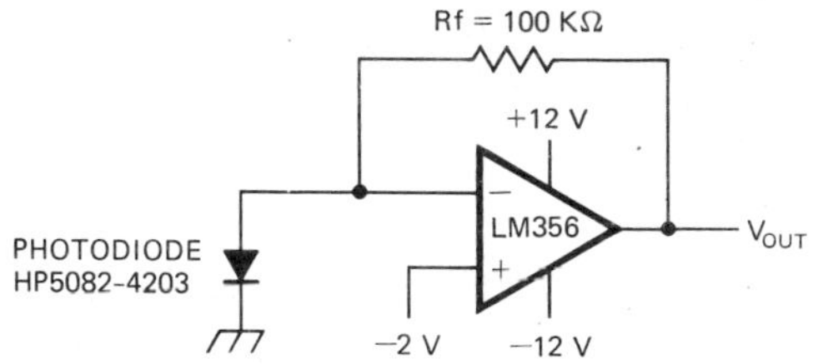

FIGURE 10-5 Photodiode circuit to measure infrared light intensity.

without any reverse bias on the diode, but with the reverse bias, the diode responds faster to changes in light. An LM356 FET input amplifier is used here because it does not require an input bias current.

A photodiode circuit such as this might be used to determine the amount of smoke being emitted from a smokestack. To do this a gallium arsenide infrared LED is put on one side of the smokestack, and the photodetector circuit put on the other. Since smoke absorbs light, the amount of light arriving at the photodetector is a measure of the amount of smoke present. An infrared LED is used here because the photodiode is most sensitive to light wavelengths in the infrared region.

Still another useful light-sensitive device is a solar cell. Common solar cells are simply large, very heavily-doped, silicon PN junctions. Light shining on the solar cell causes a reverse current to flow, just as in the photodiode. Because of the large area and the heavy doping in the solar cell, however, the current produced is milliamperes rather than microamperes. The cell functions as a light-powered battery. Solar cells can be connected in a series-parallel array to produce a solar power supply.

Light meters in cameras, photographic enlargers, and our printed-circuit-board-making machine use solar cells. The current from the solar cell is a linear function of the amount of light falling on the cell. A circuit such as that in Figure 10-5 can be used to convert the output current to a proportional voltage. Because of the larger output current we decrease R_f to a much smaller value, depending on the output current of the cell. We also connect the noninverting input of the amplifier to ground because we don't use reverse biasing with solar cells. The frequency response to light (spectral response) of solar cells has been tailored to match the output of the sun. Therefore, they are ideal in photographic applications where we want a signal proportional to the total light from the sun or from an incandescent lamp.

Temperature Sensors

Again, there are many types of temperature sensors. The two types we discuss here are: semiconductor devices, which are inexpensive and can be used to measure temperatures over the range of −55°C to 100°C, and thermocouples which can be used to measure very low temperatures and very high temperatures.

SEMICONDUCTOR TEMPERATURE SENSORS

The two main types of semiconductor temperature sensors are temperature-sensitive voltage sources and temperature-sensitive current sources. An example of the first type is the National LM35 which we show the circuit connections for in Figure 10-6*a*. The voltage output from this circuit increases by 10 mV for each degree Celsius that its temperature is increased. By connecting the output to a negative reference voltage (V_S) as shown, the sensor will give a meaningful output for a temperature range of −55 to +150°C. You adjust the output to zero volts for 0°C. The output voltage can be amplified to give the voltage range you need for a particular applica-

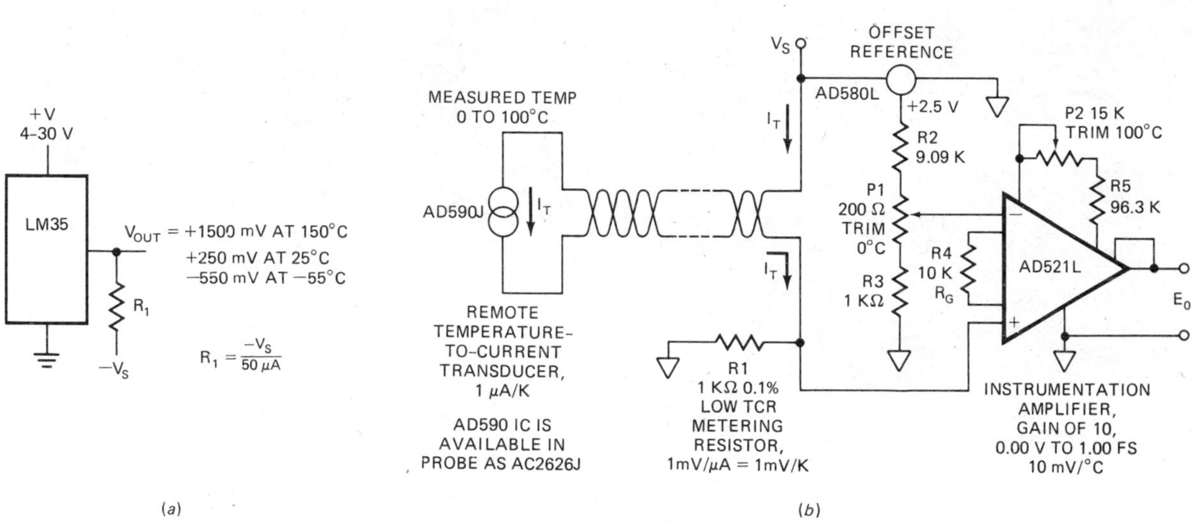

FIGURE 10-6 Semiconductor temperature-sensor circuits. *(a)* LM35 temperature-dependent voltage source. *(b)* AD590 temperature-dependent current source. *(Analog Devices Incorporated)*

tion. In a later section of this chapter we show another circuit using the LM35 temperature sensor. The accuracy of this device is about 1°C.

Another common semiconductor temperature sensor is a temperature-dependent current source such as the Analog Devices AD590. The AD590 produces a current of 1 μA/°Kelvin. Figure 10-6*b* shows a circuit which converts this current to a proportional voltage. In this circuit the current from the sensor (I_T) is passed through an approximately 1-kΩ resistor to ground. This produces a voltage which changes by 1 mV/°Kelvin. The AD580 is a precision voltage reference used to produce a reference voltage of 273.2 mV. With this voltage applied to the inverting input of the amplifier, the amplifier output will be at zero volts for 0°C. The advantage of a current-source sensor is that voltage drops in long connecting wires do not have any effect on the output value. If the gain and offset are carefully adjusted, the accuracy of the circuit in Figure 10-6*b* is ±1°C using an AD590K part.

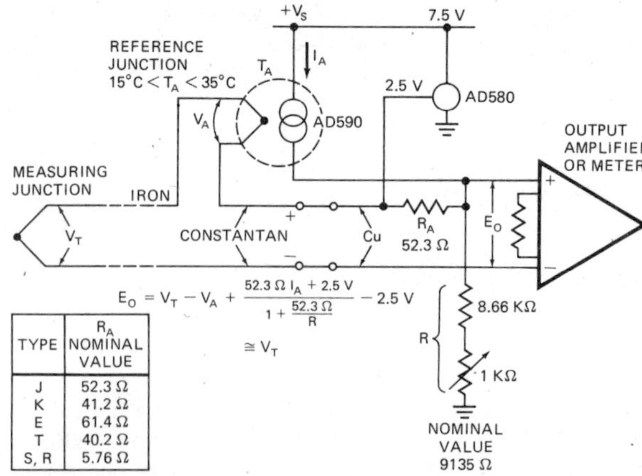

$$E_O = V_T - V_A + \frac{52.3\ \Omega\ I_A + 2.5\ V}{1 + \frac{52.3\ \Omega}{R}} - 2.5\ V \cong V_T$$

TYPE	R_A NOMINAL VALUE
J	52.3 Ω
K	41.2 Ω
E	61.4 Ω
T	40.2 Ω
S, R	5.76 Ω

FIGURE 10-7 Circuit showing amplification and cold-junction compensation for thermocouple. *(Analog Devices Incorporated)*

THERMOCOUPLES

Whenever two different metals are put in contact, a small voltage is produced between them. The voltage developed depends on the types of metals used and the temperature. Depending on the metals, the developed voltage increases between 7 μV and 75 μV for each degree Celsius increase in temperature. Different combinations of metals are useful for measuring different temperature ranges. A thermocouple junction made of iron and constantan, commonly called a type J thermocouple, has a useful temperature range of about −184 to +760°C. A junction of platinum and an alloy of platinum and 13 percent rhodium has a useful range of 0°C to about 1600°C. Thermocouples can be made small, rugged, and stable; however, they have three major problems which must be overcome.

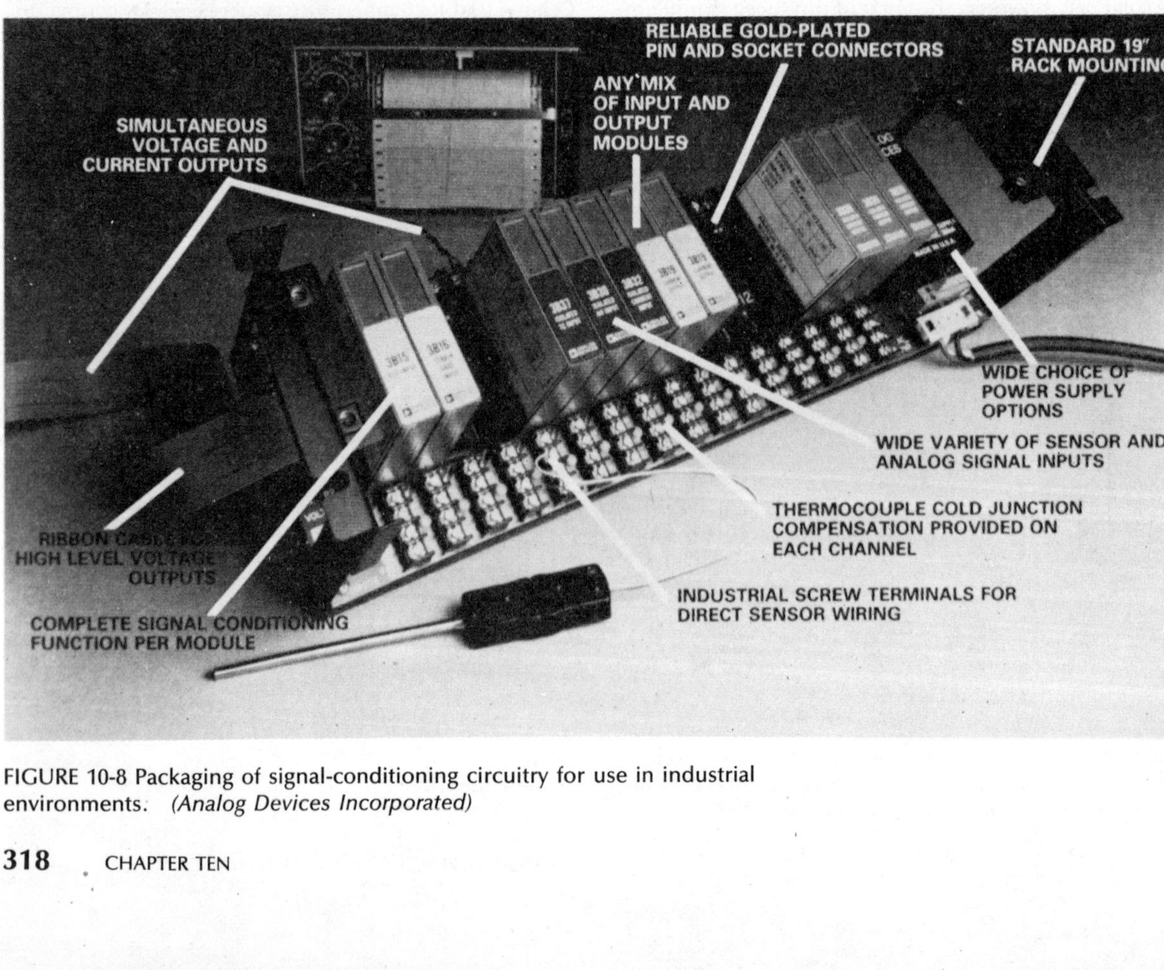

FIGURE 10-8 Packaging of signal-conditioning circuitry for use in industrial environments. *(Analog Devices Incorporated)*

First of these is the fact that the output is very small and must be amplified a great deal to bring it up into the range where it can, for example, drive an A/D converter. Second, in order to make accurate measurements, a second junction made of the same metals must be included in the circuit as a reference. Adding this second junction is referred to as *cold-junction compensation*. Figure 10-7 shows a circuit to amplify the output of a thermocouple and provide cold-junction compensation for a type J thermocouple.

The first thing to notice in the circuit is that the reference junction is connected in the reverse direction from the measuring junction. This is done so that the output connecting wires are both constantan. The thermocouples formed by connecting these wires to the copper wires going to the amplifier will then cancel out. The resultant output voltage will be the difference between the voltages across the two thermocouples. If we simply amplify the output of the two thermocouples, however, there is a problem if the temperature of both thermocouples is changing. The problem is that it is impossible to tell which thermocouple caused a change in output voltage. One cure for this is to put the reference junction in an ice bath or a small oven to hold it at a constant temperature. This solution is usually inconvenient, so instead a circuit such as that in Figure 10-7 is used to compensate electronically for changes in the temperature of the reference junction.

As we discussed in a previous section the AD590 shown here produces a current proportional to its temperature. The AD590 is attached to the reference thermocouple so that they are both at the same temperature. The current from the AD590, when passed through the resistor network, produces a voltage which compensates for changes in the reference thermocouple with temperature. The output amplifier for this circuit is a differential amplifier such as that shown in Figure 10-1*g* or the instrumentation amplifier shown in Figure 10-1*h*.

The third problem with thermocouples is that their output voltages do not change linearly with temperature. This can be corrected with analog circuitry which changes the gain of an amplifier according to the value of the signal. However, when a thermocouple is used with a microcomputer-based instrument, the correction can be easily done using a lookup table in ROM. An A/D converter converts the voltage from the thermocouple to a digital value. The digital value is then used as a pointer to a ROM location which contains the correct temperature for that reading.

For use in industrial environments, circuitry such as that in Figure 10-7 is usually packaged in durable modules and mounted on racks in metal cabinets. Figure 10-8 shows some of the Analog Devices 3B series signal-conditioning modules on a rack-mount panel. The 3B37, for example, is a thermocouple-amplifier module with built in cold-junction compensation. The silver probe in front of the unit is a common type of thermocouple. This rack unit is constructed so that you can plug in the modules you need for a given application. Modules such as these usually have both a voltage output and a current output. Sending a signal as a current has the advantages that the signal amplitude is then not affected by resistance, induced-voltage noise, or voltage drops in a long connecting line. A common range of currents used to represent analog signals in industrial environments is 4 mA to 20 mA. A current of 4 mA represents a zero output, and a current of 20 mA represents the full-scale value. The reason that the current range is offset from zero is so that a current of zero is left to represent an open circuit. The current can be converted to a proportional voltage at the receiving end by simply passing it through a resistor.

Force and Pressure Transducers

To convert force or pressure (force/area) to a proportional electrical signal, the most common methods use *strain gages* and *linear variable differential transformers* (LVDTs). Both of these methods involve moving something. This is why we refer to them as *transducers* rather than as sensors. Here's how strain gages work.

STRAIN GAGES AND LOAD CELLS

A strain gage is a small resistor whose value changes when its length is changed. It may be made of thin wire, thin foil, or semiconductor material. Figure 10-9*a* shows a simple setup for measuring force or weight with strain gages. One end of a piece of spring steel is attached to a fixed surface. A strain gage is glued on the top of the flexible bar. The force or weight to be measured is applied to the unattached end of the bar. As the applied force bends the bar, the strain gage is stretched, increasing its resistance. Since the amount that the bar is bent is directly proportional to the applied force, the change in resistance will be proportional to the applied force. If a current is passed through the strain gage, then the change in voltage across the strain gage will be proportional to the applied force.

Unfortunately, the resistance of the strain-gage elements also changes with temperature. To compensate for this problem two strain-gage elements mounted at right angles as shown in Figure 10-9*b* are often used. Both of the elements will change resistance with temperature, but only element A will change resistance appreciably with applied force. When these two elements are connected in a balanced-bridge configuration as shown in Figure 10-9*c*, any change in resistance of the elements due to temperature will have no effect on the differential output of the bridge. However, as force is applied, the resistance of the element under strain will change and produce a small differential output voltage. The full-scale differential output voltage is typically 2 or 3 mV per volt of applied voltage. For example if 10 V is applied to the bridge, the full-load output voltage will only be 20 or 30 mV. This small signal can be amplified with a differential amplifier or an instrumentation amplifier. The Analog Devices 3B16 module shown in Figure 10-8 provides a 10-V excitation voltage and amplification for the differential output signal for a strain-gage bridge.

Strain-gage bridges are used in many different forms to measure many different types of force and pressure. If

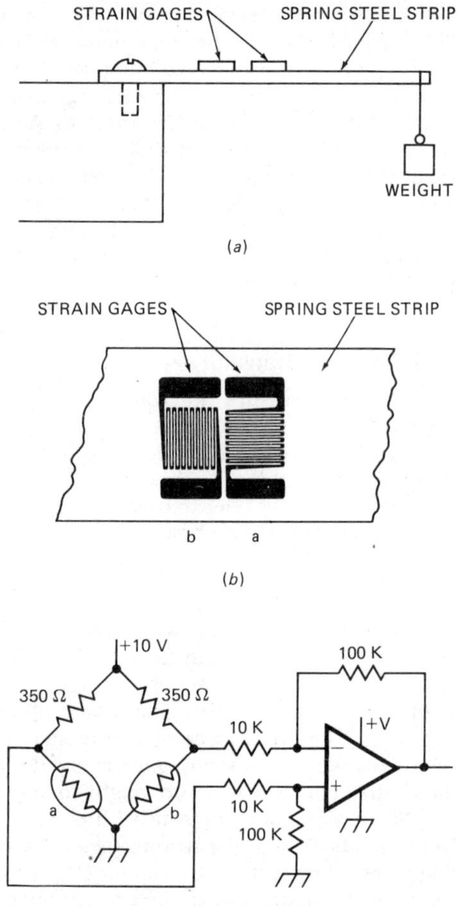

FIGURE 10-9 Strain gauges used to measure force. *(a)* Side view. *(b)* Top view (expanded). *(c)* Circuit connections.

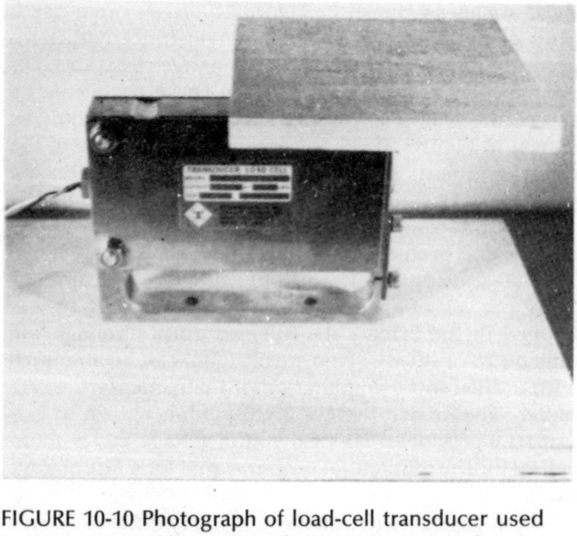

FIGURE 10-10 Photograph of load-cell transducer used to measure weight. *(Transducers, Incorporated)*

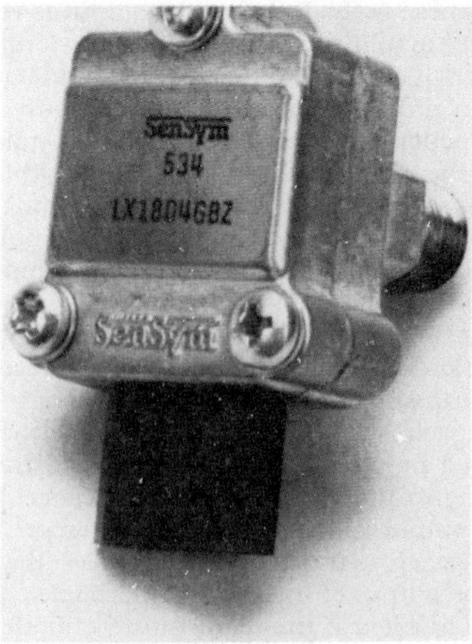

FIGURE 10-11 LX1804GBZ pressure transducer. *(Sensym, Inc.)*

the strain-gage bridge is connected to a bendable beam structure as shown in Figure 10-9*a*, the result is called a *load cell* and is used to measure weight. Figure 10-10 shows a 10-lb load cell that might be used in a microprocessor-controlled delicatessen scale or postal scale. Larger versions can be used to weigh barrels being filled, or even trucks.

If a strain-gage bridge is mounted on a movable diaphragm in a threaded housing, the output of the bridge will be proportional to the pressure applied to the diaphragm. If a vacuum is present on one side of the diaphragm, then the value read out will be a measure of the absolute pressure. If one side of the diaphragm is open then the output will be a measure of the pressure relative to atmospheric pressure. If the two sides of the diaphragm are connected to two other pressure sources, then the output will be a measure of the differential pressure between the two sides. Figure 10-11 shows a SENSYM LX1804GBZ pressure transducer which measures pressures in the range of 0 to 15 lb per square inch. A transducer such as this might be used to measure blood pressure in a microcomputer-based medical instrument.

LINEAR VARIABLE DIFFERENTIAL TRANSFORMERS

An *LVDT* is another type of transducer often used to measure force, pressure or position. Figure 10-12 shows the basic structure of an LVDT. It consists of three coils of wire wound on the same form and a movable iron core. An ac excitation signal of perhaps 20 kHz is applied to the primary. The secondaries are connected so that the voltage induced in one opposes the voltage induced in the other. If the core is centered then the induced voltages are equal and they cancel, so there is no net output voltage. If the coil is moved off center, coupling will be stronger to one secondary coil so that coil

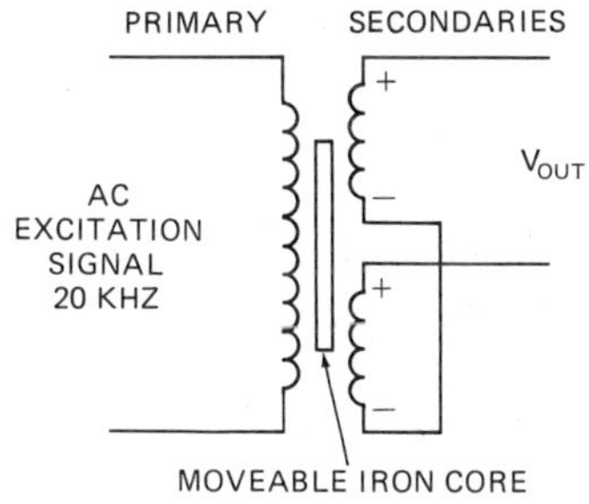

FIGURE 10-12 Linear variable differential transformer (LVDT) structure.

will produce a greater output voltage. The result will be a net output voltage. The phase relationship between the output signal and the input signal is an indication of which direction the core moved from the center position. The amplitude of the output signal is linearly proportional to how far the core moves from the center position.

An LVDT can be used directly in this form to measure displacement or position. If you add a spring so that a force is required to move the core, then the voltage out of the LVDT will be proportional to the force applied to the core. In this form the LVDT can be used in a load cell for an electronic scale. Likewise, if a spring is added and the core of the LVDT attached to a diaphragm in a threaded housing, the output from the LVDT will be proportional to the pressure exerted on the diaphragm. We do not have space here to show the ac-interface circuitry required for an LVDT.

Flow Sensors

If we are going to control the flow rate of some material in our electronics factory, we need to be able to measure it. Depending on the material, flow rate, and temperature, we use different methods.

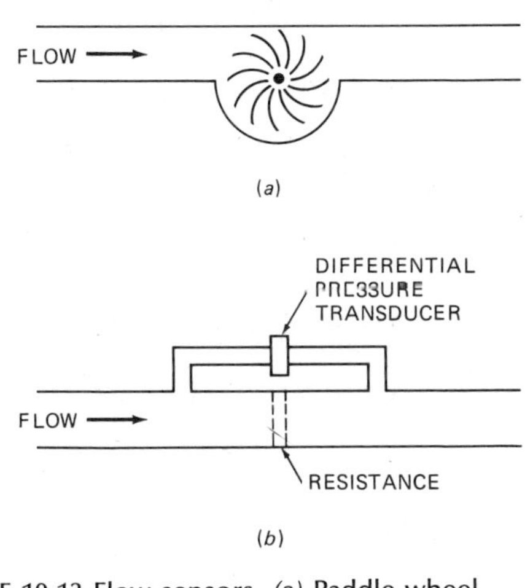

FIGURE 10-13 Flow sensors. *(a)* Paddle wheel. *(b)* Differential pressure.

One method used is to put a paddle wheel in the flow as shown in Figure 10-13*a*. The rate at which the paddle wheel turns is proportional to the rate of flow of a liquid or gas. An optical encoder can be attached to the shaft of the paddle wheel to produce digital information as to how fast the paddle wheel is turning.

A second common method of measuring flow is with a *differential pressure transducer*, as shown in Figure 10-13*b*. A wire mesh or screen is put in the pipe to create some resistance. Flow through this resistance produces a difference in pressure between the two sides of the resistance. The pressure transducer gives an output proportional to the difference in pressure between the two sides of the resistance. In the same way that the voltage across an electrical resistor is proportional to the flow of current through the resistor, the output of the pressure transducer is proportional to the flow of a liquid or gas through the pipe.

Other Sensors

As we mentioned previously, the number of different types of sensors is very large. In addition to the types we have discussed, there are sensors to measure pH, concentration of various gases, thickness of materials, and just about anything else you might want to measure. Often you can use commonly available transducers in creative ways to solve a particular application problem you have. Suppose, for example, that you need to accurately determine the level of a liquid in a large tank. To do this you could install a pressure transducer at the bottom of the tank. The pressure in a liquid is proportional to the height of the liquid in the tank, so you can easily convert a pressure reading to the desired liquid height. The point here is to check out what is available and then be creative.

D/A CONVERTER OPERATION, INTERFACING, AND APPLICATIONS

In the previous sections of this chapter we have discussed how we use sensors to get electrical signals proportional to pressure, temperature, etc. and how we use op amps to amplify and filter these electrical signals. The next logical step would be to show you how we use an A/D converter to get these signals into digital form that a microcomputer can work with. However, since D/A converters are simpler and several types of A/D converter have D/As as part of their circuitry, we will discuss D/As first.

D/A Converter Operation and Specifications

OPERATION

The purpose of a digital-to-analog converter is to convert a binary word to a proportional current or voltage. To see how this is done let's look at the simple op-amp circuit in Figure 10-14. This circuit functions as an adder. Since the noninverting input of the op amp is grounded,

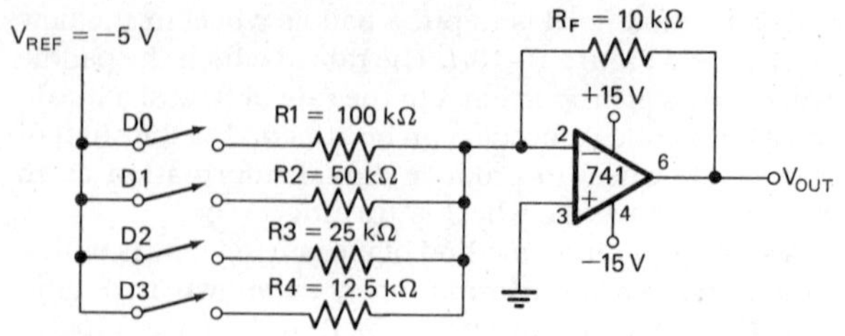

FIGURE 10-14 Simple 4-bit digital-to-analog (D/A) converter.

the op amp will work day and night to hold the inverting input also at zero volts. This point, remember, is referred to as a virtual ground or summing point. When one of the switches is closed a current will flow from the −5-V (V_{REF}) through that resistor to the summing point. The op amp will pull the current on through the feedback resistor to produce a proportional output voltage. If you close switch D0, for example, a current of 0.05 mA will flow into the summing point. In order to pull this current through the feedback resistor, the op amp must put a voltage of 0.05 mA × 10 kΩ or 0.5 V on its output. If you also close switch D1, it will send another 0.1 mA into the summing point. In order to pull the sum of the currents through the feedback resistor, the op amp has to output a voltage of 0.15 mA × 10 kΩ or 1.5 V.

The point here is that the binary-weighted resistors produce binary-weighted currents which are summed by the op amp to produce a proportional output voltage. The binary word applied to the switches produces a proportional output voltage. Technically the output voltage is "digital" because it can only have certain fixed values just as the display on a digital voltmeter can. However, the output simulates an analog signal, so we refer to it as analog. Switch D3 in Figure 10-14 represents the most significant bit, because closing it produces the largest current. Note that since V_{REF} is negative, the output will go positive as switches are closed.

As you see here, the heart of a D/A converter is a set of binary-weighted current sources which can be switched on or off according to a binary word applied to its inputs. Since these current sources are usually inside an IC, we don't need to discuss the different ways the binary-weighted currents can be produced. As shown in Figure 10-14, a simple op-amp circuit can be used to convert the sum of the currents to a proportional voltage if needed.

D/A CHARACTERISTICS AND SPECIFICATIONS

Figure 10-15 shows the circuit for an inexpensive IC D/A converter with an op-amp circuit as a current-to-voltage converter. We will use this circuit for our discussion of D/A characteristics.

The first characteristic of a D/A converter to consider is resolution. This is determined by the number of bits in the input binary word. A converter with eight binary inputs such as the one in Figure 10-15 has 2^8 or 256 possible output levels, so its resolution is 1 part in 256. As another example, a 12-bit converter has a resolution of 1 part in 2^{12} or 4096. Resolution is sometimes expressed as a percentage. The resolution of an 8-bit converter then is about 0.39 percent.

The next D/A characteristic to determine is the *full-scale output voltage*. For the converter in Figure 10-15 the current for all of the switches is supplied by V_{REF} through *R*14. The current output from pin 4 of the D/A is pulled through R_o to produce the output voltage. The formula for the output voltage is shown under the circuit in Figure 10-15. In the equation the term *A*1, for example, represents the condition of the switch for that bit. If a switch is closed, allowing a current to flow, put a 1 in that bit. If a switch is open, put a 0 in that bit. As we also show in Figure 10-15, if all of the switches are closed, the output will be (10 V)(255/256) or 9.961 V. Even though the output voltage can never actually get to 10 V, this is referred to as a *10-V output converter*. The maximum output voltage of a converter will always have a value 1 least significant bit less than the named value. As another example of this, suppose that you have a 12-bit, 10-V converter. The value of 1 LSB will be (10 V)/4096 or 2.44 mV. The highest voltage out of this converter when it is properly adjusted will then be (10.0000 − 0.0024) V or 9.9976 V.

Several different binary codes such as *straight binary*, BCD, and *offset binary* are commonly used as inputs to D/A converters. We will show examples of these codes in a later discussion of A/D converters.

The accuracy specification for a D/A converter is a comparison between the actual output and the expected output. It is specified as a percentage of the full-scale

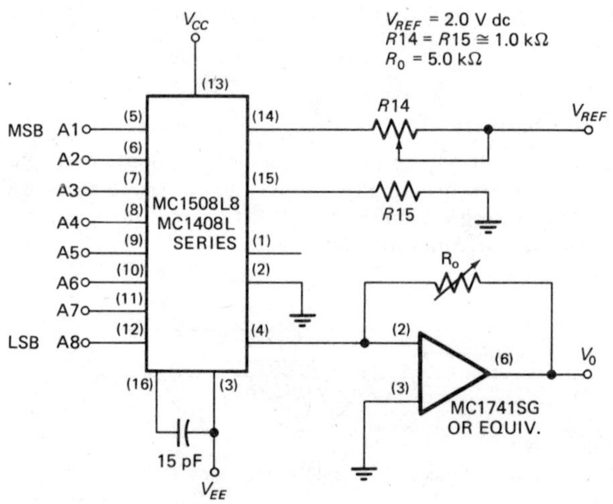

Theoretical V_0

$$V_0 = \frac{V_{REF}}{R14}(R_0)\left\{\frac{A1}{2} + \frac{A2}{4} + \frac{A3}{8} + \frac{A4}{16} + \frac{A5}{32} + \frac{A6}{64} + \frac{A7}{128} + \frac{A8}{256}\right\}$$

ADJUST V_{REF}, $R14$ OR R_0 SO THAT V_0 WITH ALL DIGITAL INPUTS AT HIGH LEVEL IS EQUAL TO 9.961 V

$$V_0 = \frac{2\text{ V}}{1\text{ k}\Omega}(5\text{ k}\Omega)\left\{\frac{1}{2} + \frac{1}{4} + \frac{1}{8} + \frac{1}{16} + \frac{1}{32} + \frac{1}{64} + \frac{1}{128} + \frac{1}{256}\right\}$$

$$= 10\text{ V}\left\{\frac{255}{256}\right\} = 9.961\text{ V}$$

FIGURE 10-15 Motorola MC1408 8-bit D/A with current-to-voltage converter.

output voltage or current. If a converter has a full-scale output of 10 V and ±0.2 percent accuracy, then the *maximum error* for any output will be 0.002 × 10.00 V or 20 mV. Ideally the maximum error for a D/A converter should be no more than ±1/2 of the value of the LSB.

Another important specification for a D/A converter is *linearity*. Linearity is a measure of how much the output ramp deviates from a straight line as the converter is stepped from no switches on to all switches on. Ideally the deviation of the output from a straight line should be no greater than ±1/2 of the value of the LSB to maintain overall accuracy. However, many D/A converters are marketed which have linearity errors greater than that. National Semiconductor, for example, markets the DAC1020, DAC1021, DAC1022 series of 10-bit-resolution converters. The linearity specification for the DAC1020 is 0.05 percent, which is appropriate for a 10-bit converter. The DAC1021 has a linearity specification of 0.10 percent and the DAC1022 has a specification of 0.20 percent. The question that may occur to you at this point is, "What good is it to have a 10-bit converter if the linearity is only equivalent to that of an 8- or 9-bit converter?" The answer to this question is that for many applications the resolution given by a 10-bit converter is needed for small output signals, but it doesn't matter if the output value is somewhat nonlinear for large signals. The price you pay for a D/A converter is proportional not only to its resolution, but also to its linearity specification.

Still another D/A specification to look for is *settling time*. When you change the binary word applied to the input of a converter, the output will change to the appropriate new value. The output, however, may overshoot the correct value and "ring" for a while before finally settling down to the correct value. The time the output takes to get within ±1/2 LSB of the final value is called settling time. As an example, the National DAC1020 10-bit converter has a typical settling time of 500 ns for a full-scale change on the output. This specification is important, because if a converter is operated at too high a frequency, it may not have time to settle to one value before it is switched to the next.

D/A Applications and Interfacing to Microcomputers

D/A converters have many applications besides those where they are used with a microcomputer. In a compact-disk audio player, for example, a 13- or 14-bit D/A converter is used to convert the binary data read off the disk by a laser to an analog audio signal. Most speech-synthesizer ICs contain an D/A converter to convert stored binary data for words into analog audio signals. Here, however, we are primarily interested in the use of a D/A converter with a microcomputer.

The inputs of the D/A circuit (A1–A8) in Figure 10-15 can connected directly to a microcomputer output port. As part of a program you can produce any desired voltage on the output of the D/A. Here's some ideas as to what you might use this circuit for.

As a first example, suppose that you want to build a microcomputer-controlled tester which determines the effect of power-supply voltage on the output voltage of some integrated-circuit amplifiers. If you connect the output of the D/A converter to the reference input of a programmable power supply, or simply add the high-current buffer circuit shown in Figure 10-16 to the output of the D/A, you have a power supply which you can vary under program control. To determine the output voltage of the IC under test as you vary its supply voltage, connect the input of an A/D converter to the IC output, and connect the output of the A/D converter to an input port of your microcomputer. You can then read in the value of the output voltage on the IC.

Another application you might use a D/A and a power buffer for is to vary the voltage supplied to a small resistive heater under program control. Also, the speed of small dc motors is proportional to the amount of current passed through them, so you could connect a small dc motor on the output of the power buffer and control the speed of the motor by the value you output to the D/A. Note that without feedback control the speed of the motor will vary if the load changes. Later we show you how to add feedback control to maintain constant motor speed under changing loads.

So far we have talked about using an 8 bit D/A with a microprocessor. Interfacing an 8-bit converter involves simply connecting the inputs of the converter to an output port, or for some D/As simply connecting the inputs to the buses as you would a port device. Now, suppose that for some application you need 12 bits of resolution, so you need to interface a 12-bit converter. If you are working with a system which has an 8-bit data bus, your first thought might be to connect the lower eight inputs of the 12-bit converter to one output port and the upper four inputs to another port. You could send the lower 8 bits with one write operation, and the upper 4 bits with another write operation. However, there is a potential problem with this approach caused by the time between the two writes. Suppose, for example, that you want to change the output of a 12-bit converter from 0000 1111 1111 to 0001 0000 0000. When you write the lower 8 bits, the output will go from 0000 1111 1111 to 0000 0000 0000. When you write the upper 4 bits, the output will then go back up to the desired 0001 0000 0000. The point here is that for the time between the two writes the output will go to an unwanted value. In

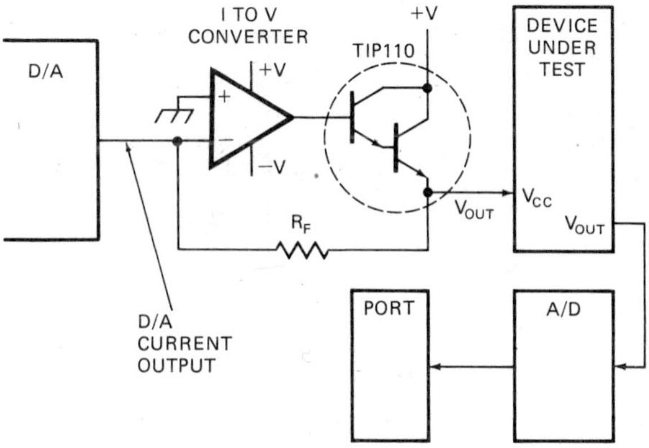

FIGURE 10-16 High-power buffer for D/A output.

many systems this could be disastrous. The cure for this problem is to put latches on the input lines. The latches can be loaded separately and then strobed together to pass all 12 bits to the D/A converter at the same time.

Many currently available D/A converters contain built-in latches to make this easier. Figure 10-17*a* shows a block diagram of the National DAC1230- and DAC1208-type 12-bit converters. Note the internal latches and the register. The DAC1230 series of parts has the upper 4 input bits connected to the lower 4 bits so that the 12 bits can be written with two write operations from an 8-bit port or data bus such as that of the 8088 microprocessor. The DAC1208 series of parts has the upper four data inputs available separately so they can be connected directly to the bus in a system which has a 16-bit data bus, as shown in Figure 10-17*a*. If, for example, you want to connect up a DAC1208 converter to an SDK-86 board, you can simply connect the DAC1208 data inputs to the lower 12 data bus lines, connect the $\overline{CS}$ input to an address decoder output, connect the $\overline{WR1}$ input to the system $\overline{WR}$ line, and tie the $\overline{WR2}$ and $\overline{XFER}$ inputs to ground. The BYTE1/$\overline{BYTE2}$ input is tied high. You then write words to the converter just as if it were a 16-bit port. The timing parameters for the DAC1208 are acceptable for an 8086 operating with a clock frequency of 5 MHz or less. For higher 8086 clock frequencies you would have to add a one-shot or other circuitry which inserts a WAIT state each time you write to the D/A. Here's a few notes about the analog connections for these devices.

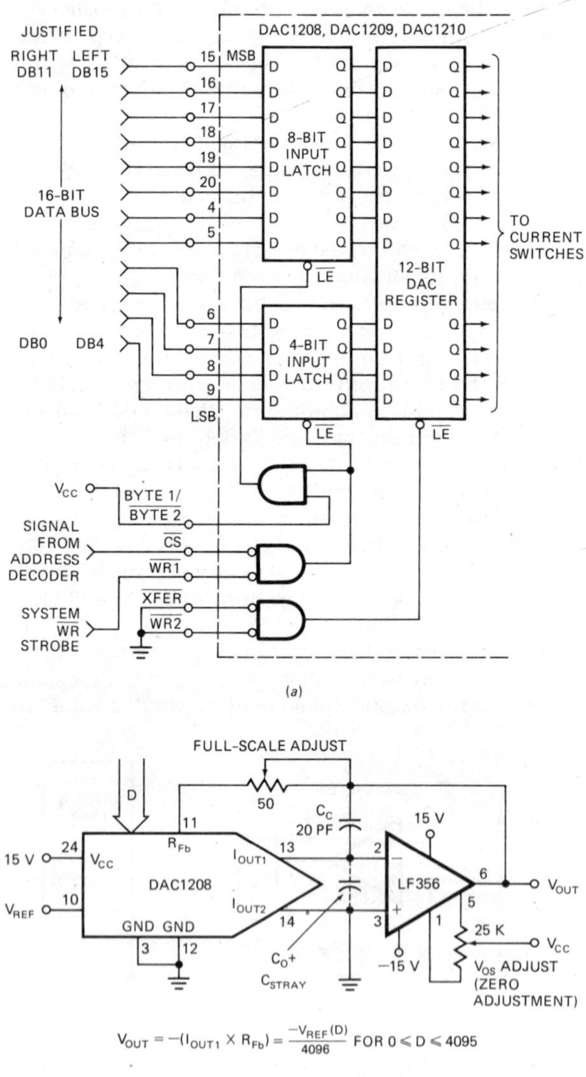

FIGURE 10-17 *(a)* National DAC1208 12-bit D/A input block diagram showing internal latches. *(b)* Analog circuit connections.

These D/A converters require a precision voltage reference. The circuit in Figure 10-17*b* uses a −10.000-V reference. The D/A converters have a current output so we use an op amp, as shown, to convert the output current to a proportional voltage. A FET input amplifier is used, because the input bias current of a bipolar input amp might affect the accuracy of the output. The DAC1208 and DAC1230 have built-in feedback resistors which match the temperature characteristics of the internal current-divider resistors, so all you have to add externally is a 50-Ω resistor for "tweaking" purposes. With a −10.000-V reference as shown, the output voltage will be equal to (the digital input word/4096) × (+10.000 V). Note that the D/A has both a digital ground and an analog ground. To avoid getting digital noise in the analog portions of the circuit, these two should be connected together only at the power supply.

A/D CONVERTER TYPES, SPECIFICATIONS, AND INTERFACING

A/D Converter Types

The function of an A/D converter is to produce a digital word which represents the magnitude of some analog voltage or current. The specifications for an A/D converter are very similar to those for a D/A converter. The resolution of an A/D converter refers to the number of bits in the output binary word. An 8-bit converter, for example, has a resolution of 1 part in 256. Accuracy and linearity specifications have the same meanings for an A/D converter as we described previously for a D/A converter. Another important specification for an A/D converter is its *conversion time*. This is simply the time it takes the converter to produce a valid output binary code for an applied input voltage. When we refer to a converter as *high-speed* we mean it has a short conversion time. There are many different ways to do an A/D conversion, but we only have space here to review three commonly used methods, which represent a wide variety of conversion times.

PARALLEL COMPARATOR A/D CONVERTER

Figure 10-18 shows a circuit for a 2-bit A/D converter using *parallel comparators*. A voltage divider sets reference voltages on the inverting inputs of each of the com-

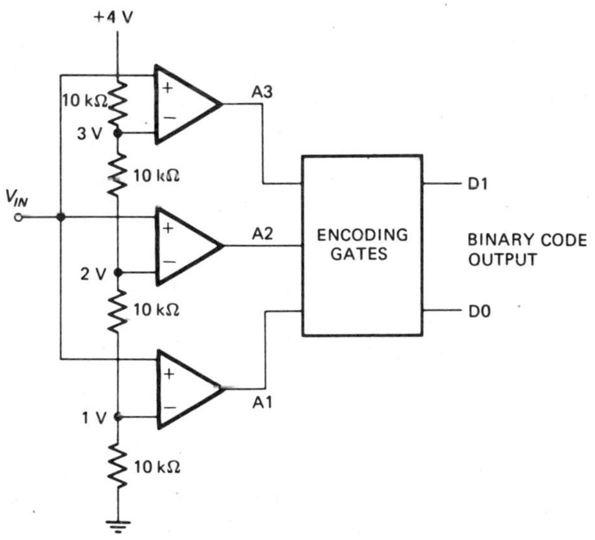

FIGURE 10-18 Parallel comparator A/D converter.

parators. The voltage at the top of the divider chain represents the full-scale value for the converter. The voltage to be converted is applied to the noninverting inputs of all of the comparators in parallel. If the input voltage on a comparator is greater than the reference voltage on the inverting input, the output of the comparator will go high. The outputs of the comparators then give us a digital representation of the voltage level of the input signal. With an input voltage of 2.6 V, for example, the outputs of comparators $A1$ and $A2$ will be high.

The major advantage of a parallel, or *flash*, A/D is its speed of conversion, which is simply the propagation delay time of the comparators. The output code from the comparators is not a standard binary code, but it can be converted with some simple logic. The major disadvantage of a flash A/D is the number of comparators needed to produce a result with a reasonable amount of resolution. The 2-bit converter in Figure 10-18 requires three comparators. To produce a converter with N bits of resolution you need $(2^N - 1)$ comparators. In other words for an 8-bit conversion you need 255 comparators, and for a 10-bit flash converter you need 1023 comparators. Single-package flash converters are available from TRW for applications where the high speed is required, but they are relatively expensive. Devices are available which can do an 8-bit conversion in 20 ns.

DUAL-SLOPE A/D CONVERTERS

Figure 10-19*a* shows a functional block diagram of a *dual-slope* A/D converter. This type of converter is often used as the heart of a digital voltmeter because it can give a large number of bits of resolution at a low cost. Here's how the converter in Figure 10-19 works.

To start, the control circuitry resets all of the counters to zero and connects the input of the integrator to the input voltage to be converted. If you assume the input voltage is positive, then this will cause the output of the integrator to ramp negative as shown in Figure 10-19*b*. As soon as the output of the integrator goes a few microvolts below ground, the comparator output will go high. The comparator output being high enables the AND gate and lets the 1-MHz clock into the counter chain. After some fixed number of counts the control circuitry switches the input of the integrator to a negative reference voltage and resets the counters all to zero. With a negative input voltage the integrator output will ramp positive as shown in Figure 10-19*b*. When the integrator output crosses zero volts, the comparator output will drop low and shut off the clock signal to the counters. The number of counts required for the integrator output to get back to zero is directly proportional to the input voltage. For the circuit shown in Figure 10-19*a*, an input signal of +1 V, for example, produces a count of 1000. Because the resistor and the capacitor on the integrator are used for both the input voltage integrate and the reference integrate, small variations in their value with temperature do not have any effect on the accuracy of the conversion.

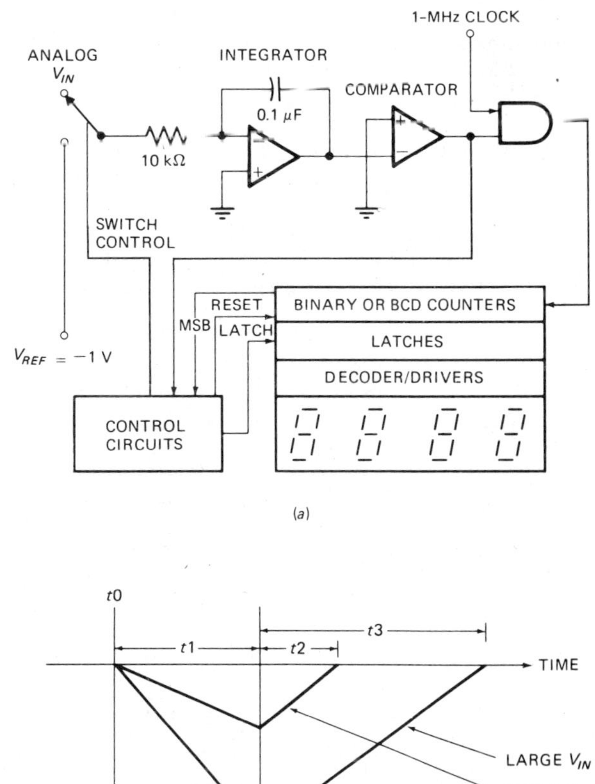

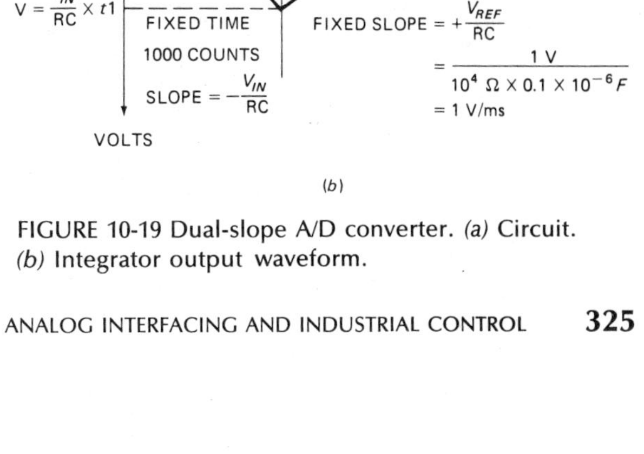

FIGURE 10-19 Dual-slope A/D converter. *(a)* Circuit. *(b)* Integrator output waveform.

Complete slope-type A/D converters are readily available in single IC packages. One example is the Intersil ICL7136 which contains all of the circuitry for a 3½-digit A/D converter and all of the interface circuitry needed to drive a 3½-digit LCD. Another example is the Intersil ICL7135 which contains all of the circuitry for a 4½-digit A/D converter and has a multiplexed BCD output. Note that, because of the usual use of this type of converter, we often express its resolution in terms of a number of digits. The full-scale reading for a 3½-digit converter is 1999, so the resolution corresponds to about 1 part in 2000. A two-chip set, the Intersil ICL8068 and ICL7104-16, contains all of the circuitry for a slope-type 16-bit binary output A/D converter.

The main disadvantage of slope-type converters is their slow speed. A 4½-digit unit may take 300 ms to do a conversion.

SUCCESSIVE-APPROXIMATION A/D CONVERTERS

Figure 10-20 shows a circuit for an 8-bit *successive-approximation* converter which uses readily available parts. The heart of this converter is a successive-approximation register (SAR) such as the MC14549 which functions as follows.

On the first clock pulse at the start of a conversion cycle the SAR outputs a high on its most-significant bit to the MC1408 D/A converter. The D/A converter and the amplifier convert this to a voltage and apply it to one input of a comparator. If this voltage is higher than the input voltage on the other input of the comparator, the comparator output will go low and tell the SAR to turn off that bit because it is too large. If the voltage from the D/A converter is less than the input voltage, then the comparator output will be high, which tells the SAR to keep that bit on. When the next clock pulse occurs, the SAR will turn on the next-most-significant bit to the D/A converter. Based on the answer this produces from the comparator, the SAR will keep or reset this bit. The SAR proceeds in this way on down to the least-significant bit, adding each bit to the total in turn and using the signal from the comparator to decide whether to keep that bit in the result. Only 8 clock pulses are needed to do the actual conversion here. When the conversion is complete the binary result is on the parallel outputs of the SAR. The SAR sends out an end-of-conversion (EOC) signal to indicate this. In the circuit in Figure 10-20 the EOC signal is used to strobe the binary result into some latches where it can be read by a microcomputer. If the EOC signal is connected to the start-conversion (SC) input as shown, then the converter will do continuous conversions. Note in the circuit in Figure 10-20 that the noninverting input of the op amp on the 1408 D/A converter is connected to −5 V instead of to ground. This shifts the analog input range from −5 V to +5 V instead of 0 V to +10 V so that sine-waves and other ac signals can be input directly to the converter to be digitized.

The National ADC1280 is a single-chip 12-bit successive-approximation converter which does a conversion in about 22 μs. Datel and Analog Devices have several 12-bit converters with conversion times about 1 μs.

Several commonly available successive-approximation A/D converters have analog multiplexers on their inputs. The National ADC0816, for example, has a 16-input multiplexer. This allows one converter to digitize any one of 16 input signals. You specify the input chan-

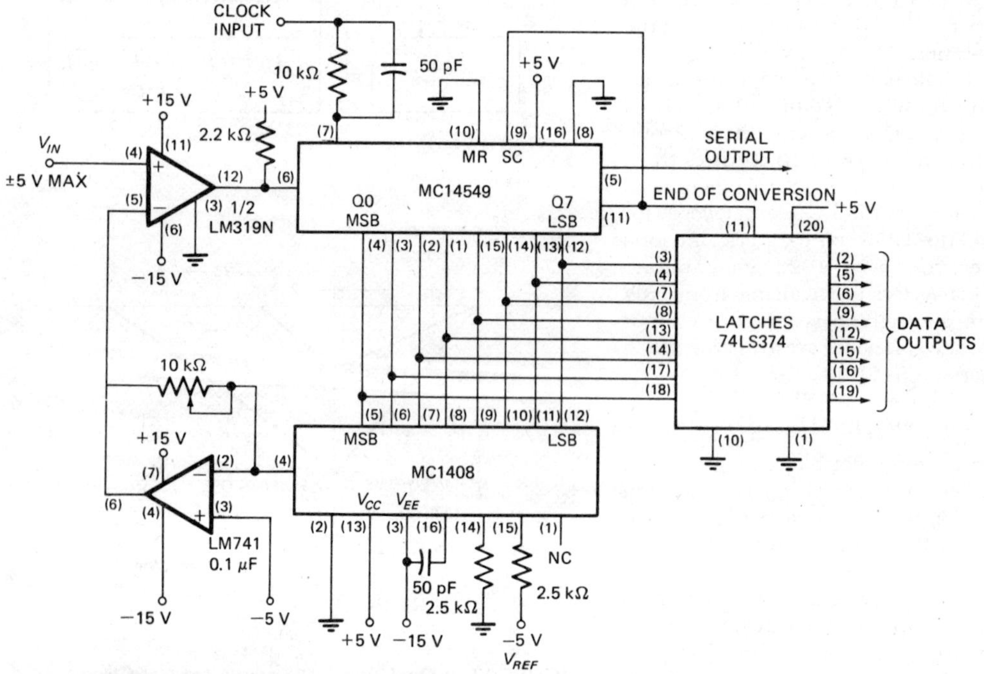

FIGURE 10-20 Successive-approximation A/D converter circuit.

nel you want to digitize with a 4-bit address you apply to the address inputs of the device. An A/D converter with a multiplexer on its inputs is often called a *data acquisition system* or DAS. Later in this chapter we show an application of a DAS in a factory control system.

Before we go on to discuss A/D interfacing, we need to make a few comments about common A/D output codes.

A/D OUTPUT CODES

For convenience in different applications, A/D converters are available with several different, somewhat confusing, output codes. The best way to make sense out of these different codes is to see them all together with representative values as shown in Figure 10-21. The values shown here are for an 8-bit converter, but you can extend them to any number of bits.

For an A/D converter with only a positive input range (*unipolar*) a straight binary code or inverted binary code is usually used. If the output of an A/D converter is going to drive a display, then it is convenient to have the output coded in BCD. For applications where the input range of the converter has both a negative and a positive range (*bipolar*) we usually use offset-binary coding. As you can see in Figure 10-21 the values of 00000000 to 11111111 are simply shifted downward so that 00000000 represents the most negative input value and 10000000 represents an input value of zero. This coding scheme has the advantage that the 2's complement representation can be produced by simply inverting the most-significant bit. Some bipolar converters output the digital value directly in 2's complement form.

UNIPOLAR BINARY CODES

VALUE	10 VOLTS FULL SCALE	BINARY (BIN)	COMPLEMENTARY BINARY (CB)	INVERTED BINARY (IB)	INVERTED COMPLEMENTARY BINARY (ICB)
+FS −1 LSB	9.9609	1111 1111	0000 0000		
+½ FS	5.0000	1000 0000	0111 1111		
+½FS −1 LSB	4.9609	0111 1111	1000 0000		
+1 LSB	0.0391	0000 0001	1111 1110		
ZERO	0.0000	0000 0000	1111 1111	0000 0000	1111 1111
−1 LSB	−0.0391			0000 0001	1111 1110
−½ FS + 1 LSB	−4.9609			0111 1111	1000 0000
−½ FS	−5.0000			1000 0000	0111 1111
− FS + 1 LSB	−9.9609			1111 1111	0000 0000

UNIPOLAR BINARY CODED DECIMAL CODES

VALUE	10 VOLTS FULL SCALE	BINARY CODED DECIMAL (BCD)	COMPLEMENTARY BINARY CODED DECIMAL (CBCD)	INVERTED BINARY CODED DECIMAL (IBCD)	INVERTED COMPLEMENTARY BINARY CODED DECIMAL (ICBCD)
+ FS −1 LSB	9.9	1001 1001	0110 0110		
+½ FS	5.0	0101 0000	1010 1111		
+1 LSB	0.1	0000 0001	1111 1110		
ZERO	0.0	0000 0000	1111 1111	0000 0000	1111 1111
−1 LSB	−0.1			0000 0001	1111 1110
−½ FS	−5.0			0101 0000	1010 1111
−FS +1 LSB	−9.9			1001 1001	0110 0110

BIPOLAR BINARY CODES

VALUE	10 VOLTS FULL SCALE RANGE	OFFSET BINARY (OB)	COMPLEMENTARY OFFSET BINARY (COB)	TWO'S COMPLEMENT (TC)
+FS	5.0000			
+FS −1 LSB	4.9609	1111 1111	0000 0000	0111 1111
+1 LSB	0.0391	1000 0001	0111 1110	0000 0001
ZERO	0.0000	1000 0000	0111 1111	0000 0000
−1 LSB	−0.0391	0111 1111	1000 0000	1111 1111
−FS +1 LSB	−4.9609	0000 0001	1111 1110	1000 0001
−FS	−5.0000	0000 0000	1111 1111	1000 0000

FIGURE 10-21 Common A/D output codes.

Interfacing Different Types of A/D Converters to Microcomputers

INTERFACING TO PARALLEL-COMPARATOR A/D CONVERTERS

In any application where a parallel comparator converter is used, the converter is most likely going to be producing digital output values much faster than a microcomputer could possibly read them in. Therefore, separate circuitry is used to bypass the microprocessor and load a set of samples from the converter directly into a series of memory locations. The microprocessor can later perform the desired operation on the samples. Bypassing the microprocessor in this way is called *direct memory access* or DMA. The basic principle of DMA is that an external controller IC tells the microprocessor to float its buses. When the microprocessor does this, the DMA controller takes control of the buses and allows data to be transferred directly from the A/D converter to successive memory locations. We discuss DMA in detail in the next chapter.

INTERFACING TO SLOPE-TYPE A/D CONVERTERS

Most of the commonly available slope-type converters were designed to drive seven-segment displays in, for example, a digital voltmeter. Therefore, they usually output data in a multiplexed BCD or seven-segment form. Figure 10-23 shows how you can connect the multiplexed BCD outputs of an inexpensive 3½-digit slope converter, the MC14433, to a microprocessor port. In the section of the chapter where Figure 10-23 is located, we use this converter as part of a microcomputer-based scale. The BCD data is output from the converter on lines Q0–Q3. A logic high is output on one of the digit strobe lines, DS1–DS4, to indicate when the BCD code for the corresponding digit is on the Q outputs. The MC14433 converter shown in Figure 10-23 outputs the BCD code for the most-significant digit, and then outputs a high on the DS1 pin. After a period of time it outputs the BCD code for the next-most-significant digit and outputs a high on the DS2 pin. After all 4 digits have been put out, the cycle repeats.

To read in the data from this converter, the principle is simply to poll the bit corresponding to a strobe line until you find it high, read in the data for that digit, and put the data in a reserved memory location for future reference. After you have read the BCD code for one digit, you poll the bit which corresponds to the strobe line for the next digit until you find it high, read the code for that digit, and put it in memory. Repeat the process until you have the data for all of the digits. The A/D converter in Figure 10-23 is connected to do continuous conversions, so you can call the procedure to read in the value from the A/D converter at any time.

Frequency counters, digital voltmeters, and other test instruments often have multiplexed BCD outputs available on their back panel. With the connections and procedure we have just described you can use these instruments to input data to your microcomputer.

INTERFACING A SUCCESSIVE-APPROXIMATION A/D CONVERTER

Successive-approximation A/D converters usually have outputs for each bit. The code output on these lines is usually straight binary or offset binary. You can simply connect the parallel outputs of the the converter to the required number of input port pins and read the converter output in under program control. In addition to the data lines, there are two other successive approximation A/D converter signal lines you need to interface to the microcomputer for the data transfer. The first of these is a START CONVERT signal which you output from the microcomputer to the A/D to tell it to do a conversion for you. The second signal is a STATUS signal which the A/D converter outputs to indicate that the conversion is complete and that the word on the outputs is valid. Here are the program steps you use to get a data sample from the converter.

First you pulse the START CONVERT high for a minimum of 100 ns. You then detect the STROBE signal going low on a polled or interrupt basis. You then read in the digitized value from the parallel outputs of the converter. In a later section of this chapter we show a detailed example of this for the ADC0808 converter.

If you are working with a personal computer such as the IBM PC, there are available a wide variety of multichannel A/D and D/A converter boards which plug directly into the bus connectors of these machines.

A MICROCOMPUTER-BASED SCALE

So far in this book we have shown you how a lot of the pieces of a microcomputer system function. Now it's time to show you how some of these pieces are put together to make a microcomputer-based instrument. The first instrument we have chosen is a "smart" scale such as you might see at the checkout stand in your local grocery store.

Overview of Smart Scale Operation

Figure 10-22 shows a block diagram of our smart scale. A load cell converts the applied weight of, for example, a bunch of carrots to a proportional electrical signal. This small signal is amplified and converted to a digital value which can be read in by the microprocessor and sent to the attached display. The user then enters the price per pound with the keyboard and this price per pound is shown on the display. When the user hits the compute key on the keyboard, the microprocessor multiplies the weight times the price per pound and shows the result on the display. After holding the price display long enough for the user to read it, the scale goes back to reading in the weight value and displaying it. To save the user from having to type the computed price into the cash register, an output from the scale could be connected directly into the cash register circuitry. A speech synthesizer, such as the Votrax SC-01A we described in Chapter 9, could be attached to verbally tell the customer the weight, price per pound, and total price.

Smart scales such as this have many applications other than weighing carrots. A modified version of this scale is used in company mail rooms to weigh packages and calculate the postage required to send them to different postal zones. The output of the scale can be connected to a postage meter which then automatically prints out the required postage sticker. Another application of smart scales is to count coins in a bank or gambling casino. For this application the user simply enters the type of coin being weighed. A conversion factor in the program then computes the total number of coins and the total dollar amount. Still another application of a scale such as this is in packaging items for sale. Suppose, for example, that we are manufacturing woodscrews, and that we want to package 100 of them per box. We can pass the boxes over the load cell on a

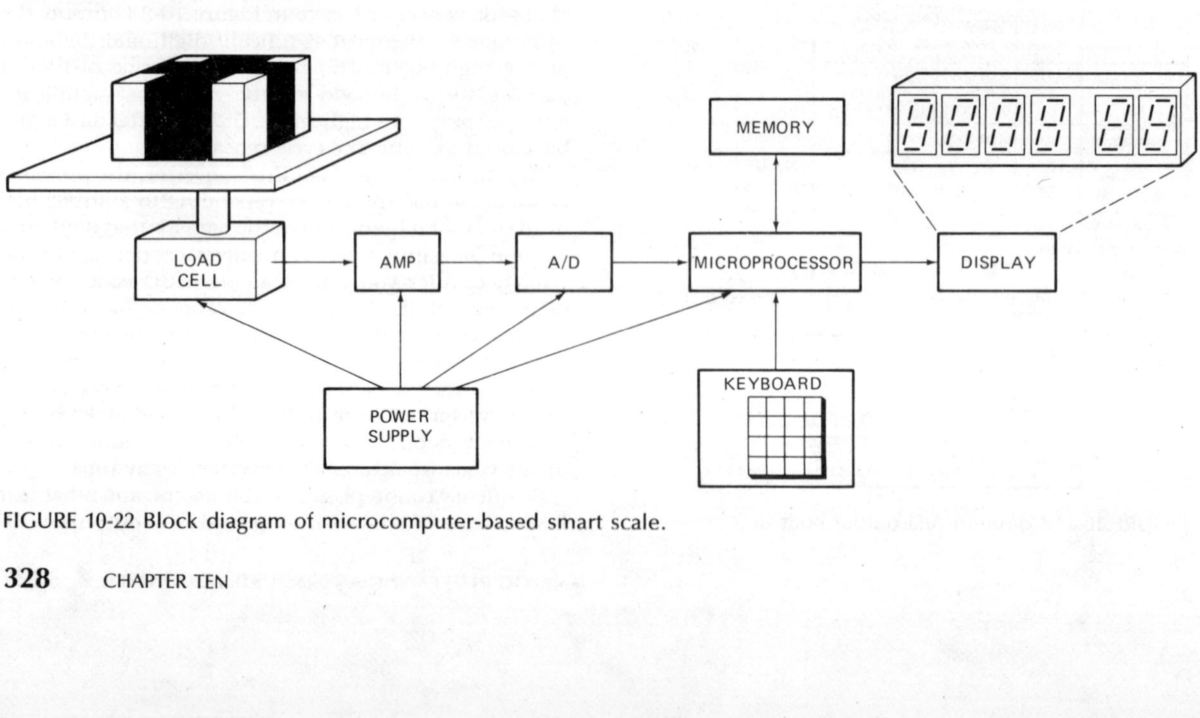

FIGURE 10-22 Block diagram of microcomputer-based smart scale.

conveyor belt and fill them from a chute until the weight, and therefore the count, reaches some entered value. The point here is that the combination of intelligence and some simple interface circuitry gives you an instrument with as many uses as your imagination can come up with.

Smart Scale Input Circuitry

Figure 10-10 shows a picture of the Transducers, Inc. Model C462-10#-10P1 strain-gage load cell we used when we built this scale. We added a piece of plywood to the top of the load cell to keep the carrots from falling off. This load cell has an accuracy of about 1 part in 1000 or 0.01 lb over the 0- to 10-lb range for which it was designed.

As shown in Figure 10-23, the load cell consists of four 350-Ω resistors connected in a bridge configuration. A stable 10.00-V excitation voltage is applied to the top of the bridge. With no load on the cell, the outputs from the bridge are at about the same voltage, 5 V. When a load is applied to the bridge, the resistance of one of the lower resistors will be changed. This produces a small differential output voltage from the bridge. The maximum differential output voltage for this 10-lb load cell is 2 mV per volt of excitation. With a 10.00 V excitation as shown, the maximum differential-output voltage is then 20 mV.

To amplify this small differential signal we use a National LM363 instrumentation amplifier. This device contains all of the circuitry shown for the instrumentation amplifier in Figure 10-1*h*. The closed-loop gain of the amplifier is programmable for fixed values of 5, 100, and 500 with jumpers on pins 2, 3, and 4. We have jumpered it for a gain of 100 so that the 20-mV maximum signal from the load cell will give a maximum voltage of 2.00 V to the A/D converter input. A precision voltage divider on the output of the amplifier divides this signal in half so that a weight of 10.00 lb produces an output voltage of 1.000 V. This scaling simplifies the display of the weight after it is read into the microprocessor. The 0.1-μF capacitor between pins 15 and 16 of the amplifier reduces the bandwidth of the amplifier to about 7.5 Hz. This removes 60 Hz and any high-frequency noise that might have been induced in the signal lines.

The MC14433 A/D converter used here is an inexpensive dual-slope device intended for use in 3½-digit digital voltmeters, etc. Because the load cell changes slowly, a fast converter isn't needed here. The voltage across an LM329 6.9-V precision reference diode is amplified by *IC*4 to produce the 10.00-V excitation voltage for the load cell, and a 2.000-V reference for the A/D. With a 2.000-V reference voltage, the full-scale input voltage for the A/D is 2.000 V. Conversion rate and multiplexing frequency for the converter are determined by an internal oscillator and *R*11. An *R*11 of 300 kΩ gives a clock frequency of 66 kHz, a multiplex frequency of 0.8 kHz, and about four conversions per second. Accuracy of the converter is ±0.05 percent and ±1 count, which is comparable to the accuracy of the load cell. In other words, the last digit of the displayed weight may be off by 1 or 2

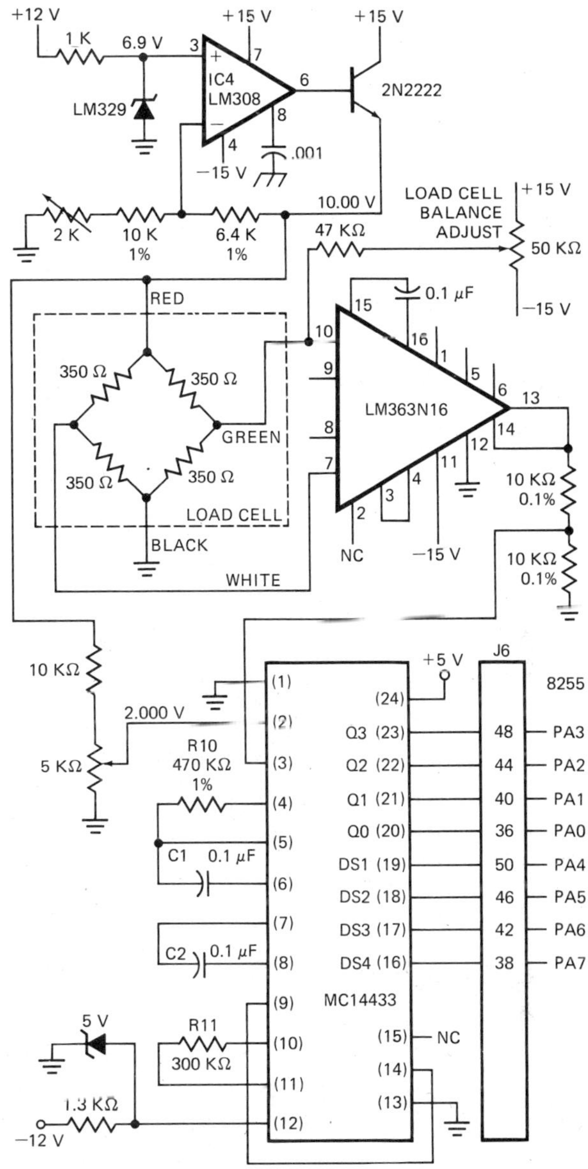

FIGURE 10-23 Circuit diagram for load-cell interface circuitry and A/D converter for smart scale.

counts. As we described in a previous section, the output from this converter is in multiplexed BCD form.

An Algorithm for the Smart Scale

Figure 10-24 shows the flowchart for our smart scale. Note that, as indicated by the double-ended boxes in the flowchart, most major parts of the program are written as procedures. The output of the A/D is in multiplexed BCD form as we described in the section on slope-converter interfacing. Therefore, each strobe has to be polled until it goes high, and then the BCD code for that digit can be read in.

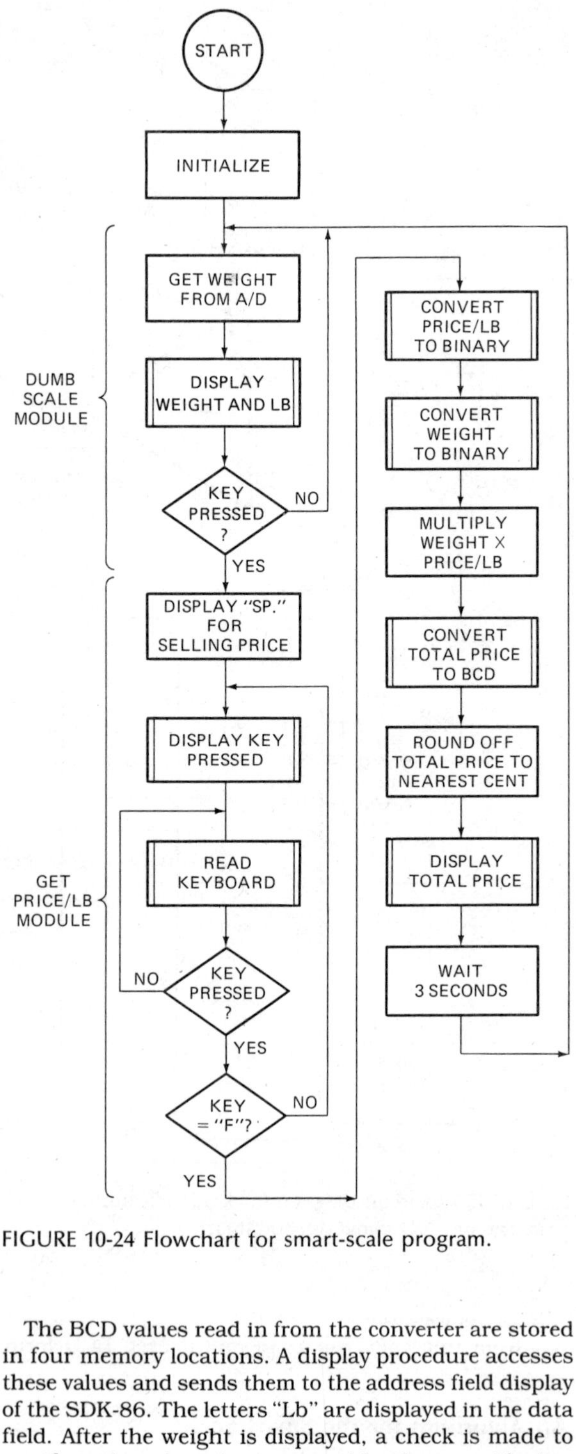

FIGURE 10-24 Flowchart for smart-scale program.

The BCD values read in from the converter are stored in four memory locations. A display procedure accesses these values and sends them to the address field display of the SDK-86. The letters "Lb" are displayed in the data field. After the weight is displayed, a check is made to see if any keys have been pressed by the user. If a key has been pressed, the letters "SP," which represents selling price, are displayed in the address field. Keycodes are read from the 8279 as entered and displayed on the data field display. Keys can be pressed until the desired price per pound shows on the display. The price per pound entered by a user is assembled in a series of memory locations. When a nonnumeric key is pressed, it is assumed that the entered price per pound is correct, and the program goes on to compute the total price.

Computing the price involves multiplying the weight in BCD form times the price per pound in BCD form. It is not easy to do a BCD × BCD multiply directly, so we took an alternate route to get there. Both the weight and the price per pound are converted to binary. The two binary numbers are then multiplied. The binary result of the multiplication is then converted to BCD, rounded off to the nearest cent, and displayed in the data field. The letters "Pr" are displayed in the address field. After a few seconds the program goes back to reading and displaying weight over and over, until a key is pressed.

The Microprocessor-based Scale Program

Figure 10-25 shows the complete program for our microprocessor-based scale. It is important for you not to be overwhelmed by a multipage program such as this. If you use the 5-minute rule, and work your way through this program one module at a time, you should pick up some more useful programming techniques and procedures for your "toolbox."

Three 4-byte buffers set up at the start of the program are used to store the unpacked BCD values of the weight, the price per pound, and the computed total price. These values will be used by the display procedure. Instead of using the display procedure we showed you in Figure 9-33, we used here a more versatile procedure which can display a few letters as well as hex digits. The SEVEN_SEG table in the data segment contains the seven-segment codes for hex digits and these letters. In the display procedure you will see how these codes are accessed.

After initializing everything the program polls the digit strobe for the most-significant digit from the A/D converter. Since this A/D converter is a 3½-digit unit, the MSD can only be a 0 or a 1. The value for this digit is sent in the third bit (bit 2) of the 4-bit digit read in. If this bit is a one, then 01 is loaded into the buffer location. If the bit is a 0, then the value which will access the seven-segment code for a blank (14H) is loaded into the buffer location. Each of the other digit strobes is then polled in turn and the values for those digits read in. When all of the BCD digits for the weight are in the WEIGHT_BUFFER, the display procedure is called to show the weight on the address field.

To use this display procedure you first load a 0 or a 1 into AL to specify data field or address field and a 1 or a 0 in AH to specify a decimal point in the middle of the display, or no decimal point. You then load BX with the offset of a buffer containing codes for the digits to be displayed. A program loop in the display procedure uses the XLAT instruction and the SEVEN_SEG table to convert these codes to the required seven-segment values and send the values to the 8279 display RAM. Note how a 1 is ORed into the seven-segment code for digit 3 when a 1 is in AH passed to the procedure. For displaying the weight, BX is simply loaded with the offset of WEIGHT_BUFFER, AL loaded with 01 to display the

```
Page ,132
;8086 PROGRAM FOR SMART SCALE
;PORTS:       Uses input port P1A - FFF9H
;PROCEDURES: READ_KEY, DISPLAY, PACK, EXPAND, CONVERT2BIN, BINCVT

DATA_HERE SEGMENT       WORD    PUBLIC

WEIGHT_BUFFER  DB  4 DUP(0)              ; Space for unpacked BCD weight
SELL_PRICE     DB  4 DUP(0)              ; Space for unpacked price/pound
PRICE_TOTAL    DB  4 DUP(0)              ; Space for total price to display
BINARY_WEIGHT  DW  0                     ; Space for converted weight
LB             DB  0BH, 10H, 14H, 14H,   ; b, L, blank, blank
S_P            DB  12H, 11H, 14H, 14H,   ; P, S, blank, blank
PR             DB  13H, 12H, 14H, 14H    ; r, P, bblank, blank

SEVEN_SEG       DB      3FH, 06H, 5BH, 4FH, 66H, 6DH, 7DH, 07H
                ;        0    1    2    3    4    5    6    7
                DB      7FH, 6FH, 77H, 7CH, 39H, 5EH, 79H, 71H
                ;        8    9    A    b    C    d    E    F
                DB      38H, 6DH, 73H, 50H, 00H, 76H
                ;        L    S    P    r  blank  H


DATA_HERE ENDS

STACK_HERE SEGMENT
                DW      40 DUP(0)
        STACK_TOP       LABEL   WORD
STACK_HERE      ENDS

CODE_HERE       SEGMENT WORD    PUBLIC
        ASSUME CS:CODE_HERE, DS:DATA_HERE, SS:STACK_HERE
;initialize data & stack segments
START:  MOV     AX, DATA_HERE
        MOV     DS, AX
        MOV     AX, STACK_HERE
        MOV     SS, AX
        MOV     SP, OFFSET STACK_TOP
;8279 initialized at power-up of SDK-86 for 8 character display, left entry
;encoded scan, 2-key lockout.
        MOV     DX, 0FFEAH              ; point at 8279 control address
        MOV     AL, 00H                 ; control word for above conditions
        OUT     DX, AL                  ; send control word
        MOV     AL, 00111000B           ; clock word for divide by 24
        OUT     DX, AL
        MOV     AL, 11000000B           ; Clear display character is all 0's
        OUT     DX, AL

;Dumb scale start
RDWT:   MOV     CX, 04H                 ; Zero out weight buffer
        MOV     BX, OFFSET WEIGHT_BUFFER
NEXT1:  MOV     BYTE PTR[BX], 00H
        INC     BX
        LOOP    NEXT1
        MOV     CX, 04H                 ; Zero out price/pound buffer
```

FIGURE 10-25 Assembly language program for smart scale.

```
        MOV     BX, OFFSET SELL_PRICE
NEXT2:  MOV     BYTE PTR[BX], 00H
        INC     BX
        LOOP    NEXT2
;Get weight from A/D converter and display
        MOV     BX, OFFSET WEIGHT_BUFFER+3      ; MSD Position in weight buffer
        MOV     DX, 0FFF9H              ; Point at A/D port
DS1:    IN      AL, DX                  ; Read byte from A/D
        AND     AL, 10H                 ; Check for MSD strobe high
        JZ      DS1                     ; Loop till high
        IN      AL, DX                  ; Read MSD data from A/D
        AND     AL, 0FH                 ; Mask strobe bits
        CMP     AL, 04H                 ; See if MSD in bit 3 is a one
        JE      LOAD1                   ; Yes, go load 01H in buffer
        MOV     BYTE PTR[BX], 14H       ; No, load code for blank
        JMP     NXTCHR
LOAD1:  MOV     BYTE PTR[BX], 01H
NXTCHR: DEC     BX                      ; Point to next buffer location
DS2:    IN      AL, DX                  ; Poll for digit 2 strobe
        AND     AL, 20H
        JZ      DS2
        IN      AL, DX                  ; Read digit 2 from A/D
        AND     AL, 0FH                 ; Mask strobe bits
        MOV     [BX], AL                ; Digit 2 BCD to buffer
        DEC     BX                      ; Point at next buffer location
DS3:    IN      AL, DX                  ; Poll for digit 3 from A/D
        AND     AL, 40H
        JZ      DS3
        IN      AL, DX                  ; Read digit 3 from A/D
        AND     AL, 0FH                 ; Mask strobe bits
        MOV     [BX],AL                 ; Digit 3 to buffer
        DEC     BX                      ; Point to next buffer location
DS4:    IN      AL, DX                  ; Poll for digit 4 (LSD)
        AND     AL, 80H
        JZ      DS4
        IN      AL, DX                  ; Read digit 4 from A/D
        AND     AL, 0FH                 ; Mask strobe bits
        MOV     [BX], AL                ; Digit 4 BCD to buffer
;Display weight on address field of SDK-86
        MOV     BX, OFFSET WEIGHT_BUFFER  ; Point at stored weight
        MOV     AL, 01H                   ; Specifies address field
        MOV     AH, 01H                 ; Specifies decimal point
        CALL    DISPLAY
        MOV     BX, OFFSET LB           ; Point at Lb string
        MOV     AL, 00                  ; Specifies data field
        MOV     AH, 00                  ; Specifies no decimal point
        CALL    DISPLAY
;Check if key has been pressed
        MOV     DX, 0FFEAH              ; Point at 8279 status address
        IN      AL, DX                  ; Read 8279 FIFO status
        AND     AL, 01H                 ; See if FIFO has keycode
        JNZ     GETKEY                  ; Yes, go read it
        JMP     RDWT                    ; No, go get weight and display
GETKEY: MOV     AL, 01000000B           ; Control word for read FIFO
        OUT     DX, AL                  ; Send to 8279
```

FIGURE 10-25 (*continued*)

```
        MOV     DX, ØFFE8H              ; Point at 8279 data address
        IN      AL, DX                  ; Read code from FIFO
        CMP     AL, Ø9H                 ; Check if legal keycode(number)
        JBE     OK                      ; Go on if below or equal 9
        JMP     RDWT                    ; Else ignore, read weight again

;Read in and display price/pound

OK:     MOV     BX, OFFSET SELL_PRICE   ; Point at price per pound buffer
        MOV     [BX], AL                ; Keycode to buffer
        MOV     AL, ØØ                  ; Specify data field for display
        MOV     AH, Ø1                  ; Specify decimal point
        CALL    DISPLAY
        MOV     BX, OFFSET S_P          ; Point at SP string
        MOV     AL, Ø1                  ; Specify address field
        MOV     AH, ØØ                  ; Specify no decimal point
        CALL    DISPLAY
NXTKEY: CALL    READ_KEY                ; Wait for next keypress
        CMP     AL, Ø9H                 ; See if more price or command
        JA      COMPUTE                 ; Go compute total price
        MOV     BX, OFFSET SELL_PRICE   ; Point at price per pound buffer
        MOV     CL, [BX+2]              ; Shift contents of buffer one
        MOV     [BX+3], CL              ; position left and insert new
        MOV     CL, [BX+1]              ; keycode
        MOV     [BX+2], CL
        MOV     CL, [BX]
        MOV     [BX+1], CL
        MOV     [BX], AL
        MOV     AL, ØØ                  ; Specify data field
        MOV     AH, Ø1                  ; Specify decimal point
        CALL    DISPLAY
        JMP     NXTKEY                  ; Keep reading and shifting keys
                                        ; until command key pressed
;compute total price

COMPUTE:
        MOV     BX, OFFSET WEIGHT_BUFFER        ; Point at weight buffer for pack
        CMP     BYTE PTR[BX+3], 14H     ; See if MSD of weight = Ø
        JNE     NOTZER
        MOV     BYTE PTR[BX+3], ØØ      ; Yes, load Ø in place of blank code
NOTZER: CALL    PACK                    ; Pack BCD weight into word
        CALL    CONVERT2BIN             ; Convert to 16 bit binary in AX
        MOV     BINARY_WEIGHT, AX       ;  and save
        MOV     BX, OFFSET SELL_PRICE   ; Point at price per pound for pack
        CALL    PACK                    ; Pack BCD price into AX for convert
        CALL    CONVERT2BIN             ; Convert price to 16-bit binary in AX
        MUL     BINARY_WEIGHT           ; Price per pound in AX x binary weight
                                        ; total price result in DX:AX
        MOV     BX, AX                  ; Prepare for convert to BCD
        CALL    BINCVT                  ; Packed BCD price result in DX:BX

;round off price to nearest cent and display

        CMP     BL, 49H                 ; Carry set if BL >49H
        MOV     AL, ØØ                  ; Clear AL, keep carry
```

FIGURE 10-25 *(continued)*

```
        ADC     AL, BH                  ; Add any carry to next digit
        DAA                             ; Keep in BCD format
        MOV     BL, AL                  ; Save lower two digits of price
        MOV     AL, 00                  ; Clear AL, save carry
        ADC     AL, DL                  ; Propagate carry to upper digits
        DAA                             ; Keep in BCD form
        MOV     AH, AL                  ; Position upper digits for EXPAND
        MOV     AL, BL                  ; Position lower digits for EXPAND
        MOV     BX, OFFSET PRICE_TOTAL  ; Point at buffer for expanded BCD
        CALL    EXPAND                  ; Unpack BCD for DISPLAY procedure
        MOV     AL, 00                  ; Display total price on data field
        MOV     AH, 01                  ;  with decimal point
        CALL    DISPLAY
        MOV     BX, OFFSET PR           ; Point at price/lb string
        MOV     AL, 01                  ; Display in address field
        MOV     AH, 00                  ; without decimal point
        CALL    DISPLAY
;delay a few seconds
        MOV     CX,0FFFFH               ; Delay a few seconds
CNTDN1: MOV     BX, 000AH
CNTDN2: DEC     BX
        JNZ     CNTDN2
        LOOP    CNTDN1
;go read next weight
        JMP     RDWT                    ; Jump back to dumb scale
;**********************************************************************
;************* procedures used in smart scale program ****************

;PROCEDURE READ_KEY
;ABSTRACT       reads the SDK-86 keyboard - procedure polls the status register
;               of the 8279 on the SDK-86 board until it finds a key pressed.
;               It then reads the keypressed code from the FIFO RAM to AL and exits
;REGISTERS:     Destroys AL - returns with character read in AL

READ_KEY        PROC    NEAR
        PUSH    DX
        MOV     DX, 0FFEAH              ; point at 8279 control address
NO_KEY: IN      AL, DX                  ; get FIFO status
        AND     AL, 00000001B           ; mask all but LSB, high if key in FIFO
        JZ      NO_KEY                  ; loop until a key is pressed
        MOV     AL, 01000000B           ; control word for read FIFO
        OUT     DX, AL                  ; send control word
        MOV     DX, 0FFE8H              ; point at 8279 data address
        IN      AL, DX                  ; read character in FIFO ram
        POP     DX
        RET
READ_KEY ENDP

;8086 PROCEDURE called DISPLAY
;ABSTRACT:  This procedure displays characters on the SDK-86 display
;           The data is sent to the procedure in the following manner:
; AL=0 for data field
; AL=1 for address field
; AH=0 for no decimal point
; AH=1 for decimal point between second & third digit
```

FIGURE 10-25 (*continued*)

```
; BX= offset of buffer containing 7-seg codes of the 4 characters to be displayed

DISPLAY PROC    NEAR
        PUSHF                           ; save flags and registers
        PUSH    AX
        PUSH    BX
        PUSH    CX
        PUSH    DX
        PUSH    SI
        MOV     DX, ØFFEAH              ; point at 8279 control address
        CMP     AL, ØØH                 ; see if data field required
        JZ      DATFLD                  ; yes, load control word for data field
        MOV     AL, 94H                 ; no, load address-field control word
        JMP     SEND                    ; go send control word
DATFLD: MOV     AL, 9ØH                 ; load control word for data field
SEND:   OUT     DX, AL                  ; send control word to 8279
        MOV     CL, Ø4H                 ; counter for number of characters
        MOV     SI, BX                  ; Free BX for use with XLAT
        MOV     BX, OFFSET SEVEN_SEG    ; pointer to seven-segment codes
        MOV     DX, ØFFE8H              ; point at 8279 display RAM
AGAIN:  MOV     AL, [SI]                ; Get character to be displayed
        XLATB                           ; translate to 7-seg code
        CMP     CL, Ø2H                 ; see if digit that gets decimal point
        JNE     MORE                    ; no, go send digit
        CMP     AH, Ø1H                 ; yes, see if decimal point specified
        JNE     MORE                    ; no, go send character
        OR      AL, 8ØH                 ; yes, OR in decimal point
MORE:   OUT     DX, AL                  ; send seven seg codeto 8279 display RAM
        INC     SI                      ; Point to next character
        LOOP    AGAIN                   ; until all four characters sent
        POP     SI
        POP     DX                      ; restore all registers and flags
        POP     CX
        POP     BX
        POP     AX
        POPF
        RET
DISPLAY ENDP

;8Ø86 PROCEDURE PACK
;ABSTRACT: This procedure converts four unpacked BCD digits pointed to by
;          BX to four packed BCD digits in AX
;DESTROYS: AX

PACK  PROC  NEAR
        PUSHF                           ; save flags and registers
        PUSH    BX
        PUSH    CX
        MOV     AL, [BX]                ; first BCD digit to AL
        MOV     CL, Ø4H                 ; counter for rotate
        ROL     BYTE PTR[BX+1], CL      ; position second BCD digit
        ADD     AL, [BX+1]              ; first 2 digits in AL
        MOV     AH, [BX+2]              ; third digit to AH
        ROL     BYTE PTR[BX+3], CL      ; position fourth digit
        ADD     AH, [BX+3]              ; second two digits now in AH
```

FIGURE 10-25 *(continued)*

```
        POP    BX
        POP    CX
        POPF
        RET
PACK  ENDP

;8086 PROCEDURE EXPAND
;ABSTRACT:      This procedure expands a packed BCD number in AX
;               to 4 unpacked BCD digits in a buffer pointed to by BX

EXPAND  PROC NEAR
        PUSHF
        PUSH   AX
        PUSH   BX
        PUSH   CX
        MOV    [BX],AL                 ; move first 2 BCD digits to buffer
        AND    BYTE PTR[BX],0FH        ; mask off upper digit
        MOV    CL, 04H                 ; counter for rotates
        ROR    AL, CL                  ; position digit 2 in low nibble
        AND    AL, 0FH                 ; mask upper nibble
        MOV    [BX+1], AL              ; digit 2 to buffer
        MOV    [BX+2], AH              ; second 2 BCD digits to buffer
        AND    BYTE PTR[BX+2],0FH      ; mask off upper digit
        ROR    AH, CL                  ; position digit 4 in low nibble
        AND    AH, 0FH                 ; mask upper nibble
        MOV    [BX+3], AH              ; digit 4 to buffer
        POP    CX
        POP    BX
        POP    AX
        POPF
        RET
EXPAND ENDP

;PROCEDURE CONVERT2BIN
;          : This procedure converts a 4 digit BCD number in
;          : the AX register into its BINARY (HEX) equivalent. It
;          : returns the result in the AX register
;SAVES     : FLAG register, BX, DX, CX, SI
;DESTROYS  : AX register
        THOU    EQU     3E8H            ; 1000 = 3E8H
CONVERT2BIN PROC     NEAR
        PUSHF                           ; save registers
        PUSH    BX
        PUSH    DX
        PUSH    CX
        PUSH    DI
        MOV     BX, AX          ; copy number into BX
        MOV     AL, AH          ; place for upper 2 digits
        MOV     BH, BL          ; place for lower 2 digits
; split up numbers so that we have one digit in each register
        MOV     CL, 04          ; nibble count for rotate
        ROR     AH, CL          ; digit 1 in correct place
        ROR     BH, CL          ; digit 3 in correct place
        AND     AX, 0F0FH
        AND     BX, 0F0FH       ; mask upper nibbles of each digit
```

FIGURE 10-25 (*continued*)

```
; copy AX into CX so that can use AX for multiplication
        MOV     CX, AX
        MOV     AX, 0000H
; now multiply each number by its place value
        MOV     AL, CH          ; multiply byte in AL * word
        MOV     DI, THOU        ; no immediate multiplication
        MUL     DI              ; digit 1 * 1000
; result in DX and AX, because BCD digit will not be greater than 9
; the result will only be in AX
; zero DX and add BL because that digit needs no multiplication for
; place value. Then add the result in AX for digit 4
        MOV     DX, 0000H
        ADD     DL, BL          ; add digit 1
        ADD     DX, AX          ; add digit 4
; continue with multiplications
        MOV     AX, 0064H       ; byte * byte result in AX
        MUL     CL              ; digit 2 * 100
        ADD     DX, AX          ; add digit 3
        MOV     AX, 000AH       ; byte * byte result in AX
        MUL     BH
        ADD     DX, AX          ; add digit 2
        MOV     AX, DX          ; put result in correct place
        POP     DI
        POP     CX
        POP     DX              ; restore registers
        POP     BX
        POPF
        RET
CONVERT2BIN     ENDP

;8086 PROCEDURE BINCVT
;ABSTRACT: Converts a 24-bit binary number in DL and BX to a
;          packed BCD equivalent in DX:BX
;INPUTS:   DL, BX - 24 BIT BINARY NUMBER
;OUTPUTS:  DX, BX - PACKED BCD RESULT
;CALLS:    CNVT1
;DESTROYS: DX and BX

BINCVT PROC NEAR
        PUSHF                   ; save registers and flags
        PUSH  AX
        PUSH  CX
        MOV   DH, 19H           ; bit counter for 24 bits
        CALL  CNVT1             ; produce 2 LS BCD digits in CH
        MOV   CL, CH            ; save in CL
        MOV   DH, 19H           ; bit counter for 24 BITS
        CALL  CNVT1             ; produce next 2 BCD digits in CH
        PUSH  CX                ; save lower four BCD digits on stack
        MOV   DH, 19H           ; bit counter for 24 bits
        CALL  CNVT1             ; produce next two BCD digits in CH
        MOV   CL, CH            ; position in CL
        MOV   DH,19H            ; set bit counter for 24 bits
        CALL  CNVT1             ; produce last two BCD digits in CH
        MOV   DX,CX             ; position 4 MS BCD DIGITS for return
        POP   BX                ; 4 LS BCD digits back from stack for RET
```

FIGURE 10-25 *(continued)*

```
        POP   CX
        POP   AX
        POPF
        RET
BINCVT  ENDP

;8086 PROCEDURE CNVT1
CNVT1   PROC  NEAR
        XOR   AL, AL          ; clear AL and carry as workspace
        MOV   CH, AL          ; clear CH
CNVT2:  XOR   AL, AL          ; clear AL and CARRY
        DEC   DH              ; decrement bit counter
        JNZ   CONTINUE        ; do all bits
        RET                   ; done if DH down to zero
CONTINUE:
        RCL   BX, 1           ; BX left one bit, MSB to carry
        RCL   DL, 1           ; MSB from BX to LSB of DL, MSB of DL to carry
        MOV   AL, CH          ; move BCD digit being built to AL
        ADC   AL, AL          ; double AL and add carry from DL shift
        DAA                   ; keep result in BCD form
        MOV   CH, AL          ; put back in CH for next time through
        JNC   CNVT2           ; no carry from DAA, continue
        ADC   BX, 0000H       ; if carry, propogate to BX and DL
        ADC   DL, 00H         ;  for future terms
        JMP   CNVT2           ; continue conversion
CNVT1   ENDP

CODE_HERE ENDS
        END START
```

FIGURE 10-25 (*continued*)

weight in the address field, and AH loaded with 01 to insert a decimal point at the appropriate place.

To display the letters Lb in the data field, BX is loaded with the offset of the string named LB, and the display procedure is called. Again, the XLAT instruction loop converts the codes from the LB string to the required seven-segment codes and sends them out to the 8279 display RAM. The codes in the string named LB represent the offsets from the start of the SEVEN_SEG table for the desired seven-segment codes. For example, the seven-segment code for a P is at offset 12H in the SEVEN_SEG table. Therefore, if you want to display a P, you put 12H in the appropriate location in the the character string in memory. The XLAT instruction will then use the value 12H to access the seven-segment code for P in the SEVEN_SEG table.

After displaying the weight, the program reads the 8279 status register to see if the operator has pressed a key to start entering a price per pound. If no key has been pressed, or if a nonnumeric key has been pressed, the program simply goes back and reads the weight again. If a number key has been pressed, the weight is removed from the address field and the letters SP for "selling price" displayed there. The number entered is put in the SELL_PRICE buffer and displayed on the rightmost digit of the data field. The program then polls the 8279 status register until another keypress is detected. If the pressed key is a numeric key, then the code(s) for the previously entered number(s) will be shifted one location in the buffer to make room for the new number. The new number is then put in the first location in the buffer so that is will be displayed in the rightmost digit of the display. In other words, previously entered numbers are continuously shifted to the left as new numbers are entered. If a mistake is made, the operator can simply enter a 0 followed by the correct price per pound.

If the pressed key is not a numeric key, then this is the signal that the displayed price per pound is correct and that the total price should now be computed. Before the weight and the price/pound can be multiplied, they must each be put in packed BCD form and converted to binary. The PACK procedure converts four unpacked BCD digits in a memory buffer pointed to by BX to a 4-digit packed result in AX. This procedure is simply some masking and moving nibbles. Note how the process is simplified by the ability to rotate the contents of a memory location. Conversion of the packed weight and the packed price per pound is done by the CONVERT2BIN procedure. The algorithm for this procedure is explained in detail in Chapter 5.

For the 8086 a single MUL instruction does the 16 ×

16 binary multiply to produce the total price. Earlier processors required a messy procedure to do this. After the multiplication the total price is in binary form, which is not the form needed for the display procedure. The procedure BINCVT is used to convert the binary total price to packed BCD form. Here's how this procedure works.

In a binary number, each bit position represents a power of 2. An 8-bit binary number, for example, can be represented as

$$b7 \times 2^7 + b6 \times 2^6 + b5 \times 2^5 + b4 \times 2^4 + b3 \times 2^3 + b2 \times 2^2 + b1 \times 2^1 + b0$$

This can be shuffled around and expressed

$$\text{as binary number} = (((((2b7 + b6)2 + b5)2 + b4)2 + b3)2 + b2)2 + b1)2 + b0$$

where b7 through b0 are the values of the binary bits. If we start with a binary number and do each operation in the nested parentheses in BCD with the aid of the DAA instruction, the result will be the BCD number equivalent to the original binary number.

The procedure in Figure 10-25 produces two BCD digits of the result at a time by calling the subprocedure CNVT1. Figure 10-26 shows a flowchart for the operation of CNVT1. The main principle here is to shift the 24-bit number left one bit position so the MSB goes into the carry flip-flop, then add this bit to twice the previous result. We use the DAA instruction to keep the result of the addition in BCD format. If the DAA produces a carry we add this carry back into the shifted 24-bit number in DL and BX so that it will be propagated into higher BCD digits. After each run of CNVT1 (24 runs of CNVT2), DL and BX will be left with a binary number which is equal to the original binary number minus the value of the two BCD digits produced. You can adapt this procedure to work with a different number of bits by simply calling CNVT1 more or fewer times, and by adjusting the count loaded into DH to be one more than the number of binary bits in the number to be converted. The count has to be one greater because of the position of the decrement in the loop. The temperature-controller procedure in Figure 10-35 shows another example of this conversion.

The least-significant 2 digits of the BCD value for the total price returned by BINCVT in BL represent tenths and hundredths of a cent. If the value of these two BCD digits is greater than 49H, then the carry produced by the compare instruction and the next two higher BCD digits in BH are added to AL. This must be done in AL, because the DAA instruction, used to keep the result in BCD format, only works on an operand in AL. Any carry from these two BCD digits is propagated on to the upper two digits of the result in DL. After this rounding off, the packed BCD for the total price is left in AX.

In order for the display procedure to be able to display this price, it must be converted to unpacked BCD form and put in four successive memory locations. Another "mask and move nibbles" procedure called EXPAND does this. The DISPLAY procedure is then called to display the total price on the data field. The DISPLAY procedure is called again to display the letters Pr in the address field.

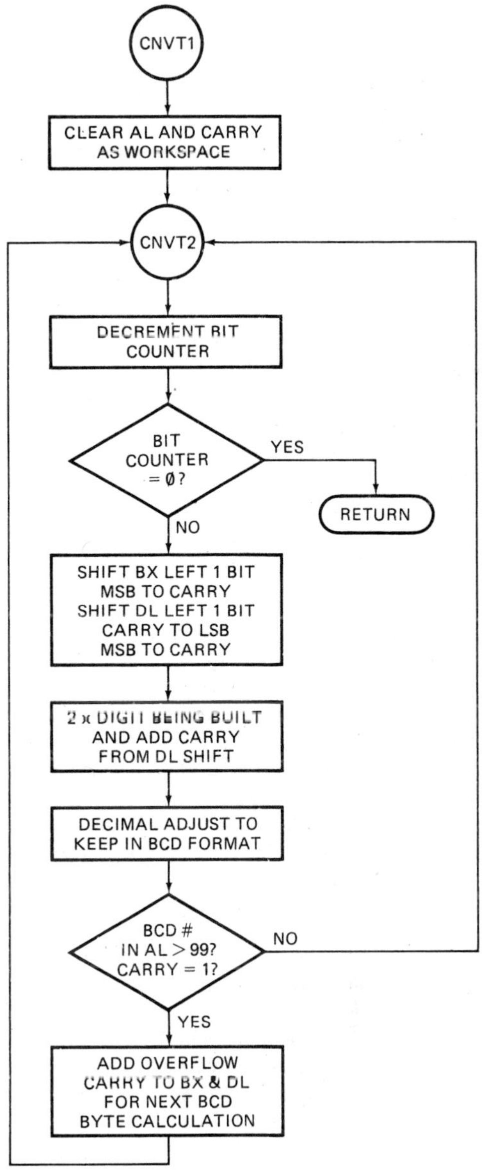

FIGURE 10-26 Flowchart for CNVT1 subprocedure.

Finally, after delaying a few seconds to give the operator time to read the price, execution returns to the "dumb scale" portion of the program and starts over.

A question that may occur to you when reading a long program such as this is, "How do you decide which parts of the program to keep in the mainline, and which parts to write as procedures?" There is no universal agreement on the answer to this question. The general guidelines we follow are to write a program section as a procedure if: it is going to be used more than once in the program; it is reusable (could be used in other programs); it is so lengthy (more than 1 page) that it clutters up the conceptual flow of the main program; or it is an essentially independent section. The disadvantage of using too many procedures is the time and overhead required for each procedure call. As you write more programs, you will arrive at a balance that feels comfortable to you. The following section shows you another long program example which was written in a highly modular manner so that it can easily be expanded. This example should further help you see when and how to use procedures.

A MICROCOMPUTER-BASED INDUSTRIAL PROCESS-CONTROL SYSTEM

Overview of Industrial Process Control

An area in which microprocessors and microcomputers have had a major impact is *industrial process control*. Process control involves first measuring system variables such as motor speed, temperature, the flow of reactants, the level of a liquid in a tank, the thickness of a material, etc. The system is then adjusted until the value of each variable is equal to a predetermined value for that variable called a *set point*. The system controller must maintain each variable as close as possible to its set point value, and it must compensate as quickly and accurately as possible for any change in the system such as an increased load on a motor. A simple example will show the traditional approach to control of a process variable and explain some of the terms used in control systems.

The circuit in Figure 10-27 shows one approach to controlling the speed of a dc motor. Attached to the shaft of the motor is a dc generator, or *tachometer*, which puts out a voltage proportional to the speed of the motor. The output voltage is typically a few volts per 1000 rpm. A fraction of the output voltage from the tachometer is fed back to the inverting input of the power amplifier driving the motor. A positive voltage is applied to the noninverting input of the amplifier as a set point. When the power is turned on, the motor accelerates

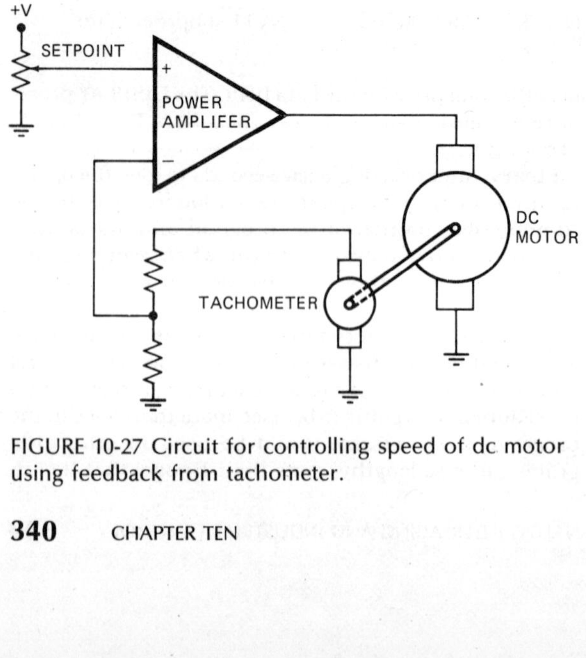

FIGURE 10-27 Circuit for controlling speed of dc motor using feedback from tachometer.

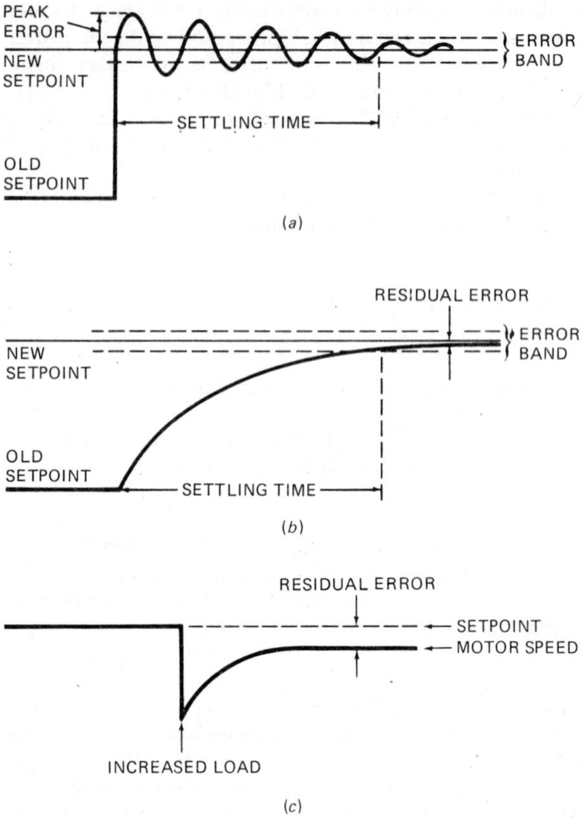

FIGURE 10-28 Overshoot and undershoot of system when setpoint or load is changed. *(a)* Overshoot. *(b)* Undershoot. *(c)* Load change.

until the voltage fed back from the tachometer to the inverting input of the amplifier is nearly equal to the set-point voltage. Using negative feedback to control a system such as this is often called *servo control*. A control loop of this type keeps the motor speed quite constant for applications where the load on the motor does not change much. Some hard-disk drive motors and high-quality phonograph turntables use this method of speed control.

For applications in which the load and/or set point changes drastically, there are several potential problems. The first of these is overshoot when you change the set point, as shown in Figure 10-28*a*. In this case the variable, motor speed for example, overshoots the new set point and bounces up and down for a while. The time it takes the bouncing to settle within a specified error range or error band is called the *settling time*. This type of response is referred to as *underdamped* and is similar to the response of a car with bad shock absorbers when it hits a bump. Figure 10-28*b* shows the opposite situation where the system is overdamped so that it takes a long time for the variable to reach the new setpoint.

Another problem of any real control system is *residual error*. Figure 10-28*c* shows the response of a control system such as the motor speed controller in Figure 10-27 when more load is added on the motor. When the increased load is first added the motor slows down so the voltage out of the tachometer decreases. This increases the voltage difference between the amplifier inputs and causes the amplifier output to increase. Increased amplifier output increases the speed of the motor and thereby the output from the tachometer. When the system reaches equilibrium, however, there is some noticeable difference between the set point and the voltage fed back from the tachometer. It is this difference or residual error which is amplified by the gain of the amplifier to produce the additional drive for the motor. For stability reasons, the gain of many control systems cannot be too high. Therefore, even if you adjust the speed of a motor, for example, to be exactly at a given speed for one load, when you change the load there will always be some residual error between the set point and the actual output.

To help solve these problems, circuits with more complex feedback are used. Figure 10-29 shows a circuit which represents the different types of feedback commonly used. First note in this circuit that the output power amplifier is an adder with four inputs. The current supplied to the summing point of the adder by the set-point input produces the basic output drive current. The other three inputs do not supply any current unless there is a difference between the set point and the feedback voltage from the tachometer. Amplifier 1 is another adder whose function is to compare the set-point voltage with the feedback voltage from the tachometer. Let's assume the two input resistors, $R1$ and $R2$, are equal. Since the set-point voltage is negative and the voltage from the tachometer is positive, there will be no net current through the feedback resistor of the amplifier if the two voltages are equal in magnitude. In other words, if the speed of the motor is at its set-point value, the output of amplifier 1 will be zero, and amplifiers 2, 3, and 4 will contribute no current to the summing junction of the power amp.

Now, suppose that you add more load on the motor, slowing it down. The tachometer voltage is no longer equal to the set-point voltage so amplifier 1 now has some output. This error signal on the output produces three types of feedback to the summing junction of the power amp.

Amplifier 1 produces simple dc feedback proportional to the difference between the set point and the tachometer output. This is exactly the same effect as the voltage divider on the tachometer output in Figure 10-27. *Proportional feedback*, as this is called, will correct for most of the effect of the increased load, but as we discussed before, there will always be some residual error.

The cure for residual error is to use some *integral feedback*. Amplifier 3 in Figure 10-29 provides this type of feedback. Remember from a previous discussion that this circuit produces a ramp on its output whenever a voltage is applied to its input. For the example here the integrator will ramp up or ramp down as long as there is any error signal present on its input. By ramping up and down just a tiny bit about the set point, the integrator can eliminate most of the residual error. Too much integral feedback, however, will cause the output to oscillate up and down.

A third type of feedback called *derivative feedback* is produced by amplifier 4 in Figure 10-29. Integral feed-

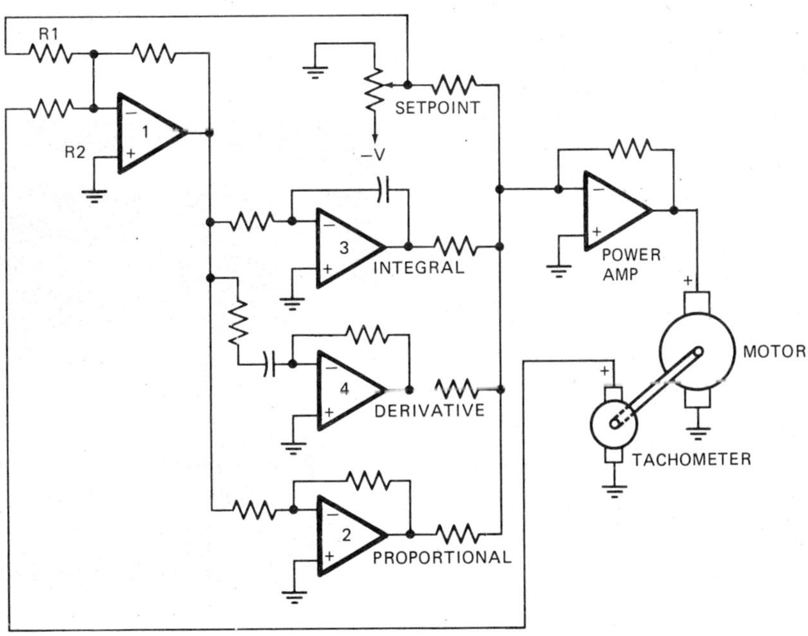

FIGURE 10-29 Circuit showing proportional, integral, and derivative feedback control.

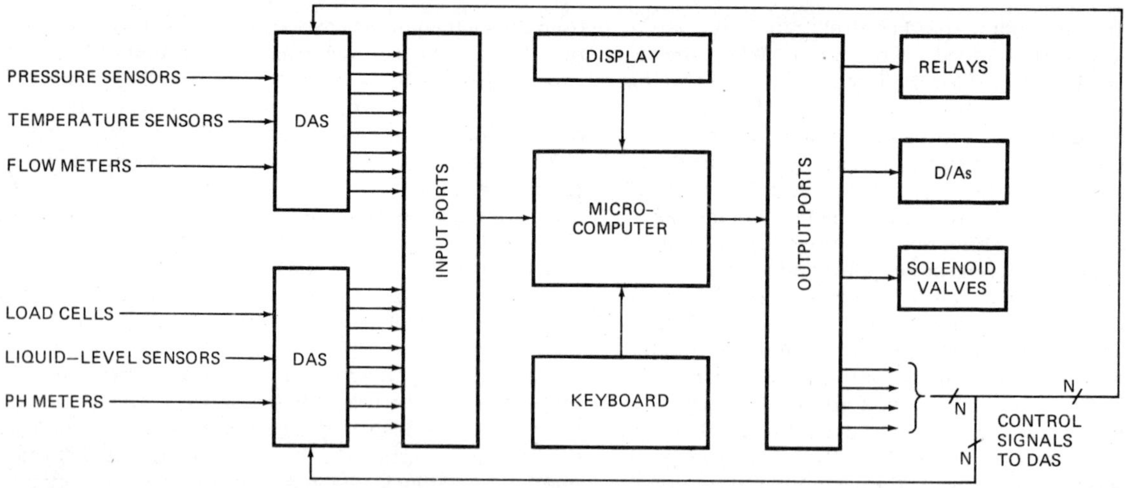

FIGURE 10-30 Block diagram of microcomputer-based process control system.

back discussed in the previous paragraph is slow because the error signal must be present for some time before the integrator has much output. Derivative feedback is a signal proportional to the rate of change of the error signal. If the load on the system is suddenly changed, the derivative amplifier circuit will give a quick shot of feedback to try and correct the error. When the error signal is first applied to the differentiator circuit, the capacitor in series with the input is not charged, so it acts like a short circuit. This initially lets a large current flow so the amplifier has a sizable output. As the capacitor charges, the current decreases, so the feedback from the differentiator decreases. Too much derivative feedback can cause the system to overshoot and oscillate.

The point here is that by using a combination of some or all of these types of feedback, a given feedback-controlled system can be adjusted for optimum response to changes in load or set point. Process control loops that use all three types of feedback are called *proportional integral-derivative* or PID control loops. Because process variables change much more slowly than the microsecond operation of a microcomputer, a microcomputer with some simple input and output circuitry can perform all of the functions of the analog circuitry in Figure 10-29 for several PID loops.

Figure 10-30 shows a block diagram of a microcomputer-based process-control system. Data acquisition systems convert the analog signals from various sensors to digital values that can be read in and processed by the microcomputer. A keyboard and display in the system allow the user to enter set point values and to read the current values of process variables. Relays, D/A converters, solenoid valves, and other actuators are used to control process variables under program direction. A programmable timer in the system determines the rate at which control loops get serviced.

Microcomputer-based process-control systems range from a small programmable controller such as the one shown in Figure 10-31, which might be used to control one or two machines on a factory floor, to a large minicomputer used to control an entire fractionating column in an oil refinery. To show you how these microcomputer-based control systems work, here's an example system you can build and experiment with.

AN 8086-BASED PROCESS-CONTROL SYSTEM

Program Overview

Figure 10-32 shows in flowchart form one way in which the program for a microcomputer-based control system with eight PID loops can operate. After power is turned on, a mainline or *executive program* initializes ports, initializes the timer, and initializes process variables to

FIGURE 10-31 Photograph of Texas Instruments' programmable controller for up to eight PID loops.

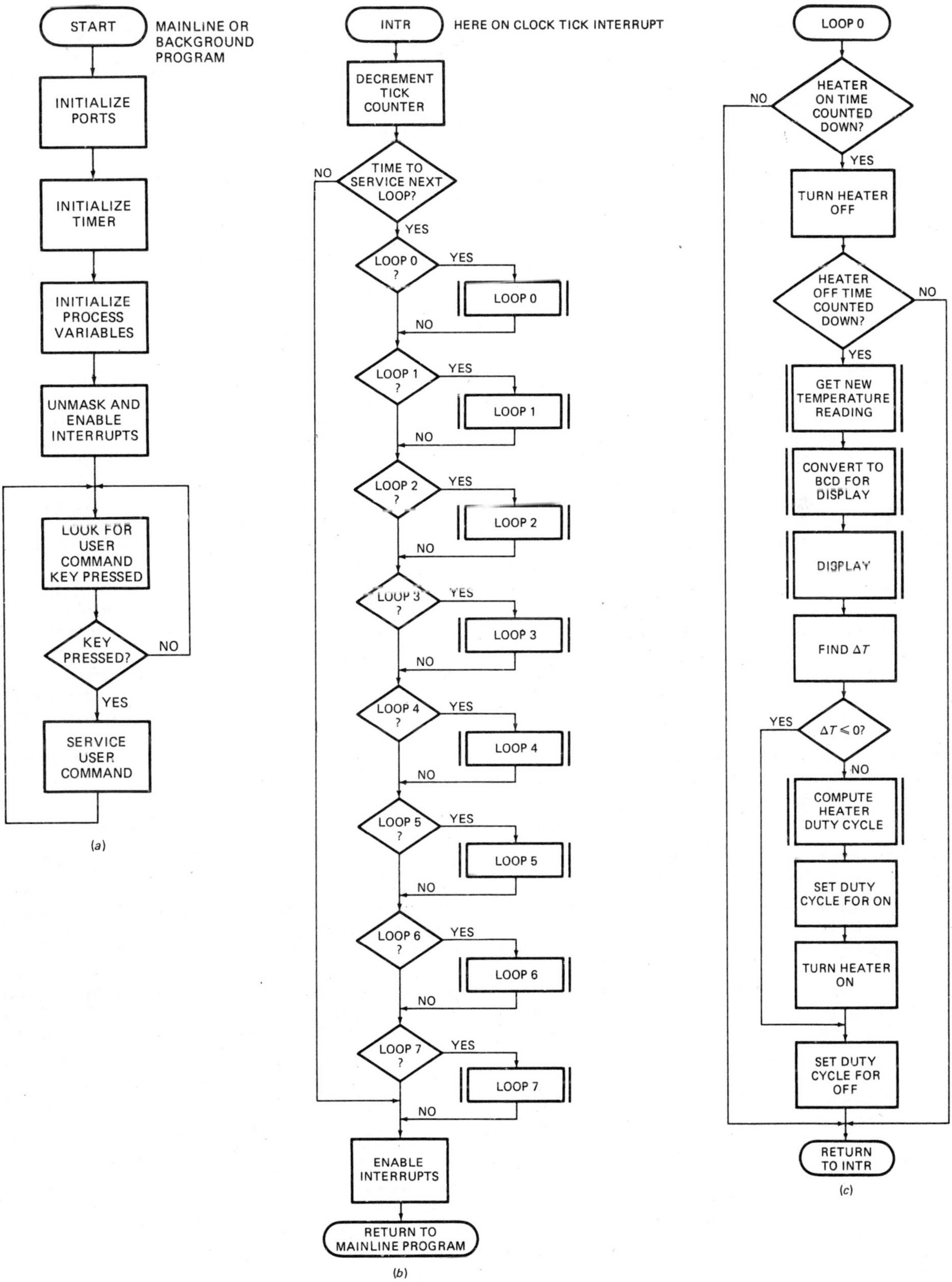

FIGURE 10-32 Flowchart for microcomputer-based process control system. *(a)* Mainline or executive. *(b)* Loop selector. *(c)* Temperature-control loop.

some starting values. The executive program then sits in a loop waiting for a user command from the keyboard or a clock "tick" from the timer. Both the keyboard and the clock are connected to interrupt inputs.

When the microcomputer receives an interrupt from the timer it goes to a procedure which determines whether it is time to service the next control loop. The interrupt procedure does this by counting interrupts in the same way as the real-time clock we described in Chapter 8. For example if you program the timer to produce a pulse every 1 ms, and you want the controller to service another loop every 20 ms, you can simply have the interrupt procedure count 20 interrupts before going on to update the next loop. Once you have counted down 20 interrupts, the program then falls into a decision structure which determines which loop is to be updated next. Every 20 ms a new loop is updated in turn, so with eight loops, each loop gets updated every 160 ms. Note that the microcomputer services each loop at a regular interval instead of simply updating all eight loops, one loop right after another. This is done so that the timing for each loop is independent of the timing for the other loops. A change in the internal timing for one loop then will not affect the timing in the other loops. This system is one type of *time-slice* system, because each loop gets a 20-ms "slice" of time.

The procedures which actually update each control loop are independent of each other. For our example system here we only have space to show the implementation of one loop, the control of the temperature of a tank of liquid in, perhaps, our printed-circuit-board-making machine. You could write other similar control loop procedures to control pH, flow, light exposure timing, motor speed, etc. Figure 10-32*c* shows the flowchart for our temperature-controller loop. We will explain how this works after we have a look at the hardware for the system.

Hardware for Control System and Temperature Controller

To build the hardware for this project we started with an SDK-86 board and added an 8254 programmable timer and an 8259A priority interrupt controller as shown in Figure 8-14. The timer is initialized to produce 1-kHz clock ticks. The 8259A provides interrupt inputs for the clock-tick interrupts and for keyboard interrupts. We built the actual temperature sensing and detecting circuitry on a separate prototyping board and connected it to some ports on the SDK-86 with a ribbon cable. Figure 10-33 shows the added circuitry.

The temperature-sensing element in the circuit is an LM35 precision Centigrade-temperature sensor. The voltage between the output pin and the ground pin of this device will be 0 V at 0°C, and will increase by 10 mV for each increase of 1°C above that. The 300 kΩ resistor connecting the output of the LM35 to −15 V allows the output to go negative for temperatures below 0°C. (If you are operating with ±12-V supplies, use a 240-kΩ resistor.) This makes the circuit able to measure temperatures over the range of −55 to +150°C. For our application here we only use the positive part of the output range, but we thought you might find this circuit useful for some of your other projects. An LM308 amplifies the signal from the sensor by 2 so that the signal uses a greater part of the input range of the A/D converter. This improves the noise immunity and resolution.

The ADC0808 A/D converter used here is an eight-input *data acquisition system*. You tell the device which input signal you want digitized with a 3-bit address you send to the ADC, ADB, and ADA inputs. This eight-input device was chosen so that other control loops could be added later. Some Schmitt-trigger inverters in a 74C14 are connected as an oscillator to produce a 300-kHz clock for the DAS. The voltage drop across an LM329 low-drift zener is buffered by an LM308 amplifier to produce a V_{CC} and a V_{REF} of 5.12 V for the A/D converter. With this reference voltage the A/D converter will have 256 steps of 20 mV each. Since the temperature sensor signal is amplified by 2, each degree Celsius of temperature change will produce an output change of 20 mV or one step on the A/D converter. This gives us a resolution of 1°C, which is about equal to the typical accuracy of the sensor. The advantage of using V_{REF} as the V_{CC} for the device is that this voltage will not have the switching noise that the digital V_{CC} line has. The control inputs and data outputs of the A/D converter are simply connected to SDK-86 ports as shown.

Figure 10-34 shows the timing waveforms and parameters for the ADC0808. Note the sequence in which control signals must be sent to the device. The 3-bit address of the desired input channel is first sent to the multiplexer inputs. After at least 50 ns the ALE input is sent high. After another 2.5 μs the START CONVERSION input is sent high and then low. Then the ALE input is brought low again. When you detect the END OF CONVERSION signal from the A/D converter going high, you can then read in the 8-bit data value which represents the temperature.

To control the power delivered to the heater we used a 25 A, 0-V turn on, solid-state relay such as the Potter-Brumfield unit described in Chapter 9. With this relay we can control a 120- or 240-V ac-powered hot plate or immersion heater. The heater is pulsed on and off under program control. The duty cycle of the pulses determines the amount of heat put out by the heater.

For very low power applications, a D/A converter and a power amplifier could be used to drive the heater. However, in high-power applications this is not very practical, because the power amplifier dissipates as much or more power than the load. For example, when driving a 5000-W heater, the amplifier will dissipate 5000 W or more. The D/A-converter approach has the added disadvantage that it cannot directly use the available ac line voltage.

The driver transistor on the input of the solid-state relay serves three purposes: it supplies the drive for the relay, isolates the port pin from the relay, and holds the relay in the off position when the power is first turned on. Port pins, remember, are in a floating state after a reset. Now that you know how the hardware is connected we can explain how the program for this system works.

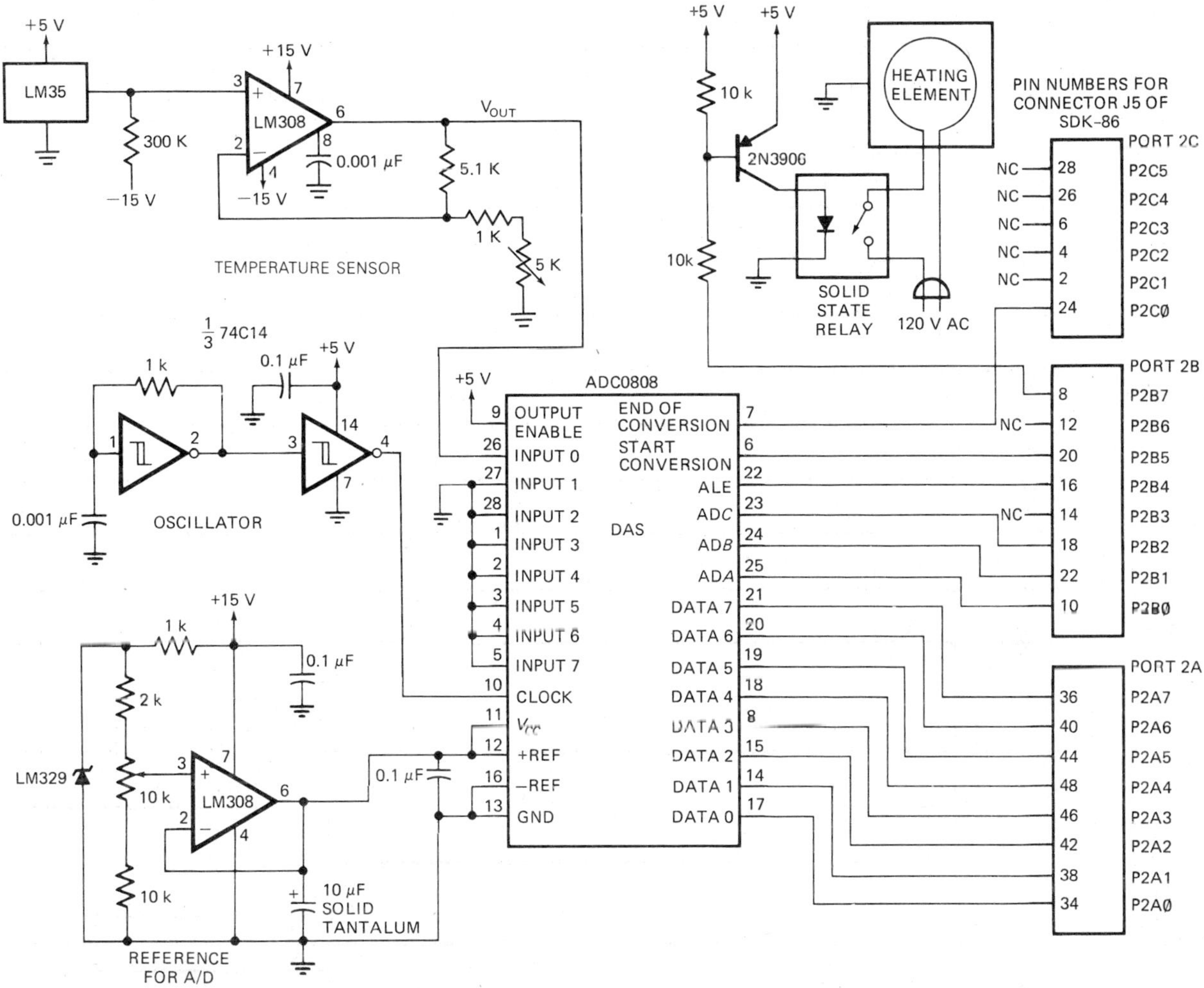

FIGURE 10-33 Temperature-sensing and heater-control circuitry for microcomputer-based controller.

The Controller System Program

THE MAINLINE OR EXECUTIVE SECTION

Figure 10-35 shows the assembly language program for our controller system. Refer to the flowchart in Figure 10-32 as you work your way through this program. The mainline or executive part of the program starts by initializing port FFFAH for output, the 8259A to receive interrupt inputs from the timer and the keyboard, and the 8254 to produce a 1-kHz square wave from its counter 0. We have described all of these operations in detail previously, so we won't dwell on them here. We also initialize here some process variables which we will explain later when they will have more meaning. After enabling the 8086 INTR input with an STI instruction, the program then enters a loop and waits for an interrupt from the user via the keyboard, or from the timer. The keyboard-interrupt procedure would normally contain a command recognizer and subprocedures to implement each of the commands, similar to the way the SDK-86 monitor program is structured. Due to severe space limitations, we do not show here the implementation of the keyboard interrupt procedure which allows the user to change set points, stop a process, or examine the value of process variables at any time.

THE CLOCK-TICK INTERRUPT HANDLER

The next part of the program to discuss is the interrupt procedure which counts clock ticks and decides which process control loop to service. At the start of this procedure we simply decrement an interrupt counter kept in a memory location. In the initialization this counter was set to 20 decimal or 14H. If the counter is not down to zero, execution is simply returned to the wait loop. If the tick counter is now down to zero, the clock tick counter is reset to 20, and one of the loop procedures is called to service the next loop. It is important that this clock tick procedure be reentrant, because if one of the loop procedures takes more than the time between clock ticks (1 ms), the procedure will be reentered before its first

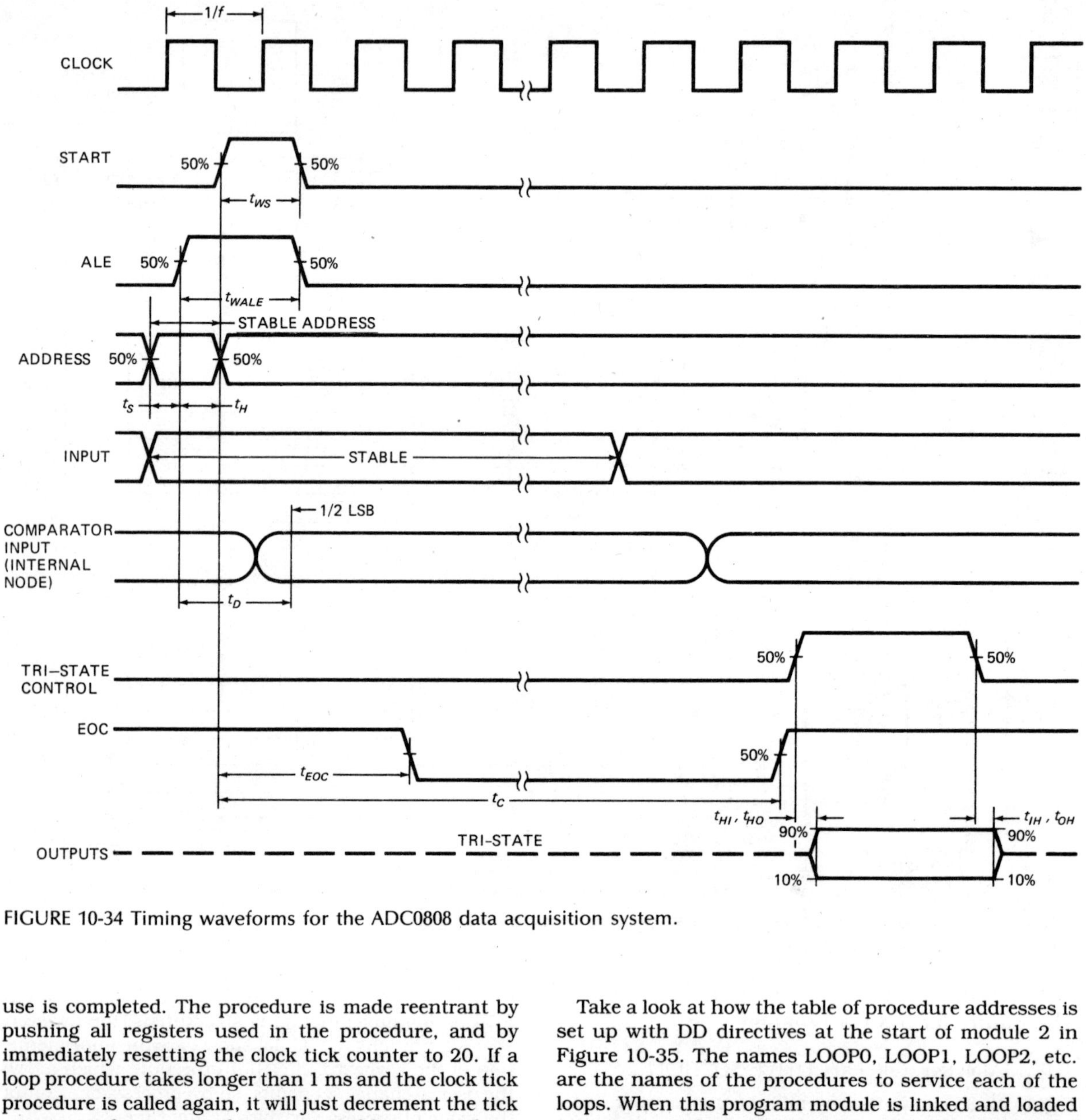

FIGURE 10-34 Timing waveforms for the ADC0808 data acquisition system.

use is completed. The procedure is made reentrant by pushing all registers used in the procedure, and by immediately resetting the clock tick counter to 20. If a loop procedure takes longer than 1 ms and the clock tick procedure is called again, it will just decrement the tick counter and return to the interrupted loop procedure.

The method used here to call the desired loop procedure is an important programming technique. It uses a *call table* to efficiently implement the CASE or nested IF—THEN—ELSE programming structure described in Chapter 3. Here's how it works.

To keep track of which loop should be serviced next, we use a variable called LOOPNUM in memory. During initialization LOOPNUM is loaded with OOH. When it is time to service the first loop, the value in LOOPNUM is loaded into BX. The CALL DWORD PTR LOOP_ADDR_TABLE[BX] instruction then gets a far call address from a table called LOOP_ADDR_TABLE in memory. BX functions as a pointer to the desired address in the table. For the first access BX is zero so the first address in the table is used.

Take a look at how the table of procedure addresses is set up with DD directives at the start of module 2 in Figure 10-35. The names LOOPO, LOOP1, LOOP2, etc. are the names of the procedures to service each of the loops. When this program module is linked and loaded into memory, the instruction pointer and code segment addresses for each of the procedures will be loaded into the table.

When execution returns from one of the loop procedures, 4 is added to LOOPNUM so that execution will go to the next loop in sequence the next time the tick counter is counted down to zero. LOOPNUM must be incremented by four because each address in the call table uses 4 bytes. If all loops have been serviced, LOOPNUM is set back to 0 so LOOPO will be serviced again. Now let's look at the actual temperature-control loop.

THE TEMPERATURE-CONTROLLER PROCEDURE

As we said previously, the amount of heat output by the heater is controlled by the duty cycle of a pulse waveform

```
page ,132
;8086 PROGRAM FOR CONTROLLER SYSTEM - MODULE 1
;
;ABSTRACT:       This program services eight process control loops on a
;                rotating basis. It is written to run on an Intel SDK-86 board.
;                Timing for the control loops is generated on an interrupt basis
;                by an on-board 8254 timer. Control-loop 0 in the program controls
;                the temperature of a water bath.
;PORTS:          Uses port P2B (FFFAH) as output
;                bits 7 = heater, bit 6,3 = not connected, bit 5 = start conversion
;                bit 4 = ALE,     bit 2,1,0 = channel address
;                Uses port P2A (FFF8H) as data input
;                Uses port P2C (FFFCH) as end-of-conversion input from A/D
;PROCEDURES:     Calls: CLOCK_TICK - interrupt service procedure
;                       KEYBOARD   - interrupt service procedure (empty)

INT_PROC_HERE   SEGMENT WORD    PUBLIC
        EXTRN   CLOCK_TICK:FAR, KEYBOARD:FAR
INT_PROC_HERE   ENDS

PUBLIC  COUNTER, TIMEHI, TIMELO, LOOPNUM, CURTEMP, SETPOINT

AINT_TABLE      SEGMENT WORD    PUBLIC
        TYPE_64         DW      2 DUP(0)        ;reserve space for clock-tick proc addr IR0
        TYPE_65         DW      2 DUP(0)        ;not used in this program - IR1
        TYPE_66         DW      2 DUP(0)        ;reserve space for keyboard proc addr - IR2
AINT_TABLE      ENDS

DATA_HERE       SEGMENT WORD    PUBLIC
        COUNTER         DB      00              ;counter for number of interrupts
        TIMEHI          DB      01              ;heater relay - time on
        TIMELO          DB      01              ;heater relay - time off
        LOOPNUM         DB      00              ;temp storage for loop counter
        CURTEMP         DB      00              ;current temperature
        SETPOINT        DB      60              ;setpoint temperature
DATA_HERE       ENDS

STACK_HERE      SEGMENT                         ; no STACK directive because using EXE2BIN
                DW      40      DUP(0)          ; so can then download code to SDK-86
        TOP_STACK       LABEL   WORD
STACK_HERE      ENDS

CODE_HERE       SEGMENT WORD    PUBLIC
        ASSUME CS:CODE_HERE, DS:AINT_TABLE, SS:STACK_HERE
;initialize stack segment register, stack pointer,data segment
        MOV     AX, STACK_HERE
        MOV     SS, AX
        MOV     SP, OFFSET TOP_STACK
        MOV     AX, AINT_TABLE
        MOV     DS, AX
;define the addresses for the interrupt service procedures
        MOV     TYPE_64+2, SEG CLOCK_TICK       ; put in clock-tick proc addr
        MOV     TYPE_64,   OFFSET CLOCK_TICK
        MOV     TYPE_66+2, SEG KEYBOARD         ; put in keyboard proc addr
        MOV     TYPE_66,   OFFSET KEYBOARD
```

FIGURE 10-35 8086 assembly language program for process control system. *(a)* Module 1—Mainline. *(b)* Module 2—interrupt service procedures. *(c)* Module 3—loop service procedures. *(d)* Module 4—utility procedures.

```
;initialize data segment
        ASSUME DS:DATA_HERE
        MOV     AX, DATA_HERE
        MOV     DS, AX
;initialize port P2B (FFFA) as output - mode 0, P2A & P2C as inputs - mode 0
        MOV     DX, 0FFFEH              ; point DX at port control addr
        MOV     AL, 10011001B           ; mode control word for above conditions
        OUT     DX, AL                  ; send control word
;initialize 8259A
        MOV     AL, 00010011B           ; edge triggered, single, ICW4
        MOV     DX, 0FF00H              ; point at 8259A control
        OUT     DX, AL                  ; send ICW1
        MOV     AL, 01000000B           ; type 64 is first 8259A type (IR0)
        MOV     DX, 0FF02H              ; point at ICW2 address
        OUT     DX, AL                  ; send ICW2
        MOV     AL, 00000001B           ; ICW4, 8086 mode
        OUT     DX, AL                  ; send ICW4
        MOV     AL, 11111110B           ; OCW1 to unmask IR0 only leave IR2 masked
        OUT     DX, AL                  ; because not used & send OCW1
;initialize 8254 counter 0 for 1-kHz output
; 8254 command word for counter 0, LSB then MSB, square wave, BCD
        MOV     AL, 00110111B
        MOV     DX, 0FF07H              ; point at 8254 control addr
        OUT     DX, AL                  ; send counter 0 command word
        MOV     AL, 58H                 ; Load LSB of count
        MOV     DX, 0FF01H              ; point at counter 0 data addr
        OUT     DX, AL                  ; send LSB of count
        MOV     AL, 24H                 ; load MSB of count
        OUT     DX, AL                  ; send MSB of count
;initialize variables
        MOV     SETPOINT, 3CH           ; initialize final temp at 60 deg
        MOV     COUNTER,  14H           ; intialize time counter
        MOV     LOOPNUM,  00H           ; start at first loop
        MOV     TIMEHI,   01H
        MOV     TIMELO,   01H
        MOV     CURTEMP,  00H
;enable interrupt input of 8086
        STI
HERE:   JMP     HERE                    ; wait for interrupt
        NOP                             ; if required, can put more
        NOP                             ; instructions here
        NOP
CODE_HERE ENDS
          END
```

(a)

FIGURE 10-35 (*continued*)

sent to the solid-state relay. The time on for the output waveform to the solid-state relay is determined by counting down a value called TIMEHI. The time off for this waveform is determined by counting down a value called TIMELO. At start-up the mainline program initializes TIMEHI and TIMELO to 01H, so that the first time the LOOPO procedure is called both of these are decremented to 0 and execution falls through to the A/D conversion procedure. This needs to be done so that we have a temperature value to use for computing the duty cycle.

The number of the A/D channel that we want to digitize is passed to the A/D conversion procedure in the AL register. The procedure then sends out this channel number to the A/D converter and generates the control wave forms shown in Figure 10-34 under software control. The binary value for the temperature is returned in AL.

```
PAGE ,132
;MODULE 2 - CONTAINS THE INTERRUPT SERVICE SUBROUTINES

PUBLIC  CLOCK_TICK, KEYBOARD

;tell assembler where to find loop addresses used in this module
DATA_HERE       SEGMENT WORD    PUBLIC
        LOOP_ADDR_TABLE DD      LOOP0
                        DD      LOOP1
                        DD      LOOP2
                        DD      LOOP3
                        DD      LOOP4
                        DD      LOOP5
                        DD      LOOP6
                        DD      LOOP7

        EXTRN   COUNTER:BYTE, LOOPNUM:BYTE
DATA_HERE       ENDS

;tell assembler where to find procedures used in this module
CODE_HERE       SEGMENT WORD    PUBLIC
        EXTRN   LOOP0:FAR, LOOP1:FAR, LOOP2:FAR, LOOP3:FAR
        EXTRN   LOOP4:FAR, LOOP5:FAR, LOOP6:FAR, LOOP7:FAR
CODE_HERE       ENDS

INT_PROC_HERE   SEGMENT WORD    PUBLIC  ;segment for interrupt service procedures
        ASSUME  CS:INT_PROC_HERE, DS:DATA_HERE

;8086 INTERRUPT PROCEDURE TO SERVICE PROCESS CONTROL LOOPS
;
;ABSTRACT:       This procedure services calls 1 of 8 process control
;                loops on a rotating basis.
;PORTS USED:     none
;PROCEDURES:     calls LOOP0,LOOP1,LOOP2,LOOP3,LOOP4,LOOP5,LOOP6,LOOP7
;REGISTERS:      saves all

CLOCK_TICK      PROC    FAR
        PUSH    AX                      ; save registers
        PUSH    BX
        PUSH    DX
        PUSH    DS                      ; save DS of interrupted program
        STI                             ; enable higher interrupts if any
        MOV     AX, DATA_HERE           ; load DS needed here
        MOV     DS, AX
        DEC     COUNTER                 ; decrement interrupt counter
        JNZ     EXIT2                   ; not zero yet , go wait
        MOV     COUNTER, 20             ; if zero, reset tick counter to 20
        MOV     BH, 00                  ; load BX with number of loop to
        MOV     BL, LOOPNUM                     ;  service
        CALL    DWORD PTR LOOP_ADDR_TABLE[BX]   ; and service that loop
        ADD     LOOPNUM, 04                     ; point at next loop address
        CMP     LOOPNUM, 20H                    ; is this the last loop?
        JNE     EXIT2                           ; no, exit
        MOV     LOOPNUM, 00                     ; yes, get back to first loop
EXIT2:  MOV     AL, 00100000B           ; OCW2 for nonspecific EOI
```

FIGURE 10-35 (*continued*)

```
        MOV     DX, ØFFØØH              ; address for OCW2
        OUT     DX, AL
        POP     DS                      ; restore registers
        POP     DX
        POP     BX
        POP     AX
        IRET
CLOCK_TICK      ENDP

;DUMMY INTERRUPT PROCEDURE TO SERVICE KEYBOARD
KEYBOARD PROC   FAR
;       :                               ; keyboard proc intructions
        MOV     AL, ØØ1ØØØØØB           ; OCW2 for non-specific EOI
        MOV     DX, ØFFØØH              ; address for OCW2
        OUT     DX, AL                  ; send OCW2 for end of interrupt
        IRET
KEYBOARD ENDP

INT_PROC_HERE   ENDS
        END
```

(*b*)

```
PAGE ,132
;MODULE 3 - CONTAINS THE PROCEDURES TO SERVICE EACH LOOP
;
DATA_HERE       SEGMENT WORD    PUBLIC
        EXTRN   TIMEHI :BYTE, TIMELO  :BYTE     ; imported into this
        EXTRN   CURTEMP:BYTE, SETPOINT:BYTE     ; module from the mainline
DATA_HERE       ENDS

PUBLIC  LOOPØ, LOOP1, LOOP2, LOOP3, LOOP4, LOOP5, LOOP6, LOOP7

CODE_HERE SEGMENT WORD PUBLIC
EXTRN   DISPLAY  : NEAR         ; These procedures can be
EXTRN   A_D_READ : NEAR         ; found in another assembly
EXTRN   BINCVT   : NEAR         ; module which will be linked
CODE_HERE ENDS                  ; with this module & other modules

CODE_HERE       SEGMENT WORD    PUBLIC
        ASSUME  CS:CODE_HERE, DS:DATA_HERE

;8Ø86 PROCEDURE TO SERVICE TEMPERATURE CONTROLLER
;
;ABSTRACT:       This procedure services the temperature controller
;REGISTERS:      Destroys none
;PORTS:          Uses P2B (FFFAH) as output port to turn heater with
;                bit 7.
;CALLS:          DISPLAY, A_D_READ, BINCVT
LOOPØ   PROC    FAR
        PUSHF                           ; save registers
        PUSH    AX
        PUSH    BX
        PUSH    CX
        PUSH    DX
        DEC     TIMEHI                  ; decrement time for heater on
```

FIGURE 10-35 (*continued*)

```
JNZ     EXIT                    ; return to interrupt procedure
MOV     TIMEHI, Ø1              ; reset time high to fall through value
MOV     DX, ØFFFAH              ; point at o/p port P2B &
MOV     AL, 8ØH                 ; turn off heater
OUT     DX, AL
DEC     TIMELO                  ; decrement time for heater off
JNZ     EXIT                    ; return to interrupt procedure
MOV     BL, ØØ                  ; load channel address (Ø)
CALL    A_D_READ                ; do A/D conversion
MOV     CURTEMP, AL             ; save current temperature
CALL    BINCVT                  ; convert to BCD
MOV     CL, AL                  ; put result in CX to display
MOV     CH, ØØ
MOV     AL, ØØ                  ; temp in data field of SDK-86
CALL    DISPLAY
MOV     AL, SETPOINT            ; get setpoint temperature
SUB     AL, CURTEMP             ; get temperature & subtract from setpoint
JBE     DONE                    ; heater off if above or equal setpoint
MOV     DL, AL                  ; save temperature difference
MOV     AX, ØØ64H               ; compute new TIMELO
DIV     DL                      ; ØØ64/error, quotient is value
MOV     TIMELO, AL              ; for new time low
        MOV     TIMEHI, Ø4              ; set time high for 4 loops on
        MOV     AL, ØØ
        MOV     DX, ØFFFAH              ; point at output port
        OUT     DX,AL                   ; turn on heater
        JMP     EXIT
DONE:   MOV     TIMEHI, Ø1H             ; fall through value for time high
        MOV     TIMELO, 7FH             ; long off value for time low
EXIT:   POP     DX                      ; loop serviced - return to
        POP     CX                      ; interrupt service procedure
        POP     BX
        POP     AX
        POPF
        RET
LOOPØ   ENDP
;DUMMY LOOPS HERE
LOOP1   PROC    FAR
;       :                               ; instructions for this loop
        RET
LOOP1   ENDP
LOOP2   PROC    FAR
;       :                               ; instructions for this loop
        RET
LOOP2   ENDP
LOOP3   PROC    FAR
;       :                               ; instructions for this loop
        RET
LOOP3   ENDP
LOOP4   PROC    FAR
;       :                               ; instructions for this loop
        RET
LOOP4   ENDP
```

FIGURE 10-35 (*continued*)

```
LOOP5   PROC   FAR
;       :                        ; instructions for this loop
        RET
LOOP5   ENDP

LOOP6   PROC   FAR
;       :                        ; instructions for this loop
        RET
LOOP6   ENDP

LOOP7   PROC   FAR
;       :                        ; instructions for this loop
        RET
LOOP7   ENDP

CODE_HERE      ENDS
        END
```

(c)

```
PAGE ,132
;MODULE 4 - CONTAINS THE SERVICE PROCEDURES NEEDED BY THE LOOP MODULES

PUBLIC  DISPLAY, A_D_READ, BINCVT ; make procedures available to other modules

DATA_HERE SEGMENT       WORD    PUBLIC
SEVEN_SEG       DB      3FH, Ø6H, 5BH, 4FH, 66H, 6DH, 7DH, Ø7H
                ;        Ø    1    2    3    4    5    6    7
                DB      7FH, 6FH, 77H, 7CH, 39H, 5EH, 79H, 71H
                ;        8    9    A    b    C    d    E    F
DATA_HERE ENDS
CODE_HERE       SEGMENT WORD    PUBLIC
                ASSUME CS:CODE_HERE, DS:DATA_HERE

;8Ø86 PROCEDURE TO DISPLAY DATA ON SDK-86 LEDs
;ABSTRACT:  This procedure will display a 4-digit hex or BCD number
;           passed in the CX register on LEDs of the SDK-86
;INPUTS:    Data in CX, control in AL.
;           AL = ØØH  data displayed in data-field of LEDs
;           AL <> 00H, data displayed in address field of LEDs.
;PORTS:     Uses none
;PROCEDURES:Uses none
;REGISTERS: saves all registers and flags

DISPLAY PROC    NEAR
        PUSHF                   ; save flags
        PUSH DS                 ; save caller's DS
        PUSH AX                 ; save registers
        PUSH BX
        PUSH CX
        PUSH DX
        MOV  BX, DATA_HERE      ; init DS as needed for procedure
        MOV  DS, BX
        MOV  DX, ØFFEAH         ; point at 8279 control address
        CMP  AL, ØØH            ; see if data field required
        JZ   DATFLD             ; yes, load control word for data field
```

FIGURE 10-35 (*continued*)

```
        MOV  AL, 94H             ; no, load address-field control word
        JMP  SEND                ; go send control word
DATFLD: MOV  AL, 90H             ; load control word for data field
SEND:   OUT  DX, AL              ; send control word to 8279
        MOV  BX, OFFSET SEVEN_SEG ; pointer to seven-segment codes
        MOV  DX, 0FFE8H          ; point at 8279 display RAM
        MOV  AL, CL              ; get low byte to be displayed
        AND  AL, 0FH             ; mask upper nibble
        XLATB                    ; translate lower nibble to 7-seg code
        OUT  DX, AL              ; send to 8279 display RAM
        MOV  AL, CL              ; get low byte again
        MOV  CL, 04              ; load rotate count
        ROL  AL, CL              ; Move upper nibble into low position
        AND  AL, 0FH             ; Mask upper nibble
        XLATB                    ; translate 2nd nibble to 7-seg code
        OUT  DX, AL              ; send to 8279 display RAM
        MOV  AL, CH              ; Get high byte to translate
        AND  AL, 0FH             ; Mask upper nibble
        XLATB                    ; Translate to 7-seg code
        OUT  DX, AL              ; send to 8279 display RAM
        MOV  AL, CH              ; get high byte to fix upper nibble
        ROL  AL, CL              ; move upper nibble into low position
        AND  AL, 0FH             ; mask upper nibble
        XLATB                    ; translate to 7-seg code
        OUT  DX, AL              ; 7-seg code to 8279 display RAM
        POP  DX                  ; restore all registers and flags
        POP  CX
        POP  BX
        POP  AX
        POP  DS
        POPF
        RET
DISPLAY ENDP

;8086 PROCEDURE TO CONTROL A/D CONVERTER
;
;PORTS:          Port P2A is input from A/D
;                Port P2B, bit 7 = heater, bit 5 = start conversion
;                          bit 4 = ALE     bits 2,1,0 = channel address
;                Port P2C  bit 0 = end of conversion
;INPUTS:         Channel address for A/D in BL
;OUTPUTS:        A/D data in AL
;REGISTERS:      DESTROYS AL & BL

A_D_READ         PROC    NEAR
        PUSHF
        PUSH     DX
        MOV      AL, 80H              ; control for heater off
        OR       AL, BL               ; combine with channel address
        MOV      DX, 0FFFAH           ; point at P2B, output port
        OUT      DX, AL               ; send
        MOV      AL, 90H              ; send ALE, keep heater on
        OR       AL, BL               ; keep channel address on
        OUT      DX, AL
        MOV      AL, 0B0H             ; send start of conversion
```

FIGURE 10-35 (*continued*)

```
        OR    AL, BL              ; keep channel address on
        OUT   DX, AL
        MOV   AL, 80H             ; turn off ALE and start
        OR    AL, BL              ; keep channel address
        OUT   DX, AL
        MOV   DX, 0FFFCH          ; point at port P2C
EOCL:   IN    AL, DX              ; wait for end of conversion
        RCR   AL, 01              ; to go low
        JC    EOCL
EOCH:   IN    AL, DX              ; wait for end of conversion
        RCR   AL, 01              ; to go high
        JNC   EOCH
        MOV   DX, 0FFF8H          ; point at port P2A
        IN    AL, DX              ; read data from A/d
        POP   DX
        POPF
        RET
A_D_READ      ENDP

;BINCVT -      Converts an 8-bit binary number in AL
;               to packed BCD equivalent in AL
;INPUTS:       AL - 8-bit binary number
;OUTPUTS:      AL - packed BCD result

BINCVT  PROC   NEAR
        PUSHF                ; save registers and flags
        PUSH  CX
        MOV   AH,09H         ; bit counter for 8 bits
        MOV   CL, AL         ; save binary in CL
        MOV   CH, 00         ; clear CH for use as buffer
CNVT2:  XOR   AL, AL         ; clear AL and carry
        DEC   AH             ; decrement bit counter
        JNZ   GO_ON          ; do all bits
        JMP   HOME           ; done if AH down to zero
GO_ON:  RCL   CL,1           ; MSB from CL to carry
        MOV   AL, CH         ; move BCD digit being built to AL
        ADC   AL, AL         ; double AL and add carry from CL shift
        DAA                  ; keep result in BCD form
        MOV   CH, AL         ; put back in CH for next time through
        JMP   CNVT2          ; continue conversion
HOME:   MOV   AL, CH         ; BCD in AL for return
        POP   CX             ; restore registers
        POPF
        RET
BINCVT  ENDP

CODE_HERE  ENDS
        END
```

(*d*)

FIGURE 10-35 (*continued*)

Upon return, the binary value of the temperature is stored in a memory location called CURTEMP for future reference. For testing purposes we wanted to display the temperature on the address field of the SDK-86 display. To do this we convert the binary value for the temperature to a BCD value using a reduced version of the binary-to-BCD procedure from the scale program earlier in this chapter, and the display routine from Chapter 9.

After displaying the current temperature, it is then compared with the set-point temperature to see if the heater needs to be turned on. If the temperature is at or above the set point, TIMEHI is loaded with the fall-through value, and TIMELO with a large number.

If the temperature is below the set point, we call a procedure, DUTY_CYCLE, which computes the correct values for TIMEHI and TIMELO based on the difference between the set point and the current temperature. A complex PID algorithm might be used for this procedure in a precision system. For our example here, however, we have used simple proportional feedback. To further simplify the calculations a fixed value of 4 was used for TIMEHI. The thinking for the value of TIMELO then goes as follows. If the difference in temperature is large, then TIMELO should be small so the heater is on for a longer duty cycle. If the difference in temperature is small, then the value of TIMELO should be large so the heater has a short duty cycle. Experimentally we found that a good first approximation for our system was (difference in temperature) × TIMELO = 100 decimal (64H). For example, if the difference in temperature is 20° (14H), then 64H/14H gives a value of 5 for TIMELO. The values for TIMEHI and TIMELO are returned in their named memory locations. Upon return to the main loop procedure we send a control word which turns on the heater. Execution then jumps to EXIT.

When execution returns to loop 0 again after 160 ms, TIMEHI will be decremented. If TIMEHI did not decrement to 0, then execution simply adjusts a few things and returns. If TIMEHI is 0 after the decrement, the heater is turned off, and TIMELO is decremented. TIMELO is then decremented every time loop 0 is serviced (every 160 ms) until TIMELO reaches 0. When TIMELO gets counted down to 0, a new A/D conversion is done, and a new feedback value for TIMELO recalculated.

An important point here is that the part of the program that determines the feedback is separate from the rest of the program so it can be easily altered without changing any of the rest of the program. All that needs to be changed in this procedure is the value of TIMEHI, the value of TIMELO, and the rate at which these change in response to a difference in temperature to produce proportional, integral, and derivative feedback control.

TEMPERATURE CONTROLLER RESPONSE

The dotted line in Figure 10-36 shows the temperature versus time response of our system with traditional thermostat control, which is often called *on-off control* or "bang-bang" control. As you can see the temperature overshoots the set point a great deal, and then oscillates around the set-point temperature. The solid line in Figure 10-36 shows the response of the system operating with our temperature-controller program. The initial overshoot was caused by the large thermal inertia of the hot plate we used. The overshoot and the residual error of about 1° could be eliminated by using a more complex feedback algorithm. This example should make you aware of the advantages of computer feedback control.

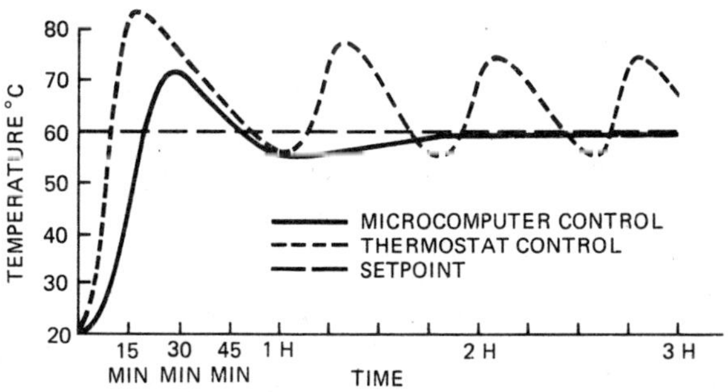

FIGURE 10-36 Temperature versus time responses for thermostat-controlled and microcomputer-controlled heaters.

Robotics

In recent years the term *robot* has become a "buzzword" in the media and in many people's minds. Science fiction movies have helped us form an image of robots as mobile, rational companions. Robots, however, have many forms, and in operation they are simply a special case of feedback control systems such as we described in the previous section. This is why we have not included a chapter dedicated just to robotics. The Rhino robot arm shown in Figure 9-43, for example, uses optical encoders to detect the position of its different joints, motors (actuators) to move each joint to a desired position, and a microcomputer to control the motors based on feedback from the sensors. In large industrial robots such as those that weld or spray-paint cars, the sensors used may also include vision, and the actuators may be hydraulic or pneumatic, but the control principle is the same. Feedback from the various sensors is used to control the output to the actuators.

Most of you have probably used some simple robots around your home without realizing it. One example is an electric garage door opener which starts to open or close when you tell it to, and then stops when a sensor indicates that it is closed or open as desired. Other examples are an automatic clothes washer, a clothes dryer, and a microwave oven with a temperature probe.

The next major section of this chapter is a discussion of how you develop the prototype of a microcomputer-based instrument such as the smart scale or the control system we discussed in the preceding sections.

DEVELOPING THE PROTOTYPE OF A MICROCOMPUTER-BASED INSTRUMENT

The first step in developing a new instrument is to very carefully define exactly what you want the instrument to do. The next step is to decide which parts of the instrument you want to do in hardware, and which parts you want to do in software. You then can decide how you want to do each of these.

For the software, you will break the overall programming job down into modules which can be individually tested and debugged as we have described previously.

For the hardware there are several different approaches you can take.

Using a Microcomputer-Prototyping Board

One approach is to use a commercially available microcomputer-prototyping board such as the SDK-86 we used for the examples in this chapter. An advantage of this approach is that it gives you the basic CPU, RAM, ROM, and ports already tested. You can then easily add any needed timers, priority interrupt controllers, and other interface circuitry. Some of the available prototyping boards such as the SDK-86 have on-board monitor programs which let you load and execute your programs. The major advantage of this approach is that it allows you to quickly get a prototype up and running to see if the instrument is feasible. If the instrument is feasible you can then design a custom hardware board which exactly fits your needs.

Computer-Aided Design Approach

Another approach to creating the needed hardware for the prototype is with a *computer-aided design* or CAD system. This system may be a large and powerful engineering workstation such as those made by Mentor Graphics, or simply an IBM PC-type computer with programs such as the PCAD system from Personal CAD Systems, Inc., Electronic Design Automation Division. The programs on these systems allow you, first of all, to easily design and draw a schematic for your hardware. You can just select the schematic symbol for a part you want to use by number from a large library of common devices in a disk file and bring it on to your CRT screen. You use a "mouse" to move the symbol into position and to draw signal lines connecting it to other symbols. You can move a device around as needed, and the connecting signal lines will follow.

When you get the schematic drawn up, you can then use another program in the CAD system to *simulate* the operation of the circuit. By simulate we mean to "run" the circuit in software. This helps you to find out if the signals are connected correctly, and if timing parameters are acceptable. If the circuit passes simulation, you can make a printout of the schematic on a printer or plotter.

The next step is to design a printed circuit board for your circuit. Another program in the CAD system will, with a little help from you, produce the artwork for the printed-circuit board. Some systems will even produce the control tape for the machine which automatically drills the required holes in the printed circuit board. The time is not too far away when the engineering workstation can be connected directly to the printed-circuit-board-making machine, the machine that gets parts from the warehouse, the machine that stuffs the parts in the printed circuit board, and the machine that does the initial functional tests on the board. This concept, incidentally, is called *computer integrated manufacturing* or CIM and seems to be where the industry is heading, but it isn't quite there yet. Therefore, you still have some work to do when you get the prototype printed circuit board back.

After you stuff the board with the required parts you can power it up and check for hot or otherwise unhappy components. If there are no apparent problems you can proceed to test the board. Probably the best tool to test the board with is an emulator.

Using an Emulator

Figure 3-12 shows a picture of an Applied Microsystems ES 1800 emulator which works with the IBM PC and other compatible computers. Several other companies make similar emulators. The hardware of an emulator consists of control circuitry, memory to store the trace data after each instruction executes, and an "umbilical" cable with a plug at the end of it. To use the emulator you remove the microprocessor from the prototype unit and insert the plug at the end of the umbilical cable in its place. The emulator contains a microprocessor which will actually run your test programs under control of the emulator. The emulator then gives you a window into the operation of the circuitry on the prototype under control of a development system or PC.

The software of the emulator is similar to a powerful monitor program or debugger program. You can use the emulator commands and the system memory to test each part of the prototype. For example, you can write a short program to test the RAM in the prototype, load this program into the system RAM, and run the program under emulator control. To help with debugging, emulators allow you to set breakpoints, examine and change the contents of registers and memory locations, and do a trace which shows the contents of registers after each instruction executes. Some emulators have an additional pod like those used on logic analyzers so that you can do a trace of the sequence of hardware signals on a group of lines to check timing.

An important point here is that, just as we stressed with building programs, the fastest way to get a prototype debugged and operating is one small part at a time. Because problems tend to interact, trying to debug too large a section at a time can be frustrating and time-consuming. Therefore, remove all but the basic CPU group ICs for your first test then keep adding, testing and debugging one section at a time. As you get a hardware section working, you can if you want write and debug the software module which uses or drives that hardware module. To give you a better idea of how to do this development process, we will briefly describe the steps we went through to develop the process-control system discussed earlier in this chapter.

A Product Development Example

For our process-control demonstration system we started with an SDK-86 board because we only wanted to make one unit, and because we did not have CAD equipment and a printed-circuit-board-making machine. For the controller we needed a timer to produce 1-kHz clock ticks and a priority interrupt controller to handle keyboard and clock tick interrupts. We added these two devices and some address decoder circuitry to the SDK-86 as shown in Figure 8-14, and proceeded to test this circuitry with an emulator. To do this we wrote a short program which wrote a byte to the starting address for the timer over and over again. We ran this program with the emulator and with a scope we checked to see if the $\overline{CS}$ input of the timer was getting asserted. It was, so we knew that the address decoding circuitry was working correctly. We then connected the 2.45-MHz PCLK signal to the clock inputs of all three timers in the 8254 and wrote the instructions needed to initialize the three timers for 1-kHz square-wave output. Even though we only need one timer here, it was very little additional work to check the other two for future reference. Hooray, the timers worked the first time, now on to the 8259A PIC.

Testing the 8259A was a little more complex because we had to provide an interrupt signal, initialize the 8259A, initialize the interrupt vector table in low RAM, and provide a location for execution to go to when the PIC received an interrupt. We used the 1-kHz clock tick from the timer as the interrupt signal to the 8259A. For 8259A initialization and the interrupt jump table initialization we used the instructions in the mainline program in Figure 10-35. For the test-interrupt procedure we actually used a real-time clock and display procedure that we developed for examples in previous chapters. We used these so that we could see if the interrupt mechanism was working correctly by watching the displays on the SDK-86 count off seconds. This again shows the advantage of writing programs as separate reusable modules. Note in the program in Figure 10-35 that we initialize the 8259A before we initialize and start the timer. When we first wrote a test program to test an 8259A and an 8254, we did this in the reverse order. When we ran the test program with the emulator, the system would only accept one interrupt and hang up. We did a trace with the emulator and found that execution was returning from the interrupt procedure to the WAIT loop in the mainline program properly, but not recognizing the next interrupt. Careful reading of the 8259A data sheet showed us that we had to initialize the 8259A *before* we started sending it interrupt signals, or it would not respond correctly to the nonspecific EOI command that we used at the end of the interrupt procedure to reset the 8269A's in-service register.

After the interrupt mechanism was working correctly, we wrote the interrupt procedure which implements the decision structure shown in Figure 10-32*b*. Initially we made all eight loops dummy loops to test the basic structure. By inserting breakpoints with the emulator we were able to see if execution was getting to each of the eight loops. When all of this was working, we went on to build and test the temperature-control section.

For the temperature-control section we first built the analog circuitry and tested it. Then we wrote a small program to read the temperature from the A/D converter and display the result on the SDK-86 displays. Initially then, the loop 0 procedure simply read in the temperature, displayed it in binary (hex) form and returned. This worked the first time, so we went on to add the binary-to-BCD conversion routine and run the result with the emulator. This was a previously written and tested module, and when added, the result worked fine.

Next we added a couple of instructions to turn the heater on during one execution of loop 0 and turn the heater off during the next time through loop 0. We then used an oscilloscope to check that the solid-state relay was getting turned on and off correctly.

Finally, we added the actual duty cycle and control instructions, and sat back waiting for the system to heat us up a big container of water for tea.

The actual development cycle will obviously be somewhat different for every instrument developed. The main points here are to develop and test both the hardware and software in small modules. To speed up the debugging process, take the time to learn to use all or most of the power of the emulator and system you are working with.

DIGITAL FILTERS

A section at the start of this chapter showed how op amps can be used to build high-pass and low-pass filter circuits. Filtering of a signal can also be done by taking samples of the signal with an A/D converter, performing mathematical operations on the samples from the A/D converter, and outputting the result to a D/A converter. This approach, referred to as a *digital filter*, can easily produce a response curve which is difficult, if not impossible, to produce with analog circuitry. This digital approach has the further advantage that the filter response can be changed under program control. Digital filters are used in speech synthesizers, satellite image-enhancement systems, and many other applications.

There are two basic types of digital filter, the *finite impulse response* or FIR type and the *infinite impulse response* or IIR type. The basic principle of a digital filter is to operate on the samples as a function of time rather than as a function of frequency as the analog filter does.

Figure 10-37*a* shows a functional diagram of the operation of an FIR-type filter. The box containing Z^{-1} represents a delay of one sample interval. Circles containing an X represent a multiplication operation, and the letters to the left of each circle represent the number that the term will be multiplied by. Y_0 represents the value of the current sample from the A/D, Y_1 represents the value of the previous sample from the A/D, and Y_2 represents the value of the sample before that. Here's how this works. The output value, V, at any time is produced by summing the (current sample × some coefficient) + (the previous sample × some coefficient) + (the sample before that × some coefficient), etc. To do all of this with a microprocessor involves simple operations of saving previous samples, multiplying, and adding. The Intel 2920 microprocessor, which was specifically designed for this type of operation, contains an A/D converter, D/A converter, and an architecture and instruction set which works with the 25-bit numbers required for accurate filter response.

Figure 10-37*b* shows a functional diagram for an IIR digital filter. Here again the blocks containing Z^{-1} rep-

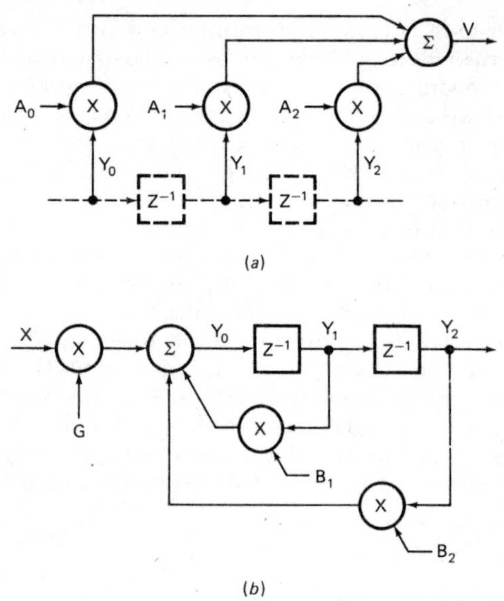

FIGURE 10-37 FIR and IIR digital filter principles. *(a)* FIR. *(b)* IIR.

resent a delay of one sample time. The value of the current sample from the A/D converter is represented by the *X* at the left of the diagram. The Y_0 point represents the output from the microprocessor to the D/A converter. Note that for an IIR filter it is this output value which is saved to be used in computing feedback terms for future samples. In the FIR type, remember, the samples from the A/D converter were saved directly for future use. The output for an IIR type is produced by summing (the current sample × a calculated coefficient) + (the previous output value × a calculated coefficient) + (the output value before that × a calculated coefficient), etc. The coefficients for both the FIR- and the IIR-type filters are usually calculated using a computer program. FIR filters are easier to design, but they may require many terms to produce a given filter response. IIR filters require fewer stages, but they have to be carefully designed so that they do not become oscillators.

A new type of filter called a *switched capacitor filter* implements digital filtering for simple filter responses without the need for the A/D and D/A converter. An example is the National MF10. In this type of filter an input signal is sampled on a capacitor. The signal is passed on to other capacitors and fractions of the outputs from these capacitors are summed to produce an analog output signal directly. Switched capacitor filters are less expensive, but they do not give the degree of programmability that the microprocessor-based filters do.

IMPORTANT TERMS AND CONCEPTS FROM THIS CHAPTER

If you do not remember any of the terms in the following list, use the index to help you find them in the chapter for review.

Op amp

Comparator

Hysteresis

Noninverting amplifier

Inverting amplifier
  Virtual ground

Gain-bandwidth product

Unity-gain bandwidth

Adder circuit—summing point

Differential amplifier
  Common-mode signal, common-mode rejection

Instrumentation amplifier

Op-amp integrator circuit
  Linear ramp
  Saturation

Op-amp differentiator

Op-amp active filters
  Low-pass filter, high-pass filter, bandpass filter
  Critical frequency or breakpoint
  Second-order low-pass filter, second-order high-pass filter

Light sensor
  Photodiode
  Solar cell

Temperature-sensitive voltage sources

Temperature-sensitive current sources

Thermocouples
  Type J thermocouple
  Cold-junction compensation

Force and pressure transducers
  Strain gage, LVDT, load cell

Flow sensors
  Paddle wheel, differential pressure transducer

D/A converters
  Binary weighted
  Resolution
  Full-scale output voltage
  Maximum error
  Linearity
  Settling time

A/D converters
  Conversion time
  Parallel-comparator A/D converter
  Dual-slope A/D converter
  Successive-approximation A/D converter
  Data acquisition system

A/D output codes
  Unipolar binary code, unipolar BCD code, biopolar binary code

Direct memory access

Set point

Servo control

Settling time, underdamped, overdamped

Residual error

Proportional-integral-derivative control loop, PID

Time slice

On-off control

Robotics

Digital filters
Finite impulse response (FIR), infinite impulse response (IIR)

Switched capacitor filter

Computer-aided design
Simulate

Computer-integrated manufacturing, CIM

Emulator

REVIEW QUESTIONS AND PROBLEMS

1. *a.* A comparator circuit such as the one in Figure 10-1*b* is powered by ±15 V, the inverting input is tied to +5 V, and the noninverting input is at +5.3 V. About what voltage will be on the output of the comparator?

   b. An amplifier circuit, such as the one in Figure 10-1*d*, has an $R1$ of 10 kΩ and an $R2$ of 190 kΩ. Calculate the closed-loop voltage gain for the circuit and the V_{out} that will be produced by a V_{in} of 0.030 V. What voltage would you measure on the inverting input? What would be the gain of the circuit if $R2 = 0\ \Omega$.

   c. An amplifier circuit, such as the one in Figure 10-1*e*, is built with an $R1$ of 15 kΩ and an R_f of 75 kΩ. Calculate the closed-loop voltage gain for the circuit and the output voltage for an input voltage of 0.73 V. What voltage will you always measure on the inverting input of this circuit?

   d. A differential amplifier, such as the one in Figure 10-1*g*, is built with $R1 = R2 = 100$ kΩ and $R_f = R3 = 1$ MΩ. V_1 is 4.9 V and $V_2 = 5.1$ V. Calculate the output voltage and polarity.

   e. Describe the main advantage of the instrumentation amplifier in Figure 10-1*h* over the simple differential amplifier in Figure 10-1*g*.

   f. If the amplifier used in the circuit in Question 1*b* has a gain-bandwidth product of 1 MHz, what will be the closed-loop bandwidth of the circuit?

2. Draw a circuit showing how a light-dependent resistor can be connected to a comparator so the output of the comparator changes state when the resistance of the LDR is 10 kΩ.

3. For the photodiode amplifier circuit in Figure 10-5, what voltage will you measure on the inverting input of the amplifier? Why is it important to use an FET input amplifier for this circuit? Which direction are electrons flowing through the photodiode?

4. In what application might you use a temperature-dependent current device such as the AD590 rather than a temperature-dependent voltage device such as the LM35?

5. Why must thermocouples be cold-junction compensated in order to make accurate measurements? How can the nonlinearity of a thermocouple be compensated for?

6. Why are strain gages usually connected in a bridge configuration? Why do you use a differential amplifier to amplify the signal from a strain-gage bridge?

7. Calculate the full-scale output voltage for the simple D/A converter in Figure 10-14.

8. What is the resolution of a 13-bit D/A converter? If the converter has a full-scale output of 10.000 V, what is the size of each step? What will be the actual maximum output voltage of this converter? What accuracy should this converter have to be consistent with its resolution.

9. Why must a 12-bit D/A converter have latches on its inputs if it is to be connected to 8-bit ports or an 8-bit data bus?

10. Describe the operation of a "flash" type A/D converter. What are its main advantages and disadvantages?

11. For the dual-slope A/D converter in Figure 10-19, what will be the displayed count for an input voltage of 2.372 V? What is the resolution of a 4½-digit slope-type A/D converter expressed in bits?

12. How many clock cycles does a 12-bit successive-approximation A/D converter take to do a conversion on a 0.1-V input signal? On a 5-V input signal? How does this compare with the number of clock cycles required for a 12-bit dual-slope type?

13. *a.* Assume the inputs of the MC1408 D/A converter in Figure 10-20 are connected to an output port on your microcomputer board and the output of the comparator is connected to bit D0 of an input port. Write the algorithm for a procedure to do an A/D conversion by outputting an incrementing count to the output port.

    b. Write an algorithm for a procedure to do the conversion by the successive approximation method. Which method will produce a faster result? If the hardware is available, write the programs for these algorithms and compare the times by watching the comparator output with an oscilloscope.

14. Show the detailed algorithm for the procedure you would use to read in the data from a multiplexed BCD output A/D converter such as the MC14433 in Figure 10-23 and assemble the value in a 16-bit register for display.

15. The data sheet for an A/D converter indicates that its output is in offset-binary code. If the converter is set up for a range of −5 to +5 V and the output code is 01011011, what input voltage does this represent? How could you convert this code to 2's complement form after you read the code into your microcomputer?

16. Write a procedure to round a 32-bit BCD number in DX:AX to a 16-bit BCD number in DX.

17. For the scale circuitry in Figure 10-23, what voltage should you measure on the inverting input of the LM308 amplifier? What voltages should you measure on the two inputs of the LM363 amplifier with no load on the scale? What voltage should you measure on the output of the LM363 with no load on the scale?

18. The section of the scale program following the label NXTKEY in Figure 10-35 moves some bytes around in memory. Rewrite this section of the program using an 8086 string instruction to do the move operations. Which version seems more efficient in this case?

19. Describe how feedback helps hold the value of some variable, such as a motor speed, constant. Refer to Figure 10-27 in your explanation.

20. What problem in a control loop does integral feedback help solve? Why is derivative feedback sometimes added to a control loop?

21. What is the major advantage of a microcomputer-controlled loop over the analog approach shown in Figure 10-29?

22. Suppose that you want to control the speed of a small dc motor, such as the one in Figure 10-27, with LOOP1 of our microcomputer-based process controller.

a. Show how you would connect the output from the motor's tachometer to the system in Figure 10-33. Also show how you would connect an 8-bit D/A to control the current to the motor.

b. Write a flowchart for the LOOP1 procedure to control the speed of the motor.

c. Describe how a lookup table could be used to determine the feedback value.

23. Describe the major difference in how the feedback is produced in an FIR digital filter and how it is produced in an IIR filter.

24. When developing a prototype, why is it very important to build, test and debug both software and hardware in small modules?

CHAPTER

11 Multiple Microprocessor Systems and Buses

The major point of the first six chapters of this book was to introduce you to structured programming and to writing 8086 assembly language programs. Chapters 7 through 10 introduced you to the hardware of an 8086 minimum-mode system, showed you how to interface a microcomputer to a wide variety of input and output devices, and finally demonstrated how all of these pieces are put together to build a microcomputer-based instrument or simple control system. The major theme of Chapters 11 through 14 is to show you how larger microcomputer systems are built and programmed. As some of the parts of this we show you how large memory banks are added to a system, how multiple processors are used in a system, how you interface to more complex peripherals such as disk drives, and how systems communicate with each other. We also discuss the system programs used to coordinate all of this.

OBJECTIVES

At the end of this chapter you should be able to:

1. Show how an 8086 is connected with a controller device for operation in its maximum mode.
2. Show how a direct memory access (DMA) controller device can be connected in an 8086 system and describe how a DMA data transfer takes place.
3. Describe how large banks of dynamic RAM can be connected in a system and how automatic error-correcting circuitry works with this memory.
4. Describe the added architectural features of the 80186 microprocessor.
5. Show how a coprocessor can be connected to an 8086 or 8088 operating in maximum mode.
6. Describe the operation of the 8087 math coprocessor, and write simple programs for the 8087.
7. Show how several CPU boards can be connected to share a common set of buses, and describe how data is transferred on this common bus.
8. Use a timing diagram to describe how control of the bus is transferred from one CPU board on the bus to another.

THE 8086 MAXIMUM MODE

Many of the circuits shown in this chapter and the following chapters use the 8086 or 8088 in its maximum mode. Therefore, we start this chapter with a discussion of maximum-mode operation.

Figure 11-1*a* shows the pin diagram of the 8086 again. You may remember from our discussion in Chapter 7 that if pin 33, the MN/$\overline{MX}$ pin is tied high, the 8086 operates in its minimum mode. In minimum mode the 8086 generates control-bus signals directly. Specifically, for pins 24–31, the 8086 in minimum mode generates the signals identified in parentheses in Figure 11-1*a*.

If the MN/$\overline{MX}$ pin is tied low, the 8086 operates in its maximum mode and pins 24–31 generate the signals named next to the pins in Figure 11-1*a*. In maximum mode the control-bus signals are sent out in coded form on the status lines, $\overline{S0}$, $\overline{S1}$, and $\overline{S2}$. As shown in Figure 11-1*b*, an external controller device such as the Intel 8288 is used to produce the required control-bus signals from these lines. Figure 11-1*b* shows the expanded names for each of the control-bus signals generated by the 8288. Note in Figure 11-1*b* that we use 8282 octal latches to demultiplex the address signals and 8286 bidirectional drivers to buffer the data bus so that it can drive a board full of devices. Figure 11-1c shows the status line codes for each type of machine cycle.

The *request/grant* pins, $\overline{RQ}/\overline{GT0}$ and $\overline{RQ}/\overline{GT1}$, are used by other devices to tell the 8086 that they want to use the address, data, and control buses. These pins are bidirectional. They operate in a similar way to that in which the HOLD and HLDA pins operate when some other device wants to borrow the buses in an 8086 minimum-mode system. We will show you how these signals work in a later section on the 8087 coprocessor. A signal can be sent out from the 8086 on the $\overline{LOCK}$ pin under program control (the LOCK prefix) to prevent some other device from taking over the bus during exe-

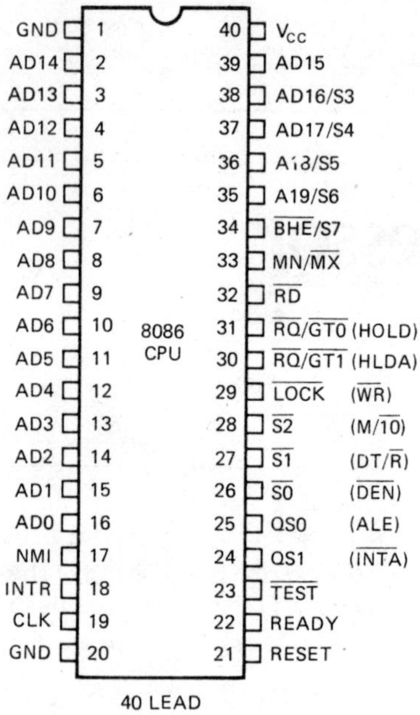

(a)

$\overline{S_2}$	$\overline{S_1}$	$\overline{S_0}$	
0 (LOW)	0	0	INTERRUPT ACKNOWLEDGE
0	0	1	READ I/O PORT
0	1	0	WRITE I/O PORT
0	1	1	HALT
1 (HIGH)	0	0	CODE ACCESS
1	0	1	READ MEMORY
1	1	0	WRITE MEMORY
1	1	1	PASSIVE

(c)

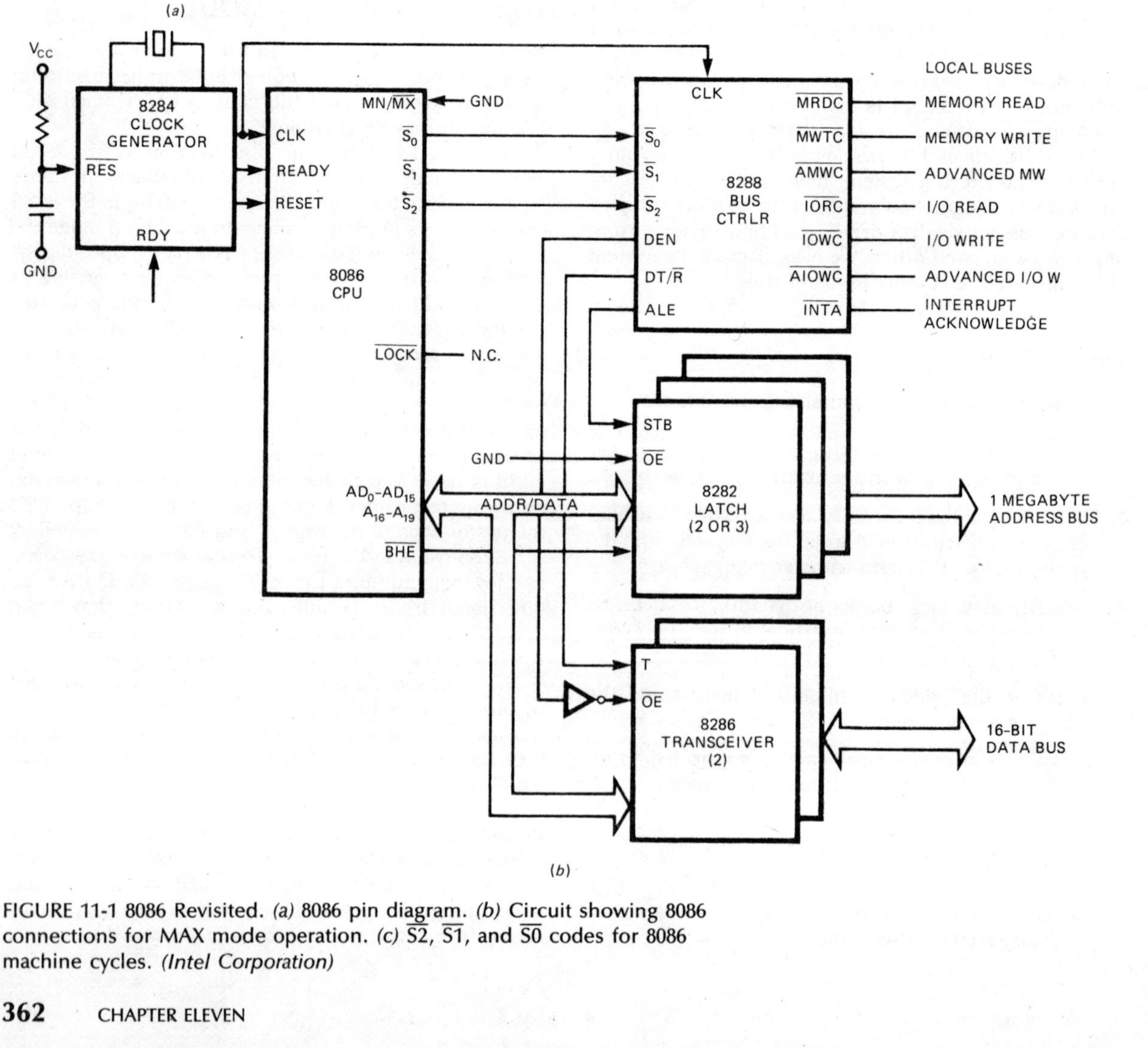

(b)

FIGURE 11-1 8086 Revisited. *(a)* 8086 pin diagram. *(b)* Circuit showing 8086 connections for MAX mode operation. *(c)* $\overline{S2}$, $\overline{S1}$, and $\overline{S0}$ codes for 8086 machine cycles. *(Intel Corporation)*

cution of a critical instruction. The queue status signals, QS1 and QS0, indicate the operation most recently done to the instruction-byte queue in the 8086 BIU. These signals allow an external device such as the 8087 math coprocessor, which we discuss later in this chapter, to monitor the 8086 queue and read the same instruction bytes as the 8086. Now we will show you some of the ways that a microprocessor can share its buses in minimum mode and in maximum mode.

DIRECT MEMORY ACCESS (DMA) DATA TRANSFER

DMA Overview

Up to this point in the book we have used program instructions to transfer data from ports to memory, or from memory to ports. For some applications, such as transferring data bytes to memory as they are read off a magnetic or optical disk, however, the data bytes are being sent from the disk faster than they can be read in with program instructions. In a case like this we use a dedicated hardware device called a *direct memory access* controller. A DMA controller temporarily borrows the address bus, data bus, and control bus from the microprocessor and transfers the data bytes directly from the port to a series of memory locations. Because the data transfer is handled totally in hardware, it is much faster than it would be if done by program instructions. A DMA controller can also transfer data from memory to a port. Some DMA devices can also do memory-to-memory transfers. Here's an example of how a common DMA controller is connected and used in an 8086 minimum-mode system.

Circuit Connections and Operation of the Intel 8237 DMA Controller

We chose the 8237 DMA controller as the example for this section because it is a commonly used device, and because it is one of the devices you will find if you start poking around inside an IBM PC or PC/AT. However, before we dig into the actual connections and operation of an 8237 circuit, let's take a look at the block diagram in Figure 11-2 to get an overview of how a DMA transfer takes place. The main point to keep in your mind here is simply that the microprocessor and the DMA controller time-share the use of the address, data, and control buses. We have tried to indicate this with the three switches in the middle of the block diagram. Here's how a transfer takes place.

When the system is first turned on, the switches are in the position where the buses are connected from the microprocessor to system memory and peripherals. We initialize all of the programmable devices in the system and go on executing our program until we need to, for example, read a file off a disk. To read a disk file we send a series of commands to the smart disk-controller device, telling it to go find and read the desired block of data from the disk. When the disk controller has the first byte of data from the disk block ready, it sends a *DMA request* (DREQ) signal to the DMA controller. If that input (channel) of the DMA controller is unmasked, the DMA controller will send a *hold-request* (HRQ) to the microprocessor HOLD input. The microprocessor will respond to this input by floating its buses and sending out an *hold-acknowledge* (HLDA) signal to the DMA controller. When the DMA controller receives the HLDA signal it will send out a control signal which throws the three bus switches to their DMA position, disconnecting

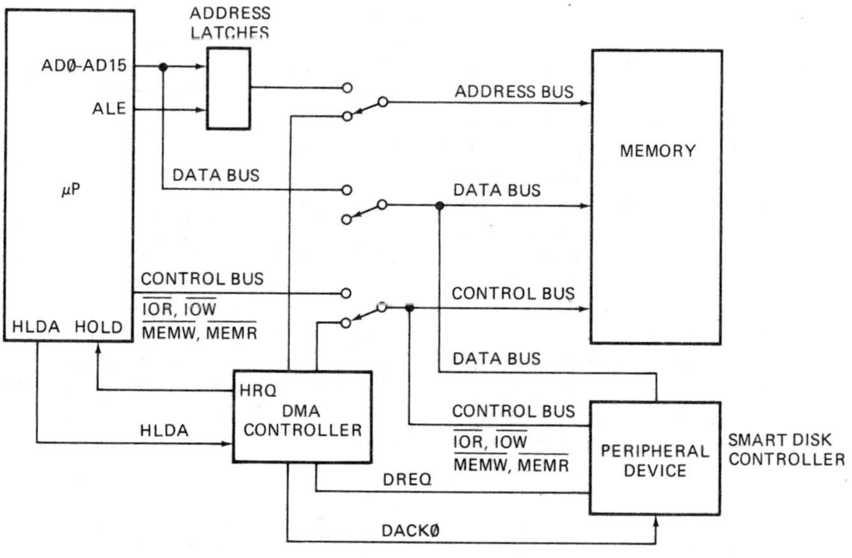

FIGURE 11-2 Block diagram showing how a DMA controller operates in a microcomputer system.

the processor from the buses. The DMA controller then outputs on the address bus the memory address where we want the byte of data from the disk controller to go. Next the DMA controller sends a *DMA-acknowledge* (DACK0) signal to the disk-controller device to tell it to get ready to output the byte. Finally, the DMA controller asserts both the $\overline{MEMW}$ and the $\overline{IOR}$ lines on the control bus. Asserting $\overline{MEMW}$ enables the addressed memory for writing data to it. Asserting $\overline{IOR}$ enables the disk controller to output the byte of data from the disk on the data bus. The byte of data will then be written to the addressed memory location.

NOTE: For this type of transfer the disk controller chip-select input does not have to be enabled by the port address decoding circuitry as it does for normal reading from and writing to registers in the device. In fact, the normal port-decoding circuitry is disabled during DMA operations to prevent the combination of $\overline{IOR}$ and the output memory address from turning on unwanted ports.

When the data transfer is complete the DMA controller unasserts its hold-request signal to the processor, and lets the processor take over the buses again until another DMA transfer is needed. The processor then continues executing from where it left off. A DMA transfer from memory to the disk controller would proceed in a similar manner except that the DMA controller would assert the *memory-read* control signal ($\overline{MEMR}$), and the *output-write* control signal ($\overline{IOW}$). DMA transfers may be done a byte at a time or in blocks.

Now, to give you more practice working your way through actual microprocessor circuits, let's look at Figure 11-3 to see some of the circuitry we might add to an 8086 system so that we can do DMA transfers to and from a disk controller. This circuitry is simply a more detailed version of the block diagram in Figure 11-2.

The first thing to do in analyzing this schematic is to identify the major devices and relate their function, where possible, to the block diagram. The 8086 and 8284 should be old friends from your exploration of the SDK-86. The 8237 is, of course, the DMA controller, and the 8272 is the floppy-disk controller. We will discuss the operation of the disk controller in the next chapter, but for now all we need to know about it is an overview of how it interacts with the 8237 as we described above. The 8282s in this circuit are octal latches with three-state outputs. They are used here to latch addresses output from either the 8086 or from the DMA controller. These devices are controlled by ALE from the 8086 and by AEN and ADSTB from the DMA controller.

When the power is first turned on, the *address-enable* signal (AEN) from the DMA controller is low. Devices *U*1, *U*2, and *U*4 are then enabled, and the ALE signal from the 8086 gets to the strobe inputs of all three devices. When the 8086 sends out an address and an ALE signal, these three devices will grab the address and send it out on the address-bus lines, A19–A0. This is just as would be done in a simpler 8086 system. Now, when the DMA controller wants to take over the bus, it asserts its AEN output high. This does several things. First, it disables device *U*1 so that address lines A7–A0 no longer come from the 8086 bus. The 8237 directly outputs the lower 8 bits of the memory address for the DMA transfer.

Secondly, AEN, going high, switches the strobe multiplexer so that the strobe for device *U*2 comes from the address strobe output of the 8237. To save pins, the 8237 outputs the upper 8 bits of the memory address for the DMA transfer on its data-bus pins and asserts its ADSTB output high to let you know that this address is present there. At the start of a DMA transfer, then, memory address bits A15–A8 will be sent out by the 8237 and latched on the outputs of *U*2.

Still another effect of AEN going high is to switch the source of address bits A19–A16 from device *U*4 to device *U*3. The DMA controller does not send out these address bits during a DMA transfer, so you have to produce them in some other way. You can either hard-wire the inputs of *U*3 to ground or +5 V to produce a fixed value for these bits, or you can connect these inputs to an output port so you can specify these address bits under program control.

Finally, AEN, going high, switches the source of the control-bus signals from the outputs of the control-bus decoder circuitry to the control-bus signal outputs of the DMA controller. This is necessary because, during a DMA transfer, the 8237 generates the required control-bus signals such as $\overline{MEMW}$ and $\overline{IOR}$. Incidentally, the NOR gate decoder circuitry in the upper right corner of the schematic is necessary to produce processor control-bus signals compatible with those from the 8237.

The final part of the circuit in Figure 11-3 to analyze is the two 8286 octal bus transceivers. The disk controller has only an 8-bit data bus output. If we connected these eight lines on the lower eight data bus lines of the 8086 system, the DMA controller could only transfer bytes to even addresses. Likewise, if we connected the disk-controller data outputs on the upper eight data lines of the 8086 system, the DMA controller could only transfer bytes to odd addresses in memory. To solve this problem, we connect the two 8286s as a switch which can route data to/from the disk controller from/to either odd or even addresses in memory. A0 determines which half of the data bus is connected to the eight data pins of the disk controller. Now let's look at the signal flow and timing for this circuit.

A DMA Transfer Timing Diagram

Figure 11-4 shows the sequence of signals that will take place for a DMA transfer in a system such as that in Figure 11-3. Keep a copy of the circuit handy as you work your way down through these waveforms. The labels we have added to each signal should help you. We will pick up where the 8237 asserts AEN high and gains control of the buses. After the 8237 gains control of the bus it sends out the lower 8 bits of the memory address on its A7–A0 pins, and the upper 8 bits of the memory address on its DB0–DB7 pins. The 8237 pulses ADSTB high to latch these address bits in the 8282, and then removes these address bits from the data bus. At about the same time the 8237 sends a DACK signal to the disk controller to tell it to get ready for a data transfer. Now

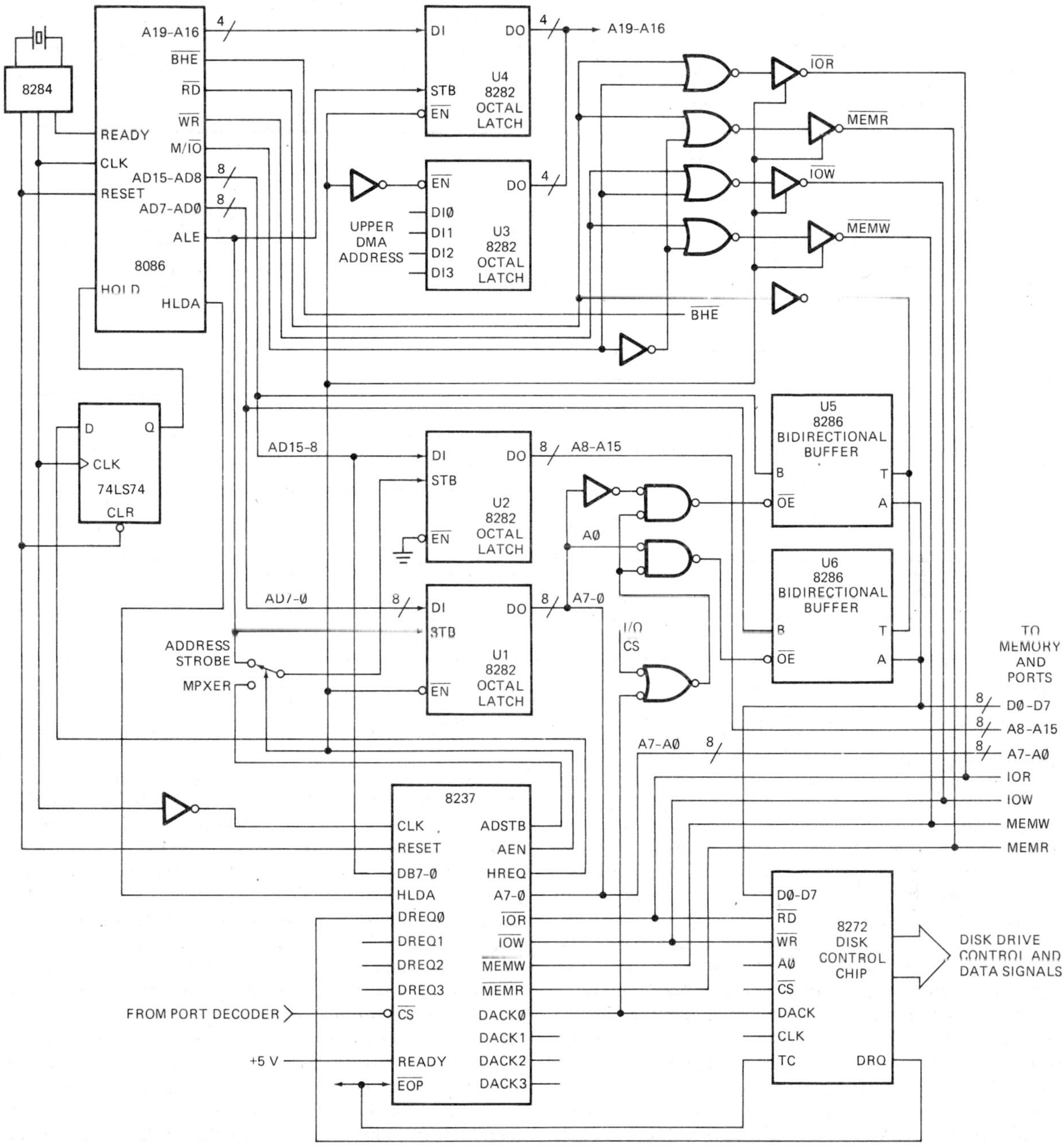

FIGURE 11-3 Schematic for 8086 system with 8237 DMA controller and 8272 floppy-disk controller.

that everything is ready the 8237 asserts two control-bus signals to enable the actual transfer. For a transfer from memory to the disk controller, it will assert $\overline{\text{MEMR}}$ and $\overline{\text{IOW}}$. For a transfer from the disk controller to memory, it will assert $\overline{\text{MEMW}}$ and $\overline{\text{IOR}}$. Note that the 8237 does not have to put out an I/O address to enable the disk controller for this transfer. When programmed in DMA mode, the disk controller needs only $\overline{\text{IOR}}$ or $\overline{\text{IOW}}$ to be asserted to enable it for the transfer. Also note that the 8237 will not output a new address on A8–A15 when a second transfer is done, unless those bits have to be changed. This saves time during multiple-byte transfers.

When the programmed number of bytes have been transferred, the DMA controller pulses its end-of-process ($\overline{\text{EOP}}$) pin low, unasserts its hold request to the 8086, and drops its AEN signal low to release the buses back to the 8086. Now that you have an idea how an 8237 is connected and operates in a system, we will give you an overview of what is involved in initializing it.

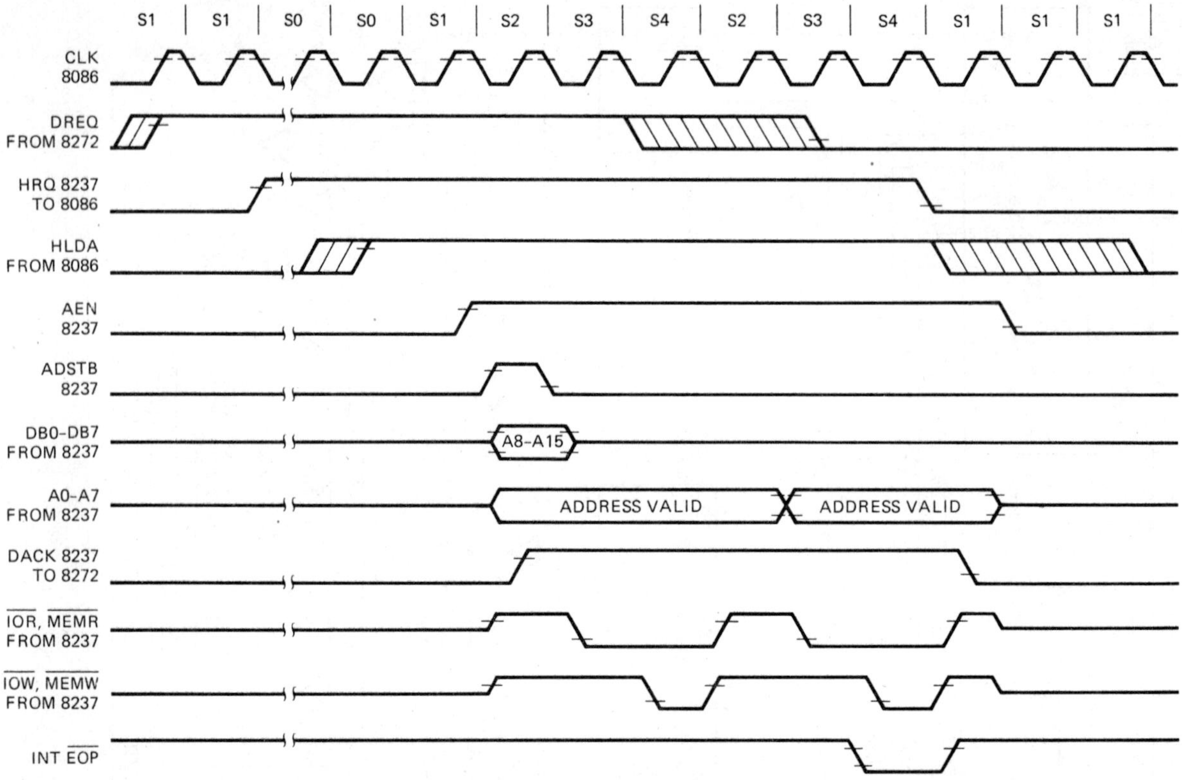

FIGURE 11-4 Timing diagram for 8237 DMA transfer. *(Intel Corporation)*

8237 Initialization Overview

Initializing an 8237 is not difficult, but it does require a fairly large number of bytes. Therefore, we do not have space here to show you a complete initialization. What we can do is give you an overview so that, hopefully, the data sheet will make more sense to you if you do have to initialize one.

As shown by the pin labels on the 8237 in Figure 11-3, the 8237 has four DMA request inputs or *channels*, as they are commonly called. To initialize an 8237 you need to send it a command word which specifies the general operation. You also need to send it mode words, starting transfer addresses, and the number of bytes to be transferred for each channel you are using. Figure 11-5*a* shows the names of the different types of registers used to hold this data in the 8237, the number of bits in each type of register, and how many registers of that type the device has. Register names with a 4 next to them have a register of that type for each channel. Now that you know about all of these registers, the next question is how do you write to or read from all of them.

The 8237 is connected in a system as a port device, so you write initialization words to it just as you would to any other port device. The lower four address bits, A3–A0, together with the input/output read signal, $\overline{\text{IOR}}$, and the input/output write signal, $\overline{\text{IOW}}$, determine which internal register you write to or read from. Figure 11-5*b* shows the internal addresses that you use when sending commands to and reading the status byte, etc. from an 8237. We'll come back to these in a minute. Figure 11-5*c* shows the internal addresses you use when you send or read addresses and counts to/from the 8237. Note that the low byte and the high byte of, for example, the base memory address are written to the same internal address in the 8237. The 8237 keeps track of which byte is being sent or read with an internal first-last flip-flop. If the flip-flop is reset then the 8237 assumes that the byte being sent or read is the least-significant byte. If the first-last flip-flop is set, the 8237 assumes the byte being sent or read is the most significant byte. The first-last flip-flop is automatically toggled after each write to a particular address so you can just write to an internal address twice to send a full 16-bit count. It is important to understand this mechanism so that you remember to keep track of the state of the first-last flip-flop as you send counts and addresses to the device. Also, at the start of the initialization before you read or write any words, it is a good idea to send the device a command which resets the first-last flip-flop. You do this by writing a byte to internal address 1100 as shown in Figure 11-5*b*. The contents of the byte written don't matter, it is the act of writing to the particular address which resets the flip-flop. Here's the order in which you might send initialization words to an 8237. Consult the data sheet in an Intel data book to get the details of each command word.

1. Output a master reset command word to internal address 1101. The actual word written doesn't matter; the command resets the first-last flip-flop.

NAME	SIZE	NUMBER
BASE ADDRESS REGISTERS	16 BITS	4
BASE WORD COUNT REGISTERS	16 BITS	4
CURRENT ADDRESS REGISTERS	16 BITS	4
CURRENT WORD COUNT REGISTERS	16 BITS	4
TEMPORARY ADDRESS REGISTER	16 BITS	1
TEMPORARY WORD COUNT REGISTER	16 BITS	1
STATUS REGISTER	8 BITS	1
COMMAND REGISTER	8 BITS	1
TEMPORARY REGISTER	8 BITS	1
MODE REGISTERS	6 BITS	4
MASK REGISTER	4 BITS	1
REQUEST REGISTER	4 BITS	1

(a)

SIGNALS						OPERATION
A3	A2	A1	A0	$\overline{IOR}$	$\overline{IOW}$	
1	0	0	0	0	1	READ STATUS REGISTER
1	0	0	0	1	0	WRITE COMMAND REGISTER
1	0	0	1	0	1	ILLEGAL
1	0	0	1	1	0	WRITE REQUEST REGISTER
1	0	1	0	0	1	ILLEGAL
1	0	1	0	1	0	WRITE SINGLE MASK REGISTER BIT
1	0	1	1	0	1	ILLEGAL
1	0	1	1	1	0	WRITE MODE REGISTER
1	1	0	0	0	1	ILLEGAL
1	1	0	0	1	0	CLEAR BYTE POINTER FLIP/FLOP
1	1	0	1	0	1	READ TEMPORARY REGISTER
1	1	0	1	1	0	MASTER CLEAR
1	1	1	0	0	1	ILLEGAL
1	1	1	0	1	0	CLEAR MASK REGISTER
1	1	1	1	0	1	ILLEGAL
1	1	1	1	1	0	WRITE ALL MASK REGISTER BITS

(b)

CHANNEL	REGISTER	OPERATION	SIGNALS							INTERNAL FLIP-FLOP	DATA BUS DB0-DB7
			$\overline{CS}$	$\overline{IOR}$	$\overline{IOW}$	A3	A2	A1	A0		
0	BASE AND CURRENT ADDRESS	WRITE	0	1	0	0	0	0	0	0	A0-A7
			0	1	0	0	0	0	0	1	A8-A15
	CURRENT ADDRESS	READ	0	0	1	0	0	0	0	0	A0-A7
			0	0	1	0	0	0	0	1	A8-A15
	BASE AND CURRENT WORD COUNT	WRITE	0	1	0	0	0	0	1	0	W0-W7
			0	1	0	0	0	0	1	1	W8-W15
	CURRENT WORD COUNT	READ	0	0	1	0	0	0	1	0	W0-W7
			0	0	1	0	0	0	1	1	W8-W15
1	BASE AND CURRENT ADDRESS	WRITE	0	1	0	0	0	1	0	0	A0-A7
			0	1	0	0	0	1	0	1	A8-A15
	CURRENT ADDRESS	READ	0	0	1	0	0	1	0	0	A0-A7
			0	0	1	0	0	1	0	1	A8-A15
	BASE AND CURRENT WORD COUNT	WRITE	0	1	0	0	0	1	1	0	W0-W7
			0	1	0	0	0	1	1	1	W8-W15
	CURRENT WORD COUNT	READ	0	0	1	0	0	1	1	0	W0-W7
			0	0	1	0	0	1	1	1	W8-W15
2	BASE AND CURRENT ADDRESS	WRITE	0	1	0	0	1	0	0	0	A0-A7
			0	1	0	0	1	0	0	1	A8-A15
	CURRENT ADDRESS	READ	0	0	1	0	1	0	0	0	A0-A7
			0	0	1	0	1	0	0	1	A8-A15
	BASE AND CURRENT WORD COUNT	WRITE	0	1	0	0	1	0	1	0	W0-W7
			0	1	0	0	1	0	1	1	W8-W15
	CURRENT WORD COUNT	READ	0	0	1	0	1	0	1	0	W0-W7
			0	0	1	0	1	0	1	1	W8-W15
3	BASE AND CURRENT ADDRESS	WRITE	0	1	0	0	1	1	0	0	A0-A7
			0	1	0	0	1	1	0	1	A8-A15
	CURRENT ADDRESS	READ	0	0	1	0	1	1	0	0	A0-A7
			0	0	1	0	1	1	0	1	A8-A15
	BASE AND CURRENT WORD COUNT	WRITE	0	1	0	0	1	1	1	0	W0-W7
			0	1	0	0	1	1	1	1	W8-W15
	CURRENT WORD COUNT	READ	0	0	1	0	1	1	1	0	W0-W7
			0	0	1	0	1	1	1	1	W8-W15

(c)

FIGURE 11-5 8237 registers and internal addresses. *(a)* Internal registers. *(b)* Internal addresses for writing commands and reading status. *(c)* Internal addresses for writing transfer addresses and counts. *(Intel Corporation)*

2. Output a master command word to internal address 1000.
3. Output a mode word for each channel you are using to internal address 1011.
4. Output the starting memory address for the transfer, one byte at a time, to the base register internal address for each channel you are using.
5. Output the number of bytes you want to transfer to the base word count internal address for each channel you are using.
6. Output clear mask command word(s) to unmask the channel(s) you are using.

Each channel of the 8237 can be programmed to transfer a single byte for each request, a block of bytes for each request, or to keep transferring bytes until it receives a wait signal on the EOP input/output. 8237s can be cascaded in a master-slave arrangement to give more input channels. As we said before, the main concept here is that the microprocessor and the DMA controller share the use of the address, data, and control buses.

As another DMA example we will now show you how the IBM PC is designed to allow peripheral boards to interface on a DMA basis.

The IBM PC Expansion Bus and DMA

To continue our evolution toward larger and larger microcomputer systems we will now take a brief look at the IBM PC, which is a multiboard system. Figure 11-6 shows a view of the component side of the main microprocessor board, often called a *motherboard*, for the IBM PC. After you find the ROM, RAM, and microprocessor on this board, note the system expansion slots in the upper left corner. These slots allow you to add the specific function boards you need in your system in addition to the basic CPU board. For example, you may want to add a disk-controller board, a serial-port board, a monochrome- or color-CRT board, a board with additional memory, an A/D-D/A board, or a board which allows your PC to function as a logic analyzer. This "open system" approach lets you easily customize the system for your application and your financial state. Now let's see how these slots connect into the basic system.

Figure 11-7 shows a block diagram for the motherboard of the IBM PC. Start on the left side of the diagram and work your way across it from the 8088 CPU and the 8259A priority-interrupt controller. The next vertical line of devices to the right consists of the address bus buffers, the data bus buffers, and the 8288 bus-controller chip. The bus-controller chip is required because the 8088 is operated in maximum mode. The buses from

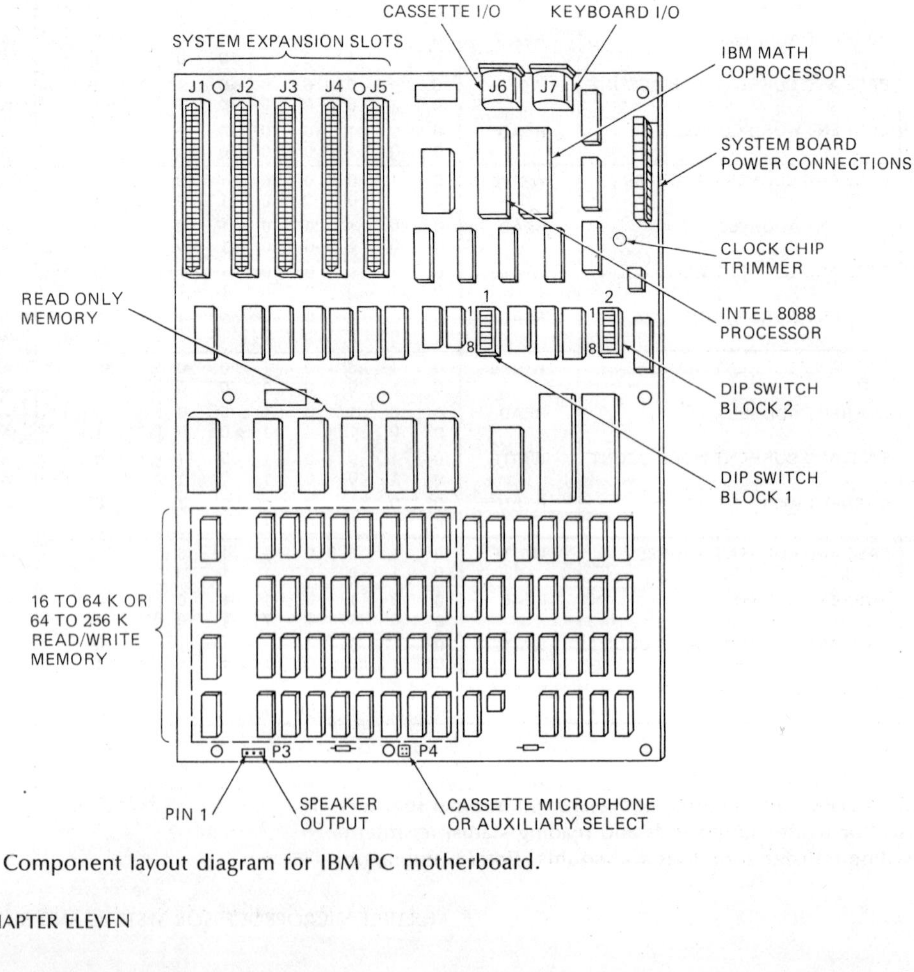

FIGURE 11-6 Component layout diagram for IBM PC motherboard.

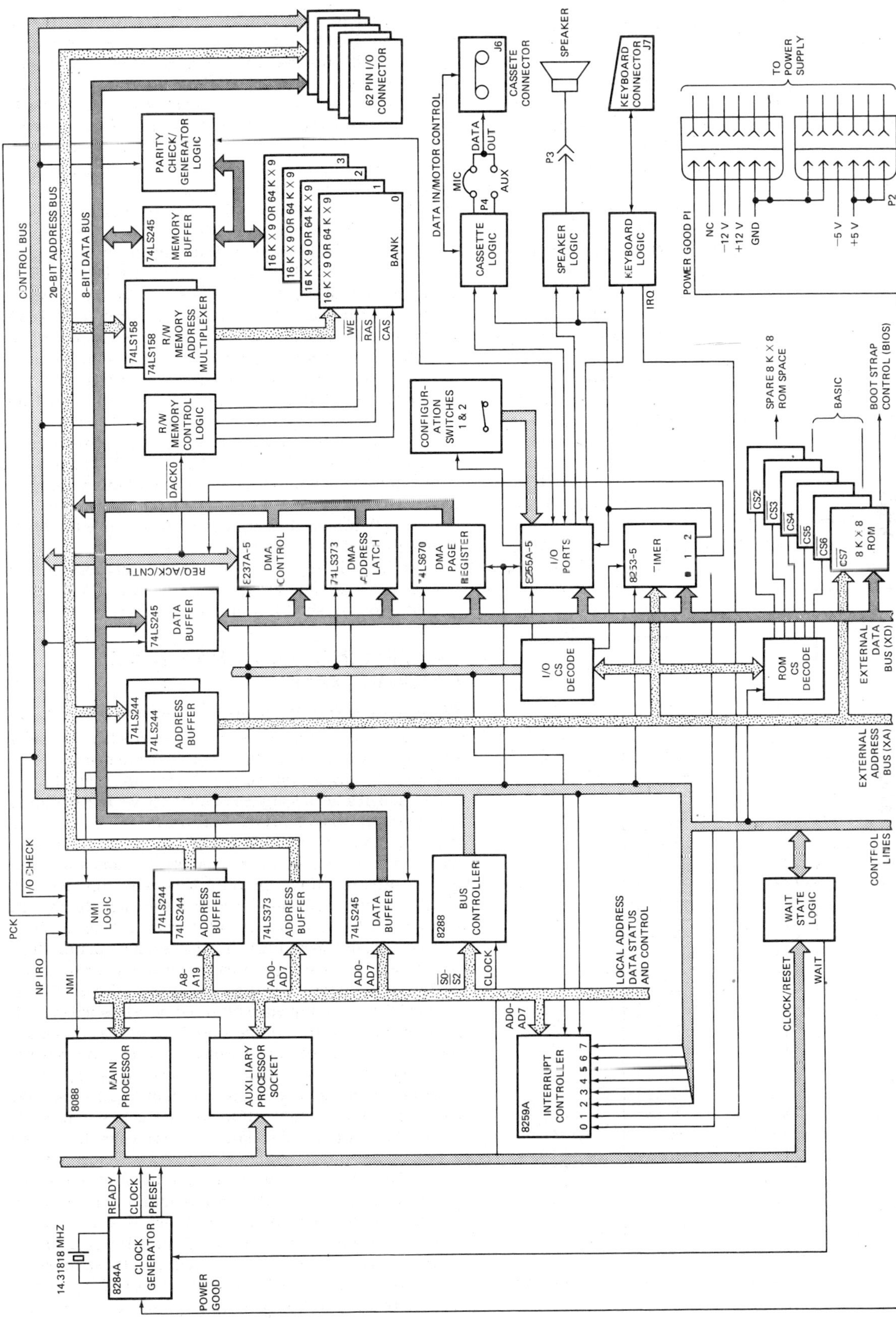

FIGURE 11-7 Block diagram of circuitry on IBM PC motherboard.

these devices go across the drawing and connect to the 62-pin peripheral board connectors. The CPU then can use these buses to communicate directly with the boards in the peripheral expansion slots. Now find the ROM in the lower right, the keyboard logic etc. in the middle right, and the dynamic RAM in the upper right. We will discuss this dynamic RAM later. Finally, take a look at the column of devices which contains the 8237A-5 DMA controller. Starting at the bottom of this column you see an 8253-5 programmable timer which is nearly identical to the 8254 we described in Chapter 8. Just above this is a familiar 8255A-5 programmable port device. Now you are left with just the three devices with DMA in their labels to ponder. The 8237A-5 is, of course, the DMA controller. The 74LS373 just under it is used to grab the upper 8 bits of the DMA address sent out on the data bus by the 8237A-5 during a transfer. This device has the same function as device *U*2 in Figure 11-3. The 74LS670 just below this is used to output bits A16–A19 of the DMA transfer address, the same function performed by *U*3 in the circuit in Figure 11-3. When you have worked your way around the diagram in Figure 11-7 and feel reasonably comfortable with it, take a look at the pin descriptions for the peripheral connectors in Figure 11-8.

The signals shown in Figure 11-8 are bused to all six peripheral connectors. Most of the signals on these connectors should be easily recognizable to you. A + in front of a signal indicates that the signal is active high, and a − indicates that the signal is active low. A0–A19 on the connectors are the 20 demultiplexed address lines, and D0–D7 are the eight data lines. IRQ2–IRQ7 are interrupt request lines which go to the 8259A priority-interrupt controller so that peripheral boards can interrupt the 8088 if necessary. Some other simple signals on the connectors are the power supply voltages; the standard ALE, $\overline{\text{MEMW}}$, $\overline{\text{MEMR}}$, $\overline{\text{IOW}}$, and $\overline{\text{IOR}}$ control-bus signals; and some clock signals.

Finally, we are down to the DMA signals on the expansion connectors. The DMA request pins DRQ1–DRQ3 allow peripheral boards to request use of the buses. A disk controller board, for example, might request a DMA transfer of a block of data from system memory. When the DMA controller gains control of the system buses, it lets the peripheral device or board know by asserting the appropriate DACK0–DACK3 signal. The AEN signal on the connectors is used to gate the DMA address on the bus as we described earlier. When the programmed number of bytes has been transferred, the T/C pin on the connector goes high to let the peripheral know that the transfer is complete. A peripheral board can assert the I/O CH RDY pin on the connector low to cause the 8088 to insert WAIT states until it is ready.

Later you will see many more examples of bus sharing. For our next example we show how dynamic RAM is connected and refreshed in a microcomputer.

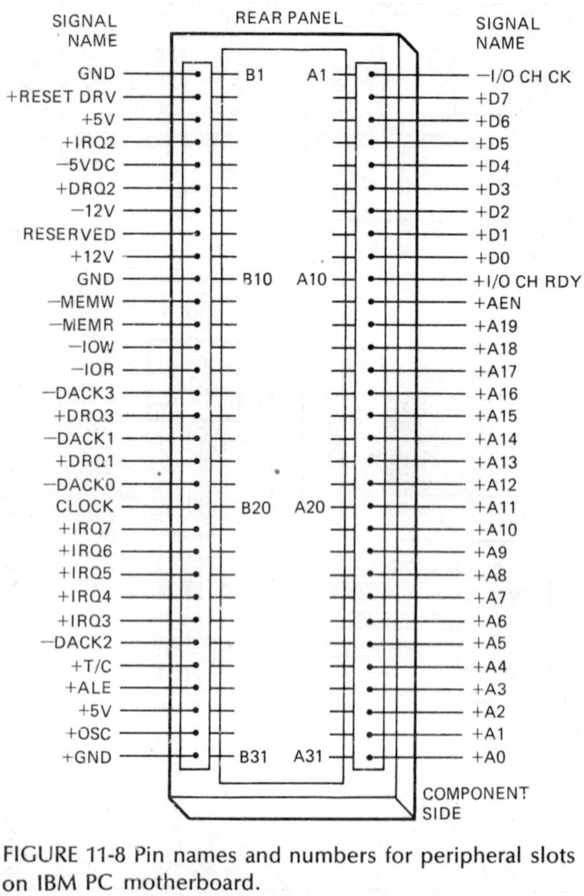

FIGURE 11-8 Pin names and numbers for peripheral slots on IBM PC motherboard.

INTERFACING AND REFRESHING DYNAMIC RAM

Review of Dynamic RAM Characteristics

For small systems such as the SDK-86 where we only need a few kilobytes of RAM we usually use static RAM devices. For larger systems where we want several hundred kilobytes or megabytes of memory we use dynamic RAMs, often called DRAMs. Here's why.

Static RAMs store each bit in an internal flip-flop which requires six or so transistors. DRAMs store bits as a charge or no charge on a tiny capacitor, so they need only one transistor to access the capacitor when you write a bit to it or read a bit from it. The result of this is that DRAMs require much less power per bit, and many more bits can be stored in a given-size chip. The cost per bit of storage is then much less. The disadvantage of DRAMs is that each bit must be refreshed every 2 ms or so because the charge stored on those tiny capacitors tends to change due to leakage. The internal refresh circuitry has to check the voltage level in each storage location; if the voltage is greater than $V_{cc}/2$ then that location is charged to V_{cc}, if the voltage is less than $V_{cc}/2$ then that location is discharged to zero volts. Let's take a look at the pin diagram and timing waveforms for a typical DRAM to see how we read, write, and refresh it.

Figure 11-9 shows the pin diagram for an Intel 51C256H CHMOS DRAM. This device is a 256K by 1 device, so it stores 262,144 words of 1 bit each in its 16-pin package. You can connect eight of these in parallel to store bytes, or 16 in parallel to store 16-bit words. Now, according to the basic rules of address decoding,

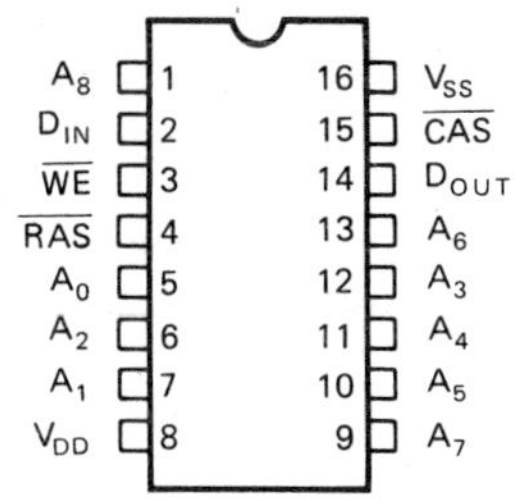

$\overline{RAS}$	ROW ADDRESS STROBE
$\overline{CAS}$	COLUMN ADDRESS STROBE
$\overline{WE}$	WRITE ENABLE
A_0-A_8	ADDRESS INPUTS
D_{IN}	DATA INPUT
D_{OUT}	DATA OUTPUT
V_{DD}	POWER (+5V)
V_{SS}	GROUND

(*a*)

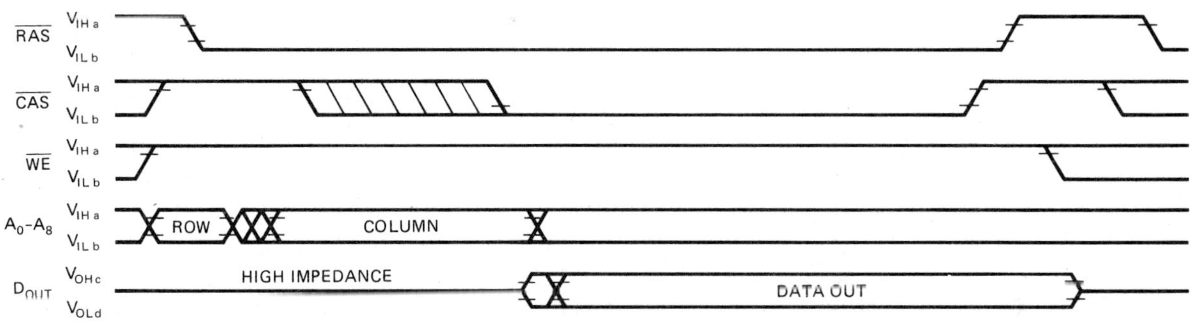

(*b*)

FIGURE 11-9 51C256H CHMOS dynamic RAM. *(a)* Pin diagram. *(b)* Read-operation timing diagram. *(Intel Corporation)*

18 address lines should be required to address one of the 2^{18} words stored in this device. The pin diagram in Figure 11-9*a*, however, shows only nine address inputs, A0–A8. The trick here is that to save pins, DRAMs usually send in the address one-half at a time. A look at the timing diagram for a read operation in Figure 11-9*b* should help you to see how this works.

To read a word from a bank of dynamic RAMs, a DRAM-controller device or other circuitry asserts the write-enable ($\overline{WE}$) pin of the DRAMs high to enable them for a read operation. It then sends the lower half of the address, called the row address, to the address inputs of the DRAMs. The controller then asserts the *row-address-strobe* ($\overline{RAS}$) input of the DRAM low to indicate a row address is present. After the proper timing interval, the controller removes the row address and outputs the upper half of the address, called the column address, to the address inputs of the DRAM. The controller then asserts the *column-address-strobe* ($\overline{CAS}$) inputs of the DRAMs low to indicate that the column address is present. After a propagation delay, the data word from the addressed memory cells will appear on the data outputs of the DRAMs.

The timing diagram for a write cycle is nearly the same except that after it sends out the column address and $\overline{CAS}$, the controller asserts the write-enable ($\overline{WE}$) input low to enable the DRAMs for writing, and asserts a signal which gates the data to be written onto the data inputs of the DRAMs.

The DRAM controller refreshes the cells in the DRAMs by sending out each of the 512 row addresses and pulsing $\overline{RAS}$ low at least every 4 ms. The refresh can be done in either a *burst* mode or in a *distributed* mode. In the burst mode all 512 rows are addressed and pulsed one right after the other every 4 ms. In the distributed mode another row is addressed and pulsed after every 4/512 ms or 7.8 μs. In a particular system you use the mode which will least interfere with the operation of the system. Now that the operation of dynamic RAMs is fresh in your mind, we will show you how you interface banks of DRAMs to an 8086.

Interfacing DRAMs to an 8086

As perhaps you can see from the preceding discussion, the main tasks you have to do to interface a bank of DRAMs to a microprocessor are:

1. To refresh each location at the proper interval.
2. To "funnel" the two halves of the address into the each device with the appropriate $\overline{RAS}$ and $\overline{CAS}$ strobes.
3. To assure that a read or write operation and a refresh operation do not take place at the same time.
4. To provide a read/write control signal to enable data into or out of the devices.

There are several ways to do these tasks.

DRAM Refreshing and Error Checking in the IBM PC

As you can see in Figure 11-7, the IBM PC 256K version has four banks of DRAMs which are 64K by 1 devices. The PC uses a dummy DMA read approach to refresh its

DRAM. Here's how it works. An 8253 timer is programmed to produce a pulse every 15 μs. This pulse is connected into one of the DMA request inputs (DREQ0) of an 8237 DMA controller which has been programmed to read from memory and write to a nonexistent port. When the 8237 DMA controller receives this pulse, it sends a hold request to the 8088 microprocessor. After the 8088 responds with a HLDA signal, the 8237 takes over the buses, sends out a memory address, sends out a memory read signal, and sends out a DMA acknowledge (DACK0) signal. The lower 8 bits of the memory address it sends out goes to the address inputs of all of the DRAMs. The DACK0 signal is used to pulse the $\overline{RAS}$ lines of all of the DRAM banks low at this time. After each DMA operation the current address register in the DMA controller will be automatically incremented or decremented, depending on how the device was programmed. In either case, the next DMA operation will refresh the next row in the DRAMs. If the 8237 is programmed for transfer of 64 Kbytes, start at address 0, increment count after DMA, and autoinitialize, the sequence of addresses sent out will refresh all 256 rows in the DRAMs over and over. One row in each of the banks is then refreshed every 15 μs. With the 4.77-MHz clock used in the IBM PC, a refresh DMA cycle takes about 820 ns every 15 μs, or about 5 percent of the processor's time.

Another point about the DRAM memory banks in the IBM PC is that each bank is 9 bits wide. Eight of these bits make up the data byte being stored, and the ninth bit is a parity bit which is used to detect errors in the stored data. A 74LS280 parity generator/checker circuit generates a parity bit for each byte and stores it in the ninth location as each byte is written to memory. When the 9 bits are read out, the parity is checked by the parity generator checker circuit. If the parity is not correct, an error signal is sent to an 8255 port pin where it can be read by the 8088 microprocessor. When you first turn on the power to an IBM PC or warm boot it by pressing the Ctrl, Alt, and Del keys at the same time, one of the self tests that it performs is to write byte patterns to all of the RAM locations and check if the byte read back and the parity of that byte are correct. If any error is found, an error message is displayed on the screen so you don't try to load and run programs in defective RAM.

A DRAM Controller IC—The Intel 8208

In high-performance systems where we want DRAM refreshing to take up a minimum amount of the processor's time, we usually use a dedicated device which handles all of the refreshing chores without tying up the microprocessor or its buses as the DMA approach does. An example of this type of device is the Intel 8208. Figure 11-10 shows, in block diagram form, how an 8208 can be connected with an 8086 in maximum mode to refresh and control 1 Mbyte of dynamic RAM. The memories here are 256K by 1 devices. As usual for an 8086, the memory array is set up as two banks. Each bank has two blocks containing 8 RAM chips, or 256 Kbytes.

In the system in Figure 11-10 the 8208 takes care of all of the refresh tasks in addition to funneling in addresses from the 8086 for read and write operations as we described previously. One important point to observe here is that the status signals, S0–S3, from the 8086 are connected directly to the control inputs of the 8208. The 8208 decodes these status signals to produce the read and write signals it needs. This means that most of the time the 8086 will be able to read a byte or word from the DRAMs without any WAIT states being required. If the 8208 happens to be in the middle of a refresh cycle when the 8086 tries to read a DRAM location, the 8208 will hold its AACK high and cause the 8086 to insert a WAIT state. The 8086 will then have to wait 1 clock cycle while the 8208 finishes its refresh cycle before it can access the DRAMs. The occasional access conflict here is arbitrated by the controller, and slows the 8086 up less than the DMA approach shown in the previous section.

Another interesting point about the 8208 is that, in order to save pins, it does not connect to the data bus to allow command words to be sent to it for initialization. If the PDI pin is tied to ground, the 8208 will initialize itself in a mode suitable for many applications. For applications where the default mode will not work, the output of a parallel-in–serial-out shift register is connected to the PDI input of the 8208. After a reset the $\overline{WE}/\overline{PCLK}$ pin outputs a series of pulses. These pulses are used to clock the desired command word from the shift register into the 8208. The desired mode word is simply hardwired on the parallel inputs of the shift register.

Battery Backup of Dynamic RAMs

In Chapter 8 we discussed the use of an 8086 NMI interrupt procedure to save program data in the case of a power failure. In the few milliseconds between the time the ac power goes off and the time the dc power drops below operating levels, the interrupt procedure copies program data to a block of static RAM which has a battery backup power supply. When the system is repowered, the saved data is copied back into the main RAM, and processing takes up where it left off. In larger systems such as the one in Figure 11-10, there may not be time enough to copy all of the important data etc. to another RAM. In this case we simply use a battery backup for the entire RAM array as shown in Figure 11-10.

In this circuit we used CHMOS DRAMs because when these devices are not being accessed for reading, writing, or refreshing, they take only microwatts of power. During battery backup of the DRAMs they must still be refreshed, so the 8208 DRAM controller is also connected to the battery power. The 8208 normally receives its required clock signal from the 8284 clock generator, but since that is a high-current device, we added a CMOS clock generator which will be switched in when the power fails. We use a nickel-cadmium or some other type of battery which can stand the continuous recharging and supply the needed current. The diodes in the circuit prevent the power supply output and the battery from fighting with each other.

In applications where the entire system must be kept running during an ac power outage, we use a *noninterruptible power supply* which contains a large battery

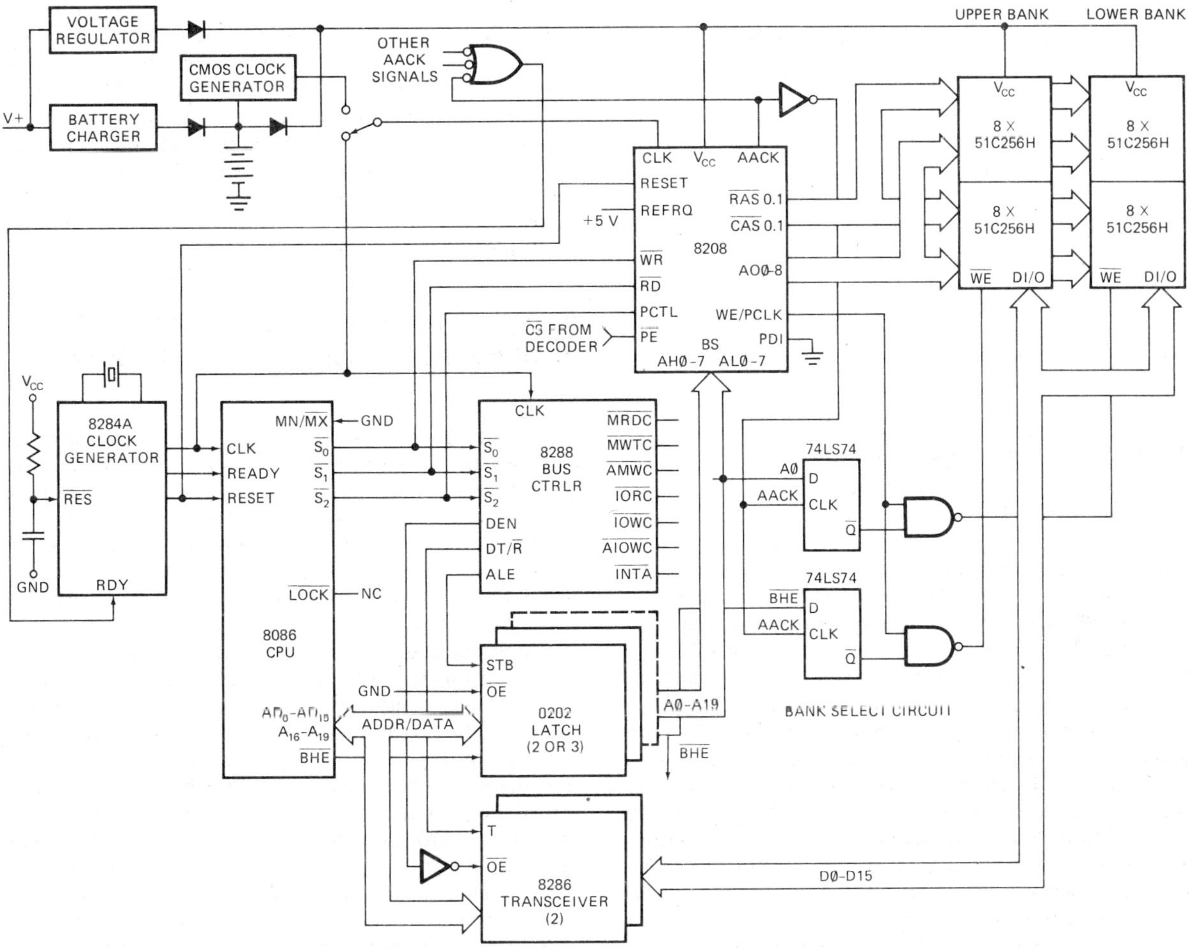

FIGURE 11-10 Circuit for refreshing dynamic RAMs with the 8208 dynamic RAM controller.

and the circuitry needed to convert the battery voltage to the voltage needed by the microcomputer.

Error Detecting and Correcting in RAM Arrays

Data read from RAMs is subject to two types of errors, *hard errors* and *soft errors*. Hard errors are caused by permanent device failure. This may be caused by a manufacturing defect or simply random breakdown in the chip. Soft errors are one-time errors caused by a noise pulse in the system or, in the case of dynamic RAMs, perhaps an alpha particle or some other radiation causing the charge to change on the tiny capacitor on which a data bit is stored. As we add larger and larger arrays of RAMs to a system, the chance of a hard or a soft error occurring increases sharply. This then increases the chance that the entire system will fail. It seems unreasonable that one fleeting alpha particle could possibly cause an entire system to fail. To prevent or at least reduce the chances of this kind of failure, we add circuitry which detects and in some cases corrects errors in the data read out from RAMs. There are several ways to do this, depending on the amount of detection and correction needed.

The simplest method for detecting an error is with a parity bit. As we described previously for the IBM PC, we do this by first determining the parity of, say, an 8-bit data word as it is being written to a memory location. We then generate a parity bit such that the overall parity of the 8 data bits plus the parity bit is, for example, always odd. The generated parity bit for each byte is written into a separate memory device in parallel with the devices containing the data byte. When the data byte and the parity bit are read from memory, we check the parity of the 9 bits together. If the overall parity of these nine is not odd as it should be, then we know that somewhere in the read/write process an error was introduced. If external hardware is being used to generate/check parity, then an output from this circuitry can be used to tell the processor that an error occurred, and that the data is not valid. The processor can then respond appropriately.

One difficulty with a simple parity check is that two errors in a data word may cancel each other. A second problem with simple parity is that it does not tell you

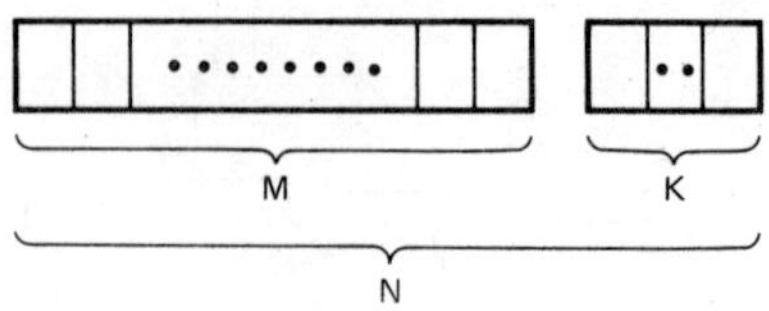

(a)

K	SINGLE CORRECT/ SINGLE DETECT ≤M≤		SINGLE CORRECT/ DOUBLE DETECT ≤M≤	
4	4	11	1	3
5	12	26	4	10
6	27	57	11	25
7	58	120	26	56
8	121	245	57	119

(b)

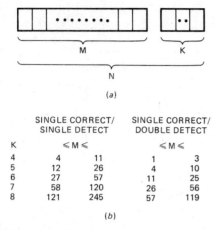

FIGURE 11-11 Error detecting/correcting codes. *(a)* Encoding bits. *(b)* Number of encoding bits for detecting/correcting.

which bit in a word is wrong so that you can correct the error. More complex error detecting/correcting codes (ECCs), often called *Hamming codes* after the man who did some of the original work in this area, permit you to detect multiple-bit errors in a word and to correct at least one bit error. Here's how they work.

As the data word is read in, several encoding bits are generated and stored in memory along with the data word. Figure 11-11*a* shows this in diagram form. The number of encoding bits, *k*, required is determined by the size of the data word, *m*, and the degree of detection/correction required. The total number of bits required for a data word, *n*, is equal to $m + k$. Figure 11-11*b* shows the number of encoding bits needed for different numbers of data bits and different degrees of detection/correction. According to these values, 5 encoding bits are required to detect and correct a single-bit error in a 16-bit data word, so the total number of bits that have to be stored for each word in this case is 21. To give you enough information to correct one error and detect 2 wrong bits in a 16-bit word would require 6 encoding bits.

When you write a word to memory the error detecting/correcting circuitry generates the required encoding bits and writes them to memory along with the data word. The encoding bits, incidentally, are not just tacked on to one end of the data word as a parity bit is. They are interspersed in the data word. When you read a data word from memory, the error detecting/correcting circuitry recalculates the encoding bits for the data word read out. It then exclusive-NORs these encoding bits with the encoding bits that were stored in memory for that data word. The word produced by this operation is known as a *syndrome word*. The encoding bits are generated in such a way that the value of this syndrome word indicates which bit, if any, is wrong in the total word of data word plus encoding bits. The erroneous bit can then be corrected by simply inverting it. Because of hardware implementation tradeoffs, there are actually several different schemes for determining the encoding bits, so if you are working with a RAM system with error correction, consult the manual for it if you need the specific coding. Several available ICs such as the Intel 8206 will automatically detect/correct single bit errors and detect two bit errors in 16-bit data words. The devices can be cascaded to work with up to 80-bit data words.

Now that we have shown you the operation of a DMA controller and the operation of the 8086 in maximum mode, we can introduce you to the 80186 and 80188 microprocessors which have DMA controllers and other peripherals built in.

PROCESSORS WITH INTEGRATED PERIPHERALS—THE 80186 AND 80188

Overview

Figure 11-12*a* shows a block diagram of the internal architecture of the Intel 80186 microprocessor. The architecture and instruction set of the 80188 are identical to those of the 80186 except that the 80188 has only an 8-bit data bus instead of the 16-bit data bus that the 80186 has. With this in mind, we will use 80186 to represent both the 80186 and the 80188 in our discussions here.

The 80186 has the same bus-interface unit and execution unit as the 8086 which we discussed previously, so there is nothing new there for you. Unlike the 8086, however, the 80186 has the clock generator built in so that all you have to add is an external crystal. Also note that the 80186 does not have a pin labeled MN/$\overline{\text{MX}}$. The 80186 is packaged in a 68-pin leadless package as shown in Figure 11-12*b*, so it has enough pins to send out both the minimum-mode type signals $\overline{\text{RD}}$ and $\overline{\text{WR}}$ and the S0–S3 status signals which can be connected to external bus-controller ICs for maximum-mode systems. Now let's look at the four peripheral chip function blocks in the 80186.

First is a priority interrupt controller which has up to four interrupt inputs, INT0, INT1, INT2/$\overline{\text{INTA0}}$, and INT3/$\overline{\text{INTA1}}$ as well as an NMI interrupt input. If the four INT inputs are programmed in their internal mode, then a signal applied to one of them will cause the 80186 to push the return address on the stack and vector directly to the start of the interrupt service procedure for that interrupt. Figure 11-14 shows the interrupt type which corresponds to each of these inputs. The INT2/$\overline{\text{INTA0}}$, and INT3/$\overline{\text{INTA1}}$ pins can be programmed to be used as interrupt inputs as we have just described, or they can be programmed to function as interrupt acknowledge outputs. This mode is used to interface with external 8259As. The interrupt request line from an external 8259A is connected to, for example, the 80186 INT0 input, and the 80186 INT2/$\overline{\text{INTA0}}$ pin is connected to the interrupt acknowledge input of the 8259A. When the 8259A receives an interrupt request, it asserts the INT0 input of the 80186. When the 8259A receives interrupt acknowledge signals from the INT2/$\overline{\text{INTA0}}$ pin, it sends the desired interrupt type to the 80186 on the data bus. This second mode is commonly referred to in the literature as *iRMX mode*, because an 80186 system must have an external 8259A and 8254 if the iRMX operating system is going to be run on it.

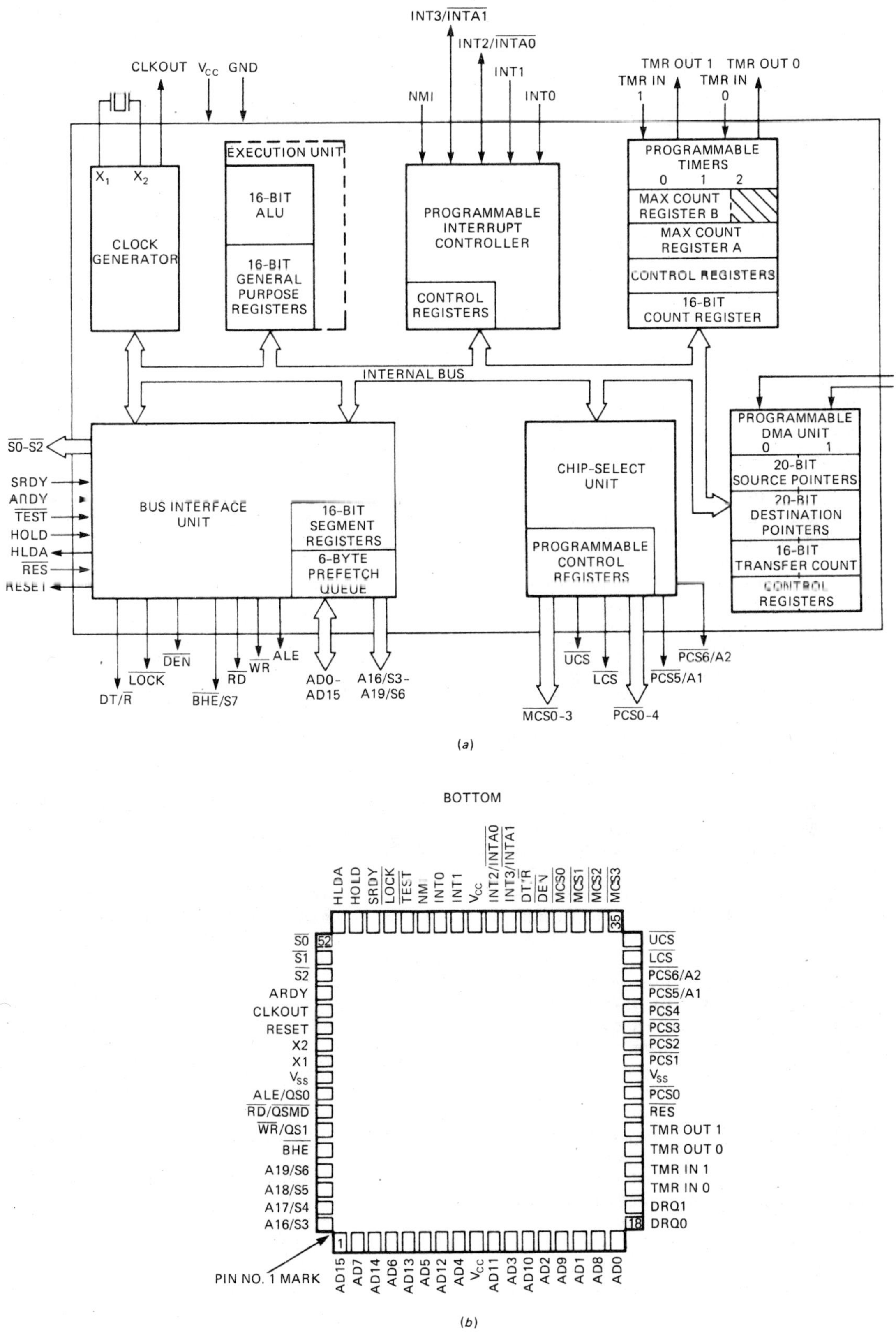

FIGURE 11-12 80186. *(a)* Internal block diagram. *(b)* Pin diagram. *(Intel Corporation)*

Next to look at in the block diagram is the built-in address decoder, referred to in the drawing as the chip-select unit. This unit can be programmed to produce an active low chip-select signal when a memory address in the specified range or a port address in a specified range is sent out. Six memory address chip-select signals are available: $\overline{LCS}$, $\overline{UCS}$, and $\overline{MCS0-3}$. The *lower-chip-select* signal, $\overline{LCS}$, will be asserted by addresses between 00000H and some address which you specify in a control word. The specified ending address can be anywhere between 1K and 256K. The highest address that will assert the *upper-chip-select* signal, $\overline{UCS}$, is fixed at FFFFFH. The lowest address for this block of memory is again programmable by some bits you put in a control word. The size of the upper memory block can be anywhere between 1K and 256K. Finally, there are four *middle-chip-select* lines, $\overline{MCS0-3}$. Each of these four is asserted by an address in a block of memory in the middle range of memory. Both the starting address and the size of the four blocks can be specified for this middle-range block. The specified size of blocks can be anywhere from 2K to 128K. The memory areas assigned to different chip selects cannot overlap, or two chip-select outputs will be asserted at the same time, possibly damaging some memory devices. The point of this built-in decoder is to select major blocks of memory. External decoders can then be used to select specific groups of memory devices. At the rate memory devices are growing in size, external decoders may soon not be needed.

In addition to producing memory chip-select signals, the 80186 can be programmed to produce up to seven peripheral chip-select signals on its $\overline{PCS0}$–$\overline{PCS4}$, $\overline{PCS5}$/A1, and $\overline{PCS6}$/A2 pins. You program a base address for these I/O addresses in a control word. $\overline{PCS0}$ will be asserted when this base address is output during an IN or an OUT instruction. The other PCS outputs will be asserted by addresses at intervals of 128 bytes above.

Now let's look at the programmable DMA unit in the 80186. As you can see from the block diagram in Figure 11-12, the DMA unit has two DMA request inputs, DRQ0 and DRQ1. These inputs allow external devices such as disk controllers, CRT controllers, etc. to request use of one of the DMA channels as we described earlier in this chapter. For each DMA channel the 80186 has a full 20-bit register to hold the address of the source of the DMA transfer, a 20-bit register to hold the destination address, and a 16-bit counter to keep track of how many words or bytes have been transferred. DMA transfers can be from memory to memory, I/O to I/O, or between I/O and memory.

Finally, let's look at the three 16-bit programmable counter/timers in the 80186. The inputs and outputs of counters 0 and 1 are available on pins of the 80816. These two counters can be used to divide down the frequency of external signals, produce programmed-width pulses, etc. just as you do with the counters in an 8254. You can also internally direct the processor clock to the input of one of these counter inputs by clearing the appropriate bit in a control word. The input of the third counter in the 80186 is internally connected to the processor clock. Because of the way the counters in the 80186 are decremented, counter 2 will be decremented

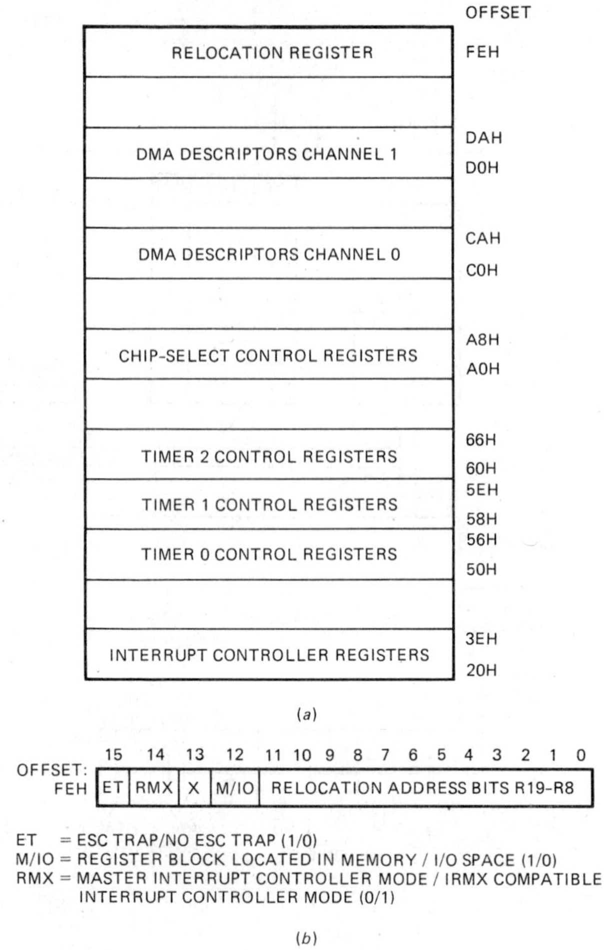

FIGURE 11-13 80186 Peripheral control block. *(a)* Control block format. *(b)* Relocation word format. *(Intel Corporation)*

every 4 processor clocks. By setting or clearing the appropriate bits in a control word, you can direct the output of counter 2 to a DMA input, an interrupt input, or the input of counter 1 and/or counter 0.

As you can see from the preceding discussion, the 80186 contains many of the peripheral chip functions needed in a medium-complexity microcomputer system. In order to use these integrated peripherals, you have to initialize them just as you do external peripherals. Note in Figure 11-12 that control registers are shown as part of each of the integrated peripherals. These 16-bit control registers are all contained in an internal 256-byte block, as shown in Figure 11-13*a*. After a reset this block will be located at I/O address FF00H. Control and status registers in this block can then be accessed with IN and OUT instructions. This peripheral control block can be relocated to some other address in the I/O space, or to an address in memory by writing the appropriate word to the relocation register in the control block. Figure 11-13*b* shows the format for the word you send to the relocation register. We do not have space here to show and explain all of the control word formats for the

80186. If you have to work with an 80186, you can find the formats for these words in the 80186 data sheet, and work out the control words you need for your particular application on a bit-by-bit basis, just as you do for the separate peripherals. You may also find Intel application note 186, *Introduction to the 80186 Microprocessor*, helpful.

The final points we want to make about the 80186 are the additional instructions and interrupts it has beyond those of the 8086. The 10 additional instructions that the 80186 has are:

ENTER —Enter a procedure
LEAVE —Leave a procedure
BOUND —Check if an array index in a register is in range of array
INS —Input string byte or string word
OUTS —Output string byte or string word
PUSHA —Push all registers on stack
POPA —Pop all registers off stack
PUSH immediate—Push immediate number on stack
IMUL destination register, source, immediate
—Immediate × source to destination
SHIFT/ROTATE destination, immediate
—Shift register or memory contents specified immediate number of times

These instructions are explained in greater detail in Chapter 6 for your reference. Now let's look at the additional built-in interrupt types the 80186 has.

Figure 11-14 shows the names, type numbers, and default priorities for the 80186 internal interrupts. The first five of these should be familiar to you from our discussions of the 8086 internal interrupts. The 80186 will do a type 5 interrupt if an array index number in a register is outside of the specified range for the array when the BOUND instruction executes. It will do a type 6 interrupt if it finds an undefined opcode in the instructions fetched from memory by the BIU. If this interrupt is unmasked, and if the 80186 finds an ESC opcode in the instructions it fetches from memory, the 80186 will do a type 7 interrupt. The additional interrupt types are dedicated to the integrated peripherals as shown.

INTERRUPT NAME	VECTOR TYPE	DEFAULT PRIORITY	RELATED INSTRUCTIONS
DIVIDE ERROR EXCEPTION	0	*1	DIV, IDIV
SINGLE STEP INTERRUPT	1	12**2	ALL
NMI	2	1	ALL
BREAKPOINT INTERRUPT	3	*1	INT
INTO DETECTED OVERFLOW EXCEPTION	4	*1	INTO
ARRAY BOUNDS EXCEPTION	5	*1	BOUND
UNUSED-OPCODE EXCEPTION	6	*1	UNDEFINED OPCODES
ESC OPCODE EXCEPTION	7	*1***	ESC OPCODES
TIMER 0 INTERRUPT	8	2A****	
TIMER 1 INTERRUPT	18	2B****	
TIMER 2 INTERRUPT	19	2C****	
RESERVED	9	3	
DMA 0 INTERRUPT	10	4	
DMA 1 INTERRUPT	11	5	
INT0 INTERRUPT	12	6	
INT1 INTERRUPT	13	7	
INT2 INTERRUPT	14	8	
INT3 INTERRUPT	15	9	

NOTES:
*1. These are generated as the result of an instruction execution.
**2. This is handled as in the 8086.
****3. All three timers constitute one source of request to the interrupt controller. The Timer interrupts all have the same default priority level with respect to all other interrupt sources. However, they have a defined priority ordering amongst themselves. (Priority 2A is higher priority than 2B.) Each Timer interrupt has a separate vector type number.
4. Default priorities for the interrupt sources are used only if the user does not program each source into a unique priority level.
***5. An escape opcode will cause a trap only if the proper bit is set in the peripheral control block relocation register.

FIGURE 11-14 80186 internal interrupt types. *(Intel Corporation)*

Now that you have an overview of the relatively sophisticated 80186 microprocessor, our next step is to show you how two or more microprocessors are used in a system. Earlier in this chapter we showed you how a DMA controller, a floppy-disk controller, and a dynamic RAM controller can share the buses of a microprocessor. These devices have some "intelligence," but they are not comparable to the microprocessor in power. In the next section we show you how two or more processors can directly share the address, data, and control buses. Processors which share the local buses in this way are referred to as *coprocessors*.

A COPROCESSOR—THE 8087 MATH COPROCESSOR

Overview

The instruction set of general-purpose processors such as the 8086 is not optimized to do complex numerical calculations, CRT graphics manipulations, or word processing. Therefore, specialized coprocessors have been developed for these applications. These coprocessors operate in parallel with an 8086-type processor on the same buses and with the same instruction-byte stream. To show you how a coprocessor works, we will use a specific example, the 8087 math coprocessor. If you have an IBM PC type of computer, you can plug an 8087 chip in it directly and run 8087 programs. You can then run the example programs, and write some of your own as you work your way through this chapter.

In many microcomputer programs such as those used for scientific research, engineering, business, and graphics, you need to make mathematical calculations such as computing the square root of a number, the tangent of a number, or the log of a number. Another common need is to do arithmetic operations on very large and very small numbers. There are several ways to do all of this. One way is to write the number-crunching part of the program in a high-level language such as FORTRAN, compile this part of the program, and link in I/O modules written in assembly language. The difficulty with this approach is that programs written in high level languages tend to run considerably slower than programs written in assembly language.

Another way is for you to write an assembly language program which uses the normal instruction set of the processor to do the arithmetic functions. Reference books which contain the algorithms for these are readily available. Our experience has shown that it is often time-consuming to get from the algorithm to a working assembly language program.

Still another approach is to buy a library of floating-point arithmetic object modules from the manufacturer of the microprocessor you are working with or from an independent software house. In your program you just declare a procedure needed from the library as external, call the procedure as required, and link the library to the object code for your program. This approach spares you the labor of writing all the procedures.

In an application where you need a calculation to be done as quickly as possible, however, all of the previous approaches have a problem. The architecture and instruction sets of general-purpose microprocessors such as the 8086 are not designed to work efficiently with mathematical manipulation. Therefore, even highly optimized number-crunching programs run slowly on these general-purpose machines. To solve this problem, special processors, which have architectures and instruction sets optimized for number crunching, have been developed. An example of this type of number-crunching processor is the Intel 8087 math processor. An 8087 is used in parallel with the main microprocessor in a system, rather than serving as a main processor itself. Therefore it is referred to as a *coprocessor*. The major principle here is that the main microprocessor, an 8088 for example, handles the general program execution, and the 8087 coprocessor handles specialized math computations. First we will show you how an 8087 is connected and functions in a system, then we will show you how to program it.

Circuit Connection for an 8087

Figure 11-15 shows the first sheet of the schematics for the 256K version of the IBM PC. We chose this schematic not only to show you how an 8087 is connected in a system with an 8088 microprocessor, but also to show you another way in which schematics for microcomputers are commonly drawn. The more schematics you work your way around, the easier it will get for you.

First in Figure 11-15, note the numbers along the left and right edges of the schematic. These numbers indicate the other sheet(s) that the signal goes to. This is an alternative approach to the zone coordinates used in the schematics in Figure 7-6. In the schematic here the zone coordinates are not needed because all of the input signal lines are extended to the left edge of the schematic, and all of the output signal lines are run to the right edge of the schematic. If you see that an output signal goes to sheet 10, then it is a simple task to scan down the left edge of sheet 10 to find that signal. The wide crosshatched strips in Figure 11-15 represent the address, data, and control buses. From the pin descriptions for the major ICs you know where these signals are produced. You can then scan along the bus to see where various signals get dropped off at other devices. On this type of schematic the buses are always expanded to individual lines where they enter or leave a schematic. Now let's look at how the 8087 is connected.

First note that the multiplexed address-data bus lines, AD0–AD7, go directly from the 8088 to the 8087. The 8088, remember, has the same instruction set as the 8086, but it only has an 8-bit data bus, so all read and writes are byte operations. The upper address lines, A8–A19, also connect directly from the 8088 to the 8087. If you look a little closer at the schematic, you should see that the status lines, $\overline{S2}$, $\overline{S1}$, and $\overline{S0}$, from the 8088 and the queue status lines, QS1 and QS0, from the 8088 also connect directly to the 8087. The 8087 receives the same clock and reset signals as the 8088. Three more connections to the 8087 that you need to pay close attention to are:

First, the request/grant signal, $\overline{RQ}/\overline{GT0}$, from the 8087 is connected to the request/grant pin, $\overline{RQ}/\overline{GT1}$, of the 8088. The way you figure this out from the schematic is to notice that the signal from the 8087 $\overline{RQ}/\overline{GT0}$ pin is labeled just $\overline{RQ}/\overline{GT}$ where it enters the crosshatched bus. Likewise, the label on the signal coming from the crosshatched bus to the $\overline{RQ}/\overline{GT1}$ pin of the 8088 is also just labeled $\overline{RQ}/\overline{GT}$. You know from the fact that the two lines have the same label, they are connected together.

Second, the BUSY signal from the 8087 is connected to the $\overline{TEST}$ input of the 8088. If the 8088 must have the result of some computation that the 8087 is doing before it can go on with its instructions, you tell the 8088 with a WAIT instruction to keep looking at its $\overline{TEST}$ pin until it finds the pin low. A low on the 8087 BUSY output indicates that the 8087 has completed the computation.

Third, the interrupt output, INT, of the 8087 is connected to the nonmaskable interrupt, NMI, input of the 8088. This connection is made so that an error condition in the 8087 can interrupt the 8088 to let it know about the error condition. The signal from the 8087 INT output actually goes through some circuitry on sheet 2 of the schematics and returns to the input labeled NMI on the left edge of Figure 11-15. We do not have room here to show and explain all of the circuitry on sheet 2. The main purposes of the circuitry between the INT output of the 8087 and the NMI input of the 8088 is to make sure that an NMI signal is not present upon reset, to make it possible to mask the NMI input, and to make it possible for other devices to cause an NMI interrupt.

A couple of pins on the 8087 that we aren't concerned with here are the bus-high-enable (BHE) and request/grant1 (RQ/GT1) pins. When the 8087 is used with an 8086, the BHE pin is connected to the system BHE line to enable the upper bank of memory. The RQ/GT1 input is available so that another coprocessor such as the 8089 I/O processor can be connected and function in parallel with the 8087.

As you can see from the preceding discussion, the 8087 is connected very tightly with the 8088. Now let's talk about how the two devices work together.

8087-8088 Cooperation

The point that we need to make about the 8087 is that it is an actual processor with its own, specialized instruction set. Instructions for the 8087 are written in a program as needed, interspersed with the 8088/8086 instructions. To you, the programmer, adding an 8087 to

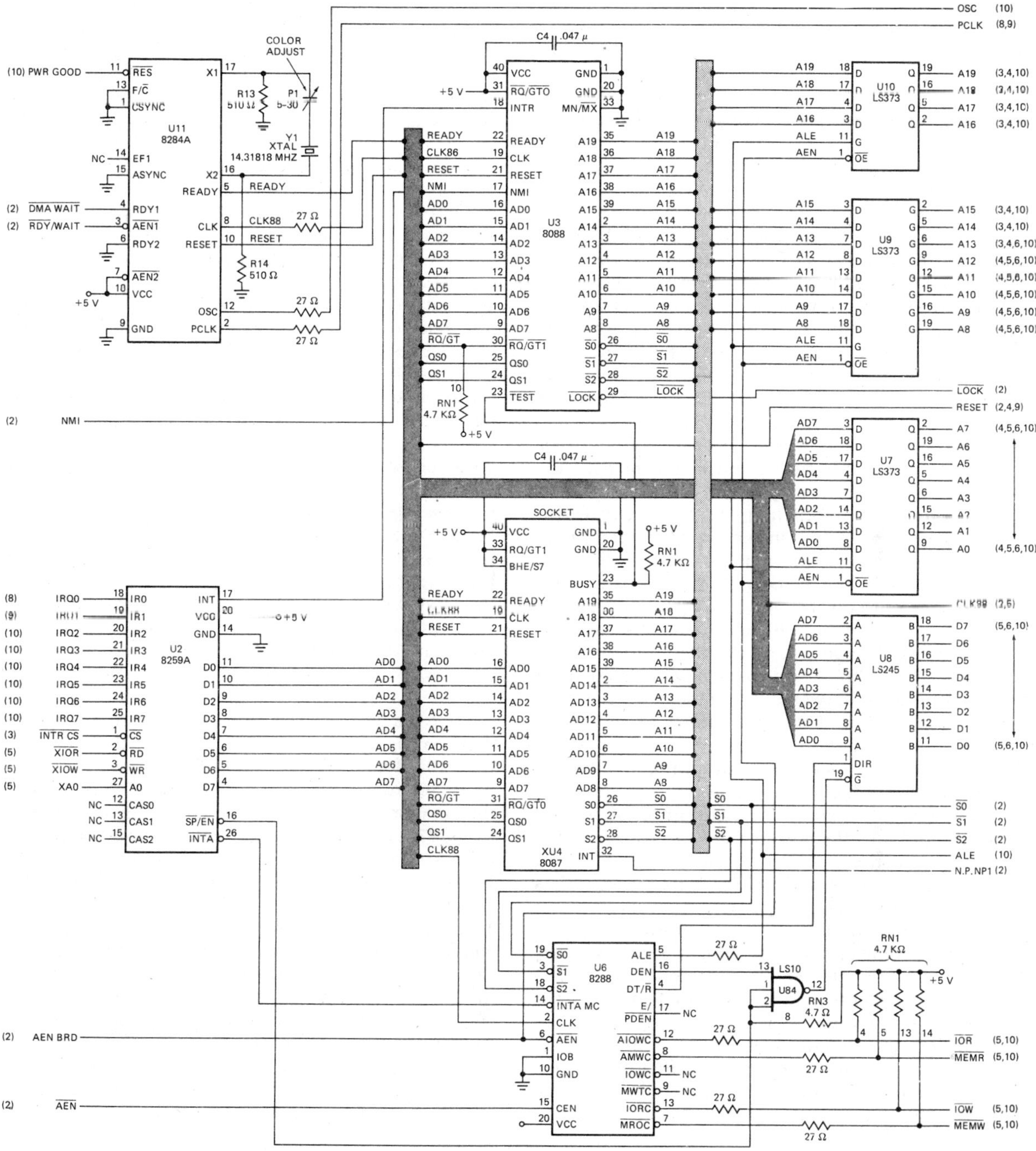

FIGURE 11-15 8088 and 8087 section of IBM PC schematic.

the system simply makes it appear that you have suddenly been given a whole new set of powerful math instructions to use in writing your programs. The opcodes for the 8087 instructions are put in memory right along with the codes for the 8086 or 8088 instructions. As the 8086 or 8088 fetches instruction bytes from memory and puts them in its queue, the 8087 also reads these instruction bytes and puts them in its internal queue.

The fact that the status lines and the queue status lines from the 8086 are connected directly to the 8087 allows the 8087 to track the 8086 or 8088 queue in this way. The 8087 decodes each instruction that comes into its queue. When it decodes an instruction from its queue and finds that it is an 8086 instruction, the 8087 simply treats the instruction as an NOP. Likewise, when the 8086 or 8088 decodes an instruction from its queue and finds that it is an 8087 instruction, the 8086 simply treats the instruction as an NOP, or in some cases reads one additional word from memory for the 8087. The point here is that each processor decodes all of the instructions in the fetched instruction byte stream, but only executes its own instructions. The first question that may occur to you is, "How do the two processors recognize 8087 instructions?" The answer is that all of the 8087 instruction codes have 11011 as the most-significant bits of their first code byte. Later we show you how to code 8087 instructions. The synchronous operation of these two processors is an example of a *tightly coupled* multiprocessor system. Now, before we get into the 8087 data types, architecture, and instruction set, let's dig a little more into how the two processors are synchronized during various operations.

One type of cooperation between the two processors that you need to know about is how the 8087 transfers data between memory and its internal registers. When the 8086 or 8088 reads an 8087 instruction that needs data from memory or wants to send data to memory, the 8086 sends out the memory address coded in the instruction and sends out the appropriate memory read or memory write signals to transfer a word of data. In the case of a memory read, the addressed word will be put on the data bus by the memory. The 8087 then simply reads in this word off the data bus. The 8086 or 8088 ignores this word. If the 8087 only needs this one word of data, it can then go on and execute its instruction, However, some 8087 instructions need to read in or write out up to 80-bit words. For these cases the 8086 outputs the address of the first data word on the address bus and outputs the appropriate control signal. The 8087 reads the data word put on the data bus by memory or writes a data word to memory on the data bus. The 8087 then grabs the 20-bit physical address that was output by the 8086 or 8088. To transfer additional words it needs to/from memory, the 8087 then takes over the buses from the 8086. To take over the bus the 8087 sends out a low-going pulse on its $\overline{RQ}/\overline{GT0}$ pin as shown in Figure 11-16. The 8086 or 8088 responds to this by sending another low-going pulse back to the $\overline{RQ}/\overline{GT0}$ pin of the 8087 and by floating its buses. The 8087 then increments the address it grabbed during the first transfer and outputs the incremented address on the address bus. When the 8087 outputs a memory read or memory write signal, another data word will be transferred to or from the 8087. The 8087 continues the process until it has transferred all of the data words required by the instruction to/from memory. When the 8087 is through using the buses for its data transfer, it sends another low-going pulse out on its $\overline{RQ}/\overline{GT0}$ pin to let the 8086 or 8088 know it can have the buses back again. This bus sharing is another example of a DMA-type operation. The $\overline{RQ}/\overline{GT0}$ line is used in a bidirectional mode here to save pins. The key point here then is that the coprocessor, by pulsing the $\overline{RQ}/\overline{GT0}$ input of the host processor, can take over the buses from the *host* or *bus master* processor to transfer data when it needs to. This is another example of a DMA operation.

The next type of synchronization between the host processor and the coprocessor that you need to know about is that required to make sure the 8086 or 8088 host does not attempt to execute the next instruction before the 8087 has completed an instruction. There are two possible problem situations here.

One problem situation is the case where the 8086 needs the data produced by execution of an 8087 instruction to carry out its next instruction. In the instruction sequence in Figure 11-17*a* for example, the 8087 must complete the FSTSW STATUS instruction before the 8086 will have the data it needs to execute the MOV AX, STATUS instruction. Without some mechanism to make the 8086 wait until the 8087 completes the FSTSW instruction, the 8086 will go on and execute the MOV AX, STATUS instruction with erroneous data. We solve this problem by connecting the 8087 BUSY output to the $\overline{TEST}$ pin of the 8086 or 8088, and putting an 8086 WAIT instruction in the program. Here's how it works.

While the 8087 is executing an instruction it asserts its BUSY pin high. When it is finished with an instruction, the 8087 will drop its BUSY pin low. Since the BUSY pin from the 8087 is connected to the $\overline{TEST}$ pin of the 8086 or 8088, the host can check this pin to see if the 8087 is done with an instruction. The 8086 instruction used to check the $\overline{TEST}$ pin is the WAIT instruction. You put the 8086 WAIT instruction in your program after the 8087 FSTSW instruction, as shown in Figure 11-17*b*. (Actually, for reasons we explain later, you should use the 8087 FWAIT instruction which does the same thing.) When the 8086 or 8088 executes the WAIT instruction, it enters an internal loop where it repeatedly checks the logic level on the $\overline{TEST}$ input. The 8086 will stay in this loop until it finds the $\overline{TEST}$ input asserted low, indicating the 8087 has completed its instruction. The 8086 will then exit the internal loop, fetch and execute its next instruction.

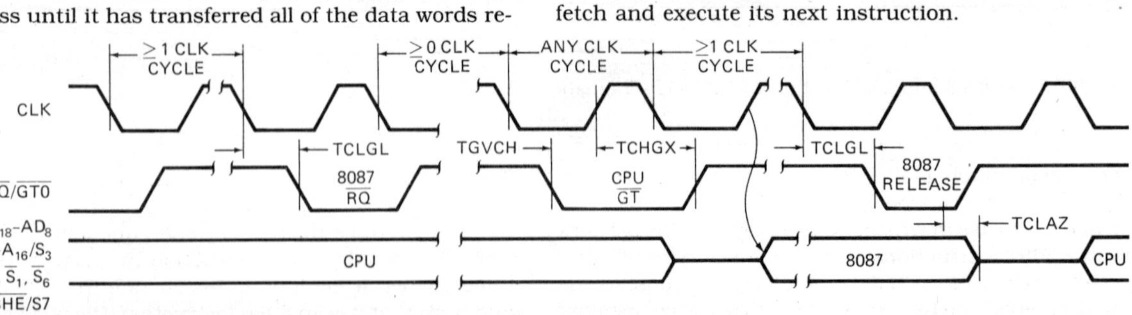

FIGURE 11-16 Signals on 8087 to 8088 $\overline{RQ}/\overline{GT}$ line during bus takeover by 8087 instruction coding formats. (Intel Corporation)

```
FSTSW  STATUS          ; copy 8087 status word to memory
MOV  AX, STATUS        ; copy status word to AX to check bits
```

(a)

```
FSTSW  STATUS          ; copy 8087 status word to memory
FWAIT                  ; wait for 8087 to finish before doing
                       ; next 8086 instruction
MOV  AX, STATUS        ; copy status word to AX to check bits
```

(b)

FIGURE 11-17 Synchronizing 8086 and 8087 instruction execution. (*a*) Code section without needed synchronization. (*b*) Code section with needed FWAIT instruction.

Another execution case where you need synchronization of the host and the coprocessor is the case where a program has several 8087 instructions in sequence. The 8087 can obviously execute only one instruction at a time, so you have to make sure that the 8087 has completed one instruction before you allow the 8086 to fetch the next 8087 instruction from memory. Here again you use the BUSY-TEST connection and the FWAIT instruction to solve the problem. If you are hand coding, you can just put the 8086 WAIT (FWAIT) instruction after each 8087 instruction to make sure that instruction is completed before going on to the next. If you are using an assembler which accepts 8087 mnemonics, the assembler will automatically insert the 8-bit code for the WAIT instruction, 10011011 binary (9BH), as the first byte of the code for the 8087 instruction. You can see an example of this in the code column of the sample program in Figure 11-24 which we discuss later. When the 8086 or 8088 fetches and decodes this code byte, it will enter the internal loop and wait for the TEST input to go low before fetching and decoding the 8087 instruction following this byte. The point here is that by putting the FWAIT instruction after an 8087 instruction in some way, you make sure that one instruction is finished before the next is started.

In the preceding sections we have shown you how two tightly coupled processors can operate essentially as one unit, sharing the same buses, memory, and instruction stream. Because the 8087 math coprocessor which we used as an example in the preceding sections is such a useful and common device, we now want to go on and show you how you can use one. We can't show you all there is to know about the 8087, because it is a fairly complex device. However we can show you enough that if you have an 8087 in your system, you can write a few simple programs for it. We will start with a discussion of the types of numbers that the 8087 is designed to work with.

8087 Data Types

Figure 11-18 shows the formats for the different types of numbers that the 8087 is designed to work with. The three general types are binary integer, packed decimal, and real. We will discuss and show examples of each type individually.

BINARY INTEGERS

The first three formats in Figure 11-18 show different-length binary integer numbers. These all have the same basic format that we have been using to represent signed binary numbers throughout the rest of the book. The most-significant bit is a sign bit which is 0 for positive numbers and 1 for negative numbers. The other 15–63 bits of the data word in these formats represent the magnitude of the number. If the number is negative, the magnitude of the number is represented in 2's complement form. Zero, remember, is considered a positive number in this format, because it has a sign bit of 0. Note also in Figure 11-18 the range of values that can be represented by each of the three integer lengths. When you put numbers in this format in memory for the 8087 to access, you put the least-significant byte in the lowest address.

PACKED DECIMAL NUMBERS

The second type of 8087 data format to look at in Figure 11-18 is the packed decimal. In this format a number is represented as a string of 18 BCD digits, packed two per byte. The most-significant bit is a sign bit which is 0 for positive numbers and 1 for negative numbers. The bits indicated with an X are don't cares. This format is handy for working with financial programs. Using this format you can represent a dollar amount as large as $9,999,999,999,999,999.99, which is probably about what the national debt will be by the year 2000. Again, when you are putting numbers of this type in memory locations for the 8087 to access, the least-significant byte goes in the lowest address.

REAL NUMBERS

Before we discuss the 8087 real number formats, we need to talk a little about real numbers in general.

So far the computations we have shown in this book have used signed integer numbers or BCD numbers. These numbers are referred to as *fixed-point* numbers because they contain no information as to the location of the decimal point or binary point in the number. The

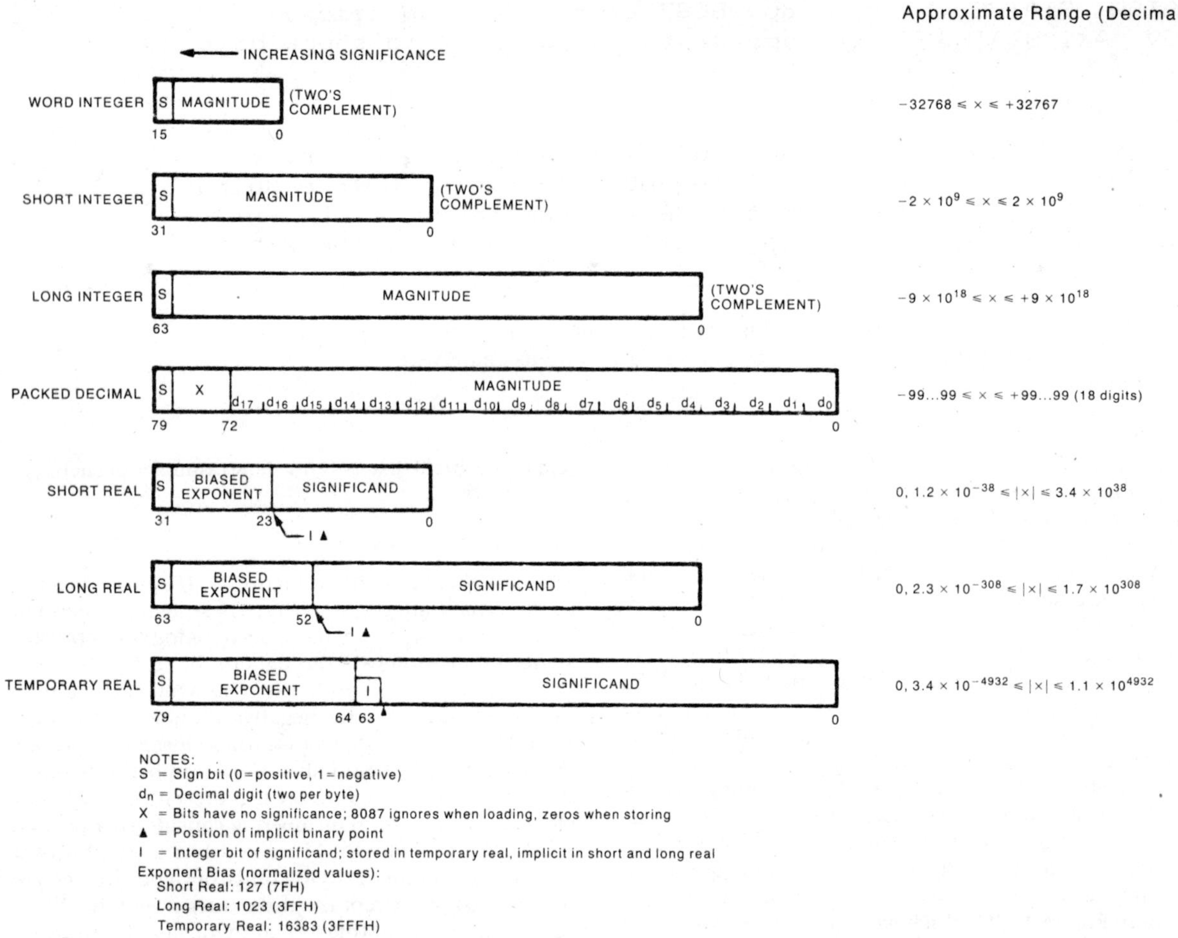

FIGURE 11-18 8087 data formats. (Intel Corporation)

decimal or binary point is always assumed to be to the right of the least-significant digit, so all numbers are represented in this form as whole numbers with no fractional part. A weight of 9.4 lb, for example, is stored in a memory location simply as 10010100 BCD or 01011110 binary. A price of $0.29 per pound is stored in a memory location as 00101001 in BCD or 00011101 in binary. When the binary representation of the weight is multiplied by the price per pound to give the total price, the result is 101010100110 binary, or 2726 decimal. To give the desired display of $2.73, the programmer must round off the result and keep track of where to put the decimal point in the result. For simple numbers such as these from the scale program in Chapter 10, it is not too difficult to do this. However, for a great many applications we need a representation that automatically keeps track of the position of the decimal or binary point for us. In other words we need to be able to represent numbers which have both an integer part and a fractional part. Such numbers are called *real numbers*, or *floating-point numbers*.

There are several different formats for representing real numbers in binary form. The basic principle of all of these, however, is to use one group of bits to represent the digits of the number, and another group of bits to represent the position of the binary point with respect to these digits. This is very similar to the way numbers are represented in scientific notation, so as a lead-in we will refresh your memory about scientific notation.

To convert the number 27,934 to scientific notation you move the decimal point four digit positions to the left and multiply the number by 10^4. The result, 2.7934×10^4, is said to be in scientific notation. As another example, you convert 0.00857 to scientific notation by moving the decimal point three digit positions to the right and multiplying by 10^{-3} to give 8.57×10^{-3}. The process of moving the decimal point to a position just to the right of the most-significant, nonzero digit is called *normalizing* the number. In these examples you can see the digit part, sometimes called the *significand* or the *mantissa*, and the *exponent* part of the representation. When you are working with a calculator or computer, the number of digits you can store for the significand part of the number determines the accuracy or *precision* of the representation. In most cases the real numbers you work with in your computer will be approximations, because to "accurately" represent a number such as π would require an infinite number of digits. The point here is that more digits give more precision, or in other words a better approximation.

The number of digits you can store for the exponent of a number determines the range of magnitudes of numbers you can store in your computer or calculator. The sign of the exponent indicates whether the magnitude of the number is greater than one or less than one. The sign of the significand or mantissa indicates whether the number itself is positive or negative. Now let's see how you represent real numbers in binary form so the 8087 can digest them.

First let's look at the short-real format shown in Figure 11-18. This format, which uses 32 bits to represent a number, is sometimes referred to as *single-precision* representation. In this format 23 bits are used to represent the magnitude of the number, 8 bits to represent the magnitude of the exponent, and 1 bit to indicate whether the number is negative or positive. The magnitude of the number is normalized so that there is only a single one to the left of the binary point. The one to the left of the binary point is not actually present in the representation, it is simply assumed to be there. This leaves more bits for representing the magnitude of the number. You can think of the binary point as being between the bits numbered 22 and 23. The exponent for this format is put in an *offset form*, which means that an offset of 127 (7FH) is added to the 2's complement value of the exponent. This is done so that the magnitude of two numbers can be compared without having to do arithmetic on the exponents first. The sign bit is 0 for positive numbers, and 1 for negative numbers. To help make this clear to you, we will show you how to convert a decimal number to this format.

We chose the number 178.625 for this example because the fractional part converts exactly, and therefore we don't have to cope with rounding at this point. The first step is to convert the decimal number to binary to give 10110010.101 as shown in Figure 11-19. Next normalize the binary number so that only a single one is to the left of the binary point and represent the number of bit positions you had to move the binary point as an exponent as shown in Figure 11-19. The result at this point is 1.0110010101E7. If you now add the bias of

```
178.625   DECIMAL
10110010.101   BINARY
1.0110010101   E7

01000011001100101010000000000000
↑        └BINARY POINT
|  BIASED          SIGNIFICAND
| EXPONENT
|
└SIGN
```

FIGURE 11-19 Converting a decimal number to short-real format.

127 (7FH) to the exponent of 7, you get the biased exponent value of 86H that you need for the short-real representation. The final line in Figure 11-19 shows the complete short-real result. For the significand you put in the binary bits to the right of the binary point. Remember, the 1 to the left of the binary point is assumed. The biased exponent value of 86H or 10000110 binary is put in as bits 23 through 30. Finally, since the number is positive, a 0 is put in bit 31 as the sign bit. The complete result is then 01000011001100101010000000000000 or 4332A000H, which is lengthy, but not difficult to produce.

The long-real format shown in Figure 11-18 uses 64 bits to represent each number. This format is often referred to as *double-precision* representation. This format is basically the same as that of the short-real, except that it allows greater range and accuracy because more bits are used for each number. For long-real, 52 bits are used to represent the magnitude of the number. Again the number is normalized so that only a single 1 is to the left of the binary point. You can think of the binary point as being between the bits numbered 51 and 52. The one to the left of the binary point is not actually put in as one of the 64 bits. For this format, 11 bits are used for the exponent, so the offset added to each exponent value is 1023 decimal or 3FFH. The most significant bit is the sign bit. Our example number of 178.625 will be represented in this long-real or double-precision format as 4066540000000000H. Note in Figure 11-18 the range of numbers that can be represented with this format. This range should be large enough for most of the problems you want to solve with an 8087.

The final format in Figure 11-18 to discuss is the temporary-real format which uses 80 bits to represent each number. This is the format that all numbers are converted to by the 8087 as it reads them in, and it is the format in which the 8087 works with numbers internally. The large number of bits used in this format reduces rounding errors in long chain calculations. To understand what this means, think of multiplying 1234×4567 in a machine that can only store the upper 4 digits of the result. The actual result of 5,635,678 will be truncated to 5,635,000. If you then divide this by 1234 to get back to the original 4567, you get instead 4566 because of the limited precision of the intermediate number.

As you can see in Figure 11-18, the temporary-real format has a sign bit, 15 bits for a biased exponent, and 64 bits for the significand. The offset or bias added to the exponent here is 16,383 decimal or 3FFFH. A major difference in the significand for this format from that for short-reals and long-reals is that the 1 to the left of the binary point after normalization is included as bit 63 in the significand. To express our example number of 178.625 in this form, then, we convert it to binary and normalize it as before to give 1.0110010101E7. This gives us the upper bits of the significand directly as 10110010101. We simply add enough 0's on the right of this to fill up the rest of the 64 bits reserved for the significand. To produce the required exponent, we add the bias value of 3FFFH to our determined value of 7. This gives 4006H or 100000000000110 binary as the value for the exponent. The sign bit is a 0 because the number is positive. Putting all of these pieces together gives4006B2A000000000000000H as the temporary-real representation of 178.625. This concludes our initial discussion of the way numbers are represented for the 8087.

The 8087 Internal Architecture

Figure 11-20 shows a block diagram of the 8087. As we described before, the 8087 connects directly on the address, data, and status lines of the 8086 or 8088 so that it can track and decode instructions fetched by the 8086 or 8088 host. The 8087 has a control-word register and a status register. Control words are sent to the 8087 by writing them to a memory location and having the 8087 execute an instruction which reads in the control word from memory. Likewise, to read the status word from an 8087 you have it execute an instruction which writes the status word to memory where you can read or check it with an 8086 instruction. Figure 11-21 shows the formats for the 8087 control and status words. Take a look at these now so you have an overview of the meaning of the various bits of these words. We will discuss the meaning of most of these bits as we work our way through the following sections.

The 8087 works internally with all numbers in the 80-bit temporary-real format which we discussed in the preceding paragraphs. To hold numbers being worked on, the 8087 has a register stack of eight, 80-bit registers labeled (0)–(7) in Figure 11-20. These registers are used as a last-in–first-out stack in the same way the 8086 uses a stack. The 8087 has a 3-bit stack pointer which holds the number of the register which is the current top-of-stack. When the 8087 is initialized, the 3-bit stack pointer in the 8087 is loaded with 000, so register 0 is then the TOS. When the 8087 reads in the first number that it is going to work on from memory, it converts the number to 80-bit temporary-real format if necessary. It then decrements the stack pointer to 111, and writes the temporary-real representation of the number in register number 111 (7). Figure 11-22*a* shows this in diagram form. As shown by the arrow in the figure, you can think of the stack as being wrapped around in a circle so that if you decrement 000 you get 111. Also from this diagram you can see that if you push more than 8 numbers on the stack they wrap around and write over previous numbers. After this write-to-stack operation, register 7 is now the TOS.

In the 8087 instructions the register that is currently the TOS is referred to as ST(0), or simply ST. The register just below this in the stack is referred to as ST(1). By the register "just below," we mean the register that the stack pointer would be pointing to if we popped one number off the stack. For the example in Figure 11-22*a*, register 000 would be ST(1) after the first push.

To help you understand this concept, Figure 11-22*b* shows another example. In this example we have pushed three numbers on the stack after initializing. Register 101 is now the TOS, so it is referred to as ST(0) or just ST. The preceding number pushed on the stack is in register 110, so it is referred to as ST(1). Likewise, the location below this in the stack is referred to as ST(2). If you draw a diagram such as that in Figure 11-22*b*, it is relatively easy to keep track of where everything is in the stack as instructions execute. In a program you can determine which register is currently the ST by simply transferring the status word to memory and checking the bits labeled ST in the status-word format in Figure 11-21*b*. Now let's have a look at the 8087 instruction set.

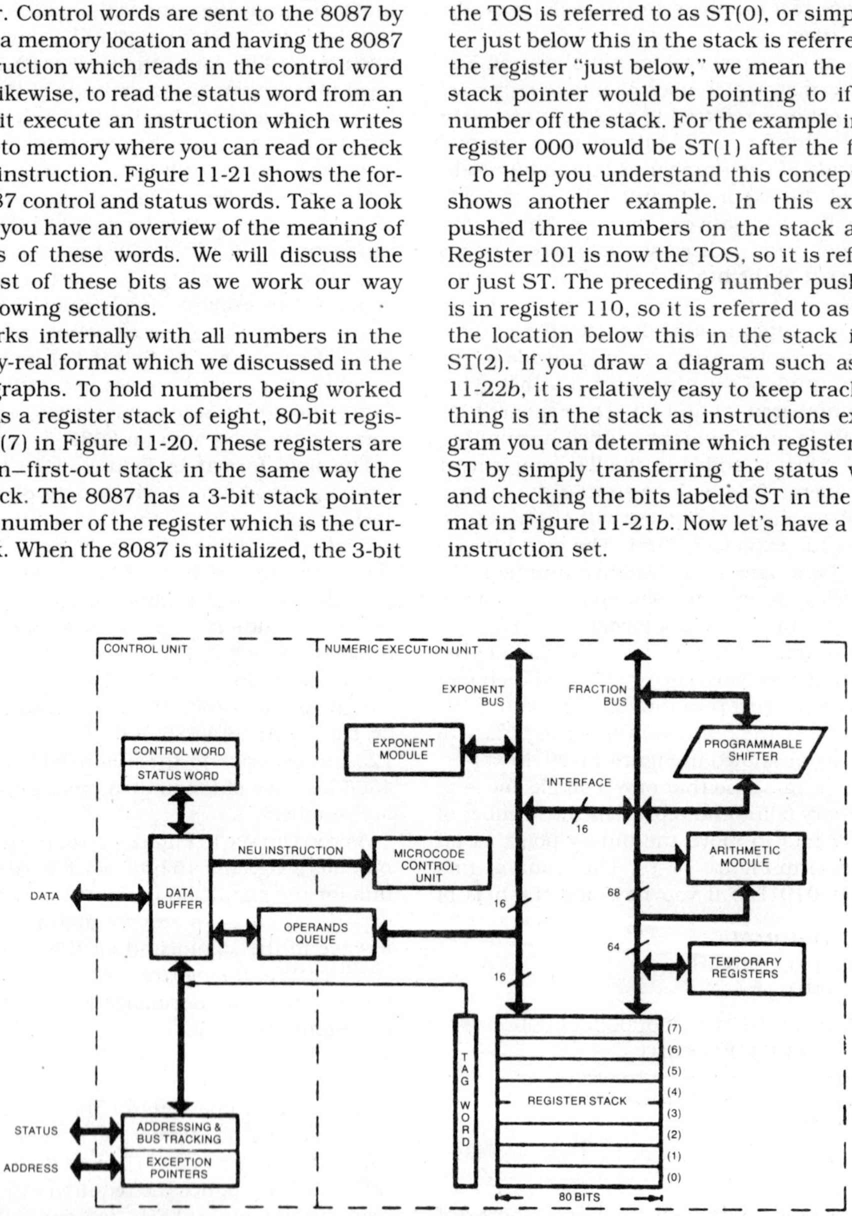

FIGURE 11-20 8087 internal block diagram. (Intel Corporation)

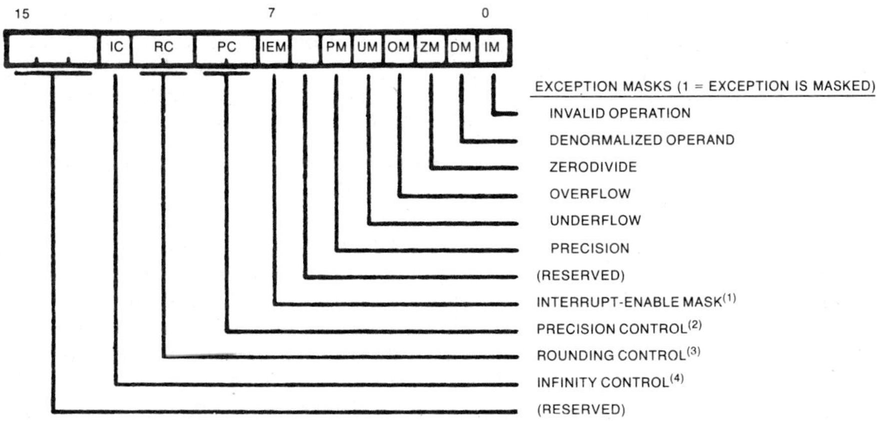

(1) Interrupt-Enable Mask:
0 = Interrupts Enabled
1 = Interrupts Disabled (Masked)*

(2) Precision Control:
00 = 24 bits
01 = (reserved)
10 = 53 bits
11 = 64 bits

(3) Rounding Control:
00 = Round to Nearest or Even*
01 = Round Down (toward −∞)
10 = Round Up (toward +∞)
11 = Chop (Truncate Toward Zero)

(4) Infinity Control:
0 = Projective*
1 = Affine

* DEFAULT AFTER FINIT

(a)

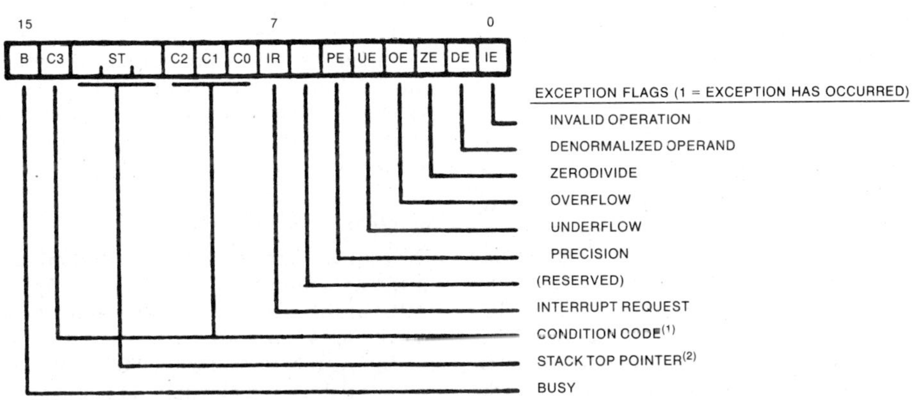

(1) See descriptions of compare, test, examine and remainder instructions in section S.7 for condition code interpretation.

(2) ST values:
000 = register 0 is stack top
001 = register 1 is stack top
.
.
.
111 = register 7 is stack top

(b)

FIGURE 11-21 8087 control and status word formats. *(a)* Control. *(b)* Status. (Intel Corporation)

8087 Instruction Set

8087 INSTRUCTION FORMATS

Before we work our way through the list of 8087 instructions we will use one simple instruction to show you how 8087 instructions are written, how they operate, and how they are coded. Instructions for the 8087 can be written as 8086 ESCAPE (ESC) instructions followed by a number as, for example, ESC 28, or they can be written using specific 8087 mnemonics. For the following discussion we will use the 8087 mnemonics. Later we will show you how to code the instructions as ESC instructions if you don't have an assembler which accepts 8087 mnemonics. The instruction we have chosen to use as an example here is the FADD instruction.

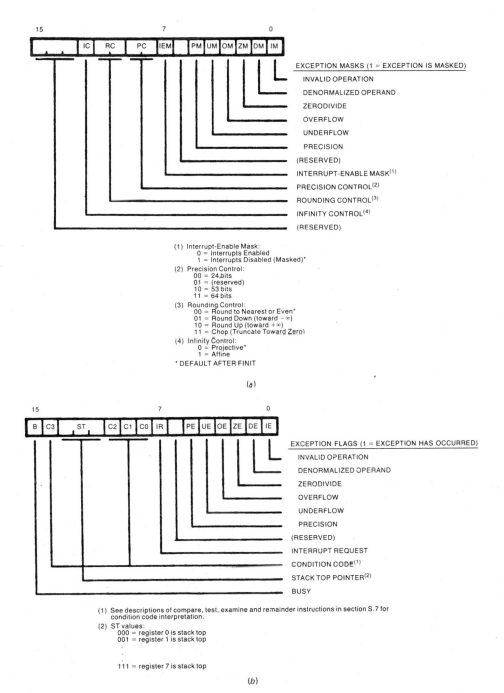

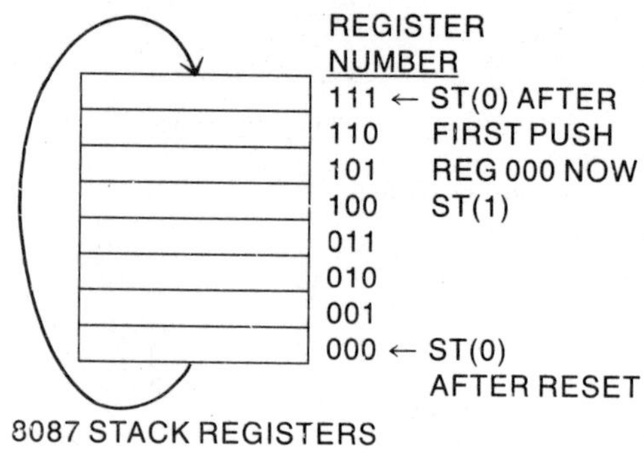

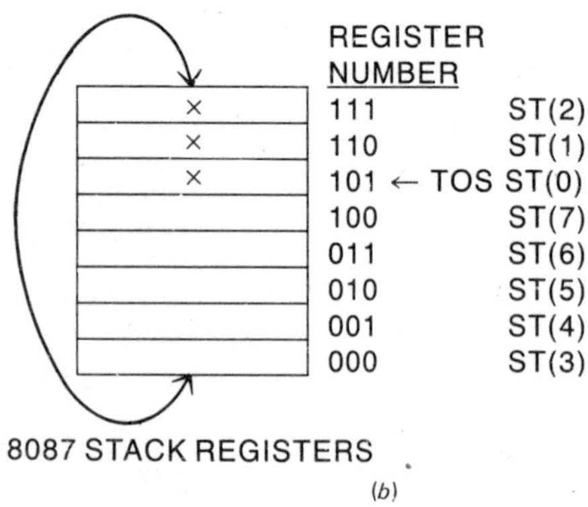

FIGURE 11-22 8087 stack operation. *(a)* Condition of stack after reset and one push. *(b)* Condition of stack after reset and three pushes.

All of the 8087 mnemonics start with an F, which stands for floating point, the form in which the 8087 works with numbers internally. If you look in the Intel data book, you will see this instruction represented as FADD //source/destination, source. This cryptic representation means that the instruction can be written in three different ways.

The // at the start indicates that the instruction can be written without any specified operands as simply FADD. In this case, when the 8087 executes the instruction it will automatically add the number at the top of the stack (ST) to the number in the next location under it in the stack, ST(1). The 8087 stack pointer will be incremented by one so that the register containing the result will be ST.

The word "source" by itself in the expression means that the instruction can be written as FADD source. The source specified here can be one of the stack elements, or a memory location. For example, the instruction FADD ST(2) will add the number from two locations below ST to the number in ST, and leave the result in ST. As another example, the instruction FADD CORRECTION_FACTOR will add a real number from the memory location named CORRECTION_FACTOR to the number in ST and leave the result in ST. The assembler will be able to determine whether the number in memory is a short-real, long-real, or temporary-real by the way that CORRECTION_FACTOR was declared. Short-reals, for example, are declared with the DD directive, long-reals with the DQ directive, and temporary-reals with the DT directive. If you want to add an integer number from memory to ST, you use an instruction such as FIADD CORRECTION_FACTOR. The I in the mnemonic tells the assembler to code the instruction so that the 8087 treats the number read in as an integer.

NOTE: The FIADD instruction only works for a source operand in memory.

The /destination, source in the representation of the FADD instruction means that you can write the instruction with both a specified source and a specified destination. The source can be one of the stack elements, or a number from memory. The destination has to be one of the stack elements. The instruction FADD ST(2), ST(1), for example, will add the number one location down from ST to the number two locations down from ST, and leave the result in ST(2). The instruction FADD ST(3), CORRECTION_FACTOR will add the real number from the memory location named CORRECTION_FACTOR to the contents of the ST(3) stack element.

Another form of the 8087 FADD instruction shown in the data book is FADDP. The P at the end of this mnemonic means to POP. When the 8087 executes this form of the FADD instruction, it will increment the stack pointer by one after it does the add operation. This is referred to as "popping the stack." The instruction FADDP ST(1), ST(4), for example, will add the number at ST(4) to the number at ST(1), and put the result in ST(1). It will then "pop the stack" or, in other words, increment the stack pointer so that what was ST(1) is now ST. This form of the instruction leaves the result at ST where it can easily be transferred to memory. Now let's see how the different forms of this instruction are coded.

CODING 8087 INSTRUCTIONS

The coding templates for all of the 8087 instructions are in the appendix for your reference. You use these templates in the same manner as you do those for the 8086 instructions. For closer reference, Figure 11-23 shows the coding templates for the 8087 FADD instructions in the form shown in the appendix and in expanded form so you can see the individual bits. Note that the figure shows coding for "8087" encoding and for "emulator" encoding. The 8087 encoding represents the codes required by the actual device. The emulator encoding represents the codes needed to call an FADD procedure from an available Intel library of 8086 procedures which perform the same functions as the 8087 instructions. The procedures in this library, written in 8086 code, run much slower, but they allow you to test an 8087 program without having an actual 8087 in the system. We will concentrate here on the codes for the actual 8087 device.

First let's look at the coding for the FADD instruction with no specified operands. This instruction, remember, will add the contents of ST to the contents of ST(1), put the results in ST(1), and then pop the stack so that the result is at ST. The first byte of the instruction code, 10011011, is the code for the 8086 WAIT instruction. Remember from our previous discussion that this in-

Type 1: Stack top and stack element

8087	10011011	11011 d 00	11000(i)
Emulator	11001101	00011 d 00	11000(i)

Type 2: Stack top and memory operand

8087	10011011	11011 m 00	mod 000 r/m
Emulator	11001101	00011 m 00	mod 000 r/m

m = 0 for short real operand; 1 for long real operand

Type 3: Pop stack

8087	10011011	11011110	11000(i)
Emulator	11001101	00011110	11000(i)

8087 Timing (clocks)	**TYPICAL**	**RANGE**
stack element and stack top	85	70-100
stack element, stack top + pop	90	75-105
short real memory and stack top	105+EA	90-120+EA
long real memory and stack top	110+EA	95-125+EA

FADD

Stack top + Stack Element

WAIT	op1	op2 + i

8087 Encoding	**Emulator Encoding**	**Execution Clocks Typical Range**	**Operation**
9B D8 C0+i	CD 18 C0+i	85 70-100	ST←ST + ST(i)
9B DC C0+i	CD 1C C0+i	85 70-100	ST(i)←ST + ST(i)

Stack top + memory operand

WAIT	op1	mod 000 r/m	addr1	addr2

8087 Encoding	**Emulator Encoding**	**Execution Clocks Typical Range**	**Operation**
9B D8 m0rm	CD 18 m0rm	105+EA (90-120)+EA	ST←ST + mem-op (short-real)
9B DC m0rm	CD 1C m0rm	110+EA (95-125)+EA	ST←ST + mem-op (long-real)

FADDP = Add Real and Pop

Stack top + Stack Element

WAIT	op1	op2 + i

8087 Encoding	**Emulator Encoding**	**Execution Clocks Typical Range**	**Operation**
9B DE C1	CD 1E C1	90 75-105	ST(1)←ST + ST(1) pop stack
9B DE C0+i	CD 1E C0+i	90 75-105	ST(i)←ST + ST(i) pop stack

FIGURE 11-23 8087 FADD coding templates. (Intel Corporation)

struction code is put here to make the 8086 and 8087 wait until it has completed a previous instruction before starting this one. The second byte shown is actually the first byte of the 8087 FADD instruction. The 5 most-significant bits, 11011, identify this as an 8087 instruction. The lower 3 bits of the first code byte and the middle 3 bits of the second code byte are the opcode for the particular 8087 instruction. The bit labeled d at the start of these 6 is a 0 if the destination for an FADD ST(N), ST(N) type instruction is ST. The d bit is a 1 if the destination stack element is one other than ST, as it is for the FADD instruction with no specified operands. For the FADD instruction with no specified operands these 6 bits will be 100 000. The two most significant (MOD) bits in the second code byte are ones, because this form of the FADD instruction does not read a number from memory. The least-significant 3 bits of the second instruction byte, represented by an i in the template, indicate which stack element other than ST is specified in the instruction. Since the simple FADD instruction uses ST(1) as a destination, 001 will be put in these bits. Putting all of this together for the FADD instruction with no specified source or destination gives 10011011 11011100 11000001 binary or 9BH DCH C1H as the code bytes.

For a little more practice with this see if you can code the 8087 instruction FADD ST, ST(2). Most of the coding for this instruction is the same as that for the previous instruction. For this one, however, the d bit is a 0 because ST is the specified destination. Also, the R/M bits are 010, because the other register involved in the addition is ST(2). The answer is 9BD8C2H. Now let's try an example which uses memory as the source of an operand for FADD.

For an FADD instruction such as FADD CORRECTION_FACTOR, which brings in one operand from memory and adds it to ST, the memory address can be specified in any of the 24 ways shown in Figure 3-8. For the memory reference form of the FADD instruction the MOD and R/M bits in the second code byte are used to specify the desired addressing mode. FADD CORRECTION_FACTOR represents direct addressing, so the MOD bits will be 00 and the R/M bits will be 110 as shown in Figure 3-8. Two additional code bytes will be used to put in the direct address, low byte first. Since we are not using any of the other stack elements other than ST for this instruction, we don't need the d bit to specify the other stack element. Instead, as shown in Figure 11-23, this bit is labeled m. A 0 in this bit is used to specify a short-real, and a 1 in this bit is used for a long-real. Assuming CORRECTION_FACTOR is declared as a long-real, the code bytes for our FADD CORRECTION_FACTOR instruction will then be 10011011 11011100 00000110 followed by the two bytes of the direct address.

So far in this section we have shown you examples of how some 8087 instructions will be coded out if they are written with 8087 mnemonics and assembled with an assembler that recognizes 8087 mnemonics. If you don't have an assembler with 8087 capability, you can code 8087 instructions as 8086 ESCAPE (ESC) instructions with the help of the 8087 instruction templates in the appendix. Your 8086 assembler should then produce the correct code for them. Here's how you do it.

The 8086 ESC instruction has two basic forms. The first of these, ESC immediate, memory, is used to produce 8087 instructions which bring an operand in from memory or send an operand out to memory. The second form of the escape instruction is ESC immediate, register. This second form is used to produce 8087 instructions which operate only on registers. We'll start with the ESC immediate, memory form.

The coding template for the ESC immediate, memory instruction is 11011xxx MODxxxR/M. The 6 x's in the template represent a 6-bit binary opcode for the desired 8087 instruction. You specify these 6 bits in the ESC instruction as an immediate number. You determine the value for this immediate number by looking at the coding template for the 8087 instructions in the appendix. For example, suppose that you want to use the ESC instruction to produce the FADD CORRECTION_FACTOR instruction. In our discussion of the coding for this instruction above, we determined that if CORRECTION_FACTOR is a long-real, then the binary code for the instruction is 10011011 11011100 00000011 followed by the direct address in the next 2 bytes. The first byte of this is the WAIT instruction. The values for the 6 bits that we are interested in are 100000. Converting this to hex gives 20H, the immediate number you need to put in the ESC instruction. You write the 8086 ESC instruction then as ESC 20H, CORRECTION_FACTOR. When the assembler reads this instruction statement, it will automatically determine the values for the MOD and the R/M bits of the instruction from the reference to the memory location CORRECTION_FACTOR. You can put an 8086 WAIT instruction before the ESC instruction if necessary.

Now we will show you how to write an ESC instruction to produce an 8087 instruction which operates only on internal registers of the 8087. The coding template for this ESC instruction is also 11011xxx MODxxxR/M. The 6 x's again represent the desired 8087 opcode. You get the values for these bits from the 8087 code templates in the appendix. As an example, suppose that you want to get the 8086 assembler to produce the correct code for the 8087 instruction FADD ST, ST(2), which was one of our examples in a previous section. According to the templates in the appendix, the code for this instruction is 10011011 11011000 11000i. The 10011011 at the start is the code for the WAIT instruction. The 6 underlined bits from this code that you use for the immediate number in the ESC instruction are 000 000. The ones in the MOD position of the last code byte indicate that only registers are involved in the operation. The three i bits in this byte must contain the number of the other stack element besides ST that is used in the instruction. For our example here we want the assembler to put in 010 or 2 for these bits, because we are using ST(2). To get the assembler to put in 11 for the MOD bits in the ESC instruction code, you simply put an 8086 register name after the immediate number in the ESC instruction. The register name you use tells the assembler what to put in the 3-bit i field of the instruction. Look at Figure 3-8 to determine the 3-bit codes for each of the 8086 registers. The code of 010 which you need here corresponds to that for the DX register. If you put all of these pieces together, the 8087 FADD ST, ST(2) can be written as the 8086 ESC 00H, DX instruction. Once you see the correspondence here, it is not too difficult to get the 8086 assembler to produce a desired 8087 instruction code. Now that you have an overview of how 8087 instructions are written and coded, we will briefly discuss the operation of the available 8087 instructions.

8087 INSTRUCTION DESCRIPTIONS

The 8087 instruction mnemonics all begin with the letter F which stands for floating point and distinguishes the 8087 instructions from 8086 instructions. We have found that if we mentally remove the F as we read the mnemonic, it makes it easier to connect the mnemonic and the operation performed by the instruction. Here we briefly describe the operation of each of the 8087 instructions so that you can use some of them to write simple programs. As you read through these instructions the first time, don't try to absorb them all, or you probably won't remember any of them. Concentrate first on the instructions you need to get operands from memory into the 8087, simple arithmetic instructions, and the instructions you need to get results copied back from the 8087 to memory where you can use them. Then work your way through the first example program in the next section. After that, read through the instructions again and pay special attention to the transcendental instructions and the load constant instructions. Finally, work your way through the second example program in the next section. We hope that you will find, as we have, that the 8087 is fun to work with. After you play around with some simple programs such as those here, you can go to the Intel literature to get more information about error handling and more complex program examples.

The instructions are grouped here in four functional groups so that you can more easily find the instruction which performs a desired operation. A section in the appendix shows the coding templates and clock cycles for each instruction.

If the 8087 detects an error condition, usually called an *exception*, while it is executing an instruction, it will set the appropriate bit in its status register. After the instruction finishes executing, the status register contents can be transferred to memory with another 8087 instruction. You can then use 8086 instructions to check the status bits and decide what action to take if an error has occurred. Figure 11-21*b* shows the format of the 8087 status word. The lowest 6 bits are the exception status bits. These bits will all be 0's if no errors have occurred. In the instruction descriptions following, we use the first letter of each exception type to indicate the status bits affected by each instruction.

If you send the 8087 a control word which unmasks the exception interrupts as shown in Figure 11-21*a*, the 8087 will also send a hardware interrupt signal to the 8086, when an error occurs. If the 8086 interrupt input is enabled, this allows the 8086 to go directly to an exception handling procedure.

DATA TRANSFER INSTRUCTIONS

Real Transfers

FLD source—Decrements the stack pointer by one and copies a real number from a stack element or memory location to the new ST. A short-real or long-real number from memory is automatically converted to temporary-real format by the 8087 before it is put in ST. Exceptions: I, D. Examples:

```
FLD ST(3)            ; Copies ST(3) to ST
FLD LONG_REAL[BX]    ; Number from memory copied
                     ; to ST
```

FST destination—Copies ST to a specified stack position or to a specified memory location. If a number is transferred to a memory location, the number and its exponent will be rounded to fit in the destination memory location. Exceptions: I, O, U, P. Examples:

```
FST ST(2)             ; Copy ST to ST(2), and
                      ; increment stack pointer
FST SHORT_REAL[BX]    ; Copy ST to memory
                      ; at SHORT_REAL[BX]
```

FSTP destination—Copies ST to a specified stack element or memory location and increments the stack pointer by one to point to the next element on the stack. This is a stack POP operation. It is identical to FST except for the effect on the stack pointer.

FXCH //destination—Exchanges the contents of ST with the contents of a specified stack element. If no destination is specified, then ST(1) used. Exception: I. Example:

```
FXCH ST(5)   ; Swap ST and ST(5)
```

Integer Transfers

FILD source—Integer load. Convert integer number from memory to temporary-real format and push on 8087 stack. Exception: I. Example:

```
FILD DWORD PTR [BX]   ; Short integer from memory
                      ; at [BX]
```

FIST destination—Integer store. Convert number from ST to integer form, and copy to memory. Exceptions: I, P. Example:

```
FIST LONG_INT   ; ST to memory locations
                ; named LONG_INT
```

FISTP destination—Integer store and pop. Identical to FIST except that stack pointer is incremented after copy.

Packed Decimal Transfers

FBLD source—Packed decimal(BCD) load. Convert number from memory to temporary-real format and push on top of 8087 stack. Exception: I. Example:

```
FBLD MONEY_DUE   ; Ten byte BCD number from
                 ; memory to ST
```

FBSTP destination—BCD store in memory and pop 8087 stack. Pops temporary-real from stack, converts to 10-byte BCD, and writes result to memory. Exception: I. Example:

```
FBSTP TAX   ; ST converted to BCD, sent to memory
```

ARITHMETIC INSTRUCTIONS

Addition

FADD //source/destination, source—Add real from specified source to real at specified destination. Source can be stack element or memory location. Destination must be a stack element. If no source or destination is specified, then ST is added to ST(1) and the stack pointer is incremented so that the result of the addition is at ST. Exceptions: I, D, O, U, P. Examples:

```
FADD ST(3), ST    ; Add ST to ST(3), result in ST(3)
FADD ST,ST(4)     ; Add ST(4) to ST, result in ST
FADD INTEREST     ; Real num from mem + ST
FADD              ; ST + ST(1), pop stack-result at ST
```

FADDP destination, source—Add ST to specified stack element and increment stack pointer by one. Exceptions: I, D, O, U, P. Example:

```
FADDP ST(1)   ; Add ST(1) to ST. Increment stack
              ; pointer so ST(1) becomes ST
```

FIADD source—Add integer from memory to ST, result in ST. Exceptions: I, D, O, P. Example:

```
FIADD CARS_SOLD   ; Integer number from
                  ; memory + ST
```

Subtraction

FSUB //source/destination, source—Subtract the real number at the specified source from the real number at the specified destination and put the result in the specified destination. Exceptions: I, D, O, U, P. Examples:

```
FSUB ST(2), ST   ; ST(2) becomes ST(2) - ST
FSUB CHARGE      ; ST becomes ST - real from memory
FSUB             ; ST becomes (ST(1) - ST)
```

FSUBP destination, source—Subtract ST from specified stack element and put result in specified stack element. Then increment stack pointer by one. Exceptions: I, D, O, U, P. Examples:

FSUBP ST(1) ; ST(1) − ST. ST(1) becomes new ST.

FISUB source—Integer from memory subtracted from ST, result in ST. Exceptions: I, D, O, P. Example:

```
FISUB CARS_SOLD   ; ST becomes ST – integer
                  ; from memory
```

Reversed Subtraction

FSUBR //source/destination, source
FSUBRP //destination, source
FISUBR source

These instructions operate the same as the FSUB instructions described above except that these instructions subtract the contents of the specified destination from the contents of the specified source and put the difference in the specified destination. Normal FSUB instructions, remember, subtract source from destination.

Multiplication

FMUL //source/destination, source—Multiply real number from source by real number from specified destination, and put result in specified stack element. See FADD instruction description for examples of specifying operands. Exceptions: I, D, O, U, P.

FMULP destination, source—Multiply real number from specified source by real number from specified destination, put result in specified stack element, and increment stack pointer by one. See FADDP instruction for examples of how to specify operands for this instruction. With no specified operands FMULP multiplies ST(1) by ST and pops stack to leave result at ST. Exceptions: I, D, O, U, P.

FIMUL source—Multiply integer from memory times ST and put result in ST. Exceptions: I, D, O, P. Example:

FIMUL DWORD PTR [BX]

Division

FDIV //source/destination, source—Divide destination real by source real, result goes in destination. See FADD formats. Exceptions: I, D, Z, O, U, P.

FDIVP destination, source—Same as FDIV, but also increment stack pointer by one after DIV. See FADDP formats. Exceptions: I, D, Z, O, U, P.

FIDIV source—Divide ST by integer from memory, result in ST. Exceptions: I, D, Z, O, U, P.

Reversed Division

FDIVR //source/destination, source
FDIVP destination, source
FIDIVR source

These three instructions are identical in format to the FDIV, FDIVP, and FIDIV instructions above except that they divide the source operand by the destination operand and put the result in the destination.

Other Arithmetic Operations

FSQRT—Contents of ST are replaced with its square root. Exceptions: I, D, P. Example: FSQRT

FSCALE—Scale the number in ST by adding an integer value in ST(1) to the exponent of the number in ST. Fast way of multiplying by integral powers of two. Exceptions: I, O, U.

FPREM—Partial remainder. The contents of ST(1) are subtracted from the contents of ST over and over again until the contents of ST are smaller than the contents of ST(1). In an 8087 program example later in this chapter we show how FPREM is used to reduce a large angle to less than $\pi/4$ so that we can use the 8087 trig functions on it. Exceptions: I, D, U. Example: FPREM

FRNDINT—Round number in ST to an integer. The *round-control* (RC) bits in the control word determine how the number will be rounded. If the RC bits are set for down or chop, a number such as 205.73 will be rounded to 205. If the RC bits are set for up or nearest, 205.73 will be rounded to 206. Exceptions: I, P.

FXTRACT—Separates the exponent and the significand parts of a temporary-real number in ST. After the instruction executes, ST contains a temporary-real representation of the significand of the number and ST(1) contains a temporary-real representation of the exponent of the number. These two could then be written separately out to memory locations. Exception: I.

FABS—Number in ST is replaced by its absolute value. Instruction simply makes sign positive. Exception: I.

FCHS—Complements the sign of the number in ST. Exception: I.

COMPARE INSTRUCTIONS

The compare instructions with COM in their mnemonic compare contents of ST with contents of specified or default source. The source may be another stack element or real number in memory. These compare instructions set the condition code bits C3, C2, and C0 of the status word shown in Figure 11-21*b* as follows:

C3	C2	C0	
0	0	0	ST > source
0	0	0	ST < source
1	0	0	ST = source
1	1	1	numbers cannot be compared

You can transfer the status word to memory with the 8087 FSTSW instruction and then use 8086 instructions to determine the results of the comparison. Here are the different compares.

FCOM //source—Compares ST with real number in another stack element or memory. Exceptions: I, D. Examples:

```
FCOM                      ; Compares ST with ST(1)
FCOM ST(3)                ; Compares ST with ST(3)
FCOM MINIMUM_PAYMENT      ; Compares ST with real
                          ; from memory
```

FCOMP //source—Identical to FCOM except that the stack pointer is incremented by one after the compare operation. Old ST(1) becomes new ST.

FCOMPP—Compare ST with ST(1) and increment stack pointer by 2 after compare. This puts the new ST above the two numbers compared. Exceptions: I, D.

FICOM source—Compares ST to a short or long integer from memory. Exceptions: I, D. Example:
FICOM MAX_ALTITUDE

FICOMP source Identical to FICOM except stack pointer is incremented by one after compare.

FTST—Compares ST with zero. Condition code bits C3, C2, and C0 in the status word are set as shown above if you assume the source in this case is zero. Exceptions: I, D.

FXAM—Tests ST to see if it is zero, infinity, unnormalized, or empty. Sets bits C3, C2, C1, and C0 to indicate result. See Intel data book for coding. Exceptions: None.

TRANSCENDENTAL (TRIGONOMETRIC AND EXPONENTIAL) INSTRUCTIONS

FPTAN—Computes the values for a ratio of Y/X for an angle in ST. The angle must be expressed in radians, and the angle must be in the range of $0 < \text{angle} < \pi/4$.

NOTE: FPTAN does not work correctly for angles of exactly 0 and $\pi/4$. You can convert an angle from degrees to radians by dividing it by 57.295779. An angle greater than $\pi/4$ can be brought into range with the 8087 FPREM instruction. The Y value replaces the angle on the stack, and the X value is pushed on the stack to become the new ST. The values for X and Y are created separately so you can use them to calculate other trig functions for the given angle as we show in an example program later in this chapter. Exceptions: I, P.

FPATAN—Computes the angle whose tangent is Y/X. The X value must be in ST, and the Y value must be in ST(1). Also, X and Y must satisfy the inequality $0 < Y < X < \infty$. The resulting angle expressed in radians replaces Y in the stack. After the operation the stack pointer is incremented so the result is then ST. Exceptions: U, P.

F2XM1—Computes the function $Y = 2^X - 1$ for an X value in ST. The result, Y, replaces X in ST. X must be in the range $0 \leq X \leq 0.5$. To produce 2^X, you can simply add one to the result from this instruction. Using some common equalities you can produce values often needed in engineering and scientific calculations. Here are some examples.

$$10^X = 2^{X(\text{LOG}_2 10)}$$

$$e^X = 2^{X(\text{LOG}_2 e)}$$

$$Y^X = 2^{X(\text{LOG}_2 Y)}$$

FYL2X—Calculates Y times the LOG to the base 2 of X or $Y(\text{LOG}_2 X)$. X must be in the range of $0 < X < \infty$ and Y must be in the range $-\infty < Y < +\infty$. X must initially be in ST and Y must be in ST(1). The result replaces Y and then the stack is popped so that the result is then at ST. This instruction can be used to compute the log of a number in any base, n, using the identity $\text{LOG}_n X = \text{LOG}_n 2(\text{LOG}_2 X)$. For a given n, $\text{LOG}_n 2$ is a constant which can easily be calculated and used as the Y value for the instruction. Exceptions: P

FYL2XP1—Computes the function Y times the LOG to the base 2 of $(X + 1)$ or $Y(\text{LOG}_2(X + 1))$. This instruction is almost identical to FYL2X except that it gives more accurate results when computing the LOG of a number very close to one. Consult the Intel manual for further detail.

INSTRUCTIONS WHICH LOAD CONSTANTS

The following instructions simply push the indicated constant on the stack. Having these commonly used constants available reduces programming effort.

FLDZ —Push 0.0 on stack
FLD1 —Push +1.0 on stack
FLDPI —Push the value of π on stack
FLD2T —Push LOG of 10 to the base 2 on stack ($\text{LOG}_2 10$)
FLDL2E —Push LOG of e to the base 2 on stack ($\text{LOG}_2 e$)
FLDLG2—Push LOG of 2 to the base 10 on stack ($\text{LOG}_{10} 2$)

PROCESSOR CONTROL INSTRUCTIONS

These instructions do not perform computations. They are used to do tasks such as initializing the 8087, enabling interrupts, writing the status word to memory, etc.

Instruction mnemonics with an N as the second character have the same function as those without the N, but they put a NOP in front of the instruction instead of putting a WAIT instruction there.

FINIT/FNINT—Initializes 8087. Disables interrupt output, sets stack pointer to register 7, sets default status.

FDISI/FNDISI—Disables the 8087 interrupt output pin so that it cannot cause an interrupt when an exception (error) occurs.

FENI/FNENI—Enables 8087 interrupt output so it can cause an interrupt when an exception occurs.

FLDCW source—Loads a status word from a named memory location into the 8087 status register. This instruction should be preceded by the FCLEX instruction to prevent a possible exception response if an exception bit in the status word is set.

FSTCW/FNSTCW destination—Copies the 8087 control word to a named memory location where you can determine its current value with 8086 instructions.

FSTSW/FNSTSW destination—Copies the 8087 status word to a named memory location. You can check various status bits with 8086 instructions and base further action on the state of these bits.

FCLEX/FNCLEX—Clears all of the 8087 exception flag bits in the status register. Unasserts BUSY and INT outputs.

FSAVE/FNSAVE destination—Copies the 8087 control word, status word, pointers, and entire register stack to a named, 94-byte area of memory. After copying all of this the FSAVE/FNSAVE instruction initializes the 8087 as if the FINIT/FNINIT instruction had been executed.

FRSTOR source—Copies a 94-byte named area of memory into the 8087 control register, status register, pointer registers, and stack registers.

FSTENV/FNSTENV destination—Copies the 8087 control register, status register, tag words, and exception pointers to a named series of memory locations. This instruction does not copy the 8087 register stack to memory as the FSAVE/FNSAVE instruction does.

FLDENV source—Loads the 8087 control register, status register, tag word, and exception pointers from a named area in memory.

FINCSTP—Increment the 8087 stack pointer by one. If the stack pointer contains 111 and it is incremented, it will point to 000.

FDECSTP—Decrement stack pointer by one. If the stack pointer contains 000 and it is decremented, it will contain 111.

FFREE destination—Changes the tag for the specified destination register to empty. See the Intel manual for a discussion of the tag word which you usually don't need to know about.

FNOP—Performs no operation. Actually copies ST to ST.

FWAIT—This instruction is actually an 8086 instruction which makes the 8086 wait until it receives a not busy signal from the 8087 to its $\overline{\text{TEST}}$ pin. This is done to make sure that neither the 8086 nor the 8087 starts the next instruction before the preceding 8087 instruction is completed.

8087 Example Programs

PYTHAGORAS REVISITED

As you may remember from back there somewhere in geometry, the pythagorean theorem states that the hypotenuse (longest side) of a right triangle squared is equal to the square of one of the other sides plus the square of the remaining side. This is commonly written as $C^2 = A^2 + B^2$. For this example program we want to solve this for the hypotenuse, C, so we take the square root of both sides of the equation to give $C = \sqrt{A^2+B^2}$. Figure 11-24 shows a simple 8087 program you can use to compute the value of C for given values of A and B. We have shown the assembler listing for the program so you can see the actual codes that are generated for the 8087 instructions. Note, for example, that each of the codes for the 8087 instructions here starts with 9BH, the code for the WAIT instruction whose function we explained before.

At the start of the program we set aside some named memory locations to store the values of the three sides of our triangle, the control word we want to send the 8087, and the status word we will read from the 8087 to check for error conditions. Remember, the only way you can pass numbers to and from the 8087 is by using 8087 instructions to read the numbers from memory locations or write the numbers to memory locations. In this section of the example program the statement SIDE_A DD 3.0 tells the assembler to set aside two words in memory to store the value of one of the sides of the triangle. The decimal point in 3.0 tells the assembler that this is a real number. The assembler then produces the short-real representation of 3.0 (40400000) and puts it in the reserved memory locations. Likewise the statement SIDE_B DD 4.0 tells the assembler to set aside two word locations and put the short-real representation of 4.0 in them. The statement HYPOTENUSE DD 0 reserves a double word space for the result of our computation. When the program is finished, these locations will contain 40A00000, the short-real representation for 5.0.

The actual code section of this program you would normally write as a procedure so that you could call it as needed. To make it simple here we have written it as a stand-alone program. We start by initializing the data segment register to point to our data in memory. We then initialize the 8087 with the FINIT instruction. The notations for the control word in Figure 11-21*a* show the default values for each part of the control word after FINIT executes. For most computations these values give the best results. However, just in case you might want to change some of these settings from their default values, we have included the instructions needed to send a new control word to the 8087. You first load the desired control word in a reserved memory location with the MOV CONTROL_WORD, 03FFH instruction, and then load this word into the 8087 with the FLDCW CONTROL_WORD instruction.

To perform the actual computation, we start in the inside of the equation and work our way outward. FLD SIDE_A brings in the value of the first side and pushes it on the 8087 stack. FMUL ST, ST(0) multiplies ST by ST

```
IBM Personal Computer MACRO Assembler   Version 2.00    Page    1-1
                                                                05-21-85
                              page ,132
                              ;8087 NUMERIC DATA PROCESSOR EXAMPLE PROGRAM
                              ;
                              ;ABSTRACT:      This program calculates the hypotenuse of a right
                              ;               triangle, given SIDE A and SIDE B
                              NAME    PYTHAG

0000                          DATA_HERE       SEGMENT WORD    PUBLIC
0000  00 00 40 40                     SIDE_A          DD      3.0     ; set aside space for Side A, short integer
0004  00 00 80 40                     SIDE_B          DD      4.0     ; set aside space for Side B, short integer
0008  00 00 00 00                     HYPOTENUSE      DD      0       ; set aside space for result, short integer
                                                                      ; 5.0 normalized = 40A00000
000C  0000                            CONTROL_WORD    DW      0       ; space for control word
000E  0000                            STATUS_WORD     DW      0       ; space for status  word
0010                          DATA_HERE       ENDS

0000                          CODE_HERE       SEGMENT WORD    PUBLIC
                                      ASSUME CS:CODE_HERE, DS:DATA_HERE
                              ;                                                               ESCAPE CODES
0000  B8 ---- R               START:  MOV     AX, DATA_HERE           ; initialize data segment
0003  8E D8                           MOV     DS, AX
0005  9B DB E3                        FINIT                           ; initialize 8087        ESC 1CH, BX

0008  C7 06 000C R 03FF               MOV     CONTROL_WORD, 03FFH     ; put control word in memory
                                                                      ; so 8087 can access it, round
                                                                      ; to even, mask interrupts
000E  9B D9 2E 000C R                 FLDCW   CONTROL_WORD            ; load control to 8087   ESC 0DH,CONTROL_WORD
0013  9B D9 06 0000 R                 FLD     SIDE_A                  ; put value of SIDE_A on
                                                                      ; stack top              ESC 08H, SIDE_A
0018  9B D8 C8                        FMUL    ST, ST(0)               ; square SIDE_A          ESC 01H, AX
001B  9B D9 06 0004 R                 FLD     SIDE_B                  ; get SIDE_B  stack top  ESC 08H, SIDE_B
0020  9B D8 C8                        FMUL    ST, ST(0)               ; square SIDE_B          ESC 01H, AX
0023  9B D8 C1                        FADD    ST, ST(1)               ; add A squared + B squared ESC 00H, CX
                                                                      ; result at top of stack
0026  9B D9 FA                        FSQRT                           ; take square root of ST
                                                                      ; result in ST           ESC 0FH, DX
0029  9B DD 3E 000E R                 FSTSW   STATUS_WORD             ; copy status word to memory
                                                                      ; where 8086 can access it  ESC 2FH, STATUS_WORD
002E  9B                              FWAIT                           ; wait until status store done
002F  A1 000E R                       MOV     AX, STATUS_WORD         ; bring status to AX to check
                                                                      ; for errors
0032  24 BF                           AND     AL, 0BFH                ; mask unneeded bit
0034  75 05                           JNZ     STOP                    ; handle error if found
0036  9B D9 1E 0008 R                 FSTP    HYPOTENUSE              ; no error, copy result from
                                                                      ; 8087 to memory          ESC 0BH, HYPOTENUSE
003B  90                      STOP:   NOP
003C                          CODE_HERE ENDS
                                      END     START
```

FIGURE 11-24 8087 program to compute the hypotenuse of a right triangle.

and puts the result in ST, so ST now has A^2. Next we bring in SIDE_B and push it on the 8087 stack with the FLD SIDE_B instruction, and square it with the MUL ST, ST(0) instruction. ST now contains B^2 and ST(1) now contains A^2. We add these together and leave the result in ST with the FADD instruction. FSQRT takes the square root of the contents of ST and leaves the results in ST. To see if the result is valid, we copy the 8087 status word to the memory location we reserved for it with the FSTSW STATUS_WORD instruction. We then use 8086 instructions to check the six exception status bits to see if anything went wrong in the square root computation. If there were no exceptions (errors), these status bits will all be 0's, and the program will copy the result from ST to the memory location named HYPOTENUSE using the FSTP HYPOTENUSE instruction. We used the POP form of this instruction so that after this instruction the stack pointer is back at the same register as it was when we started. This makes it easier to keep track of which register is ST if necessary.

For the case where our test found an error had occurred, we could have program execution go to an error handling routine instead of simply to the STOP label as we did for this simple example. We have shown the ESC form for each of the instructions to give you more examples of these.

When you feel comfortable with the preceding example, read through the 8087 instruction descriptions again, and then work your way through the next example which shows you how to use the 8087 for working with trig functions.

AN 8087 TRIG FUNCTION EXAMPLE

As our second example of an 8087 program we have chosen to show you how to compute the tangent of a given angle using the 8087 FPTAN instruction. Figure 11-25*a* shows the standard reference triangle we will be using for this example, and the definitions of sine, cosine, and

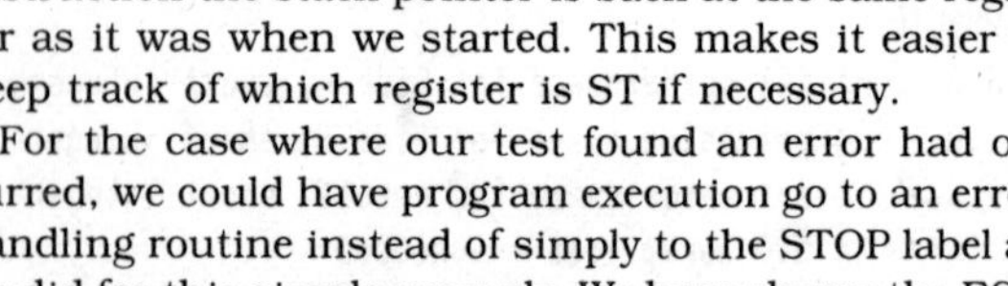

(*a*)

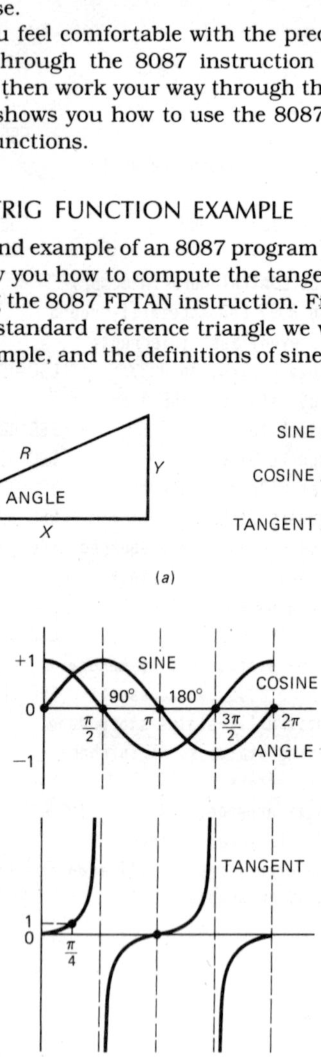

(*b*)

FIGURE 11-25 *(a)* Reference triangle for trigonometry examples. *(b)* Definitions of basic trigonometric functions.

tangent based on this triangle. The 8087 FPTAN instruction does not produce the tangent of a given angle directly, it produces two proportional values which represent Y and X. You then divide the Y value by the X value to produce the actual tangent value. To produce the sine and cosine values for the given angle, you can use these values of X and Y in the relationships shown in Figure 11-25*a*. Here are a few more general comments before we get into the actual program example.

To simplify the internal design of the 8087, the FPTAN instruction was designed to work only with an angle in the range $0 < \text{angle} < \pi/4$ radians (0 to 45°). Figure 11-25*b* shows graphically the values for the sine, cosine, and tangent for angles from 0 to 360°. As you can see there, the values for each function are repetitive. If you compute the values for these functions for the first octant (0–45° or 0–$\pi/4$ radians), you can determine their values for any other octant by using various simple trigonometric equations. The values for the sine of an angle in octant 4 (180 to 225°), for example, are the same as those in octant 0 except that their sign is changed. As another example, the tangent of an angle in octant 1 is equal to 1/(tan (90° − the angle)). The point of all of this is that before you can use the FPTAN instruction to find the Y and X values for an angle, you have to reduce the angle to within the range of 0–$\pi/4$ radians. The 8087 FPREM instruction is used to do this.

The FPREM instruction subtracts $\pi/4$ from ST over and over until the result in ST is less than $\pi/4$. After FPREM executes, ST contains the remainder from these subtractions, and bits C0, C3, and C1 of the status word contain a 3-bit binary number which indicates the number of times $\pi/4$ had to be subtracted from the original angle. The number in these status bits then tells you what octant the original angle was in. You use this octant information to tell you which formula you use to compute the desired value for the angle. For example, suppose that you are computing the sine of an angle, and the status bits tell you that the angle is in octant 4. You then know that you have to change the sign of the result produced by $Y/\sqrt{X^2+Y^2}$ to get the correct sine for the angle, because as shown in Figure 11-25*b*, sines in octant 4 are negative.

Finally, you have to check if the remainder left in ST after the FPREM instruction is zero. You need to do this because the FPTAN instruction will not work for a value of exactly 0 in ST. Now, let's look at our example program in Figure 11-26.

In order to keep this example short so you don't get lost in trigonometric manipulations, we require that the input angle be in the range of 0 to $\pi/2$ radians (0–90°). If you need to expand this example to work with other angles, use the identities from a standard trig text to help you.

In the data segment of the program in Figure 11-26 we set aside memory spaces for the status word and for the values we want to calculate. Since we set aside double word spaces with the DD directive, the 8087 assembler will convert results to short-real format before writing them to these spaces. We also declare here the value of $\pi/4$ which will be used by the FPREM instruction. We could have calculated this value with the 8087, but we

```
IBM Personal Computer MACRO Assembler   Version 2.00    Page    1-1
                                                                05-21-85
                                  page ,132
                                  ;8087 PROGRAM SECTION
                                  ;ABSTRACT       :This program section uses an 8087 to compute the tangent
                                  ;                of an angle in the range of 0<angle<pi/2 radians.
                                  ;                The angle, expressed in radians, is assumed to be in a
                                  ;                memory location named ANGLE.
                                  ;                The program also produces values for Y and X,
                                  ;                the opposite and adjacent sides of a right
                                  ;                triangle with the given angle. These Y and X values can
                                  ;                be used to compute the SINE and COSINE of the given angle.

0000                              DATA_HERE SEGMENT WORD
0000  0000                               STATUS     DW 0                          ; Reserve space for status word
0002  35 C2 68 21 A2 DA                  PI_OVER_4  DT 3FFEC90FDAA22168C235R      ; Temporary real form for Pi/4
      0F C9 FE 3F
000C  00 00 00 3F                        ANGLE      DD 0.5                        ; Angle to be processed, 0.5 Radians
0010  00 00 00 00                        X          DD 0                          ; Space for X value for angle
0014  00 00 00 00                        Y          DD 0                          ; Space for Y value for angle
                                                                                  ; Normalized results X=3FBC7A43 Y=3F4DEE7A
0018  00 00 00 00                        TANGENT    DD 0                          ; Space for Tangent of angle 0.5 radians
                                                                                  ; Tan 0.5r = 0.546; normalized = 3F0BDA7B
001C  00 00 00 00                        SINE       DD 0                          ; Space for SINE of angle
0020  00 00 00 00                        COSINE     DD 0                          ; Space for COSINE of angle
0024                              DATA_HERE ENDS

0000                              CODE_HERE SEGMENT WORD
                                         ASSUME CS:CODE_HERE, DS:DATA_HERE

0000  B8 ---- R                   START: MOV AX, DATA_HERE         ; Initialize data segment register
0003  8E D8                              MOV DS, AX
0005  9B DB E3                           FINIT                     ; Initialize 8087
0008  9B DB 2E 0002 R                    FLD     PI_OVER_4         ; PI/4 to ST from memory
000D  9B D9 06 000C R                    FLD     ANGLE             ; ANGLE to ST, PI/4 in ST(1)
0012  9B D9 F8                           FPREM                     ; Reduce angle to range 0 to pi/4,
                                                                   ; result in ST
0015  9B DD 3E 0000 R                    FSTSW    STATUS           ; Check status to determine octant
001A  9B                                 FWAIT
001B  A1 0000 R                          MOV  AX, STATUS           ; Bring to register to check
001E  F6 C4 02                           TEST AH, 00000010B        ; Bit 1 = 0 if octant 0 (0 - pi/4)
                                                                   ; Bit 1 = 1 if octant 1 (pi/4-pi/2)
0021  74 2A                              JZ       OCTANT_0         ; If octant 0, go check for 0, do Tangent
0023  9B D9 E4                           FTST                      ; Else, Test for result from FREM = 0.
                                                                   ; FPTAN will not work with value of 0
0026  9B DD 3E 0000 R                    FSTSW    STATUS           ; Status word to memory
002B  9B                                 FWAIT
002C  A1 0000 R                          MOV  AX, STATUS           ; Copy to register to check
002F  F6 C4 40                           TEST AH, 01000000B        ; Bit 6 = 1 if result from FTST = 0
0032  75 5B                              JNZ  ANGLE_45             ; Go load values for ANGLE = 45 degrees
0034  9B DE E9                           FSUB                      ; Else subtract angle from pi/4 for octant 1
0037  9B D9 F2                           FPTAN                     ; X value put in ST, Y value put in ST(1)
003A  9B D9 16 0010 R                    FST X                     ; Copy X value to memory for SIN and COS
```

```
003F  9B D9 C9                    FXCH                  ; Swap ST and ST(1) to get Y in ST
0042  9B D9 16 0014 R             FST Y                 ; Copy Y value to memory for SIN and COS
0047  9B DE F9                    FDIV                  ; Divide X value in ST(1) by Y value in ST
                                                        ; To give Tangent for octant 1 angle in ST
004A  EB 67 90                    JMP STORE_TAN
004D                       OCTANT_0:
004D  9B D9 E4                    FTST                  ; Test for result from FREM = 0. FPTAN
                                                        ; will not work with value of 0
0050  9B DD 3E 0000 R             FSTSW    STATUS       ; Status word to memory
0055  9B                          FWAIT
0056  A1 0000 R                   MOV  AX, STATUS       ; Copy to register to check
0059  F6 C4 40                    TEST AH, 01000000B    ; Bit 6 = 1 if result from FTST = 0
005C  75 19                       JNZ  ANGLE_0          ; Go load values for ANGLE = 0
005E  9B D9 F2                    FPTAN                 ; X value put in ST, Y value put in ST(1)
0061  9B D9 16 0010 R             FST X                 ; Copy X value to memory for SIN and COS
0066  9B D9 C9                    FXCH                  ; Swap ST and ST(1) to get Y in ST
0069  9B D9 16 0014 R             FST Y                 ; Copy Y value to memory for SIN and COS
006E  9B D9 C9                    FXCH                  ; Put X back in ST and Y in ST(1)
0071  9B DE F9                    FDIV                  ; Divide Y value in ST(1) by X value in ST
                                                        ; to give Tangent for octant 0 angle in ST
0074  EB 3D 90                    JMP STORE_TAN
0077                       ANGLE_0:
0077  9B D9 E8                    FLD1                  ; Load one in ST
007A  9B D9 1E 0020 R             FSTP COSINE           ; Copy to memory reserved for COSINE
007F  9B D9 EE                    FLDZ                  ; Load zero in ST
0082  9B D9 16 001C R             FST  SINE             ; Copy to memory reserved for SINE
0087  9B D9 16 0018 R             FST  TANGENT          ; Copy to memory reserved for TANGENT
008C  EB 2A 90                    JMP DONE
008F                       ANGLE_45:
008F  9B D9 E8                    FLD1                  ; Load 1 in ST
0092  9B D9 16 0018 R             FST  TANGENT          ; Tangent of 45 degrees is 1
0097  9B D9 E8                    FLD1                  ; ST and ST(1) now 1
009A  9B DE C1                    FADD                  ; ST now = 2
009D  9B DD D1                    FST ST(1)             ; ST(1) now = 2
00A0  9B D9 FA                    FSQRT                 ; ST now = square root of 2
00A3  9B D8 F1                    FDIV  ST, ST(1)       ; ST now = (square root 2)/2 = sin 45
00A6  9B D9 16 001C R             FST  SINE             ; Copy to memory reserved for SINE
00AB  9B D9 16 0020 R             FST  COSINE           ; Copy to memory reserved for COSINE
00B0  EB 06 90                    JMP  DONE
00B3                       STORE_TAN:
00B3  9B D9 16 0018 R             FST TANGENT           ; Copy computed Tangent value to memory
00B8                       DONE:
00B8  90                          NOP
00B9  90                          NOP
00BA  90                          NOP
00BB                       CODE_HERE  ENDS
                                         END START
```

FIGURE 11-26 8087 program to compute the tangent of an angle.

did it this way to show you how temporary-real numbers can be declared with a DT directive.

In the code segment of the program we first initialize the data segment register and the 8087. We then load $\pi/4$ and the angle we want to compute the values for, reduce the angle to within range with the FPREM instruction as described before, and check the status word to determine what octant the angle is in. Remember from our previous example program that you check the 8087 status word by first having the 8087 write the status word to a memory location with the FSTSW STATUS instruction. You then read this status word and check it with 8086 instructions. Since we have limited the input angle to be in either octant 0 or octant 1, we only have to check bit C1 of the status word. To do this we copy the status word to a register with the MOV AX, STATUS instruction, and check bit C1 with the TEST AH, 00000010B instruction. The 8086 TEST instruction, remember, ANDs the specified source and destination. Flags are affected, but the operands are not.

If the result of the TEST was 0, we know the angle is in octant 0, so we jump to the program section which deals with octant 0 angles. If the result of the TEST was not 0, then we know the angle is in octant 1, and process it accordingly. Let's start with how we handle an angle in octant 0.

For an octant 0 angle we first use the 8087 FTST instruction to check if the result left in ST by the FPREM instruction is zero. FTST will not change the contents of ST, but it will affect the flags in the 8087 status register. With the FSTSW STATUS instruction we then copy the 8087 status register out to memory where we can get at it. Bit 6 of the upper byte of the status word will be a 1 after FTST if the value in ST is 0. If we find that ST is zero for an octant 0 angle, we know that the angle is zero. The sine, cosine, and tangent values for 0 are known constants. Therefore, for this case we simply load these values one at a time into the 8087 ST register, and copy them to the appropriate memory locations. The cosine of 0, for example, is 1, so we simply load 1 into ST with the FLD1 instruction and copy this value to memory with the FST COSINE instruction. If we find that the result of FPREM in ST is not 0, we go and compute its tangent ratio with FPTAN.

FPTAN, remember, does not give the value for tangent directly, it gives two numbers whose ratio is equal to the tangent of the angle. We call these two values *Y* and *X* as shown in Figure 11-25*a*. This form of result is actually more useful than having just a tangent result, because you can use the *X* and *Y* values to compute the sine and cosine of the angle as shown. To save the *X* and *Y* values for these computations, we first copy *X* from ST to memory with the FST X instruction. We then exchange ST and ST(1) with the FXCH instruction. The *Y* value is now at ST. We copy *Y* to memory with the FST Y instruction, and then put *X* and *Y* back in their original position with another FXCH instruction. To get the actual tangent value we divide the *Y* in ST(1) by the *X* in ST with the FDIV instruction. The result of the division is left in ST where we can copy it to memory with an FST TANGENT instruction. Now let's look at how we handle an octant 1 angle.

Again for octant 1, the first thing we have to do is to check if the result left in ST by the FPREM instruction is 0. If it is, we know that the angle is $\pi/2$ or 45°. Again, the sine, cosine, and tangent for this angle are known constants which we can just calculate and load into the reserved memory locations. If the value left in ST by FPREM is not 0, we go on and compute the tangent ratio. The computation in this case is a little less direct.

The formula we use to compute the tangent of an angle in octant 1 is: tangent angle $= 1/(\text{tangent } ((\pi/4) - R)$, where *R* is the remainder left in ST after FPREM executes. This looks messy, but it is really quite simple because of the way the 8087 works. Since at this point in our calculations *R* is in ST and $\pi/4$ is in ST(1), we can do the needed subtraction with the FSUB instruction. The result of this subtraction will be in ST. We then use the FPTAN instruction to find the *Y* and *X* values for this result. After storing the *X* and *Y* values as we do for octant 0, we leave out the final FXCH instruction that we use there. This leaves the *X* value in ST(1) and the *Y* value in ST. Therefore, when we do the FDIV instruction, we really are dividing the *X* value by the *Y* value. This gives the reciprocal we need to satisfy our equation. The tangent value for the octant 1 angle will then be left in ST where we can copy it to memory with an FST TANGENT instruction. For your reference we have shown the short-real values produced by this program for an angle of 0.5 radians.

Hopefully, by now you have some understanding of how the 8087 works and can write some programs of your own. To debug 8087 programs, you may find it helpful to insert extra store instructions to copy intermediate results out to reserved memory locations where you can get at them to see if they are correct. The extra store instructions can be removed when the program works. If you are going to be writing programs where you need extensive 8087 sections, there are several commercially available 8087 software packages. These packages contain 8087 routines which you can simply link into your assembly or higher level language programs.

MULTIPLE BUS MICROCOMPUTER SYSTEMS

In a previous section of this chapter we discussed how a coprocessor can be connected on the local buses of a microcomputer to increase its computing power. In another section of this chapter we discussed how several different function boards can be connected on the local bus of the IBM PC to customize and extend its capabilities. The amount that you can add to a system using this approach, however, is limited. In both of these cases the coprocessor or peripheral boards borrow the local buses from the main microprocessor on a DMA basis as needed. When the coprocessor or peripheral board is using the buses, the main microprocessor is just sitting in a hold state doing nothing. Therefore, if you add too many DMA operations, you soon start to slow down the main processor. The thought might occur to you to simply add another main processor, another 8086 for example, on the local buses to increase the computing power. Since only one main processor on the local buses could be active at a time, this does not gain anything for you. There are several ways, however, that two or more standard microprocessors can be used in a system.

One major approach is to set up each microprocessor as essentially a stand-alone microcomputer with its own RAM, ROM, ports, and local buses. These separate microcomputer boards then communicate with each other and with shared resources such as a hard disk drive using a separate system bus. Figure 11-27 shows a block diagram of this type of multiprocessor system.

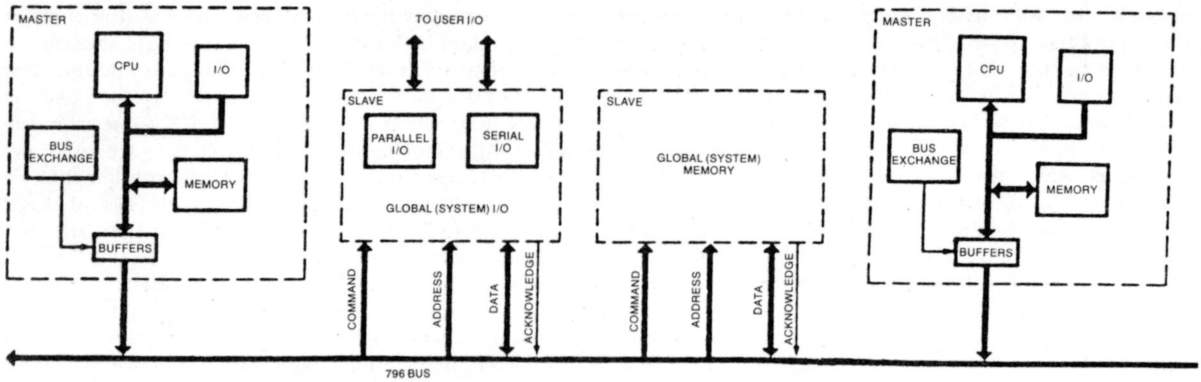

FIGURE 11-27 Multiprocessor bused system. (Intel Corporation)

The principle here is that each processor board can operate independently, fetching and executing its own instructions from its on-board memory, until it needs to access some shared resource such as the system memory, printer, or I/O board. This type of system is often referred to as a *loosely coupled* system. Each board which can take over and use the bus is called a *bus master*. A board which can only be written to or read from is called a *bus slave*.

A question that may occur to you at this point is, "What happens if two or more masters on the system bus try to use the bus at the same time?" The answer to this question is that the system must contain logic which in some way "arbitrates" the dispute and makes sure only one master at a time asserts its control signals on the bus. Later we will show you in detail some ways in which this is done. For now we will start with an overview of a commonly used system bus so you have an idea about how it operates.

The Intel Multibus—IEEE 796 Bus Standard

There are many different microcomputer system buses currently in use. Almost every microprocessor and microcomputer manufacturer has its own bus standard, so it is not possible here to give even a survey of the different ones. We have chosen to discuss the Intel Multibus because several hundred companies make several thousand different boards using this standard, and because it fits in with the devices and systems we are working with in the rest of this book. This bus standard was developed and evolved by Intel and later adopted by IEEE as Bus Standard 796.

The basic IEEE 796 standard defines the signals for two printed circuit board edge connectors, an 86-pin connector referred to as P1 and a 60-pin connector referred to as P2. Figure 11-28 shows the signal names and numbers for these connectors. Figure 11-29 shows a block diagram for part of a single-board computer that might be used as a master on the Multibus. Keep copies of these two figures handy to refer to as we discuss how all of this works.

Whenever you are confronted with a long list of pin names such as that in Figure 11-28, start with the easily recognizable groups of signal lines and work your way down to the toughies. For the Multibus 86-pin P1 connector, start with the power supply and ground connections on pins 1–12 and 75–86. This knocks off 24 pins. Next check out the 16 data lines labeled DATA*–DAT1* on pins 59–74. The * after each of these signal names indicates that these signals are active low (inverted) on the Multibus. Note in Figure 11-29 that there are three sets of buffers on the data lines. Device *A*44, on the left side, buffers the local data bus for the 8-bit peripheral devices. Buffers *A*60 and *A*61 in the right center of the diagram buffer the local data bus for the dual-port RAM. Two bidirectional inverting buffers, *A*69 and *A*89, are used to interface these local RAM data lines to the Multibus. Due to a lack of space, the on-board ROM, ROM data-bus buffers, and ROM decoder are not shown in Figure 11-29.

Next look for the address lines on the Multibus connectors. Sixteen address lines, ADRE*–ADR1*, are on pins 43–58 of the P1 connector. Four more address lines, AD10*–AD13*, are on pins 28, 30, 32, and 34. Four more address lines, ADR14*–ADR17*, are bused on pins 55–58 of the P2 connector. These 24 address lines make it possible for the bus to address up to 16 Mbytes of memory. For many systems we use simple inverting buffers to interface the on-board local bus to these address lines. In some systems, however, we want both the on-board CPU and another master on the bus to be able to access on-board RAM. In this case we use bidirectional buffers on the address lines so that addresses for RAM can be sent from the board to the Multibus, or come from the Multibus to the on-board memory. RAM which is accessible from two directions in this way is called *dual-port RAM*. Note in Figure 11-29 that devices *A*42 and *A*58 are used to buffer the local address bus to the RAM. Bidirectional devices *A*86, *A*87, and *A*88 are used to buffer address lines to and from the Multibus.

Next observe the eight interrupt request lines, INT0*–INT7*, on pins 35–42 of the P1 connector. These lines can be routed to the inputs of a processor or an 8259A on one of the master boards so that a slave board or another master board down the bus can interrupt that master when it needs attention. This is indicated by the wirewrap interrupt jumper matrix in the lower right corner of Figure 11-29.

P1

	Pin	(Component Side)		Pin	(Circuit Side)	
		Mnemonic	Description		Mnemonic	Description
Power Supplies	1	GND	Signal GND	2	GND	Signal GND
	3	+5V	+5Vdc	4	+5V	+5Vdc
	5	+5V	+5Vdc	6	+5V	+5Vdc
	7	+12V	+12Vdc	8	+12V	+12Vdc
	9		Reserved, bussed	10		Reserved, bussed
	11	GND	Signal GND	12	GND	Signal GND
Bus Controls	13	BCLK*	Bus Clock	14	INIT*	Initialize
	15	BPRN*	Bus Pri. In	16	BPRO*	Bus Pri. Out
	17	BUSY*	Bus Busy	18	BREQ*	Bus Request
	19	MRDC*	Mem Read Cmd	20	MWTC*	Mem Write Cmd
	21	IORC*	I/O Read Cmd	22	IOWC*	I/O Write Cmd
	23	XACK*	XFER Acknowledge	24	INH1*	Inhibit 1 (disable RAM)
Bus	25	LOCK*	Lock	26	INH2*	Inhibit 2 (disable PROM or ROM)
Controls and Address	27	BHEN*	Byte High Enable	28	AD10*	
	29	CBRQ*	Common Bus Request	30	AD11*	Address
	31	CCLK*	Constant CLK	32	AD12*	Bus
	33	INTA*	Intr Acknowledge	34	AD13*	
Interrupts	35	INT6*	Parallel	36	INT7*	Parallel
	37	INT4*	Interrupt	38	INT5*	Interrupt
	39	INT2*	Requests	40	INT3*	Requests
	41	INT0*		42	INT1*	
Address	43	ADRE*		44	ADRF*	
	45	ADRC*		46	ADRD*	
	47	ADRA*	Address	48	ADRB*	Address
	49	ADR8*	Bus	50	ADR9*	Bus
	51	ADR6*		52	ADR7*	
	53	ADR4*		54	ADR5*	
	55	ADR2*		56	ADR3*	
	57	ADR0*		58	ADR1*	
Data	59	DATA*		60	DATF*	
	61	DATC*		62	DATD*	
	63	DATA*	Data	64	DATB*	Data
	65	DAT8*	Bus	66	DAT9*	Bus
	67	DAT6*		68	DAT7*	
	69	DAT4*		70	DAT5*	
	71	DAT2*		72	DAT3*	
	73	DAT0*		74	DAT1*	

	Pin	(Component Side)		Pin	(Circuit Side)	
		Mnemonic	Description		Mnemonic	Description
Power Supplies	75	GND	Signal GND	76	GND	Signal GND
	77		Reserved, Bussed	78		Reserved, bussed
	79	−12V	−12Vdc	80	−12V	−12Vdc
	81	+5V	+5Vdc	82	+5V	+5Vdc
	83	+5V	+5Vdc	84	+5V	+5Vdc
	85	GND	Signal GND	86	GND	Signal GND

P2

	Pin	(Component Side)		Pin	(Circuit Side)	
		Mnemonic	Description		Mnemonic	Description
	1		Reserved, Not Bussed	2		Reserved, Not Bussed
	3		Reserved, Not Bussed	4		Reserved, Not Bussed
	5		Reserved, Not Bussed	6		Reserved, Not Bussed
	7		Reserved, Not Bussed	8		Reserved, Not Bussed
	9		Reserved, Not Bussed	10		Reserved, Not Bussed
	11		Reserved, Not Bussed	12		Reserved, Not Bussed
	13		Reserved, Not Bussed	14		Reserved, Not Bussed
	15		Reserved, Not Bussed	16		Reserved, Not Bussed
	17		Reserved, Not Bussed	18		Reserved, Not Bussed
	19		Reserved, Not Bussed	20		Reserved, Not Bussed
	21		Reserved, Not Bussed	22		Reserved, Not Bussed
	23		Reserved, Not Bussed	24		Reserved, Not Bussed
	25		Reserved, Not Bussed	26		Reserved, Not Bussed
	27		Reserved, Not Bussed	28		Reserved, Not Bussed
	29		Reserved, Not Bussed	30		Reserved, Not Bussed
	31		Reserved, Not Bussed	32		Reserved, Not Bussed
	33		Reserved, Not Bussed	34		Reserved, Not Bussed
	35		Reserved, Not Bussed	36		Reserved, Not Bussed
	37		Reserved, Not Bussed	38		Reserved, Not Bussed
	39		Reserved, Not Bussed	40		Reserved, Not Bussed
	41		Reserved, Bussed	42		Reserved, Bussed
	43		Reserved, Bussed	44		Reserved, Bussed
	45		Reserved, Bussed	46		Reserved, Bussed
	47		Reserved, Bussed	48		Reserved, Bussed
	49		Reserved, Bussed	50		Reserved, Bussed
	51		Reserved, Bussed	52		Reserved, Bussed
	53		Reserved, Bussed	54		Reserved, Bussed
Address	55	ADR16*	Address Bus	56	ADR17*	Address Bus
	57	ADR14*		58	ADR15*	
	59		Reserved, Bussed	60		Reserved, Bussed

All Reserved Pins are reserved for future use and should not be used if upwards compatibility is desired.

FIGURE 11-28 Pin assignments for bus signals on IEEE 796 (Multibus) P1 and P2 connectors. (Intel Corporation)

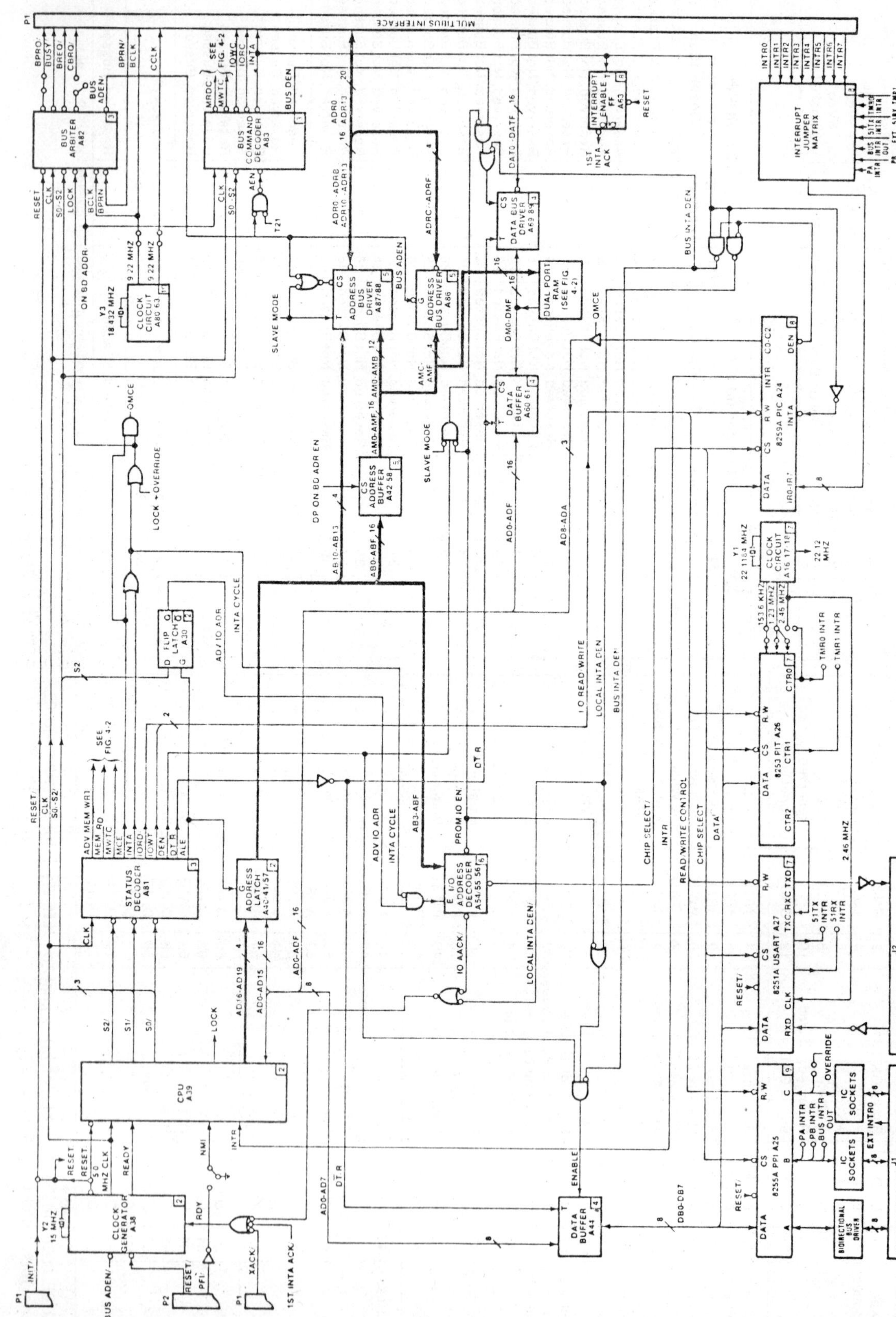

FIGURE 11-29 Block diagram of circuitry on a typical Multibus master board.

Now that we have found the address and data bus connections, let's see how the control-bus read and write signals are produced. The 8086 here is operating in MAX mode so device *A*81, an 8288 controller, is used to produce the control-bus signals for on-board operations. Another 8288, device *A*83 on the right side of the diagram, is used to produce the control-bus signals needed when this board writes to or reads from another board on the Multibus. The control signals produced by this device connect to Multibus pins 19–22 and 33. The transfer of data from, for example, a memory board on the bus to a master is nearly identical to a transfer from local memory to the CPU. The master outputs the desired address on the Multibus, and the controller outputs a memory-read command, MRDC*. The addressed location on the memory board is enabled, and puts its data on the Multibus to be returned to the CPU on the master board which has control of the bus.

The last major group of signals on the Multibus are those which transfer control of the Multibus from one master to another. The signals in this group are BCLK*, BPRN*, BPRO*, BUSY*, BREQ*, and CBRQ*. In our example system in Figure 11-29, these signals are for the most part produced and interpreted by an 8289 bus arbiter, device *A*82, in the upper right corner. In a later section we show you how these signals interact when two masters exchange control of the bus. For now we will make some comments about the few remaining signals on the bus.

The INIT* signal on pin 14 can be used to reset all of the master and slave boards on the bus. It is usually driven by circuitry on the highest priority master. The INTA* signal is produced by the 8288 bus-controller device in response to an interrupt request. An I/O board on the bus, for example, might send an interrupt request on the bus to our master board in Figure 11-29. The master board, when ready, might assert INTA* on the bus to tell an 8259A on the I/O board to send back the desired interrupt type to the master board on the system data bus. The inhibit signals, INH1* and INH2*, on pins 24 and 26, can be used to disable a block of memory in the system. You might, for example, want to have ROM in a particular address space when the system is first turned on, and later have RAM in that address space. These signals can be used to do this. Finally, note the transfer-acknowledge signal, XACK*, on pin 23 of the Multibus. In our example system in Figure 11-29 this signal is connected to the READY input logic of the 8284 in the upper left corner. This connection allows an addressed memory or peripheral on the bus to make the 8086 insert WAIT states until it has accepted or output valid data. Now, we will show you how a master can gain control of the bus.

Arbitrating and Transferring Bus Control

When two or more masters share a bus such as the Multibus, some mechanism must be used to settle the argument when two masters want to use the bus at the same time. The Multibus can use either a serial priority scheme or a parallel priority scheme to do this.

Figure 11-30*a* shows how three masters are connected for the serial priority arbitration scheme. The key here is that, in order for a master board to take over the bus, its bus priority input, BPRN/, must be asserted low. The highest priority master has its BPRN/ input tied to ground, so it can take over the Multibus anytime it needs to. If the highest priority master does not need to use the bus, it will assert its bus priority output, BPRO/, low. This will assert the BPRN/ of the next lower priority master so it can take over the bus if it needs to. If the second priority master does not need to use the bus, it asserts its BPRO/ output low. This enables the lowest priority master to use the bus if it needs to. If a low priority master is using the bus and a higher priority master needs to use the bus, the lower priority master will be allowed to finish transferring its current byte or word, and then the higher priority master will take control.

Figure 11-30*b* shows the connections for a parallel priority scheme. This scheme uses the bus request signals, BREQ/, from each master, and the BPRN/ signals to each master. Here again, the BPRN/ input of a master must be asserted low in order for that master to be able to take control of the Multibus. Here's how this scheme works. When a master needs to use the bus to transfer some data, it asserts its BREQ/ signal. This signal, along with bus-request signals from other masters, goes into the inputs of a priority encoder. The priority encoder will output a 3-bit code which represents the highest numbered input that is asserted low. The 3-bit code from the 74LS148 is connected to the select inputs of a 74LS138 one-of-eight-low decoder. The result of all of this is that the 74LS138 will assert the BPRN/ input of the highest priority master requesting service. When this master finishes its data transfer, it raises its BREQ/ signal high. The 74LS138 will then assert the BPRN/ input of the next highest priority master requesting use of the bus. Now we will show you how bus control gets transferred from one master to another in an orderly manner.

Figure 11-31 shows the signal waveforms for transfer of control of the Multibus from a lower priority master (*A*) to a higher priority master (*B*). Keep a copy of Figure 11-29 handy as we work through these waveforms with you.

The bus-control transfer process starts when the CPU on the higher priority master, *B*, outputs an address which is not decoded on that board. This off-board address causes a signal labeled ON BD ADDR/ to be sent to the system-bus-request input of the 8289 bus arbiter, telling it to attempt to take over the bus. Refer to the top right corner of Figure 11-29 to see how these signals are connected. On the waveforms in Figure 11-31 this signal is referred to as TRANSFER REQUEST/. In response to this request, the 8289 asserts its BREQ/ output low. Assuming we are using a parallel priority scheme as described before, the priority encoder-decoder will then unassert the BPRN/ input of master *A* and assert the BPRN/ input of master *B* as shown in the waveforms.

While master *A* is using the bus for a data transfer, it holds the BUSY/ line on the Multibus low. The BUSY/ line is an open-collector line which any master can pull low when it is using the bus. No other master can actually

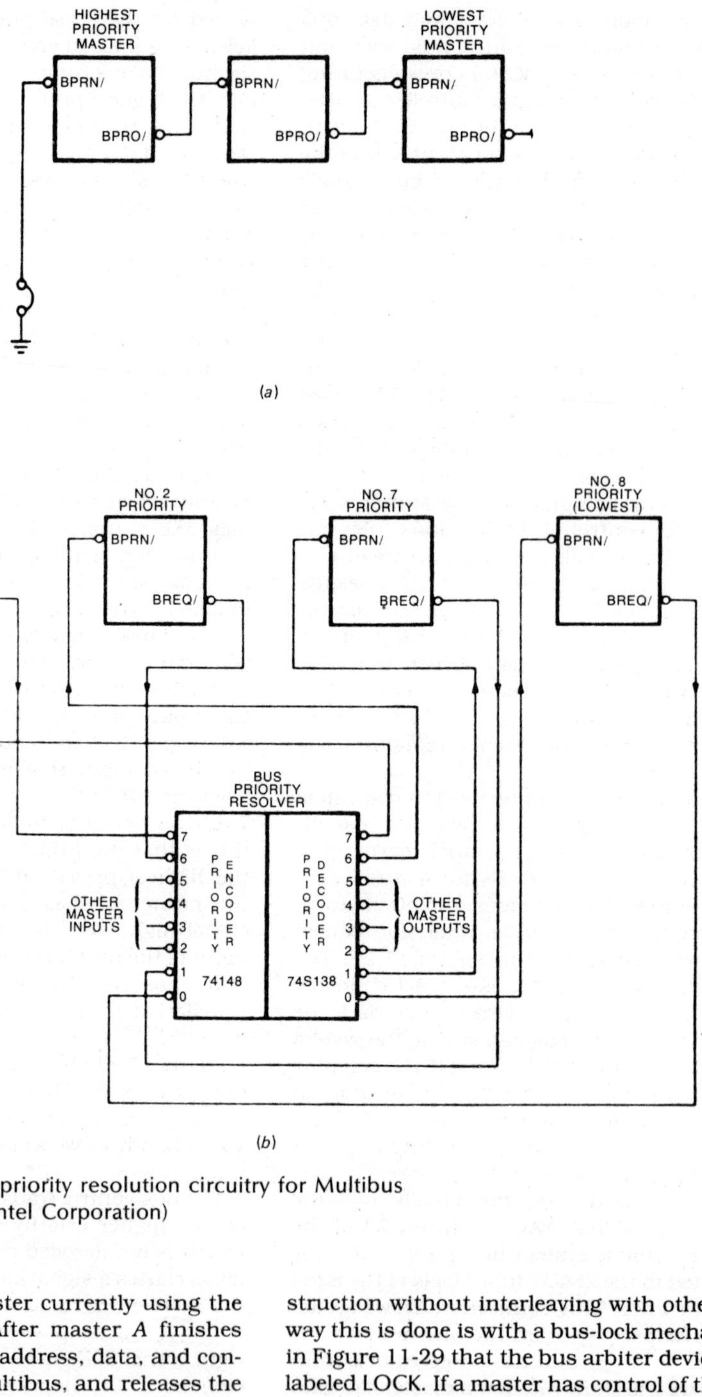

FIGURE 11-30 Serial and parallel priority resolution circuitry for Multibus systems *(a)* Serial. *(b)* Parallel. (Intel Corporation)

take over the bus until the master currently using the bus releases the BUSY/ line. After master *A* finishes transferring data it disables its address, data, and control buffers connected to the Multibus, and releases the BUSY/ line. Master *B* then pulls the BUSY/ line low to let other masters know that the bus is in use. Master *B* next enables its buffers to output address and control signals on the Multibus for its data transfer. In normal operation a master releases the bus after each byte or word data transfer so that a high priority master cannot prevent lower priority masters from ever having a turn.

In some cases, however, we want a master to be able to transfer several bytes or words of data needed for an instruction without interleaving with other masters. The way this is done is with a bus-lock mechanism. Observe in Figure 11-29 that the bus arbiter device has an input labeled LOCK. If a master has control of the bus, the bus arbiter will hold the BUSY/ line on the Multibus low as long as the LOCK input is asserted. As we described above, this prevents any other masters from taking a turn on the bus. The LOCK signal for the 8289 bus arbiter can come from one of several sources on the board, but the main source we are concerned with here is the LOCK output of the 8086.

Note in Figure 11-29 that, when the 8086 is set to operate in maximum mode, pin 29 outputs a signal

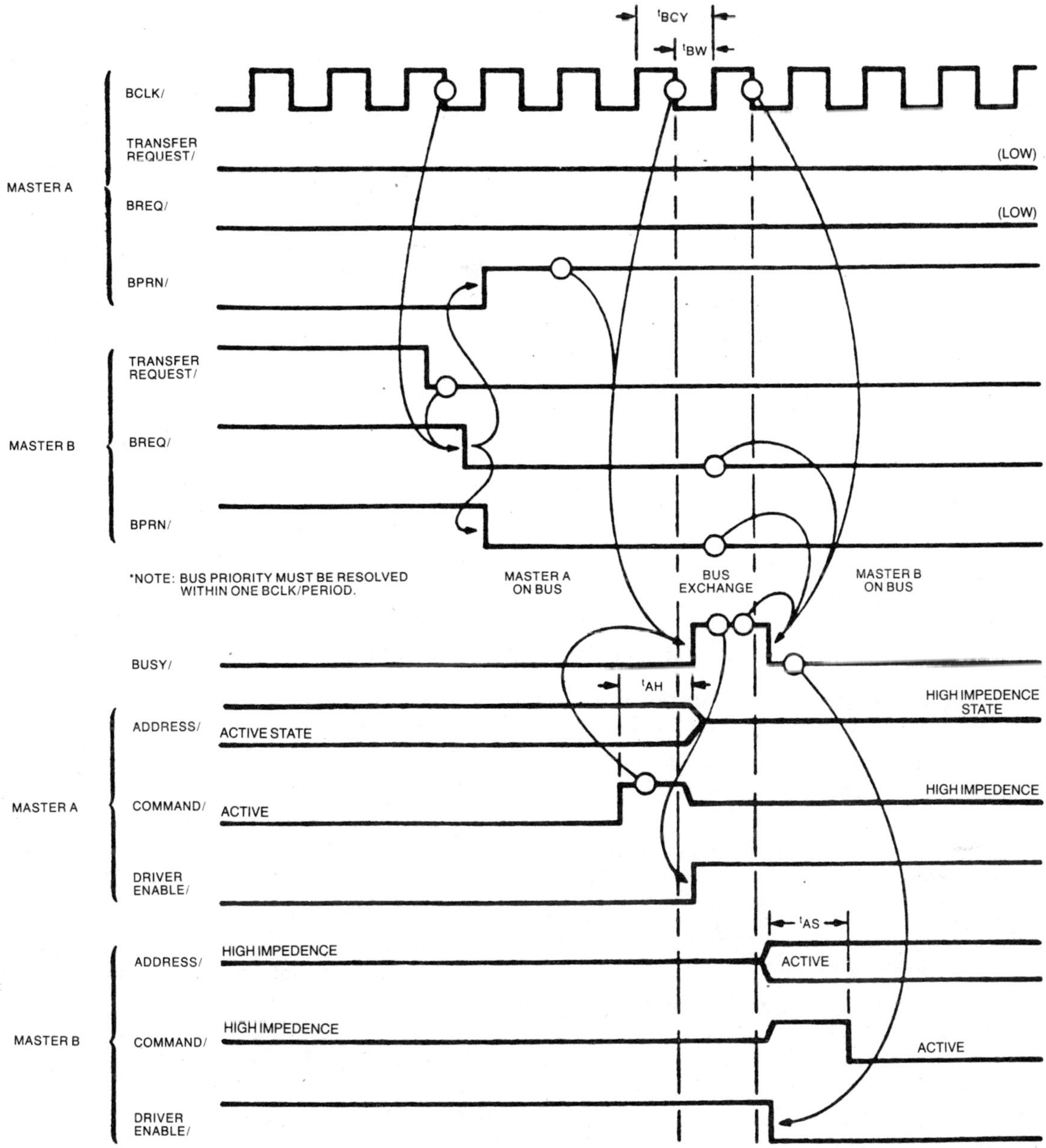

FIGURE 11-31 Signal waveforms for transfer of control of Multibus from a lower priority master to a higher priority master. (Intel Corporation)

called LOCK. The 8086 will assert this signal when it executes an instruction which has a LOCK prefix in front of it. An example of this type of instruction is LOCK XCHG AL, FLAG. When this instruction is assembled, the code for the LOCK prefix, 111000, will be put in before the code for the XCHG instruction. The XCHG instruction takes two bus cycles, one to read in the byte from the memory location named FLAG on a shared memory board, and another to copy AL to the memory location. Without the LOCK mechanism, another master on the Multibus might take over the bus between the two operations and read the old value of FLAG, rather than the new value that you are trying to put there with the XCHG AL, FLAG instruction.

This section has shown you how several processor boards can share a common bus in a harmonious manner. In Chapter 13 we show you how several complete computers can be networked together to communicate and share a common data base.

IMPORTANT TERMS AND CONCEPTS FROM THIS CHAPTER

8086 maximum mode

DMA
- DMA channel
- 8237 DMA controller

Motherboard and system expansion slots

DRAM
- Refresh—burst and distributed modes
- $\overline{RAS}$ and $\overline{CAS}$ strobes

8208 DRAM controller IC

Error detecting and correcting
- Hard and soft errors
- Parity check
- Hamming codes
- Syndrome word

80186/80188—integrated peripherals

Tightly coupled multiprocessor system

8087 math coprocessor
- data types and terms
- Word, short, and long integers
- Packed decimals
- Short-, long-, and temporary-reals
- Fixed-point numbers
- Floating-point numbers
- Normalizing
- Significand, mantissa, exponent
- Single- and double-precision representation
- Offset form

8087 control and status words

Multiple bus microprocessor systems
- loosely coupled system
- Intel Multibus—796 bus standard
- Bus master and bus slave

Dual-port RAM

Serial and parallel bus arbitration schemes

REVIEW QUESTIONS AND PROBLEMS

1. Describe how the control-bus signals are produced for an 8086 system operating in maximum mode.
2. Why is DMA data transfer faster than doing the same data transfer with program instructions?
3. Describe the series of actions that a DMA controller will do after it receives a request from a peripheral device to transfer data from the peripheral device to memory.
4. Describe how the 20-bit memory address for a DMA transfer is produced by the circuit in Figure 11-3.
5. Describe the function and operation of devices *U*5 and *U*6 in Figure 11-3.
6. Why are microcomputers such as the IBM PC designed with peripheral expansion slots instead of having functions such as a CRT controller designed into the motherboard?
7. List the series of signals that must occur to read a data word from a dynamic RAM such as the 51C256H.
8. List the major tasks that must be done when using dynamic RAM in a microcomputer system.
9. How does a dynamic RAM controller, in a system such as that in Figure 11-10, arbitrate the dispute that occurs when you attempt to read from or write to a bank of dynamic RAMs while the controller is doing a refresh cycle?
10. Describe how parity is used to check for RAM data errors in microcomputers such as the IBM PC. What is a major shortcoming of the parity method of error detection?
11. When using a Hamming code error detection/correction scheme for DRAMs, how many encoding bits must be added to detect and correct a single-bit error in a 32-bit data word?
12. List the peripheral functions integrated into the Intel 80186 microprocessor.
13. Describe the function of the relocation register in the 80186 peripheral control block.
14. How can you tell from the schematic that the 8088 in Figure 11-15 is configured in maximum mode?
15. Device *U*7 in Figure 11-15 has a signal named AEN connected to its OE input. If, in troubleshooting this system, you find that this signal is not getting asserted, on which schematic sheet would you first look to see how this signal is produced?
16. In what ways are a standard microprocessor and a coprocessor different from each other?
17. *a.* When a coprocessor and a standard processor are connected together in a system such as that in Figure 11-15, why are the $\overline{S2}$–$\overline{S0}$ status lines, the QS1–QS0 lines, the address, and the data lines of the two devices connected directly together?
    b. Where does the 8087 coprocessor in Figure 11-15 get its instructions from?
    c. How does the main processor distinguish its instructions from those for the 8087 as it fetches instructions from memory?
    d. Describe how the 8087 and 8088 work together to load a long-real data item from memory to the 8087 ST.

e. How does the 8087 in Figure 11-15 signal the 8088 that it needs to use the buses?

f. How can you prevent the 8088 in Figure 11-15 from going on with its next instruction before the 8087 has completed an instruction? What hardware connection in Figure 11-15 is part of this mechanism?

18. *a.* Given the decimal number 2435.5625, convert this number to binary, normalized binary, long-real, and temporary-real format.

b. Why are most floating-point numbers actually approximations?

19. *a.* Which 8087 stack register is ST after a reset?

b. Which 8087 stack register will be ST after one data item is read into the 8087?

c. Describe the operation that will be done by the 8087 FADD ST(2), ST(3) instruction.

d. How does the operation of the instruction FADDP ST(2), ST(3) differ from the operation of the instruction in 19c?

20. Describe the operation performed by each of the following 8087 instructions.

a. FLD TAX-RATE

b. FMUL INFLATION_FACTOR

c. FSQRT

d. FLDPI

e. FSTSW CHECK_ANSWER

f. FPTAN

21. *a.* Show the binary codes required for each of the instructions in question 20.

b. Why is 9BH, the code for the 8086 WAIT instruction, put in before most of the 8087 instructions?

c. Show the 8086 ESC instructions required to get an 8086 assembler to produce the correct codes for the instructions in question 20.

22. In the example program in Figure 11-24, why did we put a WAIT instruction after the FSTSW STATUS instruction?

23. Using the example program in Figure 11-24, write an 8087 program which computes the volume of a sphere. The formula is $V = \frac{4}{3}\pi R^3$.

24. Describe the function of the FPREM instruction in the example program in Figure 11-26.

25. Extend the example program in Figure 11-26 to calculate the sine and the cosine of the given angle in the range $0 \leq$ ANGLE $< 90°$ ($\pi/2$).

26. What are the advantages of having several microprocessors connected on a common system bus such as the Multibus? What is the major problem that has to be worked out in order for these multiple processors to exist peacefully on the common bus?

27. Name the two schemes used to determine which master on the Multibus gets control when two or more masters request use of the bus at the same time.

28. On board a master, how is the signal produced which tells the bus controller to take over the bus?

29. How can a master keep control of the bus for more than 1 byte or word access if that master is in the middle of some critical program section?

CHAPTER

12 Microcomputer System Peripherals

In the preceding chapters we discussed basic microcomputer systems and some of the programmable peripheral devices used in these systems. In this chapter we expand outward to discuss the hardware and software of system peripherals such as CRT displays, computer vision devices, disk drives, and printers.

OBJECTIVES

1. Describe how characters are produced on a CRT or an LCD screen.
2. Use BIOS calls to display a message on the CRT display of an IBM PC-compatible computer.
3. Describe how bit-mapped and vector graphic displays are produced on a CRT.
4. Describe how computer vision systems produce an image that can be stored in a digital memory.
5. Show in general terms the formats in which digital data is stored on magnetic and optical disks.
6. Describe the operation of disk controller circuitry.
7. Use DOS calls to open, read or write, and close disk files.
8. Describe the mechanism used in several common types of computer printers.
9. Describe how phoneme, formant filters, and linear predictive coding synthesizers produce human-sounding speech from a computer.
10. Describe the basic principle used in speech recognition systems.

MICROCOMPUTER DISPLAYS

Currently there are several different technologies used to display numbers, letters, and graphics for a microcomputer. The most common types are *cathode-ray tubes* (CRTs) and *liquid-crystal displays* (LCDs). In Chapter 9 we discussed the operation of alphanumeric LCD displays and a little later in this chapter we will show how large LCD screens are interfaced to a microcomputer. For now, however, we want to discuss the operation and interfacing of CRT-type displays.

Basic CRT Operation

A CRT is a large, bottle-shaped vacuum tube. The picture tube used in a TV set is an example of a CRT. An electron gun at the rear of the tube produces a beam of electrons which is directed towards the front of the tube. The inside surface of the front of the tube is coated with a phosphor substance which gives off light when it is struck by electrons. The color of the light given off is determined by the particular phosphor used. To produce color displays as in a color TV set, dots of red-, blue-, and green-producing phosphors are put on the inside of the screen in triangle patterns. Separate electron beams are focused on the dots for each color phosphor. By altering the intensity ratio of the three beams we can make the three-dot triangle appear any desired color.

The most common method of producing images on the CRT screen is to sweep the electron beam(s) back and forth across the screen. When the beam reaches the right side of the screen, it is turned off (blanked) and retraced rapidly back to the left side of the screen to start over. If the beam is slowly swept from the top of the screen to the bottom of the screen as it is swept back and forth horizontally, the entire screen appears lighted. When the beam reaches the bottom of the screen, it is blanked and rapidly retraced back to the top to start over. A display produced in this way is referred to as a *raster* display. To produce an image we turn the electron beam on or off as it sweeps across the screen. The trick here is to get the beam intensity or *video information* synchronized with the horizontal and vertical sweeping so that we get a stable display.

Black-and-white TVs in the United States use a horizontal sweep frequency of 15,750 Hz and a vertical sweep frequency of 60 Hz. One sweep of the beam from the top of the screen to the bottom is called a *field*. Sixty fields per second are then swept out. To get better pic-

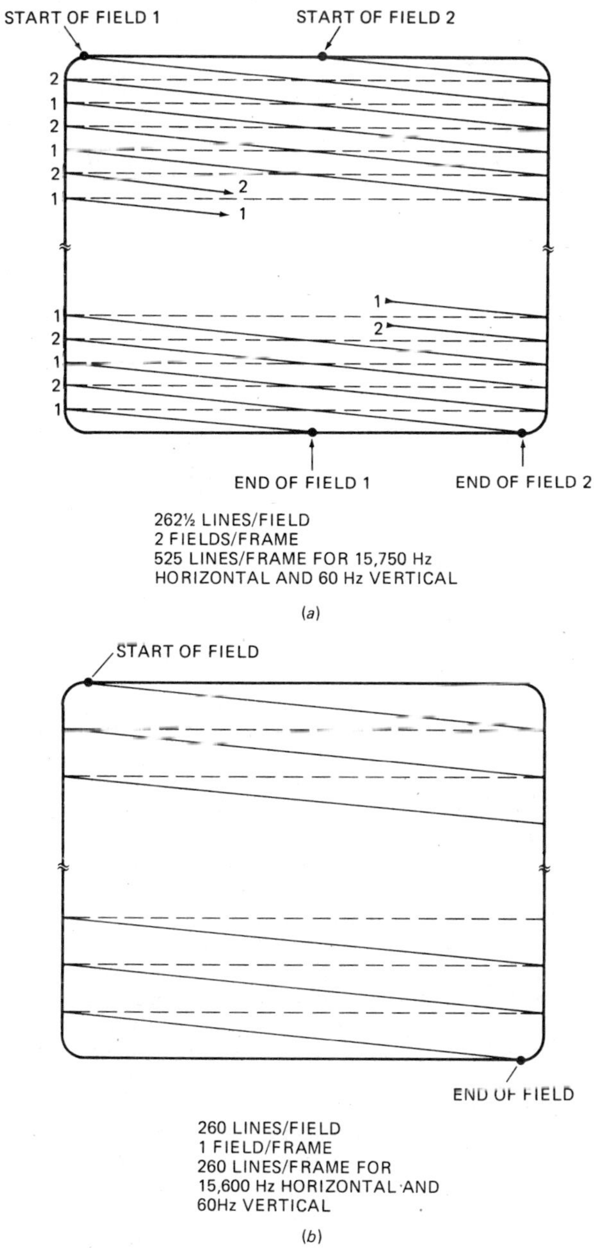

FIGURE 12-1 CRT scan patterns. *(a)* Interlaced. *(b)* Noninterlaced.

ture resolution and avoid flicker, TVs use *interlaced scanning*. As shown in Figure 12-1*a*, this means the scan lines for one field are offset and interleaved with those of of the next field. After every other field the scan lines repeat. Therefore, two fields are required to make a complete picture or frame. The frame rate is then 30 frames/second. The beam sweeps 262.5 times horizontally for each vertical sweep.

CRT units used for computer readouts usually have *noninterlaced scanning*, as shown in Figure 12-1*b*. In this case a horizontal sweep rate of 15,600 Hz and a vertical sweep rate of 60 Hz gives 260 sweep lines/field. The field rate and the frame rate are both 60 Hz in this case.

Whether the CRT you are using to display your programs is in a TV set, a video monitor, or a terminal, there are certain basic circuits required to drive the CRT. These are: the vertical oscillator to produce the vertical sweep signal for the beam, the horizontal oscillator to produce the horizontal sweep signal for the beam, and the video amplifier to control the intensity of the electron beam. A unit which contains only this basic drive circuitry is referred to as a *video monitor*. A TV set contains the basic monitor functions plus RF and audio decoding circuitry. A CRT *terminal* contains a keyboard, memory, communication circuitry, and usually a microprocessor to control all of these parts.

The basic CRT drive circuitry for a one-color, or *monochrome*, display requires three input signals to operate properly. It must have horizontal sync pulses to keep the horizontal oscillator synchronized, and vertical sync pulses to keep the vertical oscillator synchronized. Also it must have the video information for each point as the beam sweeps across the screen. All of this must be synchronized so that a particular dot of video information gets displayed at the same point on the screen during each frame. If you have seen a TV picture rolling, or a TV picture with jagged horizontal lines in it, you have seen what happens if the horizontal, vertical, and video information are not synchronized.

When transmitted to a TV set or to a video monitor, the two sync signals and the video information are usually combined into a single signal called *composite video*. Figure 12-2 shows a typical TV-type composite video signal waveform. It is hard to show in a figure, but there is one vertical sync pulse for each 262.5 horizontal sync pulses. The video information is represented by the waveform sections between horizontal sync pulses. For

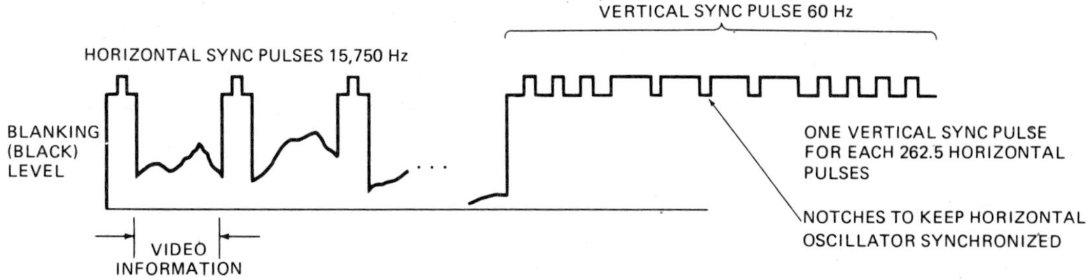

FIGURE 12-2 Composite video waveforms.

these waveforms, a more positive voltage turns the beam off. Therefore, the beam will be *blanked* during the horizontal retrace time represented by the pulse that the horizontal sync pulse sits on top of. The beam will also be blanked during the vertical retrace time. Now let's see how we generate these three signals to display characters on a CRT screen.

Creating a Page of Monochrome Characters on a CRT

Characters or graphics are generated on a CRT screen as a pattern of light and dark dots. To do this we turn the electron beam on and off as it sweeps across the screen. Figure 12-3 shows how this works. The round dots in the figure represent the beam on, and the empty, square boxes represent the beam off. With this dot matrix we can produce a reasonable approximation to any letter or symbol. The more dots used for each character, the better the representation. Common dot-matrix sizes for a character are 5 by 7, 7 by 9, and 7 by 12. The dot patterns for each character we want to display are stored in a ROM called a *character generator* ROM. Figure 12-4 shows the matrix for a Motorola MC6571 character generator. The MC6571 uses a 7 by 9 matrix for the actual character, but it has extra dot rows to leave space between rows of characters and so that lowercase letters can be dropped in the matrix to show descenders correctly. Each dot row in Figure 12-4 represents the pattern of dots for a horizontal scan line of the character. Figure 12-5 shows how the character generator is connected with some RAM, a shift register, and some counters to produce the signals required to display characters on a CRT. Here's how it works.

The ASCII or EBCDIC codes for the characters to be displayed on the screen are stored in a RAM so that they can be changed when you want to display something new on the screen. This RAM is often referred to as the *display RAM* or the *display refresh RAM*. The RAM must contain at least one byte location for each character to be displayed. A common display size is 25 rows of characters with 80 characters in each row. This display then requires about 2 Kbytes of display RAM. A character counter and a row counter are used to address the ASCII codes in this RAM.

To start the display in the upper left corner, the character counter and the row counter outputs are all 0's so

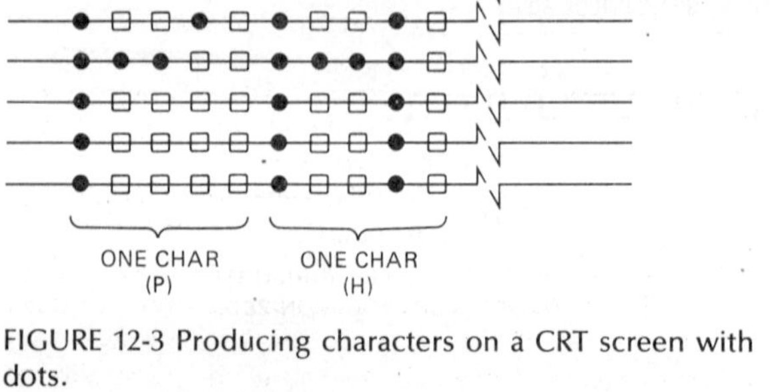

FIGURE 12-3 Producing characters on a CRT screen with dots.

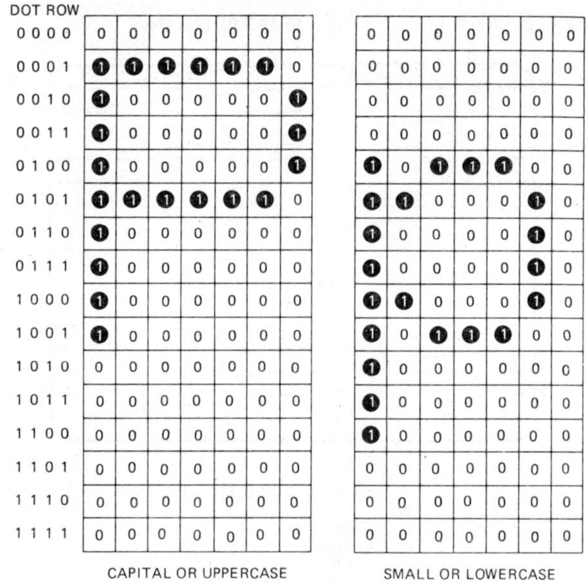

FIGURE 12-4 Dot format for Motorola MC6571 character generator ROM.

the ASCII code for the first character is addressed in the display RAM. The addressed code will be output from the ROM to the data inputs of the character generator ROM. The outputs of a dot row counter are also applied to the character generator. With these two inputs the character generator will output the 7-bit dot pattern for one dot row in the character. For the first scan across the screen the counter will output 0000 so the dot pattern output will be that for dot row 0000 of the character. The output from the character generator is in parallel form. In order to turn the beam on and off at the correct time as it sweeps across the screen, this dot pattern must be in serial form. A simple parallel-to-serial shift register is used to do this conversion. Note that the eighth data input of the shift register is tied to ground so that there is always one dark dot or *undot* between characters. The high-frequency clock used to clock this shift register is called the *dot clock* because it controls the rate at which dot information is sent out to the video amplifier.

After the dots for the first scan line of the first character are shifted out, the character counter is incremented by one. It then points to the ASCII code for the second character in the top row of characters in the display RAM. Therefore the ASCII code for this second character will be output to the character generator ROM. Since the dot line counter inputs to the ROM are still 0000, the ROM will output the dot pattern for the top scan line of the second character in the top row of characters on the screen. When all of the dots for the top scan line of this character are all shifted out, the character counter will be incremented by one again, and the process repeated for the third character in the top row of characters. The process continues until the first scan line for all 80 characters in the top row of characters is traced out.

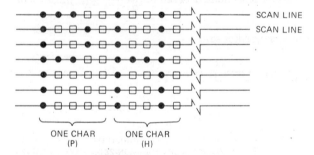

A horizontal sync pulse is then produced to cause the

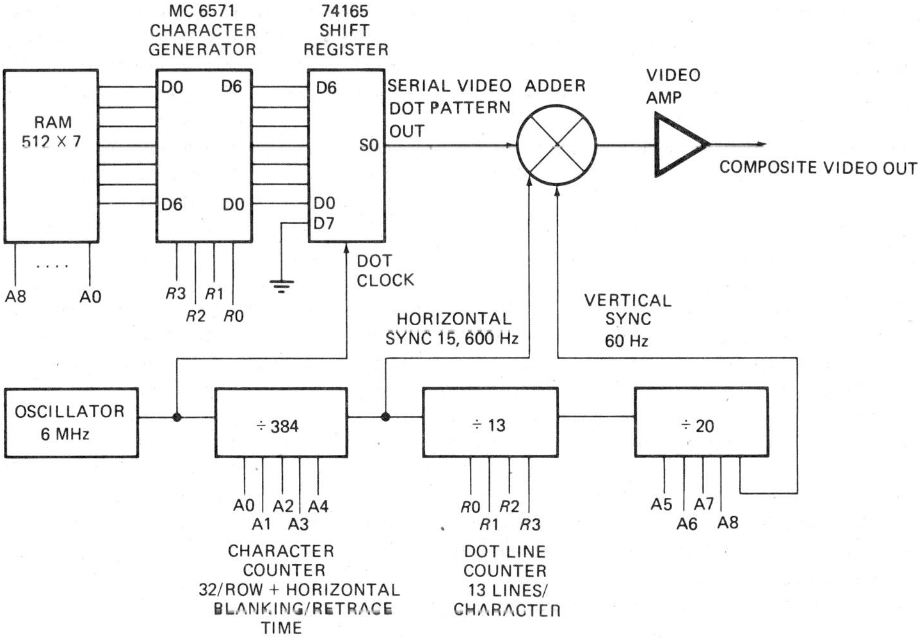

FIGURE 12-5 Block diagram of circuitry to produce dot-matrix character display on CRT.

beam to sweep back to the left side of the screen. After the beam retraces to the left, the character counter is rolled back to zero to point to the ASCII code for the first character in the row again. The dot line counter is incremented to 0001 so that the character generator will now output the dot patterns for the second scan line of each character. After the dot pattern for the second scan line of the first character in the row is shifted out to the video amplifier, the character counter is incremented to point to the ASCII code for the second character in the display RAM. The process repeats until all of the scan lines for one row of characters have been scanned.

The character row counter is then incremented by one. The outputs of the character counter and the character row counter now point to the display RAM address where the ASCII code for the first character of the second row of characters is stored. The process we described for the first row will be repeated for the second row of characters. After the second row of characters is swept out, the process will go on to the third row of characters, and then on to the fourth, and so on until all 25 rows of characters have been swept out.

When all of the character rows have been swept out, the beam is at the lower right corner of the screen. The counter circuitry then sends out a horizontal sync pulse to retrace the beam to the left side of the screen, and a vertical sync pulse to retrace the beam to the top of the screen. When the beam reaches the top left corner of the screen, the whole *screen-refresh* process that we have described will repeat. As we mentioned before, the entire screen must be scanned (refreshed) 30 to 60 times a second to avoid a blinking display. Now let's see what frequencies are involved in each major part of the circuitry.

CRT Display Timing and Frequencies

There are many different horizontal, vertical, and dot clock frequencies commonly used in raster-scan CRT displays. The horizontal sweep frequency is usually in the range of 15–30 kHz, the vertical sweep frequency is usually 50 or 60 Hz, and the dot clock frequency is usually 5–25 MHz. For our first specific example, we will use the frequencies used in the IBM PC monochrome display adapter, which we use as a circuit example in a later section.

The IBM monochrome display adapter produces a display of 25 rows of 80 characters/row. Each character is produced as a 7 by 9 matrix of dots in a 9 by 14 dot space. This means that because clear space is left around each actual character, each character actually uses 9 dot spaces horizontally, and 14 scan lines vertically. The active horizontal display area then is 9 dots/character × 80 characters/line or 720 dots per line. The active vertical display area is 25 rows × 14 scan lines/row or 350 scan lines.

Now, according to the IBM Technical Reference Manual, the monochrome adapter uses a dot clock frequency of 16.257 MHz. This means that the video shift register is shifting out 16,257,000 dots/second. The manual also indicates that the board uses a horizontal sweep frequency of 18,432 lines/second. Multiplying 16,257,000 dots per second by 1/18,432 seconds per line tells you that the board is shifting out 882 dots/line. Just above we showed you that the active display area of a line is only 720 dots. The extra 162 dot times actually present are required to give the beam time to get from the right edge of the active display to the right edge of the screen, retrace to the left edge of the screen, and sweep to the left edge of the active display area. This large number of extra dot times is necessary because most monitors have a large amount of *overscan*. This means that the beam is actually swept far off the left and right sides of the screen so that the portion of the sweep actually displayed is linear.

The manual for the display adapter indicates that the frame rate is 50 Hz. In other words the beam sweeps from the top of the screen to the bottom and back again 50 times/second. To see how many horizontal lines are in each frame, you can divide the 18,432 lines/second by 50 frames/second to give 369 scan lines/frame. As we showed above, the active vertical display area is 350 lines, so this gives 19 extra scan line times for the beam to get to the bottom of the screen, retrace to the top of the screen, and get to the start of the active display area again. Note that the dot clock, horizontal sweep frequency, and vertical sweep frequency must all be related to each other so that the display is synchronized.

Another point we need to make here concerns the bandwidth required by a video amplifier or monitor to clearly display a given number of dots per line. For our example here, the dot clock frequency is 16.257 MHz. This means that the dot shift register is shifting out 16,257,000 dots/second. If we are shifting out alternating dots and undots, then the waveform on the serial output pin of the shift register will be a square wave with a frequency of half that of the dot clock or 8.1285 MHz. In order to produce a clear display with this many dots per line, then, the video amplifier in the monitor connected to the display adapter must have a bandwidth of at least 8 MHz. In other words, the circuitry in the monitor must be able to turn on and off fast enough so that dots and undots don't smear together.

This bandwidth requirement is the reason that normal TV sets connected to computers cannot display high-resolution 80-character lines for word processing, etc. In order to filter out the sound subcarrier and the color subcarrier, the bandwidth of TV video amplifiers is limited to about 3 MHz. When using a TV as a readout device for a microcomputer, then, we usually limit the display to a smaller number of dots per character, and to 40 characters/line. A CRT monitor used for displaying characters or graphics should have a bandwidth greater than one-half the dot clock frequency.

A final point we want to make about CRT timing is how often the display-refresh RAM has to be accessed. As the circuitry scans one line of the display, it has to access a new character in RAM after each 9 dots are shifted out, assuming 9 dots horizontally per character. Dividing the dot clock frequency of 16.257 MHz by 9 dots/character tells you that characters are read from RAM at a rate of 1,806,333 characters/second, or one character every 553 ns! Next we will show you how programmable CRT display controllers are used to produce a desired display.

CRT Controller ICs and Circuits

In addition to the chain of counters shown in Figure 12-5, considerable other circuitry is needed to produce horizontal blanking pulses, vertical blanking pulses, a cursor, scrolling, and highlighting for a CRT display. Several manufacturers offer CRT controller ICs that contain different amounts of the required circuitry. The two devices we discuss here are the Intel 8275 and the Motorola 6845.

THE INTEL 8275 CRT CONTROLLER

Figure 12-6 shows, in block diagram form, how an 8275 controller is connected with other circuitry to produce

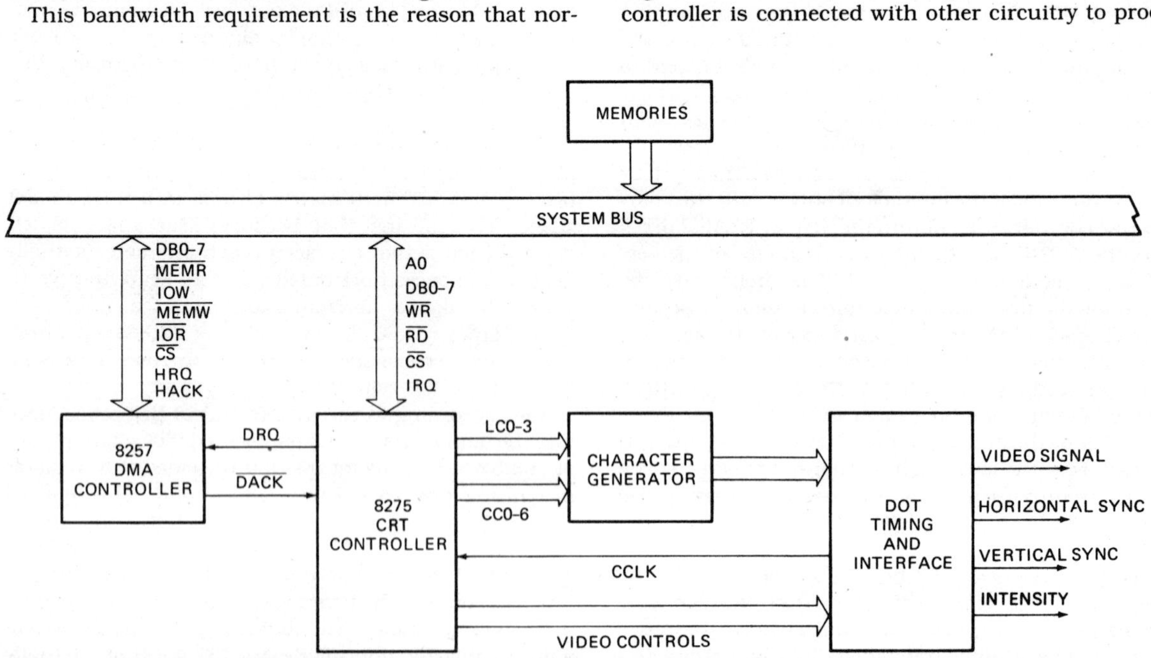

FIGURE 12-6 Block diagram showing connections of Intel 8275 CRT controller in a microcomputer system.

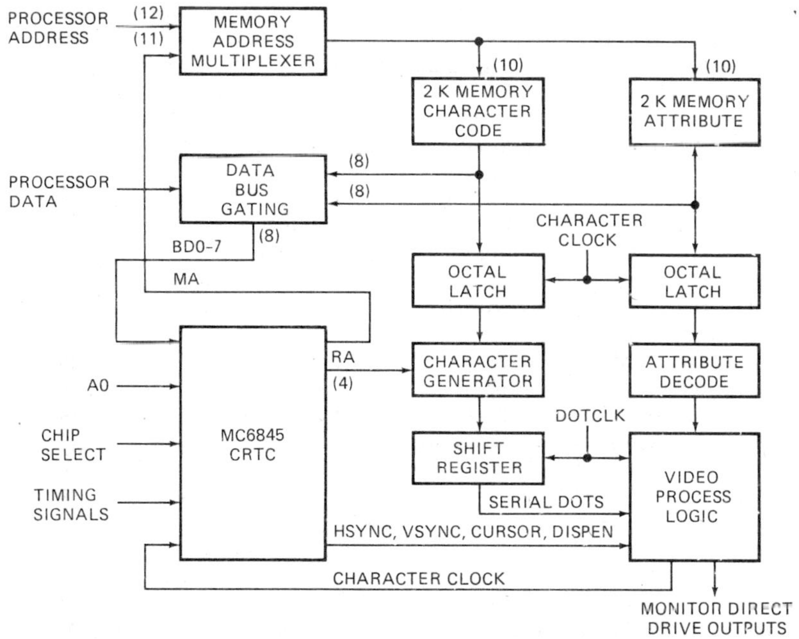

FIGURE 12-7 IBM monochrome display adapter board block diagram.

the drive signals for a CRT monitor. The 8275 contains a row counter which can be programmed for a display of 1 to 64 rows, a character counter which can be programmed for a display of 1 to 80 characters/row, and a scan line counter which can be programmed for 1 to 16 scan lines/character. The 8275 also has an 80-byte buffer to hold the ASCII characters for the row currently being displayed and an 80-byte buffer to hold the ASCII characters for the next row of characters to be displayed.

For the system in Figure 12-6 the page of characters to be displayed is stored in a buffer in the main microprocessor memory. While the 8275 is using the contents of one of its 80-byte buffers to refresh a row of characters on the screen, it fills the other 80-byte buffer from the main memory on a DMA basis. To do this it sends a DMA request signal (DREQ) to the 8257 DMA controller. The DMA controller sends a DMA request signal to, for example, the HOLD input of an 8086. When the 8086 sees the request signal, it floats its buses and sends a hold acknowledge signal. As we described in Chapter 11, the DMA controller then sends out the memory address and control signals needed to transfer the characters from memory to the 8275 buffer. The DMA approach uses only a small percentage of the microprocessor's time and, since the *display page* is located in the main memory, new characters are easily written to it.

The character generator is left out of the controller so that a ROM for any desired character set can be used. The dot clock and the dot shift register are also external because of the high frequencies involved in that part of the circuit. The 8275 produces vertical and horizontal sync signals, but external circuitry is used to massage the timing of these signals to correspond with the video information from the dot shift register. Now we will show you another CRT controller approach.

THE MOTOROLA 6845 CRT CONTROLLER

The 6845 CRT controller chip performs most of the 8275 functions discussed in the previous section, but it interfaces with the display refresh RAM in a very different way. The 6845 is used in both the monochrome display adapter board and the color/graphics monitor adapter boards for the IBM PC, so we will use some circuitry from these boards to show you how it works.

Figure 12-7 shows a block diagram for the IBM PC monochrome display adapter board. Take a look at this figure and see what parts you recognize from our previous discussions. You should quickly find the CRT controller, character generator, and dot shift register. Next, find the 2 Kbyte memory where the ASCII codes for the characters to be displayed are stored. To the right of this memory is another 2 Kbyte memory used to store an attribute code for each character. An attribute code specifies how the character is to be displayed. For example, with an underline or with increased or decreased intensity. As you may have observed, it is common practice to display a menu at reduced intensity so it does not distract from the main text on the screen.

Now observe that there is a multiplexer in series with the address lines going to the character and attribute memories. This is done so that either the CPU or the CRT controller can access the display-refresh RAM. The 6845 has 14 address outputs, so it can address up to 16 Kbyte display and attribute locations. To keep the display refreshed, the 6845 sends out the memory address for a character code and an attribute code. The character clock signal latches the code from memory for the character generator and the attribute code for the attribute decode circuitry. The character clock also increments the address counter in the 6845 to point to the next character code in memory. The next character clock transfers the next codes to the character generator and attribute decoder. The process cycles through all of the characters on the page and then repeats. Now, when you want to display some new characters on the screen, you simply have the CPU execute some instructions which write the ASCII codes for the new characters to the appropriate address in the display RAM. When the address decoding circuitry detects a display RAM address, it produces a signal which toggles the multiplexers so that the CPU has access to the display RAM. The question that probably occurs to you at this point is, "What happens if the 6845 and the CPU both want to access the display RAM at the same time?" There are several solutions to this problem. One solution is to allow the CPU to access the RAM only during horizontal and/or vertical retrace times. Another solution is to interleave 6845 accesses and CPU accesses. This is how it is done in the IBM board. The character clock signal going to the 6845 and the multiplexers allows the CPU to access the RAM during one half of the clock signal and the 6845 to access the RAM during the other half of the clock signal. If the CPU tries to access the display RAM during the controller's half of the character clock cycle, a not-ready signal from the CRT controller board will cause the processor to insert WAIT states until the half of the character clock signal when it can access the display refresh RAM.

6845 INTERNAL REGISTERS AND INITIALIZATION

Figure 12-8 shows the pin diagram and labels for the 6845. We will take a brief look at these pin functions, and then discuss the internal registers so we can show you how the device is initialized.

The function of most of the pins should be easily recognizable to you from the block diagram in Figure 12-7. Ground is on pin 1, +5 V is on pin 20, and a reset input is on pin 2. The 6845 sends out the display RAM address of the character currently being scanned on the MA0–MA13 lines. On the RA0–RA4 pins the 6845 sends out the number of the character scan line currently being scanned to the character generator. A character clock signal which changes state when it is time for the controller to access the next character in memory is connected to the 6845 CLK input. The horizontal and vertical sync output signals on pins 39 and 40 are produced by dividing down this CLK input signal. The 6845 has eight data inputs, D0–D7, which connect to the system data bus so that initialization words can be written to the device and status words read from the device, just as with any of the other peripheral devices we have discussed. The 6845 will be enabled for a read or write on its data bus when its CS input is asserted low. The R/$\overline{W}$ is asserted high for a read and low for a write. The processor clock, or a signal derived from it, is applied to the E input of the 6845 to synchronize data transfers in or out on the data lines. As seen from the processor, the 6845 has two internal addresses, a control address selected when RS is low and a data address selected when the RS input is high. We will tell you more about this after we talk briefly about the few remaining pins.

The Cursor output pin will be asserted high when the controller is displaying the cursor. This signal can be combined with signals from the attribute decoder to cause the cursor to blink or to be highlighted, depending on attributes stored for the cursor location.

The Display Enable output pin will be asserted when the 6845 is scanning the active display area of the screen. This signal can be used to produce blanking pulses during horizontal and vertical retrace times. In a system that accesses the display RAM during retrace times, this signal can be used to tell the CPU when it can access the display RAM.

When the light pen strobe input, LPSTB, is made to go from low to high, the current refresh address will be latched in two registers inside the 6845.

The 6845 has a register bank of 19 registers which are used to set and to keep track of display counts during display refreshing. Figure 12-9 shows the function of each of these registers. Even if you are not going to be programming a 6845, it is worth taking a look at this figure so you have an idea of the types of parameters you specify for a CRT controller chip such as the 6845.

The 6845 has only two internal I/O addresses which are selected by the RS input. When the RS input is low, the internal address register is selected. When the RS input is high, one of the 18 internal data registers is selected. In order to access one of the internal data registers, you first have to write the number (address) of that register to the address register with RS low, and then write the data to the 6845 with RS high. RS is usually tied to a system address line so that you just write the address word to one address, perhaps 3B4H, and the data word to another address, perhaps 3B5H.

The standard way to initialize all of these parameters for a 6845 in a system is to use a program loop of the form:

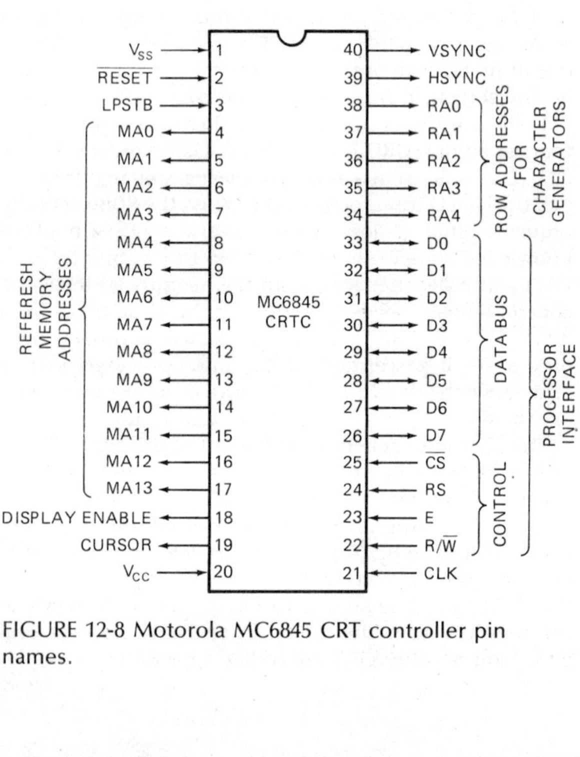

FIGURE 12-8 Motorola MC6845 CRT controller pin names.

RS	Register number	Function
0	X	Holds number of data register to write to.
1	0	Total number of horizontal character times +1, including retrace.
1	1	Number of horizontal characters displayed.
1	2	Character number when horizontal-sync pulse is produced. Determines horizontal display position.
1	3	Width of horizontal-sync pulse in character times.
1	4	Total number of vertical character rows-1, including vertical retrace.
1	5	Adjusts vertical timing to get exactly 50 or 60 Hz.
1	6	Number of vertical character rows displayed.
1	7	Vertical row number when vertical-sync pulse produced. Controls vertical position on screen.
1	8	Sets controller for interlaced or non-interlaced scanning.
1	9	Number of horizontal scan lines-1 per character row.
1	10	Starting scan line for the cursor and cursor blink rate.
1	11	Ending scan line for the cursor.
1	12	Starting address (high byte) for character to be put out after vertical retrace. Determines which character row from buffer appears at top of screen. Change this value to scroll display.
1	13	Low byte of first row starting address.
1	14	High byte of current cursor address.
1	15	Low byte of current cursor address in display RAM.
1	16	High byte of display RAM address when LPSTR occurs.
1	17	Low byte of display RAM address when LPSTR occurs.

FIGURE 12-9 MC6845 internal register functions.

REPEAT
  Output a data register number to the 6845 internal address register (RS = 0).
  Output parameter byte for that register to data register address (RS = 1).
UNTIL all required registers of the 18 are initialized.

Before we start the next section on computer graphics, let's take a brief look at how you can use the IBM PC BIOS procedures to display characters on the CRT screen.

Using the IBM PC INT 10H to Display Characters on the CRT

If you are working on an IBM PC it is quite easy to display characters on the CRT as part of your program by using the BIOS routines in ROM. In Chapter 8 we showed you how to use the BIOS INT 17H procedure to send characters to a printer. To use the BIOS procedures you load the parameters required by the procedure into registers specified for that procedure in the IBM Technical Reference Manual, and then execute the INT # instruction that accesses that procedure. You can use the BIOS INT 10H procedure for 15 different functions related to the CRT display. Some of these functions are: set display mode, set cursor position, scroll page up, scroll page down, set color palette, write dot, and write character to screen. You specify the function you want by loading the number for that function in the AH register before executing the INT 10H instruction. To write a character to the screen you simply load AH with 14 decimal, load AL with the character you want to display, and then execute the INT 10H instruction. Another BIOS procedure that you might want to use with this one is the INT 16H procedure which you can use to read characters from the keyboard. If you load AH with 0 and execute the INT 16H instruction, the ASCII code for the next key pressed on the keyboard will be left in the AL register after the procedure executes. Coupling the two INT procedures lets you read in characters from the keyboard and display them on the CRT.

RASTER SCAN CRT GRAPHICS DISPLAYS

The previous section of this chapter showed you how a monochrome display of alphanumeric characters can be produced on a CRT screen. In this section we show you how we produce a picture or graphics display. The two major methods of producing a graphics display are the *bit-mapped raster scan* approach, and the *vector graphics* approach. We'll explore the raster approach first.

Figure 12-5 shows a block diagram of some simple circuitry that can be used to create a display of characters on a CRT screen by turning the electron beam on and off as it is scanned across the screen. Characters are produced as a series of dots and undots on the screen. The ASCII codes for the page of characters to be displayed are stored in a display-refresh RAM. The dot patterns for each scan line of each character are stored in a character generator ROM. Now, suppose that we leave the character generator out of this circuit and connect the outputs of the RAM directly to the inputs of the dot shift register. And further suppose that instead of storing the ASCII codes for characters in the RAM, we store the dot patterns we want for each eight dots of a scan line in successive memory locations. When a byte is read from the RAM and loaded into the shift register, the stored dot pattern will be shifted out to the CRT beam to produce the desired pattern of dots for that section of a scan line on the screen. The next RAM byte will hold the dot pattern for the next 8 dots on a scan line, etc. Operating in this mode, each bit location in memory corresponds to a dot location on the screen. The entire screen then can be thought of as a matrix of dots. Each dot can be programmed to be on or off by putting a one or a zero in the corresponding bit location in RAM. A graphics display produced in this way is known as a *bit-mapped raster scan display*. Each dot or in some cases block of dots is called a *picture element*. Most people shorten this to *pixel* or *pel*. For our first example let's assume a pixel is 1 dot.

Now, suppose that we want a graphics display of 640 pels horizontally by 200 pels vertically. This gives a total of 200 × 640 or 128,000 dots on the screen. Since each dot corresponds to a bit location in memory, this means that we have to have at least 128,000 bits or 16 Kbytes of RAM to hold the pel information for just one display screen. Compare this with the 2 Kbytes needed for each page of an 80 by 24 character display. As we will show you a little later, producing a color graphics display with a large number of pels increases the memory requirements even further.

Now that you have a picture of a raster graphics screen as a large matrix of dots, the question that may occur to you is, "How do I draw a rocket ship or other picture on the screen?" One method is to program each of the 128,000 dots to be on or off as required to produce the desired display. This method works, but it is somewhat analogous to handprinting copies of the Bible, a very tedious process. To make your life easier, many graphics programs are now available. These programs allow you to create a complex graphics display, dump the display to a printer, store the display on a disk, or include the display in another program you are writing. These graphics programs contain graphics routines or *primitives* which allow you to draw lines, draw arcs, draw three-dimentional figures, shade in areas, set up "windows," etc. Often these programs work with a *mouse*. A mouse in this case is a device which moves a cursor around the CRT screen when you move it around on the desk next to your computer. To draw a straight line between two points, for example, you move the cursor to the point on the screen where you want one end of the line and press a button on the mouse. You then move the cursor to the point on the screen where you want the other end of the line, and press a button on the mouse again. The graphics program then computes the coordinates for the other points on the line and puts 1's in the appropriate locations in the display RAM to draw in the line. By moving the cursor around on the screen and pressing buttons on the mouse at the appropriate times, you can quickly create some elaborate graphics displays. If you have not had a chance to play with a computer that has these graphics capabilities, do not pass go, proceed directly to your nearest computer store and experiment with a graphics program on the Apple Macintosh or IBM PC.

CRT TERMINALS

Several times previously in this book we have used the term CRT terminal. You may have used a CRT terminal to communicate with a minicomputer or mainframe computer. In addition to the basic CRT drive circuitry, a terminal contains a keyboard so you can talk to it, the CRT-refresh RAM and controller to keep the display refreshed, and a UART to communicate to and from a computer. Most CRT terminals now have one or more built-in microprocessors to coordinate keyboard, display, and communications functions. A major advantage of using a microprocessor instead of dedicated logic here is that key functions and communications parameters can be changed to match a given computer by simply typing a few keystrokes. A device from National Semiconductor, the NS456, contains a microprocessor-based CRT controller, a keyboard interface, a UART, and most of the other functions needed for a graphics/character CRT terminal.

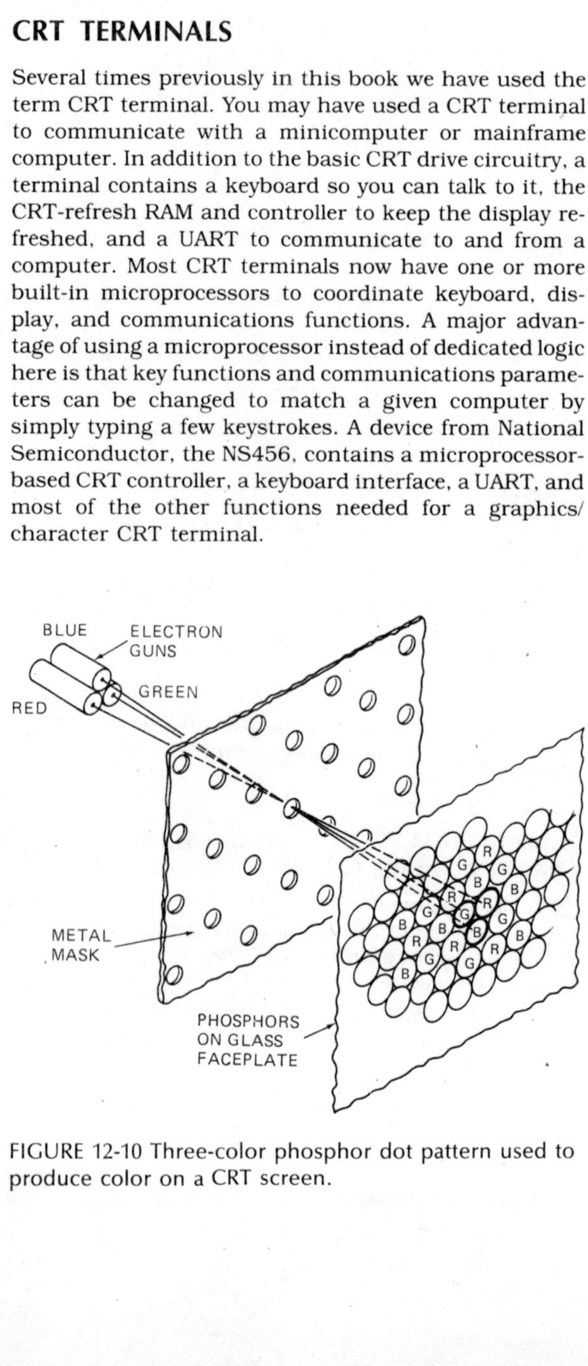

FIGURE 12-10 Three-color phosphor dot pattern used to produce color on a CRT screen.

I	R	G	B	COLOR
0	0	0	0	BLACK
0	0	0	1	BLUE
0	0	1	0	GREEN
0	0	1	1	CYAN
0	1	0	0	RED
0	1	0	1	MAGENTA
0	1	1	0	BROWN
0	1	1	1	WHITE
1	0	0	0	GRAY
1	0	0	1	LIGHT BLUE
1	0	1	0	LIGHT GREEN
1	0	1	1	LIGHT CYAN
1	1	0	0	LIGHT RED
1	1	0	1	LIGHT MAGENTA
1	1	1	0	YELLOW
1	1	1	1	HIGH INTENSITY WHITE

FIGURE 12-11 Sixteen colors produced by different combinations of red, blue, and green beams at normal and at increased intensity.

RASTER SCAN COLOR GRAPHICS

Monochrome graphics displays get boring after a while, so let's see how you can get some color in the picture.

To produce a monochrome display we coat the inside of the tube with a single phosphor which produces the desired color light when bombarded with electrons from a single electron gun at the rear of the tube. To produce a color CRT display we apply red, green, and blue phosphors to the inside of the tube, and bombard these three different phosphors with three separate electron beams. One approach is to have dots of the three phosphors in a triangular pattern as shown in Figure 12-10. The dots are close enough together so that to your eye they appear as a single dot. By changing the intensity ratio of the three beams we can make the three-part dot appear any color we want, including black and white. If all three beams are off, the dot is of course black. If the beams are turned on in the ratio of 0.30 RED, 0.59 GREEN, and 0.11 BLUE, then the dot will appear white. The overall intensity of the three beams, often represented with the letter I or the letter Y determines whether the dot will be a light or a dark shade of the color. Figure 12-11 shows 16 colors that can be produced by simply turning on or off different combinations of the red, blue, and green beams. A 1 in the I bit means that the overall intensity of the beam is increased to lighten the color as shown. If we drive the color guns and the intensity with the output of a D/A converter instead of simply on or off signals, we can produce a much wider variety of colors. A 2-bit D/A converter on each of the color signals and the intensity signal, for example, gives 256 color variations. In order to produce a display with a large number of pixels and a large number of colors, a large memory is needed. As we discuss a common color graphics adapter in the next section, we will show you some of the tradeoffs involved in this.

The IBM PC Color/Graphics Adapter Board

As a real system example here we will use the IBM PC color/graphics adapter board whose block diagram is shown in Figure 12-12.

This board again uses the Motorola MC6845 CRT controller device to do the overall display control. It produces the sequential addresses required for the display-refresh RAM, the horizontal sync pulses, and the vertical sync pulses as we described in a previous section. The 16 Kbyte display-refresh RAM is *dual-*

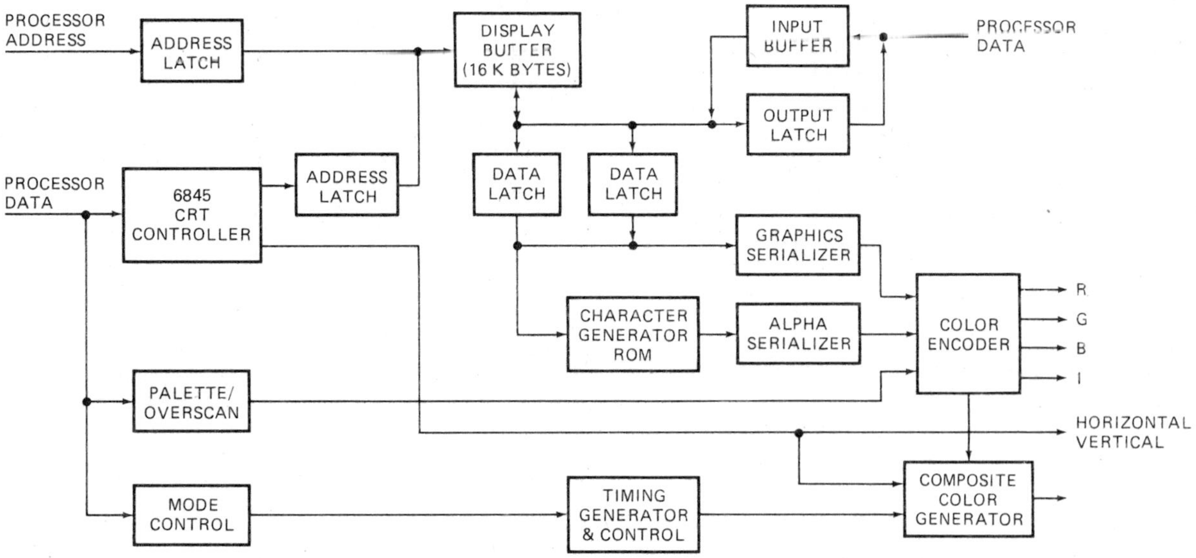

FIGURE 12-12 IBM PC color graphics adapter board block diagram.

ported. This means that it can be accessed by either the system processor or the CRT controller on a time-share basis as we also described previously. A little later we will show you how display information is stored in RAM for various display modes.

This adapter board can operate in either a character mode or a graphics mode. In the character mode it uses a character generator ROM and a single shift register (alpha serializer) to produce the serial dot information for display scan lines. When operating in a color graphics mode, the board uses separate shift registers (graphics serializer) to produce the dot information for each of the color guns and for the overall intensity.

As you can see by the signals shown in the lower right corner of Figure 12-12, the adapter board is designed to drive either of the two common types of color monitor. One type, commonly called an RGB monitor, has separate inputs for each of the required signals, red, green, blue, intensity, horizontal sync, and vertical sync. The other type of color monitor is called a *composite color monitor* because all of the required signals are combined on a single line. Color TV sets used as color monitors for computers require a composite video signal if they have a direct video input, or they require a radio-frequency signal modulated with the composite video signal if they do not have a direct video input. Later we will show you how we produce a composite color video signal from the separate signals. Now let's look at how the display information is stored in the display-refresh RAM for various display modes.

In the character or alphanumeric mode each character is represented by two bytes in the display-refresh RAM in the format shown in Figure 12-13*a*. This is the same format as the monochrome adapter board. The upper byte contains the 8-bit ASCII code for the character to be displayed. The lower byte contains an attribute code which you use to specify the character color (foreground) and the background color for the character.

DISPLAY-CHARACTER CODE BYTE								ATTRIBUTE BYTE							
7	6	5	4	3	2	1	0	7	6	5	4	3	2	1	0

(*a*)

ATTRIBUTE FUNCTION	ATTRIBUTE BYTE							
	7	6	5	4	3	2	1	0
	B	R	G	B	I	R	G	B
	FG	BACKGROUND			FOREGROUND			
NORMAL	B	0	0	0	1	1	1	1
REVERSE VIDEO	B	1	1	1	1	0	0	0
NONDISPLAY (BLACK)	B	0	0	0	1	0	0	0
NONDISPLAY (WHITE)	B	1	1	1	1	1	1	1

I = HIGHLIGHTED FOREGROUND (CHARACTER)
B = BLINKING FOREGROUND (CHARACTER)

(*b*)

FIGURE 12-13 Data storage formats for IBM color graphics board operating in alphanumeric mode. *(a)* Character byte and attribute byte. *(b)* Attribute byte format.

7	6	5	4	3	2	1	0
C1	C0	C1	C0	C1	C0	C1	C0
FIRST DISPLAY PEL		SECOND DISPLAY PEL		THIRD DISPLAY PEL		FOURTH DISPLAY PEL	

C1	C0	FUNCTION
0	0	DOT TAKES ON THE COLOR OF 1 of 16 PRESELECTED BACKGROUND COLORS
0	1	SELECTS FIRST COLOR OF PRESELECTED COLOR SET 1 OR COLOR SET 2
1	0	SELECTS SECOND COLOR OF PRESELECTED COLOR SET 1 OR COLOR SET 2
1	1	SELECTS THIRD COLOR OF PRESELECTED COLOR SET 1 OR COLOR SET 2

COLOR SET 1	COLOR SET 2
COLOR 1 IS GREEN COLOR 2 IS RED COLOR 3 IS BROWN	COLOR 1 IS CYAN COLOR 2 IS MAGENTA COLOR 3 IS WHITE

FIGURE 12-14 Data storage format for medium-resolution graphics mode of IBM PC color adapter board.

The intensity bit, I, in the attribute byte allows you to specify normal intensity or increased intensity for a character. The bit patterns used to produce different colors with the RGB and I bits are shown in Figure 12-13*b*. The B bit in the attribute byte allows you to specify that a character will be blinked. Only 4 Kbytes of the display RAM are needed to hold the character and attribute codes for an 80 character by 25 row display.

For displaying graphics, the adapter board can be operated in three different modes: low resolution, medium resolution, and high resolution. Higher resolution means more pixels in the display. We will use these three modes to show you the tradeoffs between number of colors, resolution, and memory requirements.

We often use the low-resolution mode when we are using a color TV set or a composite video monitor as a display device, because this mode requires less video amplifier bandwidth than high-resolution modes. In this low-resolution mode each pel is 2 dot times horizontally and 2 dot times vertically, so the picture is actually being made with larger blocks. The display consists of 100 rows of pels with 160 pels in each row. The total number of pels is then 16,000. The color and intensity for each pixel is specified by the I, R, G, and B bits in the lower half of a byte in the display RAM. Since 4 bits are being used to specify color and intensity, a pel can be any one of 16 colors. Since a byte is used to store the information for each pel, all 16 Kbytes of the display RAM are used to display the 100 by 160 pel display.

In the medium-resolution mode each pel is a single dot. The display consists of 200 rows of pels with 320 pels in each row, or a total of 64,000 pels. The 16 Kbytes of display refresh RAM corresponds to 16 Kbits × 8 or

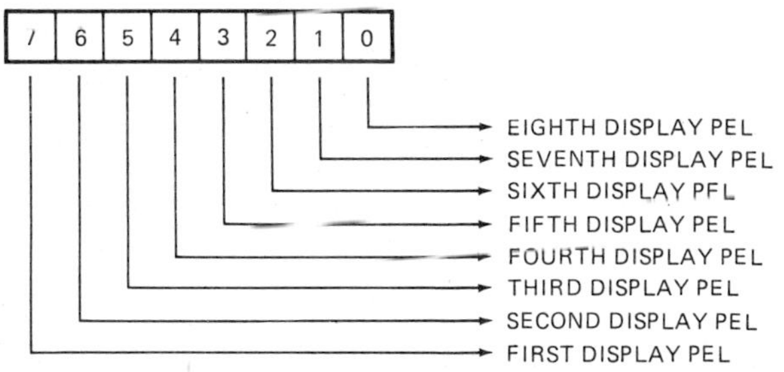

FIGURE 12-15 Data storage format for high-resolution graphics mode of IBM PC color graphics adapter board.

128 Kbits. Dividing the number of pels into the number of bits available for storage tells you that in this mode there are only 2 bits per pel available to store color information. With 2 bits we can specify only one of four colors for each pel. As you can see, increasing the resolution of the display has reduced the number of colors that can be specified with a given amount of memory. Figure 12-14 shows the format in which the pel information is stored in RAM bytes and the meaning of the bits in these bytes. The background color is selected by outputting a control byte through port 3D9H to the palette circuit shown on the left edge of Figure 12-12. Consult the IBM Technical Reference Manual for more details if you need them.

In the high-resolution graphics mode the IBM color graphics adapter board displays 200 rows of pels with 640 pels in each row, or a total of 128,000 pels. Since the 16 Kbyte refresh RAM contains 128,000 bits, this corresponds to 1 bit per pel. Therefore, you can only specify for each bit whether it is on or off. In other words, in this high-resolution mode you are limited to a black-and-white display, because there are no bits left to specify colors. Figure 12-15 shows the format in which pel data is stored in display RAM bytes for high-resolution displays. Here again we want to point out that if you want to produce color graphics displays as part of your programs, the best approach is probably to buy one of the commercially available graphics packages. These programs allow you to produce the figures you want with a mouse or with drawing instructions rather than specifying the bit values for each pixel.

As you should see by now, the limiting factor for color graphics displays is the amount of memory you are willing to devote to the display. Some high-resolution displays used in engineering work stations have a display of 1000 by 1000 pels with 16 colors. A display such as this requires about 500 Kbytes of high-speed-refresh RAM.

For each of the graphics formats above, data for a pel is read from the display RAM and converted to separate R, G, B, and I signals. These signals, along with the horizontal and vertical sync signals, can be sent directly to an RGB-type monitor. Before they can be sent to a composite video-type monitor these signals must be put together in a single signal. Here's how we do it.

Producing a Composite Color Video Signal

In order to produce a composite color signal from the R, G, B, and sync signals, we can't just add all the signals together. Instead, the approach we use is based on the NTSC standards for color television signals. Figure 12-16 shows in diagram form the somewhat complex method used to put the pieces together.

As a first step, the red, the green, and the blue signals are combined in the ratios shown to produce a signal proportional to the overall intensity or *luminance*. If horizontal and vertical sync pulses are added to this signal, the result is a monochrome composite video signal identical to that we described earlier in this chapter. This signal will produce a monochrome display on either a monochrome monitor or a color monitor.

To develop the correct color signals we pass the luminance signal through a 1.5-MHz low-pass filter and then an inverter. The filter is required to comply with FCC bandwidth rules if this signal is going to be sent out as part of a TV signal modulation. The inverted luminance signal, −Y, is then added to the red signal to produce

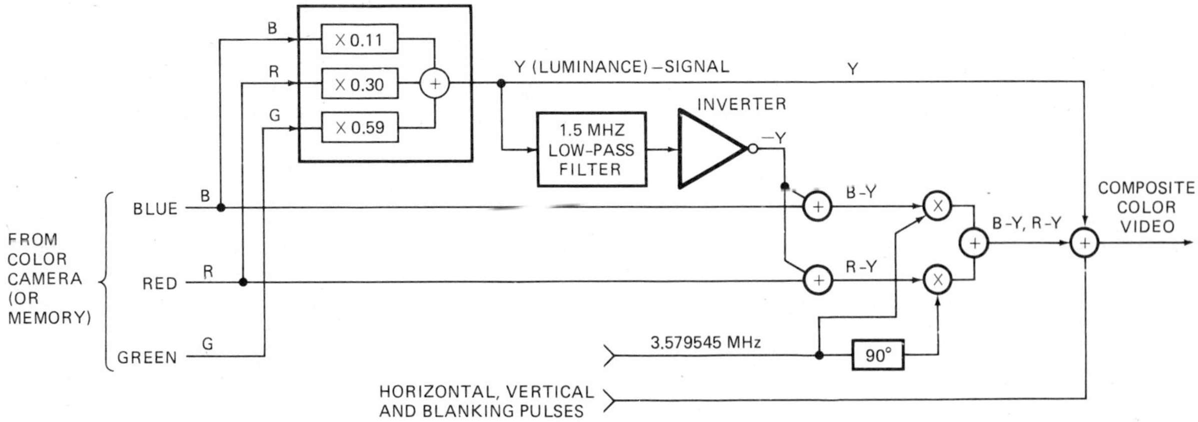

FIGURE 12-16 Block diagram of circuitry used to produce composite color video signal.

R − Y, and it is added to the blue signal to produce B − Y. The reason we do this is probably not obvious to the casual observer, but this scheme reduces the number of separate signals which have to be sent. Here's how it works. The Y, R − Y, and B − Y signals are sent as part of the color TV signal or as part of the composite video signal. In the receiver the Y signal is added to the R − Y signal to reconstruct the red signal. The Y signal is added to the B − Y signal to reconstruct the blue signal. Since the Y signal is composed of red, green, and blue, the red signal and the blue signal are subtracted from the Y signal to reconstruct the green signal. Because of all this we don't have to send a separate green signal. Now that you have an idea why we do all of this, let's continue the story of a composite color video signal.

The key to the next step is a stable 3.579545-MHz signal produced by a crystal oscillator. The B − Y signal is used to modulate this signal, and the R − Y signal is used to modulate a portion of this 3-MHz signal whose phase has been shifted by 90°. The two modulated 3.579545-MHz signals are then added together. The result is sometimes called the *chroma* signal, because it contains the color information.

Now, to produce the color composite video signal we simply add the horizontal sync pulses, the vertical sync pulses, the Y signal, and this chroma signal together as shown in Figure 12-16. When the composite video monitor receives this signal, it will separate all of the pieces again.

To produce a composite signal which can be fed into the antenna input of a color TV set, we usually use a chroma modulator device such as the Motorola MC1372 shown in Figure 12-17. This device produces the 3.579545-MHz color carrier frequency, and it produces the chroma signal from the R − Y and B − Y signals. The device also produces a radio-frequency carrier at the frequency for standard TV channel 3 or 4 and modulates this carrier signal with the Y, R − Y, B − Y, and sync information. When a color TV set receives this modulated signal, it demodulates the signal and separates the various parts. Because it has to filter out the remnants of the 3.579545-MHz color carrier frequency, the band-

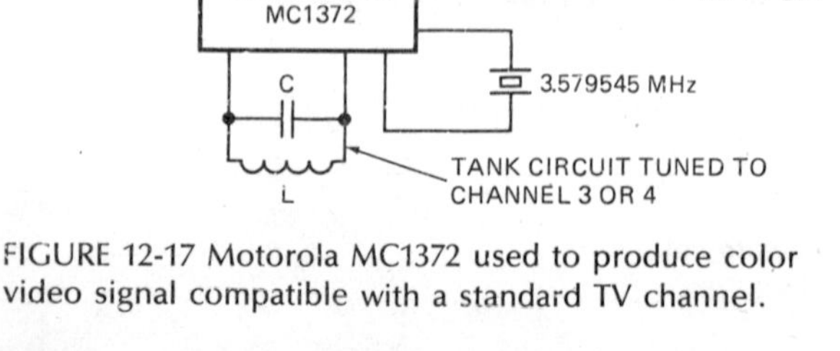

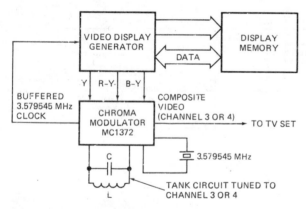

FIGURE 12-17 Motorola MC1372 used to produce color video signal compatible with a standard TV channel.

width of a composite color monitor or a color TV is limited to less than 3 MHz. As we explained in the section of the chapter on monochrome displays, this limits the resolution, and makes it difficult to display 80-character lines or detailed graphics on standard TV displays. Now that we have beat raster scan displays into the ground, we will show you how vector scan displays work.

VECTOR SCAN CRT DISPLAYS

A raster scan CRT display scans the electron beam over the entire screen and turns the beam on and off to produce a light or dark spot at each point in the scan. For certain CRT display applications such as computer-aided design workstations where the display consists mostly of background and an array of straight lines, it seems wasteful to sweep the beam back and forth over the entire screen. Also diagonal lines drawn on a raster scan display look like stair steps if you look closely at them, because of the rigid placement of the pixels on the screen.

A vector graphics scheme solves both of these problems by directly tracing out only the desired lines on the CRT. In other words, if we want a line connecting point A with point B on a vector graphics display, we simply drive the beam-deflection circuitry with a signal which causes the beam to go directly from point A to point B. If we want to move the beam from point A to point B without showing a line between the points, we can blank the beam as we move it. To draw a line on the CRT, then, we simply tell the beam how far to move and in what direction to move across the CRT. The name vector graphics comes from the fact that in physics a quantity which has magnitude and direction is called a vector.

The question that may occur to you at this point is, "How do you tell the beam where to move on the screen?" One way to direct the beam is by connecting a D/A converter to the horizontal deflection circuitry and another D/A converter to the vertical deflection circuitry. The values input to the two D/A converters then determine the position of the beam on the screen. If we use 10-bit D/A converters, we can direct the beam to one of 1024 positions horizontally and one of 1024 positions vertically. This is equivalent in resolution to a 1K by 1K raster display Color displays can be produced by using a three-beam, three-phosphor CRT and moving the three beams together as we described for the raster scan color display.

The next question that may occur to you is, "If this scheme is so simple, why don't we use it for all CRT graphics displays?" The answer is that a vector display works well where the information we want to display is mostly straight lines, but it does not work well for displays that have many curves and large shaded areas. When using a vector graphics system, we draw, for example, a circle by drawing many short vectors around in a circle. The circle is then made up of short line segments or points. The number of vectors you can draw on the screen is limited by the fact that you have to go back and redraw each vector 60 times a second to keep the display refreshed. Some current vector graphics systems can draw 150,000 short vectors 60 times a second, but for complex images you soon run out of vectors. The point here is that no one display technique or technology has all of the marbles at this point in time. Here's another display technology that has some advantages for portable instruments and computers.

ALPHANUMERIC/GRAPHICS LCD DISPLAYS

In Chapter 9 we discussed how LCDs work and how they can be used to display numbers and letters as individual digits. To make a screen-type display the liquid crystal elements are constructed in a large X-Y matrix of dots. The elements in each row are connected together, and the elements in each column are connected together. An individual element is activated by driving both the row and the column that contain that element. LCD elements cannot be turned on and off fast enough to be scanned one dot at a time in the way that we scan a CRT display. Therefore, we apply the data for one dot line of one character, or for an entire line, to the X axis of the matrix, and activate that dot row of the matrix. For a graphics display we wait a short time, then we deactivate that dot row, apply the data for the next dot row to the X axis, and activate that dot row. We continue the process until we get to the bottom of the display and then start over at the top of the screen. For large LCDs the matrix may be divided into several blocks of perhaps 40 dot lines each. Since each block of dot rows can be refreshed individually, this reduces the speed at which each liquid crystal element must be switched in order to keep the entire display refreshed. Large LCDs usually come with the multiplexing circuitry built in so that all you have to do is send the display data to the unit in the format specified by the manufacturer for that unit. We should soon see color LCDs for use with computers.

COMPUTER VISION

For many applications we need a microcomputer to be able to "see" its environment or perhaps a part that the machine it controls is working on. As part of a microcomputer-controlled security system, for example, we might want the microcomputer to "look" down a corridor to see if any intruders are present. In an automated factory application we might want a microcomputer-controlled robot to "look" in a bin of parts, recognize a specified part, pick up the part, and mount the part on an engine being assembled. There are several mechanisms we can use to allow a computer to see. The first one we will discuss uses sound waves.

Ultrasonic Vision

Bats "see" in the dark by emitting sound waves that are above the human hearing range or *ultrasonic*. A bat sends out ultrasonic pulses, and on the basis of the time it takes for echoes to return, determines how far it is from obstacles. Some Polaroid cameras use the same mechanism to determine the distance to an object being photographed. The camera then uses the distance information to automatically focus the camera lens.

The major parts of the range finder circuitry used in these cameras, including a printed circuit board, is available as a kit from Texas Instruments. With one of these kits and some simple circuitry you can add this type of vision to your microcomputer. Figure 12-18*a* shows a block diagram for the circuitry on the experimental board, and Figure 12-18*b* shows the major waveforms for one cycle of operation. A cycle starts when the VSW input is pulsed high. The transmitter section then sends out a "chirp" of 56 pulses through the transducer. The output is called a chirp because the 56 pulses step through four frequencies, 60 kHz, 57 kHz, 53 kHz, and 50 kHz to avoid absorption problems that might occur with just one frequency. This transmission is represented by the XLG signal in Figure 12-18*b*.

After the pulses are sent out, the circuitry is switched so that the transducer functions as a receiver. When the echo of the sound waves returns to the transducer it produces an analog electrical signal out of the transducer. A programmable-gain amplifier amplifies this echo and converts it to a digital pulse shown as the FLG signal in Figure 12-18*b*. The time it took the ultrasonic signal to go out to the target and return then is the time between the first rising edge of XLG and the rising edge of the FLG signal.

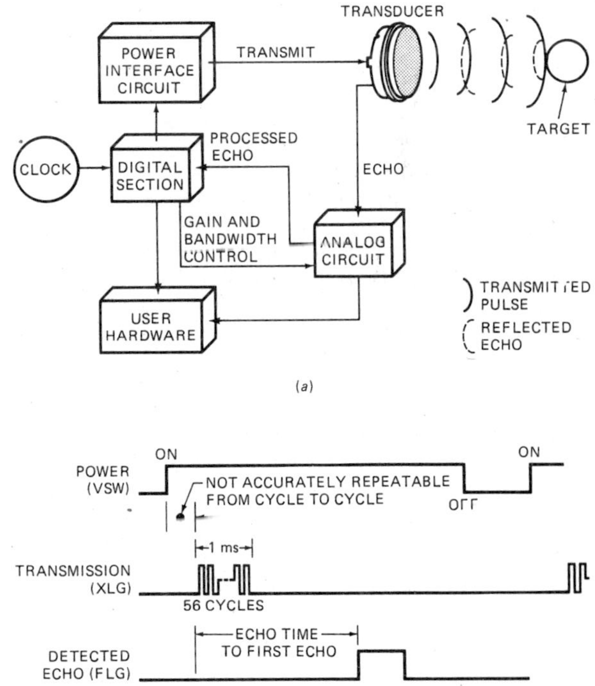

FIGURE 12-18 Polaroid ultrasonic range finder. *(a)* Block diagram of interface circuitry. *(b)* Major signal waveforms.

You can measure this time in any one of several ways. One way is to start a counter with the rising edge of XLG, and stop the counter with the rising edge of FLG. the number left in the counter then is the number of clock pulses required for the signal to go out to the target and back. To get the total time for the trip, you can multiply the number of clock pulses counted by the period of the clock pulses. Divide this time by 2 to get the actual time to the target. Since sound travels at about 1 foot in 0.888 ms, you can easily convert the transit time to an equivalent distance. An exercise in the laboratory manual written to accompany this book shows you more about all of this.

A simple ultrasonic range finder such as we have described here could be mounted in a mobile robot. By scanning the rangefinder back and forth the robot could determine a clear path through a series of obstacles, or detect when someone intrudes into its space. The range finder we described has a range of about 35 feet, and a resolution of about ⅛ inch when looking at a flat surface perpendicular to the sound waves. For applications where we need greater resolution or to recognize the shapes of objects, we use optical vision devices with our microcomputer.

Video Cameras and Computers

Cameras used in TV stations and for video recorders use a special vacuum tube called a *vidicon*. A light-sensitive coating on the inside of the face of the vidicon is swept horizontally and vertically by a beam of electrons. The beam is swept in the same way as the beam in a TV set displaying the picture will be swept. The amount of beam current that flows when the beam is at a particular spot on the vidicon is proportional to the intensity of the light that falls on that spot. The output signal from the vidicon for each scan line then is an analog signal proportional to the amount of light falling on the points along that scan line. This signal is represented by the waveform between the horizontal sync pulses in Figure 12-2. In order to get this analog video information into a digital form that a computer can store and process, we have to pass it through an A/D converter. For a color camera we need an A/D converter on each of the three color signals. Each output value from an A/D converter then represents a dot of the picture. The number of bits of resolution in the A/D converter will determine the number of intensity levels stored for each dot.

Standard video cameras and the associated digitizing circuitry are relatively expensive, so they are not cost-effective for many applications. In cases where we don't need the resolution available from a standard video camera we often use a CCD camera.

CCD Cameras

Charge-coupled devices or CCDs are constructed as long shift registers on semiconductor material. Figure 12-19 shows the structure for a CCD shift register section. As you can see, the structure consists of simply a *P*-type substrate, an insulating layer, and isolated gates. If a gate is made positive with respect to the substrate, a

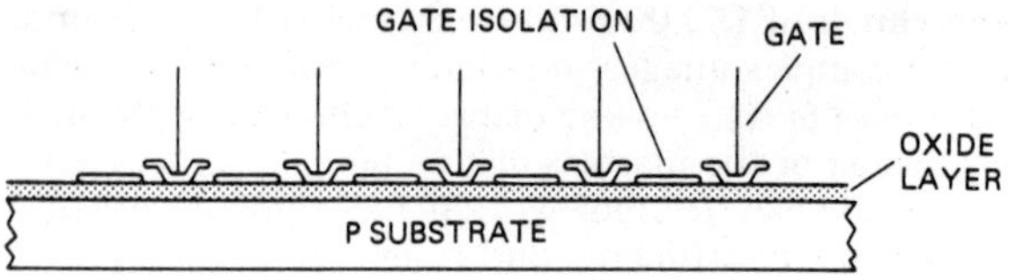

FIGURE 12-19 Basic structure of charge-coupled device used in CCD video cameras.

"potential well" is created under that gate. What this means is that, if a charge of electrons is injected into the region under the gate, the charge will be held there. By applying a sequence of clock signals to the gates, this stored charge can be shifted along to the region under the next gate. In this way a CCD can function as an analog or a digital shift register.

To make an image sensor, several hundred CCD shift registers are built in parallel on the same chip. A photodiode is doped in under every other gate. When all of the gates with photodiodes under them are made positive, potential wells are created. A camera lens is used to focus an image on the surface of the chip. Light shining on the photodiodes causes a charge proportional to the light intensity to be put in each well which has a diode. These charges can be shifted out to produce the dot-by-dot values for the scan lines of a picture. Improved performance can be gained by alternating nonlighted shift registers with the lighted ones. Information for a scan line is shifted in parallel from the lighted register to the dark, and then shifted out serially.

The video information shifted out from a CCD register is in discrete samples, but these samples are analog because the charge put in a well is simply a function of the light shining on the photodiode. To get the video information into a form that can be stored in memory and processed by a microcomputer, it must be passed through an A/D converter, or in some way converted to digital. For many robot applications and surveillance applications, a black-and-white image with no gray tones is all we need. In this case the video information from the CCD registers can simply be passed through a comparator to produce a 1 or a 0 for each dot of the image. CCD cameras have the advantages that they are smaller in size, more rugged, less expensive, and easier to interface to computer circuitry than vidicon-based cameras. Next we describe an inexpensive type of camera which produces digital video information directly.

OPTICRAM Cameras

Figure 12-20 shows a picture of the Micron Eye camera produced by Micron Technology in Boise, Idaho. This camera is relatively inexpensive, interfaces easily to common microcomputers, and has enough resolution for simple robot-type applications.

The heart of this camera is a 64 Kbit dynamic RAM with a glass cover instead of the usual metal lid. A lens on the front of the camera is used to focus the image directly onto the surface of the dynamic RAM. Here's how it works.

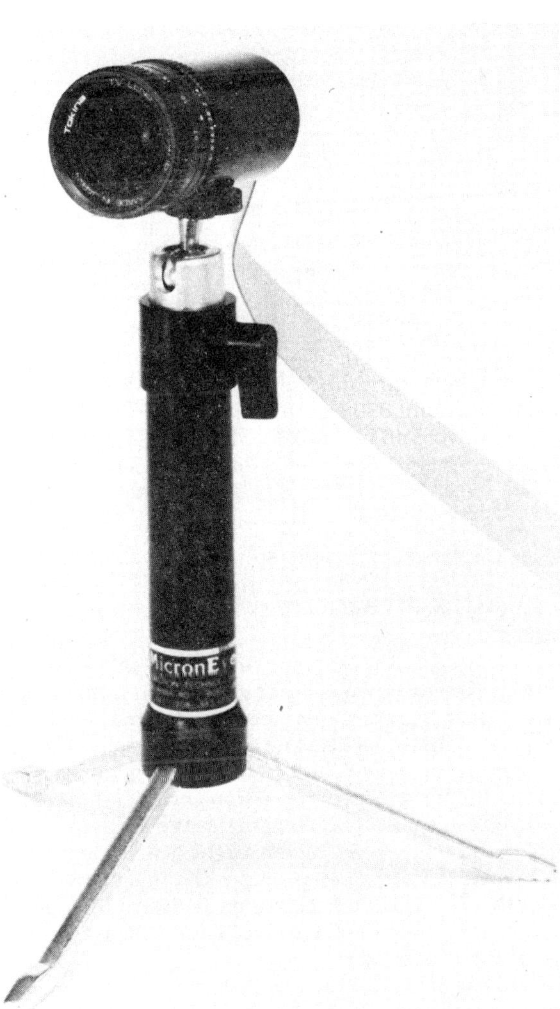

FIGURE 12-20 Micron Eye optic RAM video camera with interface board for IBM PC. *(Micron Technology Inc.)*

The 65,536 storage cells of dynamic RAM are arranged in two arrays of 128 by 256 cells each. Each cell functions as a pel. There is a dead zone of about 25 cell widths between the two arrays. If the two arrays are used together, this dead zone has to be taken into account.

Remember now that data is stored in dynamic RAMs as a charge on a tiny capacitor. Dynamic RAMs have to be refreshed because the charge gradually changes due to leakage. If you shine a light on a dynamic RAM cell, the charge changes faster than it would without the light. To use the dynamic RAM as an image sensor, then, we start by charging up all of the cells to a logic 1 level. After some amount of time we read the logic level on each cell. A cell which still contains a logic 1 represents a dark spot, and a cell which has dropped to a logic 0 represents a light spot. The logic levels can be read out of the OPTICRAM and stored directly in a microcomputer memory for processing. The sensitivity of the camera to light can be adjusted by changing the time between when you charge up all the cells and when you read out the logic levels on the cells. For brighter light conditions, use a shorter time, etc.

Available with the Micron Eye are printed circuit boards which contain circuitry to interface the camera to common microcomputers such as the IBM PC, the APPLE, and the Commodore 64. With these boards installed you can display images on the CRT screen, adjust display parameters under program control, and save images on a disk. Once you get the bit pattern for an image into memory, you can then experiment with programs which attempt to recognize the image of, for example, a square in the image.

Figure 12-21 shows an example of what a little vision can do for a robot. The Sumitomo Electric Company robot shown here can play an organ using both hands on the keys and both feet on the pedals. It can press up to 15 keys per second. The robot can play selections from memory when verbally told to do so. Using its vision it can read and play songs from standard sheet music. The robot uses seventeen 16-bit microprocessors and fifty 8-bit controllers to control all of its activities.

If you think some about what is involved in recognizing complex visual shapes, in all of their possible orientations, with a computer program, it should give you a

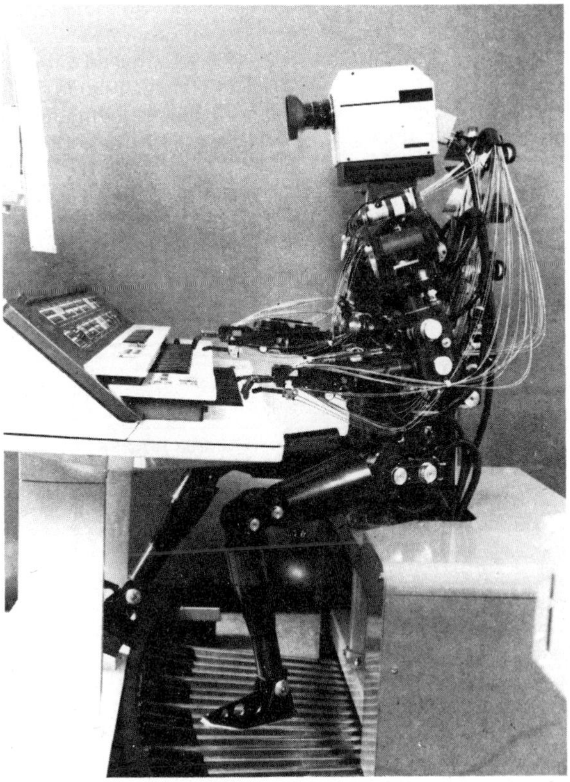

FIGURE 12-21 Organ-playing robot developed by Sumitomo Electric Company.

new appreciation for the pattern recognition capabilities of the human eye-brain system.

Another area where the human brain excels is in that of data storage. Only very recently have the devices used to store computer data approached the capacity of the human brain. In the next section we look at how some of these mass data storage systems operate, and how they are interfaced to microcomputers.

MASS DATA STORAGE SYSTEMS

Since the ROM and RAM in a computer cannot possibly hold all of the programs that we might want to run and all of the data that we might want to analyze, a computer system needs some other form of data storage which can hold massive amounts of data, is nonvolatile, can be updated, and has relatively low cost per bit of storage. The most common devices used for mass data storage are magnetic tape, floppy magnetic disks, hard magnetic disks, and optical disks. Magnetic tapes are used mostly for backup storage, because the access time to get to data stored in the middle of the tape is usually too long to be acceptable. Therefore, in our limited space here we will concentrate on the three types of disk storage.

FLOPPY DISK DATA STORAGE

Floppy Disk Overview

Figure 12-22 shows a picture of a typical floppy disk enclosed in its protective envelope. The common sizes for disks are 8, 5.25, and 3.5 inches. The disk itself is made of Mylar and coated with a magnetic material. The Mylar disk is only a few thousanths of an inch thick, thus the name floppy. When the disk is inserted in a drive unit, a spindle clamps in the large center hole and

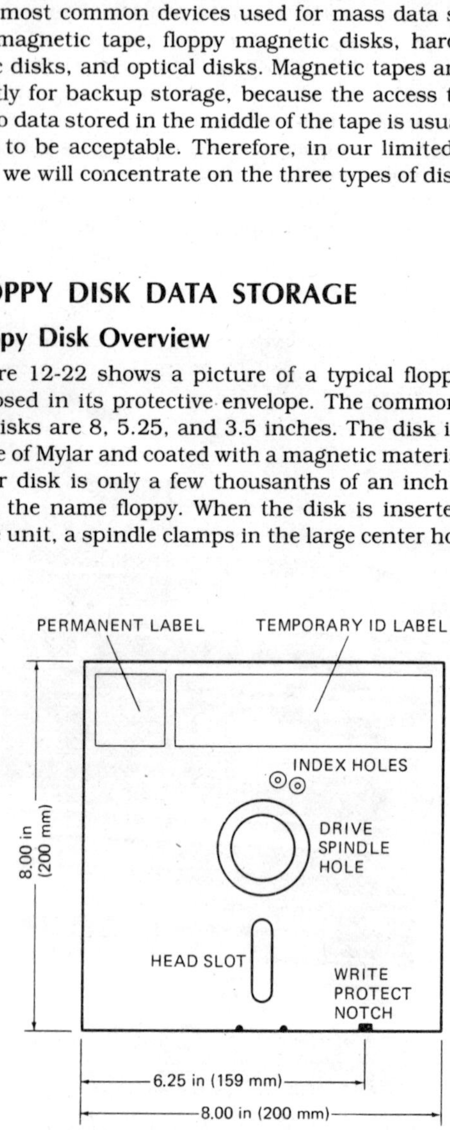

FIGURE 12-22 Floppy disk in protective envelope.

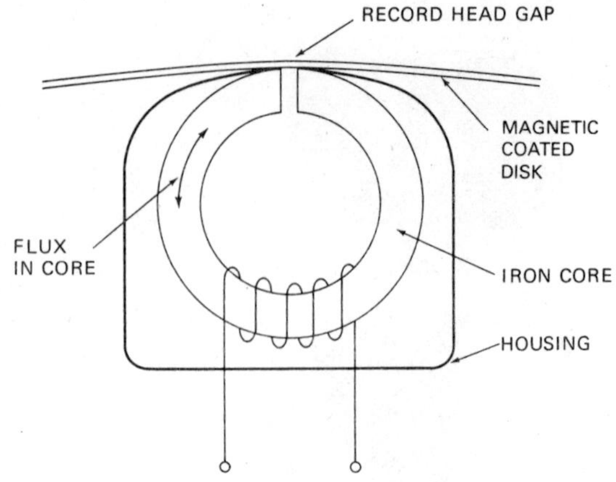

FIGURE 12-23 Magnetic disk read/write head.

spins the disk at a constant speed of perhaps 300 or 360 rpm.

Data is stored on the disk in concentric, circular tracks, rather than in a spiral track as it is on a phonograph record. A read/write head contacts the disk through the racetrack-shaped slot to read from or write to the disk. Figure 12-23 shows a diagram of a read/write head. In the write mode a current passing through the coil in the head creates a magnetic flux in the iron core of the head. A gap in the iron core allows the magnetic flux to spill out and magnetize the magnetic material on the disk. Once a region on the disk is magnetized in a particular direction, it retains that magnetism. The polarity of the magnetized region is determined by the direction of the current through the coil. We will say more about this later.

Data can be read from the disk with the same head. Whenever the polarity of the magnetism changes as the track passes over the gap in the read/write head, a small voltage, typically a few millivolts, is induced in the coil. An amplifier and comparator are used to convert this small signal to standard logic levels.

The write-protect notch in a floppy disk envelope can be used to protect stored data from being written over, as the knock-out plastic tabs on audiotape cassettes are. An LED and a phototransistor can indicate whether the notch is present and disable the write circuits if it is.

An index hole punched in the disk indicates the start of the recorded tracks. An LED and a phototransistor are used to detect when the index hole passes.

Disk Drive and Head Positioning

The motor used to spin the floppy disk is usually a dc motor whose speed is precisely controlled by negative feedback as we described in Chapter 10. In most systems this speed will be held constant at all times. Typically it takes about 250 ms for the motor to start up after a start motor command.

The most common method of positioning the read/write head over a desired track is with a stepper motor. A lead screw or a let-out–take-in steel band, such as that

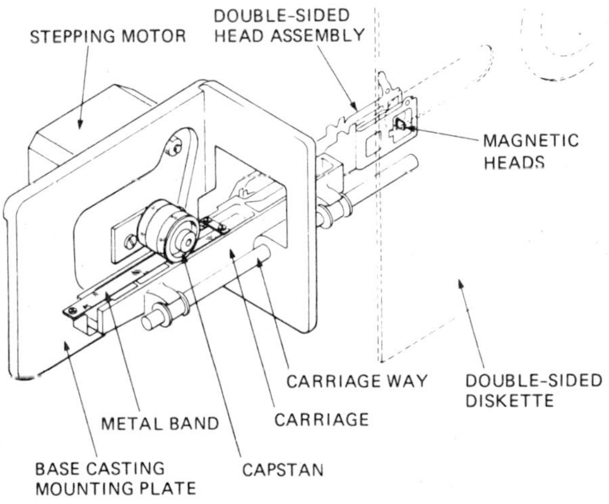

FIGURE 12-24 Head positioning mechanism for floppy disk drive unit. *(Shugart Corporation)*

shown in Figure 12-24, converts the rotary motion of the stepper motor to the linear motion needed to position the head over the desired track on the disk. As the stepper motor in Figure 12-24 rotates, the steel band is let out on one side of the motor pulley, and pulled in on the other side. This slides the head along its carriage.

To find a given track, the motor is usually stepped to move the head to track zero near the outer edge of the disk. The motor is then stepped the number of steps required to move the head to the desired track. Typically it takes a few hundred milliseconds to position the head over a desired track.

Once the desired track is found, the head must be pressed against the disk or *loaded*. Typically it takes about 50 ms to load the head and allow it time to settle against the disk.

Floppy Disk Data Formats and Error Detection

As we said previously, floppy disks come in several standard sizes. Larger disks tend to have more data tracks than smaller disks, but there is no one standard number of tracks for any size disk; 8-in disks typically have about 77 tracks/side, 5.25-in disks about 40 tracks/side, and the new 3.5-in disks in hard plastic envelopes about 80 tracks/side. Single-sided drives record data tracks on only one side of the disk. Double-sided drives use two read/write heads to store data on both sides of the disk. The data tracks on floppy disks are divided into sectors. There are two different methods of indicating the start of sectors: *hard sectoring* and *soft sectoring*. Hard-sectored 8-in disks typically have 32 additional index holes spaced equally around the disk. Each hole signals the start of a sector. The index hole photodetector is used to detect these sector holes.

Soft-sectored disks have only the one index hole which indicates the start of all of the tracks. The sector format is established by bytes stored on the track. Most newer systems use soft sectoring because it is more reliable than hard sectoring.

The actual digital data is stored on floppy disks in many different formats, so we can't begin to show you all of them. To give you a general idea, we will use an old standard, the IBM 3740 format, which is the basis of most current formats. Figure 12-25 shows how bytes are written to a track in this format.

In the 3740 format a track has three types of fields. An *index field* identifies the start of the track. *ID fields* contain the track and sector identification numbers for each of the 26 data sectors on the track. Each of the 26 sectors also contains a *data field* which consists of 128 bytes of data plus two bytes for an error-checking code. As you can see, in addition to the bytes used to store data, many bytes are used for identification, synchronization, error checking, and buffering between sectors. One type of separator used here is called a *gap*. A gap is simply a region which contains no data. Gaps are provided to separate fields, so that the information stored in one field can be changed without altering an adjacent field.

Address marks shown at several places in this format are special bytes which have an extra clock pulse recorded along with their D2 data bit. Address marks are used to identify the start of a field. The four types of address mark are: index, ID, data, and deleted data.

Two bytes at the end of each ID field and 2 bytes at the end of each data field are used to store *checksums* or *cyclic redundancy characters*. These are used to check for errors when the ID and the data are read out. A data checksum, for example, is produced by adding up all of the data bytes and keeping only the least-significant 2 bytes of the result. These 2 bytes are then recorded after the data bytes. When the data is read, it is readded and the sum is compared with the recorded checksum bytes. If the two sums are equal, then the data was probably read out correctly. If the sums do not agree, then another attempt can be made to read the data. If, after several tries, the sums still do not compare, then a disk read error can be sent out to the CRT.

Instead of using a checksum, most disk systems use a cyclic redundancy character or CRC method. There are actually several similar techniques using CRC. Here's one way to give you the idea. To produce the 2 CRC bytes the 128 data bytes are treated as a single large binary number and are divided by a constant number. The 16-bit remainder from this division is written in after the data bytes as the CRC bytes. When the data bytes and the CRC bytes are read out, the CRC bytes are subtracted from the data string. The result is divided by the original constant. Since the original remainder has already been subtracted, the remainder of the division should be zero if the data was read out correctly. Higher-quality systems usually write data to a disk and immediately read it back to see if it was written correctly. If an error is detected, then another attempt can be made. If 10 write attempts are unsuccessful, then the operator can be prompted to throw out the disk, or the write can be directed to another sector on the disk.

The IBM 3740 format we have been describing is referred to as *single density*. An 8-in disk in this format has one index track and 76 data tracks. Since each track has 26 sectors with 128 data bytes in each sector,

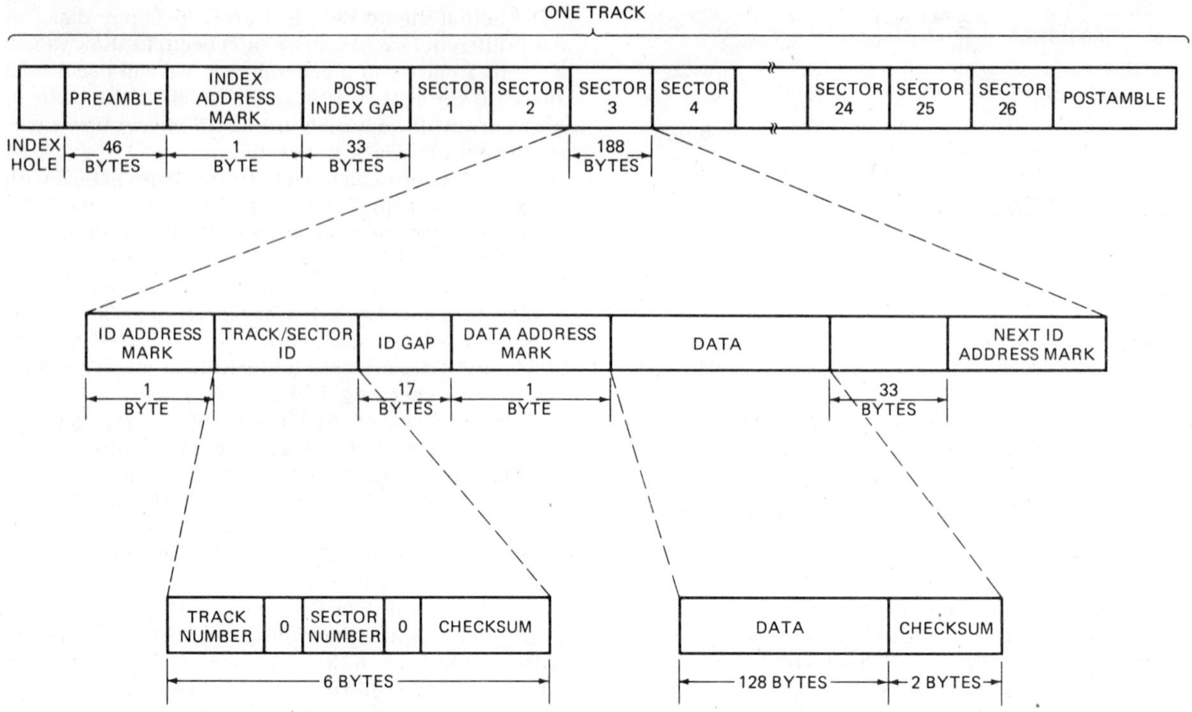

FIGURE 12-25 IBM 3740 floppy disk soft-sectored track format (single density).

the total is about 250 Kbytes. If we use both sides of the disk we get about 500 Kbytes. To increase the storage capacity even further, most systems use *double-density* recording. Double-density recording uses a different clock and data bit pattern to pack twice as many sectors in a track. Now let's look at how data bits are actually recorded on floppy disks.

Recorded Bit Formats – FM and MFM

A "one" bit is represented on magnetic disks as a change in the polarity of the magnetism on the track. A "zero" bit is represented as no change in the polarity of the magnetism. This form of recording is often called *non-return-to-zero* or NRZ recording, because the magnetic field is never zero on a recorded track. Each point on the track is always magnetized in one direction or the other. The read head produces a signal when a region where the magnetic field changes passes over it. As you read through the next section, keep in mind that what we show in the waveforms as a pulse simply represents a change in magnetic polarity on the disk.

Figure 12-26 shows how bits are stored on a disk track in single-density format. This format is often called *frequency modulation*, FM, or F2F recording. Note that there is a clock pulse, C, at the start of each bit cell in this format. These pulses represent the basic frequency. A 1 is written in a bit cell by putting in a pulse, D, between the clock pulses; a 0 is represented by no pulse between the clock pulses. Putting in the data pulses modifies the frequency, thus the name frequency modulation.

The recorded clock pulses are required to sychronize the readout circuits. The actual distance, and therefore time, between data bits read from an outer track is longer than it is for data bits read from an inner track. A circuit called a *phase-locked loop* adjusts its frequency to that of the clock pulses and produces a signal which

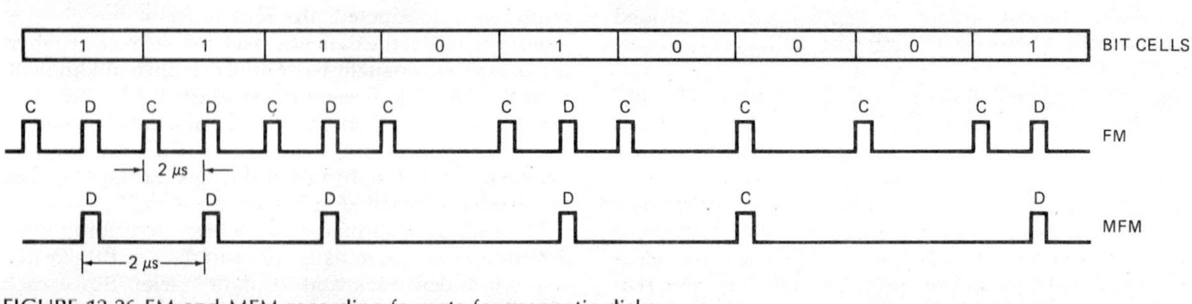

FIGURE 12-26 FM and MFM recording formats for magnetic disks.

tells the read circuit when to check for a data bit. Recording clock information along with data information not only makes it possible to accurately read data from different tracks, but it also reduces the chances of a read error caused by small changes in disk speed.

A disadvantage of standard F2F recording is that a clock pulse and the data bit are required to represent each data bit. Since bits can only be packed just so close together on a disk track without interfering with each other, this limits the amount of data that can be stored on a track in this format. To double the amount of data that we can store on a track we use the *modified frequency modulation* or MFM recording format shown as the second waveform in Figure 12-26. The basic principle of this format is that both clock pulses and "one" data pulses are used to keep the phase-locked loop and read circuitry synchronized. A clock pulse is not put in unless data pulses do not happen to come often enough in the data bytes to keep the phase-locked loop locked. Clock bits are put at the start of the bit cell and data bits are put in the middle of the bit cell time. A clock bit will only be put in, however, if the data bit in the previous cell was a 0, and the data bit in the current bit cell is also a 0. Since this format has in all cases only one pulse per bit cell, a bit cell can be half as long, or in other words, twice as many of them can be packed into a track. This is the way that double-density recording is achieved in the IBM PC and other common microcomputers. For a 5.25-in double-density recorded disk, data bits will be read out at about 250,000 *bits*/s. Incidentally, a new disk recording technology called *perpendicular* or *vertical* recording should allow 4 to 8 times as much data to be put on a given-size disk. With perpendicular recording the tiny magnetic regions are oriented perpendicular to the disk surface instead of parallel to it as they are for standard disks.

Now that we have shown you how digital data is stored on floppy disks, we will show you the circuitry required to interface a floppy disk drive to a microcomputer.

A Floppy Disk Controller – the Intel 8272A

As you can probably tell from the preceding discussion, writing data to a floppy disk and reading the data back requires coordination at several levels. One level is the motor and head drive signals. Another level is the actual writing and reading at the bit level. Still another level is interfacing with the rest of the circuitry of a microcomputer. Doing all of this coordination is a full-time job, so we use a specially designed floppy disk controller to do it. As our example device here we will use the Intel 8272A controller, which is equivalent to the NEC μPD765A controller used in the IBM PC. It is easier to find data sheets and application notes for the 8272A, if you need further information.

8272 SIGNALS AND CIRCUIT CONNECTIONS

Figure 11-3 showed you how an 8272A controller can be connected in an 8086-based microcomputer system. Also in Chapter 11 we discussed in detail how data can be transferred to and from a floppy disk controller on a DMA basis. Now we want to take a closer look at the controller itself to show you the types of signals it produces and how it is programmed.

To start, take a look at the block diagram of the 8272A in Figure 12-27. The signals along the left side of the diagram should be readily recognizable to you. The data bus lines, $\overline{RD}$, $\overline{WR}$, A0, RESET, and $\overline{CS}$ are the standard peripheral interface signals. The DRQ, $\overline{DACK}$, and INT signals are used for DMA transfer of data to and from the controller. To refresh your memory from Chapter 11, here's a review of how the DMA works. When a microcomputer program needs some data off the disk, it sends a series of command words to registers inside the controller. The controller then proceeds to read the data from the specified track and sector on the disk. When the controller reads the first byte of data from a sector, it sends a DMA request, DRQ, signal to the DMA controller. The DMA controller sends a hold request signal to the HOLD input of the CPU. The CPU floats its buses and sends a hold acknowledge signal to the DMA controller. The DMA controller then sends out the first transfer address on the bus and asserts the $\overline{DACK}$ input of the 8272 to tell it that the DMA transfer is underway. When the number of bytes specified in the DMA initialization has been transferred, the DMA controller asserts the TERMINAL COUNT input of the 8272. This causes the 8272 to assert its interrupt output signal, INT. The INT signal can be connected to a CPU or 8259A interrupt input to tell the CPU that the requested block of data has been read in from the disk to a buffer in memory. The process would proceed in a similar manner for a DMA write-to-disk operation.

Now let's work our way through the drive control signals shown in the lower right corner of the 8272 block diagram in Figure 12-27. Reading through our brief descriptions of these signals should give you a better idea of what is involved in the interfacing to the disk drive hardware. Note the direction of the arrow on each of these signals.

The READY input signal from the disk drive will be high if the drive is powered and ready to go. If, for example, you forget to close the disk drive door, the READY signal will not be asserted.

The WRITE PROTECT/TWO SIDE signal indicates whether the write protect notch is covered when the drive is in the read or write mode. When the drive is operating in track-seek mode, this signal indicates whether the drive is two-sided or one-sided.

The INDEX signal will be pulsed when the index hole in the disk passes between the LED and phototransistor detector.

The FAULT/TRACK 0 signal indicates some disk drive problem condition during a read/write operation. During a track-seek operation this signal will be asserted when the head is over track 0, the outermost track on the disk.

The DRIVE SELECT output signals, DS0 and DS1, from the controller are sent to an external decoder which uses these signals to produce an enable signal for one to four drives.

The MFM output signal will be asserted high if the controller is programmed for modified frequency modu-

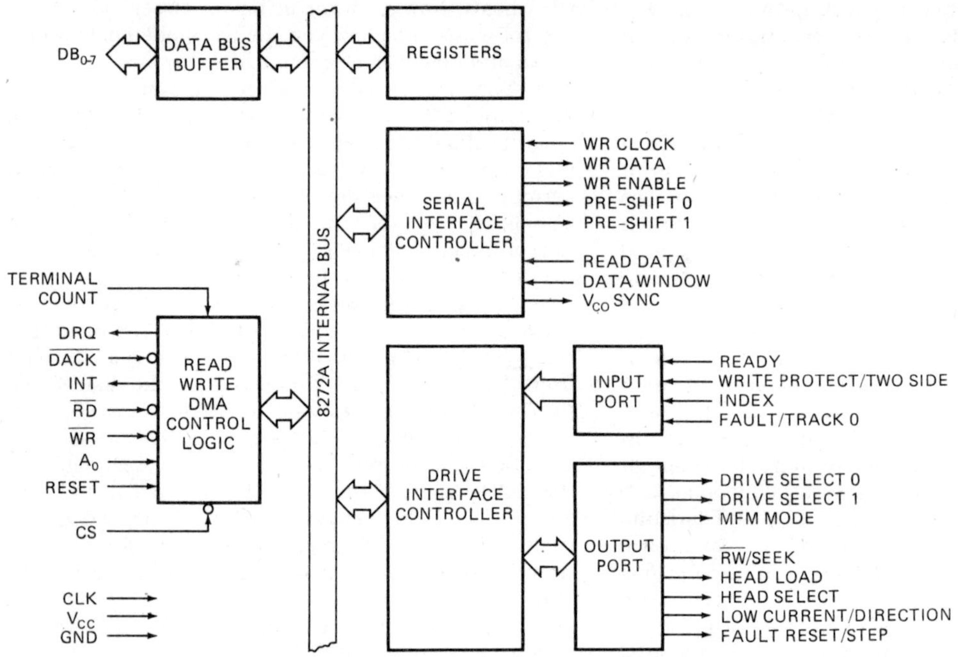

FIGURE 12-27 INTEL 8272A floppy disk controller block diagram.

lation, and low if the controller is programmed for standard frequency modulation (FM).

The $\overline{\text{RW}}$/SEEK signal is used to tell the drive to operate in read-write mode or in track-seek mode. Remember, some of the other controller signals have different meanings in the read-write mode than they do in the seek mode.

The HEAD LOAD signal is asserted by the controller to tell the drive hardware to put the read/write head in contact with the disk. When interfacing to a double-sided drive, the HEAD SELECT from the controller is used along with this signal to indicate which of the two heads should be loaded.

During write operations on inner tracks of the disk the LOW CURRENT/DIRECTION signal is asserted by the controller. Because the bits are closer together on the inner tracks, the write current must be reduced to prevent recorded bits from splattering over each other. When executing a seek-track command this signal pin is used to tell the drive whether to step outward toward the edge of the disk or inward toward the center.

The FAULT RESET/STEP output signal is used to reset the fault flip-flop after a fault has been corrected when doing a read or write command. When the controller is carrying out a track-seek command, this pin is used to output the pulses which step the head from track to track.

Now that we have led you quickly through the drive interface signals, let's take a look at the 8272A signals used to read and write the actual clock and data bits on a track. To help with this, Figure 12-28 shows a block diagram of the circuitry between these pins and the read/write head.

Remember from our discussion of FM and MFM recording that clock information is recorded on the track with the data information. We use the clock bits to tell us when to read the data bits. The V_{CO} SYNC signal from the controller tells an external phase-locked loop circuit to synchronize its frequency with that of the clock pulses being read off the disk. (In the case of MFM recording, the data bits are also part of the signal the PLL locks on). The output from the phase-locked loop circuitry is a DATA WINDOW signal. This signal is sent to the controller to tell it where to find the data pulses in the data stream coming in on the READ DATA input.

For writing pulses to the disk, the story is a little more complex. External circuitry supplies a basic WR CLOCK signal at a frequency of 500 kHz for FM and 1 MHz for

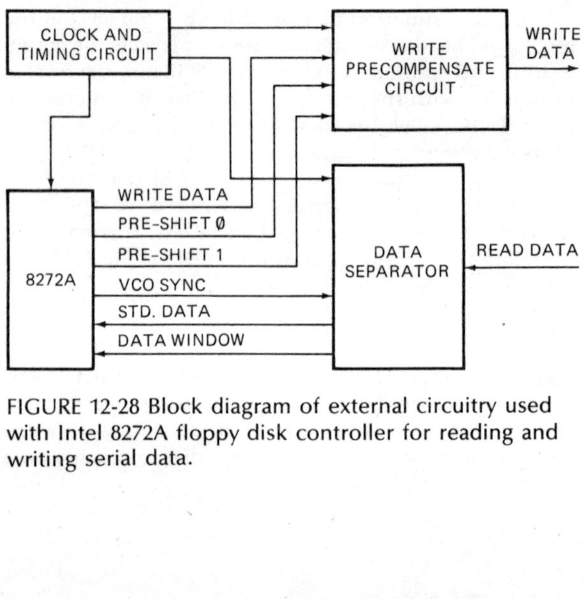

FIGURE 12-28 Block diagram of external circuitry used with Intel 8272A floppy disk controller for reading and writing serial data.

MFM recording. The 8272 outputs the serial stream of clock bits and data bits that are to be written to the disk on its WR DATA pin. During a write operation the 8272 asserts its WR ENABLE signal to turn on the external circuitry which actually sends this serial data to the read/write head. Now, data bits written in MFM on a disk will tend to shift in position as they are read out. A "one" bit, for example, will tend to shift toward an adjacent "zero" bit. This shift could cause errors in readout unless it were compensated for. The PRE-SHIFT 0 and PRE-SHIFT 1 signals from the controller go to external circuitry which shifts bits forward or backward as they are being written. The bits will then be in the correct position when read out.

8272 COMMANDS

The 8272 can execute 15 different commands. Each of these commands is sent to the data register in the controller as a series of bytes. Consult an 8272 data sheet to find the formats for these commands if you need them. After a command has been sent to the 8272, it carries out the command, and returns the results to status registers in the 8272, and/or to the data register in the 8272. To give you an overview of the commands you get to send to an 8272, we list them here with a short description for each.

SPECIFY	—Initialize head load time, head step time, DMA/non-DMA.
SENSE DRIVE STATUS	—Return drive status information.
SENSE INTERRUPT STATUS	—Poll the 8272 interrupt signal.
SEEK	—Position read/write head over specified track.
RECALIBRATE	—Position head over track 0.
FORMAT TRACK	—Write ID field, gaps, and address marks on track.
READ DATA	—Load head, read specified amount of data from sector.
READ DELETED DATA	—Read data from sectors marked as deleted.
WRITE DATA	—Load head, write data to specified sector.
WRITE DELETED DATA	—Write deleted data address mark in sector.
READ TRACK	—Load head, read all sectors on track.
READ ID	—Return first ID field found on track.
SCAN EQUAL	—Compare sector of data bytes read from disk with data bytes sent from CPU or DMA controller until strings match. Set bit in status register if match.
SCAN HIGH OR EQUAL	—Set flag if data string from disk sector greater than or equal to data string from CPU or DMA controller.
SCAN LOW OR EQUAL	—Set flag if data string from disk sector is less than or equal to data string from CPU or DMA controller.

Working out a series of commands for a disk controller such as the 8272 on a bit-by-bit basis is quite tedious and time-consuming. Fortunately, you usually don't have to do this, because in most systems, you can use higher level procedures to read from and write to a disk. In the next section we show you some of the software used to interface to disk drives.

Disk Drive Interface Software

There are several different software levels at which you can interact with a disk drive. One level is directly at the controller level. The next level up is at the BIOS level. A still higher and easier-to-use level is at the disk operating system (DOS) level. Using the IBM PC as an example we will show you in the following sections how to interface your programs with a disk drive using the BIOS approach and the DOS approach.

BIOS Level Floppy Disk Interfacing

In previous discussions we have shown you how to use IBM BIOS procedures to interface with the keyboard, the CRT, and a printer. BIOS procedures, remember, are called with the INT (type) instruction. Figure 12-29 shows the header for the BIOS INT 13H procedure which allows you to interact with disk drives. To give you an idea of what is involved in using this procedure, read through the list of parameters you must pass to it for different operations. As you can see from the header, when using this procedure, you have to specify the particular track and sector(s) that you want to read or write. You have to set up a buffer in memory and pass a pointer to the start of the buffer. Also, you have to set up a table in memory that contains the numbers of tracks and sectors you have recorded data on. The point here is that, yes, you can use this BIOS procedure to interact with a disk by loading registers with the indicated parameters and executing the INT 13H instruction. However, to use an old cliche, it is not a very user friendly way to do it. An easier way to interface your programs with the disk drive is to use DOS procedures. Here's how you do this.

Disk Operating System (DOS) Interfacing

DISK OPERATING SYSTEM OVERVIEW

First of all, let's clarify some terms for you. An *operating system* is simply a program or collection of programs which allows you to format disks, execute programs, create disk files, write data to files, read data from files,

```
SOURCE

        JMZ     K65                     ; DO ANOTHER CYCLE
        POP     AX                      ; RECOVER CONTROL
        OUT     KB_CTL.AL               ; OUTPUT THE CONTROL
        JMP     K27
;--------------------------------------------------------------------
:       ROS CHECKSUM SUBROUTINE                    :
;--------------------------------------------------------------------
ROS_CHECKSUM    PROC    HEAR            ; NEXT_ROS_MODULE
        MOV     CX,8192                 ; NUMBER OF BYTES TO ADD
ROS_CHECKSUM_CNT:                       ; ENTRY FOR OPTIONAL ROS TEST
        XDR     AL,AL
C26:
        ADD     AL,DS:[BX]
        INC     BX                      ; POINT TO NEXT BYTE
        LOOP    C26                     ; ADD ALL BYTES IN ROS MODULE
        OR      AL,AL                   ; SUM = 0?
        RET
ROS_CHECKSUM    ENDP

; -- INT 13 ---------------------------------------------------------------
; DISKETTE I/O                                                              :
;       THIS INTERFACE PROVIDES ACCESS TO THE 5 1/4" DISKETTE DRIVES        :
; INPUT                                                                     :
;       (AH) = 0  RESET DISKETTE SYSTEM                                     :
;                 HARD RESET TO NEC, PREPARE COMMAND, RECALL REQUIRED       :
;                 ON ALL DRIVES                                             :
;       (AH) = 1  READ THE STATUS OF THE SYSTEM INTO (AL)                   :
;                 DISKETTE_STATUS FROM LAST OPERATION IS USED               :
;                                                                           :
; REGISTERS FOR READ/WRITE/VERIFY/FORMAT                                    :
;       (DL)—DRIVE NUMBER (0-3 ALLOWED, VALUE CHECKED)                      :
;       (DH)—HEAD NUMBER (0-1 ALLOWED, NOT VALUE CHECKED)                   :
;       (CH)—TRACK NUMBER (0-39, NOT VALUE CHECKED)                         :
;       (CL)—SECTOR NUMBER (1-8, NOT VALUE CHECKED,                         :
;                      NOT USED FOR FORMAT)                                 :
;       (AL)—NUMBER OF SECTORS ( MAX = 8, NOT VALUE CHECKED, NOT USED       :
;                            FOR FORMAT)                                    :
;       (ES:BX)—ADDRESS OF BUFFER ( NOT REQUIRED FOR VERIFY)                :
;                                                                           :
;       (AH) = 2  READ THE DESIRED SECTORS INTO MEMORY                      :
;       (AH) = 3  WRITE THE DESIRED SECTORS FROM MEMORY                     :
;       (AH) = 4  VERIFY THE DESIRED SECTORS                                :
;       (AH) = 5  FORMAT THE DESIRED TRACK                                  :
;                 FOR THE FORMAT OPERATION, THE BUFFER POINTER (ES.BX)      :
;                 MUST POINT TO THE COLLECTION OF DESIRED ADDRESS FIELDS    :
;                 FOR THE TRACK. EACH FIELD IS COMPOSED OF 4 BYTES,         :
;                 (C,H,R,N), WHERE C = TRACK NUMBER, H = HEAD NUMBER,       :
;                 R = SECTOR NUMBER, N = NUMBER OF BYTES PER SECTOR         :
;                 (00 = 128, 01 = 256, 02 = 512, 03 = 1024). THERE MUST BE ONE :
;                 ENTRY FOR EVERY SECTOR ON THE TRACK. THIS INFORMATION     :
;                 IS USED TO FIND THE REQUESTED SECTOR DURING READ/WRITE    :
;                 ACCESS.                                                   :
;                                                                           :
; DATA VARIABLE—DISK_POINTER                                                :
;       DOUBLE WORD POINTER TO THE CURRENT SET OF DISKETTE PARAMETERS       :
; OUTPUT                                                                    :
;       AM = STATUS OF OPERATION                                            :
;                 STATUS BITS ARE DEFINED IN THE EQUATES FOR                :
;                 DISKETTE_STATUS VARIABLE IN THE DATA SEGMENT OF THIS      :
;                 MODULE.                                                   :
;       CY = 0    SUCCESSFUL OPERATION (AH = 0 ON RETURN)                   :
;       CY = 1    FAILED OPERATION (AH HAS ERROR REASON)                    :
;       FOR READ/WRITE/VERIFY                                               :
;                 DS,BX,DX,CH,CL PRESERVED                                  :
;                 AL = NUMBER OF SECTORS ACTUALLY READ                      :
;                 ***** AL MAY NOT BE CORRECT IF TIME OUT ERROR OCCURS      :
;       NOTE:     IF AN ERROR IS REPORTED BY THE DISKETTE CODE, THE         :
;                 APPROPRIATE ACTION IS TO RESET THE DISKETTE, THEN RETRY   :
;                 THE OPERATION, ON READ ACCESSES, NO MOTOR START DELAY     :
;                 IS TAKEN, SO THAT THREE RETRIES ARE REQUIRED ON READS     :
;                 TO ENSURE THAT THE PROBLEM IS NOT DUE TO MOTOR            :
;                 START-UP.                                                 :
;---------------------------------------------------------------------------
        ASSUME  CS:CODE,DS:DATA,ES:DATA
        ORS     0EC59H
DISKETTE_IO     PROC    FAR
        STI             ; INTERRUPTS BACK ON
```

FIGURE 12-29 Header for IBM BIOS INT 13H procedure for interfacing with floppy disk drives.

communicate with system peripherals such as modems and printers, etc. As we will discuss in Chapter 14, some operating systems allow several users to share a CPU on a time-share basis. The term *disk operating system* or DOS means that the operating system resides on a disk and is loaded into memory and executed when you turn on or reset the system. The term *file* in this case refers to a collection of related data accessible by name. The principle is the same as having a named file folder in an office file cabinet.

Using DOS to format disks, write files, and read files relieves you of the burden of keeping track of the individual tracks and sectors. DOS does all of this for you. Now, before we show you how to use DOS procedure calls, we will briefly show you how DOS keeps track of where it puts everything.

Boot record—variable size
First copy of file allocation table—variable size
Second copy of file allocation table—variable size
Root directory—variable size
Data area

FIGURE 12-30 IBM PC DOS format for floppy disks.

Figure 12-30 shows the "housekeeping" information that IBM PC DOS puts on the first track of a disk to do this. The basic structure for these parts is put on a disk when it is formatted with a DOS format command. As files are created and written to the disk, the relevant information for each file is put in the directory and tables.

The boot record in the first sector of the first track indicates whether the disk contains the DOS files needed to load DOS into RAM and run it. Loading DOS and running it is commonly referred to as "booting" the system.

The directory on the disk contains a 32-byte entry for each file. Let's take a quick look at the use of these bytes to get an overview of the information stored for each file.

Byte number
(decimal)

- 0-7 Filename
- 8-10 Filename extension
- 11 File attribute
  - 01H – read only
  - 02H – hidden file
  - 04H – system file
  - 08H – volume label in first 11 bytes, not filename
  - 10H – file is a subdirectory of files in lower level of hierarchical file tree
  - 20H – file has been written to and closed
- 12-21 Reserved
- 22-23 Time the file was created or last updated
- 24-25 Date the file was created or last updated
- 26-27 Starting cluster number – DOS allocates space for files in *clusters* of one or more adjacent sectors in size.
- 28-31 Size of the file in bytes

DOS uses the first *file allocation table* or FAT to keep track of which clusters on a disk are currently being used for each file, and which clusters are still available. The FAT is part of the link between a filename and the actual track and sector numbers where that file is stored. The second FAT is simply a copy of the first, included for backup purposes.

Most current microcomputer operating systems, IBM PC DOS 2.1 and later versions for example, allow you to set up a *hierarchical file structure*. In this structure you have one main or *root* directory which resides in the directory of the disk as shown in Figure 12-30. This root directory can contain the names of program or data files. The root directory can also have the names of *subdirectories* of files. Each subdirectory can also refer directly to program or data files, or it can refer to lower subdirectories. The point here is that this structure allows you to group similar files together, and to avoid going through a long list of filenames to find a particular file you need. To get to a file in a lower level directory, you simply specify the *path* to that file. The path is the series of directory names that you go through to get to that file.

USING IBM PC DOS CALLS IN YOUR PROGRAMS

As we said previously, DOS is largely a collection of procedures which you can call from your programs, similar to the way you call BIOS procedures. Many disk operating systems and earlier versions of PC DOS require you to construct a *file control block* or FCB in order to access disk files from your programs. The format of a file control block differs from system to system, but basically the FCB must contain among other things, the name of the file, the length of the file, the file attribute, and information about the blocks in the file. Version 2.0 and later versions of PC DOS simplify calling DOS file procedures by letting you refer to a file with a single 16-bit number. This number is called the *file handle* or *token*. You simply put the file handle for a file you want to access in a register, and call the DOS procedure. DOS then constructs the FCB needed to access the file. The question that may occur to you at this point is, "How do I know what the file handle is for a file I want to access on a disk?" The answer is that to get the file handle for a disk file you simply call a DOS procedure which returns the file handle in a register. You can then pass the file handle to the procedure that you want to call to access the file. PC DOS treats external devices such as printers, the keyboard, and the CRT as files for read and write operations. These devices are assigned fixed file handles by DOS as follows: 0000 – keyboard, 0001 CRT, 0002 – error output to CRT, 0003 – serial port, 0004 – printer. The point here is that file handles make it easy for you to access files. There are more examples of calling DOS procedures in Figure 13-24, but here are a few to get you started.

Each DOS function (procedure) has an identification number. To call a DOS function you put the function number in the AH register, put any parameters required by the procedure in other registers, and then execute the INT 21H instruction. For example, DOS function call 40H can be used to print a string. To use this procedure set up the registers as follows:

1. Load the function number, 40H, into the AH register.
2. Load the DS register with the segment base of the segment which contains the string.
3. Load the DX register with the offset of the start of the string.
4. Load the CX register with the number of bytes to write.

5. Load the BX register with the fixed file handle for the printer, 0004H.

Then, to call the DOS procedure, execute the INT 21H instruction. Note that the DOS function allows you to send an entire string to the printer, rather than just a single character at a time as the BIOS INT 17H does.

As another example, the DOS 0AH function will read in a string from the keyboard and put the string in a buffer pointed to by DS:DX. Characters will also be displayed on the CRT as they are entered on the keyboard. The function terminates when a carriage return is entered. To use this function, first set up a buffer in the data segment with the DB directive. The first byte of the buffer must contain the maximum number of bytes the buffer can hold. The 0AH call will return the actual number of characters read in the second byte. The function does not require you to pass it a file handle, because the file handle is implied in the function.

To leave a program and return to the DOS command level, you can use the DOS 4AH call. Load AL with 00, AH with 4CH, and execute the INT 21H instruction.

As a final example here, we will show you how DOS calls can be used to open a file, read data from a file into a buffer in memory, and close the file. Opening a file means copying the file parameters from the directory to a file control block in memory and marking the file as open. Closing a file means updating the directory information for the file and marking the file closed. To open a file and get the file handle we use DOS function call 3DH. For this call DS:DX must point to the start of an *ASCIIZ* string which contains the disk drive number, the path, and the filename. An ASCIIZ string is a string of ASCII characters which has a byte of of all 0's as its last byte. Also AL must contain an access code which indicates the type of operation that you want to perform on the file. Use an access code of 00 for read only, 01 for write only, and 02 for read and write. Again, to actually call the function, you load 3DH into AH and execute the INT 21H instruction. The handle for the opened file is returned in the BX register. The first part of Figure 12-31 shows how these pieces are put together.

To read a file we use function call 3FH. For this call BX must contain the file handle and CX the number of bytes to read from the file. DS:DX must point to the buffer location in RAM that the data from the file will be read into. To do the actual call we load 3FH into AH and do an INT 21H instruction. After the file is read, AX contains the number of bytes actually read from the file.

To close the file we load function number 3EH into AH, load the file handle into BX, and execute the INT 21H instruction. The last half of Figure 12-31 shows the instructions you can use to read and close a file. Watch for some more examples in Figure 13-24. Consult the IBM DOS Technical Reference Manual for the details of all of the available function calls.

RAM DISKS

Currently available for most microcomputers are programs which allow you to set aside an area of RAM in

```
;8086 Program fragment
;ABSTRACT:        This code shows how to use DOS functions to
;                 open a file, read the file contents into a buffer
;                 in memory, and close the file
;
; point at start of buffer containing file name
      MOV   DX, OFFSET FILE_NAME ; and move pointer over
      ADD   DX, 02H              ; string length bytes
      MOV   AL, 00               ; open file for read
      MOV   AH, 3DH              ; and get file handle
      INT   21H
      MOV   BX, AX               ; save file handle in BX
      PUSH  BX                   ; and push for future use
      MOV   CX, 2048             ; set up maximum read
; point at memory buffer reserved for disk file contents
      MOV   DX, OFFSET FILE_BUF
      MOV   AH, 3FH              ; read disk file
      INT   21H
      POP   BX                   ; get back file handle for close
      PUSH  AX                   ; save file length returned by
                                 ; 3FH function call
      MOV   AH, 3EH              ; close disk file
      INT   21H
; use the file now stored in memory
```

FIGURE 12-31 Instruction sequence to open a disk file and get file handle, read file contents to a buffer in memory, and close file using IBM PC DOS function calls.

such a way that it appears to DOS as simply another disk drive. In an IBM PC that has two actual drives, A: and B:, the installed RAM disk becomes C:. You can copy files to and from this RAM disk by name just as you would for any other drive. Here's the point of this. Suppose you are using Wordstar to edit program files. Most of the time when you execute a Wordstar command the system must go and get the code for that command from the Wordstar system disk and load it into memory before it can execute the command. This makes you spend a lot of time waiting. If you load all of the Wordstar files into the RAM disk, they can then be accessed much faster because there is no mechanical access time. The advantage of configuring the RAM as a disk drive is that the software can be accessed just as if it were on a disk.

MAGNETIC HARD DISK DATA STORAGE

The floppy disks that we discussed in the previous section have the advantage that they are relatively inexpensive and removable. The distance between tracks, and therefore the amount of data that can be stored on floppy disks, is limited to a large extent by the flexibility of the disks. The rate at which data can be read off a disk is limited by the fact that a floppy disk can only be rotated at 300 or 360 rpm. To solve these problems, we use a hard disk system like the one in Figure 12-32.

The disks in a hard disk system are made of a metal alloy, coated on both sides with a magnetic material. Hard disks are more dimensionally stable. This means that they can be spun at higher speed, and that tracks and the bits on the tracks can be put closer together. In most cases the hard disks are permanently fastened in the drive mechanism and sealed in a dust-free package, but some systems do have removable enclosed disks. Common hard disk sizes are 3.5, 5.25, 8, 10.5, 14, and 20 in. To increase the amount of storage per drive, several disks may be stacked with spacers between. A read-write head is used for each disk surface. Current technology allows 3 to 10 Mbytes per 5.25-in disk, 5 to 20 Mbytes per 8-in disk, 30 to 50 Mbytes per 10-in disk, and 40 to 100 Mbytes per 14-in disk.

FIGURE 12-32 Multiple-platter hard disk memory system.

Rigid disks are rotated at 1000 to 3600 rpm. This high speed not only makes it possible to read and write data faster, it also creates a thin cushion of air that floats the read-write head 10 μin off the disk. Unless the head *crashes*, it never touches the recorded area of the disk, so wear is minimized. When data is not being read or written, the head is retracted to a *parking zone* where no data is recorded. Hard disks must be kept in a dust-free environment, because the diameter of dust and smoke particles may be 10 times the distance the head floats off the disk. If dust does get into a hard disk system, the result will be the same as that which occurs when a plane does not fly high enough to get over some mountains. The head will crash and perhaps destroy the data stored on the disk.

Hard disk drives are often referred to as Winchesters. Legend has it that the name came from an early IBM dual-drive unit with a planned storage of 30 Mbytes/drive. The 30-30 configuration apparently reminded someone of the famous rifle, and the name stuck.

In some hard disk drives the read-write heads are positioned over the desired track by a stepper motor and a band actuator as we described for the floppy disk drive. Other hard disk drives use a *linear voice coil* mechanism to position the read-write heads. This mechanism uses feedback control, such as that we described in Chapter 10, to control the position of what is essentially a linear motor. The feedback system adjusts the position of the head over the desired track until the strength of the read signal is a maximum.

Most hard disk drives record data bits on a disk track using the MFM method we described in the floppy disk section of this chapter. As with floppy disks, there is no real standard for the format in which the data is recorded. Most systems format a track in a manner similar to that shown for floppy disks in Figure 12-25. The hard disk drive unit used in the IBM PC XT, for example, uses two double-sided hard disks with 306 tracks on each disk surface. On disk drives with more than one recording surface, tracks are often referred to as *cylinders*, because if you mentally connect same numbered tracks on the two sides of a disk or on different disks, the result is a cylinder. The cylinder number then is the same as the track number. On the PC XT hard disk, each track has 17 sectors with 512 bytes in each sector. This adds up to about 10 Mbytes of data storage. Data is read out at 5 *Mbits*/s, which is about 10 times faster than the readout rate for double-density floppy disks.

To interface a hard disk drive to a microcomputer system we use a dedicated controller device such as the Intel 82064, which operates similarly to the 8272 floppy disk controller we described previously in this chapter. An added feature of this controller is the ability to record either CRC words or error-correcting code words with each data sector.

From a software standpoint, writing files to and reading files from a hard disk is very similar to the same operations for a floppy disk. To DOS the hard disk appears for the most part as simply another drive. One difference is that a hard disk is often divided into *parti-*

tions so that groups of programs can be separated from each other. Partitions function essentially as separate disks. An operating system loaded from one partition, for example, cannot accidentally destruct another operating system stored in another partition. The only way to get to the other partition in many systems is to reboot the system into that partition.

Another term encountered in connection with hard disks is *file server*. A file server is a hard disk system which has its own CPU and operating system. The unit is usually a major part of a computer network. The function of the file server is to manage the access to and use of files stored on the disk by other systems on the network.

To prevent data loss in the event of a head crash, hard disk files are backed up on some other medium such as floppy disks or magnetic tape. The difficulty with using floppy disks for backup is the number of disks required. Backing up a 10 Mbyte hard disk with 360 Kbyte floppies requires 30 disks and considerable time shoving disks in and out. Many systems now use a high-speed magnetic tape system for backup. A typical streaming tape system, as these high-speed systems are often called, can dump or load the entire contents of a 10 Mbyte hard disk to a single tape in a few minutes. The next technology we discuss here, optical disks, can store even larger amounts of data on a single drive unit than magnetic hard disks can.

OPTICAL DISK DATA STORAGE

Optical disks are probably familiar to you from their use as laser video disks and compact audio disks. Higher quality versions of the same type of disk can be used to store very large quantities of digital data for computers. One currently available unit, the Shugart Optimem 1000, for example, stores up to a total of 1 gigabyte (1000 Mbytes) of data on one side of a single 12-in disk. This amount of storage corresponds to about 400,000 pages of text. In addition to their ability to store large amounts of data, optical disks have the advantages that they are relatively inexpensive, immune to dust, and in most cases, removable. Also, since data is written on the disk and read off the disk with the light from a tiny laser diode, the read/write head does not have to touch the disk. The laser head is held in position above the disk, so there is no disk wear, and the head cannot crash and destroy the recorded data as it can with magnetic hard disks.

The actual drive and head positioning mechanisms for optical disk drives are very similar to those for magnetic hard disk drives. A feedback system is used to precisely control the speed of the motor which rotates the disk. Some units spin the disk at a constant speed of 700 to 1200 rpm. Other systems such as those based on the compact disk (CD) audio format adjust the rotational speed of the disk so that the track passes under the head with a constant linear velocity. In this case the disk is rotated more slowly when outer tracks are read. Some optical disk systems record data in concentric tracks as magnetic disks do. The CD disk systems and some other systems record data on a single spiral track as a phonograph record does. A linear voice coil mechanism with feedback control is used to precisely position the read head over a desired track or section of the track. The head positioning must be very precise, because the tracks on an optical disk are so close together. The 24-μin-wide tracks on the Optimem 1000 disks, for example, are only 70 μin between centers. This spacing allows 40,000 tracks to be put on the disk. For the Optimem 1000 the average access time to a track is 150 ms, and data is read out at 5 Mbits/s. The disk sizes currently available in different systems are 4.72 (the compact audio disk size), 5.25, 12, and 14 in. Optical disk systems are available in three basic types: read only, write once/read, and read/write.

Read only systems allow only prerecorded disks to be read out. A disk which can only be read from is often referred to as an optical ROM or OROM. Examples of this type are the 4.7-in audio compact disks.

Write-once/read systems allow you to write data to a disk, but once the data is written, it cannot be erased or changed. Once data is written, you can read it out as many times as you want. Write-once systems are sometimes referred to by the name DRAW, which stands for *direct read after write*.

Read/write optical disk systems, as the name implies, allow you to erase recorded data and write new data on a disk. The recording materials and the recording methods are different for these different types of systems.

Disks used for read-only and write-once/read systems are coated with a substance which will be altered when a high-intensity laser beam is focused on it with a lens. The principle here is similar to using a magnifying glass to burn holes in paper as you may have done in your earlier days. In some systems the focused laser light actually produces tiny pits along a track to represent 1's. In other systems a special metal coating is applied to the disk over a plastic polymer layer. When the laser beam is focused on a spot on the metal, heat is transferred to the polymer, causing it to give off a gas. The gas given off produces a microscopic bubble at that spot on the thin metal coating to represent a stored 1. Both of these recording mechanisms are irreversible, so once written, the data can only be read. Data can be read from this type of disk using the same laser diode used for recording, but at reduced power (a system might, for example, use 25 mW for writing, but only 5 mW for reading). In some systems, such as the one in Figure 12-33, a separate laser is used for reading. The laser beam is focused on the track and a photodiode used to detect the beam reflected from the data track. A pit or bubble on the track will spread the laser beam light out so that very little of it reaches the photodiode. A spot on the track with no pit or bubble will reflect light to the photodiode. Read-only and write-once systems are less expensive than read/write systems, and for many data storage applications the inability to erase and rerecord is not a major disadvantage.

For the most common read/write optical disk system the disks are coated with an exotic metal alloy which has the required magnetic properties. The read/write head in this type of system has a laser diode and a coil of wire.

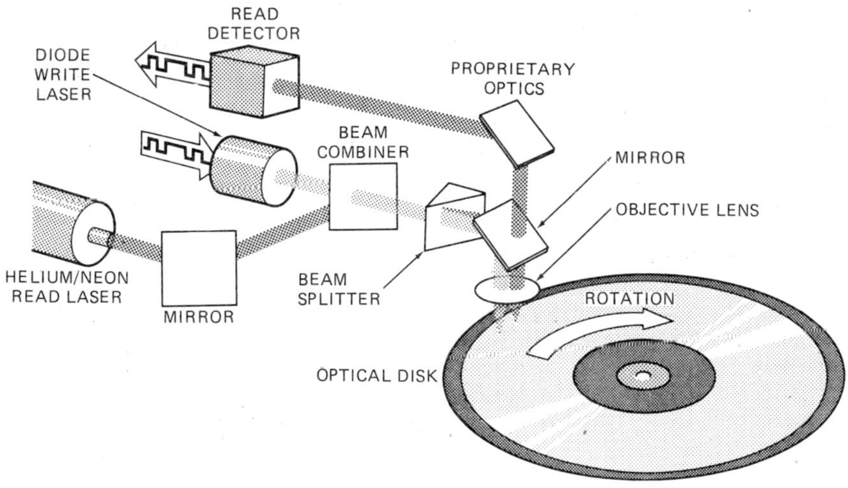

FIGURE 12-33 Read/write mechanism for optical disks.

A current is passed through the coil to produce a magnetic field perpendicular to the disk. At room temperature the applied vertical magnetic field is not strong enough to change the horizontal magnetization present on the disk. To record a 1 at a spot in a data track, a pulse of light from the laser diode is used to heat up that spot. Heating the spot makes it possible for the applied magnetic field to flip the magnetic domains around at that spot and create a tiny vertical magnet. To read data from the disk, polarized laser light is focused on the track. When the polarized light reflects from one of the tiny vertical magnets representing a 1, its plane of polarization is rotated a few degrees. Special optical circuitry can detect this shift and convert the reflections from a data track to a data stream of 1's and 0's. A bit is erased by turning off the vertical magnetic field and heating the spot corresponding to that bit with the laser. When heated with no field present, the magnetism of the spot will flip around in line with the horizontal field on the disk. Other techniques for producing read/write disks are now being researched intensely because of the promise this form of data storage has.

Data is stored on optical disks in several different formats. Figure 12-34 shows the format in which digital data is stored on the 4.7-in audio compact disks.

As shown in Figure 12-34*a*, data is stored serially in one long spiral track, starting near the center of the disk. The track is divided into blocks, each containing 2 Kbytes of actual data. Figure 12-34*b* shows the format for each block. Note that a considerable number of bytes in each block are used for header, synchronization, and error-detecting/correcting codes. Extensive error detection/correction is necessary to bring the error rate down to that of magnetic disks. The position of each block on the track is identified with coordinates of minutes, seconds, and block number. As shown in Figure 12-34*a*, a second represents 75 blocks numbered 0–74. A minute represents 60 seconds, or a total of 4500 blocks. The entire disk represents one hour or 270K blocks. Note that although data can be read out from the disk at 150

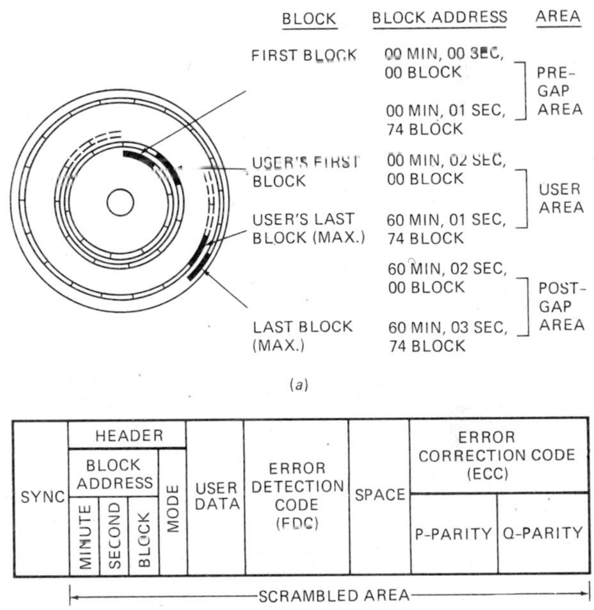

ONE BLOCK (TOTAL): 2352 BYTES

SYNC:	12 BYTES
HEADER:	4 BYTES
USER DATA:	2 KBYTES (2048 BYTES)
EDC:	4 BYTES
SPACE:	8 BYTES
ECC	
P-PARITY:	172 BYTES (REED SOLOMON CODE)
Q-PARITY:	104 BYTES (REED SOLOMON CODE)

USER DATA

1 BLOCK = 2 KBYTES (2048 BYTES)
1 SECOND = 75 BLOCKS = 150 KBYTES
1 MINUTE = 60 SECONDS = 4500 BLOCKS
1 DISK = 1 HOUR = 60 MINUTES = 270 K BLOCKS

AVERAGE DATA TRANSFER RATE (SEQUENTIAL) = 150 KBYTES/SEC.

(*b*)

FIGURE 12-34 Industry-wide data structure for audio compact disk (CD) optical disks. *(a)* Disk format. *(b)* Track format. (*Electronic Engineering Times*, March 25, 1985)

Kbytes/s (about 3 times the rate for floppy disks), the disk contains so much data that it takes an hour to read out all of the data on the disk. Also note that a large area at the start of the track and a large area at the end of the track are used as gaps. In all, about half of the total area on an optical disk is used for synchronization, identification, and error correction. This is not a big drawback because of the immense amount of data that can be stored on the disk.

There are currently available several "jukebox" optical disk systems, which contain up to 256 disks. Typically it takes only a few seconds to access a disk. The potentially low cost of a few cents per megabyte and the hundreds of gigabytes of data storage possible for optical disk systems may change the whole way our society transfers and processes information. The contents of a sizable library, for example, can be stored on a few disks. Likewise, the entire financial records of a large company may be able to be kept on a single disk. "Expert" systems for medical diagnosis or legal defense development can use a massive data base stored on disk to do a more thorough analysis. Engineering workstations can use optical disks to store drawings, graphics, or IC-mask layouts. The point here is that optical disks bring directly to your desktop computer a massive data base that previously was only available through a link to large mainframe computers, or in many cases was not available at all. Perhaps the distribution of data made possible by optical disks will reduce the need for printers which we discuss in the next section.

PRINTER MECHANISMS

Many different mechanisms and techniques are used to produce printouts or "hard" copies of programs and data. This section is intended to give you an overview of

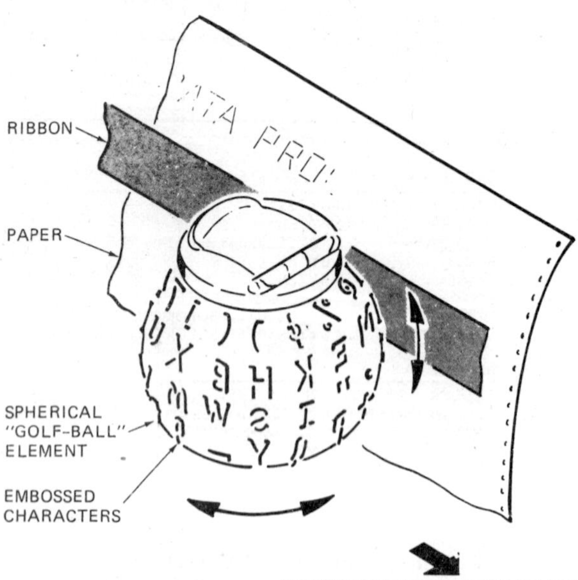

FIGURE 12-35 IBM Selectric printer mechanism. (Data Products Corporation)

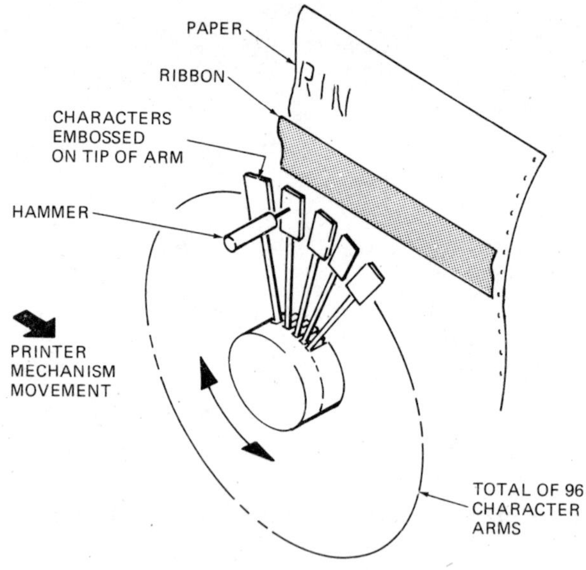

FIGURE 12-36 Daisy-wheel printer mechanism. (Data Products Corporation)

the operation and tradeoffs of some of the common printer mechanisms. We start with those that mechanically hit the paper in some way.

Formed Character Impact Printers

This category of printers function in the same way as a typewriter. In fact the unofficial standard of comparison for print quality is the print produced by the "spinning golf ball" IBM Selectric typewriter.

IBM SELECTRIC MECHANISM

To refresh your memory, Figure 12-35 shows how this works. The entire character set is present as raised type around a sphere. The bottom of the sphere is connected to the drive mechanism. By shifting the ball up or down, rotating it, and tilting it, the character to be printed can be precisely positioned over the desired spot on the paper. When the ball is hit against the ribbon, the letter is printed on the paper. The head is moved across the paper to print a string of characters. Selectric typewriters can be interfaced to computers to do printouts.

The advantages of the Selectric mechanism are the excellent print quality and the fact that the *font* can be changed by simply changing the sphere. Font is the name used to refer to the character set of a printer. The disadvantages of this mechanism are: it is mechanically complicated, noisy, and can only print about 14 characters per second (cps).

DAISY-WHEEL PRINTERS

Figure 12-36 shows a drawing of a daisy-wheel printer mechanism. Here the raised letters are attached at the ends of spokes of a wheel. To print a letter the wheel is rotated until the desired letter is in position over the paper. A solenoid-driven hammer then hits the "petal" against the ribbon to print the letter.

The advantages of the daisy-wheel mechanism are: high print quality, interchangeable fonts, and print speed up to 55 cps. Print quality is not quite as good as that produced by the spinning golf ball.

DRUM, BAND, AND CHAIN PRINTERS

A daisy-wheel produces good quality print, but for massive data output from large mini and mainframe computers, 55 cps is not nearly fast enough. For these systems *drum*, *band*, or *chain* type line printers are used. Figure 12-37 shows a diagram of how a drum type is constructed. A rapidly spinning drum has a complete raised character set constructed around the drum for each character position across the paper. To print characters, magnetically driven hammers in each character position hit the paper and ribbon against the spinning drum. An entire line of characters can be printed during each rotation of the drum. Some drum printers can print 2000 lines/min. If you assume 80 character per line, this corresponds to 2700 cps. However, print lines may be wavy, fonts are not easily changed, and the noise level is high.

In a band printer several raised character sets are constructed on a metal band which is rapidly pulled across a line position behind the paper. Each character position has a magnetically driven hammer such as those shown for the drum printer. When the desired character is under a hammer, the hammer is fired. This hits the ribbon and paper against the letter on the band and prints the character. Some band printers can print up to 2000 lines/min. Print quality is acceptable, fonts are easily changed, and the noise level is high.

Chain printers operate like band printers, except that the character sets are held in a metal or rubber chain and rotated across the paper along a print line. Another variation of this type of printer is the *train* printer which rotates metal slugs with characters on them around in a track across the paper. These mechanisms also produce print speeds up to 2000 lines/min and the font is changeable, but they are noisy and the print mechanism tends to wear out.

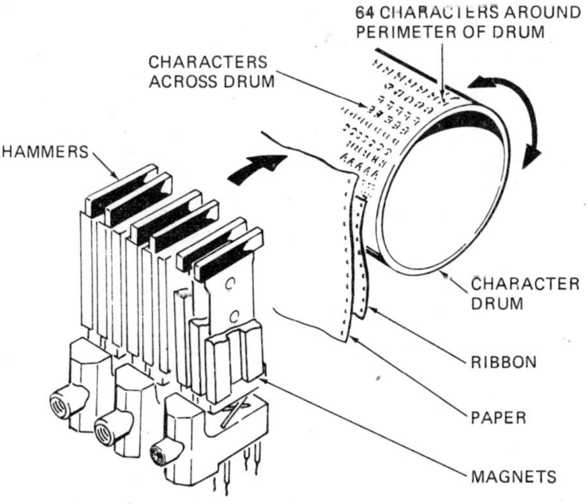

FIGURE 12-37 Drum printer mechanism. (Data Products Corporation)

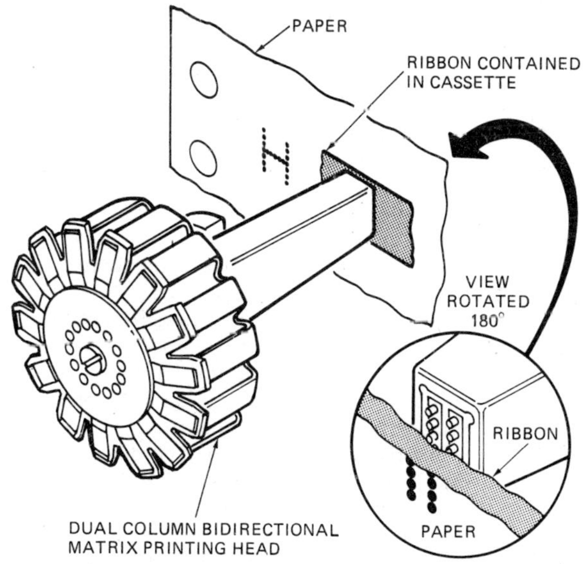

FIGURE 12-38 Impact dot-matrix printer mechanism. (Data Products Corporation)

Dot-Matrix Impact Print Mechanisms

Figure 12-38 shows an impact-type dot-matrix print head. Characters are printed as a matrix of dots. Thin print wires driven by solenoids at the rear of the print head hit the ribbon against the paper to produce dots. The print wires are arranged in a vertical column so that characters are printed out one dot column at a time as the print head is moved across a line. Early dot-matrix print heads had only seven print wires, so print quality of these units was not too good. Currently available dot-matrix printers use 9, 14, 18, or even 24 print wires in the print head. Using a large number of print wires and/or printing a line twice with the dots for the second printing offset slightly from those of the first, produces print that is difficult to tell from that of a Selectric or daisy wheel.

Unlike the formed character printers, dot-matrix printers can also print graphics. To do this the dot pattern for each column of dots is sent out to the print head solenoids as the print head is moved across the paper. The principle is similar to the way we produce bit-mapped raster graphics on a CRT screen. By using different color ribbons and making several passes across a line, some dot-matrix impact printers allow you to print color graphics. Most dot-matrix printers now contain one or more microprocessors to control all of this.

Print speeds for dot-matrix impact printers range up to 350 cps. Some units allow you to use a low-resolution mode of 200 cps for rough drafts, a medium resolution mode of 100 cps for finish copy, or 50 cps for near-letter-quality printing. A big advantage of dot-matrix impact printers is their ability to change fonts or print graphics under program control.

Dot-Matrix Thermal Print Mechanisms

Most thermal printers require paper which has a special heat-sensitive coating. When a spot on this special paper is heated, the spot turns dark. Characters or graphics are printed with a matrix of dots. There are two main print head shapes for producing the dots. For one of these the print head consists of a 5 by 7 or 7 by 9 matrix of tiny heating elements. To print a character the head is moved to a character position and the dot-sized heating elements for the desired character turned on. After a short time the heating elements are turned off and the head is moved to the next character position. Printing then is done one complete character at a time.

The second print head configuration for thermal dot-matrix printers has the heating elements along a metal bar which extends across the entire width of the paper. There is a heating element for each dot position on a print line, so this type can print an entire line of dots at a time. The metal bar removes excess heat. Characters and graphics are printed by stepping the paper through the printer one dot line at a time. A few thermal printers can print up to 400 lines/min.

Some of the newer thermal printers have the heat-sensitive material on a ribbon instead of on the paper. When a spot on the ribbon is heated, a dot of ink is transferred to the paper. This approach makes it possible to use standard paper, and by switching ribbons, to print color graphics as well as text.

The main advantage of thermal printers is their low noise. Their main disadvantages are: the special paper or ribbon is expensive, printing carbon copies is not possible, and most thermal printers with good print quality are slow.

Spark Gap Printers

These printers use a special paper that looks and feels somewhat like aluminum foil. When a spot on the paper is "zapped" with a high voltage, the outer coating at that point is burned off, exposing a dark layer underneath. Characters are printed as a matrix of dots. These printers are often used to print out movie theater tickets because they can print out as many as 2000 cps. Most of the disadvantages relate to the paper which is expensive, difficult to handle, not very durable, and does not produce very good print quality.

Laser and Other Xerographic Printers

These printers operate on the same principle as most office copy machines, such as Xerox machines. The basic approach is to first form an image of the page that is to be printed on a photosensitive drum in the machine. Powdered ink, or "toner," is then applied to the image on the drum. Next the image is electrostatically transferred from the drum to a sheet of paper. Finally the inked image on the paper is "fused," usually with heat.

In a Xerox machine the image on the photosensitive drum is simply a copy of an "original" produced with a camera lens. A more computer-compatible method of producing an image on the photosensitive drum is with a laser. Turning a laser on and off as it is swept back and forth across the drum produces an image in about the same way that an image is produced on a raster scan CRT. Figure 12-39 shows a diagram of how this is done. The rotating mirror sweeps the laser beam across the rotating drum. A modulator controlled by a microcomputer turns the laser beam on or off to produce dots. After the image is inked and transferred to the paper, the drum is cleaned and is ready for the next page.

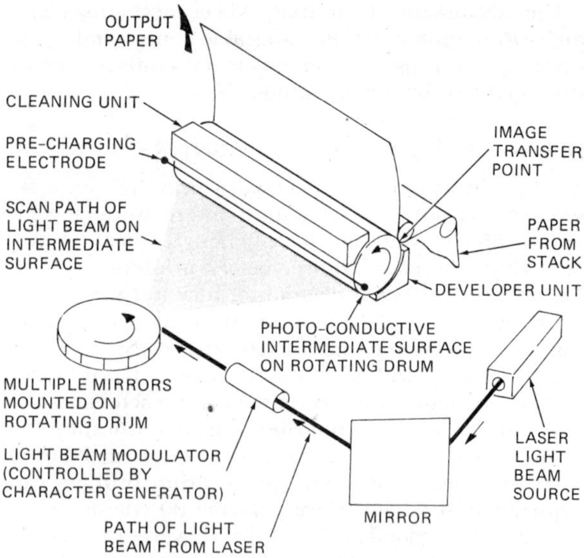

FIGURE 12-39 Laser printer mechanism. (Data Products Corporation)

An alternative to the photosensitive drum is a magnetically sensitive drum used in some units. An image is written on this magnetic drum in the same way that data is recorded on magnetic disks. Magnetized ink particles are then applied to the drum, transferred to the paper, and fused.

Laser and other xerographic printers have the advantages of very high print quality (text and graphics can easily be printed on the same page), very high print speeds (up to 20,000 lines/min), ability to use standard paper, and relatively quiet operation. They have the disadvantages that the copies "look like Xerox copies," the machines are very expensive, and the machines require a lot of maintenance.

Ink-Jet Printers

Still another type of printer that uses a dot-matrix approach to produce text and graphics is the ink jet. Early ink-jet printers used a pump and a tiny nozzle to send out a continuous stream of tiny ink globules. These ink globules were passed though an electric field which left them with an electrical charge. The stream of charged ink globules was then electrostatically deflected to produce characters on the paper in the same way that the electron beam is deflected to produce an image on a CRT screen. Excess ink was deflected to a gutter and returned to the ink reservoir. Ink-jet printers are relatively quiet, and some of these electrostatically deflected ink-jet printers can print up to 45,000 lines/min. Several disadvantages, however, prevented them from being used more widely. They tend to be messy and difficult to keep working well. Print quality at high speeds is poor and multiple copies are not possible.

Newer ink-jet printers use a variety of approaches to solve these problems. Some, such as the HP Thinkjet, use ink cartridges which contain a column of tiny heaters. When one of these tiny heaters is pulsed on, it caused a drop of ink to explode onto the paper. Others, such as the IBM Quietwriter, for example, use an electric current to explode microscopic ink bubbles from a special ribbon directly onto the paper. These last two approaches are really hybrids of thermal and ink-jet technologies. They can produce very near letter-quality print at speeds comparable to those of slower dot-matrix impact printers. A disadvantage of some ink-jet printers is that they require special paper for best results.

SPEECH SYNTHESIS AND RECOGNITION WITH A COMPUTER

In a great many cases it is very convenient for a computer to communicate verbally with a user. Some examples of the use of computer-created speech are talking games, talking cash registers, and text-to-speech machines used by blind people. Other examples are medical monitor systems that give verbal warnings and directions when some emergency condition exists. This use demonstrates some of the major advantages of speech readout. The verbal signal attracts more attention than a simple alarm, and the user does not have to search through a series of readouts to determine the problem.

Adding speech recognition circuitry to a computer so that it can interpret verbal commands from a user also makes the computer much easier to use. The pilot of a rocket ship or space shuttle, for example, can operate some controls verbally while operating other controls manually. (It probably won't be too long before we eliminate the verbal/manual link and control the whole ship directly from the brain, but that is another story, perhaps in the next book.) Voice entry systems are also useful for handicapped programmers and other computer users. We will first describe for you the different methods used to create speech with a computer, and then describe speech recognition methods.

Speech Synthesis Methods

There are several common methods of producing speech from a computer. The tradeoffs between the different methods are speech quality and the number of bits that must be stored/sent for each word. In other words, the higher the speech quality you want, the more bits you have to store in memory to represent each word, and the faster you have to send bits to the synthesizer circuitry. All of the common methods of speech synthesis fall into two general categories: waveform modification, and direct digitization. In order to explain how the waveform modification approaches work we need to talk briefly about how humans produce sounds.

WAVEFORM MODIFICATION SPEECH SYNTHESIS

Some sounds, called voiced sounds, are produced by vibration of the vocal cords as air passes from the lungs. The frequency of vibration or *pitch*, the position of the tongue, the shape of the mouth, and the position of the lips determine the actual sound produced. The vowels A and E are examples of voiced sounds. Another type of sound, called unvoiced sounds, in speech are produced by modifying the position of the tongue and the shape of the mouth as a constant stream of air comes from the lungs. The letter S is an example of this type of sound. A third type of sound, the nasal sounds called *fricatives*, consist of a mixture of voiced and unvoiced sounds. In electronic terms then the human vocal system consists of a variable-frequency signal generator as the source for voiced sounds, a "white" noise signal source for unvoiced sounds, and a series of filters which modify the outputs from the two signal sources to produce the desired sounds. Figure 12-40 shows this in block diagram form.

The three main approaches to implementing this model electronically are *linear predictive coding* or LPC, *formant*, and *phoneme*. These methods differ mostly in the type of filter used, and in how often the filter characteristics are updated.

LPC synthesizers, such as that in the Texas Instruments "Speak and Spell," use a digital filter such as we described in Chapter 10 to modify the signals from a pulse and a white noise source. For this type of filter the parameters that must be sent from the microcomputer are the coefficients for the filter and the pitch for the pulse source. Remember from the discussion in Chapter 10 that for a digital filter, the current output value is computed or "predicted" as the sum of the current input value and portions of previous input values. A high-quality LPC synthesizer may require as many as 16 Kbits/s. One difficulty with most LPC devices has been that complex computer equipment and programs had to be used to analyze a spoken word and determine the series of coefficients required to produce that word. Usually the IC manufacturer did this for a fee, and produced a ROM with the parameters for a particular vocabulary. The General Instruments SP1000, however, has now simplified this process somewhat.

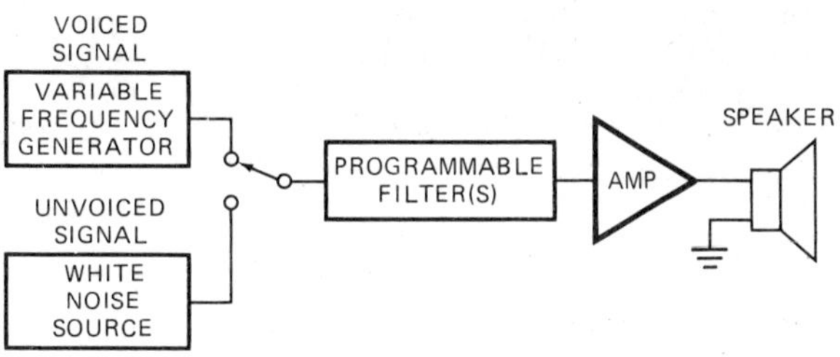

FIGURE 12-40 Electronic model of human vocal tract.

The SP1000 can function an LPC speech processor, an LPC speech recognizer, and an LPC speech synthesizer under the control of a microcomputer. In learn mode the device generates LPC coefficients for spoken words. The microcomputer reads these coefficients from the SP1000, and stores them in memory. To operate in recognition mode the SP1000 is used to generate the coefficients for the unknown word. These coefficients are then compared with those of known words in memory to identify the unknown word. For use as a speech synthesizer the SP1000 is switched to talk mode and the coefficients for the desired word are sent to it by the microcomputer. Consult the General Instruments data sheet for more information about this interesting device.

The formant approach uses several resonant or *formant* filters to massage the signals from a variable-frequency signal source and a white-noise source. Figure 12-41 shows how the frequencies of these formant filters might be arranged for a male and for a female voice. For this type of system the parameters that must be sent from the computer are the pitch of the variable-frequency signal, the center frequency for each formant filter, and the bandwidth of each formant filter. The data rate for direct formant synthesis is only about 1 Kbit/s, but the parameters must again be determined with complex equipment. It is then not easy to develop a custom vocabulary for a specific application. A phoneme approach solves this problem and requires a still lower data rate, at the expense of lower speech quality.

Phonemes are fragments of words. An example of a phoneme speech synthesizer is the Votrax SC-01 which we described in Chapter 9. In the case of the SC-01 you get it to sound one of its 64 phonemes by sending it a 6-bit binary code from a computer port. When the SC-01 finishes sending the phoneme, it asserts a REQUEST signal which indicates that it is ready for the next phoneme. Words are produced by sending a series of phoneme codes. In addition to the 6-bit phoneme code, an additional 2 bits can be sent to specify rising, falling, or flat inflection for each phoneme. Inside the SC-01A, the 6-bit phoneme code is used to control the characteristics of some formant filters as described in the previous paragraph. Since only one code is sent out for a relatively long period of speech, the required bit rate is only about 70 bits/s. However, the long period between codes gives less control over waveform details, and therefore sound quality. A phoneme synthesizer has a mechanical sound. One big advantage of phoneme synthesizers is that you can make up any message you want by simply putting together a sequence of phoneme codes. Another example of a phoneme synthesizer is the SSI263 from Silicon Systems, Inc.

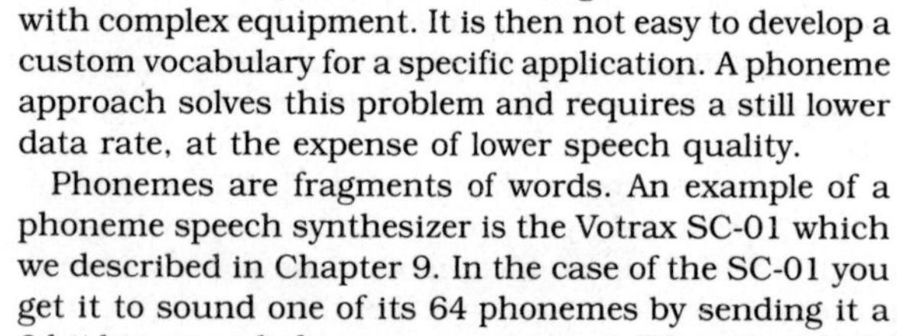

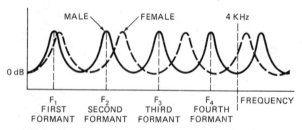

FIGURE 12-41 Filter responses for formant speech synthesizer.

DIRECT DIGITIZATION SPEECH SYNTHESIS

This method produces the highest-quality speech, because it is essentially just a playback of digitally recorded speech. To start, the word you want the computer to speak is spoken clearly into a microphone. The output voltage from the microphone is amplified and applied to the input of perhaps a 12-bit A/D converter. One approach at this point might be to simply store the A/D samples for the word in a ROM and read the values out to a D/A converter when you want the computer to speak the word. The difficulty with this approach is that, if the samples are taken often enough to produce good speech quality, a lot of memory is required to store the samples for a word. To reduce the amount of memory required, several speech compression algorithms are used. These algorithms are too complex to discuss here, but the basic principles involve storing repeated waveforms only once, taking advantage of symmetry in waveforms, and not storing values for silent periods. To further reduce the memory required for direct digital speech, some systems use differential or *delta* modulation. In these systems only a 3-bit or 4-bit code, representing how much a sample has changed from the last sample, is stored in memory instead of storing the complete 12-bit value. Since audio signals change slowly, this is very acceptable. Even with compression, however, direct digital speech requires considerable memory and a bit rate as high as 64 Kbits/s. The OKI Semiconductor MSM5218RS is an example of a device which functions in this way. In record mode it can be used with an A/D converter to produce the differential codes for a spoken word. In play mode the device produces speech from applied codes using an internal 10-bit D/A converter. Another example of a direct digital system is the National Semiconductor *Digitalker*. For further information, consult the data sheets for these devices.

Speech Recognition

Speech recognition is considerably more difficult than speech synthesis. The process is similar to trying to recognize human faces with a computer vision system. The first step in speech recognition is to train the system, or in other words produce templates for each of the words that the system needs to recognize, and store these templates in memory. To produce a template for a word, the intended user speaks the word several times into a microphone connected to the system. The system then determines several parameters or *features* for each repetition of the word and averages them to produce the actual template.

Different systems extract different parameters to form the template. Figure 12-42 shows a block diagram for

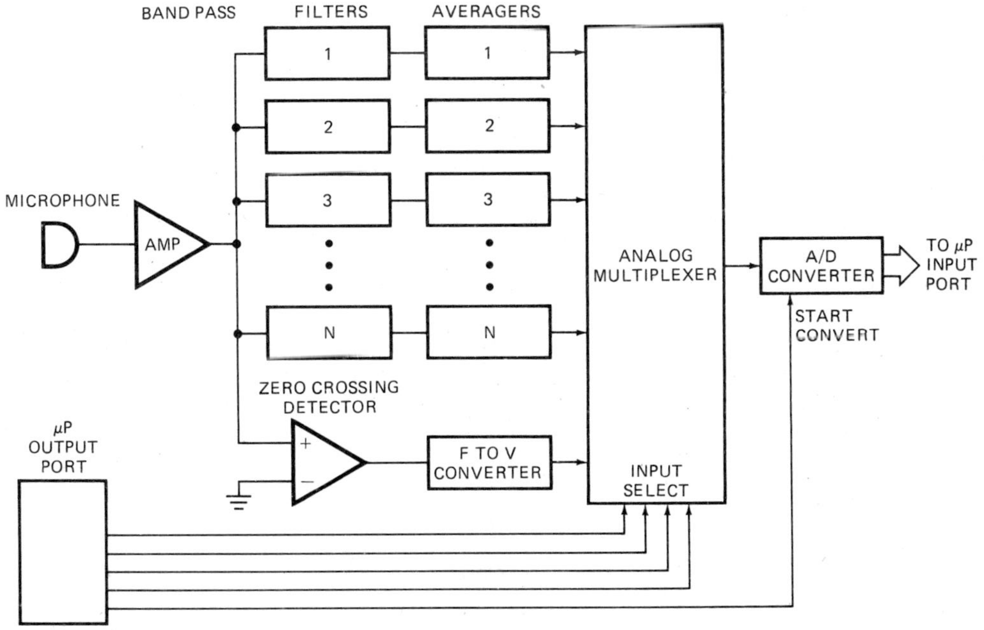

FIGURE 12-42 Block diagram of one type of speech recognition system.

one of the most common methods. This method uses a set of formant filters with their center frequencies adjusted to match those of the average speaker. The output amplitudes of these formant filters are averaged to produce a signal proportional to the energy in each of the frequency bands. Also used are one or more zero-crossing detectors to give basic frequency information. The pulse train from the zero-crossing detector is converted to a proportional voltage, so it can be digitized along with the outputs from the formant averagers.

Now, when a word is spoken, samples of each of the features are taken and digitized at evenly spaced intervals of 10–20 ms during the duration of the word. The features are stored in memory. If this is a training run, this set of samples will be averaged with others to form the template for the word. If this is a recognition run, this set of features will be compared with the templates stored in memory. The best match is assumed to be the correct word. Currently none of the available voice recognition systems is 100 percent accurate. The most accurate systems are those that only work with the speaker who trained them and those that only work with isolated words. However, considerable progress is being made in this area. The VPC 2000 from VOTAN Inc., for example, is a speech recognition unit which plugs into IBM PC-compatible computers and can recognize continuous phrases. It also has a built in voice-activated telephone dialing and answering service. Another PC-compatible unit, the VocaLink from Interstate Voice Products, permits the programming of up to 240 spoken commands to control standard PC software such as word processors and business programs. Perhaps the HAL 9000 is not too far away.

IMPORTANT TERMS AND CONCEPTS FROM THIS CHAPTER

CRT operation
- raster display
- field
- interlaced scanning
- frame

Video monitor

CRT terminal

Horizontal and vertical sync pulses

Composite video

Character generator

Display refresh RAM
- Dot, undot
- dot clock
- overscan

Display page

Attribute code

Bit-mapped raster scan CRT graphics display
- picture element—pixel—pel

Mouse

Composite color monitor

Luminance signal

Chroma signal

Vector-scan CRT displays

Alphanumeric/graphics liquid crystal displays (LCDs)

Computer vision
- ultrasonic vision
- video cameras—vidicon
- CCD cameras
- OPTICRAM cameras

Floppy disks
- hard and soft sectoring
- index holes
- index, ID, and data fields, gaps, address marks

Checksums

Cyclic redundency character

Single and double density

FM and MFM recording

File allocation table

Hierarchical file structure

Root directory and subdirectory

File control block

File handle

ASCIIZ

RAM disk

Hard disk systems
- cylinders
- partitions

Optical disk systems
- OROM
- DRAW

Printer mechanisms
- IBM Selectric
- daisy wheel
- drum, band, and chain
- dot-matrix impact and dot-matrix thermal
- spark gap
- laser and xerographic
- ink jet

Speech synthesis
- pitch, unvoiced sounds, and fricatives
- linear predictive coding, formant, phoneme
- direct digitization

Speech recognition

REVIEW QUESTIONS AND PROBLEMS

1. With the help of a simple drawing explain how a noninterlaced raster is produced on a CRT.

2. Use a simple drawing to help you describe how a display of the letter X is produced on a noninterlaced raster-scan CRT display.

3. Refer to Figure 12-5 to help you answer the following questions.
   a. What is the purpose of the RAM in this circuit?
   b. At what point(s) in displaying a frame do the address inputs of this RAM get changed?
   c. At what point(s) in displaying a frame do the R0–R3 address inputs of the character generator ROM get changed?
   d. What is the purpose of the shift register on the output of the character generator ROM?
   e. Why is one input of the shift register tied to ground?
   f. At what point(s) in displaying a frame are horizontal sync pulses produced?
   g. At what point(s) in displaying a frame are vertical sync pulses produced?
   h. List the three components of a composite video signal.

4. A CRT display is designed to display 24 character rows with 72 characters in each row. The system uses a 7 by 9 character generator in a 9 by 12 dot matrix. Assuming a 60-Hz noninterlaced frame rate, three additional character times for horizontal overscan, and 120 additional scan lines for vertical overscan, answer the following questions.
   a. Total number of character times/row
   b. Total number of scan lines/frame
   c. Horizontal frequency (number of lines/second)
   d. Dot-clock frequency (dots/second)
   e. Minimum bandwidth required for video amplifier
   f. Time between RAM accesses

5. The IBM PC color adapter board uses a 14-MHz dot clock frequency, a 15.750-kHz horizontal scan rate, and a 60-Hz frame rate. Characters are produced in an 8 by 8 dot matrix. There are 80 characters/row and 25 rows/frame.
   a. What is the total number of dot times per scan line?
   b. How many dot times then are left for horizontal overscan?
   c. What is the total number of scan lines per frame including overscan?
   d. How many scan lines then are left for vertical overscan?

6. Describe how a DMA controller is used with a CRT controller such as the 8275 to keep a CRT display refreshed.

7. How does the CRT display system in Figure 12-7 arbitrate the dispute that occurs when the 6845 CRT controller and the microprocessor both want to access the display RAM at the same time.

8. Write a program which uses the IBM BIOS procedures to read a string of characters entered from

the keyboard, put the key codes in a buffer in memory, and display the characters for the pressed keys on the CRT.

9. How much memory is required to store the pel data for a bit-mapped display of 640 by 480?

10. What is the difference between a CRT monitor and a CRT terminal?

11. Describe how three electron beams are used to produce all possible colors on a color CRT screen.

12. How much memory is required to store the pel data for a 512 by 512 display where each pel can be any one of 16 colors?

13. Describe how a composite color video signal is produced from the red, blue, green, and sync signals. Include in your answer the function of the 3.579545-MHz signal.

14. Describe how a vector graphics CRT display system produces a display of a triangle on the screen. What is the major problem with the vector approach to CRT graphics?

15. The inputs of an 8-bit D/A converter are connected to port FFF8H of a microcomputer and the output of the D/A converter is connected to the *X* axis of an oscilloscope. The inputs of another 8-bit D/A converter are connected to port FFFAH of a microcomputer, and the output of this D/A is connected to the *Y* axis of the oscilloscope. Write a program which uses these D/A converters to display a square on the screen of the oscilloscope. Then modify the program so that the square enlarges after each 100 refreshes.

16. Describe the methods used by CCD and OPTICRAM cameras to produce visual images which can be stored in computer memory.

17. How is the read/write head for a disk drive moved into position over a specified track?

18. What additional information besides the actual data is recorded on each track of a soft-sectored floppy disk? Describe the purpose of the CRC bytes included with each block of data recorded on the disk.

19. Why must clock bits be recorded along with data bits on floppy disks? Under what conditions will a clock pulse be inserted in a bit cell when recording data on a disk in MFM format?

20. List the major types of information contained in the directory of a magnetic disk formatted by a DOS. If a data file requires several clusters on a disk, how does a DOS keep track of where the pieces of the file are located.

21. What is meant by the term *hierarchical file structure*? What is a major advantage of this type of file structure?

22. Write a program which uses the IBM PC DOS function calls to read in a string containing your name from the keyboard to a buffer in memory, and sends the string to a printer. Remember to use the DOS 4CH function call to return to DOS at the end of the program.

23. Explain why magnetic hard disks can store much more data than floppy disks, and why data can be written or read out much faster from hard disks.

24. Why must hard disks be operated in a dust-free environment?

25. Two terms often encountered in hard disk system manuals are *cylinder* and *partition*. Define and tell the difference between these two terms.

26. Describe how stored data is read from optical disks. What advantages does this readout method have over that used for hard magnetic disks?

27. Describe how data bits are recorded in magneto-optic read/write optical disk systems and in DRAW optical disk systems.

28. A human brain can store about 10^{10} bits of data and has an access time in the order of about a second. Compare these parameters with those of an optical disk system such as the Optimem 1000.

29. Describe the operation of the print mechanism for each of the following types of printer. Also give an advantage and a disadvantage for each type.
    a. Spinning golf ball
    b. Daisy wheel
    c. Drum
    d. Chain or band
    e. Dot matrix
    f. Thermal
    g. Laser
    h. Ink jet

30. Draw a block diagram of a waveform modification type of speech systhesizer. Describe the operation of the LPC, formant, and phoneme types of speech synthesizer that use this model.

31. What are the major differences between an LPC speech synthesizer and a formant speech synthesizer?

32. Describe the operation of a direct digitization speech synthesizer. What is the major advantage and the major disadvantage of this type?

CHAPTER

13 Data Communication and Networks

In Chapter 2 we discussed "computerizing" an electronics factory. What this means is that computers are integrated into all of the operations of the factory, and that each person in the company has access to a computer. The company may have a large centrally located mainframe computer, several supermicrocomputers that serve groups of users, individual computer engineering workstations, and portable computers spread around the world with the salespeople. In order for all of these computers to work together, they must be able to communicate with each other in an organized manner. In this chapter we show you some of the devices, signal standards, and systems used for communication with and between computers.

OBJECTIVES

At the end of this chapter you should be able to:

1. Show and describe the meaning of the bits in the format used for sending asynchronous serial data.
2. Describe the use of the major signals in the RS-232C standard.
3. Show how to connect RS-232C equipment directly or with a "null-modem" connection.
4. Initialize a common UART for transmitting serial data in a specified format.
5. Use the IBM PC BIOS and DOS procedures to send and receive serial data.
6. Describe several voltage, current, and light (fiber-optic) signal methods used to transmit data.
7. Describe the three types of modulation commonly used by modems.
8. Show the formats for a byte-oriented protocol and for a bit-oriented protocol used in synchronous serial data transmission.
9. Draw diagrams to show the common computer network configurations.
10. Describe how data is transmitted on an Ethernet system.
11. Describe how data is transmitted in a token-passing ring system.
12. Show the major signal groups for the GPIB (IEEE 488) bus, describe how bus control is managed, and how data is transferred on a handshake basis for the GPIB.

ASYNCHRONOUS SERIAL DATA COMMUNICATION

Introduction and Overview

Serial data communication is a somewhat difficult subject to approach, because you need pieces of information from several different topics in order for each part of the subject to really make sense. To make this approach easier, we will first give an overview of how all the pieces fit together and then describe the details of each piece later in specific sections. A problem with this subject is that it contains a great many terms and acronyms. To help you absorb all of these, you may want to make a glossary of terms as you work your way through the chapter. Within a microcomputer data is transferred in parallel, because that is the fastest way to do it. For transferring data over long distances, however, parallel data transmission requires too many wires. Therefore, data to be sent long distances is usually converted from parallel form to serial form so that it can be sent on a single wire or pair of wires. Serial data received from a distant source is converted to parallel form so that it can easily be transferred on the microcomputer buses. Three terms often encountered in literature on serial data systems are *simplex*, *half-duplex*, and *full-duplex*. A simplex data line can transmit data only in one direction. An earthquake sensor sending data back from Mount St. Helens or a commercial radio station are examples of simplex transmission. Half-duplex transmission means that communication can take place in either direction between two systems, but can only occur in one direction at a time. An example of half-duplex transmission is a two-way radio system, where one user always listens while the other talks because the receiver circuitry is turned off during transmit. The term full-duplex means that each system can send and receive data at the same time. A normal phone conversation is an example of a full-duplex operation.

Serial data can be sent *synchronously* or *asynchronously*. For synchronous transmission, data is sent in blocks at a constant rate. The start and end of a block are identified with specific bytes or bit patterns. We discuss synchronous data formats in a later section of this chapter. For asynchronous transmission, each data character has a bit which identifies its start and 1 or 2 bits which identify its end. Since each character is individually identified, characters can be sent at any time (asynchronously), in the same way that a person types on a keyboard at different rates.

Figure 13-1 shows the bit format often used for transmitting asynchronous serial data. When no data is being sent, the signal line is in a constant high or *marking* state. The beginning of a data character is indicated by the line going low for 1 bit time. This bit is called a *start* bit. The data bits are then sent out on the line one after the other. Note that the least-significant bit is sent out first. Depending on the system, the data word may consist of 5, 6, 7, or 8 bits. Following the data bits is a parity bit which, as we explained in Chapter 11, is used to check for errors in received data. Some systems do not insert or look for a parity bit. After the data bits and the parity bit, the signal line is returned high for at least 1 bit time to identify the end of the character. This always-high bit is referred to as a *stop bit*. Some systems use 2 stop bits. For future reference note that the efficiency of this format is low, because 10 or 11 bit times are required to transmit a 7-bit data word such as an ASCII character.

The term *baud rate* is used to indicate the rate at which serial data is being transferred. Baud rate is defined as 1/(the time for a bit cell). If a bit time is 3.33 ms, for example, the baud rate is 1/(3.33 ms), or 300 Bd. There is an almost unavoidable, but incorrect, tendency to refer to this as 300 bits/s. In some cases, the two do correspond, but in other cases 2 to 4 actual data bits are encoded within one transmitted bit time, so data bits per second and baud do not correspond. Commonly used baud rates are 110, 300, 1200, 2400, 4800, 9600, and 19,200 Bd.

To interface a microcomputer with serial data lines the data must be converted to and from serial form. A parallel-in–serial-out shift register and a serial-in–parallel-out shift register can be used to do this. Also needed, for some cases of serial data transfer, is handshaking circuitry to make sure that a transmitter does not send data faster than it can be read in by the receiving system. There are available several programmable LSI devices which contain most of the circuitry needed for serial communication. A device such as the National INS8250, which can only do asynchronous communication, is often referred to as a *universal asynchronous receiver-transmitter* or *UART*. A device such as the Intel 8251A, which can be programmed to do either asynchronous or synchronous communication, is often called a *universal synchronous-asynchronous receiver-transmitter* or *USART*.

Once the data is converted to serial form it must in some way be sent from the transmitting UART to the receiving UART. There are several ways in which serial data is commonly sent. One method is to use a current to represent a 1 in the signal line and no current to represent a 0. We discuss this *current loop* approach in a later section. Another approach is to add line drivers on the output of the UART to produce a sturdy voltage signal. The range of each of these methods, however, is limited to a few thousand feet.

For sending serial data over long distances the standard telephone system is a convenient path, because the wiring and connections are already in place. Standard phone lines, often referred to as *switched lines* because any two points can be connected together through a series of switches, have a bandwidth of only about 300 to 3000 Hz. Therefore, for several reasons, digital signals of the form shown in Figure 13-1 cannot be sent directly over standard phone lines. (NOTE: Phone lines capable of carrying digital data directly can be leased, but these are somewhat costly, and limited to the specific destination of the line.)

The solution to this problem is to convert the digital signals to audio-frequency tones, which are in the frequency range that the phone lines can transmit. The device used to do this conversion and to convert transmitted tones back to digital information is called a *modem*. The term is a contraction of modulator-demodulator. In a later section of this chapter we discuss the operation of some common types of modems. For now, take a look at Figure 13-2 which shows how two modems can be connected to allow a remote terminal to communicate with a distant mainframe computer over a phone line. Modems and other equipment used to send serial data over long distances are known as *data communication equipment* or DCE. The terminals and

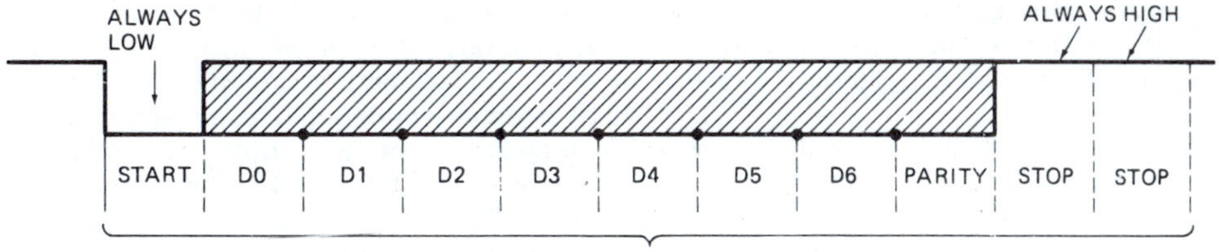

FIGURE 13-1 Bit format used for sending asynchronous serial data.

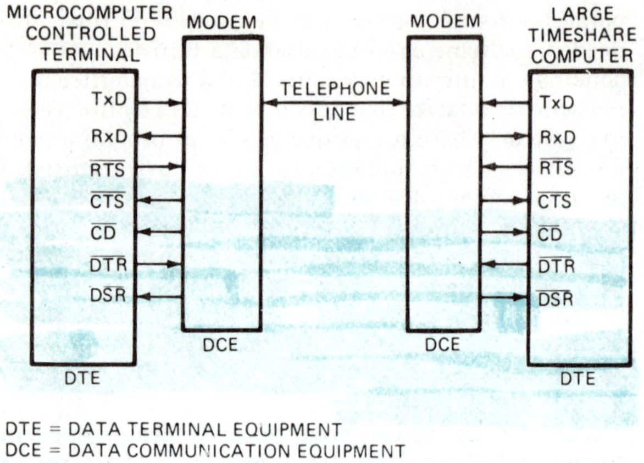

FIGURE 13-2 Digital data transmission using modems and standard phone lines.

computers that are sending or receiving the serial data are referred to as *data terminal equipment* or DTE. The signal names shown in Figure 13-2 are part of a serial data communications standard called RS-232C, which we discuss in detail in a later section. For now you just need enough of an overview so that the signals and initialization of the 8251A described in the next section make sense to you. Note the direction arrowheads on each of these signals. Here is a sequence of signals that might occur when a user at a terminal wants to send some data to the computer.

After the terminal power is turned on and the terminal runs any self-checks, it asserts the *data-terminal-ready* ($\overline{DTR}$) signal to tell the modem it is ready. When it is powered up and ready to transmit or receive data, the modem will assert the *data-set-ready* ($\overline{DSR}$) signal to the terminal. Under manual control or terminal control the modem then dials up the computer.

If the computer is available, it will send back a specified tone. Now, when the terminal has a character actually ready to send, it will assert a *request-to-send* ($\overline{RTS}$) signal to the modem. The modem will then assert its *carrier-detect* ($\overline{CD}$) signal to the terminal to indicate that it has established contact with the computer. When the modem is fully ready to transmit data, it asserts the *clear-to-send* ($\overline{CTS}$) signal back to the terminal. The terminal then sends serial data characters to the modem. When the terminal has sent all the characters it needs to, it makes its $\overline{RTS}$ signal high. This causes the modem to unassert its $\overline{CTS}$ signal and stop transmitting. As we show later, a similar handshake occurs between the modem and the computer at the other end of the data link. The important point at this time is that a set of handshake signals are defined for transferring serial data to and from a modem.

Now that you have an overview of asynchronous serial data, modems, and handshaking, we will describe the operation of a typical device used to interface a microcomputer to a modem or other device which requires serial data.

A Serial Interface Device—The Intel 8251A

Since the 8251A is used as the serial port on SDK-86 boards, on the IBM PC synchronous communication board, and on many other boards, we will use it as an example here.

SIGNALS AND SYSTEM CONNECTIONS

Figure 13-3 shows a block diagram and the pin descriptions for the 8251A. Figure 7-6, sheet 9, shows how an 8251A is connected on the SDK-86 board. Keep copies of these two handy as you work your way through the following discussion.

As shown in the SDK-86 schematic, the eight parallel lines, D7–D0, connect to the system data bus so that data words and control/status words can be transferred to and from the device. The *chip-select* ($\overline{CS}$) input is connected to an address decoder so the device is enabled when addressed. The 8251A has two internal addresses, a control address which is selected when the C/$\overline{D}$ input is high, and a data address which is selected when the C/$\overline{D}$ input is low. For the SDK-86 the control/status address is FFF2H and the data read/write address is FFF0H. The RESET, $\overline{RD}$, and $\overline{WR}$ lines are connected to the system signals with the same names. The clock input of the 8251A is usually connected to the system clock to synchronize internal operations with system operations. In the case of the SDK-86 the clock input is connected to the 2.45-MHz PCLK signal because it is related to the system clock, but at a frequency the 8251A can handle.

The signal labeled TxD on the upper right corner of the 8251A block diagram is the actual *serial-data* output. The pin labeled RxD is the *serial-data* input. The additional circuitry connected to the TxD pin on the SDK-86 board is needed to convert the TTL logic levels from the 8251A to current loop or RS-232C signals. The circuitry connected to the RxD pin performs the opposite conversion. We will discuss current loop and RS-232C signal standards a little later.

The shift registers in the UART require clocks to shift the serial data in and out. $\overline{TxC}$ is the *transmit shift-register clock* input, and $\overline{RxC}$ is the *receive shift-register clock* input. Usually these two inputs are tied together so they are driven by the same clock frequency. Look at Figure 7-6, sheet 9, to see how a variety of clock signals are produced from a 74LS393 counter. A wirewrap jumper is installed to select the desired $\overline{TxC}$ and $\overline{RxC}$. The frequency of the applied clock signal must be 1, 16, or 64 times the transmit and receive baud rate, depending on the mode in which the 8251A is initialized. Using a clock frequency higher than the baud rate allows the receive shift register to be clocked at the center of a bit time rather than at a transition. This reduces the chance of noise at a transition causing a read error.

The 8251A is *double-buffered*. This means that one character can be loaded into a holding buffer while another character is being shifted out of the actual transmit shift register. The TxRDY output from the 8251A will go high when the holding buffer is empty and another character can be sent from the CPU. The TxEMPTY pin

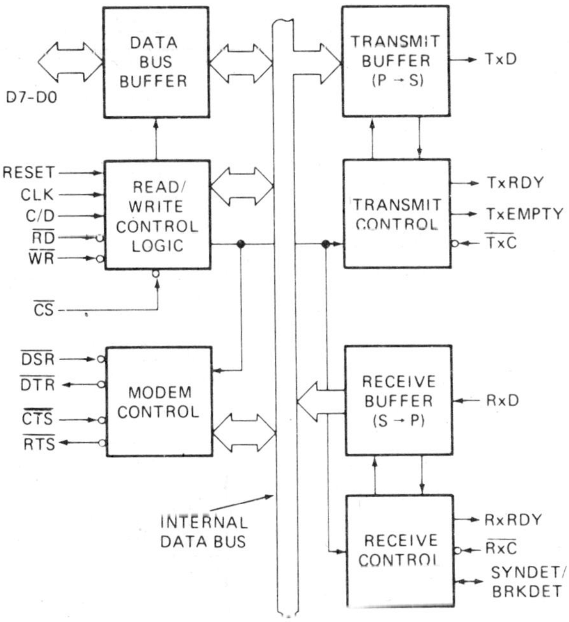

(a)

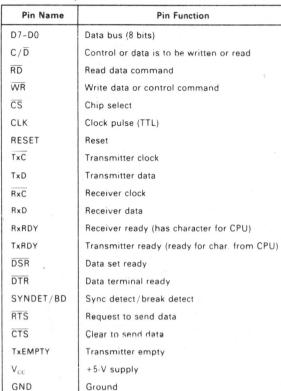

8251 Pin Functions

Pin Name	Pin Function
D7-D0	Data bus (8 bits)
C/$\overline{D}$	Control or data is to be written or read
$\overline{RD}$	Read data command
$\overline{WR}$	Write data or control command
$\overline{CS}$	Chip select
CLK	Clock pulse (TTL)
RESET	Reset
$\overline{TxC}$	Transmitter clock
TxD	Transmitter data
$\overline{RxC}$	Receiver clock
RxD	Receiver data
RxRDY	Receiver ready (has character for CPU)
TxRDY	Transmitter ready (ready for char. from CPU)
$\overline{DSR}$	Data set ready
$\overline{DTR}$	Data terminal ready
SYNDET/BD	Sync detect/break detect
$\overline{RTS}$	Request to send data
$\overline{CTS}$	Clear to send data
TxEMPTY	Transmitter empty
V_{CC}	+5-V supply
GND	Ground

(b)

FIGURE 13-3 Block diagram and pin descriptions for the Intel 8251A USART. *(a)* Block diagram. *(b)* Pin descriptions. *(Intel Corp.)*

on the 8251A will go high when both the holding buffer and the transmit shift register are empty. The RxRDY pin of the 8251A will go high when a character has been shifted into the receiver buffer and is ready to be read out by the CPU. Incidentally, if a character is not read out before another character is shifted in, the first character will be overwritten and lost.

The *sync-detect/break-detect* (SYNDET/BD) pin has two uses. When the device is operating in asynchronous mode, which we are interested in here, this pin will go high if the serial data input line, RxD, stays low for more than 2 character times. This signal then indicates an intentional break in data transmission, or a break in the signal line. When programmed for synchronous data transmission this pin will go high when the 8251A finds a specified sync character(s) in the incoming string of data bits.

The four signals connected to the box labeled MODEM CONTROL in the 8251A block diagram are handshake signals which we described in a previous section.

INITIALIZING AN 8251A

To initialize an 8251A you must send first a mode word and then a command word to the control register address for the device. Figure 13-4 shows the formats for these words and for the 8251A status word which is read from the same address. Baud rate factor, specified by the two least-significant bits of the mode word, is the ratio between the clock signal applied to the $\overline{TxC}$-$\overline{RxC}$ inputs and the desired baud rate. For example, if you want to use a $\overline{TxC}$ of 19,200 Hz and transmit data at 1200 Bd, the baud rate factor is 19,200/1200 or 16 × . If bits D0 and D1 are both made 0's, the 8251A is programmed for synchronous data transfer. In this case the baud rate will be the same as the applied $\overline{TxC}$ and $\overline{RxC}$. The other three combinations for these 2 bits represent asynchronous transfer. A baud rate factor of one can only be used for asynchronous transfer if the transmitting system and the receiving system both use the same $\overline{TxC}$ and $\overline{RxC}$. The character length specified by bits D2 and D3 in the mode word includes only the actual data bits, not the start bit, parity bit or stop bit(s). If parity is disabled, no parity bit is inserted in the transmitted bit string. If the 8251A is programmed for 5, 6, or 7 data bits, the extra bits in the data character byte read from the device will be 0's.

After you send a mode word to an 8251A, you must then send it a command word. A 1 in the least-signifi-

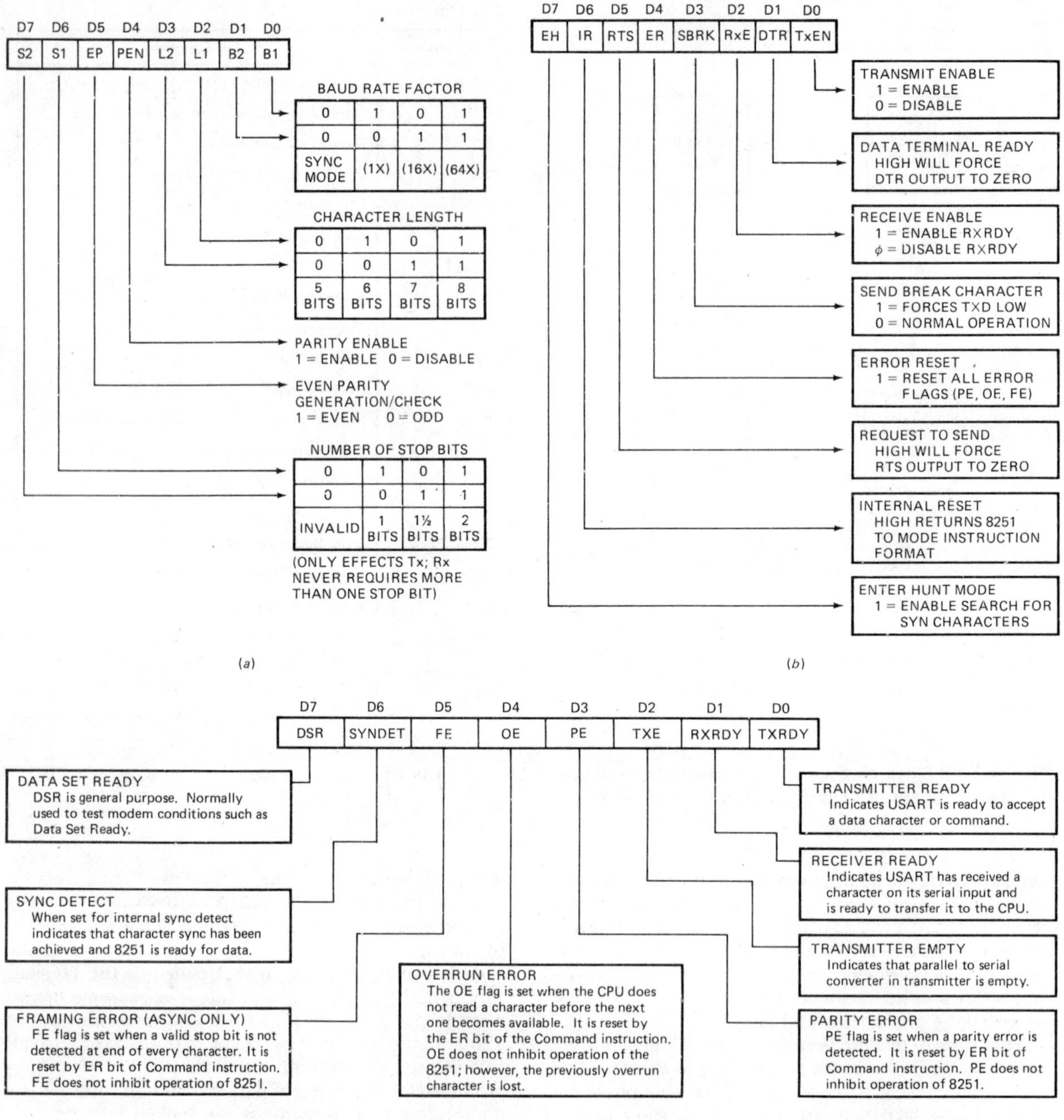

FIGURE 13-4 Format of 8251A mode, command, and status words. *(a)* Mode word. *(b)* Command word. *(c)* Status word. *(Intel Corp.)*

cant bit of the command word enables the transmitter section of the 8251A and the TxRDY output. When enabled, the 8251A TxRDY output will be asserted high if the $\overline{CTS}$ input has been asserted low, and the transmitter holding buffer is ready for another character from the CPU. The TxRDY signal can be connected to an interrupt input on the CPU or an 8259A, so that characters to be transmitted can be sent to the 8251A on an interrupt basis. When a character is written to the 8251A data address, the TxRDY signal will go low and remain low until the holding buffer is again ready for another character. Putting a 1 in bit D1 of the command word will cause the $\overline{DTR}$ output of the 8251A to be asserted low. As we explained before, this signal is used to tell a modem that a terminal or computer is operational. A 1 in bit D2 of the command word enables the RxRDY output pin of the 8251A. If enabled, the RxRDY pin will go high when the 8251A has a character in its receiver buffer ready to be read. This signal can be connected to an interrupt input so that characters can be read in on an interrupt basis. The RxRDY output is reset when a character is read from the 8251A.

Putting a 1 in bit D3 of the command word causes the 8251A to output a character of all 0's, which is called a break character. A break character is sometimes used to indicate the end of a block of transmitted data. Sending a command word with a 1 in bit D4 causes the 8251A to reset the parity, overrun, and framing error flags in the 8251A status register. The meanings of these flags are explained in Figure 13-4c. A 1 in bit D5 of the command word will cause the 8251A to assert its request-to-send ($\overline{RTS}$) output low. This signal, remember, is sent to a modem to ask whether the modem and the receiving system are ready for a data character to be sent.

Putting a 1 in bit D6 of the command word causes the 8251A to be internally reset when the command word is sent. After a software reset command is sent in this way, a new mode word must be sent. Later we will show you how this is used.

The D7 bit in the command word is only used when the device is operating in synchronous mode. A command word with a 1 in this bit position tells the 8251A to look for specified sync character(s) in a stream of bits being shifted in. If the 8251A finds the specified sync character(s), it will assert its SYNDET/BD pin high. We will discuss this more in the synchronous data communication section of this chapter.

Figure 13-5 shows an example of the instruction sequence you can use to initialize an 8251A. This sequence is somewhat lengthy for two reasons. First, the

```
;8086 Instructions to initialize the 8251A on an SDK-86 board
        MOV   DX, OFFF2H          ; point at command register address
        MOV   AL, 00H             ; send 0's to guarantee device is in
        OUT   DX, AL              ; the command instruction format before
        MOV   CX, 2               ; the RESET command is issued
DO:     LOOP  DO                  ; and delay after sending each command
        OUT   DX, AL              ; instruction.
        MOV   CX, 2
D1:     LOOP  D1
        OUT   DX, AL
        MOV   CX, 2
D2:     LOOP  D2
        MOV   AL, 40H             ; Send internal reset command to
        OUT   DX, AL              ; return device to idle state
        MOV   CX, 2               ; Load delay constant
D3:     LOOP  D3                  ; and delay
        MOV   AL, 11001110B       ; Load mode control word and send it

; 1 1 0 0 1 1 1 0         Mode Word
;  \ \ \ \ \ \ \_\______baud rate factor of 16x
;   \ \ \ \ \_\_________character length of 8 bits
;    \ \ \_\____________parity disabled
;     \_\_______________2 stop bits

        OUT   DX, AL
        MOV   CX, 2               ; and delay
D4:     LOOP  D4
        MOV   AL, 00110111B       ; Load command word and send it
        OUT   DX, AL

; 0 0 1 1 0 1 1 1       Command Word
; \ \ \ \ \ \ \ \_____Transmit enable
;  \ \ \ \ \ \ \______Data terminal ready, DTR will output 0
;   \ \ \ \ \ \_______Receive enable
;    \ \ \ \ \________Normal operation
;     \ \ \ \_________Reset all error flags
;      \ \ \__________RST output 0, request to send
;       \ \___________Do not return to mode instruction form
;        \____________Disable hunt mode
```

FIGURE 13-5 Instruction sequence for 8251A initialization.

8251A does not always respond correctly to a hardware reset on power-up. Therefore, a series of software commands must be sent to the device to make sure it is reset properly before the desired mode and command words are sent. The device is put into a known state by writing 3 bytes of all 0's to the 8251A control register address, and then it is reset by sending a control word with a 1 in bit D6. After this reset sequence the desired mode and control words can be sent to the 8251A. The 8251A distinguishes a command word from a mode word by the order they are sent to the device. After reset, a mode word must be sent to the command address. Any words sent to the command address after the mode word will be treated as command words until the device is reset.

The second factor which lengthens this initialization is the *write-recovery* time T_{RV} of the 8251A. According to the data sheet the 8251A requires a worst case recovery time of 16 cycles of the clock signal connected to the CLK input. On the SDK-86 board the PCLK signal, which is the same as the processor clock frequency, is connected to the CLK input of the 8251A. Therefore, for the SDK-86 board, the required write-recovery time corresponds to 16 processor clock cycles. What all this means is that you have to delay this many clock cycles between successive initialization byte writes to the 8251A. A simple way to produce the required delay and some extra is to load CX with 0002 and count it down with the LOOP instruction. The MOV CX, 0002 instruction takes 4 clock cycles, the first execution of the LOOP instruction takes 17 clock cycles, and the last execution of the LOOP instruction takes 5 cycles. The 8 cycles required for the OUT instruction, which writes the control words, also count as part of the time between writes, so the sum of all these is more than enough. When writing data characters to an 8251A you don't have to worry about this recovery time, because a new character will not be written to the 8251A until the previous character has been shifted out. This shifting, of course, requires much more time than T_{RV}.

The comments in Figure 13-5 explain the meanings of the bits in the mode and control words used in this example. Once the 8251A is initialized as shown, new control words can be sent at any time to, for example, reset the error flags. Now let's look at how characters are sent to and read from an 8251A.

SENDING AND RECEIVING CHARACTERS WITH AN 8251A

Data characters can be sent to and read from the 8251A on an interrupt basis or on a polled basis. To send characters on an interrupt basis the TxRDY pin of the 8251A is connected to an interrupt input on the processor or an 8259A priority interrupt controller. The transmitter and the TxRDY output are enabled by putting a 1 in bit D1 of the control word sent to the 8251A during initialization. When the $\overline{CTS}$ input of the 8251A is asserted low, and the 8251A buffer is ready for a character, the TxRDY pin will go high. If the processor and 8259A interrupt path is enabled, the processor will go to an interrupt service procedure which writes a data character to the 8251A data address. Writing the data character causes the 8251A to reset its TxRDY output until the buffer is again ready to receive a character. A counter can be used to keep track of how many characters have been sent.

In a similar manner characters can be read from an 8251A on an interrupt basis. In this case the RxRDY output of the 8251A is connected to an interrupt input of the processor or an 8259A, and this output is enabled by putting a 1 in bit D2 of the command word sent during initialization. When a character has been shifted into the 8251A, and the character is in the receiver buffer ready to be read, the RxRDY pin will go high. If the interrupt chain through the 8259A and the processor is enabled, the processor will go to an interrupt procedure which reads in the data character. Reading a data character from the 8251A causes it to reset the RxRDY output signal. This signal will stay low until another character is ready to be read.

To send characters to an 8251A on a polled basis, the 8251A status register is read and checked over and over until the TxRDY bit (D0) is found to be a 1. In some systems you also want to check bit D7 of the status register to make sure the $\overline{DSR}$ input of the 8251A has been asserted by a signal from, for example, a modem. When the required bit(s) of the status register are all high, a data character is then written to the 8251A. Figure 13-6*a* shows the instruction sequence needed to do this. Note that the status register has the same internal address as the control register. Also note that both an AND and a CMP operation must be done to determine when the two desired bits are both high. Writing a data character to the 8251A resets the TxRDY bit in the status register.

Reading a character from the 8251A on a polled basis is a similar process, except that the RxRDY bit (D1) of the status register is polled to determine when a character is ready to be read. When bit D1 is found high, a character is read in from the 8251A data address. Figure 13-6*b* shows the instruction sequence for this. Status register bits D3, D4, and D5 can be checked to see if a parity error, overrun error, or framing error has occurred. If an error has occurred, a message to retransmit the data can be sent to the transmitting system.

The next step in our journey into serial data communications is to discuss the signal standards used to connect the serial inputs and outputs of UARTS to modems and other serial devices.

SERIAL DATA TRANSMISSION METHODS AND STANDARDS

Aside from drum beats in the jungle, one of the earliest forms of serial data communication was the telegraph. In a telegraph pressing a key at one end of a signal line causes a current to flow through the line. When this current reaches the receiving end of the line, it activates a solenoid (sounder) which produces a sound. Letters and numbers are sent using the familiar Morse code or some other convenient code. After a hundred years or so the telegraph key and sounder evolved into the teletypewriter. A teletypewriter terminal has a typewriter-style keyboard so that the user can simply press a key to send a desired letter or number. A teletype terminal also has a print mechanism which prints out characters as they

```
; Instructions for transmitting data using an SDK-86 8251A
; using polling method

        MOV   DX, OFFF2H          ; point at control register address
TEST1:  IN    AL, DX              ; read status
        AND   AL, 10000001B       ; and check status of
;                \______\________data set ready & transmit ready
        CMP   AL, 10000001B       ; is it ready?
        JNE   TEST1               ; continue to poll if not ready
        MOV   DX, OFFF0H          ; otherwise point at data address
        MOV   AL, DATA_TO_SEND    ; load data to send
        OUT   DX, AL              ; and send it
```

(a)

```
; Instructions for receiving data with an SDK-86 8251A
; using polling method

        MOV   DX, OFFF2H          ; point at control register address
TEST2:  IN    AL, DX              ; read status
        AND   AL, 00000010B       ; and check status of RxRdy
        JZ    TEST2               ; continue to poll if not ready
        MOV   DX, OFFF0H          ; otherwise point at data address
        IN    AL, DX              ; get data
```

(b)

FIGURE 13-6 Instruction sequences for transmitting and receiving with an 8251A on a polled basis. *(a)* Transmit. *(b)* Receive.

are received. Most teletypes use a current to represent a 1 and no current to represent a 0. We start this section by briefly describing the old current-loop standards, and then go on to newer methods.

20- and 60-mA Current Loops

In teletypewriters or other current signal systems some manufacturers use a nominal current of 20 mA to represent a 1, or mark, and no current to represent a 0. Other manufacturers use a nominal current of 60 mA to represent a 1 and no current to represent a 0.

NOTE: The actual current in a specific system may be considerably different from the nominal value.

Sheet 9 of Figure 7-6 shows circuitry which can be used to interface current type signals with the TTL input and output of an 8251A USART on the SDK-86 board. With the jumpers in place as shown, a high on the TxD output of the 8251A will produce a low on the PNP transistor. This will turn the transistor on and cause a positive current to flow out the TTY TX line. Inside a teletypewriter this current flows through an electromagnet and back to the TTY TX RET. To help you visualize this, think of a coil of wire connected between pins 13 and 25 of J7 in the drawing. The current then flows on to −12 V through *R*2 to complete the path. To send a data bit, the teletypewriter opens or closes a switch in a current path. The current for this path in the SDK-86 circuitry is supplied from +5 V through *R*10 to the TTY RX RET line. Think of the key mechanism of the teletypewriter as a simple switch connected between pins 24 and 12 of J7 on the circuit. When the switch is closed the current flows back on the TTY RX line and through *R*3 to −12 V. The current flowing through *R*3 will produce a legal TTL high logic level on the input of the 74LS14 inverter. This high signal passes through two inverters and produces a high on the RxD input of the 8251A.

The circuit as shown in Figure 7-6 is set up for full-duplex operation because the transmit circuit and the receive circuit are independent of each other. In this case, a character sent to the SDK-86 will not be printed out on the teletypewriter until it is echoed back from the SDK-86. This is sometimes referred to as *echoplex* mode. If jumper W17 is installed, the system will operate in half-duplex mode. In this mode a character will be printed out directly as it is typed on the keyboard. However, data cannot be sent and received at the same time because typed characters will be interspersed with characters sent from the SDK-86. We will leave it to you to trace the current paths for this half-duplex mode. Now we will go on to a more common signal standard which uses voltages.

RS-232C Serial Data Standard

OVERVIEW

In the 1960s as the use of time-share computer terminals became more widespread, modems were developed so that terminals could use phone lines to communicate with distant computers. As we stated earlier, modems and other devices used to send serial data are often referred to as *data communication equipment* or DCE. The terminals or computers that are sending or receiving the data are referred to as *data terminal equipment* or DTE. In response to the need for signal and handshake standards between DTE and DCE, the Electronic Industries Association (EIA) developed EIA standard *RS-232C*. This standard describes the functions of 25 signal and handshake pins for serial data transfer. It also describes the voltage levels, impedance levels, rise and fall times, maximum bit rate, and maximum capacitance for these signal lines. Before we work our way through the 25 pin functions, we will take a brief look at some of the other hardware aspects of RS-232C.

RS-232C specifies 25 signal pins and it specifies that the DTE connector should be a male, and the DCE connector should be a female. A specific connector is not given, but the most commonly used connectors are the DB-25P male and the DB-25S female shown in Figure 13-7. When you are wiring up these connectors, it is important to note the order in which the pins are numbered.

The voltage levels for all RS-232C signals are as follows. A logic high, or mark, is a voltage between −3 V and −15 V under load (−25 V no load). A logic low or space is a voltage between +3 V and +15 V under load (+25 V no load). Voltages such as ±12 V are commonly used.

RS-232C TO TTL INTERFACING

Obviously a USART such as the 8251A is not directly compatible with these signal levels. Sheet 9 of the SDK-86 schematics in Figure 7-6 shows one way to interface TTL signals of the 8251A to RS-232C signal levels. If the jumpers shown are removed and the jumpers shown in the jumper table under CRT are inserted, the circuit will produce and accept RS-232C signals. (NOTE: This is the jumpering needed to prepare the SDK-86 board for downloading programs from an IBM PC or other computer.) Here's an example of how it works.

With a jumper between the points numbered 7 and 8, a high on the TxD output of the 8251A produces a high on the base of the transistor, which turns it off. With points numbered 9 and 10 jumpered, the CR TX line will then be pulled to −12 V which is a legal high or marking condition for RS-232C. A low on the TxD output of the 8251A will turn on the transistor and pull the CR TX line to +5 V, which is a legal low or space condition for RS-232C.

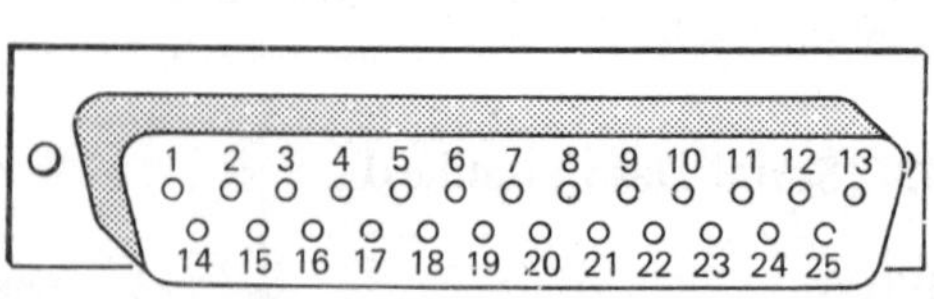

FIGURE 13-7 DB-25P connector often used for RS-232C connections.

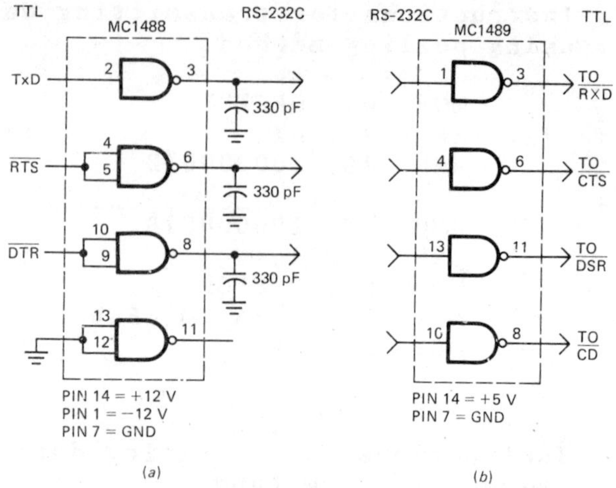

FIGURE 13-8 TTL to RS-232C and RS-232C to TTL signal conversion. *(a)* MC1488 used to convert TTL to RS-232C. *(b)* MC1489 used to convert RS-232C to TTL.

Another, more standard way to interface between RS-232C and TTL levels is with MC1488 quad TTL-to-RS-232C drivers and MC1489 quad RS-232C-to-TTL receivers shown in Figure 13-8. The MC1488s require + and − supplies, but the MC1489s require only +5 V. Note the capacitor to ground on the outputs of the MC1488 drivers. To reduce cross talk between adjacent wires the rise and fall times for RS-232C signals are limited to 30 V/μs. Also note that the RS-232C handshake signals such as $\overline{RTS}$ are active low. Therefore, if one of these signals is asserted, you will find a positive voltage on the actual RS-232C signal line when you check it during troubleshooting. Now let's look at the RS-232C pin descriptions.

RS-232C SIGNAL PINS

Figure 13-9 shows the signal names, signal direction, and a brief description for each of the 25 pins defined for RS-232C. For most applications only a few of these pins are used, so don't get overwhelmed. Here are a few additional notes about these signals.

First note that the signal direction is specified with respect to the DCE. This convention is part of the standard. We have found it very helpful to put arrowheads on all signal lines as shown in Figure 13-2 when we are drawing circuits for connecting RS-232C equipment.

Next observe that there is both a chassis ground (pin 1) and a signal ground (pin 7). To prevent large ac-induced ground currents in the signal ground these two should be connected together only at the power supply in the terminal or the computer.

The TxD, RxD, and handshake signals shown with common names in Figure 13-9 are the ones most often used for simple systems. We gave an overview of their use in the introduction to this section of the chapter, and will discuss them further in a later section of the chapter on modems. These signals control what is called the *primary* or *forward* communications channel of the

PIN NUMBER	COMMON NAME	RS-232-C NAME	DESCRIPTION	SIGNAL DIRECTION ON DCE
1		AA	PROTECTIVE GROUND	–
2	TXD	BA	TRANSMITTED DATA	IN
3	RXD	BB	RECEIVED DATA	OUT
4	$\overline{RTS}$	CA	REQUEST TO SEND	IN
5	$\overline{CTS}$	CB	CLEAR TO SEND	OUT
6	$\overline{DSR}$	CC	DATA SET READY	OUT
7	GND	AB	SIGNAL GROUND (COMMON RETURN)	–
8	$\overline{CD}$	CF	RECEIVED LINE SIGNAL DETECTOR	OUT
9		–	(RESERVED FOR DATA SET TESTING)	–
10		–	(RESERVED FOR DATA SET TESTING)	–
11			UNASSIGNED	–
12		SCF	SECONDARY REC'D. LINE SIG. DETECTOR	OUT
13		SCB	SECONDARY CLEAR TO SEND	OUT
14		SBA	SECONDARY TRANSMITTED DATA	IN
15		DB	TRANSMISSION SIGNAL ELEMENT TIMING (DCE SOURCE)	OUT
16		SBB	SECONDARY RECEIVED DATA	OUT
17		DD	RECEIVER SIGNAL ELEMENT TIMING (DCE SOURCE)	OUT
18			UNASSIGNED	–
19		SCA	SECONDARY REQUEST TO SEND	IN
20	$\overline{DTR}$	CD	DATA TERMINAL READY	IN
21		CG	SIGNAL QUALITY DETECTOR	OUT
22		CE	RING INDICATOR	OUT
23		CH/CI	DATA SIGNAL RATE SELECTOR (DTE/DCE SOURCE)	IN/OUT
24		DA	TRANSMIT SIGNAL ELEMENT TIMING (DTE SOURCE)	IN
25			UNASSIGNED	–

FIGURE 13-9 RS-232C pin names and signal descriptions.

modem. Some modems allow communication over a *secondary* or *backward* channel which operates in the reverse direction from the forward channel and at a much lower baud rate. Pins 12, 13, 14, 16, and 19 are the data and handshake lines for this backward channel.

Pins 15, 17, 21, and 24 are used for synchronous data communication. We will tell you a little more about these in the section of the chapter on modems. Next we want to show you some of the tricks in connecting up RS-232C "compatible" equipment.

CONNECTING UP RS-232C EQUIPMENT

A major point we need to make right now is that you can seldom just connect together two pieces of equipment, described by their manufacturers as RS-232C compatible, and expect them to work the first time. There are several reasons for this. To give you an idea of one of the reasons, suppose that you want to connect the terminal in Figure 13-2 directly to the computer rather than through the modem-modem link. The terminal and the computer probably both have DB-25 type connectors so that, other than a possible male-female mismatch, you might think you could just plug the terminal cable directly into the computer. To see why this doesn't work, hold your fingers over the modems in Figure 13-2 and refer to the pin numbers for the RS-232C signals in Figure 13-9. As you should see, both the terminal and the computer are trying to output data (TxD) from their number 2 pins to the same line. Likewise, they are both trying to input data (RxD) from the same line on their number 3 pins. The same problem exists with the handshake signals. RS-232C drivers are designed so that connecting the lines together in this way will not destroy anything, but connecting outputs together is not a productive relationship. A solution to this problem is to make an adapter with two connectors so that the signals cross over as shown in Figure 13-10*a*. This crossover connection is often called a *null modem*. We have again put arrowheads on the signals in Figure 13-10*a* to help you keep track of the direction for each. As you can see in the figure, the TxD from the terminal now sends data to the RxD input of the computer. Likewise, the TxD from the computer now sends data to the RxD input of the computer as desired. The handshake signals also are crossed over so that each handshake output signal is connected to the corresponding input signal.

A second reason that you can't just plug RS-232C compatible equipment together and expect it to work is that a partial implementation of RS-232C is often used to communicate with printers, plotters, and other computer peripherals besides modems. These other peripherals may be configured as DCE or as DTE. Also, they may use all, some, or none of the handshake signals. As an example of this, suppose that we want to connect the RS-232C port on the IBM PC asynchronous communication board to the serial port on the SDK-86 so that we can download object-code programs.

The IBM PC asynchronous board is configured as DTE, so TxD is on pin 2, RxD on pin 3, $\overline{RTS}$ on pin 4, $\overline{CTS}$ on pin 5, $\overline{DTR}$ on pin 20, $\overline{DSR}$ on pin 6, and carrier detect ($\overline{CD}$) on pin 8. In order for the IBM board to be able to transmit and receive, its $\overline{CTS}$, $\overline{DSR}$, and $\overline{CD}$ inputs must be asserted. The BIOS software asserts the $\overline{DTR}$ and $\overline{RTS}$ outputs. Now take another look at sheet 9 of the SDK-86 schematics in Figure 7-6 to see how the data and handshake signals are connected there.

For communicating with RS-232C type equipment, the SDK-86 board is jumpered as shown in the jumper

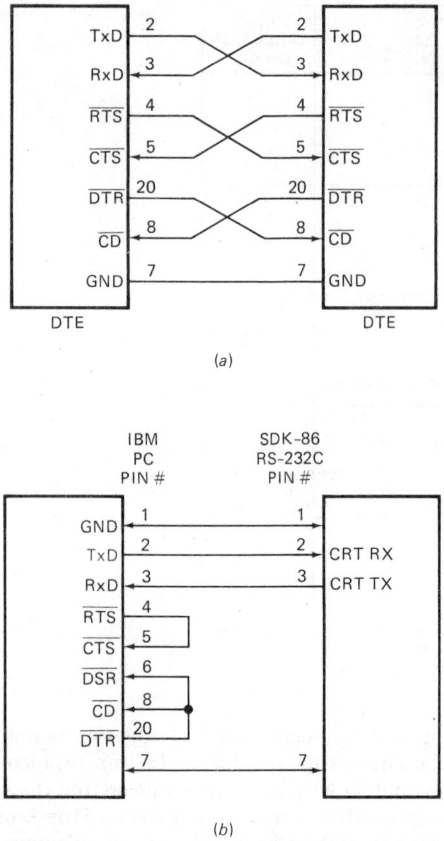

FIGURE 13-10 Nonmodem RS-232C connections. *(a)* Null modem for connecting two RS-232C data-terminal-type devices. *(b)* IBM PC serial port to SDK-86 RS-232C connection.

table column labeled "stand alone CRT." The output data on CRT TX then connects to pin 3 of connector J7, a DB-25S–type connector. This corresponds to the RxD on the IBM connector, so no crossover is needed. Likewise, the CRT RX of the SDK corresponds to the TxD of the IBM board, so this is also a straight-through connection. The handshake signals here are another story.

The $\overline{\text{RTS}}$ of the SDK-86 is simply looped into the $\overline{\text{CTS}}$ so $\overline{\text{CTS}}$ will automatically be asserted when $\overline{\text{RTS}}$ is asserted by the 8251A. Therefore, neither of these signals is available for external handshaking. The $\overline{\text{DTR}}$ output of the 8251A on the SDK board is used for a teletypewriter function, and does not connect to the normal RS-232C $\overline{\text{DTR}}$ pin number, so it is not available either. The $\overline{\text{DSR}}$ input of the 8251A is connected to the RxD input so that it will be asserted when a start bit comes in on the serial data line, but this line is also not available for handshaking with external devices. The problem here then is that the SDK-86 is not set up to supply the handshake signals needed by the IBM PC serial board. Figure 13-10*b* shows the connections we make to solve this problem so the PC can talk to the SDK-86. The PC $\overline{\text{RTS}}$ line on pin 4 is jumpered on the connector to its $\overline{\text{CTS}}$ line on pin 5, so that $\overline{\text{CTS}}$ will automatically be asserted when $\overline{\text{RTS}}$ is asserted. Pins 6, 8, and 20 are also jumpered together on the connector so that when the PC asserts its $\overline{\text{DTR}}$ output on pin 20, the $\overline{\text{DSR}}$ input and the $\overline{\text{CD}}$ input will automatically be asserted. These connections do not provide for any hardware handshaking. They are necessary just to get the PC and the SDK-86 to talk to each other.

The point here is that whenever you have to connect RS-232C "compatible" devices such as terminals, serial printers, etc., get the schematic for each and work your way through the connections one pin at a time. Make sure that an output on one device goes to the appropriate input on the other device. Sometimes you have to look at the actual drivers and receivers on the schematic to determine which pins on the connector are outputs and which are inputs. This is necessary because some manufacturers label an output pin connected to pin 3 as RxD, indicating that this signal goes to the RxD input of the receiving system.

RS-422A, RS-423A, and RS-449

A major problem with RS-232C is that it can only transmit data reliably about 50 ft [16.4 m] at its maximum rate of 20,000 Bd. If longer lines are used, the transmission rate has to be drastically reduced. This limitation is caused by the open signal lines with a single common ground that are used for RS-232C.

A newer standard, *RS-422A* specifies that each signal will be sent differentially over two adjacent wires in a ribbon cable or a twisted pair of wires as shown in Figure 13-11*a*. Differential signals are produced by differential line drivers such as the MC3487 and translated back to TTL levels by differential line receivers such as the MC3486. Data rates for this standard are 10 MBd for a distance of 50 ft [16.4 m] or 100,000 Bd for a distance of 4000 ft [1220 m]. The higher transmission rate of RS-422A is possible because the differential lines are terminated by resistors so they act as transmission lines instead of simply open wires. A further advantage of differential signals is that any common-mode electrical noise induced in the two lines will be rejected by the differential line receiver. For RS-422A a logic high or mark is indicated by the B signal line being more positive than the A signal line. A logic low or space is indicated by the A signal line being more positive than the B signal line. The voltage difference between the two lines must be greater than 0.4 V, but not greater than 12 V. The common-mode voltage on the signal lines must be in the range of −7 V to +7 V.

Another EIA standard intended to solve the speed and distance problems of RS-232C is *RS-423A*. This standard specifies a low impedance single-ended signal instead of the differential signal of RS-422A. The low-impedance signal can be sent over 50-Ω coaxial cable and terminated at the receiving end to prevent reflections. Figure 13-11*b* shows how an MC3487 driver and MC3486 receiver can be connected to produce the required signals. A logic high in this standard is represented by the A line being between 4 and 6 V negative with respect to the B line (ground), and a logic low is

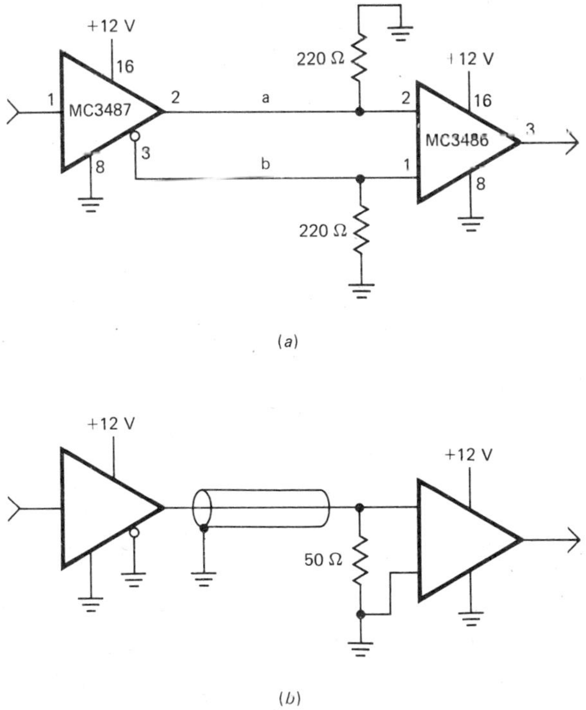

FIGURE 13-11 RS-422A and RS-423A drivers and receivers. *(a)* MC3487 driver and MC3486 receiver used for RS-422A differential signals. *(b)* MC3487 driver and MC3486 receiver used for RS-423A signal on coax cable.

represented by the A line being 4–6 V positive with respect to ground.

The RS-422A and RS-423A standards do not specify connector pin numbers or handshake signals the way that RS-232C does. An additional EIA standard called *RS-449* does this for the two. RS-449 specifies 37 signal pins on a main connector and 9 additional pins on an optional connector. The signals on these connectors are a superset of the RS-232C signals so adapters can be used to interface RS-232C equipment with RS-449 equipment. Still another EIA standard, *RS-366*, incorporates signals for automatic telephone dialing with modems.

Telephone Circuits and Systems

A large amount of the communication between users and computers and between different computers takes place over telephone lines in some form. In this section we give you an introduction to the terminology and operation of phone lines, and then discuss how different types of modems send and receive data over telephone lines.

In the case where digital data needs to be transferred only between two fixed points, *broadband* lines can be leased from telephone or other companies. Depending on the type of line leased, digital data can be directly sent and received at rates from 10 Kbits/s to several megabits per second on these lines. However, in cases where digital data needs to be transferred to many different locations, or where the amount of data to be transferred does not warrant the cost of a leased line, standard (switched) phone lines are used. As we indicated in a previous section, the bandwidth of standard phone lines is limited to 300 Hz–3 kHz, so modems must be used to convert digital data to tones that it can be sent over standard telephone lines, and to convert the tones back to digital data at the receiving end of the phone line. Before we can explain how modems interact with phone lines, we need to look at some telephone circuitry.

Figure 13-12 shows some generic circuitry for *plain old telephone service* or POTS, that many of us are still stuck with. A POTS system uses a rotary dial and electromechanical ringer. Newer systems use *dual-tone multifrequency* (DTMF or touch-tone) dialing and electronic ringers, but the line connections we are interested in here are the same.

The circuitry to the left of the first vertical line in Figure 13-12 is contained in the telephone. The circuitry between the two dotted vertical lines may be located in a *private branch exchange*, or PBX, if the phone is part of a multiphone system in a large building. It may be located in a centralized building if the phone is that of a single subscriber connected directly to the phone lines. Note that for a simple system such as this, only two wires are required to connect the phone to the PBX or local exchange. To send and receive the signals over long distances, however, the two-wire signal must be converted into separate send and receive signals. This must be done so that amplifiers or *repeaters* can be put in each signal path. The conversions back and forth between two-wire signals and four-wire signals is done by a device called a *hybrid coil*. Now let's see how this works and pick up some more terminology.

Let's assume that the phone handset is in its cradle, or *on-hook*, to start. At this point, then, switches *S1* and *S2* are open, and *S3* is closed. To ring up the phone, an ac voltage of 48 V or 96 V rms is sent from the local exchange or PBX to the phone. This voltage activates the ringer mechanism. When the phone handset is lifted off-hook, switches *S1* and *S2* close, and switch *S3* opens. Closing *S1* and *S2* causes a direct current to flow from the 48-V dc supply. Circuitry in the PBX or in the local exchange detects this direct current and turns off the ringing signal within about 200 ms. The process is referred to as a *ring trip*. Incidentally, the ring and tip names on the signal lines in Figure 13-12 refer to the parts of the old-fashioned phone plug connected to that signal line. Closing *S1* and *S2* also allows the voice signals to get in and out of the phone. As you talk into the phone, the induction coil feeds back part of the transmitted voice signal to the receiver so you can tell how loud you are talking. Now let's see what happens when you make a call.

Again when the handset is lifted off-hook a direct current is produced in the wires to the PBX or local exchange. The PBX or local exchange returns a dial tone to let you know that it is alive and ready to serve you. As you turn the rotary dial to dial a number, switches *S4* and *S5* open and close as the dial passes each number.

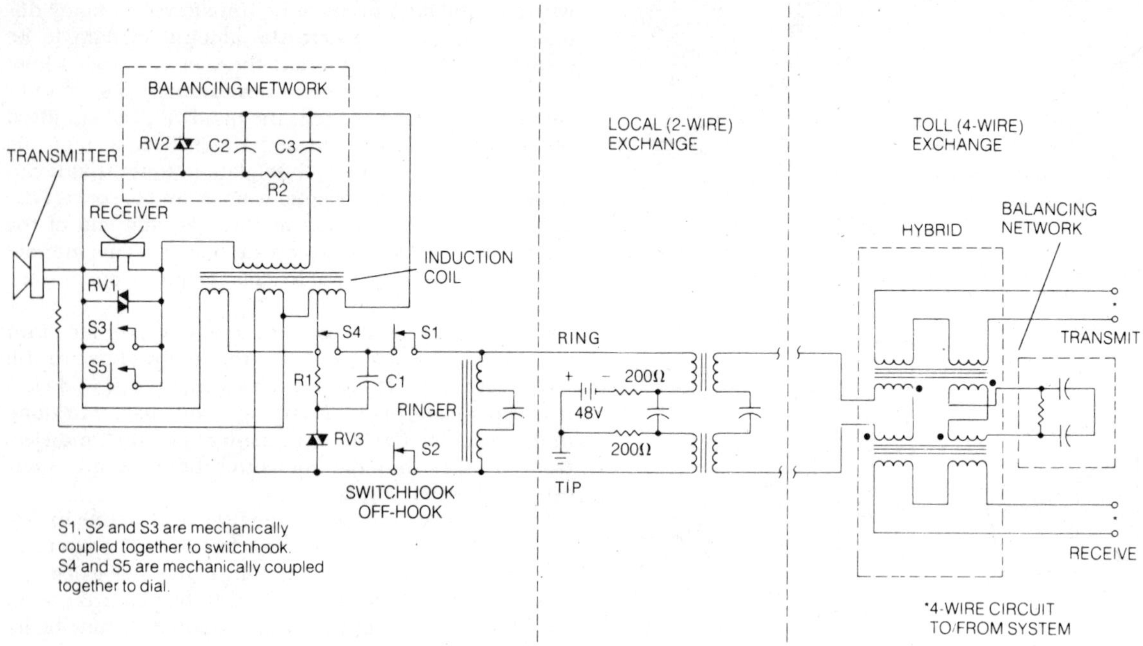

FIGURE 13-12 Circuitry and line connections for standard rotary-dial telephone. *(Texas Instruments, Inc.)*

This produces a series of pulses equal to the desired number. Switching circuitry in the local exchange uses the series of pulses to start finding a path to the dialed phone. *Dual-tone frequency-modulation* or DTFM telephones produce a mixture of two tones for each number button pushed. Circuitry in the PBX or the local exchange decodes these tones to get the required number information. In either case when all the desired numbers have been entered, the local exchange attempts to complete the connection. If the dialed unit is unavailable, the local exchange returns a busy signal to your phone.

An important point here is that any circuit or system that is going to be connected to standard phone lines must be approved by the FCC. This regulation is intended to prevent untested devices from damaging the phone system or creating a shock hazard.

The next topic we want to discuss here is one way that analog phone signals are converted to digital form so that they can be more efficiently sent over long distances.

CODECs, TDM, and PCM

Because digital signals have much better noise immunity than analog signals, analog signals are often converted to digital signals with an A/D converter for transmission over long distances. A D/A converter at the destination uses the received binary codes to reconstruct a replica of the original analog signal. Sending analog signals such as phone signals as a series of binary codes is called *pulse-code modulation* or PCM. The A/D converter that produces the binary codes in this application is usually called a *coder* and the D/A converter that reconstructs the analog signal from the pulse codes is referred to as a *decoder*. Since both a coder and a decoder are needed for two-way communication, they are often packaged in the same IC. This combined coder and decoder is called a *codec*. Common examples of codecs are the Intel 2910A and the Intel 2911A-1. These devices each contain a sample-and-hold circuit on the analog input, an 8-bit A/D converter, an 8-bit D/A converter, and appropriate control circuitry.

Normal A/D converters are linear, which means that the steps are the same size over the full range of the converters. The A/D converters used in codecs are nonlinear. They have small steps for small signals and large steps for large signals. In other words, for signals near the zero point of the A/D converter, it only takes a small change in the signal to change the code on the output of the A/D. For a signal near the full scale of the converter, a large change in the input signal is required to produce a change in the output binary code. This nonlinearity of the A/D converter is said to *compress* the signal, because it reduces the dynamic range of the signal. Compression in this way greatly improves the accuracy for small signals where it is needed, without going to a converter with more bits of resolution. The D/A in the codec is nonlinear in the reverse manner, so that when the binary pulse codes are converted to analog, the result is *expanded* to duplicate the original waveform. A codec which has this intentional nonlinearity is often referred to as a *compander* or a *companding codec*. The two most common ways of changing the size of the steps as the signal gets larger are called the μ *LAW*, and the *A LAW*. Consult the Intel 2910A data sheet for more infor-

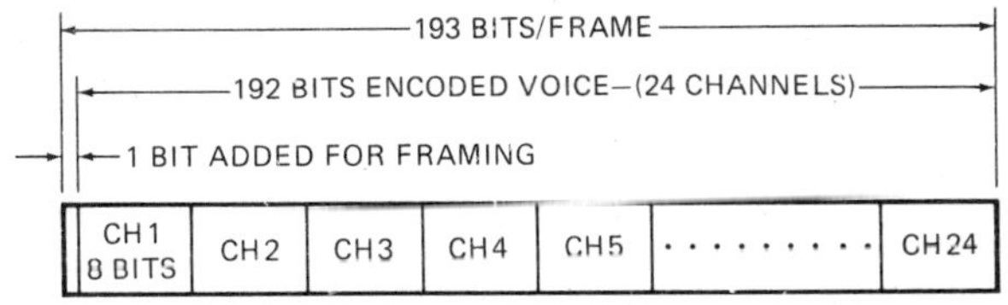

FIGURE 13-13 Frame format for time-domain-multiplexed data from codec.

mation about a μ-LAW device, and the Intel 2911A-1 data sheet for more information about an A-LAW device.

In most systems the output of the codec A/D is not simply sent on a wire by itself, it is multiplexed with the outputs of many other codecs in a manner known as *time-division multiplexing*, TDM. There are several different formats used. A simple one will give you the idea of how it's done.

One of the first TDM systems was the T1 or DS-1 system which multiplexes 24 PCM voice channels onto a single wire. For this system an 8-bit codec on each channel samples and digitizes the input signal at an 8-kHz rate. The 8-bit codes from the codecs are sent to a multiplexer which sends them out serially, one after the other. One set of bits from each of the 24 codecs plus a framing bit is referred to as a frame. Figure 13-13 shows the format of a frame for this system. The framing bit at the start of each frame toggles after each frame is sent. It is used to keep the receiver and the transmitter synchronized and for keeping track of how many frames have been sent. After it sends the framing bit, the multiplexer sends out the 8-bit code from the first codec, then sends out the 8-bit code from the next codec, and so on until the codes for all 24 have been sent out. At specified intervals the multiplexer sends out a frame which contains synchronization information and signaling information. This does not seriously affect the quality of the transmitted data.

Since the multiplexer is sending out 193-bit frames at a rate of 8000 per second, the data rate on the wire is 193 × 8000, or 1.544 Mbits/s. A newer system, known as T4M or DS-4 multiplexes 4032 channels onto a single coaxial cable or optic fiber. The bit rate for this system is 274.176 Mbits/s.

Telephone companies transmit long-distance phone signals over high-speed digital lines, and all local phone service may eventually be converted to a wideband digital system known as the *integrated services digital network* or ISDN. For now, however, the bandwidth of each standard user channel is still only about 4 kHz. Until this "weak link" is removed we still have to use modems to communicate with computers over standard phone lines. The next section shows how modems operate.

Modems

MODULATION METHODS

As we described in a previous section, a modulator-demodulator, or modem, sends digital 1's and 0's over standard phone lines as modulated tones. The frequency of the tones is within the bandpass of the lines. The three major forms of modulation used are *amplitude*, *frequency-shift keying* (FSK), and *phase-shift keying* (PSK).

To produce amplitude modulation, a single-frequency tone of perhaps 387 Hz is turned on to represent a 1 and turned off to represent a 0 as shown in Figure 13-14*a*. Amplitude modulation is only used for very low speed reverse channel transmission because of its poor noise rejection characteristics. A temporary change in line resistance, for example, might drop the amplitude of the 1 signal below the threshold of the detector, and the data would appear to be all 0's.

Frequency-shift keying uses one tone to represent a 0 and another tone to represent a 1 as shown in Figure 13-14*b*. In order to allow full-duplex communication, four different frequencies are often used. An old standard, the Bell 103A, 300-Bd FSK modem, for example, uses 2025 Hz for a 0 and 2225 Hz for a 1 in one direction, and 1070 Hz for a 0 and 1270 Hz for a 1 in the other direction. Another standard, the Bell 202 modem, permits half-duplex communication at 1200 Bd. The 202 uses 1200 Hz to represent a 0 and 1700 Hz to represent a 1 for the main channel. Different versions of the 202 may also have either a 5 bit/s amplitude-modulated back channel, or a 150 bit/s FSK back channel which uses 387 Hz for a 0 and 487 Hz for a 1.

LSI has made it possible to build an FSK modem with very few parts. Figure 13-15 shows a circuit diagram for a modem which uses the Advanced Micro Devices Am 7910 device. The 7910 can be programmed to operate in a Bell 103 compatible mode, Bell 202 compatible mode, or in a mode compatible with one of several other standards. It uses A/Ds, D/As, and the digital filter techniques we described in Chapter 10 to synthesize and filter all of the data signals. The circuit in Figure 13-15 is connected to a terminal or microcomputer through a stand-

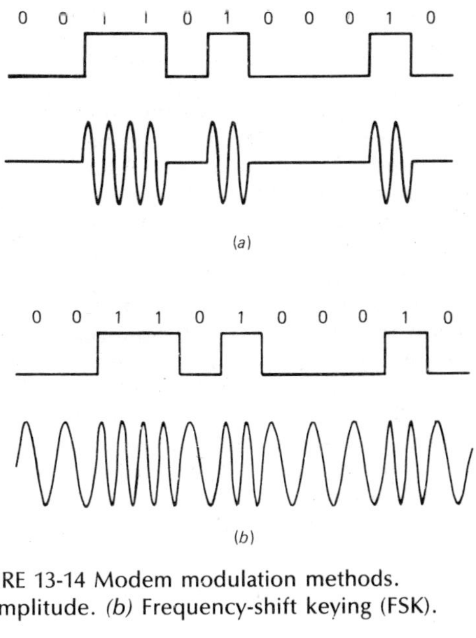

FIGURE 13-14 Modem modulation methods. *(a)* Amplitude. *(b)* Frequency-shift keying (FSK).

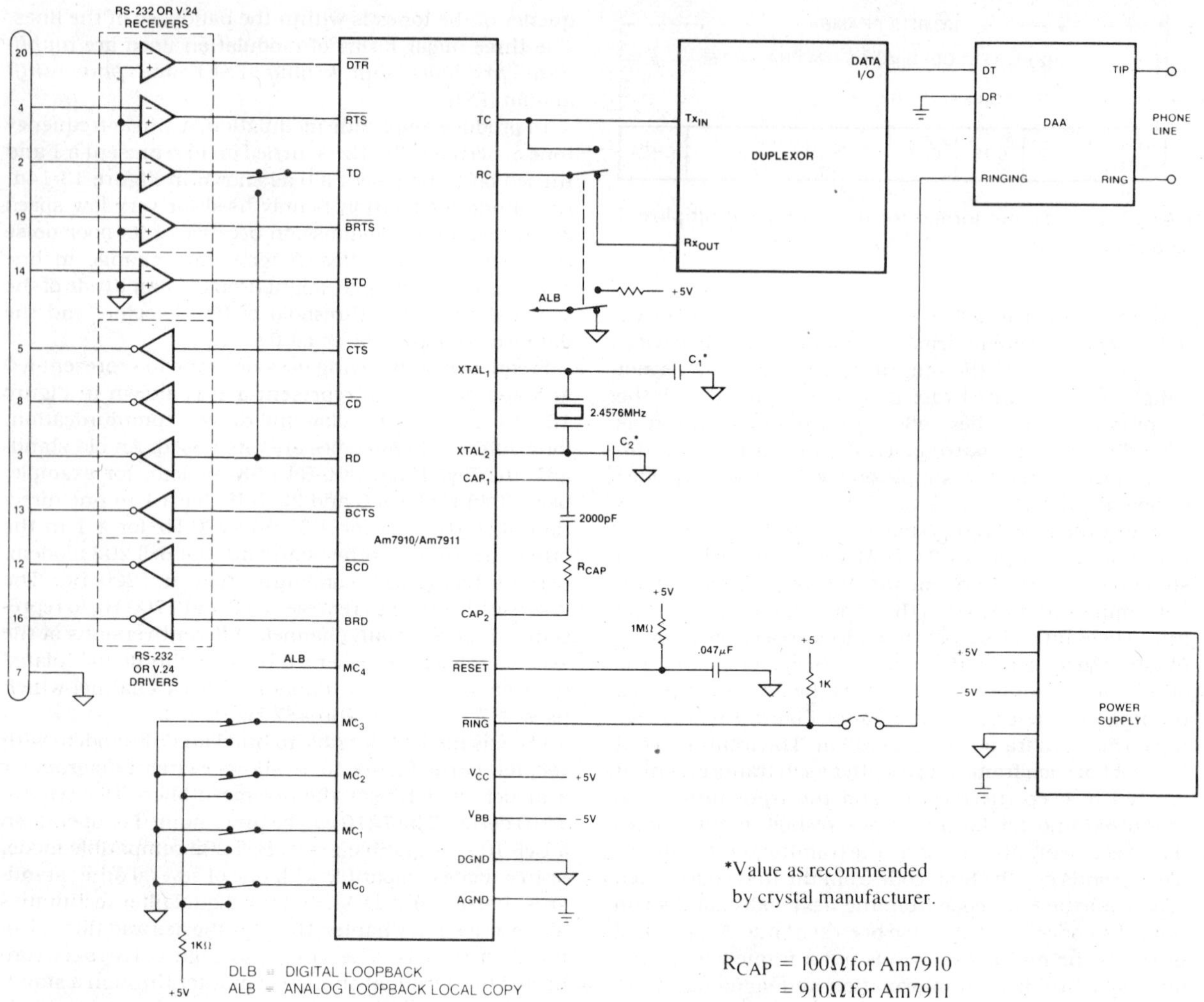

FIGURE 13-15 FSK modem circuit using AM7910 modem chip. *(Advanced Micro Devices)*

ard RS-232C interface. Signal names on the 7910 which start with a B are the back channel signals. The 7910 inputs labeled MC0–MC4 are used to program the operating mode for the device. These can be connected to manual switches as shown, or in a microcomputer system where the data and handshake signals are connected directly to the UART, these lines could be connected to a port so that the operating mode could be changed under program control.

FSK-modulated data is sent out from the 7910 on the pin labeled TC and received on the pin labeled RC. The duplexer puts the transmitted signal on the common signal line and taps off the received signal from the common signal line. Remember from a previous discussion that data is sent and received on the same wire for a standard two-wire phone service. The box labeled DAA in Figure 13-15 is the *data access arrangement* circuitry which actually interfaces the signals with the phone lines. It is this circuitry that must conform to the provisions of FCC rules section 68. Several integrated packages are available which contain a duplexer and sophisticated DAA circuitry. One of these is the CH1810 from Cermetek Microelectronics, Inc. Note in Figure 13-15 that the DAA circuitry provides a RINGING signal to the 7910 when the modem is being called.

Two important features of the modem circuit in Figure 13-15 are the *analog loopback* (ALB) and the *digital loopback* (DLB) which are used for testing. The analog loop mode is used to test the modem locally. When the ALB switch in the middle of Figure 13-15 is in the up position, the FSK-modulated data will be routed back into the FSK input. A software test procedure can then compare the incoming data with the transmitted data to see if the modem is a 7910 and the connecting circuitry is working correctly. The digital loopback is used to check the operation of the modem from some remote location. When the DLB switch in Figure 13-15 is in the down position, data received from the phone line will be retransmitted back to the sending system. The sending system can then compare the sent and returned data to see if the data link is operating correctly.

On a standard two-wire phone line FSK modulation is

limited to half-duplex operation at 1200 Bd. (With a four-line service, full-duplex operation at 1200 bits/s is possible.) For higher bit rates and full-duplex operation on standard phone service, phase-shift modulation is used.

PHASE-SHIFT MODULATION VARIATIONS

In the simplest form of *phase-shift modulation*, the phase of a constant-frequency sine-wave carrier of perhaps 1700 Hz is shifted by 180° to represent a change in the data from a 1 to a 0 or from a 0 to a 1. Figure 13-16*a* shows an example of this. As the digital data changes from a 0 to a 1, near the left edge of the figure, the phase of the signal is shifted by 180°. When the data changes from a 1 to a 0, the phase of the carrier is again shifted by 180°. For the next section of the digital data where the data stays 0 for 3 bit times, the phase of the carrier is not changed. Likewise, in a later section of the waveform where the data remains at a one level for 2 bit times, the phase of the carrier is not changed. The phase of the carrier then is only shifted by 180° when the data line changes from a 1 to a 0 or from a 0 to a 1.

The simple phase-shift modulation shown in Figure 13-16*a* has no real advantage over FSK as far as maximum bit rate etc. are concerned. However, by using additional phase angles besides 180°, 2 or 3 bits can be sent in 1 baud time. Two bits sent in 1 baud time are called *dibits*, and 3 bits sent in 1 baud time are called *tribits*. Here's how dibits and tribits are encoded.

For dibit encoding, each pair of bits in the data stream is treated together, and referred to as a dibit. The value of these two bits determines the amount that the phase of the carrier will be shifted. Figure 13-16*b* shows the angle that the phase of the carrier will be shifted for the four possible values of a dibit. If, for example, the value of 2 bits taken together is 00, the phase of the carrier will be shifted 90° to represent that dibit. The trick here is that the phase of the carrier only has to shift once for each two transmitted bits. Remember from a previous discussion that baud rate is the rate at which the carrier is changing. In this case it is not the same as the number of bits per second. Bell 212A type modems use this scheme to transmit 1200 bits/s at a baud rate of only 600 Bd. Two carrier frequencies, 1200 Hz and 2400 Hz, are used to permit full-duplex operation.

For tribit encoding, the data stream is divided into groups of three bits each, called tribits. Figure 13-16c shows one possible set of angles that the phase of the carrier might be shifted to represent the eight different values that the tribit can have. The Bell 208 modem uses this tribit scheme to transmit data at 4800 bits/s.

Another similar scheme which makes it possible to transmit data at 9600 bits/s over conditioned standard phone lines encodes 4 bits in each baud time and is called *quaternary amplitude modulation* (QAM). This scheme uses eight different angles to represent three of the bits in each group of four, and two different amplitudes to represent the fourth bit in each quadbit. A baud rate of 2400 Bd and 4 bits/Bd produces a data rate of 9600 bits/s.

Dibit and tribit phase-shift modulation permits higher data rates on phone lines, but detecting this type of phase-encoded data presents some unique problems. We will use a dibit example to describe these problems and how they are solved. Remember from our previous discussion that, in a dibit system, the value of a dibit is represented by shifting the phase of a carrier signal some specified number of degrees from a reference phase. In order to detect the amount of phase shift, the receiver and the transmitter must be using the same reference phase. This would be easy if we could just run another wire to carry a synchronizing clock signal. Since this is not easily done, the synchronizing signal must in some way be included with the data. The carrier signal itself cannot be used directly, because we want to measure the phase of that signal.

The solution to this problem is to use transitions in the transmitted signal to synchronize a phase-locked loop oscillator in the receiver. In order for this to work, two factors must be included in the transmitted data. First of all, the system must be operated synchronously rather than asynchronously, so that data, sync, or null characters are always being received by the receiver. Secondly, the transmitted data must have enough transitions at regular intervals to keep the phase-locked loop locked in the desired phase. The transmitted data from the USART may not have enough transitions in it to satisfy this second condition, so a special circuit called a *scrambler* in the transmitter puts any required extra transitions in the signal. The scrambler usually consists of a shift register with feedback. The output from the

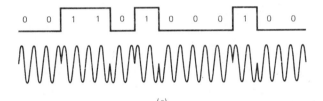

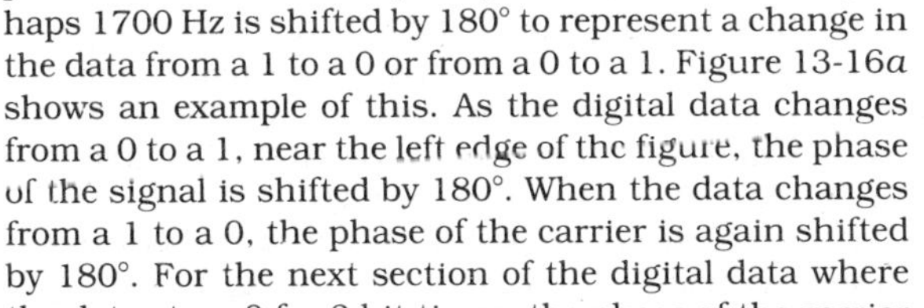

(*a*)

GREY CODE DIBIT VALUE	DEGREES OF PHASE SHIFT
0 0	0
0 1	90
1 1	180
1 0	270

(*b*)

GRAY CODE DIBIT VALUE	DEGREES OF PHASE SHIFT
0 0 1	22.5
0 0 0	67.5
0 1 0	112.5
0 1 1	157.5
1 1 1	202.5
1 1 0	247.5
1 0 0	292.5
1 0 1	337.5

(*c*)

FIGURE 13-16 Phase-shift modulation (PSK). *(a)* Waveforms for simple phase-shift modulation. *(b)* One common set of phase shifts used to represent four possible dibit combinations. *(c)* One common set of phase shifts used to represent the eight possible tribit combinations.

scrambler is then used to modulate the phase of the carrier. When the carrier signal reaches the receiver, the signal is demodulated to produce a signal of 1's and 0's. This signal is then passed through a descrambler which reverses the scrambling process and outputs the original data.

MODEM HANDSHAKING

Many of the currently available modems, such as the Hayes Smartmodem 1200 that we use with our IBM PC, contain a dedicated microprocessor. The built-in intelligence allows these units to automatically dial a specified number with either tones or pulses, and redial the number if it is found busy or doesn't answer. When a smart modem makes contact with another modem, it will automatically try to set its transmit circuitry to match the baud rate of the other modem. It can be set to automatically answer a call after a programmed number of rings so that you can access your computer from a remote location. Some CRT terminals now come with a smart modem which allows the user to establish voice contact and then switch over to digital data communication.

After a modem dials up another modem, a series of handshake signals takes place. The handshake signals may be generated by hardware in the modem or by software in the system connected to the modem. The handshake sequence is similar for most of the Bell compatible modems, so we will use the Bell 202 as an example.

Figure 13-17 shows the data and handshake waveforms for a 202 modem as produced by the AM7910 single-chip modem described in the FSK section previously. Keep a copy of the circuit diagram in Figure 13-15 handy as you work your way through the waveforms here.

The modem which makes a call is usually referred to as the *originate* modem, and the modem which receives the call is usually referred to as the *answer* modem. In the following discussion we will use the terms *calling modem* and *called modem*, respectively, to agree with the labels on the waveforms in Figure 13-17.

At the left side of the waveforms, a call is being made from one modem to another. Assuming that the $\overline{\text{DTR}}$ of the called modem is asserted, the ringing signal on the line will cause the DAA circuitry to assert the *ringing input* ($\overline{\text{RI}}$) of the 7910. In response to this the 7910 will send out a silent period of about 2 seconds to accommodate billing signals, and then it will send out an answer tone of 2025 Hz to the calling modem for 2 seconds. If the $\overline{\text{DTR}}$ and the $\overline{\text{RTS}}$ of the calling modem are asserted indicating that data is ready to be sent, the calling modem then puts a tone of 2225 Hz (mark) on the line

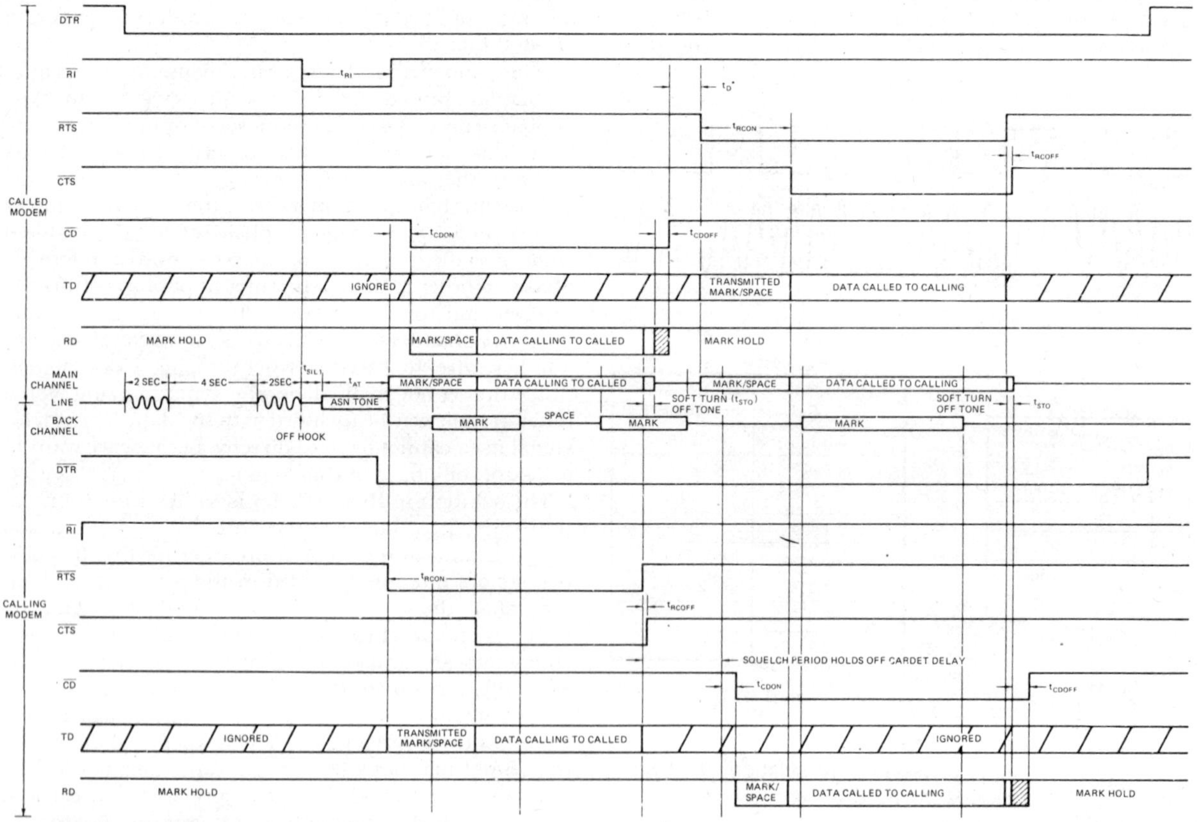

FIGURE 13-17 Handshake signal sequence for Bell type 202 FSK modem using AM7910 modem chip.

for 8 ms to let the called modem know that contact is complete. In response to this mark the called modem asserts its *carrier detect* output ($\overline{CD}$) to enable the receiving UART. The calling modem then sends data until its $\overline{RTS}$ input is released by the computer or terminal sending the data. While it is receiving data on the main channel, the called modem can send data to the calling modem on the 5 bit/s back channel. Releasing $\overline{RTS}$ causes the modem to release $\overline{CTS}$ to the sending computer, and remove the carrier from the line. The called modem senses the loss of the carrier and unasserts its carrier detect, $\overline{CD}$ signal.

Now, if the called system is to send some data back to the calling system on the main channel, it asserts the $\overline{RTS}$ input to its modem. The called modem sends a marking tone to the calling modem for 8 ms. The calling modem asserts its $\overline{CD}$ output to its UART. The called modem then sends data to the calling modem on the main channel until its $\overline{RTS}$ input is unasserted by the called system, indicating no more data to send. While the called modem is transmitting on the main channel, the calling modem can transmit over the back channel if necessary. For a full-duplex system the handshake is similar, but the data rates are equal in both directions.

MODEM STANDARDS

Two organizations are responsible for most of the current standards regarding modems. In the United States most modems follow one of the Bell Telephone standards. Examples of these de facto standards are the Bell types 103, 202, 208, and 212A which we have used as examples in this section. Throughout much of the rest of the world modems follow one of the standards of the *Comité Consultatif International Téléphonique et Télégraphique* (CCITT), which is part of the International Telecommunications Union. CCITT standards which relate to modems start with a V. Examples are the V.26, which is a 2400 bit/s modem, and the V.27, which is a 4800 bit/s modem. In the next section we discuss a means of data communication that may make modems obsolete.

Fiber-Optic Data Communication

INTRODUCTION

All of the data communication methods we have discussed so far use metallic conductors. The systems we describe here transfer data through very thin glass or plastic fibers with a beam of light and have no conducting electrical path. Therefore, they are called *fiber-optic* systems. Figure 13-18 shows the connections for a basic fiber-optic data link you can build and experiment with.

The light source here is a simple infrared LED. Higher performance systems use an *infrared injection laser diode* (ILD) or some other laser driven by a high speed, high-current driver.

NOTE: If you are working on a fiber-optic system you should never look directly into the end of the fiber to see

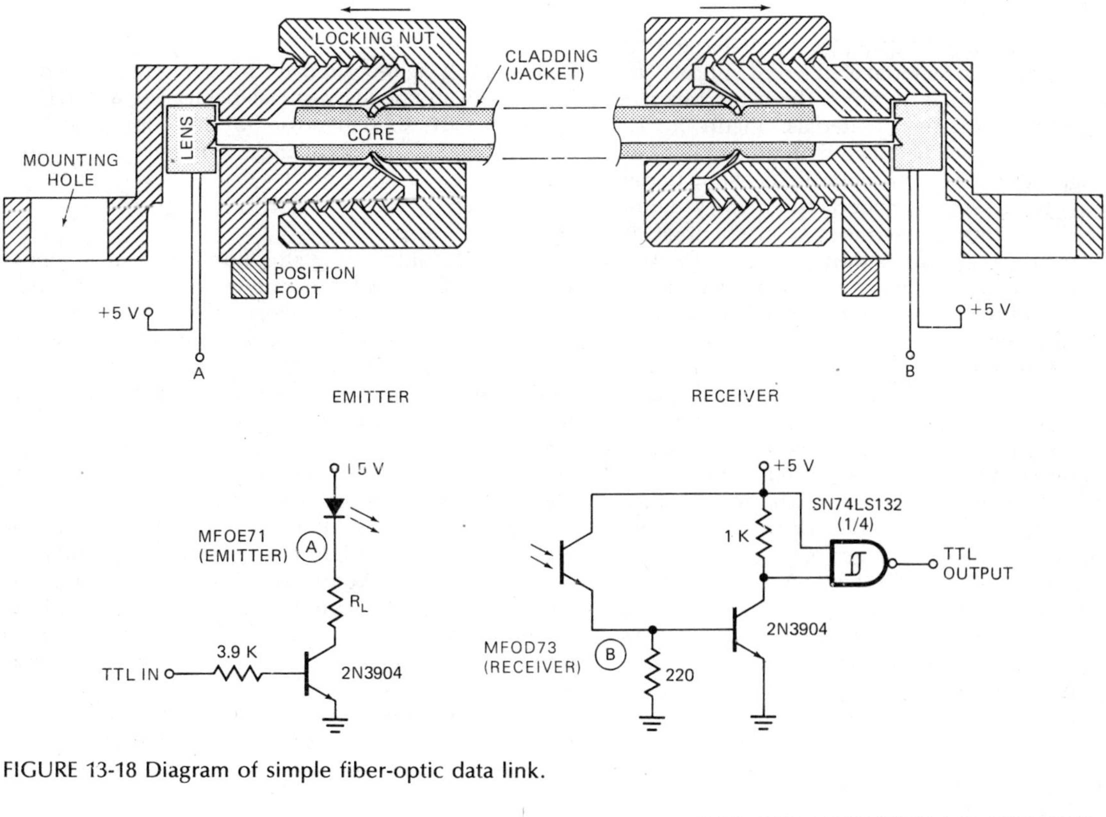

FIGURE 13-18 Diagram of simple fiber-optic data link.

if the light source is working, because the light beam from some laser diodes is powerful enough to cause permanent eye damage. Use a light meter, or point the cable at a nonreflecting surface to see if the light source is working.

Digital data is sent over the fiber by turning the light beam on for a 1 and off for a 0. Data rates for some currently available systems are as high as a gigabit per second. Current systems use one of three light wavelengths, 0.85 μm, 1.3 μm, or 1.500 μm. These wavelengths are used because the absorption of light by the optical fibers is minimum at these wavelengths.

The fibers used are made of special plastic or glass. Depending on the desired operating mode, bandwidth, and transmitting distance, different diameter fibers are used. Fiber diameters used range from 2–1000 μm. As shown in Figure 13-19*e*, the fiber-optic cable consists of three parts. The optical-fiber core is surrounded by a *cladding* material which is also transparent to light. We will explain the function of this cladding later. The cable is enclosed in a sheath which protects the cladding and does not allow external light to enter.

To convert the light signal back into an electrical signal at the receiving end, Darlington photodetectors such as the MFOD73 shown in Figure 13-18, p-i-n FET devices, or avalanche photodiodes (APDs) are used. APDs are more sensitive and operate at higher frequencies, but the circuitry for them is more complex. A Schmitt trigger is usually used on the output of the detector to "square up" the output pulses. Now that you have an overview of an optical-fiber link, we will briefly discuss some of the optics involved.

THE OPTICS OF FIBERS

The path of a beam of light going from a material with one optical density to a material of different optical density depends on the angle at which the beam hits the boundary between the two materials. Figure 13-19 shows the path that will be taken by beams of light at various angles going from an optically dense material such as glass to a less dense material such as a vacuum or air. If the beam hits the boundary at a right angle, it will go straight through as shown in Figure 13-19*a*. When a beam hits the boundary at a small angle away from the perpendicular or *normal*, it will be bent away from the normal when it goes from the more dense to the less dense, as shown in Figure 13-19*b*. A light beam going in the other direction would follow the same path. A quantity called the *index of refraction* is used to describe the amount that the light beam will be bent. Using the angle identifications shown in Figure 13-19*b*, the index of refraction, n, is defined as the (sine of angle B)/(sine of angle A). A typical value for the index of refraction of glass is 1.5. The larger the index of refraction, the more the beam will be bent when it goes from one material to another.

Figure 13-19*c* shows a unique situation that occurs when a beam going from a dense material to a less dense material hits the boundary at a special angle called the *critical angle*. The beam will be bent so that it travels parallel to the boundary after it enters the less dense material.

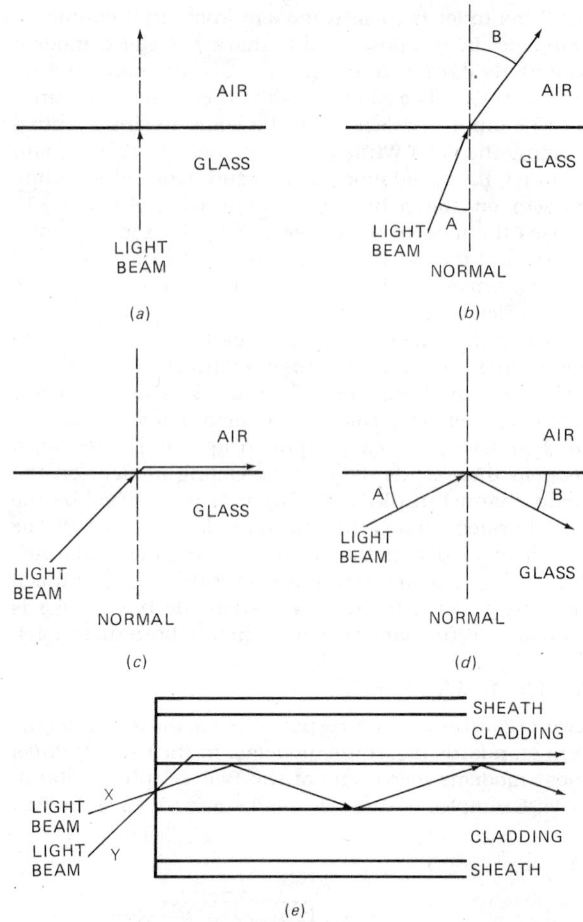

FIGURE 13-19 Light-beam paths for different angles of incidence with boundary of material leaving a lower optical density. *(a)* Right angle. *(b)* Angle less than critical angle. *(c)* At critical angle. *(d)* Angle greater than critical angle. *(e)* Angle greater than critical angle in optical fiber.

A still more interesting situation is shown in Figure 13-19*d*. If the beam hits the boundary at an angle greater than the critical angle, it will be totally reflected from the boundary at the same angle on the other side of the normal. This is somewhat like skipping stones across water. In this case the light beam will not leave the more dense material.

To see how all of this relates to optical fibers, take a look at the cross-sectional drawing of an optical fiber in Figure 13-19*e*. If a beam of light enters the fiber parallel to the axis of the fiber, it will simply travel through the fiber. If the beam enters the fiber so that it hits the glass-cladding layer boundary at the critical angle, it will travel through the fiber optic cable in the cladding layer as shown for beam Y in Figure 13-19*e*. However, if the beam enters the cable so that it hits the glass-cladding layer boundary at an angle greater than the critical angle, it will bounce back and forth between the walls of the fiber as shown for beam X in Figure 13-19*e*. The glass or plastic used for fiber-optic cables has very low absorption, so the beam can bounce back and forth along the fiber for several feet or miles without excessive attenuation.

MULTIMODE AND SINGLE-MODE FIBERS

If an optical fiber has a diameter many times larger than the wavelength of the light being used, then beams which enter the fiber at different angles will arrive at the other end of the fiber at slightly different times. The different angles of entry for the beams are referred to as *modes*. A fiber with a diameter large enough to allow beams with several different entry angles to propagate through it is called a *multimode* fiber. Since multimode fibers are larger, they are easier to manufacture, are easier to manually work with, and can use inexpensive LED drivers. However, the phase difference between the output beams in multimode fibers causes problems at high data rates. One partial solution to this problem is to dope the glass of the fiber so that the index of refraction decreases toward the outside of the fiber. Light beams travel faster in the region where the index of refraction is lower, so beams which take a longer path back and forth through the faster outer regions tend to arrive at the end of the fiber at the same time as those that take a shorter path through the slower center region.

A better solution to the phase problems of the multimode fiber is to use a fiber that has a diameter only a few times the wavelength of the light being transmitted. Only beams very nearly parallel to the axis of the fiber can be transmitted then. This is referred to as *single-mode* operation. Single-mode systems currently available can transmit data a distance of over 30 km at a rate of nearly 1 Gbit/s. An experimental system developed by AT&T multiplexes 10 slightly different wavelength laser beams onto one single-mode fiber. The system can transmit data at an effective rate of 20 Gbit/s over a distance of 68 km without amplification.

One of the main problems with single-mode fibers is the difficulty in making low-loss connections with the tiny fibers. Another difficulty is that in order to get enough light energy into the tiny fiber, relatively expensive laser diodes or other lasers rather than inexpensive LEDs must be used.

FIBER-OPTIC CABLE USES

Fiber-optic transmission has the advantages that the signal lines are much smaller than the equivalent electrical signal lines, the signal lines are immune to electrical noise, and signals can be sent much longer distances without repeater amplifiers. A large number of fibers can be put in a single cable. One of the first major uses of fiber-optic transmission systems has been for carrying large numbers of phone conversations across oceans and throughout cities. The specifications of currently available fiber-optic systems are impressive, but relatively primitive as compared to the possiblities shown by laboratory research. In the future it is possible that the high data rate of fiber-optic transmission may make picture phones a household reality, replace TV cables, replace satellite communication for many applications, replace modems, and provide extensive computer networking.

ASYNCHRONOUS COMMUNICATION SOFTWARE ON THE IBM PC

In a previous section of this chapter we discussed how asynchronous serial data can be sent or received with an 8251A on a polled or an interrupt basis. Any serial communication at some point has to get down to that level of hardware interaction. However, when you are working in a microcomputer system which has a DOS and BIOS available, you can often add serial communication capability to a program without getting down to this hardware level. To help you see how to decide which software level to use for a given application, we will show you how we developed a simple terminal emulator program and a program which downloads object code files from the IBM PC to an SDK-86 board.

A Terminal Emulator Using DOS Function Calls and BIOS Calls

As a first step in developing the SDK download program we decided to write a simple terminal emulator program. A terminal emulator program, when run, makes the PC act like a dumb CRT terminal. Characters typed on the keyboard are sent out on an RS-232C line to a modem or some other RS-232C compatible equipment, and characters coming into the PC on an RS-232C line are displayed on the CRT.

Whenever you want to write a system program such as this, you should first see what DOS function calls are available to do all or part of the job for you. There are several reasons for this approach. First, DOS function calls are usually very easy to use because they do not

```
INIT COM1
REPEAT
     IF CHARACTER READY IN UART THEN
          READ CHARACTER
          SEND TO CRT
     IF KEYPRESSED ON IBM KEYBOARD THEN
          READ KEY
          SEND TO SERIAL PORT
UNTIL FOREVER
```

FIGURE 13-20 Algorithm for simple terminal emulator program.

require you to have extensive knowledge of the hardware details. Second, programs written at the DOS level are much more likely to run correctly on another "compatible" system. If you are going to be writing system programs for the IBM PC, you should get a copy of the *DOS Technical Reference Manual.*

If you need some operation that is not provided by a DOS function call, then the next step is to check the available BIOS procedures in the *IBM PC Technical Reference Manual.* Finally, if neither DOS or BIOS has the functions you need, you invoke the 5-minute rule, and then dig into the Technical Reference Manual to find the pieces you need to write the functions yourself. First, we will show you an example using DOS and BIOS calls, and then an example which manipulates the hardware directly to obtain greater performance.

Figure 13-20 shows the algorithm for our terminal emulator program. Let's see which parts of the algorithm can be done with DOS, and which must be done with BIOS.

The relevant DOS function calls are:

Function Call 2—The character in DL is sent to the CRT and the cursor is advanced one position.

Function Call 3—Waits for a character to be received in the serial port, then returns character in AL.

Function Call 4—The character in DL is output to the serial port.

Function Call 8—Waits for a key to be pressed on the keyboard, and then returns the ASCII code for the key in AL.

Function call 2 looks useful for sending a character to the CRT, and function call 4 looks useful for sending a character to the serial port. However, the keyboard call and the serial-read call will not work because they both sit in loops waiting for input. In other words, if you call function 8, execution will not return from that function until a key is pressed on the keyboard. If execution is in the function 8 loop, characters coming into the serial port will not be read. What is needed here are procedures which allow polling to go back and forth between the keyboard and the serial port receiver. Also needed is the least painful way to initialize the serial port. Let's see what BIOS has to offer.

Figure 13-21*a* shows the header for the IBM PC BIOS, INT 14H procedure. This procedure will do one of four functions, depending on the value passed to it in the AH register. If AH = 0 when the procedure is called, the byte in AL is used to initialize the serial port device as shown. If AH = 1, then the character in AL will be sent out from the serial port. Likewise, if AH = 2, then a character will be read in from the serial port and left in AL. Finally, if AH = 3 when the procedure is called, the status of the serial port will be returned in AH and AL. The first of these four options solves the initialization problem. The last (AH = 3) supplies most of the solution for the problem of determining when the UART has a character

```
RS232_IO
    THIS ROUTINE PROVIDES BYTE STREAM I/O TO THE COMMUNICATIONS
    PORT ACCORDING TO THE PARAMETERS:
    (AH) = 0  INITIALIZE THE COMMUNICATIONS PORT
              (AL) HAS PARAMETERS FOR INITIALIZATION

    7       6        5       4        3       2          1       0
    ----- BAUD RATE --      -PARITY--       STOPBIT     --WORD LENGTH--
        000-110             X0-NONE          0-1            10-7 BITS
        001-150             01-ODD           1-2            11-8 BITS
        010-300             11-EVEN
        011-600
        100-1200
        101-2400
        110-4800
        111-9600

    ON RETURN, CONDITIONS SET AS IN CALL TO COMMO STATUS (AH = 3)
    (AH) = 1  SEND THE CHARACTER IN (AL) OVER THE COMMO LINE
              (AL) REGISTER IS PRESERVED
              ON EXIT, BIT 7 OF AH IS SET IF THE ROUTINE WAS UNABLE
                  TO TRANSMIT THE BYTE OF DATA OVER THE LINE.
                  IF BIT 7 OF AH IS NOT SET, THE REMAINDER OF AH
                  IS SET AS IN A STATUS REQUEST, REFLECTING THE
                  CURRENT STATUS OF THE LINE.
    (AH) = 2  RECEIVE A CHARACTER IN (AL) FROM COMMO LINE BEFORE
                  RETURNING TO CALLER
              ON EXIT, AH HAS THE CURRENT LINE STATUS, AS SET BY
                  THE STATUS ROUTINE, EXCEPT THAT THE ONLY BITS
                  LEFT ON ARE THE ERROR BITS (7,4,3,2,1)
                  IF AH HAS BIT 7 ON (TIME OUT) THE REMAINING
                  BITS ARE NOT PREDICTABLE.
                  THUS, AH IS NON ZERO ONLY WHEN AN ERROR
                  OCCURRED.
    (AH) = 3  RETURN THE COMMO PORT STATUS IN (AX)
              AH CONTAINS THE LINE STATUS
              BIT 7 = TIME OUT
              BIT 6 = TRANS SHIFT REGISTER EMPTY
              BIT 5 = TRANS HOLDING REGISTER EMPTY
              BIT 4 = BREAK DETECT
              BIT 3 = FRAMING ERROR
              BIT 2 = PARITY ERROR
              BIT 1 = OVERRUN ERROR
              BIT 0 = DATA READY
              AL CONTAINS THE MODEM STATUS
              BIT 7 = RECEIVED LINE SIGNAL DETECT
              BIT 6 = RING INDICATOR
              BIT 5 = DATA SET READY
              BIT 4 = CLEAR TO SEND
              BIT 3 = DELTA RECEIVE LINE SIGNAL DETECT
              BIT 2 = TRAILING EDGE RING DETECTOR
              BIT 1 = DELTA DATA SET READY
              BIT 0 = DELTA CLEAR TO SEND

    (DX) = PARAMETER INDICATING WHICH RS232 CARD (0,1 ALLOWED)

DATA AREA RS232_BASE CONTAINS THE BASE ADDRESS OF THE 8250 ON THE
    CARD LOCATION 400H CONTAINS UP TO 4 RS232 ADDRESSES POSSIBLE
    DATA AREA LABEL RS232_TIM_OUT (BYTE) CONTAINS OUTER LOOP COUNT
    VALUE FOR TIMEOUT (DEFAULT = 1)
OUTPUT
    AX MODIFIED ACCORDING TO PARMS OF CALL
    ALL OTHERS UNCHANGED
```

(*a*)

```
KEYBOARD I/O
    THESE ROUTINES PROVIDE KEYBOARD SUPPORT
INPUT
    (AH) = 0  READ THE NEXT ASCII CHARACTER STRUCK FROM THE KEYBOARD
              RETURN THE RESULT IN (AL), SCAN CODE IN (AH)
    (AH) = 1  SET THE Z FLAG TO INDICATE IF AN ASCII CHARACTER IS
              AVAILABLE TO BE READ.
              (ZF) = 1—NO CODE AVAILABLE
              (ZF) = 0—CODE IS AVAILABLE
              IF ZF = 0, THE NEXT CHARACTER IN THE BUFFER TO BE READ
              IS IN AX, AND THE ENTRY REMAINS IN THE BUFFER
    (AH) = 2  RETURN THE CURRENT SHIFT STATUS IN AL REGISTER
              THE BIT SETTINGS FOR THIS CODE ARE INDICATED IN
              THE EQUATES FOR KB_FLAG
OUTPUT
    AS NOTED ABOVE, ONLY AX AND FLAGS CHANGED
    ALL REGISTERS PRESERVED
```

(*b*)

FIGURE 13-21 Header for IBM PC BIOS calls. *(a)* INT 14 serial communication procedure. *(b)* INT 16 keyboard access procedure.

```
;TERMINAL EMULATOR PROGRAM FOR SDK-86
; This program sends characters entered on the IBM PC to the COM1
; serial port at 600 baud, and displays characters received from the
; COM1 serial port on the CRT.
PAGE ,132

STACK_HERE        SEGMENT STACK
                  DW 100 DUP(0)
        STACK_TOP LABEL WORD
STACK_HERE ENDS

CODE_HERE         SEGMENT
        ASSUME CS:CODE_HERE, SS:STACK_HERE

START:  MOV   AX, STACK_HERE      ; Initialize stack segment
        MOV   SS, AX
        MOV   SP, OFFSET STACK_TOP; Initialize stack pointer
        MOV   AH, 00              ; Initialize COM1
        MOV   DX, 0000            ; Point at COM1
        MOV   AL, 01100111B       ; 600 Bd, no parity, 2 stop,8-bit
        INT   14H                 ; via BIOS INT 14H
        STI                       ; Enable interrupts
CHKAGN: MOV   DX, 0000            ; Point at COM1
        MOV   AH, 03              ; Check for character from SDK
        INT   14H
        TEST  AH, 01H             ; See if char waiting in UART
        JNZ   RDCHAR              ; If char, read it
        JMP   KYBD                ; else, go look for keypress
RDCHAR: MOV   AH, 02              ; Read character
        INT   14H                 ; from UART to AL
        MOV   DL, AL              ; Character to DL for DOS call
        MOV   AH, 02H             ; DOS call number for CRT display
        INT   21H                 ; Do DOS call
KYBD:
        MOV   AH, 1               ; Check if key has been pressed
        INT   16H                 ;  using BIOS call
        JNZ   RDKY                ; If keypress, read key code
        JMP   CHKAGN              ; else look for more from SDK
RDKY:   MOV   AH, 0               ; Read key code
        INT   16H                 ; using BIOS call
        MOV   DX, 0000H           ; Point at COM1 serial port
        MOV   AH, 01
        INT   14H                 ; Send character to UART with BIOS
        JMP   CHKAGN              ; Go look for another char from UART
                                  ; or from keyboard

CODE_HERE ENDS
END START
```

FIGURE 13-22 Simple 300/600-Bd terminal emulator program.

ready to be read. Bit 0 of the status byte returned in AH will be set if the UART contains a character ready to be read. If a character is ready, it can be read in and sent to the CRT. If no character is present, the program can go check to see if a key on the keyboard has been pressed.

Figure 13-21*b* shows the header for the IBM PC BIOS, INT 16H procedure which accesses the keyboard. As you can see in the figure, this procedure supplies three different functions, depending on the value passed to it in AH. AH = 0 returns the code for a pressed key in AL. AH = 1 returns the zero flag = 0 if a key has been pressed and the code is available to be read. AH = 2

```
INITIALIZE EVERYTHING
   REPEAT
     IF KEYPRESSED THEN
        READ KEY
        IF KEY = Q THEN
             QUIT
        ELSE IF KEY = L THEN
               DOWNLOAD BINARY FILE FORM DISK TO SDK-86
             ELSE SEND CHAR FOR PRESSED KEY TO SDK-86
     IF UART BUFFER HAS CHARACTER THEN
        SEND CHARACTER TO CRT
  UNTIL QUIT
```

(a)

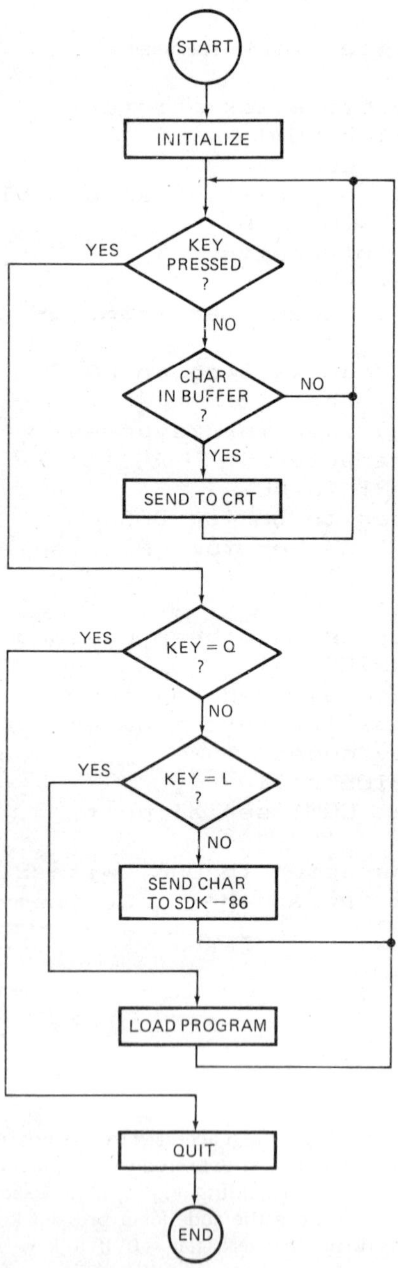

(b)

FIGURE 13-23 Alogorithm for download program. *(a)* Pseudocode. *(b)* Flowchart.

causes the procedure to return the shift status in AL. Bit D6 of this byte will be set if the shift key is depressed. Calling the INT 16 procedure with AH = 1 solves the problem of polling the keyboard without sitting in a loop the way the DOS function call does. The zero flag can simply be checked upon return from the INT 16 procedure, and if no key is ready, execution can go check the UART again. If a key code is ready, it can be read in with a DOS call or another INT 16 call and sent to the UART.

Figure 13-22 shows a simple terminal emulator program which uses the procedures we have described. The program follows the algorithm almost line by line. Remember from previous chapter examples that BIOS procedures are called directly by an INT (number) instruction, and DOS calls are done by putting the function number in AH and doing an INT 21H instruction.

The program in Figure 13-22 works well at 300 Bd or 600 Bd. You can connect the serial port on an IBM-compatible computer to the serial port of an SDK-86 board as shown in Figure 13-10*b* and use this program to communicate with the board. However, if you try to use the program at 1200 or 2400 Bd, the first character of each line of characters received from the SDK-86 or other source will be lost. It took some work to figure out why this is the case, because with a processor as fast as the 8088 in the IBM PC, even 4800-Bd communication should be nc problem. The problem is in the procedure which sends a character to the CRT. After a carriage return is sent to the CRT, the display on the screen is scrolled up one line. To avoid flicker, however, the screen is not scrolled until the next frame update. Since the frame rate for the monochrome display is 50 Hz, the return from the display procedure may take as long as 20 ms. One or more characters that come in during this time will be lost. The next section shows how we solved this problem in a high-speed download program.

IBM PC to SDK-86 Download Program

The main purpose of the program described in this section is to allow the binary codes for programs developed on an IBM PC compatible computer to be downloaded through an RS-232C link to an SDK-86 board. The program also functions as a dumb terminal so that downloaded programs can be run, memory contents displayed, and registers examined by typing the appropriate keys on the computer keyboard. Figure 13-23 shows the overall algorithm for the program. The main difference between this algorithm and the one for the dumb terminal in Figure 13-20 is the addition of the actions when the letter Q or the letter L is typed. However, we implemented the algorithm in a different way in order to solve the speed problem described in the previous section and to show you some very important programming techniques.

Figure 13-24 shows the complete program. The data segment declared at the start of the program contains buffers, flags, and the messages used in the program. We will refer to these as needed throughout the discussion. The mainline part of the program which follows this is only a little over a page long and consists mostly of various initializations. Four procedures, SERIAL_IN, CHK_N_DISPLAY, XMIT, and LOAD_IT, are used to do most of the work. After we give an overview, we will explain in detail how each of these five parts work.

OVERVIEW

The procedure SERIAL_IN reads in characters received from the SDK-86 by the IBM serial board on an interrupt basis and puts the characters in a buffer. The procedure CHK_N_DISPLAY, when called, checks the buffer to see if it contains any characters. If the buffer contains a character, the procedure sends it to the CRT. If the buffer is empty, the procedure simply returns. Now, take a look at the mainline section of the program starting at the label THERE, to see how these two procedures, together with XMIT and LOAD_IT implement the algorithm in Figure 13-23.

After interrupts are enabled, the BIOS INT 16 procedure is used to see if a key on the PC has been pressed. If a key has been pressed, then the BIOS INT 16 procedure is called again to read in the code for the pressed key. An IF—THEN—ELSE decision structure is then used to quit the program and return to DOS, go download a binary file, or simply send the character to the SDK-86. If the first call of the BIOS INT 16 procedure does not indicate that a key has been pressed, then the CHK_N_DISPLAY procedure is called. If the UART interrupt procedure, SERIAL_IN, has read in a character from the SDK and put it in the buffer, this procedure will send the character to the CRT and return to the mainline. If there is no character in the buffer, CHK_N_DISPLAY will simply return to the mainline. Once back in the mainline, execution loops back to again see if a key on the PC has been pressed. The interrupt and buffer approach used to read and hold the characters coming in from the SDK solves the timing problem we described for the program in Figure 13-22. Here's how.

The problem with operating the program in Figure 13-22 at over 600 Bd is that characters which come into the UART while the INT 10H procedure is scrolling the CRT display are missed. The program in Figure 13-24 reads characters from the UART on an interrupt basis. Even if the PC is in the middle of the INT 10H procedure or some other procedure when the UART has a character ready, the interrupt procedure will read the character and put it in the buffer. When execution loops back around to the CHK_N_DISPLAY procedure again, the character will then be read from the buffer and sent to the display. The XMIT procedure is used at several places in the program to send characters to the UART for transmission to the SDK.

The LOAD_IT procedure, which is called when the user presses the L key on the PC, prompts the user to enter the name of a binary file, converts the binary file to a form the SDK can digest, and sends the result to the SDK through the UART. Now let's look at the details of the initialization and the four procedures.

INITIALIZATION

The UART used on the IBM asynchronous communications adapter board is an INS8250. If the board is configured as system serial port COM1, the interrupt output from this device is connected to the IR4 input of an 8259A priority interrupt controller on the main PC board of the IBM computer. The major part of the initialization here involves getting the 8250 initialized and setting up the interrupt mechanism. Remember from previous discussions that when an 8259A receives an interrupt on an IR input that is unmasked, it sends a specified interrupt type to the processor. The 8259A is initialized by the BIOS so that type 8 will be sent for an IR0 input. Therefore, for an IR4 signal from the UART, the 8259A will send type 0CH to the processor. The processor multiplies the type number by 4 and goes to that address to get the starting address of the service procedure for that interrupt. The recommended way of putting the IP and CS values for the procedure in these absolute memory locations is with the DOS function call 25H. To use this call, the interrupt type number is put in AL, the CS value in the DS register, and the IP value in the DX register as shown in Figure 13-24.

The 8259A itself is mostly initialized by the BIOS when the system is turned on. However, since the UART is connected to IR4 of the 8259A, that input has to be unmasked. To do this the current contents of the 8259A interrupt mask register are read in from address 21H. The bit corresponding to IR4 is then ANDed with a 0 to unmask the interrupt, and the result sent back to the interrupt mask register. Using this approach saves the system environment. It is important to do this rather than just sending out a control word directly, so that you don't disable other system functions. In this system, for example, the system clock tick is connected to IR0 and the keyboard is connected to IR1, so these would be disabled if you accidentally put 1's in these bits of the control word.

Initializing the 8250 UART is next. Figure 13-25 shows the internal addresses and the bit formats for the control words we need here. The first part of the initialization is the baud rate, parity, and stop bits. Since this step requires several words to be sent, we simply used the BIOS INT 14H procedure to do it. (NOTE: We initialize the 8250 here for 4800 Bd, so the baud-rate jumper on the SDK-86 must be set for this baud rate.)

The next task we need to do here is enable the desired interrupt circuitry in the 8250. In order to do this, the DLAB bit of the line control word must first be made a 0. Note that this is done by reading in the line control

```
;TERMINAL EMULATOR AND DOWNLOAD PROGRAM FOR SDK-86
;This program allows an IBM PC compatible computer to operate as a 4800-Bd
; dumb terminal for use with an SDK-86 board. The program also allows
; binary files to be downloaded from a disk in the PC to RAM in the SDK-86
; through the COM1 serial port on the PC.
PAGE ,132

DATA_HERE  SEGMENT
SIGN_ON         DB 'SDKDMP PROGRAM',0DH,0AH
                DB '  BY DOUGLAS V. HALL',0DH,0AH
                DB 'COPYRIGHT - McGraw-Hill Book Co.,1986',0DH,0AH
                DB 0DH,0AH
                DB 'Press RESET key on SDK-86 to get Monitor prompt.'
                DB 0DH,0AH,0DH,0AH,24H
QUEUE           DB 1000 DUP(0)  ; Circular buffer
HEAD_POINTER    DW 0            ; Next character to read out of queue
TAIL_POINTER    DW 0            ; Next location to put char in queue
CHAR_COUNTER    DB 0            ; Used to count number of char sent to CRT

TIME_OUT_MESS   DB ' Transmit Timeout - Check Hardware ',     0DH, 0AH
PROMPT          DB ' Please type in binary filename',          0DH, 0AH
                DB ' Filename format is d:path filename.bin', 0DH, 0AH
                DB  24H         ; String terminator

FILE_NAME       DB  40 DUP(0)   ; Space for user entered filename

ERR_MESS_POINTERS  DW   0       ; Dummy, no ERR_MESS0
                DW   OFFSET ERR_MESS1
                DW   OFFSET ERR_MESS2
                DW   OFFSET ERR_MESS3

ERR_MESS1       DB 'INVALID FUNCTION NUMBER', 0DH, 0AH, 24H
ERR_MESS2       DB 'FILE NOT FOUND', 0DH, 0AH, 24H
ERR_MESS3       DB 'PATH NOT FOUND', 0DH, 0AH, 24H

FILE_BUF        DB 2048 DUP(0)  ; Buffer for bin file read from disk

HEADER          DB 53H,20H,30H,30H,30H,30H,3AH,30H
                DB 31H,30H,30H,2CH                ; SDK-86 Substitute Command

DATA_HERE  ENDS

STACK_HERE SEGMENT STACK
                DW 400H DUP(0)
     STACK_TOP  LABEL WORD
STACK_HERE ENDS

CODE_HERE SEGMENT
       ASSUME CS:CODE_HERE, DS:DATA_HERE, SS:STACK_HERE

START:  MOV  AX, STACK_HERE      ; Initialize stack segment
        MOV  SS, AX
        MOV  SP, OFFSET STACK_TOP; Initialize stack pointer
        MOV  AL, 0CH             ; Initialize interrupt vector for UART
        MOV  BX, SEG SERIAL_IN   ; using DOS function call 25H
```

FIGURE 13-24 Program to download object code programs from IBM PC to SDK-86 and allow PC to function as a "dumb" terminal for the SDK-86.

```
        MOV  DS, BX
        MOV  DX, OFFSET SERIAL_IN
        MOV  AH, 25H
        INT  21H
        MOV  AX, DATA_HERE      ; Initialize DS register
        MOV  DS, AX
        MOV  HEAD_POINTER, 0    ; Initialize circular buffer pointers
        MOV  TAIL_POINTER, 0
;unmask 8259A IR4
        IN  AL, 21H             ; Read 8259A IMR
        AND  AL, 0ECH           ; Unmask IR4
        OUT  21H, AL
;initialize 8250 UART baud rate,etc.
        MOV  AH, 00             ; Initialize COM1
        MOV  DX, 0000           ; Point at COM1
        MOV  AL, 11000111B      ; 4800 Bd,No parity,2 stop,8-bit
        INT  14H                ;  via BIOS INT 14H
;enable 8250 RxRDY interrupt
        MOV  DX, 03FBH          ; Point at 8250 line control port
        IN   AL, DX             ; Read in line control word
        AND  AL, 7FH            ; Set DLAB = 0
        OUT  DX, AL             ; Send line control word back out
        MOV  AL, 01H            ; Value to enable RxRDY interrupt
        MOV  DX, 03F9H          ; Point at interrupt enable register
        OUT  DX, AL             ; Enable RxRDY interrupt
        MOV  AL, 0BH            ; Assert 8250 OUT2, RTS, DTR byte
        MOV  DX, 03FCH          ; Point at modem control reg in 8250
        OUT  DX, AL             ; Send to 8250
        STI                     ; Enable 8086 interrupts
;main program starts here
        MOV  DX, OFFSET SIGN_ON ; Send sign on message to CRT
        MOV  AH, 09H            ;  with DOS call
        MOV  BH, 0
        INT 21H
;look for response from SDK
THERE:  MOV  AH, 1              ; Check if key has been pressed
        INT  16H
        JNZ  RDKY               ; If keypress, go read key code
        CALL CHK_N_DISPLAY      ; See if char in circular buffer from
                                ; UART and send it to CRT if there
        JMP  THERE              ; Go look for keypress or char from UART
RDKY:   MOV  AH, 0              ; Read key code
        INT  16H
        CMP  AL, 51H            ; See if quit commnad
        JNE  NXCHK              ; No,  go check if load command
        JMP  QUIT               ; Yes, go exit to DOS
NXCHK:  CMP  AL, 4CH            ; Check if load command
        JNE  SENDIT             ; No, go send char to SDK
        CALL LOAD_IT            ; Yes, go load file
        JMP  THERE              ; Go wait for next command
SENDIT: CALL XMIT               ; Char not Q or L, send char to SDK-86
        JMP  THERE              ; Go wait for next command
QUIT:   IN   AL, 21H
        OR   AL, 10H            ; Mask UART interrupt
        OUT  21H, AL            ;  to prevent it from disrupting DOS
```

Figure 13-24 (*continued*)

```
        MOV  AL, 0              ; Exit to DOS using DOS function call 4CH
        MOV  AH, 4CH
        INT  21H
        NOP

SERIAL_IN PROC FAR
        STI                     ; Interrupts back on for keyboard, etc
        PUSH AX
        PUSH BX
        PUSH DX
        PUSH DI
        PUSH DS                 ; Save DS of interrupted program
        MOV  AX, DATA_HERE      ; Install DS needed here
        MOV  DS, AX
        MOV  DX, 03F8H          ; Receiver buffer address for 8250
        IN   AL, DX             ; Read character
        MOV  DI, TAIL_POINTER   ; Get current tail pointer value
        INC  DI                 ; Point to next storage location
        CMP  DI, 1000           ; Compare with queue length to see if time
                                ;  to wrap around
        JNE  FULCHK             ; No, go check if queue full
        MOV  DI, 00             ; Yes, set DI for wraparound to start
FULCHK: CMP  DI, HEAD_POINTER   ; Check for full queue
        JE   NO_MORE            ; Full, return without writing char
        MOV  BX, TAIL_POINTER   ; Not full, point at location to put char
        MOV  QUEUE[BX], AL      ; Character to circular buffer
        MOV  TAIL_POINTER, DI   ; Tail pointer to next location
NO_MORE:
        MOV  AL, 20H            ; Non-specific End Of Interrupt command
        OUT  20H, AL            ;  to 8259A
        POP  DS
        POP  DI
        POP  DX
        POP  BX
        POP  AX
        IRET
SERIAL_IN  ENDP

CHK_N_DISPLAY PROC NEAR
        PUSH BX
        IN   AL, 21H
        OR   AL, 10H            ; Disable 8259A IR4 during critical region
        OUT  21H, AL            ;  by masking bit 4 of int mask register
        MOV  DI, HEAD_POINTER
        CMP  DI, TAIL_POINTER   ; Is queue empty ?
        JE   NOCHAR             ; Yes, just return
        MOV  AL, QUEUE[DI]      ; No, get char from queue to AL
        INC  DI                 ; Point DI at next byte in queue
        CMP  DI, 1000           ; See if time to wrap pointer around
        JNE  OK                 ; No, go on
        MOV  DI, 00             ; Yes, wrap pointer around
OK:     MOV  HEAD_POINTER, DI   ; Store new pointer value
        PUSH AX                 ; Save character in AL on stack
        IN   AL, 21H
        AND  AL, 0ECH           ; Enable IR4 interrupt so new char in 8250
```

Figure 13-24 (*continued*)

```
        OUT  21H, AL           ;   can interrupt INT 10H
        POP  AX                ; Get character back from stack
        MOV  AH, 14            ; Use BIOS INT 10H procedure to send to CRT
        MOV  BH, 0
        INT  10H               ; Send char to CRT
        DEC  CHAR_COUNTER      ; Decrement count of char to be sent to CRT
NOCHAR: IN   AL, 21H
        AND  AL, 0ECH          ; End of critical region. Enable IR4 by
                               ;  unmasking bit 4 in IMR of 8259A so new
        OUT  21H, AL           ;  char in UART can interrupt
        POP  BX
        RET
CHK_N_DISPLAY ENDP

;Send character in AL to COM1 serial port after checking if
; handshake signals asserted
; INPUTS:  character in AL
; OUTPUTS: character in AL, CY flag set if xmitter not ready

XMIT PROC NEAR
        PUSH BX
        PUSH CX
        PUSH DX
        PUSH AX                ; Save char
        MOV  CX,0
RECHK:  MOV  DX, 03FEH         ; Check CTS and DSR asserted
        IN   AL, DX
        AND  AL, 30H
        CMP  AL, 30H
        JNE  NOT_READY
        MOV  DX, 03FDH         ; Check if transmitter buffer ready
        IN   AL, DX
        AND  AL, 20H
        JZ   NOT_READY
        POP  AX                ; Get character back
        MOV  DX, 03F8H         ; Send to UART
        OUT  DX, AL
        CLC                    ; Clear carry to indicate char sent
        JMP  DONE
NOT_READY:
        LOOP RECHK             ; Check for status ready CX times
        MOV  DX, OFFSET TIME_OUT_MESS   ; If still not ready
        MOV  AH, 09H                    ;  send error message
        MOV  BH, 0
        INT  21H
        STC                    ; Set carry to indicate xmitter not ready
        POP  AX                ; Restore registers
DONE:   POP  DX
        POP  CX
        POP  BX
        RET
XMIT    ENDP

;LOAD_IT - Down-load procedure
```

Figure 13-24 (*continued*)

```
; Prompts user to enter name of binary file, then reads file from disk
;  to a buffer in memory. The file is converted to the ASCII character form
;  required by the SDK-86, and sent to an SDK-86 board via the COM1 serial
;  port. Replies from the SDK-86 are displayed on the CRT as received.

LOAD_IT PROC NEAR
        PUSH BX
        MOV  DX, OFFSET PROMPT   ; Send message to CRT telling user
        MOV  AH, 09H             ;  to enter filename with DOS function
        MOV  BH, 0               ;  call 09H
        INT  21H                 ;
        MOV  DX, OFFSET FILE_NAME; Point at filename buffer
        MOV  FILE_NAME, 40       ; Make first byte of buffer = max chars
        MOV  AH, 0AH             ; DOS function call number to
        INT  21H                 ;  read in filename from keyboard
        MOV  BL, FILE_NAME+1     ; Get length of file name from buffer
        ADD  BL, 02              ; Add 2 to reach carriage return at end
        MOV  BH, 00              ; BX now has offset of CR at end of file name
        MOV  FILE_NAME[BX], 00   ; Replace 0DH at end of file name with 00H
        MOV  DX, OFFSET FILE_NAME; Point at start of file name buffer
        ADD  DX, 02H             ; Move pointer over string length bytes
        MOV  AL, 0               ; Open file for read
        MOV  AH, 3DH             ;  and get file handle with DOS 3DH call
        INT  21H
; Check for file error
        JNC  FILEOK              ; No carry, file opened properly
        ROL  AX, 1               ; Multiply error code in AX by 2
        MOV  BX, AX                     ; Copy to BX for use as pointer
        MOV  DX, ERR_MESS_POINTERS[BX]  ; Get pointer to desired error
        MOV  BH, 0                      ; message from table to DX
        MOV  AH, 09H             ; Use DOS function call 09 to send
        INT  21H                 ;  error message to CRT
        MOV  DL, 0DH             ; Send carriage return to CRT
        MOV  AH, 02H             ;  with DOS function call 02H
        INT  21H
        JMP  EXIT1               ; Return to look for next command from user
; Read binary file from disk to buffer in memory
FILEOK: MOV  BX, AX              ; File handle to BX
        PUSH BX                  ; Save file handle for file close
        MOV  CX, 2048            ; Set maximum number of char to read
        MOV  DX, OFFSET FILE_BUF ; Point at buffer to store char read in
        MOV  AH, 3FH             ; Read disk file
        INT  21H
        POP  BX                  ; Get back file handle for close
        PUSH AX                  ; Save length of file returned by 3FH
        MOV  AH, 3EH             ; Close disk file with DOS 3EH call
        INT  21H
;SEND SUBSTITUTE COMMAND TO SDK86
        MOV  BX, OFFSET HEADER   ; Point at string buffer
        MOV  CX, 000CH           ; Number of characters to send to SDK
NEXT1:  MOV  AL, [BX]            ; Get a character to be sent
TRY2:   CALL XMIT                ; Character to COM1
        JC   EXIT1
        INC  BX                  ; Pointer to next location in buffer
        LOOP NEXT1               ; Loop until all sent
        MOV  CHAR_COUNTER, 11H   ; Number of characters to send to CRT
```

Figure 13-24 (*continued*)

```
MORE1:  CALL CHK_N_DISPLAY       ; Characters to CRT
        DEC  CHAR_COUNTER        ; See if all sent
        JNZ  MORE1
;SEND FIRST CODE BYTE TO SDK86
        POP  AX                  ; Get back length of file
        MOV  CX, AX              ; Use CX as counter
        MOV  BX, OFFSET FILE_BUF ; Point at start of object code file
NXTCHR: MOV  DL, [BX]            ; Get a character from file buffer
        AND  DL, 0F0H            ; Mask lower nibble
        ROR  DL, 1               ; Move to lower nibble position
        ROR  DL, 1
        ROR  DL, 1
        ROR  DL, 1
        CMP  DL, 0AH             ; Convert to ASCII
        JAE  ADD37
        ADD  DL, 30H
        JMP  SEND1
ADD37:  ADD  DL, 37H
SEND1:  MOV  AL, DL              ; Position character for send
        CALL XMIT                ; Send upper nibble of byte to SDK
        JC   EXIT1
        MOV  DL, [BX]            ; Get char again and make ASCII for
        AND  DL, 0FH             ;  low nibble
        CMP  DL, 0AH
        JAE  ADDHI
        ADD  DL, 30H
        JMP  SEND2
ADDHI:  ADD  DL, 37H
SEND2:  MOV  AL, DL              ; Position for send
        CALL XMIT                ; Send ASCII for low nibble to SDK
        JC   EXIT1
        DEC  CX                  ; Check if all bytes sent
        JZ   EXIT                ; Yes, done, send carriage return
        MOV  AL, 2CH             ; Else load comma
        CALL XMIT                ; Send to SDK
        JC   EXIT1
        MOV  CHAR_COUNTER, 0EH   ; Number of characters to send to CRT
MORE2:  CALL CHK_N_DISPLAY       ; SDK echo message to CRT
        CMP  CHAR_COUNTER, 0     ; See if all of message sent to CRT
        JNE  MORE2
        INC  BX                  ; Point to next byte in binary file
        JMP  NXTCHR              ; Send next code byte
EXIT:   MOV  AL, 0DH             ; Load carriage return
        CALL XMIT                ; Send to SDK
EXIT1:  POP BX
        RET                      ; Go look for SDK answer and next command
LOAD_IT  ENDP

CODE_HERE ENDS
END START
```

Figure 13-24 (*continued*)

word, resetting the desired bit, and sending the word out again. This preserves the previous state of the rest of the bits in the line control register. As shown in Figure 13-25*b*, with DLAB = 0, a control word which enables the enable-receive line status interrupt can be sent to the interrupt enable register at address 03F9H. As shown in Figure 13-25*b*, the 8250 has four different conditions which can be enabled to assert the interrupt output when true. In cases where multiple interrupts are used, the interrupt identification register can be

I/O DECODE (IN HEX)		REGISTER SELECTED	DLAB STATE
PRIMARY ADAPTER	ALTERNATE ADAPTER		
3F8	2F8	TX BUFFER	DLAB=0 (WRITE)
3F8	2F8	RX BUFFER	DLAB=0 (READ)
3F8	2F8	DIVISOR LATCH LSB	DLAB=1
3F9	2F9	DIVISOR LATCH MSB	DLAB=1
3F9	2F9	INTERRUPT ENABLE REGISTER	DLAB=X
3FA	2FA	INTERRUPT IDENTIFICATION REGISTERS	DLAB=X
3FB	2FB	LINE CONTROL REGISTER	DLAB=X
3FC	2FC	MODEM CONTROL REGISTER	DLAB=X
3FD	2FD	LINE STATUS REGISTER	DLAB=X
3FE	2FE	MODEM STATUS REGISTER	DLAB=X

(a)

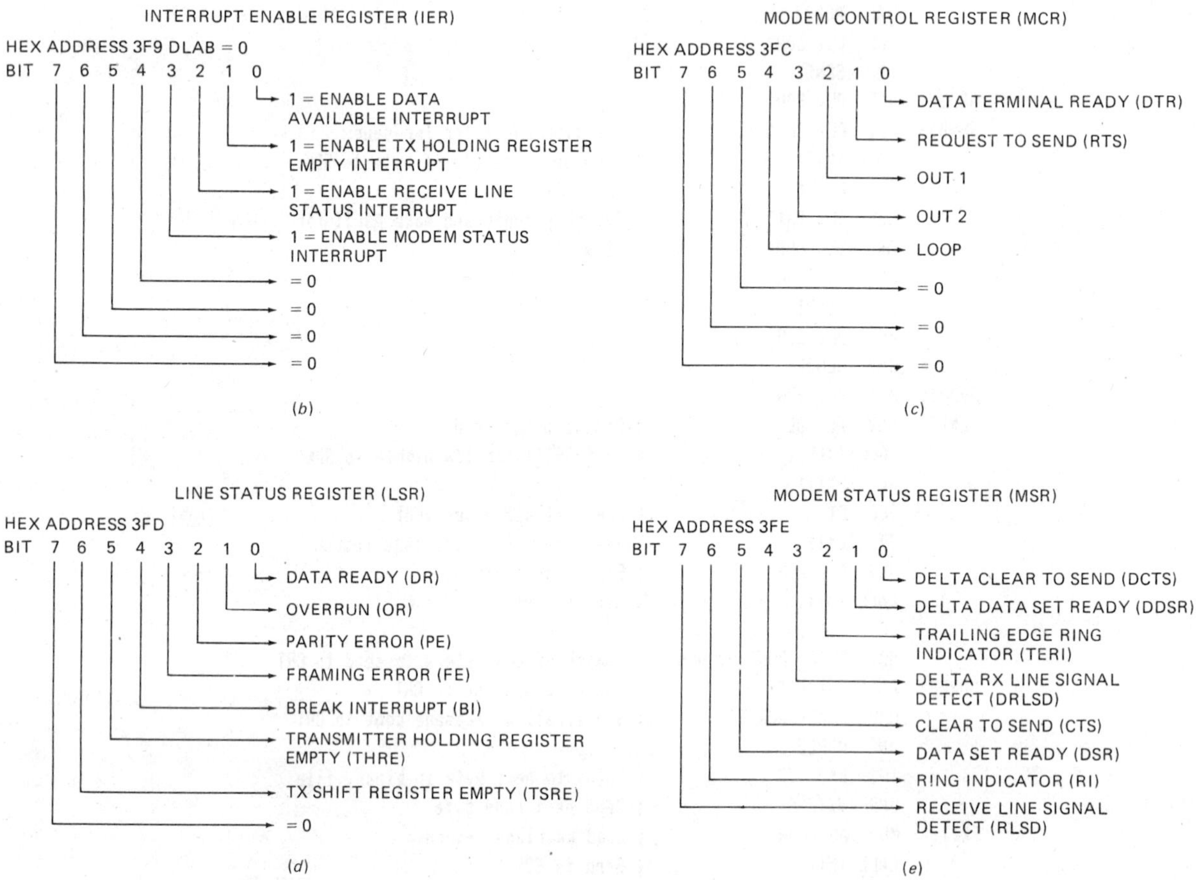

FIGURE 13-25 8250 internal addresses, registers, and control words. *(a)* System addresses. *(b)* Interrupt enable register. *(c)* Modem control register. *(d)* Line status register. *(e)* Modem status register.

read to determine the source of an interrupt. For this application we are only using the enable receive line status interrupt, so a 1 is put in that bit. The final step in the 8250 initialization is to assert the $\overline{RTS}$, $\overline{DTR}$, and $\overline{OUT2}$ output signals. As shown by the circuit connections in Figure 13-10*a*, asserting $\overline{RTS}$ is necessary to assert the $\overline{CTS}$ input so the UART can transmit. Likewise, asserting $\overline{DTR}$ is necessary to assert the $\overline{DSR}$ and $\overline{CD}$ inputs of the UART. The $\overline{OUT2}$ signal from the 8250 must be asserted in order to enable a three-state buffer which is in series with the interrupt signal from the 8250 to the 8259A.

When you are working out an initialization sequence such as this, read the data sheet carefully, and check out the actual hardware circuitry for the system you are working on. We missed the $\overline{OUT2}$ connection the first time through, but a second look at the schematic for the communications board showed that it was necessary to assert this signal. Now let's see how the procedure which reads characters from the UART works.

THE SERIAL_IN PROCEDURE

The purpose of this interrupt procedure is to read characters in from the UART and put them in a buffer. Note that since this is an interrupt procedure which can occur at any time, it is important to save the DS register of the interrupted program and load the DS register with DATA_HERE, the value needed for this procedure.

The buffer used here is a special type of queue called a *circular buffer*. Figure 13 26 attempts to show how this works. One pointer, called the TAIL_POINTER, is used to keep track of where the next byte is to be written to the buffer. Another pointer called the HEAD_POINTER is used to keep track of where the next character to be read from the buffer is located. The buffer is circular because, when the tail pointer reaches the highest location in the memory space set aside for the buffer, it is "wrapped around" to the beginning of the buffer again. The head pointer follows the tail pointer around the circle as characters are read from the buffer. Two checks are made on the tail pointer before a character is written to the buffer.

First the tail pointer is brought into a register and incremented. This incremented value is then compared with the maximum number of bytes the buffer can hold. If the values are equal, the pointer is at the highest address in the buffer, so the register is reset to zero. After the current character is put in the buffer, this value will be loaded into TAIL_POINTER to wrap around to the lowest address in the buffer again.

Second, a check is made to see if the incremented value of the tail pointer is equal to the head pointer. If the two are equal, it means that the current byte can be written, but that the next byte would be written over the byte at the head of the queue. In this case we simply return to the interrupted program without writing the current character into the buffer. Actually this wastes a byte of buffer space, but it is necessary to do this so that the pointers have different values for this buffer-full condition than they do for the buffer-empty condition.

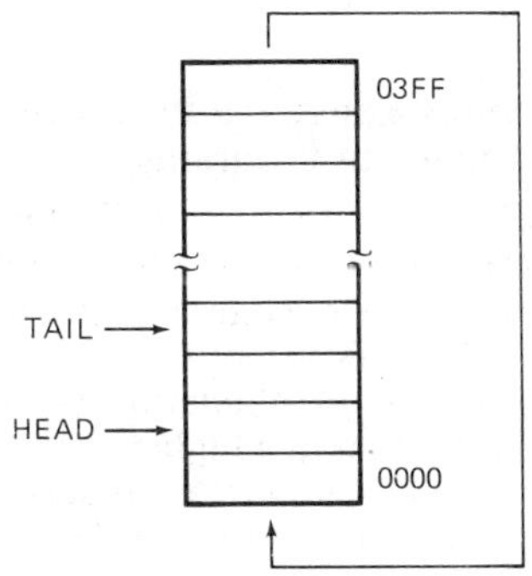

FIGURE 13-26 Diagram showing how circular buffer pointers wrap around at the top of allocated buffer space.

The buffer-empty condition is indicated when the head pointer is equal to the tail pointer. If the buffer is not full, the character read in from the UART is written to the buffer, and the pointer to the next available location in the buffer is transferred from the register to the memory location called TAIL_POINTER. Finally, before returning, an end-of-interrupt command must be sent to the 8259A to reset bit 4 of the interrupt mask register.

To summarize the operation of a circular buffer, then, bytes are put in at the tail pointer location and read out from the head pointer location. The buffer is considered full when the tail pointer reaches one less than the head pointer. An empty buffer is indicated by head pointer equal to tail pointer.

THE CHK_N_DISPLAY PROCEDURE

The main purpose of this procedure is to read a character from the circular buffer and send it to the CRT with the BIOS INT 10H procedure. In order to make sure the procedure operates correctly under all conditions, however, we mask the IR4 interrupt in the 8259A right at the start so that an interrupt from the UART cannot call the SERIAL_IN procedure while CHK_N_DISPLAY is using the head and tail pointers. This is necessary to prevent the SERIAL_IN procedure from altering the values of the pointers in the middle of CHK_N_DISPLAY's use of them and causing the CHK_N_DISPLAY procedure to make the wrong decisions about whether the buffer is empty, for example. The group of instructions which you need to protect from interruption is called a *critical region*. It is important to keep critical regions as short as possible so that interrupts need not be masked for unnecessarily long times. Note that we masked the IR4 interrupt input of the 8259A rather than disable the processor interrupt. This was done so that the keyboard and the timer interrupts, which have nothing to do with the critical region in this procedure, can keep running.

Once the critical region is safe, a check is made to see if there are any characters in the buffer. If not, the 8259A IR4 input is unmasked, and execution returned to the calling program. If a character is available in the buffer, the character is read out and the head pointer updated to point to the next available character. If the pointer is at the top of the space allotted for the buffer, the pointer will be wrapped around to the start of the buffer again. As soon as the character is read out from the buffer and the pointers updated, an interrupt from the UART cannot do any damage, so we unmask IR4. The BIOS INT 10H procedure is then used to send the character to the CRT. If a UART interrupt occurs during the INT 10H procedure, the SERIAL_IN procedure will read the character from the UART and return execution to the INT 10H procedure. This short interruption produces no noticeable effect on the operation of the INT 10H procedure, and it makes sure no characters from the UART are missed. After the INT 10H procedure finishes, a character-sent counter is decremented and execution is returned to the calling program. This counter counts the number of characters actually sent to the CRT rather than just the number of times the CHK_N_DISPLAY procedure is called. This allows the procedure to be called over and over again until a given number of characters are sent to the CRT.

THE XMIT PROCEDURE

After first checking to see if the handshake signals and the transmitter buffer are ready, this procedure sends a character to the 8250 UART. The status of the $\overline{DSR}$ input and the $\overline{CTS}$ input are available as bits 5 and 4 respectively, of the modem status word of the 8250. For an IBM PC with the serial board configured as COM1, the address of this register will be 03FEH. The transmitter-buffer-ready status bit of the 8250 is available as bit 5 of the line status register at address 03FDH. This bit will be high when a character can be sent to the internal buffer for transmission. Rather than having the program hang in a loop checking the status signals forever, if any one of them is not ready, we send an error message and exit after a specified number of tries. The CX register is loaded with the desired number of tries, and counted down after each loop through the status check. After the error message is sent to the CRT with the DOS 09H function call, the carry flag is set to indicate that a character could not be sent. Execution is then returned to the section of the program from which XMIT was called. After each call of XMIT in the program, a JC instruction is used to send execution back to look for the next user command if the transmitter was not ready for some reason.

THE LOAD_IT PROCEDURE

In response to a filename entered by the user, this procedure reads a binary file from disk to a buffer in memory, converts each byte to the form needed for sending to the SDK, and sends the result to the SDK-86. SDK responses are displayed on the CRT.

At the start of the procedure, DOS function call 09H is used to send a prompt message to the user on the CRT. DOS function call 0AH is then used to read in the filename entered by the user on the computer keyboard. To use this call the DX register must point to the start of a buffer in memory where the characters are to be put. The first byte of the buffer must contain the maximum number of characters that you want to be read in. The function call terminates when the user enters a carriage return. The second byte of the buffer then holds the number of characters actually read in, not including the carriage return.

The next task in the procedure is to open the named binary file for reading with the DOS 3DH function call. Before this can be done, however, the carriage return at the end of the filename in the buffer must be changed to 00H. To do this we get the number of characters read from the second byte in the buffer, and use it to construct a pointer to the carriage return in the buffer. That location is then loaded with 0's.

If an error is detected while trying to open the file, the DOS 3DH function call will return the carry flag set, and an error number in AX. We multiply the error number by 2 and use it as a pointer to a table which contains the offsets of the error messages for some of the errors. As a refresher on the use of address tables, work your way through how the pointer gets to DX for the 09H function call which is used to display the error message. If the file was opened correctly, then the binary file is read in from a disk with the DOS function call 3FH, and the file is closed with the DOS function call 3EH as we described in Chapter 12.

The next section of the procedure sends the substitute command to the SDK-86 to get it ready to receive the binary file. The SERIAL_IN interrupt procedure will read the echo of the command sent back from the SDK and put it in the circular buffer. To display this echo on the CRT, we load CX with the number of characters in the echo message, and enter a loop which calls the CHK_N_DISPLAY procedure over and over until all of the SDK response is sent to the CRT. Now comes the bit-fiddling section of the procedure.

The SDK-86 requires that each nibble of a program code byte be sent in as the corresponding ASCII character. The code byte 3AH for example must be sent as 33H (ASCII 3), followed by 41H (ASCII A). You can work your way through this section to see how the conversion is done if you need to use it in some other program. After the ASCII characters for each code byte are sent, the ASCII code for a comma is sent as required by the SDK-86. The SERIAL_IN procedure reads in the SDK reply and puts it in the circular buffer. A loop containing a call to the CHK_N_DISPLAY procedure is used to send the SDK reply to the CRT. After the ASCII codes for the final byte are sent to the SDK, a carriage return is sent to the SDK to terminate the substitute command. Execution then returns to the mainline program and then to the section of the program which waits for the user to enter another command.

CONCLUSION

This program was written to do a specific job and to demonstrate some important programming concepts. Space limitation prevented us from making the program as "friendly" as we would have liked it to be. Instructions could be added, for example, to allow the desired baud rate to be entered by the user from the keyboard. Hopefully you can use some of the techniques shown here in your own programs. In the next section we show you how to call assembly language procedures from higher level language programs.

Calling Assembly Language Procedures from High Level Language Programs

Programs which need to do a lot of bit-fiddling and hardware manipulation are usually written in assembly language because this level gives direct hardware control. However, business, scientific, and other programs which involve mostly manipulating large amounts of data are usually written in a higher level language such as BASIC, Pascal, or C. For programs such as communications programs, which involve both types of operation, the main program is usually written in a high level language, and assembly language procedures are called to do the bit fiddling as needed. The intent of this section is to give you an overview of how to write and call these procedures. However, we first have to briefly discuss the two ways that high level language programs are converted into machine code and executed, interpreting and compiling.

Figure 13-27*a* shows in flowchart form how an interpreter program executes a high level language program. The interpreter reads a high level language statement of the source program, translates that statement to machine code and, if it doesn't need information from another instruction, executes the code for that statement immediately. It then reads the next high level language source statement, translates it, and executes it. BASIC programs are often executed in this way.

The advantage of using an interpreter is that if an error is found, you can just correct the source program and immediately rerun it. The major disadvantage of the interpreter approach is that an interpreted program runs 5 to 25 times slower than the same program will run after being compiled. The reason is that with an interpreter each statement must be translated to machine code every time the program is run. In other words, the translation time is part of the execution time.

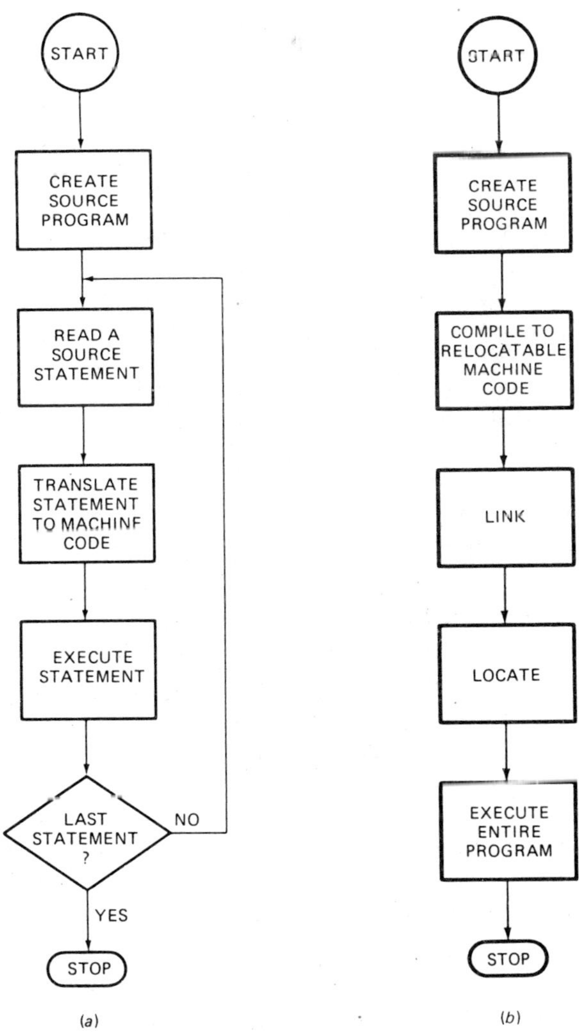

FIGURE 13-27 Comparison of compiler and interpreter operation. *(a)* Interpreter. *(b)* Compiler.

Figure 13-27*b* shows how a compiler fits into the translation-execution process. A compiler program reads through the entire high-level language source program, and in two or more passes through it, translates it all to a relocatable machine code version. Before the program can be run, however, this relocatable object code version must be linked with any other required object modules from the system library, a user library, or assembly language procedures. The output file from the linker is then located, which means that it is given absolute addresses so that it can be loaded into memory. Finally the located program is loaded into memory. Some systems, incidentally, combine two or more of the link, locate, and load functions in a single program. Once the located program is loaded into memory, the entire program can be run without any further translation. Therefore, it will run much faster than it would if executed by an interpreter. The major disadvantage of the compiler approach is that when an error is found, it usually must be corrected in the source program and the entire compile-load sequence repeated.

Calling assembly language procedures from an interpreted high level language is quite messy, especially in the case of multiple procedures, because of the way the interpreter uses memory. If you are working with IBM PC Microsoft BASIC, consult Appendix C of the *IBM BASIC Reference Manual* to see how to do it for that language. We recommend the compiler approach for most hybrid programs, because of the obvious execution speed advantage and the ease with which assembly language modules can be called.

Calling assembly language procedures in compiled programs is much simpler, because the object modules produced by the assembler can be simply linked with object modules produced by the compiler and object modules from libraries. The major task when using assembler-created modules with compiler-created modules is to make sure the modules can find and communicate with each other. Since common Pascal, C, and BASIC compilers use similar conventions to do this, we will use the Microsoft BASIC compiler conventions as an example here.

The BASIC call statement has the format CALL numvar (var1, var2, var3, . . ., varN). For compiled BASIC programs numvar is the name of the assembly language procedure being called. When the BASIC program and the assembly language module are linked, the linker will establish the required connection between the call and the named procedure. Var1, var2, var3, etc. represent the names of variables which are being passed to or from the assembly language procedure. Numeric variables or string variables can be passed to or from an assembly language procedure.

Figure 13-28*a* shows a simple BASIC program which inputs a line of characters from the keyboard, calls an assembly language procedure to make sure all letters in the string are upper case, and then outputs the resultant string to the CRT. The program terminates when

```
10 REM This program inputs a string of characters from the keyboard,
11 REM   converts any lowercase letters to upper case, and outputs
12 REM   the resulting string to the CRT
13 REM  Procedures called: UCASE
14
100 LINE INPUT Q$
110 CALL UCASE(Q$)
120 PRINT Q$
130 IF Q$<>"END" THEN 100
140 END
```

(a)

```
; This procedure is intended to be called from an interpreted
;   or compiled BASIC program.
; The procedure converts any lower case ASCII characters in an
;   input string to upper case and returns the result as the same string.
        PUBLIC  UCASE
;Segment name 'CODE' required for complier compatibility
CODE    SEGMENT   PUBLIC  'CODE'
        UCASE PROC FAR
                ASSUME CS:CODE
        PUSH BP                 ; Save old BP
        MOV  BP,SP              ; Set up second stack pointer
        MOV  BX,[BP+6]          ; Pointer to string descriptor from stack
                                ;  add 2 to displacement for each word
                                ;  pushed on stack in addition to BP
        MOV  CX,[BX]            ; Length of string from memory to CX as counter
;       MOV  CL,[BX]            ; For IBM interpreted BASIC string length 1byte
        MOV  BX,[BX+2]          ; Offset of start of string from descriptor
                                ;  displacement in instruction=1 for
                                ;  IBM interpreted BASIC
NEXT:   MOV  AL,[BX]            ; Get byte from string
        CMP  AL,61H             ; Check if lower case
        JB   OK
        CMP  AL,7AH
        JA   OK                 ; No, leave as is
        SUB  AL,20H             ; Yes, convert to upper case
OK:     MOV  [BX],AL            ; Put back in string
        INC  BX                 ; Increment pointer to next char
        LOOP NEXT               ; Repeat until all string done
        POP  BP                 ; Restore old BP
        RET  2                  ; Return and increment stack pointer over
                                ;  string descriptor passed on stack
UCASE   ENDP
CODE    ENDS
        END
```

(b)

FIGURE 13-28 Integration of high-level language and assembly language programs. *(a)* BASIC calling program. *(b)* Assembly language procedure called.

END followed by a carriage return is entered. The CALL instruction here calls a procedure called UCASE and passes the "hooks" needed for the procedure to access the string named Q$. Before we can discuss the actual assembly language procedure, we need to show you how BASIC passes these parameters to the procedure on the stack.

The left side of Figure 13-29 shows the contents of the top few locations of the stack after the CALL executes. The top 2 words on the stack contain the IP and the CS

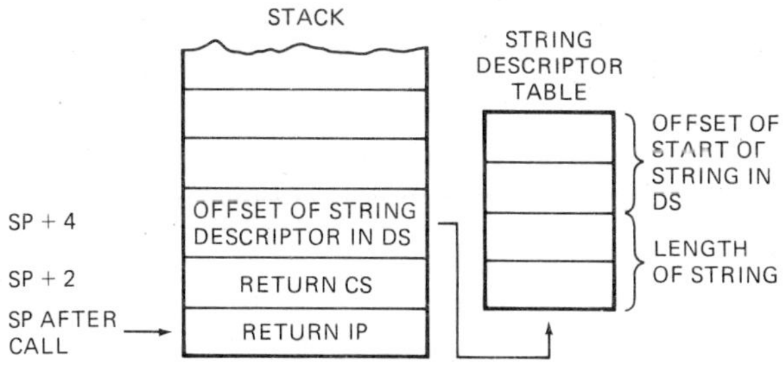

FIGURE 13-29 Stack after assembly language procedure call showing pointer to string variable descriptor table passed by calling program.

of the return address. The next word in the stack contains the offset (in the data segment) of a 4-byte *descriptor* for the string passed to the procedure. As shown in the right side of Figure 13-29 the lower 2 bytes of the descriptor contain the length of the string. The upper 2 bytes of the descriptor contain the offset of the actual string in the data segment. In other words, the offset passed on the stack for the string is a pointer to a table which contains the length of the string and the actual location of the string. When a simple nonstring variable is passed to a procedure, the offset in the stack points directly to a word location which contains the offset of the actual variable in the data segment. Once you know where the offset for a variable or a descriptor is stored in the stack, it is a simple matter to access the variable in your assembly language program.

Figure 13-28*b* shows how this is done. The BP register is first saved on the stack, and then the value in the stack pointer copied to it. Since a value in the BP register is added to the stack segment base to produce a physical address, BP can now function as a second pointer into the stack. After BP is pushed on the stack, the offset of the string descriptor that we need will be 6 bytes up in the stack. The instruction MOV BX, [BP + 6] then brings the string descriptor offset from the stack to BX. The MOV CX, [BX] instruction then uses this offset to bring the length of the string into CX for use as a counter. The MOV BX, [BX + 2] instruction then brings the offset of the start of the string from the descriptor table to BX where it can be used to access the elements in the string. The next section of the program runs through a simple check and fix-if-necessary loop until all of the string elements have been processed. In this case, the results of the processing are passed back to the calling program in the same string as the data was passed in. Finally, the initial BP is restored and execution returned to the calling program. Note that the RET 2 instruction is used to increment the stack pointer after the return. This is done to move the stack pointer over the variable offset that was passed to the procedure on the stack. If this is not done to balance the stack, the stack may grow downward in memory over the program. For each variable passed to the procedure, add 2 to the return number. It is not necessary to push the flags and the general-purpose registers used in the procedure, because this is done as part of the call statement. The registers and flags are restored upon return if the correct number is used with the RET instruction. If you are not sure they are pushed by the call statement in a particular high level language, it won't hurt anything to push and pop them in the procedure.

Combining assembly language procedures with high level language programs is a powerful programming technique. Hopefully, this section has shown you that it is not a difficult one if done in a compiler environment.

SYNCHRONOUS SERIAL DATA COMMUNICATION AND PROTOCOLS

Introduction

Most of the discussion of serial data transfer up to this point in the chapter has been about asynchronous transmission. For asynchronous serial transmission a start bit is used to identify the beginning of each data character, and at least one stop bit is used to identify the end of each data character. The transmitter and the receiver are effectively synchronized on a character-by-character basis. Since a total of 10 bits must be sent for each 8-bit data character, 20 percent of the transmission time is wasted. A more efficient method of transferring serial data is to synchronize the transmitter and the receiver, and then send a large block of data characters one after the other. No start or stop bits are then needed for individual data characters, because the receiver automatically knows that every 8 bits received after synchronization represents a data character. When a block of data is not being sent through a synchronous data link, the line is held in a marking condition. To indicate the start of a transmission the transmitter sends out one or more unique characters called *sync characters*, or a unique bit pattern called a *flag*, depending on the system being used. The receiver uses the sync characters or the flag to synchronize its internal clock with that of the receiver. The receiver then shifts in the data characters following and converts them to parallel form so they can be read in by a computer. Higher-speed data links or digital communication channels usually use synchronous transmission.

Now, remember from a previous section that a hardware level set of handshake signals is required to transmit asynchronous or synchronous digital data over phone lines with modems. In addition to this handshaking, a higher level of coordination, or *protocol*, is required between transmitter and receiver to assure the orderly transfer of data. A protocol in this case is an agreed set of rules concerning the form in which the data is to be sent. There are many different serial data protocols. The two most common that we discuss here are the IBM *binary synchronous communications protocol*, or BISYNC, and the *high-level data link control protocol*, or HDLC.

Binary Synchronous Communication Protocol—BISYNC

BISYNC is a referred to as a *byte-controlled protocol* (BCP), because specified ASCII or EBCDIC characters (bytes) are used to indicate the start of a message and to handshake between the transmitter and the receiver. Incidentally, even in a full-duplex system, BISYNC only allows data transfer in one direction at a time.

Figure 13-30 shows the general message format for BISYNC. For our first cycle through this we will assume that the transmitter has received a message from the receiver that it is ready to receive a transmission. If no message is being sent, the line is in an "idle" condition with a continuous high on the line. To indicate the start of a message, the transmitting system sends 2 or more previously agreed upon sync characters. For example, a sync character might be the ASCII 16H. As we said before, the receiver uses these sync characters to synchronize its clock with that of the transmitter. A header may then be sent if desired. The header contents are usually defined for a specific system and may include information about the type, priority, and destination of the message that follows. The start of the header is indicated with a special character called *start-of-header* (SOH), which in ASCII is represented by 01H.

After the header, if present, the beginning of the text portion of the message is indicated by another special character called *start-of-text* (STX), which in ASCII is represented by 02H. To indicate the end of the text portion of the message, an *end-of-text* (ETX) character or an *end-of-block* (ETB) character is sent. The text portion may contain 128 or 256 characters (different systems use different-size blocks of text). Immediately following the ETX, character 1 or 2 block check characters (BCC) are sent. For systems using ASCII, the BCC is a single byte which represents complex parity information computed for the text of the message. For systems using EBCDIC a 16-bit *cyclic redundancy check* is performed on the text part of the message and the 16-bit result sent as 2 BCCs. The point of these BCCs is that the receiving system can recompute the value for them from the received data and compare the results with the BCCs sent from the transmitter. If the BCCs are not equal, the receiver can send a message to the transmitter, telling it to send the message again. Now let's look at how messages are used for data transfer handshaking between the transmitter and the receiver.

To start let's assume that we have a remote "smart" terminal connected to a computer with a half-duplex connection. Further, let's assume that the computer is in the receive mode. Now, when the brain in the terminal determines that it has a block of data to send to the computer, it first sends a message with the text containing only the single character ENQ (ASCII 05H), which stands for *enquiry*. The terminal then switches to receive mode to await the reply from the computer. The computer reads the ENQ message, and, if it is not ready

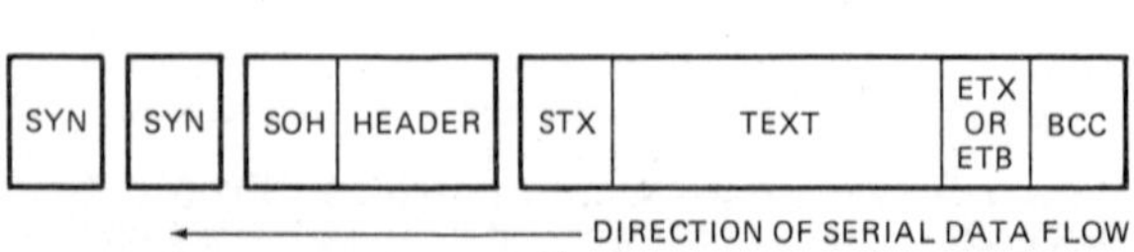

FIGURE 13-30 General message format for BISYNC.

to receive data, it sends back a text message containing the single character for *negative acknowledge*, NAK (ASCII 15H). If the receiver is ready, it sends a message containing the *affirmative acknowledge*, ACK, character (ASCII 06H). In either case, the computer then switches to receive mode to await the next message from the terminal. If the terminal received a NAK, it may give up, or it may wait a while and try again. If the terminal received an ACK, it will send a message containing a block of text and ending with a BCC character(s). After sending the message, the terminal switches to receive mode and awaits a reply from the computer as to whether the message was received correctly. The computer meanwhile computes the BCC for the received block of data and compares it with the BCC sent with the message. If the two BCCs are not equal, the computer sends a NAK message to the terminal. This tells the terminal to send the message again, because it was not received correctly. If the two BCCs are equal, then the computer sends an ACK message to the terminal, which tells it to send the next message or block of text. In a system where multiple blocks of data are being transferred, an ACK 0 message is usually sent for one block, an ACK 1 message sent for the next, and an ACK 0 again sent for the next. The alternating ACK messages are a further help in error checking. In either case, after the message is sent the computer switches to receive mode to await a response from the terminal.

One major problem with this protocol is that the transmitter must stop after each block of data is transferred, and wait for an ACK or NAK signal from the receiver. Due to the wait and line turnaround times, the actual data transfer rate may be only half of the theoretical rate predicted by the physical bit rate of the data link. The HDLC protocol discussed in a later section greatly reduces this problem. Next we want to return to the Intel 8251A USART and give you a brief look at how it is used for BISYNC communication.

Using the Intel 8251A USART for BISYNC Communication

As shown in Figure 13-5, we initialize an 8251A by first getting its attention, sending it a mode word, and then sending it a command word. To initialize the 8251A for synchronous communication, 0's are put in the least-significant 2 bits of the mode word. The rest of the bits in the mode word then have the meanings shown in Figure 13-31. Most of the bit functions should be reasonably clear from the descriptions in the figure, but a couple need a little more explanation.

Bit 6 of the mode word specifies the SYNDET pin on the 8251A to be an input or an output. The pin is programmed to function as an input if external circuitry is used to detect the sync character in the data bit stream. When programmed as an output, this pin will go high when the 8251A has found one or more sync characters in the data bit stream.

Bit 7 of the mode word is used to specify whether 1 sync character, or a sequence of 2 different sync characters is to be looked for at the start of a message.

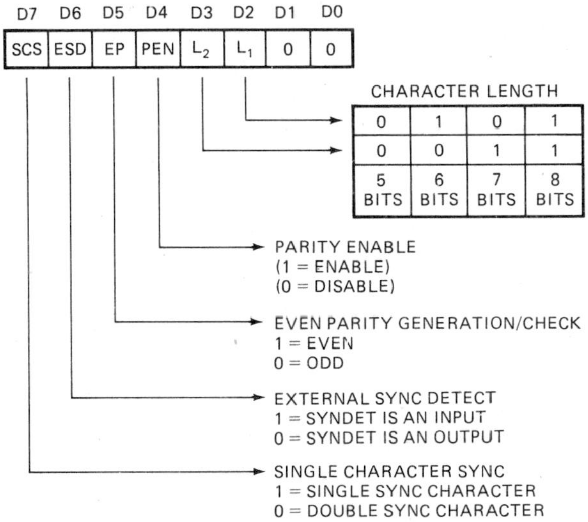

NOTE: IN EXTERNAL SYNC MODE, PROGRAMMING DOUBLE CHARACTER SYNC WILL AFFECT ONLY THE Tx.

FIGURE 13-31 8251A mode word for synchronous communication.

To initialize an 8251A for synchronous operation:

1. Send a series of nulls and a software reset command to the control address as shown at the start of Figure 13-5.
2. Send a mode word based on the format in Figure 13-31 to the control address.
3. Send the desired sync character for that particular system to the control address of the 8251A.
4. If a second sync character is needed, send it to the control address.
5. Finally, send a command word to the control address to enable the transmitter, enable the receiver, and enable the device to look for sync characters in the data bit stream coming in on the RxD input.

The format for the command word is shown in Figure 13-32. Now, let's examine how the 8251A participates in a synchronous data transfer. As you work your way through this section, try to keep separate in your mind the parts of the process that are done by the 8251A and the parts that are done by software at one end of the link or the other.

To start, let's assume the 8251A is in a terminal which has blocks of data to send to a computer as we described earlier in this section. Further assume that the computer is in receive mode waiting for a transmission from the terminal, and that the 8251A in the terminal has been initialized and is sending out a continuous high on the TxD line.

An I/O driver routine in the terminal will start the transfer process by sending a sync character(s), SOH character, header characters, STX character, ENQ character, ETX character, and BCC byte to the 8251A, one after the other. The 8251A sends the characters out in synchronous serial format (no start and stop bits). If, for some reason such as a high-priority interrupt, the CPU stops sending characters while a message is being sent, the 8251A will automatically insert sync characters until the flow of data characters from the CPU resumes.

After the ENQ message has been sent, the CPU in the terminal awaits a reply from the computer through the RxD input of the 8251A. If the 8251A has been programmed to enter hunt mode by sending it a control word with a 1 in bit 7, it will continuously shift in bits from the RxD line and check after each shift if the character in the receive buffer is a sync character. When it finds a sync character, the 8251A asserts the SYNDET pin high, exits the hunt mode, and starts the normal data read operation. When the 8251A has a valid data character in its receiver buffer, the RxRDY pin will be asserted, and the RxRDY bit in the status register will be set. Characters can then be read in by the CPU on a polled or an interrupt basis.

When the CPU has read in the entire message, it can determine whether the message was a NAK or an ACK. If the message was an ACK, the CPU can then send the actual data message sequence of characters to the 8251A. Handshake and data messages will be sent back and forth until all of the desired block of data has been

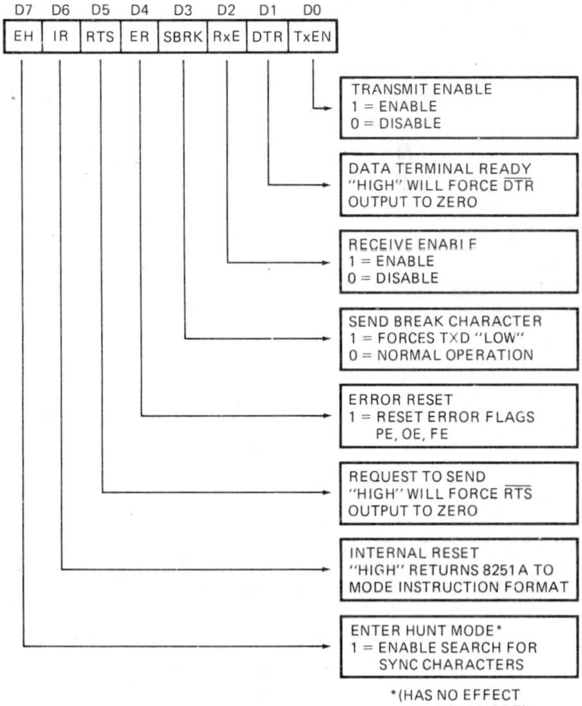

NOTE: ERROR RESET MUST BE PERFORMED WHENEVER RxENABLE AND ENTER HUNT ARE PROGRAMMED.

FIGURE 13-32 8251A command-word format for synchronous operation.

sent to the computer. In the next section we discuss another protocol used for synchronous serial data transfer.

High-level Data Link Control (HDLC) and Synchronous Data Link Control (SDLC) Protocols

The BISYNC protocol, which we discussed in the previous section only works in half-duplex mode, has difficulty accurately transmitting pure binary data such as object code for programs, and is not easily adapted to serving multiple units sharing a common data link. In an attempt to solve these problems, the *International Standards Organization* (ISO) proposed the *high level data link control protocol* (HDLC) and IBM developed the *synchronous data link control protocol* (SDLC). The standards are so nearly identical that, for the discussion here, we will treat them together under the name HDLC and indicate any significant differences as needed.

As we said previously, BISYNC is referred to as a byte-controlled protocol because character codes or bytes such as SOH, STX, and ETX are used to mark off parts of a transmitted message or act as control messages. HDLC is referred to as a *bit-oriented protocol* (BOP) because messages are treated simply as a string of bits, rather than a string of characters. The group of bits which make up a message is referred to as a *frame*. The three types of frames used are: information or *I frames*, supervisory control sequences or *S frames*, and command/response or *U frames*. The three types of frames all have the same basic format.

Figure 13-33*a* shows the format of an HDLC frame. Each part of the frame is referred to as a *field*. A frame starts and ends with a specific bit pattern, 01111110, called a *flag* or *flag field*. When no data is being sent, the line idles with all 1's, or continuous flags. Immediately after the flag field is an 8-bit address field which contains the address of the destination unit for a control or information frame, and the source of the response for a response frame.

Figure 13-33*b* shows the meaning of the bits in the 8-bit control field for each of the three types of frames. We don't have the space or the desire to explain here the meaning of all of these. A little later we will, however, explain the use of the Ns and Nr bits in the control byte for an information frame.

The information field, which is only present in information frames, can have any number of bits in HDLC protocol, but in SDLC the number of bits has to be a multiple of 8. In some systems as many as 10,000 or 20,000 information bits may be sent per frame. Now, the question may occur to you, "What happens if the data contains the flag bit pattern, 01111110?" The answer to this question is that a special hardware circuit modifies the bit stream between flags so that there are never more than 5 ones in sequence. To do this the circuit monitors the data stream and automatically stuffs in a zero after any series of 5 ones. A complementary circuit in the receiver removes the extra zeros. This scheme allows any type of characters or data to be sent without the problems BISYNC has in this area.

The next field in a frame is the 16-bit *frame check sequence* (FCS). This is a cyclic redundancy word derived from all of the bits between the beginning and end flags, but not including 0's inserted to prevent false flag bytes. This CRC value can be recomputed by the receiving system to check for errors.

Finally, a frame is terminated by another flag byte. The ending flag for one frame may be the starting flag for another frame.

In order to describe the HDLC data-transfer process, we first need to define a couple of terms. HDLC is used for communication between two or more systems on a data link. One of the systems or *stations* on the link will always be set up as a controller for the link. This station is called the *primary station*. Other stations on the link are referred to as *secondary stations*.

Now, suppose that a primary station, a computer for example, wants to send several frames of information to a secondary station such as another computer or terminal. Here's how a transfer might take place.

The primary station starts by sending an S frame containing the address of the desired secondary station and a control word which inquires if the receiver is ready. The secondary station then sends an S frame which contains the address of the primary station and a con-

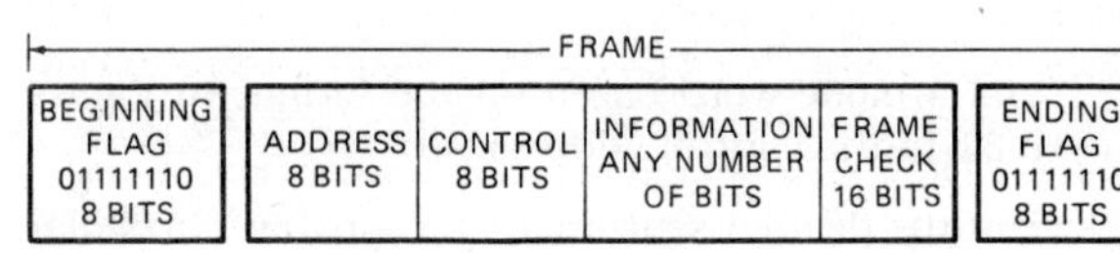

(*a*)

	BITS IN CONTROL FIELD							
HDLC FRAME FORMAT	7	6	5	4	3	2	1	Ø
I-FRAME (INFORMATION TRANSFER COMMANDS/RESPONSES)	Nr	Nr	Nr	P/F	Ns	Ns	Ns	Ø
S-FRAME (SUPERVISORY COMMANDS/RESPONSES)	Nr	Nr	Nr	P/F	S	S	Ø	1
U-FRAME (UNNUMBERED COMMANDS/RESPONSES)	M	M	M	P/F	M	M	1	1

SENDING ORDER - BIT Ø FIRST, BIT 7 LAST

NS THE TRANSMITTING STATION SEND SEQUENCE NUMBER, BIT 2 IS THE LOW-ORDER BIT.

P/F THE POLL BIT FOR PRIMARY STATION TRANSMISSIONS, AND THE FINAL BIT FOR SECONDARY STATION TRANSMISSIONS.

Nr THE TRANSMITTING STATION RECEIVE SEQUENCE NUMBER, BIT 6 IS THE LOW-ORDER BIT.

S THE SUPERVISORY FUNCTION BITS

M THE MODIFIER FUNCTION BITS

(*b*)

FIGURE 13-33 *(a)* Format of HDLC frame. *(b)* Meaning of bits in 8-bit control field for a frame.

trol word which indicates its ready status. If the secondary station receiver was ready, the primary station then sends a sequence of information frames. The information frames contain the address of the secondary station, a control word, a block of information, and the FCS words. For all but the last frame of a sequence of information frames, the P/F bit in the control byte will be a 0. The 3 Ns bits in the control byte will contain the number of the frame in the sequence.

Now, as the secondary station receives each information frame, it reads the data into memory and computes the frame check sequence for the frame. For each frame in a sequence that the secondary station receives correctly, it increments an internal counter. When the primary station sends the last frame in a sequence of up to seven frames, it makes the P/F bit in the control byte a 1. This is a signal to the secondary station that the primary station wants a response as to how many frames were received correctly. The secondary station responds with an S frame. The Nr bits in the control word of this S frame contain the sequence number of the last frame that was received correctly plus 1. In other words, Ns represents the number of the next expected frame. The primary station compares Ns − 1 with the number of frames sent in the sequence. If the two numbers do not agree, the primary station knows that it must retransmit some frames, because they were not all received correctly. The Nr number tells the primary station which frame number to start the retransmission from. For example, if Nr is 3, the primary station will retransmit the sequence of frames starting with frame 3. If the sequence of frames was received correctly, another series of frames can be sent if desired. Actually, since HDLC operates in full duplex, the receiving station can be queried after each frame is sent to see if the previous frame was received correctly. A similar series of actions takes place when a secondary station transmits to a primary station or to another secondary station.

One advantage of this HDLC scheme is that the transmitter does not have to stop after every short message for an acknowledge as it does in BISYNC protocol. True, several frames may have to be resent in case of an error, but in low error rate systems, this is the exception. As we will show in the next section, HDLC is used along with some higher level protocols to allow communication between a wide variety of systems.

One final point to make here is how HDLC protocol is implemented with a microcomputer. At the basic hardware level, a standard USART cannot be used because of the need to stuff and strip 0 bits. Instead, specially designed parts such as the Intel 8273 HDLC/SDLC protocol controller are used. Devices such as this automatically stuff and strip the required 0 bits, generate and check frame-check sequence words, and produce the interface signals for RS-232C. The devices interface directly to microcomputer buses.

The actual control of which station uses the data link at a particular time and the formatting of frames is done by the system software. The next section discusses how several systems can be connected together or "networked" so they can communicate with each other.

LOCAL AREA NETWORKS

Introduction

The objective of this section is to show you how several computers can be connected together to communicate with each other and to share common peripherals such as printers and large disk drives. We will start with simple cases and progress to the type of network that might be used in the computerized electronics factory we described in an earlier chapter.

To communicate between a single terminal and a nearby computer, a simple RS 232C connection is sufficient. If the computer is distant, then a modem and phone line or a leased phone line is used, depending on the required baud rate. Now, for a more difficult case, suppose that we have in a university building 100 terminals that need to communicate with a distant computer. We could use 100 phone lines with modems, but this seems quite inefficient. One solution to this problem is to run wires from all of the terminals to a central point in the building, and then use a multiplexer or *data concentrator* of some type to send all of the communications over one wideband line. Either time-domain multiplexing or frequency-division multiplexing can be used. A demultiplexer at the other end of the line reconstructs the original signals.

As another example of computer communication, suppose that we have several computers in one building or a complex of buildings that need to communicate with each other. Our computerized electronics factory is an example of this situation. What is needed in this case is a high-speed network, commonly called a *local area network* or *LAN*, connecting the computers together. We start our discussion of LANs by showing you some of the basic connection configurations.

LAN Connection Configurations

The different ways of physically connecting devices on a network with each other are commonly referred to as *topologies*. Figure 13-34 shows the five most common topologies and some other pertinent data about each, such as examples of commercially available systems which use each type.

In a *star topology* network, a central controller coordinates all communication between devices on the network. The most familiar example of how this works is probably a private automatic branch exchange, or PABX, phone system. In a PABX all calls from one phone on the system to another or to an outside phone are routed through a central switchboard. The new digital PABX systems allow direct communication between computers within a building at baud rates up to perhaps 100 kBd.

In the *loop topology*, one device acts as a controller. If a device wants to communicate with one or more other devices on the loop, it sends a request to the controller. If the loop is not in use, the controller enables the one device to output, and the other device(s) to receive. The GPIB or IEEE488 bus described in the last section of this chapter is an example of this topology.

TOPOLOGY	TYPICAL PROTOCOLS	TYPICAL NO. OF NODES	TYPICAL SYSTEMS
STAR	RS-232C OR COMPUTER	TENS	PABX, COMPUTER μC CLUSTERS
LOOP	SDLC	TENS	IBM 3600/3700, μC CLUSTERS
COMMON BUS	CSMA/CD OR CSMA WITH ACKNOWLEDGMENT	TENS TO HUNDREDS PER SEGMENT	ETHERNET, NET/ONE, OMNINET, Z-NET μC CLUSTERS
RING	SDLC HDLC (TOKEN PASSING)	TENS TO HUNDREDS PER CHANNEL	PRIMENET, DOMAIN, OMNILINK μC CLUSTERS
OTHER SERVICES BROADBAND BUS	CSMA/CD RS-232C & OTHERS PER CHANNEL	TWO TO HUNDREDS PER CHANNEL	WANGNET, LOCALNET M/A-COM

• TERMINAL
DISTRIBUTED CONTROL
(C) LOCAL CONTROLLER
((C)) MULTINETWORK CONTROLLER
(FDM) FREQUENCY DIVISION MULTIPLEX

FIGURE 13-34 Summary of common computer network topologies.

In the *common-bus topology*, control of the bus is spread among all of the devices on the bus. The connection in this type of system is simply a wire (usually but not always a coaxial cable) which any number of devices can be tapped into. Any device can take over the bus to transmit data. Data is transmitted in fixed-length blocks called *packets*. Two devices are prevented from transmitting at the same time by a scheme called *carrier sense, multiple access with collision detection*, or CSMA/CD. This is discussed in detail in a later section on Ethernet.

In a *ring network*, the control is also distributed among all of the devices on the network. Each device on the ring functions as a repeater, which means that it simply takes in the data stream and passes the data stream on to the next device on the ring if it is not the intended receiver for the data. Data always circulates around the ring in one direction. Any device can transmit on the ring. A *token* is one common way used to prevent two or more devices from transmitting at the same time. A token is a specific lone byte such as 01111111 which is circulated around the ring when no device is transmitting. A device must possess the token in order to transmit. When a device needs to transmit, it removes the token from the bus, thus preventing any other devices from transmitting. After transmitting one or more packets of data, the transmitting device puts the token back on the ring so another device can grab it and transmit. We discuss this more in a later section.

The final topology we want to discuss here is the *tree* structured network which often uses broadband transmission. Before we can really explain this one, we need to introduce you to a couple of terms commonly used with networks. In some networks such as Ethernet, data is transmitted directly as digital signals at a rate of 10M bits/s. With this type of signal, only one device can transmit at a time. This form of data transmission is often referred to as *baseband* transmission, because only one basic frequency is used. The other common form of data transmission on a network is referred to as *broadband* transmission. Broadband transmission is based on a frequency-division multiplexing scheme such as that used for community antenna television (CATV) systems. The radio-frequency spectrum is divided up into 6-MHz-bandwidth channels.

A single device or group of devices can be assigned one channel for transmitting and another for receiving. Each channel or pair of channels is considered a branch on the tree. Special modems are used to convert digital signals to and from the modulated radio-frequency signals required. The multiple channels and the 6-MHz bandwidth of the channels in a broadband network allow voice, data, and video signals to be transmitted at the same time throughout the network. This is an advantage over baseband systems which can only transmit one digital data signal at a time, but the broadband system is much more expensive.

Network Protocols

In order for different systems on a network to communicate effectively with each other, a series of rules or protocols must be agreed upon and followed by all of the devices on the network. The International Standards Organization, in an attempt to bring some order to the chaos of network communication, has developed a standard called the *open systems interconnection* (OSI) model. This model is not a rigid standard. It is a seven-layer hierarchy of protocols as shown in Figure 13-35. This layered approach structures the design tasks and makes it possible to change, for example, the actual hardware used to transmit the data without changing the other layers. We will use a common network operation, electronic mail, to explain to you the function of the upper-layers model.

Electronic mail allows a user on one system on a network to send a message to another user on the same system or on another system. The message is actually sent to a "mailbox" in a hard-disk file. Each user on the network periodically checks a personal mailbox to see if it contains any messages. If any messages are present, they can be read out and then deleted from the mailbox.

The *application layer* of the OSI protocol specifies the general operation of network services such as electronic mail, access to common data bases, and access to common peripherals. For our example, this layer of the protocol dictates how you go about invoking the electronic mail function of the network.

The *presentation layer* of the OSI protocol governs the programs which convert messages to the code and format that will be understood by the receiver. For our electronic mail message this layer might involve translating a message from ASCII to EBCDIC, or perhaps from English to French.

LAYER	NUMBER	FUNCTION
APPLICATION	7	SELECTS APPROPRIATE SERVICE FOR APPLICATIONS
PRESENTATION	6	PROVIDES CODE CONVERSION, DATA REFORMATTING
SESSION	5	COORDINATES INTERACTION BETWEEN END-APPLICATION PROCESSES
TRANSPORT	4	PROVIDES END-TO-END DATA INTEGRITY AND QUALITY OF SERVICE
NETWORK	3	SWITCHES AND ROUTES INFORMATION
DATA LINK	2	TRANSFERS UNITS OF INFORMATION TO OTHER END OF PHYSICAL LINK
PHYSICAL	1	TRANSMITS BIT STREAM TO MEDIUM

FIGURE 13-35 ISO open system interconnect model for network communications.

The *session layer* of the OSI protocol establishes and terminates logical connections on the network. This layer is responsible for opening and closing named files, and for translating a user name into a physical network address. Electronic mail allows you to specify the intended receiver of a message by name. It is the responsibility of this layer of the protocol to make the connection between the name and the network address of the named receiver.

The *transport layer* of the protocol is responsible for making sure a message is transmitted and received correctly. An example of the operation of this protocol layer is the ACK or NAK handshake used in BISYNC transmission after the receiver has checked to see if the data was received correctly. For electronic mail, the message can be written to the addressed mailbox and then read back to make sure it was sent correctly.

The *network layer* of the protocol is only used in multichannel systems. It is responsible for finding a path through the network to the desired receiver by switching between channels. The function of this layer is similar to the function of postal mail routing which finds a route to get a letter from your house to the addressed destination.

The *data link layer* of the OSI model includes the framing of the data in terms of packet size and address information, the means used to check errors (parity or CRC), the handshake signals between the transmitter and the receiver, etc. The HDLC protocol described earlier in this chapter is an example of the type of transmission factors involved in this layer.

The *physical layer* of the OSI model is the lowest level. This layer is used to specify the connectors, cables, voltage levels, bit rates, modulation methods, etc. RS-232C is an example of a standard which falls in this layer of the model.

Now we will take a more detailed look at the operation of a very widespread "common-bus" network, Ethernet. Ethernet is a trademark of Xerox Corporation.

Ethernet

The *Ethernet network standard* was originally developed by Xerox Corporation. Later Xerox, DEC, and Intel worked on defining the standard sufficiently so that commercial products for implementing the standard were possible. It has now been adopted, with slight changes, as the IEEE802.3 standard.

Physically, Ethernet is implemented in a common-bus topology with a single 50-Ω coaxial cable. Data is sent over the cable using baseband transmission at 10 Mbits/s. Data bits are encoded using Manchester coding as shown in Figure 13-36. The advantage of this coding is that each bit cell contains a signal transition. A system that wants to transmit data on the network first checks for these transitions to see if the network is currently busy. If the system detects no transitions, then it can go ahead and transmit on the network.

Figure 13-37 shows how a very simple Ethernet is set up. The backbone of the system is the coaxial cable. Terminations are put on each end of the cable to prevent signal reflections. Each unit is connected into the cable with a simple tee-type tap. A transmitter-receiver, or *transceiver*, sends out data on the coax, receives data from the coax, and detects any attempt to transmit while the coax is already in use. The transceiver is connected to an interface board with a 15-pin connector and four twisted-wire pairs. The transceiver cable can be as long as 15 m. The interface board, as the name implies, performs most of the work of getting data on and off the network in the correct form. The *interface board* assembles and disassembles data frames, sends out source and destination addresses, detects transmission errors, and prevents transmission while some other unit on the network is transmitting.

The method used by a unit to gain access to the network is *CSMA/CD*. Before a unit attempts to transmit on the network, it looks at the coax to see if a carrier (Manchester code transitions) is present. If a carrier is present, the unit waits some random length of time and then tries again. When the unit finds no carrier on the line, it starts transmitting. While it is transmitting, it

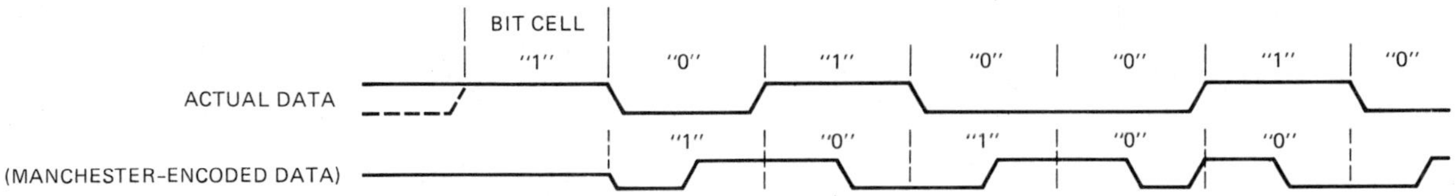

FIGURE 13-36 Manchester encoding used for Ethernet data bit stream. Note that encoded data has a transition at center of each data bit cell time.

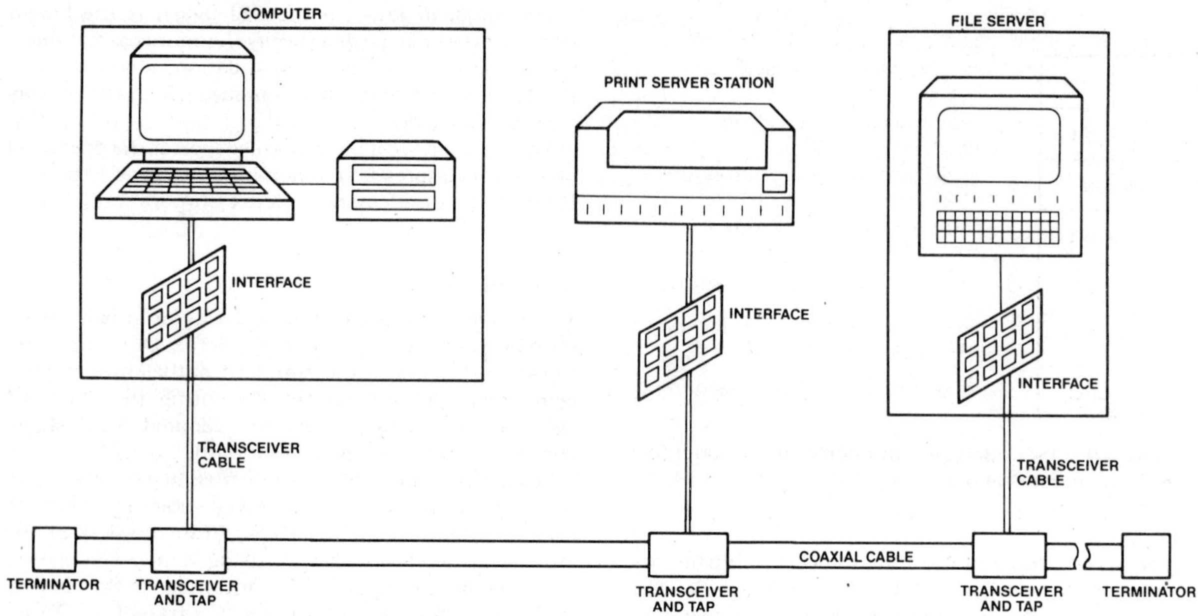

FIGURE 13-37 Very simple Ethernet system.

also monitors the line to make sure no other unit is transmitting at the same time. The question may occur to you at this point, "If a unit cannot start transmitting until it finds no carrier on the coax, how can another unit be transmitting at the same time?" The answer to this question involves propagation delay. Since transceivers can be as much as 2500 m apart, it may take as long as 23 μs for data transmitted from one unit to reach another unit. In other words, one unit may start transmitting before the signal from a transmitter that started earlier reaches it. Two units transmitting at the same time is referred to as a *collision*. When a unit detects a collision, it will keep transmitting long enough that all transmitting stations detect that a collision has occurred. All of the units then stop transmitting and try again after a random period of time. The term "multiple access" in the CSMA/CD name means that any unit on the network can attempt to transmit. No central controller decides which unit has use of the network at a particular time. Access is gained by any unit using the mechanism we have just described. The maximum number of units that can be connected on a single Ethernet is 1024. For further information about how an interface board is built, consult the data sheets for the Intel 82586 LAN coprocessor, and the data sheet for the Intel 82501 Ethernet serial interface.

Incidentally, a file server such as the one shown in Figure 13-37 is a "smart" hard-disk system which manages file access requests by other systems on the network. A print server is a "smart" printer which queues up print requests from other systems on the network.

Token-Passing Rings

Token-passing ring networks solve the multiple access problem in an entirely different way from the CSMA/CD approach described for Ethernet. As the name implies, systems are connected in series around a ring. Data always travels in one direction around the ring. Unlike the passive taps used in an Ethernet system, however, each active station on a token ring receives data, examines it to see if the data is addressed to it, and retransmits the data to the next station on the ring. A bypass relay is used to shunt data around defective or inactive units. Data is sent around the ring at perhaps 4–5Mbits/s in HDLC or SDLC frames which we described in an earlier section of the chapter on synchronous transmission. Any station on the network can use the network.

A token is a byte of data with an agreed-upon, unique bit pattern such as 01111111. If no station is transmitting, this token is circulated continuously around the ring. When a station needs to transmit, it withdraws the not-busy token, changes it to a busy token of perhaps 01111110, and sends the busy token on around the ring. The transmitting unit then sends a frame of data around the ring to the intended receiver(s). When the transmitting station receives the busy token back again, it reads it in, and sends out the not-busy token again. The transmitting station also pulls the transmitted data off the ring as it returns, so it can't circulate around again. As soon as a transmitting station releases the not-busy token again, the next station on the loop can grab the token and transmit on the network. The first station that transmitted cannot transmit again until the not-busy token works its way around the ring. This gives all units on the network a chance to transmit in a "round-robin" manner. (NOTE: Some token ring networks use tokens with priority bits so that one station can transmit again if necessary before a lower-priority station gets a turn.)

Two questions occurred to us the first time we read about token-passing rings; perhaps these same two questions may have occurred to you. The first question is, "How does a station on the network tell the bit pattern for a token from the same bit pattern in the data frame?" The answer to this question is bit-stuffing, the same technique that is used to prevent the flag bit pattern from being present in the data section of an HDLC frame. A hardware circuit in the transmitter alters the data stream so that certain bit patterns are not present. Another hardware circuit in the receiver reconstructs the original data.

The second question is, "What happens if the not-busy token somehow gets lost going around the ring?" A couple of different approaches are used to solve this problem. One approach uses a timer in each station. When a station has a frame to transmit, it starts a timer. If the station does not detect a token in the data stream before the timer counts down, it assumes that the token was lost, and sends out a new token. Another approach used by IBM sets up one station as a network monitor. If this station does not detect a token within a prescribed time, it clears any leftover data from the ring and sends out a new not-busy token.

Token-passing ring networks have the disadvantage that more complex hardware is required where each station connects to the network. The receive and transmit circuitry at the connection, however, acts as a repeater which maintains signal quality throughout the network. Since signals only travel in one direction around the ring, this topology is ideally suited for fiber-optic transmission, which allows high data rates and long distances between repeaters. IEEE802.5 standard describes a token-passing ring network. Texas Instruments is currently offering the TI TMS380 chip set which implements a token-passing ring microcomputer interface.

In the next section we discuss a somewhat different type of network which is used to connect instruments with a computer to form an integrated test and measurement system.

The GPIB, HPIB, IEEE488 Bus

The *general-purpose interface bus* (GPIB), also known as the *Hewlett-Packard interface bus* and the *IEEE488 bus* is not intended for use as a computer network in the same way that the Ethernet and token rings are used. It was developed by Hewlett-Packard to interface smart test instruments with a computer.

The standard describes three types of devices that can be connected on the GPIB. First is a *listener*, which can receive data from other instruments or from the controller. Examples of listeners are printers, display devices, programmable power supplies, and programmable signal generators. The second type of device defined is a *talker*, which can send data to other instruments. Examples of talkers are tape readers, digital voltmeters, frequency counters, and other measuring equipment. A device can be both a talker and a listener. The third type of device on the bus is a *controller*, which determines who talks and who listens on the bus.

Physically the bus consists of a 24-wire cable with a connector such as that shown in Figure 13-38*a* on each end. Actually, each end of the cable has both a male connector and a female connector, so that cables can daisy-chain from one unit to the next on the bus. Instruments intended for use on a GPIB usually have some switches which allow you to select the 4-bit address that the instrument will have on the bus. Standard TTL signal voltage levels are used.

As shown in Figure 13-38*b*, the GPIB has eight bidirectional data lines. These lines are used to transfer data, addresses, commands, and status bytes among as many as 8 or 10 instruments.

The GPIB also has five bus management lines which function basically as follows. The *interface clear* line (IFC), when asserted by the controller, resets all devices on the bus to a starting state. It is essentially a system reset. The *attention* (ATN) line, when asserted (low), indicates that the controller is putting a universal command or an address-command such as "listen" on the data bus. When the ATN line is high, the data lines contain data or a status byte. *Service request* (SRQ) is similar to an interrupt. Any device that needs to transfer data on the bus asserts the SRQ line low. The controller then polls all the devices to determine which one needs service. When asserted by the system controller, the *remote enable* (REN) signal allows an instrument to be controlled directly by the controller rather than by its front panel switches. The *end or identify* (EOI) signal is usually asserted by a talker to indicate that the transfer of a block of data is complete.

Finally, the bus has three handshake lines that coordinate the transfer of data bytes on the data bus. These are *data valid* (DAV), *not ready for data* (NRFD), and *not data accepted* (NDAC). These handshake signals allow devices with very different data rates to be connected together in a system. A little later we will show you how this handshake works. First we will give you an overview of general bus operation.

Upon power-up the controller takes control of the bus and sends out an IFC signal to set all instruments on the bus to a known state. The controller then proceeds to use the bus to perform the desired series of measurements or tests. To do this the controller sends out a series of commands with the ATN line asserted low. Figure 13-38*c* shows the formats for the combination command-address codes that a controller can send to talkers and listeners. Bit 8 of these words is a don't-care, bits 7 and 6 specify which command is being sent, and bits 5 through 1 give the address of the talker or listener to which the command is being sent. For example, to enable (address) a device at address 04 as a talker, the controller simply asserts the ATN line low and sends out a command-address byte of X1000100 on the data bus. A listener is enabled by sending out a command-address byte of $X01A_5A_4A_3A_2A_1$, where the lower 5 bits contain the address that the listener has been given in the system. When a data transfer is complete, all listeners are turned off by the controller sending an unlisten command, X0111111. The controller turns off the talker by sending an untalk command, X1011111. *Universal commands* sent by the controller with bits 7, 6, and 5 all 0's will go to all listeners and talkers. The lower 4 bits of these words specify one of 16 universal commands.

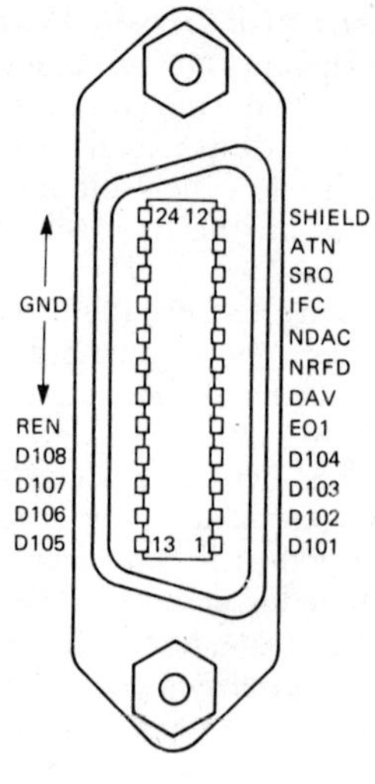

(a)

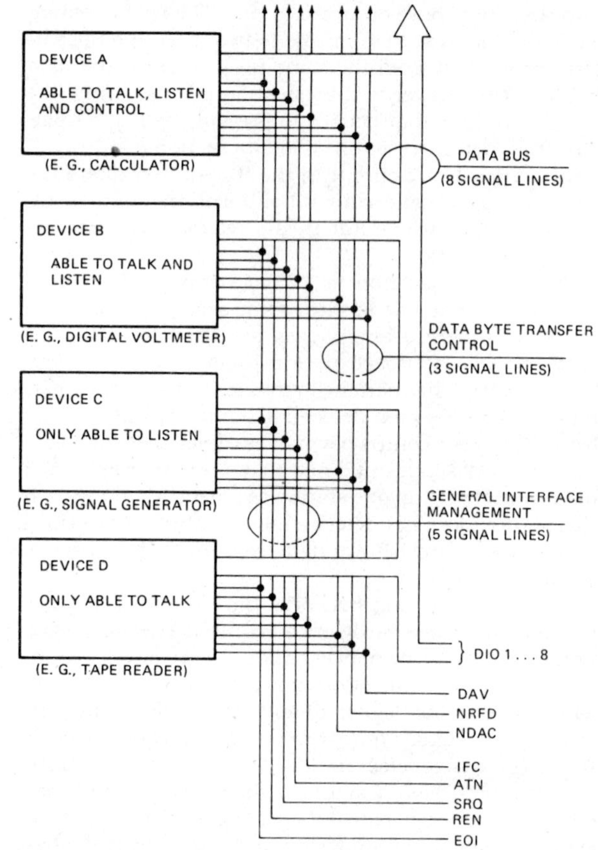

(b)

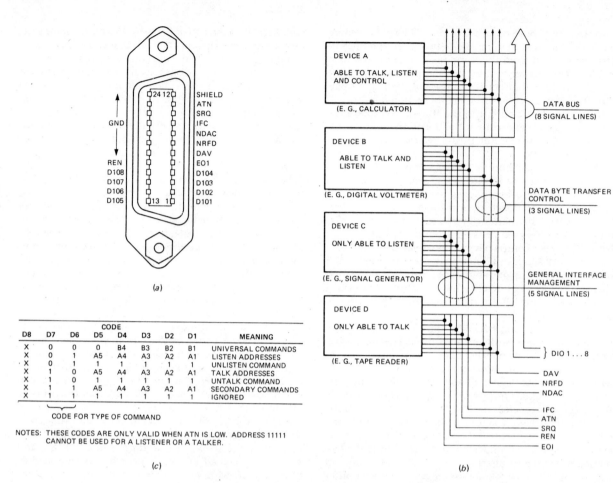

			CODE					
D8	D7	D6	D5	D4	D3	D2	D1	MEANING
X	0	0	0	B4	B3	B2	B1	UNIVERSAL COMMANDS
X	0	1	A5	A4	A3	A2	A1	LISTEN ADDRESSES
X	0	1	1	1	1	1	1	UNLISTEN COMMAND
X	1	0	A5	A4	A3	A2	A1	TALK ADDRESSES
X	1	0	1	1	1	1	1	UNTALK COMMAND
X	1	1	A5	A4	A3	A2	A1	SECONDARY COMMANDS
X	1	1	1	1	1	1	1	IGNORED

CODE FOR TYPE OF COMMAND (D7, D6)

NOTES: THESE CODES ARE ONLY VALID WHEN ATN IS LOW. ADDRESS 11111 CANNOT BE USED FOR A LISTENER OR A TALKER.

(c)

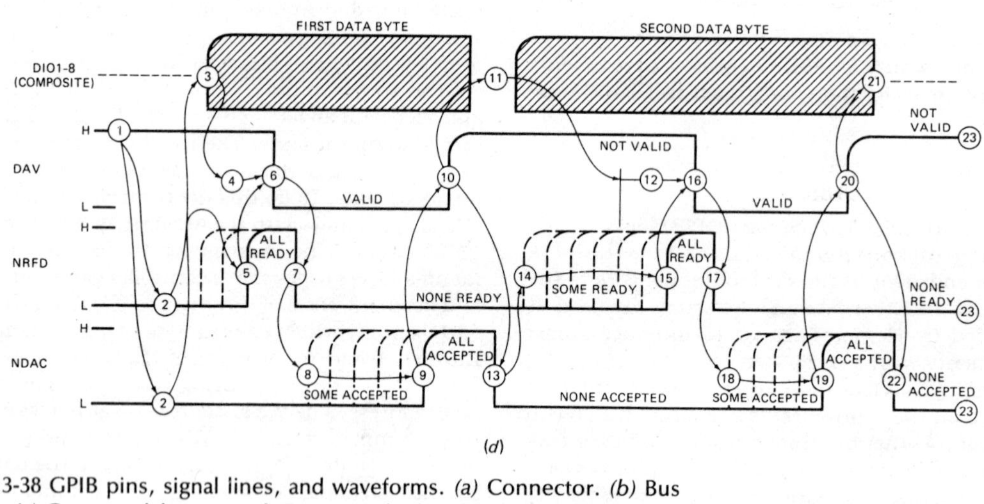

(d)

FIGURE 13-38 GPIB pins, signal lines, and waveforms. *(a)* Connector. *(b)* Bus structure. *(c)* Command formats. *(d)* Data transfer handshake waveforms.

Periodically while it is using the bus, the controller checks the SRQ line for a service request. If the SRQ line is low, the controller polls each device on the bus one after another (serial), or all at once (parallel), until it finds the device requesting service. A talker such as a DVM, for example, might be indicating that it has completed a series of conversions and has some data to send to a listener such as a chart recorder. When the controller determines the source of the SRQ, it asserts the ATN line low and sends listener address commands to each listener that is to receive the data and a talk address command to the talker that requested service. The controller then raises the ATN line high, and data is transferred directly from the talker to the listeners using a double-handshake-signal sequence.

Figure 13-38*d* shows the sequence of signals on the handshake lines for a transfer of data from a talker to several listeners. The DAV, NRFD, and NDAC lines are all open-collector. Therefore, any listener can hold NRFD low to indicate that it is not ready for data, or hold NDAC low to indicate that it has not yet accepted a data byte. The sequence proceeds as follows. When all listeners have released the NRFD line (5 in Figure 13-38*d*), indicating that they are ready (not not-ready), the talker asserts the DAV line low to indicate that a valid data byte is on the bus. The addressed listeners then all pull NRFD low and start accepting the data. When the slowest listener has accepted the data, the NDAC line will be released high (9 in Figure 13-38*d*). The talker senses NDAC becoming high and unasserts its DAV signal. The listeners all pull NDAC low again, and the sequence is repeated until the talker has sent all of the data bytes it has to send. The rate of data transfer is determined by the rate at which the slowest listener can accept the data.

When the data transfer is complete, the talker pulls the EOI line in the management group low to tell the controller that the transfer is complete. The controller then takes control again and sends an untalk command to the talker. It also sends an unlisten command to turn off the listeners, and continues to use the bus according to its internal program.

A standard microprocessor bus can be interfaced to the GPIB with dedicated devices such as the Intel 8291 GPIB talker-listener, and 8292 GPIB controller. The importance of the GPIB is that it allows a microcomputer to be connected with several test instruments to form an integrated test system.

IMPORTANT TERMS AND CONCEPTS FROM THIS CHAPTER

Serial data communication
- simplex, half-duplex, full-duplex
- synchronous, asynchronous
- marking state, spacing state
- start bit, stop bit
- baud rate

UART, USART, DTE, DCE

20- and 60-mA current loops

RS-232C, RS-422A, RS-423A, RS-449 serial data standards

Telephone circuits and systems
- leased and switched lines
- POTS, DTMF, PBX
- repeater, hybrid-coil, ring trip

CODECs, TDM, and PCM

Modems
- FSK, PSK
- analog and digital loopback (ALB and DLB)
- dibit, tribit
- quaternary amplitude modulation (QAM)

Fiber-optic data communication
- index of refraction, critical angle
- Multimode and single-mode fibers

Circular buffer

Critical region

Compiler and interpreter

Descriptor

Binary synchronous communications protocol (BISYNC)
- byte-controlled protocol (BCP)
- cyclic redundancy check

HDLC,SDLC protocols
- bit-oriented protocol (BOP)
- frame, field, flag
- frame check sequence (FCS)

Local area network (LAN)

Star, loop, ring, common-bus, broadband-bus (tree) topologies
- token
- baseband and broadband transmission

Electronic mail

Open system interconnection model (OSI)
- presentation, session, transport, network, data link, physical layers

Ethernet
- transceiver
- collision
- CSMA/CD

Token-passing rings

GPIB, HPIB, IEEE 488 bus standard
- listener, talker, controller

REVIEW QUESTIONS AND PROBLEMS

1. Draw a diagram showing the bit format used for asynchronous serial data. Label the start, stop, and parity bits. Use numbers to indicate the order of the data bits.
2. A terminal is transmitting asynchronous serial data at 1200 Bd. What is the bit time? Assuming 7 data bits, a parity bit, and 1 stop bit, how long does it take to transmit one character?

3. What is the main difference between a UART and a USART?

4. Define the term modem and explain why a modem is required to send digital data over standard switched phone lines.

5. Describe the functions of the $\overline{DSR}$, $\overline{DTR}$, $\overline{RTS}$, $\overline{CTS}$, TxD, and RxD signals exchanged between a terminal and a modem.

6. What frequency transmit clock (TxC) is required by an 8251A in order for it to transmit data at 4800 Bd with a baud rate factor of 16.

7. *a.* Show the bit pattern for the mode word and the command word that must be sent to an 8251A to initialize the device as follows: baud rate factor of 64, 7 bits/character, even parity, 1 stop bit, transmit interrupt enabled, receive interrupt enabled, $\overline{DTR}$ and $\overline{RTS}$ asserted, error flags reset, no hunt mode, no break character.
   b. Show the sequence of instructions required to initialize an 8251A at addresses 80H and 81H with the mode and command words you worked out in part *a*.
   c. Show the sequence of instructions that can be used to poll this 8251A to determine when the receiver buffer has a character ready to be read.
   d. How can you determine whether a character received by an 8251A contains a parity error?
   e. What frequency transmit and receive clock will this 8251A require in order to send data at 2400 Bd?
   f. What other way besides polling does the 8251A provide for determining when a character can be sent to the device for transmission? Describe the additional hardware connections required for this method.

8. Draw a flowchart showing how asynchronous serial data can be sent from a port line using a software routine.

9. Give the signal voltage ranges for a logic high and for a logic low in the RS-232C standard.

10. Describe the problem that occurs when you attempt to connect together two RS-232C devices that are both configured as DTE. Draw a diagram which shows how this problem can be solved.

11. Why are the two ground pins on an RS-232C connector not just jumpered together?

12. What symptom will you observe if the wire connected to pin 5 of an RS-232C terminal is broken?

13. Explain why systems which use the RS-422A or RS-423A signal standards can transmit data over longer distances and at higher baud rates than RS-232C systems.

14. Describe the function of the hybrid coil in Figure 13-12.

15. Describe the operation of a codec. Why are codecs designed with nonlinear response?

16. How does an FSK modem represent digital 1's and 0's in the signal it sends out on a phone line? How does an FSK modem perform full-duplex communication over standard phone lines?

17. Draw a waveform to show the signal that a simple phase-shift keying (PSK) modem will send out to represent the binary data 011010100.

18. *a.* Draw a diagram which shows the construction of a fiber-optic cable, and label each part.
   b. Identify two types of devices which are used to produce the light beam for a fiber-optic cable and two devices which are commonly used to detect the light at the receiving end of the fiber.
   c. Why should you never look into the end of a fiber optic cable to see if light is getting through?
   d. Describe the difference between a multimode fiber and a single-mode fiber. Give a major advantage and a major disadvantage of each type.
   e. What are the major advantages of fiber-optic cables over metallic conductors?

19. Using IBM PC BIOS and DOS calls, write an assembly language program which reads characters from the keyboard and puts them in a buffer until a carriage return is entered. The characters should be displayed on the CRT as entered. When a carriage return is entered, the contents of the buffer should be sent out the COM1 serial port.

20. The SDK-86 will only accept uppercase letters as commands. The SDK-86 download program in Figure 13-24 would be friendlier if you did not have to remember to press the caps lock key on the IBM. Write an assembly language routine that will convert a letter entered in lowercase to uppercase without affecting entered uppercase letters or numbers. Show where you would insert this section of code in the program in Figure 13-24.

21. Describe the operation of a circular buffer. Include in your answer the function of the tail pointer, the head pointer, and how the buffer-full and buffer-empty conditions are detected.

22. Why is it necessary to disable the UART interrupt input of the 8259A during part of the CHK_N_DISPLAY procedure in Figure 13-24?

23. Why, when changing a bit in a control word or interrupt mask word, should you not alter the other bits in the word? Show the assembly language instructions you would use to unmask IR5 of an 8259A at base address 80H without changing the interrupt status of any other bits.

24. Describe the major difference between the way that an interpreter translates a high level language program and the way that a compiler does the translation. Give a major advantage and a major disadvantage of each approach.

25. Show how the CALL statement in Figure 13-28*a* would be modified to pass two additional parame-

ters, A2 and B3, to the assembly language procedure. Show the assembly language instructions that you would add to the procedure to bring the actual value of A2 into the CX register. Assume the order of the passed parameters is Q$, A2, B3. Describe the function of the string descriptor table during a procedure call.

26. Why is synchronous serial data communication much more efficient than asynchronous communication?

27. If an 8251A is being used in synchronous mode for a BISYNC data link, what additional initialization word(s) must be sent to the device. How does the 8251A detect the start of a message? How does the 8251A indicate that is has found the start of a message? How does the receiving station in a BISYNC link indicate that it found an error in the received data?

28. How is the start of a message frame indicated in a bit-oriented protocol such as HDLC? How does an HDLC system prevent the flag bit pattern from appearing in the data part of the message? How does the receiver in an HDLC system tell the transmitter that an error was found in a transmitted frame?

29. Draw simple diagrams which show the five common network topologies. For each topology identify one commercially available system which uses it.

30. What is the difference between a baseband network and a broadband network?

31. List the seven layers of the ISO open systems model. Which of these layers is responsible for assembling messages into frames or packets? Which layer is responsible for making sure the message was transmitted and received correctly?

32. *a.* Describe the topology, physical connections, and signal type used in Ethernet.
b. Describe the method used by a unit on an Ethernet to gain access to the network for transmitting a message.
c. What response will a transmitting station make if it finds that another station starts transmitting after it starts? What is the term used to refer to this condition?

33. Describe the method used by a unit on a token-passing ring to take control of the network for transmitting a message frame. What is the advantage of this scheme over the method used in Ethernet?

34. How can a token ring network recover if the token is lost while being passed around the ring?

35. *a.* For what purpose was the GPIB designed?
b. Give the names for the three types of devices which the GPIB defines.
c. List and briefly describe the function of the three signal groups of the GPIB.
d. Describe the sequence of handshake signals that take place when a talker on a GPIB transfers data to several listeners. How does this handshake scheme make it possible for talkers and listeners with very different data rates to operate correctly on the bus?

CHAPTER

14 Operating Systems, the 80286 Microprocessor, and the Future

As we told you in an earlier chapter, a general-purpose operating system in its simplest form is a program which allows a user to create, print, copy, delete, display, and in other ways work with files. It also allows a user to load and execute other programs. The operating system insulates the user from needing to know the intricate hardware details of the system in order to use it. Up to this point in the book we have only referred to single-user operating systems such as the IBM PC DOS. To round out the book we now want to give you an overview of multiuser/multitasking operating systems, and an introduction to the 80286 microprocessor. The 80286 (used in the IBM PC/AT and its clones) has advanced features which make it suitable as the CPU in a multitasking system. Finally, in this chapter we discuss a few directions in which microcomputer evolution seems to be heading.

OBJECTIVES

At the conclusion of this chapter you should be able to:

1. Describe the difference between time-slice scheduling and preemptive priority-based scheduling.
2. Define the terms blocked, task queue, deadlock, deadly embrace, critical region, semaphore, kernel, memory management unit, and virtual memory.
3. Describe two methods that can be used to protect a critical region of code.
4. Show with assembly language instructions how a semaphore can be used to accomplish mutual exclusion.
5. Describe the major features of the UNIX operating system and define the terms kernel, pipe, and shell.
6. List and describe the types of "objects" used in the RMX 86 operating system.
7. List and describe the states that an RMX task can be in.
8. Describe the mechanism used to schedule tasks in RMX 86.
9. List some of the differences between UNIX and RMX 86.
10. Draw a block diagram of the internal structure of the 80286.
11. List the major hardware and software features that the 80286 microprocessor has beyond those of the 8086.
12. Show how the 80286 constructs physical addresses in its real address mode and in its protected virtual address mode.
13. Describe how the 80286 uses descriptor tables and call gates to control memory access.
14. Define the term "demand-paged virtual memory" and describe briefly how the 80386 produces a physical address in paged mode.

OPERATING SYSTEM CONCEPTS AND TERMS

Multiuser/Multitasking Operating System Overview

Newer 16-bit and 32-bit microprocessors are designed to be used as the CPU in multiuser/multitasking microcomputer systems. Therefore, to understand how these processors operate, you need to understand some of the terms and concepts of operating systems.

In Chapter 2 we discussed how several terminals can be connected to a single CPU and operated on a time-share basis. An operating system which coordinates the actions of a time-share system such as this is referred to as a *multiuser* operating system. The basic principle of a time-share system is that the CPU services one terminal for a few milliseconds, then services the next for a few milliseconds, and so on until all of the terminals have had a turn. It cycles through all of the terminals over and over, fast enough that each user seems to have the complete attention of the CPU. The program or section of a program which services each user is referred to as a *task* or *process*. A multiuser operating system then, can also be referred to as *multitasking*, but this term is more often used when referring to real-time industrial-control operating systems. With the addition of a user interface, the factory controller program in Figure 10-35 would be an example of a very simple real-time multitasking operating system.

The multiple tasks that are to be executed by a CPU must in some way be scheduled so that they execute properly. This part of the operating system is called the *scheduler*, *dispatcher*, or *supervisor*. There are several different methods of scheduling tasks, but we are mainly interested in two of them.

The first method is the *time-slice* method which we discussed previously. In this approach the CPU executes one task for perhaps 20 ms, then switches to the next task. After all tasks have had their turn, execution returns to the first. The UNIX operating system, which we discuss in detail later, uses this scheduling approach for a multiple-user system. The advantage of the time-slice approach in a multiuser system is that all users are serviced at approximately equal time intervals. As more users are added, however, each user gets serviced less often, so each user's program takes longer to execute. This is referred to as *system degradation*. In industrial-control operating systems this variable time between services is often not acceptable, so a different scheduling method is used.

The second scheduling method we are interested in is *preemptive priority-based scheduling*. In this approach an executing low priority task can be interrupted by a higher priority task. When the high priority task finishes executing, execution returns to the low priority task. This approach is well suited to some control applications because it allows the most important tasks to be done first. Priority interrupt controllers such as the 8259A are often used to set up and manage the task service requests. The Intel RMX 86 operating system, which we discuss later, uses priority-based scheduling.

In addition to scheduling, multitasking operating systems have several other considerations which have to be taken into account. The next section discusses some of these.

Problems Encountered in Building Multitasking Operating Systems

There are a great many operating system variations, and many different ways of solving various problems in an operating system. What we have tried to do in this section is use simple enough examples to illustrate the basic problems without getting lost in all of the possible variations.

PRESERVING THE ENVIRONMENT

The first problem to be solved in a multitasking system is to preserve the registers, data, and return address (environment) of each task when execution is switched to another task. This is necessary so that the task can be restarted correctly. The usual way of preserving the environment is to keep it on a stack. Often the operating system keeps a separate stack for each task. Current processors such as the 80186 and 80286 have the ENTER and LEAVE instructions to make it easy to save and restore the environment. Any procedures used in a multitasking system have to be reentrant.

ACCESSING RESOURCES

The second problem encountered in a multitasking system is assuring that tasks have orderly access to resources such as printers, disk drives, etc. As one example of this, suppose that a user at a terminal needs to read a file from a hard disk and print it on the system printer. Obviously the file cannot be read in from the disk and printed in one of the 20-ms time slices allotted to the terminal service, so several provisions must be made to gain access to the resources and hang on to them long enough to get the job done properly. A flag or *semaphore* in memory is used to indicate whether the disk drive is in use by another task or not. Likewise, another semaphore is used to indicate whether the printer is in use. If a task cannot access a resource because it is busy, the task is said to be *blocked*. Now, rather than making the user type in a print command over and over until the disk drive and the printer are available, most operating systems of this type set up queues of tasks waiting for each resource. When one task finishes with a resource, it resets the semaphore for that resource. The next task in the queue can then set the semaphore, to indicate the resource is busy, and use the resource.

In order to keep track of the state of a task, a block of data called a *process control block*, *process header*, or *process descriptor* is set up by the operating system for each task. Part of the information contained in the process control block is the progress of the read disk and print job. To simplify the disk and printer queues, all that needs to be put in these queues are pointers to the process control blocks of tasks that are waiting for access. This is similar to the way a pointer to a string descriptor table is passed to a procedure, rather than passing the string itself, as shown in Figure 13-29. Incidentally, most systems use a separate I/O processor to actually handle disks, printers, and other slow resources, so that these do not load down the main processor.

Another problem situation in a multitasking system can occur when two tasks need the same two resources, for example, a disk drive and a printer. Suppose that one task gains access to the disk drive and sets its semaphore to indicate that the disk drive is busy at the end of its time slice. The next task finds the disk busy, so its request goes on the queue. However, suppose that the second task finds the printer not busy, so it sets the printer semaphore to indicate it has control of the printer, and goes on about its business. When execution returns to the first task, it will try to access the printer so it has both the disk drive and the printer it needs. However, it finds the printer busy, so its request is put on the printer queue. The situation here is that each task controls a resource that the other needs in order to proceed. Therefore, neither can proceed. This condition is called *deadlock* or *deadly embrace*. The problem can be solved in a number of ways. One way is to link the printer and the disk drive together under one semaphore so that the two resources are accessed with a single action. Another more practical approach is to set up a hierarchy among the tasks, so that if deadlock occurs, the higher priority task can gain access to all of the resources it needs.

Still another interesting problem can occur in a multitasking operating system when two or more users attempt to read and change the contents of some memory locations at the same time. As an example, suppose that an airline ticket reservation system is operating on a time-slice basis. Now, further suppose that one user examines the memory location which represents a seat on a plane and finds the seat empty, just before the end of its time slice. Another user on the system can then, in its time slice, examine the same memory location, find it empty, mark it full, and print out a reservation confirmation on the CRT. When execution returns to the first user, it has already checked the seat during its previous time slice, so it marks the seat full, and prints out a reservation confirmation on the CRT. The two people assigned to the same seat may make nasty remarks about computers unless this problem is solved.

The section of a program where the value of a variable is being examined and changed must be protected from access by other tasks until the operation is complete. The section of code which must be protected is called a *critical region*. A technique called *mutual exclusion* is used to prevent two tasks from accessing a critical region at the same time. In the CHK_N_DISPLAY procedure in Figure 13-24 we showed one way in which a critical region can be protected from an interrupt procedure by simply masking the interrupt. In a time-slice system, however, a semaphore is used to provide mutual exclusion.

Figure 14-1 shows how this can be done with 8086 assembly language instructions. The instruction sequence is the same for each task. If task 1 needs to enter a critical section of code, it first loads the semaphore value for critical-region-busy into AL. The single instruction, XCHG AL, SEMAPHORE, then swaps the byte in AL with the byte in the memory location named SEMAPHORE. It is important to do this in one instruction so that the time-slice mechanism cannot switch to another task halfway through the exchange and cause our airline problem.

After the semaphore is read in Figure 14-1, it is compared with the busy value. If the critical region is busy, execution will remain in a wait loop for as many time slices as are required for the critical region to become free. If the semaphore value is a 0, indicating not busy, then execution enters the critical region. The XCHG instruction has already set the semaphore to indicate the critical region is busy. After execution of the critical region finishes, the MOV SEMAPHORE,00 instruction resets the semaphore to indicate that the critical region is

```
;Instructions for accessing critical region of code protected by
; semaphore - User 1

      MOV  AL, 01            ; Load semaphore value for region busy
HOLD: XCHG AL, SEMAPHORE     ; Swap and set semaphore
      CMP  AL, 01            ; Check if region busy
      JE   HOLD              ; Yes, loop until not busy
                             ; No, enter critical region of code
;Instructions which access critical region inserted here.
      MOV  SEMAPHORE, 00     ; Reset semaphore to make critical
                             ;  region available to other users.


;Instructions for accessing critical region of code protected by
; semaphore - User 2

      MOV  AL, 01            ; Load semaphore value for region busy
HOLD: XCHG AL, SEMAPHORE     ; swap and set semaphore
      CMP  AL , 01           ; Check if region busy
      JE   HOLD              ; Yes, loop until not busy
                             ; No, enter critical region of code
;Instructions which access critical region inserted here.
      MOV  SEMAPHORE, 00     ; Reset semaphore to make critical
                             ;  region available to other users.
```

FIGURE 14-1 8086 assembly language instruction sequences showing how a semaphore can be used to provide mutual exclusion for a critical region.

no longer busy. Task 2 can then swap the semaphore and access the critical region when needed. The semaphore functions in the same way as the "occupied" sign on a restroom of a plane or train. If you mentally try interrupting each sequence of instructions at different points, you should see that there is no condition where both tasks can get into the critical region at the same time.

In an Intel multibus system a problem could still result if the semaphore were located on a different board from the CPU. The XCHG instruction takes 2 bus cycles, so a task on another master could take over the bus and access the semaphore between the 2 bus cycles. This problem is solved by putting the LOCK prefix in front of the XCHG instruction. The LOCK prefix causes the 8086 to assert a hardware signal which can be used to maintain control of the bus as we described in Chapter 11. In the next section we look at some other ways that an operating system must protect various data and code areas that it uses.

The Need for Protection

Most single-user operating systems do little to prevent user programs from "tromping on" the code or data areas of the operating system. The usual results of this and Murphy's law are that an incorrect address in a user program will cause it to write over critical sections of the operating system. The system then "locks up," and the only way to get control again is to reboot the system. In a multitasking system this is intolerable, so several methods are used to protect the operating system.

The major method is to construct the operating system in two or more *layers*. Figure 14-2*a* shows an "onionskin" diagram for a two-layer operating system. The basic principle here is that the inner circle represents the code and data areas used by the operating system. The outer layer represents the code and data areas of user programs or tasks that are being run under control of the operating system. The inner layer is protected because user programs can only access operating system resources through very specific mechanisms rather than a simple, accidental call or jump. The Motorola MC68000 family of microprocessors is designed to accommodate a two-level structure such as this. The MC68000 has two modes of operation, user and supervisory. Certain privileged instructions which affect the operating system can only be executed when the processor is in supervisory mode.

The AT&T UNIX™ operating system, which we discuss in the next major section of the chapter, is an example of a three-layer operating system. Figure 14-2*b* shows the three layers for UNIX. The innermost layer, or *kernel*, contains the major operating system functions such as the scheduler. The middle layer or *shell* contains the command line interpreter, which translates user-entered commands to a sequence of kernel operations. The shell level is the user-interface level. The outer layer contains application programs such as data base management programs. It also contains utilities such as editors, compilers, etc. which programmers can use to write more application programs.

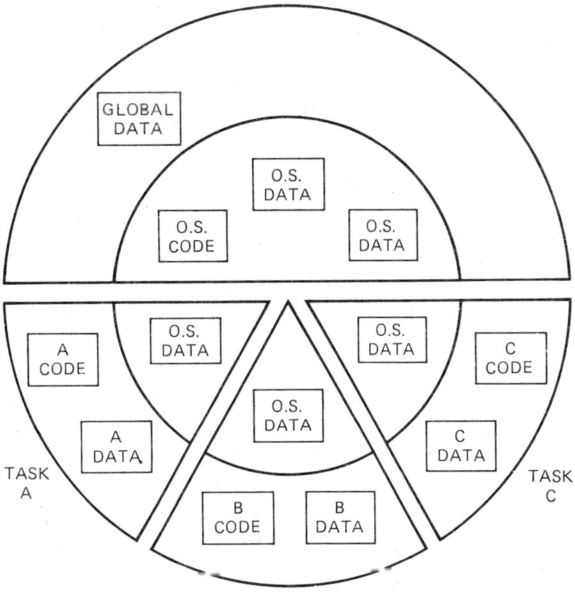

(*a*)

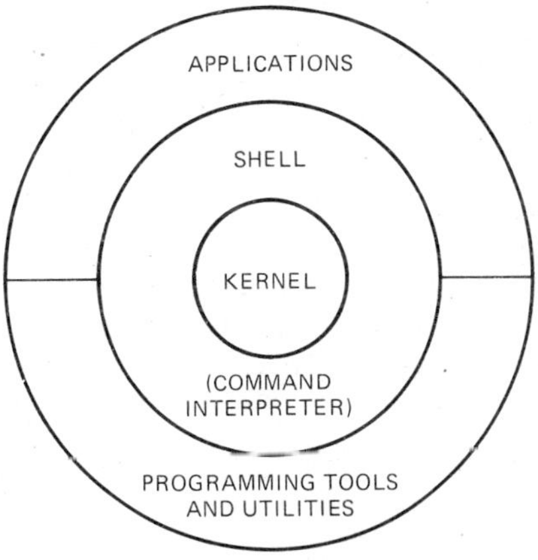

(*b*)

FIGURE 14-2 "Onionskin" diagrams for multitasking operating systems. *(a)* Operating system with two levels of protection. *(b)* Operating system with three levels, e.g., UNIX.

Other systems use even more levels of protection. The Intel 80286 processor has designed into its hardware a mechanism which allows up to four levels of protection to be built into an operating system running on it.

In addition to protecting itself from being tromped on by executing tasks, an operating system should provide some way of protecting tasks from each other. Throughout the rest of this chapter we will be showing you how protection layers are actually implemented, and how tasks can be protected from each other. To start this, we need to introduce you to the concepts of memory management.

Memory Management

There are two major reasons why memory must be specifically managed in a multitasking operating system. The first reason is that the physical memory is usually not large enough to hold all of the operating system and all of the application programs that are being executed by the multiple users. The second reason is to make sure that executing tasks do not access protected areas of memory. Memory management can be done totally by the operating system or with the aid of hardware called a *memory management unit* or MMU. Before we get into the operation of an MMU, we want to give you a little background on methods used to solve the limited memory problem.

A common problem, especially in older, single-user systems, is that the physical memory is not large enough to hold, for example, an assembler and the program being assembled. The traditional solution to this problem is to write the assembler in modules, and use an *overlay* scheme. When the assembler is invoked, the executive module of the assembler is loaded into memory, and reserves an additional memory space called the *overlay area*. The assembler then reads through the source program. When it reaches a point where it needs a particular module, it reads that module, referred to as an *overlay*, from the disk into the overlay area reserved in memory. When the assembler reaches a point where it needs another overlay, it reads the overlay from the disk, and loads it into the same overlay area in memory. The overlay approach is commonly used and works well for specific cases such as the assembler example we used here, but it is not flexible enough for multitasking systems.

Another approach traditionally used to expand the available memory in a microcomputer is *bank switching*. A system which has only 16 address lines can directly address only 64 Kbytes of memory. As shown in Figure 14-3, however, the addition of some simple selection hardware allows the system to access up to eight memory "banks" of 64 Kbytes each. The hardware is configured so that when the power is turned on, the system is using bank 0. To switch to bank 1, a byte which turns off bank 0 and turns on bank 1 is output to the selection port. Execution then proceeds in bank 1. In practice, some system-dependent tricks are often necessary to get execution smoothly from one bank to another, but the approach does help overcome the memory limits designed into the processor.

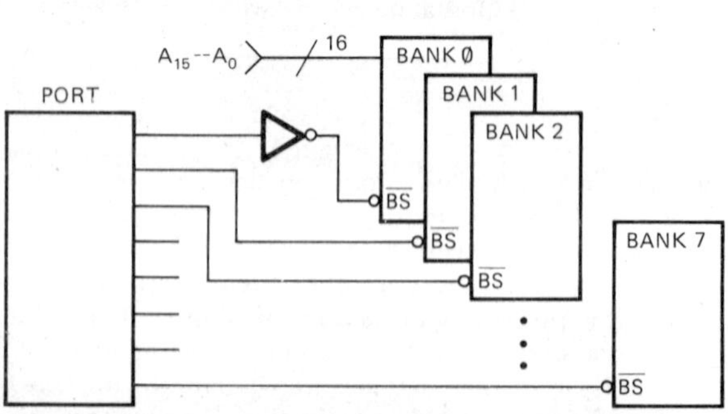

FIGURE 14-3 Block diagram showing how microcomputer memory can be expanded with bank switching.

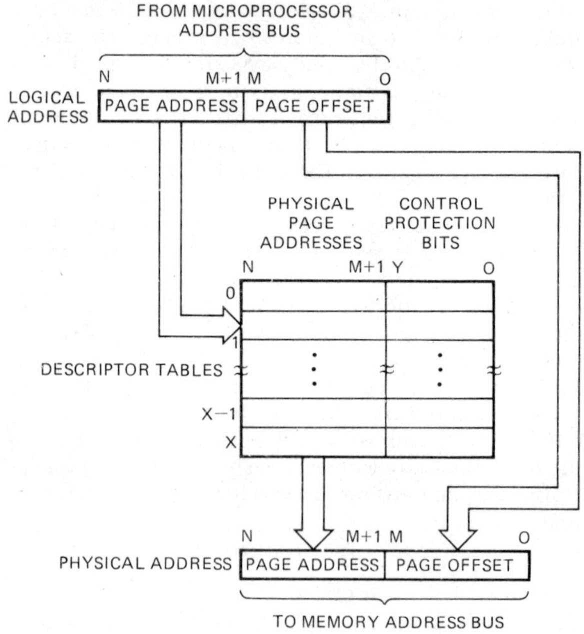

FIGURE 14-4 Block diagram showing operation of a memory management unit.

To use bank switching in a multiuser system, each user's program might be assigned to a bank. The difficulty with this is that a copy of the operating system kernel must be kept in each bank, the actual memory available for each user is still limited to 64 Kbytes, and users cannot easily share code or data. Thus memory is not very efficiently used. Also, protection is not as easily implemented as it is in the MMU approaches we discuss next.

MMUs

To reduce the burden of memory management on the operating system, most microprocessor manufacturers now have hardware memory management devices available. The device may be built into the processor as it is in the 80286, or be available separately. In either case the MMU is functionally positioned between the processor and the actual memory as shown in Figure 14-4. The major function of the MMU is to translate logical program addresses to physical memory addresses. We will explain how it does this, but first let's clarify what is meant, in this case, by logical address and physical address.

When you write an assembly language program, you usually refer to addresses by name. The addresses you work with in a program are called *logical addresses*, because they indicate the logical positions of code and data. An example of this is the 8086 instruction, JNZ NEXT. The label NEXT represents a logical address that execution will go to if the zero flag is not set. When an

8086 program is assembled, each logical address is represented with a 16-bit offset and a 16-bit segment base. The 8086 BIU then produces the actual physical memory address by simply adding these two parts together as explained many times previously.

When a program is assembled or compiled to run on a system with an MMU, each logical address is also represented by two components, but the components function differently. In a segment-oriented system the upper component is referred to as a *segment selector*, and the lower component is referred to as the *offset*. In a page-oriented system the upper component is referred to as the *page address*, and the lower component is referred to as the *page offset*.

In the segment case the MMU uses the upper component, a segment selector for example, to access a *descriptor* in a table of descriptors in memory, rather than just adding it to the offset in the lower component. A descriptor is a series of memory locations that contains the physical base address for a segment, the privilege level of the segment, and some control bits. As an example, let's assume the selector has 14 address bits and 2 privilege-level bits. The 14 address bits in the selector can select any one of 16,384 descriptors in the descriptor table. If the offset component of the logical address has 16 bits, then each segment can contain 64 Kbytes. Since each descriptor points to a segment, the logical address space for our example system here is 65,536 bytes/segment × 16,384 segments, or about 1 gigabyte. What this means is that the operating system can function as if a gigabyte of memory were available. Now let's see how this relates to the actual semiconductor memory.

Physically the MMU may have perhaps 24 address lines so it can only address 16 Mbytes of physical memory. The question that may immediately come to mind here is, "How can the operating system function as if there were a gigabyte of memory, when the maximum physical memory that the system can have is 16 Mbytes?" The answer to this question is that the physical memory, whatever its actual size, is simply a holding place for the segments currently being used by the operating system or user programs.

When the MMU receives a logical address from the CPU, it checks to see if that segment is currently in the physical memory. If the segment is present in physical memory, the MMU adds the offset component of the address to the segment base component of the address from the segment descriptor to form the physical address. It then outputs the physical address to memory on the memory address bus. If the MMU finds that the segment specified by the logical address is not in memory, it sends an interrupt signal to the CPU. In response to the interrupt, the CPU reads the desired code or data segment from a disk or other secondary storage, and loads it into the physical memory. The MMU then computes and outputs the physical address as described above. The operation is semiautomatic, so other than a slight delay, the operating system or other program is not aware that the segment had to be loaded. The gigabyte of logical address space that is available to programs is called *virtual memory* and the logical address in this type of system is usually called the *virtual address*. The term virtual refers to something that appears to be present but actually isn't.

When the CPU or smart MMU wants to load a segment from secondary storage into physical memory, it must first make space for it in the physical memory. Depending on the system, it may do this by compacting the segments already present and changing the descriptors to point to the new physical locations, or by swapping the segment being brought in with one currently in physical memory. To help in deciding which segment to swap back to memory, most systems use some bits in the descriptor to keep track of how many times the sector has been used. A low-use segment is the most likely candidate to swap back to memory. Most systems also have a *dirty bit* in each descriptor. This bit will be set if the contents of a segment have been changed. If the dirty bit is set, a segment must be swapped back to secondary storage if its space is needed. If the dirty bit is not set, then the segment has not been altered. The copy of the segment in secondary storage is still correct, so the segment can just be overwritten. This eliminates one write-to-disk operation.

Another term often found in MMU data sheets is the term *hit rate*. Hit rate refers to the percentage of the time that the segment required at a particular time is present in the physical memory. In a well-structured system the hit rate may be 85–90 percent.

The use of a descriptor table to translate logical addresses to physical addresses has another major advantage besides making virtual memory possible. The selector component of each address contains 1 or 2 bits which represent the privilege level of the program section requesting access to a segment. The descriptor for each segment also contains 1 or 2 bits which represent the privilege level of that segment. When an executing program attempts to access a segment, the MMU can compare the privilege level in the selector with the privilege level in the descriptor. If the selector has the same or greater privilege, then the MMU allows the access. If the selector privilege is lower, the MMU can send an interrupt signal to the CPU indicating a privilege-level violation. The indirect method of producing physical addresses then gives a method of providing privilege levels and protecting program sections such as the operating system kernel.

In most segment-oriented systems the segments swapped in and out of physical memory are quite large. The disadvantages of these large segments are the time required to load them and the compaction that often must be done to make space for a segment in physical memory. The method described next helps solve this problem.

The second major memory management approach currently used is called *demand-paged virtual memory*. In this approach the virtual memory is mapped as fixed-length pages of perhaps 4 Kbytes in length. The two components of the virtual address are called the page address and the page offset. The page offset, as the name implies, contains the offset of a desired byte within a page. The page address is used as a pointer to a descriptor table just as the selector is in the segmentation approach. The descriptors function in about the same way here that they do in the segmentation scheme. When a demanded page is found to be not present in the physical memory, the MMU or the CPU swaps it in. The typically smaller and fixed length of the pages makes the swapping operation much easier.

Before we summarize and go on to the next topic, we need to explain one more term commonly used with MMUs. For some MMUs the descriptor table is stored in a part of the main physical memory. Other MMUs have a built-in, high-speed memory called a *cache* (pronounced "cash"). The descriptors for the currently used segments or pages are kept in the cache memory so that they can be accessed much more quickly than they could if they were in the main memory. The descriptors for pages not currently being used are kept in a table in main memory. If the descriptor for a required page is not present in the cache, then it is read in from the descriptor table in main memory. The descriptor is then used to read in the required page.

To summarize then, MMUs translate logical program addresses to physical addresses with an indirect method through a descriptor table. This indirect approach makes possible a virtual address space much larger than the physical address space. The indirect approach also makes it possible to protect a memory segment or page from access by a program section with a lower privilege level. You will meet all of these concepts again in a later section which describes the operation of the 80286 microprocessor. First, however, we want to give you overviews of UNIX, a common multiuser operating system, and RMX 86, a common real-time multitasking operating system.

THE UNIX OPERATING SYSTEM

The purpose of this section is to show you the structure, terminology, and overall operation of the UNIX operating system, so you can see how it relates to multiuser microcomputer systems. If you are going to be working with UNIX, there are available several books which show with step-by-step examples how to use it.

History

In 1969 Ken Thompson, a researcher at Bell Laboratories, decided to write some system programs that would make it easier to develop other programs. Over the next few years, with the help of another researcher, Dennis Richie, these programs evolved into a powerful multiuser operating system. The original versions were written in assembly language for a DEC PDP-7 minicomputer, but when the value of the operating system became obvious, there was a strong desire to write versions for other machines. Adapting an assembly language program to run on another machine with a different CPU means rewriting the whole thing. To help solve this portability problem, Dennis Richie developed a high level language called C. This language has much of the capability of assembly language to work with hardware and twiddle bits, but it also allows a programmer to write high level language structured programs. Adapting a high level language program to run on a different machine involves rewriting the I/O sections as needed by the hardware of the new machine, and compiling the high level language program to the machine code for the new machine. By 1972 a version of UNIX written in C was operating successfully on the DEC PDP-11 computer.

In the following years Western Electric, a parent company of Bell Laboratories, licensed the source code of UNIX to several universities where it underwent further evolution. A commonly available enhanced version was developed at the University of California at Berkeley. The evolution also continued at Bell Labs. In 1979 version 7 was released, and later versions III and V were released by Western Electric.

Unfortunately, the basic structure of UNIX is easy to understand and alter. Therefore, each group using UNIX tended to extend and modify it to fit its specific needs or prejudices. Furthermore, due to licensing difficulties with Western Electric, commercial companies developing UNIX-like operating systems developed their own proprietary versions. The result of all of this is that there are many different versions of UNIX-type operating systems in use. Hopefully, the current efforts to work out a standard will be successful.

UNIX Operating System Structure

As shown in Figure 14-2*b*, the UNIX operating system consists of three layers. The innermost, most privileged layer, or kernel, contains a process scheduler, a hierarchal file structure, and mechanisms for processes to communicate with each other. The middle layer of the operating system, called the shell, is the layer that a user interfaces with. This layer contains the command interpreter which decodes and carries out user-entered commands. The outermost layer contains programming tools such as editors, assemblers, compilers, debuggers, etc., and application programs such as an accounting package. Let's take a closer look at how each of these layers function, and how they operate together.

OPERATIONS OF THE KERNEL

The UNIX operating system was designed to allow several users to share a CPU on a time-slice basis. Each user program is referred to as a *process*. One of the major functions of the kernel is to schedule and service the needs of processes. To do this the kernel keeps two tables in memory.

One of these, the *process table*, contains information about the state of each process. Among other things the entry in the process table contains the location of the process in memory, the length of the process, the identification number of the process, the identification of the user, and whether the process is active or blocked.

The second type of table maintained by the kernel is called a *user table*, or a *per-process segment*. The user table contains pointers to the data, files, and directories currently being used by the process.

When a user or process is added to the system, the kernel creates a process table entry and a user table for that process. The length of the process table is fixed for each system, so only this number of processes can be present in the system at one time. A process can create a subprocess, called a *child process*. When a child process is created, an entry is made in the process table for it, and a user table created for it. When any process is removed from the system, its process table entry and user table are removed to make room for another process.

At any given instant in time, only one process can actually be running since there is only one CPU to run processes on. All of the other processes are *suspended*. Processes essentially compete with each other for service. The scheduler in the kernel determines which process is to be run at a given time. The scheduling mechanism works as follows.

An external clock signal interrupts the CPU 50 or 60 times a second to produce the basic time slices. The interrupt procedure which services this clock interrupt checks the process table entries for each process to determine which process should be run next. The decision as to which process to run is based on several factors. The first factor is whether the process is ready to run or blocked. A process may be blocked or *put to sleep* if it has to wait for: an input or output operation to complete, a child process to complete, a signal from some other process, an external interrupt signal, or some fixed amount of time before continuing. A sleeping process will not be given a turn until the waited-for event occurs and the process is marked as active (ready to run).

A second factor used to determine which active process should be serviced next is how recently each process has been serviced. An active process which has recently had a turn will have a lower priority than an active process which has not recently had a turn because it just became unblocked or was just *swapped in* from disk. Because there is usually not room enough in memory to store all of the suspended processes, some of them are *swapped out* to secondary storage such as a disk. The scheduler decides which processes to swap so that all processes get serviced as needed.

A second major function of the UNIX kernel is to maintain the system file structure. Unix uses a hierarchical file structure as shown in Figure 14-5. This structure is sometimes called a tree structure because it looks like an inverted tree. The highest level in the hierarchy is the root directory. This directory contains the names of system files and the names of subdirectories. A directory in UNIX is simply a file which contains the name of the directory above it in the hierarchy, and names of files or directories below it in the hierarchy. The directory above a file or directory is often referred to as the *parent* directory.

A user logged on to the system is given a directory, under a directory labeled usr in Figure 14-5, under the parent directory. The user can then create subdirectories or files under this directory. To refer to other files or directories in the system, a user specifies the directory path to it. For example, a user whose directory is at point 2 in Figure 14-5 can refer to the file at point 8 as

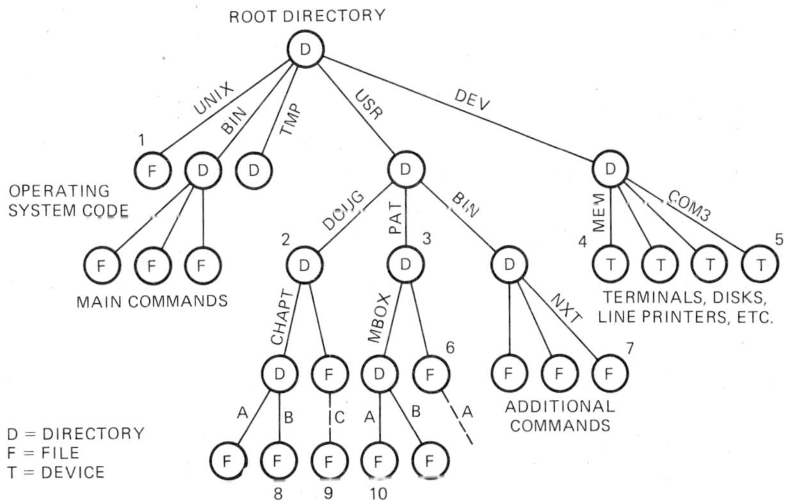

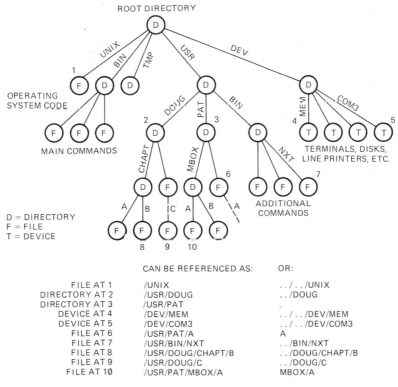

	CAN BE REFERENCED AS:	OR:
FILE AT 1	/UNIX	../../UNIX
DIRECTORY AT 2	/USR/DOUG	../DOUG
DIRECTORY AT 3	/USR/PAT	.
DEVICE AT 4	/DEV/MEM	../../DEV/MEM
DEVICE AT 5	/DEV/COM3	../../DEV/COM3
FILE AT 6	/USR/PAT/A	A
FILE AT 7	/USR/BIN/NXT	../BIN/NXT
FILE AT 8	/USR/DOUG/CHAPT/B	../DOUG/CHAPT/B
FILE AT 9	/USR/DOUG/C	../DOUG/C
FILE AT 10	/USR/PAT/MBOX/A	MBOX/A

FIGURE 14-5 UNIX hierarchial file structure.

/usr/doug/chapt/b. The name of a parent directory is often represented simply by two periods, so ../doug/chap/b can also be used to refer to the file at point 8 in, for example, a copy command. Figure 14-5 shows some other examples of how files, directories, and input/output devices are referred to in this type file structure.

In addition to names, the directory entry for each file and subdirectory contains a 2-byte *inode* number. The inode number identifies the position of the inode for that file or directory in a table of inodes kept by the operating system kernel. An inode is similar to a file control block, which we discussed in chapter 12. It contains the type of the file, the length of the file, the location of the file, the identification number of the owner, and the times the file was created, modified, and last accessed. The kernel uses inodes to manipulate files, but normally a user only has to be concerned with the file names.

Still another function of the kernel is to provide a means of communication between processes. The two methods it provides are *signals* and *pipes*. Signals are software interrupts generated by one process to tell another process to stop what it is doing, respond to the signal, and then go on with what it was doing. Signals can also be generated by user commands such as an abort command, or by processing errors such as a divide by zero error.

A pipe is a mechanism for passing the output data from one program directly to another program as input. We will discuss how a pipe is used later. Now that you have an overview of some of the kernel functions, let's take a look at some of the shell functions.

THE UNIX SHELL

As we said before, the shell layer of UNIX is the level at which a user usually interacts with the system. The shell executes user commands and programs. It calls kernel procedures as needed to do this. The UNIX command shell has some interesting features that we want to acquaint you with.

The first feature of the shell to discuss is how it handles I/O. At the user level, UNIX essentially treats I/O devices as files in a directory called dev as shown in Figure 14-5. A modem connected to the system at point 5 in the system, for example, can be referred to simply as /dev/com3. Devices are opened, read from or written to, and closed just as other files are. When a process is created, it has three files already open for use. The three are referred to as *standard input*, *standard output*, and *error output*. Standard input usually means the keyboard on the user's terminal. Standard output and error output usually mean the CRT on the user's terminal.

What this means is that when a user enters a command, which requires input, the input will be taken from the keyboard unless otherwise specified. Likewise a command that produces output data will send it to the user's CRT unless some other destination is indicated. The UNIX command ls, for example, will send a simple list of the user's directory to the CRT on the user's terminal. However, input data or output data can be *redirected* to other devices or files. The < and the > symbols are used to indicate redirection. For a user at point 3 in Figure 14-5, the command ls /usr/doug > /dev/com3, for example, reads the directory of /usr/doug and sends it to the device named com3, instead of to the user's CRT. The command sort −d < /usr/pat > lpr will alphabetically sort the directory /user/pat and send the result to the line printer. Note that UNIX commands are entered in lowercase letters.

Another feature of the UNIX shell which we mentioned previously is the pipe command. The pipe command allows output data from one program to be passed directly as input data to another program. The unique feature of a pipe is the way that the data is passed between the two. In most other systems data is passed from one program to another through files. One program processes some data and puts the results in a file. When the first program is done, a second program may access the file, further process the data, and put its results in another file which can be accessed by another program or command. The command ls −l > myfile, for example, might be used to produce a long listing (including subdirectories etc.) of a user's directory and put it in a file called myfile. The command sort −d < myfile might then be used to sort the directory listing in the file in alphabetical order, and display the result on the user's CRT. The pipe command makes it possible to do both the list and the sort operations without the need for an intermediate file to pass the data between the two commands. The single command ls −l ¦ sort −d can be used to do this. The dashed vertical line ¦ in the middle of the command indicates that the two commands are to be piped together. When a UNIX user issues a command to pipe two programs together, the kernel makes a connection so that the output from the first program is fed directly to the second program *as it is produced*. The pipe feature is often used with programs called *filters*. A filter is a program which simply performs some operation on a stream of input data and outputs the results. Some common types of filters are programs that format data into columns, sort data in various ways, and translate from one file format to another. As another example of how this is used, suppose that a user on the system shown in Figure 14-5 at point 2 wants to sort a directory alphabetically and send the result to the line printer. The simple command sort −d < /usr/doug ¦ lpr will do this. The designated output for the lpr command is the line printer, so no redirection is needed.

Another useful feature of the UNIX command shell allows a user to execute two commands concurrently. As an example of how this capability might be used, suppose that a programmer wants to assemble and print the listing of one program module, while editing a second module. The terms *foreground* and *background* are often used to describe the way in which the two processes are executing. For our example here, the compiling and printing will be done in the background while editing is done in the foreground. A command followed by an & (ampersand) indicates that the command should be carried out in the background mode. The sort and print command from the previous paragraph, for example, can be run as a background process by simply entering the command sort −d < /usr/doug ¦ lpr &.

The kernel actually does the background command by

creating a new process for it. The initial user process is referred to as the parent process and the new process is referred to as the child process. The parent process may be put to sleep until the child process finishes, or the parent process and the child process can compete for time slices and execute concurrently as we described in the example above.

The UNIX shell also provides a simple way to execute a series of commands over and over again. The commands to be executed are simply written into a named file using the editor. The resulting file is called a *shell file*. The shell file can be executed with the single command sh followed by the name that was given to the shell file.

One final feature of the UNIX shell and kernel that we want to describe is *spooling*. Spooling is a mechanism that allows users to send files off to get printed without worrying about whether the printer is available at that particular moment. Incidentally, the term spool stands for *simultaneous peripheral operation on line*. Here's how it works. A user sends a file off to the spooler with the lpr command. The file to be printed and another short file containing information about the file are put in a dedicated directory called /usr/lpd. Writing a file in this directory causes a special printer program called the *printer demon* to start running and print the file. The printer demon program does this by stealing small amounts of time between other operations. If the printer is busy, then the print request is queued up behind other print requests and eventually gets printed. The main point here is that while all of this printing is going on, users can go on editing, compiling, or executing other programs. Now we will take a brief look at the programs and utilities that are included in the outermost UNIX layer.

The UNIX Utilities/Application Layer

Utilities are software tools used to develop, write, compile, debug, and document programs. Because UNIX has been around for so many years, there are a great many utilities available for it. Among these are several powerful editors, programs which format text for typesetting machines, compilers for many high level languages, and a host of debuggers. For just about any function that a programmer or writer might want, there is probably a UNIX utility which does it. There are also a large number of application programs which will run under UNIX. Application programs, in contrast to utilities, are self-contained or "canned" programs. Examples of application programs are accounting packages, data base management packages, and computer-aided engineering design packages.

Some of the advantages of UNIX are its portability to new systems, the shell features we described previously, and the large number of utilities available for it. However, UNIX has various shortcomings, some of which have been remedied in later versions and in newer UNIX-like operating systems. One problem is that a user can load down the system with multiple background programs, fill a disk with files, and even crash the system. Also, a full UNIX system requires 8–10 Mbytes of disk space, which makes it more difficult to implement on small systems such as personal computers. Another major problem is that the basic time-slice approach of UNIX, which works well for a time-share system, responds too slowly for many real-time control applications. For these applications, an operating system such as Intel's RMX 86, which we describe in the next section, is used.

THE INTEL RMX 86 OPERATING SYSTEM

The UNIX operating system, described in the preceding section, is designed to allow several users to develop programs or run application programs on a time-share basis. UNIX and similar operating systems are usually sold to users as complete packages which can simply be configured to the hardware of a particular system and run. The time-slice approach of UNIX works well for a multiuser time-share system, but it does not respond fast enough and does not have a suitable priority setup for many real-time control systems. Several companies offer operating systems more suitable to the needs of real-time control systems. One example is the Intel RMX 86 operating system.

RMX 86 is a "building block" operating system. It is primarily intended to assist OEMs (original equipment manufacturers) in building custom control systems to sell to end users. Therefore, RMX 86 consists of a group of highly structured functional modules and utilities from which a system designer can chose the required functions. The purpose of this section of the chapter is to introduce you to the structure, terminology, and scheduling used in this common operating system.

RMX 86 Structure

Figure 14-6 shows an onionskin diagram of the basic structure of RMX 86. At the center is the *nucleus*, which corresponds to the kernel in UNIX. The nucleus consists mostly of a few dozen procedures which system developers can call as needed to implement a desired end user application program. This is indicated in Figure 14-6 by the fact that the user application section of the diagram extends all the way to the nucleus. The nucleus is the only software module required in a system. All of the other modules shown in Figure 14-6 are optional.

The *basic input/output system* contains device drivers to interface the system to disk drives, UARTS, keyboards, multiple CRT terminals, parallel printers, and other devices. The *extended input/output system*, or *EIOS*, contains higher level I/O routines which include built-in buffering. The application loader allows user programs to be loaded from disk into memory to be run. The *human interface* part of the system corresponds roughly to the shell in a UNIX system. It decodes and carries out user-entered commands. The basic human interface comes with commands for working with disk files, but other commands can be added as needed for a particular application. The final piece of the puzzle

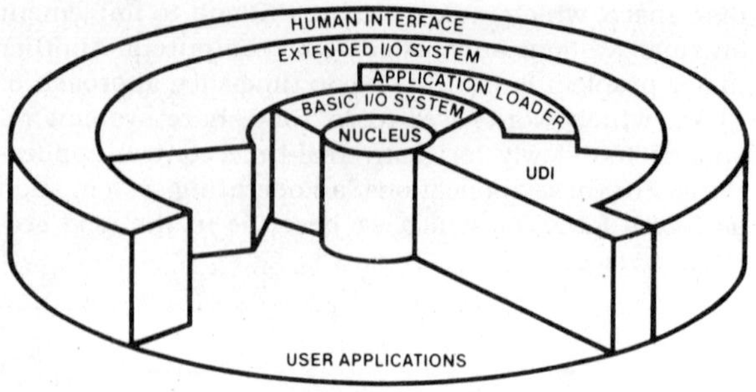

FIGURE 14-6 "Onionskin" diagram of Intel RMX 86 operating system.

shown in Figure 14-6 is the *universal development interface* or *UDI*. This software module, when added to the basic system, allows program development tools such as editors, assemblers, compilers, linkers, etc. to be loaded and run. Other software modules are also available. The point here is that software modules can be included, added to, or left out to produce a wide variety of custom operating systems. Now let's look a little closer to see how RMX 86 provides for multitasking.

RMX 86 Objects

The basic building blocks for RMX 86 programs are called *objects*. Objects are program structures which are created and manipulated by calls to procedures in the nucleus. The major object types are tasks, jobs, segments, mailboxes, regions, and semaphores. We will briefly describe each of these types, and then show how they are used.

Tasks in RMX 86 are equivalent to processes described previously. Tasks are the only active type of object. As a task executes, it manipulates the other types of objects by calling procedures in the nucleus. Tasks compete with each other for CPU time. Tasks are scheduled for execution on a preemptive, priority basis. We will talk about this more later, but basically what it means is that if several tasks are ready to run, the task which has been assigned the highest priority will be run first.

A *job* in RMX 86 is a logical environment in which tasks and other objects reside. A job usually corresponds to an application. The system initially has one job called the *root job* and a task that can be used to create other jobs. Tasks use system calls to create jobs. When a job is created, it is given a memory pool. From this memory pool tasks can create child jobs and other objects as needed. Figure 14-7 shows a simple diagram to illustrate this hierarchy.

A *segment* in RMX 86 is a contiguous block of memory, 16 to 64 Kbytes in size. When a task requests a segment, the requested memory is taken from the memory pool of the job which contains that task.

A *mailbox* is an object used to pass objects from one task to another. The object being passed through a mailbox is usually referred to as a message. What is actually passed through the mailbox is not the object itself, but a 16-bit *token* which represents the object. When each object is created, it is assigned a unique 16-bit number called a token. This approach is similar to the "file handle" approach used in some PC DOS function calls. Tasks can create mailboxes, delete mailboxes, send message tokens to mailboxes, and receive message tokens from mailboxes. A mailbox has two queues, one for tasks that are waiting to receive a message (object), and the other for objects that have not yet been received by their destination task. If a task attempts to receive a message, and there is no message in the mailbox, the task may be *put to sleep* for a while to wait for the message. This mechanism can be used to make one task wait at a mailbox until another task is finished before it starts.

A *region* in RMX 86 is one mechanism that can be used to prevent two or more tasks from accessing shared data at the same time. A task can create a region, delete a region, receive control, or send control of a region. If a task has received control of a region (has the token for it), no other task can access the region, until the task "sends control" of it (releases the token for it) back to the operating system. A region then is usually used to provide mutual exclusion for a collection of data shared by two tasks. To provide mutual exclusion for a single variable or protect a critical region of code, a simpler way is to use a semaphore.

A *semaphore* in its simplest case is simply a 1-bit flag used to indicate that a resource is busy. Figure 14-1 shows an example of how a simple semaphore can be set up to do this. In RMX 86 a semaphore is a *counter*. A task can create a semaphore, delete a semaphore, send units to the semaphore, and receive units from a semaphore. For a simple case such as that shown in Figure 14-1, a semaphore can be created and sent one unit. When a task wants to access the variable protected by the semaphore, it is made to receive one unit from the semaphore. If the variable is not busy, then the semaphore will contain one unit. The task can receive that unit and access the variable. If any other tasks that access that variable are made to receive one unit before accessing the variable, then once one task has received the unit, no other tasks can access it. This is because there is no unit in the semaphore for the other tasks to receive. When a task has finished with a shared variable

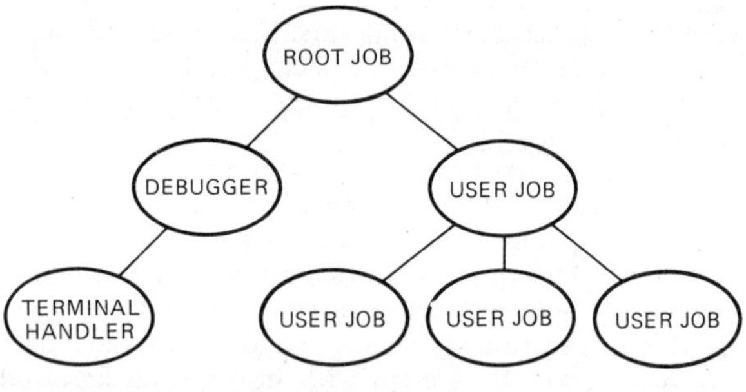

FIGURE 14-7 RMX 86 hierarchy of jobs.

or critical region, it sends one unit to the semaphore to release the variable so other tasks can access it. All of this is the same principle as the example in Figure 14-1 described with a different vocabulary. The fact that a semaphore can have values other than 0 or 1 allows it to be used to synchronize two tasks. For example, a task can be written to send one unit to a semaphore each time it executes. A second task can be made to wait at the semaphore until it is able to receive a specified number of units from the semaphore.

In addition to the defined object types, a programmer can create custom objects. Now that we have given you an overview of the types of objects that RMX 86 works with, we will describe how RMX 86 handles task execution.

RMX 86 Task Execution

Real-time control systems usually must respond to asynchronous requests for service in a manner that makes sure the most important request gets serviced first. In an RMX 86 system each service for a request is set up as a task. When an RMX 86 task is created, it is assigned a priority number between 0 and 255. The lower the number the higher the priority. Numbers between 0 and 127 are used for interrupt tasks. In addition to the built-in 8086 interrupt types, RMX 86 supports a single or several cascaded 8259A priority interrupt controllers for multiple hardware interrupts. Priority numbers between 128 and 255 are used for software tasks. RMX 86 schedules the execution of tasks on a preemptive priority basis. This means that if two or more tasks are ready to run, the task which has been assigned the highest priority will be executed first. This task will execute until it finishes, or until it reaches a point where it needs some resource which is not yet available. If a higher priority task becomes ready while a task is executing, the executing task will be preempted (put to sleep), and the higher priority task will be executed.

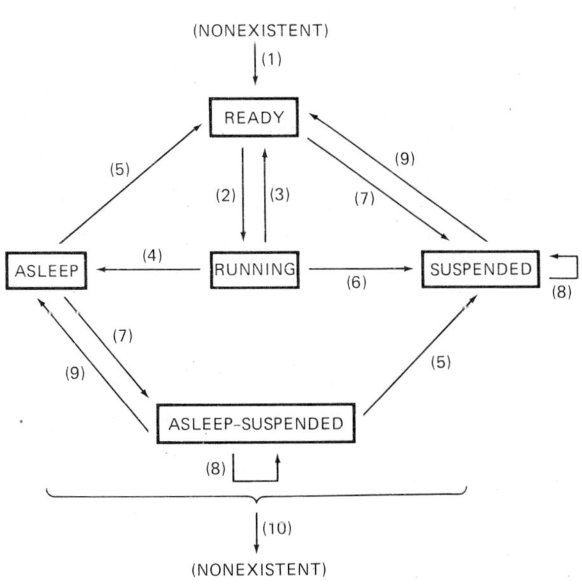

FIGURE 14-8 RMX 86 task state diagram.

Figure 14-8 shows an RMX 86 *task state diagram* which is often used to summarize the different states a task can be in, and the conditions necessary to go from one state to another. As we work our way around through the numbers in the figure, try to develop some intuitive feel for how it all works.

1. When a task is created, it is placed in the ready state. A task is created, remember, by a call to a procedure in the nucleus.
2. A task enters the running state when it has a higher priority than any other ready task, or it has been waiting longer than another ready task of the same priority.
3. A task is returned to the ready state when a task with a higher priority becomes ready and preempts it.
4. A task goes from the running state to the asleep state if:
   a. The task puts itself to sleep with a sleep system call. A task can put itself to sleep for a specified amount of time and then return to the ready state.
   b. The task must wait for a semaphore, a message, or a region in order to proceed.
5. Note that there are two number 5's on Figure 14-8. A task will go from the asleep state to the ready state or from the asleep-suspended state to the suspended state if:
   a. The time specified in the sleep call has expired.
   b. The semaphore, message, or region that the task was waiting for becomes available.
   c. The task's waiting time limit expires without the object that the task was waiting for becoming available.
6. A task goes from the running state to the suspended state when it does a suspend-task system call or a wait-for-interrupt system call.
7. A task in the ready state will be suspended when another task suspends it by calling the suspend-task procedure from the nucleus.
8. A task remains in a suspended state or an asleep-suspended state until the resume-task nucleus procedure has been called as many times as the suspend-task procedure was called for the task.
9. A task in the suspended state will return to the ready state, and a task in the asleep-suspended state will return to the asleep state when the resume-task system call has been done as many times for the task as the suspend-task system call. Another case where a task may exit from a sus-

pended state is when an interrupt that the task was waiting for occurs.

10. A task can be deleted with the delete-task system call.

A question that might occur to you at this point is, "If RMX 86 tasks are executed on a priority basis, how can a multiuser capability be included in an RMX 86 based system?" The answer to this question is that a clock tick can be used to produce an interrupt every 20 ms or so. The interrupt service procedure for that interrupt can then cycle around to a different terminal after each interrupt.

The final point we want to make here about RMX 86 is how a designer goes about using it to develop a custom system program. The design steps usually follow a sequence such as:

1. Define the system requirements.
2. Break the overall system into logical jobs.
3. Carefully define the functions of each job.
4. Determine the data structures needed for each job.
5. Determine whether jobs need to communicate or share resources.
6. Break down each job into tasks.
7. Write the algorithms for each task, including any needs for shared resources, synchronization, or communication between tasks.
8. Write the system initialization modules which set up the jobs, tasks, segments, regions, and semaphores using nucleus calls.
9. Write and test the program code for each task using system calls to define and manipulate "objects" as needed.
10. Integrate and test the completed system.

In the next section of this chapter we introduce you to the Intel 80286 microprocessor, which was designed to be used as the CPU in a multitasking system with virtual memory capability.

THE INTEL 80286 MICROPROCESSOR

We started this chapter with an introduction to some of the needs of multitasking/multiuser operating systems such as protection, mutual exclusion, and virtual memory capability. Later sections gave brief overviews of UNIX, a common multiuser operating system, and of RMX 86, a common real-time multitasking operating system. The Intel 80286 microprocessor was designed to serve as the CPU in a multitasking microcomputer system such as those we have described. The IBM PC/AT and its clones, as well as several other common systems capable of multitasking operation, use the 80286 as their CPU.

After a brief introduction to the internal architecture, signals, and hardware connections of the 80286, we will show you how memory management, task switching, and protection are done with the features built into the device.

80286 Architecture, Signals, and System Connections

Figure 14-9 shows an internal block diagram for the 80286. Note that the 80286 contains four separate processing units.

The *bus unit* (BU) in the device performs all memory and I/O reads and writes. When the BU is not using the buses for other operations, it prefetches instruction bytes and puts them in a 6-byte prefetch queue. When a

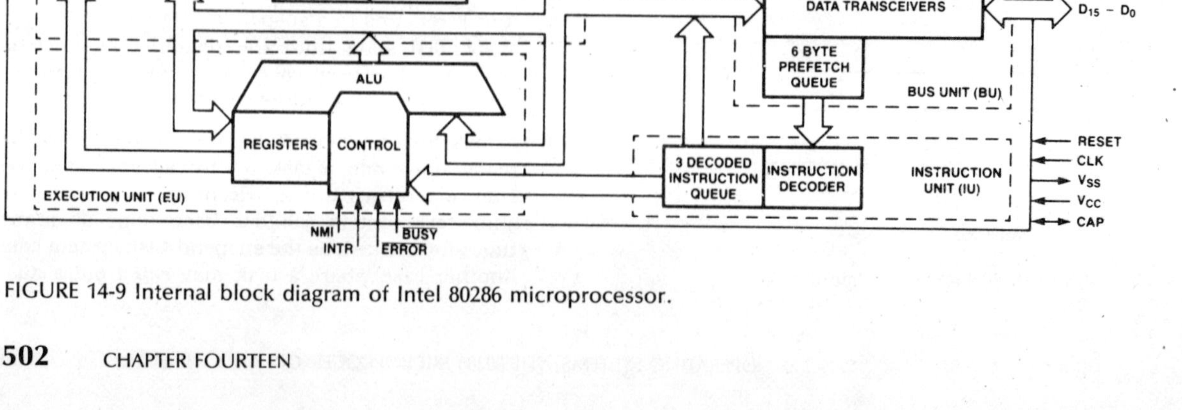

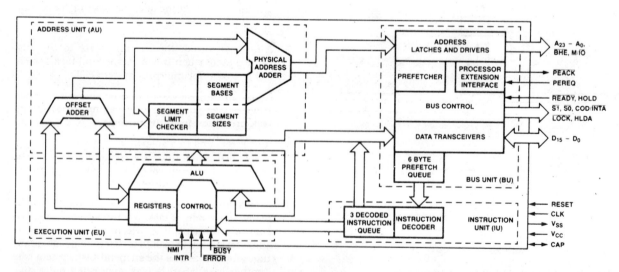

FIGURE 14-9 Internal block diagram of Intel 80286 microprocessor.

jump or call instruction executes, the BU will dump the queue and start filling it from the destination address. The BU also controls transfer of data to and from processor extension devices such as the 80287 math coprocessor.

The *instruction unit* (IU) fully decodes up to three prefetched instructions and holds them in a queue where the execution unit can access them. This is a further example of using pipelining to speed up the operation of a processor.

The *execution unit* (EU) sequentially executes instructions it receives from the instruction unit. It contains a set of index, pointer, and general-purpose registers identical to those of the 8086. In addition to a flag register like that of the 8086, the 80286 also has a 16-bit machine status word register which we will discuss later. The execution unit directs the BU to access memory or I/O operands as needed.

The *address unit* (AU) computes the physical addresses that will be sent out to memory or I/O by the BU. The 80286 can operate in one of two memory addressing modes, *real address mode* or *protected virtual address mode*. When programmed to operate in the real address mode, the address unit computes addresses with a segment base and an offset component just as the 8086 does. The familiar CS, DS, SS, and ES registers are used to hold the base addresses for the segments currently in use. The instruction pointer holds the offset of the currently addressed code byte in the code segment, and the stack pointer register holds the offset of the top of the stack. The maximum physical address space in this mode is 1 Mbyte, just as it is for the 8086.

When programmed to operate in *protected virtual address mode* (PVAM), the address unit functions as a complete memory management unit (MMU). In this address mode the 80286 uses all 24 address lines to access up to 16 Mbytes of physical memory. It is this mode which provides the protection and virtual memory capability desirable in multitasking operating systems. We will discuss this mode in detail in a later section.

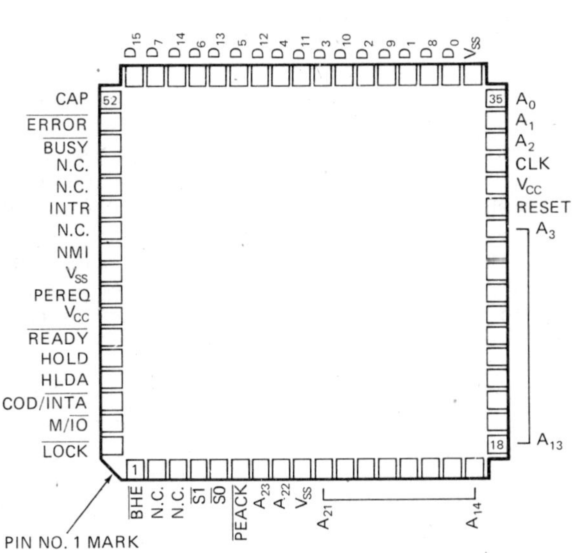

FIGURE 14-10 Pin diagram for 80286 microprocessor.

The 80286 is packaged in a 68-pin ceramic flat-pack as shown in Figure 14-10. Figure 14-11 shows how an 80286 is connected with some other components to form a simple system. Keep these figures handy as we work our way around the pins of the 80286. Many of the signals of the 80286 should be familiar to you from our discussion of the 8086 signals in Chapter 7.

An external 82284 clock generator is required to produce a single-phase clock signal for the 80286. The 82284 also synchronizes RESET and READY signals with the clock signal for the 80286.

The 80286 has a 16-bit data bus and a 24-bit nonmultiplexed address bus. Having the address lines available directly speeds up processing and simplifies the hardware. The 24-bit address bus allows the processor to access 16 Mbytes of physical memory. External buffers are used on both the address and the data bus.

Memory for the 80286 is set up as an odd bank and an even bank, just as it is for the 8086. The even bank will be enabled when A0 is low, and the odd bank will be enabled when BHE is low. To access an aligned word, both A0 and BHE will be low.

From a control standpoint, the 80286 functions similarly to an 8086 operating in maximum mode. Status signals S0, S1, and M/IO are decoded by an external 82288 bus controller to produce the control bus, read, write, and interrupt acknowledge signals.

Some other familiar pins on the 80286 should be the HOLD input which is used to request use of the buses for DMA operations, and the hold acknowledge (HLDA) output which is used to tell the DMA controller that the buses are available. An 8259A priority interrupt controller can be connected to the 80286 INTR interrupt input to funnel in external hardware interrupts. The interrupt acknowledge output (INTA) from the 82288 tells the 8259 to send the desired interrupt type to the processor on the data bus. The nonmaskable interrupt (NMI) input, the READY input, and the LOCK output pins function the same as they do with the 8086.

After the RESET input has been held high for the required time and then made low, the 80286 will begin executing in the real address mode at address FFFFF0H. The internal registers will be initialized as follows: flag word—0002H, machine status word—FFF0H, IP—FFF0H, CS register—F000H, DS register—0000, ES register—0000H, SS register—0000.

A new pin function on the 80286 is the CAP pin. In order to operate at maximum speed, the substrates of the MOS devices in the 80286 must be biased with a negative voltage. This negative voltage is produced from the +5-V supply by a bias generator on the 80286. The external capacitor connected to the CAP pin filters this bias voltage. Note the polarity of the capacitor.

A second new pin here is the COD/INTA pin. This is a type of status signal output which can be used with M/IO to produce early control bus signals. COD/INTA will be asserted low for interrupt acknowledge, memory data read, or memory data write signals. It will be high for I/O read, I/O write, or memory instruction read machine cycles.

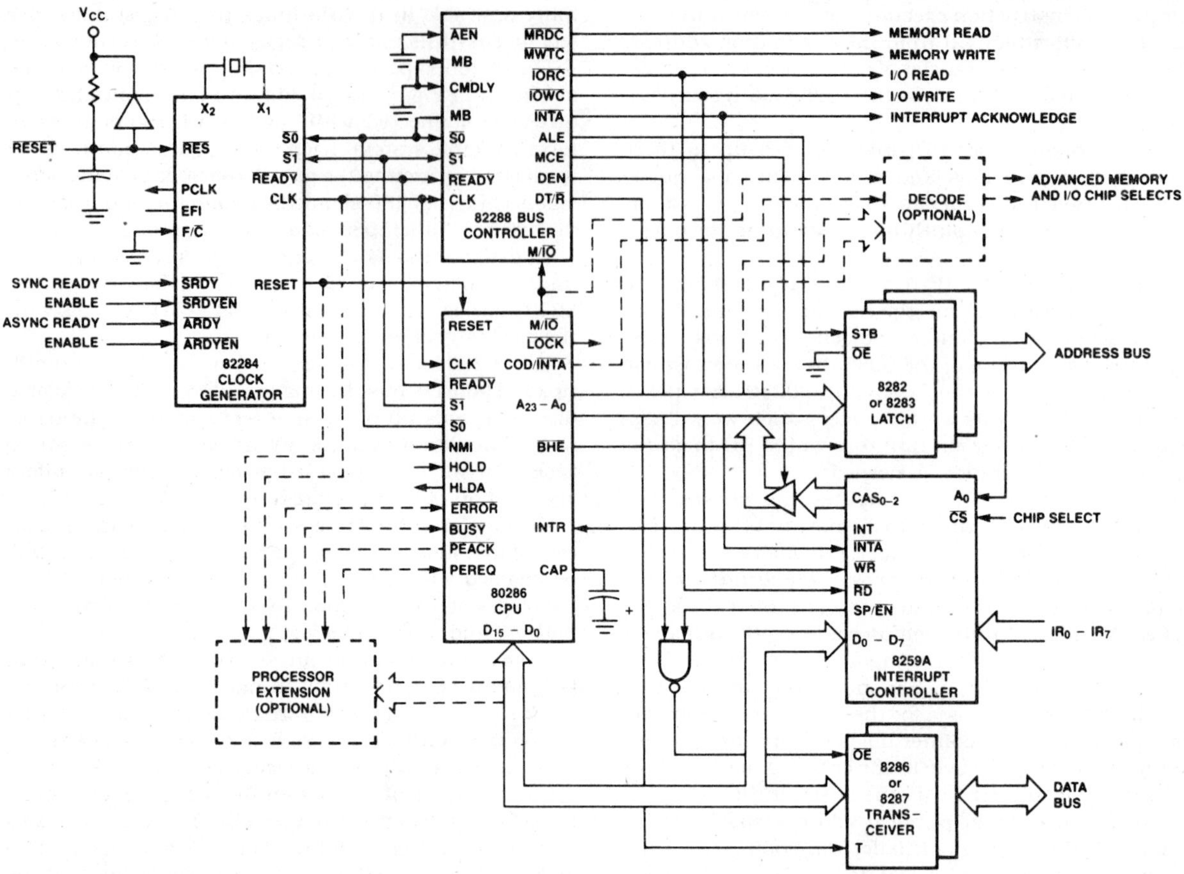

FIGURE 14-11 Circuit connections for simple 80286 system. *(Intel)*

The final four signal pins we need to discuss here are used to interface with processor extensions (coprocessors) such as the 80287 math coprocessor. The *processor extension request* (PEREQ) input pin will be asserted by a coprocessor to tell the 80286 to perform a data transfer to or from memory for it. When the 80286 gets around to do the transfer, it asserts the *processor extension acknowledge* ($\overline{\text{PEACK}}$) signal to the coprocessor to let it know the data transfer has started. Data transfers are done through the 80286 in this way so that the coprocessor uses the protection and virtual memory capability of the MMU in the 80286. The $\overline{\text{BUSY}}$ signal input on the 80286 functions the same as it does on the 8086. When the 80286 executes the WAIT instruction, it will remain in a WAIT loop until it finds the $\overline{\text{BUSY}}$ signal from the coprocessor high. If a coprocessor finds some error during processing, it will assert the $\overline{\text{ERROR}}$ input of the 80286. This will cause the 80286 to automatically do a type 16H interrupt call. An interrupt service procedure can be written to make the desired response to the error condition.

The machine cycle waveforms for the 80286 are very similar to those of the 8086 that we showed and discussed in earlier chapters. You should be able to work your way through them in the Intel 80286 data sheets if you need that type of information. In the limited space we have for the remainder of this section, we want to concentrate on the operation of the 80286 in its real address mode and in its protected virtual address mode.

The 80286 Real Address Mode

After an 80286 is reset, it starts executing in its real address mode. This mode is referred to as real because physical memory addresses are produced directly by adding an offset to a segment base just as the 8086 does. In this mode the 80286 can address up to 1 Mbyte of physical memory. As you will see later, the virtual address mode computes addresses in a very different way. In the real address mode, an 80286 functions essentially as a "souped up" 8086. Here are some of the ways in which the 80286 is enhanced when operating in real address mode.

The 80286 will directly execute 8086 machine code programs with only minor modifications to them. However, due to the extensive pipelining and other hardware improvements, the 80286 will execute most programs several times faster.

The instruction set of the 80286, in real address mode, is a "superset" of the 8086 instructions as illustrated by Figure 14-12. The 80286 in real address mode executes all of the 8086 instructions; the additional in-

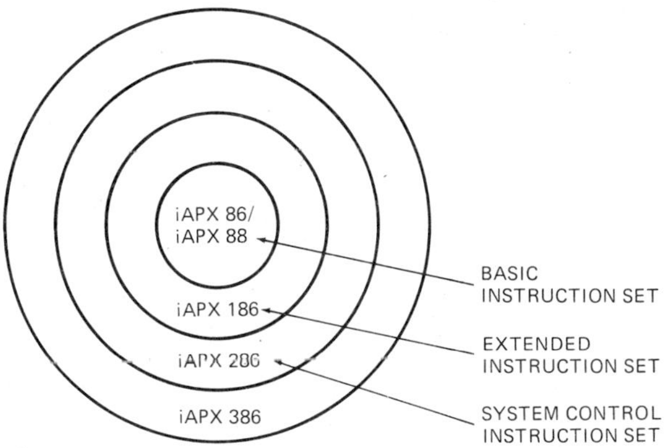

FIGURE 14-12 Relationship of 80386, 80286, 80186, and 8086 instruction set.

structions of the 80186 such as ENTER, LEAVE, and BOUND; and a few instructions used to switch the 80286 from real address mode to virtual address mode. In a later section we will show the additional instructions the 80286 can execute in its virtual address mode.

The 80286 has several additional built-in interrupt types as shown in Figure 14-13. For your reference we have included here a brief explanation of the cause of each of the interrupt types; some of these may not make much sense until we dig a little deeper into the operation of the 80286. As with the 8086, when an interrupt occurs, the 80286 multiplies the interrupt type number by 4 and goes to the resulting address in the interrupt vector table to get the CS and IP values for the interrupt procedure. In real address mode the interrupt vector table is in the first 1 Kbyte of memory. As we show later, the interrupt vector table has no fixed physical address in memory in the virtual address mode.

FUNCTION	INTERRUPT NUMBER
DIVIDE ERROR EXCEPTION	0
SINGLE STEP INTERRUPT	1
NMI INTERRUPT	2
BREAKPOINT INTERRUPT	3
INTO DETECTED OVERFLOW EXCEPTION	4
BOUND RANGE EXCEEDED EXCEPTION	5
INVALID OPCODE EXCEPTION	6
PROCESSOR EXTENSION NOT AVAILABLE EXCEPTION	7
INTERRUPT TABLE LIMIT TOO SMALL	8
PROCESSOR EXTENSION SEGMENT OVERRUN INTERRUPT	9
INVALID TASK STATE SEGMENT	10
SEGMENT NOT PRESENT	11
STACK SEGMENT OVERRUN OR NOT PRESENT	12
SEGMENT OVERRUN EXCEPTION	13
RESERVED	14,15
PROCESSOR EXTENSION ERROR INTERRUPT	16
RESERVED	17-31
USER DEFINED	32-255

FIGURE 14-13 80286 interrupt types.

The 80286 was designed to be upward-compatible from the 8086 and, except for the integrated peripherals, from the 80186, so that the huge amount of software developed for these could be easily transported to the 80286. Previously debugged modules can then be integrated with new program modules written to take advantage of the advanced features of the 80286. Let's take a look at how some of these advanced features work.

The 80286 Protected Virtual Address Mode

The major features of the *protected virtual address mode*, PVAM, of the 80286 that we want to discuss in this section are memory management, protection, task switching, and interrupt processing. We will work our way through these in order, but first we want to make some general comments about how an 80286 is switched to PVAM when an operating system running on it is booted.

After the 80286 is reset, it initially operates in real address mode. This mode is usually used to initialize peripheral devices, load the main part of the operating system from disk into memory, load some registers, enable interrupts, and enter the PVAM. The PVAM is entered by setting the protection enable bit of the *machine status word* (MSW) in the 80286. Figure 14-14*a* shows the format for the MSW. Bit 0 of this word is the

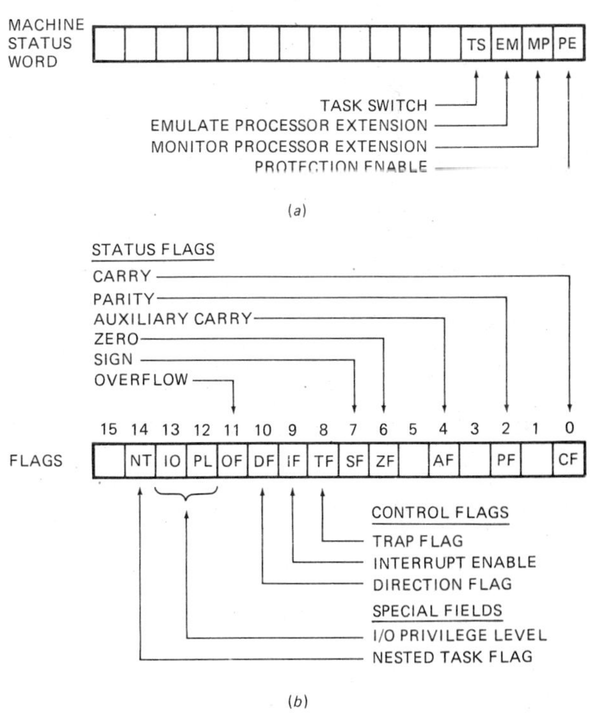

FIGURE 14-14 80286 machine status word and flag word. *(a)* Machine status word. *(b)* Flag word.

protection enable bit we are interested in here. Bits 1, 2, and 3 are used, for the most part, to indicate whether a processor extension (coprocessor) is present or not. Bits are changed in the MSW by loading the desired word in a register or memory location and executing the *load machine status word* (LMSW) instruction. Once the PVAM is entered by executing the LMSW instruction, the only way to get back to real address mode is by resetting the system. The 80286 was designed this way so that a "clever" programmer could not switch the system back into real address mode to defeat the protection schemes in PVAM. The final step to get the 80286 operating in PVAM is to execute an intersegment jump to the start of the main system program. This jump is necessary to flush the queue because the queue functions differently in PVAM. Now that we have the 80286 in PVAM, let's see how it manages memory and does all of those other wondrous things.

The 80286 Memory Management Scheme

Just as with the 8086, the basic building blocks of memory management in PVAM are logical segments. The segments probably should be referred to as virtual segments, because they may not all be present in physical memory at the same time. Unlike 8086 segments, however, segments for the 80286 can be any length from 1 byte to 64 Kbytes. A size or *limit* is given to each segment when it is created. Defining the length of segments in this way makes more efficient use of memory,

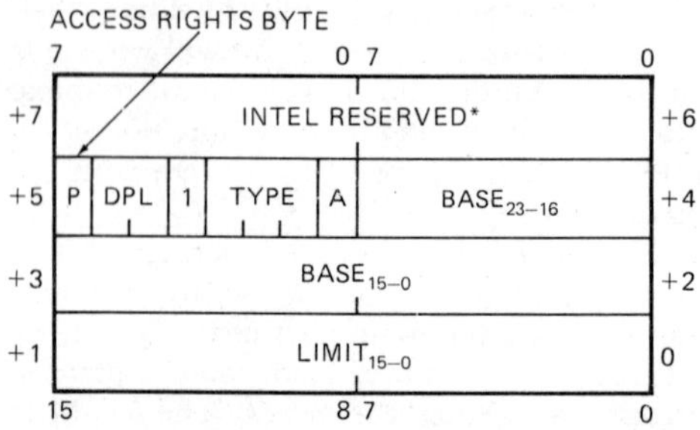

P = PRESENT
DPL = DESCRIPTOR PRIVILEGE LEVEL
TYPE = SEGMENT TYPE AND ACCESS INFORMATION
A = ACCESSED

*MUST BE SET TO 0 FOR COMPATIBILITY WITH iAPZ 386

(*a*)

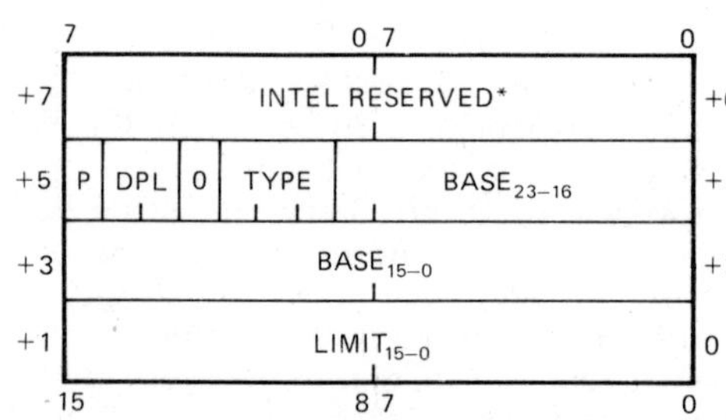

P = PRESENT
DPL = DESCRIPTOR PRIVELEGE LEVEL
TYPE = TYPE OF SPECIAL DESCRIPTOR (INCLUDES CONTROL AND SYSTEM SEGMENTS)

0 = INVALID DESCRIPTOR
1 = AVAILABLE TASK STATE SEGMENT
2 = LDT DESCRIPTOR
3 = BUSY TASK STATE SEGMENT
4-7 = CONTROL DESCRIPTOR
8→F = INVALID DESCRIPTOR

*MUST BE SET TO 0 FOR COMPATIBILITY WITH iAPX 386

(*b*)

BIT	NAME	DESCRIPTION
7	PRESENT	1 MEANS PRESENT AND ADDRESSABLE IN REAL MEMORY; 0 MEANS NOT PRESENT. SEE SECTION 11.3.
6, 5	DPL	2-BIT DESCRIPTOR PRIVILEGE LEVEL, 0 TO 3.
4	SEGMENT	1 MEANS SEGMENT DESCRIPTOR; 0 MEANS CONTROL DESCRIPTOR.
FOR SEGMENT = 1, THE REMAINING BITS HAVE THE FOLLOWING MEANINGS:		
3	EXECUTABLE	1 MEANS CODE, 0 MEANS DATA.
2	C OR ED	IF CODE, CONFORMING: 1 MEANS YES, 0 NO. IF DATA, EXPAND DOWN: 1 YES, 0 NO—NORMAL CASE.
1	R OR W	IF CODE, READABLE: 1 MEANS READABLE, 0 NOT. IF DATA, WRITABLE: 1 MEANS WRITABLE, 0 NOT.
0	ACCESSED	1 IF SEGMENT DESCRIPTOR HAS BEEN ACCESSED, 0 IF NOT.

(*c*)

FIGURE 14-15 Descriptor and access byte formats. *(a)* Code or data segment descriptor format. *(b)* Control descriptor format. *(c)* Access byte format for code and data segment descriptors.

and makes it possible for the 80286 to check if a memory access accidentally goes outside of the bounds of a segment. The segments currently being used by a task or program are kept in physical memory. In PVAM all 24 address lines are active, so 16 Mbytes of physical memory can be accessed. Segments not currently being used are swapped out from physical memory to secondary storage such as a hard disk, and then swapped back into physical memory as needed. When you write programs for the 80286 in PVAM, you refer to segments by the names you give them, just as you do in 8086 programs. For example, if you want to use a segment named ACCOUNTS_IN as the current data segment, you can do this with the simple instructions: MOV AX, ACCOUNTS_IN, and MOV DS,AX. When the 80286 in PVAM executes these two instructions, it will automatically point at the segment ACCOUNTS_IN as the current data segment. If the segment ACCOUNTS_IN is not currently present in physical memory, an interrupt will be produced. The service routine for this interrupt will load the segment from disk to physical memory and return execution to the interrupted program. To summarize, then, the MMU translates the logical (virtual) addresses for segments used in programs to the actual physical addresses of the segments in memory. The MMU also has a mechanism for detecting if a named segment is not present in physical memory, and interrupting the program to get the segment into physical memory.

DESCRIPTORS

When a program is assembled and made ready for execution in PVAM, a unique, 8-byte quantity called a *descriptor* is produced for each segment. Figure 14-15*a* and Figure 14-15*b* show the formats for the two major types of descriptors, segment and system control. For now we will concentrate on the characteristics of segment descriptors.

The least-significant 2 bytes of a descriptor (bytes 0 and 1) contain the length, or *limit*, of the segment in bytes. This limit value is used by the MMU to produce an interrupt if an attempt is made to access a location beyond the end of a segment. Bytes 2, 3, and 4 of a descriptor contain the 24-bit base address where the segment is, or will be located in the 16 Mbyte physical address space. Byte 5 of a descriptor, the access byte, contains information about the privilege level, access, and type of the segment. To give you an idea of the kind of information contained in the access byte of a descriptor, Figure 14-15*c* summarizes the meanings of the bits in the access bytes of code segment and data segment descriptors. For now, just skim through the descriptions to get an overview. We will say more about the meanings of these bits later. Bytes 6 and 7 of a descriptor are filled with 0's for the 80286. These bytes were included in the descriptor format for upward compatibility with the 80386 microprocessor.

DESCRIPTOR TABLES

The descriptors for an 80286 system are kept in tables in memory and read into the processor as needed. There

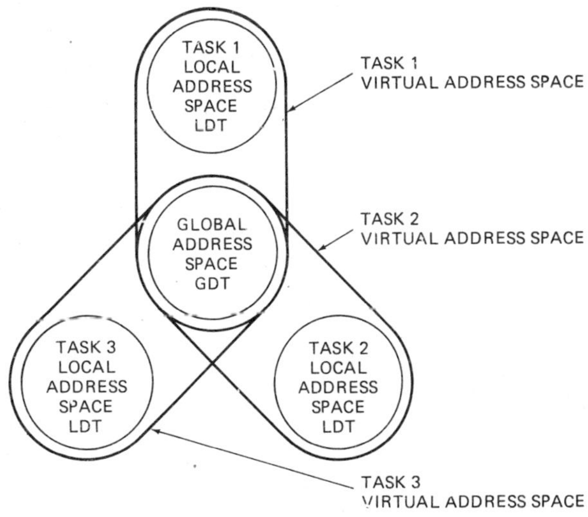

FIGURE 14-16 Diagram showing how tasks can share memory mapped by the global descriptor table, but be isolated from each other by having their own local descriptor tables. *(Intel)*

are two major categories of descriptor table, global and local. A system can have only one global descriptor table. The *global descriptor table* contains, among other things, the segment descriptors for the operating system segments and the descriptors for segments which need to be accessed by all user tasks. A *local descriptor table* is set up in the system for each task or closely related group of tasks. Figure 14-16 shows, in diagram form, how this works. Tasks share a global descriptor table and the memory area defined by the descriptors in it. Each task has its own local descriptor table and memory area defined by the descriptors in it. Because each task has its own local descriptor table, tasks are protected from each other. A four-level protection scheme which we discuss a little later can be used to protect the operating system kernel descriptors in the global descriptor table from unauthorized access by user tasks.

ACCESSING SEGMENTS

Before going on, let's review where we are at in our discussion of PVAM. The major building block of programs in PVAM are segments. Each segment has associated with it an 8-byte descriptor which contains the length, starting address, and access rights for that segment. Segment descriptors for programs are kept in a local descriptor table or in the global descriptor table in memory. To complete the general picture of how segment descriptors are used, we simply need to show you how the 80286 keeps track of where the descriptor tables are in memory, how it accesses a descriptor that corresponds to a named segment, and how it produces the physical address for a word or byte in a segment in memory.

The 80286 keeps the base addresses and limits for the descriptor tables currently in use in internal registers.

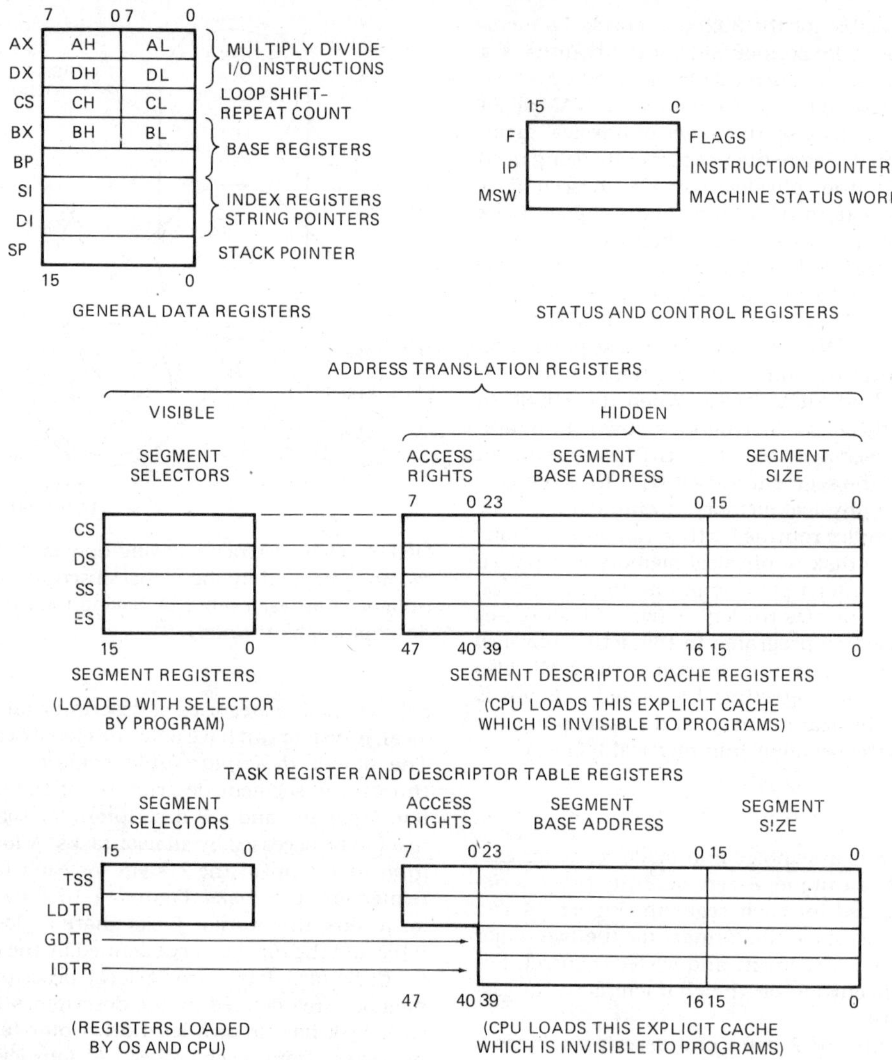

FIGURE 14-17 80286 complete register set.

Figure 14-17 shows the complete register set of the 80286. The *global descriptor table register* (GDTR) contains the 24-bit base address and limit for the table containing the global address space descriptors. This register is initialized with the LGDT instruction when the system is booted.

The *local descriptor table register* (LDTR) in the 80286 contains the base address and limit of the local descriptor table for the task currently being executed. The LLDT instruction is used to load this register when the system is booted. The LLDT instruction can only be executed by programs executing at the highest privilege level. Therefore, unless a task is operating at the highest privilege level, it cannot intentionally or maliciously access the local descriptor table of another task. Task switching is often handled by the operating system kernel which, of course, operates at the highest priority level.

Each local descriptor table is actually a named segment which has its own unique descriptor. The descriptors for the local descriptor tables in the system are kept in the global descriptor table. This sounds more complex than it really is, so let's see if we can clarify it. For example, when the operating system does a task switch, the new local descriptor table descriptor is read from the global descriptor table and loaded into the LDTR register in the 80286. As the new task executes, it then uses the descriptors in the local descriptor table pointed to by the LDT register to access the segments it needs. In later sections we will explain the function of the *task register* (TSS or TR) and the *interrupt descriptor table register* (IDTR) shown in Figure 14-17. The next step in our explanation here is to show how a descriptor in a descriptor table is accessed.

SELECTORS

When a program is assembled and prepared for execution on an 8086 or on an 80286 operating in real address mode, each named segment is given a 16-bit base

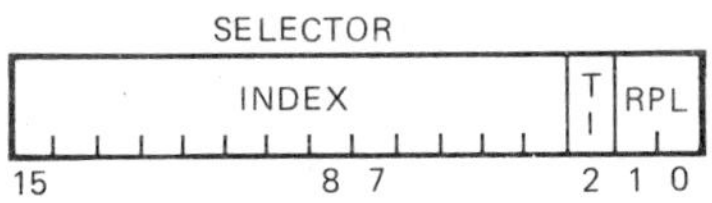

BITS	NAME	FUNCTION
1-0	REQUESTED PRIVILEGE LEVEL (RPL)	INDICATES SELECTOR PRIVILEGE LEVEL DESIRED
2	TABLE INDICATOR (TI)	TI = 0 USE GLOBAL DESCRIPTOR TABLE (GDT) TI = 1 USE LOCAL DESCRIPTOR TABLE (LDT)
15-3	INDEX	SELECT DESCRIPTOR FROM TABLE

FIGURE 14-18 Segment selector format.

address. Offsets in program instructions are added to this segment base address to produce a physical address, as we have described many times before. When a program is assembled for execution on an 80286 in PVAM, instead of being directly assigned a base address, each segment is assigned a 16-bit *selector*. Figure 14-18 shows the format for segment selectors. The upper 13 bits of a selector contain the number of the descriptor in a descriptor table. This part of the selector is referred to as an index because the value in these bits, when internally multiplied by 8, points at the descriptor for that segment in a descriptor table. (The index value is multiplied by 8 because each descriptor requires 8 bytes in the descriptor table.) If the *table indicator* bit (bit 2) of the selector is a 0, then the upper 13 bits will index a segment descriptor in the global descriptor table. If the TI bit of the selector is a 1, then the upper 13 bits of the selector will index a segment descriptor in a local descriptor table. This is a form of indirect addressing. The selector points to a descriptor location in a descriptor table, and the descriptor at that location contains the actual base address and other information about the desired segment. This may seem to be a roundabout way to get to the start of a segment, but it is this indirect approach which allows several forms of protection to be built into the 80286. The least-significant 2 bits of a selector, the *requested privilege level* or RPL bits, are part of the built-in protection. We will explain 80286 protection mechanisms after we finish discussing the basic addressing scheme.

ADDRESS TRANSLATION REGISTERS AND PHYSICAL ADDRESSES

For an 8086 or a 80286 operating in real address mode the base addresses for the currently used segments are kept in the CS, DS, SS, and ES registers in the processor, where they can be used to produce physical addresses. For the 80286 operating in PVAM, descriptors must be brought in from descriptor tables in memory to registers in the 80286, where they can be used for producing and checking physical addresses. Figure 14-17 shows the *segment address translation registers* which hold the descriptors for the currently used segments. These are usually referred to simply as segment registers with names the same as those of the 8086. However, the registers have an added 48-bit hidden part, and they are used differently.

When a segment register is loaded in a PVAM program, the selector for that segment is loaded into the the selector portion of the indicated address translation (segment) register. For example, the instructions MOV AX,ACCOUNTS_IN and MOV DS,AX will load the selector for the segment ACCOUNTS_IN into the selector portion of the DS address translation register. When the selector is loaded into the DS register, the 80286 will automatically get the descriptor pointed to in a descriptor table by that selector and put the lower 6 bytes of the descriptor in the 48-bit hidden descriptor part of the DS address translation register. This part of the DS address translation register is referred to as hidden because it cannot be directly accessed by program instructions. If the TI bit of the selector is a 0, then the index part of the selector will be multiplied by 8 and added to the global descriptor table base address in the global descriptor table register to produce the physical address where the descriptor is located. If the TI bit in the selector is a one, then the index part of the selector will be multiplied by 8 and added to the local descriptor base address held in the local descriptor table register to produce the physical address of the DS descriptor in the local descriptor table. The left side of Figure 14-19 shows how this works. The main points are that the descriptor table registers hold the 24-bit physical starting addresses for the global and local descriptor tables. The 13-bit index part of a selector is multiplied by 8 and added to one of these bases to produce the physical address of the descriptor which corresponds to that selector. When a selector is loaded into the visible part of the segment register, the 80286 automatically computes the physical address of the corresponding descriptor, and loads that descriptor into the hidden part of the segment register. The next point to consider is how the physical address for a code or data byte within a segment is produced.

Each virtual memory address for an 80286 operating in PVAM is represented by two components, a 16-bit selector and a 16-bit offset, similar to the way that 8086 addresses are represented as two 16-bit components. As we explained in the previous paragraphs, however, the selector points to a descriptor for the segment which contains that address. When the selector for a segment is loaded into the visible part of an address translation register such as DS, the lower 6 bytes of the segment descriptor will be loaded into the hidden part of the address translation register. As shown in Figure 14-19, the 24-bit physical base address for the segment is part of the descriptor that is loaded in. The actual physical address of an addressed byte or word is produced by adding the 16-bit offset part of the original virtual address directly to the segment base address in the hidden part of the address translation register. The right side of Figure 14-19 shows in diagram form how this works. NOTE: The 24-bit segment base in the address translation register is NOT shifted left four bit positions before the add as is done when producing physical addresses in an 8086.

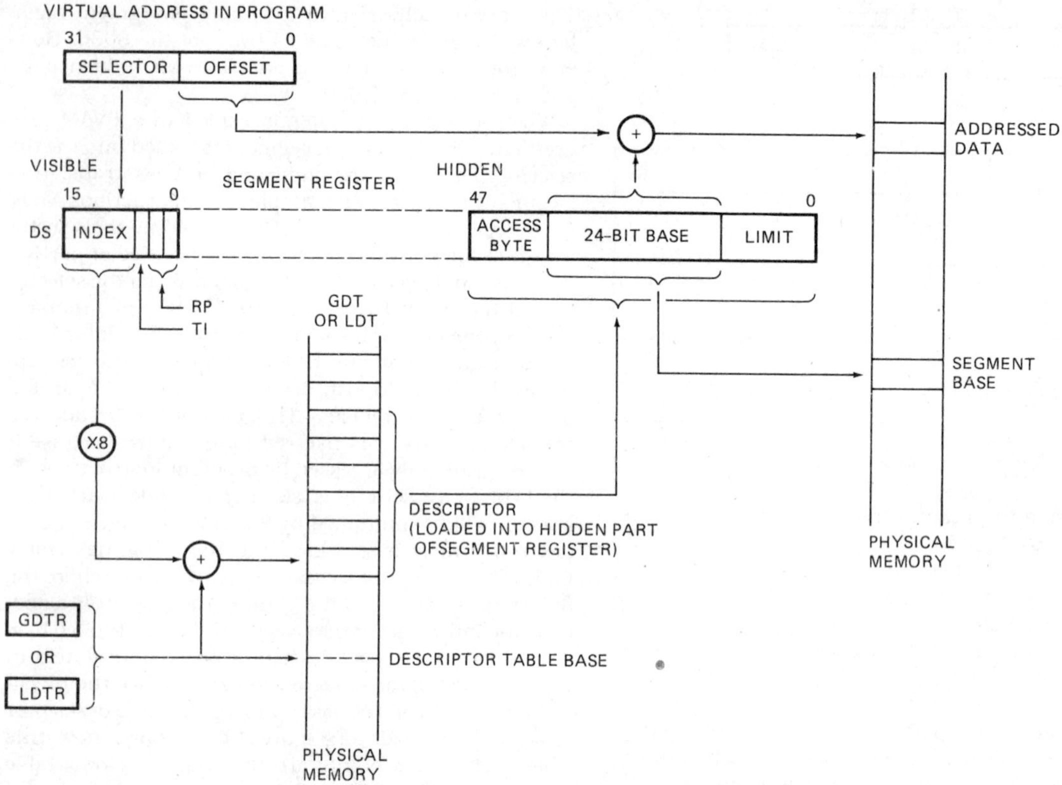

FIGURE 14-19 80286 translation of 32-bit virtual address to 24-bit physical address.

To summarize all of this, then, each address in PVAM is represented by a 32-bit virtual address consisting of a 16-bit selector and a 16-bit offset. To access a segment, the selector for that segment is loaded into the visible part of the appropriate segment register in the 80286. The 80286 then automatically multiplies the index value in the upper 13 bits of the selector by 8 and adds the result to a descriptor table base address in its GDT register or its LDT register. The 80286 then fetches the segment descriptor from the resultant address in a descriptor table and loads it into the hidden part of the segment register. The 24-bit segment base address, loaded in as part of the descriptor, is added directly to the offset part of the virtual address to produce the physical address of the desired byte or word. In a system which always runs the same program, the physical base addresses in descriptors are fixed by program development tools when the program is built. In a general-purpose system which runs many different programs, the physical base addresses in descriptors may be changed by the operating system when a program is loaded into memory to be run. This is done so that the program can be loaded into available memory without disturbing programs or tasks already present. In a dynamic system such as this, the selectors do not need to be changed, only the base addresses in the descriptors. This is one advantage of addressing segments indirectly through selectors and descriptors. In the next section we discuss how this indirect access to memory makes it possible to establish privilege levels and protection.

80286 Protection Mechanisms

The 80286 operating in PVAM has mechanisms to protect system software from user programs, protect user tasks from each other, and generally protect regions of memory from accidental access. All of these mechanisms depend in some way on the contents of descriptors. The overall principle here is that information in segment descriptors is used to check every attempt to access memory to see if the attempt is valid. If the attempt is valid, then the access is allowed. If the access is not valid, then an error interrupt is produced. Let's take another look at the descriptor formats in Figure 14-15*a* and the access rights byte definition in Figure 14-15*c*, so we can discuss how some of this protection works. First we will discuss some of the ways in which regions of memory are protected from accidental access.

When an attempt is made to load a segment selector into a segment register, the 80286 automatically makes several checks. First of all, it checks to see if the descriptor table indexed by the selector contains a valid descriptor for that selector. If a valid descriptor is not present, then an interrupt is produced. If a valid descriptor is present, the limit, base, and access rights byte of the descriptor are loaded into the hidden part of the segment register. The 80286 then checks to see if the segment for that descriptor is present in physical memory. If it is not present, as indicated by the P bit in the access rights byte of the descriptor, an interrupt is produced. The service procedure for this interrupt can swap the segment into physical memory and return execution to the interrupted program. The 80286 also checks to see if the segment descriptor is of the right type to be loaded into the specified segment register. The descriptor for a read-only data segment, for example, cannot be loaded into the SS register, because a stack must be able to be written to. A descriptor for a code segment which does not allow reading cannot be loaded into the DS register.

After a segment selector and descriptor are loaded into a segment register, further checks are made each time a location in the actual segment is accessed. An attempt to write to a code segment or a read-only data segment, for example, will cause an error. Furthermore, the limit value contained in the segment descriptor is used to check that an address produced by program instructions does not fall outside the limit defined for the segment. For example, an effective address produced by a memory reference in an instruction might, because of some error, give an offset which exceeds the limit for a segment.

User tasks are protected from each other in an 80286 PVAM system by the fact that one task cannot access descriptors in the local descriptor table for another task. When execution is switched from one task to another, the LDT register is loaded with the base address of the local descriptor table for the new task by instructions operating at a high privilege level. Instructions in the user task, operating at a lower privilege level, cannot change the contents of the LDT register.

System software, such as the operating system kernel, is protected from corruption in several ways. One way we have already mentioned is that code segments cannot be written to. The second and most important way is with privilege levels. Figure 14-20 illustrates how an 80286 PVAM system can be set up with four privilege levels. As you can see, the operating system kernel is assigned a privilege level of 0, which is the highest privilege level. Privilege level 1 might contain system services such as BIOS procedures and file handling procedures. Custom utilities and user application programs operate at lower (higher-numbered) privilege levels. The privilege level of a code or data segment is inserted as bits 5 and 6 of the access byte of the segment descriptor, when the program is built. These 2 bits are referred to as the *descriptor privilege level* or DPL. When a task executes, it executes at a specified privilege level. The privilege level of the executing task is contained in the least-significant 2 bits of the selector currently in the CS register (see Figure 14-18). This privilege level is referred to as the *current privilege level* or CPL. When an attempt is made to access, for example, a data segment by loading its selector and descriptor into the DS register, the DPL from the descriptor is compared with the CPL. If the DPL is lower (a higher number) or the same as the the CPL, the descriptor will be loaded and the access allowed. If the task attempts to access a segment with a higher DPL (lower number), an error interrupt will be produced. The point here then is that a task cannot directly access a segment which has a higher DPL.

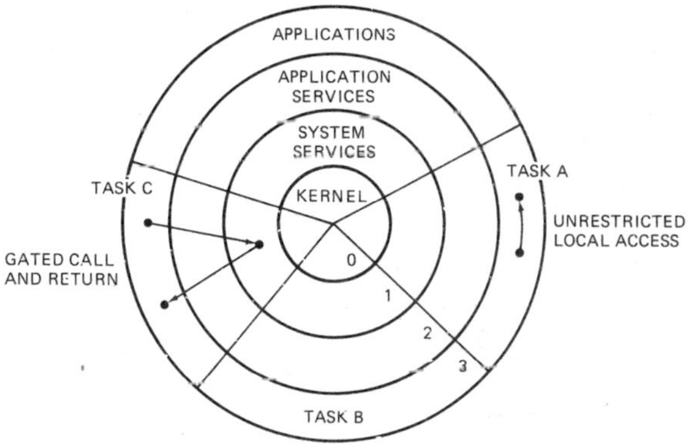

FIGURE 14-20 80286 use of privilege levels in PVAM.

The question that might come to mind at this point is, "If a task cannot access a segment with a higher DPL, how can user programs access operating system kernel, BIOS, or utilities procedures in segments which have higher DPLs?" The answer to this is that a procedure in a segment at a higher privilege level cannot be called directly, but it can be called through a special structure called a *gate*. There are four types of gates: call, trap, interrupt, and task. Figure 14-21 shows the format for gate descriptors. For now, we will just describe how a call gate operates.

As you can see in Figure 14-21, a gate is simply a special type of descriptor. Gate descriptors can be put in the GDT or a LDT, just as segment and other descriptors are. When a program does a call to a far procedure, the selector for a call gate is loaded into the CS register, and the call gate descriptor is loaded into the hidden part of the CS register. If you compare the call gate descriptor in Figure 14-21 with the segment descriptor in Figure 14-15*a*, you will see that the call gate descriptor does not directly contain the 24-bit base address of the segment containing the procedure called. The call gate descriptor instead contains a selector which points to the segment containing the called procedure. If the call is determined to be valid, then the selector from the call gate and the corresponding segment descriptor will be loaded into the CS register. The called procedure will then execute. In other words, the call is done indirectly through the call gate descriptor rather than directly through a descriptor to the segment containing the procedure.

This indirect approach permits another level of checking before access to the procedure in the higher privileged segment is allowed. The privilege level of the calling program is compared with the privilege level of that specified in the call gate. If the privilege level of the calling program is lower than the privilege level specified in the call gate, the access will not be allowed. If, for example, the DPL in the call gate descriptor is 2, a level 2 program can use the call gate to call a privilege level 1 procedure, but a level 3 program cannot. Another advantage of the indirect call gate approach is that user

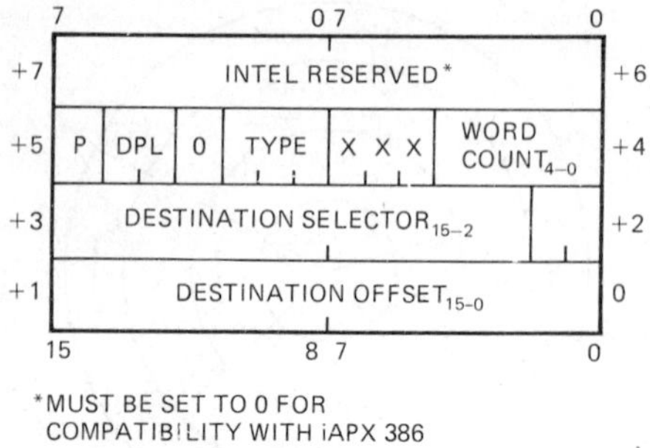

NAME	VALUE	DESCRIPTION
TYPE	4 5 6 7	CALL GATE TASK GATE INTERRUPT GATE TRAP GATE
P	0 1	DESCRIPTOR CONTENTS ARE NOT VALID DESCRIPTOR CONTENTS ARE VALID
DPL	0-3	DESCRIPTOR PRIVILEGE LEVEL
WORD COUNT	0-31	NUMBER OF WORDS TO COPY FROM CALLERS STACK TO CALLED PROCEEDURES STACK. ONLY USED WITH CALL GATE.
DESTINATION SELECTOR	16-bit SELECTOR	SELECTOR TO THE TARGET CODE SEGMENT (CALL, INTERRUPT OR TRAP GATE) SELECTOR TO THE TARGET TASK STATE SEGMENT (TASK GATE)
DESTINATION OFFSET	16-bit OFFSET	ENTRY POINT WITHIN THE TARGET CODE SEGMENT

FIGURE 14-21 80286 gate descriptor format and gate descriptor access byte format.

programs cannot accidentally enter higher privileged segments at just any old point. If they are going to enter at all, they must enter at the specific points contained in the call gate descriptors. In the next section we discuss how the 80286 does task switching and uses task gates.

Task Switching and Task Gates

Each task in a PVAM system has a 22-word *task state segment* (TSS) associated with it. A TSS holds copies of all registers and flags, the selector for the task's LDT, and a link to the TSS of the previously executing task. Descriptors for each task state segment are kept in the global descriptor table. A *task register* (TR) in the 80286 holds the selector and the task state segment descriptor for the currently executing task. The *load task register* (LTR) instruction can be used to initialize the task register to the task state segment for a particular task. During a task switch the task register is automatically loaded with the selector and descriptor for the new task.

A task switch may be done in any one of four ways:

1. A long jump or call instruction contains a selector which points at a task state segment descriptor. The call instruction is used if a return to the previously executing task is desired. A jump instruction is used if a return is not desired.
2. An IRET instruction is executed with the NT bit in the flag word equal 1. The return address is contained in the TSS.
3. The selector in a long jump or call instruction points to a task gate. In this case the selector for the destination TSS is in the task gate. The indirect mechanism here is similar to that we described above for call gates, and has the same advantages regarding privilege levels and protection.
4. An interrupt occurs, and the interrupt vector points to a task gate descriptor.

The point here is that once the task state segments and the rest of the structure are set up, normal program instructions such as CALL, JUMP, and IRET are used to accomplish a task switch. The next point we want to discuss here is how the 80286 handles interrupts in PVAM.

80286 Interrupt Handling in PVAM

An 8086 or an 80286 operating in real address mode uses the first 1 Kbyte of memory as an interrupt vector table. The CS and the IP values for up to 256 interrupt types are put in this table as we explained in Chapter 8. When operating in PVAM the 80286 can also handle up to 256 interrupts but, in order to provide protection, it is done with descriptors. An *interrupt descriptor table* (IDT) is set up in memory. The base address, access byte, and limit for the interrupt descriptor table are held in a special *interrupt descriptor table register* in the 80286. The interrupt descriptor table can be located anywhere in memory. When an interrupt occurs, the interrupt type is multiplied by 8 and used as a selector to point to the desired descriptor in the interrupt descriptor table.

The descriptors in the interrupt descriptor table are gates. They can be interrupt gates, trap gates, or task gates. Interrupt and trap gates are used to handle interrupts with procedures in the same task. A task gate is used if the interrupt handling procedure is in another task. The only difference between an interrupt gate and a trap gate is that an interrupt gate is used to go to a procedure with interrupts disabled, and a trap gate is used to get to an interrupt procedure with interrupts enabled. Interrupt gates and trap gates have the same format as the call gates described in a previous section. Incidentally, in 80286 data books the term *exception* is often used for interrupts which are caused by errors during execution of an instruction. The term interrupt is reserved for interrupts caused by some event external to the executing program.

In summary, then, the PVAM interrupt response proceeds as follows. The interrupt descriptor table register in the 80286 holds the base address of the interrupt descriptor table in memory. When an interrupt occurs, the interrupt type is multiplied by 8 and added to the

IDT register to index a gate descriptor in the IDT. The gate descriptor contains a selector for the segment which holds the interrupt service procedure, and the offset of the procedure in that segment. If the access is valid, the selector from the gate descriptor is loaded into the visible part of the CS register, and the segment descriptor pointed to by the selector is loaded into the hidden part of the CS register. This is essentially the same indirect process that we described for the operation of call gates.

As you should see by now, the key to understanding the 80286 PVAM is to think of segments, descriptors to define the characteristics of the segments, selectors to point at desired segment descriptors, and 80286 registers to hold the selectors and descriptors for the descriptor tables and segments currently in use. Now that you have an overview of how the 80286 operates in PVAM, we want to briefly introduce you to the 80286 instructions which are used for PVAM.

80286 Instructions for PVAM

The following brief descriptions are intended to introduce you to the instructions that the 80286 has beyond those of the 80186. For further details, consult Intel's 80286 Assembly Language Programming Manual.

CTS—Clear task switched flag in machine status word

LGDT—Load global descriptor table register from memory

SGDT—Store global descriptor table register contents in memory

LIDT—Load interrupt descriptor table register from memory

LLDT—Load selector and associated descriptor into LDTR

SLDT—Store selector from LDTR in specified register or memory

LTR—Load task register with selector and descriptor for TSS

STR—Store selector from task register in register or memory

LMSW—Load machine status register from register or memory

SMSW—Store machine status word in register or memory

LAR—Load access rights byte of descriptor into register or memory

LSL—Load segment limit from descriptor into register or memory

ARPL—Adjust requested privilege level of selector (down only)

VERR—Determine if segment pointed to by selector is readable

VERW—Determine if segment pointed to be selector is writable

In the remaining sections of this chapter we show you some of the directions microprocessor evolution is heading beyond the 80286.

NEW DIRECTIONS

Microprocessor evolution has been proceeding very rapidly in the last few years, and the rate of evolution seems to be increasing. Throughout this book we have tried to point out some of the directions this evolution has been taking. One area of evolution has been from batch-processing computer systems to time-share and multitasking systems. Another direction has been to distributed processing systems linked together in networks such as we described in Chapter 13. Also, the development of optical disk storage makes available, at each user's desk, more data than was previously available at many large mainframe computers. The overall direction of evolution is toward microcomputers with greater screen resolution, more memory capability, larger data words, and higher processing speeds. We will use the remainder of this chapter to introduce you to some developing areas: the Intel 80386 32-bit microprocessor, parallel processing, RISC machines, and optical computers.

The Intel 80386 32-bit Microprocessor

The 80386 microprocessor is a logical extension of the 80286 we discussed in the preceding section to a 32-bit machine. Remember that the 80286 in PVAM can address up to 8192 virtual memory segments of 64 Kbytes each. The 80286 then has a virtual address space of 1 gigabyte. Virtual memory segments are swapped in from disk storage to as much as 16 Mbytes of physical memory, addressed with 24 address lines. In order to operate with maximum efficiency, a processor working with virtual memory should have an addressing capability equivalent to the size of the secondary storage. As the capacity of optical disk storage units passes the 10-gigabyte mark, processors with greater addressing range than the 80286 are needed. Also, in the eternal quest for speed, processors which use more pipelining and work with wider data words are needed. Here's how the 80386 attempts to answer these needs.

First of all, the 80386 is more highly pipelined than the 80286. Figure 14-22 shows a block diagram of the functional processors within the 80386. Instruction fetching, instruction decoding, instruction execution and memory management are all carried out in parallel. As shown in Figure 14-12, the 80386 instruction set is a superset of that of the other members of the 8086 family. Figure 14-23 shows the register set of the 80386. All of the general-purpose registers can be used for bytes, words, or long words. Long registers are referred to in instructions by simply putting an L in front of the register name, as in the instruction MOV LAX, 7FF0845BH.

Secondly, the 80386 has a 32-bit data bus and produces a full 32-bit nonmultiplexed address bus. This

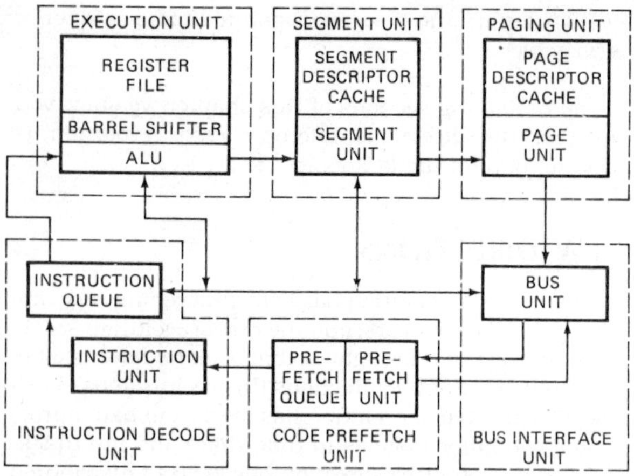

FIGURE 14-22 Intel 80386 internal block diagram showing separate processing units.

32-bit address bus allows it to address 2^{32}, or about 4 gigabytes of physical memory. The 80386 memory management unit allows it to address 2^{46}, or about 64 terabytes of virtual memory. The 80386 can be operated in one of two memory management modes, *paged mode* and *nonpaged mode*. In nonpaged mode the MMU operates very similarly to that of the 80286. Virtual addresses are represented with a selector component and an offset component as they are with the 80286. The selector component is used to index a descriptor in a descriptor table. The descriptor contains the 32-bit physical base address for the desired segment. The offset part of the virtual address is added to the base address from the descriptor to produce the actual physical address as we described in previous sections. The offset part of a virtual address can be 16 or 32 bits, so segments can be as large as 4 gigabytes, rather than being limited to 64 Kbytes as they are for the 80286. In nonpaged mode, named segments are swapped into physical memory as needed. For very large programs, a pure segment approach is not efficient because the whole segment has to be swapped in, even if only a very small part of the segment is needed. To solve this problem, the 80386 can be operated in paged mode.

When operated in paged mode, the 80386 switches the paging unit in after the segment unit. The paging unit allows memory pages of perhaps 4 Kbytes each to be swapped in and out from disk. The principle here is that only the pages that are actually needed from a large segment have to be swapped in. This type of operation is often called *demand-paged virtual memory*. The 80386 paging unit inserts another layer of address computation after adding a base address from a descriptor and an offset. Figure 14-24 shows how this works.

The 32-bit address produced by adding a base address from a descriptor and an offset is referred to as a *linear address*. Components of this linear address are used as pointers to two levels of paging tables. The first level is the page table directory which contains the base addresses of all of the page tables for the system. The second level is the actual page tables. Page tables contain the addresses of each page in the system.

To summarize the operation of an 80386 in paging mode, then, task and memory protection are provided by descriptor tables and the segment unit, just as they are for an 80286 operating in PVAM. The paging unit allows small, fixed-length pages to be swapped into physical memory as needed. In systems with very large segments this is more efficient than swapping in an entire segment. When coupled with optical disk storage, devices such as the 80386 will make it possible to have much of the power of a mainframe computer on your desk. In the next section we discuss how several proces-

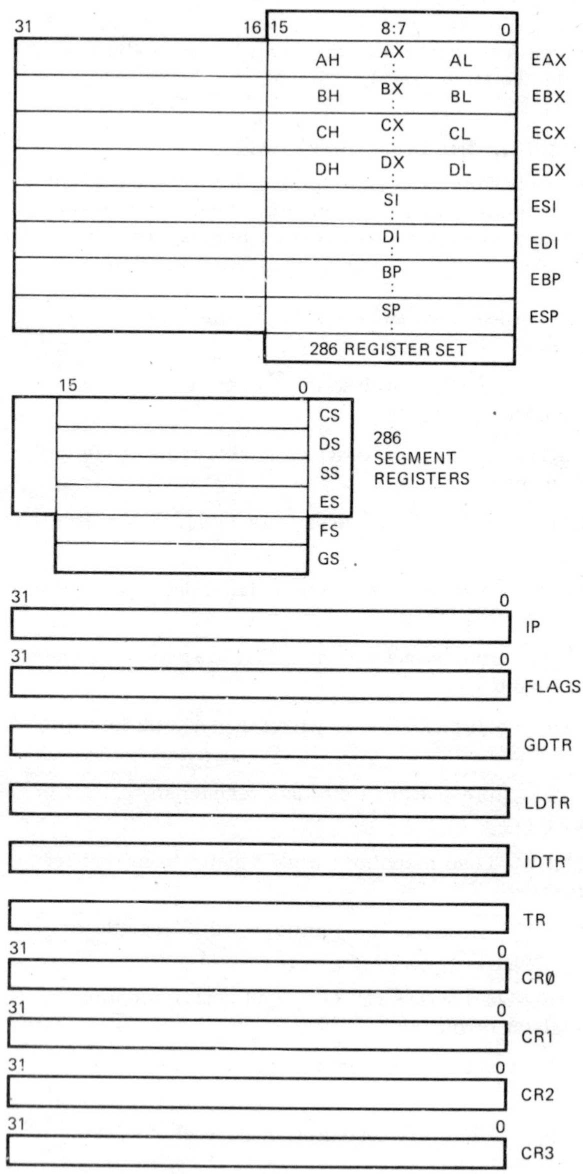

FIGURE 14-23 Intel 80386 registers.

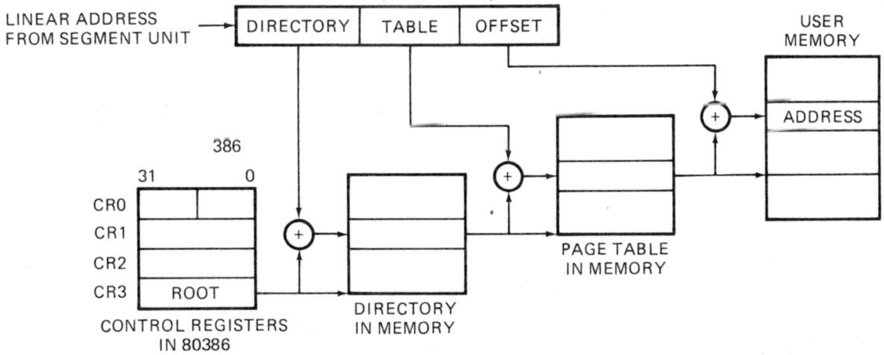

FIGURE 14-24 Linear address from 80386 segment unit is converted to 32-bit physical address by two-level paging mechanism. *(Intel)*

sors such as the 80286 or 80386 can be connected in parallel to produce a "supercomputer."

Parallel Processing

Some computer jobs such as analyzing weather data, modelling the response of complex drugs, or creating the graphics for high-tech science fiction movies require a type of computer commonly called a *supercomputer*. These supercomputers typically work with 64-bit data words, address large amounts of memory, and execute hundreds of millions of instructions per second. The processing speed of these supercomputers is usually expressed in millions of instructions per second (MIPS) or in millions of floating-point operations per second (megaflops). An example of a floating-point operation is adding together two numbers expressed in floating-point form. One current supercomputer, the X-MP2 from Cray Research, Inc. is capable of about 500 megaflops. Depending on configuration, the X-MP2 costs between $9 million and $12 million. The high price of supercomputers is caused by the fact that in order to achieve their great speed, they have to use large quantities of expensive, state-of-the-art discrete components. Less expensive LSI components are not nearly fast enough for a supercomputer with a traditional one- or two-processor architecture. One solution to this problem is to build a system using many LSI processors which operate in parallel or *concurrently*. Each processor can then work on a part of the overall problem that the computer is analyzing.

There are several different ways of connecting processors in parallel. In Chapter 11 we discussed how several microprocessors can operate in parallel on an Intel Multibus. The difficulty with a simple bus structure such as this is that processors compete for shared resources such as memory. If one processor is using the bus, others must wait. This slows down the overall processing speed. One of the more efficient multiprocessor architectures is the hypercube topology developed by Seitz and Fox at Caltech. A diagram of this topology is shown in Figure 14-25. Each node in the system consists of a complete processing unit with the ability to communicate with other units. The number of nodes can be expanded to give the power and speed needed to handle the problem the computer is being used to solve. Each processor unit is typically connected to its nearest neighbors as shown.

Intel has produced the iPSC family of commercial products based on the hypercube topology. The three currently available versions have 32, 64, and 128 nodes. Figure 14-26 shows the components contained on the processor board for each node. Each node is a complete microcomputer with an 80286 processor, 80287 math coprocessor, 500 Kbytes of RAM, 64 Kbytes of ROM, and interface circuitry. The procesor board also has an Intel 82586 Ethernet coprocessor to control communications with other nodes. Each processor has seven 10 Mbit/s lines to communicate with other processors and one 10 Mbit/s line to communicate with a central controller. The Intel systems use an Intel 286/310 minicomputer as the central controller for the hypercube. The advantage of this structure is that each processor has enough

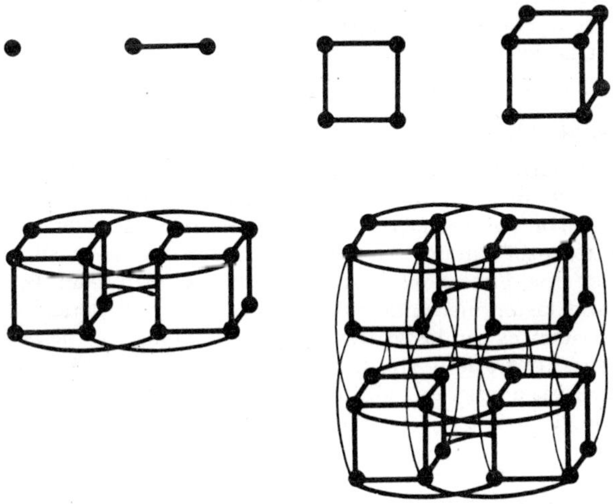

FIGURE 14-25 Hypercube multiprocessor topologies for 1 to 32 nodes.

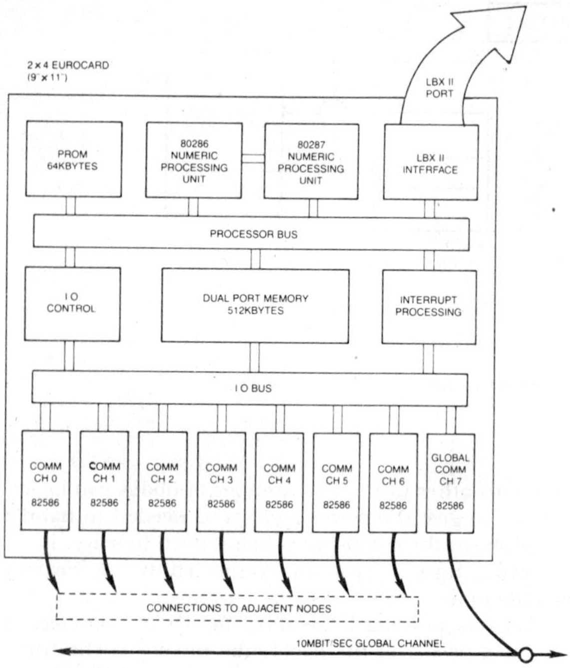

FIGURE 14-26 Block diagram of Intel iPSC hypercube node processor board.

memory to operate independently, and communication between processors can take any one of several routes, instead of being limited to a single bus. Current systems operate at 2–10 Mflops, which puts them in the low end of the supercomputer range. However, because common LSI components are used, the cost is much less than that of an equivalent single-processor supercomputer. Adding more nodes should produce faster systems in the future because parallel processors eliminate much of the "bottleneck" caused by a single serial processor. Another method currently being developed to speed up the operation of processors is to streamline their instruction set.

RISC Machines

The term RISC stands for *reduced instruction set computer*. By designing a microprocessor instruction set with only simple logical and arithmetic instructions, the processor can operate faster. There are several reasons for this. First of all, fewer instructions mean a simpler and faster instruction decoder. Secondly, instruction sequences can be written to most efficiently do the desired operation. The tradeoff here, of course, is that writing a program requires more work on the part of the programmer.

Optical Computers

So far in this book the microcomputer devices we have discussed use electrical currents or voltage levels to represent logic levels. In the final section of this chapter we want to introduce you to experimental computers which use light beams to represent logic levels and switch logic devices on and off. The basic principle is to let a light beam represent a logic 1 and no beam represent a logic 0 just as is done in simple fiber-optic digital signal transmission systems. Logic gates in an optical computer transmit a beam when switched on and block the light beam when switched off. The logic gates themselves are switched on and off by a light beam shining on them. In other words, logic levels are represented by light or no light, logic gate switches are controlled by the presence or absence of a light beam, and the connecting links between optical logic gates are light beams.

One advantage of optical logic gates and computers is that signals can easily be sent to many elements in parallel. This may lead to their use in parallel processor systems. Another major advantage of optical logic gates is their switching speeds. Even though current optical logic gates are quite primitive, switching speeds are in the picosecond range. Optics researchers believe that switching speeds of a few femtoseconds (10^{-15} seconds) are possible. Optical computers may be able to run with clocks of several hundred MHz. A major disadvantage of current optical devices is the relatively large amount of power they require. Hopefully, further research will realize the potential of this technology.

EPILOGUE

This book has only been able to show you a small view of where microcomputer electronics is and where it seems to be evolving. Hopefully it has given you enough of a start that you can proceed on your own and enjoy playing a part in the continuing evolution. As you are faced with learning some new and seemingly difficult material, remember the 5-minute rule and the old saying "Grapevines and people bear the best fruit on new growth."

IMPORTANT TERMS AND CONCEPTS FROM THIS CHAPTER

Multiuser, multitasking

Scheduler

Time-slice and preemptive priority-based scheduling

Semaphore

Process control block

Deadlock

Critical region

Application programs

Utilities

Memory management

Overlay

Bank switching

Descriptor table

Virtual memory

Demand-paged virtual memory

Cache

UNIX

Kernel
- process and user tables
- child process
- parent process
- inode number
- signal
- pipe

Shell
- standard input and output, error output
- redirection
- spooling

RMX 86
- nucleus
- modules
- basic I/O systems
- extended I/O systems
- human interface
- universal development interface

Objects
- tasks, jobs, segments, mailboxes, regions, semaphores

Task states

80286
- bus, instruction, execution, and address units
- real address mode
- protected virtual address mode (PVAM)
- descriptor
- global and local descriptor tables
- selector
- segment address translation register (segment register)
- privilege level
- descriptor and current privilege levels
- task state segment
- interrupt, trap, and task gates

80386
- paged and nonpaged mode
- demand-paged virtual memory

Parallel processing
- parallel or concurrent operation
- hypercube topology

RISC machine

Optical computer

5-minute rule

Grapevine

REVIEW QUESTIONS AND PROBLEMS

1. List and briefly describe the two types of scheduling commonly used in multiuser/multitasking operating systems.
2. Suppose that two users in a time-share computer system each want to print out a file. How can the system be prevented from printing lines from one file between lines of the other file?
3. Define the term "deadlock" and describe one way it can be prevented.
4. Define the term "critical region" and show with 8086 assembly language instructions how a semaphore can be used to protect a critical region.
5. The UNIX operating system is set up as a "three-layer" operating system. What is the major reason why it is configured as layers? Identify and describe the function of each of the three layers.
6. Describe how an overlay scheme is used to run programs such as compilers which are too large to be loaded into physical memory all at once.
7. Define the term "virtual memory," and use Figure 14-4 to help you briefly describe how a logical address is converted to a physical address by a memory management unit using a descriptor table. What action will the MMU take if it finds that a requested segment is not present in physical memory? What is another major advantage of the indirect addressing provided by descriptor tables, besides the ability to address a large amount of virtual memory?
8. How does a UNIX scheduler determine which active process to service next?
9. Define the term "hierarchial file structure" as used in the UNIX operating system. What is the advantage of this type file structure over a simple list type?
10. In a UNIX system, input or output can be "redirected." Explain briefly what this means.
11. What is meant by the term "piping" in a UNIX system. What symbol is used to indicate a pipe?
12. A programmer was heard to say that she "sent the file off to the print spooler before going to lunch." What did she mean by this statement?
13. For what types of applications was the RMX 86 operating system designed? Compare the scheduling method of RMX 86 with that of UNIX.

14. List the four major processing units in an 80286 microprocessor and briefly describe the function of each.

15. The data sheet for a computer which uses the 80286 as its CPU indicates that the 80286 is operated in its "real address mode." What does this mean?

16. How is an 80286 switched from real address mode to protected virtual address mode operation? How can it be switched back to real address mode operation?

17. Explain the term "virtual memory." How much virtual memory can an 80286 address? How much physical memory can an 80286 address?

18. Why is the length of the segment included in the descriptor for the segment? How does the 80286 keep track of where the global descriptor table and the currently used local descriptor table are located in memory?

19. Give the names of the two parts of an 80286 PVAM virtual address. Using Figure 14-19, describe how a 32-bit virtual address for a data segment location in a task's local memory is translated to the actual 24-bit physical address. How would this translation be different if the memory location were in the global memory area?

20. How are tasks in an 80286 system protected from each other?

21. How can operating system kernel procedures and data be protected from access by application programs in an 80286 system?

22. In an 80286 system a task operating at a level 2 privilege can in a special way call a procedure at a higher privilege level. Describe briefly the mechanism that is used to make this access.

23. The 80286 maintains a task state segment for each active task in a system. How are these task state segments accessed?

24. List three major advances that the 80386 microprocessor has over the 80286.

25. What are the major advantages of using parallel processors such as is done in the Intel hypercube, instead of using a single fast processor.

26. What factor makes optical computers an inviting technology?

BIBLIOGRAPHY

Because of the technical level of this book, the major sources of further information on the topics discussed are manufacturers' data books, application notes, and articles in current engineering periodicals. With the foundation provided here, you should be able to comfortably read these materials. Listed below, by chapter, are some materials you may find helpful. Following the chapter listings is a list of periodicals which we have found to be particularly helpful in keeping up with the latest advances in microcomputer evolution and applications. Most of these periodicals have articles which review the basic principles of a particular area of electronics and then discuss the latest developments in that area.

Chapter 1

Hall, Douglas V., *Microprocessors and Digital Systems*, 2d ed., McGraw-Hill, Inc., New York, 1983.

Chapters 2–6

Mick, John, and Jim Brick, *Bit-Slice Microprocessor Design*, McGraw-Hill, Inc., New York, 1980.

Myers, Glendford J., *Digital System Design with LSI Bit-Slice Logic*, John Wiley and Sons, New York, 1980.

A History of Microprocessor Development at Intel, Intel Article Reprint/AR-173, Intel Corporation, Santa Clara, Calif.

8086/8087/8088 Macro Assembly Language Reference Manual for 8086-Based Development Systems, Intel Corporation, Santa Clara, Calif., 1980.

Microsystem Components Handbook Microprocessors and Peripherals, vols. 1 and 2, Intel Corporation, Santa Clara, Calif., 1985.

iAPX 86/88, 186/188 User's Manual Programmer's Reference, Intel Corporation, Santa Clara, Calif. 1983.

Staff of Microsoft, *Macro Assembler*, IBM Corp., 1981.

Chapters 7 and 8

Microsystem Components Handbook Microprocessors and Peripherals, vols. 1 and 2, Intel Corporation, Santa Clara, Calif., 1985.

SDK-86 MCS-86 System Design Kit User's Guide, Intel Corporation, Santa Clara, Calif., 1981.

8086 System Design, Application Note, Intel Corporation/AP-67, Intel Corporation, Santa Clara, Calif.

Chapter 9

Microsystem Components Handbook Microprocessors and Peripherals, vols. 1 and 2, Intel Corporation, Santa Clara, Calif. 1985.

Technical Reference Manual, IBM Corp., 1983.

SC-01 Speech Synthesizer Data Sheet, Votrax, Troy, Mich., 1980.

Dorf, Richard C., *Robotics and Automated Manufacturing*, Reston Publishing Company, Reston, Va., 1983.

Hall Effect Manual, 2d ed., Helipot Division of Beckman Instruments, Inc., Fullerton, Calif. 1964.

AMF Potter & Brumfield Catalog, Potter & Brumfield, Princeton, Ind., 1984.

Optoelectronics Designer's Catalog, Hewlett-Packard, Palo Alto, Calif., 1985.

Interfacing Liquid Crystal Displays in Digital Systems, Application Note AN-B, Beckman Instruments, Inc., Scottsdale, Ariz., 1980.

Hot Ideas in CMOS Data Book, GE-Intersil, Cupertino, Calif., 1984.

Optoelectronics Device Data Book, DL118R1, Motorola Semiconductor Products Inc., Phoenix, Ariz. 1983.

Power Mosfet Selector Guide and Cross Reference, SG56R4, Motorola Semiconductor Products Inc., Phoenix, Ariz., 1983.

Sandhu, H.S., *Hands-On-Introduction to ROBOTICS—The Manual for the XR-Series Robots*, Rhino Robots, Champaign, Ill., 1983.

Slo-Syn DC Stepping Motors Catalog, DCM1078, Superior Electric Company, Bristol, Conn., 1979.

Auslander, David M., and Paul Sagues, *Microprocessors for Measurement and Control*, Osborne/McGraw-Hill, Berkeley, Calif., 1981.

Allocca, John A., and Allen Stuart, *Transducers Theory and Applications*, Reston Publishing Company, Inc., Reston, Va., 1984.

Seippel, Robert G., *Transducers, Sensors, and Detectors*, Reston Publishing Company, Inc., Reston, Va., 1983.

Johnson, Curtis D., *Process Control Instrumentation Technology*, 2d ed., John Wiley & Sons, New York, 1982.

Hunter, Ronald P., *Automated Process Control Systems Concepts and Hardware*. Prentice-Hall, Englewood Cliffs, N.J., 1978.

Sheingold, Daniel H. (ed), *Transducer Interfacing Handbook —A Guide to Analog Signal Conditioning*, Analog Devices, Inc., Norwood, Mass. 1981.

Analog Devices 3B Industrial Control Series Data Sheet, Analog Devices, Inc., Norwood, Mass., 1981.

Chapter 11

8086/8087/8088 Macro Assembly Language Reference Manual for 8086-Based Development Systems, Intel Corporation, Santa Clara, Calif. 1980.

iAPX 86/88, 186/188 User's Manual Programmer's Reference, Intel Corporation, Santa Clara, Calif., 1983.

Error Detecting and Correcting Codes, Application Note AP-46, Intel Corporation, Santa Clara, Calif., 1979.

Introduction to the 80186 Microprocessor, Application Note AP-186 Intel Corporation, Santa Clara, Calif., 1983.

Multibus Handbook, Intel Corporation, Santa Clara, Calif., 1983.

Multibus Interfacing, Application Note AP-28A, Intel Corporation, Santa Clara, Calif., 1979.

Getting Started With the Numeric Data Processor, Application Note AP-113, Intel Corporation, Santa Clara, Calif., 1981.

Chapter 12

Raster Graphics Handbook, Conrac Corporation, Covina, Calif., 1980.

Lesea, Austin, and Rodnay Zaks, *Microprocessor Interfacing Techniques*, 2d ed., Sybex Inc., Berkeley, Calif., 1977.

CRT Terminal Design Using the Intel 8275 and 8279, Application Note AP-32, Intel Corporation, Santa Clara, Calif., 1977.

An Intelligent Data Base System Using the 8272, Application Note AP-116, Intel Corporation, Santa Clara, Calif., 1981.

Intel SBC 202 Double Density Diskette Controller Hardware Reference Manual, Intel Corporation, Santa Clara, Calif., 1977.

Chapter 13

LAN Components User's Manual, Intel Corporation, Santa Clara, Calif., 1984.

Data Acquisition Telecommunications Local Area Networks 1983 Data Book, Advanced Micro Devices, Inc., Sunnyvale, Calif., 1983.

Data Communications—A User's Handbook, Racal-Vadic, Sunnyvale, Calif.

Stallings, William, *Local Networks—An Introduction*, Macmillan, New York, 1984.

Friend, George E., et al., *Understanding Data Communications*, Texas Instruments, Dallas, 1984.

Fike, John L., and George E. Friend, *Understanding Telephone Electronics*, Texas Instruments, Dallas, 1984.

McNamara, John E., *Technical Aspects of Data Communication*, Digital Equipment Corporation, Maynard, Mass., 1977.

Smartmodem 1200 Hardware Reference Manual, Hayes Microcomputer Products, Inc., Norcross, Ga., 1983.

AM7911 FSK Modem Product Specification, Advanced Micro Devices, Sunnyvale, Calif., 1983.

EIA Standard RS-422, Electrical Characteristics of Balanced Voltage Digital Interface Circuits, Electronic Industries Association, Engineering Department, Washington, 1975.

EIA Standard RS-232-C, Interface Between Data Terminal Equipment and Data Communication Equipment Employing Serial Binary Data Interchange, Electronic Industries Association, Engineering Department, Washington, 1969.

Using the 8273 SDLC/HDLC Protocol Controller, Application Note AP-36, Intel Corporation, Santa Clara, Calif., 1978.

Essentials of Data Communications, Tektronix Inc., Beaverton, Ore., 1978.

Designer's Guide to Fiber Optics, EDN Magazine, Cahners Publishing Company, Boston, 1978.

Fiber Optic Applications in Electrical Substations, IEEE 83TH0104 PWR, IEEE Service Center, Piscataway, N.J., 1983.

Improving Measurements in Engineering and Manufacturing - HP-1B, Hewlett-Packard, Palo Alto, Calif., 1976.

Using the 8292 GPIB Controller, Application Note AP-66, Intel Corporation, Santa Clara, Calif., 1980.

The AM7990 Family Ethernet Node, Advanced Micro Devices, Sunnyvale, Calif., 1983.

Chapter 14

iRMX 86 Introduction and Operator's Reference Manual, Intel Corporation, Santa Clara, Calif., 1984.

Kaisler, Stephen H., *The Design of Operating Systems for Small Computer Systems*, John Wiley & Sons, New York, 1983.

Multiprogramming with the iAPX 88 and iAPX 86 Microsystems, Application Note AP-106, Intel Corporation, Santa Clara, Calif., 1980.

ASM286 Assembly Language Reference Manual, Intel Corporation, Santa Clara, Calif., 1983.

Gauthier, Richard: *Using the UNIX System*, Reston Publishing Company, Inc., Reston, Va., 1981.

Lumuto, Ann Nicols and Nico Lomuto: *A UNIX Primer*, Prentice-Hall, Englewood Cliffs, N.J., 1983.

Christian, Kaare, *The UNIX Operating System*, John Wiley & Sons, New York, 1983.

Introduction to the iAPX 286, Intel Corporation, Santa Clara, Calif., 1985.

iAPX 286 Operating Systems Writer's Guide, Intel Corporation, Santa Clara, Calif., 1983.

iAPX 286 Hardware Reference Manual, Intel Corporation, Santa Clara, Calif., 1983.

iAPX 286 Programmer's Reference Manual, Intel Corporation, Santa Clara, Calif., 1983.

iAPX 386 High Performance 32-bit Microcomputer Product Preview, Advance Information, Intel Corporation, Santa Clara, Calif.

Periodicals

BYTE. ISSN 0360-5280. Byte Publications, Inc., 70 Main Street, Peterborough, New Hampshire 03458.

EDN. ISSN 0012-7515. Cahners Publishing Co., Boston, Mass.

Electronic Design. USPS-172-080. Hayden Publishing Co., Rochelle Park, N.J.

Electronics. ISSN 0013-5070, McGraw-Hill, Inc., New York.

Instruments and Control Systems. ISSN 0164-0089. Chilton Company, Radnor, Pa.

Mini-Micro Systems. ISSN 0364-9342. Cahners Publishing Co. Boston, Mass.

Electronic Engineering Times. ISSN 0192-1541. Electronic Engineering Times, Manhasset, N.Y.

intel

iAPX 86/10
16-BIT HMOS MICROPROCESSOR

8086/8086-2/8086-1

- **Direct Addressing Capability 1 MByte of Memory**
- **Architecture Designed for Powerful Assembly Language and Efficient High Level Languages.**
- **14 Word, by 16-Bit Register Set with Symmetrical Operations**
- **24 Operand Addressing Modes**
- **Bit, Byte, Word, and Block Operations**
- **8 and 16-Bit Signed and Unsigned Arithmetic in Binary or Decimal Including Multiply and Divide**
- **Range of Clock Rates:**
  5 MHz for 8086,
  8 MHz for 8086-2,
  10 MHz for 8086-1
- **MULTIBUS™ System Compatible Interface**
- **Available in EXPRESS**
  - Standard Temperature Range
  - Extended Temperature Range

The Intel iAPX 86/10 high performance 16-bit CPU is available in three clock rates: 5, 8 and 10 MHz. The CPU is implemented in N-Channel, depletion load, silicon gate technology (HMOS), and packaged in a 40-pin CerDIP package. The iAPX 86/10 operates in both single processor and multiple processor configurations to achieve high performance levels.

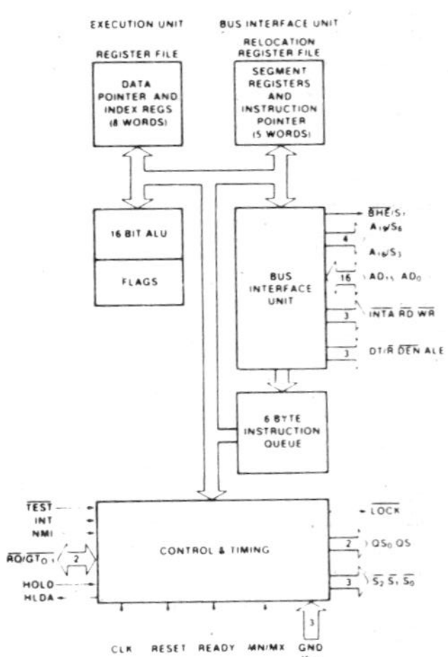

Figure 1. iAPX 86/10 CPU Block Diagram

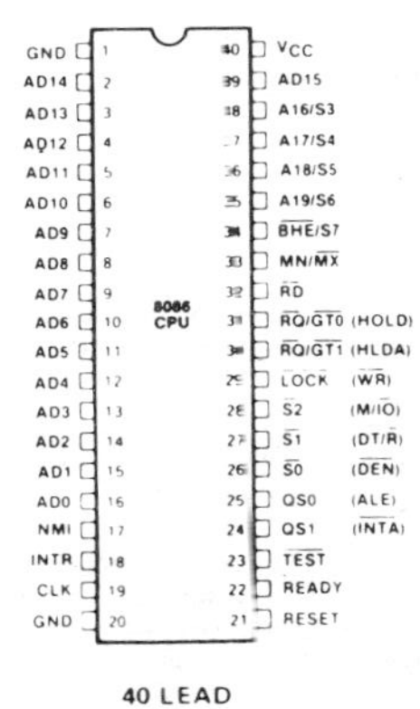

Figure 2. iAPX 86/10 Pin Configuration

Table 1. Pin Description

The following pin function descriptions are for iAPX 86 systems in either minimum or maximum mode. The "Local Bus" in these descriptions is the direct multiplexed bus interface connection to the 8086 (without regard to additional bus buffers).

Symbol	Pin No.	Type	Name and Function
AD_{15}-AD_0	2-16, 39	I/O	**Address Data Bus:** These lines constitute the time multiplexed memory/IO address (T_1) and data (T_2, T_3, T_W, T_4) bus. A_0 is analogous to $\overline{BHE}$ for the lower byte of the data bus, pins D_7-D_0. It is LOW during T_1 when a byte is to be transferred on the lower portion of the bus in memory or I/O operations. Eight-bit oriented devices tied to the lower half would normally use A_0 to condition chip select functions. (See $\overline{BHE}$.) These lines are active HIGH and float to 3-state OFF during interrupt acknowledge and local bus "hold acknowledge."
A_{19}/S_6, A_{18}/S_5, A_{17}/S_4, A_{16}/S_3	35-38	O	**Address/Status:** During T_1 these are the four most significant address lines for memory operations. During I/O operations these lines are LOW. During memory and I/O operations, status information is available on these lines during T_2, T_3, T_W, and T_4. The status of the interrupt enable FLAG bit (S_5) is updated at the beginning of each CLK cycle. A_{17}/S_4 and A_{16}/S_3 are encoded as shown. A_{17}/S_4 / A_{16}/S_3 / Characteristics: 0 (LOW) / 0 / Alternate Data 0 / 1 / Stack 1 (HIGH) / 0 / Code or None 1 / 1 / Data S_6 is 0 (LOW) This information indicates which relocation register is presently being used for data accessing. These lines float to 3-state OFF during local bus "hold acknowledge."
$\overline{BHE}/S_7$	34	O	**Bus High Enable/Status:** During T_1 the bus high enable signal ($\overline{BHE}$) should be used to enable data onto the most significant half of the data bus, pins D_{15}-D_8. Eight-bit oriented devices tied to the upper half of the bus would normally use $\overline{BHE}$ to condition chip select functions. $\overline{BHE}$ is LOW during T_1 for read, write, and interrupt acknowledge cycles when a byte is to be transferred on the high portion of the bus. The S_7 status information is available during T_2, T_3, and T_4. The signal is active LOW, and floats to 3-state OFF in "hold." It is LOW during T_1 for the first interrupt acknowledge cycle. $\overline{BHE}$ / A_0 / Characteristics: 0 / 0 / Whole word 0 / 1 / Upper byte from/to odd address 1 / 0 / Lower byte from/to even address 1 / 1 / None
$\overline{RD}$	32	O	**Read:** Read strobe indicates that the processor is performing a memory of I/O read cycle, depending on the state of the S_2 pin. This signal is used to read devices which reside on the 8086 local bus. $\overline{RD}$ is active LOW during T_2, T_3 and T_W of any read cycle, and is guaranteed to remain HIGH in T_2 until the 8086 local bus has floated. This signal floats to 3-state OFF in "hold acknowledge."
READY	22	I	**READY:** is the acknowledgement from the addressed memory or I/O device that it will complete the data transfer. The READY signal from memory/IO is synchronized by the 8284A Clock Generator to form READY. This signal is active HIGH. The 8086 READY input is not synchronized. Correct operation is not guaranteed if the setup and hold times are not met.
INTR	18	I	**Interrupt Request:** is a level triggered input which is sampled during the last clock cycle of each instruction to determine if the processor should enter into an interrupt acknowledge operation. A subroutine is vectored to via an interrupt vector lookup table located in system memory. It can be internally masked by software resetting the interrupt enable bit. INTR is internally synchronized. This signal is active HIGH.
$\overline{TEST}$	23	I	**$\overline{TEST}$:** input is examined by the "Wait" instruction. If the $\overline{TEST}$ input is LOW execution continues, otherwise the processor waits in an "Idle" state. This input is synchronized internally during each clock cycle on the leading edge of CLK.

Table 1. Pin Description (Continued)

Symbol	Pin No.	Type	Name and Function
NMI	17	I	**Non-maskable interrupt:** an edge triggered input which causes a type 2 interrupt. A subroutine is vectored to via an interrupt vector lookup table located in system memory. NMI is not maskable internally by software. A transition from a LOW to HIGH initiates the interrupt at the end of the current instruction. This input is internally synchronized.
RESET	21	I	**Reset:** causes the processor to immediately terminate its present activity. The signal must be active HIGH for at least four clock cycles. It restarts execution, as described in the Instruction Set description, when RESET returns LOW. RESET is internally synchronized.
CLK	19	I	**Clock:** provides the basic timing for the processor and bus controller. It is asymmetric with a 33% duty cycle to provide optimized internal timing.
V_{CC}	40		V_{CC}: +5V power supply pin.
GND	1, 20		**Ground**
MN/$\overline{MX}$	33	I	**Minimum/Maximum:** indicates what mode the processor is to operate in. The two modes are discussed in the following sections.

The following pin function descriptions are for the 8086/8288 system in maximum mode (i.e., MN/$\overline{MX}$ = V_{SS}). Only the pin functions which are unique to maximum mode are described; all other pin functions are as described above.

Symbol	Pin No.	Type	Name and Function
$\overline{S_2}$, $\overline{S_1}$, $\overline{S_0}$	26-28	O	**Status:** active during T_4, T_1, and T_2 and is returned to the passive state (1,1,1) during T_3 or during T_W when READY is HIGH. This status is used by the 8288 Bus Controller to generate all memory and I/O access control signals. Any change by $\overline{S_2}$, $\overline{S_1}$, or $\overline{S_0}$ during T_4 is used to indicate the beginning of a bus cycle, and the return to the passive state in T_3 or T_W is used to indicate the end of a bus cycle. These signals float to 3-state OFF in "hold acknowledge." These status lines are encoded as shown.

$\overline{S_2}$	$\overline{S_1}$	$\overline{S_0}$	Characteristics
0 (LOW)	0	0	Interrupt Acknowledge
0	0	1	Read I/O Port
0	1	0	Write I/O Port
0	1	1	Halt
1 (HIGH)	0	0	Code Access
1	0	1	Read Memory
1	1	0	Write Memory
1	1	1	Passive

Symbol	Pin No.	Type	Name and Function
$\overline{RQ}/\overline{GT_0}$, $\overline{RQ}/\overline{GT_1}$	30, 31	I/O	**Request/Grant:** pins are used by other local bus masters to force the processor to release the local bus at the end of the processor's current bus cycle. Each pin is bidirectional with $\overline{RQ}/\overline{GT_0}$ having higher priority than $\overline{RQ}/\overline{GT_1}$. $\overline{RQ}/\overline{GT}$ has an internal pull-up resistor so may be left unconnected. The request/grant sequence is as follows (see Figure 9): 1. A pulse of 1 CLK wide from another local bus master indicates a local bus request ("hold") to the 8086 (pulse 1). 2. During a T_4 or T_1 clock cycle, a pulse 1 CLK wide from the 8086 to the requesting master (pulse 2), indicates that the 8086 has allowed the local bus to float and that it will enter the "hold acknowledge" state at the next CLK. The CPU's bus interface unit is disconnected logically from the local bus during "hold acknowledge." 3. A pulse 1 CLK wide from the requesting master indicates to the 8086 (pulse 3) that the "hold" request is about to end and that the 8086 can reclaim the local bus at the next CLK. Each master-master exchange of the local bus is a sequence of 3 pulses. There must be one dead CLK cycle after each bus exchange. Pulses are active LOW. If the request is made while the CPU is performing a memory cycle, it will release the local bus during T_4 of the cycle when all the following conditions are met: 1. Request occurs on or before T_2. 2. Current cycle is not the low byte of a word (on an odd address). 3. Current cycle is not the first acknowledge of an interrupt acknowledge sequence. 4. A locked instruction is not currently executing.

Table 1. Pin Description (Continued)

Symbol	Pin No.	Type	Name and Function
			If the local bus is idle when the request is made the two possible events will follow: 1. Local bus will be released during the next clock. 2. A memory cycle will start within 3 clocks. Now the four rules for a currently active memory cycle apply with condition number 1 already satisfied.
$\overline{LOCK}$	29	O	$\overline{LOCK}$: output indicates that other system bus masters are not to gain control of the system bus while $\overline{LOCK}$ is active LOW. The $\overline{LOCK}$ signal is activated by the "LOCK" prefix instruction and remains active until the completion of the next instruction. This signal is active LOW, and floats to 3-state OFF in "hold acknowledge."
QS_1, QS_0	24, 25	O	**Queue Status:** The queue status is valid during the CLK cycle after which the queue operation is performed. QS_1 and QS_0 provide status to allow external tracking of the internal 8086 instruction queue.

QS_1	QS_0	CHARACTERISTICS
0 (LOW)	0	No Operation
0	1	First Byte of Op Code from Queue
1 (HIGH)	0	Empty the Queue
1	1	Subsequent Byte from Queue

The following pin function descriptions are for the 8086 in minimum mode (i.e., MN/$\overline{MX}$ = V_{CC}). Only the pin functions which are unique to minimum mode are described; all other pin functions are as described above.

Symbol	Pin No.	Type	Name and Function
M/$\overline{IO}$	28	O	**Status line:** logically equivalent to S_2 in the maximum mode. It is used to distinguish a memory access from an I/O access. M/$\overline{IO}$ becomes valid in the T_4 preceding a bus cycle and remains valid until the final T_4 of the cycle (M = HIGH, IO = LOW). M/$\overline{IO}$ floats to 3-state OFF in local bus "hold acknowledge."
$\overline{WR}$	29	O	**Write:** indicates that the processor is performing a write memory or write I/O cycle, depending on the state of the M/$\overline{IO}$ signal. $\overline{WR}$ is active for T_2, T_3 and T_W of any write cycle. It is active LOW, and floats to 3-state OFF in local bus "hold acknowledge."
$\overline{INTA}$	24	O	$\overline{INTA}$ is used as a read strobe for interrupt acknowledge cycles. It is active LOW during T_2, T_3 and T_W of each interrupt acknowledge cycle.
ALE	25	O	**Address Latch Enable:** provided by the processor to latch the address into the 8282/8283 address latch. It is a HIGH pulse active during T_1 of any bus cycle. Note that ALE is never floated.
DT/$\overline{R}$	27	O	**Data Transmit/Receive:** needed in minimum system that desires to use an 8286/8287 data bus transceiver. It is used to control the direction of data flow through the transceiver. Logically DT/$\overline{R}$ is equivalent to $\overline{S_1}$ in the maximum mode, and its timing is the same as for M/$\overline{IO}$. (T = HIGH, R = LOW.) This signal floats to 3-state OFF in local bus "hold acknowledge."
$\overline{DEN}$	26	O	**Data Enable:** provided as an output enable for the 8286/8287 in a minimum system which uses the transceiver. $\overline{DEN}$ is active LOW during each memory and I/O access and for INTA cycles. For a read or $\overline{INTA}$ cycle it is active from the middle of T_2 until the middle of T_4, while for a write cycle it is active from the beginning of T_2 until the middle of T_4. $\overline{DEN}$ floats to 3-state OFF in local bus "hold acknowledge."
HOLD, HLDA	31, 30	I/O	**HOLD:** indicates that another master is requesting a local bus "hold." To be acknowledged, HOLD must be active HIGH. The processor receiving the "hold" request will issue HLDA (HIGH) as an acknowledgement in the middle of a T_1 clock cycle. Simultaneous with the issuance of HLDA the processor will float the local bus and control lines. After HOLD is detected as being LOW, the processor will LOWer the HLDA, and when the processor needs to run another cycle, it will again drive the local bus and control lines. The same rules as for $\overline{RQ}/\overline{GT}$ apply regarding when the local bus will be released. HOLD is not an asynchronous input. External synchronization should be provided if the system cannot otherwise guarantee the setup time.

FUNCTIONAL DESCRIPTION

GENERAL OPERATION

The internal functions of the iAPX 86/10 processor are partitioned logically into two processing units. The first is the Bus Interface Unit (BIU) and the second is the Execution Unit (EU) as shown in the block diagram of Figure 1.

These units can interact directly but for the most part perform as separate asynchronous operational processors. The bus interface unit provides the functions related to instruction fetching and queuing, operand fetch and store, and address relocation. This unit also provides the basic bus control. The overlap of instruction pre-fetching provided by this unit serves to increase processor performance through improved bus bandwidth utilization. Up to 6 bytes of the instruction stream can be queued while waiting for decoding and execution.

The instruction stream queuing mechanism allows the BIU to keep the memory utilized very efficiently. Whenever there is space for at least 2 bytes in the queue, the BIU will attempt a word fetch memory cycle. This greatly reduces "dead time" on the memory bus. The queue acts as a First-In-First-Out (FIFO) buffer, from which the EU extracts instruction bytes as required. If the queue is empty (following a branch instruction, for example), the first byte into the queue immediately becomes available to the EU.

The execution unit receives pre-fetched instructions from the BIU queue and provides un-relocated operand addresses to the BIU. Memory operands are passed through the BIU for processing by the EU, which passes results to the BIU for storage. See the Instruction Set description for further register set and architectural descriptions.

MEMORY ORGANIZATION

The processor provides a 20-bit address to memory which locates the byte being referenced. The memory is organized as a linear array of up to 1 million bytes, addressed as 00000(H) to FFFFF(H). The memory is logically divided into code, data, extra data, and stack segments of up to 64K bytes each, with each segment falling on 16-byte boundaries. (See Figure 3a.)

All memory references are made relative to base addresses contained in high speed segment registers. The segment types were chosen based on the addressing needs of programs. The segment register to be selected is automatically chosen according to the rules of the following table. All information in one segment type share the same logical attributes (e.g. code or data). By structuring memory into relocatable areas of similar characteristics and by automatically selecting segment registers, programs are shorter, faster, and more structured.

Word (16-bit) operands can be located on even or odd address boundaries and are thus not constrained to even boundaries as is the case in many 16-bit computers. For address and data operands, the least significant byte of the word is stored in the lower valued address location and the most significant byte in the next higher address location. The BIU automatically performs the proper number of memory accesses, one if the word operand is on an even byte boundary and two if it is on an odd byte boundary. Except for the performance penalty, this double access is transparent to the software. This performance penalty does not occur for instruction fetches, only word operands.

Physically, the memory is organized as a high bank (D_{15}-D_8) and a low bank (D_7-D_0) of 512K 8-bit bytes addressed in parallel by the processor's address lines

A_{19} - A_1. Byte data with even addresses is transferred on the D_7-D_0 bus lines while odd addressed byte data (A_0 HIGH) is transferred on the D_{15}-D_8 bus lines. The processor provides two enable signals, $\overline{BHE}$ and A_0, to selectively allow reading from or writing into either an odd byte location, even byte location, or both. The instruction stream is fetched from memory as words and is addressed internally by the processor to the byte level as necessary.

Memory Reference Need	Segment Register Used	Segment Selection Rule
Instructions	CODE (CS)	Automatic with all instruction prefetch.
Stack	STACK (SS)	All stack pushes and pops. Memory references relative to BP base register except data references.
Local Data	DATA (DS)	Data references when: relative to stack, destination of string operation, or explicitly overridden.
External (Global) Data	EXTRA (ES)	Destination of string operations: Explicitly selected using a segment override.

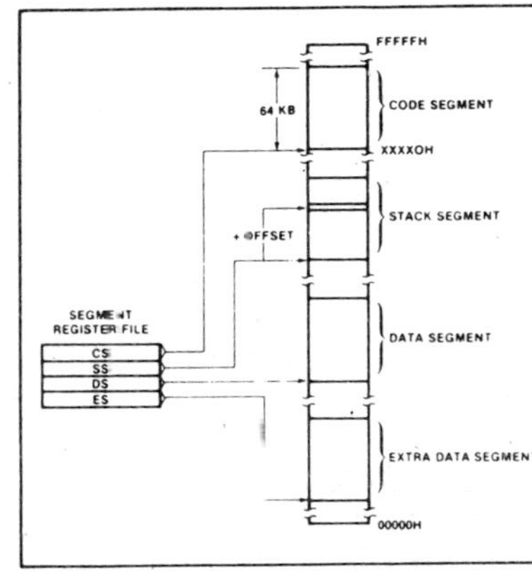

Figure 3a. Memory Organization

In referencing word data the BIU requires one or two memory cycles depending on whether or not the starting byte of the word is on an even or odd address, respectively. Consequently, in referencing word operands performance can be optimized by locating data on even address boundaries. This is an especially useful technique for using the stack, since odd address references to the stack may adversely affect the context switching time for interrupt processing or task multiplexing.

Certain locations in memory are reserved for specific CPU operations (see Figure 3b.) Locations from address FFFF0H through FFFFFH are reserved for operations including a jump to the initial program loading routine. Following $\overline{RESET}$, the CPU will always begin execution at location FFFF0H where the jump must be. Locations 00000H through 003FFH are reserved for interrupt operations. Each of the 256 possible interrupt types has its service routine pointed to by a 4-byte pointer element consisting of a 16-bit segment address and a 16-bit offset address. The pointer elements are assumed to have been stored at the respective places in reserved memory prior to occurrence of interrupts.

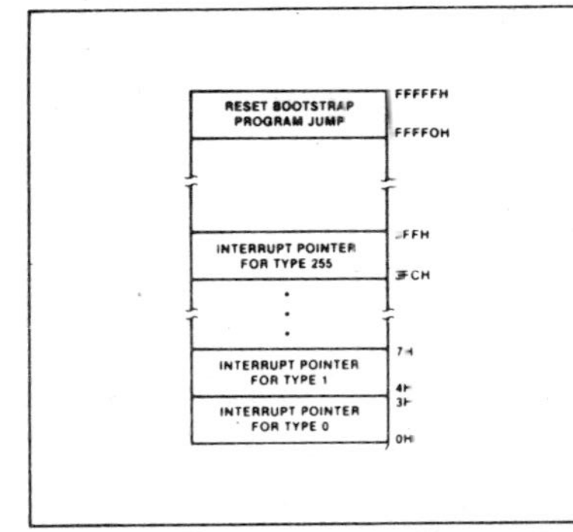

Figure 3b. Reserved Memory Locations

MINIMUM AND MAXIMUM MODES

The requirements for supporting minimum and maximum iAPX 86/10 systems are sufficiently different that they cannot be done efficiently with 40 uniquely defined pins. Consequently, the 8086 is equipped with a strap pin (MN/$\overline{MX}$) which defines the system configuration. The definition of a certain subset of the pins changes dependent on the condition of the strap pin. When MN/$\overline{MX}$ pin is strapped to GND, the 8086 treats pins 24 through 31 in maximum mode. An 8288 bus controller interprets status information coded into $\overline{S_0}$,$\overline{S_1}$,$\overline{S_2}$ to generate bus timing and control signals compatible with the MULTIBUS™ architecture. When the MN/$\overline{MX}$ pin is strapped to V_{CC}, the 8086 generates bus control signals itself on pins 24 through 31, as shown in parentheses in Figure 2. Examples of minimum mode and maximum mode systems are shown in Figure 4.

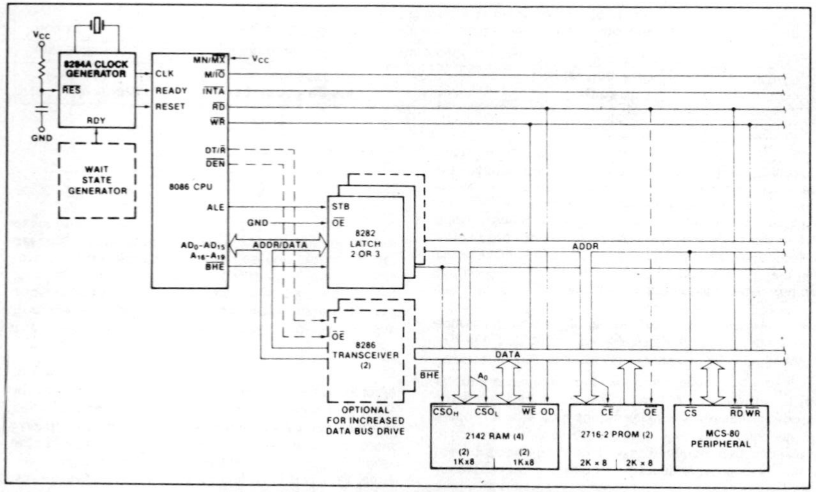

Figure 4a. Minimum Mode iAPX 86/10 Typical Configuration

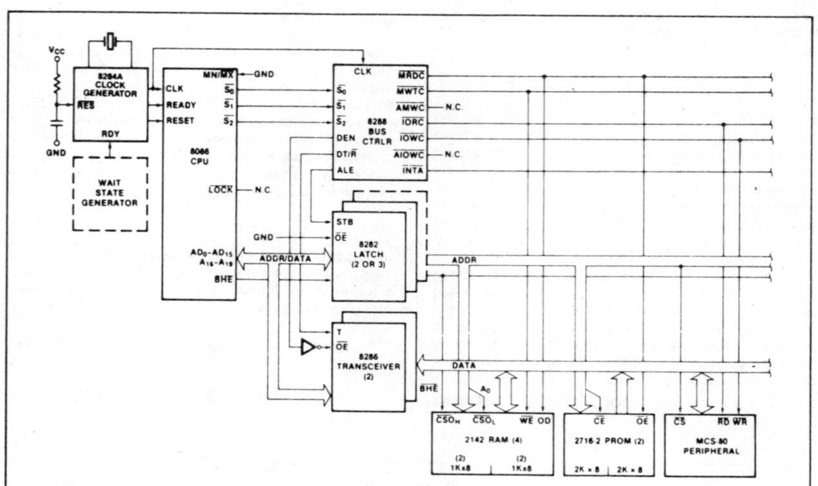

Figure 4b. Maximum Mode iAPX 86/10 Typical Configuration

BUS OPERATION

The 86/10 has a combined address and data bus commonly referred to as a time multiplexed bus. This technique provides the most efficient use of pins on the processor while permitting the use of a standard 40-lead package. This "local bus" can be buffered directly and used throughout the system with address latching provided on memory and I/O modules. In addition, the bus can also be demultiplexed at the processor with a single set of address latches if a standard non-multiplexed bus is desired for the system.

Each processor bus cycle consists of at least four CLK cycles. These are referred to as T_1, T_2, T_3 and T_4 (see Figure 5). The address is emitted from the processor during T_1 and data transfer occurs on the bus during T_3 and T_4. T_2 is used primarily for changing the direction of the bus during read operations. In the event that a "NOT READY" indication is given by the addressed device, "Wait" states (T_W) are inserted between T_3 and T_4. Each inserted "Wait" state is of the same duration as a CLK cycle. Periods can occur between 8086 bus cycles. These are referred to as "Idle" states (T_I) or inactive CLK cycles. The processor uses these cycles for internal housekeeping.

During T_1 of any bus cycle the ALE (Address Latch Enable) signal is emitted (by either the processor or the 8288 bus controller, depending on the MN/$\overline{MX}$ strap). At the trailing edge of this pulse, a valid address and certain status information for the cycle may be latched.

Status bits $\overline{S_0}$, $\overline{S_1}$, and $\overline{S_2}$ are used, in maximum mode, by the bus controller to identify the type of bus transaction according to the following table:

$\overline{S_2}$	$\overline{S_1}$	$\overline{S_0}$	CHARACTERISTICS
0 (LOW)	0	0	Interrupt Acknowledge
0	0	1	Read I/O
0	1	0	Write I/O
0	1	1	Halt
1 (HIGH)	0	0	Instruction Fetch
1	0	1	Read Data from Memory
1	1	0	Write Data to Memory
1	1	1	Passive (no bus cycle)

Status bits S_3 through S_7 are multiplexed with high-order address bits and the $\overline{BHE}$ signal, and are therefore valid during T_2 through T_4. S_3 and S_4 indicate which segment register (see Instruction Set description) was used for this bus cycle in forming the address, according to the following table:

S_4	S_3	CHARACTERISTICS
0 (LOW)	0	Alternate Data (extra segment)
0	1	Stack
1 (HIGH)	0	Code or None
1	1	Data

S_5 is a reflection of the PSW interrupt enable bit. S_6=0 and S_7 is a spare status bit.

I/O ADDRESSING

In the 86/10, I/O operations can address up to a maximum of 64K I/O byte registers or 32K I/O word registers. The I/O address appears in the same format as the memory address on bus lines A_{15}-A_0. The address lines A_{19}-A_{16} are zero in I/O operations. The variable I/O instructions which use register DX as a pointer have full address capability while the direct I/O instructions directly address one or two of the 256 I/O byte locations in page 0 of the I/O address space.

I/O ports are addressed in the same manner as memory locations. Even addressed bytes are transferred on the D_7-D_0 bus lines and odd addressed bytes on D_{15}-D_8. Care must be taken to assure that each register within an 8-bit peripheral located on the lower portion of the bus be addressed as even.

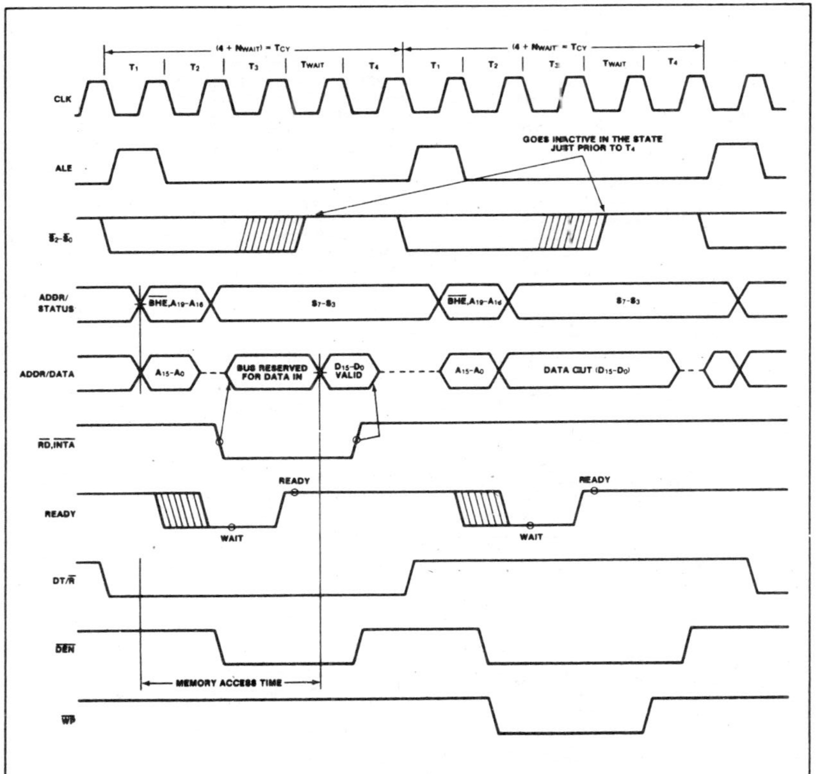

Figure 5. Basic System Timing

EXTERNAL INTERFACE

PROCESSOR RESET AND INITIALIZATION

Processor initialization or start up is accomplished with activation (HIGH) of the RESET pin. The 8086 RESET is required to be HIGH for greater than 4 CLK cycles. The 8086 will terminate operations on the high-going edge of RESET and will remain dormant as long as RESET is HIGH. The low-going transition of RESET triggers an internal reset sequence for approximately 10 CLK cycles. After this interval the 8086 operates normally beginning with the instruction in absolute location FFFF0H (see Figure 3B). The details of this operation are specified in the Instruction Set description of the MCS-86 Family User's Manual. The RESET input is internally synchronized to the processor clock. At initialization the HIGH-to-LOW transition of RESET must occur no sooner than 50 μs after power-up, to allow complete initialization of the 8086.

NMI may not be asserted prior to the 2nd CLK cycle following the end of RESET.

INTERRUPT OPERATIONS

Interrupt operations fall into two classes; software or hardware initiated. The software initiated interrupts and software aspects of hardware interrupts are specified in the Instruction Set description. Hardware interrupts can be classified as non-maskable or maskable.

Interrupts result in a transfer of control to a new program location. A 256-element table containing address pointers to the interrupt service program locations resides in absolute locations 0 through 3FFH (see Figure 3b), which are reserved for this purpose. Each element in the table is 4 bytes in size and corresponds to an interrupt "type". An interrupting device supplies an 8-bit type number, during the interrupt acknowledge sequence, which is used to "vector" through the appropriate element to the new interrupt service program location.

NON-MASKABLE INTERRUPT (NMI)

The processor provides a single non-maskable interrupt pin (NMI) which has higher priority than the maskable interrupt request pin (INTR). A typical use would be to activate a power failure routine. The NMI is edge-triggered on a LOW-to-HIGH transition. The activation of this pin causes a type 2 interrupt. (See Instruction Set description.)

NMI is required to have a duration in the HIGH state of greater than two CLK cycles, but is not required to be synchronized to the clock. Any high-going transition of NMI is latched on-chip and will be serviced at the end of the current instruction or between whole moves of a block-type instruction. Worst case response to NMI would be for multiply, divide, and variable shift instructions. There is no specification on the occurrence of the low-going edge; it may occur before, during, or after the servicing of NMI. Another high-going edge triggers another response if it occurs after the start of the NMI procedure. The signal must be free of logical spikes in general and be free of bounces on the low-going edge to avoid triggering extraneous responses.

MASKABLE INTERRUPT (INTR)

The 86/10 provides a single interrupt request input (INTR) which can be masked internally by software with the resetting of the interrupt enable FLAG status bit. The interrupt request signal is level triggered. It is internally synchronized during each clock cycle on the high-going edge of CLK. To be responded to, INTR must be present (HIGH) during the clock period preceding the end of the current instruction or the end of a whole move for a block-type instruction. During the interrupt response sequence further interrupts are disabled. The enable bit is reset as part of the response to any interrupt (INTR, NMI, software interrupt or single-step), although the

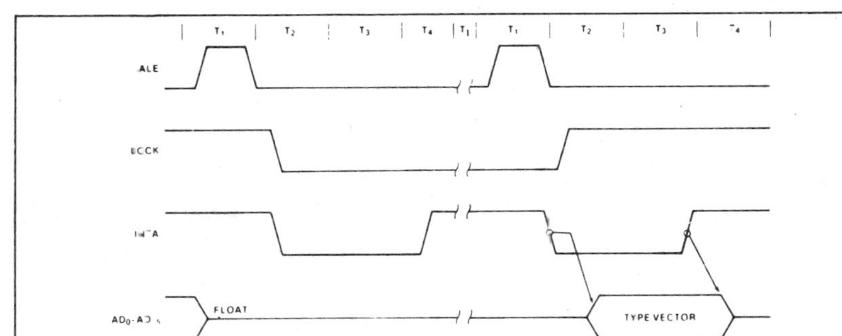

Figure 6. Interrupt Acknowledge Sequence

FLAGS register which is automatically pushed onto the stack reflects the state of the processor prior to the interrupt. Until the old FLAGS register is restored the enable bit will be zero unless specifically set by an instruction.

During the response sequence (figure 6) the processor executes two successive (back-to-back) interrupt acknowledge cycles. The 8086 emits the LOCK signal from T_2 of the first bus cycle until T_2 of the second. A local bus "hold" request will not be honored until the end of the second bus cycle. In the second bus cycle a byte is fetched from the external interrupt system (e.g., 8259A PIC) which identifies the source (type) of the interrupt. This byte is multiplied by four and used as a pointer into the interrupt vector lookup table. An INTR signal left HIGH will be continually responded to within the limitations of the enable bit and sample period. The INTERRUPT RETURN instruction includes a FLAGS pop which returns the status of the original interrupt enable bit when it restores the FLAGS.

HALT

When a software "HALT" instruction is executed the processor indicates that it is entering the "HALT" state in one of two ways depending upon which mode is strapped. In minimum mode, the processor issues one ALE with no qualifying bus control signals. In Maximum Mode, the processor issues appropriate HALT status on $\overline{S_2}\overline{S_1}\overline{S_0}$ and the 8288 bus controller issues one ALE. The 8086 will not leave the "HALT" state when a local bus "hold" is entered while in "HALT". In this case, the processor reissues the HALT indicator. An interrupt request or RESET will force the 8086 out of the "HALT" state.

READ/MODIFY/WRITE (SEMAPHORE) OPERATIONS VIA LOCK

The $\overline{LOCK}$ status information is provided by the processor when directly consecutive bus cycles are required during the execution of an instruction. This provides the processor with the capability of performing read/modify/write operations on memory (via the Exchange Register With Memory instruction, for example) without the possibility of another system bus master receiving intervening memory cycles. This is useful in multi-processor system configurations to accomplish "test and set lock" operations. The $\overline{LOCK}$ signal is activated (forced LOW) in the clock cycle following the one in which the software "LOCK" prefix instruction is decoded by the EU. It is deactivated at the end of the last bus cycle of the instruction following the "LOCK" prefix instruction. While $\overline{LOCK}$ is active a request on a RQ/GT pin will be recorded and then honored at the end of the LOCK.

EXTERNAL SYNCHRONIZATION VIA TEST

As an alternative to the interrupts and general I/O capabilities, the 8086 provides a single software-testable input known as the $\overline{TEST}$ signal. At any time the program may execute a WAIT instruction. If at that time the $\overline{TEST}$ signal is inactive (HIGH), program execution becomes suspended while the processor waits for $\overline{TEST}$ to become active. It must remain active for at least 5 CLK cycles. The WAIT instruction is re-executed repeatedly until that time. This activity does not consume bus cycles. The processor remains in an idle state while waiting. All 8086 drivers go to 3-state OFF if bus "Hold" is entered. If interrupts are enabled, they may occur while the processor is waiting. When this occurs the processor fetches the WAIT instruction one extra time, processes the interrupt, and then re-fetches and re-executes the WAIT instruction upon returning from the interrupt.

BASIC SYSTEM TIMING

Typical system configurations for the processor operating in minimum mode and in maximum mode are shown in Figures 4a and 4b, respectively. In minimum mode, the MN/$\overline{MX}$ pin is strapped to V_{CC} and the processor emits bus control signals in a manner similar to the 8085. In maximum mode, the MN/$\overline{MX}$ pin is strapped to V_{SS} and the processor emits coded status information which the 8288 bus controller uses to generate MULTIBUS compatible bus control signals. Figure 5 illustrates the signal timing relationships.

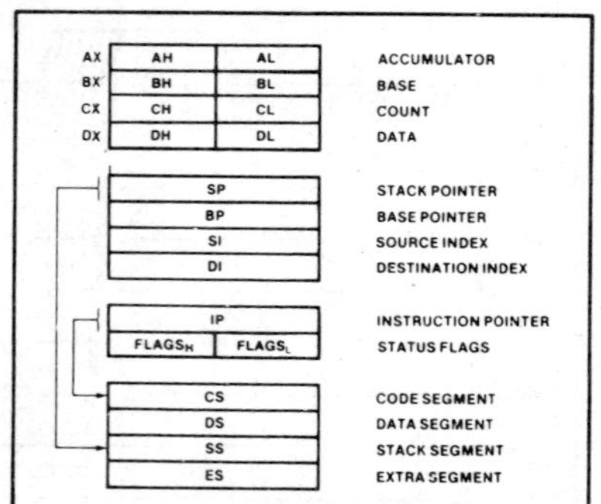

Figure 7. iAPX 86/10 Register Model

SYSTEM TIMING — MINIMUM SYSTEM

The read cycle begins in T_1 with the assertion of the Address Latch Enable (ALE) signal. The trailing (low-going) edge of this signal is used to latch the address information, which is valid on the local bus at this time, into the 8282/8283 latch. The $\overline{BHE}$ and A_0 signals address the low, high, or both bytes. From T_1 to T_4 the M/$\overline{IO}$ signal indicates a memory or I/O operation. At T_2 the address is removed from the local bus and the bus goes to a high impedance state. The read control signal is also asserted at T_2. The read ($\overline{RD}$) signal causes the addressed device to enable its data bus drivers to the local bus. Some time later valid data will be available on the bus and the addressed device will drive the READY line HIGH. When the processor returns the read signal to a HIGH level, the addressed device will again 3-state its bus drivers. If a transceiver (8286/8287) is required to buffer the 8086 local bus, signals DT/$\overline{R}$ and $\overline{DEN}$ are provided by the 8086.

A write cycle also begins with the assertion of ALE and the emission of the address. The M/$\overline{IO}$ signal is again asserted to indicate a memory or I/O write operation. In the T_2 immediately following the address emission the processor emits the data to be written into the addressed location. This data remains valid until the middle of T_4. During T_2, T_3, and T_W the processor asserts the write control signal. The write ($\overline{WR}$) signal becomes active at the beginning of T_2 as opposed to the read which is delayed somewhat into T_2 to provide time for the bus to float.

The $\overline{BHE}$ and A_0 signals are used to select the proper byte(s) of the memory/IO word to be read or written according to the following table:

$\overline{BHE}$	A0	CHARACTERISTICS
0	0	Whole word
0	1	Upper byte from/ to odd address
1	0	Lower byte from/ to even address
1	1	None

I/O ports are addressed in the same manner as memory location. Even addressed bytes are transferred on the D_7–D_0 bus lines and odd addressed bytes on D_{15}–D_8.

The basic difference between the interrupt acknowledge cycle and a read cycle is that the interrupt acknowledge signal ($\overline{INTA}$) is asserted in place of the read ($\overline{RD}$) signal and the address bus is floated. (See Figure 6.) In the second of two successive INTA cycles, a byte of information is read from bus lines D_7–D_0 as supplied by the interrupt system logic (i.e., 8259A Priority Interrupt Controller). This byte identifies the source (type) of the interrupt. It is multiplied by four and used as a pointer into an interrupt vector lookup table, as described earlier.

BUS TIMING—MEDIUM SIZE SYSTEMS

For medium size systems the MN/$\overline{MX}$ pin is connected to V_{SS} and the 8288 Bus Controller is added to the system as well as an 8282/8283 latch for latching the system address, and a 8286/8287 transceiver to allow for bus loading greater than the 8086 is capable of handling. Signals ALE, DEN, and DT/$\overline{R}$ are generated by the 8288 instead of the processor in this configuration although their timing remains relatively the same. The 8086 status outputs ($\overline{S_2}$, $\overline{S_1}$, and $\overline{S_0}$) provide type-of-cycle information and become 8288 inputs. This bus cycle information specifies read (code, data, or I/O), write (data or I/O), interrupt acknowledge, or software halt. The 8288 thus issues control signals specifying memory read or write, I/O read or write, or interrupt acknowledge. The 8288 provides two types of write strobes, normal and advanced, to be applied as required. The normal write strobes have data valid at the leading edge of write. The advanced write strobes have the same timing as read strobes, and hence data isn't valid at the leading edge of write. The 8286/8287 transceiver receives the usual T and OE inputs from the 8288's DT/$\overline{R}$ and DEN.

The pointer into the interrupt vector table, which is passed during the second INTA cycle, can derive from an 8259A located on either the local bus or the system bus. If the master 8259A Priority Interrupt Controller is positioned on the local bus, a TTL gate is required to disable the 8286/8287 transceiver when reading from the master 8259A during the interrupt acknowledge sequence and software "poll".

ABSOLUTE MAXIMUM RATINGS*

Ambient Temperature Under Bias ....0°C to 70°C
Storage Temperature .... −65°C to +150°C
Voltage on Any Pin with
Respect to Ground .... −1.0 to +7V
Power Dissipation .... 2.5 Watt

**NOTICE: Stresses above those listed under "Absolute Maximum Ratings" may cause permanent damage to the device. This is a stress rating only and functional operation of the device at these or any other conditions above those indicated in the operational sections of this specification is not implied. Exposure to absolute maximum rating conditions for extended periods may affect device reliability.*

D.C. CHARACTERISTICS

(8086: T_A = 0°C to 70°C, V_{CC} = 5V ± 10%)
(8086-1: T_A = 0°C to 70°C, V_{CC} = 5V ± 5%)
(8086-2: T_A = 0°C to 70°C, V_{CC} = 5V ± 5%)

Symbol	Parameter	Min.	Max.	Units	Test Conditions
V_{IL}	Input Low Voltage	−0.5	+0.8	V	
V_{IH}	Input High Voltage	2.0	V_{CC}+0.5	V	
V_{OL}	Output Low Voltage		0.45	V	I_{OL}=2.5 mA
V_{OH}	Output High Voltage	2.4		V	I_{OH} = −400 μA
I_{CC}	Power Supply Current: 8086 8086-1 8086-2		340 360 350	mA	T_A = 25°C
I_{LI}	Input Leakage Current		±10	μA	0V ≤ V_{IN} ≤ V_{CC}
I_{LO}	Output Leakage Current		±10	μA	0.45V ≤ V_{OUT} ≤ V_{CC}
V_{CL}	Clock Input Low Voltage	−0.5	+0.6	V	
V_{CH}	Clock Input High Voltage	3.9	V_{CC}+1.0	V	
C_{IN}	Capacitance of Input Buffer (All input except AD_0-AD_{15}, $\overline{RQ}/\overline{GT}$)		15	pF	fc = 1 MHz
C_{IO}	Capacitance of I/O Buffer (AD_0-AD_{15}, $\overline{RQ}/\overline{GT}$)		15	pF	fc = 1 MHz

A.C. CHARACTERISTICS

(8086: T_A = 0°C to 70°C, V_{CC} = 5V ± 10%)
(8086-1: T_A = 0°C to 70°C, V_{CC} = 5V ± 5%)
(8086-2: T_A = 0°C to 70°C, V_{CC} = 5V ± 5%)

MINIMUM COMPLEXITY SYSTEM TIMING REQUIREMENTS

Symbol	Parameter	8086		8086-1 (Preliminary)		8086-2		Units	Test Conditions
		Min.	Max.	Min.	Max.	Min.	Max.		
TCLCL	CLK Cycle Period	200	500	100	500	125	500	ns	
TCLCH	CLK Low Time	118		53		68		ns	
TCHCL	CLK High Time	69		39		44		ns	
TCH1CH2	CLK Rise Time		10		10		10	ns	From 1.0V to 3.5V
TCL2CL1	CLK Fall Time		10		10		10	ns	From 3.5V to 1.0V
TDVCL	Data in Setup Time	30		5		20		ns	
TCLDX	Data in Hold Time	10		10		10		ns	
TR1VCL	RDY Setup Time into 8284A (See Notes 1, 2)	35		35		35		ns	
TCLR1X	RDY Hold Time into 8284A (See Notes 1, 2)	0		0		0		ns	
TRYHCH	READY Setup Time into 8086	118		53		68		ns	
TCHRYX	READY Hold Time into 8086	30		20		20		ns	
TRYLCL	READY Inactive to CLK (See Note 3)	−8		−10		−8		ns	
THVCH	HOLD Setup Time	35		20		20		ns	
TINVCH	INTR, NMI, $\overline{TEST}$ Setup Time (See Note 2)	30		15		15		ns	
TILIH	Input Rise Time (Except CLK)		20		20		20	ns	From 0.8V to 2.0V
TIHIL	Input Fall Time (Except CLK)		12		12		12	ns	From 2.0V to 0.8V

A.C. CHARACTERISTICS (Continued)

TIMING RESPONSES

Symbol	Parameter	8086		8086-1 (Preliminary)		8086-2		Units	Test Conditions
		Min.	Max.	Min.	Max.	Min.	Max.		
TCLAV	Address Valid Delay	10	110	10	50	10	60	ns	
TCLAX	Address Hold Time	10		10		10		ns	
TCLAZ	Address Float Delay	TCLAX	80	10	40	TCLAX	50	ns	
TLHLL	ALE Width	TCLCH−20		TCLCH−10		TCLCH-10		ns	
TCLLH	ALE Active Delay		80		40		50	ns	
TCHLL	ALE Inactive Delay		85		45		55	ns	
TLLAX	Address Hold Time to ALE Inactive	TCHCL−10		TCHCL−10		TCHCL−10		ns	
TCLDV	Data Valid Delay	10	110	10	50	10	60	ns	*C_L = 20-100 pF for all 8086 Outputs (In addition to 8086 self-load)
TCHDX	Data Hold Time	10		10		10		ns	
TWHDX	Data Hold Time After WR	TCLCH−30		TCLCH−25		TCLCH−30		ns	
TCVCTV	Control Active Delay 1	10	110	10	50	10	70	ns	
TCHCTV	Control Active Delay 2	10	110	10	45	10	60	ns	
TCVCTX	Control Inactive Delay	10	110	10	50	10	70	ns	
TAZRL	Address Float to READ Active	0		0		0		ns	
TCLRL	$\overline{RD}$ Active Delay	10	165	10	70	10	100	ns	
TCLRH	$\overline{RD}$ Inactive Delay	10	150	10	60	10	80	ns	
TRHAV	$\overline{RD}$ Inactive to Next Address Active	TCLCL−45		TCLCL−35		TCLCL−40		ns	
TCLHAV	HLDA Valid Delay	10	160	10	60	10	100	ns	
TRLRH	$\overline{RD}$ Width	2TCLCL−75		2TCLCL−40		2TCLCL−50		ns	
TWLWH	$\overline{WR}$ Width	2TCLCL−60		2TCLCL−35		2TCLCL−40		ns	
TAVAL	Address Valid to ALE Low	TCLCH−60		TCLCH−35		TCLCH−40		ns	
TOLOH	Output Rise Time		20		20		20	ns	From 0.8V to 2.0V
TOHOL	Output Fall Time		12		12		12	ns	From 2.0V to 0.8V

NOTES:
1. Signal at 8284A shown for reference only.
2. Setup requirement for asynchronous signal only to guarantee recognition at next CLK.
3. Applies only to T2 state. (8 ns into T3).

A.C. TESTING INPUT, OUTPUT WAVEFORM

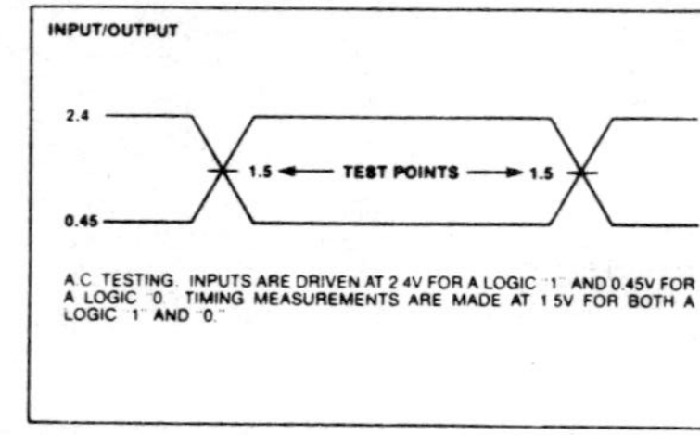

A.C. TESTING LOAD CIRCUIT

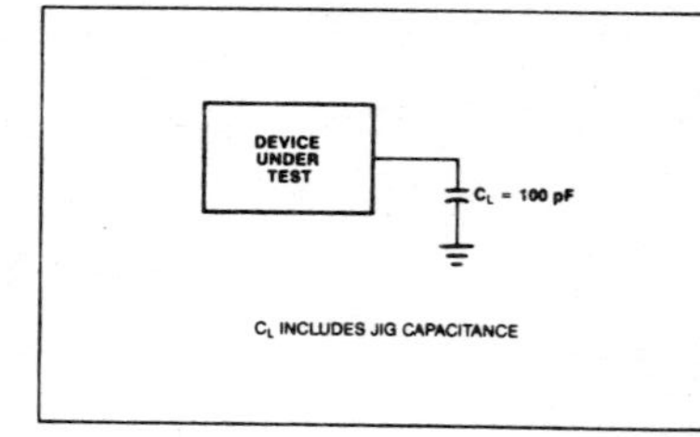

WAVEFORMS

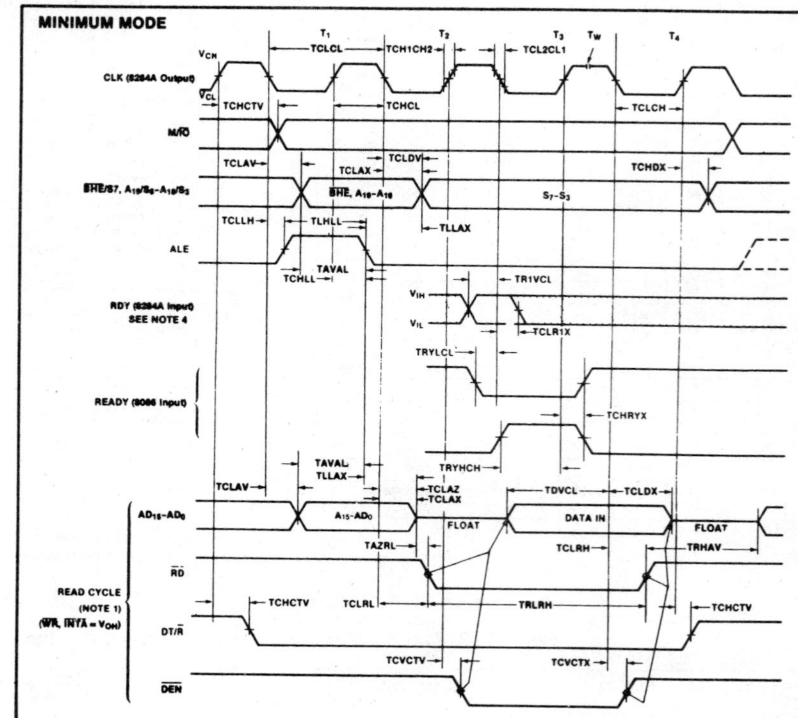

WAVEFORMS (Continued)

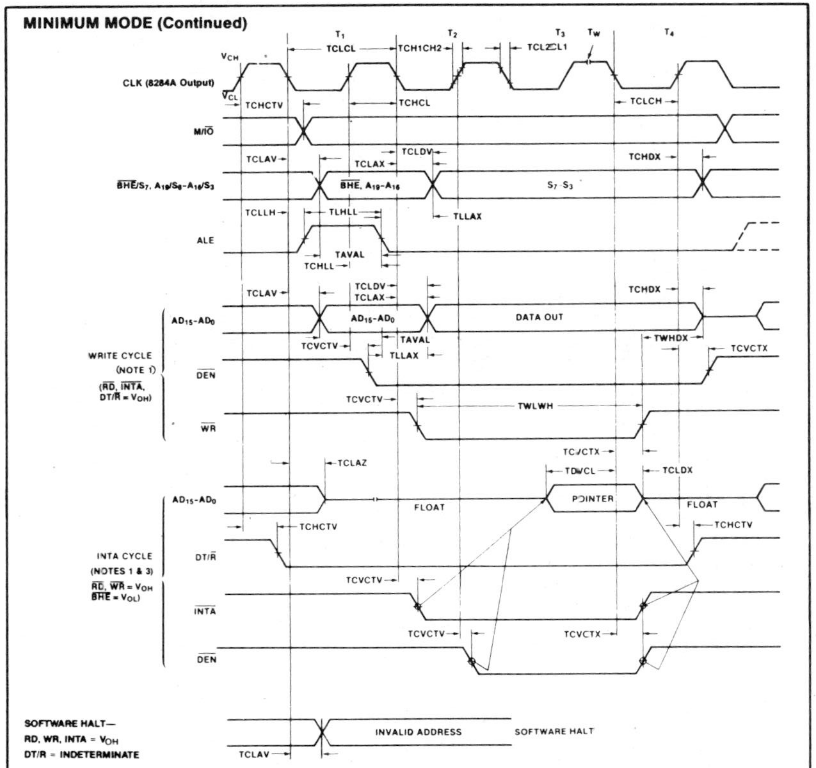

NOTES:
1. All signals switch between V_{OH} and V_{OL} unless otherwise specified.
2. RDY is sampled near the end of T_2, T_3, T_W to determine if T_W machines states are to be inserted.
3. Two INTA cycles run back-to-back. The 8086 LOCAL ADDR/DATA BUS is floating during both INTA cycles. Control signals shown for second INTA cycle.
4. Signals at 8284A are shown for reference only.
5. All timing measurements are made at 1.5V unless otherwise noted.

A.C. CHARACTERISTICS

MAX MODE SYSTEM (USING 8288 BUS CONTROLLER) TIMING REQUIREMENTS

Symbol	Parameter	8086		8086-1 (Preliminary)		8086-2 (Preliminary)		Units	Test Conditions
		Min.	Max.	Min.	Max.	Min.	Max.		
TCLCL	CLK Cycle Period	200	500	100	500	125	500	ns	
TCLCH	CLK Low Time	118		53		68		ns	
TCHCL	CLK High Time	69		39		44		ns	
TCH1CH2	CLK Rise Time		10		10		10	ns	From 1.0V to 3.5V
TCL2CL1	CLK Fall Time		10		10		10	ns	From 3.5V to 1.0V
TDVCL	Data in Setup Time	30		5		20		ns	
TCLDX	Data In Hold Time	10		10		10		ns	
TR1VCL	RDY Setup Time into 8284A (See Notes 1, 2)	35		35		35		ns	
TCLR1X	RDY Hold Time into 8284A (See Notes 1, 2)	0		0		0		ns	
TRYHCH	READY Setup Time into 8086	118		53		68		ns	
TCHRYX	READY Hold Time into 8086	30		20		20		ns	
TRYLCL	READY Inactive to CLK (See Note 4)	−8		−10		−8		ns	
TINVCH	Setup Time for Recognition (INTR, NMI, TEST) (See Note 2)	30		15		15		ns	
TGVCH	RQ/GT Setup Time	30		12		15		ns	
TCHGX	RQ Hold Time into 8086	40		20		30		ns	
TILIH	Input Rise Time (Except CLK)		20		20		20	ns	From 0.8V to 2.0V
TIHIL	Input Fall Time (Except CLK)		12		12		12	ns	From 2.0V to 0.8V

NOTES:
1. Signal at 8284A or 8288 shown for reference only.
2. Setup requirement for asynchronous signal only to guarantee recognition at next CLK.
3. Applies only to T3 and wait states.
4. Applies only to T2 state (8 ns into T3).

A.C. CHARACTERISTICS (Continued)

TIMING RESPONSES

Symbol	Parameter	8086		8086-1 (Preliminary)		8086-2 (Preliminary)		Units	Test Conditions
		Min.	Max.	Min.	Max.	Min.	Max.		
TCLML	Command Active Delay (See Note 1)	10	35	10	35	10	35	ns	
TCLMH	Command Inactive Delay (See Note 1)	10	35	10	35	10	35	ns	
TRYHSH	READY Active to Status Passive (See Note 3)		110		45		65	ns	
TCHSV	Status Active Delay	10	110	10	45	10	60	ns	
TCLSH	Status Inactive Delay	10	130	10	55	10	70	ns	
TCLAV	Address Valid Delay	10	110	10	50	10	60	ns	
TCLAX	Address Hold Time	10		10		10		ns	
TCLAZ	Address Float Delay	TCLAX	80	10	40	TCLAX	50	ns	
TSVLH	Status Valid to ALE High (See Note 1)		15		15		15	ns	
TSVMCH	Status Valid to MCE High (See Note 1)		15		15		15	ns	
TCLLH	CLK Low to ALE Valid (See Note 1)		15		15		15	ns	
TCLMCH	CLK Low to MCE High (See Note 1)		15		15		15	ns	
TCHLL	ALE Inactive Delay (See Note 1)		15		15		15	ns	C_L = 20-100 pF for all 8086 Outputs (In addition to 8086 self-load)
TCLMCL	MCE Inactive Delay (See Note 1)		15		15		15	ns	
TCLDV	Data Valid Delay	10	110	10	50	10	60	ns	
TCHDX	Data Hold Time	10		10		10		ns	
TCVNV	Control Active Delay (See Note 1)	5	45	5	45	5	45	ns	
TCVNX	Control Inactive Delay (See Note 1)	10	45	10	45	10	45	ns	
TAZRL	Address Float to Read Active	0		0		0		ns	
TCLRL	RD Active Delay	10	165	10	70	10	100	ns	
TCLRH	RD Inactive Delay	10	150	10	60	10	80	ns	
TRHAV	RD Inactive to Next Address Active	TCLCL−45		TCLCL−35		TCLCL−40		ns	
TCHDTL	Direction Control Active Delay (See Note 1)		50		50		50	ns	
TCHDTH	Direction Control Inactive Delay (See Note 1)		30		30		30	ns	
TCLGL	GT Active Delay	0	85	0	45	0	50	ns	
TCLGH	GT Inactive Delay	0	85	0	45	0	50	ns	
TRLRH	RD Width	2TCLCL−75		2TCLCL−40		2TCLCL−50		ns	
TOLOH	Output Rise Time		20		20		20	ns	From 0.8V to 2.0V
TOHOL	Output Fall Time		12		12		12	ns	From 2.0V to 0.8V

WAVEFORMS

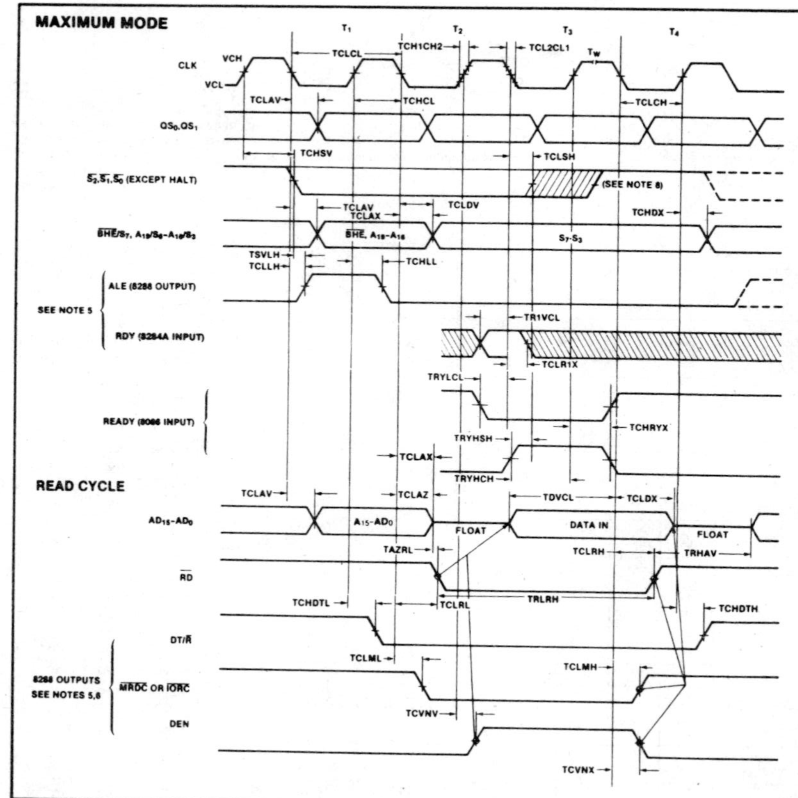

WAVEFORMS (Continued)

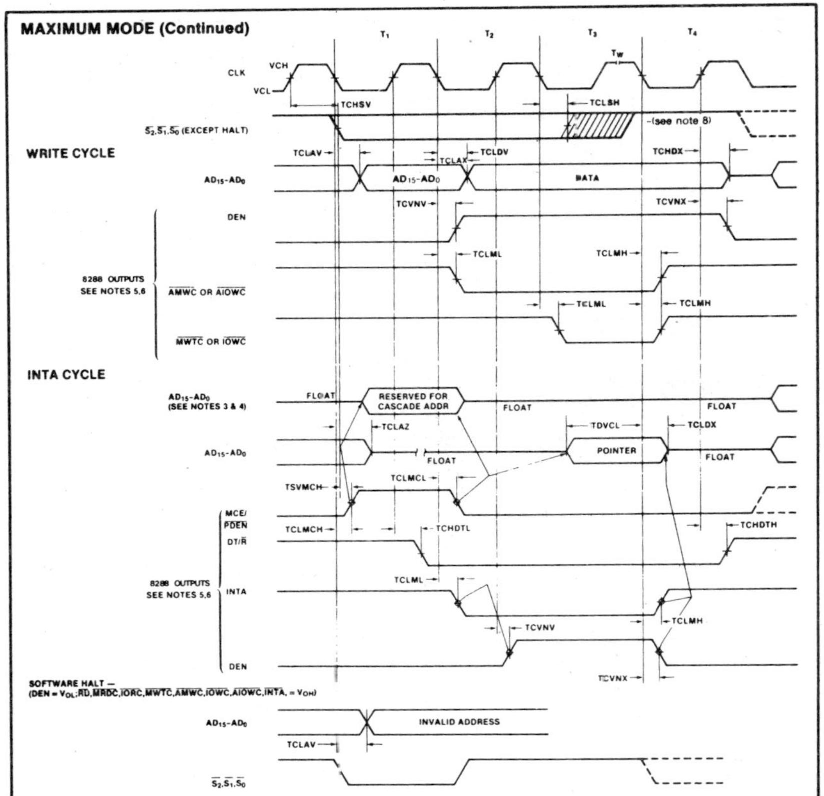

NOTES:

1. All signals switch between V_{OH} and V_{OL} unless otherwise specified.
2. RDY is sampled near the end of T_2, T_3, T_W to determine if T_W machines states are to be inserted.
3. Cascade address is valid between first and second INTA cycle.
4. Two INTA cycles run back-to-back. The 8086 LOCAL ADDR/DATA BUS is floating during both INTA cycles. Control for pointer address is shown for second INTA cycle.
5. Signals at 8284A or 8288 are shown for reference only.
6. The issuance of the 8288 command and control signals ($\overline{MRDC}$, $\overline{MWTC}$, $\overline{AMWC}$, $\overline{IORC}$, $\overline{IOWC}$, $\overline{AIOWC}$, $\overline{INTA}$ and DEN) lags the active high 8288 CEN.
7. All timing measurements are made at 1.5V unless otherwise noted.
8. Status inactive in state just prior to T_4.

WAVEFORMS (Continued)

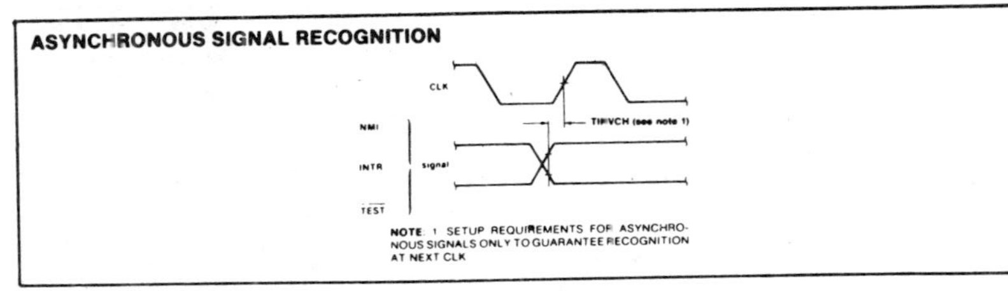

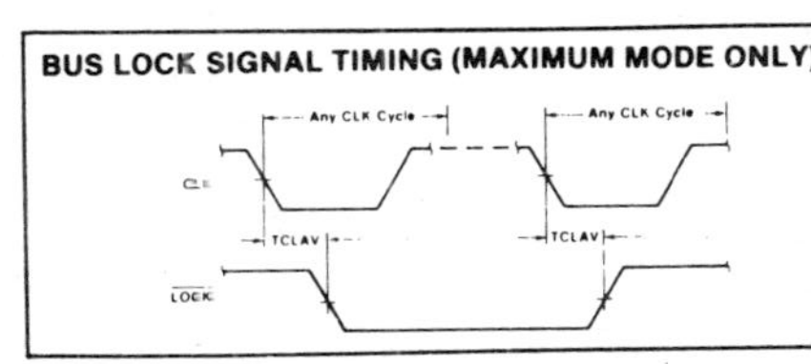

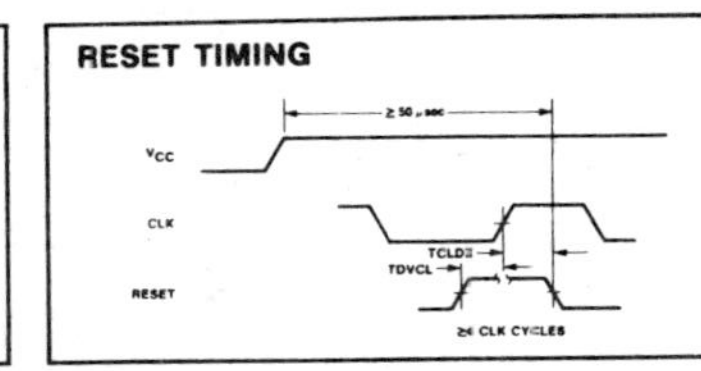

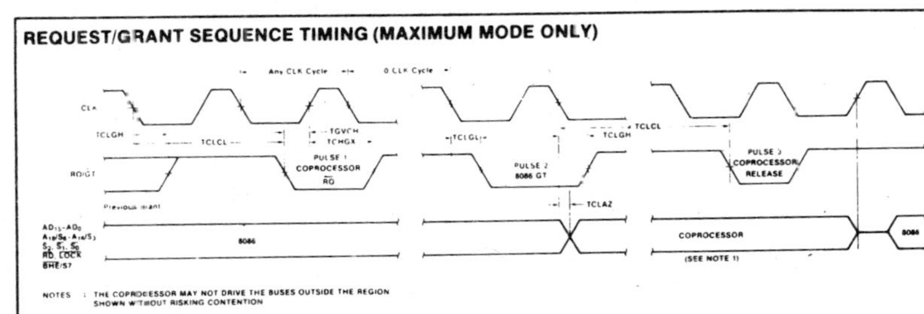

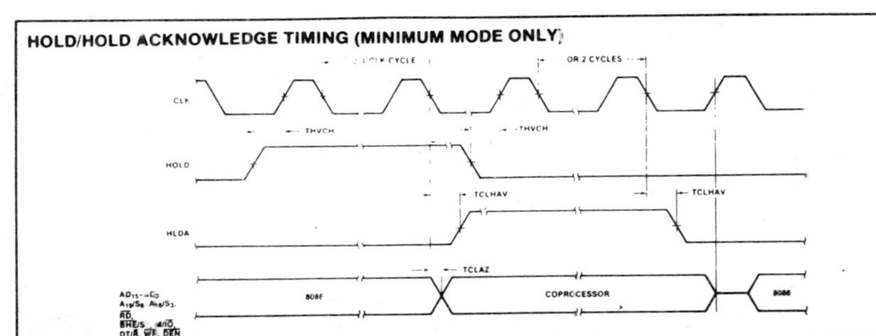

Table 2. Instruction Set Summary

Instruction	7 6 5 4 3 2 1 0	7 6 5 4 3 2 1 0	7 6 5 4 3 2 1 0	7 6 5 4 3 2 1 0
DATA TRANSFER				
MOV - **Move**:				
Register/memory to/from register	1 0 0 0 1 0 d w	mod reg r/m		
Immediate to register/memory	1 1 0 0 0 1 1 w	mod 0 0 0 r/m	data	data if w 1
Immediate to register	1 0 1 1 w reg	data	data if w 1	
Memory to accumulator	1 0 1 0 0 0 0 w	addr-low	addr-high	
Accumulator to memory	1 0 1 0 0 0 1 w	addr-low	addr-high	
Register/memory to segment register	1 0 0 0 1 1 1 0	mod 0 reg r/m		
Segment register to register/memory	1 0 0 0 1 1 0 0	mod 0 reg r/m		
PUSH - **Push**:				
Register/memory	1 1 1 1 1 1 1 1	mod 1 1 0 r/m		
Register	0 1 0 1 0 reg			
Segment register	0 0 0 reg 1 1 0			
POP - **Pop**:				
Register/memory	1 0 0 0 1 1 1 1	mod 0 0 0 r/m		
Register	0 1 0 1 1 reg			
Segment register	0 0 0 reg 1 1 1			
XCHG - **Exchange**:				
Register/memory with register	1 0 0 0 0 1 1 w	mod reg r/m		
Register with accumulator	1 0 0 1 0 reg			
IN=Input from:				
Fixed port	1 1 1 0 0 1 0 w	port		
Variable port	1 1 1 0 1 1 0 w			
OUT = Output to:				
Fixed port	1 1 1 0 0 1 1 w	port		
Variable port	1 1 1 0 1 1 1 w			
XLAT - Translate byte to AL	1 1 0 1 0 1 1 1			
LEA=Load EA to register	1 0 0 0 1 1 0 1	mod reg r/m		
LDS-Load pointer to DS	1 1 0 0 0 1 0 1	mod reg r/m		
LES=Load pointer to ES	1 1 0 0 0 1 0 0	mod reg r/m		
LAHF-Load AH with flags	1 0 0 1 1 1 1 1			
SAHF-Store AH into flags	1 0 0 1 1 1 1 0			
PUSHF=Push flags	1 0 0 1 1 1 0 0			
POPF=Pop flags	1 0 0 1 1 1 0 1			
ARITHMETIC				
ADD = **Add**:				
Reg./memory with register to either	0 0 0 0 0 0 d w	mod reg r/m		
Immediate to register/memory	1 0 0 0 0 0 s w	mod 0 0 0 r/m	data	data if s w 01
Immediate to accumulator	0 0 0 0 0 1 0 w	data	data if w 1	
ADC = **Add with carry**:				
Reg./memory with register to either	0 0 0 1 0 0 d w	mod reg r/m		
Immediate to register/memory	1 0 0 0 0 0 s w	mod 0 1 0 r/m	data	data if s w=01
Immediate to accumulator	0 0 0 1 0 1 0 w	data	data if w 1	
INC = **Increment**:				
Register/memory	1 1 1 1 1 1 1 w	mod 0 0 0 r/m		
Register	0 1 0 0 0 reg			
AAA=ASCII adjust for add	0 0 1 1 0 1 1 1			
DAA=Decimal adjust for add	0 0 1 0 0 1 1 1			
SUB = **Subtract**:				
Reg./memory and register to either	0 0 1 0 1 0 d w	mod reg r/m		
Immediate from register/memory	1 0 0 0 0 0 s w	mod 1 0 1 r/m	data	data if s w=01
Immediate from accumulator	0 0 1 0 1 1 0 w	data	data if w 1	
SBB = **Subtract with borrow**				
Reg./memory and register to either	0 0 0 1 1 0 d w	mod reg r/m		
Immediate from register/memory	1 0 0 0 0 0 s w	mod 0 1 1 r/m	data	data if s w=01
Immediate from accumulator	0 0 0 1 1 1 0 w	data	data if w=1	

Instruction	7 6 5 4 3 2 1 0	7 6 5 4 3 2 1 0	7 6 5 4 3 2 1 0	7 6 5 4 3 2 1 0
DEC **Decrement**:				
Register/memory	1 1 1 1 1 1 1 w	mod 0 0 1 r/m		
Register	0 1 0 0 1 reg			
NEG Change sign	1 1 1 1 0 1 1 w	mod 0 1 1 r/m		
CMP **Compare**:				
Register/memory and register	0 0 1 1 1 0 d w	mod reg r/m		
Immediate with register/memory	1 0 0 0 0 0 s w	mod 1 1 1 r/m	data	data if s w 01
Immediate with accumulator	0 0 1 1 1 1 0 w	data	data if w 1	
AAS ASCII adjust for subtract	0 0 1 1 1 1 1 1			
DAS Decimal adjust for subtract	0 0 1 0 1 1 1 1			
MUL Multiply (unsigned)	1 1 1 1 0 1 1 w	mod 1 0 0 r/m		
IMUL Integer multiply (signed)	1 1 1 1 0 1 1 w	mod 1 0 1 r/m		
AAM ASCII adjust for multiply	1 1 0 1 0 1 0 0	0 0 0 0 1 0 1 0		
DIV Divide (unsigned)	1 1 1 1 0 1 1 w	mod 1 1 0 r/m		
IDIV Integer divide (signed)	1 1 1 1 0 1 1 w	mod 1 1 1 r/m		
AAD ASCII adjust for divide	1 1 0 1 0 1 0 1	0 0 0 0 1 0 1 0		
CBW Convert byte to word	1 0 0 1 1 0 0 0			
CWD Convert word to double word	1 0 0 1 1 0 0 1			
LOGIC				
NOT Invert	1 1 1 1 0 1 1 w	mod 0 1 0 r/m		
SHL/SAL Shift logical/arithmetic left	1 1 0 1 0 0 v w	mod 1 0 0 r/m		
SHR Shift logical right	1 1 0 1 0 0 v w	mod 1 0 1 r/m		
SAR Shift arithmetic right	1 1 0 1 0 0 v w	mod 1 1 1 r/m		
ROL Rotate left	1 1 0 1 0 0 v w	mod 0 0 0 r/m		
ROR Rotate right	1 1 0 1 0 0 v w	mod 0 0 1 r/m		
RCL Rotate through carry flag left	1 1 0 1 0 0 v w	mod 0 1 0 r/m		
RCR Rotate through carry right	1 1 0 1 0 0 v w	mod 0 1 1 r/m		
AND **And**:				
Reg./memory and register to either	0 0 1 0 0 0 d w	mod reg r/m		
Immediate to register/memory	1 0 0 0 0 0 0 w	mod 1 0 0 r/m	data	data if w 1
Immediate to accumulator	0 0 1 0 0 1 0 w	data	data if w 1	
TEST **And function to flags, no result**:				
Register/memory and register	1 0 0 0 0 1 0 w	mod reg r/m		
Immediate data and register/memory	1 1 1 1 0 1 1 w	mod 0 0 0 r/m	data	data if w 1
Immediate data and accumulator	1 0 1 0 1 0 0 w	data	data if w 1	
OR **Or**:				
Reg./memory and register to either	0 0 0 0 1 0 d w	mod reg r/m		
Immediate to register/memory	1 0 0 0 0 0 0 w	mod 0 0 1 r/m	data	data if w 1
Immediate to accumulator	0 0 0 0 1 1 0 w	data	data if w 1	
XOR **Exclusive or**:				
Reg./memory and register to either	0 0 1 1 0 0 d w	mod reg r/m		
Immediate to register/memory	1 0 0 0 0 0 0 w	mod 1 1 0 r/m	data	data if w 1
Immediate to accumulator	0 0 1 1 0 1 0 w	data	data if w 1	
STRING MANIPULATION				
REP=Repeat	1 1 1 1 0 0 1 z			
MOVS=Move byte/word	1 0 1 0 0 1 0 w			
CMPS=Compare byte/word	1 0 1 0 0 1 1 w			
SCAS=Scan byte/word	1 0 1 0 1 1 1 w			
LODS=Load byte/wd to AL/AX	1 0 1 0 1 1 0 w			
STOS=Stor byte/wd from AL/A	1 0 1 0 1 0 1 w			

Mnemonics ©Intel, 1978

Table 2. Instruction Set Summary (Continued)

Instruction	7 6 5 4 3 2 1 0	7 6 5 4 3 2 1 0	7 6 5 4 3 2 1 0
CONTROL TRANSFER			
CALL = **Call**:			
Direct within segment	1 1 1 0 1 0 0 0	disp-low	disp-high
Indirect within segment	1 1 1 1 1 1 1 1	mod 0 1 0 r/m	
Direct intersegment	1 0 0 1 1 0 1 0	offset-low	offset-high
		seg-low	seg-high
Indirect intersegment	1 1 1 1 1 1 1 1	mod 0 1 1 r/m	
JMP = **Unconditional Jump**:			
Direct within segment	1 1 1 0 1 0 0 1	disp-low	disp-high
Direct within segment-short	1 1 1 0 1 0 1 1	disp	
Indirect within segment	1 1 1 1 1 1 1 1	mod 1 0 0 r/m	
Direct intersegment	1 1 1 0 1 0 1 0	offset-low	offset-high
		seg-low	seg-high
Indirect intersegment	1 1 1 1 1 1 1 1	mod 1 0 1 r/m	
RET = **Return from CALL**:			
Within segment	1 1 0 0 0 0 1 1		
Within seg. adding immed to SP	1 1 0 0 0 0 1 0	data-low	data-high
Intersegment	1 1 0 0 1 0 1 1		
Intersegment, adding immediate to SP	1 1 0 0 1 0 1 0	data-low	data-high
JE/JZ=Jump on equal/zero	0 1 1 1 0 1 0 0	disp	
JL/JNGE=Jump on less/not greater or equal	0 1 1 1 1 1 0 0	disp	
JLE/JNG=Jump on less or equal/not greater	0 1 1 1 1 1 1 0	disp	
JB/JNAE=Jump on below/not above or equal	0 1 1 1 0 0 1 0	disp	
JBE/JNA=Jump on below or equal/not above	0 1 1 1 0 1 1 0	disp	
JP/JPE=Jump on parity/parity even	0 1 1 1 1 0 1 0	disp	
JO=Jump on overflow	0 1 1 1 0 0 0 0	disp	
JS=Jump on sign	0 1 1 1 1 0 0 0	disp	
JNE/JNZ=Jump on not equal/not zero	0 1 1 1 0 1 0 1	disp	
JNL/JGE=Jump on not less/greater or equal	0 1 1 1 1 1 0 1	disp	
JNLE/JG=Jump on not less or equal/greater	0 1 1 1 1 1 1 1	disp	

Instruction	7 6 5 4 3 2 1 0	7 6 5 4 3 2 1 0
JNB/JAE Jump on not below/above or equal	0 1 1 1 0 0 1 1	disp
JNBE/JA Jump on not below or equal/above	0 1 1 1 0 1 1 1	disp
JNP/JPO Jump on not par/par odd	0 1 1 1 1 0 1 1	disp
JNO Jump on not overflow	0 1 1 1 0 0 0 1	disp
JNS Jump on not sign	0 1 1 1 1 0 0 1	disp
LOOP Loop CX times	1 1 1 0 0 0 1 0	disp
LOOPZ/LOOPE Loop while zero/equal	1 1 1 0 0 0 0 1	disp
LOOPNZ/LOOPNE Loop while not zero/equal	1 1 1 0 0 0 0 0	disp
JCXZ Jump on CX zero	1 1 1 0 0 0 1 1	disp
INT **Interrupt**		
Type specified	1 1 0 0 1 1 0 1	type
Type 3	1 1 0 0 1 1 0 0	
INTO Interrupt on overflow	1 1 0 0 1 1 1 0	
IRET Interrupt return	1 1 0 0 1 1 1 1	
PROCESSOR CONTROL		
CLC Clear carry	1 1 1 1 1 0 0 0	
CMC Complement carry	1 1 1 1 0 1 0 1	
STC Set carry	1 1 1 1 1 0 0 1	
CLD Clear direction	1 1 1 1 1 1 0 0	
STD Set direction	1 1 1 1 1 1 0 1	
CLI Clear interrupt	1 1 1 1 1 0 1 0	
STI Set interrupt	1 1 1 1 1 0 1 1	
HLT Halt	1 1 1 1 0 1 0 0	
WAIT Wait*	1 0 0 1 1 0 1 1	
ESC Escape (to external device)	1 1 0 1 1 x x x	mod x x x r/m
LOCK Bus lock prefix	1 1 1 1 0 0 0 0	

Footnotes:

AL = 8-bit accumulator
AX = 16-bit accumulator
CX = Count register
DS = Data segment
ES = Extra segment
Above/below refers to unsigned value.
Greater = more positive;
Less = less positive (more negative) signed values
if d = 1 then "to" reg; if d = 0 then "from" reg
if w = 1 then word instruction; if w = 0 then byte instruction

if mod = 11 then r/m is treated as a REG field
if mod = 00 then DISP = 0*, disp-low and disp-high are absent
if mod = 01 then DISP = disp-low sign-extended to 16 bits, disp-high is absent
if mod = 10 then DISP = disp-high: disp-low

if r/m = 000 then EA = (BX) + (SI) + DISP
if r/m = 001 then EA = (BX) + (DI) + DISP
if r/m = 010 then EA = (BP) + (SI) + DISP
if r/m = 011 then EA = (BP) + (DI) + DISP
if r/m = 100 then EA = (SI) + DISP
if r/m = 101 then EA = (DI) + DISP
if r/m = 110 then EA = (BP) + DISP*
if r/m = 111 then EA = (BX) + DISP
DISP follows 2nd byte of instruction (before data if required)

*except if mod = 00 and r/m = 110 then EA = disp-high: disp-low.

if s:w = 01 then 16 bits of immediate data form the operand
if s:w = 11 then an immediate data byte is sign extended to form the 16-bit operand
if v = 0 then "count" = 1; if v = 1 then "count" in (CL)
x = don't care
z is used for string primitives for comparison with ZF FLAG.

SEGMENT OVERRIDE PREFIX

0 0 1 reg 1 1 0

REG is assigned according to the following table:

16-Bit (w = 1)		8-Bit (w = 0)		Segment	
000	AX	000	AL	00	ES
001	CX	001	CL	01	CS
010	DX	010	DL	10	SS
011	BX	011	BL	11	DS
100	SP	100	AH		
101	BP	101	CH		
110	SI	110	DH		
111	DI	111	BH		

Instructions which reference the flag register file as a 16-bit object use the symbol FLAGS to represent the file:

FLAGS = X:X:X:X:(OF):(DF):(IF):(TF):(SF):(ZF):X:(AF):X:(PF):X:(CF)

Mnemonics ©Intel, 1978

APPENDIX B

8086/8088 Instructions

Notes for 8086/8088 Instructions

The individual instruction descriptions are shown by a format box such as the following:

Opcode	m/op/r/m	– –	– –	Data	– –

These are byte-wise representations of the object code generated by the assembler and are interpreted as follows:

- Opcode is the 8-bit opcode for the instruction. The actual opcode generated is defined in the "Opcode" column of the instruction table that follows each format box.
- m/op/r/m is the byte that specifies the operands of the instruction. It contains a 2-bit mode field (m), a 3-bit register field (op), and a 3-bit register or memory (r/m) field.
- Dashed blank boxes following the m/op/r/m box are for any displacement required by the mode field.
- Data is for a byte of immediate data.
- A dashed blank box following a Data box is used whenever the immediate operand is a word quantity.

Operand Summary

"reg" field Bit Assignments:

Word Operand	Byte Operand	Segment
000 AX	000 AL	00 ES
001 CX	001 CL	01 CS
010 DX	010 DL	10 SS
011 BX	011 BL	11 DS
100 SP	100 AH	
101 BP	101 CH	
110 SI	110 DH	
111 DI	111 BH	

Second Instruction Byte Summary

mod xxx r/m

mod	Displacement
00	DISP = 0*, disp-low and disp-high are absent
01	DISP = disp-low sign-extended to 16-bits, disp-high is absent
10	DISP = disp-high: disp-low
11	r/m is treated as a "reg" field

r/m	Operand Address
000	(BX) + (SI) + DISP
001	(BX) + (DI) + DISP
010	(BP) + (SI) + DISP
011	(BP) + (DI) + DISP
100	(SI) + DISP
101	(DI) + DISP
110	(BP) + DISP*
111	(BX) + DISP

DISP follows 2nd byte of instruction (before data if required).

*except if mod = 00 and r/m = 110 then EA = disp-high: disp-low.

Flags

AF: AUXILIARY CARRY — BCD
CF: CARRY FLAG
DF: DIRECTION FLAG (STRINGS)
IF: INTERRUPT ENABLE FLAG
OF: OVERFLOW FLAG (CF SF)
PF: PARITY FLAG
SF: SIGN FLAG
TF: TRAP (SINGLE STEP FLAG)
ZF: ZERO FLAG

Instructions that reference the flag register file as a 16-bit object use the symbol FLAGS to represent the file:

15							8								0
X	X	X	X	OF	DF	IF	TF	SF	ZF	X	AF	X	PF	X	CF

X = Don't Care

Segment Override Prefix

0 0 1 reg 1 1 0

Timing: 2 clocks

Use of Segment Override

Operand Register	Default	With Override Prefix
IP (code address)	CS	Never
SP (stack address)	SS	Never
BP (stack address or stack marker)	SS	BP + DS or ES, or CS
SI or DI (not incl. strings)	DS	ES, SS, or CS
SI (implicit source addr for strings)	DS	ES, SS, or CS
DI (implicit dest addr for strings)	ES	Never

Operand Address (EA) Timing (Clocks):

Add 4 clocks for word operands at ODD ADDRESSES.
Immed Offset = 6
Base (BX, BP, SI, DI) = 5
Base + DISP = 9
Base + Index (BP + DI, BX + SI) = 7
Base + Index (BP + SI, BX + DI) = 8
Base + Index (BP + DI, BX + SI) + DISP = 11
Base + Index (BP + SI, BX + DI) + DISP = 12

AAA = ASCII Adjust for Addition

Opcode

Opcode	Clocks	Operation
37	4	adjust AL, flags, AH

AAD = ASCII Adjust for Division

Long——Opcode

Opcode	Clocks	Operation
D5,0A	60	Adjust AL, AH prior to division

AAM = ASCII Adjust for Multiplication

Long——Opcode

Opcode	Clocks	Operation
D4,0A	83	Adjust AL, AH after multiplication

AAS = ASCII Adjust for Subtraction

Opcode

Opcode	Clocks	Operation
3F	4	adjust AL, flags, AH

ADC = Integer Add with Carry

Memory/Reg + Reg

Opcode	mod reg r/m	– –	– –

	Opcode	Clocks	Operation
Byte	12	3	Reg8 ← CF + Reg 8 + Reg8
	12	9+EA	Reg8 ← CF + Reg8 + Mem8
	10	16+EA	Mem8 ← CF + Mem8 + Reg8
Word	13	3	Reg16 ← CF + Reg16 + Reg16
	13	9+EA	Reg16 ← CF + Reg16 + Mem16
	11	16+EA	Mem16 ← CF + Mem16 + Reg16

Immed to AX/AL

Opcode	Data	– –

	Opcode	Clocks	Operation
Byte	14	4	AL ← CF + AL + Immed8
Word	15	4	AX ← CF + AX + Immed16

Immed to Memory/Reg

Opcode	mod 010 r/m			Data	

	Opcode	Clocks	Operation
Byte	80	4	Reg8 ← CF + Reg8 + Immed8
	80	17+EA	Mem8 ← CF + Mem8 + Immed8
Word	81	4	Reg16 ← CF + Reg16 + Immed16
	81	17+EA	Mem16 ← CF + Mem16 + Immed16
	83	4	Reg16 ← CF + Reg16 + Immed8
	83	17+EA	Mem16 ← CF + Mem16 + Immed8

ADD = Integer Addition

Memory/Reg + Reg

Opcode	mod reg r/m		

	Opcode	Clocks	Operation
Byte	02	3	Reg8 ← Reg8 + Reg8
	02	9+EA	Reg8 ← Reg8 + Mem8
	00	16+EA	Mem8 ← Mem8 + Reg8
Word	03	3	Reg16 ← Reg16 + Reg16
	03	9+EA	Reg16 ← Reg16 + Mem16
	01	16+EA	Mem16 ← Mem16 + Reg16

Immed to AX/AL

Opcode	Data	

Opcode	Clocks	Operation
04	4	AL ← AL + Immed8
05	4	AX ← AX + Immed16

Immed to Memory/Reg

Opcode	mod 000 r/m			Data	

	Opcode	Clocks	Operation
Byte	80	4	Reg8 ← Reg8 + Immed8
	80	17+EA	Mem8 ← Mem8 + Immed8
Word	81	4	Reg16 ← Reg16 + Immed16
	81	17+EA	Mem16 ← Mem16 + Immed16
	83	4	Reg16 ← Reg16 + Immed8
	83	17+EA	Mem16 ← Mem16 + Immed8

AND = Logical AND

Memory/Reg with Reg

Opcode	mod reg r/m		

	Opcode	Clocks	Operation
Byte	22	3	Reg8 ← Reg8 AND Reg8
	22	9+EA	Reg8 ← Reg8 AND Mem8
	20	16+EA	Mem8 ← Mem8 AND Reg8
Word	23	3	Reg16 ← Reg16 AND Reg16
	23	9+EA	Reg16 ← Reg16 AND Mem16
	21	16+EA	Mem16 ← Mem16 AND Reg16

Immed to AX/AL

Opcode	Data	

	Opcode	Clocks	Operation
Byte	24	4	AL ← AL AND Immed8
Word	25	4	AX ← AX AND Immed16

Immed to Memory/Reg

Opcode	mod 100 r/m			Data	

	Opcode	Clocks	Operation
Byte	80	4	Reg8 ← Reg8 AND Immed8
	80	17+EA	Mem8 ← Mem8 AND Immed8
Word	81	4	Reg16 ← Reg16 AND Immed16
	81	17+EA	Mem16 ← Mem16 AND Immed16

CALL = Call

Within segment or group, IP relative

Opcode	DispL	DispH

Opcode	Clocks	Operation
E8	19	IP ← IP + Disp16—(SP) ← return link

Within segment or group, Indirect

Opcode	mod 010 r/m		

Opcode	Clocks	Operation
FF	16	IP ← Reg16—(SP) ← return link
FF	21+EA	IP ← Mem16—(SP) ← return link
FF	21+EA	IP ← Mem16—(SP) ← return link

Inter-segment or group, Direct

Opcode	offset	offset	segbase	segbase	segbase

Opcode	Clocks	Operation
9A	28	CS ← segbase IP ← offset

Inter-segment or group, Indirect

Opcode	mod 011 r/m		

Opcode	Clocks	Operation
FF	37+EA	CS ← segbase IP ← offset

CBW = Convert Byte to Word

Opcode

Opcode	Clocks	Operation
98	2	convert byte in AL to word in AX

CLC = Clear Carry Flag

Opcode

Opcode	Clocks	Operation
F8	2	clear the carry flag

CLD = Clear Direction Flag

Opcode

Opcode	Clocks	Operation
FC	2	clear direction flag

CLI = Clear Interrupt Enable Flag

Opcode	Clocks	Operation
FA	2	clear interrupt flag

CMC = Complement Carry Flag

Opcode

Opcode	Clocks	Operation
F5	2	complement carry flag

CMP = Compare Two Operands

Memory/Reg with Reg

Opcode	mod reg r/m		

	Opcode	Clocks	Operation
Byte	38	3	flags ← Reg8 - Reg8
	38	9+EA	flags ← Reg8 - Mem8
	3A	9+EA	flags ← Mem8 - Reg8
Word	39	3	flags ← Reg16 - Reg16
	39	9+EA	flags ← Reg16 - Mem16
	3B	9+EA	flags ← Mem16 - Reg16

Immed to AX/AL

Opcode	Data	

	Opcode	Clocks	Operation
Byte	3C	4	flags AL - Immed8
Word	3D	4	flags AX - Immed16

Immed to Memory/Reg

`Opcode | mod 111 r/m | [ ] [ ] | Data | [ ]`

	Opcode	Clocks	Operation
Byte	80	4	flags ← Reg8 - Immed8
	80	10+EA	flags ← Mem8 - Immed8
Word	81	4	flags ← Reg16 - Immed16
	81	10+EA	flags ← Mem16 - Immed16
	83	4	flags ← Reg16 - Immed8
	83	10+EA	flags ← Mem16 - Immed8

CWD — Convert Word to Doubleword

`Opcode`

Opcode	Clocks	Operation
99	5	convert word in AX to doubleword in DX:AX

DAA = Decimal Adjust for Addition

`Opcode`

Opcode	Clocks	Operation
27	4	adjust AL, flags, AH

DAS = Decimal Adjust for Subtraction

`Opcode`

Opcode	Clocks	Operation
2F	4	adjust AL, flags, AH

DEC = Decrement by 1

Word Register

`Opcode + reg`

Opcode	Clocks	Operation
48+reg	2	Reg16 ← Reg16 - 1

Memory/Byte Register

`Opcode | mod 001 r/m | [ ] [ ]`

	Opcode	Clocks	Operation
Byte	FE	3	Reg8 ← Reg8 - 1
	FE	15+EA	Mem8 ← Mem8 - 1
Word	FF	15+EA	Mem16 ← Mem16 - 1

DIV = Unsigned Division

Memory/Reg with AX or DX:AX

`Opcode | mod 110 r/m | [ ] [ ]`

	Opcode	Clocks	Operation
Byte	F6	80-90	AH,AL ← AX / Reg8
	F6	(86-96)+EA	AH,AL ← AX / Mem8
Word	F7	144-162	DX,AX ← DX:AX / Reg16
	F7	(150-168)+EA	DX,AX ← DX:AX / Mem16

ESC = Escape

`Opcode + i | mod xxx r/m | [ ] [ ]`

Opcode	Clocks	Operation
D8+i	8+EA	data bus ← (EA)
D8+i	2	data bus ← (EA)

HLT = Halt

`Opcode`

Opcode	Clocks	Operation
F4	2	halt operation

IDIV = Signed Division

Memory/Reg with AX or DX:AX

`Opcode | mod 111 r/m | [ ] [ ]`

	Opcode	Clocks	Operation
Byte	F6	101-112	AH,AL ← AX / Reg8
	F6	(107-118)+EA	AH,AL ← AX / Mem8
Word	F7	165-184	DX,AX ← DX:AX / Reg16
	F7	(171-190)+EA	DX,AX ← DX:AX / Mem16

IMUL = Signed Multiplication

Memory/Reg with AL or AX

`Opcode | mod 101 r/m | [ ] [ ]`

	Opcode	Clocks	Operation
Byte	F6	80-98	AX ← AL*Reg8
	F6	(86-104)+EA	AX ← AL*Mem8
Word	F7	128-154	DX:AX ← AX*Reg16
	F7	(134-160)+EA	DX:AX ← AX*Mem16

IN = Input Byte, Word

Fixed port

`Opcode | Port`

	Opcode	Clocks	Operation
Byte	E4	10	AL ← Port8
	E5	10	AX ← Port8

Variable port

`Opcode`

	Opcode	Clocks	Operation
Word	EC	8	AL ← Port16(in DX)
	ED	8	AX ← Port16(in DX)

INC = Increment by 1

Word Register

`Opcode + reg`

Opcode	Clocks	Operation
40+reg	2	Reg16 ← Reg16 + 1

Memory/Byte Register

`Opcode | mod 000 r/m | [ ] [ ]`

	Opcode	Clocks	Operation
Byte	FE	3	Reg8 ← Reg8 + 1
	FE	15+EA	Mem8 ← Mem8 + 1
Word	FF	15+EA	Mem16 ← Mem16 + 1

INT / INTO = Interrupt

`Opcode | type`

Opcode	Clocks	Operation
CC	52	Interrupt 3
CD	51	Interrupt 'type'
CE	53 or 4	Interrupt4 if FLAGS.OF=1, else NOP

IRET = Return from Interrupt

`Opcode`

Opcode	Clocks	Operation
CF	24	Return from interrupt

Jcond = Jump on Condition

Operation

```
if condition is true then do;
  sign-extend displacement to 16 bits;
  IP ← IP + sign-extended displacement;
  end if;
```

Format

`Opcode | Disp`

Opcode	Clocks	Operation	*cond* =
77	16 or 4	jump if above	JA
73	16 or 4	jump if above or equal	JAE
72	16 or 4	jump if below	JB
76	16 or 4	jump if below or equal	JBE
72	16 or 4	jump if carry set	JC
74	16 or 4	jump if equal	JE
7F	16 or 4	jump if greater	JG
7D	16 or 4	jump if greater or equal	JGE
7C	16 or 4	jump if less	JL
7E	16 or 4	jump if less or equal	JLE
76	16 or 4	jump if not above	JNA
72	16 or 4	jump if neither above nor equal	JNAE
73	16 or 4	jump if not below	JNB
77	16 or 4	jump if neither below nor equal	JNBE
73	16 or 4	jump if no carry	JNC
75	16 or 4	jump if not equal	JNE
7E	16 or 4	jump if not greater	JNG
7C	16 or 4	jump if neither greater nor equal	JNGE
7D	16 or 4	jump if not less	JNL
7F	16 or 4	jump if neither less nor equal	JNLE
71	16 or 4	jump if no overflow	JNO
7B	16 or 4	jump if no parity	JNP
79	16 or 4	jump if positive	JNS
75	16 or 4	jump if not zero	JNZ
70	16 or 4	jump if overflow	JO
7A	16 or 4	jump if parity	JP
7A	16 or 4	jump if parity even	JPE
7B	16 or 4	jump if parity odd	JPO
78	16 or 4	jump if sign	JS
74	18 or 6	jump if zero	JZ
E3	18 or 6	jump if CX is zero (does not test flags)	JCXZ

JMP = Jump

Within segment or group, IP relative

Opcode	DispL	DispH

Opcode	Clocks	Operation
E9	15	IP ← IP + Disp16
EB	15	IP ← IP + Disp8 (Disp8 sign-extended)

Within segment or group, Indirect

Opcode	mod 100 r/m		

Opcode	Clocks	Operation
FF	11	IP ← Reg16
FF	18+EA	IP ← Mem16
FF	18+EA	IP ← Mem16

Inter-segment or group, Direct

Opcode	offset	offset	segbase	segbase

Opcode	Clocks	Operation
EA	15	CS ← segbase IP ← offset

Inter-segment or group, Indirect

Opcode	mod 101 r/m		

Opcode	Clocks	Operation
FF	24+EA	CS ← segbase IP ← offset

LAHF = Load AH from Flags

Opcode

Opcode	Clocks	Operation
9F	4	copy low byte of flags word to AH

LDS/LES = Load Pointer to DS/ES and Register

Opcode	mod reg r/m		

Opcode	Clocks	Operation
C4	16+EA	dword pointer at EA goes to reg16 (1st word) and ES (2nd word)
C5	16+EA	dword pointer at EA goes to reg16 (1st word) and DS (2nd word)

LEA = Load Effective Address

Opcode	mod reg r/m		

Opcode	Clocks	Operation
8D	2+EA	Reg16 ← EA

LOCK = Assert Bus Lock

Opcode

Opcode	Clocks	Operation
F0	2	assert the bus lock next instruction

LOOP*xx* = Loop Control

Opcode	Disp

Opcode	Clocks	Operation	*xx* =
E1	18 or 6	dec CX; loop if equal and CX not 0	LOOPE
E0	19 or 5	dec CX; loop if not equal and CX not 0	LOOPNE
E1	18 or 6	dec CX; loop if zero and CX not 0	LOOPZ
E0	19 or 5	dec CX; loop if not zero and CX not 0	LOOPNZ
E2	17 or 5	dec CX; loop if CX not 0	LOOP

MOV = Move Data

Memory/Reg to or from Reg

Opcode	mod reg r/m		

	Opcode	Clocks	Operation
Byte	88	9+EA	Mem8 ← Reg8
	88	2	Reg8 ← Reg8
	8A	8+EA	Reg8 ← Mem8
Word	89	9+EA	Mem16 ← Reg16
	89	2	Reg16 ← Reg16
	8B	8+EA	Reg16 ← Mem16

Direct-Addressed Memory to or from AX/AL

Opcode	AddrL	AddrH

	Opcode	Clocks	Operation
Byte	A0	10	AL ← Mem8
	A2	10	Mem8 ← AL
Word	A1	10	AX ← Mem16
	A3	10	Mem16 ← AX

Immed to Reg

Opcode	Data	

	Opcode	Clocks	Operation
Byte	B0+reg	4	Reg 8 ← Immed8
Word	B8+reg	4	Reg16 ← Immed16

Immed to Memory/Reg

Opcode	mod 000 r/m			Data	

Opcode	Clocks	Operation
C6	4	Reg8 ← Immed8
C6	10+EA	Mem8 ← Immed8
C7	4	Reg16 ← Immed16
C7	10+EA	Mem16 ← Immed16

Memory/Reg to or from SReg

Opcode	mod sreg r/m		

	Opcode	Clocks	Operation
Word	8C	9+EA	Mem16 ← SReg
	8C	2	Reg16 ← SReg
	8E	8+EA	SReg ← Mem16
	8E	2	SReg ← Reg16

MUL = Unsigned Multiplication

Memory/Reg with AL or AX

Opcode	mod 100 r/m		

	Opcode	Clocks	Operation
Byte	F6	70-77	AX ← AL*Reg8
	F6	(76-83)+EA	AX ← AL*Mem8
Word	F7	118-133	DX:AX ← AX*Reg16
	F7	(124-139)+EA	DX:AX ← AX*Mem16

NEG = Negate an Integer

Memory/Reg

Opcode	mod 011 r/m		

Opcode	Clocks	Operation
F6	3	Reg8←00H - Reg 8
F7	3	Reg16←0000H - Reg16
F6	16+EA	Mem8←00H - Mem8
F7	16+EA	Mem16←0000H - Mem16

NOP = No Operation

Opcode

Opcode	Clocks	Operation
90	3	no operation

NOT = Form One's Complement

Memory/Reg

Opcode	mod 010 r/m		

	Opcode	Clocks	Operation
Byte	F6	3	Reg8←0FFH - Reg8
	F6	16+EA	Mem8←0FFH - Mem8
Word	F7	3	Reg16←0FFFFH - Reg16
	F7	16+EA	Mem16←0FFFFH - Mem16

OR = Logical Inclusive OR

Memory/Reg with Reg

Opcode	mod reg r/m		

	Opcode	Clocks	Operation
Byte	0A	3	Reg8←Reg8 OR Reg8
	0A	9+EA	Reg8←Reg8 OR Mem8
	08	16+EA	Mem8←Mem8 OR Reg8
Word	0B	3	Reg16←Reg16 OR Reg 16
	0B	9+EA	Reg16←Reg16 OR Mem16
	09	16+EA	Mem16←Mem16 OR Reg16

Immed to AX/AL

Opcode	Data	

Opcode	Clocks	Operation
0C	4	AL←AL OR Immed8
0D	4	AX←AX OR Immed16

Immed to Memory/Reg

Opcode	mod 001 r/m			Data	

	Opcode	Clocks	Operation
Byte	80	4	Reg8←Reg8 OR Immed8
	80	17+EA	Mem8←Mem8 OR Immed8
Word	81	4	Reg16←Reg16 OR Immed16
	81	17+EA	Mem16←Mem16 OR Immed16

OUT = Output Byte, Word

Fixed port

Opcode	Port

	Opcode	Clocks	Operation
Byte	E6	10	Port8←AL
	E7	10	Port8←AX

Variable port

Opcode

	Opcode	Clocks	Operation
Word	EE	8	Port16 (in DX)←AL
	EF	8	Port16 (in DX)←AX

POP = Pop a Word from the Stack

Word Memory

Opcode	mod 000 r/m		

Opcode	Clocks	Operation
8F	17+EA	Mem16←(SP)++

Word Register

Opcode + reg

Opcode	Clocks	Operation
58+reg	8	Reg16←(SP)++

Segment Register

Opcode + SReg

Opcode	Clocks	Operation
07+SReg	8	SReg←(SP)++

POPF = Pop the TOS into the Flags

Opcode

Opcode	Clocks	Operation
9D	8	FLAGS←(SP)++

PUSH = Push a Word onto the Stack

Memory/Reg

Opcode	mod 110 r/m		

Opcode	Clocks	Operation
FF	16+EA	—(SP)←Mem16

Word Register

Opcode + reg

Opcode	Clocks	Operation
50+reg	11	—(SP)←Reg16

Segment Register

Opcode + SReg

Opcode	Clocks	Operation
06+SReg	10	—(SP)←SReg

PUSHF = Push the Flags to the Stack

Opcode

Opcode	Clocks	Operation
9C	10	—(SP)←FLAGS

RCL = Rotate Left Through Carry

Memory or Reg by 1

Opcode	mod 010 r/m		

	Opcode	Clocks	Operation
Byte	D0	2	rotate Reg 8 by 1
	D0	15+EA	rotate Mem8 by 1
Word	D1	2	rotate Reg 16 by 1
	D1	15+EA	rotate Mem16 by 1

Memory or Reg by count in CL

Opcode	mod 010 r/m		

	Opcode	Clocks	Operation
Byte	D2	8+4/bit	rotate Reg8 by CL
	D2	20+EA+4/bit	rotate Mem8 by CL
Word	D3	8+4/bit	rotate Reg16 by CL
	D3	20+EA+4/bit	rotate Mem16 by CL

RCR = Rotate Right Through Carry

Memory or Reg by 1

Opcode	mod 011 r/m		

	Opcode	Clocks	Operation
Byte	D0	2	rotate Reg8 by 1
	D0	15+EA	rotate Mem8 by 1
Word	D1	2	rotate Reg16 by 1
	D1	15+EA	rotate Mem16 by 1

Memory or Reg by count in CL

Opcode	mod 011 r/m		

	Opcode	Clocks	Operation
Byte	D2	8+4/bit	rotate Reg8 by CL
	D2	20+EA+4/bit	rotate Mem8 by CL
Word	D3	8+4/bit	rotate Reg16 by CL
	D3	20+EA+4/bit	rotate Mem16 by CL

REP*x* = Repeat Prefix

Opcode

Opcode	Clocks	Operation	REP*x* =
F3	2	repeat next instruction until CX=0	REP
F3	2	repeat next instruction until CX=0 or ZF=1	REPE REPZ
F2	2	repeat next instruction until CX=0 or ZF=0	REPNE REPNZ

RET = Return from Subroutine

Opcode	

Opcode	Clocks	Operation
C3	8	intra-segment return
CB	18	inter-segment return

Return and add constant to SP

Opcode	DataL	DataH

Opcode	Clocks	Operation
C2	12	intra-segment ret and add
CA	17	inter-segment ret and add

ROL = Rotate Left

Memory or Reg by 1

Opcode	mod 010 r/m		

	Opcode	Clocks	Operation
Byte	D0	2	rotate Reg8 by 1
	D0	15+EA	rotate Mem8 by 1
Word	D1	2	rotate Reg16 by 1
	D1	15+EA	rotate Mem16 by 1

Memory or Reg by count in CL

Opcode	mod 010 r/m		

	Opcode	Clocks	Operation
Byte	D2	8+4/bit	rotate Reg8 by CL
	D2	20+Ea+4/bit	rotate Mem8 by CL
Word	D3	8+4/bit	rotate Reg16 by CL
	D3	20+EA+4/bit	rotate Mem16 by CL

ROR = Rotate Right

Memory or Reg by 1

Opcode	mod 011 r/m		

	Opcode	Clocks	Operation
Byte	D0	2	rotate Reg8 by 1
	D0	15+EA	rotate Mem8 by 1
Word	D1	2	rotate Reg16 by 1
	D1	15+EA	rotate Mem16 by 1

Memory or Reg by count in CL

Opcode	mod 011 r/m		

	Opcode	Clocks	Operation
Byte	D2	8+4/bit	rotate Reg8 by CL
	D2	20+EA+4/bit	rotate Mem8 by CL
	D3	8+4/bit	rotate Reg16 by CL
	D3	20+EA+4/bit	rotate Mem16 by CL

SAHF = Store AH in Flags

Opcode	

Opcode	Clocks	Operation
9E	4	copy AH to low byte of flags word

SAL/SHL = Arithmetic/Logical Left Shift

Memory or Reg by 1

Opcode	mod 100 r/m		

	Opcode	Clocks	Operation
Byte	D0	2	shift Reg8 by 1
	D0	15+EA	shift Mem8 by 1
Word	D1	2	shift Reg16 by 1
	D1	15+EA	shift Mem16 by 1

Memory or Reg by count in CL

Opcode	mod 100 r/m		

	Opcode	Clocks	Operation
Byte	D2	8+4/bit	shift Reg8 by CL
	D2	20+EA+4/bit	shift Mem8 by CL
Word	D3	8+4/bit	shift Reg16 by CL
	D3	20+EA+4/bit	shift Mem16 by CL

SAR = Arithmetic Right Shift

Memory or Reg by 1

Opcode	mod 111 r/m		

	Opcode	Clocks	Operation
Byte	D0	2	shift Reg8 by 1
	D0	15+EA	shift Mem8 by 1
Word	D1	2	shift Reg16 by 1
	D1	15+EA	shift Mem16 by 1

Memory or Reg by count in CL

Opcode	mod 111 r/m		

	Opcode	Clocks	Operation
Byte	D2	8+4/bit	shift Reg8 by CL
	D2	20+EA+4/bit	shift Mem8 by CL
Word	D3	8+4/bit	shift Reg16 by CL
	D3	20+EA+4/bit	shift Mem16 by CL

SBB = Integer Subtraction with Borrow

Memory/Reg with Reg

Opcode	mod reg r/m		

	Opcode	Clocks	Operation
Byte	1A	3	Reg8 ← Reg8 - Reg8 - CF
	1A	9+EA	Reg8 ← Reg8 - Mem8 - CF
	18	16+EA	Mem8 ← Mem8 - Reg8 - CF
Word	1B	3	Reg16 ← Reg16 - Reg16 - CF
	1B	9+EA	Reg16 ← Reg16 - Mem16 - CF
	19	16+EA	Mem16 ← Mem16 - Reg16 - CF

Immed from AX/AL

Opcode	Data	

Opcode	Clocks	Operation
1C	4	AL ← AL - Immed8 - CF
1D	4	AX ← AX - Immed16 - CF

Immed from Memory/Reg

Opcode	mod 011 r/m		Data	

Opcode	Clocks	Operation
80	4	Reg8 ← Reg8 - Immed8 - CF
80	17+EA	Mem8 ← Mem8 - Immed8 - CF
81	4	Reg16 ← Reg16 - Immed16 - CF
81	17+EA	Mem16 ← Mem16 - Immed16 - CF
83	4	Reg16 ← Reg16 - Immed8 - CF
83	17+EA	Mem16 ← Mem16 - Immed8 - CF (Immed8 is sign-extended before subtract)

SHR = Logical Right Shift

Memory or Reg by 1

Opcode	mod 101 r/m		

	Opcode	Clocks	Operation
Byte	D0	2	shift Reg8 by 1
	D0	15+EA	shift Mem8 by 1
Word	D1	2	shift Reg16 by 1
	D1	15+EA	shift Mem16 by 1

Memory or Reg by count in CL

Opcode	mod 101 r/m		

	Opcode	Clocks	Operation
Byte	D2	8+4/bit	shift Reg8 by CL
	D2	20+Ea+4/bit	shift Mem8 by CL
Word	D3	8+4/bit	shift Reg16 by CL
	D3	20+EA+4/bit	shift Mem16 by CL

STC = Set Carry Flag

Opcode	

Opcode	Clocks	Operation
F9	2	set the carry flag

STD = Set Direction Flags

Opcode	

Opcode	Clocks	Operation
FD	2	set direction flag

STI = Set Interrupt Enable Flag

Opcode	

Opcode	Clocks	Operation
FB	2	set interrupt flag

***String* = String Operations**

Opcode	

Opcode	Clocks	Operation	*String* =
A6	22	flags ← (SI) - (DI)	CMPS

A7	22	flags ← (SI) - (DI)	CMPS
A4	18	(DI) ← (SI)	MOVS
A5	18	(DI) ← (SI)	MOVS
AE	15	flags ← (DI) - AL	SCAS
AF	15	flags ← (DI) - AX	SCAS
AC	12	AL ← (SI)	LODS
AD	12	AX ← (SI)	LODS
AA	11	(DI) ← AL	STOS
AB	11	(DI) ← AX	STOS

SUB = Integer Subtraction

Memory/Reg with Reg

Opcode	mod reg r/m		

	Opcode	Clocks	Operation
Byte	2A	3	Reg8 Reg8 - Reg8
	2A	9+EA	Reg8 Reg8 - Mem8
	28	16+EA	Mem8 Mem8 - Reg8
Word	2B	3	Reg16 Reg16 - Reg16
	2B	9+EA	Reg16 Reg16 - Mem16
	29	16+EA	Mem16 Mem16 - Reg16

Immed to AX/AL

Opcode	Data	

	Opcode	Clocks	Operation
Byte	2C	4	AL ← AL - Immed8
Word	2D	4	AX ← AX - Immed16

Immed to Memory/Reg

Opcode	mod 101 r/m			Data	

	Opcode	Clocks	Operation
Byte	80	4	Reg8 ← Reg8 - Immed8
	80	17+EA	Mem8 ← Mem8 - Immed8
Word	81	4	Reg16 ← Reg16 - Immed16
	81	17+EA	Mem16 ← Mem16 - Immed16
	83	4	Reg16 ← Reg16 - Immed8
	83	17+EA	Mem16 ← Mem16 - Immed8

TEST = Logical Compare

Memory/Reg with Reg

Opcode	mod reg r/m		

	Opcode	Clocks	Operation
Byte	84	3	flags ← Reg8 AND Reg8
	84	9+EA	flags ← Reg8 AND Mem8
Word	85	3	flags ← Reg16 AND Reg16
	85	9+EA	flags ← Reg16 AND Mem16

Immed to AX/AL

Opcode	Data	

	Opcode	Clocks	Operation
Byte	A8	4	flags ← AL AND Immed8
Word	A9	4	flags ← AX AND Immed16

Immed to Memory/Reg

Opcode	mod 000 r/m			Data	

	Opcode	Clocks	Operation
Byte	F6	5	flags ← Reg8 AND Immed8
	F6	11+EA	flags ← Mem8 AND Immed8
Word	F7	5	flags ← Reg16 AND Immed16
	F7	11+EA	flags ← Mem16 AND Immed16

WAIT = Wait While TEST Pin Not Asserted

Opcode

Opcode	Clocks	Operation
9B	3+5n	none

XCHG = Exchange Memory/Register with Register

Memory/Reg with Reg

Opcode	mod reg r/m		

	Opcode	Clocks	Operation
Byte	86	4	Reg8 ↔ Reg8
	86	17+EA	Mem8 ↔ Mem8
Word	87	4	Reg16 ↔ Reg16
	87	17+EA	Mem16 ↔ Mem16

Word Register with AX

Opcode + Reg

Opcode	Clocks	Operation
90+Reg	3	AX ↔ Reg16

XLAT XLATB = Table Look-up Translation

Opcode

Opcode	Clocks	Operation
D7	11	replace AL with table entry

XOR = Logical Exclusive OR

Memory/Reg with Reg

Opcode	mod reg r/m		

	Opcode	Clocks	Operation
Byte	32	3	Reg8 ← Reg8 XOR Reg8
	32	9+EA	Reg8 ← Reg8 XOR Mem8
	30	16+EA	Mem8 ← Mem8 XOR Reg8
Word	33	3	Reg16 ← Reg16 XOR Reg16
	33	9+EA	Reg16 ← Reg16 XOR Mem16
	31	16+EA	Mem16 ← Mem16 XOR Reg16

Immed to AX/AL

Opcode	Data	

	Opcode	Clocks	Operation
	34	4	AL ← AL XOR Immed8
	35	4	AX ← AX XOR Immed16

Immed to Memory/Reg

Opcode	mod 110 r/m			Data	

	Opcode	Clocks	Operation
Byte	80	4	Reg8 ← Reg8 XOR Immed8
	80	17+EA	Mem8 ← Mem8 XOR Immed8
Word	81	4	Reg16 ← Reg16 XOR Immed16
	81	17+EA	Mem16 ← Mem16 XOR Immed16

186 INSTRUCTIONS

Notes for iAPX 186 Instructions

These instructions can be used only if the MOD186 control is specified. When MOD186 is specified, clocks for all instructions are as stated under "Clocks for MOD186 Operation."

BOUND = Check Array Against Bounds

Opcode	ModRM		

Opcode	Operation
62	if Reg16<Mem16 at EA, or Reg16>Mem16 at EA+2 then INTERRUPT 5

ENTER = High Level Procedure Entry

Opcode	DataL	DataH	Level

Opcode	Operation
C8	build new stack frame

IMUL = Signed Multiplication

Mem/Reg* Immediate to Reg

Opcode	ModRM			Data	

Opcode	Operation
6B	Reg 16 ← Reg 16 * Immed 8
6B	Reg 16 ← Reg 16 * Immed 8
6B	Reg 16 ← Mem 16 * Immed 8
69	Reg 16 ← Reg 16 * Immed 16
69	Reg 16 ← Reg 16 * Immed 16
69	Reg 16 ← Mem 16 * Immed 16

LEAVE = High Level Procedure Exit

Opcode

Opcode	Operation
C9	release current stack frame and return to prior frame.

POPA = Pop All Registers

Opcode

Opcode	Operation
61	restore registers from stack

PUSH = Push a Word onto the Stack

Word Immediate

Opcode	Data		

Opcode	Operation
6A	—(SP) ← Immed8 (sign extended)
68	—(SP) ← Immed16

PUSHA = Push All Registers

Opcode

Opcode	Operation
60	save registers on the stack

RCL = Rotate Left Through Carry

Mem or Reg by Immed8

Opcode	ModRM*				count

*—(Reg field = 011)

Opcode	Operation
C0	rotate Reg8 by Immed8
C0	rotate Mem8 by Immed8
C1	rotate Reg16 by Immed8
C1	rotate Mem16 by Immed8

RCR = Rotate Right Through Carry

Mem or Reg by Immed8

Opcode	ModRM*				count

*—(Reg field = 011)

Opcode	Operation
C0	rotate Reg8 by Immed8
C0	rotate Mem8 by Immed8
C1	rotate Reg16 by Immed8
C1	rotate Mem16 by Immed8

ROL = Rotate Left

Mem or Reg by Immed8

Opcode	ModRM*				count

*—(Reg field = 000)

Opcode	Operation
C0	rotate Reg8 by Immed8
C0	rotate Mem8 by Immed8
C1	rotate Reg16 bv Immed8
C1	rotate Mem16 by Immed8

ROR = Rotate Right

Mem or Reg by Immed8

Opcode	ModRM*				count

*—(Reg field = 001)

Opcode	Operation
C0	rotate Reg8 by Immed8
C0	rotate Mem8 by Immed8
C1	rotate Reg16 by Immed8
C1	rotate Mem16 by Immed8

SAL/SHL = Arithmetic/Logical Left Shift

Mem or Reg by immediate count

Opcode	ModRM*				count

*—(Reg field = 100)

Opcode	Operation
C0	rotate Reg8 by Immed8
C0	rotate Mem8 by Immed8
C1	rotate Reg16 by Immed8
C1	rotate Mem16 by Immed8

SAR = Arithmetic Right Shift

Mem or Reg by Immed8

Opcode	ModRM*				count

*—(Reg field = 111)

Opcode	Operation
C0	rotate Reg8 by Immed8
C0	rotate Mem8 by Immed8
C1	rotate Reg16 by Immed8
C1	rotate Mem16 by Immed8

SHR = Logical Right Shift

Mem or Reg by Immed8

Opcode	ModRM*				count

*—(Reg field = 101)

Opcode	Operation
C0	rotate Reg8 by Immed8
C0	rotate Mem8 by Immed8
C1	rotate Reg16 by Immed8
C1	rotate Mem16 by Immed8

String = String Operations (INS/OUTS)

Opcode

Opcode	Clocks	Operation
6E	INS	(DI) ← port(DX)
6F	INS	(DI) ← port(DX:DX+1)
6C	OUTS	port(DX) ← (SI)
6D	OUTS	port(DX:DX+1) ← (SI)

8087 INSTRUCTIONS

Notes for 8087 Instructions

The individual instruction descriptions are shown by a format box such as the following:

WAIT	op1	m/op/r/m	addr1	addr2

These are the byte-wise representations of the object code generated by the assembler and are interpreted as follows:

- WAIT is an 8086 wait instruction, NOP or emulator instruction.
- op1 is the opcode, possibly taking two bytes.
- m/op/r/m byte (middle 3-bits is part of the opcode).
- addr1 and addr2 are offsets of either 8 or 16 bits.

For integer functions, m = 0 for short-integer memory operand; 1 for word-integer memory operand.
For real functions, m = 0 for short-real memory operand; 1 for longreal memory operand.
i = stack element index.
If mod = 00 then DISP = 0, disp-lo and disp-hi are absent.
If mod = 01 then DISP = disp-lo sign-extended to 16 bits, disp-hi is absent.
If mod = 10 then DISP = disp-hi; disp-lo.
If mod = 11 then r/m is treated as an ST(i) field.

If r/m = 000 then EA = (BX)+(SI)+DISP
If r/m = 001 then EA = (BX)+(DI)+DISP
If r/m = 010 then EA = (BP)+(SI)+DISP
If r/m = 011 then EA = (BP)+(DI)+DISP
If r/m = 100 then EA = (SI)+DISP
If r/m = 101 then EA = (DI)+DISP
If r/m = 110 then EA (BP)+DISP*
If r/m = 111 then EA = (BX)+DISP
*Except if mod = 000 and r/m = 110 then EA = disp-hi; disp-lo.

ST(0) = Current stack top
ST(i) = i^{th} register below stack top
d = Destination
0 — Destination is ST(0)
1 — Destination is ST(i)
P = Pop
0 — No pop
1 — Pop ST(0)
R = Reverse
0 — Destination (op) source
1 — Source (op) destination

For FSQRT:	$-0 \leq ST(0) \leq +\infty$
For FSCALE:	$-2^{15} \leq ST(1) < +2^{15}$ and ST(1) integer
For F2XM1:	$0 \leq ST(0) \leq 2^{-1}$
For FYL2X:	$0 \leq ST(0) < \infty$ $-\infty < ST(1) < +\infty$
For FYL2XP1:	$0 < \lvert ST(0) \rvert < (2-\sqrt{2})/2$ $-\infty \leq ST(1) < \infty$
For FPTAN:	$0 \leq ST(0) < \pi/4$
For FPATAN:	$0 \leq ST(0) < ST(1) < +\infty$

F2XMI = Compute 2^x — 1

WAIT	op1	op2

8087 Encoding	Emulator Encoding	Execution Clocks Typical Range	Operation
9B D9 F0	CD 19 F0	500 310-630	ST ← 2ST-1

FABS = Absolute Value

WAIT	op1	op2

8087 Encoding	Emulator Encoding	Execution Clocks Typical Range	Operation
9B D9 E1	CD 19 E1	14 10-17	ST ← |ST|

FADD = Add Real

Stack top + Stack element

WAIT	op1	op2 + i

8087 Encoding	Emulator Encoding	Execution Clocks Typical Range	Operation
9B D8 C0+i	CD 18 C0+i	85 70-100	ST ← ST + ST(i)
9B DC C0+i	CD 1C C0+i	85 70-100	ST(i) ← ST + ST(i)

Stack top + memory operand

WAIT	op1	mod 000 r/m	addr1	addr2

8087 Encoding	Emulator Encoding	Execution Clocks Typical Range	Operation
9B D8 m0rm	CD 18 m0rm	105+EA (90-120)+EA	ST ← ST + mem-op (short-real)
9B DC m0rm	CD 1C m0rm	110+EA (95-125)+EA	ST ← ST + mem-op (long-real)

FADDP = Add Real and Pop

Stack top + Stack Element

WAIT	op1	op2 + i

8087 Encoding	Emulator Encoding	Execution Clocks Typical Range	Operation
9B DE C1	CD 1E C1	90 75-105	ST(1) ← ST + ST(1) pop stack
9B DE C0+i	CD 1E C0+i	90 75-105	ST(i) ← ST + ST(i) pop stack

FBLD = Packed Decimal (BCD) Load

WAIT	op1	mod 100 r/m	addr1	addr2

8087 Encoding	Emulator Encoding	Execution Clocks Typical Range	Operation
9B DF m4rm	CD 1F m4rm	300+EA (290-310)+EA	push stack ST ← mem-op

FBSTP = Packed Decimal (BCD) Store and Pop

WAIT	op1	mod 110 r/m	addr1	addr2

8087 Encoding	Emulator Encoding	Execution Clocks Typical Range	Operation
9B DF m6rm	CD 1F m6rm	530+EA (520-540)+EA	mem-op ← ST pop stack

FCHS = Change Sign

WAIT	op1	op2

8087 Encoding	Emulator Encoding	Execution Clocks Typical Range	Operation
9B D9 E0	CD 19 E0	15 10-17	ST ← —ST

FCLEX / FNCLEX = Clear Exceptions

WAIT	op1	op2

8087 Encoding	Emulator Encoding	Execution Clocks Typical Range	Operation
9B DB E2	CD 1B E2	5 2-8	clear 8087 exceptions
90 DB E2	CD 1B E2	5 2-8	clear 8087 exceptions (no wait)

FCOM = Compare Real

Compare Stack top and Stack element

WAIT	op1	op2 + i

8087 Encoding	Emulator Encoding	Execution Clocks Typical Range	Operation
9B D8 D1	CD 18 D1	45 40-50	ST — ST(1)
9B D8 D0+i	CD 18 D0+i	45 40-50	ST — ST(i)

Compare Stack top and memory operands

WAIT	op1	mod 010 r/m	addr1	addr2

8087 Encoding	Emulator Encoding	Execution Clocks Typical Range	Operation
9B D8 m2rm	CD 18 m2rm	65+EA (60-70)+EA	ST — memop (short-real)
9B DC m2rm	CD 1C m2rm	70+EA (65-75)+EA	ST — memop (long-real)

FCOMP = Compare Real and Pop

Compare Stack top and Stack element and pop

WAIT	op1	op2 + i

8087 Encoding	Emulator Encoding	Execution Clocks Typical Range	Operation
9B D8 D9	CD 18 D9	47 42-52	ST — ST(1) pop stack
9B D8 D8+i	CD 18 D8+i	47 42-52	ST — ST(i) pop stack

Compare Stack top and memory operand and pop

WAIT	op1	mod 011 r/m	addr1	addr2

8087 Encoding	Emulator Encoding	Execution Clocks Typical Range	Operation
9B D8 m3rm	CD 18 m3rm	68+EA (63-73)+EA	ST — mem-op pop stack (short-real)
9B DC m3rm	CD 1C m3rm	72+EA (67-77)+EA	ST — mem-op pop stack (long-real)

FCOMPP = Compare Real and Pop Twice

WAIT	op1	op2

8087 Encoding	Emulator Encoding	Execution Clocks Typical Range	Operation
9B DE D9	CD 1E D9	50 45-55	ST — ST(1) pop stack pop stack

FDECSTP = Decrement Stack Pointer

WAIT	op1	op2

8087 Encoding	Emulator Encoding	Execution Clocks Typical Range	Operation
9B D9 F6	CD 19 F6	9 6-12	stack pointer ← stack pointer − 1

FDISI / FNDISI = Disable Interrupts

WAIT	op1	op2

8087 Encoding	Emulator Encoding	Execution Clocks Typical Range	Operation
9B DB E1	CD 1B E1	5 2-8	Set 8087 interrupt mask
90 DB E1	CD 1B E1	5 2-8	Set 8087 interrupt mask (no wait)

FDIV = Divide Real

Stack top and Stack element

WAIT	op1	op2 + i

8087 Encoding	Emulator Encoding	Execution Clocks Typical Range	Operation
9B D8 F0+i	CD 18 F0+i	198 193-203	ST ← ST/ST(i)

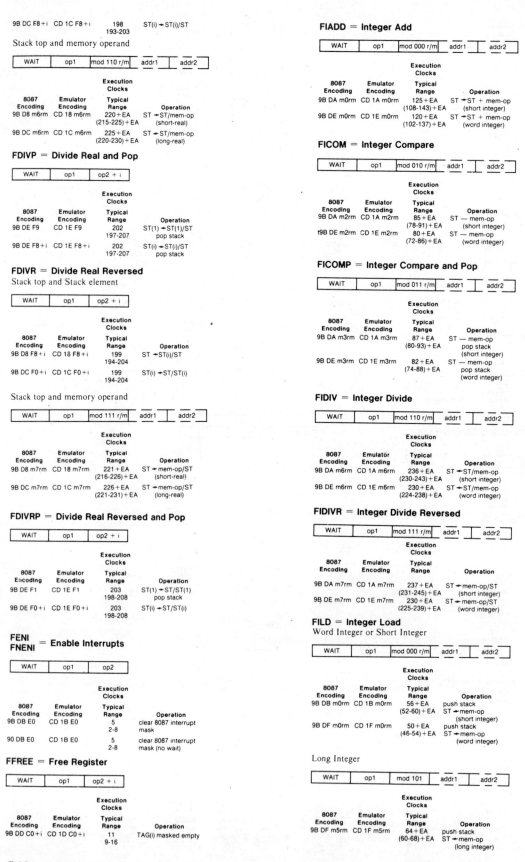

8087 Encoding	Emulator Encoding	Execution Clocks Typical Range	Operation
9B DC F8+i	CD 1C F8+i	198 193-203	ST(i) ← ST(i)/ST

Stack top and memory operand

WAIT	op1	mod 110 r/m	addr1	addr2

8087 Encoding	Emulator Encoding	Execution Clocks Typical Range	Operation
9B D8 m6rm	CD 18 m6rm	220+EA (215-225)+EA	ST ← ST/mem-op (short-real)
9B DC m6rm	CD 1C m6rm	225+EA (220-230)+EA	ST ← ST/mem-op (long-real)

FDIVP = Divide Real and Pop

WAIT	op1	op2 + i

8087 Encoding	Emulator Encoding	Execution Clocks Typical Range	Operation
9B DE F9	CD 1E F9	202 197-207	ST(1) ← ST(1)/ST pop stack
9B DE F8+i	CD 1E F8+i	202 197-207	ST(i) ← ST(i)/ST pop stack

FDIVR = Divide Real Reversed

Stack top and Stack element

WAIT	op1	op2 + i

8087 Encoding	Emulator Encoding	Execution Clocks Typical Range	Operation
9B D8 F8+i	CD 18 F8+i	199 194-204	ST ← ST(i)/ST
9B DC F0+i	CD 1C F0+i	199 194-204	ST(i) ← ST/ST(i)

Stack top and memory operand

WAIT	op1	mod 111 r/m	addr1	addr2

8087 Encoding	Emulator Encoding	Execution Clocks Typical Range	Operation
9B D8 m7rm	CD 18 m7rm	221+EA (216-226)+EA	ST ← mem-op/ST (short-real)
9B DC m7rm	CD 1C m7rm	226+EA (221-231)+EA	ST ← mem-op/ST (long-real)

FDIVRP = Divide Real Reversed and Pop

WAIT	op1	op2 + i

8087 Encoding	Emulator Encoding	Execution Clocks Typical Range	Operation
9B DE F1	CD 1E F1	203 198-208	ST(1) ← ST/ST(1) pop stack
9B DE F0+i	CD 1E F0+i	203 198-208	ST(i) ← ST/ST(i)

FENI / FNENI = Enable Interrupts

WAIT	op1	op2

8087 Encoding	Emulator Encoding	Execution Clocks Typical Range	Operation
9B DB E0	CD 1B E0	5 2-8	clear 8087 interrupt mask
90 DB E0	CD 1B E0	5 2-8	clear 8087 interrupt mask (no wait)

FFREE = Free Register

WAIT	op1	op2 + i

8087 Encoding	Emulator Encoding	Execution Clocks Typical Range	Operation
9B DD C0+i	CD 1D C0+i	11 9-16	TAG(i) masked empty

FIADD = Integer Add

WAIT	op1	mod 000 r/m	addr1	addr2

8087 Encoding	Emulator Encoding	Execution Clocks Typical Range	Operation
9B DA m0rm	CD 1A m0rm	125+EA (108-143)+EA	ST ← ST + mem-op (short integer)
9B DE m0rm	CD 1E m0rm	120+EA (102-137)+EA	ST ← ST + mem-op (word integer)

FICOM = Integer Compare

WAIT	op1	mod 010 r/m	addr1	addr2

8087 Encoding	Emulator Encoding	Execution Clocks Typical Range	Operation
9B DA m2rm	CD 1A m2rm	85+EA (78-91)+EA	ST — mem-op (short integer)
t9B DE m2rm	CD 1E m2rm	80+EA (72-86)+EA	ST — mem-op (word integer)

FICOMP = Integer Compare and Pop

WAIT	op1	mod 011 r/m	addr1	addr2

8087 Encoding	Emulator Encoding	Execution Clocks Typical Range	Operation
9B DA m3rm	CD 1A m3rm	87+EA (80-93)+EA	ST — mem-op pop stack (short integer)
9B DE m3rm	CD 1E m3rm	82+EA (74-88)+EA	ST — mem-op pop stack (word integer)

FIDIV = Integer Divide

WAIT	op1	mod 110 r/m	addr1	addr2

8087 Encoding	Emulator Encoding	Execution Clocks Typical Range	Operation
9B DA m6rm	CD 1A m6rm	236+EA (230-243)+EA	ST ← ST/mem-op (short integer)
9B DE m6rm	CD 1E m6rm	230+EA (224-238)+EA	ST ← ST/mem-op (word integer)

FIDIVR = Integer Divide Reversed

WAIT	op1	mod 111 r/m	addr1	addr2

8087 Encoding	Emulator Encoding	Execution Clocks Typical Range	Operation
9B DA m7rm	CD 1A m7rm	237+EA (231-245)+EA	ST ← mem-op/ST (short integer)
9B DE m7rm	CD 1E m7rm	230+EA (225-239)+EA	ST ← mem-op/ST (word integer)

FILD = Integer Load

Word Integer or Short Integer

WAIT	op1	mod 000 r/m	addr1	addr2

8087 Encoding	Emulator Encoding	Execution Clocks Typical Range	Operation
9B DB m0rm	CD 1B m0rm	56+EA (52-60)+EA	push stack ST ← mem-op (short integer)
9B DF m0rm	CD 1F m0rm	50+EA (46-54)+EA	push stack ST ← mem-op (word integer)

Long Integer

WAIT	op1	mod 101	addr1	addr2

8087 Encoding	Emulator Encoding	Execution Clocks Typical Range	Operation
9B DF m5rm	CD 1F m5rm	64+EA (60-68)+EA	push stack ST ← mem-op (long integer)

FIMUL = Integer Multiply

WAIT	op1	mod001 r/m	addr1	addr2

8087 Encoding	Emulator Encoding	Execution Clocks Typical Range	Operation
9B DA m1rm	CD 1A m1rm	136 + EA (130-144)+EA	ST ← ST · mem-op (short integer)
9B DE m1rm	CD 1E m1m	130 + EA (124-138)+EA	ST ← ST · mem-op (word integer)

FINCSTP = Increment Stack Pointer

WAIT	op1	op2

8087 Encoding	Emulator Encoding	Execution Clocks Typical Range	Operation
9B D9 F7	CD 19 F7	9 6-12	stack pointer ← stack pointer + 1

FINIT FNINIT = Initialize Processor

WAIT	op1	op2

8087 Encoding	Emulator Encoding	Execution Clocks Typical Range	Operation
9B DB E3	CD 1B E3	5 2-8	initialize 8087
90 DB E3	CD 1B E3	5 2-8	initialize 8087 (no wait)

FIST = Integer Store

WAIT	op1	mod 010 r/m	addr1	addr2

8087 Encoding	Emulator Encoding	Execution Clocks Typical Range	Operation
9B DB m2rm	CD 1B m2rm	88 + EA (82-92) + EA	mem-op ← ST (short integer)
9B DF m2rm	CD 1F m2rm	86 + EA (80-90) + EA	mem-op ← ST (word integer)

FISTP = Integer Store and Pop

Short Integer or Word Integer

WAIT	op1	mod 011 r/m	addr1	addr2

8087 Encoding	Emulator Encoding	Execution Clocks Typical Range	Operation
9B DB m3rm	CD 1B m3rm	90 + EA (84-94) + EA	mem-op ← ST pop stack (short integer)
9B DF m3rm	CD 1F m3rm	88 + EA (82-92) + EA	mem-op ← ST pop stack (word integer)

Long Integer

WAIT	op1	mod 111	addr1	addr2

8087 Encoding	Emulator Encoding	Execution Clocks Typical Range	Operation
9B DF m7rm	CD 1F m7rm	100 + EA (94-105) + EA	mem-op ← ST pop stack (long integer)

FISUB = Integer Subtract

WAIT	op1	mod 100 r/m	addr1	addr2

8087 Encoding	Emulator Encoding	Execution Clocks Typical Range	Operation
9B DA m4rm	CD 1A m4rm	125 + EA (108-143) + EA	ST ← ST — mem-op (short integer)
9B DE m4rm	CD 1E m4rm	120 + EA (102-137) + EA	ST ← ST — mem-op (word integer)

FISUBR = Integer Subtract Reversed

WAIT	op1	mod 101 r/m	addr1	addr2

8087 Encoding	Emulator Encoding	Execution Clocks Typical Range	Operation
9B DA m5rm	CD 1A m5rm	125 + EA (109-144) + EA	ST ← mem-op — ST (short integer)
9B DE m5rm	CD 1E m5rm	120 + EA (103-139) + EA	ST ← mem-op — ST (word integer)

FLD = Load Real

Stack element to Stack top

WAIT	op1	op2 + i

8087 Encoding	Emulator Encoding	Execution Clocks Typical Range	Operation
9B D9 C0 + i	CD 19 C0 + i	20 17-22	T_1 ← ST(i) push stack ST ← T_1

Memory operand to Stack top
Short Integer or Long Integer

WAIT	op1	mod 000 r/m	addr1	addr2

8087 Encoding	Emulator Encoding	Execution Clocks Typical Range	Operation
9B D9 m0rm	CD 19 m0rm	43 + EA (38-56) + EA	push stack ST ← mem-op (short integer)
9B DD m0rm	CD 1D m0rm	46 + EA (40-60) + EA	push stack ST ← mem-op (long integer)

Temp Real

WAIT	op1	mod 101	addr1	addr2

8087 Encoding	Emulator Encoding	Execution Clocks Typical Range	Operation
9B DB m5rm	CD 1B m5rm	57 + EA (53-65) + EA	push stack ST ← mem-op (temp real)

FLD1 = Load + 1.0

WAIT	op1	op2

8087 Encoding	Emulator Encoding	Execution Clocks Typical Range	Operation
9B D9 E8	CD 19 E8	18 15-21	push stack ST ← 1.0

FLDCW = Load Control Word

WAIT	op1	mod 101 r/m	addr1	addr2

8087 Encoding	Emulator Encoding	Execution Clocks Typical Range	Operation
9B D9 m5rm	CD 19 m5rm	10 + EA (7-14) + EA	processor control word ← mem-op

FLDENV = Load Environment

WAIT	op1	mod 100 r/m	addr1	addr2

8087 Encoding	Emulator Encoding	Execution Clocks Typical Range	Operation
9B D9 m4rm	CD 19 m4rm	40 + EA (35-45) + EA	8087 environment ← mem-op

FLDL2E = Load Log_2e

WAIT	op1	op2

8087 Encoding	Emulator Encoding	Execution Clocks Typical Range	Operation
9B D9 EA	CD 19 EA	18 15-21	push stack ST ← log_2e

FLDL2T = Load $Log_2 10$

WAIT	op1	op2

8087 Encoding	Emulator Encoding	Execution Clocks Typical Range	Operation
9B D9 E9	CD 19 E9	19 16-22	push stack ST $\leftarrow \log_2 10$

FLDLG2 = Load $\text{Log}_{10}2$

WAIT	op1	op2

8087 Encoding	Emulator Encoding	Execution Clocks Typical Range	Operation
9B D9 EC	CD 19 EC	21 18-24	push stack ST $\leftarrow \log_{10} 2$

FLDPI = Load π

WAIT	op1	op2

8087 Encoding	Emulator Encoding	Execution Clocks Typical Range	Operation
9B D9 EB	CD 19 EB	19 16-22	push stack ST $\leftarrow \pi$

FLDZ = Load + 0.0

WAIT	op1	op2

8087 Encoding	Emulator Encoding	Execution Clocks Typical Range	Operation
9B D9 EE	CD 19 EE	14 11-17	push stack ST $\leftarrow$ 0.0

FMUL = Multiply Real

Stack top and Stack element

WAIT	op1	op2 + i

8087 Encoding	Emulator Encoding	Execution Clocks Typical Range	Operation
9B D8 C8+i	CD 18 C8+i	138 130-145	ST $\leftarrow$ ST · ST(i)
9B DC C8+i	CD 1C C8+i	138 130-145	ST(i) $\leftarrow$ ST(i) — ST

Stack top and memory operand

WAIT	op1	mod 001 r/m	addr1	addr2

8087 Encoding	Emulator Encoding	Execution Clocks Typical Range	Operation
9B D8 m1rm	CD 18 m1rm	118+EA (110-125)+EA	ST $\leftarrow$ ST · mem-op (short real)
9B DC m1rm	CD 1C m1rm	161+EA (154-168)+EA	ST $\leftarrow$ ST · mem-op (long real)

FMULP = Multiply Real and Pop

WAIT	op1	op2 + i

8087 Encoding	Emulator Encoding	Execution Clocks Typical Range	Operation
9B DE C9 +i	CD 1E C9+i	142 134-148	ST(i) $\leftarrow$ ST(i) · ST pop stack

FNOP = No Operation

WAIT	op1	op2

8087 Encoding	Emulator Encoding	Execution Clocks Typical Range	Operation
9B D9 D0	CD 19 D0	13 10-16	ST $\leftarrow$ ST

FPATAN = Partial Arctangent

WAIT	op1	op2

8087 Encoding	Emulator Encoding	Execution Clocks Typical Range	Operation
9B D9 F3	CD 19 F3	650 250-800	$T_1 \leftarrow$ arctan (ST(1)/ST) pop stack ST $\leftarrow T_1$

FPREM = Partial Remainder

WAIT	op1	op2

8087 Encoding	Emulator Encoding	Execution Clocks Typical Range	Operation
9B D9 F8	CD 19 F8	125 15-190	ST $\leftarrow$ REPEAT (ST — ST(1))

FPTAN = Partial Tangent

WAIT	op1	op2

8087 Encoding	Emulator Encoding	Execution Clocks Typical Range	Operation
9B D9 F2	CD 19 F2	450 30-540	Y/X $\leftarrow$ TAN (ST) ST $\leftarrow$ Y push stack ST $\leftarrow$ X

FRNDINT = Round to Integer

WAIT	op1	op2

8087 Encoding	Emulator Encoding	Execution Clocks Typical Range	Operation
9B D9 FC	CD 19 FC	45 16-50	ST $\leftarrow$ nearest integer (ST)

FRSTOR = Restore Saved State

WAIT	op1	mod 100 r/m	addr1	addr2

8087 Encoding	Emulator Encoding	Execution Clocks Typical Range	Operation
9B DD m4rm	CD 1D m4rm	202+EA (197-207)+EA	8087 state $\leftarrow$ mem-op

FSAVE / FNSAVE = Save State

WAIT	op1	mod 110 r/m	addr1	addr2

8087 Encoding	Emulator Encoding	Execution Clocks Typical Range	Operation
9B DD m6rm	CD 1D m6rm	202+EA (197-207)+EA	mem-op $\leftarrow$ 8087 state
90 DD m6rm	CD 1D m6rm	202+EA (197-207)+EA	mem-op $\leftarrow$ 8087 state (no wait)

FSCALE = Scale

WAIT	op1	op2

8087 Encoding	Emulator Encoding	Execution Clocks Typical Range	Operation
9B D9 FD	CD 19 FD	35 32-38	ST $\leftarrow$ ST $\cdot 2^{ST(1)}$

FSQRT = Square Root

WAIT	op1	op2

8087 Encoding	Emulator Encoding	Execution Clocks Typical Range	Operation
9B D9 FA	CD 19 FA	183 180-186	ST $\leftarrow \sqrt{ST}$

FST = Store Real

Stack top to Stack element

WAIT	op1	op2 + i

8087 Encoding	Emulator Encoding	Execution Clocks Typical Range	Operation
9B DD D0+i	CD 1D D0+i	18 15-22	ST(i) ← ST

Stack top to memory operand

WAIT	op1	mod 010 r/m	addr1	addr2

8087 Encoding	Emulator Encoding	Execution Clocks Typical Range	Operation
9B D9 m2rm	CD 19 m2rm	87+EA (84-90)+EA	mem-op ← ST (short-real)
9B DD m2rm	CD 1D m2rm	100+EA (96-104)+EA	mem-op ← ST (long-real)

FSTCW
FNSTCW = Store Control Word

WAIT	op1	mod 111 r/m	addr1	addr2

8087 Encoding	Emulator Encoding	Execution Clocks Typical Range	Operation
9B D9 m7rm	CD 19 m7rm	15+EA (12-18)+EA	mem-op ← processor control word
90 D9 m7rm	CD 19 m7rm	15+EA (12-18)+EA	mem-op ← processor control word (no wait)

FSTENV
FNSTENV = Store Environment

WAIT	op1	mod 110 r/m	addr1	addr2

8087 Encoding	Emulator Encoding	Execution Clocks Typical Range	Operation
9B D9 m6rm	CD 19 m6rm	45+EA (40-50)+EA	mem-op ← 8087 environment
90 D9 r6rm	CD 19 m6rm	45+EA (40-50)+EA	mem-op ← 8087 environment (no wait)

FSTP = Store Real and Pop

Stack top to Stack element

WAIT	op1	op2 + i

8087 Encoding	Emulator Encoding	Execution Clocks Typical Range	Operation
9B DD D8+i	CD 1D D8+i	20 17-24	ST(i) ← ST pop stack

Stack top to memory operand

WAIT	op1	mod 011 r/m	addr1	addr2

Long Real or Short Real

8087 Encoding	Emulator Encoding	Execution Clocks Typical Range	Operation
9B D9 m3rm	CD 19 m3rm	89+EA (86-92)+EA	mem-op ← ST pop stack (short-real)
9B DB m3rm	CD 1B m3rm	102+EA (98-106)+EA	mem-op ← ST pop stack (long-real)

Temp Real

WAIT	op1	mod 111 r/m	disp-lo	disp-hi

8087 Encoding	Emulator Encoding	Execution Clocks Typical Range	Operation
9B DD m7rm	CD 1D m7rm	55+EA (52-58)+EA	mem-op ← ST pop stack (temp-real)

FSTSW
FNSTSW = Store Status Word

WAIT	op1	mod 111 r/m	addr1	addr2

8087 Encoding	Emulator Encoding	Execution Clocks Typical Range	Operation
9B DD m7rm	CD 1D m7rm	15+EA (12-18)+EA	mem-op ← 8087 status word
90 DD m7rm	CD 1D m7rm	15+EA (12-18)+EA	mem-op ← 8087 status word (no wait)

FSUB = Subtract Real

Stack top and Stack element

WAIT	op1	op2 + i

8087 Encoding	Emulator Encoding	Execution Clocks Typical Range	Operation
9B D8 E0+i	CD 18 E0+i	85 70-100	ST ← ST — ST(i)
9B DC E8+i	CD 1C E8+i	85 70-100	ST(i) ← ST(i) — ST

Stack top and memory operand

WAIT	op1	mod 100 r/m	addr1	addr2

8087 Encoding	Emulator Encoding	Execution Clocks Typical Range	Operation
9B D8 m4rm	CD 18 m4rm	105+EA (90-120)+EA	ST ← ST — mem-op (short-real)
9B DC m4rm	CD 1C m4rm	110+EA (95-125)+EA	ST ← ST — mem-op (long-real)

FSUBP = Subtract Real and Pop

WAIT	op1	op2 + i

8087 Encoding	Emulator Encoding	Execution Clocks Typical Range	Operation
9B DE E9	CD 1E E9	90 75-105	ST(1) ← ST(1) — ST pop stack
9B DE E8+i	CD 1E E8+i	90 75-105	ST(i) ← ST(i) — ST pop stack

FSUBR = Subtract Real Reversed

Stack top and Stack element

WAIT	op1	op2 + i

8087 Encoding	Emulator Encoding	Execution Clocks Typical Range	Operation
9B D8 E8+i	CD D8 E8+i	87 70-100	ST ← ST(i) — ST
9B DC E0+i	CD 1C E0+i	87 70-100	ST(i) ← ST — ST(i)

Stack top and memory operand

WAIT	op1	mod 101 r/m	addr1	addr2

8087 Encoding	Emulator Encoding	Execution Clocks Typical Range	Operation
9B D8 m5rm	CD 18 m5rm	105+EA (90-120)+EA	ST ← mem-op — ST (short-real)
9B DC m5rm	CD 1C m5rm	110+EA (95-125)+EA	ST ← mem-op — ST (long-real)

FSUBRP = Subtract Real Reversed and Pop

WAIT	op1	op2 + i

8087 Encoding	Emulator Encoding	Execution Clocks Typical Range	Operation

8087 Encoding	Emulator Encoding	Execution Clocks Typical Range	Operation
9B DE E1	CD 1E E1	90 75-105	ST(1) ← ST — ST(1) pop stack
9B DE E0+i	CD 1E E0+i	90 75-105	ST(i) ← ST — ST(i) pop stack

FTST = Test Stack Top Against + 0.0

WAIT	op1	op2

8087 Encoding	Emulator Encoding	Execution Clocks Typical Range	Operation
9B D9 E4	CD 19 E4	42 38-48	ST ← ST — 0.0

FWAIT = (CPU) Wait While 8087 Is Busy

WAIT

8087 Encoding	Emulator Encoding	Execution Clocks Typical Range	Operation
9B	90	3+5n 3+5n	8086 wait instruction

FXAM = Examine Stack Top

WAIT	op1	op2

8087 Encoding	Emulator Encoding	Execution Clocks Typical Range	Operation
9B D9 E5	CD 19 E5	17 12-23	set condition code

FXCH = Exchange Registers

WAIT	op1	op2 + i

8087 Encoding	Emulator Encoding	Execution Clocks Typical Range	Operation
9B D9 C8	CD 19 C8	12 10-15	T_1 ← ST(1) ST(1) ← ST ST ← T_1
9B D9 C8+i	CD 19 C8+i	12 10-15	T_1 ← ST(i) ST(i) ← ST ST ← T_1

FXTRACT = Extract Exponent and Significand

WAIT	op1	op2

8087 Encoding	Emulator Encoding	Execution Clocks Typical Range	Operation
9B D9 F4	CD 19 F4	50 27-55	T_1 ← exponent (ST) T_2 ← significand (ST) ST ← T_1 push stack ST ← T_2

FYL2X = Compute Y · Log_2 X

WAIT	op1	op2

8087 Encoding	Emulator Encoding	Execution Clocks Typical Range	Operation
9B D9 F1	CD 19 F1	950 900-1100	T_1 ← ST(1) · $\log_2$ (ST) pop stack ST ← T_1

FYL2XP1 = Compute Y · Log_2 (X+1)

WAIT	op1	op2

8087 Encoding	Emulator Encoding	Execution Clocks Typical Range	Operation
9B D9 F9	CD 19 F9	850 700-1000	T_1 ← ST + 1 T_2 ← ST(1) · $\log_2 T_1$ pop stack ST ← T_2

INDEX

APPENDIX C THE SC-01 SPEECH SYNTHESIZER

General description of the SC-01 chip

The SC-01 Speech Synthesizer is a completely self-contained solid state device. This single chip phonetically synthesizes continuous speech, of unlimited vocabulary, from low data rate inputs.

Speech is synthesized by combining phonemes (the building blocks of speech) in the appropriate sequence. The SC-01 Speech Synthesizer contains 64 different phonemes which are accessed by a 6-bit code. It is the proper sequential combination of these phoneme codes that creates continuous speech.

Phoneme description

Table 1 lists the 64 phonemes produced by the SC-01. Each sound is represented by its VOTRAX® phoneme code and is accompanied by its phoneme symbol and an example. The underlined segments of the example word demonstrate the phoneme use, i.e., sound to be pronounced.

Table 2 provides the phoneme sequences used to produce vowels in the group called dipthongs, (2 vowel sounds in sequence, identified as a single sound, e.g., the long "i" vowel).

TABLE 2: DIPTHONG CHART

Phoneme Combination	Key Words
A1-AY-Y	fate, maid
AH1-EH3-Y	find, wide
UH3-AH2-Y	fight, white
AH1-I3-UH3-L	file, smile
01-UH3-Y	foy, boy
01-I3-UH3-L	foil, spoil
AH1-02-U1	found, cow
UH3-AH2-U1	foust, house
O1-U1	float, note
Y1-IU-U1	few, you, music
AY-I1	fear, beer

/T/ must precede /CH/ to produce CH sound.
/D/ must precede /J/ to produce J sound.

TABLE 1: PHONEME CHART

Phoneme Code	Phoneme Symbol	Duration (ms)	Example Word
ØØ	EH3	59	jacket
Ø1	EH2	71	enlist
Ø2	EH1	121	heavy
Ø3	PAØ	47	no sound
Ø4	DT	47	butter
Ø5	A2	71	made
Ø6	A1	103	made
Ø7	ZH	90	azure
Ø8	AH2	71	honest
Ø9	13	55	inhibit
ØA	12	80	inhibit
ØB	11	121	inhibit
ØC	M	103	mat
ØD	N	80	sun
ØE	B	71	bag
ØF	V	71	van
1Ø	CH*	71	chip
11	SH	121	shop
12	Z	71	zoo
13	AW1	146	lawful
14	NG	121	thing
15	AH1	146	father
16	001	103	looking
17	00	185	book
18	L	103	land
19	K	80	trick
1A	J*	47	judge
1B	H	71	hello
1C	G	71	get
1D	F	103	fast
1E	D	55	paid
1F	S	90	pass
2Ø	A	185	day
21	AY	65	day
22	Y1	80	yard
23	UH3	47	mission
24	AH	250	mop
25	P	103	past
26	O	185	cold
27	I	185	pin
28	U	185	move
29	Y	103	any
2A	T	71	tap
2B	R	90	red
2C	E	185	meet
2D	W	80	win
2E	AE	185	dad
2F	AE1	103	after
30	AW2	90	salty
31	UH2	71	about
32	UH1	103	uncle
33	UH	185	cup
34	02	80	for
35	01	121	aboard
36	IU	59	you
37	U1	90	you
38	THV	80	the
39	TH	71	thin
3A	ER	146	bird
3B	EH	185	get
3C	E1	121	be
3D	AW	253	call
3E	PA1	185	no sound
3F	STOP	47	no sound

PHONETIC PROGRAMS FOR THE SC-01 SPEECH SYNTHESIZER

A A1, AY, Y
abort UH1, B, 02, 02, R, T
active AE1, EH3, K, T, I1, V
age A1, AY, Y, D, J
alarm UH1, L, AH1, R, M
alert UH1, L, ER, R, T
alpha AE1, AW2, L, F, UH1
an AE1, EH3, N
angle AE1, EH3, NG, G, UH3, L
april A1, Y, P, R, UH2, L
ASCII AE1, EH3, S, K, Y
attempt UH1, T, EH1, EH3, M, P, T
august AW2, AW2, G, EH2, S, T
average AE1, EH3, V, R, I1, D, J
B B, E1, Y
bad B, AE1, AE1, D
base B, A1, AY, Y, S
beep B, E1, Y, P
begin B, Y, G, I1, I3, N
beta B, A2, A2, AY, T, UH2
binary B, AH1, Y, N, EH3, EH3, ER, Y
birthday B, ER, R, TH, D, A1, I3, Y
black B, L, AE1, EH3, K
blue B, L, IU, U1, U1
brown B, R, AH1, UH3, U1, N
bug B, UH1, UH2, G
button B, UH1, UH3, T, EH3, N
bye B, AH1, EH3, I3, Y
byte B, UH3, AH2, Y, T
C S, E1, Y
call K, AW2, AW1, L
caution K, AW2, AW1, SH, UH3, N
centigrade S, EH1, N, T, I3, G, R, A1, Y, D
circuit S, R, R, K, I2, T
clock K, L, AH1, UH3, K
computer K, UH1, M, P, Y1, IU, U1, T, ER
connect K, UH1, N, EH1, EH3, K, T
D D, E1, Y
danger D, A1, AY, Y, N, D, J, ER
dead D, EH1, EH3, F
december D, Y, S, EH1, EH3, M, B, ER
degree D, Y, G, R, E1, Y
destroy D, Y, S, T, R, 01, UH3, I3, AY
digital D, I1, D, J, I3, T, UH3, L
directory D, ER, EH1, EH3, K, T, ER, Y
disconnect D, I1, S, K, UH1, N, EH1, EH3, KT
doctor D, AH1, UH3, K, T, ER
E E1, Y
eight A2, A2, Y, T
eighth A2, A2, Y, DT, DT, TH
eighty A2, A2, Y, T, Y
electronic EH3, L, EH1, K, T, R, AH1, N, I2, K
eleven EH1, L, EH1, EH3, V, I2, N
engineer EH1, N, D, J, I2, N, AY, I1, R
enter EH1, EH3, N, T, ER
exponent EH1, K, PAO, S, P, O2, O2, N, EH3, N, T
F EH1, EH2, F
fahrenheit F, EH1, R, I2, N, H, UH3, AH2, Y, T
farad F, EH3, EH3, ER, AE1, EH3, D
february F, EH1, B, Y1, IU, W, EH1, R, Y
feet F, E1, Y, T
female F, AY, Y, M, A1, AY, UH3, L
fifteen F, I1, I3, F, T, E1, Y, N
fifth F, I1, I3, F, TH
fifty F, I1, I3, F, T, Y
fire F, AH1, EH3, AY, R
first F, ER, R, S, T
forty F, O2, O2, R, T, Y
four F, O1, O2, R
fourth F, O1, O2, R, TH
friday F, R, AH1, EH3, Y, D, A1, I3, Y
G D, J, E1, Y
glitch G, L, I1, I3, T, CH
go G, OO1, O1, U1
good G, OO1, OO1, D
H A1, AY, Y, T, CH
half H, AE1, EH3, F
hello H, EH1, UH3, L, UH3, O1, U1
help H, EH1, EH3, L, P
henry H, EH1, EH3, N, R, Y
hertz H, R, R, T, S
high H, AH1, EH3, Y
home H, O1, U1, M
human H, Y1, IU, U1, U1, M, EH2, N
hundred H, UH1, UH2, N, D, R, I3, D
I AH1, EH3, I3, Y
inch I1, I3, N, T, CH
input I1, I3, N, P, OO1, OO1, T
install I1, I3, N, S, T, AW, L
insufficient I1, N, S, UH2, F, I1, SH, EH3, N, T
is I1, I3, Z
it I1, I3, T
J D, J, EH3, A1, AY, Y
january D, J, AE1, EH3, N, Y1, U1, EH3, EH3, ER, Y
july D, J, UH1, L, AH1, EH3, Y
june D, J, IU, U1, U1, N
K K, EH3, A1, AY, Y
kilo K, E1, AY, L, UH3, O2, U1
L EH1, EH3, UH3, L
late L, A1, AY, Y, T
linear L, I2, I3, N, AY, Y, ER
load L, UH3, O1, U1, D
M EH1, EH2, M
machine M, UH2, SH, E1, Y, N
male M, A2, A2, AY, UH3, L
march M, AH1, R, T, CH
mark M, AH1, R, K
mega M, EH1, EH3, G, UH2, UH3
memory M, EH1, EH3, M, ER, Y
meter M, E1, Y, T, ER
micro M, UH3, AH2, AY, K, R, O1, U1
mile M, AH1, EH3, I3, UH3, L
million M, I1, I3, L, Y, UH3, N
mini M, I2, I2, N, Y
minus M, AH1, Y, N, EH3, S
minute M, I1, N, EH3, T
monday M, UH3, UH1, N, D, A1, I3, Y
money M, UH3, UH1, N, AY, Y
Mr. M, I1, S, T, ER
Mrs. M, I1, S, I2, Z
Ms. M, I1, I3, Z
multiply M, UH1, L, T, I3, P, L, AH1, Y
N EH1, EH2, N
nano N, AE1, EH3, N, O1, U1
negative N, EH1, G, EH3, T, I1, V
nine N, AH1, EH3, Y, N
ninety N, AH1, EH3, Y, N, T, Y
nineth N, AH1, Y, N, DT, TH
no N, OO1, O1, U1
november N, O1, U1, V, EH1, EH3, M, B, ER
O O2, O1, U1
october AH1, UH3, K, T, O1, U1, B, ER
ohm O2, O2, U1, M
one W, UH1, UH2, N
P P, E1, Y
percent P, ER, S, EH1, EH3, N, T
pico P, E1, Y, K, O2, U1
plus P, L, UH1, UH2, S
police P, UH1, L, AY, Y, S
please P, L, E1, Y, Z
pop P, AH1, UH3, P
positive P, AH1, UH3, Z, I1, T, I1, V
print P, R, I1, I3, N, T
proceed P, R, O1, S, E1, Y, D
process P, R, AH1, UH3, S, EH1, EH3, S
PROM P, R, AH1, UH3, M
pull P, OO1, OO1, L
push P, OO1, IU, SH
Q K, Y1, IU, U1, U1
quarter K, W, O1, R, T, ER
quick K, W, I1, I3, K
quit K, W, I1, I3, T
R AH1, UH2, ER
record R, E1, K, O2, O2, R, D
red R, EH1, EH3, D
reject R, E1, D, J, EH1, EH3, K, T
repair R, E1, P, EH2, EH2, R
repeat R, E1, P, E1, AY, T
replace R, E1, P, L, A1, AY, Y, S
root R, U1, U1, T
run R, UH1, UH3, N
S EH1, EH2, S
safe S, A1, AY, Y, F
saturday S, AE1, EH3, T, ER, D, A1, Y
second S, EH1, EH3, K, UH1, N, D
security S, EH1, EH3, K, Y, ER, I1, T, Y
september S, EH1, EH3, P, T, EH1, EH3, M, B, ER
serial S, I1, R, Y, UH3, L
seven S, EH1, EH3, V, I2, N
seventh S, EH1, EH3, V, I2, N, DT, TH
seventy S, EH1, V, I2, N, D, Y
six S, I1, I3, K, PAO, S
sixth S, I1, I3, K, PAO, S, TH
sixty S, I1, I3, K, PAO, T, Y
smoke S, M, O1, U1, K
space S, P, A1, AY, Y, S
square S, K, W, EH1, R
stack S, T, AE1, EH3, K
subtract S, UH1, UH2, D, T, R, AE1, EH3, K, T
suggest S, UH1, UH2, G, D, J, EH1, EH3, S, T
sum S, UH1, UH2, M
sunday S, UH1, UH2, N, D, A1, I3, Y
syntax S, I1, N, T, AE1, EH3, K, PAO, S
system S, I1, S, T, UH3, M
T T, E1, AY, Y
tangent T, AE1, EH3, N, D, J, EH3, N, T
temperature T, EH1, EH3, M, P, ER, UH1, T, CH, ER
ten T, EH1, EH3, N
terminal T, ER, M, EH3, N, UH2, L
test T, EH1, EH3, S, T
third TH, ER, R, D
thirteen TH, ER, T, T, E1, Y, N
thirty TH, ER, R, D, Y
three TH, R, E1, Y
thursday TH, ER, R, Z, D, A1, I3, Y
too (two) T, IU, U1, U1
trace T, R, A1, AY, Y, S
transistor T, R, AE1, N, Z, I1, S, T, ER
trouble T, R, UH3, UH1, B, UH3, L
tuesday T, IU, U1, Z, D, A1, Y
twelve T, W, EH1, EH3, UH3, L, V
twenty T, W, EH1, EH3, N, T, Y
U Y1, IU, U1, U1
urgent R, R, D, J, I3, N, T
V V, E1, AY, Y
valid V, AE1, UH3, L, I1, D
very V, EH1, R, Y
volt V, O2, O2, L, T
W D, UH1, B, UH3, L, Y1, IU, U1
want W, AH1, UH3, N, T
was W, UH1, UH3, Z
water W, AH1, UH3, T, ER
we W, E1, Y
wednesday W, EH1, N, Z, D, A1, I3, Y
whiskey W, I1, I3, S, K, AY, Y
X EH1, EH2, K, PAO, S
Y W, AH1, EH3, I3, Y
yellow Y1, EH1, EH3, L, O1, U1
yes Y1, EH3, EH1, S
you (use "U" program)
your Y, 02, 02, R
Z Z, E1, Y
zap Z, AE1, EH3, P
zero Z, AY, I1, R, O1, U1

Prefixes

con... K, UH1, N
dis... D, I1, S
en... EH1, N
in... I1, N
non... N, AH1, UH3, N
pre... P, R, E1
re... R, E1
un... UH1, N